Complete Solutions Manual

Calculus of a Single Variable
Early Transcendental Functions
Volume 2 (Chapters 7-10)

SIXTH EDITION

Ron Larson
The Pennsylvania State University,
The Behrend College

Bruce Edwards
University of Florida

Prepared by

Bruce Edwards
University of Florida

CENGAGE Learning

Australia • Brazil • Mexico • Singapore • United Kingdom • United States

CENGAGE
Learning

For product information and technology assistance, contact us at **Cengage Learning Customer & Sales Support, 1-800-354-9706**.

For permission to use material from this text or product, submit all requests online at **www.cengage.com/permissions**
Further permissions questions can be emailed to **permissionrequest@cengage.com**.

ISBN-13: 978-1-285-77482-4
ISBN-10: 1-285-77482-5

Cengage Learning
200 First Stamford Place, 4th Floor
Stamford, CT 06902
USA

Cengage Learning is a leading provider of customized learning solutions with office locations around the globe, including Singapore, the United Kingdom, Australia, Mexico, Brazil, and Japan. Locate your local office at: **www.cengage.com/global**.

Cengage Learning products are represented in Canada by Nelson Education, Ltd.

To learn more about Cengage Learning Solutions, visit **www.cengage.com**.

Purchase any of our products at your local college store or at our preferred online store **www.cengagebrain.com**.

Printed in the United States of America
1 2 3 4 5 6 7 17 16 15 14 13

CONTENTS

Chapter 7 Applications of Integration **614**

Chapter 8 Integration Techniques, L'Hôpital's Rule, and Improper Integrals.............. **708**

Chapter 9 Infinite Series **854**

Chapter 10 Conics, Parametric Equations, and Polar Coordinates................................ **984**

Appendix C.1.. **1092**

Appendix C.2.. **1097**

Appendix C.3.. **1106**

CHAPTER 7
Applications of Integration

Section 7.1 Area of a Region Between Two Curves **615**

Section 7.2 Volume: The Disk Method **630**

Section 7.3 Volume: The Shell Method **646**

Section 7.4 Arc Length and Surfaces of Revolution **658**

Section 7.5 Work **671**

Section 7.6 Moments, Centers of Mass, and Centroids **677**

Section 7.7 Fluid Pressure and Fluid Force **690**

Review Exercises **695**

Problem Solving **703**

CHAPTER 7
Applications of Integration

Section 7.1 Area of a Region Between Two Curves

1. $A = \int_0^6 \left[0 - \left(x^2 - 6x\right)\right] dx = -\int_0^6 \left(x^2 - 6x\right) dx$

2. $A = \int_{-2}^{2} \left[(2x + 5) - \left(x^2 + 2x + 1\right)\right] dx$

$= \int_{-2}^{2} \left(-x^2 + 4\right) dx$

3. $A = \int_0^3 \left[\left(-x^2 + 2x + 3\right) - \left(x^2 - 4x + 3\right)\right] dx$

$= \int_0^3 \left(-2x^2 + 6x\right) dx$

4. $A = \int_0^1 \left(x^2 - x^3\right) dx$

5. $A = 2\int_{-1}^{0} 3\left(x^3 - x\right) dx = 6\int_{-1}^{0} \left(x^3 - x\right) dx$

or $-6\int_0^1 \left(x^3 - x\right) dx$

6. $A = 2\int_0^1 \left[(x - 1)^3 - (x - 1)\right] dx$

7. $\int_0^4 \left[(x + 1) - \frac{x}{2}\right] dx$

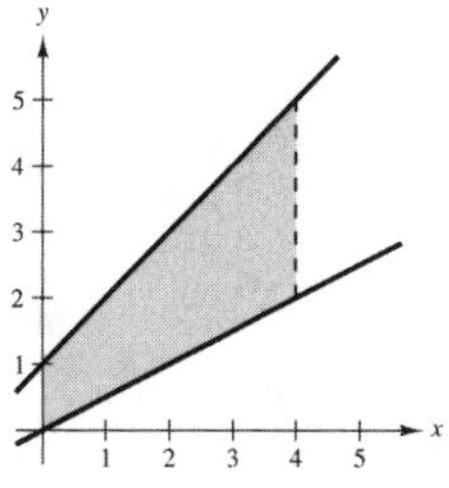

8. $\int_{-1}^{1} \left[\left(2 - x^2\right) - x^2\right] dx$

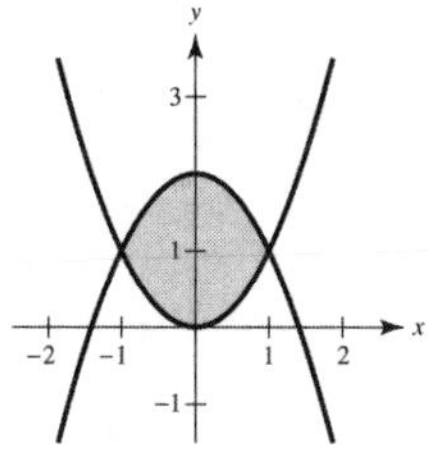

9. $\int_2^3 \left[\left(\frac{x^3}{3} - x\right) - \frac{x}{3}\right] dx$

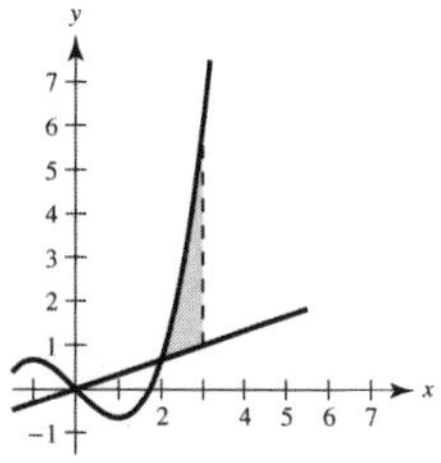

10. $\int_{-\pi/4}^{\pi/4} \left(\sec^2 x - \cos x\right) dx$

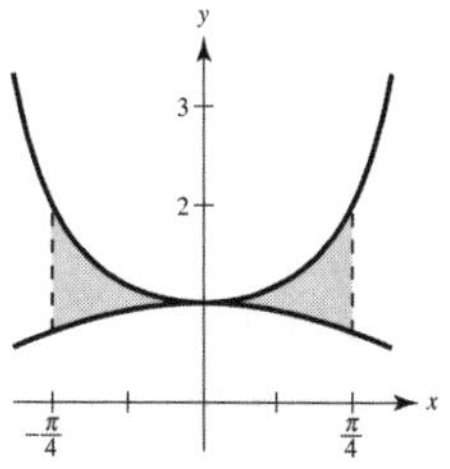

11. $\int_{-2}^{1} \left[(2 - y) - y^2\right] dy$

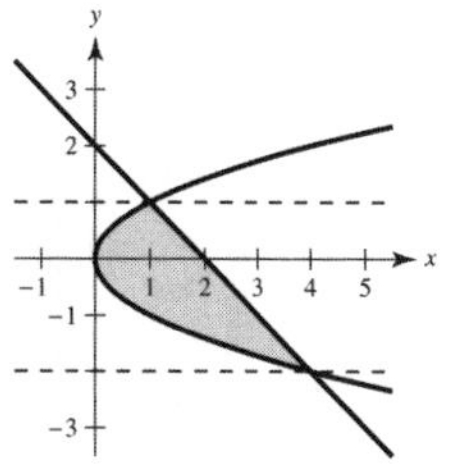

12. $\int_0^4 \left(2\sqrt{y} - y\right) dy$

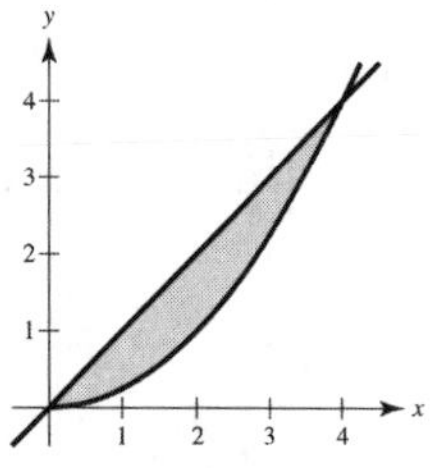

13. $f(x) = x + 1$

$g(x) = (x - 1)^2$

$A \simeq 4$

Matches (d)

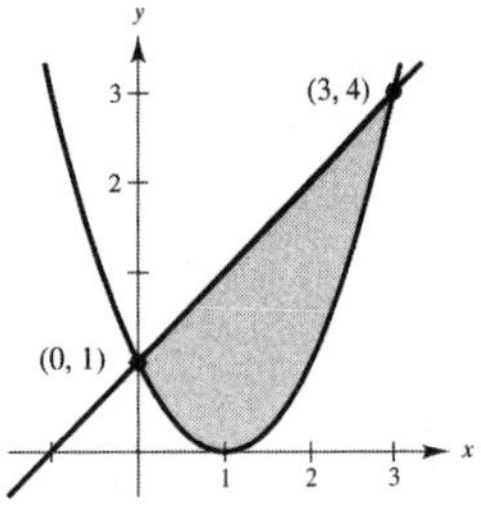

14. $f(x) = 2 - \frac{1}{2}x$

$g(x) = 2 - \sqrt{x}$

$A \approx 1$

Matches (a)

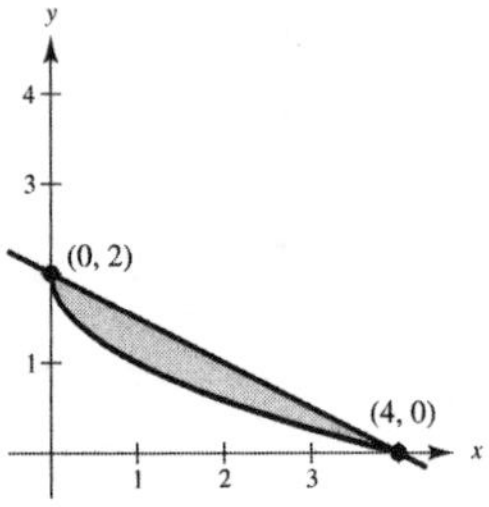

15. (a)

$$x = 4 - y^2$$
$$x = y - 2$$
$$4 - y^2 = y - 2$$
$$y^2 + y - 6 = 0$$
$$(y + 3)(y - 2) = 0$$

Intersection points: $(0, 2)$ and $(-5, -3)$

$$A = \int_{-5}^{0}\left[(x + 2) + \sqrt{4 - x}\, dx + \int_{0}^{4} 2\sqrt{4 - x}\, dx\right.$$
$$= \frac{61}{6} + \frac{32}{3} = \frac{125}{6}$$

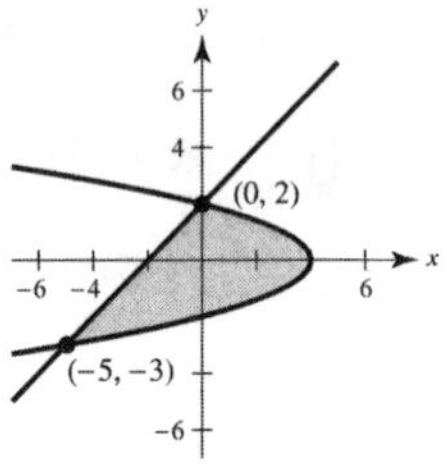

(b) $A = \int_{-3}^{2}\left[(4 - y^2) - (y - 2)\right] dy = \frac{125}{6}$

(c) The second method is simpler. Explanations will vary.

16. (a) $y = x^2$ and $y = 6 - x$

$x^2 = 6 - x \Rightarrow x^2 + x - 6 = 0 \Rightarrow (x + 3)(x - 2) = 0$

Intersection points: $(2, 4)$ and $(-3, 9)$

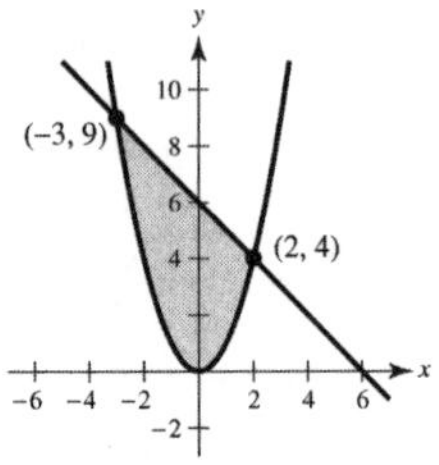

$$A = \int_{-3}^{2}\left[(6 - x) - x^2\right] dx = \frac{125}{6}$$

(b) $A = \int_{0}^{4} 2\sqrt{y}\, dy + \int_{4}^{9}\left[(6 - y) + \sqrt{y}\, dy\right. = \frac{32}{3} + \frac{61}{6} = \frac{125}{6}$

(c) The first method is simpler. Explanations will vary.

17.

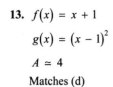

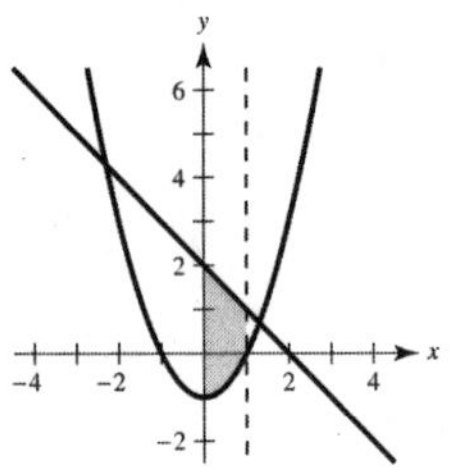

$$A = \int_{0}^{1}\left[(-x + 2) - (x^2 - 1)\right] dx$$
$$= \int_{0}^{1}(-x^2 - x + 3)\, dx$$
$$= \left[\frac{-x^3}{3} - \frac{x^2}{2} + 3x\right]_{0}^{1}$$
$$= \left(-\frac{1}{3} - \frac{1}{2} + 3\right) - 0 = \frac{13}{6}$$

18.

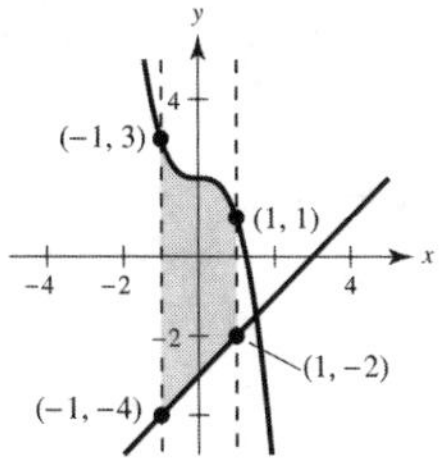

$$A = \int_{-1}^{1} \left[\left(-x^3 + 2\right) - (x - 3)\right] dx$$
$$= \int_{-1}^{1} \left(-x^3 - x + 5\right) dx$$
$$= \left[\frac{-x^4}{4} - \frac{x^2}{2} + 5x\right]_{-1}^{1}$$
$$= \left(-\frac{1}{4} - \frac{1}{2} + 5\right) - \left(-\frac{1}{4} - \frac{1}{2} - 5\right) = 10$$

19. The points of intersection are given by:

$$x^2 + 2x = x + 2$$
$$x^2 + x - 2 = 0$$
$$(x + 2)(x - 1) = 0 \text{ when } x = -2, 1$$

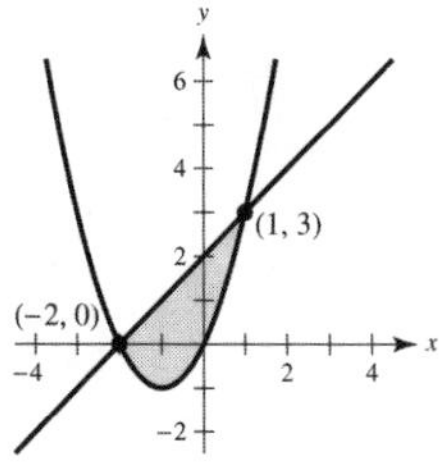

$$A = \int_{-2}^{1} \left[g(x) - f(x)\right] dx$$
$$= \int_{-2}^{1} \left[(x + 2) - \left(x^2 + 2x\right)\right] dx$$
$$= \left[\frac{-x^3}{3} - \frac{x^2}{2} + 2x\right]_{-2}^{1}$$
$$= \left(-\frac{1}{3} - \frac{1}{2} + 2\right) - \left(\frac{8}{3} - 2 - 4\right) = \frac{9}{2}$$

20. The points of intersection are given by:

$$-x^2 + 3x + 1 = -x + 1$$
$$-x^2 + 4x = 0$$
$$x(4 - x) = 0 \text{ when } x = 0, 4$$

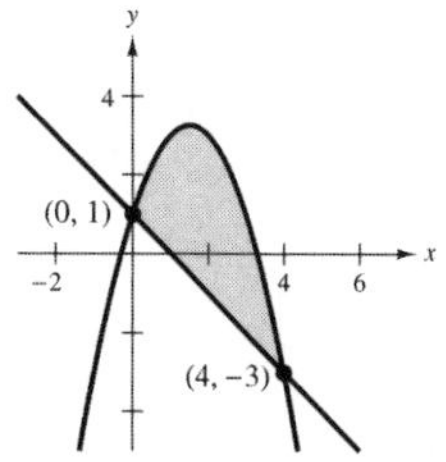

$$A = \int_0^4 \left[\left(-x^2 + 3x + 1\right) - (1 - x)\right] dx$$
$$= \int_0^4 \left(-x^2 + 4x\right) dx$$
$$= \left[\frac{-x^3}{3} + 2x^2\right]_0^4$$
$$= -\frac{64}{3} + 32 = \frac{32}{3}$$

21. The points of intersection are given by:

$x = 2 - x$ and $x = 0$ and $2 - x = 0$

$x = 1$ $\quad x = 0$ $\quad x = 2$

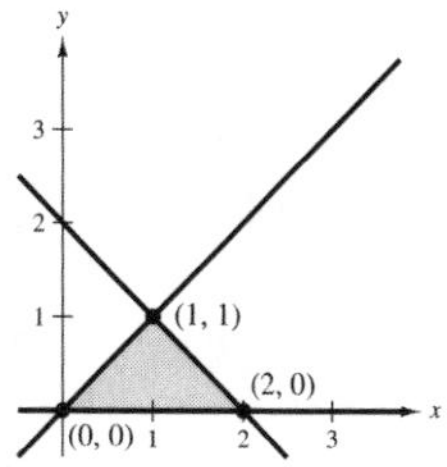

$$A = \int_0^1 \left[(2 - y) - (y)\right] dy = \left[2y - y^2\right]_0^1 = 1$$

Note that if you integrate with respect to x, you need two integrals. Also, note that the region is a triangle.

22.

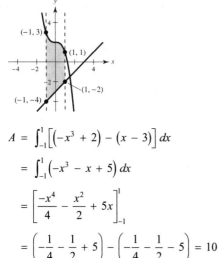

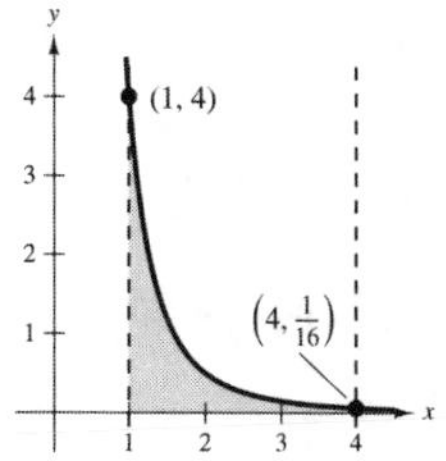

$$A = \int_1^4 \frac{4}{x^3}\, dx = \int_1^4 4x^{-3}\, dx$$
$$= \left[-2x^{-2}\right]_1^4$$
$$= \left[\frac{-2}{x^2}\right]_1^4$$
$$= -\frac{2}{16} + 2 = \frac{15}{8}$$

23. The points of intersection are given by:

$$\sqrt{x} + 3 = \frac{1}{2}x + 3$$
$$\sqrt{x} = \frac{1}{2}x$$
$$x = \frac{x^2}{4} \quad \text{when } x = 0, 4$$

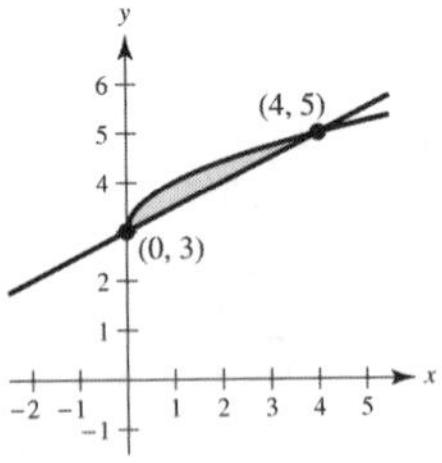

$$A = \int_0^4 \left[\left(\sqrt{x} + 3\right) - \left(\frac{1}{2}x + 3\right)\right] dx$$
$$= \left[\frac{2}{3}x^{3/2} - \frac{x^2}{4}\right]_0^4 = \frac{16}{3} - 4 = \frac{4}{3}$$

24. The points of intersection are given by:

$$\sqrt[3]{x-1} = x - 1$$
$$x - 1 = (x-1)^3 = x^3 - 3x^2 + 3x - 1$$
$$x^3 - 3x^2 + 2x = 0$$
$$x\left(x^2 - 3x + 2\right) = 0$$
$$x(x-2)(x-1) = 0 \quad \text{when } x = 0, 1, 2$$

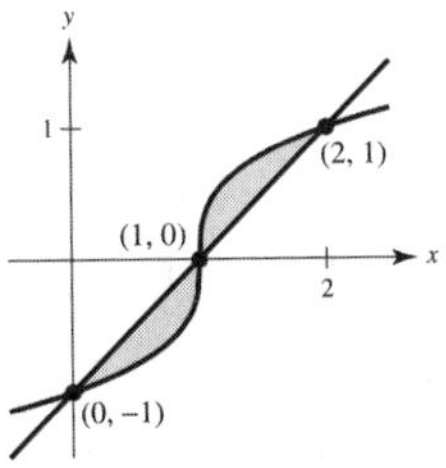

$$A = 2\int_0^1 \left[(x-1) - \sqrt[3]{x-1}\right] dx$$
$$= 2\left[\frac{x^2}{2} - x - \frac{3}{4}(x-1)^{4/3}\right]_0^1$$
$$= 2\left[\left(\frac{1}{2} - 1 - 0\right) - \left(-\frac{3}{4}\right)\right] = \frac{1}{2}$$

25. The points of intersection are given by:

$$y^2 = y + 2$$
$$(y-2)(y+1) = 0 \quad \text{when } y = -1, 2$$

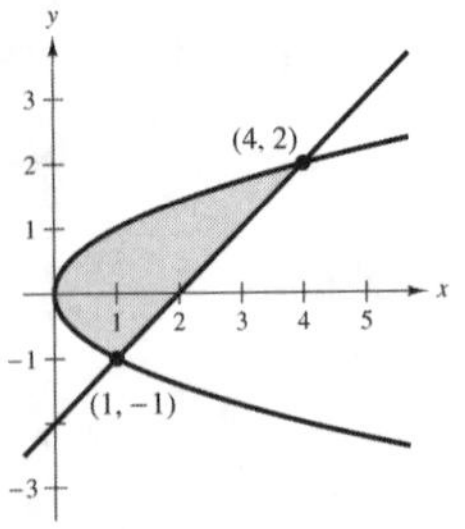

$$A = \int_{-1}^{2} \left[g(y) - f(y)\right] dy$$
$$= \int_{-1}^{2} \left[(y+2) - y^2\right] dy$$
$$= \left[2y + \frac{y^2}{2} - \frac{y^3}{3}\right]_{-1}^{2} = \frac{9}{2}$$

26. The points of intersection are given by:

$$2y - y^2 = -y$$
$$y(y-3) = 0 \quad \text{when } y = 0, 3$$

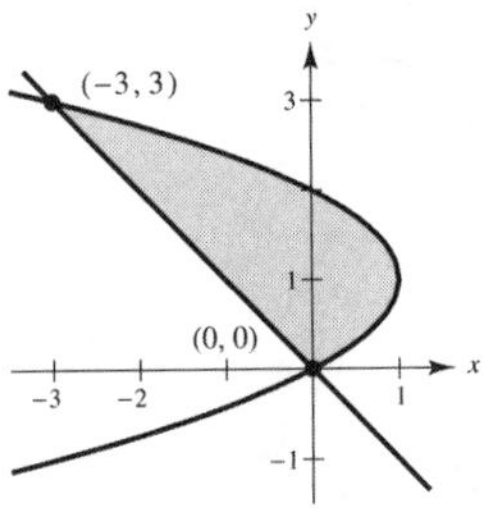

$$A = \int_0^3 \left[f(y) - g(y)\right] dy$$
$$= \int_0^3 \left[\left(2y - y^2\right) - (-y)\right] dy$$
$$= \int_0^3 \left(3y - y^2\right) dy$$
$$= \left[\tfrac{3}{2}y^2 - \tfrac{1}{3}y^3\right]_0^3 = \tfrac{9}{2}$$

27.

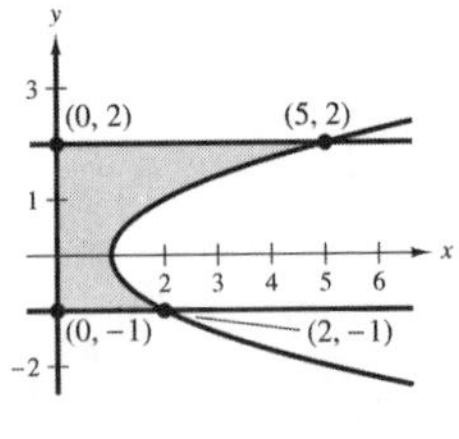

$$A = \int_{-1}^{2} \left[f(y) - g(y)\right] dy$$
$$= \int_{-1}^{2} \left[\left(y^2 + 1\right) - 0\right] dy$$
$$= \left[\frac{y^3}{3} + y\right]_{-1}^{2} = 6$$

28.

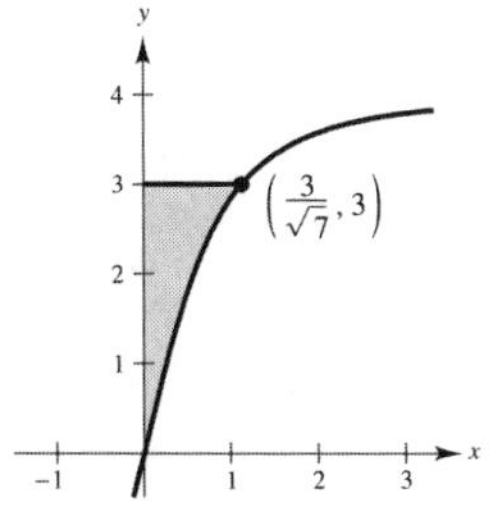

$$A = \int_0^3 [f(y) - g(y)]\,dy$$

$$= \int_0^3 \left[\frac{y}{\sqrt{16 - y^2}} - 0\right] dy$$

$$= -\frac{1}{2}\int_0^3 (16 - y^2)^{-1/2}(-2y)\,dy$$

$$= \left[-\sqrt{16 - y^2}\right]_0^3 = 4 - \sqrt{7} \approx 1.354$$

29. $y = \dfrac{10}{x} \Rightarrow x = \dfrac{10}{y}$

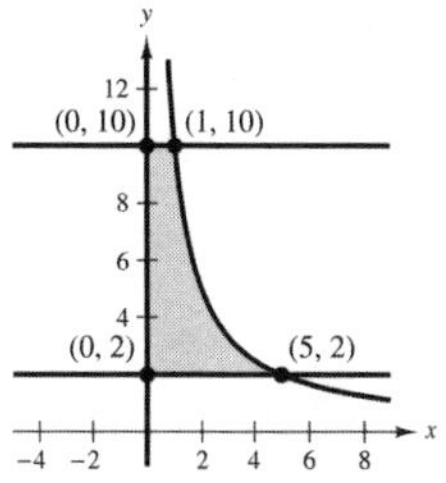

$$A = \int_2^{10} \frac{10}{y}\,dy$$

$$= \left[10 \ln y\right]_2^{10}$$

$$= 10(\ln 10 - \ln 2)$$

$$= 10 \ln 5 \approx 16.0944$$

30. The point of intersection is given by:

$$\frac{4}{2 - x} = 4$$

$$\frac{4}{2 - x} - 4 = 0 \quad \text{when } x = 1$$

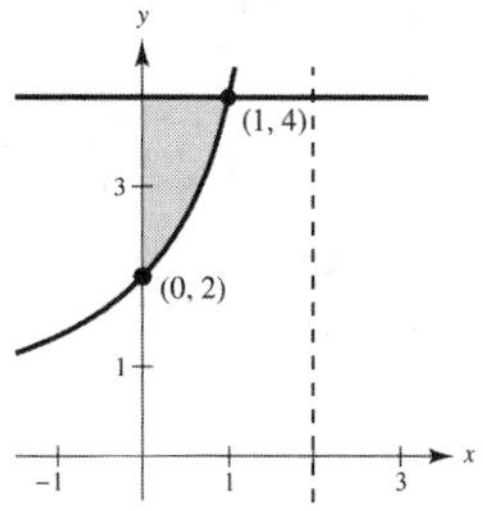

$$A = \int_0^1 \left(4 - \frac{4}{2 - x}\right) dx$$

$$= \left[4x + 4 \ln|2 - x|\right]_0^1$$

$$= 4 - 4 \ln 2$$

$$\approx 1.227$$

31. (a)

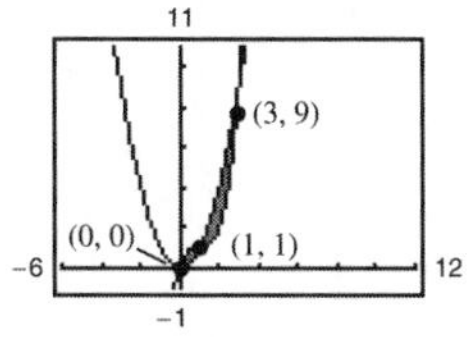

(b) The points of intersection are given by:

$$x^3 - 3x^2 + 3x = x^2$$

$$x(x - 1)(x - 3) = 0 \quad \text{when } x = 0, 1, 3$$

$$A = \int_0^1 [f(x) - g(x)]\,dx + \int_1^3 [g(x) - f(x)]\,dx$$

$$= \int_0^1 \left[(x^3 - 3x^2 + 3x) - x^2\right] dx + \int_1^3 \left[x^2 - (x^3 - 3x^2 + 3x)\right] dx$$

$$= \int_0^1 (x^3 - 4x^2 + 3x)\,dx + \int_1^3 (-x^3 + 4x^2 - 3x)\,dx = \left[\frac{x^4}{4} - \frac{4}{3}x^3 + \frac{3}{2}x^2\right]_0^1 + \left[\frac{-x^4}{4} + \frac{4}{3}x^3 - \frac{3}{2}x^2\right]_1^3 = \frac{5}{12} + \frac{8}{3} = \frac{37}{12}$$

(c) Numerical approximation: $0.417 + 2.667 \approx 3.083$

32. (a)

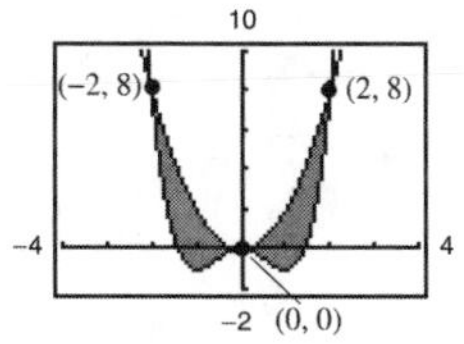

(b) The points of intersection are given by:

$$x^4 - 2x^2 = 2x^2$$

$$x^2(x^2 - 4) = 0 \quad \text{when } x = 0, \pm 2$$

$$A = 2\int_0^2 \left[2x^2 - (x^4 - 2x^2)\right] dx = 2\int_0^2 (4x^2 - x^4)\,dx = 2\left[\frac{4x^3}{3} - \frac{x^5}{5}\right]_0^2 = \frac{128}{15}$$

(c) Numerical approximation: 8.533

33. (a) $f(x) = x^4 - 4x^2, \quad g(x) = x^2 - 4$

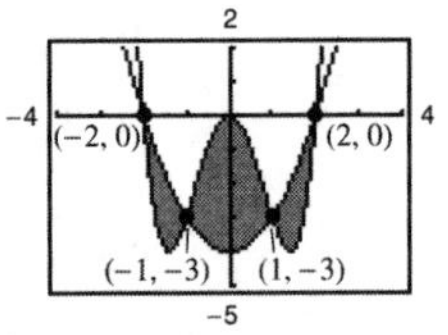

(b) The points of intersection are given by:

$$x^4 - 4x^2 = x^2 - 4$$
$$x^4 - 5x^2 + 4 = 0$$
$$(x^2 - 4)(x^2 - 1) = 0 \quad \text{when } x = \pm 2, \pm 1$$

By symmetry:

$$A = 2\int_0^1 \left[(x^4 - 4x^2) - (x^2 - 4)\right] dx + 2\int_1^2 \left[(x^2 - 4) - (x^4 - 4x^2)\right] dx$$
$$= 2\int_0^1 (x^4 - 5x^2 + 4)\, dx + 2\int_1^2 (-x^4 + 5x^2 - 4)\, dx$$
$$= 2\left[\frac{x^5}{5} - \frac{5x^3}{3} + 4x\right]_0^1 + 2\left[-\frac{x^5}{5} + \frac{5x^3}{3} - 4x\right]_1^2$$
$$= 2\left[\frac{1}{5} - \frac{5}{3} + 4\right] + 2\left[\left(-\frac{32}{5} + \frac{40}{3} - 8\right) - \left(-\frac{1}{5} + \frac{5}{3} - 4\right)\right] = 8$$

(c) Numerical approximation:

$5.067 + 2.933 = 8.0$

34. (a)

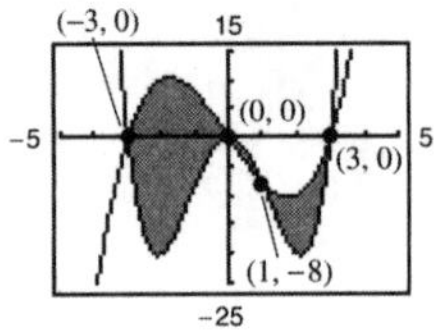

(b) The points of intersection are given by:

$$x^4 - 9x^2 = x^3 - 9x$$
$$x^4 - x^3 - 9x^2 + 9x = 0$$
$$x(x - 3)(x - 1)(x + 3) = 0 \quad \text{when } x = -3, 0, 1, 3$$

$$A = \int_{-3}^0 \left[(x^3 - 9x) - (x^4 - 9x^2)\right] dx + \int_0^1 \left[(x^4 - 9x^2) - (x^3 - 9x)\right] dx + \int_1^3 \left[(x^3 - 9x) - (x^4 - 9x^2)\right] dx$$
$$= \left[\frac{x^4}{4} - \frac{9x^2}{2} - \frac{x^5}{5} + 3x^3\right]_{-3}^0 + \left[\frac{x^5}{5} - 3x^3 - \frac{x^4}{4} + \frac{9x^2}{2}\right]_0^1 + \left[\frac{x^4}{4} - \frac{9x^2}{2} - \frac{x^5}{5} + 3x^3\right]_1^3$$
$$= \frac{1053}{20} + \frac{29}{20} + \frac{68}{5} = \frac{677}{10}$$

(c) Numerical approximation: 67.7

35. (a)

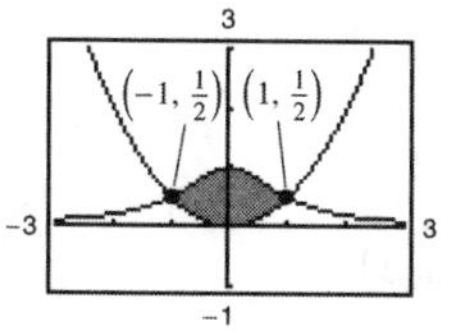

(b) The points of intersection are given by:

$$\frac{1}{1 + x^2} = \frac{x^2}{2}$$

$$x^4 + x^2 - 2 = 0$$

$$\left(x^2 + 2\right)\left(x^2 - 1\right) = 0 \quad \text{when } x = \pm 1$$

$$A = 2\int_0^1 \left[f(x) - g(x)\right] dx$$

$$= 2\int_0^1 \left[\frac{1}{1 + x^2} - \frac{x^2}{2}\right] dx$$

$$= 2\left[\arctan x - \frac{x^3}{6}\right]_0^1$$

$$= 2\left(\frac{\pi}{4} - \frac{1}{6}\right) = \frac{\pi}{2} - \frac{1}{3} \approx 1.237$$

(c) Numerical approximation: 1.237

36. (a)

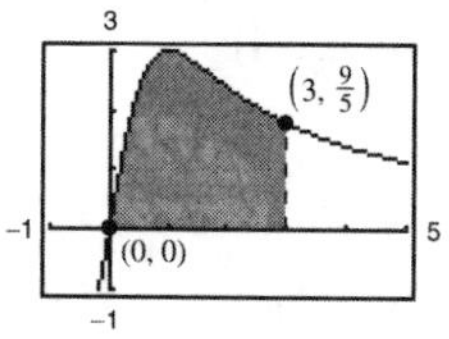

(b) $A = \displaystyle\int_0^3 \left[\frac{6x}{x^2 + 1} - 0\right] dx$

$$= \left[3 \ln\left(x^2 + 1\right)\right]_0^3$$

$$= 3 \ln 10$$

$$\approx 6.908$$

(c) Numerical approximation: 6.908

37. $A = \displaystyle\int_0^{2\pi} \left[(2 - \cos x) - \cos x\right] dx$

$$= 2\int_0^{2\pi} (1 - \cos x)\, dx$$

$$= 2\left[x - \sin x\right]_0^{2\pi} = 4\pi \approx 12.566$$

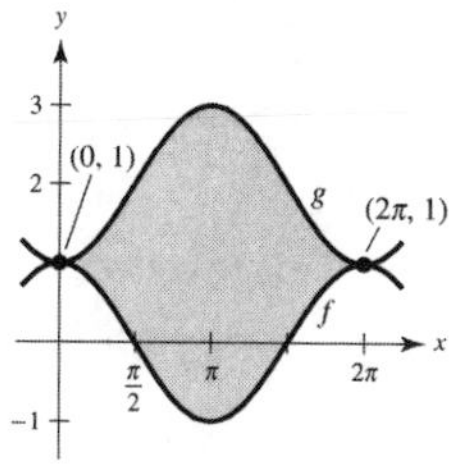

38. $A = \displaystyle\int_{-\pi/2}^{\pi/6} (\cos 2x - \sin x)\, dx$

$$= \left[\frac{1}{2}\sin 2x + \cos x\right]_{-\pi/2}^{\pi/6}$$

$$= \left(\frac{\sqrt{3}}{4} + \frac{\sqrt{3}}{2}\right) - (0) = \frac{3\sqrt{3}}{4} \approx 1.299$$

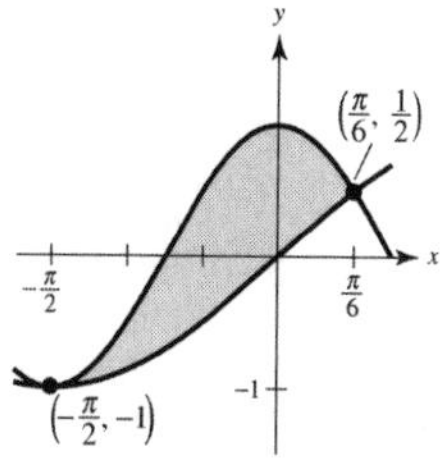

39. $A = 2\displaystyle\int_0^{\pi/3} \left[f(x) - g(x)\right] dx$

$$= 2\int_0^{\pi/3} (2\sin x - \tan x)\, dx$$

$$= 2\left[-2\cos x + \ln\left|\cos x\right|\right]_0^{\pi/3} = 2(1 - \ln 2) \approx 0.614$$

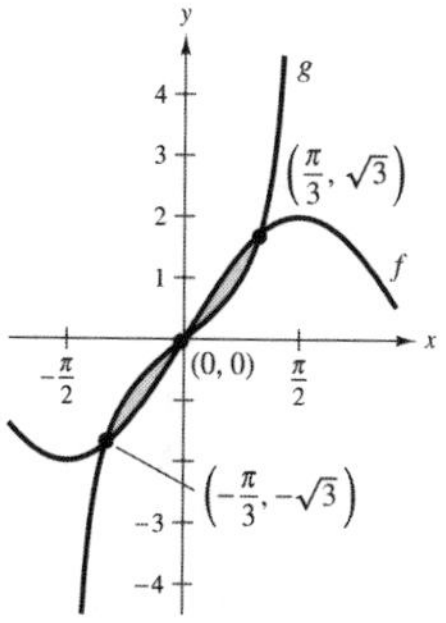

40. $A = \displaystyle\int_0^1 \left[\left(\sqrt{2} - 4\right)x + 4 - \sec\frac{\pi x}{4}\tan\frac{\pi x}{4}\right] dx$

$$= \left[\frac{\sqrt{2} - 4}{2}x^2 + 4x - \frac{4}{\pi}\sec\frac{\pi x}{4}\right]_0^1$$

$$= \left(\frac{\sqrt{2} - 4}{2} + 4 - \frac{4}{\pi}\sqrt{2}\right) - \left(-\frac{4}{\pi}\right)$$

$$= \frac{\sqrt{2}}{2} + 2 + \frac{4}{\pi}\left(1 - \sqrt{2}\right) \approx 2.1797$$

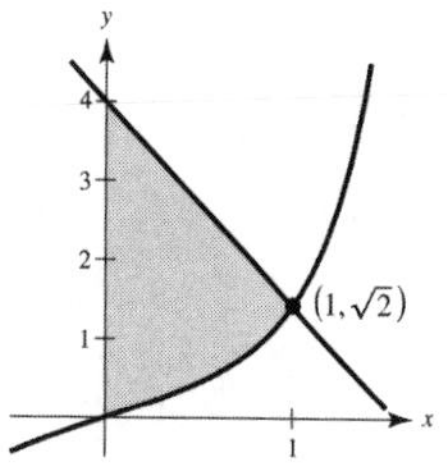

41. $A = \int_0^1 \left[xe^{-x^2} - 0 \right] dx$

$= \left[-\frac{1}{2}e^{-x^2} \right]_0^1 = \frac{1}{2}\left(1 - \frac{1}{e}\right) \approx 0.316$

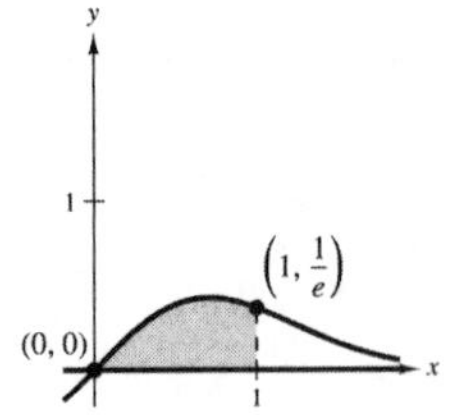

42. From the graph, f and g intersect at $x = 0$ and $x = 2$.

$A = \int_0^2 \left[\left(\frac{3}{2}x + 1 \right) - 2^x \right] dx$

$= \left[\frac{3x^2}{4} + x - \frac{2^x}{\ln 2} \right]_0^2$

$= \left(3 + 2 - \frac{4}{\ln 2} \right) + \frac{1}{\ln 2}$

$= 5 - \frac{3}{\ln 2} \approx 0.672$

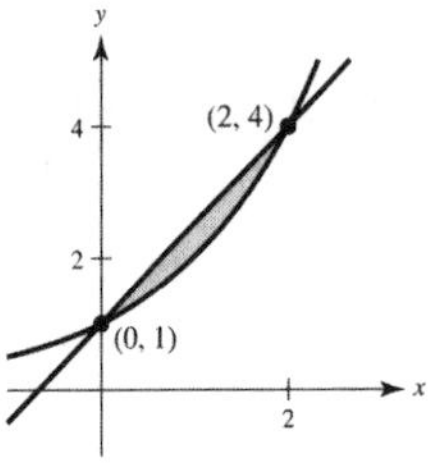

43. (a)

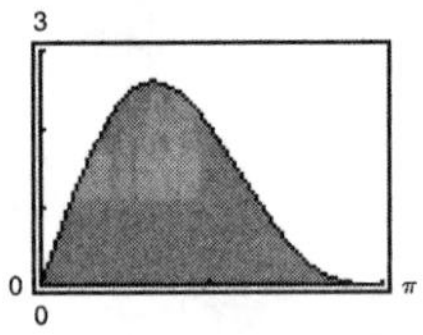

(b) $A = \int_0^{\pi} (2 \sin x + \sin 2x)\, dx$

$= \left[-2 \cos x - \frac{1}{2} \cos 2x \right]_0^{\pi}$

$= \left(2 - \frac{1}{2}\right) - \left(-2 - \frac{1}{2}\right) = 4$

(c) Numerical approximation: 4.0

44. (a)

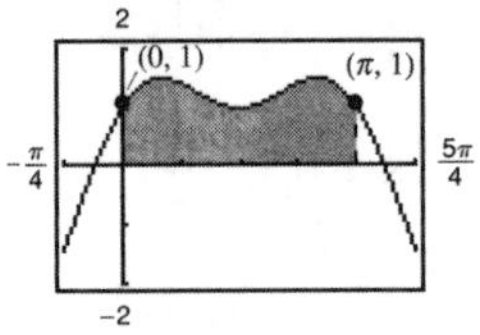

(b) $A = \int_0^{\pi} (2 \sin x + \cos 2x)\, dx$

$= \left[-2 \cos x + \frac{1}{2} \sin 2x \right]_0^{\pi} = 4$

(c) Numerical approximation: 4

45. (a)

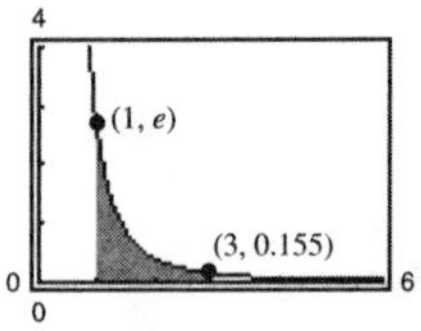

(b) $A = \int_1^3 \frac{1}{x^2} e^{1/x}\, dx$

$= \left[-e^{1/x} \right]_1^3$

$= e - e^{1/3}$

(c) Numerical approximation: 1.323

46. (a)

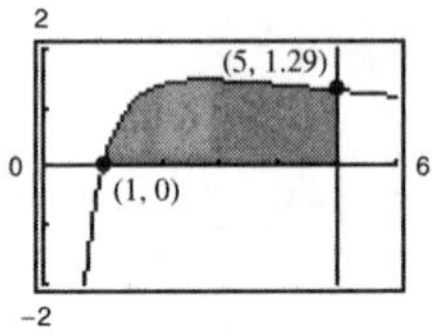

(b) $A = \int_1^5 \frac{4 \ln x}{x}\, dx$

$= \left[2(\ln x)^2 \right]_1^5$

$= 2(\ln 5)^2$

(c) Numerical approximation: 5.181

47. (a)

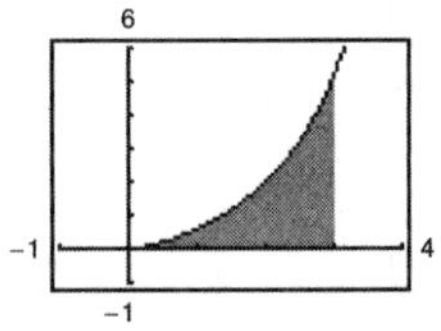

(b) The integral

$A = \int_0^3 \sqrt{\frac{x^3}{4 - x}}\, dx$

does not have an elementary antiderivative.

(c) $A \approx 4.7721$

48. (a)

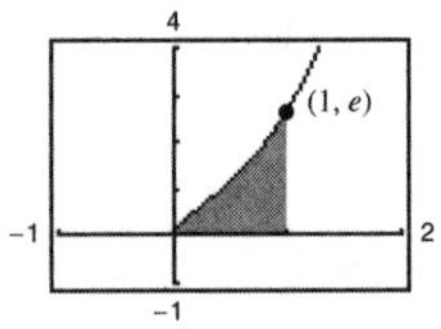

(b) The integral

$A = \int_0^1 \sqrt{x} e^x\, dx$

does not have an elementary antiderivative.

(c) 1.2556

49. (a)

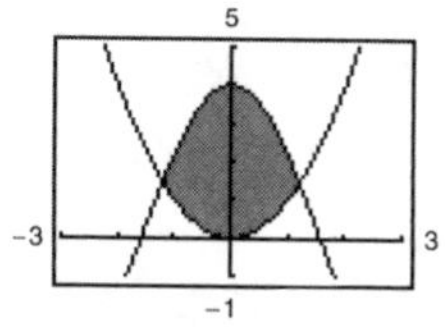

(b) The intersection points are difficult to determine by hand.

(c) Area $= \int_{-c}^{c}\left[4\cos x - x^2\right]dx \approx 6.3043$ where $c \approx 1.201538$.

50. (a)

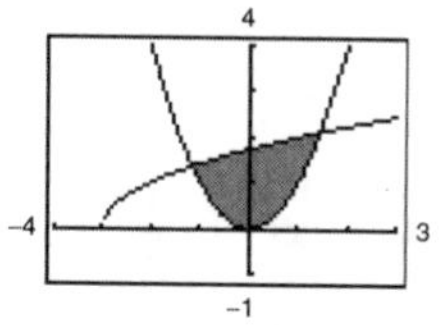

(b) The intersection points are difficult to determine.

(c) Intersection points: $(-1.164035, 1.3549778)$ and $(1.4526269, 2.1101248)$

$$A = \int_{-1.164035}^{1.4526269}\left[\sqrt{3+x} - x^2\right]dx \approx 3.0578$$

51. $F(x) = \int_0^x\left(\frac{1}{2}t + 1\right)dt = \left[\frac{t^2}{4} + t\right]_0^x = \frac{x^2}{4} + x$

(a) $F(0) = 0$

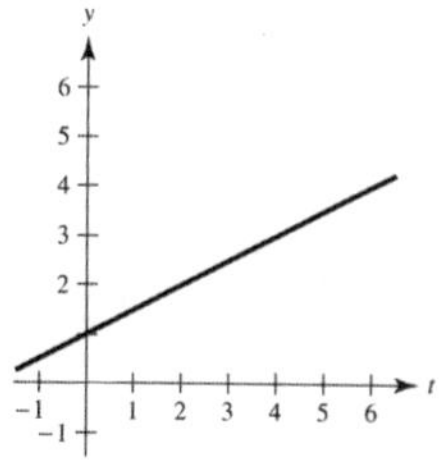

(b) $F(2) = \frac{2^2}{4} + 2 = 3$

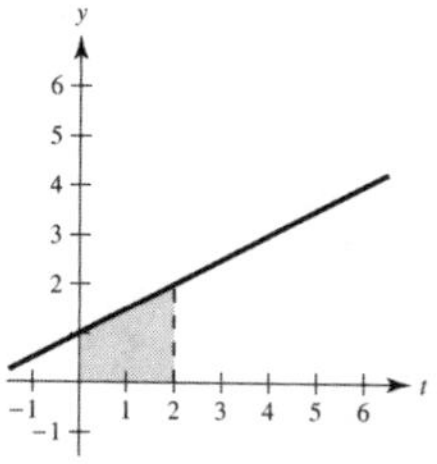

(c) $F(6) = \frac{6^2}{4} + 6 = 15$

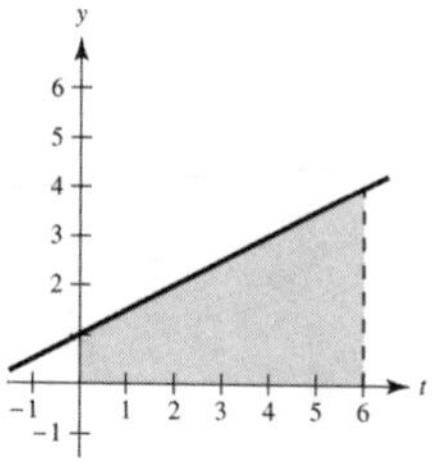

52. $F(x) = \int_0^x\left(\frac{1}{2}t^2 + 2\right)dt = \left[\frac{1}{6}t^3 + 2t\right]_0^x = \frac{x^3}{6} + 2x$

(a) $F(0) = 0$

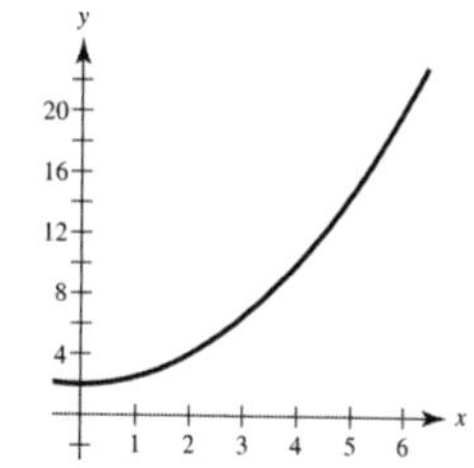

(b) $F(4) = \frac{4^3}{6} + 2(4) = \frac{56}{3}$

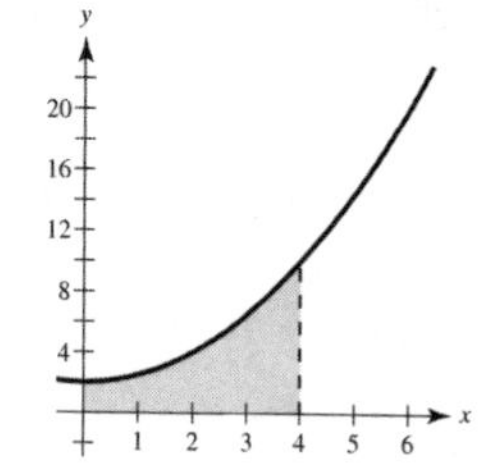

(c) $F(6) = 36 + 12 = 48$

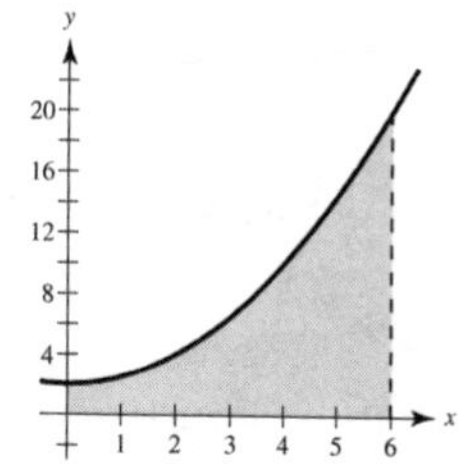

53. $F(\alpha) = \int_{-1}^{\alpha} \cos\frac{\pi\theta}{2}\,d\theta = \left[\frac{2}{\pi}\sin\frac{\pi\theta}{2}\right]_{-1}^{\alpha} = \frac{2}{\pi}\sin\frac{\pi\alpha}{2} + \frac{2}{\pi}$

(a) $F(-1) = 0$

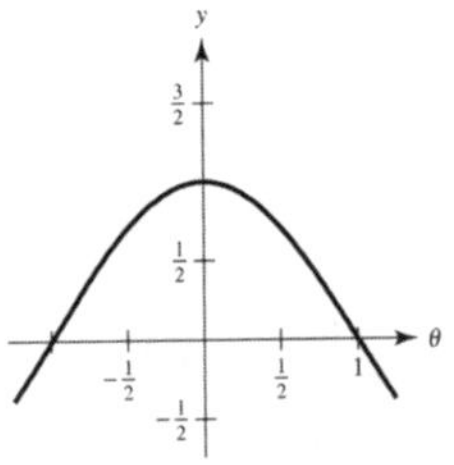

(b) $F(0) = \frac{2}{\pi} \approx 0.6366$

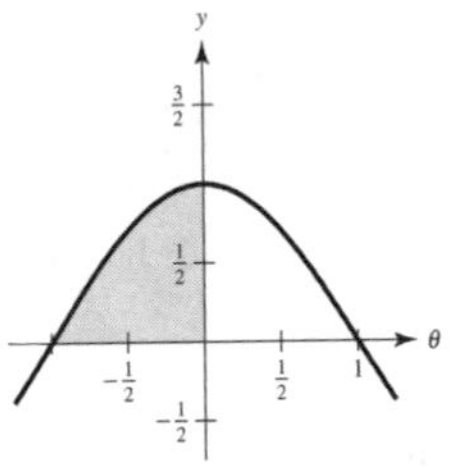

(c) $F\left(\frac{1}{2}\right) = \frac{2 + \sqrt{2}}{\pi} \approx 1.0868$

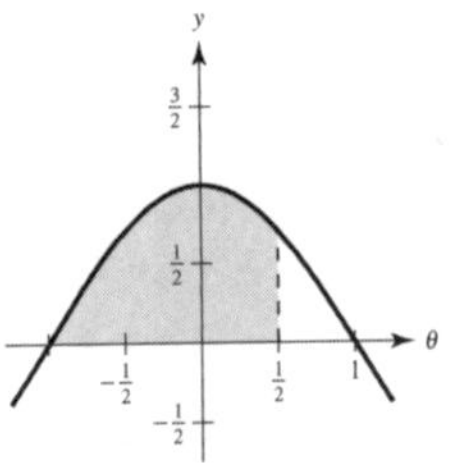

54. $F(y) = \int_{-1}^{y} 4e^{x/2}\,dx = \left[8e^{x/2}\right]_{-1}^{y} = 8e^{y/2} - 8e^{-1/2}$

(a) $F(-1) = 0$

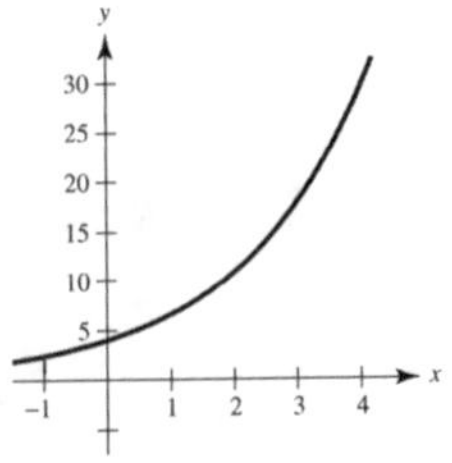

(b) $F(0) = 8 - 8e^{-1/2} \approx 3.1478$

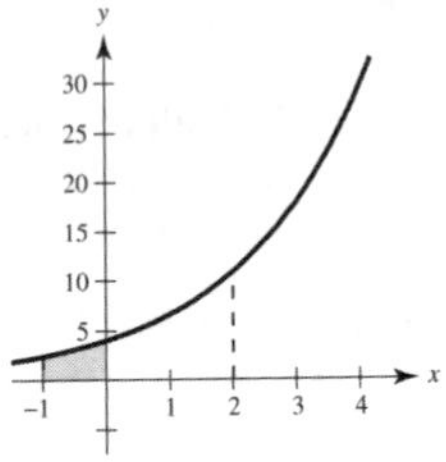

(c) $F(4) = 8e^2 - 8e^{-1/2} \approx 54.2602$

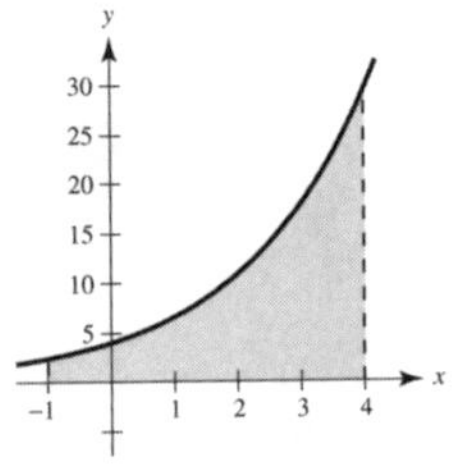

55. $A = \int_2^4 \left[\left(\frac{9}{2}x - 12\right) - (x - 5)\right]dx + \int_4^6 \left[\left(-\frac{5}{2}x + 16\right) - (x - 5)\right]dx$

$= \int_2^4 \left(\frac{7}{2}x - 7\right)dx + \int_4^6 \left(-\frac{7}{2}x + 21\right)dx = \left[\frac{7}{4}x^2 - 7x\right]_2^4 + \left[-\frac{7}{4}x^2 + 21x\right]_4^6 = 7 + 7 = 14$

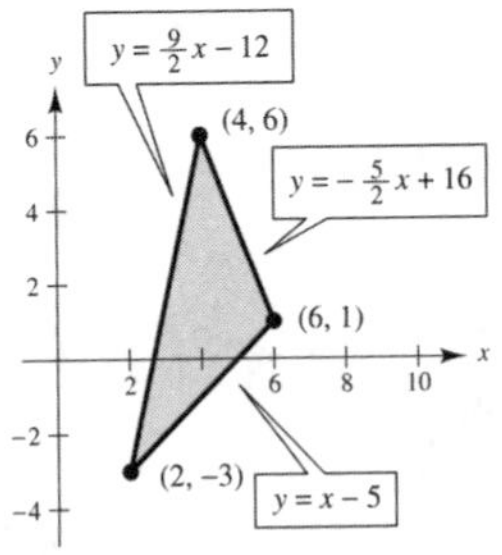

56. $A = \int_0^4 \frac{3}{4}x\,dx + \int_4^6 \left(9 - \frac{3}{2}x\right)dx$

$= \left[\frac{3x^2}{8}\right]_0^4 + \left[9x - \frac{3x^2}{4}\right]_4^6$

$= 6 + (54 - 27) - (36 - 12)$

$= 6 + 3 = 9$

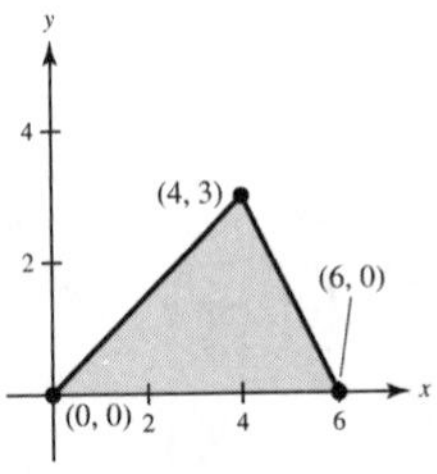

57.

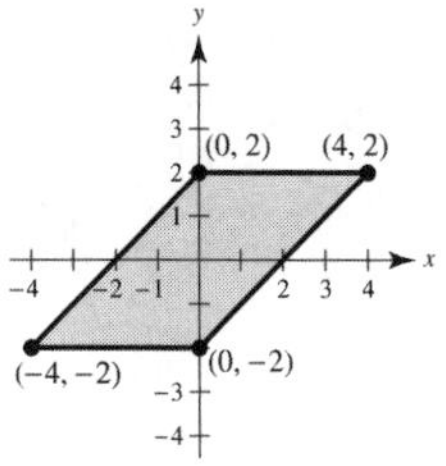

Left boundary line: $y = x + 2 \Leftrightarrow x = y - 2$

Right boundary line: $y = x - 2 \Leftrightarrow x = y + 2$

$$A = \int_{-2}^{2} \left[(y + 2) - (y - 2)\right] dy$$
$$= \int_{-2}^{2} 4\, dy = \left[4y\right]_{-2}^{2} = 8 - (-8) = 16$$

58. $A = \int_0^1 \left[2x - (-3x)\right] dx + \int_1^3 \left[(-2x + 4) - \left(\frac{1}{2}x - \frac{7}{2}\right)\right] dx$

$$= \int_0^1 5x\, dx + \int_1^3 \left(-\frac{5}{2}x + \frac{15}{2}\right) dx$$
$$= \left[\frac{5x^2}{2}\right]_0^1 + \left[-\frac{5x^2}{4} + \frac{15}{2}x\right]_1^3$$
$$= \frac{5}{2} + \left(-\frac{45}{4} + \frac{45}{2} + \frac{5}{4} - \frac{15}{2}\right)$$
$$= \frac{15}{2}$$

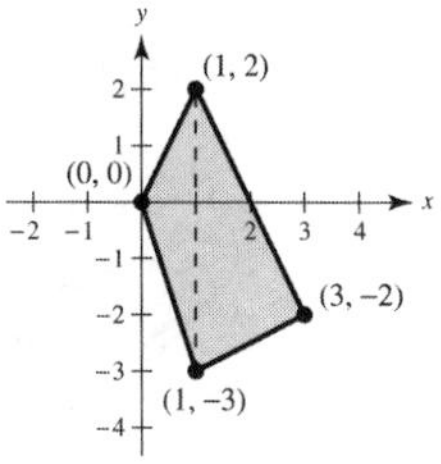

59. Answers will vary. *Sample answer*: If you let $\Delta x = 6$ and $n = 10$, $b - a = 10(6) = 60$.

(a) Area $\approx \frac{60}{2(10)}\left[0 + 2(14) + 2(14) + 2(12) + 2(12) + 2(15) + 2(20) + 2(23) + 2(25) + 2(26) + 0\right] = 3[322] = 966 \text{ ft}^2$

(b) Area $\approx \frac{60}{3(10)}\left[0 + 4(14) + 2(14) + 4(12) + 2(12) + 4(15) + 2(20) + 4(23) + 2(25) + 4(26) + 0\right] = 2[502] = 1004 \text{ ft}^2$

60. Answers will vary. *Sample answer*: $\Delta x = 4$, $n = 8$, $b - a = (8)(4) = 32$

(a) Area $\approx \frac{32}{2(8)}\left[0 + 2(11) + 2(13.5) + 2(14.2) + 2(14) + 2(14.2) + 2(15) + 2(13.5) + 0\right]$

$= 2[190.8]$

$= 381.6 \text{ mi}^2$

(b) Area $\approx \frac{32}{3(8)}\left[0 + 4(11) + 2(13.5) + 4(14.2) + 2(14) + 4(14.2) + 2(15) + 4(13.5) + 0\right]$

$= \frac{4}{3}[296.6]$

$= 395.5 \text{ mi}^2$

61. $f(x) = x^3$

$f'(x) = 3x^2$

At $(1, 1)$, $f'(1) = 3$.

Tangent line: $y - 1 = 3(x - 1)$ or $y = 3x - 2$

The tangent line intersects $f(x) = x^3$ at $x = -2$.

$$A = \int_{-2}^{1} \left[x^3 - (3x - 2)\right] dx = \left[\frac{x^4}{4} - \frac{3x^2}{2} + 2x\right]_{-2}^{1} = \frac{27}{4}$$

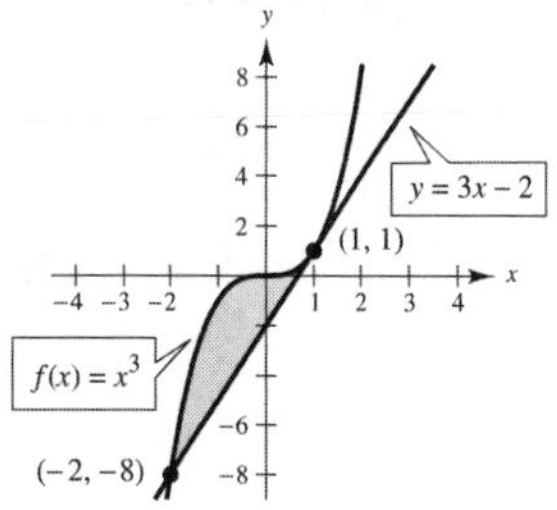

62. $y = x^3 - 2x, \quad (-1, 1)$

$y' = 3x^2 - 2$

$y'(-1) = 3 - 2 = 1$

Tangent line: $y - 1 = 1(x + 1) \Rightarrow y = x + 2$

Intersection points: $(-1, 1)$ and $(2, 4)$

$$A = \int_{-1}^{2} \left[(x + 2) - (x^3 - 2x)\right] dx = \int_{-1}^{2} \left(-x^3 + 3x + 2\right) dx$$

$$= \left[-\frac{x^4}{4} + \frac{3x^2}{2} + 2x\right]_{-1}^{2} = \left[(-4 + 6 + 4) - \left(-\frac{1}{4} + \frac{3}{2} - 2\right)\right] = \frac{27}{4}$$

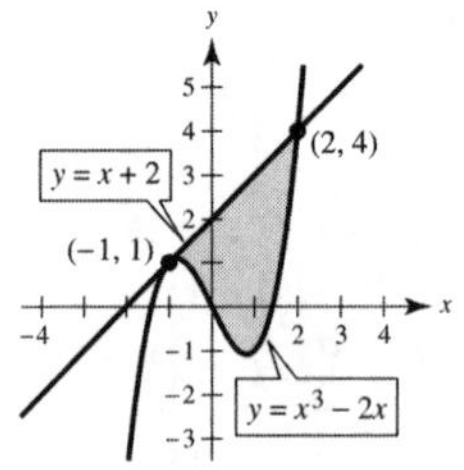

63. $f(x) = \dfrac{1}{x^2 + 1}$

$f'(x) = -\dfrac{2x}{\left(x^2 + 1\right)^2}$

At $\left(1, \frac{1}{2}\right)$, $f'(1) = -\frac{1}{2}$.

Tangent line: $y - \frac{1}{2} = -\frac{1}{2}(x - 1)$ or $y = -\frac{1}{2}x + 1$

The tangent line intersects $f(x) = \dfrac{1}{x^2 + 1}$ at $x = 0$.

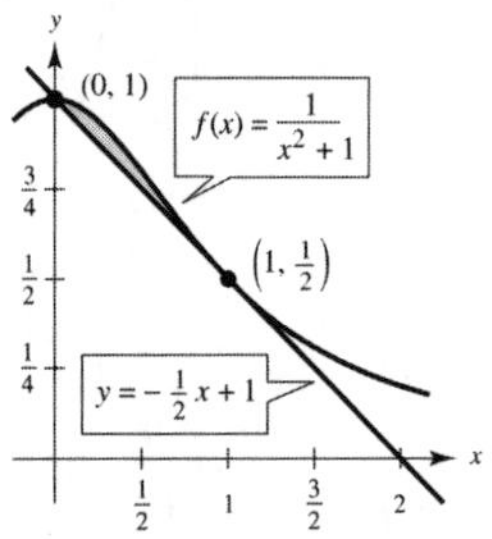

$$A = \int_0^1 \left[\frac{1}{x^2 + 1} - \left(-\frac{1}{2}x + 1\right)\right] dx = \left[\arctan x + \frac{x^2}{4} - x\right]_0^1 = \frac{\pi - 3}{4} \approx 0.0354$$

64. $y = \dfrac{2}{1 + 4x^2}, \quad \left(\dfrac{1}{2}, 1\right)$

$y' = \dfrac{-16x}{\left(1 + 4x^2\right)^2}$

$y'\left(\dfrac{1}{2}\right) = \dfrac{-8}{2^2} = -2$

Tangent line: $y - 1 = -2\left(x - \dfrac{1}{2}\right)$

$y = -2x + 2$

Intersection points: $\left(\dfrac{1}{2}, 1\right), (0, 2)$

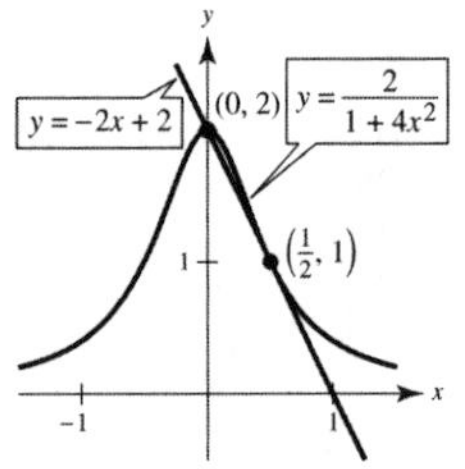

$$A = \int_0^{1/2} \left[\frac{2}{1 + 4x^2} - (-2x + 2)\right] dx = \left[\arctan(2x) + x^2 - 2x\right]_0^{1/2} = \arctan(1) + \frac{1}{4} - 1 = \frac{\pi}{4} - \frac{3}{4} \approx 0.0354$$

65. $x^4 - 2x^2 + 1 \le 1 - x^2$ on $[-1, 1]$

$$A = \int_{-1}^{1} \left[\left(1 - x^2\right) - \left(x^4 - 2x^2 + 1\right)\right] dx$$
$$= \int_{-1}^{1} \left(x^2 - x^4\right) dx$$
$$= \left[\frac{x^3}{3} - \frac{x^5}{5}\right]_{-1}^{1} = \frac{4}{15}$$

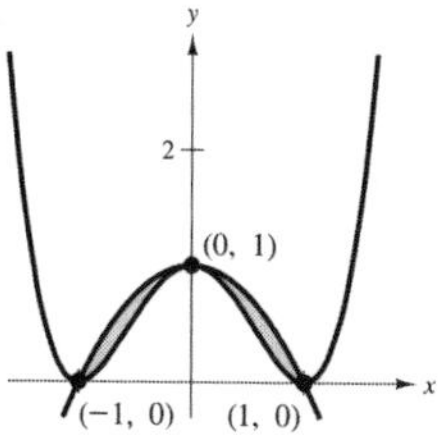

You can use a single integral because $x^4 - 2x^2 + 1 \le 1 - x^2$ on $[-1, 1]$.

66. $x^3 \ge x$ on $[-1, 0]$, $x^3 \le x$ on $[0, 1]$

Both functions symmetric to origin.

$$\int_{-1}^{0} \left(x^3 - x\right) dx = -\int_{0}^{1} \left(x^3 - x\right) dx$$

Thus, $\int_{-1}^{1} \left(x^3 - x\right) dx = 0.$

$$A = 2\int_{0}^{1} \left(x - x^3\right) dx = 2\left[\frac{x^2}{2} - \frac{x^4}{4}\right]_{0}^{1} = \frac{1}{2}$$

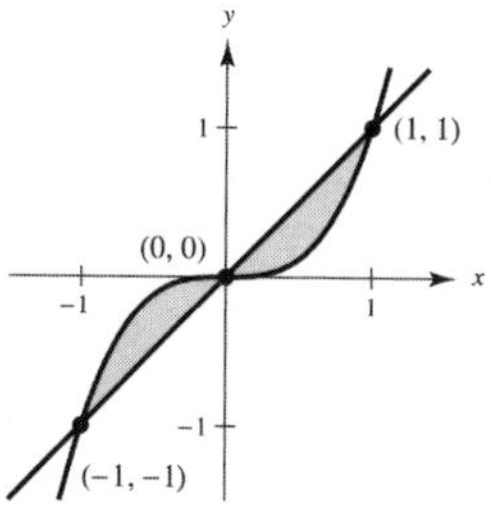

67. (a) $\int_{0}^{5} \left[v_1(t) - v_2(t)\right] dt = 10$ means that Car 1 traveled 10 more meters than Car 2 on the interval $0 \le t \le 5$.

$\int_{0}^{10} \left[v_1(t) - v_2(t)\right] dt = 30$ means that Car 1 traveled 30 more meters than Car 2 on the interval $0 \le t \le 10$.

$\int_{20}^{30} \left[v_1(t) - v_2(t)\right] dt = -5$ means that Car 2 traveled 5 more meters than Car 1 on the interval $20 \le t \le 30$.

(b) No, it is not possible because you do not know the initial distance between the cars.

(c) At $t = 10$, Car 1 is ahead by 30 meters.

(d) At $t = 20$, Car 1 is ahead of Car 2 by 13 meters. From part (a), at $t = 30$, Car 1 is ahead by $13 - 5 = 8$ meters.

68. (a) The area between the two curves represents the difference between the accumulated deficit under the two plans.

(b) Proposal 2 is better because the cumulative deficit (the area under the curve) is less.

69.
$$A = \int_{-3}^{3} \left(9 - x^2\right) dx = 36$$
$$\int_{-\sqrt{9-b}}^{\sqrt{9-b}} \left[\left(9 - x^2\right) - b\right] dx = 18$$
$$\int_{0}^{\sqrt{9-b}} \left[(9 - b) - x^2\right] dx = 9$$
$$\left[(9 - b)x - \frac{x^3}{3}\right]_{0}^{\sqrt{9-b}} = 9$$
$$\frac{2}{3}(9 - b)^{3/2} = 9$$
$$(9 - b)^{3/2} = \frac{27}{2}$$
$$9 - b = \frac{9}{\sqrt[3]{4}}$$
$$b = 9 - \frac{9}{\sqrt[3]{4}} \approx 3.330$$

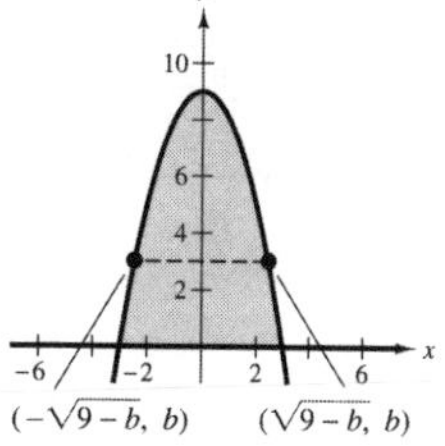

70. $A = 2\int_0^9 (9 - x)\,dx = 2\left[9x - \frac{x^2}{2}\right]_0^9 = 81$

$$2\int_0^{9-b} \left[(9 - x) - b\right] dx = \frac{81}{2}$$

$$2\int_0^{9-b} \left[(9 - b) - x\right] dx = \frac{81}{2}$$

$$2\left[(9 - b)x - \frac{x^2}{2}\right]_0^{9-b} = \frac{81}{2}$$

$$(9 - b)(9 - b) = \frac{81}{2}$$

$$9 - b = \frac{9}{\sqrt{2}}$$

$$b = 9 - \frac{9}{\sqrt{2}} \approx 2.636$$

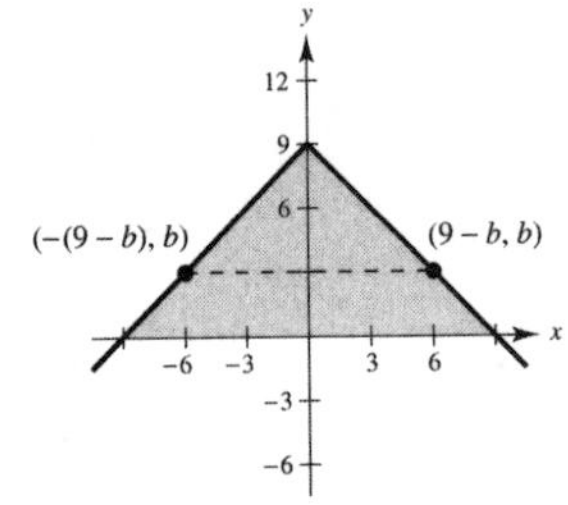

71. Area of triangle OAB is $\frac{1}{2}(4)(4) = 8$.

$$4 = \int_0^a (4 - x)\,dx = \left[4x - \frac{x^2}{2}\right]_0^a = 4a - \frac{a^2}{2}$$

$a^2 - 8a + 8 = 0$

$a = 4 \pm 2\sqrt{2}$

Because $0 < a < 4$, select $a = 4 - 2\sqrt{2} \approx 1.172$.

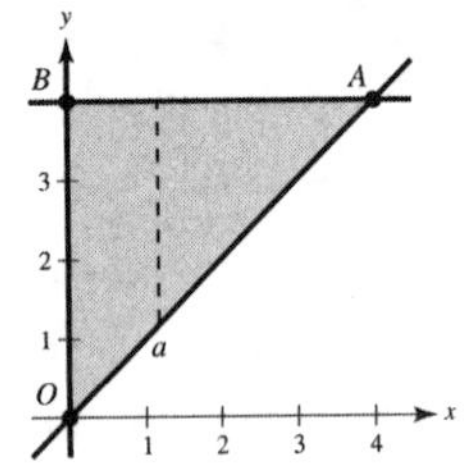

72. Total area $= \int_{-2}^{2} \left(4 - y^2\right) dy = 2\int_0^2 \left(4 - y^2\right) dy$

$$= 2\left[4y - \frac{y^3}{3}\right]_0^2 = 2\left[8 - \frac{8}{3}\right] = \frac{32}{3}$$

$$\frac{16}{3} = 2\int_a^4 \sqrt{4 - x}\,dx = -\frac{4}{3}(4 - x)^{3/2}\Big]_a^4 = \frac{4}{3}(4 - a)^{3/2}$$

$4 = (4 - a)^{3/2}$

$4^{2/3} = 4 - a$

$a = 4 - 4^{2/3} \approx 1.48$

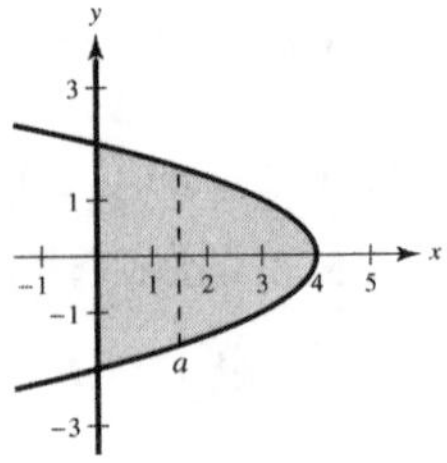

73. $\lim\limits_{\|\Delta\| \to 0} \sum\limits_{i=1}^{n} \left(x_i - x_i^2\right) \Delta x$

where $x_i = \frac{i}{n}$ and $\Delta x = \frac{1}{n}$ is the same as

$$\int_0^1 \left(x - x^2\right) dx = \left[\frac{x^2}{2} - \frac{x^3}{3}\right]_0^1 = \frac{1}{6}.$$

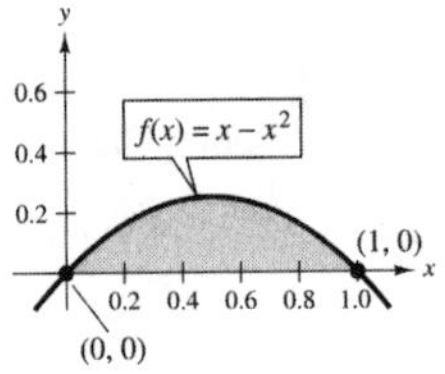

74. $\lim\limits_{\|\Delta\| \to 0} \sum\limits_{i=1}^{n} \left(4 - x_i^2\right) \Delta x$

where $x_i = -2 + \frac{4i}{n}$ and $\Delta x = \frac{4}{n}$ is the same as

$$\int_{-2}^{2} \left(4 - x^2\right) dx = \left[4x - \frac{x^3}{3}\right]_{-2}^{2} = \frac{32}{3}.$$

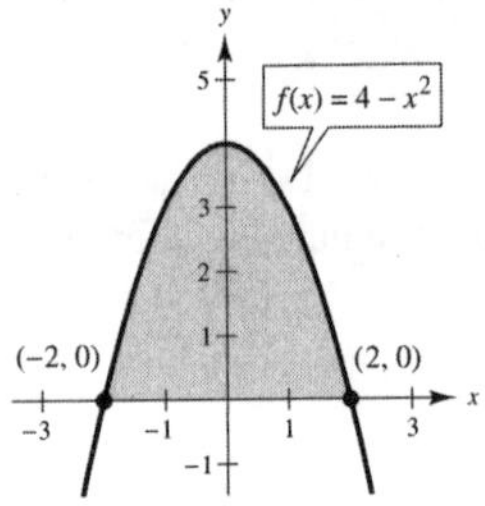

75. R_1 projects the greater revenue because the area under the curve is greater.

$$\int_{15}^{20} \left[(7.21 + 0.58t) - (7.21 + 0.45t)\right] dt$$

$$= \int_{15}^{20} 0.13t\,dt = \left[\frac{0.13t^2}{2}\right]_{15}^{20} = \$11.375 \text{ billion}$$

76. R_2 projects the greater revenue because the area under the curve is greater.

$$\int_{15}^{20}\left[\left(7.21 + 0.26t + 0.02t^2\right) - \left(7.21 + 0.1t + 0.01t^2\right)\right] dt$$

$$= \int_{15}^{20}\left(0.01t^2 + 0.16t\right) dt$$

$$= \left[\frac{0.01t^3}{3} + \frac{0.16t^2}{2}\right]_{15}^{20} \approx \$29.417 \text{ billion}$$

77. (a) $y_1 = 0.0124x^2 - 0.385x + 7.85$

(b)

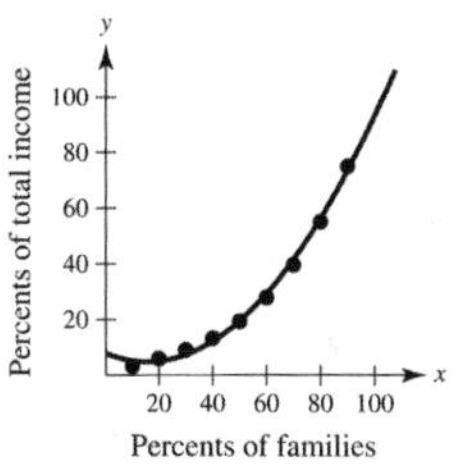

(c)

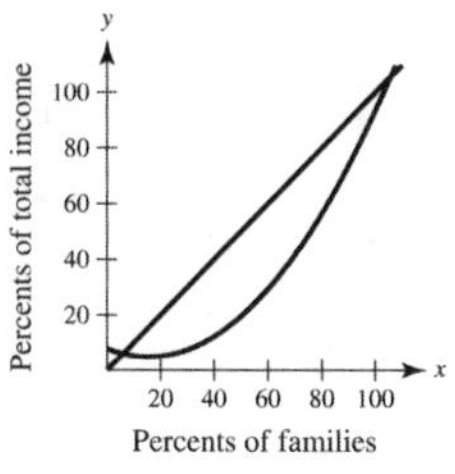

(d) Income inequality $= \int_0^{100}\left[x - y_1\right] dx \approx 2006.7$

78. 5%: $P_1 = 15.9e^{0.05t}$ (in millions)

3.5%: $P_2 = 15.9e^{0.035t}$ (in millions)

Difference in profits over 5 years:

$$\int_0^5 \left(P_1 - P_2\right) dt = \int_0^5 15.9\left(e^{0.05t} - e^{0.035t}\right) dt = 15.9\left[\frac{e^{0.05t}}{0.05} - \frac{e^{0.035t}}{0.035}\right]_0^5 \approx \$3.44 \text{ million}$$

79. (a) $A = 2\left[\int_0^5\left(1 - \frac{1}{3}\sqrt{5 - x}\right) dx + \int_5^{5.5}(1 - 0)\, dx\right]$

$$= 2\left(\left[x + \frac{2}{9}(5 - x)^{3/2}\right]_0^5 + \left[x\right]_5^{5.5}\right)$$

$$= 2\left(5 - \frac{10\sqrt{5}}{9} + 5.5 - 5\right) \approx 6.031 \text{ m}^2$$

(b) $V = 2A \approx 2(6.031) \approx 12.062 \text{ m}^3$

(c) $5000\,V \approx 5000(12.062) = 60{,}310$ pounds

80. The curves intersect at the point where the slope of y_2 equals that of y_1, 1.

$$y_2 = 0.08x^2 + k \Rightarrow y_2' = 0.16x = 1 \Rightarrow x = \frac{1}{0.16} = 6.25$$

(a) The value of k is given by

$$y_1 = y_2$$
$$6.25 = (0.08)(6.25)^2 + k$$
$$k = 3.125.$$

(b) Area $= 2\int_0^{6.25}\left(y_2 - y_1\right) dx$

$$= 2\int_0^{6.25}\left(0.08x^2 + 3.125 - x\right) dx$$

$$= 2\left[\frac{0.08x^3}{3} + 3.125x - \frac{x^2}{2}\right]_0^{6.25}$$

$$= 2(6.510417) \approx 13.02083$$

81. Line: $y = \dfrac{-3}{7\pi}x$

$$A = \int_0^{7\pi/6}\left[\sin x + \frac{3x}{7\pi}\right]dx$$
$$= \left[-\cos x + \frac{3x^2}{14\pi}\right]_0^{7\pi/6}$$
$$= \frac{\sqrt{3}}{2} + \frac{7\pi}{24} + 1$$
$$\approx 2.7823$$

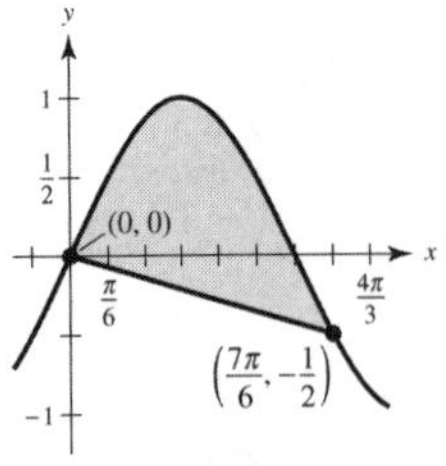

82. $A = 4\int_0^a b\sqrt{1 - \dfrac{x^2}{a^2}}\,dx = \dfrac{4b}{a}\int_0^a \sqrt{a^2 - x^2}\,dx$

$\int_0^a \sqrt{a^2 - x^2}\,dx$ is the area of $\dfrac{1}{4}$ of a circle $= \dfrac{\pi a^2}{4}$.

So, $A = \dfrac{4b}{a}\left(\dfrac{\pi a^2}{4}\right) = \pi ab.$

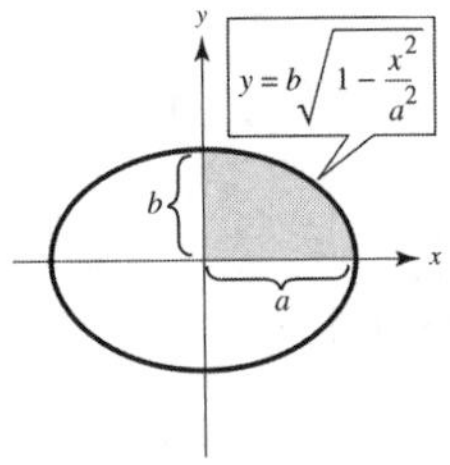

83. True. The region has been shifted C units upward (if $C > 0$), or C units downward (if $C < 0$).

84. True. This is a property of integrals.

85. False. Let $f(x) = x$ and $g(x) = 2x - x^2$, f and g intersect at $(1, 1)$, the midpoint of $[0, 2]$, but

$$\int_a^b\left[f(x) - g(x)\right]dx = \int_0^2\left[x - \left(2x - x^2\right)\right]dx = \tfrac{2}{3} \neq 0.$$

86. True. The area under $f(x)$ between 0 and 1 is $\frac{1}{6}$. The curves intersect at $x = \frac{1}{2}^{1/3}$, and the area between $y = \left(1 - \frac{1}{2}^{1/3}\right)x$ and f on the interval $\left[0, \frac{1}{2}^{1/3}\right]$ is $\frac{1}{12}$.

87. You want to find c such that:

$$\int_0^b\left[\left(2x - 3x^3\right) - c\right]dx = 0$$
$$\left[x^2 - \tfrac{3}{4}x^4 - cx\right]_0^b = 0$$
$$b^2 - \tfrac{3}{4}b^4 - cb = 0$$

But, $c = 2b - 3b^3$ because (b, c) is on the graph.

$$b^2 - \tfrac{3}{4}b^4 - \left(2b - 3b^3\right)b = 0$$
$$4 - 3b^2 - 8 + 12b^2 = 0$$
$$9b^2 = 4$$
$$b = \tfrac{2}{3}$$
$$c = \tfrac{4}{9}$$

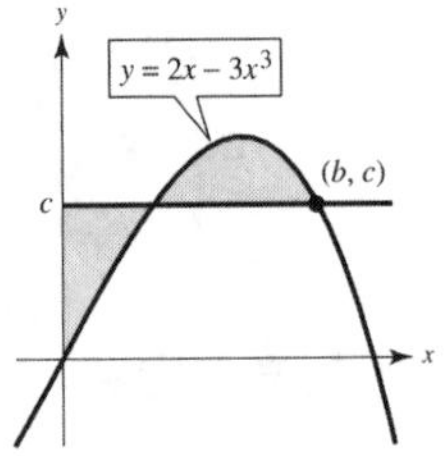

Section 7.2 Volume: The Disk Method

1. $V = \pi\int_0^1(-x + 1)^2\,dx = \pi\int_0^1\left(x^2 - 2x + 1\right)dx = \pi\left[\dfrac{x^3}{3} - x^2 + x\right]_0^1 = \dfrac{\pi}{3}$

2. $V = \pi\int_0^2\left(4 - x^2\right)^2\,dx = \pi\int_0^2\left(x^4 - 8x^2 + 16\right)dx = \pi\left[\dfrac{x^5}{5} - \dfrac{8x^3}{3} + 16x\right]_0^2 = \dfrac{256\pi}{15}$

3. $V = \pi\int_1^4\left(\sqrt{x}\right)^2\,dx = \pi\int_1^4 x\,dx = \pi\left[\dfrac{x^2}{2}\right]_1^4 = \dfrac{15\pi}{2}$

4. $V = \pi\int_0^3\left(\sqrt{9 - x^2}\right)^2\,dx = \pi\int_0^3\left(9 - x^2\right)dx = \pi\left[9x - \dfrac{x^3}{3}\right]_0^3 = 18\pi$

5. $V = \pi \int_0^1 \left[\left(x^2\right)^2 - \left(x^5\right)^2 \right] dx$

$= \pi \int_0^1 \left(x^4 - x^{10}\right) dx$

$= \pi \left[\frac{x^5}{5} - \frac{x^{11}}{11} \right]_0^1$

$= \pi \left(\frac{1}{5} - \frac{1}{11} \right) = \frac{6\pi}{55}$

6. $2 = 4 - \frac{x^2}{4}$

$8 = 16 - x^2$

$x^2 = 8$

$x = \pm 2\sqrt{2}$

$V = \pi \int_{-2\sqrt{2}}^{2\sqrt{2}} \left[\left(4 - \frac{x^2}{4}\right)^2 - (2)^2 \right] dx$

$= 2\pi \int_0^{2\sqrt{2}} \left[\frac{x^4}{16} - 2x^2 + 12 \right] dx$

$= 2\pi \left[\frac{x^5}{80} - \frac{2x^3}{3} + 12x \right]_0^{2\sqrt{2}}$

$= 2\pi \left[\frac{128\sqrt{2}}{80} - \frac{32\sqrt{2}}{3} + 24\sqrt{2} \right]$

$= \frac{448\sqrt{2}}{15}\pi \approx 132.69$

7. $y = x^2 \Rightarrow x = \sqrt{y}$

$V = \pi \int_0^4 \left(\sqrt{y}\right)^2 dy = \pi \int_0^4 y \, dy$

$= \pi \left[\frac{y^2}{2} \right]_0^4 = 8\pi$

8. $y = \sqrt{16 - x^2} \Rightarrow x = \sqrt{16 - y^2}$

$V = \pi \int_0^4 \left(\sqrt{16 - y^2}\right)^2 dy = \pi \int_0^4 \left(16 - y^2\right) dy$

$= \pi \left[16y - \frac{y^3}{3} \right]_0^4 = \frac{128\pi}{3}$

9. $y = x^{2/3} \Rightarrow x = y^{3/2}$

$V = \pi \int_0^1 \left(y^{3/2}\right)^2 dy = \pi \int_0^1 y^3 \, dy = \pi \left[\frac{y^4}{4} \right]_0^1 = \frac{\pi}{4}$

10. $V = \pi \int_1^4 \left(-y^2 + 4y\right)^2 dy = \pi \int_1^4 \left(y^4 - 8y^3 + 16y^2\right) dy$

$= \pi \left[\frac{y^5}{5} - 2y^4 + \frac{16y^3}{3} \right]_1^4 = \frac{459\pi}{15} = \frac{153\pi}{5}$

11. $y = \sqrt{x}, y = 0, x = 3$

(a) $R(x) = \sqrt{x}, r(x) = 0$

$V = \pi \int_0^3 \left(\sqrt{x}\right)^2 dx = \pi \int_0^3 x \, dx = \pi \left[\frac{x^2}{2} \right]_0^3 = \frac{9\pi}{2}$

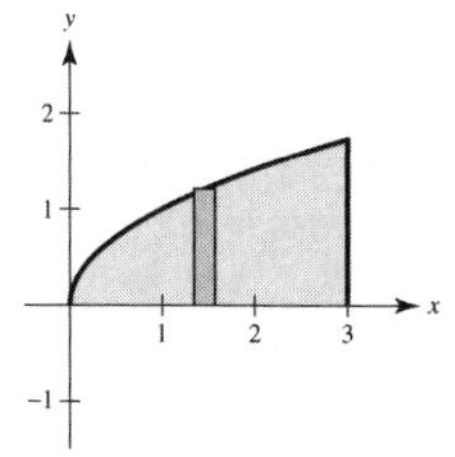

(b) $R(y) = 3, r(y) = y^2$

$V = \pi \int_0^{\sqrt{3}} \left[3^2 - \left(y^2\right)^2 \right] dy = \pi \int_0^{\sqrt{3}} \left(9 - y^4\right) dy = \pi \left[9y - \frac{y^5}{5} \right]_0^{\sqrt{3}} = \pi \left[9\sqrt{3} - \frac{9}{5}\sqrt{3} \right] = \frac{36\sqrt{3}\pi}{5}$

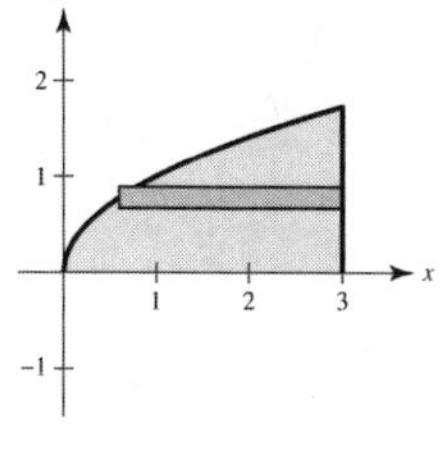

(c) $R(y) = 3 - y^2, r(y) = 0$

$$V = \pi \int_0^{\sqrt{3}} \left(3 - y^2\right)^2 dy = \pi \int_0^{\sqrt{3}} \left(9 - 6y^2 + y^4\right) dy$$

$$= \pi\left[9y - 2y^3 + \frac{y^5}{5}\right]_0^{\sqrt{3}} = \pi\left[9\sqrt{3} - 6\sqrt{3} + \frac{9\sqrt{3}}{5}\right]$$

$$= \frac{24\sqrt{3}\pi}{5}$$

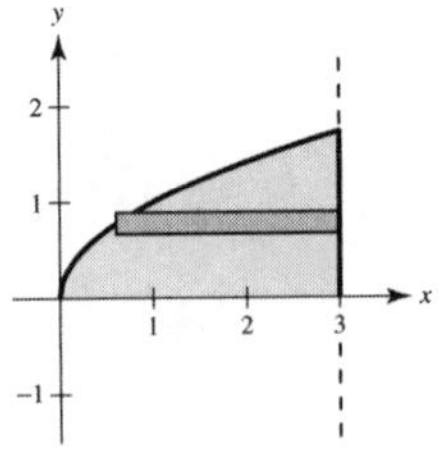

(d) $R(y) = 3 + \left(3 - y^2\right) = 6 - y^2, r(y) = 3$

$$V = \pi \int_0^{\sqrt{3}} \left[\left(6 - y^2\right)^2 - 3^2\right] dy = \pi \int_0^{\sqrt{3}} \left(y^4 - 12y^2 + 27\right) dy$$

$$= \pi\left[\frac{y^5}{5} - 4y^3 + 27y\right]_0^{\sqrt{3}} = \pi\left[\frac{9\sqrt{3}}{5} - 12\sqrt{3} + 27\sqrt{3}\right]$$

$$= \frac{84\sqrt{3}\pi}{5}$$

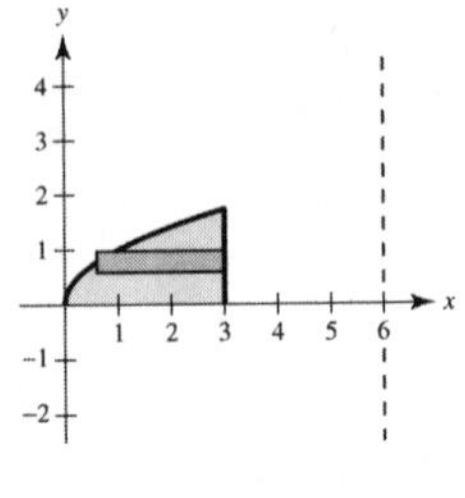

12. $y = 2x^2, y = 0, x = 2$

(a) $R(y) = 2, r(y) = \sqrt{y/2}$

$$V = \pi \int_0^8 \left(4 - \frac{y}{2}\right) dy = \pi\left[4y - \frac{y^2}{4}\right]_0^8 = 16\pi$$

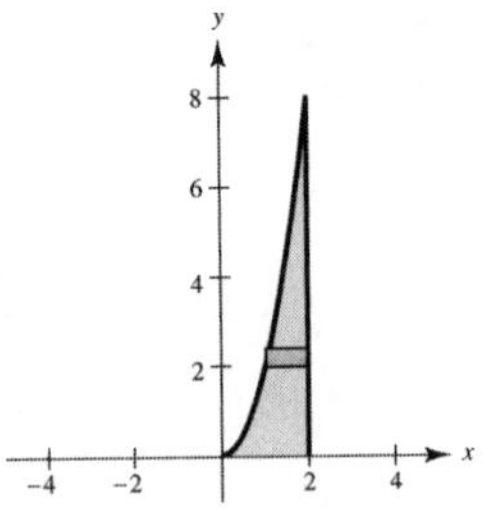

(b) $R(x) = 2x^2, r(x) = 0$

$$V = \pi \int_0^2 4x^4\, dx = \pi\left[\frac{4x^5}{5}\right]_0^2 = \frac{128\pi}{5}$$

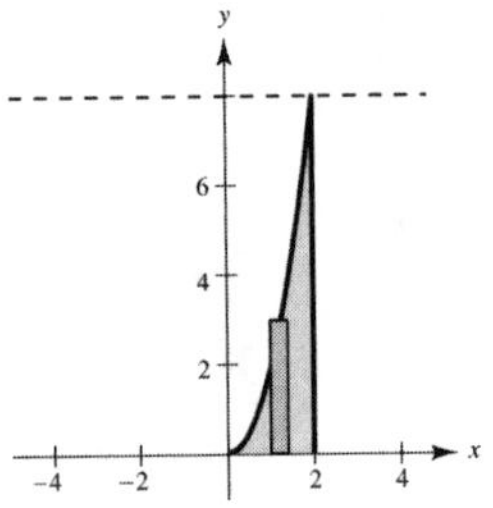

(c) $R(x) = 8, r(x) = 8 - 2x^2$

$$V = \pi \int_0^2 \left[64 - \left(64 - 32x^2 + 4x^4\right)\right] dx$$

$$= \pi \int_0^2 \left(32x^2 - 4x^4\right) dx = 4\pi \int_0^2 \left(8x^2 - x^4\right) dx$$

$$= 4\pi\left[\frac{8}{3}x^3 - \frac{1}{5}x^5\right]_0^2$$

$$= \frac{896\pi}{15}$$

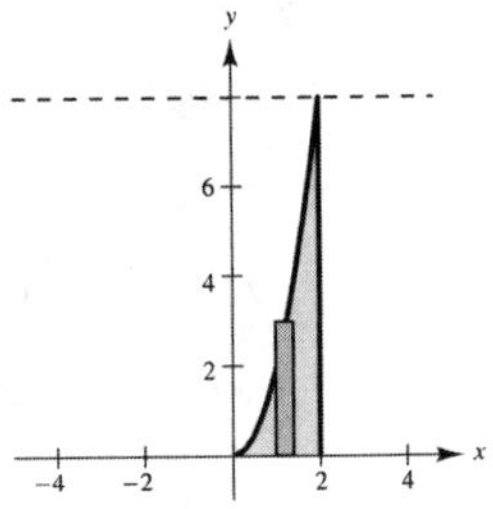

(d) $R(y) = 2 - \sqrt{y/2}, r(y) = 0$

$$V = \pi \int_0^8 \left(2 - \sqrt{\frac{y}{2}}\right)^2 dy$$

$$= \pi \int_0^8 \left(4 - 4\sqrt{\frac{y}{2}} + \frac{y}{2}\right) dy$$

$$= \pi\left[4y - \frac{4\sqrt{2}}{3}y^{3/2} + \frac{y^2}{4}\right]_0^8 = \frac{16\pi}{3}$$

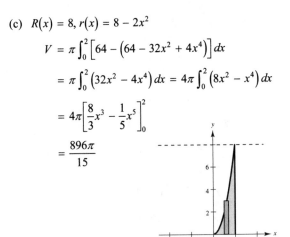

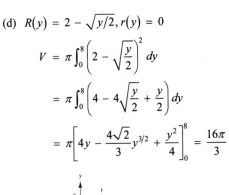

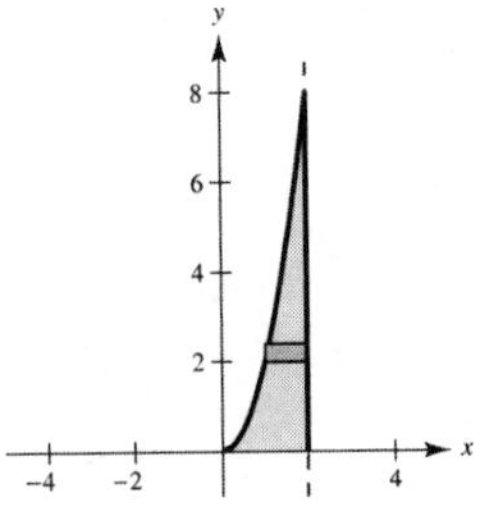

13. $y = x^2$, $y = 4x - x^2$ intersect at $(0, 0)$ and $(2, 4)$.

(a) $R(x) = 4x - x^2$, $r(x) = x^2$

$$V = \pi \int_0^2 \left[\left(4x - x^2\right)^2 - x^4 \right] dx$$

$$= \pi \int_0^2 \left(16x^2 - 8x^3\right) dx$$

$$= \pi \left[\frac{16}{3}x^3 - 2x^4 \right]_0^2 = \frac{32\pi}{3}$$

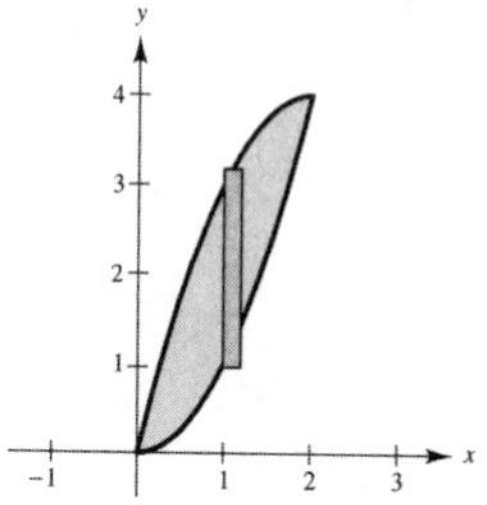

(b) $R(x) = 6 - x^2$, $r(x) = 6 - \left(4x - x^2\right)$

$$V = \pi \int_0^2 \left[\left(6 - x^2\right)^2 - \left(6 - 4x + x^2\right)^2 \right] dx$$

$$= 8\pi \int_0^2 \left(x^3 - 5x^2 + 6x\right) dx$$

$$= 8\pi \left[\frac{x^4}{4} - \frac{5}{3}x^3 + 3x^2 \right]_0^2 = \frac{64\pi}{3}$$

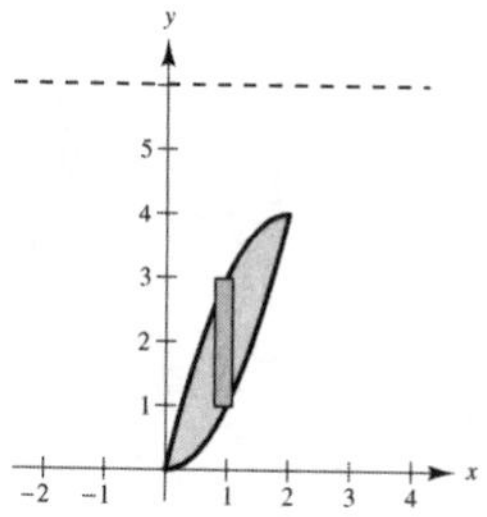

14. $y = 4 + 2x - x^2$, $y = 4 - x$ intersect at $(0, 4)$ and $(3, 1)$.

(a) $R(x) = 4 + 2x - x^2$, $r(x) = 4 - x$

$$V = \pi \int_0^3 \left[\left(4 + 2x - x^2\right)^2 - \left(4 - x\right)^2 \right] dx$$

$$= \pi \int_0^3 \left(x^4 - 4x^3 - 5x^2 + 24x\right) dx$$

$$= \pi \left[\frac{x^5}{5} - x^4 - \frac{5x^3}{3} + 12x^2 \right]_0^3 = \frac{153\pi}{5}$$

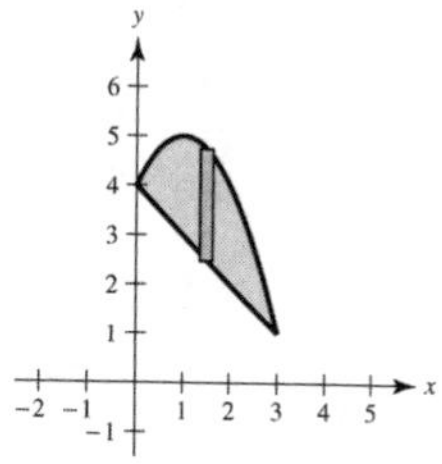

(b) $R(x) = \left(4 + 2x - x^2\right) - 1$, $r(x) = (4 - x) - 1$

$$V = \pi \int_0^3 \left[\left(3 + 2x - x^2\right)^2 - \left(3 - x\right)^2 \right] dx$$

$$= \pi \int_0^3 \left(x^4 - 4x^3 - 3x^2 + 18x\right) dx$$

$$= \pi \left[\frac{x^5}{5} - x^4 - x^3 + 9x^2 \right]_0^3 = \frac{108\pi}{5}$$

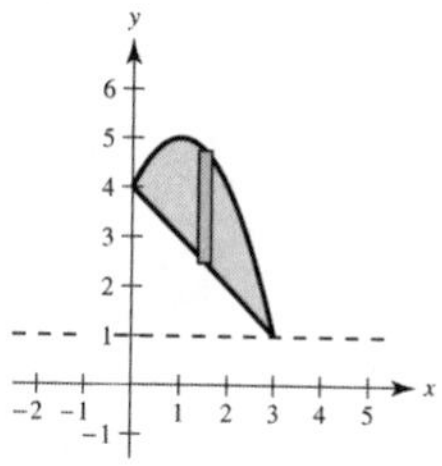

15. $R(x) = 4 - x$, $r(x) = 1$

$$V = \pi \int_0^3 \left[(4 - x)^2 - (1)^2 \right] dx$$

$$= \pi \int_0^3 \left(x^2 - 8x + 15\right) dx$$

$$= \pi \left[\frac{x^3}{3} - 4x^2 + 15x \right]_0^3$$

$$= 18\pi$$

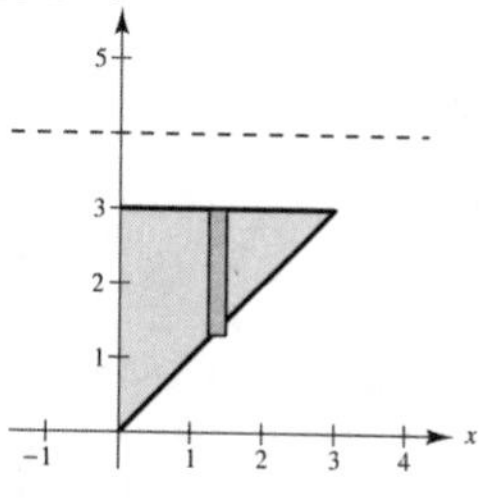

16. $R(x) = 4 - \dfrac{x^3}{2}, r(x) = 0$

$$V = \pi \int_0^1 \left(4 - \frac{x^3}{2}\right)^2 dx$$
$$= \pi \int_0^2 \left[16 - 4x^3 + \frac{x^6}{4}\right] dx$$
$$= \pi \left[16x - x^4 + \frac{x^7}{28}\right]_0^2$$
$$= \pi\left(32 - 16 + \frac{128}{28}\right)$$
$$= \frac{144}{7}\pi$$

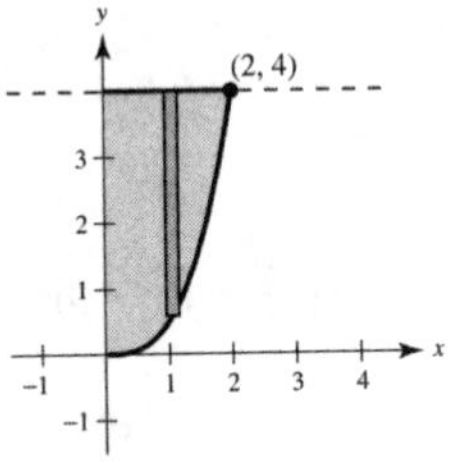

17. $R(x) = 4, r(x) = 4 - \dfrac{3}{1 + x}$

$$V = \pi \int_0^3 \left[4^2 - \left(4 - \frac{3}{1 + x}\right)^2\right] dx$$
$$= \pi \int_0^3 \left[\frac{24}{1 + x} - \frac{9}{(1 + x)^2}\right] dx$$
$$= \pi\left[24 \ln|1 + x| + \frac{9}{1 + x}\right]_0^3$$
$$= \pi\left[\left(24 \ln 4 + \frac{9}{4}\right) - 9\right]$$
$$= \left(48 \ln 2 - \frac{27}{4}\right)\pi \approx 83.318$$

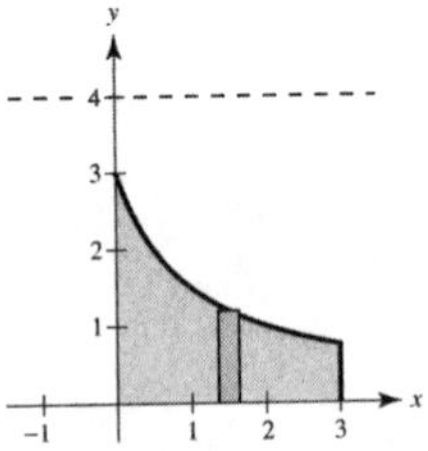

18. $R(x) = 4, r(x) = 4 - \sec x$

$$V = \pi \int_0^{\pi/3} \left[(4)^2 - (4 - \sec x)^2\right] dx$$
$$= \pi \int_0^{\pi/3} \left(8 \sec x - \sec^2 x\right) dx$$
$$= \pi\left[8 \ln|\sec x + \tan x| - \tan x\right]_0^{\pi/3}$$
$$= \pi\left[\left(8 \ln\left|2 + \sqrt{3}\right| - \sqrt{3}\right) - \left(8 \ln|1 + 0| - 0\right)\right]$$
$$= \pi\left[8 \ln\left(2 + \sqrt{3}\right) - \sqrt{3}\right] \approx 27.66$$

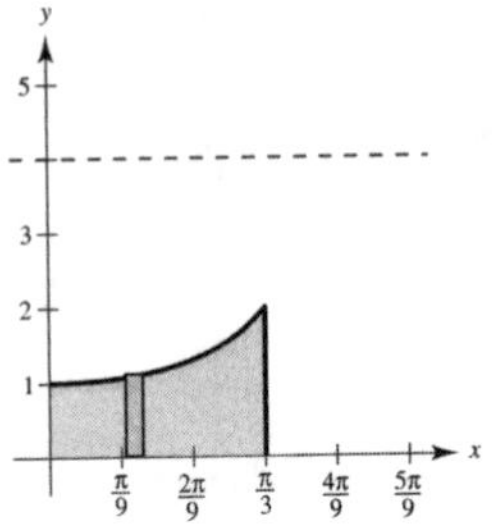

19. $R(y) = 5 - y, r(y) = 0$

$$V = \pi \int_0^4 (5 - y)^2 dy$$
$$= \pi \int_0^4 \left(25 - 10y + y^2\right) dy$$
$$= \pi\left[25y - 5y^2 + \frac{y^3}{3}\right]_0^4$$
$$= \pi\left[100 - 80 + \frac{64}{3}\right]$$
$$= \frac{124\pi}{3}$$

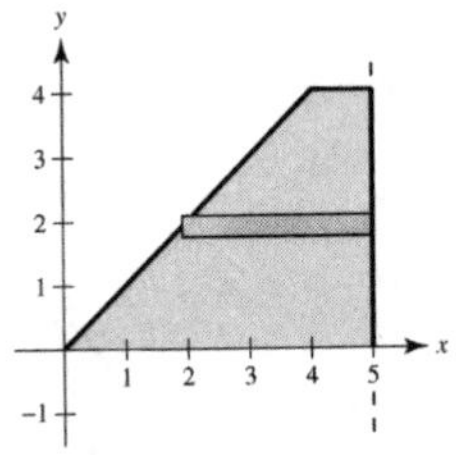

20. $y = 3 - x, x = 3 - y$
$R(y) = 5, r(y) = 5 - (3 - y) = 2 + y$

$$V = \pi \int_0^2 \left[5^2 - (2 + y)^2\right] dy$$
$$= \pi \int_0^2 \left(-y^2 - 4y + 21\right) dy$$
$$= \pi\left[\frac{-y^3}{3} - 2y^2 + 21y\right]_0^2 = \frac{94\pi}{3}$$

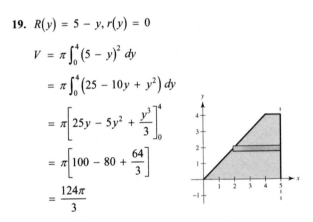

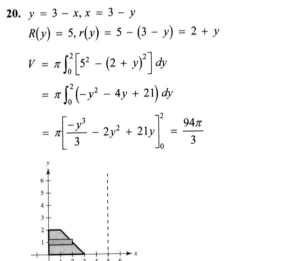

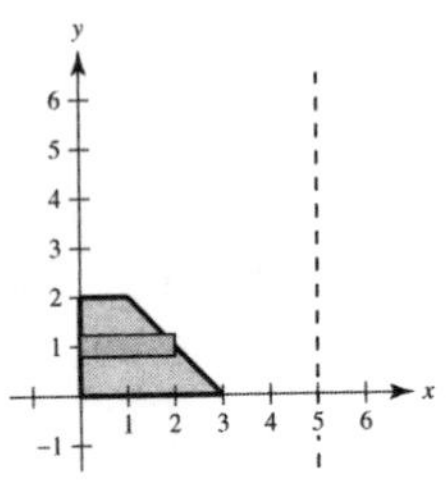

21. $R(y) = 5 - y^2, r(y) = 1$

$$V = \pi\int_{-2}^{2}\left[\left(5 - y^2\right)^2 - 1\right]dy$$
$$= 2\pi\int_{0}^{2}\left[y^4 - 10y^2 + 24\right]dy$$
$$= 2\pi\left[\frac{y^5}{5} - \frac{10y^3}{3} + 24y\right]_0^2$$
$$= 2\pi\left[\frac{32}{5} - \frac{80}{3} + 48\right] = \frac{832\pi}{15}$$

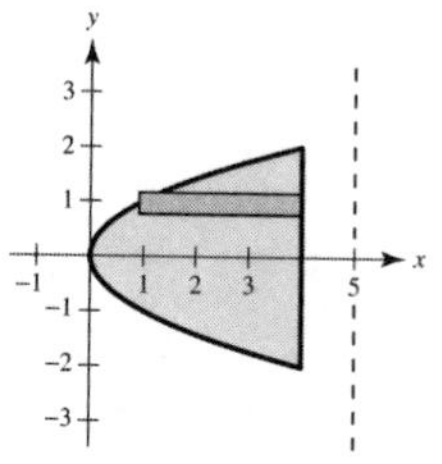

22. $xy = 3, x = \dfrac{3}{y}$

$R(y) = 5 - \dfrac{3}{y}, r(y) = 0$

$$V = \pi\int_{1}^{4}\left(5 - \frac{3}{y}\right)^2 dy$$
$$= \pi\int_{1}^{4}\left(25 + \frac{9}{y^2} - \frac{30}{y}\right)dy$$
$$= \pi\left[25y - \frac{9}{y} - 30\ln y\right]_1^4$$
$$= \pi\left[\left(100 - \frac{9}{4} - 30\ln 4\right) - (25 - 9)\right]$$
$$= \pi\left[\frac{327}{4} - 30\ln 4\right] \approx 126.17$$

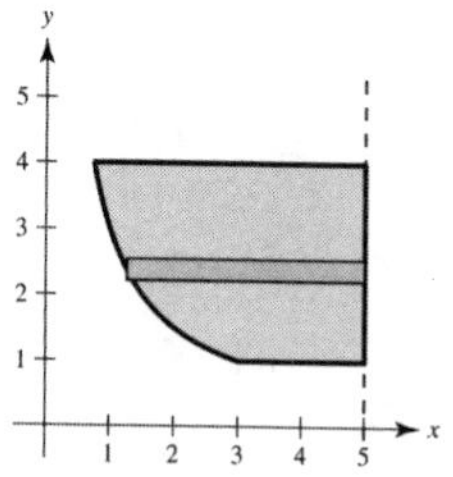

23. $R(x) = \dfrac{1}{\sqrt{x+1}}, r(x) = 0$

$$V = \pi\int_{0}^{4}\left(\frac{1}{\sqrt{x+1}}\right)^2 dx$$
$$= \pi\int_{0}^{4}\frac{1}{x+1}\,dx = \pi\Big[\ln|x+1|\Big]_0^4 = \pi\ln 5$$

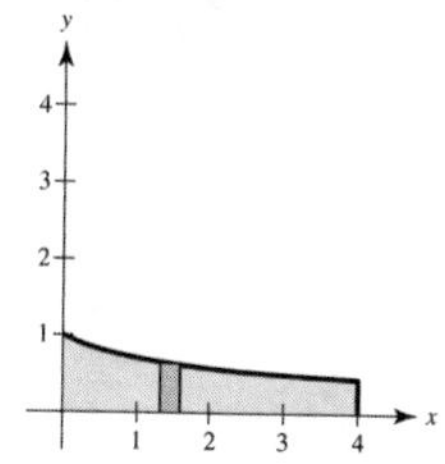

24. $R(x) = x\sqrt{4 - x^2}, r(x) = 0$

$$V = 2\pi\int_{0}^{2}\left(x\sqrt{4 - x^2}\right)^2 dx$$
$$= 2\pi\int_{0}^{2}\left(4x^2 - x^4\right)dx$$
$$= 2\pi\left[\frac{4x^3}{3} - \frac{x^5}{5}\right]_0^2$$
$$= 2\pi\left[\frac{32}{3} - \frac{32}{5}\right] = \frac{128\pi}{15}$$

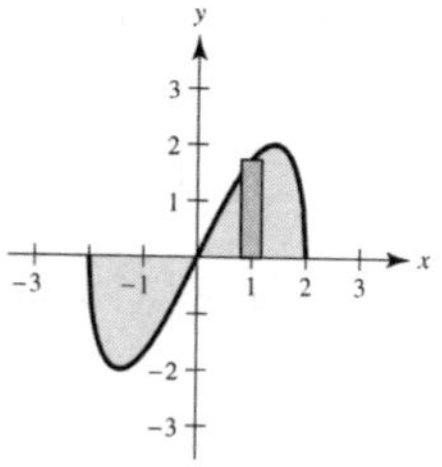

25. $R(x) = \dfrac{1}{x}, r(x) = 0$

$$V = \pi\int_{1}^{3}\left(\frac{1}{x}\right)^2 dx$$
$$= \pi\left[-\frac{1}{x}\right]_1^3$$
$$= \pi\left[-\frac{1}{3} + 1\right] = \frac{2}{3}\pi$$

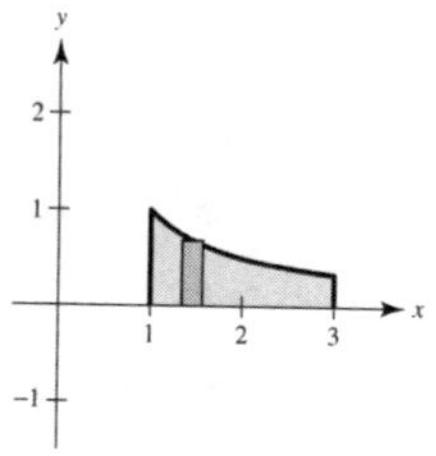

26. $R(x) = \dfrac{2}{x+1}, r(x) = 0$

$$V = \pi\int_{0}^{6}\left(\frac{2}{x+1}\right)^2 dx$$
$$= 4\pi\int_{0}^{6}(x+1)^{-2}\,dx$$
$$= 4\pi\left[\frac{-1}{x+1}\right]_0^6$$
$$= 4\pi\left[-\frac{1}{7} + 1\right] = \frac{24\pi}{7}$$

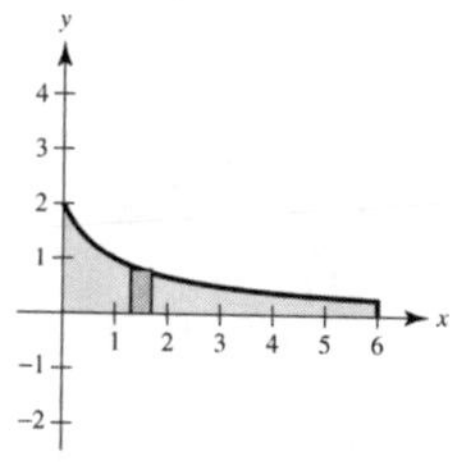

27. $R(x) = e^{-x}, r(x) = 0$

$$V = \pi \int_0^1 \left(e^{-x}\right)^2 dx$$
$$= \pi \int_0^1 e^{-2x} dx$$
$$= \left[-\frac{\pi}{2} e^{-2x}\right]_0^1$$
$$= \frac{\pi}{2}\left(1 - e^{-2}\right) \approx 1.358$$

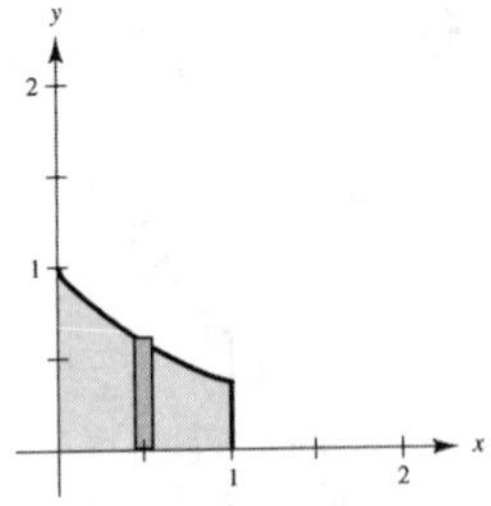

28. $R(x) = e^{x/4}, r(x) = 0$

$$V = \pi \int_0^6 \left(e^{x/4}\right)^2 dx$$
$$= \pi \int_0^6 e^{x/2} dx$$
$$= \pi\left[2e^{x/2}\right]_0^6$$
$$= \pi\left(2e^3 - 2\right) \approx 119.92$$

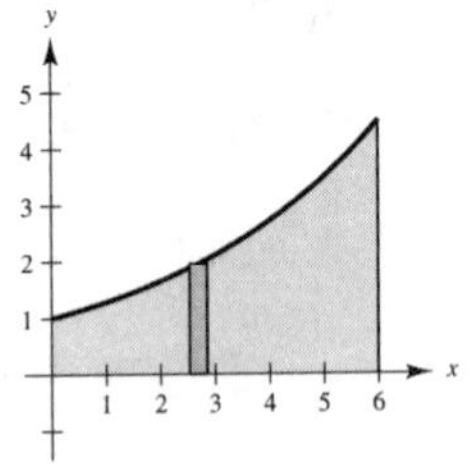

29. $x^2 + 1 = -x^2 + 2x + 5$

$$2x^2 - 2x - 4 = 0$$
$$x^2 - x - 2 = 0$$
$$(x - 2)(x + 1) = 0$$

The curves intersect at $(-1, 2)$ and $(2, 5)$.

$$V = \pi \int_0^2 \left[\left(5 + 2x - x^2\right)^2 - \left(x^2 + 1\right)^2\right] dx + \pi \int_2^3 \left[\left(x^2 + 1\right)^2 - \left(5 + 2x - x^2\right)^2\right] dx$$
$$= \pi \int_0^2 \left(-4x^3 - 8x^2 + 20x + 24\right) dx + \pi \int_2^3 \left(4x^3 + 8x^2 - 20x - 24\right) dx$$
$$= \pi\left[-x^4 - \frac{8}{3}x^3 + 10x^2 + 24x\right]_0^2 + \pi\left[x^4 + \frac{8}{3}x^3 - 10x^2 - 24x\right]_2^3$$
$$= \pi\frac{152}{3} + \pi\frac{125}{3} = \frac{277\pi}{3}$$

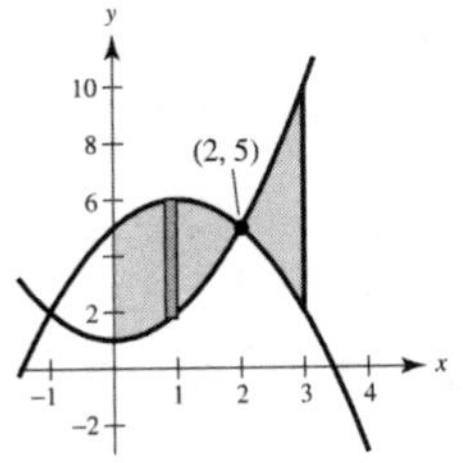

30. $\sqrt{x} = -\frac{1}{2}x + 4$

$$x = \frac{1}{4}x^2 - 4x + 16$$
$$0 = x^2 - 20x + 64$$
$$0 = (x - 4)(x - 16)$$

The curves intersect at $(4, 2)$. (Note $x = 16$ is an extraneous root.)

$$V = \pi \int_0^4 \left[\left(4 - \frac{1}{2}x\right)^2 - \left(\sqrt{x}\right)^2\right] dx + \pi \int_4^8 \left[\left(\sqrt{x}\right)^2 - \left(4 - \frac{1}{2}x\right)^2\right] dx$$
$$= \pi \int_0^4 \left(\frac{x^2}{4} - 5x + 16\right) dx + \pi \int_4^8 \left(-\frac{x^2}{4} + 5x - 16\right) dx$$
$$= \pi\left[\frac{x^3}{12} - \frac{5x^2}{2} + 16x\right]_0^4 + \pi\left[-\frac{x^3}{12} + \frac{5x^2}{2} - 16x\right]_4^8$$
$$= \frac{88}{3}\pi + \frac{56}{3}\pi = 48\pi$$

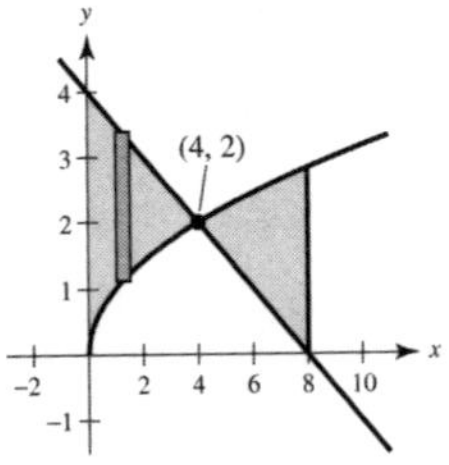

31. $y = 6 - 3x \Rightarrow x = \frac{1}{3}(6 - y)$

$$V = \pi\int_0^6 \left[\frac{1}{3}(6 - y)\right]^2 dy$$
$$= \frac{\pi}{9}\int_0^6 \left[36 - 12y + y^2\right] dy$$
$$= \frac{\pi}{9}\left[36y - 6y^2 + \frac{y^3}{3}\right]_0^6$$
$$= \frac{\pi}{9}\left[216 - 216 + \frac{216}{3}\right]$$
$$= 8\pi = \frac{1}{3}\pi r^2 h, \text{ Volume of cone}$$

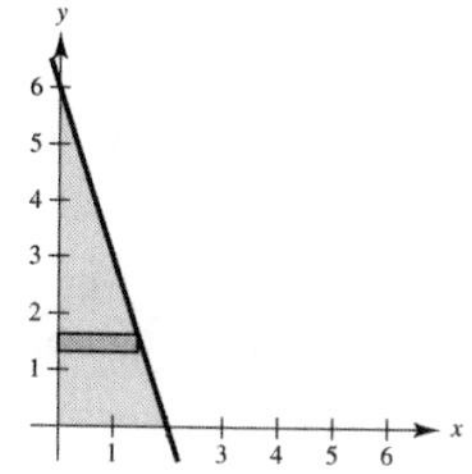

32. $y = 9 - x^2, y = 0, x = 2, x = 3$

$x = \sqrt{9 - y}$

$$V = \pi\int_0^5 \left[\left(\sqrt{9 - y}\right)^2 - 2\right] dy$$
$$= \pi\int_0^5 (5 - y)\, dy$$
$$= \pi\left[5y - \frac{y^2}{2}\right]_0^5 = \pi\left(25 - \frac{25}{2}\right) = \frac{25\pi}{2}$$

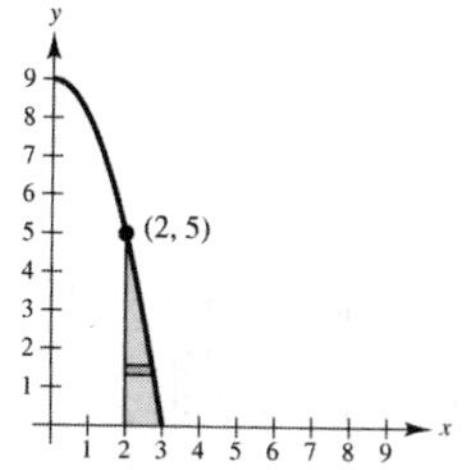

33. $V = \pi\int_0^\pi (\sin x)^2\, dx$

$$= \pi\int_0^\pi \frac{1 - \cos 2x}{2}\, dx$$
$$= \frac{\pi}{2}\left[x - \frac{1}{2}\sin 2x\right]_0^\pi = \frac{\pi}{2}[\pi] = \frac{\pi^2}{2}$$

Numerical approximation: 4.9348

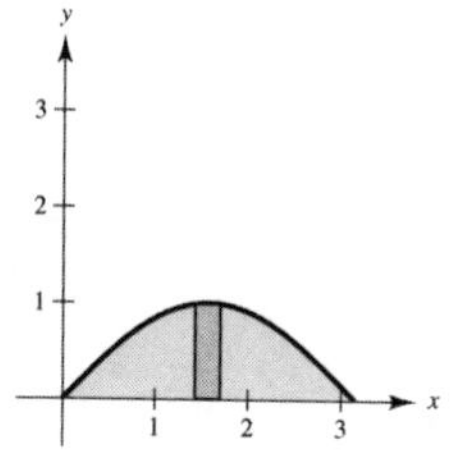

34. $V = \pi\int_0^{\pi/4} \cos^2 2x\, dx$

$$= \pi\int_0^{\pi/4} \frac{1 + \cos 4x}{2}\, dx$$
$$= \frac{\pi}{2}\left[x + \frac{\sin 4x}{4}\right]_0^{\pi/4}$$
$$= \frac{\pi}{2}\left[\frac{\pi}{4}\right] = \frac{\pi^2}{8}$$

Numerical approximation: 1.2337

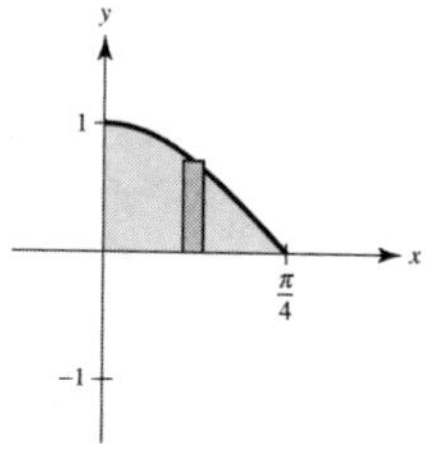

35. $V = \pi\int_1^2 \left(e^{x-1}\right)^2 dx$

$$= \pi\int_1^2 e^{2x-2}\, dx$$
$$= \left.\frac{\pi}{2}e^{2x-2}\right]_1^2$$
$$= \frac{\pi}{2}\left(e^2 - 1\right)$$

Numerical approximation: 10.0359

36. $V = \pi\int_{-1}^2 \left[e^{x/2} + e^{-x/2}\right]^2 dx$

$$= \pi\int_{-1}^2 \left[e^x + e^{-x} + 2\right] dx$$
$$= \pi\left[e^x - e^{-x} + 2x\right]_{-1}^2$$
$$= \pi\left[\left(e^2 - e^{-2} + 4\right) - \left(e^{-1} - e - 2\right)\right]$$
$$= \pi\left(e^2 + e + 6 - e^{-2} - e^{-1}\right)$$

Numerical approximation: 49.0218

37. $V = \pi\int_0^2 \left[e^{-x^2}\right]^2 dx \approx 1.9686$

38. $V = \pi\int_1^3 [\ln x]^2\, dx \approx 3.2332$

39. $V = \pi\int_0^5 \left[2\arctan(0.2x)\right]^2 dx$

≈ 15.4115

40. $x^2 = \sqrt{2x}$

$x^4 = 2x$

$x^3 = 2$

$x = 2^{1/3} \approx 1.2599$

$$V = \pi\int_0^{2^{1/3}}\left[\left(\sqrt{2x}\right)^2 - \left(x^2\right)^2\right]dx \approx 2.9922$$

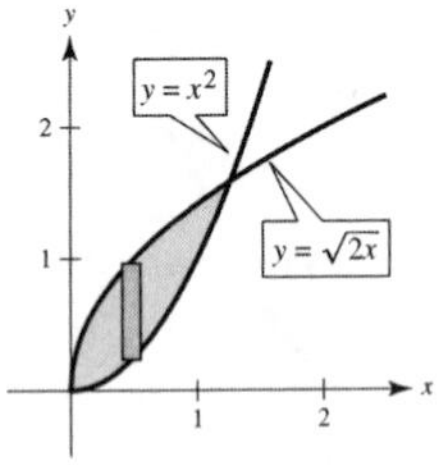

41. $V = \pi\int_0^1 y^2\,dy = \pi\dfrac{y^3}{3}\Big]_0^1 = \dfrac{\pi}{3}$

42. $V = \pi\int_0^1\left[1^2 - (1-y)^2\right]dy$

$= \pi\int_0^1\left[2y - y^2\right]dy$

$= \pi\left[y^2 - \dfrac{y^3}{3}\right]_0^1$

$= \pi\left(1 - \dfrac{1}{3}\right) = \dfrac{2}{3}\pi$

43. $V = \pi\int_0^1\left(x^2 - x^4\right)dx$

$= \pi\left[\dfrac{x^3}{3} - \dfrac{x^5}{5}\right]_0^1$

$= \pi\left(\dfrac{1}{3} - \dfrac{1}{5}\right)$

$= \dfrac{2\pi}{15}$

44. $V = \pi\int_0^1\left[\left(1-x^2\right)^2 - (1-x)^2\right]dx$

$= \pi\int_0^1\left[1 - 2x^2 + x^4 - 1 + 2x - x^2\right]dx$

$= \pi\int_0^1\left[2x - 3x^2 + x^4\right]dx$

$= \pi\left[x^2 - x^3 + \dfrac{x^5}{5}\right]_0^1$

$= \pi\left(\dfrac{1}{5}\right) = \dfrac{\pi}{5}$

45. $V = \pi\int_0^1(1-y)\,dy$

$= \pi\left[y - \dfrac{y^2}{2}\right]_0^1 = \pi\left(1 - \dfrac{1}{2}\right) = \dfrac{\pi}{2}$

46. $V = \pi\int_0^1\left(1 - \sqrt{y}\right)^2 dy$

$= \pi\int_0^1\left(1 - 2\sqrt{y} + y\right)dy$

$= \pi\left[y - \dfrac{4}{3}y^{3/2} + \dfrac{y^2}{2}\right]_0^1$

$= \pi\left(1 - \dfrac{4}{3} + \dfrac{1}{2}\right)$

$= \dfrac{\pi}{6}$

47. $V = \pi\int_0^1\left(y - y^2\right)dy$

$= \pi\left[\dfrac{y^2}{2} - \dfrac{y^3}{3}\right]_0^1 = \pi\left(\dfrac{1}{2} - \dfrac{1}{3}\right) = \dfrac{\pi}{6}$

48. $V = \pi\int_0^1\left[(1-y)^2 - \left(1 - \sqrt{y}\right)^2\right]dy$

$= \pi\int_0^1\left[1 - 2y + y^2 - 1 + 2\sqrt{y} - y\right]dy$

$= \pi\int_0^1\left[2\sqrt{y} - 3y + y^2\right]dy$

$= \pi\left[\dfrac{4}{3}y^{3/2} - \dfrac{3y^2}{2} + \dfrac{y^3}{3}\right]_0^1$

$= \pi\left(\dfrac{4}{3} - \dfrac{3}{2} + \dfrac{1}{3}\right)$

$= \dfrac{\pi}{6}$

49. $\pi\int_0^{\pi/2}\sin^2 x\,dx$ represents the volume of the solid generated by revolving the region bounded by $y = \sin x$, $y = 0$, $x = 0$, $x = \pi/2$ about the x-axis.

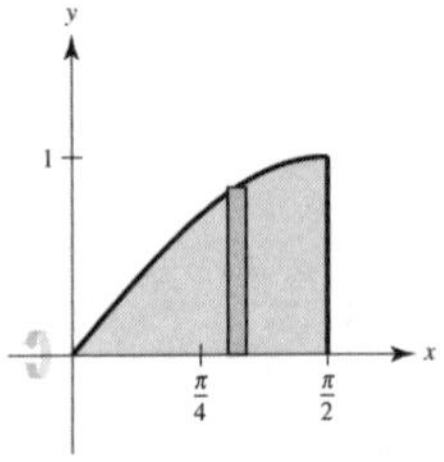

50. $\pi \int_2^4 y^4 \, dy$ represents the volume of the solid generated by revolving the region bounded by $x = y^2, x = 0, y = 2, y = 4$ about the y-axis.

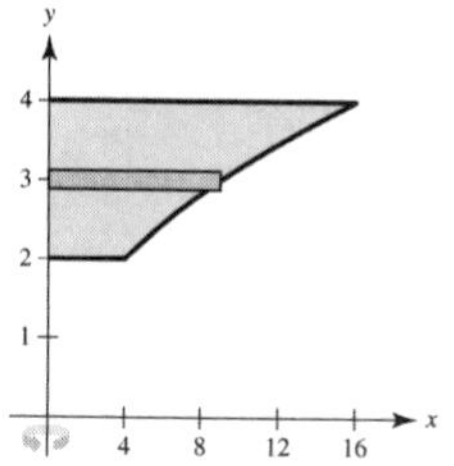

51.

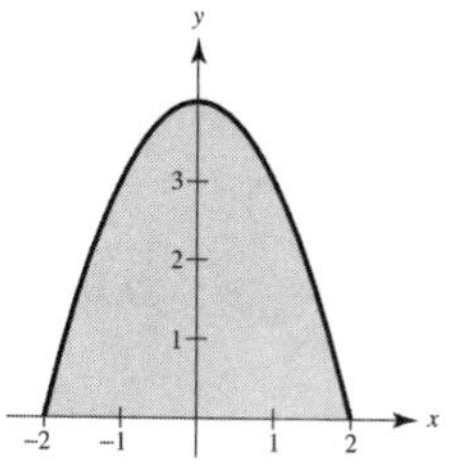

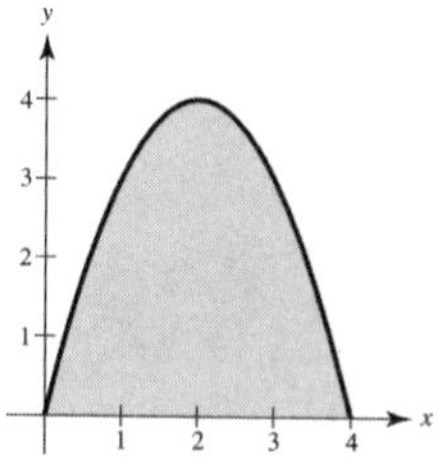

The volumes are the same because the solid has been translated horizontally. $\left(4x - x^2 = 4 - (x - 2)^2\right)$

52.

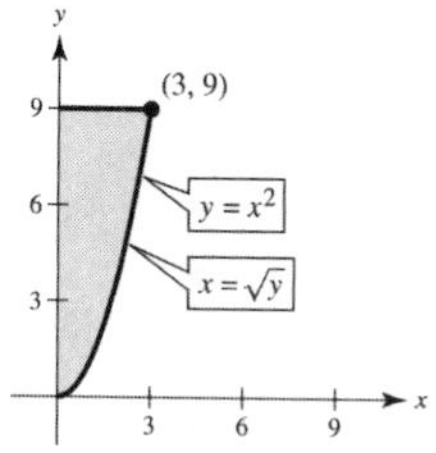

(a) Around x-axis:

$$V = \pi \int_0^3 \left[9^2 - \left(x^2\right)^2\right] dx = \tfrac{972}{5}\pi = 194.4\pi$$

(b) Around y-axis:

$$V = \pi \int_0^9 \left(\sqrt{y}\right)^2 dy = \tfrac{81}{2}\pi = 40.5\pi$$

(c) Around $x = 3$:

$$V = \pi\left(3^2\right)9 - \int_0^9 \pi\left(\sqrt{y} - 3\right)^2 dy = 81\pi - \frac{27}{2}\pi$$

$$= \frac{135\pi}{2} \approx 67.5\pi$$

So, $b < c < a$.

53. (a) True. Answers will vary.

(b) False. Answers will vary.

54. (a) Matches (ii) because the axis of rotation is vertical, and this is the washer method.

(b) Matches (iv) because the axis of rotation is horizontal, and this is the washer method.

(c) Matches (i) because the axis of rotation is horizontal.

(d) Matches (iii) because the axis of rotation is vertical.

55. $V = \pi \int_0^4 \left(\sqrt{x}\right)^2 dx = \pi \int_0^4 x \, dx = \left[\frac{\pi x^2}{2}\right]_0^4 = 8\pi$

Let $0 < c < 4$ and set

$$\pi \int_0^c x \, dx = \left[\frac{\pi x^2}{2}\right]_0^c = \frac{\pi c^2}{2} = 4\pi.$$

$$c^2 = 8$$

$$c = \sqrt{8} = 2\sqrt{2}$$

So, when $x = 2\sqrt{2}$, the solid is divided into two parts of equal volume.

56. Set $\pi \int_0^c x \, dx = \frac{8\pi}{3}$ (one third of the volume).

Then $\frac{\pi c^2}{2} = \frac{8\pi}{3}, c^2 = \frac{16}{3}, c = \frac{4}{\sqrt{3}} = \frac{4\sqrt{3}}{3}$.

To find the other value, set $\pi \int_0^d x \, dx = \frac{16\pi}{3}$ (two thirds of the volume).

Then $\frac{\pi d^2}{2} = \frac{16\pi}{3}, d^2 = \frac{32}{3}, d = \frac{\sqrt{32}}{\sqrt{3}} = \frac{4\sqrt{6}}{3}$.

The x-values that divide the solid into three parts of equal volume are $x = \left(4\sqrt{3}\right)/3$ and $x = \left(4\sqrt{6}\right)/3$.

57. $V = \pi \int_{-\sqrt{R^2-r^2}}^{\sqrt{R^2-r^2}} \left[\left(\sqrt{R^2 - x^2}\right)^2 - r^2\right] dx$

$$= 2\pi \int_0^{\sqrt{R^2-r^2}} \left(R^2 - r^2 - x^2\right) dx$$

$$= 2\pi\left[\left(R^2 - r^2\right)x - \frac{x^3}{3}\right]_0^{\sqrt{R^2-r^2}}$$

$$= 2\pi\left[\left(R^2 - r^2\right)^{3/2} - \frac{\left(R^2 - r^2\right)^{3/2}}{3}\right] = \frac{4}{3}\pi\left(R^2 - r^2\right)^{3/2}$$

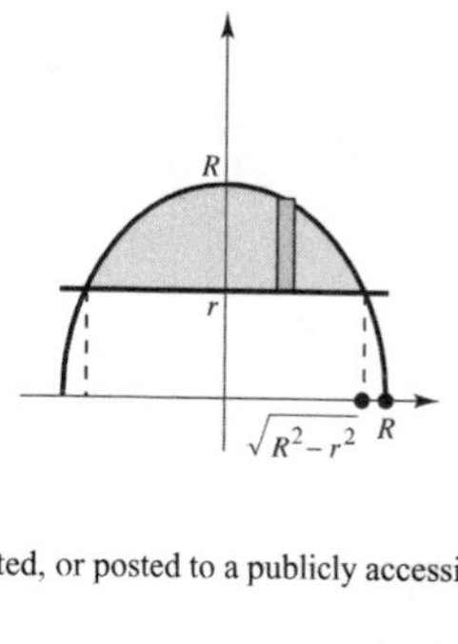

58. Let $R = 6$ in the previous Exercise.

$$\frac{4}{3}\pi\left(36 - r^2\right)^{3/2} = \frac{1}{2}\left(\frac{4}{3}\right)\pi(6)^3$$
$$\left(36 - r^2\right)^{3/2} = 108$$
$$36 - r^2 = (108)^{2/3}$$
$$r^2 = 36 - 108^{2/3}$$
$$r = \sqrt{36 - 108^{2/3}} \approx 3.65$$

59. $R(x) = \dfrac{r}{h}x,\ r(x) = 0$

$$V = \pi\int_0^h \frac{r^2}{h^2}x^2\,dx = \left[\frac{r^2\pi}{3h^2}x^3\right]_0^h = \frac{r^2\pi}{3h^2}h^3 = \frac{1}{3}\pi r^2 h$$

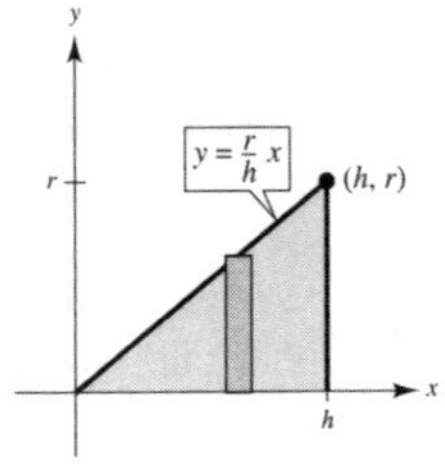

60. $R(x) = \sqrt{r^2 - x^2},\ r(x) = 0$

$$V = \pi\int_{-r}^{r}\left(r^2 - x^2\right)dx$$
$$= 2\pi\int_0^r\left(r^2 - x^2\right)dx$$
$$= 2\pi\left[r^2x - \frac{1}{3}x^3\right]_0^r$$
$$= 2\pi\left(r^3 - \frac{1}{3}r^3\right) = \frac{4}{3}\pi r^3$$

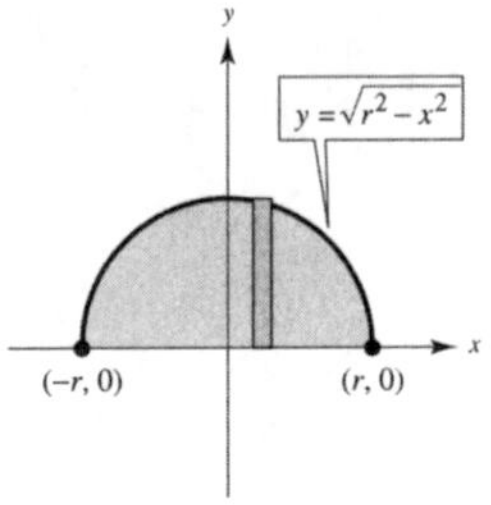

61. $x = r - \dfrac{r}{H}y = r\left(1 - \dfrac{y}{H}\right),\ R(y) = r\left(1 - \dfrac{y}{H}\right),\ r(y) = 0$

$$V = \pi\int_0^h\left[r\left(1 - \frac{y}{H}\right)\right]^2 dy = \pi r^2\int_0^h\left(1 - \frac{2}{H}y + \frac{1}{H^2}y^2\right)dy$$
$$= \pi r^2\left[y - \frac{1}{H}y^2 + \frac{1}{3H^2}y^3\right]_0^h$$
$$= \pi r^2\left(h - \frac{h^2}{H} + \frac{h^3}{3H^2}\right) = \pi r^2 h\left(1 - \frac{h}{H} + \frac{h^2}{3H^2}\right)$$

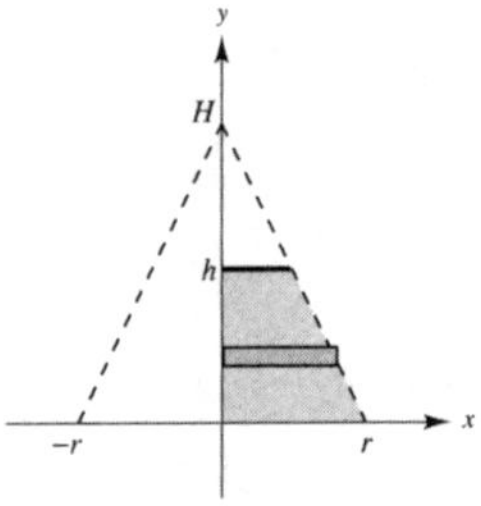

62. $x = \sqrt{r^2 - y^2},\ R(y) = \sqrt{r^2 - y^2},\ r(y) = 0$

$$V = \pi\int_h^r\left(\sqrt{r^2 - y^2}\right)^2 dy$$
$$= \pi\int_h^r\left(r^2 - y^2\right)dy$$
$$= \pi\left[r^2y - \frac{y^3}{3}\right]_h^r$$
$$= \pi\left[\left(r^3 - \frac{r^3}{3}\right) - \left(r^2h - \frac{h^3}{3}\right)\right]$$
$$= \pi\left(\frac{2r^3}{3} - r^2h + \frac{h^3}{3}\right) = \frac{\pi}{3}\left(2r^3 - 3r^2h + h^3\right)$$

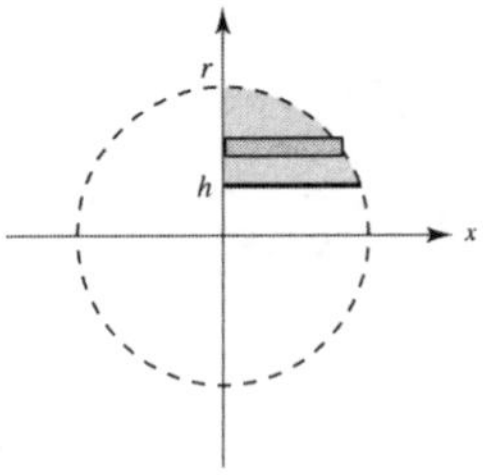

63.

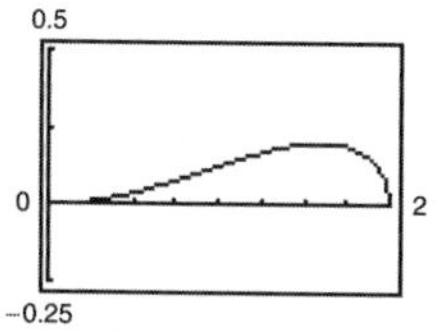

$$V = \pi\int_0^2\left(\frac{1}{8}x^2\sqrt{2-x}\right)^2 dx = \frac{\pi}{64}\int_0^2 x^4(2-x)\,dx = \frac{\pi}{64}\left[\frac{2x^5}{5} - \frac{x^6}{6}\right]_0^2 = \frac{\pi}{30}\text{ m}^3$$

64. $y = \begin{cases}\sqrt{0.1x^3 - 2.2x^2 + 10.9x + 22.2}, & 0 \le x \le 11.5\\ 2.95, & 11.5 < x \le 15\end{cases}$

$$V = \pi\int_0^{11.5}\left(\sqrt{0.1x^3 - 2.2x^2 + 10.9x + 22.2}\right)^2 dx + \pi\int_{11.5}^{15} 2.95^2\,dx$$

$$= \pi\left[\frac{0.1x^4}{4} - \frac{2.2x^3}{3} + \frac{10.9x^2}{2} + 22.2x\right]_0^{11.5} + \pi\left[2.95^2 x\right]_{11.5}^{15}$$

$\approx$ 1031.9016 cubic centimeters

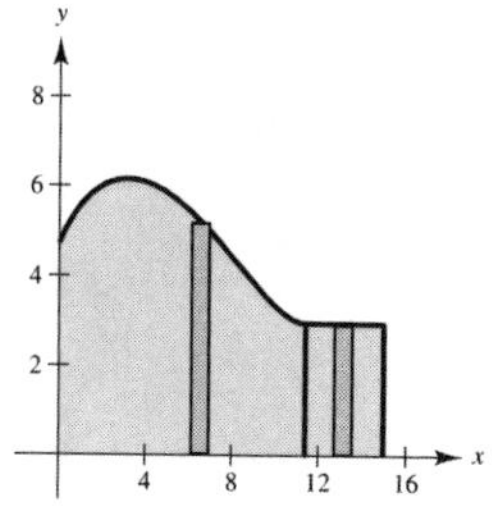

65. (a) $R(x) = \frac{3}{5}\sqrt{25 - x^2},\ r(x) = 0$

$$V = \frac{9\pi}{25}\int_{-5}^{5}\left(25 - x^2\right)dx = \frac{18\pi}{25}\int_0^5\left(25 - x^2\right)dx = \frac{18\pi}{25}\left[25x - \frac{x^3}{3}\right]_0^5 = 60\pi$$

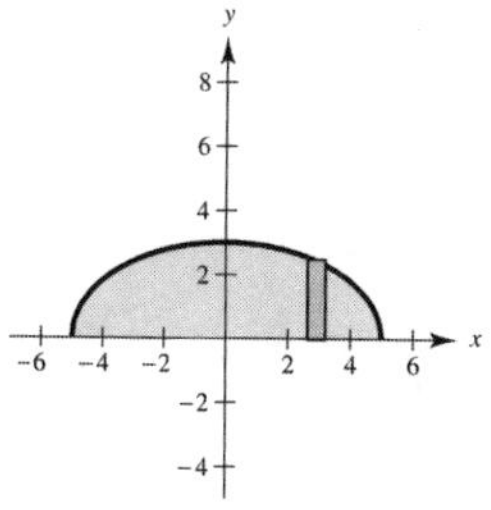

(b) $R(y) = \frac{5}{3}\sqrt{9 - y^2},\ r(y) = 0,\ x \ge 0$

$$V = \frac{25\pi}{9}\int_0^3\left(9 - y^2\right)dy = \frac{25\pi}{9}\left[9y - \frac{y^3}{3}\right]_0^3 = 50\pi$$

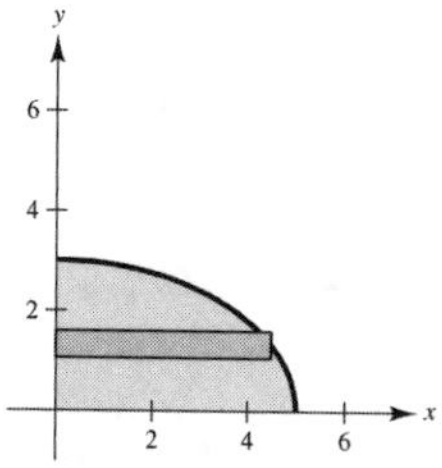

66. Total volume: $V = \dfrac{4\pi(50)^3}{3} = \dfrac{500{,}000\pi}{3}$ ft^3

Volume of water in the tank:

$$\pi\int_{-50}^{y_0}\left(\sqrt{2500 - y^2}\right)^2 dy = \pi\int_{-50}^{y_0}\left(2500 - y^2\right) dy = \pi\left[2500y - \frac{y^3}{3}\right]_{-50}^{y_0} = \pi\left(2500y_0 - \frac{{y_0}^3}{3} + \frac{250{,}000}{3}\right)$$

When the tank is one-fourth of its capacity:

$$\frac{1}{4}\left(\frac{500{,}000\pi}{3}\right) = \pi\left(2500y_0 - \frac{{y_0}^3}{3} + \frac{250{,}000}{3}\right)$$

$$125{,}000 = 7500y_0 - {y_0}^3 + 250{,}000$$

$${y_0}^3 - 7500y_0 - 125{,}000 = 0$$

$$y_0 \approx -17.36$$

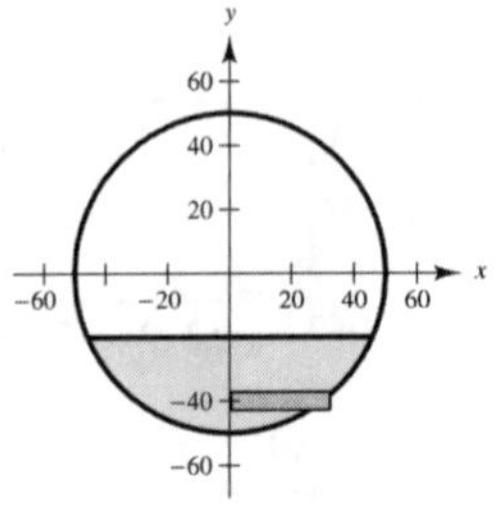

Depth: $-17.36 - (-50) = 32.64$ feet

When the tank is three-fourths of its capacity the depth is $100 - 32.64 = 67.36$ feet.

67. (a) First find where $y = b$ intersects the parabola:

$$b = 4 - \frac{x^2}{4}$$

$$x^2 = 16 - 4b = 4(4 - b)$$

$$x = 2\sqrt{4 - b}$$

$$V = \int_0^{2\sqrt{4-b}} \pi\left[4 - \frac{x^2}{4} - b\right]^2 dx + \int_{2\sqrt{4-b}}^{4} \pi\left[b - 4 + \frac{x^2}{4}\right]^2 dx$$

$$= \int_0^4 \pi\left[4 - \frac{x^2}{4} - b\right]^2 dx$$

$$= \pi\int_0^4\left[\frac{x^4}{16} - 2x^2 + \frac{bx^2}{2} + b^2 - 8b + 16\right] dx$$

$$= \pi\left[\frac{x^5}{80} - \frac{2x^3}{3} + \frac{bx^3}{6} + b^2x - 8bx + 16x\right]_0^4$$

$$= \pi\left(\frac{64}{5} - \frac{128}{3} + \frac{32}{3}b + 4b^2 - 32b + 64\right) = \pi\left(4b^2 - \frac{64}{3}b + \frac{512}{15}\right)$$

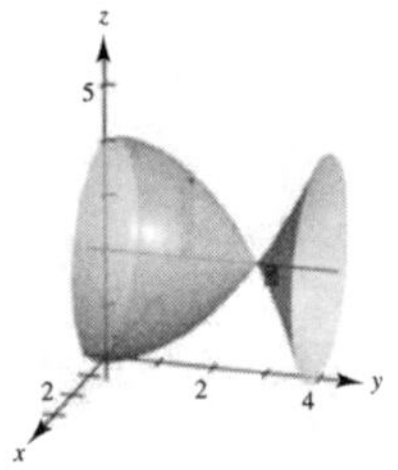

(b) Graph of $V(b) = \pi\left(4b^2 - \dfrac{64}{3}b + \dfrac{512}{15}\right)$

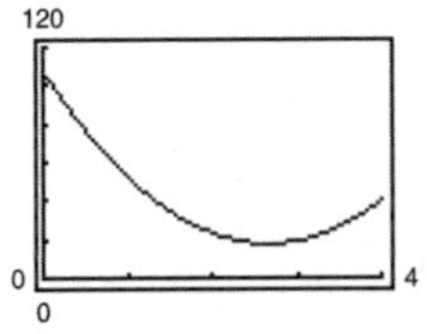

Minimum volume is 17.87 for $b = 2.67$.

(c) $V'(b) = \pi\left(8b - \dfrac{64}{3}\right) = 0 \Rightarrow b = \dfrac{64/3}{8} = \dfrac{8}{3} = 2\dfrac{2}{3}$

$V''(b) = 8\pi > 0 \Rightarrow b = \dfrac{8}{3}$ is a relative minimum.

68. (a) $V = \int_0^{10} \pi\left[f(x)\right]^2\,dx$

Simpson's Rule: $b - a = 10 - 0 = 10,\ n = 10$

$$V \approx \frac{\pi}{3}\left[(2.1)^2 + 4(1.9)^2 + 2(2.1)^2 + 4(2.35)^2 + 2(2.6)^2 + 4(2.85)^2 + 2(2.9)^2 + 4(2.7)^2 + 2(2.45)^2 + 4(2.2)^2 + (2.3)^2\right]$$

$$\approx \frac{\pi}{3}(178.405) \approx 186.83\text{ cm}^3$$

(b) $f(x) = 0.00249x^4 - 0.0529x^3 + 0.3314x^2 - 0.4999x + 2.112$

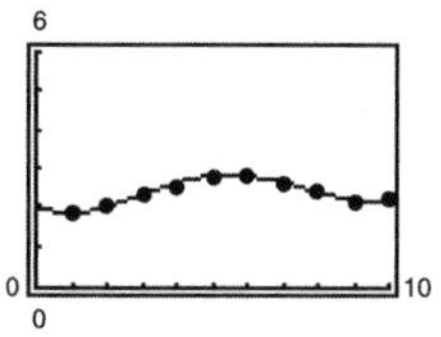

(c) $V \approx \int_0^{10} \pi f(x)^2\,dx \approx 186.35\text{ cm}^3$

69. (a) $\pi\int_0^h r^2\,dx$ (ii)

is the volume of a right circular cylinder with radius r and height h.

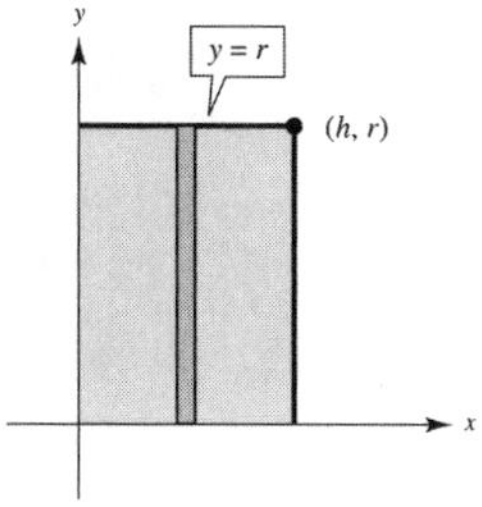

(b) $\pi\int_{-b}^{b}\left(a\sqrt{1 - \frac{x^2}{b^2}}\right)^2 dx$ (iv)

is the volume of an ellipsoid with axes $2a$ and $2b$.

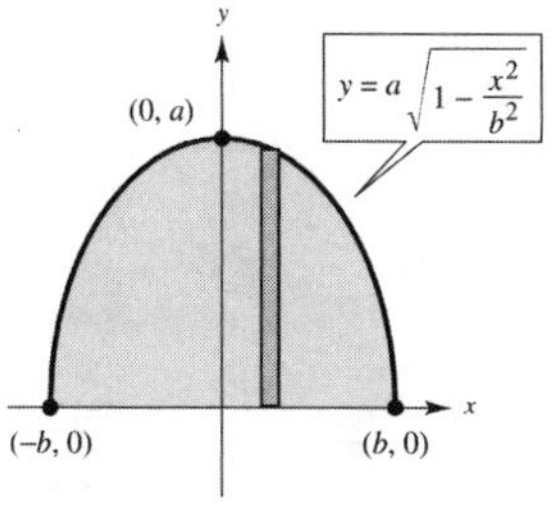

(c) $\pi\int_{-r}^{r}\left(\sqrt{r^2 - x^2}\right)^2 dx$ (iii)

is the volume of a sphere with radius r.

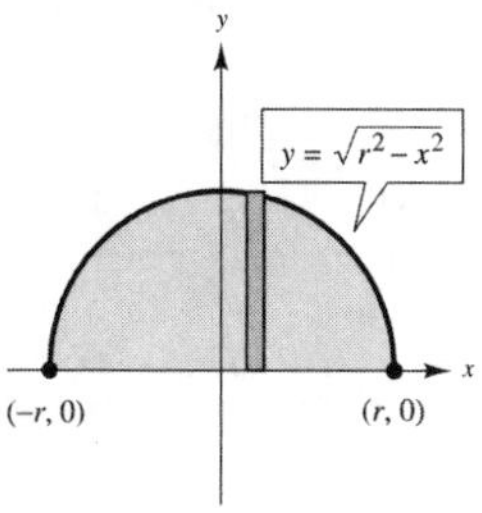

(d) $\pi\int_0^h\left(\frac{rx}{h}\right)^2 dx$ (i)

is the volume of a right circular cone with the radius of the base as r and height h.

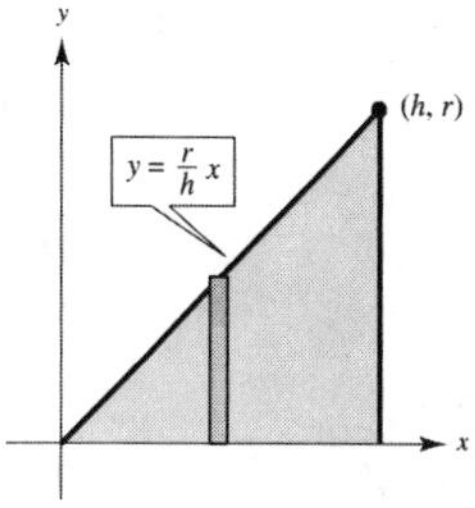

(e) $\pi\int_{-r}^{r}\left[\left(R + \sqrt{r^2 - x^2}\right)^2 - \left(R - \sqrt{r^2 - x^2}\right)^2\right] dx$ (v)

is the volume of a torus with the radius of its circular cross section as r and the distance from the axis of the torus to the center of its cross section as R.

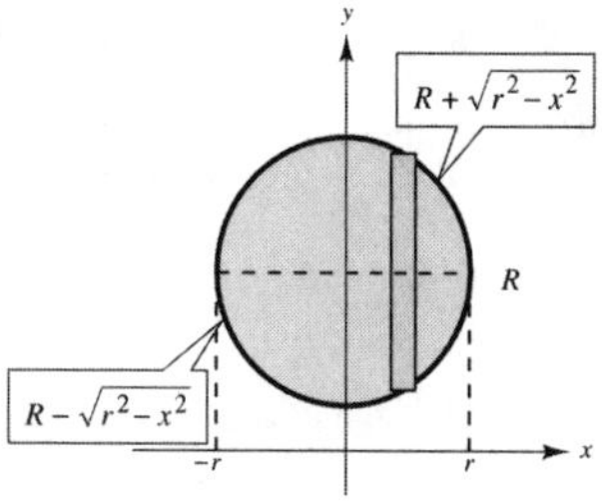

70. Let $A_1(x)$ and $A_2(x)$ equal the areas of the cross sections of the two solids for $a \le x \le b$. Because $A_1(x) = A_2(x)$, you have

$$V_1 = \int_a^b A_1(x)\,dx = \int_a^b A_2(x)\,dx = V_2.$$

So, the volumes are the same.

71.

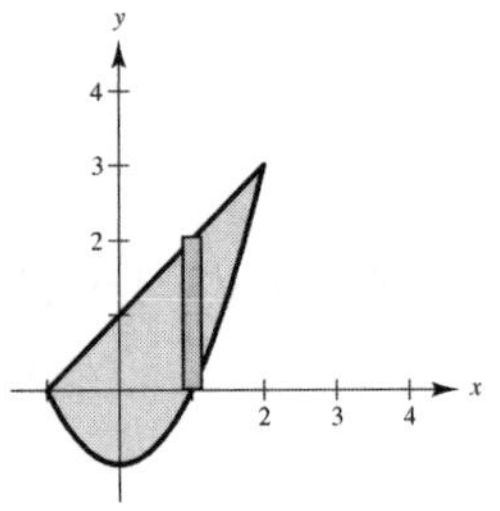

Base of cross section $= (x + 1) - (x^2 - 1) = 2 + x - x^2$

(a) $A(x) = b^2 = (2 + x - x^2)^2 = 4 + 4x - 3x^2 - 2x^3 + x^4$

$$V = \int_{-1}^{2} (4 + 4x - 3x^2 - 2x^3 + x^4)\, dx = \left[4x + 2x^2 - x^3 - \frac{1}{2}x^4 + \frac{1}{5}x^5\right]_{-1}^{2} = \frac{81}{10}$$

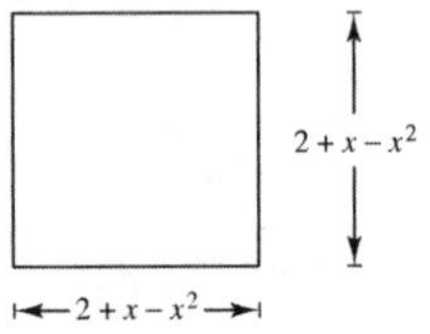

(b) $A(x) = bh = (2 + x - x^2)1$

$$V = \int_{-1}^{2} (2 + x - x^2)\, dx = \left[2x + \frac{x^2}{2} - \frac{x^3}{3}\right]_{-1}^{2} = \frac{9}{2}$$

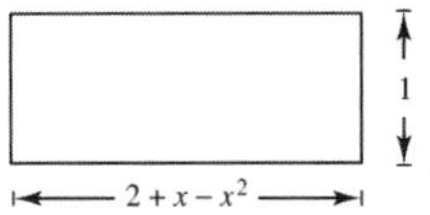

72.

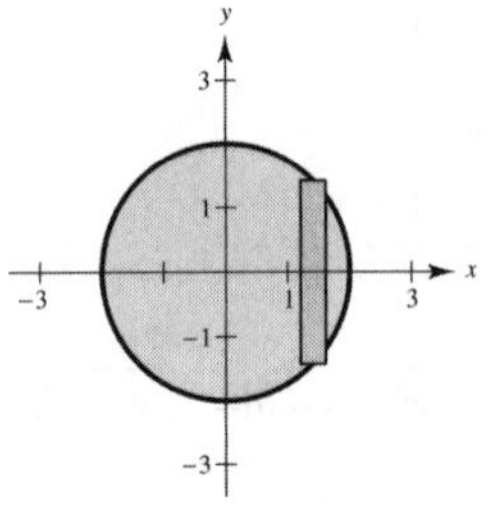

Base of cross section $= 2\sqrt{4 - x^2}$

(a) $A(x) = b^2 = \left(2\sqrt{4 - x^2}\right)^2$

$$V = \int_{-2}^{2} 4(4 - x^2)\, dx$$

$$= 4\left[4x - \frac{x^3}{3}\right]_{-2}^{2}$$

$$= \frac{128}{3}$$

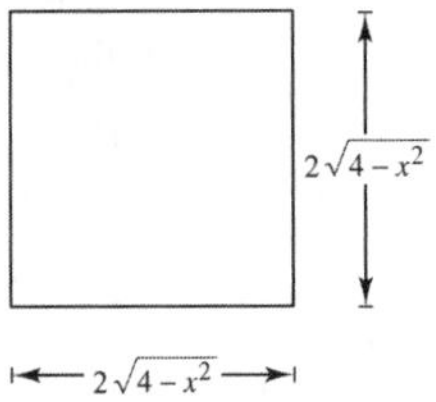

(b) $A(x) = \frac{1}{2}bh = \frac{1}{2}\left(2\sqrt{4 - x^2}\right)\left(\sqrt{3}\sqrt{4 - x^2}\right) = \sqrt{3}(4 - x^2)$

$$V = \sqrt{3}\int_{-2}^{2} (4 - x^2)\, dx$$

$$= \sqrt{3}\left[4x - \frac{x^3}{3}\right]_{-2}^{2}$$

$$= \frac{32\sqrt{3}}{3}$$

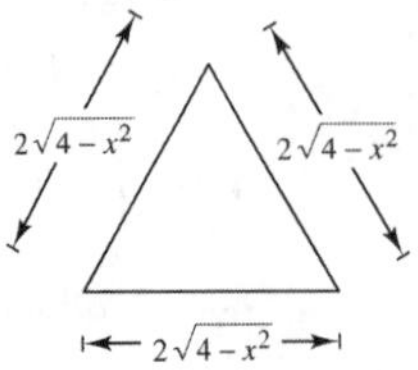

(c) $A(x) = \frac{1}{2}\pi r^2 = \frac{\pi}{2}\left(\sqrt{4 - x^2}\right)^2 = \frac{\pi}{2}\left(4 - x^2\right)$

$$V = \frac{\pi}{2}\int_{-2}^{2}\left(4 - x^2\right)dx$$

$$= \frac{\pi}{2}\left[4x - \frac{x^3}{3}\right]_{-2}^{2} = \frac{16\pi}{3}$$

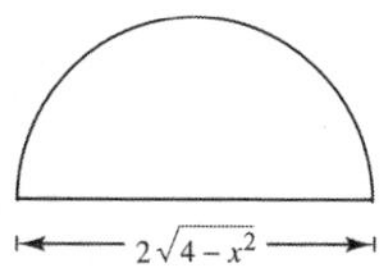

(d) $A(x) = \frac{1}{2}bh = \frac{1}{2}\left(2\sqrt{4 - x^2}\right)\left(\sqrt{4 - x^2}\right) = 4 - x^2$

$$V = \int_{-2}^{2}\left(4 - x^2\right)dx$$

$$= \left[4x - \frac{x^3}{3}\right]_{-2}^{2} = \frac{32}{3}$$

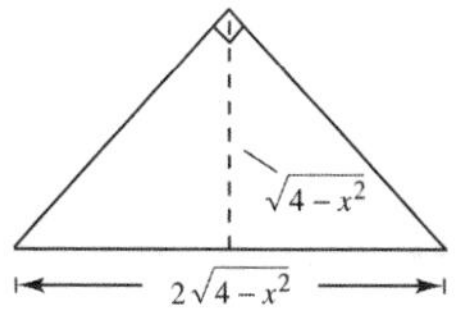

73. The cross sections are squares. By symmetry, you can set up an integral for an eighth of the volume and multiply by 8.

$$A(y) = b^2 = \left(\sqrt{r^2 - y^2}\right)^2$$

$$V = 8\int_0^r\left(r^2 - y^2\right)dy$$

$$= 8\left[r^2y - \tfrac{1}{3}y^3\right]_0^r$$

$$= \tfrac{16}{3}r^3$$

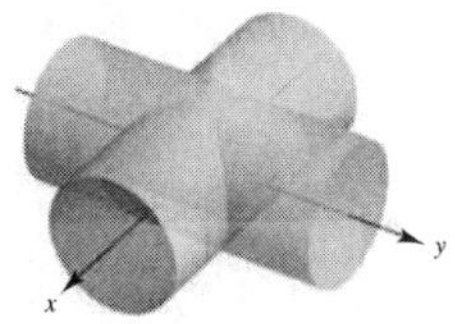

74. (a) When $a = 1$: $|x| + |y| = 1$ represents a square.

When $a = 2$: $|x|^2 + |y|^2 = 1$ represents a circle.

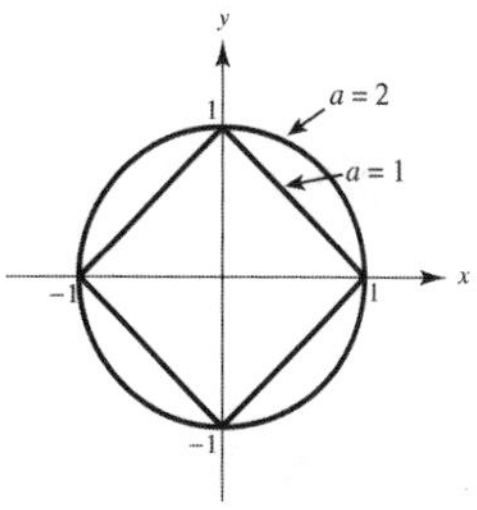

(b) $|y| = \left(1 - |x|^a\right)^{1/a}$

$$A = 2\int_{-1}^{1}\left(1 - |x|^a\right)^{1/a}dx = 4\int_0^1\left(1 - x^a\right)^{1/a}dx$$

To approximate the volume of the solid, from n slices, each of whose area is approximated by the integral above. Then sum the volumes of these n slices.

75. (a) Because the cross sections are isosceles right triangles:

$$A(x) = \frac{1}{2}bh = \frac{1}{2}\left(\sqrt{r^2 - y^2}\right)\left(\sqrt{r^2 - y^2}\right) = \frac{1}{2}\left(r^2 - y^2\right)$$

$$V = \frac{1}{2}\int_{-r}^{r}\left(r^2 - y^2\right)dy = \int_0^r\left(r^2 - y^2\right)dy = \left[r^2y - \frac{y^3}{3}\right]_0^r = \frac{2}{3}r^3$$

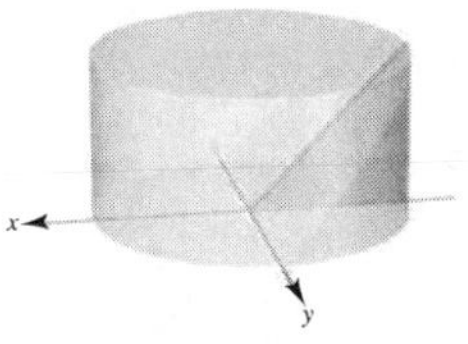

(b) $A(x) = \frac{1}{2}bh = \frac{1}{2}\sqrt{r^2 - y^2}\left(\sqrt{r^2 - y^2}\tan\theta\right) = \frac{\tan\theta}{2}\left(r^2 - y^2\right)$

$$V = \frac{\tan\theta}{2}\int_{-r}^{r}\left(r^2 - y^2\right)dy = \tan\theta\int_0^r\left(r^2 - y^2\right)dy = \tan\theta\left[r^2y - \frac{y^3}{3}\right]_0^r = \frac{2}{3}r^3\tan\theta$$

As $\theta \to 90°$, $V \to \infty$.

76. (a) $(x - R)^2 + y^2 = r^2$

$$x = R \pm \sqrt{r^2 - y^2}$$

$$V = 2\pi \int_0^r \left(\left[R + \sqrt{r^2 - y^2} \right]^2 - \left[R - \sqrt{r^2 - y^2} \right]^2 \right) dy = 2\pi \int_0^r 4R\sqrt{r^2 - y^2}\, dy = 8\pi R \int_0^r \sqrt{r^2 - y^2}\, dy$$

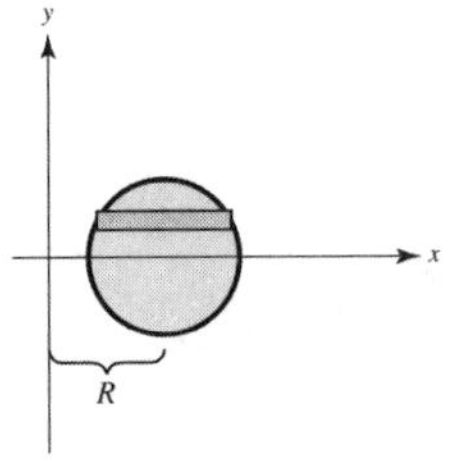

(b) $\int_0^r \sqrt{r^2 - y^2}\, dy$ is one-quarter of the area of a circle of radius r, $\frac{1}{4}\pi r^2$.

$$V = 8\pi R\left(\tfrac{1}{4}\pi r^2\right) = 2\pi^2 r^2 R$$

Section 7.3 Volume: The Shell Method

1. $p(x) = x, h(x) = x$

$$V = 2\pi \int_0^2 x(x)\, dx = \left[\frac{2\pi x^3}{3} \right]_0^2 = \frac{16\pi}{3}$$

2. $p(x) = x, h(x) = 1 - x$

$$V = 2\pi \int_0^1 x(1 - x)\, dx$$

$$= 2\pi \int_0^1 (x - x^2)\, dx = 2\pi \left[\frac{x^2}{2} - \frac{x^3}{3} \right]_0^1 = \frac{\pi}{3}$$

3. $p(x) = x, h(x) = \sqrt{x}$

$$V = 2\pi \int_0^4 x\sqrt{x}\, dx = 2\pi \int_0^4 x^{3/2}\, dx = \left[\frac{4\pi}{5} x^{5/2} \right]_0^4 = \frac{128\pi}{5}$$

4. $p(x) = x, h(x) = 3 - \left(\frac{1}{2}x^2 + 1\right) = 2 - \frac{1}{2}x^2$

$$V = 2\pi \int_0^2 x\left(2 - \frac{1}{2}x^2\right) dx$$

$$= 2\pi \left[x^2 - \frac{x^4}{8} \right]_0^2 = 2\pi(4 - 2) = 4\pi$$

5. $p(x) = x, h(x) = \frac{1}{4}x^2$

$$V = 2\pi \int_0^4 x\left(\frac{1}{4}x^2\right) dx$$

$$= \frac{\pi}{2}\left[\frac{x^4}{4} \right]_0^4$$

$$= 32\pi$$

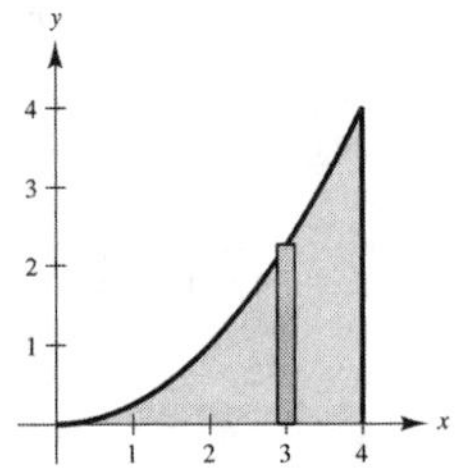

6. $p(x) = x, h(x) = \frac{1}{2}x^3$

$$V = 2\pi \int_0^3 x\left(\frac{x^3}{2}\right) dx$$

$$= \pi \left[\frac{x^5}{5} \right]_0^3$$

$$= \frac{243\pi}{5}$$

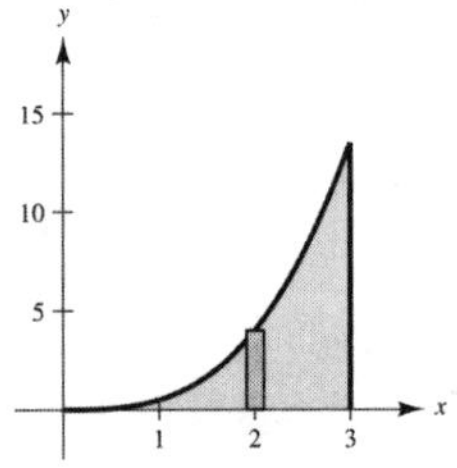

7. $p(x) = x, h(x) = (4x - x^2) - x^2 = 4x - 2x^2$

$$V = 2\pi \int_0^2 x(4x - 2x^2)\, dx$$

$$= 4\pi \int_0^2 (2x^2 - x^3)\, dx$$

$$= 4\pi \left[\frac{2}{3}x^3 - \frac{1}{4}x^4 \right]_0^2 = \frac{16\pi}{3}$$

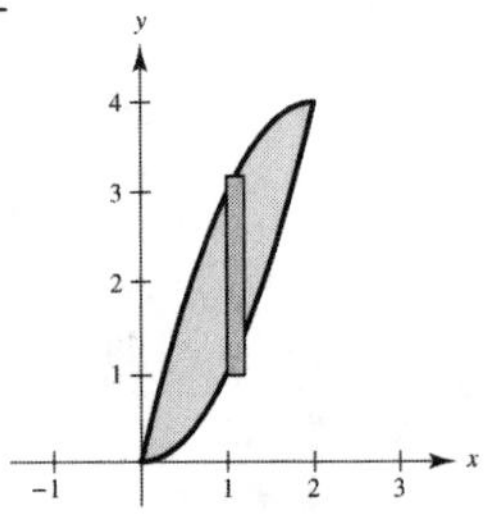

8. $p(x) = x, h(x) = 9 - x^2$

$$V = 2\pi\int_0^3 x(9 - x^2)\,dx$$
$$= 2\pi\left[\frac{9x^2}{2} - \frac{x^4}{4}\right]_0^3$$
$$= 2\pi\left[\frac{81}{2} - \frac{81}{4}\right] = \frac{81\pi}{2}$$

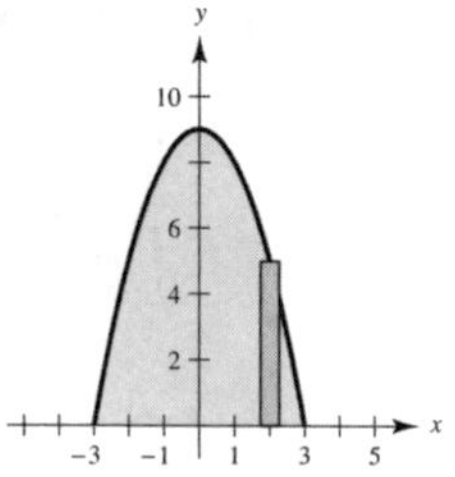

9. $p(x) = x$

$$h(x) = 4 - (4x - x^2)$$
$$= x^2 - 4x + 4$$
$$V = 2\pi\int_0^2 x(x^2 - 4x + 4)\,dx$$
$$V = 2\pi\int_0^2 (x^3 - 4x^2 + 4x)\,dx$$
$$= 2\pi\left[\frac{x^4}{4} - \frac{4}{3}x^3 + 2x^2\right]_0^2$$
$$= \frac{8\pi}{3}$$

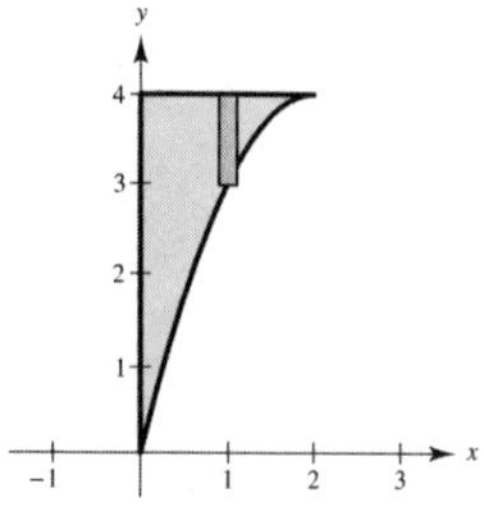

10. $p(x) = x, h(x) = 8 - x^{3/2}$

$$V = 2\pi\int_0^4 x(8 - x^{3/2})\,dx$$
$$= 2\pi\left[4x^2 - \frac{2}{7}x^{7/2}\right]_0^4$$
$$= 2\pi\left[64 - \frac{2}{7}(128)\right] = \frac{384\pi}{7}$$

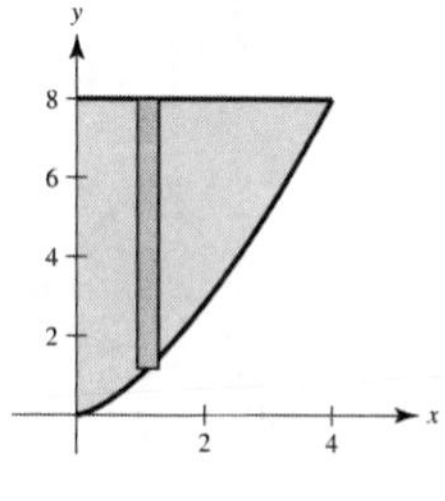

11. $p(x) = x, h(x) = \sqrt{x - 2}$

$$V = 2\pi\int_2^4 x\sqrt{x - 2}\,dx$$

Let $u = x - 2, x = u + 2, du = dx$.

When $x = 2, u = 0$.

When $x = 4, u = 2$.

$$V = 2\pi\int_0^2 (u + 2)u^{1/2}\,du$$
$$= 2\pi\left[\frac{2}{5}u^{5/2} + \frac{4}{3}u^{3/2}\right]_0^2$$
$$= 2\pi\left[\frac{2}{5}(2)^{5/2} + \frac{4}{3}(2)^{3/2}\right]$$
$$= 2\pi\sqrt{2}\left[\frac{2}{5}(4) + \frac{4}{3}(2)\right] = \frac{128\sqrt{2}\pi}{15}$$

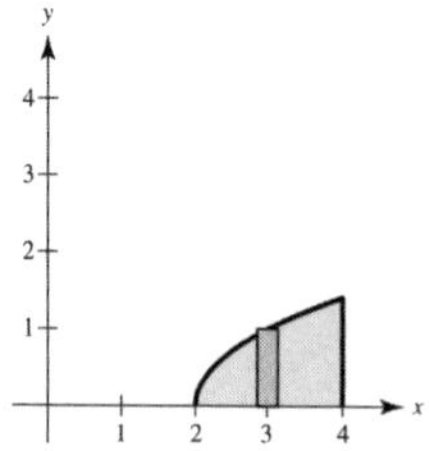

12. $p(x) = x, h(x) = 1 - x^2$

$$V = 2\pi\int_0^1 x(1 - x^2)\,dx$$
$$= 2\pi\left[\frac{x^2}{2} - \frac{x^4}{4}\right]_0^1$$
$$= 2\pi\left(\frac{1}{2} - \frac{1}{4}\right) = \frac{\pi}{2}$$

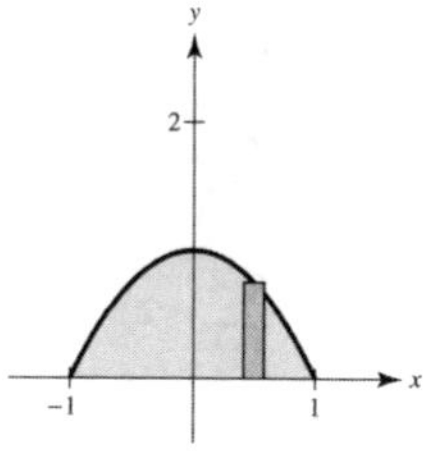

13. $p(x) = x, h(x) = \frac{1}{\sqrt{2\pi}}e^{-x^2/2}$

$$V = 2\pi\int_0^1 x\left(\frac{1}{\sqrt{2\pi}}e^{-x^2/2}\right)dx$$
$$= \sqrt{2\pi}\int_0^1 e^{-x^2/2}x\,dx$$
$$= \left[-\sqrt{2\pi}e^{-x^2/2}\right]_0^1$$
$$= \sqrt{2\pi}\left(1 - \frac{1}{\sqrt{e}}\right)$$
$$\approx 0.986$$

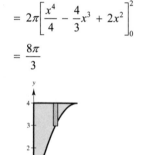

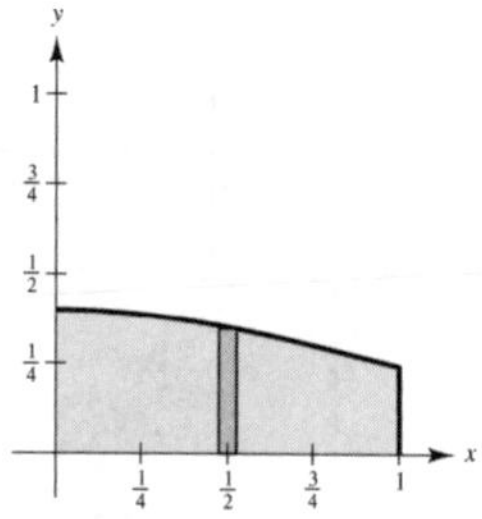

14. $p(x) = x,\ h(x) = \dfrac{\sin x}{x}$

$$V = 2\pi \int_0^{\pi} x\left[\frac{\sin x}{x}\right] dx$$
$$= 2\pi \int_0^{\pi} \sin x\, dx$$
$$= \left[-2\pi \cos x\right]_0^{\pi} = 4\pi$$

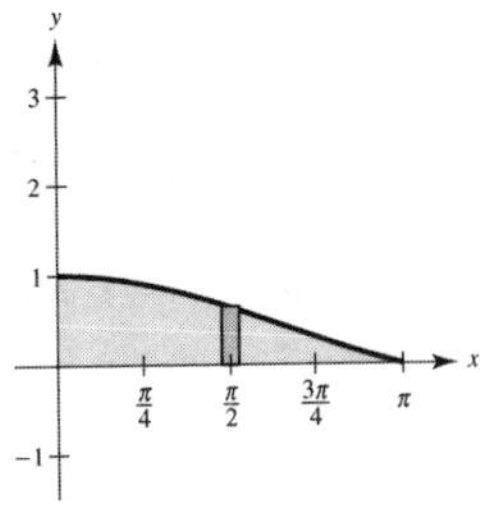

15. $p(y) = y,\ h(y) = 2 - y$

$$V = 2\pi \int_0^2 y(2 - y)\, dy$$
$$= 2\pi \int_0^2 \left(2y - y^2\right) dy$$
$$= 2\pi\left[y^2 - \frac{y^3}{3}\right]_0^2 = \frac{8\pi}{3}$$

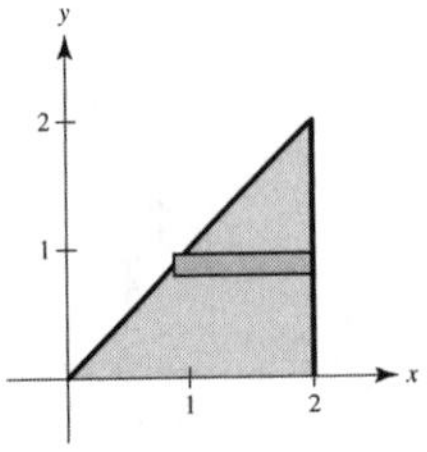

16. $p(y) = -y$ $\qquad$ (So, $p(y) \ge 0$ on $[-2, 0]$)

$h(y) = 3 - (1 - y) = 2 + y$

$$V = 2\pi \int_{-2}^0 (-y)(2 + y)\, dy$$
$$= 2\pi \int_{-2}^0 \left[-2y - y^2\right] dy$$
$$= 2\pi\left[-y^2 - \frac{y^3}{3}\right]_{-2}^0$$
$$= 2\pi\left[0 - \left(-4 + \frac{8}{3}\right)\right]$$
$$= 2\pi \frac{4}{3}$$
$$= \frac{8\pi}{3}$$

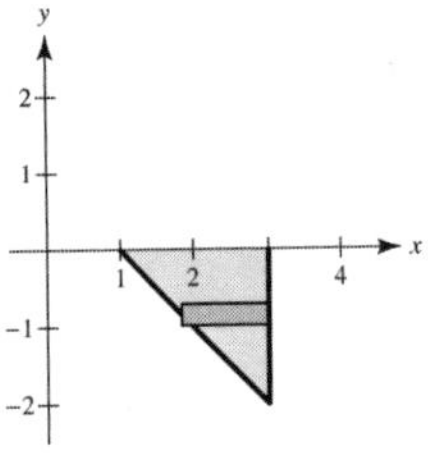

17. $p(y) = y$ and $h(y) = 1$ if $0 \le y < \dfrac{1}{2}$.

$p(y) = y$ and $h(y) = \dfrac{1}{y} - 1$ if $\dfrac{1}{2} \le y \le 1$.

$$V = 2\pi \int_0^{1/2} y\, dy + 2\pi \int_{1/2}^1 (1 - y)\, dy$$
$$= 2\pi\left[\frac{y^2}{2}\right]_0^{1/2} + 2\pi\left[y - \frac{y^2}{2}\right]_{1/2}^1 = \frac{\pi}{4} + \frac{\pi}{4} = \frac{\pi}{2}$$

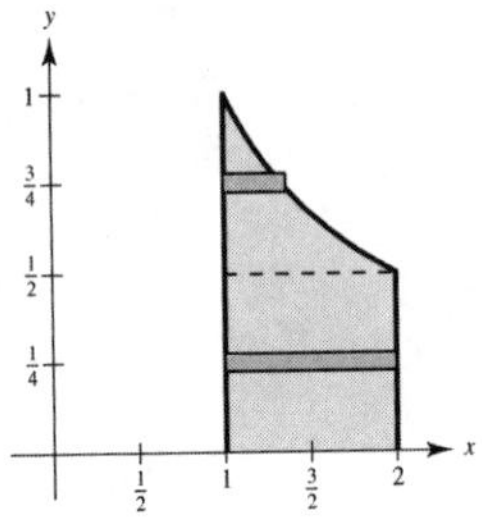

18. $p(y) = y,\ h(y) = 16 - y^2$

$$V = 2\pi \int_0^4 y\left(16 - y^2\right) dy$$
$$= 2\pi \int_0^4 \left(16y - y^3\right) dy$$
$$= 2\pi\left[8y^2 - \frac{y^4}{4}\right]_0^4$$
$$= 2\pi(128 - 64)$$
$$= 128\pi$$

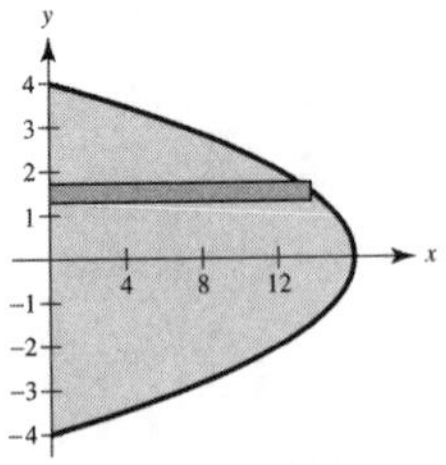

19. $p(y) = y,\ h(y) = \sqrt[3]{y}$

$$V = 2\pi \int_0^8 y\sqrt[3]{y}\, dy$$
$$= 2\pi \int_0^8 y^{4/3}\, dy$$
$$= \left[2\pi\left(\frac{3}{7}\right)y^{7/3}\right]_0^8$$
$$= \frac{6\pi}{7}\left(2^7\right) = \frac{768\pi}{7}$$

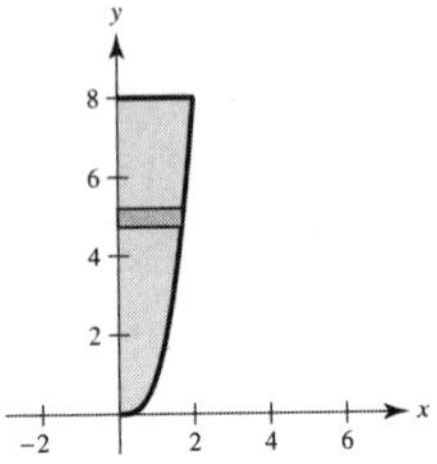

20. $y = 4x^2,\ x = \dfrac{\sqrt{y}}{2}$

$p(y) = y,\ h(y) = \dfrac{\sqrt{y}}{2}$

$$V = 2\pi \int_0^4 y\left(\frac{\sqrt{y}}{2}\right) dy$$
$$= \pi \int_0^4 y^{3/2}\, dy$$
$$= \pi\left[\frac{2}{5}y^{5/2}\right]_0^4$$
$$= \frac{64\pi}{5}$$

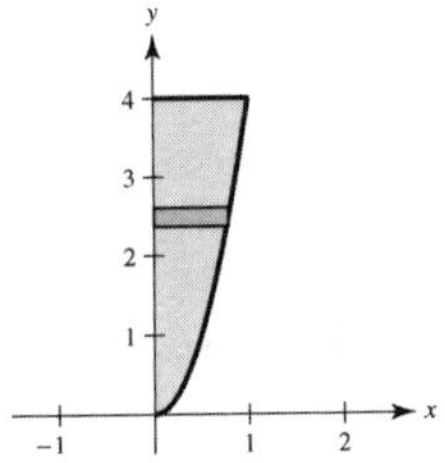

21. $p(y) = y,\ h(y) = (4 - y) - (y) = 4 - 2y$

$$V = 2\pi \int_0^2 y(4 - 2y)\, dy$$
$$= 2\pi \int_0^2 \left(4y - 2y^2\right) dy$$
$$= 2\pi\left[2y^2 - \frac{2}{3}y^3\right]_0^2$$
$$= 2\pi\left(8 - \frac{16}{3}\right) = \frac{16\pi}{3}$$

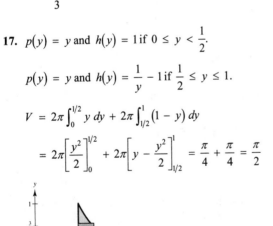

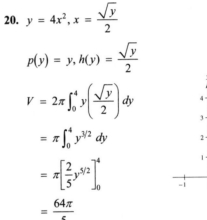

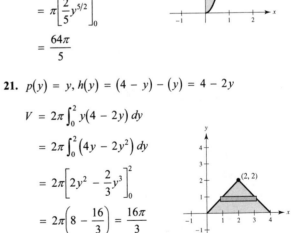

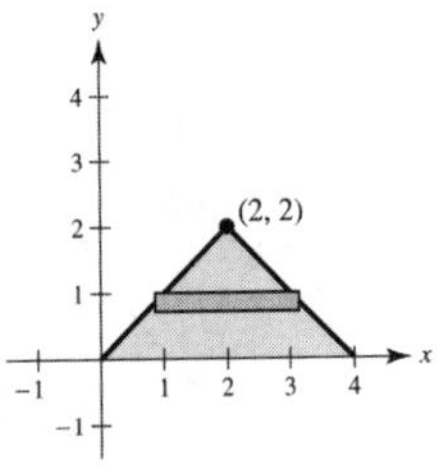

22. $p(y) = y, h(y) = y - (y^2 - 2) = 2 + y - y^2$

$$V = 2\pi\int_0^2 y(2 + y - y^2)\,dy$$
$$= 2\pi\int_0^2 (2y + y^2 - y^3)\,dy$$
$$= 2\pi\left[y^2 + \frac{y^3}{3} - \frac{y^4}{4}\right]_0^2$$
$$= 2\pi\left(4 + \frac{8}{3} - 4\right) = \frac{16\pi}{3}$$

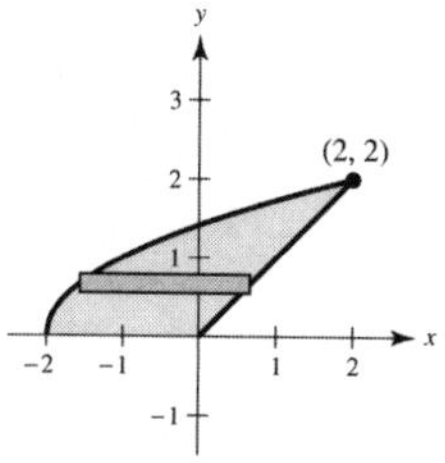

23. $p(x) = 4 - x, h(x) = 2x - x^2$

$$V = 2\pi\int_0^2 (4 - x)(2x - x^2)\,dx$$
$$= 2\pi\int_0^2 (8x - 6x^2 + x^3)\,dx$$
$$= 2\pi\left[4x^2 - 2x^3 + \frac{x^4}{4}\right]_0^2$$
$$= 2\pi[16 - 16 + 4] = 8\pi$$

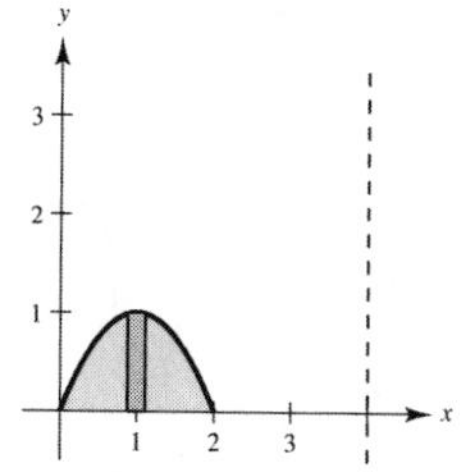

24. $p(x) = 6 - x, h(x) = \sqrt{x}$

$$V = 2\pi\int_0^4 (6 - x)\sqrt{x}\,dx$$
$$= 2\pi\int_0^4 (6x^{1/2} - x^{3/2})\,dx$$
$$= 2\pi\left[4x^{3/2} - \frac{2}{5}x^{5/2}\right]_0^4 = \frac{192\pi}{5}$$

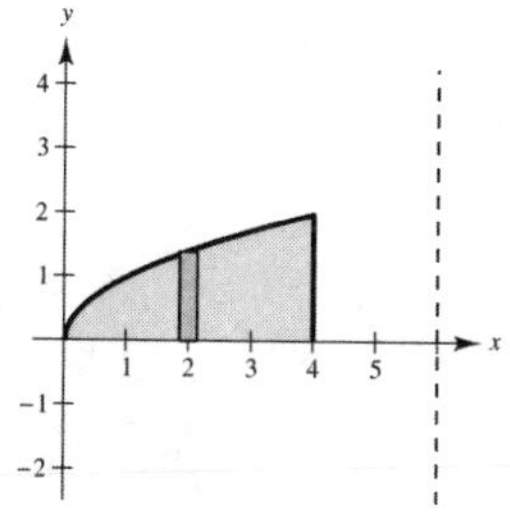

25. $p(x) = 4 - x, h(x) = 4x - x^2 - x^2 = 4x - 2x^2$

$$V = 2\pi\int_0^2 (4 - x)(4x - 2x^2)\,dx$$
$$= 2\pi(2)\int_0^2 (x^3 - 6x^2 + 8x)\,dx$$
$$= 4\pi\left[\frac{x^4}{4} - 2x^3 + 4x^2\right]_0^2 = 16\pi$$

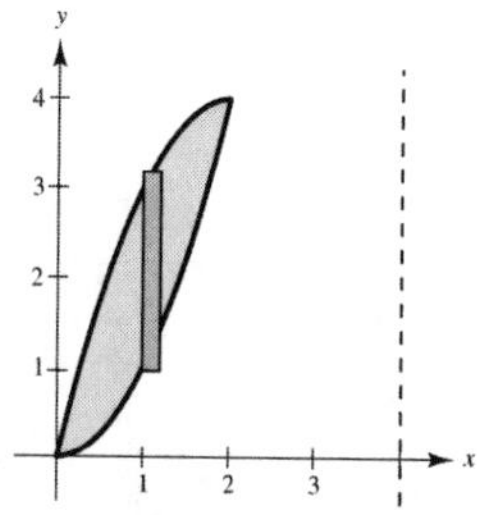

26. $p(x) = 3 - x, h(x) = (6x - x^2) - \frac{1}{3}x^3$

$$V = 2\pi\int_0^3 (3 - x)\left(6x - x^2 - \frac{x^3}{3}\right)dx$$
$$= 2\pi\int_0^3 \left(\frac{x^4}{3} - 9x^2 + 18x\right)dx$$
$$= 2\pi\left[\frac{x^5}{15} - 3x^3 + 9x^2\right]_0^3$$
$$= \frac{162\pi}{5}$$

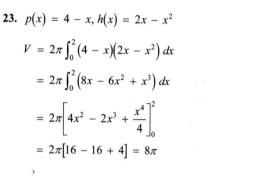

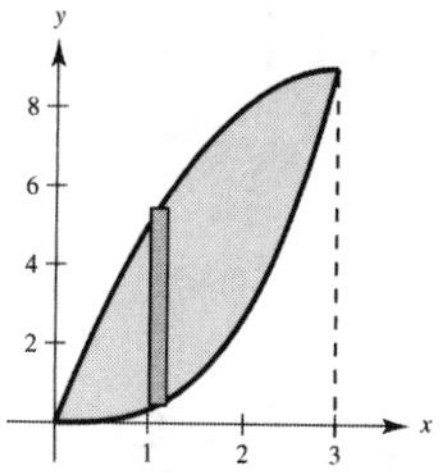

27. The shell method would be easier:

$$V = 2\pi\int_0^4 \left[4 - (y - 2)^2\right]y\,dy$$

Using the disk method:

$$V = \pi\int_0^4 \left[\left(2 + \sqrt{4 - x}\right)^2 - \left(2 - \sqrt{4 - x}\right)^2\right]dx$$

$\left[\textbf{Note: } V = \frac{128\pi}{3}\right]$

28. The shell method is easier: $V = 2\pi\int_0^{\ln 4} x(4 - e^x)\,dx$

Using the disk method, $x = \ln(4 - y)$ and $V = \pi\int_0^3 (\ln(4 - y))^2\,dy$. $\left[\textbf{Note: } V = \pi\left[8(\ln 2)^2 - 8\ln 2 + 3\right]\right]$

29. (a) **Disk**

$R(x) = x^3, r(x) = 0$

$$V = \pi\int_0^2 x^6\,dx = \pi\left[\frac{x^7}{7}\right]_0^2 = \frac{128\pi}{7}$$

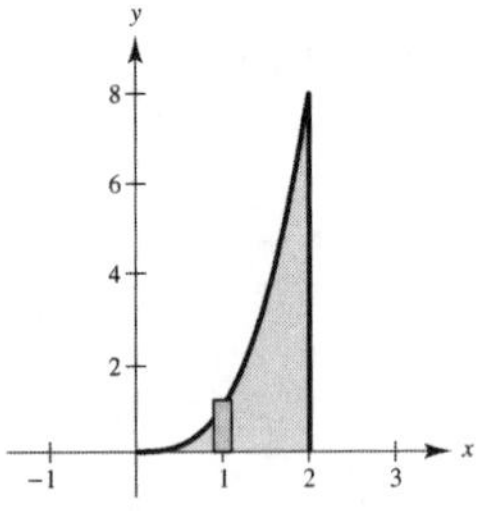

(b) **Shell**

$p(x) = x, h(x) = x^3$

$$V = 2\pi\int_0^2 x^4\,dx = 2\pi\left[\frac{x^5}{5}\right]_0^2 = \frac{64\pi}{5}$$

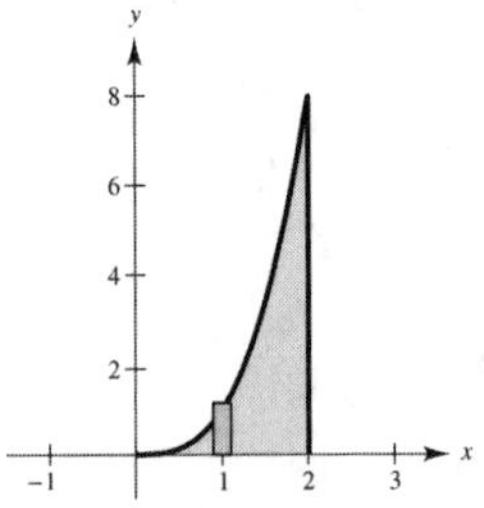

(c) **Shell**

$p(x) = 4 - x, h(x) = x^3$

$$V = 2\pi\int_0^2 (4 - x)x^3\,dx$$

$$= 2\pi\int_0^2 \left(4x^3 - x^4\right)dx$$

$$= 2\pi\left[x^4 - \frac{1}{5}x^5\right]_0^2 = \frac{96\pi}{5}$$

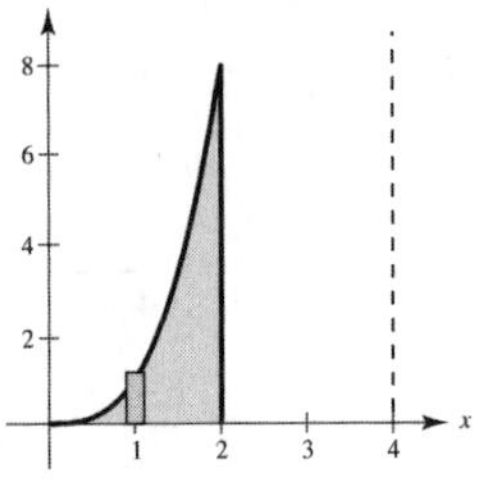

30. (a) **Disk**

$R(x) = \dfrac{10}{x^2}, r(x) = 0$

$$V = \pi\int_1^5 \left(\frac{10}{x^2}\right)^2 dx$$

$$= 100\pi\int_1^5 x^{-4}\,dx$$

$$= 100\pi\left[\frac{x^{-3}}{-3}\right]_1^5$$

$$= -\frac{100\pi}{3}\left(\frac{1}{125} - 1\right) = \frac{496}{15}\pi$$

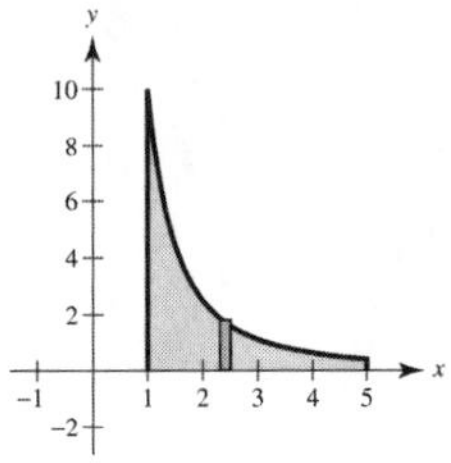

(b) **Shell**

$R(x) = x, r(x) = 0$

$$V = 2\pi\int_1^5 x\left(\frac{10}{x^2}\right)dx$$

$$= 20\pi\int_1^5 \frac{1}{x}\,dx$$

$$= 20\pi\Big[\ln|x|\Big]_1^5 = 20\pi\ln 5$$

(c) **Disk**

$R(x) = 10, r(x) = 10 - \dfrac{10}{x^2}$

$$V = \pi\int_1^5 \left[10^2 - \left(10 - \frac{10}{x^2}\right)^2\right]dx$$

$$= \pi\left[\frac{100}{3x^3} - \frac{200}{x}\right]_1^5 = \frac{1904}{15}\pi$$

31. (a) **Shell**

$$p(y) = y,\ h(y) = \left(a^{1/2} - y^{1/2}\right)^2$$

$$V = 2\pi \int_0^a y\left(a - 2a^{1/2}y^{1/2} + y\right) dy$$

$$= 2\pi \int_0^a \left(ay - 2a^{1/2}y^{3/2} + y^2\right) dy$$

$$= 2\pi\left[\frac{a}{2}y^2 - \frac{4a^{1/2}}{5}y^{5/2} + \frac{y^3}{3}\right]_0^a$$

$$= 2\pi\left(\frac{a^3}{2} - \frac{4a^3}{5} + \frac{a^3}{3}\right) = \frac{\pi a^3}{15}$$

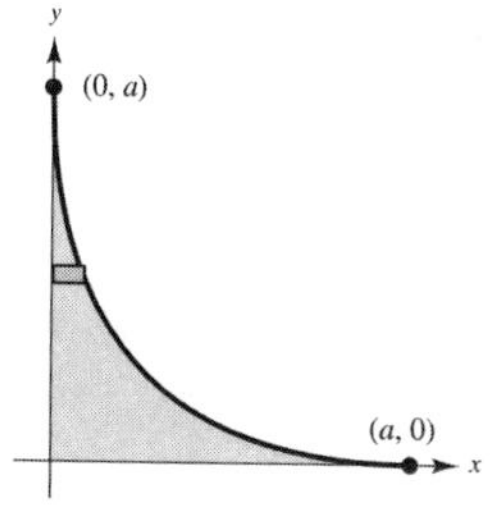

(b) Same as part (a) by symmetry

(c) **Shell**

$$p(x) = a - x,\ h(x) = \left(a^{1/2} - x^{1/2}\right)^2$$

$$V = 2\pi \int_0^a (a - x)\left(a^{1/2} - x^{1/2}\right)^2 dx$$

$$= 2\pi \int_0^a \left(a^2 - 2a^{3/2}x^{1/2} + 2a^{1/2}x^{3/2} - x^2\right) dx$$

$$= 2\pi\left[a^2x - \frac{4}{3}a^{3/2}x^{3/2} + \frac{4}{5}a^{1/2}x^{5/2} - \frac{1}{3}x^3\right]_0^a$$

$$= \frac{4\pi a^3}{15}$$

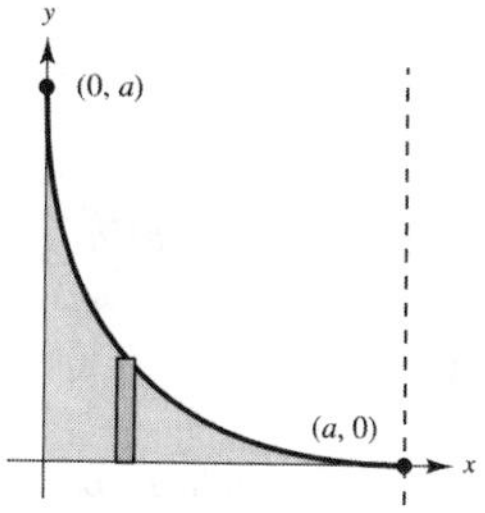

32. (a) **Disk**

$$R(x) = \left(a^{2/3} - x^{2/3}\right)^{3/2},\ r(x) = 0$$

$$V = \pi \int_{-a}^a \left(a^{2/3} - x^{2/3}\right)^3 dx$$

$$= 2\pi \int_0^a \left(a^2 - 3a^{4/3}x^{2/3} + 3a^{2/3}x^{4/3} - x^2\right) dx$$

$$= 2\pi\left[a^2x - \frac{9}{5}a^{4/3}x^{5/3} + \frac{9}{7}a^{2/3}x^{7/3} - \frac{1}{3}x^3\right]_0^a$$

$$= 2\pi\left(a^3 - \frac{9}{5}a^3 + \frac{9}{7}a^3 - \frac{1}{3}a^3\right) = \frac{32\pi a^3}{105}$$

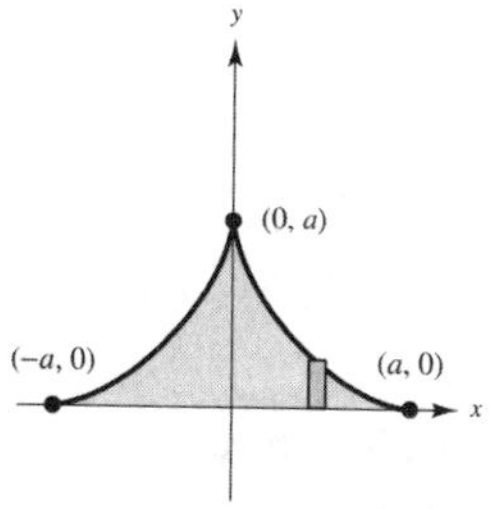

(b) Same as part (a) by symmetry

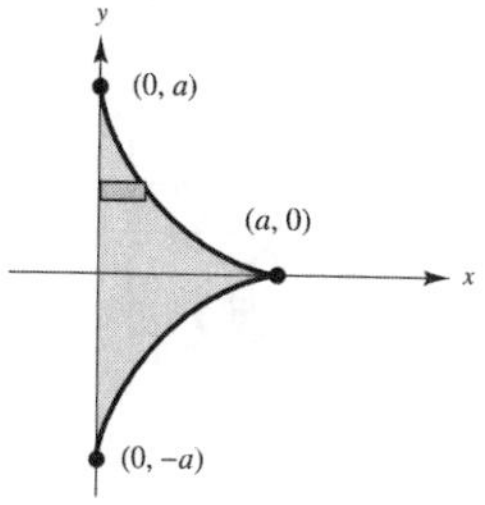

33. (a)

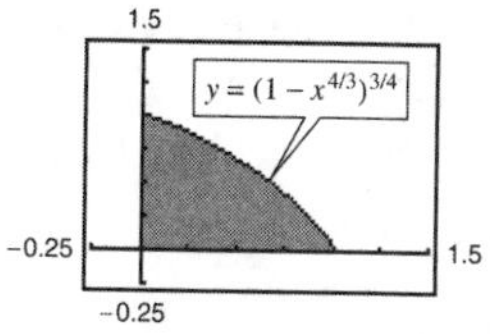

(b) $x^{4/3} + y^{4/3} = 1,\ x = 0,\ y = 0$

$$y = \left(1 - x^{4/3}\right)^{3/4}$$

$$V = 2\pi \int_0^1 x\left(1 - x^{4/3}\right)^{3/4} dx \approx 1.5056$$

34. (a)

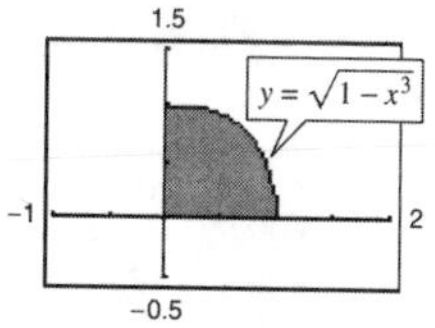

(b) $V = 2\pi \int_0^1 x\sqrt{1 - x^3}\, dx \approx 2.3222$

35. (a)

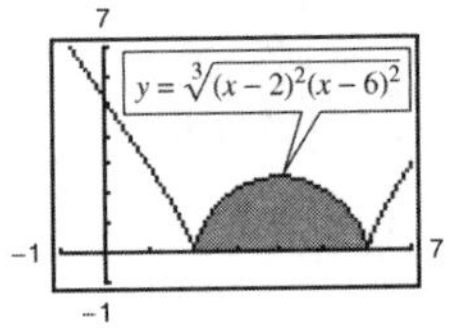

(b) $V = 2\pi\int_2^6 x\sqrt[3]{(x-2)^2(x-6)^2}\,dx \approx 187.249$

36. (a)

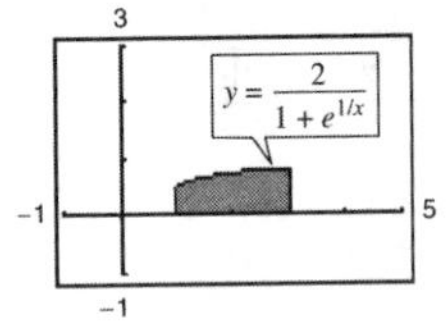

(b) $V = 2\pi\int_1^3 \dfrac{2x}{1+e^{1/x}}\,dx \approx 19.0162$

37. Answers will vary.

(a) The rectangles would be vertical.

(b) The rectangles would be horizontal.

38. (a) radius = k

height = b

(b) radius = b

height = k

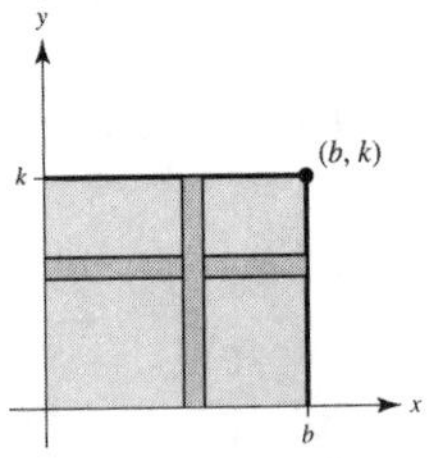

39. $\pi\int_1^5 (x-1)\,dx = \pi\int_1^5 \left(\sqrt{x-1}\right)^2 dx$

This integral represents the volume of the solid generated by revolving the region bounded by $y = \sqrt{x-1}$, $y = 0$, and $x = 5$ about the x-axis by using the disk method.

$2\pi\int_0^2 y\left[5 - \left(y^2+1\right)\right] dy$

represents this same volume by using the shell method.

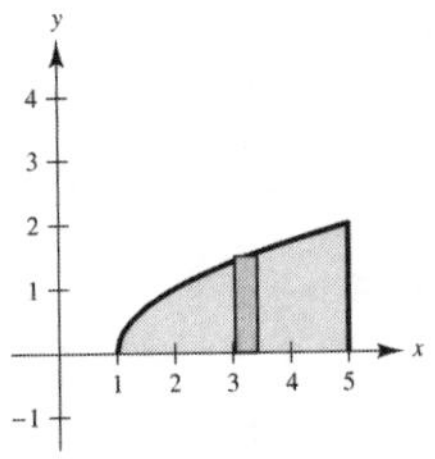

Disk method

40. $2\pi\int_0^4 x\left(\dfrac{x}{2}\right) dx$

This integral represents the volume of the solid generated by revolving the region bounded by $y = x/2$, $y = 0$, and $x = 4$ about the y-axis by using the shell method.

$\pi\int_0^2 \left[16 - (2y)^2\right] dy = \pi\int_0^2 \left[(4)^2 - (2y)^2\right] dy$

represents this same volume by using the disk method.

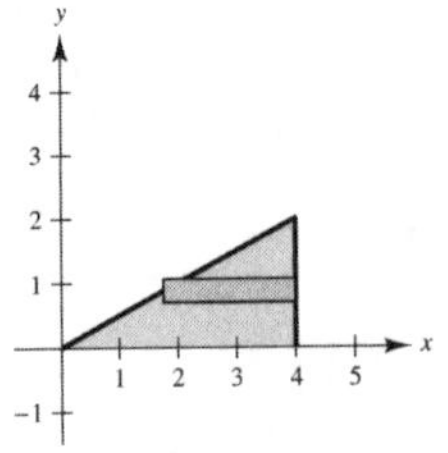

Disk method

41.

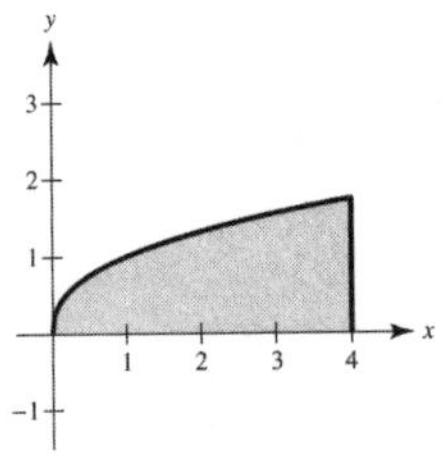

(a) Around x-axis: $V = \pi\int_0^4 \left(x^{2/5}\right)^2 dx = \left[\pi\dfrac{5}{9}x^{9/5}\right]_0^4$

$= \dfrac{5}{9}\pi(4)^{9/5} \approx 6.7365\pi$

(b) Around y-axis: $V = 2\pi\int_0^4 x\left(x^{2/5}\right) dx$

$= \left[2\pi\dfrac{5}{12}x^{12/5}\right]_0^4 \approx 23.2147\pi$

(c) Around $x = 4$:

$V = 2\pi\int_0^4 (4-x)x^{2/5}\,dx \approx 16.5819\pi$

So, $(a) < (c) < (b)$.

42. (a) The figure will be a circle of radius AB and center A.

(b) The figure will be a circular cylinder of radius AB.

(c) Disk method: $V = \pi\int_0^3 \left[g(y)\right]^2 dy$

Shell method: $V = 2\pi\int_0^{2.45} x\,f(x)\,dx$

43. $2\pi \int_0^2 x^3\,dx = 2\pi \int_0^2 x(x^2)\,dx$

(a) Plane region bounded by

$y = x^2, y = 0, x = 0, x = 2$

(b) Revolved about the y-axis

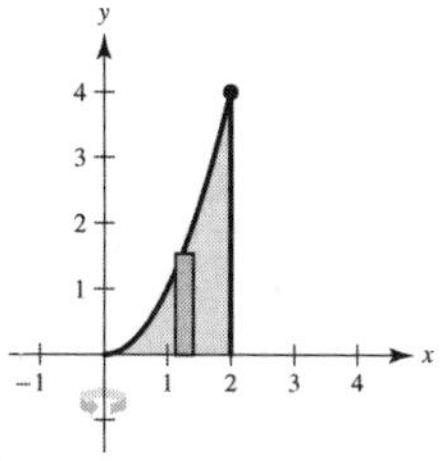

Other answers possible.

44. $2\pi \int_0^1 (y - y^{3/2})\,dy = 2\pi \int_0^1 y(1 - \sqrt{y})\,dy$

(a) Plane region bounded by $x = \sqrt{y}, x = 1, y = 0$

(b) Revolved about the x-axis

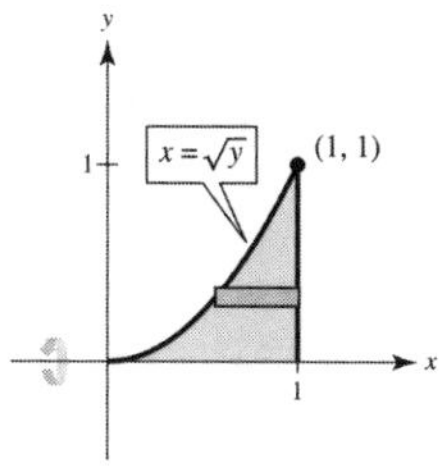

Other answers possible.

45. $2\pi \int_0^6 (y + 2)\sqrt{6 - y}\,dy$

(a) Plane region bounded by

$x = \sqrt{6 - y}, x = 0, y = 0$

(b) Revolved around line $y = -2$

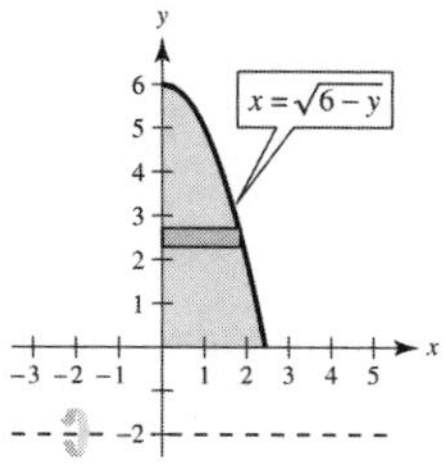

Other answers possible.

46. $2\pi \int_0^1 (4 - x)e^x\,dx$

(a) Plane region bounded by

$y = e^x, y = 0, x = 0, x = 1$

(b) Revolved about the line $x = 4$

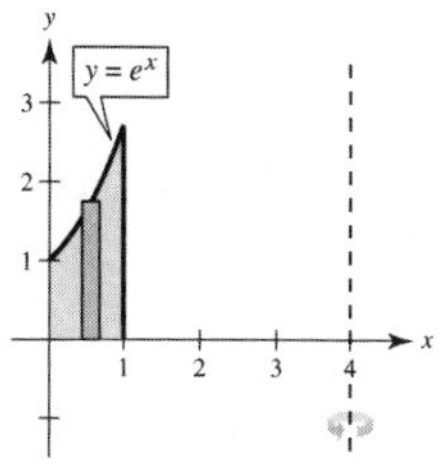

47. $p(x) = x, h(x) = 2 - \frac{1}{2}x^2$

$$V = 2\pi \int_0^2 x\left(2 - \tfrac{1}{2}x^2\right) dx$$

$$= 2\pi \int_0^2 \left(2x - \tfrac{1}{2}x^3\right) dx$$

$$= 2\pi\left[x^2 - \tfrac{1}{8}x^4\right]_0^2 = 4\pi \text{ (total volume)}$$

Now find x_0 such that:

$$\pi = 2\pi \int_0^{x_0} \left(2x - \tfrac{1}{2}x^3\right) dx$$

$$1 = 2\left[x^2 - \tfrac{1}{8}x^4\right]_0^{x_0}$$

$$1 = 2x_0^2 - \tfrac{1}{4}x_0^4$$

$$x_0^4 - 8x_0^2 + 4 = 0$$

$$x_0^2 = 4 \pm 2\sqrt{3} \quad \text{(Quadratic Formula)}$$

Take $x_0 = \sqrt{4 - 2\sqrt{3}} \approx 0.73205$, because the other root is too large.

Diameter: $2\sqrt{4 - 2\sqrt{3}} \approx 1.464$

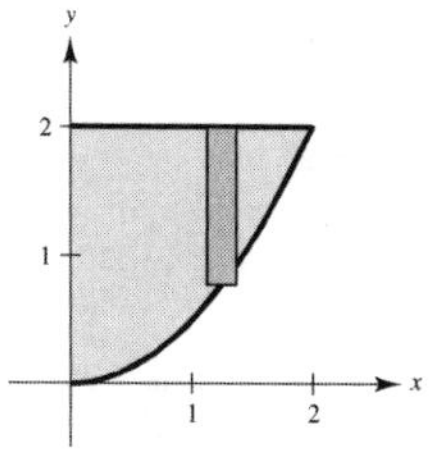

48. Total volume of the hemisphere is $\frac{1}{2}\left(\frac{4}{3}\right)\pi r^3 = \frac{2}{3}\pi(3)^3 = 18\pi$. By the Shell Method, $p(x) = x$, $h(x) = \sqrt{9 - x^2}$. Find x_0 such that:

$$6\pi = 2\pi\int_0^{x_0} x\sqrt{9 - x^2}\,dx$$

$$6 = -\int_0^{x_0}\left(9 - x^2\right)^{1/2}(-2x)\,dx$$

$$= \left[-\tfrac{2}{3}\left(9 - x^2\right)^{3/2}\right]_0^{x_0} = 18 - \tfrac{2}{3}\left(9 - x_0^2\right)^{3/2}$$

$$\left(9 - x_0^2\right)^{3/2} = 18$$

$$x_0 = \sqrt{9 - 18^{2/3}} \approx 1.460$$

Diameter: $2\sqrt{9 - 18^{2/3}} \approx 2.920$

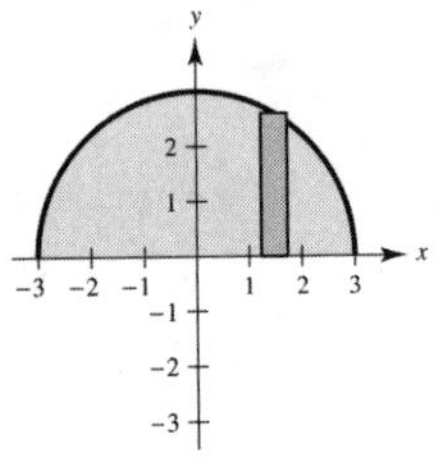

49. $$V = 4\pi\int_{-1}^{1}(2 - x)\sqrt{1 - x^2}\,dx$$

$$= 8\pi\int_{-1}^{1}\sqrt{1 - x^2}\,dx - 4\pi\int_{-1}^{1}x\sqrt{1 - x^2}\,dx$$

$$= 8\pi\left(\frac{\pi}{2}\right) + 2\pi\int_{-1}^{1}x\left(1 - x^2\right)^{1/2}(-2)\,dx$$

$$= 4\pi^2 + \left[2\pi\left(\frac{2}{3}\right)\left(1 - x^2\right)^{3/2}\right]_{-1}^{1} = 4\pi^2$$

50. $$V = 4\pi\int_{-r}^{r}(R - x)\sqrt{r^2 - x^2}\,dx$$

$$= 4\pi R\int_{-r}^{r}\sqrt{r^2 - x^2}\,dx - 4\pi\int_{-r}^{r}x\sqrt{r^2 - x^2}\,dx$$

$$= 4\pi R\left(\frac{\pi r^2}{2}\right) + \left[2\pi\left(\frac{2}{3}\right)\left(r^2 - x^2\right)^{3/2}\right]_{-r}^{r}$$

$$= 2\pi^2 r^2 R$$

51. (a) $$\frac{d}{dx}\left[\sin x - x\cos x + C\right] = \cos x + x\sin x - \cos x$$

$$= x\sin x$$

So, $\int x\sin x\,dx = \sin x - x\cos x + C$.

(b) (i) $p(x) = x$, $h(x) = \sin x$

$$V = 2\pi\int_0^{\pi/2} x\sin x\,dx$$

$$= 2\pi\left[\sin x - x\cos x\right]_0^{\pi/2}$$

$$= 2\pi\left[(1 - 0) - 0\right] = 2\pi$$

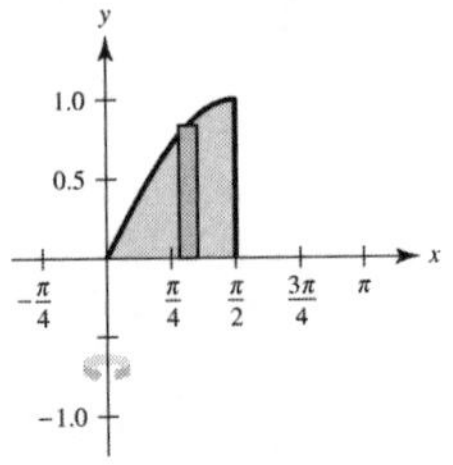

(ii) $p(x) = x$, $h(x) = 2\sin x - (-\sin x) = 3\sin x$

$$V = 2\pi\int_0^{\pi} x(3\sin x)\,dx$$

$$= 6\pi\int_0^{\pi} x\sin x\,dx$$

$$= 6\pi\left[\sin x - x\cos x\right]_0^{\pi}$$

$$= 6\pi(\pi) = 6\pi^2$$

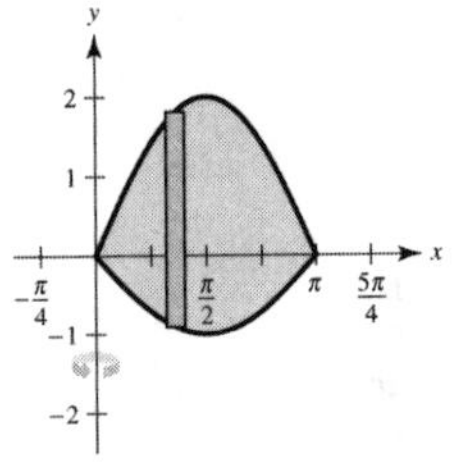

52. (a) $\frac{d}{dx}\left[\cos x + x\sin x + C\right] = -\sin x + \sin x + x\cos x$

$= x\cos x$

Hence, $\int x\cos x\,dx = \cos x + x\sin x + C.$

(b) (i) $x^2 = \cos x \Rightarrow x \approx \pm 0.8241$

$$V \approx 2(2\pi)\int_0^{0.8241} x\left[\cos x - x^2\right]dx$$

$$= 4\pi\left[\cos x + x\sin x - \frac{x^4}{4}\right]_0^{0.8241} \approx 2.1205$$

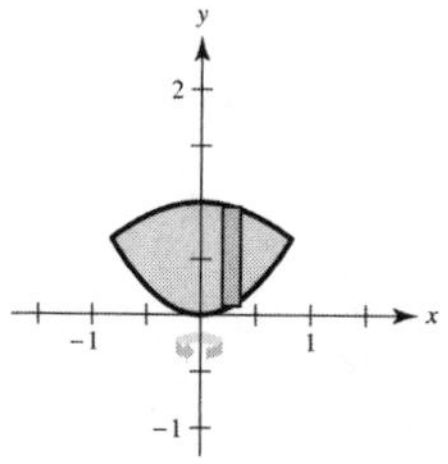

(ii) $4\cos x = (x-2)^2 \Rightarrow x = 0, 1.5110$

$$V \approx 2\pi\int_0^{1.511} x\left[4\cos x - (x-2)^2\right]dx$$

$$= 2\pi\int_0^{1.511}\left[4\cos x + 4x\sin x - \frac{(x-2)^3}{3}\right]_0^{1.511}$$

$= 6.2993$

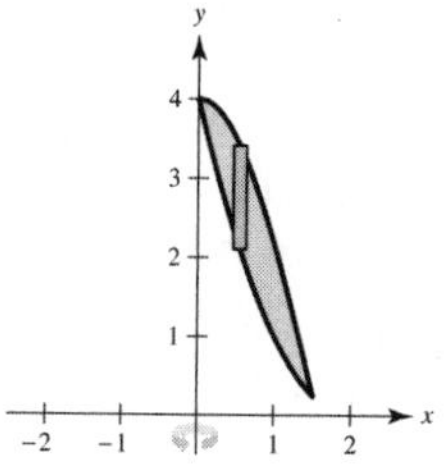

53. Disk Method

$R(y) = \sqrt{r^2 - y^2}$

$r(y) = 0$

$$V = \pi\int_{r-h}^{r}\left(r^2 - y^2\right)dy$$

$$= \pi\left[r^2y - \frac{y^3}{3}\right]_{r-h}^{r} = \frac{1}{3}\pi h^2(3r - h)$$

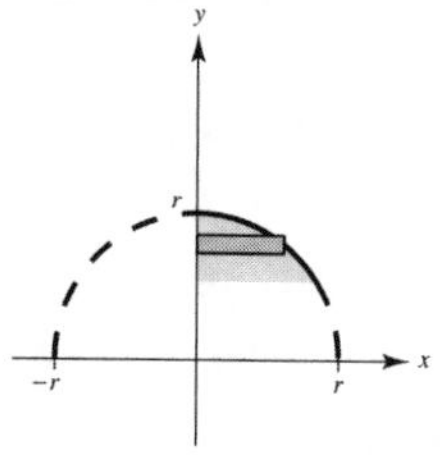

54. $\frac{x^2}{a^2} + \frac{y^2}{b^2} = 1$

$\frac{y^2}{b^2} = 1 - \frac{x^2}{a^2}$

$y = \pm b\sqrt{1 - \frac{x^2}{a^2}}$

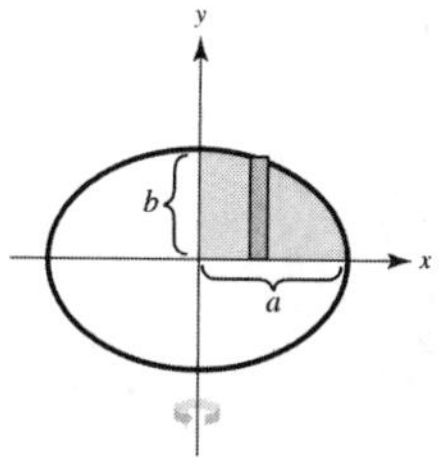

$p(x) = x,\ h(x) = b\sqrt{1 - \frac{x^2}{a^2}}$

$$V = 2(2\pi)\int_0^a xb\sqrt{1 - \frac{x^2}{a^2}}\,dx$$

$$= \frac{4\pi b}{a}\int_0^a \sqrt{a^2 - x^2}\;x\,dx$$

$$= \frac{4\pi b}{a}\left[\left(\frac{-\left(a^2 - x^2\right)^{3/2}}{3}\right)\right]_0^a$$

$$= \frac{4\pi b}{3a}a^3 = \frac{4}{3}\pi a^2 b$$

If the region is revolved about the x-axis, then by symmetry the volume would be $V = \frac{4}{3}\pi ab^2$.

Note: If $a = b$, then volume is that of a sphere.

55. (a) Area of region $= \int_0^b \left[ab^n - ax^n\right] dx$

$$= \left[ab^n x - a\frac{x^{n+1}}{n+1}\right]_0^b$$

$$= ab^{n+1} - a\frac{b^{n+1}}{n+1}$$

$$= ab^{n+1}\left(1 - \frac{1}{n+1}\right)$$

$$= ab^{n+1}\left(\frac{n}{n+1}\right)$$

$$R_1(n) = \frac{ab^{n+1}\left[n/(n+1)\right]}{\left(ab^n\right)b} = \frac{n}{n+1}$$

(b) $\lim_{n\to\infty} R_1(n) = \lim_{n\to\infty} \frac{n}{n+1} = 1$

$\lim_{n\to\infty} \left(ab^n\right)b = \infty$

(c) **Disk Method:**

$$V = 2\pi\int_0^b x\left(ab^n - ax^n\right) dx$$

$$= 2\pi a\int_0^b \left(xb^n - x^{n+1}\right) dx$$

$$= 2\pi a\left[\frac{b^n}{2}x^2 - \frac{x^{n+2}}{n+2}\right]_0^b$$

$$= 2\pi a\left[\frac{b^{n+2}}{2} - \frac{b^{n+2}}{n+2}\right] = \pi ab^{n+2}\left(\frac{n}{n+2}\right)$$

$$R_2(n) = \frac{\pi ab^{n+2}\left[n/(n+2)\right]}{\left(\pi b^2\right)\left(ab^n\right)} = \left(\frac{n}{n+2}\right)$$

(d) $\lim_{n\to\infty} R_2(n) = \lim_{n\to\infty}\left(\frac{n}{n+2}\right) = 1$

$\lim_{n\to\infty} \left(\pi b^2\right)\left(ab^n\right) = \infty$

(e) As $n \to \infty$, the graph approaches the line $x = b$.

56. (a) $2\pi\int_0^r hx\left(1 - \frac{x}{r}\right) dx$ (ii)

is the volume of a right circular cone with the radius of the base as r and height h.

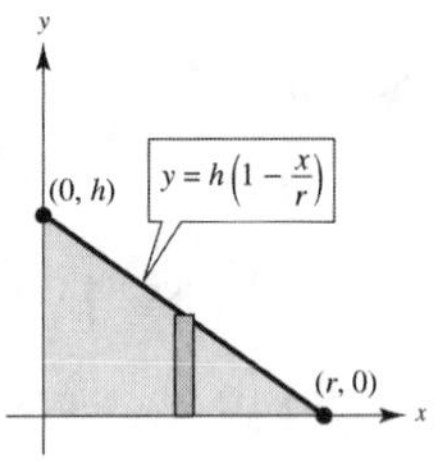

(b) $2\pi\int_{-r}^r (R - x)\left(2\sqrt{r^2 - x^2}\right) dx$ (v)

is the volume of a torus with the radius of its circular cross section as r and the distance from the axis of the torus to the center of its cross section as R.

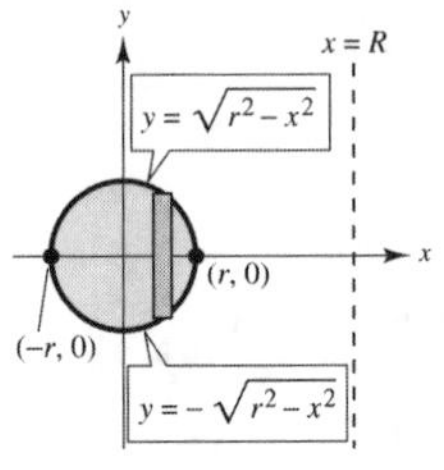

(c) $2\pi\int_0^r 2x\sqrt{r^2 - x^2}\, dx$ (iii)

is the volume of a sphere with radius r.

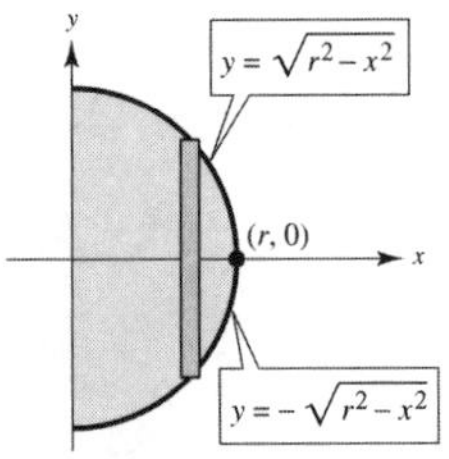

(d) $2\pi\int_0^r hx\, dx$ (i)

is the volume of a right circular cylinder with a radius of r and a height of h.

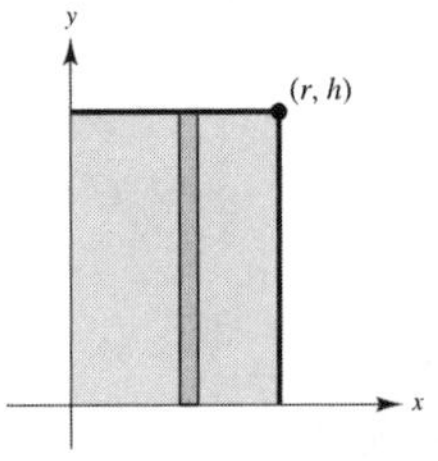

(e) $2\pi\int_0^b 2ax\sqrt{1 - \left(x^2/b^2\right)}\, dx$ (iv)

is the volume of an ellipsoid with axes $2a$ and $2b$.

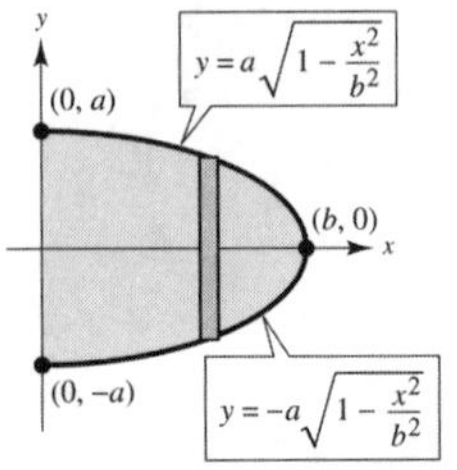

57. (a) $V = 2\pi\int_0^4 xf(x)\,dx = \frac{2\pi(40)}{3(4)}\left[0 + 4(10)(45) + 2(20)(40) + 4(30)(20) + 0\right] = \frac{20\pi}{3}(5800) \approx 121{,}475 \text{ ft}^3$

(b) Top line: $y - 50 = \frac{40 - 50}{20 - 0}(x - 0) = -\frac{1}{2}x \Rightarrow y = -\frac{1}{2}x + 50$

Bottom line: $y - 40 = \frac{0 - 40}{40 - 20}(x - 20) = -2(x - 20) \Rightarrow y = -2x + 80$

$$V = 2\pi\int_0^{20} x\left(-\frac{1}{2}x + 50\right)dx + 2\pi\int_{20}^{40} x(-2x + 80)\,dx$$

$$= 2\pi\int_0^{20}\left(-\frac{1}{2}x^2 + 50x\right)dx + 2\pi\int_{20}^{40}\left(-2x^2 + 80x\right)dx$$

$$= 2\pi\left[-\frac{x^3}{6} + 25x^2\right]_0^{20} + 2\pi\left[-\frac{2x^3}{3} + 40x^2\right]_{20}^{40} = 2\pi\left(\frac{26{,}000}{3}\right) + 2\pi\left(\frac{32{,}000}{3}\right) \approx 121{,}475 \text{ ft}^3$$

(Note that Simpson's Rule is exact for this problem.)

58. (a) $V = 2\pi\int_0^{200} xf(x)\,dx$

$$\approx \frac{2\pi(200)}{3(8)}\left[0 + 4(25)(19) + 2(50)(19) + 4(75)(17) + 2(100)15 + 4(125)(14) + 2(150)(10) + 4(175)(6) + 0\right]$$

$$\approx 1{,}366{,}593 \text{ ft}^3$$

(b) $d = -0.000561x^2 + 0.0189x + 19.39$

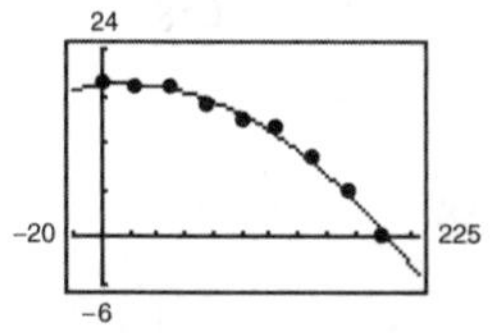

(c) $V \approx 2\pi\int_0^{200} xd(x)\,dx \approx 2\pi(213{,}800) = 1{,}343{,}345 \text{ ft}^3$

(d) Number of gallons $\approx V(7.48) = 10{,}048{,}221$ gal

59. $V_1 = \pi\int_{1/4}^{c}\frac{1}{x^2}\,dx = \pi\left[-\frac{1}{x}\right]_{1/4}^{c} = \pi\left[-\frac{1}{c} + 4\right] = \frac{4c - 1}{c}\pi$

$V_2 = \left[2\pi\int_{1/4}^{c} x\left(\frac{1}{x}\right)dx = 2\pi x\right]_{1/4}^{c} = 2\pi\left(c - \frac{1}{4}\right)$

$V_1 = V_2 \Rightarrow \frac{4c - 1}{c}\pi = 2\pi\left(c - \frac{1}{4}\right)$

$4c - 1 = 2c\left(c - \frac{1}{4}\right)$

$4c^2 - 9c + 2 = 0$

$(4c - 1)(c - 2) = 0$

$c = 2 \quad \left(c = \frac{1}{4} \text{ yields no volume.}\right)$

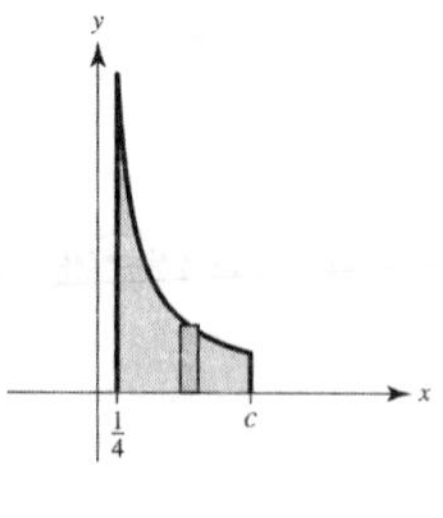

60. (a) $p(x) = x, h(x) = r^2 - x^2$

Shell method:

$$V = 2\pi \int_k^r x(r^2 - x^2)\,dx$$
$$= -\pi \int_k^r (r^2 - x^2)(-2x)\,dx$$
$$= -\pi\left[\frac{(r^2 - x^2)^2}{2}\right]_k^r$$
$$= -\pi\left[0 - \frac{(r^2 - k^2)^2}{2}\right] = \frac{\pi}{2}(r^2 - k^2)^2$$

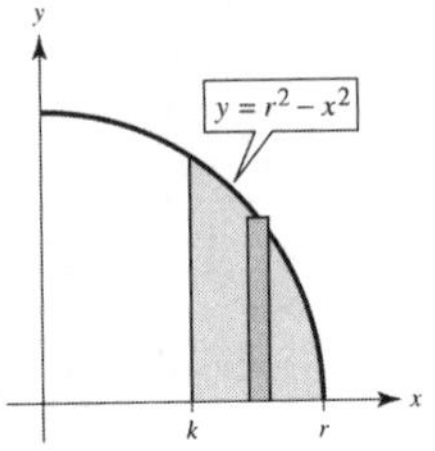

(b) $y = r^2 - x^2$

$x = \sqrt{r^2 - y}$

Disk method:

$$V = \pi \int_0^{r^2 - k^2} \left[\left(\sqrt{r^2 - y}\right)^2 - k^2\right] dy$$
$$= \pi \int_0^{r^2 - k^2} \left[r^2 - y - k^2\right] dy$$
$$= \pi\left[(r^2 - k^2)y - \frac{y^2}{2}\right]_0^{r^2 - k^2}$$
$$= \pi\left[(r^2 - k^2)^2 - \frac{(r^2 - k^2)^2}{2}\right] = \frac{\pi}{2}(r^2 - k^2)^2$$

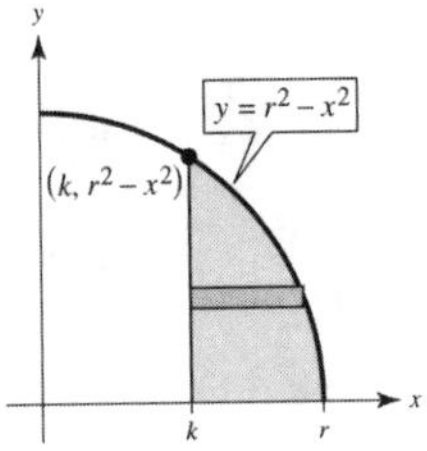

61. $y^2 = x(4 - x)^2, \quad 0 \le x \le 4$

$y_1 = \sqrt{x(4 - x)^2} = (4 - x)\sqrt{x}$

$y_2 = -\sqrt{x(4 - x)^2} = -(4 - x)\sqrt{x}$

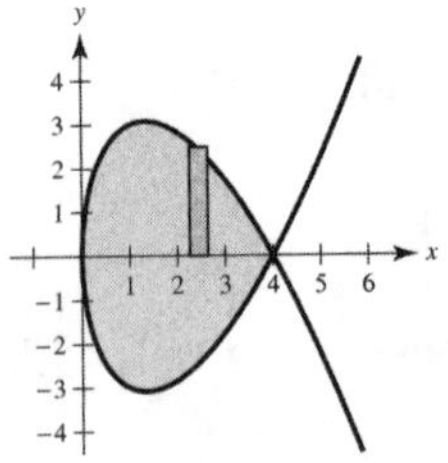

(a) $V = \pi \int_0^4 x(4 - x)^2\,dx$

$$= \pi \int_0^4 (x^3 - 8x^2 + 16x)\,dx$$
$$= \pi\left[\frac{x^4}{4} - \frac{8x^3}{3} + 8x^2\right]_0^4 = \frac{64\pi}{3}$$

(b) $V = 4\pi \int_0^4 x(4 - x)\sqrt{x}\,dx$

$$= 4\pi \int_0^4 \left(4x^{3/2} - x^{5/2}\right) dx$$
$$= 4\pi\left[\frac{8}{5}x^{5/2} - \frac{2}{7}x^{7/2}\right]_0^4 = \frac{2048\pi}{35}$$

(c) $V = 4\pi \int_0^4 (4 - x)(4 - x)\sqrt{x}\,dx$

$$= 4\pi \int_0^4 \left(16\sqrt{x} - 8x^{3/2} + x^{5/2}\right) dx$$
$$= 4\pi\left[\frac{32}{3}x^{3/2} - \frac{16}{5}x^{5/2} + \frac{2}{7}x^{7/2}\right]_0^4 = \frac{8192\pi}{105}$$

Section 7.4 Arc Length and Surfaces of Revolution

1. $(0, 0), (8, 15)$

(a) $d = \sqrt{(8 - 0)^2 + (15 - 0)^2}$

$$= \sqrt{64 + 225}$$
$$= \sqrt{289} = 17$$

(b) $y = \frac{15}{8}x$

$y' = \frac{15}{8}$

$$s = \int_0^8 \sqrt{1 + \left(\tfrac{15}{8}\right)^2}\,dx = \int_0^8 \tfrac{17}{8}\,dx = \left[\tfrac{17}{8}x\right]_0^8 = 17$$

2. $(1, 2), (7, 10)$

(a) $d = \sqrt{(7-1)^2 + (10-2)^2} = 10$

(b) $y = \frac{4}{3}x + \frac{2}{3}$

$y' = \frac{4}{3}$

$s = \int_1^7 \sqrt{1 + \left(\frac{4}{3}\right)^2}\, dx = \left[\frac{5}{3}x\right]_1^7 = 10$

3. $y = \frac{2}{3}\left(x^2 + 1\right)^{3/2}$

$y' = \left(x^2 + 1\right)^{1/2}(2x), \quad 0 \le x \le 1$

$1 + (y')^2 = 1 + 4x^2\left(x^2 + 1\right)$

$= 4x^4 + 4x^2 + 1 = \left(2x^2 + 1\right)^2$

$s = \int_0^1 \sqrt{1 + (y')^2}\, dx$

$= \int_0^1 \left(2x^2 + 1\right) dx = \left[\frac{2x^3}{3} + x\right]_0^1 = \frac{5}{3}$

4. $y = \frac{x^3}{6} + \frac{1}{2x}$

$y' = \frac{x^2}{2} - \frac{1}{2x^2} = \frac{1}{2}\left(x^2 - \frac{1}{x^2}\right), \quad 1 \le x \le 2$

$1 + (y')^2 = 1 + \frac{1}{4}\left(x^4 - 2 + \frac{1}{x^4}\right)$

$= \frac{1}{4}\left(x^4 + 2 + \frac{1}{x^4}\right)$

$= \left[\frac{1}{2}\left(x^2 + \frac{1}{x^2}\right)\right]^2$

$s = \int_1^2 \sqrt{1 + (y')^2}\, dx = \int_1^2 \frac{1}{2}\left(x^2 + \frac{1}{x^2}\right) dx$

$= \frac{1}{2}\left[\frac{x^3}{3} - \frac{1}{x}\right]_1^2$

$= \frac{1}{2}\left[\left(\frac{8}{3} - \frac{1}{2}\right) - \left(\frac{1}{3} - 1\right)\right]$

$= \frac{17}{12}$

5. $y = \frac{2}{3}x^{3/2} + 1$

$y' = x^{1/2}, \quad 0 \le x \le 1$

$s = \int_0^1 \sqrt{1 + x}\, dx$

$= \left[\frac{2}{3}(1 + x)^{3/2}\right]_0^1 = \frac{2}{3}\left(\sqrt{8} - 1\right) \approx 1.219$

6. $y = 2x^{3/2} + 3$

$y' = 3x^{1/2}, \quad 0 \le x \le 9$

$s = \int_0^9 \sqrt{1 + 9x}\, dx$

$= \left[\frac{2}{27}(1 + 9x)^{3/2}\right]_0^9 = \frac{2}{27}\left(82^{3/2} - 1\right) \approx 54.929$

7. $y = \frac{3}{2}x^{2/3}$

$y' = \frac{1}{x^{1/3}}, \quad 1 \le x \le 8$

$s = \int_1^8 \sqrt{1 + \left(\frac{1}{x^{1/3}}\right)^2}\, dx$

$= \int_1^8 \sqrt{\frac{x^{2/3} + 1}{x^{2/3}}}\, dx$

$= \frac{3}{2}\int_1^8 \sqrt{x^{2/3} + 1}\left(\frac{2}{3x^{1/3}}\right) dx$

$= \frac{3}{2}\left[\frac{2}{3}\left(x^{2/3} + 1\right)^{3/2}\right]_1^8$

$= 5\sqrt{5} - 2\sqrt{2} \approx 8.352$

8. $y = \frac{x^4}{8} + \frac{1}{4x^2}$

$y' = \frac{1}{2}x^3 - \frac{1}{2x^3}, \quad 1 \le x \le 3$

$1 + (y')^2 = \left(\frac{1}{2}x^3 + \frac{1}{2x^3}\right)^2, \quad [1, 3]$

$s = \int_a^b \sqrt{1 + (y')^2}\, dx$

$= \int_1^3 \left(\frac{1}{2}x^3 + \frac{1}{2x^3}\right) dx$

$= \left[\frac{1}{8}x^4 - \frac{1}{4x^2}\right]_1^3$

$= \frac{92}{9} \approx 10.222$

9. $y = \frac{x^5}{10} + \frac{1}{6x^3}, \quad 2 \le x \le 5$

$$y' = \frac{x^4}{2} - \frac{1}{2x^4} = \frac{1}{2}\left(x^4 - \frac{1}{x^4}\right)$$

$$1 + (y')^2 = 1 + \frac{1}{4}\left(x^4 - \frac{1}{x^4}\right)^2 = 1 + \frac{1}{4}\left(x^8 - 2 + \frac{1}{x^8}\right)$$

$$= \frac{1}{4}\left(x^8 + 2 + \frac{1}{x^8}\right) = \frac{1}{4}\left(x^4 + \frac{1}{x^4}\right)^2$$

$$s = \int_2^5 \sqrt{1 + (y')^2}\, dx = \int_2^5 \frac{1}{2}\left(x^4 + \frac{1}{x^4}\right) dx$$

$$= \frac{1}{2}\left[\frac{x^5}{5} - \frac{1}{3x^3}\right]_2^5 = \frac{1}{2}\left[\left(625 - \frac{1}{375}\right) - \left(\frac{32}{5} - \frac{1}{24}\right)\right]$$

$$= \frac{618639}{2000} \approx 309.320$$

10. $y = \frac{3}{2}x^{2/3} + 4$

$$y' = x^{-1/3}, \quad 1 \le x \le 27$$

$$s = \int_1^{27} \sqrt{1 + \left(\frac{1}{x^{1/3}}\right)^2}\, dx$$

$$= \int_1^{27} \sqrt{\frac{x^{2/3} + 1}{x^{2/3}}}\, dx$$

$$= \frac{3}{2}\int_1^{27} \sqrt{x^{2/3} + 1}\left(\frac{2}{3x^{1/3}}\right) dx$$

$$= \left[\frac{3}{2} \cdot \frac{2}{3}\left(x^{2/3} + 1\right)^{3/2}\right]_1^{27}$$

$$= 10^{3/2} - 2^{3/2} \approx 28.794$$

11. $y = \ln(\sin x), \quad \left[\frac{\pi}{4}, \frac{3\pi}{4}\right]$

$$y' = \frac{1}{\sin x}\cos x = \cot x$$

$$1 + (y')^2 = 1 + \cot^2 x = \csc^2 x$$

$$s = \int_{\pi/4}^{3\pi/4} \csc x\, dx$$

$$= \Big[\ln\left|\csc x - \cot x\right|\Big]_{\pi/4}^{3\pi/4}$$

$$= \ln\left(\sqrt{2} + 1\right) - \ln\left(\sqrt{2} - 1\right) \approx 1.763$$

12. $y = \ln(\cos x), \quad 0 \le x \le \frac{\pi}{3}$

$$y' = \frac{-\sin x}{\cos x} = -\tan x$$

$$1 + (y')^2 = 1 + \tan^2 x = \sec^2 x$$

$$s = \int_0^{\pi/3} \sqrt{\sec^2 x}\, dx$$

$$= \int_0^{\pi/3} \sec x\, dx$$

$$= \ln\left|\sec x + \tan x\right|\Big]_0^{\pi/3}$$

$$= \ln\left(2 + \sqrt{3}\right) \approx 1.3170$$

13. $y = \frac{1}{2}\left(e^x + e^{-x}\right)$

$$y' = \frac{1}{2}\left(e^x - e^{-x}\right), \quad [0, 2]$$

$$1 + (y')^2 = \left[\frac{1}{2}\left(e^x + e^{-x}\right)\right]^2, \quad [0, 2]$$

$$s = \int_0^2 \sqrt{\left[\frac{1}{2}\left(e^x + e^{-x}\right)\right]^2}\, dx$$

$$= \frac{1}{2}\int_0^2 \left(e^x + e^{-x}\right) dx$$

$$= \frac{1}{2}\left[e^x - e^{-x}\right]_0^2 = \frac{1}{2}\left(e^2 - \frac{1}{e^2}\right) \approx 3.627$$

14.
$$y = \ln\left(\frac{e^x + 1}{e^x - 1}\right) = \ln(e^x + 1) - \ln(e^x - 1)$$
$$\frac{dy}{dx} = \frac{e^x}{e^x + 1} - \frac{e^x}{e^x - 1} = \frac{-2e^x}{e^{2x} - 1} = \frac{2e^x}{1 - e^{2x}}$$
$$1 + \left(\frac{dy}{dx}\right)^2 = 1 + \frac{4e^{2x}}{1 - 2e^{2x} + e^{4x}}$$
$$= \frac{1 + 2e^{2x} + e^{4x}}{\left(1 - e^{2x}\right)^2} = \left(\frac{1 + e^{2x}}{1 - e^{2x}}\right)^2$$
$$s = \int_a^b \sqrt{1 + \left(\frac{dy}{dx}\right)^2}\, dx = \int_{\ln 2}^{\ln 3} \frac{1 + e^{2x}}{e^{2x} - 1}\, dx$$
$$= \int_{\ln 2}^{\ln 3} \frac{e^x + e^{-x}}{e^x - e^{-x}}\, dx = \int_{\ln 2}^{\ln 3} \coth x\, dx$$
$$= \ln(\sinh(x))\Big]_{\ln 2}^{\ln 3} = \ln\left(\frac{4}{3}\right) - \ln\left(\frac{3}{4}\right)$$
$$= \ln\left(\frac{4/3}{3/4}\right) = \ln\frac{16}{9} - 2\ln\left(\frac{4}{3}\right) \approx 0.57536$$

15. $x = \frac{1}{3}\left(y^2 + 2\right)^{3/2}, \quad 0 \le y \le 4$
$$\frac{dx}{dy} = y\left(y^2 + 2\right)^{1/2}$$
$$s = \int_0^4 \sqrt{1 + y^2\left(y^2 + 2\right)}\, dy$$
$$= \int_0^4 \sqrt{y^4 + 2y^2 + 1}\, dy$$
$$= \int_0^4 \left(y^2 + 1\right) dy$$
$$= \left[\frac{y^3}{3} + y\right]_0^4 = \frac{64}{3} + 4 = \frac{76}{3}$$

16.
$$x = \frac{1}{3}\sqrt{y}(y - 3), \quad 1 \le y \le 4$$
$$x = \frac{1}{3}\left(y^{3/2} - 3y^{1/2}\right)$$
$$\frac{dx}{dy} = \frac{1}{2}y^{1/2} - \frac{1}{2}y^{-1/2}$$
$$1 + \left(\frac{dx}{dy}\right)^2 = 1 + \frac{1}{4}y + \frac{1}{4}y^{-1} - \frac{1}{2}$$
$$= \frac{1}{4}\left(y + 2 + y^{-1}\right) = \frac{1}{4}\left(\sqrt{y} + \frac{1}{\sqrt{y}}\right)^2$$
$$s = \int_1^4 \frac{1}{2}\left(\sqrt{y} + \frac{1}{\sqrt{y}}\right) dy$$
$$= \left[\frac{1}{2}\left(\frac{2}{3}y^{3/2} + 2y^{1/2}\right)\right]_1^4$$
$$= \frac{1}{2}\left(\frac{16}{3} + 4\right) - \frac{1}{2}\left(\frac{2}{3} + 2\right) = \frac{10}{3}$$

17. (a) $y = 4 - x^2, \quad 0 \le x \le 2$

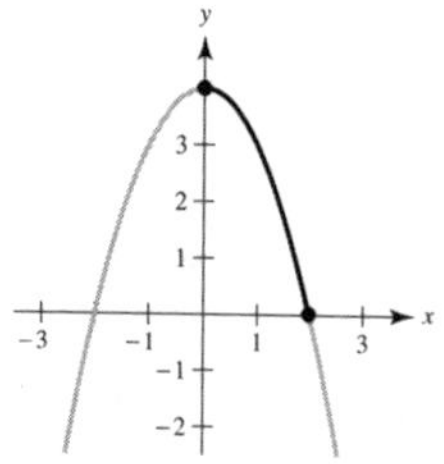

(b)
$$y' = -2x$$
$$1 + (y')^2 = 1 + 4x^2$$
$$L = \int_0^2 \sqrt{1 + 4x^2}\, dx$$

(c) $L \approx 4.647$

18. (a) $y = x^2 + x - 2, \quad -2 \le x \le 1$

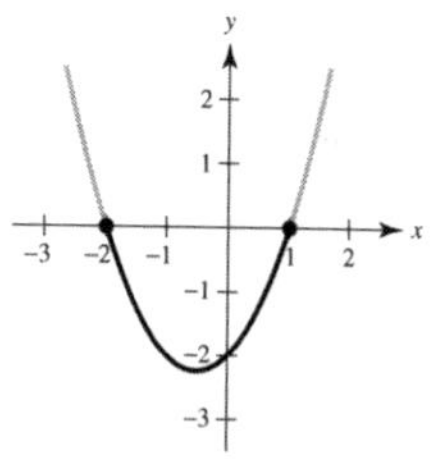

(b)
$$y' = 2x + 1$$
$$1 + (y')^2 = 1 + 4x^2 + 4x + 1$$
$$L = \int_{-2}^1 \sqrt{2 + 4x + 4x^2}\, dx$$

(c) $L \approx 5.653$

19. (a) $y = \frac{1}{x}, \quad 1 \le x \le 3$

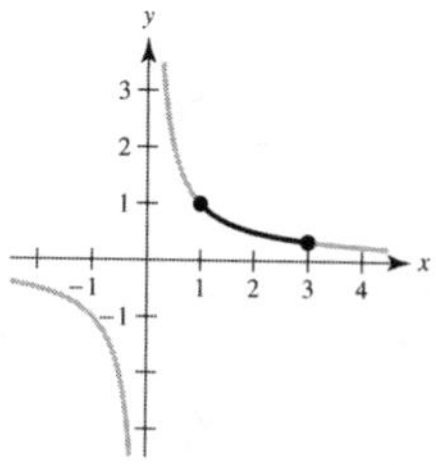

(b)
$$y' = -\frac{1}{x^2}$$
$$1 + (y')^2 = 1 + \frac{1}{x^4}$$
$$L = \int_1^3 \sqrt{1 + \frac{1}{x^4}}\, dx$$

(c) $L \approx 2.147$

20. (a) $y = \dfrac{1}{1 + x}, \quad 0 \le x \le 1$

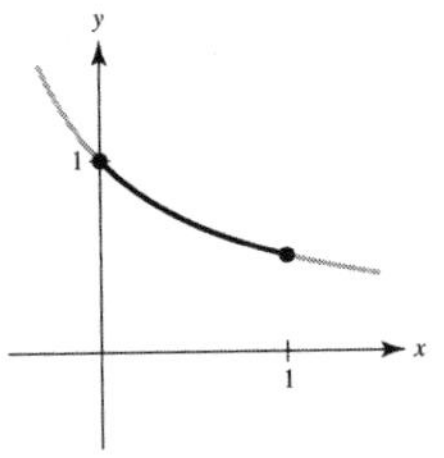

(b) $$y' = -\frac{1}{(1 + x)^2}$$
$$1 + (y')^2 = 1 + \frac{1}{(1 + x)^4}$$
$$L = \int_0^1 \sqrt{1 + \frac{1}{(1 + x)^4}}\, dx$$

(c) $L \approx 1.132$

21. (a) $y = \sin x, \quad 0 \le x \le \pi$

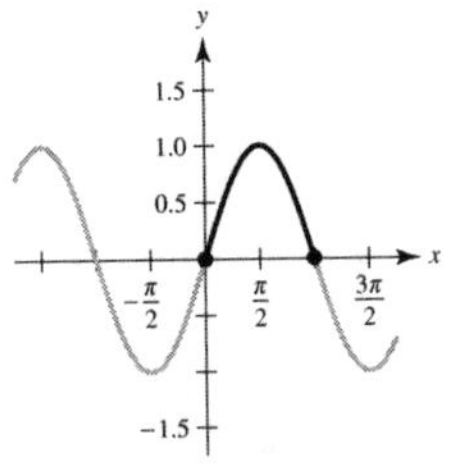

(b) $$y' = \cos x$$
$$1 + (y')^2 = 1 + \cos^2 x$$
$$L = \int_0^{\pi} \sqrt{1 + \cos^2 x}\, dx$$

(c) $L \approx 3.820$

22. (a) $y = \cos x, \quad -\dfrac{\pi}{2} \le x \le \dfrac{\pi}{2}$

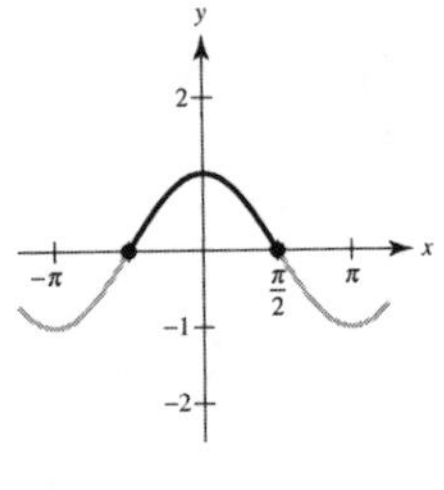

(b) $$y' = -\sin x$$
$$1 + (y')^2 = 1 + \sin^2 x$$
$$L = \int_{-\pi/2}^{\pi/2} \sqrt{1 + \sin^2 x}\, dx$$

(c) 3.820

23. (a) $x = e^{-y}, \quad 0 \le y \le 2$

$y = -\ln x$

$1 \ge x \ge e^{-2} \approx 0.135$

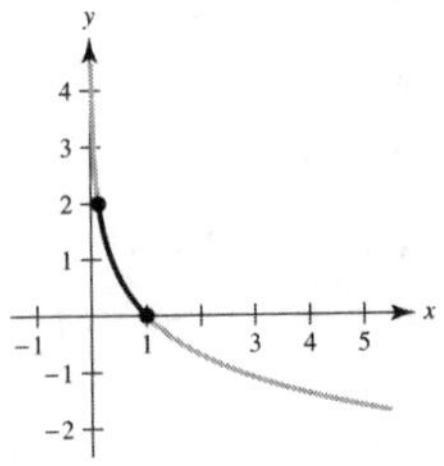

(b) $$y' = -\frac{1}{x}$$
$$1 + (y')^2 = 1 + \frac{1}{x^2}$$
$$L = \int_{e^{-2}}^{1} \sqrt{1 + \frac{1}{x^2}}\, dx$$

(c) $L \approx 2.221$

Alternatively, you can do all the computations with respect to y.

(a) $x = e^{-y}, \quad 0 \le y \le 2$

(b) $$\frac{dx}{dy} = -e^{-y}$$
$$1 + \left(\frac{dx}{dy}\right)^2 = 1 + e^{-2y}$$
$$L = \int_0^2 \sqrt{1 + e^{-2y}}\, dy$$

(c) $L \approx 2.221$

24. (a) $y = \ln x, \quad 1 \le x \le 5$

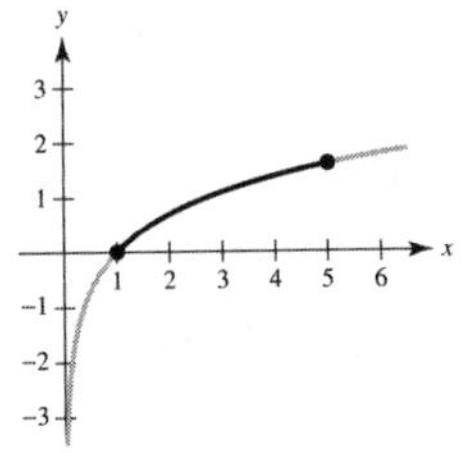

(b) $$y' = \frac{1}{x}$$
$$1 + (y')^2 = 1 + \frac{1}{x^2}$$
$$L = \int_1^5 \sqrt{1 + \frac{1}{x^2}}\, dx$$

(c) $L \approx 4.367$

25. (a) $y = 2 \arctan x, \quad 0 \le x \le 1$

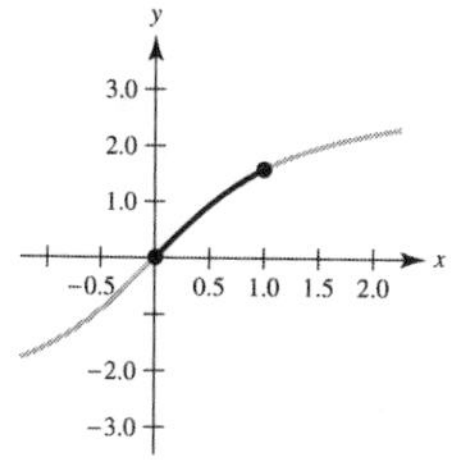

(b) $y' = \dfrac{2}{1 + x^2}$

$$L = \int_0^1 \sqrt{1 + \frac{4}{\left(1 + x^2\right)^2}}\, dx$$

(c) $L \approx 1.871$

26. (a) $x = \sqrt{36 - y^2}, \quad 0 \le y \le 3$

$y = \sqrt{36 - x^2}, \quad 3\sqrt{3} \le x \le 6$

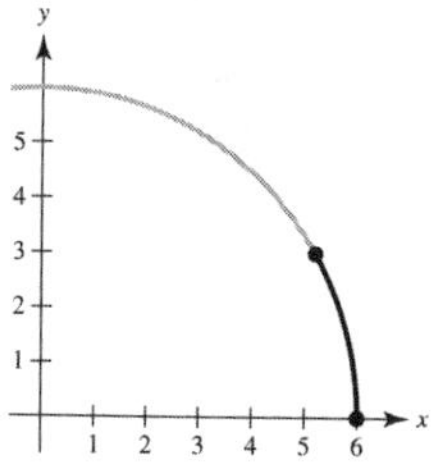

(b) $\dfrac{dx}{dy} = \dfrac{1}{2}\left(36 - y^2\right)^{-1/2}(-2y) = \dfrac{-y}{\sqrt{36 - y^2}}$

$$L = \int_0^3 \sqrt{1 + \frac{y^2}{36 - y^2}}\, dy = \int_0^3 \frac{6}{\sqrt{36 - y^2}}\, dy$$

(c) $L \approx 3.142 \quad (\pi)$

27. $\displaystyle\int_0^2 \sqrt{1 + \left[\frac{d}{dx}\left(\frac{5}{x^2 + 1}\right)\right]^2}\, dx$

$s \approx 5$

Matches (b)

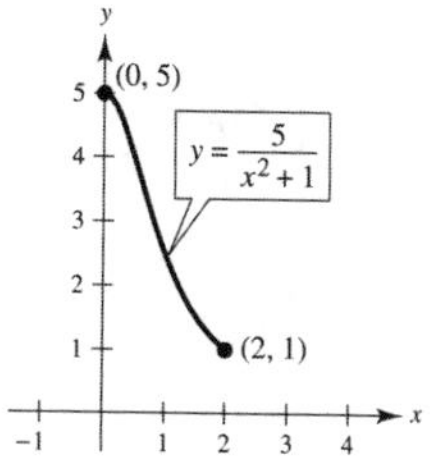

28. $\displaystyle\int_0^{\pi/4} \sqrt{1 + \left[\frac{d}{dx}(\tan x)\right]^2}\, dx$

$s \approx 1$

Matches (e)

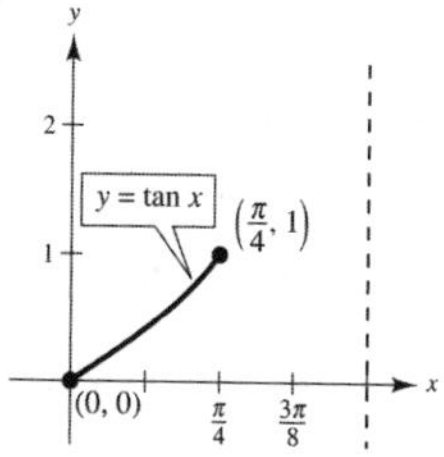

29. $y = x^3, \quad [0, 4]$

(a) $d = \sqrt{(4 - 0)^2 + (64 - 0)^2} \approx 64.125$

(b) $d = \sqrt{(1 - 0)^2 + (1 - 0)^2} + \sqrt{(2 - 1)^2 + (8 - 1)^2} + \sqrt{(3 - 2)^2 + (27 - 8)^2} + \sqrt{(4 - 3)^2 + (64 - 27)^2} \approx 64.525$

(c) $s = \displaystyle\int_0^4 \sqrt{1 + \left(3x^2\right)^2}\, dx = \int_0^4 \sqrt{1 + 9x^4}\, dx \approx 64.666$ (Simpson's Rule, $n = 10$)

(d) 64.672

30. $f(x) = \left(x^2 - 4\right)^2, \quad [0, 4]$

(a) $d = \sqrt{(4 - 0)^2 + (144 - 16)^2} \approx 128.062$

(b) $d = \sqrt{(1 - 0)^2 + (9 - 16)^2} + \sqrt{(2 - 1)^2 + (0 - 9)^2} + \sqrt{(3 - 2)^2 + (25 - 0)^2} + \sqrt{(4 - 3)^2 + (144 - 25)^2} \approx 160.151$

(c) $s = \displaystyle\int_0^4 \sqrt{1 + \left[4x\left(x^2 - 4\right)\right]^2}\, dx \approx 159.087$

(d) 160.287

31. $y = 20\cosh\dfrac{x}{20}, \quad -20 \le x \le 20$

$y' = \sinh\dfrac{x}{20}$

$1 + (y')^2 = 1 + \sinh^2\dfrac{x}{20} = \cosh^2\dfrac{x}{20}$

$$L = \int_{-20}^{20} \cosh\frac{x}{20}\,dx = 2\int_0^{20}\cosh\frac{x}{20}\,dx = \left[2(20)\sinh\frac{x}{20}\right]_0^{20} = 40\sinh(1) \approx 47.008 \text{ m}$$

32. $y = 31 - 10\left(e^{x/20} + e^{-x/20}\right)$

$y' = -\dfrac{1}{2}\left(e^{x/20} - e^{-x/20}\right)$

$1 + (y')^2 = 1 + \dfrac{1}{4}\left(e^{x/10} - 2 + e^{-x/10}\right) = \left[\dfrac{1}{2}\left(e^{x/20} + e^{-x/20}\right)\right]^2$

$$s = \int_{-20}^{20}\sqrt{\left[\frac{1}{2}\left(e^{x/20} + e^{-x/20}\right)\right]^2}\,dx = \frac{1}{2}\int_{-20}^{20}\left(e^{x/20} + e^{-x/20}\right)dx = \left[10\left(e^{x/20} - e^{-x/20}\right)\right]_{-20}^{20} = 20\left(e - \frac{1}{e}\right) \approx 47 \text{ ft}$$

So, there are $100(47) = 4700$ square feet of roofing on the barn.

33. $y = 693.8597 - 68.7672\cosh 0.0100333x$

$y' = -0.6899619478\sinh 0.0100333x$

$$s = \int_{-299.2239}^{299.2239}\sqrt{1 + (-0.6899619478\sinh 0.0100333x)^2}\,dx \approx 1480$$

(Use Simpson's Rule with $n = 100$ or a graphing utility.)

34. $x^{2/3} + y^{2/3} = 4$

$y^{2/3} = 4 - x^{2/3}$

$y = \left(4 - x^{2/3}\right)^{3/2}$

$y' = \dfrac{3}{2}\left(4 - x^{2/3}\right)^{1/2}\left(-\dfrac{2}{3}x^{-1/3}\right) = \dfrac{-\left(4 - x^{2/3}\right)^{1/2}}{x^{1/3}}$

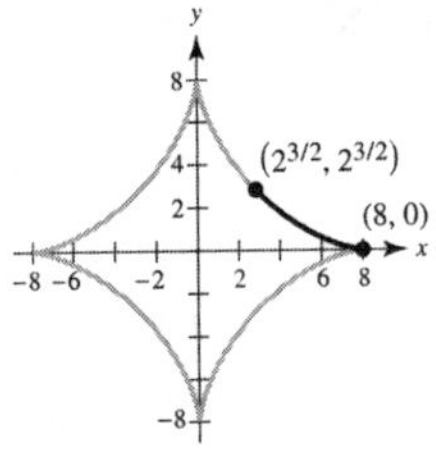

In order to avoid division by 0, compute the arc length for $2^{3/2} \le x \le 8$, and multiply the answer by 8, as indicated in the figure.

$1 + (y')^2 = 1 + \dfrac{4 - x^{2/3}}{x^{2/3}}, \quad 2^{3/2} \le x \le 8$

$= \dfrac{4}{x^{2/3}}$

$s = 8\displaystyle\int_{2^{3/2}}^{8}\sqrt{\frac{4}{x^{2/3}}}\,dx$

$= 16\displaystyle\int_{2^{3/2}}^{8} x^{-1/3}\,dx$

$= 16\left[\dfrac{3}{2}x^{2/3}\right]_{2^{3/2}}^{8}$

$= 24(4 - 2) = 48$

35.
$$y = \sqrt{9 - x^2}$$
$$y' = \frac{-x}{\sqrt{9 - x^2}}$$
$$1 + (y')^2 = \frac{9}{9 - x^2}$$
$$s = \int_0^2 \sqrt{\frac{9}{9 - x^2}}\,dx = \int_0^2 \frac{3}{\sqrt{9 - x^2}}\,dx$$
$$= \left[3 \arcsin \frac{x}{3}\right]_0^2 = 3\left(\arcsin \frac{2}{3} - \arcsin 0\right)$$
$$= 3 \arcsin \frac{2}{3} \approx 2.1892$$

36.
$$y = \sqrt{25 - x^2}$$
$$y' = \frac{-x}{\sqrt{25 - x^2}}$$
$$1 + (y')^2 = \frac{25}{25 - x^2}$$
$$s = \int_{-3}^4 \sqrt{\frac{25}{25 - x^2}}\,dx = \int_{-3}^4 \frac{5}{\sqrt{25 - x^2}}\,dx$$
$$= \left[5 \arcsin \frac{x}{5}\right]_{-3}^4 = 5\left[\arcsin \frac{4}{5} - \arcsin\left(-\frac{3}{5}\right)\right]$$
$$\approx 7.8540$$
$$\frac{1}{4}\left[2\pi(5)\right] \approx 7.8540 = s$$

37. $y = \dfrac{x^3}{3}$
$$y' = x^2, \quad [0, 3]$$
$$S = 2\pi \int_0^3 \frac{x^3}{3}\sqrt{1 + x^4}\,dx$$
$$= \frac{\pi}{6}\int_0^3 \left(1 + x^4\right)^{1/2}\left(4x^3\right)dx$$
$$= \left[\frac{\pi}{9}\left(1 + x^4\right)^{3/2}\right]_0^3$$
$$= \frac{\pi}{9}\left(82\sqrt{82} - 1\right) \approx 258.85$$

38. $y = 2\sqrt{x}$
$$y' = \frac{1}{\sqrt{x}}, \quad [4, 9]$$
$$S = 2\pi \int_4^9 2\sqrt{x}\sqrt{1 + \frac{1}{x}}\,dx$$
$$= 4\pi \int_4^9 \sqrt{x + 1}\,dx$$
$$= \left[\frac{8}{3}\pi(x + 1)^{3/2}\right]_4^9$$
$$= \frac{8\pi}{3}\left(10^{3/2} - 5^{3/2}\right) \approx 171.258$$

39.
$$y = \frac{x^3}{6} + \frac{1}{2x}$$
$$y' = \frac{x^2}{2} - \frac{1}{2x^2}$$
$$1 + (y')^2 = \left(\frac{x^2}{2} + \frac{1}{2x^2}\right)^2, \quad [1, 2]$$
$$S = 2\pi \int_1^2 \left(\frac{x^3}{6} + \frac{1}{2x}\right)\left(\frac{x^2}{2} + \frac{1}{2x^2}\right)dx$$
$$= 2\pi \int_1^2 \left(\frac{x^5}{12} + \frac{x}{3} + \frac{1}{4x^3}\right)dx$$
$$= 2\pi\left[\frac{x^6}{72} + \frac{x^2}{6} - \frac{1}{8x^2}\right]_1^2 = \frac{47\pi}{16}$$

40.
$$y = 3x$$
$$y' = 3$$
$$1 + (y')^2 = 10, \quad [0, 3]$$
$$S = 2\pi \int_0^3 3x\sqrt{10}\,dx$$
$$= 6\pi\sqrt{10}\left[\frac{x^2}{2}\right]_0^3$$
$$= 27\sqrt{10}\pi$$

41.
$$y = \sqrt{4 - x^2}$$
$$y' = \frac{1}{2}\left(4 - x^2\right)^{-1/2}(-2x) = \frac{-x}{\sqrt{4 - x^2}}, \quad -1 \le x \le 1$$
$$1 + (y')^2 = 1 + \frac{x^2}{4 - x^2} = \frac{4}{4 - x^2}$$
$$S = 2\pi \int_{-1}^1 \sqrt{4 - x^2} \cdot \sqrt{\frac{4}{4 - x^2}}\,dx$$
$$= 4\pi \int_{-1}^1 dx = 4\pi\left[x\right]_{-1}^1 = 8\pi$$

42. $y = \sqrt{9 - x^2}, \quad -2 \le x \le 2$
$$y' = \frac{1}{2}\left(9 - x^2\right)^{-1/2}(-2x) = \frac{-x}{\sqrt{9 - x^2}}$$
$$1 + (y')^2 + 1 + \frac{x^2}{\left(9 - x^2\right)} = \frac{9}{9 - x^2}$$
$$S = 2\pi \int_{-2}^2 \sqrt{9 - x^2}\,\frac{3}{\sqrt{9 - x^2}}\,dx = 2\pi \int_{-2}^2 3\,dx$$
$$= 2\pi\left[3x\right]_{-2}^2 = 24\pi$$

43. $y = \sqrt[3]{x} + 2$

$$y' = \frac{1}{3x^{2/3}}, \quad [1, 8]$$

$$S = 2\pi \int_1^8 x\sqrt{1 + \frac{1}{9x^{4/3}}}\, dx$$

$$= \frac{2\pi}{3} \int_1^8 x^{1/3}\sqrt{9x^{4/3} + 1}\, dx$$

$$= \frac{\pi}{18} \int_1^8 \left(9x^{4/3} + 1\right)^{1/2}\left(12x^{1/3}\right) dx$$

$$= \left[\frac{\pi}{27}\left(9x^{4/3} + 1\right)^{3/2}\right]_1^8$$

$$= \frac{\pi}{27}\left(145\sqrt{145} - 10\sqrt{10}\right) \approx 199.48$$

44. $y = 9 - x^2, \quad [0, 3]$

$$y' = -2x$$

$$S = 2\pi \int_0^3 x\sqrt{1 + 4x^2}\, dx$$

$$= \frac{\pi}{4} \int_0^3 \left(1 + 4x^2\right)^{1/2}(8x)\, dx$$

$$= \left[\frac{\pi}{6}\left(1 + 4x^2\right)^{3/2}\right]_0^3 = \frac{\pi}{6}\left(37^{3/2} - 1\right) \approx 117.319$$

45. $y = 1 - \dfrac{x^2}{4}$

$$y' = -\frac{x}{2}, \quad 0 \le x \le 2$$

$$1 + \left(y'\right)^2 = 1 + \frac{x^2}{4} = \frac{4 + x^2}{4}$$

$$S = 2\pi \int_0^2 x\sqrt{\frac{4 + x^2}{4}}\, dx$$

$$= \pi \int_0^2 x\sqrt{4 + x^2}\, dx$$

$$= \frac{1}{2}\pi \int_0^2 \left(4 + x^2\right)^{1/2}(2x)\, dx$$

$$= \frac{1}{2}\pi\left[\frac{2}{3}\left(4 + x^2\right)^{3/2}\right]_0^2$$

$$= \frac{\pi}{3}\left(8^{3/2} - 4^{3/2}\right)$$

$$= \frac{\pi}{3}\left(16\sqrt{2} - 8\right) \approx 15.318$$

46. $y = \dfrac{x}{2} + 3$

$$y' = \frac{1}{2}$$

$$1 + \left(y'\right)^2 = \frac{5}{4}, 1 \le x \le 5$$

$$S = 2\pi \int_1^5 x\sqrt{\frac{5}{4}}\, dx$$

$$= \sqrt{5}\pi\left[\frac{x^2}{2}\right]_1^5$$

$$= \sqrt{5}\pi\left(\frac{25}{2} - \frac{1}{2}\right) = 12\sqrt{5}\pi$$

47. $y = \sin x$

$$y' = \cos x, \quad [0, \pi]$$

$$S = 2\pi \int_0^\pi \sin x\sqrt{1 + \cos^2 x}\, dx \approx 14.4236$$

48. $y = \ln x$

$$y' = \frac{1}{x}$$

$$1 + \left(y'\right)^2 = \frac{x^2 + 1}{x^2}, \quad [1, e]$$

$$S = 2\pi \int_1^e x\sqrt{\frac{x^2 + 1}{x^2}}\, dx = 2\pi \int_1^e \sqrt{x^2 + 1}\, dx$$

$$\approx 22.943$$

49. A rectifiable curve is one that has a finite arc length.

50. The precalculus formula is the distance formula between two points. The representative element is

$$\sqrt{\left(\Delta x_i\right)^2 + \left(\Delta y_i\right)^2} = \sqrt{1 + \left(\frac{\Delta y_i}{\Delta x_i}\right)^2}\, \Delta xi.$$

51. The precalculus formula is the surface area formula for the lateral surface of the frustum of a right circular cone. The formula is $S = 2\pi rL$, where $r = \frac{1}{2}\left(r_1 + r_2\right)$, which is the average radius of the frustum, and L is the length of a line segment on the frustum. The representative element is

$$2\pi f\left(d_i\right)\sqrt{\Delta x_i^2 + \Delta y_i^2} = 2\pi f\left(d_i\right)\sqrt{1 + \left(\frac{\Delta y_i}{\Delta x_i}\right)^2}\, \Delta x_i.$$

52. The surface of revolution given by f_1 will be larger. $r(x)$ is larger for f_1.

53. (a)

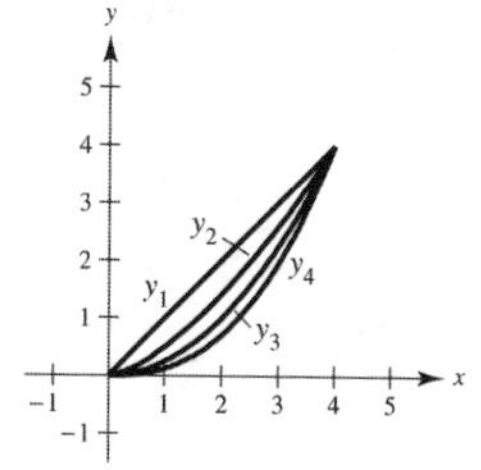

(b) y_1, y_2, y_3, y_4

(c) $y_1' = 1, \quad s_1 = \int_0^4 \sqrt{2}\,dx \approx 5.657$

$y_2' = \frac{3}{4}x^{1/2}, \quad s_2 = \int_0^4 \sqrt{1 + \frac{9x}{16}}\,dx \approx 5.759$

$y_3' = \frac{1}{2}x, \quad s_3 = \int_0^4 \sqrt{1 + \frac{x^2}{4}}\,dx \approx 5.916$

$y_4' = \frac{5}{16}x^{3/2}, \quad s_4 = \int_0^4 \sqrt{1 + \frac{25}{256}x^3}\,dx \approx 6.063$

54. (a) Area of circle with radius L: $A = \pi L^2$

Area of sector with central angle θ (in radians):

$$S = \frac{\theta}{2\pi}A = \frac{\theta}{2\pi}\left(\pi L^2\right) = \frac{1}{2}L^2\theta$$

(b) Let s be the arc length of the sector, which is the circumference of the base of the cone. Here, $s = L\theta = 2\pi r$, and you have

$$S = \frac{1}{2}L^2\theta = \frac{1}{2}L^2\left(\frac{s}{L}\right) = \frac{1}{2}Ls = \frac{1}{2}L(2\pi r) = \pi rL.$$

(c) The lateral surface area of the frustum is the difference of the large cone and the small one.

$$S = \pi r_2(L + L_1) - \pi r_1 L_1$$
$$= \pi r_2 L + \pi L_1(r_2 - r_1)$$

By similar triangles,

$$\frac{L + L_1}{r_2} = \frac{L_1}{r_1} \Rightarrow Lr_1 = L_1(r_2 - r_1). \text{ So,}$$

$$S = \pi r_2 L + \pi L_1(r_2 - r_1) = \pi r_2 L + \pi L r_1$$
$$= \pi L(r_1 + r_2).$$

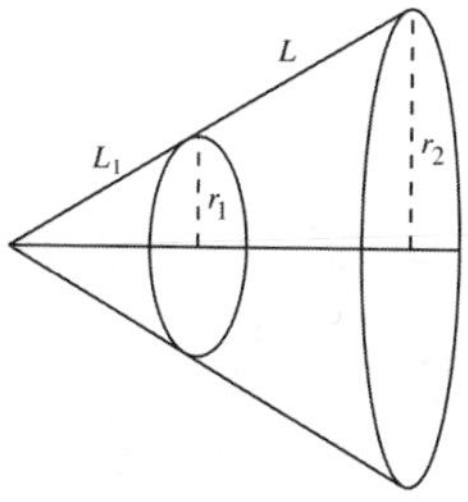

55. $y = \frac{3x}{4}, \quad y' = \frac{3}{4}$

$$1 + (y')^2 = 1 + \frac{9}{16} = 25/16$$

$$S = 2\pi \int_0^4 x\sqrt{\frac{25}{16}}\,dx = \frac{5\pi}{2}\left[\frac{x^2}{2}\right]_0^4 = 20\pi$$

56. $y = \frac{hx}{r}$

$$y' = \frac{h}{r}$$

$$1 + (y')^2 = \frac{r^2 + h^2}{r^2}$$

$$S = 2\pi \int_0^r x\sqrt{\frac{r^2 + h^2}{r^2}}\,dx$$

$$= \left[\frac{2\pi\sqrt{r^2 + h^2}}{r}\left(\frac{x^2}{2}\right)\right]_0^r = \pi r\sqrt{r^2 + h^2}$$

57. $y = \sqrt{9 - x^2}$

$$y' = \frac{-x}{\sqrt{9 - x^2}}$$

$$\sqrt{1 + (y')^2} = \frac{3}{\sqrt{9 - x^2}}$$

$$S = 2\pi \int_0^2 \frac{3x}{\sqrt{9 - x^2}}\,dx$$

$$= -3\pi \int_0^2 \frac{-2x}{\sqrt{9 - x^2}}\,dx$$

$$= \left[-6\pi\sqrt{9 - x^2}\right]_0^2$$

$$= 6\pi\left(3 - \sqrt{5}\right) \approx 14.40$$

See figure in Exercise 58.

58. From Exercise 57 you have:

$$S = 2\pi \int_0^a \frac{rx}{\sqrt{r^2 - x^2}}\,dx$$

$$= -r\pi \int_0^a \frac{-2x\,dx}{\sqrt{r^2 - x^2}}$$

$$= \left[-2r\pi\sqrt{r^2 - x^2}\right]_0^a$$

$$= 2r^2\pi - 2r\pi\sqrt{r^2 - a^2}$$

$$= 2r\pi\left(r - \sqrt{r^2 - a^2}\right)$$

$= 2\pi rh$ (where h is the height of the zone)

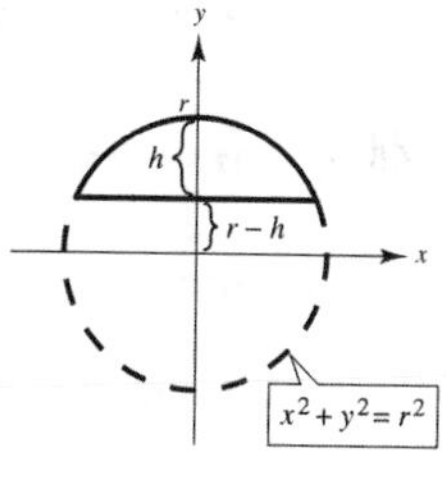

59. (a) Approximate the volume by summing six disks of thickness 3 and circumference C_i equal to the average of the given circumferences:

$$V \approx \sum_{i=1}^{6} \pi r_i^2(3) = \sum_{i=1}^{6} \pi\left(\frac{C_i}{2\pi}\right)^2(3) = \frac{3}{4\pi}\sum_{i=1}^{6} C_i^2$$

$$= \frac{3}{4\pi}\left[\left(\frac{50 + 65.5}{2}\right)^2 + \left(\frac{65.5 + 70}{2}\right)^2 + \left(\frac{70 + 66}{2}\right)^2 + \left(\frac{66 + 58}{2}\right)^2 + \left(\frac{58 + 51}{2}\right)^2 + \left(\frac{51 + 48}{2}\right)^2\right]$$

$$= \frac{3}{4\pi}\left[57.75^2 + 67.75^2 + 68^2 + 62^2 + 54.5^2 + 49.5^2\right] = \frac{3}{4\pi}(21813.625) = 5207.62 \text{ in.}^3$$

(b) The lateral surface area of a frustum of a right circular cone is $\pi s(R + r)$. For the first frustum:

$$S_1 \approx \pi\left[3^2 + \left(\frac{65.5 - 50}{2\pi}\right)^2\right]^{1/2}\left[\frac{50}{2\pi} + \frac{65.5}{2\pi}\right]$$

$$= \left(\frac{50 + 65.5}{2}\right)\left[9 + \left(\frac{65.5 - 50}{2\pi}\right)^2\right]^{1/2}.$$

Adding the six frustums together:

$$S \approx \left(\frac{50 + 65.5}{2}\right)\left[9 + \left(\frac{15.5}{2\pi}\right)^2\right]^{1/2} + \left(\frac{65.5 + 70}{2}\right)\left[9 + \left(\frac{4.5}{2\pi}\right)^2\right]^{1/2}$$

$$+ \left(\frac{70 + 66}{2}\right)\left[9 + \left(\frac{4}{2\pi}\right)^2\right]^{1/2} + \left(\frac{66 + 58}{2}\right)\left[9 + \left(\frac{8}{2\pi}\right)^2\right]^{1/2}$$

$$+ \left(\frac{58 + 51}{2}\right)\left[9 + \left(\frac{7}{2\pi}\right)^2\right]^{1/2} + \left(\frac{51 + 48}{2}\right)\left[9 + \left(\frac{3}{2\pi}\right)^2\right]^{1/2}$$

$$\approx 224.30 + 208.96 + 208.54 + 202.06 + 174.41 + 150.37 = 1168.64$$

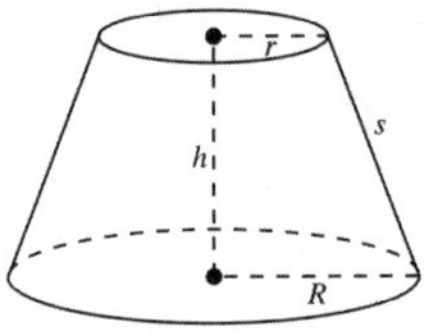

(c) $r = 0.00401y^3 - 0.1416y^2 + 1.232y + 7.943$

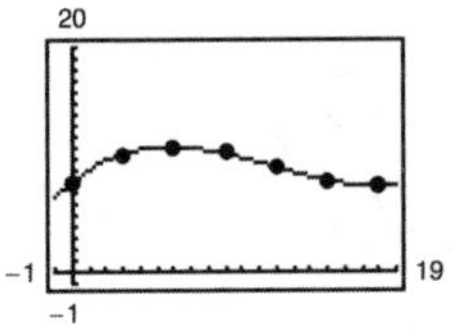

(d) $V = \int_0^{18} \pi r^2 dy \approx 5275.9 \text{ in.}^3$

$S = \int_0^{18} 2\pi r(y)\sqrt{1 + r'(y)^2}\, dy \approx 1179.5 \text{ in.}^2$

60. (a) $y = f(x) = 0.0000001953x^4 - 0.0001804x^3 + 0.0496x^2 - 4.8323x + 536.9270$

(b) Area $= \int_0^{400} f(x)\, dx \approx 131{,}734.5 \text{ ft}^2 \approx 3.0$ acres $\left(1 \text{ acre} = 43{,}560 \text{ ft}^2\right)$

(Answers will vary.)

(c) $L = \int_0^{400} \sqrt{1 + f'(x)^2}\, dx \approx 794.9$ ft

(Answers will vary.)

61. (a) $V = \pi \int_1^b \frac{1}{x^2}\,dx = \left[-\frac{\pi}{x}\right]_1^b = \pi\left(1 - \frac{1}{b}\right)$

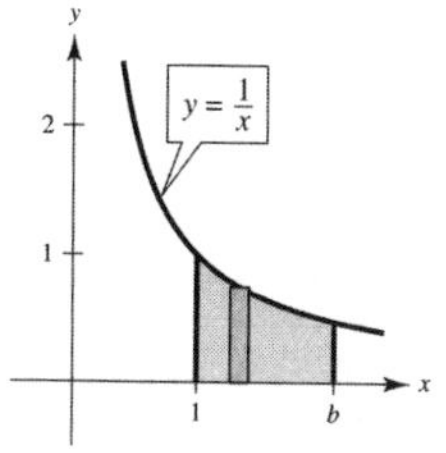

(b) $S = 2\pi \int_1^b \frac{1}{x}\sqrt{1 + \left(-\frac{1}{x^2}\right)^2}\,dx$

$= 2\pi \int_1^b \frac{1}{x}\sqrt{1 + \frac{1}{x^4}}\,dx$

$= 2\pi \int_1^b \frac{\sqrt{x^4 + 1}}{x^3}\,dx$

(c) $\lim_{b\to\infty} V = \lim_{b\to\infty} \pi\left(1 - \frac{1}{b}\right) = \pi$

(d) Because

$\frac{\sqrt{x^4 + 1}}{x^3} > \frac{\sqrt{x^4}}{x^3} = \frac{1}{x} > 0$ on $[1, b]$,

you have

$\int_1^b \frac{\sqrt{x^4 + 1}}{x^3}\,dx > \int_1^b \frac{1}{x}\,dx = \left[\ln x\right]_1^b = \ln b$

and $\lim_{b\to\infty} \ln b \to \infty$. So,

$\lim_{b\to\infty} 2\pi \int_1^b \frac{\sqrt{x^4 + 1}}{x^3}\,dx = \infty.$

62. (a) $\frac{x^2}{9} + \frac{y^2}{4} = 1$

Ellipse: $y_1 = 2\sqrt{1 - \frac{x^2}{9}}$

$y_2 = -2\sqrt{1 - \frac{x^2}{9}}$

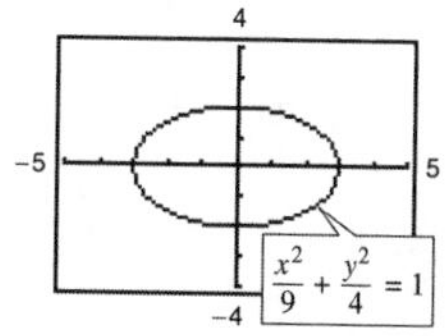

(b) $y = 2\sqrt{1 - \frac{x^2}{9}}, \quad 0 \le x \le 3$

$y' = 2\left(\frac{1}{2}\right)\left(1 - \frac{x^2}{9}\right)^{-1/2}\left(\frac{-2x}{9}\right)$

$= \frac{-2x}{9\sqrt{1 - (x^2/9)}} = \frac{-2x}{3\sqrt{9 - x^2}}$

$L = \int_0^3 \sqrt{1 + \frac{4x^2}{81 - 9x^2}}\,dx$

(c) You cannot evaluate this definite integral, because the integrand is not defined at $x = 3$. Simpson's Rule will not work for the same reason. Also, the integrand does not have an elementary antiderivative.

63. $y = \frac{1}{3}\left(x^{3/2} - 3x^{1/2} + 2\right)$

When $x = 0$, $y = \frac{2}{3}$. So, the fleeing object has traveled $\frac{2}{3}$ unit when it is caught.

$y' = \frac{1}{3}\left(\frac{3}{2}x^{1/2} - \frac{3}{2}x^{-1/2}\right) = \left(\frac{1}{2}\right)\frac{x - 1}{x^{1/2}}$

$1 + (y')^2 = 1 + \frac{(x - 1)^2}{4x} = \frac{(x + 1)^2}{4x}$

$s = \int_0^1 \frac{x + 1}{2x^{1/2}}\,dx = \frac{1}{2}\int_0^1 \left(x^{1/2} + x^{-1/2}\right)dx$

$= \frac{1}{2}\left[\frac{2}{3}x^{3/2} + 2x^{1/2}\right]_0^1 = \frac{4}{3} = 2\left(\frac{2}{3}\right)$

The pursuer has traveled twice the distance that the fleeing object has traveled when it is caught.

64.

$$y = \frac{1}{3}x^{1/2} - x^{3/2}$$

$$y' = \frac{1}{6}x^{-1/2} - \frac{3}{2}x^{1/2} = \frac{1}{6}\left(x^{-1/2} - 9x^{1/2}\right)$$

$$1 + (y')^2 = 1 + \frac{1}{36}\left(x^{-1} - 18 + 81x\right) = \frac{1}{36}\left(x^{-1/2} + 9x^{1/2}\right)^2$$

$$S = 2\pi\int_0^{1/3}\left(\frac{1}{3}x^{1/2} - x^{3/2}\right)\sqrt{\frac{1}{36}\left(x^{-1/2} + 9x^{1/2}\right)^2}\,dx = \frac{2\pi}{6}\int_0^{1/3}\left(\frac{1}{3}x^{1/2} - x^{3/2}\right)\left(x^{-1/2} + 9x^{1/2}\right)dx$$

$$= \frac{\pi}{3}\int_0^{1/3}\left(\frac{1}{3} + 2x - 9x^2\right)dx = \frac{\pi}{3}\left[\frac{1}{3}x + x^2 - 3x^3\right]_0^{1/3} = \frac{\pi}{27}\text{ ft}^2 \approx 0.1164\text{ ft}^2 \approx 16.8\text{ in.}^2$$

Amount of glass needed: $V = \frac{\pi}{27}\left(\frac{0.015}{12}\right) \approx 0.00015\text{ ft}^3 \approx 0.25\text{ in.}^3$

65. $x^{2/3} + y^{2/3} = 4$

$$y^{2/3} = 4 - x^{2/3}$$

$$y = \left(4 - x^{2/3}\right)^{3/2}, \quad 0 \le x \le 8$$

$$y' = \frac{3}{2}\left(4 - x^{2/3}\right)^{1/2}\left(-\frac{2}{3}x^{-1/3}\right) = \frac{-\left(4 - x^{2/3}\right)^{1/2}}{x^{1/3}}$$

$$1 + (y')^2 = 1 + \frac{4 - x^{2/3}}{x^{2/3}} = \frac{4}{x^{2/3}}$$

$$S = 2\pi\int_0^8\left(4 - x^{2/3}\right)^{3/2}\sqrt{\frac{4}{x^{2/3}}}\,dx = 4\pi\int_0^8\frac{\left(4 - x^{2/3}\right)^{3/2}}{x^{1/3}}\,dx = \left[-\frac{12\pi}{5}\left(4 - x^{2/3}\right)^{5/2}\right]_0^8 = \frac{384\pi}{5}$$

[Surface area of portion above the x-axis]

66.

$$y^2 = \frac{1}{12}x(4 - x)^2, \quad 0 \le x \le 4$$

$$y = \frac{(4 - x)\sqrt{x}}{\sqrt{12}}$$

$$y' = \frac{(4 - 3x)\sqrt{3}}{12\sqrt{x}}$$

$$1 + (y')^2 = 1 + \frac{(4 - 3x)^2}{48x}$$

$$= \frac{48x + 16 - 24x + 9x^2}{48x} = \frac{(4 + 3x)^2}{48x}, \quad x \ne 0$$

$$S = 2\pi\int_0^4\frac{(4 - x)\sqrt{x}}{\sqrt{12}}\cdot\frac{(4 + 3x)}{\sqrt{48x}}\,dx$$

$$= 2\pi\int_0^4\frac{(4 - x)(4 + 3x)}{24}\,dx$$

$$= \frac{\pi}{12}\int_0^4\left(16 + 8x - 3x^2\right)dx = \frac{\pi}{12}\left[16x + 4x^2 - x^3\right]_0^4 = \frac{\pi}{12}(64 + 64 - 64) = \frac{16\pi}{3}$$

67. $y = kx^2,\ y' = 2kx$

$$1 + (y')^2 = 1 + 4k^2x^2$$

$$h = kw^2 \Rightarrow k = \frac{h}{w^2} \Rightarrow 1 + (y') = 1 + \frac{4h^2}{w^4}x^2$$

By symmetry, $C = 2\int_0^w \sqrt{1 + \frac{4h^2}{w^4}x^2}\,dx.$

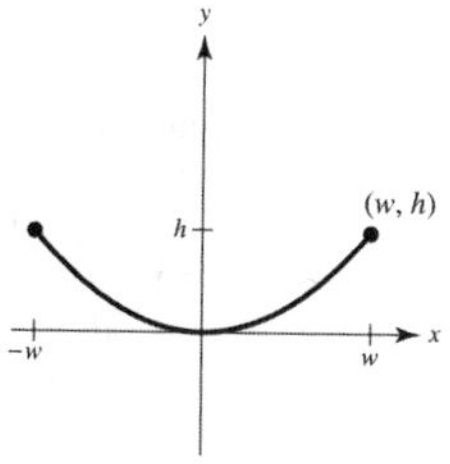

68. $C = 2\int_0^w \sqrt{1 + \frac{4h^2}{w^4}x^2}\,dx$

$$= 2\int_0^{700} \sqrt{1 + \frac{4(155)^2x^2}{700^4}}\,dx = 1444.5 \text{ m}$$

69.

$$y = f(x) = \cosh x$$

$$y' = \sinh x$$

$$1 + (y')^2 = 1 + \sinh^2 x = \cosh^2 x$$

$$\text{Area} = \int_0^t \cosh x\,dx = [\sinh x]_0^t = \sinh t$$

$$\text{Arc length} = \int_0^t \sqrt{1 + (y')^2}\,dx$$

$$= \int_0^t \cosh x\,dx = \sinh x\Big]_0^t$$

$$= \sinh t.$$

Another curve with this property is $g(x) = 1$.

$$\text{Area} = \int_0^t dx = t$$

$$\text{Arc length} = t$$

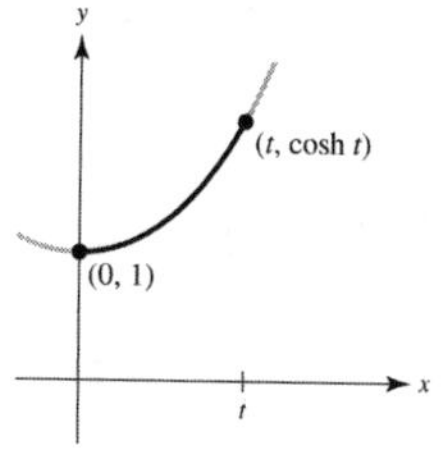

70. Let (x_0, y_0) be the point on the graph of $y^2 = x^3$ where the tangent line makes an angle of 45° with the x-axis.

$$y = x^{3/2}$$

$$y' = \tfrac{3}{2}x^{1/2} = 1$$

$$x_0 = \tfrac{4}{9}$$

$$L = \int_0^{4/9} \sqrt{1 + \tfrac{9}{4}x}\,dx$$

$$= \tfrac{8}{27}\left(2\sqrt{2} - 1\right)$$

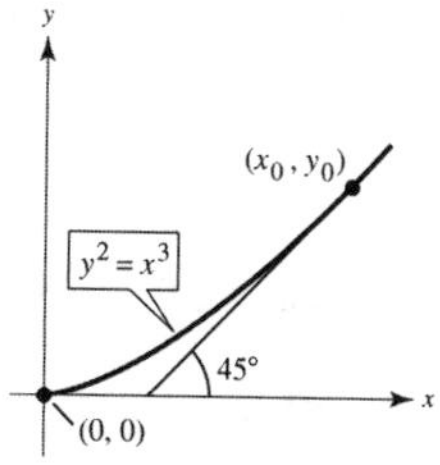

Section 7.5 Work

1. $W = Fd = 1200(40) = 48{,}000$ ft-lb

2. $W = Fd = 2500(6) = 15{,}000$ ft-lb

3. $W = Fd = (112)(8) = 896$ joules (Newton-meters)

4. $W = Fd = [9(2000)]\left[\tfrac{1}{2}(5280)\right] = 47{,}520{,}000$ ft-lb

5. $F(x) = kx$

$$5 = k(3)$$

$$k = \frac{5}{3}$$

$$F(x) = \frac{5}{3}x$$

$$W = \int_0^7 F(x)\,dx = \int_0^7 \frac{5}{3}x\,dx = \left[\frac{5}{6}x^2\right]_0^7 = \frac{245}{6} \text{ in.-lb}$$

$$\approx 40.833 \text{ in.-lb} \approx 3.403 \text{ ft-lb}$$

6. $F(x) = kx$

$$250 = k(30) \Rightarrow k = \frac{25}{3}$$

$$W = \int_{20}^{50} F(x)\,dx$$

$$= \int_{20}^{50} \frac{25}{3}x\,dx = \frac{25x^2}{6}\Big]_{20}^{50}$$

$$= 8750 \text{ n-cm}$$

$$= 87.5 \text{ joules or Nm}$$

7. $F(x) = kx$

$$20 = k(9)$$

$$k = \tfrac{20}{9}$$

$$W = \int_0^{12} \tfrac{20}{9}x\,dx = \left[\tfrac{10}{9}x^2\right]_0^{12} = 160 \text{ in.-lb} = \tfrac{40}{3} \text{ ft-lb}$$

8. $F(x) = kx$

$15 = k(1) = k$

$W = 2\int_0^4 15x\,dx = \left[15x^2\right]_0^4 = 240$ ft-lb

9. $W = 18 = \int_0^{1/3} kx\,dx = \left[\frac{kx^2}{2}\right]_0^{1/3} = \frac{k}{18} \Rightarrow k = 324$

$W = \int_{1/3}^{7/12} 324x\,dx = \left[162x^2\right]_{1/3}^{7/12} = 37.125$ ft-lb

$\left[\textbf{Note: } 4 \text{ inches} = \frac{1}{3} \text{ foot}\right]$

10. $W = 7.5 = \int_0^{1/6} kx\,dx = \left[\frac{kx^2}{2}\right]_0^{1/6} = \frac{k}{72} \Rightarrow k = 540$

$W = \int_{1/6}^{5/24} 540x\,dx = \left[270x^2\right]_{1/6}^{5/24} = 4.21875$ ft-lb

11. Assume that Earth has a radius of 4000 miles.

$F(x) = \frac{k}{x^2}$

$5 = \frac{k}{(4000)^2}$

$k = 80{,}000{,}000$

$F(x) = \frac{80{,}000{,}000}{x^2}$

(a) $W = \int_{4000}^{4100} \frac{80{,}000{,}000}{x^2}\,dx = \left[\frac{-80{,}000{,}000}{x}\right]_{4000}^{4100}$

≈ 487.8 mi-tons $\approx 5.15 \times 10^9$ ft-lb

(b) $W = \int_{4000}^{4300} \frac{80{,}000{,}000}{x^2}\,dx$

≈ 1395.3 mi-ton $\approx 1.47 \times 10^{10}$ ft-ton

12. $W = \int_{4000}^{h} \frac{80{,}000{,}000}{x^2}\,dx$

$= \left[-\frac{80{,}000{,}000}{x}\right]_{4000}^{h}$

$= \frac{-80{,}000{,}000}{h} + 20{,}000$

$\lim_{h\to\infty} W = 20{,}000$ mi-ton $\approx 2.1 \times 10^{11}$ ft-lb

13. Assume that Earth has a radius of 4000 miles.

$F(x) = \frac{k}{x^2}$

$10 = \frac{k}{(4000)^2}$

$k = 160{,}000{,}000$

$F(x) = \frac{160{,}000{,}000}{x^2}$

(a) $W = \int_{4000}^{15{,}000} \frac{160{,}000{,}000}{x^2}\,dx = \left[-\frac{160{,}000{,}000}{x}\right]_{4000}^{15{,}000} \approx -10{,}666.667 + 40{,}000$

$= 29{,}333.333$ mi-ton

$\approx 2.93 \times 10^4$ mi-ton

$\approx 3.10 \times 10^{11}$ ft-lb

(b) $W = \int_{4000}^{26{,}000} \frac{160{,}000{,}000}{x^2}\,dx = \left[-\frac{160{,}000{,}000}{x}\right]_{4000}^{26{,}000} \approx -6{,}153.846 + 40{,}000$

$= 33{,}846.154$ mi-ton

$\approx 3.38 \times 10^4$ mi-ton

$\approx 3.57 \times 10^{11}$ ft-lb

14. Weight on surface of moon: $\frac{1}{6}(12) = 2$ tons

Weight varies inversely as the square of distance from the center of the moon. Therefore:

$$F(x) = \frac{k}{x^2}$$

$$2 = \frac{k}{(1100)^2}$$

$$k = 2.42 \times 10^6$$

$$W = \int_{1100}^{1150} \frac{2.42 \times 10^6}{x^2}\,dx = \left[\frac{-2.42 \times 10^6}{x}\right]_{1100}^{1150} = 2.42 \times 10^6\left(\frac{1}{1100} - \frac{1}{1150}\right)$$

$$\approx 95.652 \text{ mi-ton} \approx 1.01 \times 10^9 \text{ ft-lb}$$

15. Weight of each layer: $62.4(20)\,\Delta y$

Distance: $4 - y$

(a) $W = \int_2^4 62.4(20)(4 - y)\,dy = \left[4992y - 624y^2\right]_2^4 = 2496$ ft-lb

(b) $W = \int_0^4 62.4(20)(4 - y)\,dy = \left[4992y - 624y^2\right]_0^4 = 9984$ ft-lb

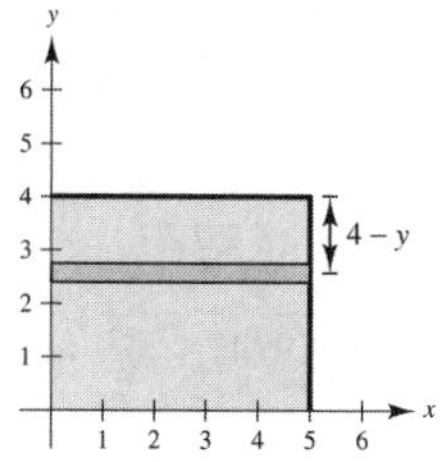

16. The bottom half had to be pumped a greater distance than the top half.

17. Volume of disk: $\pi(2)^2\,\Delta y = 4\pi\,\Delta y$

Weight of disk of water: $9800(4\pi)\,\Delta y$

Distance the disk of water is moved: $5 - y$

$$W = \int_0^4 (5 - y)(9800)4\pi\,dy = 39{,}200\pi\int_0^4 (5 - y)\,dy$$

$$= 39{,}200\pi\left[5y - \frac{y^2}{2}\right]_0^4$$

$$= 39{,}200\pi(12) = 470{,}400\pi \text{ newton–meters}$$

18. Volume of disk: $4\pi\,\Delta y$

Weight of disk: $9800(4\pi)\,\Delta y$

Distance the disk of water is moved: y

$$W = \int_{10}^{12} y(9800)(4\pi)\,dy = 39{,}200\pi\left[\frac{y^2}{2}\right]_{10}^{12}$$

$$= 39{,}200\pi(22)$$

$$= 862{,}400\pi \text{ newton–meters}$$

19. Volume of disk: $\pi\left(\frac{2}{3}y\right)^2\,\Delta y$

Weight of disk: $62.4\pi\left(\frac{2}{3}y\right)^2\,\Delta y$

Distance: $6 - y$

$$W = \frac{4(62.4)\pi}{9}\int_0^6 (6 - y)y^2\,dy$$

$$= \frac{4}{9}(62.4)\pi\left[2y^3 - \frac{1}{4}y^4\right]_0^6$$

$$= 2995.2\pi \text{ ft-lb}$$

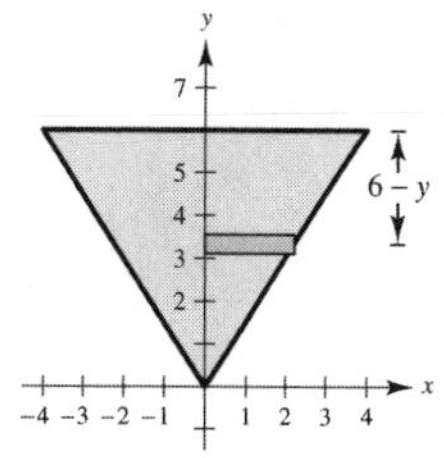

20. Volume of disk: $\pi\left(\frac{2}{3}y\right)^2 \Delta y$

Weight of disk: $62.4\pi\left(\frac{2}{3}y\right)^2 \Delta y$

Distance: y

(a) $W = \frac{4}{9}(62.4)\pi \int_0^2 y^3 dy$

$= \left[\frac{4}{9}(62.4)\pi\left(\frac{1}{4}y^4\right)\right]_0^2 \approx 110.9\pi \text{ ft} \cdot \text{lb}$

(b) $W = \frac{4}{9}(62.4)\pi \int_4^6 y^3 dy$

$= \left[\frac{4}{9}(62.4)\pi\left(\frac{1}{4}y^4\right)\right]_4^6 \approx 7210.7\pi \text{ ft-lb}$

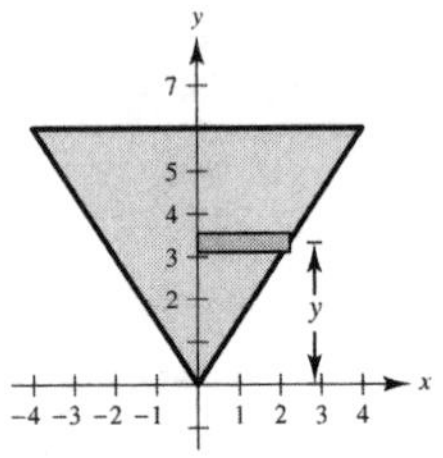

21. Volume of disk: $\pi\left(\sqrt{36 - y^2}\right)^2 \Delta y$

Weight of disk: $62.4\pi\left(36 - y^2\right) \Delta y$

Distance: y

$W = 62.4\pi \int_0^6 y\left(36 - y^2\right) dy$

$= 62.4\pi \int_0^6 \left(36y - y^3\right) dy = 62.4\pi\left[18y^2 - \frac{1}{4}y^4\right]_0^6$

$= 20{,}217.6\pi \text{ ft-lb}$

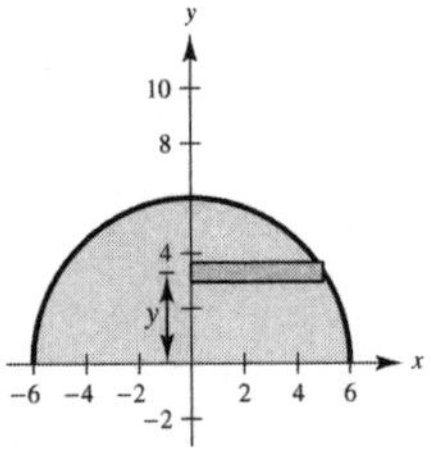

22. Volume of each layer: $\frac{y + 3}{3}(3)\, \Delta y = (y + 3)\, \Delta y$

Weight of each layer: $53.1(y + 3)\, \Delta y$

Distance: $6 - y$

$W = \int_0^3 53.1(6 - y)(y + 3)\, dy$

$= 53.1 \int_0^3 \left(18 + 3y - y^2\right) dy$

$= 53.1\left[18y + \frac{3y^2}{2} - \frac{y^3}{3}\right]_0^3$

$= 53.1\left(\frac{117}{2}\right)$

$= 3106.35 \text{ ft-lb}$

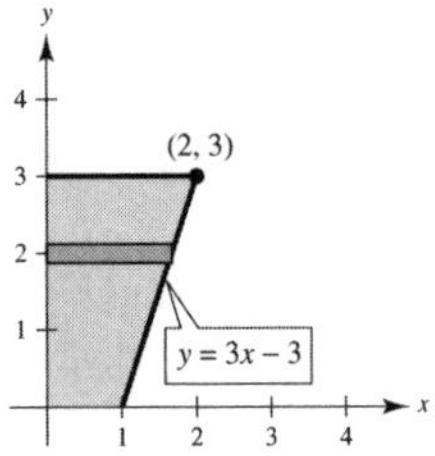

23. Volume of layer: $V = lwh = 4(2)\sqrt{(9/4) - y^2}\, \Delta y$

Weight of layer: $W = 42(8)\sqrt{(9/4) - y^2}\, \Delta y$

Distance: $\frac{13}{2} - y$

$W = \int_{-1.5}^{1.5} 42(8)\sqrt{\frac{9}{4} - y^2}\left(\frac{13}{2} - y\right) dy$

$= 336\left[\frac{13}{2}\int_{-1.5}^{1.5} \sqrt{\frac{9}{4} - y^2}\, dy - \int_{-1.5}^{1.5} \sqrt{\frac{9}{4} - y^2}\, y\, dy\right]$

The second integral is zero because the integrand is odd and the limits of integration are symmetric to the origin. The first integral represents the area of a semicircle of radius $\frac{3}{2}$. So, the work is

$W = 336\left(\frac{13}{2}\right)\pi\left(\frac{3}{2}\right)^2\left(\frac{1}{2}\right) = 2457\pi \text{ ft-lb}.$

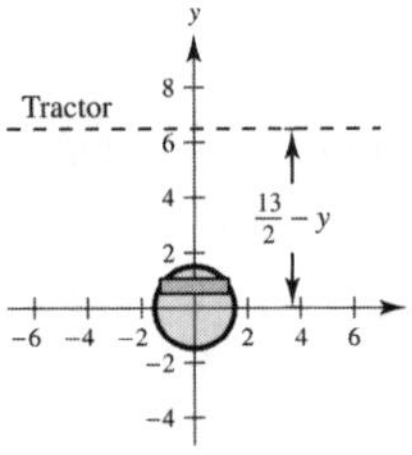

24. Volume of layer: $V = 12(2)\sqrt{(25/4) - y^2}\,\Delta y$

Weight of layer: $W = 42(24)\sqrt{(25/4) - y^2}\,\Delta y$

Distance: $\frac{19}{2} - y$

$$W = \int_{-2.5}^{2.5} 42(24)\sqrt{\tfrac{25}{4} - y^2}\left(\tfrac{19}{2} - y\right) dy = 1008\left[\frac{19}{2}\int_{-2.5}^{2.5}\sqrt{\tfrac{25}{4} - y^2}\,dy + \int_{-2.5}^{2.5}\sqrt{\tfrac{25}{4} - y^2}(-y)\,dy\right]$$

The second integral is zero because the integrand is odd and the limits of integration are symmetric to the origin. The first integral represents the area of a semicircle of radius $\frac{5}{2}$. So, the work is

$$W = 1008\left(\tfrac{19}{2}\right)\pi\left(\tfrac{5}{2}\right)^2\left(\tfrac{1}{2}\right) = 29{,}925\pi \text{ ft-lb} \approx 94{,}012.16 \text{ ft-lb}.$$

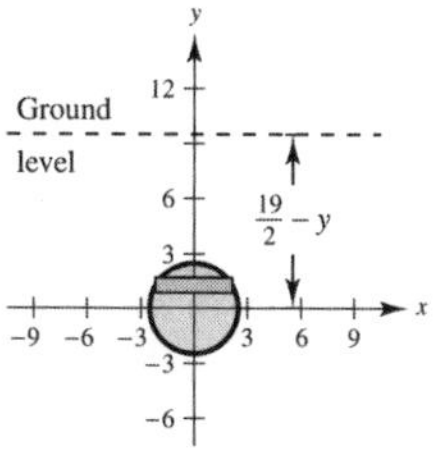

25. Weight of section of chain: $3\,\Delta y$

Distance: $20 - y$. $\quad \Delta W = (\text{force increment})(\text{distance}) = (3\,\Delta y)(20 - y)$

$$W = \int_0^{20}(20 - y)3\,dy = 3\left[20y - \frac{y^2}{2}\right]_0^{20} = 3\left[400 - \frac{400}{2}\right] = 600 \text{ ft-lb}$$

26. The lower $\frac{2}{3}(20)$ feet of chain are raised with a constant force

$$W_1 = 3\left(\frac{2}{3}(20)\right)\left(\frac{20}{3}\right) = \frac{800}{3} \text{ ft-lb}$$

The top $\frac{1}{3}(20)$ feet are raised with a variable force.

Weight of section: $3\,\Delta y$

Distance: $\frac{1}{3}(20) - y$

$$W_2 = \int_0^{20/3} 3\left(\frac{20}{3} - y\right) dy = 3\left[\frac{20}{3}y - \frac{y^2}{2}\right]_0^{20/3} = \frac{200}{3} \text{ ft-lb}$$

$$W = W_1 + W_2 = \frac{800}{3} + \frac{200}{3} = \frac{1000}{3} \text{ ft-lb}$$

27. The lower 10 feet of fence are raised 10 feet with a constant force.

$$W_1 = 3(10)(10) = 300 \text{ ft-lb}$$

The top 10 feet are raised with a variable force.

Weight of section: $3\,\Delta y$

Distance: $10 - y$

$$W_2 = \int_0^{10} 3(10 - y)\,dy = 3\left[10y - \frac{y^2}{2}\right]_0^{10} = 150 \text{ ft-lb}$$

$$W = W_1 + W_2 = 300 + 150 = 450 \text{ ft-lb}$$

28. From Exercise 25, the work required to lift the chain is 600 ft-lb.

The work required to lift the 500-pound load is $500(20) = 10{,}000$ ft-lb.

The total is $600 + 10{,}000 = 10{,}600$ ft-lb.

29. Weight of section of chain: $3\,\Delta y$

Distance: $15 - 2y$

$$W = 3\int_0^{7.5}(15 - 2y)\,dy = \left[-\tfrac{3}{4}(15 - 2y)^2\right]_0^{7.5} = \tfrac{3}{4}(15)^2 = 168.75 \text{ ft-lb}$$

30. $W = 3\int_0^6 (12 - 2y)\,dy = \left[-\frac{3}{4}(12 - 2y)^2\right]_0^6$

$= \frac{3}{4}(12)^2 = 108$ ft-lb

31. If an object is moved a distance D in the direction of an applied constant force F, then the work W done by the force is defined as force times distance, $W = FD$.

32. $W = \int_a^b F(x)\,dx$ is the work done by a force F moving an object along a straight line from $x = a$ to $x = b$.

33. (a) requires more work. In part (b) no work is done because the books are not moved:
W = force × distance

34. Because the work equals the area under the force function, you have (c) < (d) < (a) < (b).

35. (a) $W = \int_0^9 6\,dx = 54$ ft-lb

(b) $W = \int_0^7 20\,dx + \int_7^9 (-10x + 90)\,dx = 140 + 20$

$= 160$ ft-lb

(c) $W = \int_0^9 \frac{1}{27}x^2\,dx = \left.\frac{x^3}{81}\right]_0^9 = 9$ ft-lb

(d) $W = \int_0^9 \sqrt{x}\,dx = \left.\frac{2}{3}x^{3/2}\right]_0^9 = \frac{2}{3}(27) = 18$ ft-lb

36. $F(x) = \frac{k}{(2 - x)^2}$

$W = \int_{-2}^1 \frac{k}{(2 - x)^2}\,dx = \left[\frac{k}{2 - x}\right]_{-2}^1 = k\left(1 - \frac{1}{4}\right) = \frac{3k}{4}$ (units of work)

37. $p = \frac{k}{V}$

$1000 = \frac{k}{2}$

$k = 2000$

$W = \int_2^3 \frac{2000}{V}\,dV$

$= \left[2000 \ln|V|\right]_2^3 = 2000 \ln\left(\frac{3}{2}\right) \approx 810.93$ ft-lb

38. $p = \frac{k}{V}$

$2500 = \frac{k}{1} \Rightarrow k = 2500$

$W = \int_1^3 \frac{2500}{V}\,dV = \left[2500 \ln V\right]_1^3 = 2500 \ln 3$

≈ 2746.53 ft-lb

39. $W = \int_0^5 1000\left[1.8 - \ln(x + 1)\right]dx \approx 3249.44$ ft-lb

40. $W = \int_0^4 \left(\frac{e^{x^2} - 1}{100}\right)dx \approx 11{,}494$ ft-lb

41. $W = \int_0^5 100x\sqrt{125 - x^3}\,dx \approx 10{,}330.3$ ft-lb

42. $W = \int_0^2 1000 \sinh x\,dx \approx 2762.2$ ft-lb

43. (a) $W = FD = (8000\pi)(2) = 16{,}000\pi$ ft · lb

(b) $W \approx \frac{2 - 0}{3(6)}\left[0 + 4(20{,}000) + 2(22{,}000) + 4(15{,}000) + 2(10{,}000) + 4(5000) + 0\right] \approx 24888.889$ ft-lb

(c) $F(x) = -16{,}261.36x^4 + 85{,}295.45x^3 - 157{,}738.64x^2 + 104{,}386.36x - 32.4675$

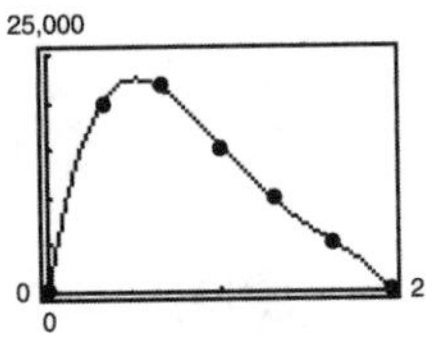

(d) $F(x)$ is a maximum when $x \approx 0.524$ feet.

(e) $W = \int_0^2 F(x)\,dx \approx 25{,}180.5$ ft-lb

Section 7.6 Moments, Centers of Mass, and Centroids

1. $\bar{x} = \dfrac{7(-5) + 3(0) + 5(3)}{7 + 3 + 5} = \dfrac{-20}{15} = -\dfrac{4}{3}$

2. $\bar{x} = \dfrac{7(-3) + 4(-2) + 3(5) + 8(4)}{7 + 4 + 3 + 8} = \dfrac{9}{11}$

3. $\bar{x} = \dfrac{1(6) + 3(10) + 2(3) + 9(2) + 5(4)}{1 + 3 + 2 + 9 + 5} = \dfrac{80}{20} = 4$

4. $\bar{x} = \dfrac{8(-2) + 5(6) + 5(0) + 12(3) + 2(-5)}{8 + 5 + 5 + 12 + 2} = \dfrac{40}{32} = \dfrac{5}{4}$

5. (a) Add 4 to each x-value because each point is translated to the right 4 units.

$$\bar{x} = \frac{1(10) + 3(14) + 2(7) + 9(6) + 5(8)}{1 + 3 + 2 + 9 + 5} = \frac{160}{20} = 8$$

Note: From Exercise 3, $4 + 4 = 8$.

(b) Subtract 2 from each x-value because each point is translated 2 units to the left.

$$\bar{x} = \frac{8(-4) + 5(4) + 5(-2) + 12(1) + 2(-7)}{8 + 5 + 5 + 12 + 2} = \frac{-24}{32} = -\frac{3}{4}$$

Note: From Exercise 4, $\dfrac{5}{4} - 2 = -\dfrac{3}{4}$.

6. The center of mass is translated k units as well.

7.
$$\begin{aligned} 48x &= 72(L - x) = 72(10 - x) \\ 48x &= 720 - 72x \\ 120x &= 720 \\ x &= 6 \text{ ft} \end{aligned}$$

8.
$$\begin{aligned} 200x &= 600(5 - x) \quad \text{(person is on the left)} \\ 200x &= 3000 - 600x \\ 800x &= 3000 \\ x &= \tfrac{15}{4} = 3\tfrac{3}{4} \text{ ft} \end{aligned}$$

9. $\bar{x} = \dfrac{5(2) + 1(-3) + 3(1)}{5 + 1 + 3} = \dfrac{10}{9}$

$\bar{y} = \dfrac{5(2) + 1(1) + 3(-4)}{5 + 1 + 3} = -\dfrac{1}{9}$

$(\bar{x}, \bar{y}) = \left(\dfrac{10}{9}, -\dfrac{1}{9}\right)$

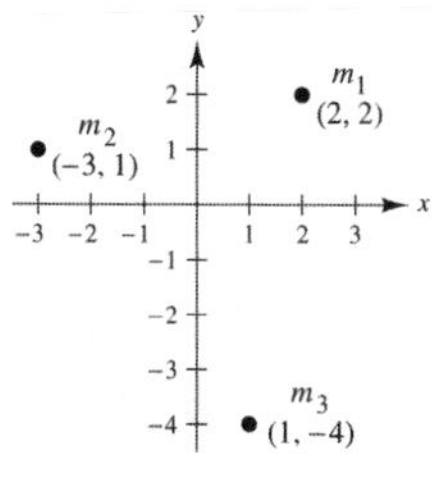

10. $\bar{x} = \dfrac{10(1) + 2(5) + 5(-4)}{10 + 2 + 5} = 0$

$\bar{y} = \dfrac{10(-1) + 2(5) + 5(0)}{10 + 2 + 5} = 0$

$(\bar{x}, \bar{y}) = (0, 0)$

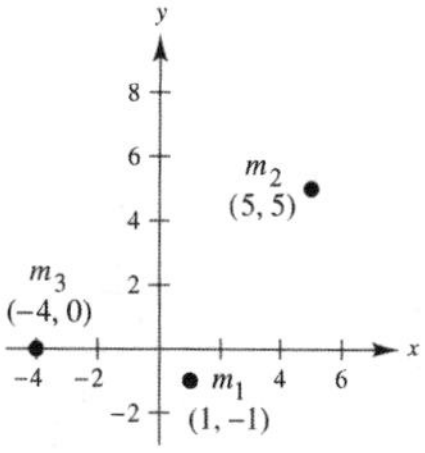

11. $\bar{x} = \dfrac{12(2) + 6(-1) + (9/2)(6) + 15(2)}{12 + 6 + (9/2) + 15} = \dfrac{75}{37.5} = 2$

$\bar{y} = \dfrac{12(3) + 6(5) + (9/2)(8) + 15(-2)}{12 + 6 + (9/2) + 15} = \dfrac{72}{37.5} = \dfrac{48}{25}$

$(\bar{x}, \bar{y}) = \left(2, \dfrac{48}{25}\right)$

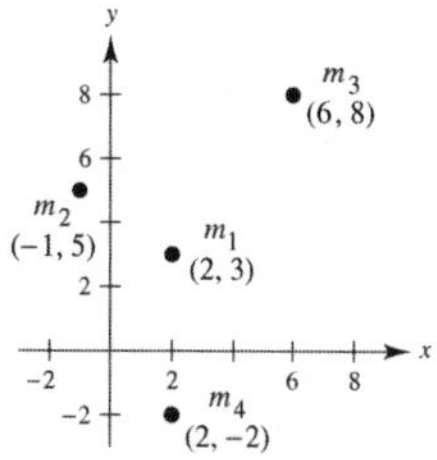

12. $\bar{x} = \dfrac{3(-2) + 4(5) + 2(7) + 1(0) + 6(-3)}{3 + 4 + 2 + 1 + 6} = \dfrac{5}{8}$

$\bar{y} = \dfrac{3(-3) + 4(5) + 2(1) + 1(0) + 6(0)}{3 + 4 + 2 + 1 + 6} = \dfrac{13}{16}$

$(\bar{x}, \bar{y}) = \left(\dfrac{5}{8}, \dfrac{13}{16}\right)$

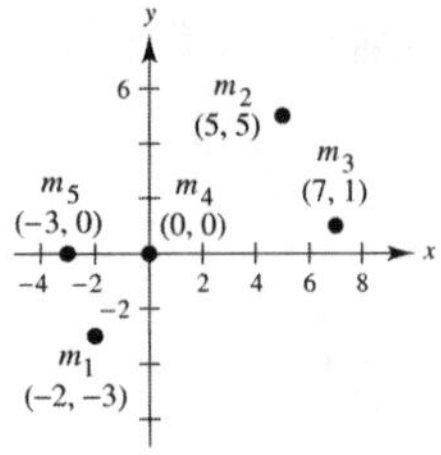

13. $m = \rho\int_0^2 \dfrac{x}{2}\,dx = \left[\rho\dfrac{x^2}{4}\right]_0^2 = \rho$

$M_x = \rho\int_0^2 \dfrac{1}{2}\left(\dfrac{x}{2}\right)^2 dx$

$= \dfrac{\rho}{8}\left[\dfrac{x^3}{3}\right]_0^2 = \dfrac{\rho}{3}$

$\bar{y} = \dfrac{M_x}{m} = \dfrac{\rho/3}{\rho} = \dfrac{1}{3}$

$M_y = \rho\int_0^2 x\left(\dfrac{x}{2}\right) dx = \dfrac{\rho}{2}\left[\dfrac{x^3}{3}\right]_0^2 = \dfrac{4}{3}\rho$

$\bar{x} = \dfrac{M_y}{m} = \dfrac{4/3\rho}{\rho} = \dfrac{4}{3}$

$(\bar{x}, \bar{y}) = \left(\dfrac{4}{3}, \dfrac{1}{3}\right)$

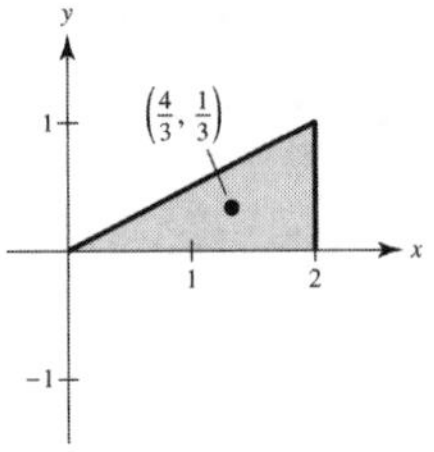

14. $m = \rho\int_0^6 (6 - x)\,dx = \rho\left[6x - \dfrac{x^2}{2}\right]_0^6 = 18\rho$

$M_x = \rho\int_0^6 \dfrac{1}{2}(6 - x)^2\,dx = \dfrac{\rho}{2}\int_0^6 \left(36 - 12x + x^2\right) dx$

$= \dfrac{\rho}{2}\left[36x - 6x^2 + \dfrac{x^3}{3}\right]_0^6$

$= \dfrac{\rho}{2}[72] = 36\rho$

$\bar{y} = \dfrac{M_x}{m} = \dfrac{36\rho}{18\rho} = 2$

$M_y = \rho\int_0^6 x(6 - x)\,dx = \rho\int_0^6 \left(6x - x^2\right) dx$

$= \rho\left[3x^2 - \dfrac{x^3}{3}\right]_0^6 = 36\rho$

$\bar{x} = \dfrac{M_y}{m} = \dfrac{36\rho}{18\rho} = 2$

$(\bar{x}, \bar{y}) = (2, 2)$

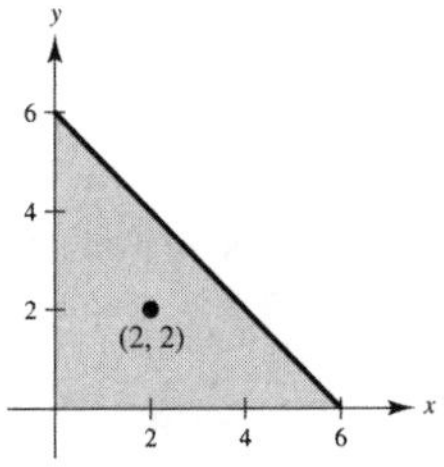

15. $m = \rho\int_0^4 \sqrt{x}\,dx = \left[\dfrac{2\rho}{3}x^{3/2}\right]_0^4 = \dfrac{16\rho}{3}$

$M_x = \rho\int_0^4 \dfrac{\sqrt{x}}{2}\left(\sqrt{x}\right) dx = \left[\rho\dfrac{x^2}{4}\right]_0^4 = 4\rho$

$\bar{y} = \dfrac{M_x}{m} = 4\rho\left(\dfrac{3}{16\rho}\right) = \dfrac{3}{4}$

$M_y = \rho\int_0^4 x\sqrt{x}\,dx = \left[\rho\dfrac{2}{5}x^{5/2}\right]_0^4 = \dfrac{64\rho}{5}$

$\bar{x} = \dfrac{M_y}{m} = \dfrac{64\rho}{5}\left(\dfrac{3}{16\rho}\right) = \dfrac{12}{5}$

$(\bar{x}, \bar{y}) = \left(\dfrac{12}{5}, \dfrac{3}{4}\right)$

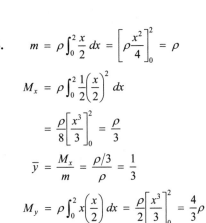

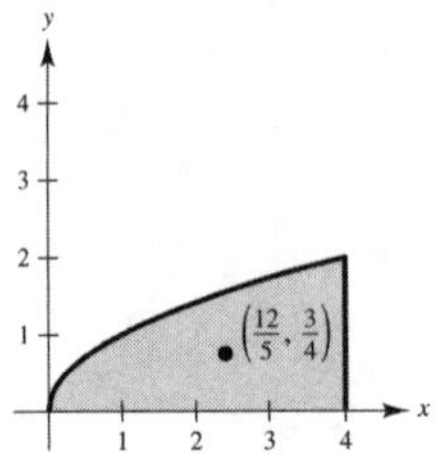

16. $$m = \rho\int_0^2 \frac{x^2}{2}\,dx = \rho\left[\frac{x^3}{6}\right]_0^2 = \frac{4}{3}\rho$$

$$M_x = \rho\int_0^2 \frac{1}{2}\left(\frac{x^2}{2}\right)^2 dx = \frac{\rho}{8}\left[\frac{x^5}{5}\right]_0^2 = \frac{4}{5}\rho$$

$$\bar{y} = \frac{M_x}{m} = \frac{4/5\rho}{4/3\rho} = \frac{3}{5}$$

$$M_y = \rho\int_0^2 x\left(\frac{x^2}{2}\right) dx = \frac{\rho}{2}\left[\frac{x^4}{4}\right]_0^2 = 2\rho$$

$$\bar{x} = \frac{M_y}{m} = \frac{2\rho}{4/3\rho} = \frac{3}{2}$$

$$(\bar{x}, \bar{y}) = \left(\frac{3}{2}, \frac{3}{5}\right)$$

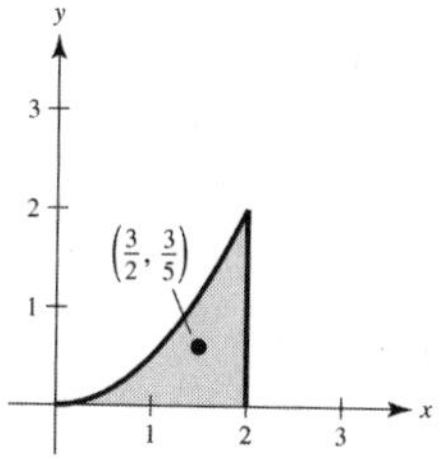

17. $$m = \rho\int_0^1 \left(x^2 - x^3\right) dx = \rho\left[\frac{x^3}{3} - \frac{x^4}{4}\right]_0^1 = \frac{\rho}{12}$$

$$M_x = \rho\int_0^1 \frac{\left(x^2 + x^3\right)}{2}\left(x^2 - x^3\right) dx = \frac{\rho}{2}\int_0^1 \left(x^4 - x^6\right) dx = \frac{\rho}{2}\left[\frac{x^5}{5} - \frac{x^7}{7}\right]_0^1 = \frac{\rho}{35}$$

$$\bar{y} = \frac{M_x}{m} = \frac{\rho}{35}\left(\frac{12}{\rho}\right) = \frac{12}{35}$$

$$M_y = \rho\int_0^1 x\left(x^2 - x^3\right) dx = \rho\int_0^1 \left(x^3 - x^4\right) dx = \rho\left[\frac{x^4}{4} - \frac{x^5}{5}\right]_0^1 = \frac{\rho}{20}$$

$$\bar{x} = \frac{M_y}{m} = \frac{\rho}{20}\left(\frac{12}{\rho}\right) = \frac{3}{5}$$

$$(\bar{x}, \bar{y}) = \left(\frac{3}{5}, \frac{12}{35}\right)$$

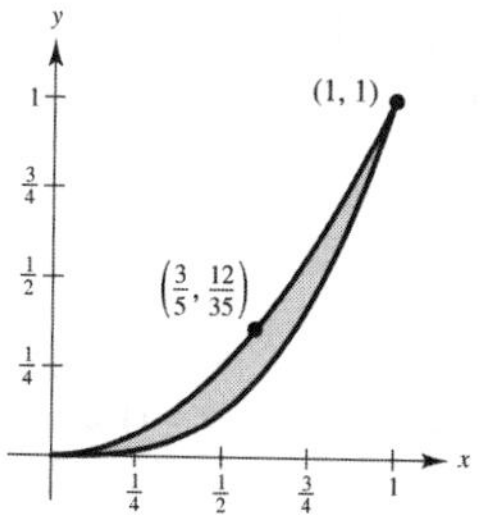

18. $$m = \rho\int_0^4 \left(\sqrt{x} - \frac{x}{2}\right) dx = \rho\left[\frac{2}{3}x^{3/2} - \frac{x^2}{4}\right]_0^4 = \rho\left[\frac{16}{3} - 4\right] = \frac{4}{3}\rho$$

$$M_x = \rho\int_0^4 \frac{1}{2}\left(\sqrt{x} + \frac{x}{2}\right)\left(\sqrt{x} - \frac{x}{2}\right) dx = \frac{1}{2}\rho\int_0^4 \left(x - \frac{x^2}{4}\right) dx = \frac{\rho}{2}\left[\frac{x^2}{2} - \frac{x^3}{12}\right]_0^4 = \frac{\rho}{2}\left[8 - \frac{16}{3}\right] = \frac{4}{3}\rho$$

$$\bar{y} = \frac{M_x}{m} = \frac{4/3\rho}{4/3\rho} = 1$$

$$M_y = \rho\int_0^4 x\left(\sqrt{x} - \frac{x}{2}\right) dx = \rho\left[\frac{2}{5}x^{5/2} - \frac{x^3}{6}\right]_0^4 = \rho\left[\frac{64}{5} - \frac{32}{3}\right] = \frac{32}{15}\rho$$

$$\bar{x} = \frac{M_y}{m} = \frac{32/15\rho}{4/3\rho} = \frac{8}{5}$$

$$(\bar{x}, \bar{y}) = (8/5, 1)$$

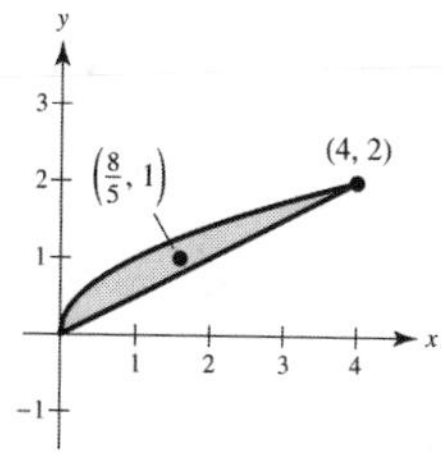

19. $m = \rho\int_0^3\left[\left(-x^2 + 4x + 2\right) - (x + 2)\right]dx = -\rho\left[\frac{x^3}{3} + \frac{3x^2}{2}\right]_0^3 = \frac{9\rho}{2}$

$$M_x = \rho\int_0^3\left[\frac{\left(-x^2 + 4x + 2\right) + (x + 2)}{2}\right]\left[\left(-x^2 + 4x + 2\right) - (x + 2)\right]dx$$

$$= \frac{\rho}{2}\int_0^3\left(-x^2 + 5x + 4\right)\left(-x^2 + 3x\right)dx = \frac{\rho}{2}\int_0^3\left(x^4 - 8x^3 + 11x^2 + 12x\right)dx = \frac{\rho}{2}\left[\frac{x^5}{5} - 2x^4 + \frac{11x^3}{3} + 6x^2\right]_0^3 = \frac{99\rho}{5}$$

$$\bar{y} = \frac{M_x}{m} = \frac{99\rho}{5}\left(\frac{2}{9\rho}\right) = \frac{22}{5}$$

$$M_y = \rho\int_0^3 x\left[\left(-x^2 + 4x - 2\right) - (x + 2)\right]dx = \rho\int_0^3\left(-x^3 + 3x^2\right)dx = \rho\left[-\frac{x^4}{4} + x^3\right]_0^3 = \frac{27\rho}{4}$$

$$\bar{x} = \frac{M_y}{m} = \frac{27\rho}{4}\left(\frac{2}{9\rho}\right) = \frac{3}{2}$$

$$(\bar{x}, \bar{y}) = \left(\frac{3}{2}, \frac{22}{5}\right)$$

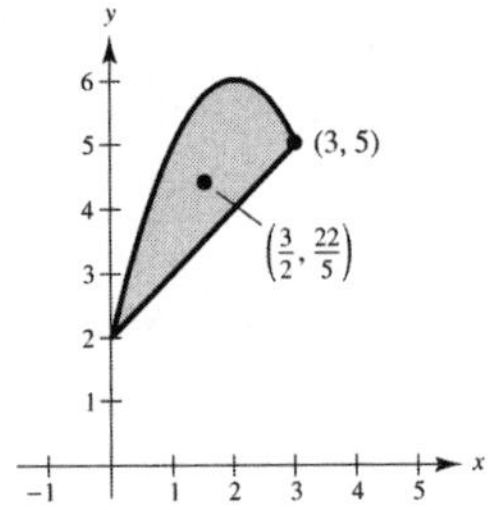

20. $m = \rho\int_0^9\left[\left(\sqrt{x} + 1\right) - \left(\frac{1}{3}x + 1\right)\right]dx = \rho\int_0^9\left(\sqrt{x} - \frac{1}{3}x\right)dx = \rho\left[\frac{2}{3}x^{3/2} - \frac{x^2}{6}\right]_0^9 = \rho\left(18 - \frac{27}{2}\right) = \frac{9}{2}\rho$

$$M_x = \rho\int_0^9\frac{\sqrt{x} + 1 + (1/3)x + 1}{2}\left(\sqrt{x} + 1 - \frac{1}{3}x - 1\right)dx = \frac{\rho}{2}\int_0^9\left(\sqrt{x} + \frac{1}{3}x + 2\right)\left(\sqrt{x} - \frac{1}{3}x\right)dx$$

$$= \frac{\rho}{2}\int_0^9\left(x - \frac{1}{3}x^{3/2} + \frac{1}{3}x^{3/2} - \frac{1}{9}x^2 + 2\sqrt{x} - \frac{2}{3}x\right)dx = \frac{\rho}{2}\int_0^9\left(\frac{1}{3}x - \frac{1}{9}x^2 + 2\sqrt{x}\right)dx$$

$$= \frac{\rho}{2}\left[\frac{x^2}{6} - \frac{x^3}{27} + \frac{4}{3}x^{3/2}\right]_0^9 = \frac{\rho}{2}\left[\frac{27}{2} - 27 + 36\right] = \frac{45}{4}\rho$$

$$M_y = \rho\int_0^9 x\left[\sqrt{x} + 1 - \frac{1}{3}x - 1\right]dx = \rho\int_0^9\left(x^{3/2} - \frac{1}{3}x^2\right)dx = \rho\left[\frac{2}{5}x^{5/2} - \frac{1}{9}x^3\right]_0^9 = \rho\left[\frac{486}{5} - 81\right] = \frac{81}{5}\rho$$

$$\bar{x} = \frac{M_y}{m} = \frac{(81/5)\rho}{(9/2)\rho} = \frac{18}{5};\ \bar{y} = \frac{M_x}{m} = \frac{(45/4)\rho}{(9/2)\rho} = \frac{5}{2}$$

$$(\bar{x}, \bar{y}) = \left(\frac{18}{5}, \frac{5}{2}\right)$$

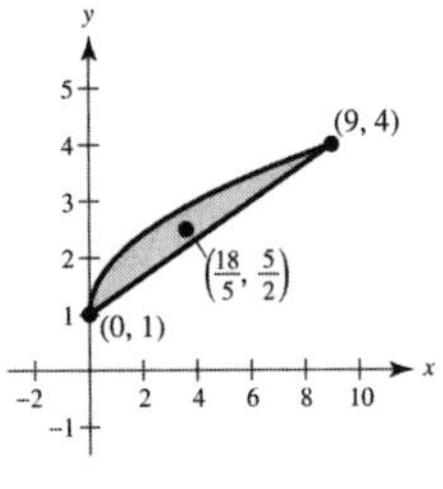

21. $m = \rho\int_0^8 x^{2/3}\,dx = \rho\left[\frac{3}{5}x^{5/3}\right]_0^8 = \frac{96\rho}{5}$

$M_x = \rho\int_0^8 \frac{x^{2/3}}{2}\left(x^{2/3}\right)dx = \frac{\rho}{2}\left[\frac{3}{7}x^{7/3}\right]_0^8 = \frac{192\rho}{7}$

$\bar{y} = \frac{M_x}{m} = \frac{192\rho}{7}\left(\frac{5}{96\rho}\right) = \frac{10}{7}$

$M_y = \rho\int_0^8 x\left(x^{2/3}\right)dx = \rho\left[\frac{3}{8}x^{8/3}\right]_0^8 = 96\rho$

$\bar{x} = \frac{M_y}{m} = 96\rho\left(\frac{5}{96\rho}\right) = 5$

$(\bar{x}, \bar{y}) = \left(5, \frac{10}{7}\right)$

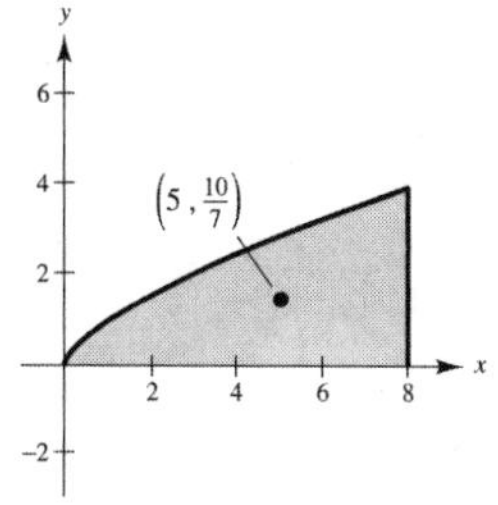

22. $m = 2\rho\int_0^8\left(4 - x^{2/3}\right)dx = 2\rho\left[4x - \frac{3}{5}x^{5/3}\right]_0^8 = \frac{128\rho}{5}$

By symmetry, M_y and $\bar{x} = 0$.

$M_x = 2\rho\int_0^8\left(\frac{4 + x^{2/3}}{2}\right)\left(4 - x^{2/3}\right)dx = \rho\left[16x - \frac{3}{7}x^{7/3}\right]_0^8 = \frac{512\rho}{7}$

$\bar{y} = \frac{512\rho}{7}\left(\frac{5}{128\rho}\right) = \frac{20}{7}$

$(\bar{x}, \bar{y}) = \left(0, \frac{20}{7}\right)$

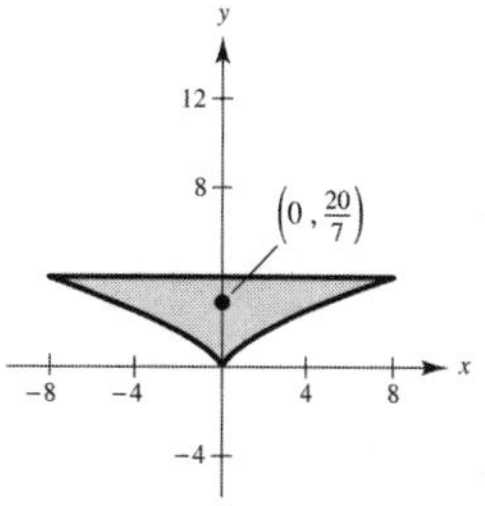

23. $m = 2\rho\int_0^2\left(4 - y^2\right)dy = 2\rho\left[4y - \frac{y^3}{3}\right]_0^2 = \frac{32\rho}{3}$

$M_y = 2\rho\int_0^2\left(\frac{4 - y^2}{2}\right)\left(4 - y^2\right)dy = \rho\left[16y - \frac{8}{3}y^3 + \frac{y^5}{5}\right]_0^2 = \frac{256\rho}{15}$

$\bar{x} = \frac{M_y}{m} = \frac{256\rho}{15}\left(\frac{3}{32\rho}\right) = \frac{8}{5}$

By symmetry, M_x and $\bar{y} = 0$.

$(\bar{x}, \bar{y}) = \left(\frac{8}{5}, 0\right)$

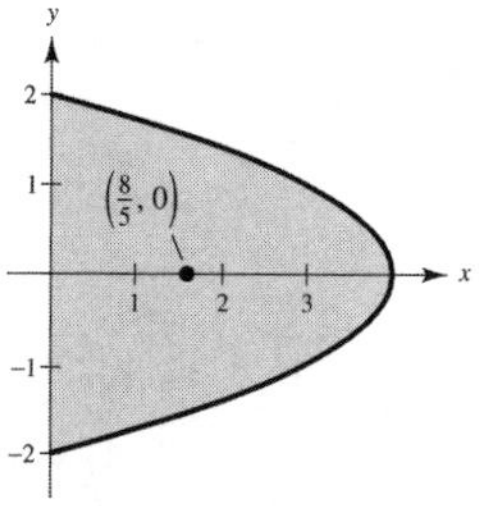

24. $m = \rho\int_0^3\left(3y - y^2\right)dy = \rho\left[\frac{3y^2}{2} - \frac{y^3}{3}\right]_0^2 = \frac{9}{2}\rho$

$M_y = \rho\int_0^3 \frac{1}{2}\left(3y - y^2\right)dy = \frac{\rho}{2}\int_0^3\left(9y^2 - 6y^3 + y^4\right)dy$

$= \frac{\rho}{2}\left[3y^3 - \frac{3y^4}{2} + \frac{y^5}{5}\right]_0^3 = \frac{81}{20}\rho$

$\bar{x} = \frac{M_y}{m} = \frac{81/20\rho}{9/2\rho} = \frac{9}{10}$

$M_x = \rho\int_0^3 y\left(3y - y^2\right)dy = \rho\left[y^3 - \frac{y^4}{4}\right]_0^3 = \frac{27}{4}\rho$

$\bar{y} = \frac{M_x}{m} = \frac{27/4\rho}{9/2\rho} = \frac{3}{2}$

$(\bar{x}, \bar{y}) = \left(\frac{9}{10}, \frac{3}{2}\right)$

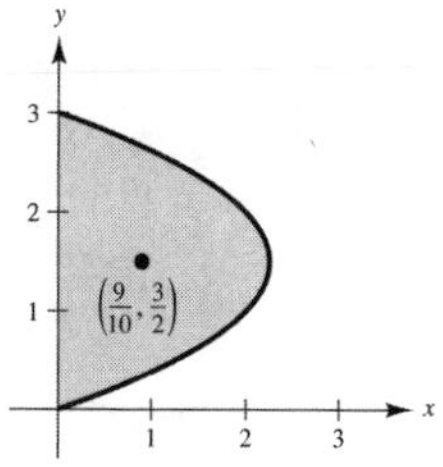

25. $m = \rho\int_0^3\left[\left(2y - y^2\right) - (-y)\right]dy = \rho\left[\frac{3y^2}{2} - \frac{y^3}{3}\right]_0^3 = \frac{9\rho}{2}$

$M_y = \rho\int_0^3 \frac{\left[\left(2y - y^2\right) + (-y)\right]}{2}\left[\left(2y - y^2\right) - (-y)\right]dy = \frac{\rho}{2}\int_0^3\left(y - y^2\right)\left(3y - y^2\right)dy$

$= \frac{\rho}{2}\int_0^3\left(y^4 - 4y^3 + 3y^2\right)dy = \frac{\rho}{2}\left[\frac{y^5}{5} - y^4 + y^3\right]_0^3 = -\frac{27\rho}{10}$

$\bar{x} = \frac{M_y}{m} = -\frac{27\rho}{10}\left(\frac{2}{9\rho}\right) = -\frac{3}{5}$

$M_x = \rho\int_0^3 y\left[\left(2y - y^2\right) - (-y)\right]dy = \rho\int_0^3\left(3y^2 - y^3\right)dy = \rho\left[y^3 - \frac{y^4}{4}\right]_0^3 = \frac{27\rho}{4}$

$\bar{y} = \frac{M_x}{m} = \frac{27\rho}{4}\left(\frac{2}{9\rho}\right) = \frac{3}{2}$

$(\bar{x}, \bar{y}) = \left(-\frac{3}{5}, \frac{3}{2}\right)$

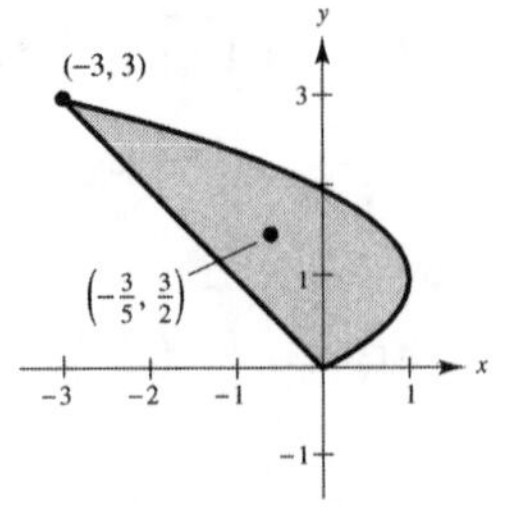

26. $m = \rho\int_{-1}^2\left[(y + 2) - y^2\right]dy = \rho\left[\frac{y^2}{2} + 2y - \frac{y^3}{3}\right]_{-1}^2 = \frac{9\rho}{2}$

$M_y = \rho\int_{-1}^2 \frac{\left[(y + 2) + y^2\right]}{2}\left[(y + 2) - y^2\right]dy = \frac{\rho}{2}\int_{-1}^2\left[(y + 2)^2 - y^4\right]dy = \frac{\rho}{2}\left[\frac{(y + 2)^3}{3} - \frac{y^5}{5}\right]_{-1}^2 = \frac{36\rho}{5}$

$\bar{x} = \frac{M_y}{m} = \frac{36\rho}{5}\left(\frac{2}{9\rho}\right) = \frac{8}{5}$

$M_x = \rho\int_{-1}^2 y\left[(y + 2) - y^2\right]dy = \rho\int_{-1}^2\left(2y + y^2 - y^3\right)dy = \rho\left[y^2 + \frac{y^3}{3} - \frac{y^4}{4}\right]_{-1}^2 = \frac{9\rho}{4}$

$\bar{y} = \frac{M_x}{m} = \frac{9\rho}{4}\left(\frac{2}{9\rho}\right) = \frac{1}{2}$

$(\bar{x}, \bar{y}) = \left(\frac{8}{5}, \frac{1}{2}\right)$

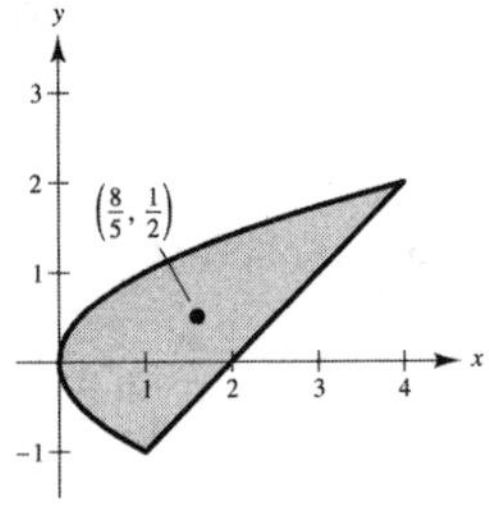

27. $m = \rho\int_0^5 10x\sqrt{125 - x^3}\,dx \approx 1033.0\rho$

$M_x = \rho\int_0^5\left(\frac{10x\sqrt{125 - x^3}}{2}\right)\left(10x\sqrt{125 - x^3}\right)dx = 50\rho\int_0^5 x^2\left(125 - x^3\right)dx = \frac{3{,}124{,}375\rho}{24} \approx 130{,}208\rho$

$M_y = \rho\int_0^5 10x^2\sqrt{125 - x^3}\,dx = -\frac{10\rho}{3}\int_0^5\sqrt{125 - x^3}\left(-3x^2\right)dx = \frac{12{,}500\sqrt{5}\rho}{9} \approx 3105.6\rho$

$\bar{x} = \frac{M_y}{m} \approx 3.0$

$\bar{y} = \frac{M_x}{m} \approx 126.0$

Therefore, the centroid is $(3.0, 126.0)$.

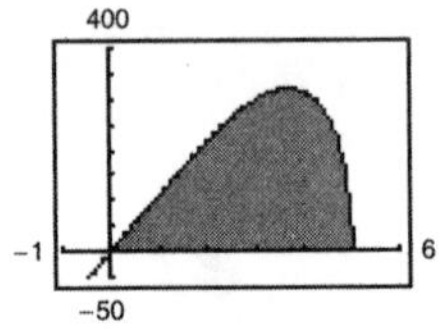

28. $m = \rho \int_0^4 xe^{-x/2}\, dx \approx 2.3760\rho$

$$M_x = \rho \int_0^4 \left(\frac{xe^{-x/2}}{2}\right)\left(xe^{-x/2}\right) dx$$

$$= \frac{\rho}{2}\int_0^4 x^2 e^{-x}\, dx \approx 0.7619\rho$$

$$M_y = \rho \int_0^4 x^2 e^{-x/2}\, dx \approx 5.1732\rho$$

$$\bar{x} = \frac{M_y}{m} \approx 2.2$$

$$\bar{y} = \frac{M_x}{m} \approx 0.3$$

Therefore, the centroid is $(2.2, 0.3)$.

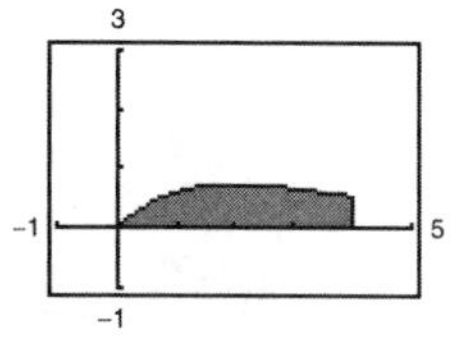

29. $m = \rho \int_{-20}^{20} 5\sqrt[3]{400 - x^2}\, dx \approx 1239.76\rho$

$$M_x = \rho \int_{-20}^{20} \frac{5\sqrt[3]{400 - x^2}}{2}\left(5\sqrt[3]{400 - x^2}\right) dx$$

$$= \frac{25\rho}{2}\int_{-20}^{20} \left(400 - x^2\right)^{2/3} dx \approx 20064.27$$

$$\bar{y} = \frac{M_x}{m} \approx 16.18$$

$\bar{x} = 0$ by symmetry. Therefore, the centroid is $(0, 16.2)$.

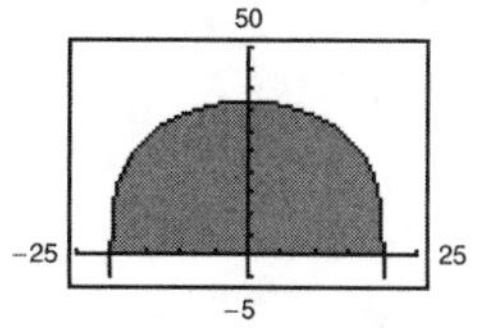

30. $m = \rho \int_{-2}^{2} \frac{8}{x^2 + 4}\, dx \approx 6.2832\rho$

$$M_x = \rho \int_{-2}^{2} \frac{1}{2}\left(\frac{8}{x^2 + 4}\right)\left(\frac{8}{x^2 + 4}\right) dx$$

$$= 32\rho \int_{-2}^{2} \frac{1}{\left(x^2 + 4\right)^2}\, dx \approx 5.14149\rho$$

$$\bar{y} = \frac{M_x}{m} \approx 0.8$$

$\bar{x} = 0$ by symmetry. Therefore, the centroid is $(0, 0.8)$.

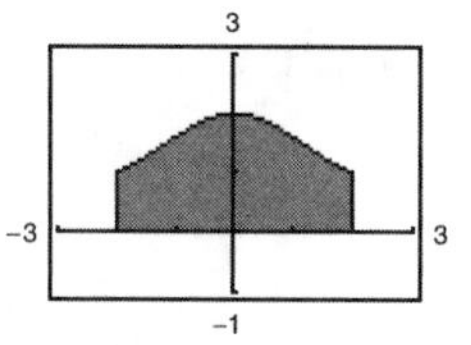

31. Centroids of the given regions: $(1, 0)$ and $(3, 0)$

Area: $A = 4 + \pi$

$$\bar{x} = \frac{4(1) + \pi(3)}{4 + \pi} = \frac{4 + 3\pi}{4 + \pi}$$

$$\bar{y} = \frac{4(0) + \pi(0)}{4 + \pi} = 0$$

$$(\bar{x}, \bar{y}) = \left(\frac{4 + 3\pi}{4 + \pi}, 0\right) \approx (1.88, 0)$$

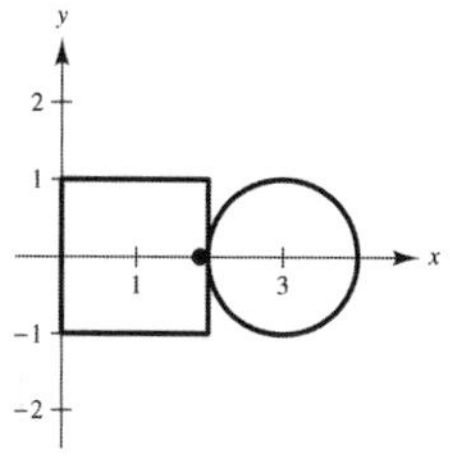

32. Centroids of the given regions: $\left(\frac{1}{2}, \frac{3}{2}\right)$, $\left(2, \frac{1}{2}\right)$, and $\left(\frac{7}{2}, 1\right)$

Area: $A = 3 + 2 + 2 = 7$

$$\bar{x} = \frac{3(1/2) + 2(2) + 2(7/2)}{7} = \frac{25/2}{7} = \frac{25}{14}$$

$$\bar{y} = \frac{3(3/2) + 2(1/2) + 2(1)}{7} = \frac{15/2}{7} = \frac{15}{14}$$

$$(\bar{x}, \bar{y}) = \left(\frac{25}{14}, \frac{15}{14}\right)$$

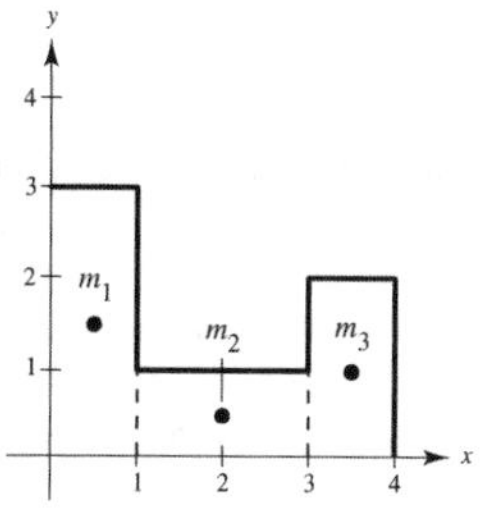

33. Centroids of the given regions: $\left(0, \frac{3}{2}\right)$, $(0, 5)$, and $\left(0, \frac{15}{2}\right)$

Area: $A = 15 + 12 + 7 = 34$

$$\bar{x} = \frac{15(0) + 12(0) + 7(0)}{34} = 0$$

$$\bar{y} = \frac{15(3/2) + 12(5) + 7(15/2)}{34} = \frac{135}{34}$$

$$(\bar{x}, \bar{y}) = \left(0, \frac{135}{34}\right) \approx (0, 3.97)$$

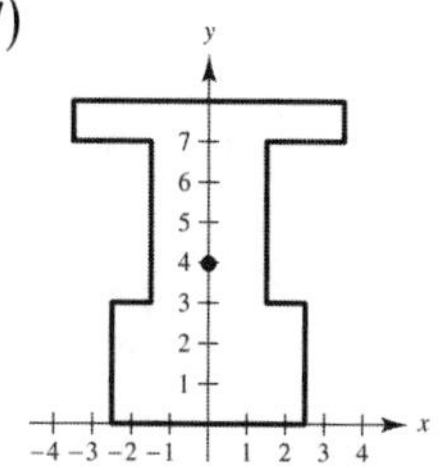

34. $m_1 = \frac{7}{8}(2) = \frac{7}{4}, P_1 = \left(0, \frac{7}{16}\right)$

$m_2 = \frac{7}{8}\left(6 - \frac{7}{8}\right) = \frac{287}{64}, P_2 = \left(0, \frac{55}{16}\right)$

By symmetry, $\bar{x} = 0$.

$$\bar{y} = \frac{(7/4)(7/16) + (287/64)(55/16)}{(7/4) + (287/64)} = \frac{16{,}569}{6384} = \frac{789}{304}$$

$$(\bar{x}, \bar{y}) = \left(0, \frac{789}{304}\right) \approx (0, 2.595)$$

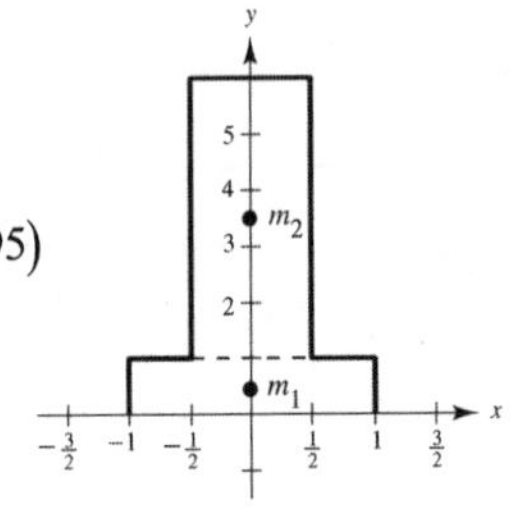

35. Centroids of the given regions: $(1, 0)$ and $(3, 0)$

Mass: $4 + 2\pi$

$$\bar{x} = \frac{4(1) + 2\pi(3)}{4 + 2\pi} = \frac{2 + 3\pi}{2 + \pi}$$

$$\bar{y} = 0$$

$$(\bar{x}, \bar{y}) = \left(\frac{2 + 3\pi}{2 + \pi}, 0\right) \approx (2.22, 0)$$

36. Centroids of the given regions: $(3, 0)$ and $(1, 0)$

Mass: $8 + \pi$

$$\bar{y} = 0$$

$$\bar{x} = \frac{8(1) + \pi(3)}{8 + \pi} = \frac{8 + 3\pi}{8 + \pi}$$

$$(\bar{x}, \bar{y}) = \left(\frac{8 + 3\pi}{8 + \pi}, 0\right) \approx (1.56, 0)$$

37. $r = 5$ is distance between center of circle and y-axis.

$A \approx \pi(4)^2 = 16\pi$ is the area of circle. So,

$V = 2\pi rA = 2\pi(5)(16\pi) = 160\pi^2 \approx 1579.14.$

38. $V = 2\pi rA = 2\pi(3)(4\pi) = 24\pi^2$

39. $A = \frac{1}{2}(4)(4) = 8$

$$\bar{y} = \left(\frac{1}{8}\right)\frac{1}{2}\int_0^4 (4 + x)(4 - x)\,dx = \frac{1}{16}\left[16x - \frac{x^3}{3}\right]_0^4 = \frac{8}{3}$$

$$r = \bar{y} = \frac{8}{3}$$

$$V = 2\pi rA = 2\pi\left(\frac{8}{3}\right)(8) = \frac{128\pi}{3} \approx 134.04$$

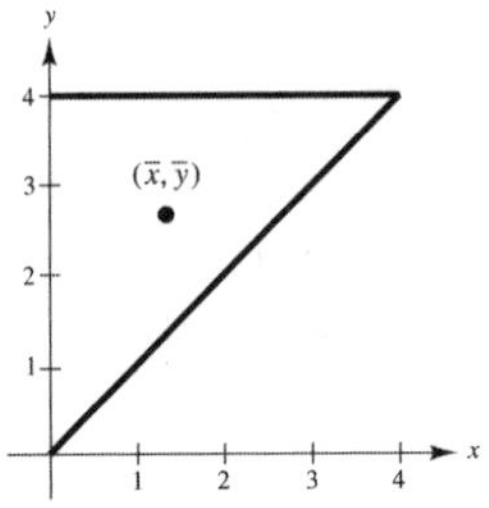

40. $A = \int_2^6 2\sqrt{x - 2}\,dx = \frac{4}{3}(x - 2)^{3/2}\Big]_2^6 = \frac{32}{3}$

$$M_y = \int_2^6 (x)2\sqrt{x - 2}\,dx = 2\int_2^6 x\sqrt{x - 2}\,dx$$

Let $u = x - 2, x = u + 2, du = dx$:

$$M_y = 2\int_0^4 (u + 2)\sqrt{u}\,du = 2\int_0^4 \left(u^{3/2} + 2u^{1/2}\right)du = 2\left[\frac{2}{5}u^{5/2} + \frac{4}{3}u^{3/2}\right]_0^4 = 2\left(\frac{64}{5} + \frac{32}{3}\right) = \frac{704}{15}$$

$$\bar{x} = \frac{M_y}{A} = \frac{704/15}{32/3} = \frac{22}{5}$$

$$r = \bar{x} = \frac{22}{5}$$

$$V = 2\pi rA = 2\pi\left(\frac{22}{5}\right)\left(\frac{32}{3}\right) = \frac{1408\pi}{15} \approx 294.89$$

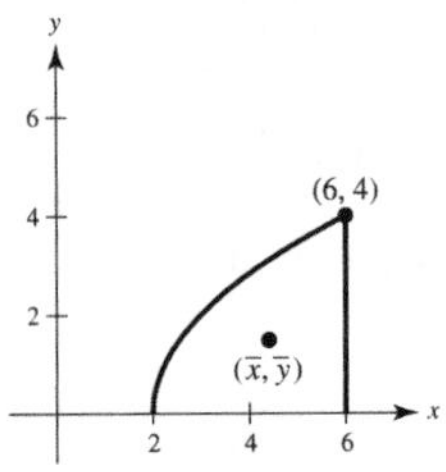

41. The center of mass $(\bar{x}, \bar{y})$ is $\bar{x} = M_y/m$ and $\bar{y} = M_x/m$, where:

1. $m = m_1 + m_2 + \cdots + m_n$ is the total mass of the system.

2. $M_y = m_1x_1 + m_2x_2 + \cdots + m_nx_n$ is the moment about the y-axis.

3. $M_x = m_1y_1 + m_2y_2 + \cdots + m_ny_n$ is the moment about the x-axis.

42. A planar lamina is a thin flat plate of constant density. The center of mass $(\bar{x}, \bar{y})$ is the balancing point on the lamina.

43. Let R be a region in a plane and let L be a line such that L does not intersect the interior of R. If r is the distance between the centroid of R and L, then the volume V of the solid of revolution formed by revolving R about L is $V = 2\pi rA$ where A is the area of R.

44. (a) Yes. The region is shifted upward two units.

$$(\bar{x}, \bar{y}) = (1.2, 1.4 + 2) = (1.2, 3.4)$$

(b) Yes. The region is shifted to the right two units.

$$(\bar{x}, \bar{y}) = (1.2 + 2, 1.4) = (3.2, 1.4)$$

(c) Yes. The region is reflected in the x-axis.

$$(\bar{x}, \bar{y}) = (1.2, -1.4)$$

(d) Not possible

45.

$$A = \frac{1}{2}(2a)c = ac$$

$$\frac{1}{A} = \frac{1}{ac}$$

$$\bar{x} = \left(\frac{1}{ac}\right)\frac{1}{2}\int_0^c \left[\left(\frac{b-a}{c}y + a\right)^2 - \left(\frac{b+a}{c}y - a\right)^2\right] dy$$

$$= \frac{1}{2ac}\int_0^c \left[\frac{4ab}{c}y - \frac{4ab}{c^2}y^2\right] dy = \frac{1}{2ac}\left[\frac{2ab}{c}y^2 - \frac{4ab}{3c^2}y^3\right]_0^c = \frac{1}{2ac}\left(\frac{2}{3}abc\right) = \frac{b}{3}$$

$$\bar{y} = \frac{1}{ac}\int_0^c y\left[\left(\frac{b-a}{c}y + a\right) - \left(\frac{b+a}{c}y - a\right)\right] dy$$

$$= \frac{1}{ac}\int_0^c y\left(-\frac{2a}{c}y + 2a\right) dy = \frac{2}{c}\int_0^c \left(y - \frac{y^2}{c}\right) dy = \frac{2}{c}\left[\frac{y^2}{2} - \frac{y^3}{3c}\right]_0^c = \frac{c}{3}$$

$$(\bar{x}, \bar{y}) = \left(\frac{b}{3}, \frac{c}{3}\right)$$

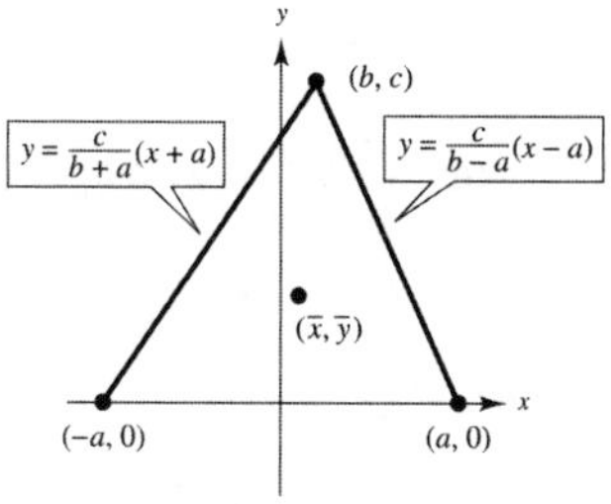

From elementary geometry, $(b/3, c/3)$ is the point of intersection of the medians.

46.

$$A = bh = ac$$

$$\frac{1}{A} = \frac{1}{ac}$$

$$\bar{x} = \frac{1}{ac}\frac{1}{2}\int_0^c \left[\left(\frac{b}{c}y + a\right)^2 - \left(\frac{b}{c}y\right)^2\right] dy$$

$$= \frac{1}{2ac}\int_0^c \left(\frac{2ab}{c}y + a^2\right) dy$$

$$= \frac{1}{2ac}\left[\frac{ab}{c}y^2 + a^2y\right]_0^c$$

$$= \frac{1}{2ac}\left[abc + a^2c\right] = \frac{1}{2}(b + a)$$

$$\bar{y} = \frac{1}{ac}\int_0^c y\left[\left(\frac{b}{c}y + a\right) - \left(\frac{b}{c}y\right)\right] dy = \left[\frac{1}{c}\frac{y^2}{2}\right]_0^c = \frac{c}{2}$$

$$(\bar{x}, \bar{y}) = \left(\frac{b+a}{2}, \frac{c}{2}\right)$$

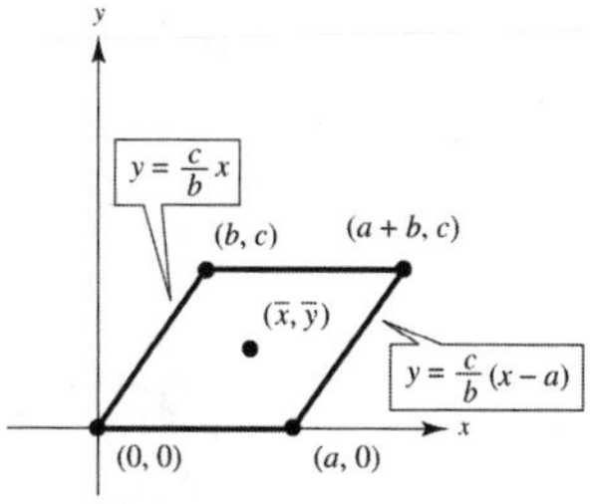

This is the point of intersection of the diagonals.

47. $A = \frac{c}{2}(a+b)$

$\frac{1}{A} = \frac{2}{c(a+b)}$

$$\bar{x} = \frac{2}{c(a+b)}\int_0^c x\left(\frac{b-a}{c}x + a\right)dx = \frac{2}{c(a+b)}\int_0^c \left(\frac{b-a}{c}x^2 + ax\right)dx = \frac{2}{c(a+b)}\left[\frac{b-a}{c}\frac{x^3}{3} + \frac{ax^2}{2}\right]_0^c$$

$$= \frac{2}{c(a+b)}\left[\frac{(b-a)c^2}{3} + \frac{ac^2}{2}\right] = \frac{2}{c(a+b)}\left[\frac{2bc^2 - 2ac^2 + 3ac^2}{6}\right] = \frac{c(2b+a)}{3(a+b)} = \frac{(a+2b)c}{3(a+b)}$$

$$\bar{y} = \frac{2}{c(a+b)}\frac{1}{2}\int_0^c \left(\frac{b-a}{c}x + a\right)^2 dx = \frac{1}{c(a+b)}\int_0^c \left[\left(\frac{b-a}{c}\right)^2 x^2 + \frac{2a(b-a)}{c}x + a^2\right]dx$$

$$= \frac{1}{c(a+b)}\left[\left(\frac{b-a}{c}\right)^2\frac{x^3}{3} + \frac{2a(b-a)}{c}\frac{x^2}{2} + a^2x\right]_0^c = \frac{1}{c(a+b)}\left[\frac{(b-a)^2c}{3} + ac(b-a) + a^2c\right]$$

$$= \frac{1}{3c(a+b)}\left[(b^2 - 2ab + a^2)c + 3ac(b-a) + 3a^2c\right]$$

$$= \frac{1}{3(a+b)}\left[b^2 - 2ab + a^2 + 3ab - 3a^2 + 3a^2\right] = \frac{a^2 + ab + b^2}{3(a+b)}$$

So, $(\bar{x}, \bar{y}) = \left(\frac{(a+2b)c}{3(a+b)}, \frac{a^2+ab+b^2}{3(a+b)}\right)$.

The one line passes through $\left(0, \frac{a}{2}\right)$ and $\left(c, \frac{b}{2}\right)$. Its equation is $y = \frac{b-a}{2c}x + \frac{a}{2}$. The other line passes through $(0, -b)$ and $(c, a+b)$. Its equation is $y = \frac{a+2b}{c}x - b$. $(\bar{x}, \bar{y})$ is the point of intersection of these two lines.

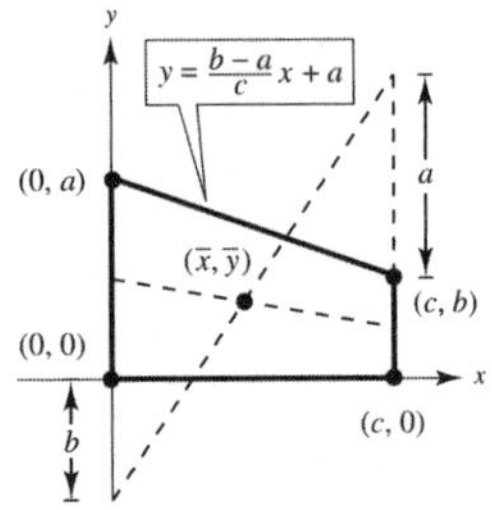

48. $\bar{x} = 0$ by symmetry.

$A = \frac{1}{2}\pi r^2$

$\frac{1}{A} = \frac{2}{\pi r^2}$

$$\bar{y} = \frac{2}{\pi r^2}\frac{1}{2}\int_{-r}^{r}\left(\sqrt{r^2 - x^2}\right)^2 dx$$

$$= \frac{1}{\pi r^2}\left[r^2x - \frac{x^3}{3}\right]_{-r}^{r} = \frac{1}{\pi r^2}\left(\frac{4r^3}{3}\right) = \frac{4r}{3\pi}$$

$(\bar{x}, \bar{y}) = \left(0, \frac{4r}{3\pi}\right)$

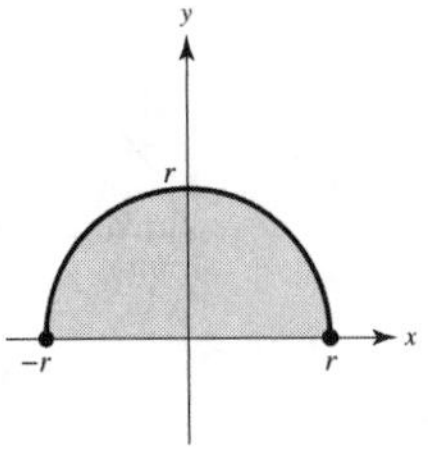

49. $\bar{x} = 0$ by symmetry.

$A = \frac{1}{2}\pi ab$

$\frac{1}{A} = \frac{2}{\pi ab}$

$$\bar{y} = \frac{2}{\pi ab}\frac{1}{2}\int_{-a}^{a}\left(\frac{b}{a}\sqrt{a^2 - x^2}\right)^2 dx$$

$$= \frac{1}{\pi ab}\left(\frac{b^2}{a^2}\right)\left[a^2x - \frac{x^3}{3}\right]_{-a}^{a} = \frac{b}{\pi a^3}\left(\frac{4a^3}{3}\right) = \frac{4b}{3\pi}$$

$(\bar{x}, \bar{y}) = \left(0, \frac{4b}{3\pi}\right)$

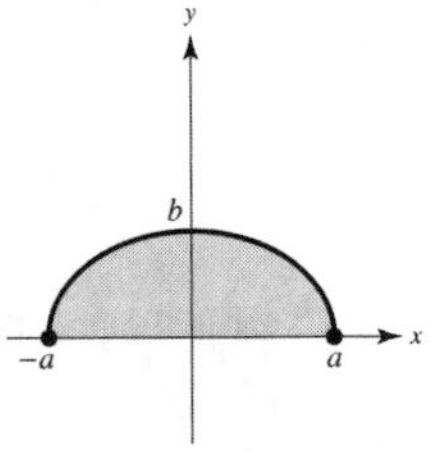

50. $A = \int_0^1 \left[1 - \left(2x - x^2\right)\right] dx = \frac{1}{3}$

$\frac{1}{A} = 3$

$\bar{x} = 3\int_0^1 x\left[1 - \left(2x - x^2\right)\right] dx = 3\int_0^1 \left[x - 2x^2 + x^3\right] dx = 3\left[\frac{x^2}{2} - \frac{2}{3}x^3 + \frac{x^4}{4}\right]_0^1 = \frac{1}{4}$

$\bar{y} = 3\int_0^1 \frac{\left[1 + \left(2x - x^2\right)\right]}{2}\left[1 - \left(2x - x^2\right)\right] dx = \frac{3}{2}\int_0^1 \left[1 - \left(2x - x^2\right)^2\right] dx$

$= \frac{3}{2}\int_0^1 \left(1 - 4x^2 + 4x^3 - x^4\right) dx = \frac{3}{2}\left[x - \frac{4}{3}x^3 + x^4 - \frac{x^5}{5}\right]_0^1 = \frac{7}{10}$

$(\bar{x}, \bar{y}) = \left(\frac{1}{4}, \frac{7}{10}\right)$

51. (a)

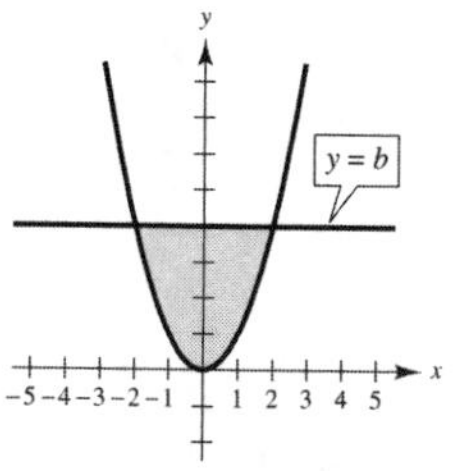

(b) $\bar{x} = 0$ by symmetry.

(c) $M_y = \int_{-\sqrt{b}}^{\sqrt{b}} x\left(b - x^2\right) dx = 0$ because $bx - x^3$ is odd.

(d) $\bar{y} > \frac{b}{2}$ because there is more area above $y = \frac{b}{2}$ than below.

(e) $M_x = \int_{-\sqrt{b}}^{\sqrt{b}} \frac{\left(b + x^2\right)\left(b - x^2\right)}{2} dx = \int_{-\sqrt{b}}^{\sqrt{b}} \frac{b^2 - x^4}{2} dx = \frac{1}{2}\left[b^2x - \frac{x^5}{5}\right]_{-\sqrt{b}}^{\sqrt{b}} = b^2\sqrt{b} - \frac{b^2\sqrt{b}}{5} = \frac{4b^2\sqrt{b}}{5}$

$A = \int_{-\sqrt{b}}^{\sqrt{b}} \left(b - x^2\right) dx = \left[bx - \frac{x^3}{3}\right]_{-\sqrt{b}}^{\sqrt{b}} = \left(b\sqrt{b} - \frac{b\sqrt{b}}{3}\right)2 = 4\frac{b\sqrt{b}}{3}$

$\bar{y} = \frac{M_x}{A} = \frac{4b^2\sqrt{b}/5}{4b\sqrt{b}/3} = \frac{3}{5}b$

52. (a)

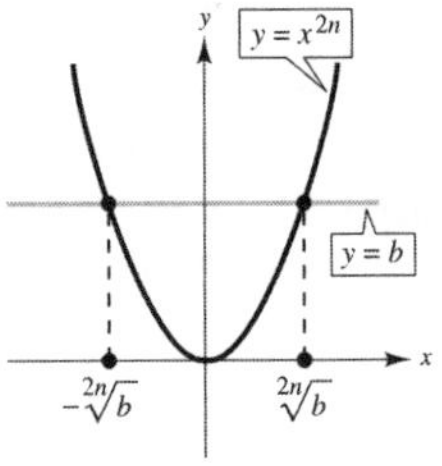

(b) $M_y = 0$ by symmetry.

$$M_y = \int_{-\sqrt[2n]{b}}^{\sqrt[2n]{b}} x\left(b - x^{2n}\right) dx = 0$$

because $bx - x^{2n+1}$ is an odd function.

(c) $\bar{y} > \dfrac{b}{2}$ because there is more area above $y = \dfrac{b}{2}$ than below.

(d) $$M_x = \int_{-\sqrt[2n]{b}}^{\sqrt[2n]{b}} \frac{\left(b + x^{2n}\right)\left(b - x^{2n}\right)}{2} dx = \int_{-\sqrt[2n]{b}}^{\sqrt[2n]{b}} \frac{1}{2}\left(b^2 - x^{4n}\right) dx$$

$$= \frac{1}{2}\left(b^2 x - \frac{x^{4n+1}}{4n+1}\right)\Bigg]_{-\sqrt[2n]{b}}^{\sqrt[2n]{b}} = b^2 b^{1/2n} - \frac{b^{(4n+1)/2n}}{4n+1} = \frac{4n}{4n+1} b^{(4n+1)/2n}$$

$$A = \int_{-\sqrt[2n]{b}}^{\sqrt[2n]{b}} \left(b - x^{2n}\right) dx = 2\left[bx - \frac{x^{2n+1}}{2n+1}\right]_0^{\sqrt[2n]{b}} = 2\left[b \cdot b^{1/2n} - \frac{b^{(2n+1)/2n}}{2n+1}\right] = \frac{4n}{2n+1} b^{(2n+1)/2n}$$

$$\bar{y} = \frac{M_x}{A} = \frac{4nb^{(4n+1)/2n}/(4n+1)}{4nb^{(24n+1)/2n}/(2n+1)} = \frac{2n+1}{4n+1} b$$

(e)

n	1	2	3	4
$\bar{y}$	$\frac{3}{5}b$	$\frac{5}{9}b$	$\frac{7}{13}b$	$\frac{9}{17}b$

(f) $$\lim_{n\to\infty} \bar{y} = \lim_{n\to\infty} \frac{2n+1}{4n+1} b = \frac{1}{2} b$$

(g) As $n \to \infty$, the figure gets narrower.

53. (a) $\bar{x} = 0$ by symmetry.

$$A = 2\int_0^{40} f(x)\, dx = \frac{2(40)}{3(4)}\left[30 + 4(29) + 2(26) + 4(20) + 0\right] = \frac{20}{3}(278) = \frac{5560}{3}$$

$$M_x = \int_{-40}^{40} \frac{f(x)^2}{2} dx = \frac{40}{3(4)}\left[30^2 + 4(29)^2 + 2(26)^2 + 4(20)^2 + 0\right] = \frac{10}{3}(7216) = \frac{72{,}160}{3}$$

$$\bar{y} = \frac{M_x}{A} = \frac{72{,}160/3}{5560/3} = \frac{72{,}160}{5560} \approx 12.98$$

$(\bar{x}, \bar{y}) = (0, 12.98)$

(b) $y = \left(-1.02 \times 10^{-5}\right)x^4 - 0.0019x^2 + 29.28$ (Use nine data points.)

(c) $$\bar{y} = \frac{M_x}{A} \approx \frac{23{,}697.68}{1843.54} \approx 12.85$$

$(\bar{x}, \bar{y}) = (0, 12.85)$

54. Let $f(x)$ be the top curve, given by $l + d$. The bottom curve is $d(x)$.

x	0	0.5	1.0	1.5	2
f	2.0	1.93	1.73	1.32	0
d	0.50	0.48	0.43	0.33	0

(a) $\text{Area} = 2\int_0^2 \left[f(x) - d(x)\right] dx$

$$\approx 2\frac{2}{3(4)}\left[1.50 + 4(1.45) + 2(1.30) + 4(.99) + 0\right] = \frac{1}{3}[13.86] = 4.62$$

$$M_x = \int_{-2}^{2} \frac{f(x) + d(x)}{2}\left(f(x) - d(x)\right) dx$$

$$= \int_0^2 \left[f(x)^2 - d(x)^2\right] dx$$

$$= \frac{2}{3(4)}\left[3.75 + 4(3.4945) + 2(2.808) + 4(1.6335) + 0\right] = \frac{1}{6}[29.878] = 4.9797$$

$$\bar{y} = \frac{M_x}{A} = \frac{4.9797}{4.62} = 1.078$$

$(\bar{x}, \bar{y}) = (0, 1.078)$

(b) $f(x) = -0.1061x^4 - 0.06126x^2 + 1.9527$

$d(x) = -0.02648x^4 - 0.01497x^2 + .4862$

(c) $\bar{y} = \frac{M_x}{A} \approx \frac{4.9133}{4.59998} = 1.068$

$(\bar{x}, \bar{y}) = (0, 1.068)$

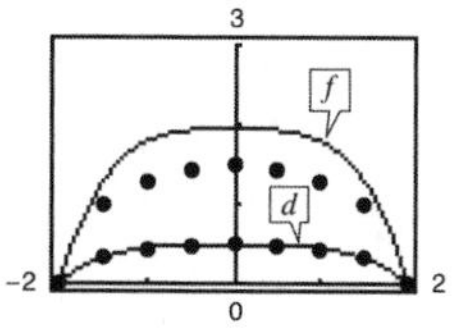

55. The surface area of the sphere is $S = 4\pi r^2$. The arc length of C is $s = \pi r$. The distance traveled by the centroid is

$$d = \frac{S}{s} = \frac{4\pi r^2}{\pi r} = 4r.$$

This distance is also the circumference of the circle of radius y.

$d = 2\pi y$

So, $2\pi y = 4r$ and you have $y = 2r/\pi$. Therefore, the centroid of the semicircle $y = \sqrt{r^2 - x^2}$ is $(0, 2r/\pi)$.

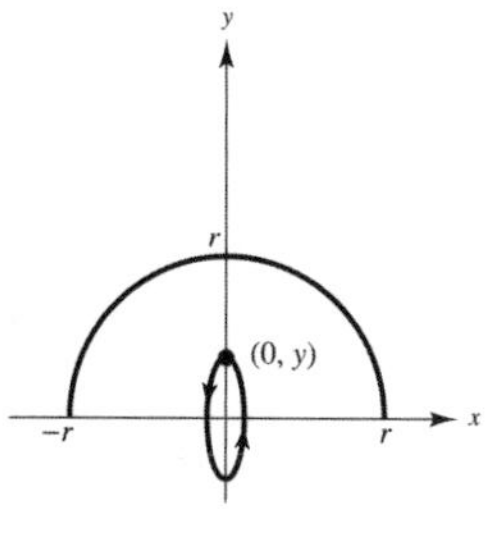

56. The centroid of the circle is $(1, 0)$. The distance traveled by the centroid is 2π. The arc length of the circle is also 2π. Therefore, $S = (2\pi)(2\pi) = 4\pi^2$.

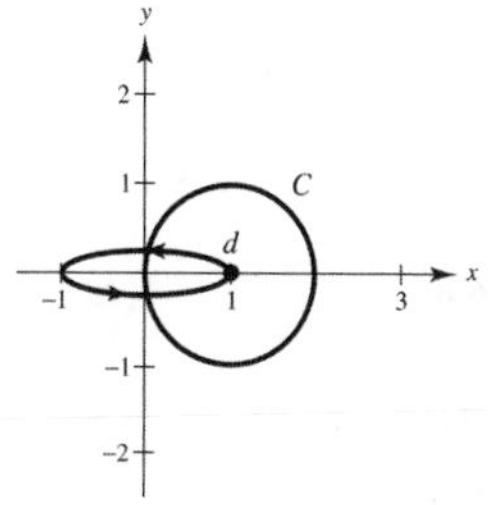

57. $A = \int_0^1 x^n\,dx = \left[\dfrac{x^{n+1}}{n+1}\right]_0^1 = \dfrac{1}{n+1}$

$m = \rho A = \dfrac{\rho}{n+1}$

$M_x = \dfrac{\rho}{2}\int_0^1 \left(x^n\right)^2\,dx = \left[\dfrac{\rho}{2}\cdot\dfrac{x^{2n+1}}{2n+1}\right]_0^1 = \dfrac{\rho}{2(2n+1)}$

$M_y = \rho\int_0^1 x\left(x^n\right)\,dx = \left[\rho\cdot\dfrac{x^{n+2}}{n+2}\right]_0^1 = \dfrac{\rho}{n+2}$

$\bar{x} = \dfrac{M_y}{m} = \dfrac{n+1}{n+2}$

$\bar{y} = \dfrac{M_x}{m} = \dfrac{n+1}{2(2n+1)} = \dfrac{n+1}{4n+2}$

Centroid: $\left(\dfrac{n+1}{n+2}, \dfrac{n+1}{4n+2}\right)$

As $n \to \infty$, $(\bar{x}, \bar{y}) \to \left(1, \dfrac{1}{4}\right)$. The graph approaches the x-axis and the line $x = 1$ as $n \to \infty$.

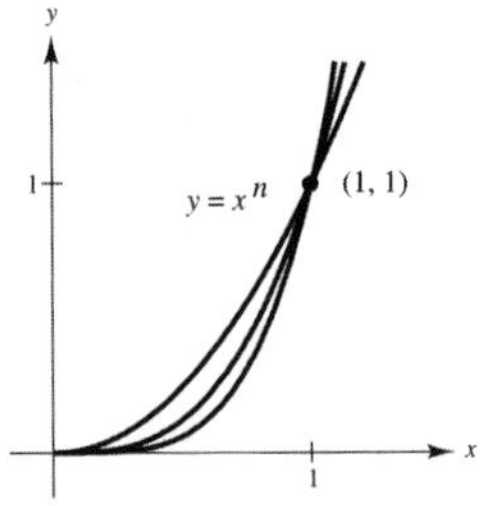

58. Let T be the shaded triangle with vertices $(-1, 4)$, $(1, 4)$, and $(0, 3)$. Let U be the large triangle with vertices $(-4, 4)$, $(4, 4)$, and $(0, 0)$. V consists of the region U minus the region T.

Centroid of T: $\left(0, \frac{11}{3}\right)$; Area $= 1$

Centroid of U: $\left(0, \frac{8}{3}\right)$; Area $= 16$

Area: $V = 16 - 1 = 15$

$\bar{x} = 0$ by symmetry.

$15\bar{y} + 1\left(\frac{11}{3}\right) = 16\left(\frac{8}{3}\right)$

$15\bar{y} = \frac{117}{3}$

$\bar{y} = \frac{13}{5}$

$(\bar{x}, \bar{y}) = \left(0, \frac{13}{5}\right)$

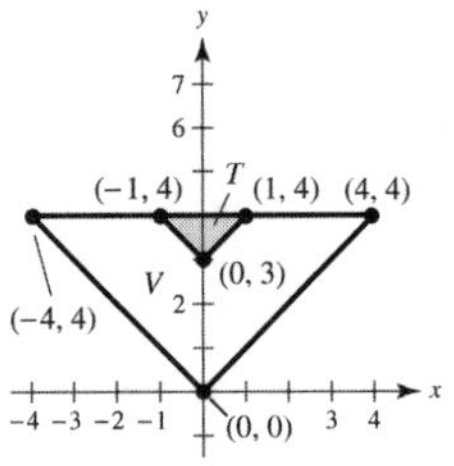

Section 7.7 Fluid Pressure and Fluid Force

1. $F = PA = [62.4(8)]3 = 1497.6$ lb

2. $F = PA = [62.4(8)]8 = 3993.6$ lb

3. $F = PA = [62.4(8)]10 = 4992$ lb

4. $F = PA = [62.4(8)]25 = 12{,}480$ lb

5. $F = 62.4(h+2)(6) - (62.4)(h)(6)$
$= 62.4(2)(6) = 748.8$ lb

6. $F = 62.4(h+4)(48) - (62.4)(h)(48)$
$= 62.4(4)(48) = 11{,}980.8$ lb

7. $h(y) = 3 - y$

$L(y) = 4$

$F = 62.4\int_0^3 (3-y)(4)\,dy$

$= 249.6\int_0^3 (3-y)\,dy$

$= 249.6\left[3y - \dfrac{y^2}{2}\right]_0^3 = 1123.2$ lb

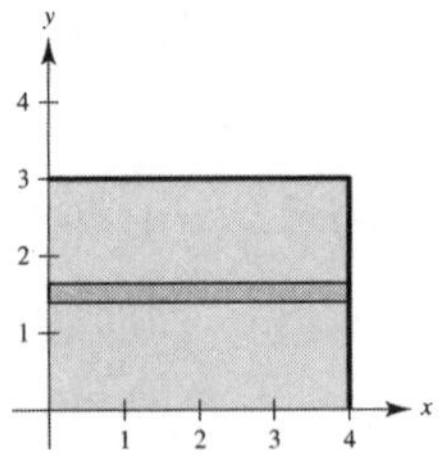

8. $h(y) = 3 - y$

$L(y) = \frac{4}{3}y$

$$F = 62.4\int_0^3 (3 - y)\left(\frac{4}{3}y\right) dy$$

$$= \frac{4}{3}(62.4)\int_0^3 \left(3y - y^2\right) dy$$

$$= \frac{4}{3}(62.4)\left[\frac{3y^2}{2} - \frac{y^3}{3}\right]_0^3 = 374.4 \text{ lb}$$

Force is one-third that of Exercise 7.

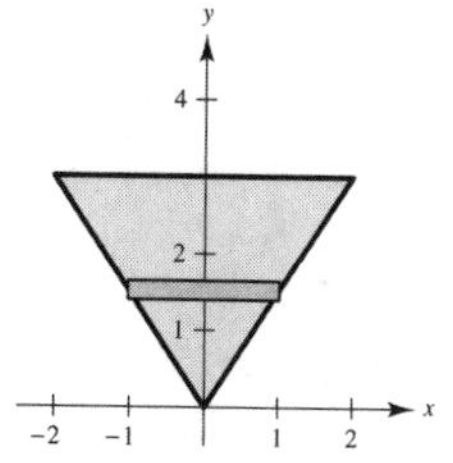

9. $h(y) = 3 - y$

$L(y) = 2\left(\frac{y}{3} + 1\right)$

$$F = 2(62.4)\int_0^3 (3 - y)\left(\frac{y}{3} + 1\right) dy$$

$$= 124.8\int_0^3 \left(3 - \frac{y^2}{3}\right) dy$$

$$= 124.8\left[3y - \frac{y^3}{9}\right]_0^3 = 748.8 \text{ lb}$$

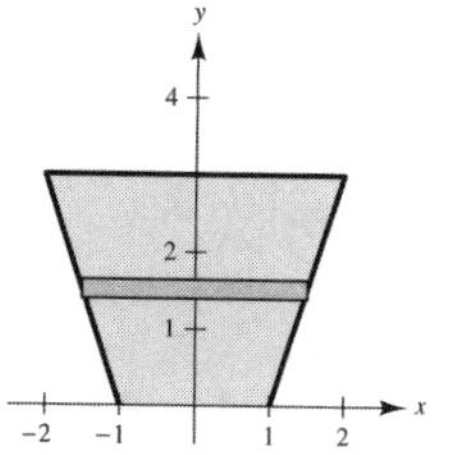

10. $h(y) = -y$

$L(y) = 2\sqrt{4 - y^2}$

$$F = 62.4\int_{-2}^0 (-y)(2)\sqrt{4 - y^2}\, dy$$

$$= \left[62.4\left(\tfrac{2}{3}\right)\left(4 - y^2\right)^{3/2}\right]_{-2}^0 = 332.8 \text{ lb}$$

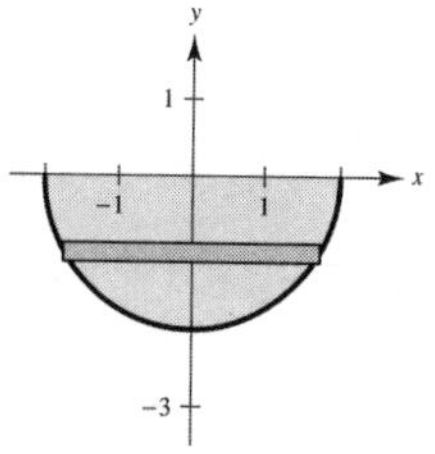

11. $h(y) = 4 - y$

$L(y) = 2\sqrt{y}$

$$F = 2(62.4)\int_0^4 (4 - y)\sqrt{y}\, dy$$

$$= 124.8\int_0^4 \left(4y^{1/2} - y^{3/2}\right) dy$$

$$= 124.8\left[\frac{8y^{3/2}}{3} - \frac{2y^{5/2}}{5}\right]_0^4 = 1064.96 \text{ lb}$$

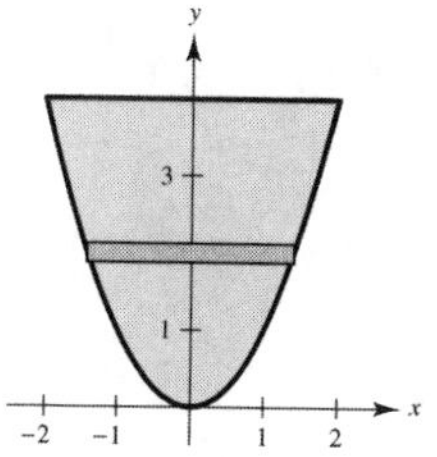

12. $h(y) = -y$

$L(y) = \frac{4}{3}\sqrt{9 - y^2}$

$$F = 62.4\int_{-3}^0 (-y)\tfrac{4}{3}\sqrt{9 - y^2}\, dy$$

$$= 62.4\left(\tfrac{2}{3}\right)\int_{-3}^0 \left(9 - y^2\right)^{1/2}(-2y)\, dy$$

$$= \left[62.4\left(\tfrac{4}{9}\right)\left(9 - y^2\right)^{3/2}\right]_{-3}^0 = 748.8 \text{ lb}$$

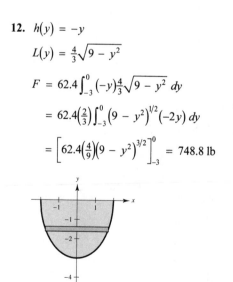

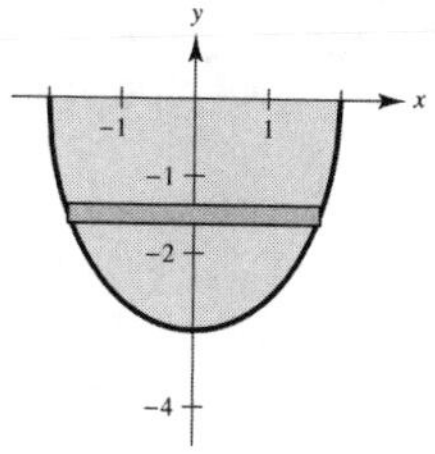

13. $h(y) = 4 - y$

$L(y) = 2$

$$F = 9800\int_0^2 2(4 - y)\,dy$$

$$= 9800\Big[8y - y^2\Big]_0^2 = 117{,}600 \text{ newtons}$$

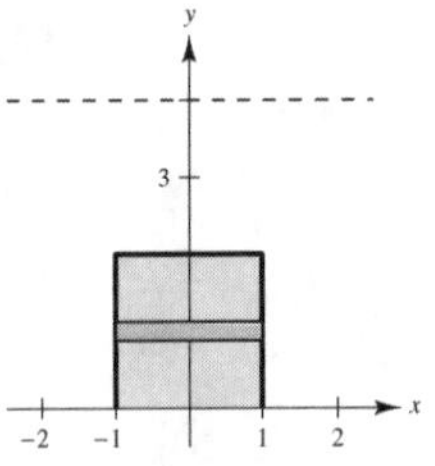

14. $h(y) = \left(1 + 3\sqrt{2}\right) - y$

$L_1(y) = 2y$ (lower part)

$L_2(y) = 2\left(3\sqrt{2} - y\right)$ (upper part)

$$F = 2(9800)\left[\int_0^{3\sqrt{2}/2}\left(1 + 3\sqrt{2} - y\right)y\,dy + \int_{3\sqrt{2}/2}^{3\sqrt{2}}\left(1 + 3\sqrt{2} - y\right)\left(3\sqrt{2} - y\right)dy\right]$$

$$= 19{,}600\left[\left[\frac{y^2}{2} - 3\sqrt{2}y - \frac{y^3}{3}\right]_0^{3\sqrt{2}/2} + \left[3\sqrt{2}y + 18y + \frac{y^3}{3} - \frac{6\sqrt{2} + 1}{2}y\right]_{3\sqrt{2}/2}^{3\sqrt{2}}\right]$$

$$= 19{,}600\left[\frac{9\left(2\sqrt{2} + 1\right)}{4} + \frac{9\left(\sqrt{2} + 1\right)}{4}\right] = 44{,}100\left(3\sqrt{2} + 2\right) \text{ newtons}$$

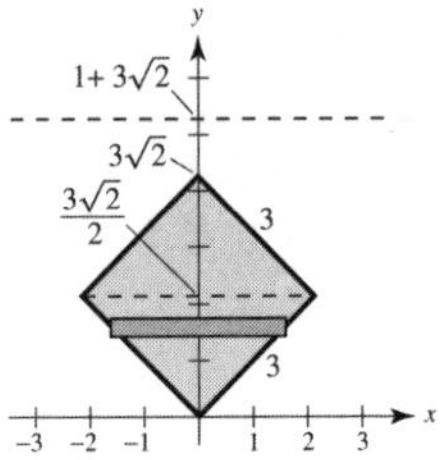

15. $h(y) = 12 - y$

$$L(y) = 6 - \frac{2y}{3}$$

$$F = 9800\int_0^9 (12 - y)\left(6 - \frac{2y}{3}\right)dy = 9800\left[72y - 7y^2 + \frac{2y^3}{9}\right]_0^9 = 2{,}381{,}400 \text{ newtons}$$

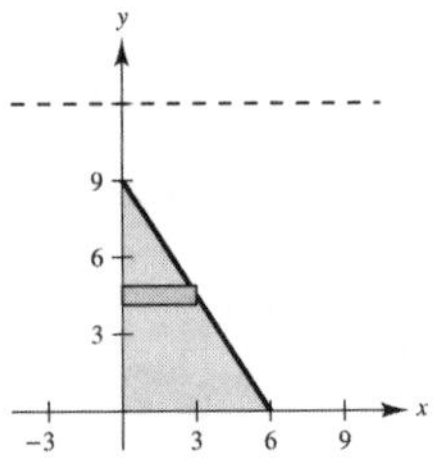

16. $h(y) = 6 - y$

$L(y) = 1$

$$F = 9800\int_0^5 1(6 - y)\,dy$$

$$= 9800\left[6y - \frac{y^2}{2}\right]_0^5$$

$= 171{,}500$ newtons

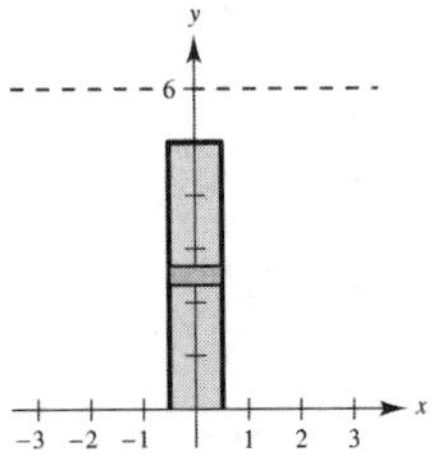

17. $h(y) = 2 - y$

$L(y) = 10$

$$F = 140.7\int_0^2 (2 - y)(10)\,dy$$

$$= 1407\int_0^2 (2 - y)\,dy$$

$$= 1407\left[2y - \frac{y^2}{2}\right]_0^2 = 2814 \text{ lb}$$

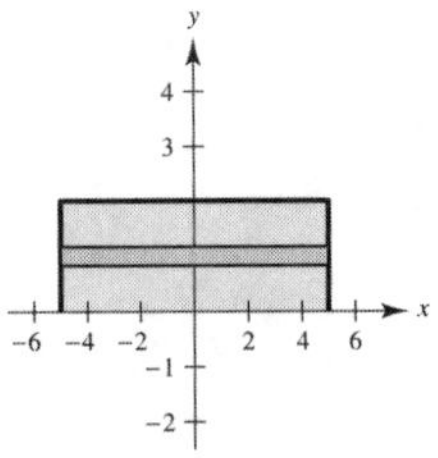

18. $h(y) = -y$

$L(y) = 2\left(\frac{4}{3}\sqrt{9 - y^2}\right)$

$$F = 140.7\int_{-3}^{0}(-y)(2)\left(\frac{4}{3}\sqrt{9 - y^2}\right)dy$$

$$= \frac{(140.7)(4)}{3}\int_{-3}^{0}\sqrt{9 - y^2}(-2y)\,dy$$

$$= \left[\frac{(140.7)(4)}{3}\left(\frac{2}{3}\right)\left(9 - y^2\right)^{3/2}\right]_{-3}^{0}$$

$$= 3376.8 \text{ lb}$$

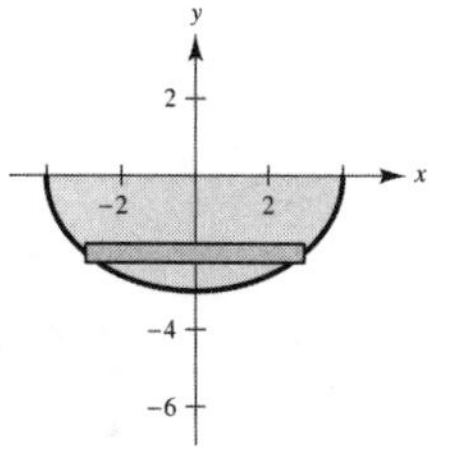

19. $h(y) = 4 - y$

$L(y) = 6$

$$F = 140.7\int_{0}^{4}(4 - y)(6)\,dy$$

$$= 844.2\int_{0}^{4}(4 - y)\,dy$$

$$= 844.2\left[4y - \frac{y^2}{2}\right]_{0}^{4} = 6753.6 \text{ lb}$$

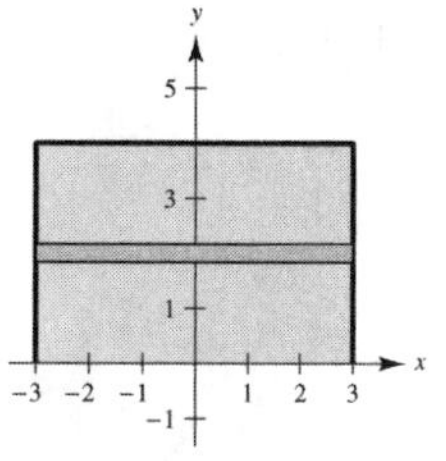

20. $h(y) = -y$

$L(y) = 5 + \frac{5}{3}y$

$$F = 140.7\int_{-3}^{0}(-y)\left(5 + \frac{5}{3}y\right)dy$$

$$= 140.7\int_{-3}^{0}\left(-5y - \frac{5}{3}y^2\right)dy$$

$$= 140.7\left[-\frac{5}{2}y^2 - \frac{5}{9}y^3\right]_{-3}^{0}$$

$$= 140.7\left[\frac{45}{2} - 15\right]$$

$$= 1055.25 \text{ lb}$$

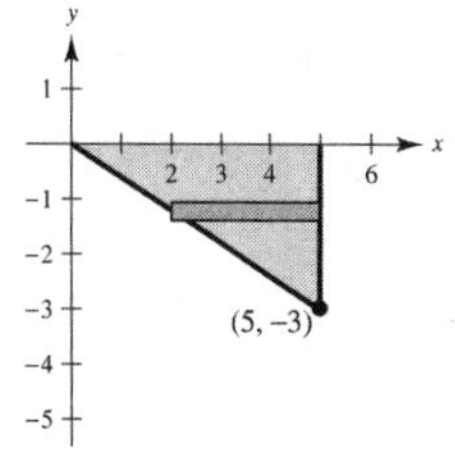

21. $h(y) = -y$

$L(y) = 2\left(\frac{1}{2}\right)\sqrt{9 - 4y^2}$

$$F = 42\int_{-3/2}^{0}(-y)\sqrt{9 - 4y^2}\,dy$$

$$= \frac{42}{8}\int_{-3/2}^{0}\left(9 - 4y^2\right)^{1/2}(-8y)\,dy$$

$$= \left[\left(\frac{21}{4}\right)\left(\frac{2}{3}\right)\left(9 - 4y^2\right)^{3/2}\right]_{-3/2}^{0} = 94.5 \text{ lb}$$

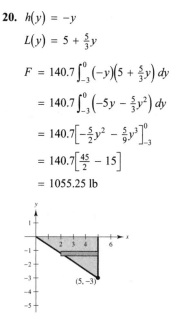

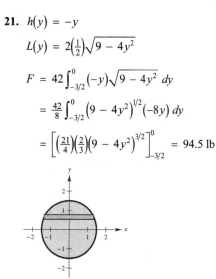

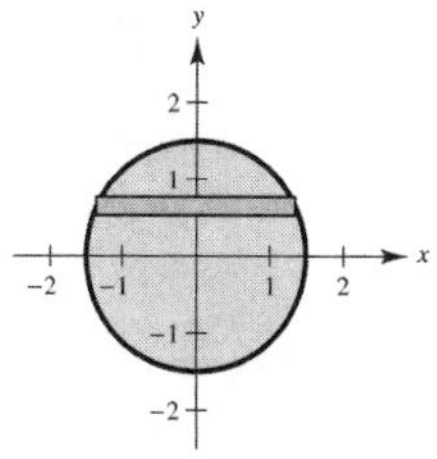

22. $h(y) = \frac{3}{2} - y$

$L(y) = 2\left(\frac{1}{2}\right)\sqrt{9 - 4y^2}$

$$F = 42\int_{-3/2}^{3/2}\left(\frac{3}{2} - y\right)\sqrt{9 - 4y^2}\,dy = 63\int_{-3/2}^{3/2}\sqrt{9 - 4y^2}\,dy + \frac{21}{4}\int_{-3/2}^{3/2}\sqrt{9 - 4y^2}(-8y)\,dy$$

The second integral is zero because it is an odd function and the limits of integration are symmetric to the origin. The first integral is twice the area of a semicircle of radius $\frac{3}{2}$.

$$\left(\sqrt{9 - 4y^2} = 2\sqrt{(9/4) - y^2}\right)$$

So, the force is $63\left(\frac{9}{4}\pi\right) = 141.75\pi \approx 445.32$ lb.

23. $h(y) = k - y$

$L(y) = 2\sqrt{r^2 - y^2}$

$$F = w\int_{-r}^{r}(k - y)\sqrt{r^2 - y^2}(2)\,dy = w\left[2k\int_{-r}^{r}\sqrt{r^2 - y^2}\,dy + \int_{-r}^{r}\sqrt{r^2 - y^2}(-2y)\,dy\right]$$

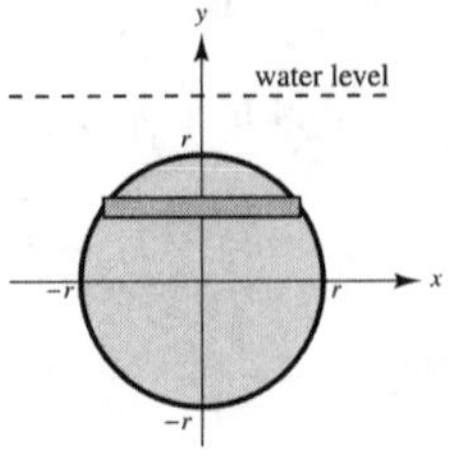

The second integral is zero because its integrand is odd and the limits of integration are symmetric to the origin. The first integral is the area of a semicircle with radius r.

$$F = w\left[(2k)\frac{\pi r^2}{2} + 0\right] = wk\pi r^2$$

24. (a) $F = wk\pi r^2 = (62.4)(7)(\pi 2^2) = 1747.2\pi$ lb

(b) $F = wk\pi r^2 = (62.4)(5)(\pi 3^2) = 2808\pi$ lb

25. $h(y) = k - y$

$L(y) = b$

$$F = w\int_{-h/2}^{h/2}(k - y)b\,dy$$

$$= wb\left[ky - \frac{y^2}{2}\right]_{-h/2}^{h/2} = wb(hk) = wkhb$$

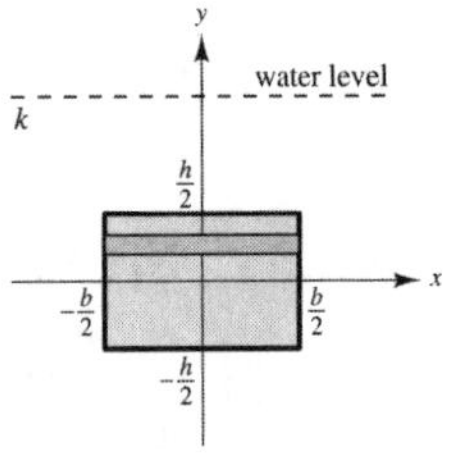

26. (a) $F = wkhb$

$= (62.4)\left(\frac{11}{2}\right)(3)(5) = 5148$ lb

(b) $F = wkhb$

$= (62.4)\left(\frac{17}{2}\right)(5)(10) = 26{,}520$ lb

27. From Exercise 25:

$F = 64(15)(1)(1) = 960$ lb

28. From Exercise 23:

$F = 64(15)\pi\left(\frac{1}{2}\right)^2 \approx 753.98$ lb

29. $h(y) = 4 - y$

$$F = 62.4\int_0^4(4 - y)L(y)\,dy$$

Using Simpson's Rule with $n = 8$ you have:

$$F \approx 62.4\left(\frac{4 - 0}{3(8)}\right)\left[0 + 4(3.5)(3) + 2(3)(5) + 4(2.5)(8) + 2(2)(9) + 4(1.5)(10) + 2(1)(10.25) + 4(0.5)(10.5) + 0\right]$$

$= 3010.8$ lb

30. $h(y) = 3 - y$

Solving $y = 5x^2/(x^2 + 4)$ for x, you obtain

$x = \sqrt{4y/(5 - y)}$.

$$L(y) = 2\sqrt{\frac{4y}{5 - y}}$$

$$F = 62.4(2)\int_0^3(3 - y)\sqrt{\frac{4y}{5 - y}}\,dy$$

$$= 2(124.8)\int_0^3(3 - y)\sqrt{\frac{y}{5 - y}}\,dy \approx 546.265 \text{ lb}$$

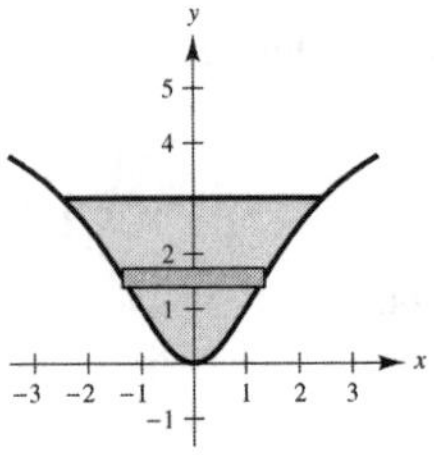

31. If the fluid force is one-half of 1123.2 lb, and the height of the water is b, then

$$h(y) = b - y$$
$$L(y) = 4$$

$$F = 62.4\int_0^b (b - y)(4)\,dy = \frac{1}{2}(1123.2)$$
$$\int_0^b (b - y)\,dy = 2.25$$
$$\left[by - \frac{y^2}{2}\right]_0^b = 2.25$$
$$b^2 - \frac{b^2}{2} = 2.25$$
$$b^2 = 4.5 \Rightarrow b \approx 2.12 \text{ ft.}$$

The pressure increases with increasing depth.

32. (a) Fluid pressure is the force per unit of area exerted by a fluid over the surface of a body.

(b) $F = Fw = w\int_c^d h(y)L(y)\,dy$, see page 498.

33. You use horizontal representative rectangles because you are measuring total force against a region between two depths.

34. The left window experiences the greater fluid force because its centroid is lower.

Review Exercises for Chapter 7

1. $A = \int_{-2}^{2}\left[\left(6 - \frac{x^2}{2}\right) - \frac{3}{4}x\right]dx$

$= \left[6x - \frac{x^3}{6} - \frac{3x^2}{8}\right]_{-2}^{2}$

$= \left(12 - \frac{4}{3} - \frac{3}{2}\right) - \left(-12 + \frac{4}{3} - \frac{3}{2}\right) = \frac{64}{3}$

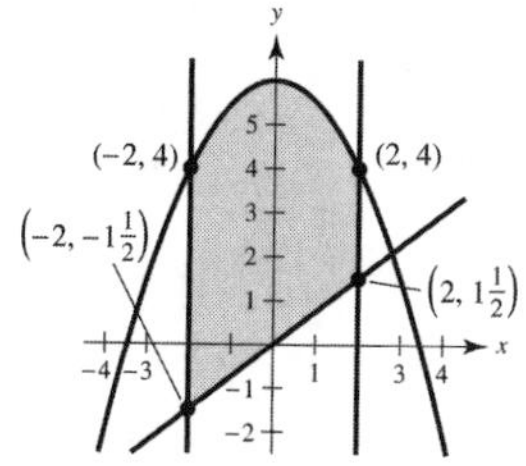

2. $A = \int_{1/2}^{5}\left(4 - \frac{1}{x^2}\right)dx = \left[4x + \frac{1}{x}\right]_{1/2}^{5} = \frac{81}{5}$

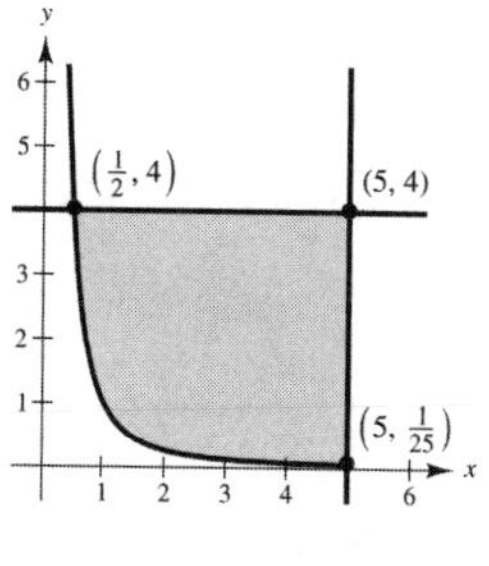

3. $A = \int_{-1}^{1}\frac{1}{x^2 + 1}\,dx = \left[\arctan x\right]_{-1}^{1} = \frac{\pi}{4} - \left(-\frac{\pi}{4}\right) = \frac{\pi}{2}$

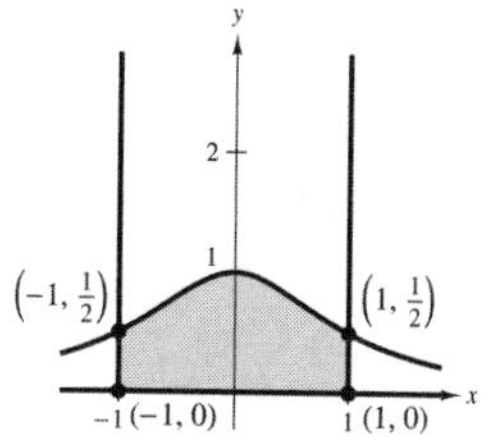

4. $A = \int_0^1\left[\left(y^2 - 2y\right) - (-1)\right]dy$

$= \int_0^1\left(y^2 - 2y + 1\right)dy$

$= \int_0^1 (y - 1)^2\,dy = \left[\frac{(y - 1)^3}{3}\right]_0^1 = \frac{1}{3}$

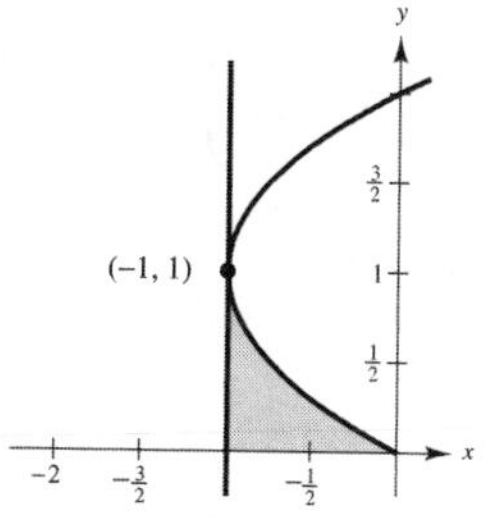

5. $A = 2\int_0^1 (x - x^3)\, dx = 2\left[\frac{1}{2}x^2 - \frac{1}{4}x^4\right]_0^1 = \frac{1}{2}$

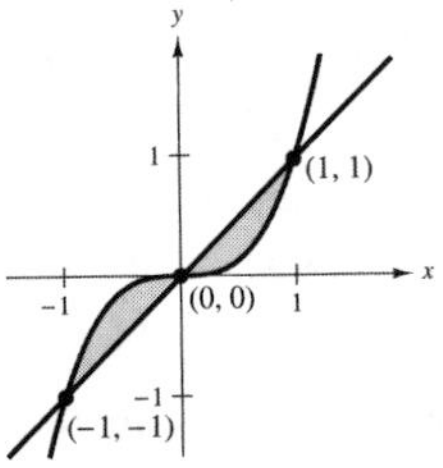

6. $A = \int_{-1}^{2} \left[(y + 3) - (y^2 + 1)\right] dy$

$= \int_{-1}^{2} (2 + y - y^2)\, dy = \left[2y + \frac{1}{2}y^2 - \frac{1}{3}y^3\right]_{-1}^{2} = \frac{9}{2}$

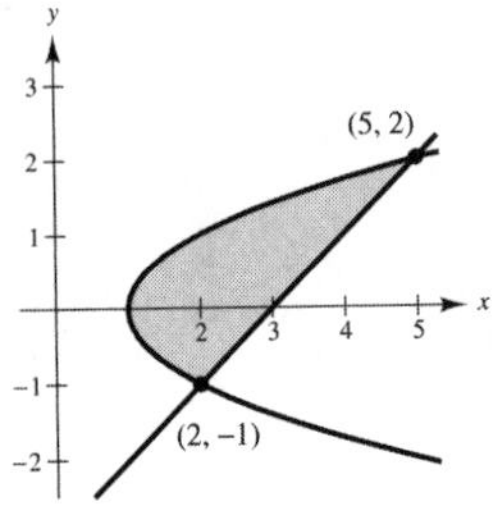

7. $A = \int_0^2 (e^2 - e^x)\, dx = \left[xe^2 - e^x\right]_0^2 = e^2 + 1$

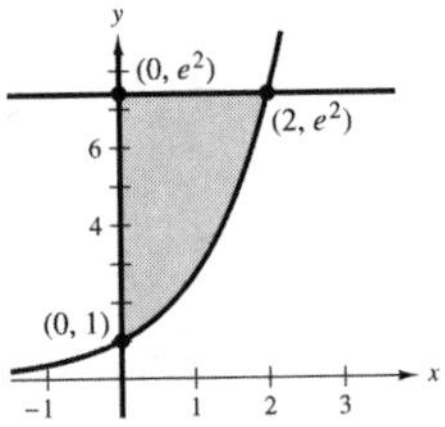

8. $A = 2\int_{\pi/6}^{\pi/2} (2 - \csc x)\, dx$

$= 2\left[2x - \ln\left|\csc x - \cot x\right|\right]_{\pi/6}^{\pi/2}$

$= 2\left(\left[\pi - 0\right] - \left[\frac{\pi}{3} - \ln\left(2 - \sqrt{3}\right)\right]\right)$

$= 2\left[\frac{2\pi}{3} + \ln\left(2 - \sqrt{3}\right)\right] \approx 1.555$

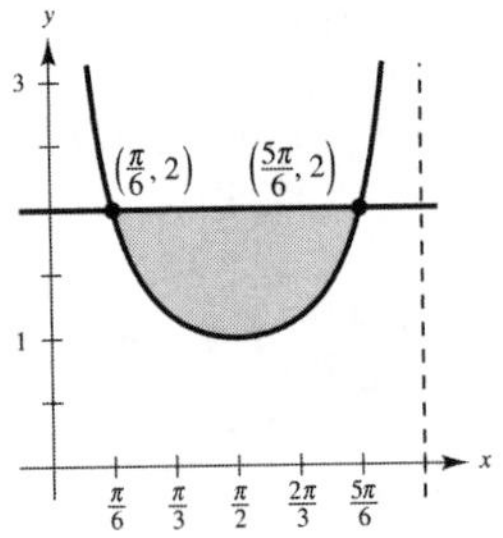

9. $A = \int_{\pi/4}^{5\pi/4} (\sin x - \cos x)\, dx$

$= \left[-\cos x - \sin x\right]_{\pi/4}^{5\pi/4}$

$= \left(\frac{1}{\sqrt{2}} + \frac{1}{\sqrt{2}}\right) - \left(-\frac{1}{\sqrt{2}} - \frac{1}{\sqrt{2}}\right)$

$= \frac{4}{\sqrt{2}} = 2\sqrt{2}$

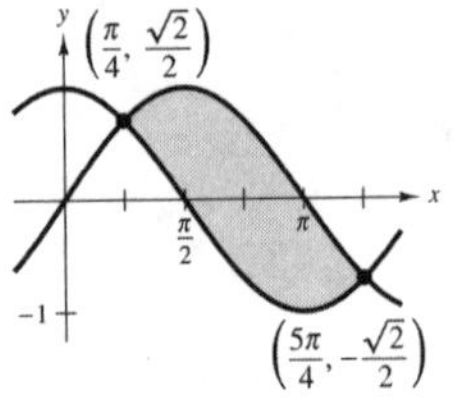

10. $A = \int_{\pi/3}^{5\pi/3} \left(\frac{1}{2} - \cos y\right) dy + \int_{5\pi/3}^{7\pi/3} \left(\cos y - \frac{1}{2}\right) dy$

$= \left[\frac{y}{2} - \sin y\right]_{\pi/3}^{5\pi/3} + \left[\sin y - \frac{y}{2}\right]_{5\pi/3}^{7\pi/3}$

$= \frac{\pi}{3} + 2\sqrt{3}$

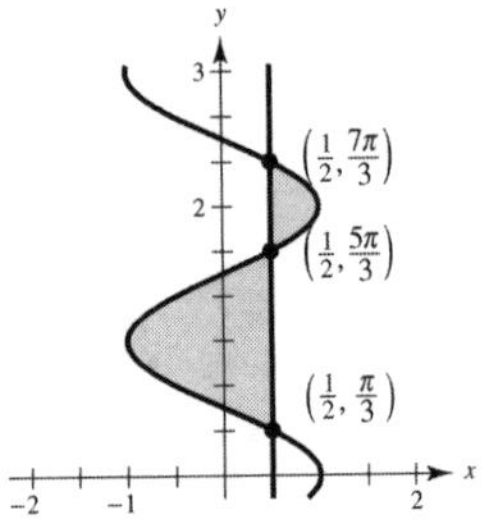

11. Points of intersection:

$x^2 - 8x + 3 = 3 + 8x - x^2$

$2x^2 - 16x = 0 \quad \text{when} \quad x = 0, 8$

$A = \int_0^8 \left[(3 + 8x - x^2) - (x^2 - 8x + 3)\right] dx$

$= \int_0^8 (16x - 2x^2)\, dx$

$= \left[8x^2 - \frac{2}{3}x^3\right]_0^8 = \frac{512}{3} \approx 170.667$

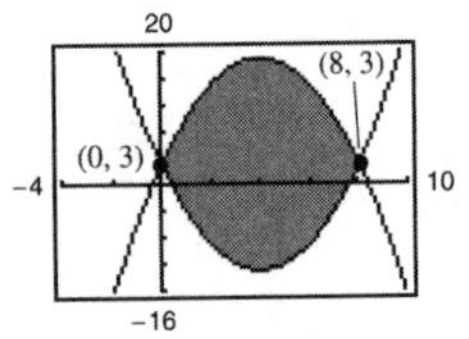

12. Point of intersection:

$$x^3 - x^2 + 4x - 3 = 0 \Rightarrow x \approx 0.783.$$

$$A \approx \int_0^{0.783} \left(3 - 4x + x^2 - x^3\right) dx$$

$$= \left[3x - 2x^2 + \frac{1}{3}x^3 - \frac{1}{4}x^4\right]_0^{0.783}$$

$$\approx 1.189$$

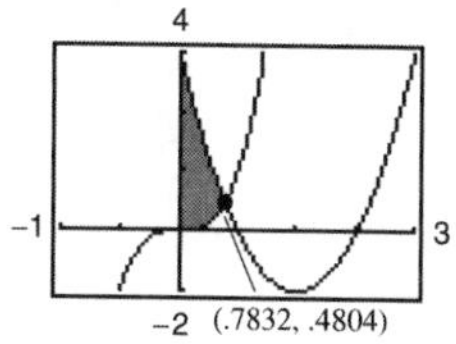

13. $y = \left(1 - \sqrt{x}\right)^2$

$$A = \int_0^1 \left(1 - \sqrt{x}\right)^2 dx$$

$$= \int_0^1 \left(1 - 2x^{1/2} + x\right) dx$$

$$= \left[x - \frac{4}{3}x^{3/2} + \frac{1}{2}x^2\right]_0^1 = \frac{1}{6} \approx 0.1667$$

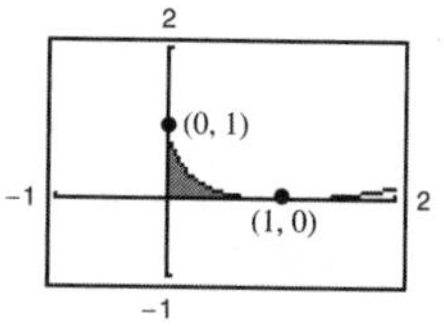

14. Points of intersection:

$$x^4 - 2x^2 = 2x^2$$

$$x^4 - 4x^2 = 0 \quad \text{when} \quad x = 0, \pm 2$$

$$A = 2\int_0^2 \left[2x^2 - \left(x^4 - 2x^2\right)\right] dx$$

$$= 2\int_0^2 \left(4x^2 - x^4\right) dx$$

$$= 2\left[\frac{4}{3}x^3 - \frac{1}{5}x^5\right]_0^2 = \frac{128}{15} \approx 8.5333$$

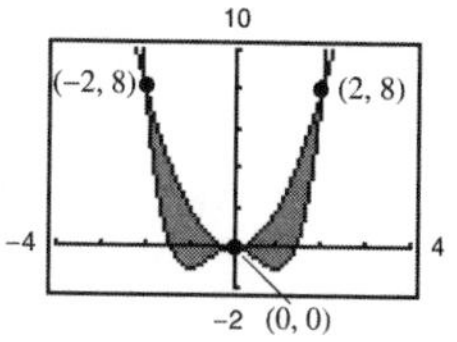

15. (a) Trapezoidal: Area $\approx \dfrac{160}{2(8)}\left[0 + 2(50) + 2(54) + 2(82) + 2(82) + 2(73) + 2(75) + 2(80) + 0\right] = 9920 \text{ ft}^2$

(b) Simpson's: Area $\approx \dfrac{160}{3(8)}\left[0 + 4(50) + 2(54) + 4(82) + 2(82) + 4(73) + 2(75) + 4(80) + 0\right] = 10{,}413\frac{1}{3} \text{ ft}^2$

16. $\displaystyle\int_{15}^{20} \left(6.4 + 0.2t + 0.01t^2\right) dt = \left[6.4t + \frac{0.2t^2}{2} + \frac{0.01t^3}{3}\right]_{15}^{20}$

$$\approx \$64.917 \text{ billion}$$

$$\int_{15}^{20} (8.4 + 0.35t)\, dt = \left[8.4t + \frac{0.35t^2}{2}\right]_{15}^{20}$$

$$\approx \$72.625 \text{ billion}$$

The second model projects the greater revenue.

The difference is about $\$72.625 - \$64.917 \approx \$7.71$ billion.

17. (a) **Disk**

$$V = \pi \int_0^3 x^2\,dx = \left[\frac{\pi x^3}{3}\right]_0^3 = 9\pi$$

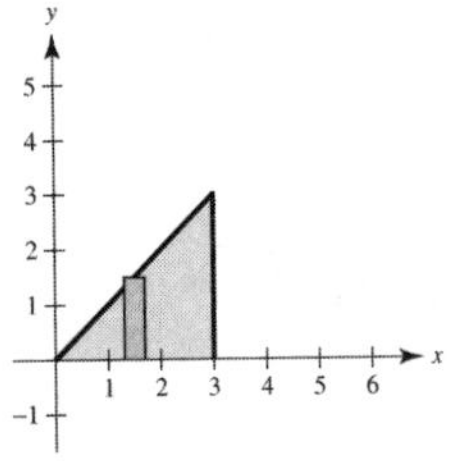

(b) **Shell**

$$V = 2\pi \int_0^3 x(x)\,dx = 2\pi\left[\frac{x^3}{3}\right]_0^3 = 18\pi$$

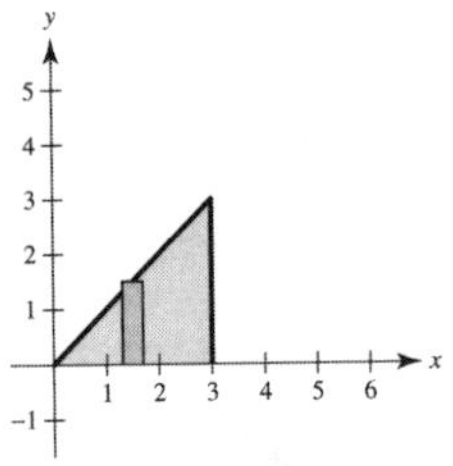

(c) **Shell**

$$V = 2\pi \int_0^3 (3 - x)x\,dx = 2\pi\left[\frac{3x^2}{2} - \frac{x^3}{3}\right]_0^3 = 9\pi$$

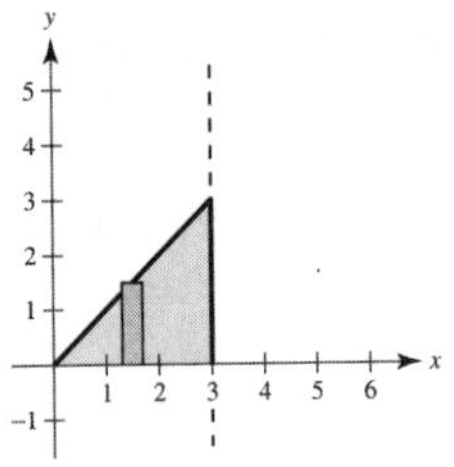

(d) **Shell**

$$V = 2\pi \int_0^3 (6 - x)x\,dx = 2\pi\left[3x^2 - \frac{x^3}{3}\right]_0^3 = 36\pi$$

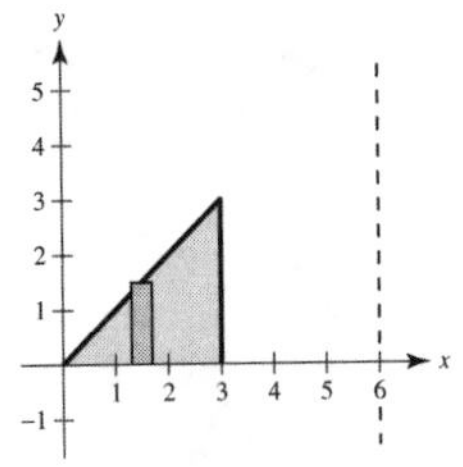

18. (a) **Shell**

$$V = 2\pi \int_0^2 y^3\,dy = \left[\frac{\pi}{2}y^4\right]_0^2 = 8\pi$$

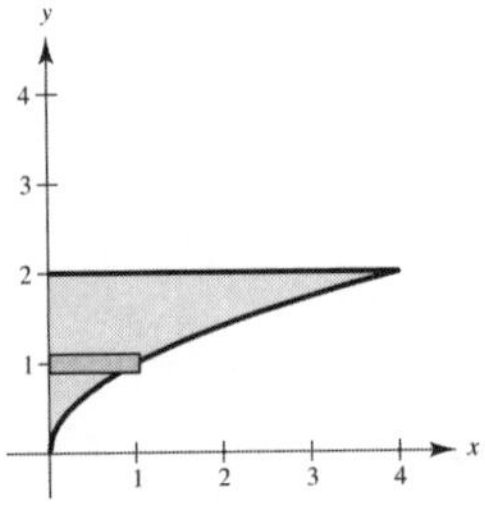

(b) **Shell**

$$V = 2\pi \int_0^2 (2 - y)y^2\,dy$$

$$= 2\pi \int_0^2 \left(2y^2 - y^3\right)dy = 2\pi\left[\frac{2}{3}y^3 - \frac{1}{4}y^4\right]_0^2 = \frac{8\pi}{3}$$

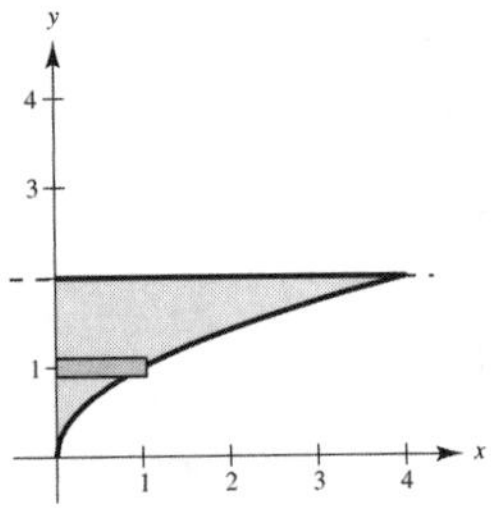

(c) **Disk**

$$V = \pi \int_0^2 y^4\,dy = \left[\frac{\pi}{5}y^5\right]_0^2 = \frac{32\pi}{5}$$

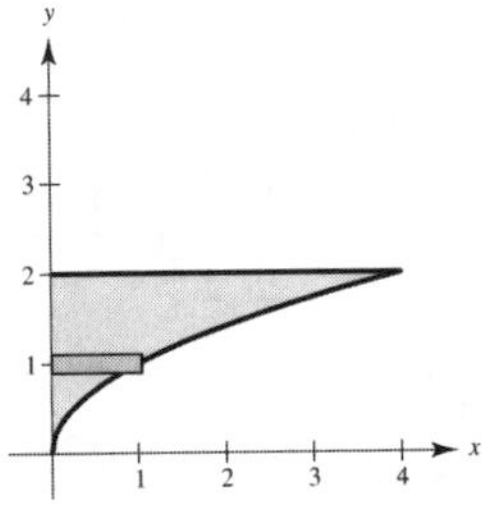

(d) **Disk**

$$V = \pi \int_0^2 \left[\left(y^2 + 1\right)^2 - 1^2\right]dy$$

$$= \pi \int_0^2 \left(y^4 + 2y^2\right)dy = \pi\left[\frac{1}{5}y^5 + \frac{2}{3}y^3\right]_0^2 = \frac{176\pi}{15}$$

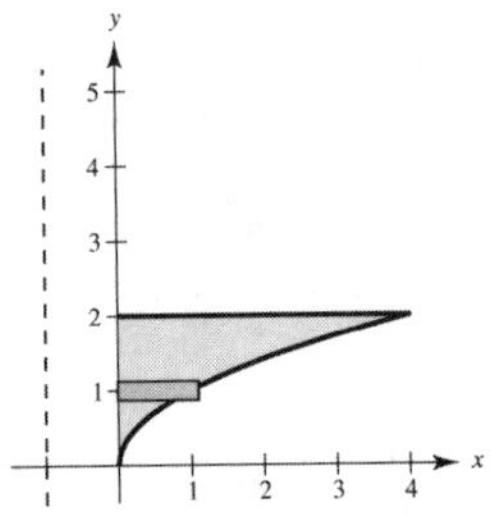

19. Shell

$$V = 2\pi\int_0^1 \frac{x}{x^4+1}\,dx = \pi\int_0^1 \frac{(2x)}{(x^2)^2+1}\,dx = \left[\pi \arctan(x^2)\right]_0^1 = \pi\left(\frac{\pi}{4} - 0\right) = \frac{\pi^2}{4}$$

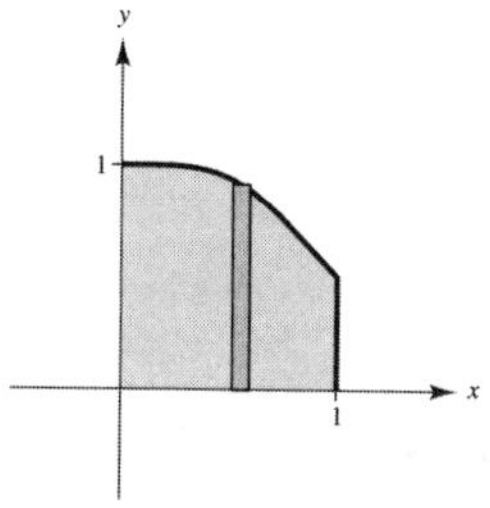

20. Disk

$$V = 2\pi\int_0^1 \left[\frac{1}{\sqrt{1+x^2}}\right]^2 dx = \left[2\pi \arctan x\right]_0^1 = 2\pi\left(\frac{\pi}{4} - 0\right) = \frac{\pi^2}{2}$$

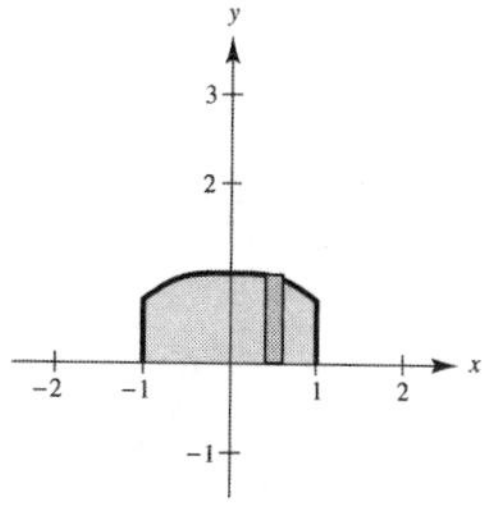

21. Shell

$$V = 2\pi\int_2^5 x\left(\frac{1}{x^2}\right) dx$$
$$= 2\pi\int_2^5 \frac{1}{x}\,dx$$
$$= \left[2\pi \ln|x|\right]_2^5$$
$$= 2\pi(\ln 5 - \ln 2)$$
$$= 2\pi \ln\left(\frac{5}{2}\right)$$

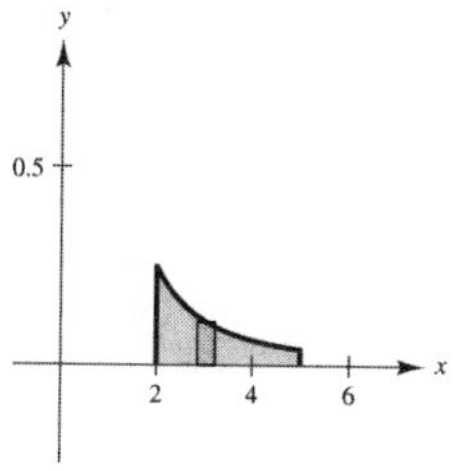

22. Disk

$$V = \pi\int_0^1 \left(e^{-x}\right)^2 dx$$
$$= \pi\int_0^1 e^{-2x}\,dx$$
$$= \left[-\frac{\pi}{2}e^{-2x}\right]_0^1$$
$$= \left(\frac{-\pi}{2e^2} + \frac{\pi}{2}\right) = \frac{\pi}{2}\left(1 - \frac{1}{e^2}\right)$$

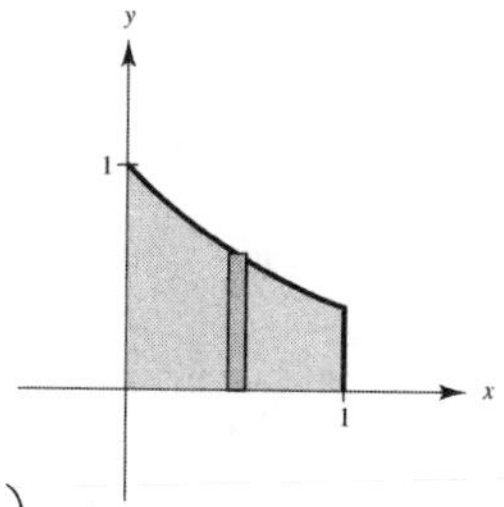

23. The volume of the spheroid is given by:

$$V = 4\pi\int_0^4 x\left(\tfrac{3}{4}\right)\sqrt{16 - x^2}\,dx$$
$$= \left[3\pi\left(-\tfrac{1}{2}\right)\left(\tfrac{2}{3}\right)\left(16 - x^2\right)^{3/2}\right]_0^4$$
$$= 64\pi$$
$$\tfrac{1}{4}V = 16\pi$$

Disk:
$$\pi\int_{-3}^{y_0} \tfrac{16}{9}\left(9 - y^2\right) dy = 16\pi$$
$$\tfrac{1}{9}\int_{-3}^{y_0} \left(9 - y^2\right) dy = 1$$
$$\left[9y - \tfrac{1}{3}y^3\right]_{-3}^{y_0} = 9$$
$$\left(9y_0 - \tfrac{1}{3}y_0^3\right) - (-27 + 9) = 9$$
$$y_0^3 - 27y_0 - 27 = 0$$

By Newton's Method, $y_0 \approx -1.042$ and the depth of the gasoline is $3 - 1.042 = 1.958$ feet.

24.

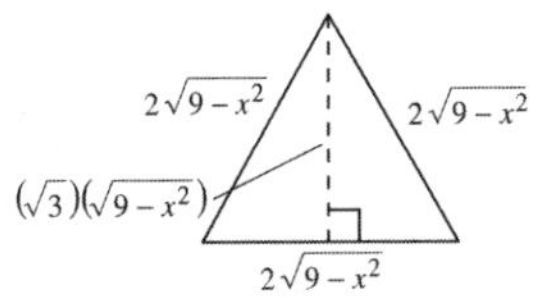

$$A(x) = \frac{1}{2}bh = \frac{1}{2}\left(2\sqrt{9-x^2}\right)\left(\sqrt{3}\sqrt{9-x^2}\right)$$
$$= \sqrt{3}\left(9 - x^2\right)$$
$$V = \sqrt{3}\int_{-3}^{3}\left(9 - x^2\right) dx = \sqrt{3}\left[9x - \frac{x^3}{3}\right]_{-3}^{3}$$
$$= \sqrt{3}\left[(27 - 9) - (-27 + 9)\right] = 36\sqrt{3}$$

25.
$$f(x) = \frac{4}{5}x^{5/4}$$
$$f'(x) = x^{1/4}$$
$$1 + \left[f'(x)\right]^2 = 1 + \sqrt{x}$$
$$u = 1 + \sqrt{x}$$
$$x = (u - 1)^2$$
$$dx = 2(u - 1)\, du$$
$$s = \int_0^4 \sqrt{1 + \sqrt{x}}\, dx = 2\int_1^3 \sqrt{u}(u - 1)\, du$$
$$= 2\int_1^3 \left(u^{3/2} - u^{1/2}\right) du$$
$$= 2\left[\frac{2}{5}u^{5/2} - \frac{2}{3}u^{3/2}\right]_1^3 = \frac{4}{15}\left[u^{3/2}(3u - 5)\right]_1^3$$
$$= \frac{8}{15}\left(1 + 6\sqrt{3}\right) = 6.076$$

26.
$$y = \frac{x^3}{6} + \frac{1}{2x}$$
$$y' = \frac{1}{2}x^2 - \frac{1}{2x^2}$$
$$1 + (y')^2 = \left(\frac{1}{2}x^2 + \frac{1}{2x^2}\right)^2$$
$$s = \int_1^3 \left(\frac{1}{2}x^2 + \frac{1}{2x^2}\right) dx = \left[\frac{1}{6}x^3 - \frac{1}{2x}\right]_1^3 = \frac{14}{3}$$

27. $y = 300 \cosh\left(\dfrac{x}{2000}\right) - 280, -2000 \le x \le 2000$
$$y' = \frac{3}{20}\sinh\left(\frac{x}{2000}\right)$$
$$s = \int_{-2000}^{2000} \sqrt{1 + \left[\frac{3}{20}\sinh\left(\frac{x}{2000}\right)\right]^2}\, dx$$
$$= \frac{1}{20}\int_{-2000}^{2000} \sqrt{400 + 9\sinh^2\left(\frac{x}{2000}\right)}\, dx$$
$= 4018.2$ ft (by Simpson's Rule or graphing utility)

28. This integral represents the arc length of the curve $f(x) = \frac{4}{x+1}$ between $x = 0$ and $x = 1$.

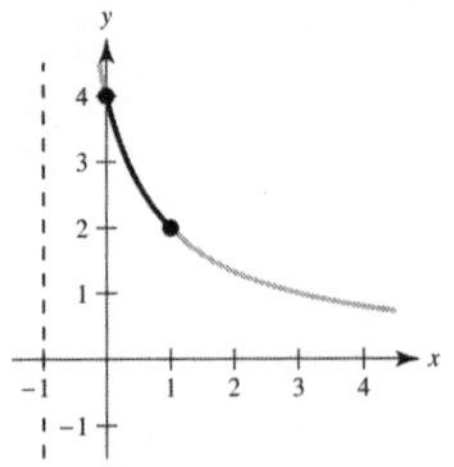

The length is a little more than 2. So, (c) is the best approximation.

29.
$$y = \frac{3}{4}x$$
$$y' = \frac{3}{4}$$
$$1 + (y')^2 = \frac{25}{16}$$
$$S = 2\pi\int_0^4 \left(\frac{3}{4}x\right)\sqrt{\frac{25}{16}}\, dx = \left[\left(\frac{15\pi}{8}\right)\frac{x^2}{2}\right]_0^4 = 15\pi$$

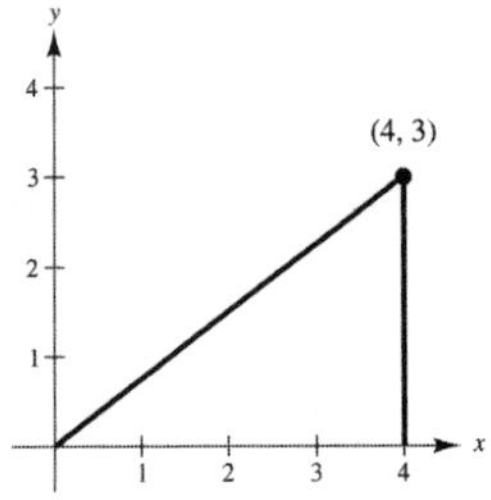

30. $y = 2\sqrt{x},\ y' = \dfrac{1}{\sqrt{x}}$
$$1 + (y')^2 = 1 + \frac{1}{x} = \frac{x + 1}{x}$$
$$S = 2\pi\int_3^8 2\sqrt{x}\sqrt{\frac{x + 1}{x}}\, dx = 4\pi\int_3^8 \sqrt{x + 1}\, dx$$
$$= 4\pi\left[\frac{2}{3}(x + 1)^{3/2}\right]_3^8 = \frac{152\pi}{3}$$

31. $F = kx$
$$5 = k(1)$$
$$F = 5x$$
$$W = \int_0^5 5x\, dx = \frac{5x^2}{2}\Bigg]_0^5 = \frac{125}{2} \text{ in.-lb} \approx 5.21 \text{ ft-lb}$$

32. $F = kx$
$$50 = k(1) \Rightarrow k = 50$$
$$W = \int_0^{10} 50x\, dx = \left[25x^2\right]_0^{10} = 2500 \text{ in.-lb} \approx 208.3 \text{ ft-lb}$$

33. Volume of disk: $\pi\left(\frac{1}{3}\right)^2 \Delta y \quad \left[\text{diameter} = \frac{2}{3} \text{ ft}\right]$

Weight of disk: $62.4\pi\left(\frac{1}{3}\right)^2 \Delta y$

Distance: $190 - y$

$$W = \frac{62.4\pi}{9}\int_0^{165}(190 - y)\,dy$$
$$= \frac{62.4\pi}{9}\left[190y - \frac{y^2}{2}\right]_0^{165}$$
$$= \frac{62.4\pi}{9}\left[\frac{35{,}475}{2}\right] = 122{,}980\pi \text{ ft-lb}$$
$$\approx 193.2 \text{ foot-tons}$$

34. $\rho = \frac{k}{V}$

$$800 = \frac{k}{2}$$
$$k = 1600$$
$$W = \int_2^3 \frac{1600}{V}\,dV$$
$$= \left[1600 \ln|V|\right]_2^3$$
$$= 1600 \ln\left(\frac{3}{2}\right) \approx 648.74 \text{ ft-lb}$$

35. Weight of section of chain: $4\,\Delta x$

Distance moved: $10 - x$

$$W = 4\int_0^{10}(10 - x)\,dx = 4\left[10x - \frac{x^2}{2}\right]_0^{10}$$
$$= 200 \text{ ft-lb}$$

36. (a) Weight of section of cable: $5\,\Delta x$

Distance: $200 - x$

$$W = 5\int_0^{200}(200 - x)\,dx$$
$$= 5\left[200x - \frac{x^2}{2}\right]_0^{200}$$
$$= 100{,}000 \text{ ft-lb}$$

(b) Work to move 300 pounds 200 feet vertically:

$300(200) = 60{,}000$ ft-lb.

Total work: $100{,}000 + 60{,}000 = 160{,}000$ ft-lb

37. $W = \int_a^b F(x)\,dx$

$$80 = \int_0^4 ax^2\,dx = \left[\frac{ax^3}{3}\right]_0^4 = \frac{64}{3}a$$
$$a = \frac{3(80)}{64} = \frac{15}{4} = 3.75$$

38. $W = \int_a^b F(x)\,dx$

$$F(x) = \begin{cases} -(2/9)x + 6, & 0 \le x \le 9 \\ -(4/3)x + 16, & 9 \le x \le 12 \end{cases}$$
$$W = \int_0^9\left(-\tfrac{2}{9}x + 6\right)dx + \int_9^{12}\left(-\tfrac{4}{3}x + 16\right)dx$$
$$= \left[-\tfrac{1}{9}x^2 + 6x\right]_0^9 + \left[-\tfrac{2}{3}x^2 + 16x\right]_9^{12}$$
$$= (-9 + 54) + (-96 + 192 + 54 - 144)$$
$$= 51 \text{ ft-lb}$$

39. $\bar{x} = \frac{8(-1) + 12(2) + 6(5) + 14(7)}{8 + 12 + 6 + 14} = \frac{144}{40} = \frac{18}{5} = 3.6$

40.
$$\bar{x} = \frac{3(2) + 2(-3) + 6(4) + 9(6)}{3 + 2 + 6 + 9} = \frac{78}{20} = \frac{39}{10}$$
$$\bar{y} = \frac{3(1) + 2(2) + 6(-1) + 9(5)}{3 + 2 + 6 + 9} = \frac{46}{20} = \frac{23}{10}$$
$$(\bar{x}, \bar{y}) = \left(\frac{39}{10}, \frac{23}{10}\right)$$

41. $A = \int_{-1}^{3}\left[(2x + 3) - x^2\right] dx = \left[x^2 + 3x - \frac{1}{3}x^3\right]_{-1}^{3} = \frac{32}{3}$

$$\frac{1}{A} = \frac{3}{32}$$

$$\bar{x} = \frac{3}{32}\int_{-1}^{3} x\left(2x + 3 - x^2\right) dx = \frac{3}{32}\int_{-1}^{3}\left(3x + 2x^2 - x^3\right) dx = \frac{3}{32}\left[\frac{3}{2}x^2 + \frac{2}{3}x^3 - \frac{1}{4}x^4\right]_{-1}^{3} = 1$$

$$\bar{y} = \left(\frac{3}{32}\right)\frac{1}{2}\int_{-1}^{3}\left[(2x + 3)^2 - x^4\right] dx = \frac{3}{64}\int_{-1}^{3}\left(9 + 12x + 4x^2 - x^4\right) dx$$

$$= \frac{3}{64}\left[9x + 6x^2 + \frac{4}{3}x^3 - \frac{1}{5}x^5\right]_{-1}^{3} = \frac{17}{5}$$

$$(\bar{x}, \bar{y}) = \left(1, \frac{17}{5}\right)$$

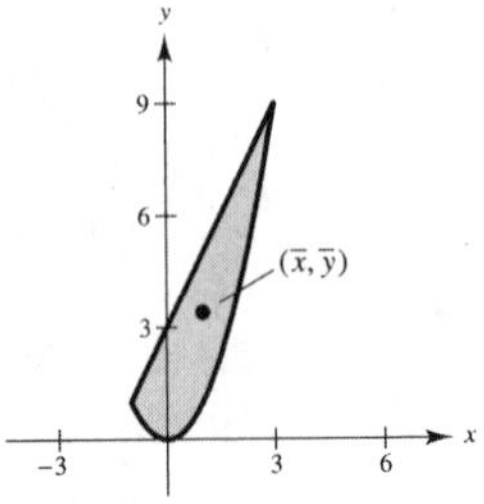

42. $A = \int_{0}^{8}\left(x^{2/3} - \frac{1}{2}x\right) dx = \left[\frac{3}{5}x^{5/3} - \frac{1}{4}x^2\right]_{0}^{8} = \frac{16}{5}$

$$\frac{1}{A} = \frac{5}{16}$$

$$\bar{x} = \frac{5}{16}\int_{0}^{8} x\left(x^{2/3} - \frac{1}{2}x\right) dx = \frac{5}{16}\left[\frac{3}{8}x^{8/3} - \frac{1}{6}x^3\right]_{0}^{8} = \frac{10}{3}$$

$$\bar{y} = \left(\frac{5}{16}\right)\frac{1}{2}\int_{0}^{8}\left(x^{4/3} - \frac{1}{4}x^2\right) dx = \frac{1}{2}\left(\frac{5}{16}\right)\left[\frac{3}{7}x^{7/3} - \frac{1}{12}x^3\right]_{0}^{8} = \frac{40}{21}$$

$$(\bar{x}, \bar{y}) = \left(\frac{10}{3}, \frac{40}{21}\right)$$

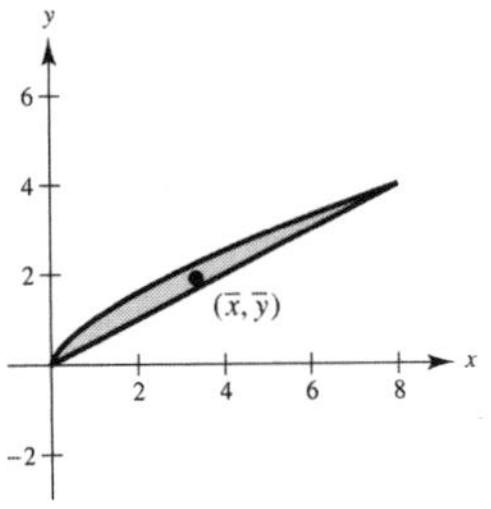

43. $\bar{y} = 0$ by symmetry.

For the trapezoid:

$$m = \left[(4)(6) - (1)(6)\right]\rho = 18\rho$$

$$M_y = \rho\int_{0}^{6} x\left[\left(\frac{1}{6}x + 1\right) - \left(-\frac{1}{6}x - 1\right)\right] dx = \rho\int_{0}^{6}\left(\frac{1}{3}x^2 + 2x\right) dx = \rho\left[\frac{x^3}{9} + x^2\right]_{0}^{6} = 60\rho$$

For the semicircle:

$$m = \left(\frac{1}{2}\right)(\pi)(2)^2\rho = 2\pi\rho$$

$$M_y = \rho\int_{6}^{8} x\left[\sqrt{4 - (x - 6)^2} - \left(-\sqrt{4 - (x - 6)^2}\right)\right] dx = 2\rho\int_{6}^{8} x\sqrt{4 - (x - 6)^2}\, dx$$

Let $u = x - 6$, then $x = u + 6$ and $dx = du$. When $x = 6, u = 0$. When $x = 8, u = 2$.

$$M_y = 2\rho\int_{0}^{2}(u + 6)\sqrt{4 - u^2}\, du = 2\rho\int_{0}^{2} u\sqrt{4 - u^2}\, du + 12\rho\int_{0}^{2}\sqrt{4 - u^2}\, du$$

$$= 2\rho\left[\left(-\frac{1}{2}\right)\left(\frac{2}{3}\right)\left(4 - u^2\right)^{3/2}\right]_{0}^{2} + 12\rho\left[\frac{\pi(2)^2}{4}\right] = \frac{16\rho}{3} + 12\pi\rho = \frac{4\rho(4 + 9\pi)}{3}$$

So, you have: $\bar{x}(18\rho + 2\pi\rho) = 60\rho + \frac{4\rho(4 + 9\pi)}{3}$

$$\bar{x} = \frac{180\rho + 4\rho(4 + 9\pi)}{3}\cdot\frac{1}{2\rho(9 + \pi)} = \frac{2(9\pi + 49)}{3(\pi + 9)}$$

The centroid of the blade is $\left(\frac{2(9\pi + 49)}{3(\pi + 9)}, 0\right)$.

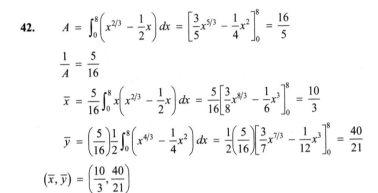

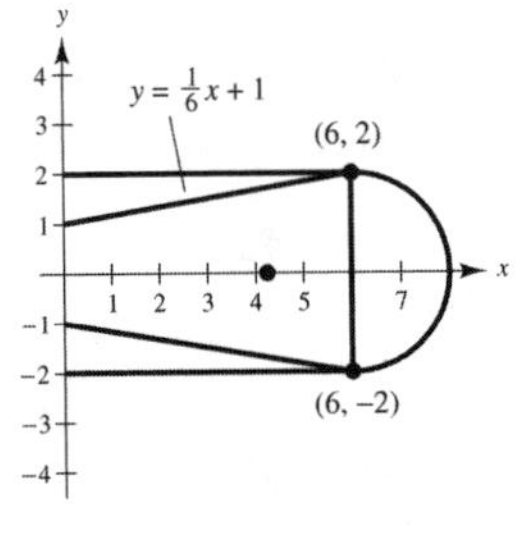

44. $r = 4$ is the distance between the center of the circle and the y-axis.

$A = \pi(2)^2 = 4\pi$ is the area of the circle. So,

$V = 2\pi rA = 2\pi(4)(4\pi) = 32\pi^2.$

45. $h(y) = 9 - y$

$L(y) = 4 - \frac{4}{3}y$

$$F = 64\int_0^3 (9 - y)\left(4 - \frac{4}{3}y\right) dy$$

$$= 64\int_0^3 \left(36 - 16y + \frac{4}{3}y^2\right) dy$$

$$= 64\left[36y - 8y^2 + \frac{4}{9}y^3\right]_0^3$$

$$= 64\left[36(3) - 8(9) + 4(3)\right] = 64(48)$$

$$= 3072 \text{ lb}$$

46. $h(y) = 5 - y$

$L(y) = 7$

$$F = 140.7\int_0^5 (5 - y)(7)\, dy$$

$$= 140.7(7)\left[5y - \frac{y^2}{2}\right]_0^5$$

$$= 984.9\left(25 - \frac{25}{2}\right)$$

$$= 12{,}311.25 \text{ lb}$$

47. Wall at shallow end:

$$F = 62.4\int_0^5 y(20)\, dy = \left[(1248)\frac{y^2}{2}\right]_0^5 = 15{,}600 \text{ lb}$$

Wall at deep end:

$$F = 62.4\int_0^{10} y(20)\, dy = \left[(624)y^2\right]_0^{10} = 62{,}400 \text{ lb}$$

Side wall:

$$F_1 = 62.4\int_0^5 y(40)\, dy = \left[(1248)y^2\right]_0^5 = 31{,}200 \text{ lb}$$

$$F_2 = 62.4\int_0^5 (10 - y)8y\, dy = 62.4\int_0^5 \left(80y - 8y^2\right) dy$$

$$= 62.4(8)\left[5y^2 - \frac{y^3}{3}\right]_0^5 = 41{,}600 \text{ lb}$$

$$F = F_1 + F_2 = 72{,}800 \text{ lb}$$

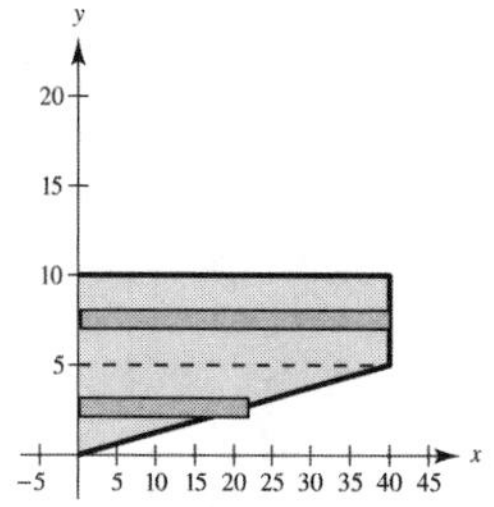

Problem Solving for Chapter 7

1. $T = \frac{1}{2}c(c^2) = \frac{1}{2}c^3$

$$R = \int_0^c \left(cx - x^2\right) dx = \left[\frac{cx^2}{2} - \frac{x^3}{3}\right]_0^c = \frac{c^3}{2} - \frac{c^3}{3} = \frac{c^3}{6}$$

$$\lim_{c \to 0^+} \frac{T}{R} = \lim_{c \to 0^+} \frac{\frac{1}{2}c^3}{\frac{1}{6}c^3} = 3$$

2. (a) By symmetry, $M_x = 0$ for L

(b) Because

$(M_y \text{ for } L) + (M_y \text{ for } A) = (M_y \text{ for } B)$,

you have

$(M_y \text{ for } L) = (M_y \text{ for } B) - (M_y \text{ for } A)$

(c) M_y for $B = 0$, because B is a circle at the origin

For A, $\bar{x} = \frac{M_y}{\text{Area}} \Rightarrow M_y = r\left(\pi r^2\right) = \pi r^3$

So, $(M_y \text{ for } L) = 0 - \pi r^3 = -\pi r^3$

(d) $\bar{y} = 0$ by symmetry.

$$\bar{x} = \frac{M_y \text{ of } L}{\text{Area of } L} = \frac{-\pi r^3}{4\pi r^2 - \pi r^2} = -\frac{r}{3}$$

$$(\bar{x}, \bar{y}) = \left(-\frac{r}{3}, 0\right)$$

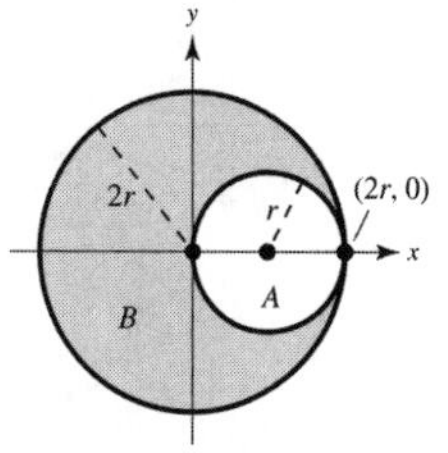

3. $R = \int_0^1 x(1-x)\,dx = \left[\frac{x^2}{2} - \frac{x^3}{3}\right]_0^1 = \frac{1}{2} - \frac{1}{3} = \frac{1}{6}$

Let (c, mc) be the intersection of the line and the parabola.

Then, $mc = c(1-c) \Rightarrow m = 1 - c$ or $c = 1 - m$.

$$\frac{1}{2}\left(\frac{1}{6}\right) = \int_0^{1-m}\left(x - x^2 - mx\right)dx$$

$$\frac{1}{12} = \left[\frac{x^2}{2} - \frac{x^3}{3} - m\frac{x^2}{2}\right]_0^{1-m}$$

$$= \frac{(1-m)^2}{2} - \frac{(1-m)^3}{3} - m\frac{(1-m)^2}{2}$$

$$1 = 6(1-m)^2 - 4(1-m)^3 - 6m(1-m)^2$$

$$= (1-m)^2\left(6 - 4(1-m) - 6m\right)$$

$$= (1-m)^2(2-2m)$$

$$\frac{1}{2} = (1-m)^3$$

$$\left(\frac{1}{2}\right)^{1/3} = 1 - m$$

$$m = 1 - \left(\frac{1}{2}\right)^{1/3} \approx 0.2063$$

So, $y = 0.2063x$.

4. $V = 2(2\pi)\int_{\sqrt{r^2-(h^2/4)}}^{r} x\sqrt{r^2 - x^2}\,dx$

$$= -2\pi\left[\frac{2}{3}\left(r^2 - x^2\right)^{3/2}\right]_{\sqrt{r^2-(h^2/4)}}^{r}$$

$$= \frac{-4\pi}{3}\left[-\frac{h^3}{8}\right] = \frac{\pi h^3}{6} \text{ which does not depend on } r$$

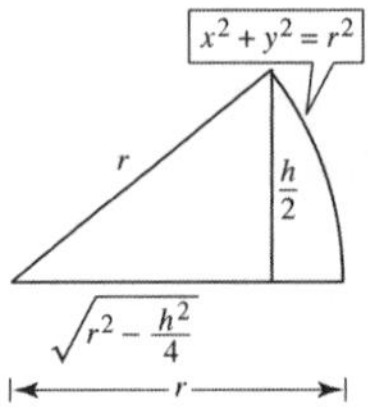

5. $8y^2 = x^2\left(1 - x^2\right)$

$$y = \pm\frac{|x|\sqrt{1-x^2}}{2\sqrt{2}}$$

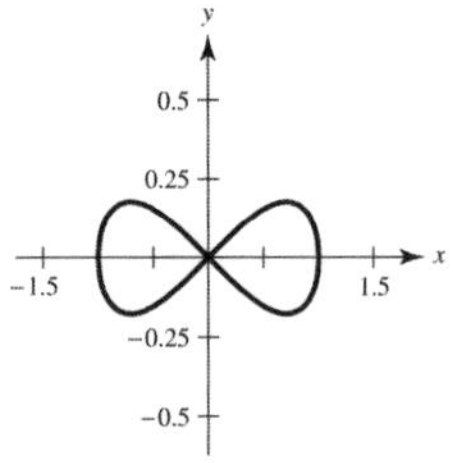

For $x > 0$, $y' = \dfrac{1 - 2x^2}{2\sqrt{2}\sqrt{1-x^2}}$.

$$S = 2(2\pi)\int_0^1 x\sqrt{1 + \left(\frac{1-2x^2}{2\sqrt{2}\sqrt{1-x^2}}\right)^2}\,dx$$

$$= \frac{5\sqrt{2}\pi}{3}$$

6. (a) $\frac{1}{2}V = \int_0^1\left[\pi\left(2 + \sqrt{1-y^2}\right)^2 - \pi\left(2 - \sqrt{1-y^2}\right)^2\right]dy$

$$= \pi\int_0^1\left[\left(4 + 4\sqrt{1-y^2} + \left(1-y^2\right)\right) - \left(4 - 4\sqrt{1-y^2} + \left(1-y^2\right)\right)\right]dy$$

$$= 8\pi\int_0^1\sqrt{1-y^2}\,dy \quad \text{(Integral represents 1/4 (area of circle))}$$

$$= 8\pi\left(\frac{\pi}{4}\right) = 2\pi^2 \Rightarrow V = 4\pi^2$$

(b) $(x-R)^2 + y^2 = r^2 \Rightarrow x = R \pm\sqrt{r^2 - y^2}$

$$\frac{1}{2}V = \int_0^r\left[\pi\left(R + \sqrt{r^2-y^2}\right)^2 - \pi\left(R - \sqrt{r^2-y^2}\right)^2\right]dy = \pi\int_0^r 4R\sqrt{r^2-y^2}\,dy = \pi(4R)\frac{1}{4}\pi r^2 = \pi^2 r^2 R$$

$$V = 2\pi^2 r^2 R$$

7. By the Theorem of Pappus,

$$V = 2\pi rA$$
$$= 2\pi\left[d + \frac{1}{2}\sqrt{w^2 + l^2}\right]lw.$$

8. (a) Tangent at A: $y = x^3,\ y' = 3x^2$

$$y - 1 = 3(x - 1)$$
$$y = 3x - 2$$

To find point B:

$$x^3 = 3x - 2$$
$$x^3 - 3x + 2 = 0$$
$$(x - 1)^2(x + 2) = 0 \Rightarrow B = (-2, -8)$$

Tangent at B: $y = x^3,\ y' = 3x^2$

$$y + 8 = 12(x + 2)$$
$$y = 12x + 16$$

To find point C:

$$x^3 = 12x + 16$$
$$x^3 - 12x - 16 = 0$$
$$(x + 2)^2(x - 4) = 0 \Rightarrow C = (4, 64)$$

$$\text{Area of } R = \int_{-2}^{1}\left(x^3 - 3x + 2\right)dx = \frac{27}{4}$$

$$\text{Area of } S = \int_{-2}^{4}\left(12x + 16 - x^3\right)dx = 108$$

$$\text{Area of } S = 16(\text{area of } R) \quad \left[\frac{\text{area } S}{\text{area } R} = 16\right]$$

(b) Tangent at $A(a, a^3)$: $y - a^3 = 3a^2(x - a)$

$$y = 3a^2x - 2a^3$$

To find point B: $x^3 - 3a^2x + 2a^3 = 0$

$$(x - a)^2(x + 2a) = 0$$
$$\Rightarrow B = \left(-2a, -8a^3\right)$$

Tangent at B: $y + 8a^3 = 12a^2(x + 2a)$

$$y = 12a^2x + 16a^3$$

To find point C: $x^3 - 12a^2x - 16a^3 = 0$

$$(x + 2a)^2(x - 4a) = 0$$
$$\Rightarrow C = \left(4a, 64a^3\right)$$

$$\text{Area of } R = \int_{-2a}^{a}\left[x^3 - 3a^2x + 2a^3\right]dx = \frac{27}{4}a^4$$

$$\text{Area of } S = \int_{-2a}^{4a}\left[12a^2x + 16a^3 - x^3\right]dx = 108a^4$$

$$\text{Area of } S = 16(\text{area of } R)$$

9.

$$f'(x)^2 = e^x$$
$$f'(x) = e^{x/2}$$
$$f(x) = 2e^{x/2} + C$$
$$f(0) = 0 \Rightarrow C = -2$$
$$f(x) = 2e^{x/2} - 2$$

10. $s(x) = \int_a^x \sqrt{1 + f'(t)^2}\,dt$

(a) $s'(x) = \dfrac{ds}{dx} = \sqrt{1 + f'(x)^2}$

(b)

$$ds = \sqrt{1 + f'(x)^2}\,dx$$
$$(ds)^2 = \left[1 + f'(x)^2\right](dx)^2$$
$$= \left[1 + \left(\frac{dy}{dx}\right)^2\right](dx)^2 = (dx)^2 + (dy)^2$$

(c) $s(x) = \int_1^x \sqrt{1 + \left(\frac{3}{2}t^{1/2}\right)^2}\,dt = \int_1^x \sqrt{1 + \frac{9}{4}t}\,dt$

(d)

$$s(2) = \int_1^2 \sqrt{1 + \frac{9}{4}t}\,dt$$
$$= \left[\frac{8}{27}\left(1 + \frac{9}{4}t\right)^{3/2}\right]_1^2$$
$$= \frac{22}{27}\sqrt{22} - \frac{13}{27}\sqrt{13} \approx 2.0858$$

This is the length of the curve $y = x^{3/2}$ from $x = 1$ to $x = 2$.

11. Let ρ_f be the density of the fluid and ρ_0 the density of the iceberg. The buoyant force is

$$F = \rho_f g \int_{-h}^{0} A(y)\,dy$$

where $A(y)$ is a typical cross section and g is the acceleration due to gravity. The weight of the object is

$$W = \rho_0 g \int_{-h}^{L-h} A(y)\,dy.$$

$$F = W$$

$$\rho_f g \int_{-h}^{0} A(y)\,dy = \rho_0 g \int_{-h}^{L-h} A(y)\,dy$$

$$\frac{\rho_0}{\rho_f} = \frac{\text{submerged volume}}{\text{total volume}}$$

$$= \frac{0.92 \times 10^3}{1.03 \times 10^3} = 0.893 \text{ or } 89.3\%$$

12. (a) $\bar{y} = 0$ by symmetry

$$M_y = \int_1^6 x\left(\frac{1}{x^3} - \left(-\frac{1}{x^3}\right)\right) dx = \int_1^6 \frac{2}{x^2}\, dx = \left[-2\frac{1}{x}\right]_1^6 = \frac{5}{3}$$

$$m = 2\int_1^6 \frac{1}{x^3}\, dx = \left[-\frac{1}{x^2}\right]_1^6 = \frac{35}{36}$$

$$\bar{x} = \frac{5/3}{35/36} = \frac{12}{7} \qquad (\bar{x}, \bar{y}) = \left(\frac{12}{7}, 0\right)$$

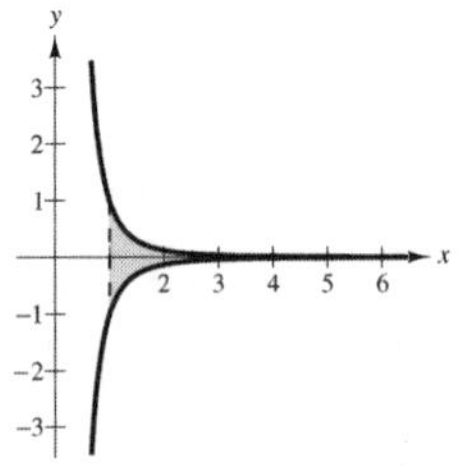

(b) $$m = 2\int_1^b \frac{1}{x^3}\, dx = \frac{b^2 - 1}{b^2}$$

$$M_y = 2\int_1^6 \frac{1}{x^2}\, dx = \frac{2(b - 1)}{b}$$

$$\bar{x} = \frac{2(b - 1)/b}{(b^2 - 1)/b^2} = \frac{2b}{b + 1} \qquad (\bar{x}, \bar{y}) = \left(\frac{2b}{b + 1}, 0\right)$$

(c) $$\lim_{b\to\infty} \bar{x} = \lim_{b\to\infty} \frac{2b}{b + 1} = 2 \qquad (\bar{x}, \bar{y}) = (2, 0)$$

13. (a) $\bar{y} = 0$ by symmetry

$$M_y = 2\int_1^6 x\frac{1}{x^4}\, dx = 2\int_1^6 \frac{1}{x^3}\, dx = \frac{35}{36}$$

$$m = 2\int_1^6 \frac{1}{x^4}\, dx = \frac{215}{324}$$

$$\bar{x} = \frac{35/36}{215/324} = \frac{63}{43} \qquad (\bar{x}, \bar{y}) = \left(\frac{63}{43}, 0\right)$$

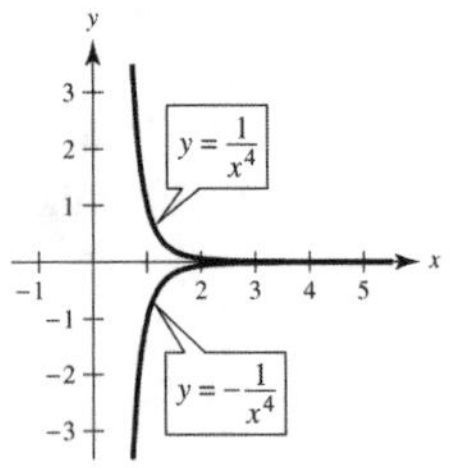

(b) $$M_y = 2\int_1^b \frac{1}{x^3}\, dx = \frac{b^2 - 1}{b^2}$$

$$m = 2\int_1^b \frac{1}{x^4}\, dx = \frac{2(b^3 - 1)}{3b^3}$$

$$\bar{x} = \frac{(b^2 - 1)/b^2}{2(b^3 - 1)/3b^3} = \frac{3b(b + 1)}{2(b^2 + b + 1)} \qquad (\bar{x}, \bar{y}) = \left(\frac{3b(b + 1)}{2(b^2 + b + 1)}, 0\right)$$

(c) $$\lim_{b\to\infty} \bar{x} = \frac{3b(b + 1)}{2(b^2 + b + 1)} = \frac{3}{2} \qquad (\bar{x}, \bar{y}) = \left(\frac{3}{2}, 0\right)$$

14. (a) $W = \text{area} = 2 + 4 + 6 = 12$

(b) $W = \text{area} = 3 + (1 + 1) + 2 + \frac{1}{2} = 7\frac{1}{2}$

15. Point of equilibrium: $50 - 0.5x = 0.125x$

$x = 80, p = 10$

$(P_0, x_0) = (10, 80)$

Consumer surplus $= \int_0^{80} [(50 - 0.5x) - 10]\, dx = 1600$

Producer surplus $= \int_0^{80} (10 - 0.125x)\, dx = 400$

16. Point of equilibrium: $1000 - 0.4x^2 = 42x$

$x = 20, p = 840$

$(P_0, x_0) = (840, 20)$

Consumer surplus $= \int_0^{20} [(1000 - 0.4x^2) - 840]\, dx$

$= 2133.33$

Producer surplus $= \int_0^{20} (840 - 42x)\, dx = 8400$

17. Use Exercise 25, Section 7.7, which gives $F = wkhb$ for a rectangle plate.

Wall at shallow end

From Exercise 25: $F = 62.4(2)(4)(20) = 9984$ lb

Wall at deep end

From Exercise 25: $F = 62.4(4)(8)(20) = 39{,}936$ lb

Side wall

From Exercise 25: $F_1 = 62.4(2)(4)(40) = 19{,}968$ lb

$F_2 = 62.4 \int_0^4 (8 - y)(10y)\, dy$

$= 624 \int_0^4 (8y - y^2)\, dy = 624 \left[4y^2 - \frac{y^3}{3}\right]_0^4$

$= 26{,}624$ lb

Total force: $F_1 + F_2 = 46{,}592$ lb

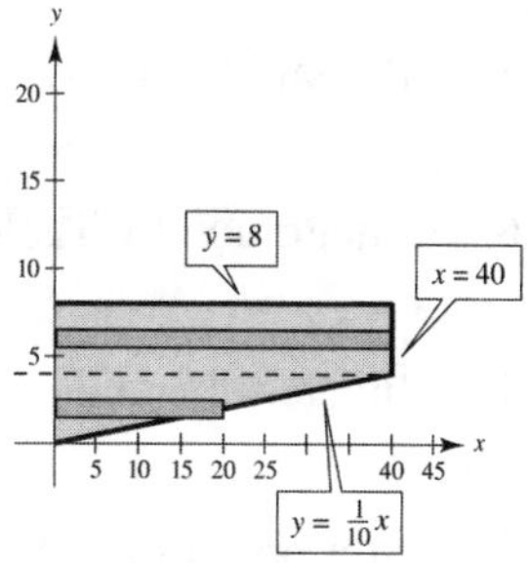

CHAPTER 8
Integration Techniques, L'Hôpital's Rule, and Improper Integrals

Section 8.1 Basic Integration Rules 709

Section 8.2 Integration by Parts 722

Section 8.3 Trigonometric Integrals 742

Section 8.4 Trigonometric Substitution 756

Section 8.5 Partial Fractions 778

Section 8.6 Integration by Tables and Other Integration Techniques 791

Section 8.7 Indeterminate Forms and L'Hôpital's Rule 802

Section 8.8 Improper Integrals 817

Review Exercises 830

Problem Solving 844

CHAPTER 8
Integration Techniques, L'Hôpital's Rule, and Improper Integrals

Section 8.1 Basic Integration Rules

1. (a) $\dfrac{d}{dx}\left[2\sqrt{x^2+1}+C\right] = 2\left(\dfrac{1}{2}\right)\left(x^2+1\right)^{-1/2}(2x) = \dfrac{2x}{\sqrt{x^2+1}}$

(b) $\dfrac{d}{dx}\left[\sqrt{x^2+1}+C\right] = \dfrac{1}{2}\left(x^2+1\right)^{-1/2}(2x) = \dfrac{x}{\sqrt{x^2+1}}$

(c) $\dfrac{d}{dx}\left[\dfrac{1}{2}\sqrt{x^2+1}+C\right] = \dfrac{1}{2}\left(\dfrac{1}{2}\right)\left(x^2+1\right)^{-1/2}(2x) = \dfrac{x}{2\sqrt{x^2+1}}$

(d) $\dfrac{d}{dx}\left[\ln\left(x^2+1\right)+C\right] = \dfrac{2x}{x^2+1}$

$\displaystyle\int \frac{x}{\sqrt{x^2+1}}\,dx$ matches (b).

2. (a) $\dfrac{d}{dx}\left[\ln\sqrt{x^2+1}+C\right] = \dfrac{1}{2}\left(\dfrac{2x}{x^2+1}\right) = \dfrac{x}{x^2+1}$

(b) $\dfrac{d}{dx}\left[\dfrac{2x}{\left(x^2+1\right)^2}+C\right] = \dfrac{\left(x^2+1\right)^2(2)-(2x)(2)\left(x^2+1\right)(2x)}{\left(x^2+1\right)^4} = \dfrac{2\left(1-3x^2\right)}{\left(x^2+1\right)^3}$

(c) $\dfrac{d}{dx}\left[\arctan x + C\right] = \dfrac{1}{1+x^2}$

(d) $\dfrac{d}{dx}\left[\ln\left(x^2+1\right)+C\right] = \dfrac{2x}{x^2+1}$

$\displaystyle\int \frac{x}{x^2+1}\,dx$ matches (a).

3. (a) $\dfrac{d}{dx}\left[\ln\sqrt{x^2+1}+C\right] = \dfrac{1}{2}\left(\dfrac{2x}{x^2+1}\right) = \dfrac{x}{x^2+1}$

(b) $\dfrac{d}{dx}\left[\dfrac{2x}{\left(x^2+1\right)^2}+C\right] = \dfrac{\left(x^2+1\right)^2(2)-(2x)(2)\left(x^2+1\right)(2x)}{\left(x^2+1\right)^4} = \dfrac{2\left(1-3x^2\right)}{\left(x^2+1\right)^3}$

(c) $\dfrac{d}{dx}\left[\arctan x + C\right] = \dfrac{1}{1+x^2}$

(d) $\dfrac{d}{dx}\left[\ln\left(x^2+1\right)+C\right] = \dfrac{2x}{x^2+1}$

$\displaystyle\int \frac{1}{x^2+1}\,dx$ matches (c).

4. (a) $\frac{d}{dx}\left[2x\sin\left(x^2+1\right)+C\right)\right] = 2x\left[\cos\left(x^2+1\right)(2x)\right] + 2\sin\left(x^2+1\right) = 2\left[2x^2\cos\left(x^2+1\right)+\sin\left(x^2+1\right)\right]$

(b) $\frac{d}{dx}\left[-\frac{1}{2}\sin\left(x^2+1\right)+C\right] = -\frac{1}{2}\cos\left(x^2+1\right)(2x) = -x\cos\left(x^2+1\right)$

(c) $\frac{d}{dx}\left[\frac{1}{2}\sin\left(x^2+1\right)+C\right] = \frac{1}{2}\cos\left(x^2+1\right)(2x) = x\cos\left(x^2+1\right)$

(d) $\frac{d}{dx}\left[-2x\sin\left(x^2+1\right)+C\right] = -2x\left[\cos\left(x^2+1\right)(2x)\right] - 2\sin\left(x^2+1\right) = -2\left[2x^2\cos\left(x^2+1\right)+\sin\left(x^2+1\right)\right]$

$\int x\cos\left(x^2+1\right)dx$ matches (c).

5. $\int(5x-3)^4\,dx$

$u = 5x-3,\ du = 5\,dx,\ n = 4$

Use $\int u^n\,du$.

6. $\int\frac{2t+1}{t^2+t-4}\,dt$

$u = t^2+t-4,\ du = (2t+1)\,dt$

Use $\int\frac{du}{u}$.

7. $\int\frac{1}{\sqrt{x}\left(1-2\sqrt{x}\right)}\,dx$

$u = 1-2\sqrt{x},\ du = -\frac{1}{\sqrt{x}}dx$

Use $\int\frac{du}{u}$.

8. $\int\frac{2}{(2t-1)^2+4}\,dt$

$u = 2t-1,\ du = 2\,dt,\ a = 2$

Use $\int\frac{du}{u^2+a^2}$.

9. $\int\frac{3}{\sqrt{1-t^2}}\,dt$

$u = t,\ du = dt,\ a = 1$

Use $\int\frac{du}{\sqrt{a^2-u^2}}$.

10. $\int\frac{-2x}{\sqrt{x^2-4}}\,dx$

$u = x^2-4,\ du = 2x\,dx,\ n = -\frac{1}{2}$

Use $\int u^n\,du$.

11. $\int t\sin t^2\,dt$

$u = t^2,\ du = 2t\,dt$

Use $\int\sin u\,du$.

12. $\int\sec 5x\tan 5x\,dx$

$u = 5x,\ du = 5\,dx$

Use $\int\sec u\tan u\,du$.

13. $\int(\cos x)e^{\sin x}\,dx$

$u = \sin x,\ du = \cos x\,dx$

Use $\int e^u\,du$.

14. $\int\frac{1}{x\sqrt{x^2-4}}\,dx$

$u = x,\ du = dx,\ a = 2$

Use $\int\frac{du}{u\sqrt{u^2-a^2}}$.

15. Let $u = x-5,\ du = dx$.

$\int 14(x-5)^6\,dx = 14\int(x-5)^6\,dx = 2(x-5)^7 + C$

16. Let $u = t+6,\ du = dt$.

$$\int\frac{5}{(t+6)^3}\,dt = 5\int(t+6)^{-3}\,dt$$
$$= 5\cdot\frac{(t+6)^{-2}}{-2} + C$$
$$= \frac{-5}{2(t+6)^2} + C$$

17. Let $u = z-10,\ du = dz$.

$\int\frac{7}{(z-10)^7}\,dz = 7\int(z-10)^{-7}\,dz = -\frac{7}{6(z-10)^6} + C$

18. Let $u = t^4 + 1$, $du = 4t^3\, dt$.

$$\int t^3\sqrt{t^4 + 1}\, dt = \frac{1}{4}\int (t^4 + 1)^{1/2}(4t^3)\, dt$$
$$= \frac{1}{4} \cdot \frac{(t^4 + 1)^{3/2}}{(3/2)} + C$$
$$= \frac{1}{6}(t^4 + 1)^{3/2} + C$$

19. $$\int \left[v + \frac{1}{(3v - 1)^3}\right] dv = \int v\, dv + \frac{1}{3}\int (3v - 1)^{-3}(3)\, dv$$
$$= \frac{1}{2}v^2 - \frac{1}{6(3v - 1)^2} + C$$

20. $$\int \left[4x - \frac{2}{(2x + 3)^2}\right] dx = \int 4x\, dx - \int 2(2x + 3)^{-2}\, dx$$
$$= 2x^2 - \frac{(2x + 3)^{-1}}{-1} + C$$
$$= 2x^2 + \frac{1}{2x + 3} + C$$

21. Let $u = -t^3 + 9t + 1$,

$du = (-3t^2 + 9)\, dt = -3(t^2 - 3)\, dt$.

$$\int \frac{t^2 - 3}{-t^3 + 9t + 1}\, dt = -\frac{1}{3}\int \frac{-3(t^2 - 3)}{-t^3 + 9t + 1}\, dt$$
$$= -\frac{1}{3}\ln\left|-t^3 + 9t + 1\right| + C$$

22. Let $u = 3x^2 + 6x$, $du = (6x + 6)\, dx = 6(x + 1)\, dx$.

$$\int \frac{x + 1}{\sqrt{3x^2 + 6x}}\, dx = \frac{1}{6}\int (3x^2 + 6x)^{-1/2} 6(x + 1)\, dx$$
$$= \frac{1}{6} \cdot \frac{(3x^2 + 6x)^{1/2}}{(1/2)} + C$$
$$= \frac{1}{3}\sqrt{3x^2 + 6x} + C$$

23. $$\int \frac{x^2}{x - 1}\, dx = \int (x + 1)\, dx + \int \frac{1}{x - 1}\, dx$$
$$= \frac{1}{2}x^2 + x + \ln|x - 1| + C$$

24. $$\int \frac{3x}{x + 4}\, dx = \int \left(3 - \frac{12}{x + 4}\right) dx$$
$$= 3x - 12\ln|x + 4| + C$$

25. Let $u = 1 + e^x$, $du = e^x\, dx$.

$$\int \frac{e^x}{1 + e^x}\, dx = \ln(1 + e^x) + C$$

26. $$\int \left(\frac{1}{2x + 5} - \frac{1}{2x - 5}\right) dx = \frac{1}{2}\int \frac{1}{2x + 5}(2)\, dx - \frac{1}{2}\int \frac{1}{2x - 5}(2)\, dx$$
$$= \frac{1}{2}\ln|2x + 5| - \frac{1}{2}\ln|2x - 5| + C$$
$$= \frac{1}{2}\ln\left|\frac{2x + 5}{2x - 5}\right| + C$$

27. $\int(5+4x^2)^2\,dx = \int(25+40x^2+16x^4)\,dx$

$= 25x + \frac{40}{3}x^3 + \frac{16}{5}x^5 + C$

$= \frac{x}{15}(48x^5 + 200x^3 + 375) + C$

28. $\int x\left(3+\frac{2}{x}\right)^2 dx = \int\left(9x + 12 + \frac{4}{x}\right)dx$

$= \frac{9}{2}x^2 + 12x + 4\ln|x| + C$

29. Let $u = 2\pi x^2$, $du = 4\pi x\,dx$.

$\int x(\cos 2\pi x^2)\,dx = \frac{1}{4\pi}\int(\cos 2\pi x^2)(4\pi x)\,dx$

$= \frac{1}{4\pi}\sin 2\pi x^2 + C$

30. Let $u = \pi x$, $du = \pi\,dx$.

$\int \csc \pi x \cot \pi x\,dx = \frac{1}{\pi}\int(\csc \pi x)(\cot \pi x)\,\pi\,dx$

$= -\frac{1}{\pi}\csc \pi x + C$

31. Let $u = \cos x$, $du = -\sin x\,dx$.

$\int \frac{\sin x}{\sqrt{\cos x}}\,dx = -\int(\cos x)^{-1/2}(-\sin x)\,dx$

$= -2\sqrt{\cos x} + C$

32. Let $u = \cot x$, $du = -\csc^2 x\,dx$.

$\int \csc^2 x e^{\cot x}\,dx = -\int e^{\cot x}(-\csc^2 x)\,dx = -e^{\cot x} + C$

33. Let $u = 1 + e^x$, $du = e^x\,dx$.

$\int \frac{2}{e^{-x}+1}\,dx = 2\int\left(\frac{2}{e^{-x}+1}\right)\left(\frac{e^x}{e^x}\right)dx$

$= 2\int \frac{e^x}{1+e^x}\,dx = 2\ln(1+e^x) + C$

34. $\int \frac{2}{7e^x+4}\,dx = 2\int \frac{1}{7e^x+4}\frac{e^{-x}}{e^{-x}}\,dx = 2\int \frac{e^{-x}}{7+4e^{-x}}\,dx$

$= 2\left(-\frac{1}{4}\right)\int \frac{1}{(7+4e^{-x})}(-4e^{-x})\,dx$

$= -\frac{1}{2}\ln|7+4e^{-x}| + C$

35. $\int \frac{\ln x^2}{x}\,dx = 2\int(\ln x)\frac{1}{x}\,dx$

$= 2\frac{(\ln x)^2}{2} + C = (\ln x)^2 + C$

36. Let $u = \ln(\cos x)$, $du = \frac{-\sin x}{\cos x}\,dx = -\tan x\,dx$.

$\int(\tan x)(\ln \cos x)\,dx = -\int(\ln \cos x)(-\tan x)\,dx$

$= \frac{-[\ln(\cos x)]^2}{2} + C$

37. $\int \frac{1+\cos\alpha}{\sin\alpha}\,d\alpha = \int \csc\alpha\,d\alpha + \int \cot\alpha\,d\alpha$

$= -\ln|\csc\alpha + \cot\alpha| + \ln|\sin\alpha| + C$

38. $\frac{1}{\cos\theta - 1} = \frac{1}{\cos\theta - 1}\cdot\frac{\cos\theta+1}{\cos\theta+1} = \frac{\cos\theta+1}{\cos^2\theta - 1}$

$= \frac{\cos\theta+1}{-\sin^2\theta} = -\csc\theta\cdot\cot\theta - \csc^2\theta$

$\int \frac{1}{\cos\theta - 1}\,d\theta = \int(-\csc\theta\cot\theta - \csc^2\theta)\,d\theta$

$= \csc\theta + \cot\theta + C$

$= \frac{1}{\sin\theta} + \frac{\cos\theta}{\sin\theta} + C$

$= \frac{1+\cos\theta}{\sin\theta} + C$

39. Let $u = 4t + 1$, $du = 4\,dt$.

$\int \frac{-1}{\sqrt{1-(4t+1)^2}}\,dt = -\frac{1}{4}\int \frac{4}{\sqrt{1-(4t+1)^2}}\,dt$

$= -\frac{1}{4}\arcsin(4t+1) + C$

40. Let $u = 2x$, $du = 2dx$, $a = 5$.

$$\int \frac{1}{25 + 4x^2}\,dx = \frac{1}{2}\int \frac{1}{5^2 + (2x)^2}(2)\,dx$$
$$= \frac{1}{10}\arctan\frac{2x}{5} + C$$

41. Let $u = \cos\left(\frac{2}{t}\right)$, $du = \frac{2\sin(2/t)}{t^2}\,dt$.

$$\int \frac{\tan(2/t)}{t^2}\,dt = \frac{1}{2}\int \frac{1}{\cos(2/t)}\left[\frac{2\sin(2/t)}{t^2}\right]dt$$
$$= \frac{1}{2}\ln\left|\cos\left(\frac{2}{t}\right)\right| + C$$

42. Let $u = \frac{1}{t}$, $du = \frac{-1}{t^2}\,dt$.

$$\int \frac{e^{1/t}}{t^2}\,dt = -\int e^{1/t}\left(\frac{-1}{t^2}\right)dt = -e^{1/t} + C$$

43. Note: $10x - x^2 = 25 - \left(25 - 10x + x^2\right)$
$$= 25 - (5 - x)^2$$

$$\int \frac{6}{\sqrt{10x - x^2}}\,dx = 6\int \frac{1}{\sqrt{25 - (5 - x)^2}}\,dx$$
$$= -6\int \frac{-1}{\sqrt{5^2 - (5 - x)^2}}\,dx$$
$$= -6\arcsin\frac{(5 - x)}{5} + C$$
$$= 6\arcsin\left(\frac{x - 5}{5}\right) + C$$

44. $$\int \frac{1}{(x - 1)\sqrt{4x^2 - 8x + 3}}\,dx = \int \frac{2}{[2(x - 1)]\sqrt{[2(x - 1)]^2 - 1}}\,dx = \operatorname{arcsec}\left|2(x - 1)\right| + C$$

45. $$\int \frac{4}{4x^2 + 4x + 65}\,dx = \int \frac{1}{\left[x + (1/2)\right]^2 + 16}\,dx$$
$$= \frac{1}{4}\arctan\left[\frac{x + (1/2)}{4}\right] + C$$
$$= \frac{1}{4}\arctan\left(\frac{2x + 1}{8}\right) + C$$

46. $$\int \frac{1}{x^2 - 4x + 9}\,dx = \int \frac{1}{x^2 - 4x + 4 + 5}\,dx$$
$$= \int \frac{1}{(x - 2)^2 + \left(\sqrt{5}\right)^2}\,dx$$
$$= \frac{1}{\sqrt{5}}\arctan\left(\frac{x - 2}{\sqrt{5}}\right) + C$$
$$= \frac{\sqrt{5}}{5}\arctan\left(\frac{\sqrt{5}}{5}(x - 2)\right) + C$$

47. $\frac{ds}{dt} = \frac{t}{\sqrt{1 - t^4}}$, $\left(0, -\frac{1}{2}\right)$

(a)

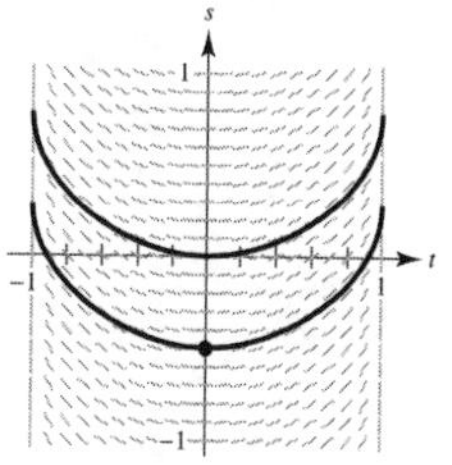

(b) $u = t^2$, $du = 2t\,dt$

$$\int \frac{t}{\sqrt{1 - t^4}}\,dt = \frac{1}{2}\int \frac{2t}{\sqrt{1 - (t^2)^2}}\,dt$$
$$= \frac{1}{2}\arcsin t^2 + C$$

$\left(0, -\frac{1}{2}\right)$: $-\frac{1}{2} = \frac{1}{2}\arcsin 0 + C \Rightarrow C = -\frac{1}{2}$

$$s = \frac{1}{2}\arcsin t^2 - \frac{1}{2}$$

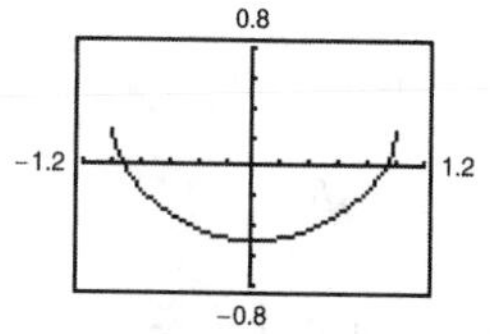

48. (a) $\dfrac{dy}{dx} = \dfrac{1}{\sqrt{4x - x^2}}, \quad \left(2, \dfrac{1}{2}\right)$

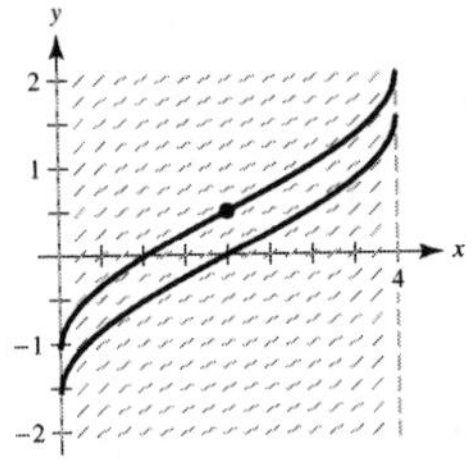

(b) $$y = \int \frac{1}{\sqrt{4x - x^2}}\,dx$$

$$= \int \frac{1}{\sqrt{4 - \left(x^2 - 4x + 4\right)}}\,dx$$

$$= \int \frac{1}{\sqrt{4 - (x - 2)^2}}\,dx = \arcsin\left(\frac{x - 2}{2}\right) + C$$

$$\left(2, \frac{1}{2}\right): \frac{1}{2} = \arcsin(0) + C \Rightarrow C = \frac{1}{2}$$

$$y = \arcsin\left(\frac{x - 2}{2}\right) + \frac{1}{2}$$

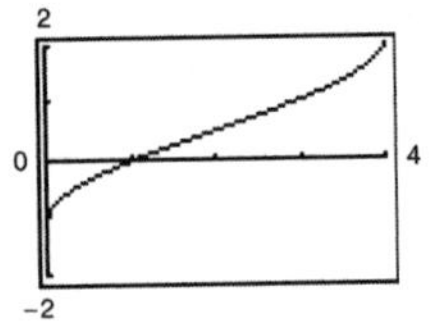

49.

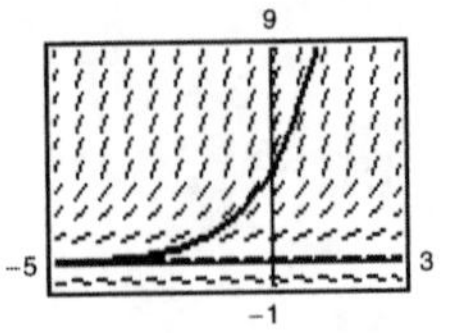

$y = 4e^{0.8x}$

50.

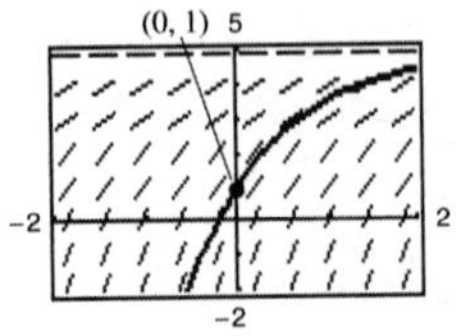

$y = 5 - 4e^{-x}$

51. $$\frac{dy}{dx} = \left(e^x + 5\right)^2 = e^{2x} + 10e^x + 25$$

$$y = \int\left(e^{2x} + 10e^x + 25\right)dx$$

$$= \frac{1}{2}e^{2x} + 10e^x + 25x + C$$

52. $$\frac{dy}{dx} = \left(4 - e^{2x}\right)^2 = 16 - 8e^{2x} + e^{4x}$$

$$y = \int\left(16 - 8e^{2x} + e^{4x}\right)dx$$

$$= 16x - 4e^{2x} + \frac{1}{4}e^{4x} + C$$

53. $$\frac{dr}{dt} = \frac{10e^t}{\sqrt{1 - e^{2t}}}$$

$$r = \int \frac{10e^t}{\sqrt{1 - \left(e^t\right)^2}}\,dt$$

$$= 10 \arcsin\left(e^t\right) + C$$

54. $$\frac{dr}{dt} = \frac{\left(1 + e^t\right)^2}{e^{3t}} = \frac{1 + 2e^t + e^{2t}}{e^{3t}} = e^{-3t} + 2e^{-2t} + e^{-t}$$

$$r = \int\left(e^{-3t} + 2e^{-2t} + e^{-t}\right)dt$$

$$= -\frac{1}{3}e^{-3t} - e^{-2t} - e^{-t} + C$$

55. $$\frac{dy}{dx} = \frac{\sec^2 x}{4 + \tan^2 x}$$

Let $u = \tan x$, $du = \sec^2 x\,dx$.

$$y = \int \frac{\sec^2 x}{4 + \tan^2 x}\,dx = \frac{1}{2}\arctan\left(\frac{\tan x}{2}\right) + C$$

56. $$y' = \frac{1}{x\sqrt{4x^2 - 9}}$$

Let $u = 2x$, $du = 2dx$, $a = 3$.

$$y = \int \frac{1}{x\sqrt{4x^2 - 9}}\,dx = \int \frac{1}{(2x)\sqrt{(2x)^2 - 3^2}}(2)\,dx$$

$$= \frac{1}{3}\operatorname{arcsec}\frac{|2x|}{3} + C$$

57. Let $u = 2x$, $du = 2\,dx$.

$$\int_0^{\pi/4} \cos 2x\,dx = \frac{1}{2}\int_0^{\pi/4} \cos 2x(2)\,dx$$
$$= \left[\frac{1}{2}\sin 2x\right]_0^{\pi/4} = \frac{1}{2}$$

58. Let $u = \sin t$, $du = \cos t\,dt$.

$$\int_0^{\pi} \sin^2 t\cos t\,dt = \left[\frac{1}{3}\sin^3 t\right]_0^{\pi} = 0$$

59. Let $u = -x^2$, $du = -2x\,dx$.

$$\int_0^1 xe^{-x^2}\,dx = -\frac{1}{2}\int_0^1 e^{-x^2}(-2x)\,dx = \left[-\frac{1}{2}e^{-x^2}\right]_0^1$$
$$= \frac{1}{2}\left(1 - e^{-1}\right) \approx 0.316$$

60. Let $u = 1 - \ln x$, $du = \dfrac{-1}{x}\,dx$.

$$\int_1^e \frac{1 - \ln x}{x}\,dx = -\int_1^e (1 - \ln x)\left(\frac{-1}{x}\right)dx$$
$$= \left[-\frac{1}{2}(1 - \ln x)^2\right]_1^e = \frac{1}{2}$$

61. Let $u = x^2 + 36$, $du = 2x\,dx$.

$$\int_0^8 \frac{2x}{\sqrt{x^2 + 36}}\,dx = \int_0^8 \left(x^2 + 36\right)^{-1/2}(2x)\,dx$$
$$= 2\left[\left(x^2 + 36\right)^{1/2}\right]_0^8 = 8$$

62. $$\int_1^3 \frac{2x^2 + 3x - 2}{x}\,dx = \int_1^3 \left(2x + 3 - \frac{2}{x}\right)dx$$
$$= \left[x^2 + 3x - 2\ln|x|\right]_1^3$$
$$= (9 + 9 - 2\ln 3) - (1 + 3 - 0)$$
$$= 14 - 2\ln 3$$

63. Let $u = 3x$, $du = 3\,dx$.

$$\int_0^{2/\sqrt{3}} \frac{1}{4 + 9x^2}\,dx = \frac{1}{3}\int_0^{2/\sqrt{3}} \frac{3}{4 + (3x)^2}\,dx$$
$$= \left[\frac{1}{6}\arctan\left(\frac{3x}{2}\right)\right]_0^{2/\sqrt{3}}$$
$$= \frac{\pi}{18} \approx 0.175$$

64. $$\int_0^7 \frac{1}{\sqrt{100 - x^2}}\,dx = \left[\arcsin\left(\frac{x}{10}\right)\right]_0^7 = \arcsin\left(\frac{7}{10}\right)$$

65. $$A = \int_0^{3/2} (-4x + 6)^{3/2}\,dx$$
$$= -\frac{1}{4}\int_0^{3/2} (6 - 4x)^{3/2}(-4)\,dx$$
$$= -\frac{1}{4}\left[\frac{2}{5}(6 - 4x)^{5/2}\right]_0^{3/2}$$
$$= -\frac{1}{10}\left(0 - 6^{5/2}\right)$$
$$= \frac{18}{5}\sqrt{6} \approx 8.8182$$

66. $$A = \int_0^5 \frac{3x + 2}{x^2 + 9}\,dx$$
$$= \int_0^5 \frac{3x}{x^2 + 9}\,dx + \int_0^5 \frac{2}{x^2 + 9}\,dx$$
$$= \left[\frac{3}{2}\ln\left|x^2 + 9\right| + \frac{2}{3}\arctan\left(\frac{x}{3}\right)\right]_0^5$$
$$= \frac{3}{2}\ln(34) + \frac{2}{3}\arctan\left(\frac{5}{3}\right) - \frac{3}{2}\ln 9$$
$$= \frac{3}{2}\ln\left(\frac{34}{9}\right) + \frac{2}{3}\arctan\left(\frac{5}{3}\right)$$
$$\approx 2.6806$$

67. $$y^2 = x^2\left(1 - x^2\right)$$
$$y = \pm\sqrt{x^2\left(1 - x^2\right)}$$
$$A = 4\int_0^1 x\sqrt{1 - x^2}\,dx$$
$$= -2\int_0^1 \left(1 - x^2\right)^{1/2}(-2x)\,dx$$
$$= -\frac{4}{3}\left[(1 - x)^{3/2}\right]_0^1$$
$$= -\frac{4}{3}(0 - 1) = \frac{4}{3}$$

68. $$A = \int_0^{\pi/2} \sin 2x\,dx = -\frac{1}{2}\left[\cos 2x\right]_0^{\pi/2} = -\frac{1}{2}(-1 - 1) = 1$$

69. $\displaystyle\int \frac{1}{x^2 + 4x + 13}\,dx = \frac{1}{3}\arctan\left(\frac{x + 2}{3}\right) + C$

The antiderivatives are vertical translations of each other.

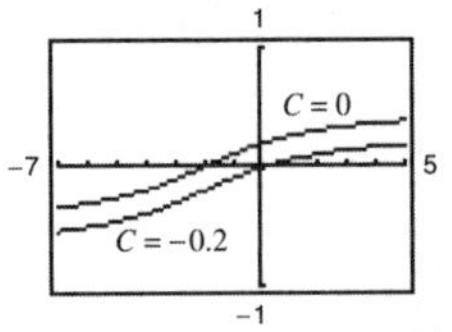

70. $\displaystyle\int \frac{x - 2}{x^2 + 4x + 13}\,dx = \frac{1}{2}\ln\left(x^2 + 4x + 13\right) - \frac{4}{3}\arctan\left(\frac{x + 2}{3}\right) + C$

The antiderivatives are vertical translations of each other.

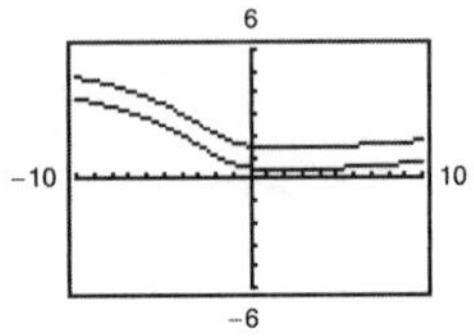

71. $\displaystyle\int \frac{1}{1 + \sin\theta}\,d\theta = \tan\theta - \sec\theta + C \quad \left(\text{or } \frac{-2}{1 + \tan(\theta/2)}\right)$

The antiderivatives are vertical translations of each other.

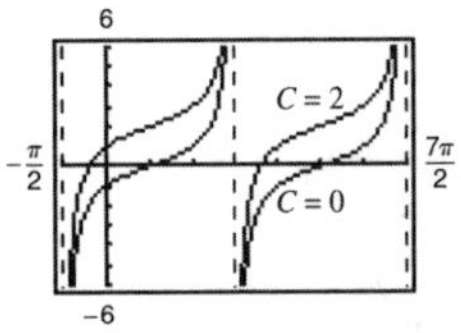

72. $\displaystyle\int \left(\frac{e^x + e^{-x}}{2}\right)^3 dx = \frac{1}{24}\left[e^{3x} + 9e^x - 9e^{-x} - e^{-3x}\right] + C$

The antiderivatives are vertical translations of each other.

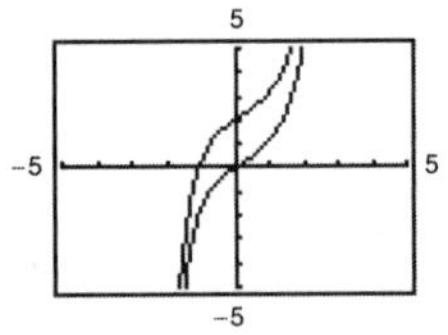

73. Power Rule: $\displaystyle\int u^n\,du = \frac{u^{n+1}}{n + 1} + C, \quad n \neq -1$

$u = x^2 + 1, n = 3$

74. $\displaystyle\int \sec u \tan u\,du = \sec u + C$

75. Log Rule: $\displaystyle\int \frac{du}{u} = \ln|u| + C, \quad u = x^2 + 1$

76. Arctan Rule: $\displaystyle\int \frac{du}{a^2 + u^2} = \frac{1}{a}\arctan\left(\frac{u}{a}\right) + C$

77. $\sin x + \cos x = a\sin(x + b)$

$\sin x + \cos x = a\sin x\cos b + a\cos x\sin b$

$\sin x + \cos x = (a\cos b)\sin x + (a\sin b)\cos x$

Equate coefficients of like terms to obtain the following.

$1 = a\cos b$ and $1 = a\sin b$

So, $a = 1/\cos b$. Now, substitute for a in $1 = a\sin b$.

$$1 = \left(\frac{1}{\cos b}\right)\sin b$$

$$1 = \tan b \Rightarrow b = \frac{\pi}{4}$$

Because $b = \dfrac{\pi}{4}$, $a = \dfrac{1}{\cos(\pi/4)} = \sqrt{2}$. So,

$$\sin x + \cos x = \sqrt{2}\sin\left(x + \frac{\pi}{4}\right).$$

$$\int\frac{dx}{\sin x + \cos x} = \int\frac{dx}{\sqrt{2}\sin(x + (\pi/4))}$$
$$= \frac{1}{\sqrt{2}}\int\csc\left(x + \frac{\pi}{4}\right)dx$$
$$= -\frac{1}{\sqrt{2}}\ln\left|\csc\left(x + \frac{\pi}{4}\right) + \cot\left(x + \frac{\pi}{4}\right)\right| + C$$

78. $$\frac{\sin x}{\cos x} + \frac{\cos x}{1 + \sin x} = \frac{\sin x(1 + \sin x) + \cos^2 x}{\cos x(1 + \sin x)}$$
$$= \frac{\sin x + \sin^2 x + \cos^2 x}{\cos x(1 + \sin x)}$$
$$= \frac{\sin x + 1}{\cos x(1 + \sin x)}$$
$$= \frac{1}{\cos x} = \sec x$$

So,

$$\int\sec x\,dx = \int\left[\frac{\sin x}{\cos x} + \frac{\cos x}{1 + \sin x}\right]dx$$
$$= -\ln|\cos x| + \ln|1 + \sin x| + C$$
$$= \ln\left|\frac{1 + \sin x}{\cos x}\right| + C$$
$$= \ln|\sec x + \tan x| + C$$

79. $$\int_0^{1/a}\left(x - ax^2\right)dx = \left[\frac{1}{2}x^2 - \frac{a}{3}x^3\right]_0^{1/a} = \frac{1}{6a^2}$$

Let $\dfrac{1}{6a^2} = \dfrac{2}{3}$, $12a^2 = 3$, $a = \dfrac{1}{2}$.

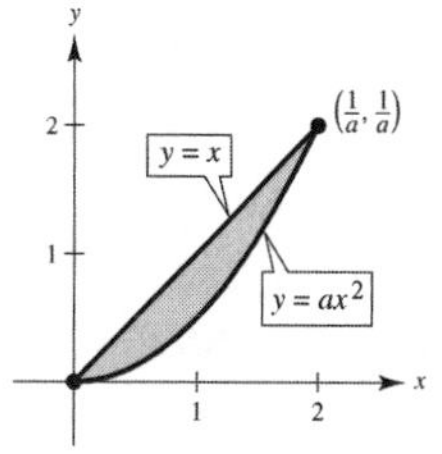

80. No. When $u = x^2$, it does not follow that $x = \sqrt{u}$ because x is negative on $[-1, 0)$.

81. (a) They are equivalent because

$e^{x+C_1} = e^x \cdot e^{C_1} = Ce^x$, $C = e^{C_1}$.

(b) They differ by a constant.

$\sec^2 x + C_1 = \left(\tan^2 x + 1\right) + C_1 = \tan^2 x + C$

82. $\int_0^5 f(x)\,dx < 0$ because there is more area below the x-axis than above.

83. $$\int_0^2 \frac{4x}{x^2 + 1}\,dx \approx 3$$

Matches (a).

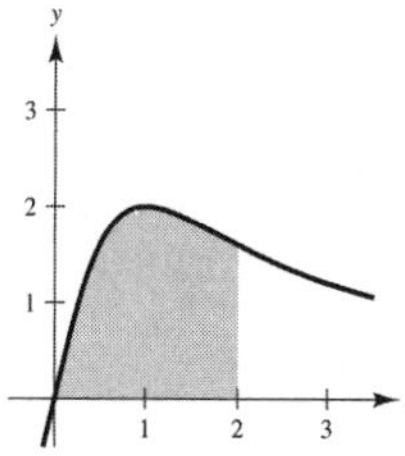

84. $$\int_0^2 \frac{4}{x^2 + 1}\,dx \approx 4$$

Matches (d).

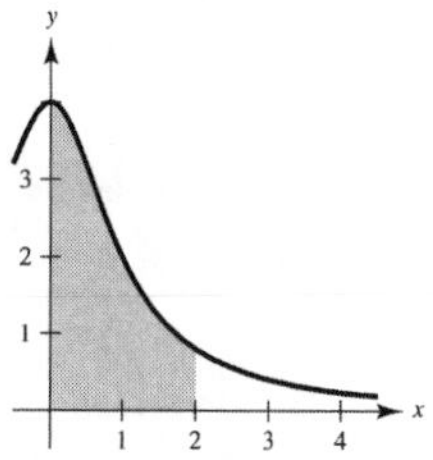

85. (a) $y = 2\pi x^2, \quad 0 \le x \le 2$

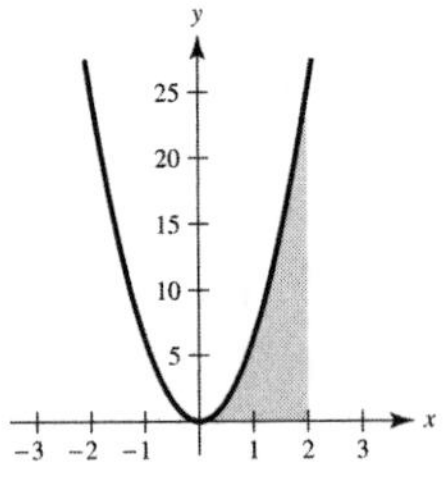

(b) $y = \sqrt{2}x, \quad 0 \le x \le 2$

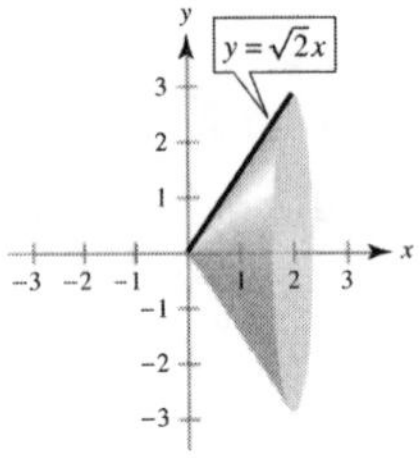

(c) $y = x, \quad 0 \le x \le 2$

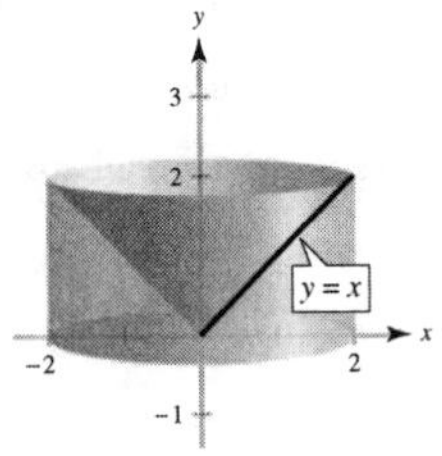

86. (a) $x = \pi y, \quad 0 \le y \le 4$

$y = \dfrac{1}{\pi}x, \quad 0 \le x \le 4\pi$

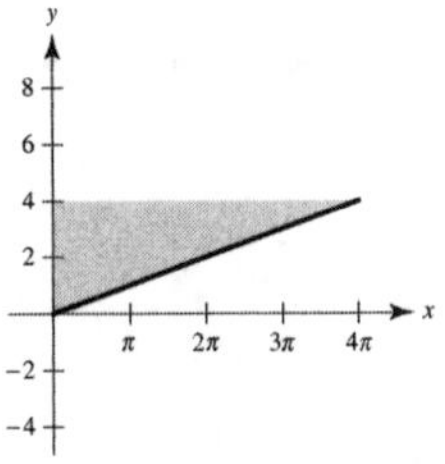

(b) $x = \sqrt{y}, \quad 0 \le y \le 4$

$y = x^2, \quad 0 \le x \le 2$

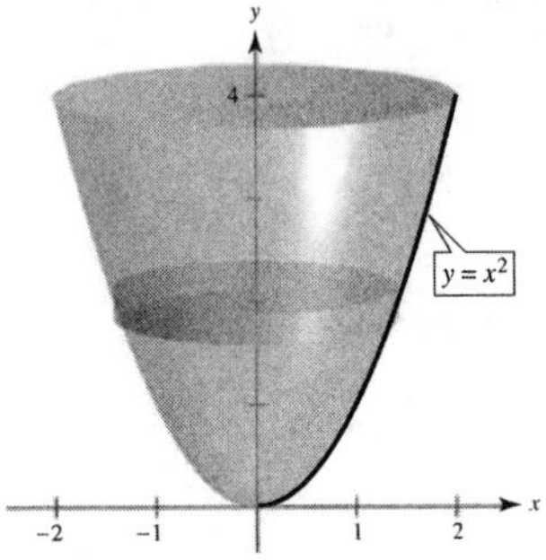

(c) $x = \dfrac{1}{2}, \quad 0 \le y \le 4$

$2\pi \int_0^4 y\left(\dfrac{1}{2}\right) dy$

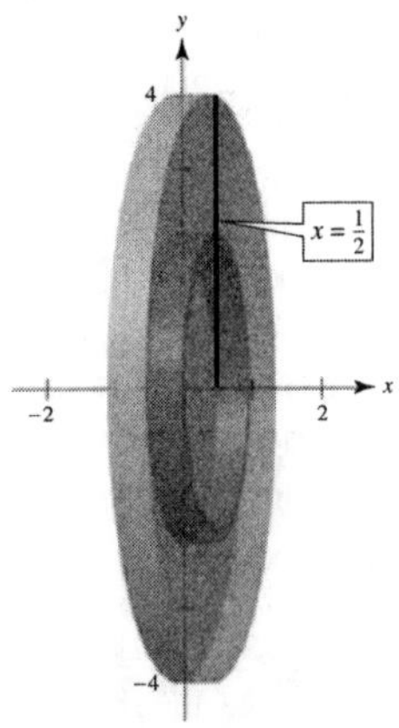

87. (a) **Shell Method:**

Let $u = -x^2, du = -2x\,dx$.

$$V = 2\pi \int_0^1 xe^{-x^2}\,dx$$
$$= -\pi \int_0^1 e^{-x^2}(-2x)\,dx$$
$$= \left[-\pi e^{-x^2}\right]_0^1$$
$$= \pi\left(1 - e^{-1}\right) \approx 1.986$$

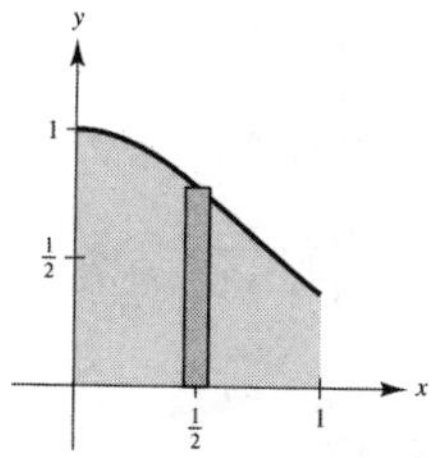

(b) **Shell Method:**

$$V = 2\pi \int_0^b xe^{-x^2}\,dx$$
$$= \left[-\pi e^{-x^2}\right]_0^b$$
$$= \pi\left(1 - e^{-b^2}\right) = \frac{4}{3}$$
$$e^{-b^2} = \frac{3\pi - 4}{3\pi}$$
$$b = \sqrt{\ln\left(\frac{3\pi}{3\pi - 4}\right)} \approx 0.743$$

88.

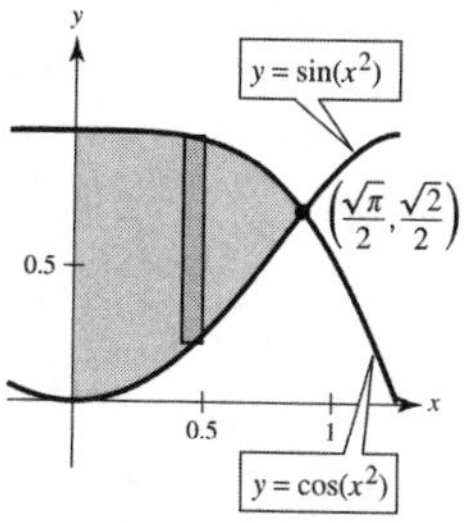

Shell Method:

$$V = 2\pi \int_0^{\sqrt{\pi}/2} x\left(\cos\left(x^2\right) - \sin\left(x^2\right)\right)dx$$
$$= \pi\left[\sin\left(x^2\right) + \cos\left(x^2\right)\right]_0^{\sqrt{\pi}/2}$$
$$= \pi\left[\left(\frac{\sqrt{2}}{2} + \frac{\sqrt{2}}{2}\right) - (0 + 1)\right]$$
$$= \pi\left(\sqrt{2} - 1\right)$$

89. $y = f(x) = \ln(\sin x)$

$$f'(x) = \frac{\cos x}{\sin x}$$
$$s = \int_{\pi/4}^{\pi/2} \sqrt{1 + \frac{\cos^2 x}{\sin^2 x}}\,dx = \int_{\pi/4}^{\pi/2} \sqrt{\frac{\sin^2 x + \cos^2 x}{\sin^2 x}}\,dx$$
$$= \int_{\pi/4}^{\pi/2} \frac{1}{\sin x}\,dx = \int_{\pi/4}^{\pi/2} \csc x\,dx$$
$$= \left[-\ln\left|\csc x + \cot x\right|\right]_{\pi/4}^{\pi/2}$$
$$= -\ln(1) + \ln\left(\sqrt{2} + 1\right)$$
$$= \ln\left(\sqrt{2} + 1\right) \approx 0.8814$$

90. $y = \ln(\cos x), \quad 0 \le x \le \pi/3$

$$y' = \frac{-\sin x}{\cos x} = -\tan x$$
$$1 + (y')^2 = 1 + \tan^2 x = \sec^2 x$$
$$s = \int_0^{\pi/3} \sqrt{1 + (y')^2}\,dx = \int_0^{\pi/3} \sec x\,dx$$
$$= \left[\ln\left|\sec x + \tan x\right|\right]_0^{\pi/3}$$
$$= \ln\left(2 + \sqrt{3}\right) - \ln(1) = \ln\left(2 + \sqrt{3}\right) \approx 1.317$$

91. $y = 2\sqrt{x}$

$$y' = \frac{1}{\sqrt{x}}$$
$$1 + (y')^2 = 1 + \frac{1}{x} = \frac{x + 1}{x}$$
$$S = 2\pi \int_0^9 2\sqrt{x}\sqrt{\frac{x + 1}{x}}\,dx$$
$$= 2\pi \int_0^9 2\sqrt{x + 1}\,dx$$
$$= \left[4\pi\left(\frac{2}{3}\right)(x + 1)^{3/2}\right]_0^9 = \frac{8\pi}{3}\left(10\sqrt{10} - 1\right) \approx 256.545$$

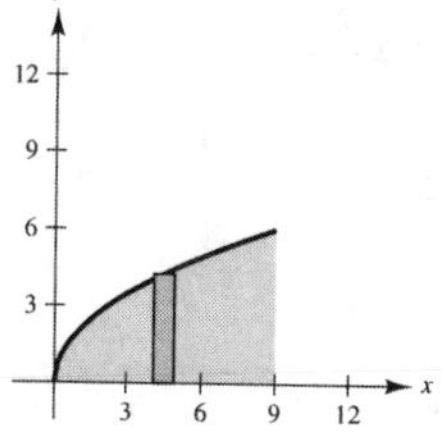

92. $A = \int_0^4 \frac{5}{\sqrt{25 - x^2}}\,dx = \left[5 \arcsin \frac{x}{5}\right]_0^4 = 5 \arcsin \frac{4}{5}$

$$\bar{x} = \frac{1}{A}\int_0^4 x\left(\frac{5}{\sqrt{25 - x^2}}\right)dx$$

$$= \frac{1}{5 \arcsin(4/5)}\left(-\frac{5}{2}\right)\int_0^4 \left(25 - x^2\right)^{-1/2}(-2x)\,dx$$

$$= \frac{1}{5 \arcsin(4/5)}(-5)\left[\left(25 - x^2\right)^{1/2}\right]_0^4$$

$$= -\frac{1}{\arcsin(4/5)}[3 - 5] = \frac{2}{\arcsin(4/5)} \approx 2.157$$

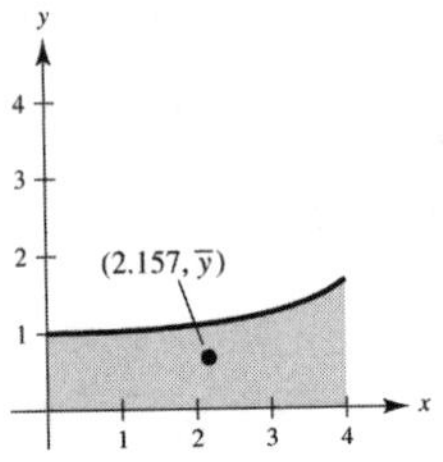

93. Average value $= \frac{1}{b - a}\int_a^b f(x)\,dx$

$$= \frac{1}{3 - (-3)}\int_{-3}^{3} \frac{1}{1 + x^2}\,dx$$

$$= \frac{1}{6}\left[\arctan(x)\right]_{-3}^{3}$$

$$= \frac{1}{6}\left[\arctan(3) - \arctan(-3)\right]$$

$$= \frac{1}{3}\arctan(3) \approx 0.4163$$

94. Average value $= \frac{1}{b - a}\int_a^b f(x)\,dx$

$$= \frac{1}{(\pi/n) - 0}\int_0^{\pi/n} \sin(nx)\,dx$$

$$= \frac{n}{\pi}\left[\frac{-1}{n}\cos(nx)\right]_0^{\pi/n}$$

$$= -\frac{1}{\pi}\left[\cos(\pi) - \cos(0)\right] = \frac{2}{\pi}$$

95.

$$y = \tan(\pi x)$$

$$y' = \pi \sec^2(\pi x)$$

$$1 + (y')^2 = 1 + \pi^2 \sec^4(\pi x)$$

$$s = \int_0^{1/4} \sqrt{1 + \pi^2 \sec^4(\pi x)}\,dx \approx 1.0320$$

96.

$$y = x^{2/3}$$

$$y' = \frac{2}{3x^{1/3}}$$

$$1 + (y')^2 = 1 + \frac{4}{9x^{2/3}}$$

$$s = \int_1^8 \sqrt{1 + \frac{4}{9x^{2/3}}}\,dx \approx 7.6337$$

97. (a) $\int \cos^3 x\,dx = \int\left(1 - \sin^2 x\right)\cos x\,dx = \sin x - \frac{\sin^3 x}{3} + C = \frac{1}{3}\sin x\left(\cos^2 x + 2\right) + C$

(b) $\int \cos^5 x\,dx = \int\left(1 - \sin^2 x\right)^2 \cos x\,dx = \int\left(1 - 2\sin^2 x + \sin^4 x\right)\cos x\,dx$

$$= \sin x - \frac{2}{3}\sin^3 x + \frac{\sin^5 x}{5} + C = \frac{1}{15}\sin x\left(3\cos^4 x + 4\cos^2 x + 8\right) + C$$

(c) $\int \cos^7 x\,dx = \int\left(1 - \sin^2 x\right)^3 \cos x\,dx$

$$= \int\left(1 - 3\sin^2 x + 3\sin^4 x - \sin^6 x\right)\cos x\,dx$$

$$= \sin x - \sin^3 x + \frac{3}{5}\sin^5 x - \frac{1}{7}\sin^7 x + C$$

$$= \frac{1}{35}\sin x\left(5\cos^6 x + 6\cos^4 x + 8\cos^2 x + 16\right) + C$$

(d) $\int \cos^{15} x\,dx = \int\left(1 - \sin^2 x\right)^7 \cos x\,dx$

You would expand $\left(1 - \sin^2 x\right)^7$.

98. (a) $\int \tan^3 x\,dx = \int(\sec^2 x - 1)\tan x\,dx$

$= \int \sec^2 x \tan x\,dx - \int \tan x\,dx$

$= \dfrac{\tan^2 x}{2} - \int \tan x\,dx$

$\int \tan^3 x\,dx = \dfrac{\tan^2 x}{2} + \ln|\cos x| + C$

(b) $\int \tan^5 x\,dx = \int(\sec^2 x - 1)\tan^3 x\,dx$

$= \dfrac{\tan^4 x}{4} - \int \tan^3 x\,dx$

(c) $\int \tan^{2k+1} x\,dx = \int(\sec^2 x - 1)\tan^{2k-1} x\,dx$

$= \dfrac{\tan^{2k} x}{2k} - \int \tan^{2k-1} x\,dx$

(d) You would use these formulas recursively.

99. Let $f(x) = \dfrac{1}{2}\left(x\sqrt{x^2+1} + \ln\left|x + \sqrt{x^2+1}\right|\right) + C.$

$$f'(x) = \frac{1}{2}\left(x\frac{1}{2}(x^2+1)^{-1/2}(2x) + \sqrt{x^2+1} + \frac{1}{x+\sqrt{x^2+1}}\left(1 + \frac{1}{2}(x^2+1)^{-1/2}(2x)\right)\right)$$

$$= \frac{1}{2}\left(\frac{x^2}{\sqrt{x^2+1}} + \sqrt{x^2+1} + \frac{1}{x+\sqrt{x^2+1}}\left(1 + \frac{x}{\sqrt{x^2+1}}\right)\right)$$

$$= \frac{1}{2}\left(\frac{x^2 + (x^2+1)}{\sqrt{x^2+1}} + \frac{1}{x+\sqrt{x^2+1}}\left(\frac{\sqrt{x^2+1}+x}{\sqrt{x^2+1}}\right)\right)$$

$$= \frac{1}{2}\left(\frac{2x^2+1}{\sqrt{x^2+1}} + \frac{1}{\sqrt{x^2+1}}\right) = \frac{1}{2}\left(\frac{2(x^2+1)}{\sqrt{x^2+1}}\right) = \sqrt{x^2+1}$$

So, $\int\sqrt{x^2+1}\,dx = \dfrac{1}{2}\left(x\sqrt{x^2+1} + \ln\left|x+\sqrt{x^2+1}\right|\right) + C.$

Let $g(x) = \dfrac{1}{2}\left(x\sqrt{x^2+1} + \operatorname{arcsinh}(x)\right).$

$$g'(x) = \frac{1}{2}\left(x\frac{1}{2}(x^2+1)^{-1/2}(2x) + \sqrt{x^2+1} + \frac{1}{\sqrt{x^2+1}}\right)$$

$$= \frac{1}{2}\left(\frac{x^2}{\sqrt{x^2+1}} + \sqrt{x^2+1} + \frac{1}{\sqrt{x^2+1}}\right)$$

$$= \frac{1}{2}\left(\frac{x^2 + (x^2+1) + 1}{\sqrt{x^2+1}}\right)$$

$$= \frac{1}{2}\left(\frac{2(x^2+1)}{\sqrt{x^2+1}}\right) = \sqrt{x^2+1}$$

So, $\int\sqrt{x^2+1}\,dx = \dfrac{1}{2}\left(x\sqrt{x^2+1} + \operatorname{arcsinh}(x)\right) + C.$

100. Let $I = \int_2^4 \frac{\sqrt{\ln(9-x)}}{\sqrt{\ln(9-x)} + \sqrt{\ln(x+3)}}\,dx.$

I is defined and continuous on $[2, 4]$. Note the symmetry: as x goes from 2 to 4, $9 - x$ goes from 7 to 5 and $x + 3$ goes from 5 to 7. So, let $y = 6 - x$, $dy = -dx$.

$$I = \int_4^2 \frac{\sqrt{\ln(3+y)}}{\sqrt{\ln(3+y)} + \sqrt{\ln(9-y)}}(-dy) = \int_2^4 \frac{\sqrt{\ln(3+y)}}{\sqrt{\ln(3+y)} + \sqrt{\ln(9-y)}}\,dy$$

Adding:

$$2I = \int_2^4 \frac{\sqrt{\ln(9-x)}}{\sqrt{\ln(9-x)} + \sqrt{\ln(x+3)}}\,dx + \int_2^4 \frac{\sqrt{\ln(3+x)}}{\sqrt{\ln(3+x)} + \sqrt{\ln(9-x)}}\,dx = \int_2^4 dx = 2 \Rightarrow I = 1$$

You can easily check this result numerically.

Section 8.2 Integration by Parts

1. $\int xe^{2x}\,dx$

$u = x,\ dv = e^{2x}\,dx$

2. $\int x^2e^{2x}\,dx$

$u = x^2,\ dv = e^{2x}\,dx$

3. $\int(\ln x)^2\,dx$

$u = (\ln x)^2,\ dv = dx$

4. $\int \ln 4x\,dx$

$u = \ln 4x,\ dv = dx$

5. $\int x\sec^2 x\,dx$

$u = x,\ dv = \sec^2 x\,dx$

6. $\int x^2\cos x\,dx$

$u = x^2,\ dv = \cos x\,dx$

7. $dv = x^3\,dx \Rightarrow v = \int x^3\,dx = \frac{x^4}{4}$

$u = \ln x \quad \Rightarrow du = \frac{1}{x}\,dx$

$$\begin{aligned}\int x^3\ln x\,dx &= uv - \int v\,du\\ &= (\ln x)\frac{x^4}{4} - \int\left(\frac{x^4}{4}\right)\frac{1}{x}\,dx\\ &= \frac{x^4}{4}\ln x - \frac{1}{4}\int x^3\,dx\\ &= \frac{x^4}{4}\ln x - \frac{1}{16}x^4 + C\\ &= \frac{1}{16}x^4(4\ln x - 1) + C\end{aligned}$$

8. $dv = e^x\,dx \quad \Rightarrow v = \int e^x\,dx = e^x$

$u = 4x + 7 \Rightarrow du = 4dx$

$$\begin{aligned}\int(4x+7)e^x\,dx &= uv - \int v\,du\\ &= (4x+7)e^x - \int e^x\,4\,dx\\ &= (4x+7)e^x - 4e^x + C\\ &= (4x+3)e^x + C\end{aligned}$$

9. $dv = \sin 3x\,dx \Rightarrow v = \int\sin 3x\,dx = -\frac{1}{3}\cos 3x$

$u = x \quad \Rightarrow du = dx$

$$\begin{aligned}\int x\sin 3x\,dx &= uv - \int v\,du\\ &= x\left(-\frac{1}{3}\cos 3x\right) - \int -\frac{1}{3}\cos 3x\,dx\\ &= -\frac{x}{3}\cos 3x + \frac{1}{9}\sin 3x + C\end{aligned}$$

10. $dv = \cos 4x\,dx \Rightarrow v = \int\cos 4x\,dx = \frac{1}{4}\sin 4x$

$u = x \quad \Rightarrow du = dx$

$$\begin{aligned}\int x\cos 4x\,dx &= uv - \int v\,du\\ &= x\left(\frac{1}{4}\sin 4x\right) - \int\frac{1}{4}\sin 4x\,dx\\ &= \frac{x}{4}\sin 4x + \frac{1}{16}\cos 4x + C\end{aligned}$$

11. $dv = e^{-4x}\,dx \Rightarrow v = \int e^{-4x}\,dx = -\frac{1}{4}e^{-4x}$

$u = x \qquad \Rightarrow du = dx$

$$\int xe^{-4x}\,dx = x\left(-\frac{1}{4}e^{-4x}\right) - \int -\frac{1}{4}e^{-4x}\,dx$$

$$= -\frac{x}{4}e^{-4x} - \frac{1}{16}e^{-4x} + C$$

$$= -\frac{1}{16e^{4x}}(1 + 4x) + C$$

12. $dv = e^{-2x}\,dx \Rightarrow v = \int e^{-2x}\,dx = -\frac{1}{2}e^{-2x}$

$u = 5x \Rightarrow du = 5dx$

$$\int \frac{5x}{e^{2x}}\,dx = \int 5xe^{-2x}\,dx$$

$$= (5x)\left(-\frac{1}{2}e^{-2x}\right) - \int\left(-\frac{1}{2}e^{-2x}\right)5dx$$

$$= -\frac{5}{2}xe^{-2x} + \frac{5}{2}\int e^{-2x}\,dx$$

$$= -\frac{5}{2}xe^{-2x} - \frac{5}{4}e^{-2x} + C$$

$$= -\frac{5}{4}e^{-2x}(2x + 1) + C$$

13. Use integration by parts three times.

(1) $dv = e^x\,dx \Rightarrow \quad v = \int e^x\,dx = e^x$

$u = x^3 \qquad \Rightarrow du = 3x^2\,dx$

(2) $dv = e^x\,dx \Rightarrow \quad v = \int e^x\,dx = e^x$

$u = x^2 \qquad \Rightarrow du = 2x\,dx$

(3) $dv = e^x\,dx \Rightarrow \quad v = \int e^x\,dx = e^x$

$u = x \qquad \Rightarrow du = dx$

$$\int x^3e^x\,dx = x^3e^x - 3\int x^2e^x\,dx = x^3e^x - 3x^2e^x + 6\int xe^x\,dx$$

$$= x^3e^x - 3x^2e^x + 6xe^x - 6e^x + C = e^x\left(x^3 - 3x^2 + 6x - 6\right) + C$$

14. $\displaystyle\int \frac{e^{1/t}}{t^2}\,dt = -\int e^{1/t}\left(\frac{-1}{t^2}\right)dt = -e^{1/t} + C$

15. $dv = t\,dt \qquad \Rightarrow \quad v = \int t\,dt = \frac{t^2}{2}$

$u = \ln(t + 1) \Rightarrow du = \frac{1}{t + 1}\,dt$

$$\int t\ln(t + 1)\,dt = \frac{t^2}{2}\ln(t + 1) - \frac{1}{2}\int\frac{t^2}{t + 1}\,dt$$

$$= \frac{t^2}{2}\ln(t + 1) - \frac{1}{2}\int\left(t - 1 + \frac{1}{t + 1}\right)dt$$

$$= \frac{t^2}{2}\ln(t + 1) - \frac{1}{2}\left[\frac{t^2}{2} - t + \ln(t + 1)\right] + C$$

$$= \frac{1}{4}\left[2\left(t^2 - 1\right)\ln\left|t + 1\right| - t^2 + 2t\right] + C$$

16. $dv = x^5\,dx \quad \Rightarrow \quad v = \int x^5\,dx = \frac{1}{6}x^6$

$u = \ln 3x \quad \Rightarrow \quad du = \frac{1}{x}\,dx$

$$\int x^5\ln 3x\,dx = \frac{x^6}{6}\ln 3x - \int\frac{x^6}{6}\left(\frac{1}{x}\right)dx$$

$$= \frac{x^6}{6}\ln 3x - \frac{x^6}{36} + C$$

17. Let $u = \ln x$, $du = \frac{1}{x}\,dx$.

$$\int\frac{(\ln x)^2}{x}\,dx = \int(\ln x)^2\left(\frac{1}{x}\right)dx = \frac{(\ln x)^3}{3} + C$$

18. $dv = x^{-3}\,dx \Rightarrow v = \int x^{-3}\,dx = -\frac{1}{2}x^{-2}$

$u = \ln x \Rightarrow du = \frac{1}{x}\,dx$

$$\int \frac{\ln x}{x^3}\,dx = -\frac{1}{2}x^{-2}\ln x - \int\left(-\frac{1}{2}x^{-2}\right)\frac{1}{x}\,dx$$
$$= -\frac{1}{2x^2}\ln x + \frac{1}{2}\int x^{-3}\,dx$$
$$= -\frac{1}{2x^2}\ln x + \left(\frac{1}{2}\right)\frac{x^{-2}}{-2} + C$$
$$= -\frac{1}{2x^2}\ln x - \frac{1}{4x^2} + C$$

19. $dv = \frac{1}{(2x+1)^2}\,dx \Rightarrow v = \int (2x+1)^{-2}\,dx = -\frac{1}{2(2x+1)}$

$u = xe^{2x} \Rightarrow du = \left(2xe^{2x} + e^{2x}\right)dx = e^{2x}(2x+1)\,dx$

$$\int \frac{xe^{2x}}{(2x+1)^2}\,dx = -\frac{xe^{2x}}{2(2x+1)} + \int \frac{e^{2x}}{2}\,dx$$
$$= \frac{-xe^{2x}}{2(2x+1)} + \frac{e^{2x}}{4} + C = \frac{e^{2x}}{4(2x+1)} + C$$

20. $dv = \frac{x}{\left(x^2+1\right)^2}\,dx \Rightarrow v = \int \left(x^2+1\right)^{-2} x\,dx = -\frac{1}{2\left(x^2+1\right)}$

$u = x^2e^{x^2} \Rightarrow du = \left(2x^3e^{x^2} + 2xe^{x^2}\right)dx = 2xe^{x^2}\left(x^2+1\right)dx$

$$\int \frac{x^3e^{x^2}}{\left(x^2+1\right)^2}\,dx = -\frac{x^2e^{x^2}}{2\left(x^2+1\right)} + \int xe^{x^2}\,dx = -\frac{x^2e^{x^2}}{2\left(x^2+1\right)} + \frac{e^{x^2}}{2} + C = \frac{e^{x^2}}{2\left(x^2+1\right)} + C$$

21. $dv = \sqrt{x-5}\,dx \Rightarrow v = \int (x-5)^{1/2}\,dx = \frac{2}{3}(x-5)^{3/2}$

$u = x \Rightarrow du = dx$

$$\int x\sqrt{x-5}\,dx = x\tfrac{2}{3}(x-5)^{3/2} - \int \tfrac{2}{3}(x-5)^{3/2}\,dx$$
$$= \tfrac{2}{3}x(x-5)^{3/2} - \tfrac{4}{15}(x-5)^{5/2} + C$$
$$= \tfrac{2}{15}(x-5)^{3/2}\left(5x - 2(x-5)\right) + C$$
$$= \tfrac{2}{15}(x-5)^{3/2}(3x+10) + C$$

22. $dv = (6x+1)^{-1/2}\,dx \Rightarrow v = \int (6x+1)^{-1/2}\,dx = \frac{1}{3}(6x+1)^{1/2}$

$u = x \Rightarrow du = dx$

$$\int \frac{x}{\sqrt{6x+1}}\,dx = \frac{x\sqrt{6x+1}}{3} - \int \frac{\sqrt{6x+1}}{3}\,dx$$
$$= \frac{x\sqrt{6x+1}}{3} - \frac{(6x+1)^{3/2}}{27} + C$$
$$= \frac{\sqrt{6x+1}}{27}\left[9x - (6x+1)\right] + C$$
$$= \frac{\sqrt{6x+1}}{27}(3x-1) + C$$

23. $dv = \cos x\,dx \Rightarrow v = \int \cos x\,dx = \sin x$

$u = x \Rightarrow du = dx$

$$\int x\cos x\,dx = x\sin x - \int \sin x\,dx = x\sin x + \cos x + C$$

24. $u = t,\ du = dt,\ dv = \csc t\cot t\,dt,\ v = -\csc t$

$$\int t\csc t\cot t\,dt = -t\csc t + \int \csc t\,dt = -t\csc t - \ln\left|\csc t + \cot t\right| + C$$

25. Use integration by parts three times.

(1) $u = x^3,\ du = 3x^2\,dx,\ dv = \sin x\,dx,\ v = -\cos x$

$$\int x^3 \sin dx = -x^3 \cos x + 3\int x^2 \cos x\,dx$$

(2) $u = x^2,\ du = 2x\,dx,\ dv = \cos x\,dx,\ v = \sin x$

$$\int x^3 \sin x\,dx = -x^3 \cos x + 3\left(x^2 \sin x - 2\int x \sin x\,dx\right) = -x^3 \cos x + 3x^2 \sin x - 6\int x \sin x\,dx$$

(3) $u = x,\ du = dx,\ dv = \sin x\,dx,\ v = -\cos x$

$$\begin{aligned}\int x^3 \sin x\,dx &= -x^3 \cos x + 3x^2 \sin x - 6\left(-x \cos x + \int \cos x\,dx\right)\\ &= -x^3 \cos x + 3x^2 \sin x + 6x \cos x - 6 \sin x + C\\ &= \left(6x - x^3\right) \cos x + \left(3x^2 - 6\right) \sin x + C\end{aligned}$$

26. Use integration by parts twice.

(1) $u = x^2,\ du = 2x\,dx,\ dv = \cos x\,dx,\ v = \sin x$

$$\int x^2 \cos x\,dx = x^2 \sin x - 2\int x \sin x\,dx$$

(2) $u = x,\ du = dx,\ dv = \sin x\,dx,\ v = -\cos x$

$$\int x^2 \cos x\,dx = x^2 \sin x - 2\left(-x \cos x + \int \cos x\,dx\right) = x^2 \sin x + 2x \cos x - 2 \sin x + C$$

27. $dv = dx \quad \Rightarrow \quad v = \int dx = x$

$u = \arctan x \quad \Rightarrow \quad du = \dfrac{1}{1 + x^2}\,dx$

$$\begin{aligned}\int \arctan x\,dx &= x \arctan x - \int \frac{x}{1 + x^2}\,dx\\ &= x \arctan x - \frac{1}{2} \ln\left(1 + x^2\right) + C\end{aligned}$$

28. $dv = dx \quad \Rightarrow \quad v = \int dx = x$

$u = \arccos x \quad \Rightarrow \quad du = -\dfrac{1}{\sqrt{1 - x^2}}\,dx$

$$\begin{aligned}4\int \arccos x\,dx &= 4\left(x \arccos x + \int \frac{x}{\sqrt{1 - x^2}}\,dx\right)\\ &= 4\left(x \arccos x - \sqrt{1 - x^2}\right) + C\end{aligned}$$

29. Use integration by parts twice.

(1) $dv = e^{-3x}\,dx \quad \Rightarrow \quad v = \int e^{-3x}\,dx = -\frac{1}{3}e^{-3x}$

$u = \sin 5x \quad \Rightarrow \quad du = 5 \cos 5x\,dx$

$$\int e^{-3x} \sin 5x\,dx = \sin 5x\left(-\tfrac{1}{3}e^{-3x}\right) - \int\left(-\tfrac{1}{3}e^{-3x}\right) 5 \cos x\,dx = -\tfrac{1}{3}e^{-3x} \sin 5x + \tfrac{5}{3}\int e^{-3x} \cos 5x\,dx$$

(2) $dv = e^{-3x}\,dx \quad \Rightarrow \quad v = \int e^{-3x}\,dx = -\frac{1}{3}e^{-3x}$

$u = \cos 5x \quad \Rightarrow \quad du = -5 \sin 5x\,dx$

$$\begin{aligned}\int e^{-3x} \sin 5x\,dx &= -\tfrac{1}{3}e^{-3x} \sin 5x + \tfrac{5}{3}\left[\left(-\tfrac{1}{3}e^{-3x} \cos 5x - \int\left(-\tfrac{1}{3}e^{-3x}\right)(-5 \sin 5x)\right) dx\right]\\ &= -\tfrac{1}{3}e^{-3x} \sin 5x - \tfrac{5}{9}e^{-3x} \cos 5x - \tfrac{25}{9}\int e^{-3x} \sin 5x\,dx\\ \left(1 + \tfrac{25}{9}\right)\int e^{-3x} \sin 5x\,dx &= -\tfrac{1}{3}e^{-3x} \sin 5x - \tfrac{5}{9}e^{-3x} \cos 5x\\ \int e^{-3x} \sin 5x\,dx &= \tfrac{9}{34}\left(-\tfrac{1}{3}e^{-3x} \sin 5x - \tfrac{5}{9}e^{-3x} \cos 5x\right) + C = -\tfrac{3}{34}e^{-3x} \sin 5x - \tfrac{5}{34}e^{-3x} \cos 5x + C\end{aligned}$$

30. Use integration by parts twice.

(1) $dv = e^{4x}\,dx \quad \Rightarrow \quad v = \int e^{4x}\,dx = \frac{1}{4}e^{4x}$

$u = \cos 2x \quad \Rightarrow \quad du = -2 \sin 2x\,dx$

(2) $dv = e^{4x}\,dx \Rightarrow v = \int e^{4x}\,dx = \frac{1}{4}e^{4x}$

$u = \sin 2x \Rightarrow du = 2\cos 2x\,dx$

$$\int e^{4x}\cos 2x\,dx = \frac{1}{4}e^{4x}\cos 2x + \frac{1}{2}\left[\frac{1}{4}e^{4x}\sin 2x - \int \frac{1}{4}e^{4x}(2\cos 2x)\,dx\right]$$

$$= \frac{1}{4}e^{4x}\cos 2x + \frac{1}{8}e^{4x}\sin 2x - \frac{1}{4}\int e^{4x}\cos 2x\,dx + C$$

$$\left(1 + \frac{1}{4}\right)\int e^{4x}\cos 2x\,dx = \frac{1}{4}e^{4x}\cos 2x + \frac{1}{8}e^{4x}\sin 2x + C$$

$$\int e^{4x}\cos 2x\,dx = \frac{1}{5}e^{4x}\cos 2x + \frac{1}{10}e^{4x}\sin 2x + C$$

31. $dv = dx \Rightarrow v = x$

$u = \ln x \Rightarrow du = \frac{1}{x}\,dx$

$y' = \ln x$

$$y = \int \ln x\,dx = x\ln x - \int x\left(\frac{1}{x}\right)dx = x\ln x - x + C = x(-1 + \ln x) + C$$

32. $dv = dx \Rightarrow v = \int dx = x$

$$u = \arctan\frac{x}{2} \Rightarrow du = \frac{1}{1 + (x/2)^2}\left(\frac{1}{2}\right)dx = \frac{2}{4 + x^2}\,dx$$

$$y = \int \arctan\frac{x}{2}\,dx = x\arctan\frac{x}{2} - \int\frac{2x}{4 + x^2}\,dx = x\arctan\frac{x}{2} - \ln\left(4 + x^2\right) + C$$

33. Use integration by parts twice.

(1) $dv = \frac{1}{\sqrt{3 + 5t}}\,dt \Rightarrow v = \int(3 + 5t)^{-1/2}\,dt = \frac{2}{5}(3 + 5t)^{1/2}$

$u = t^2 \Rightarrow du = 2t\,dt$

$$\int\frac{t^2}{\sqrt{3 + 5t}}\,dt = \frac{2}{5}t^2(3 + 5t)^{1/2} - \int\frac{2}{5}(3 + 5t)^{1/2}2t\,dt$$

$$= \frac{2}{5}t^2(3 + 5t)^{1/2} - \frac{4}{5}\int t(3 + 5t)^{1/2}\,dt$$

(2) $dv = (3 + 5t)^{1/2}\,dt \Rightarrow v = \int(3 + 5t)^{1/2}\,dt = \frac{2}{15}(3 + 5t)^{3/2}$

$u = t \Rightarrow du = dt$

$$\int\frac{t^2}{\sqrt{3 + 5t}}\,dt = \frac{2}{5}t^2(3 + 5t)^{1/2} - \frac{4}{5}\left[\frac{2}{15}t(3 + 5t)^{3/2} - \int\frac{2}{15}(3 + 5t)^{3/2}\,dt\right]$$

$$= \frac{2}{5}t^2(3 + 5t)^{1/2} - \frac{8}{75}t(3 + 5t)^{3/2} + \frac{8}{75}\int(3 + 5t)^{3/2}\,dt$$

$$= \frac{2}{5}t^2(3 + 5t)^{1/2} - \frac{8}{75}t(3 + 5t)^{3/2} + \frac{16}{1875}(3 + 5t)^{5/2} + C$$

$$= \frac{2}{1875}\sqrt{3 + 5t}\left(375t^2 - 100t(3 + 5t) + 8(3 + 5t)^2\right) + C$$

$$= \frac{2}{625}\sqrt{3 + 5t}\left(25t^2 - 20t + 24\right) + C$$

34. Use integration by parts twice.

(1) $dv = \sqrt{x-3}\,dx \Rightarrow v = \int (x-3)^{1/2}\,dx = \frac{2}{3}(x-3)^{3/2}$

$u = x^2 \Rightarrow du = 2x\,dx$

$$\int x^2\sqrt{x-3}\,dx = \frac{2}{3}x^2(x-3)^{3/2} - \int \frac{2}{3}(x-3)^{3/2}2x\,dx$$
$$= \frac{2}{3}x^2(x-3)^{3/2} - \frac{4}{3}\int (x-3)^{3/2}\,x\,dx$$

(2) $dv = (x-3)^{3/2}dx \Rightarrow v = \int (x-3)^{3/2}\,dx = \frac{2}{5}(x-3)^{5/2}$

$u = x \Rightarrow du = dx$

$$\int x^2\sqrt{x-3}\,dx = \frac{2}{3}x^2(x-3)^{3/2} - \frac{4}{3}\left[\frac{2}{5}x(x-3)^{5/2} - \int \frac{2}{5}(x-3)^{5/2}\,dx\right]$$
$$= \frac{2}{3}x^2(x-3)^{3/2} - \frac{8}{15}x(x-3)^{5/2} + \frac{8}{15}\left[\frac{2}{7}(x-3)^{7/2}\right] + C$$
$$= \frac{2}{35}(x-3)^{3/2}\left(5x^2 + 12x + 24\right) + C$$

35. (a)

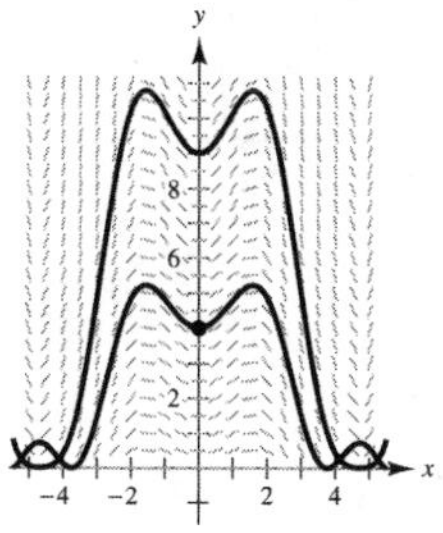

(b) $\dfrac{dy}{dx} = x\sqrt{y}\cos x, \quad (0, 4)$

$$\int \frac{dy}{\sqrt{y}} = \int x\cos x\,dx$$

$\int y^{-1/2}\,dy = \int x\cos x\,dx \qquad (u = x, du = dx, dv = \cos x\,dx, v = \sin x)$

$2y^{1/2} = x\sin x - \int \sin x\,dx = x\sin x + \cos x + C$

$(0, 4)$: $2(4)^{1/2} = 0 + 1 + C \Rightarrow C = 3$

$2\sqrt{y} = x\sin x + \cos x + 3$

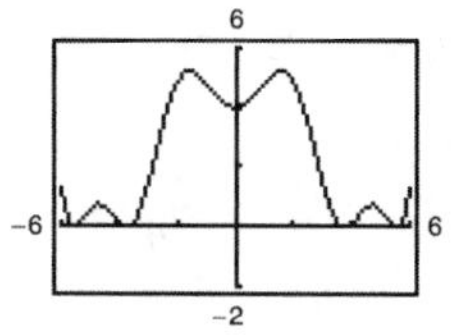

36. (a)

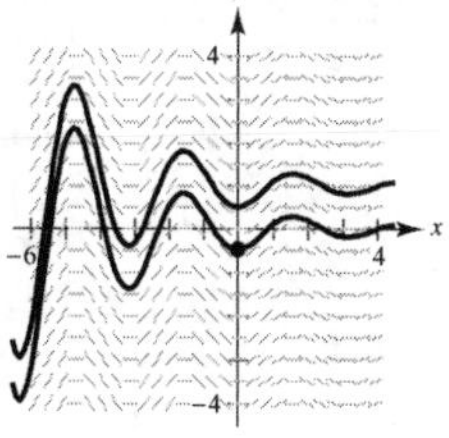

(b) $\frac{dy}{dx} = e^{-x/3} \sin 2x, \quad \left(0, -\frac{18}{37}\right)$

$y = \int e^{-x/3} \sin 2x \, dx$

Use integration by parts twice.

(1) $u = \sin 2x, du = 2 \cos 2x$

$dv = e^{-x/3} \, dx, v = -3e^{-x/3}$

$\int e^{-x/3} \sin 2x \, dx = -3e^{-x/3} \sin 2x + \int 6e^{-x/3} \cos 2x \, dx$

(2) $u = \cos 2x, du = -2 \sin 2x$

$dv = e^{-x/3} \, dx, v = -3e^{-x/3}$

$$\int e^{-x/3} \sin 2x \, dx = -3e^{-x/3} \sin 2x + 6\left(-3e^{-x/3} \cos 2x - \int 6e^{-x/3} \sin 2x \, dx\right) + C$$

$$37 \int e^{-x/3} \sin 2x \, dx = -3e^{-x/3} \sin 2x - 18e^{-x/3} \cos 2x + C$$

$$y = \int e^{-x/3} \sin 2x \, dx = \frac{1}{37}\left(-3e^{-x/3} \sin 2x - 18e^{-x/3} \cos 2x\right) + C$$

$$\left(0, \frac{-18}{37}\right): \frac{-18}{37} = \frac{1}{37}[0 - 18] + C \Rightarrow C = 0$$

$$y = \frac{-1}{37}\left(3e^{-x/3} \sin 2x + 18e^{-x/3} \cos 2x\right)$$

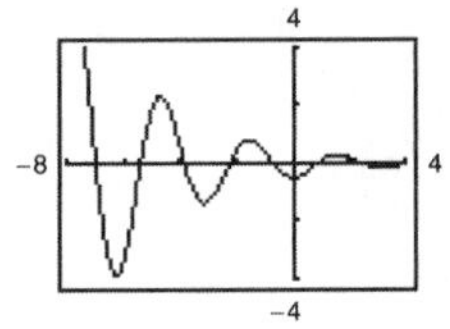

37. $\frac{dy}{dx} = \frac{x}{y} e^{x/8}, y(0) = 2$

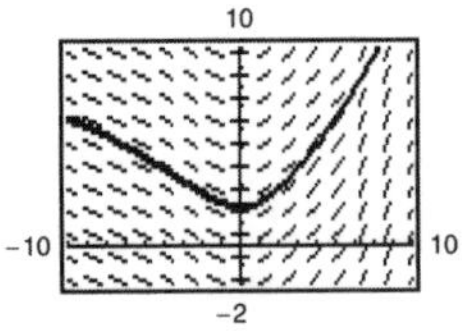

38. $\frac{dy}{dx} = \frac{x}{y} \sin x, y(0) = 4$

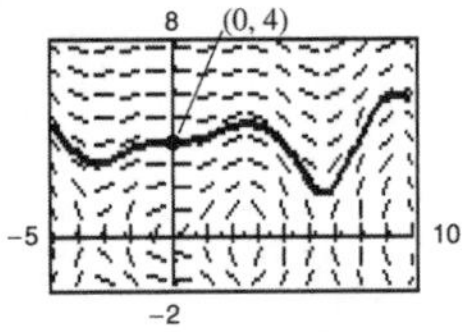

39. $u = x, du = dx, dv = e^{x/2} \, dx, v = 2e^{x/2}$

$$\int xe^{x/2} \, dx = 2xe^{x/2} - \int 2e^{x/2} \, dx$$
$$= 2xe^{x/2} - 4e^{x/2} + C$$

So,

$$\int_0^3 xe^{x/2} \, dx = \left[2xe^{x/2} - 4e^{x/2}\right]_0^3$$
$$= \left(6e^{3/2} - 4e^{3/2}\right) - (-4)$$
$$= 4 + 2e^{3/2} \approx 12.963$$

40. Use integration by parts twice.

(1) $u = x^2, du = 2x \, dx, dv = e^{-2x} \, dx,$

$v = -\frac{1}{2}e^{-2x}$

$$\int x^2 e^{-2x} \, dx = -\frac{1}{2}x^2 e^{-2x} - \int \left(-\frac{1}{2}e^{-2x}\right) 2x \, dx$$
$$= -\frac{1}{2}x^2 e^{-2x} + \int xe^{-2x} \, dx$$

(2) $u = x, du = dx, dv = e^{-2x} \, dx, v = -\frac{1}{2}e^{-2x}$

$$\int x^2 e^{-2x} \, dx = -\frac{1}{2}x^2 e^{-2x} + \left(-\frac{1}{2}xe^{-2x} - \int -\frac{1}{2}e^{-2x} \, dx\right)$$
$$= -\frac{1}{2}x^2 e^{-2x} - \frac{1}{2}xe^{-2x} - \frac{1}{4}e^{-2x} + C$$
$$= e^{-2x}\left(-\frac{1}{2}x^2 - \frac{1}{2}x - \frac{1}{4}\right)$$

So,

$$\int_0^2 x^2 e^{-2x} \, dx = \left[e^{-2x}\left(-\frac{1}{2}x^2 - \frac{1}{2}x - \frac{1}{4}\right)\right]_0^2$$
$$= e^{-4}\left(-2 - 1 - \frac{1}{4}\right) - \left(-\frac{1}{4}\right)$$
$$= \frac{-13}{4e^4} + \frac{1}{4} \approx 0.190$$

41. $u = x,\ du = dx,\ dv = \cos 2x\,dx,\ v = \frac{1}{2}\sin 2x$

$$\int x\cos 2x\,dx = \frac{1}{2}x\sin 2x - \int\frac{1}{2}\sin 2x\,dx = \frac{1}{2}x\sin 2x + \frac{1}{4}\cos 2x + C$$

So,

$$\int_0^{\pi/4} x\cos 2x\,dx = \left[\frac{1}{2}x\sin 2x + \frac{1}{4}\cos 2x\right]_0^{\pi/4} = \left(\frac{\pi}{8}(1) + 0\right) - \left(0 + \frac{1}{4}\right) = \frac{\pi}{8} - \frac{1}{4} \approx 0.143$$

42. $dv = \sin 2x\,dx \Rightarrow v = \int\sin 2x\,dx = -\frac{1}{2}\cos 2x$

$u = x \Rightarrow du = dx$

$$\int x\sin 2x\,dx = \frac{-1}{2}x\cos 2x + \frac{1}{2}\int\cos 2x\,dx = \frac{-1}{2}x\cos 2x + \frac{1}{4}\sin 2x + C = \frac{1}{4}(\sin 2x - 2x\cos 2x) + C$$

So,

$$\int_0^{\pi} x\sin 2x\,dx = \left[\frac{1}{4}(\sin 2x - 2x\cos 2x)\right]_0^{\pi} = -\frac{\pi}{2}.$$

43. $u = \arccos x,\ du = -\frac{1}{\sqrt{1-x^2}}\,dx,\ dv = dx,\ v = x$

$$\int\arccos x\,dx = x\arccos x + \int\frac{x}{\sqrt{1-x^2}}\,dx = x\arccos x - \sqrt{1-x^2} + C$$

So,

$$\int_0^{1/2}\arccos x = \left[x\arccos x - \sqrt{1-x^2}\right]_0^{1/2} = \frac{1}{2}\arccos\left(\frac{1}{2}\right) - \sqrt{\frac{3}{4}} + 1 = \frac{\pi}{6} - \frac{\sqrt{3}}{2} + 1 \approx 0.658.$$

44. $dv = x\,dx \Rightarrow v = \int x\,dx = \frac{x^2}{2}$

$u = \arcsin x^2 \Rightarrow du = \frac{2x}{\sqrt{1-x^4}}\,dx$

$$\int x\arcsin x^2\,dx = \frac{x^2}{2}\arcsin x^2 - \int\frac{x^3}{\sqrt{1-x^4}}\,dx = \frac{x^2}{2}\arcsin x^2 + \frac{1}{4}(2)\left(1-x^4\right)^{1/2} + C = \frac{1}{2}\left(x^2\arcsin x^2 + \sqrt{1-x^4}\right) + C$$

So, $\int_0^1 x\arcsin x^2\,dx = \frac{1}{2}\left[x^2\arcsin x^2 + \sqrt{1-x^4}\right]_0^1 = \frac{1}{4}(\pi - 2).$

45. Use integration by parts twice.

(1) $dv = e^x\,dx \Rightarrow v = \int e^x\,dx = e^x$ (2) $dv = e^x\,dx \Rightarrow v = \int e^x\,dx = e^x$

$u = \sin x \Rightarrow du = \cos x\,dx$ $u = \cos x \Rightarrow du = -\sin x\,dx$

$$\int e^x\sin x\,dx = e^x\sin x - \int e^x\cos x\,dx = e^x\sin x - e^x\cos x - \int e^x\sin x\,dx$$

$$2\int e^x\sin x\,dx = e^x(\sin x - \cos x)$$

$$\int e^x\sin x\,dx = \frac{e^x}{2}(\sin x - \cos x) + C$$

So, $\int_0^1 e^x\sin x\,dx = \left[\frac{e^x}{2}(\sin x - \cos x)\right]_0^1 = \frac{e}{2}(\sin 1 - \cos 1) + \frac{1}{2} = \frac{e(\sin 1 - \cos 1) + 1}{2} \approx 0.909.$

46. $u = \ln(4 + x^2),\ du = \dfrac{2x}{4 + x^2}\,dx,\ dv = dx,\ v = x$

$$\int \ln(4 + x^2)\,dx = x\ln(4 + x^2) - \int \frac{2x^2}{4 + x^2}\,dx$$
$$= x\ln(4 + x^2) - 2\int\left(1 - \frac{4}{4 + x^2}\right)dx$$
$$= x\ln(4 + x^2) - 2\left(x - \frac{4}{2}\arctan\frac{x}{2}\right) + C$$
$$= x\ln(4 + x^2) - 2x + 4\arctan\frac{x}{2} + C$$

So, $\displaystyle\int_0^1 \ln(4 + x^2)\,dx = \left[x\ln(4 + x^2) - 2x + 4\arctan\frac{x}{2}\right]_0^1 = \left(\ln 5 - 2 + 4\arctan\left(\frac{1}{2}\right)\right) \approx 1.464.$

47. $dv = x\,dx,\ v = \dfrac{x^2}{2},\ u = \operatorname{arcsec} x,\ du = \dfrac{1}{x\sqrt{x^2 - 1}}\,dx$

$$\int x \operatorname{arcsec} x\,dx = \frac{x^2}{2}\operatorname{arcsec} x - \int \frac{x^2/2}{x\sqrt{x^2 - 1}}\,dx = \frac{x^2}{2}\operatorname{arcsec} x - \frac{1}{4}\int \frac{2x}{\sqrt{x^2 - 1}}\,dx = \frac{x^2}{2}\operatorname{arcsec} x - \frac{1}{2}\sqrt{x^2 - 1} + C$$

So,

$$\int_2^4 x \operatorname{arcsec} x\,dx = \left[\frac{x^2}{2}\operatorname{arcsec} x - \frac{1}{2}\sqrt{x^2 - 1}\right]_2^4 = \left(8\operatorname{arcsec} 4 - \frac{\sqrt{15}}{2}\right) - \left(\frac{2\pi}{3} - \frac{\sqrt{3}}{2}\right) = 8\operatorname{arcsec} 4 - \frac{\sqrt{15}}{2} + \frac{\sqrt{3}}{2} - \frac{2\pi}{3} \approx 7.380.$$

48. $u = x,\ du = dx,\ dv = \sec^2 2x\,dx,\ v = \dfrac{1}{2}\tan 2x$

$$\int x\sec^2 2x\,dx = \frac{1}{2}x\tan 2x - \int \frac{1}{2}\tan 2x\,dx = \frac{1}{2}x\tan 2x + \frac{1}{4}\ln\left|\cos 2x\right| + C$$

So,

$$\int_0^{\pi/8} x\sec^2 2x\,dx = \left[\frac{1}{2}x\tan 2x + \frac{1}{4}\ln\left|\cos 2x\right|\right]_0^{\pi/8} = \frac{\pi}{16}(1) + \frac{1}{4}\ln\left(\frac{\sqrt{2}}{2}\right) = \frac{\pi}{16} - \frac{1}{8}\ln(2) \approx 0.1097.$$

49. $\displaystyle\int x^2e^{2x}\,dx = x^2\left(\tfrac{1}{2}e^{2x}\right) - (2x)\left(\tfrac{1}{4}e^{2x}\right) + 2\left(\tfrac{1}{8}e^{2x}\right) + C$

$$= \tfrac{1}{2}x^2e^{2x} - \tfrac{1}{2}xe^{2x} + \tfrac{1}{4}e^{2x} + C$$
$$= \tfrac{1}{4}e^{2x}\left(2x^2 - 2x + 1\right) + C$$

Alternate signs	u and its derivatives	v' and its antiderivatives
+	x^2	e^{2x}
−	$2x$	$\frac{1}{2}e^{2x}$
+	2	$\frac{1}{4}e^{2x}$
−	0	$\frac{1}{8}e^{2x}$

50. $\int x^3 e^{-2x}\,dx = x^3\left(-\frac{1}{2}e^{-2x}\right) - 3x^2\left(\frac{1}{4}e^{-2x}\right) + 6x\left(-\frac{1}{8}e^{-2x}\right) - 6\left(\frac{1}{16}e^{-2x}\right) + C$

$= -\frac{1}{8}e^{-2x}\left(4x^3 + 6x^2 + 6x + 3\right) + C$

Alternate signs	u and its derivatives	v' and its antiderivatives
+	x^3	e^{-2x}
−	$3x^2$	$-\frac{1}{2}e^{-2x}$
+	$6x$	$\frac{1}{4}e^{-2x}$
−	6	$-\frac{1}{8}e^{-2x}$
+	0	$\frac{1}{16}e^{-2x}$

51. $\int x^3 \sin x\,dx = x^3(-\cos x) - 3x^2(-\sin x) + 6x\cos x - 6\sin x + C$

$= -x^3\cos x + 3x^2\sin x + 6x\cos x - 6\sin x + C$

$= \left(3x^2 - 6\right)\sin x - \left(x^3 - 6x\right)\cos x + C$

Alternate signs	u and its derivatives	v' and its antiderivatives
+	x^3	$\sin x$
−	$3x^2$	$-\cos x$
+	$6x$	$-\sin x$
−	6	$\cos x$
+	0	$\sin x$

52. $\int x^3 \cos 2x\,dx = x^3\left(\frac{1}{2}\sin 2x\right) - 3x^2\left(-\frac{1}{4}\cos 2x\right) + 6x\left(-\frac{1}{8}\sin 2x\right) - 6\left(\frac{1}{16}\cos 2x\right) + C$

$= \frac{1}{2}x^3\sin 2x + \frac{3}{4}x^2\cos 2x - \frac{3}{4}x\sin 2x - \frac{3}{8}\cos 2x + C$

$= \frac{1}{8}\left(4x^3\sin 2x + 6x^2\cos 2x - 6x\sin 2x - 3\cos 2x\right) + C$

Alternate signs	u and its derivatives	v' and its antiderivatives
+	x^3	$\cos 2x$
−	$3x^2$	$\frac{1}{2}\sin 2x$
+	$6x$	$-\frac{1}{4}\cos 2x$
−	6	$-\frac{1}{8}\sin 2x$
+	0	$\frac{1}{16}\cos 2x$

53. $\int x\sec^2 x\,dx = x\tan x + \ln\left|\cos x\right| + C$

Alternate signs	u and its derivatives	v' and its antiderivatives
+	x	$\sec^2 x$
−	1	$\tan x$
+	0	$-\ln\left|\cos x\right|$

54. $\int x^2(x-2)^{3/2}\,dx = \frac{2}{5}x^2(x-2)^{5/2} - \frac{8}{35}x(x-2)^{7/2} + \frac{16}{315}(x-2)^{9/2} + C = \frac{2}{315}(x-2)^{5/2}(35x^2 + 40x + 32) + C$

Alternate signs	u and its derivatives	v' and its antiderivatives
+	x^2	$(x-2)^{3/2}$
−	$2x$	$\frac{2}{5}(x-2)^{5/2}$
+	2	$\frac{4}{35}(x-2)^{7/2}$
−	0	$\frac{8}{315}(x-2)^{9/2}$

55. $u = \sqrt{x} \Rightarrow u^2 = x \Rightarrow 2u\,du = dx$

$$\int \sin\sqrt{x}\,dx = \int \sin u(2u\,du) = 2\int u \sin u\,du$$

Integration by parts:

$w = u,\ dw = du,\ dv = \sin u\,du,\ v = -\cos u$

$$\begin{aligned}2\int u\sin u\,du &= 2\left(-u\cos u + \int \cos u\,du\right)\\ &= 2(-u\cos u + \sin u) + C\\ &= 2\left(-\sqrt{x}\cos\sqrt{x} + \sin\sqrt{x}\right) + C\end{aligned}$$

56. $u = x^2,\ du = 2x\,dx$

$$\int 2x^3\cos(x^2)\,dx = \int x^2\cos(x^2)(2x)\,dx = \int u\cos u\,du$$

Integration by parts:

$w = u,\ dw = du,\ dv = \cos u\,du,\ v = \sin u$

$$\begin{aligned}\int u\cos u\,du &= u\sin u - \int \sin u\,du\\ &= u\sin u + \cos u + C\\ &= x^2\sin(x^2) + \cos(x^2) + C\end{aligned}$$

57. $u = x^2,\ du = 2x\,dx$

$$\int x^5 e^{x^2}\,dx = \frac{1}{2}\int e^{x^2} x^4\, 2x\,dx = \frac{1}{2}\int e^u u^2\,du$$

Integration by parts twice.

(1) $w = u^2,\ dw = 2u\,du,\ dv = e^u\,du,\ v = e^u$

$$\begin{aligned}\frac{1}{2}\int e^u u^2\,du &= \frac{1}{2}\left[u^2e^u - \int 2ue^u\,du\right]\\ &= \frac{1}{2}u^2e^u - \int ue^u\,du\end{aligned}$$

(2) $w = u,\ dw = du,\ dv = e^u\,du,\ v = e^u$

$$\begin{aligned}\frac{1}{2}\int e^u u^2\,du &= \frac{1}{2}u^2e^u - \left(ue^u - \int e^u\,du\right)\\ &= \frac{1}{2}u^2e^u - ue^u + e^u + C\\ &= \frac{1}{2}x^4e^{x^2} - x^2e^{x^2} + e^{x^2} + C\\ &= \frac{e^{x^2}}{2}\left(x^4 - 2x^2 + 2\right) + C\end{aligned}$$

58. Let $u = \sqrt{2x},\ u^2 = 2x,\ 2u\,du = 2dx.$

$$\int e^{\sqrt{2x}}\,dx = \int e^u(u\,du)$$

Now use integration by parts.

$dv = e^u\,du \Rightarrow v = \int e^u\,du = e^u$

$w = u \Rightarrow dw = du$

$$\begin{aligned}\int e^{\sqrt{2x}}\,dx &= ue^u - \int e^u\,du\\ &= ue^u - e^u + C\\ &= \sqrt{2x}\,e^{\sqrt{2x}} - e^{\sqrt{2x}} + C\end{aligned}$$

59. (a) Integration by parts is based on the Product Rule.

(b) Answers will vary. *Sample answer:* You want dv to be the most complicated portion of the integrand.

60. In order for the integration by parts technique to be efficient, you want dv to be the most complicated portion of the integrand and you want u to be the portion of the integrand whose derivative is a function simpler than u. Suppose you let $u = \sin x$ and $dv = x\,dx$. Then $du = \cos x\,dx$ and $v = x^2/2$. So

$$\int x\sin x\,dx = uv - \int v\,du = \frac{x^2}{2}\sin x - \int \frac{x^2}{2}\cos x\,dx,$$

which is a more complicated integral than the original one.

61. (a) No

Substitution

(b) Yes

$u = \ln x,\ dv = x\,dx$

(c) Yes

$u = x^2,\ dv = e^{-3x}\,dx$

(d) No

Substitution

(e) Yes. Let $u = x$ and

$$dv = \frac{1}{\sqrt{x+1}}\,dx.$$

(Substitution also works. Let $u = \sqrt{x+1}$.)

(f) No

Substitution

62. (a) The slope of f at $x = 2$ is approximately 1.4 because $f'(2) \approx 1.4$.

(b) $f' < 0$ on $(0, 1) \Rightarrow f$ is decreasing on $(0, 1)$.

$f' > 0$ on $(1, \infty) \Rightarrow f$ is increasing on $(1, \infty)$.

63. (a) $dv = \dfrac{x}{\sqrt{4 + x^2}}\,dx \quad \Rightarrow \quad v = \int (4 + x^2)^{-1/2} x\,dx = \sqrt{4 + x^2}$

$u = x^2 \quad \Rightarrow \quad du = 2x\,dx$

$$\int \frac{x^3}{\sqrt{4 + x^2}}\,dx = x^2\sqrt{4 + x^2} - 2\int x\sqrt{4 + x^2}\,dx$$

$$= x^2\sqrt{4 + x^2} - \frac{2}{3}(4 + x^2)^{3/2} + C = \frac{1}{3}\sqrt{4 + x^2}(x^2 - 8) + C$$

(b) $u = 4 + x^2 \Rightarrow x^2 = u - 4$ and $2x\,dx = du \Rightarrow x\,dx = \dfrac{1}{2}du$

$$\int \frac{x^3}{\sqrt{4 + x^2}}\,dx = \int \frac{x^2}{\sqrt{4 + x^2}}x\,dx = \int \left(\frac{u - 4}{\sqrt{u}}\right)\frac{1}{2}\,du$$

$$= \frac{1}{2}\int \left(u^{1/2} - 4u^{-1/2}\right)du = \frac{1}{2}\left(\frac{2}{3}u^{3/2} - 8u^{1/2}\right) + C$$

$$= \frac{1}{3}u^{1/2}(u - 12) + C$$

$$= \frac{1}{3}\sqrt{4 + x^2}\left[(4 + x^2) - 12\right] + C = \frac{1}{3}\sqrt{4 + x^2}(x^2 - 8) + C$$

64. (a) $dv = \sqrt{4 - x}\,dx \quad \Rightarrow \quad v = \int (4 - x)^{1/2}\,dx$

$= -\frac{2}{3}(4 - x)^{3/2}$

$u = x \quad \Rightarrow \quad du = dx$

$$\int x\sqrt{4 - x}\,dx = -\tfrac{2}{3}x(4 - x)^{3/2} + \tfrac{2}{3}\int (4 - x)^{3/2}\,dx$$

$$= -\tfrac{2}{3}x(4 - x)^{3/2} - \tfrac{4}{15}(4 - x)^{5/2} + C$$

$$= -\tfrac{2}{15}(4 - x)^{3/2}\left[5x + 2(4 - x)\right] + C = -\tfrac{2}{15}(4 - x)^{3/2}(3x + 8) + C$$

(b) $u = 4 - x \Rightarrow x = 4 - u$ and $dx = -du$

$$\int x\sqrt{4 - x}\,dx = -\int (4 - u)\sqrt{u}\,du$$

$$= -\int \left(4u^{1/2} - u^{3/2}\right)du$$

$$= -\tfrac{8}{3}u^{3/2} + \tfrac{2}{5}u^{5/2} + C$$

$$= -\tfrac{2}{15}u^{3/2}(20 - 3u) + C$$

$$= -\tfrac{2}{15}(4 - x)^{3/2}\left[20 - 3(4 - x)\right] + C$$

$$= -\tfrac{2}{15}(4 - x)^{3/2}(3x + 8) + C$$

65. $n = 0$: $\int \ln x \, dx = x(\ln x - 1) + C$

$n = 1$: $\int x \ln x \, dx = \frac{x^2}{4}(2 \ln x - 1) + C$

$n = 2$: $\int x^2 \ln x \, dx = \frac{x^3}{9}(3 \ln x - 1) + C$

$n = 3$: $\int x^3 \ln x \, dx = \frac{x^4}{16}(4 \ln x - 1) + C$

$n = 4$: $\int x^4 \ln x \, dx = \frac{x^5}{25}(5 \ln x - 1) + C$

In general, $\int x^n \ln x \, dx = \frac{x^{n+1}}{(n+1)^2}\left[(n+1)\ln x - 1\right] + C$.

66. $n = 0$: $\int e^x \, dx = e^x + C$

$n = 1$: $\int xe^x \, dx = xe^x - e^x + C = xe^x - \int e^x \, dx$

$n = 2$: $\int x^2e^x \, dx = x^2e^x - 2xe^x + 2e^x + C = x^2e^x - 2\int xe^x \, dx$

$n = 3$: $\int x^3e^x \, dx = x^3e^x - 3x^2e^x + 6xe^x - 6e^x + C = x^3e^x - 3\int x^2e^x \, dx$

$n = 4$: $\int x^4e^x \, dx = x^4e^x - 4x^3e^x + 12x^2e^x - 24xe^x + 24e^x + C = x^4e^x - 4\int x^3e^x \, dx$

In general, $\int x^ne^x \, dx = x^ne^x - n\int x^{n-1}e^x \, dx$.

67. $dv = \sin x \, dx \Rightarrow v = -\cos x$

$u = x^n \Rightarrow du = nx^{n-1} \, dx$

$\int x^n \sin x \, dx = -x^n \cos x + n\int x^{n-1} \cos x \, dx$

68. $dv = \cos x \, dx \Rightarrow v = \sin x$

$u = x^n \Rightarrow du = nx^{n-1} \, dx$

$\int x^n \cos x \, dx = x^n \sin x - n\int x^{n-1} \sin x \, dx$

69. $dv = x^n \, dx \Rightarrow v = \frac{x^{n+1}}{n+1}$

$u = \ln x \Rightarrow du = \frac{1}{x} \, dx$

$$\begin{aligned}\int x^n \ln x \, dx &= \frac{x^{n+1}}{n+1} \ln x - \int \frac{x^n}{n+1} \, dx \\ &= \frac{x^{n+1}}{n+1} \ln x - \frac{x^{n+1}}{(n+1)^2} + C \\ &= \frac{x^{n+1}}{(n+1)^2}\left[(n+1)\ln x - 1\right] + C\end{aligned}$$

70. $dv = e^{ax} \, dx \Rightarrow v = \frac{1}{a}e^{ax}$

$u = x^n \Rightarrow du = nx^{n-1}dx$

$\int x^n e^{ax} \, dx = \frac{x^ne^{ax}}{a} - \frac{n}{a}\int x^{n-1} e^{ax} \, dx$

71. Use integration by parts twice.

(1) $dv = e^{ax} \, dx \Rightarrow v = \frac{1}{a}e^{ax}$ (2) $dv = e^{ax} \, dx \Rightarrow v = \frac{1}{a}e^{ax}$

$u = \sin bx \Rightarrow du = b \cos bx \, dx$ $u = \cos bx \Rightarrow du = -b \sin bx \, dx$

$$\begin{aligned}\int e^{ax} \sin bx \, dx &= \frac{e^{ax} \sin bx}{a} - \frac{b}{a}\int e^{ax} \cos bx \, dx \\ &= \frac{e^{ax} \sin bx}{a} - \frac{b}{a}\left(\frac{e^{ax} \cos bx}{a} + \frac{b}{a}\int e^{ax} \sin bx \, dx\right) = \frac{e^{ax} \sin bx}{a} - \frac{b}{a^2} e^{ax} \cos bx - \frac{b^2}{a^2}\int e^{ax} \sin bx \, dx\end{aligned}$$

Therefore, $\left(1 + \frac{b^2}{a^2}\right)\int e^{ax} \sin bx \, dx = \frac{e^{ax}(a \sin bx - b \cos bx)}{a^2}$

$$\int e^{ax} \sin bx \, dx = \frac{e^{ax}(a \sin bx - b \cos bx)}{a^2 + b^2} + C.$$

72. Use integration by parts twice.

(1) $dv = e^{ax}\,dx \Rightarrow v = \frac{1}{a}e^{ax}$ (2) $dv = e^{ax}\,dx \Rightarrow v = \frac{1}{a}e^{ax}$

$u = \cos bx \Rightarrow du = -b\sin bx$ $u = \sin bx \Rightarrow du = b\cos bx$

$$\int e^{ax}\cos bx\,dx = \frac{e^{ax}\cos bx}{a} + \frac{b}{a}\int e^{ax}\sin bx\,dx = \frac{e^{ax}\cos bx}{a} + \frac{b}{a}\left(\frac{e^{ax}\sin bx}{a} - \frac{b}{a}\int e^{ax}\cos bx\,dx\right)$$

$$= \frac{e^{ax}\cos bx}{a} + \frac{be^{ax}\sin bx}{a^2} - \frac{b^2}{a^2}\int e^{ax}\cos bx\,dx$$

Therefore, $\left(1 + \frac{b^2}{a^2}\right)\int e^{ax}\cos bx\,dx = \frac{e^{ax}(a\cos bx + b\sin bx)}{a^2}$

$$\int e^{ax}\cos bx\,dx = \frac{e^{ax}(a\cos bx + b\sin bx)}{a^2 + b^2} + C.$$

73. $n = 2$ (Use formula in Exercise 67.)

$$\int x^2\sin x\,dx = -x^2\cos x + 2\int x\cos x\,dx$$

$$= -x^2\cos x + 2\left[x\sin x - \int\sin x\,dx\right]$$ (Use formula in Exercise 68; $(n = 1)$.)

$$= -x^2\cos x + 2x\sin x + 2\cos x + C$$

74. $n = 2$ (Use formula in Exercise 68.)

$$\int x^2\cos x\,dx = x^2\sin x - 2\int x\sin x\,dx,$$ (Use formula in Exercise 67.) $(n = 1)$

$$= x^2\sin x - 2\left(-x\cos x + \int\cos x\,dx\right) = x^2\sin x + 2x\cos x - 2\sin x + C$$

75. $n = 5$ (Use formula in Exercise 69.)

$$\int x^5\ln x\,dx = \frac{x^6}{6^2}(-1 + 6\ln x) + C = \frac{x^6}{36}(-1 + 6\ln x) + C$$

76. $n = 3, a = 2$ (Use formula in Exercise 70 three times.)

$$\int x^3e^{2x}\,dx = \frac{x^3e^{2x}}{2} - \frac{3}{2}\int x^2e^{2x}\,dx, \quad (n = 3, a = 2)$$

$$= \frac{x^3e^{2x}}{2} - \frac{3}{2}\left[\frac{x^2e^{2x}}{2} - \int xe^{2x}\,dx\right], \quad (n = 2, a = 2)$$

$$= \frac{x^3e^{2x}}{2} - \frac{3x^2e^{2x}}{4} + \frac{3}{2}\left[\frac{xe^{2x}}{2} - \frac{1}{2}\int e^{2x}\,dx\right]$$

$$= \frac{x^3e^{2x}}{2} - \frac{3x^2e^{2x}}{4} + \frac{3xe^{2x}}{4} - \frac{3e^{2x}}{8} + C, \quad (n = 1, a = 2)$$

$$= \frac{e^{2x}}{8}(4x^3 - 6x^2 + 6x - 3) + C$$

77. $a = -3, b = 4$ (Use formula in Exercise 71.)

$$\int e^{-3x}\sin 4x\,dx = \frac{e^{-3x}(-3\sin 4x - 4\cos 4x)}{(-3)^2 + (4)^2} + C$$

$$= \frac{-e^{-3x}(3\sin 4x + 4\cos 4x)}{25} + C$$

78. $a = 2, b = 3$ (Use formula in Exercise 72.)

$$\int e^{2x}\cos 3x\,dx = \frac{e^{2x}(2\cos 3x + 3\sin 3x)}{13} + C$$

79.

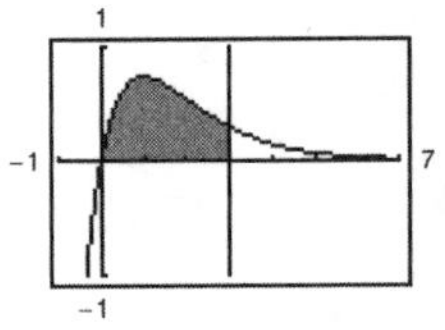

$$dv = e^{-x}\,dx \Rightarrow v = \int e^{-x}\,dx = -e^{-x}$$
$$u = 2x \Rightarrow du = 2\,dx$$
$$\int 2xe^{-x}dx = 2x\left(-e^{-x}\right) - \int -2e^{-x}\,dx$$
$$= -2xe^{-x} - 2e^{-x} + C$$
$$A = \int_0^3 2xe^{-x}\,dx = \left[-2xe^{-x} - 2e^{-x}\right]_0^3$$
$$= \left(-6e^{-3} - 2e^{-3}\right) - (-2)$$
$$= 2 - 8e^{-3} \approx 1.602$$

80.

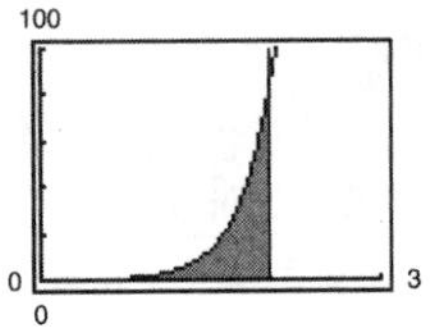

$$A = \int_0^2 \frac{1}{10}xe^{3x}\,dx = \frac{1}{10}\int_0^2 xe^{3x}\,dx$$
$$dv = e^{3x}\,dx \Rightarrow v = \int e^{3x}\,dx = \frac{1}{3}e^{3x}$$
$$u = x \Rightarrow du = dx$$
$$\frac{1}{10}\int xe^{3x}\,dx = \frac{1}{10}\left[\frac{x}{3}e^{3x} - \int \frac{1}{3}e^{3x}\,dx\right]$$
$$= \frac{x}{30}e^{3x} - \frac{1}{90}e^{3x} + C$$
$$A = \left[\frac{x}{30}e^{3x} - \frac{1}{90}e^{3x}\right]_0^2$$
$$= \left(\frac{1}{15}e^6 - \frac{1}{90}e^6\right) + \frac{1}{90}$$
$$= \frac{1}{90}\left(5e^6 + 1\right) \approx 22.424$$

81.

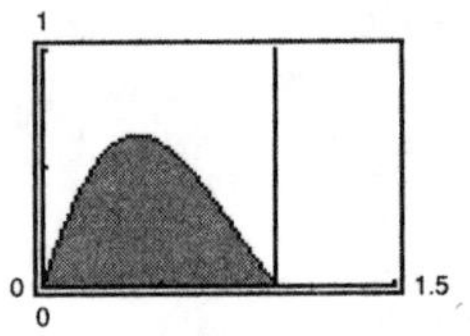

$$A = \int_0^1 e^{-x}\sin(\pi x)\,dx$$
$$= \left[\frac{e^{-x}(-\sin \pi x - \pi \cos \pi x)}{1 + \pi^2}\right]_0^1$$
$$= \frac{1}{1 + \pi^2}\left(\frac{\pi}{e} + \pi\right)$$
$$= \frac{\pi}{1 + \pi^2}\left(\frac{1}{e} + 1\right)$$
$$\approx 0.395 \quad \text{(See Exercise 71.)}$$

82.

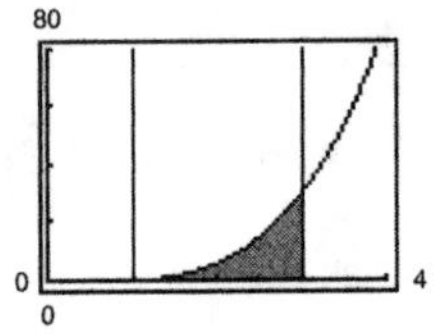

$$dv = x^3\,dx \Rightarrow v = \int x^3\,dx = \frac{x^4}{4}$$
$$u = \ln x \Rightarrow du = \frac{1}{x}\,dx$$
$$\int x^3 \ln x\,dx = \frac{x^4}{4}\ln x - \int \frac{x^4}{4}\left(\frac{1}{x}\,dx\right)$$
$$= \frac{x^4}{4}\ln x - \int \frac{x^3}{4}\,dx$$
$$= \frac{x^4}{4}\ln x - \frac{x^4}{16} + C$$
$$A = \int_1^3 x^3 \ln x\,dx = \left[\frac{x^4}{4}\ln x - \frac{x^4}{16}\right]_1^3$$
$$= \left(\frac{81}{4}\ln 3 - \frac{81}{16}\right) + \frac{1}{16}$$
$$= \frac{81}{4}\ln 3 - 5 \approx 17.247$$

83. (a) $dv = dx \Rightarrow v = x$

$$u = \ln x \Rightarrow du = \frac{1}{x}\,dx$$
$$A = \int_1^e \ln x\,dx = \left[x \ln x - x\right]_1^e = 1 \qquad \text{(Use integration by parts once.)}$$

(b) $R(x) = \ln x, r(x) = 0$

$$V = \pi \int_1^e (\ln x)^2\, dx$$

$$= \pi\left[x(\ln x)^2 - 2x \ln x + 2x\right]_1^e \qquad \text{(Use integration by parts twice, see Exercise 3.)}$$

$$= \pi(e - 2) \approx 2.257$$

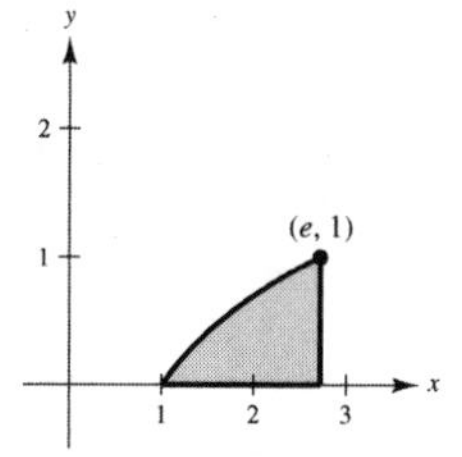

(c) $p(x) = x, h(x) = \ln x$

$$V = 2\pi \int_1^e x \ln x\, dx = 2\pi\left[\frac{x^2}{4}(-1 + 2\ln x)\right]_1^e$$

$$= \frac{(e^2 + 1)\pi}{2} \approx 13.177 \text{ (See Exercise 91.)}$$

(d) $$\bar{x} = \frac{\int_1^e x \ln x\, dx}{1} = \frac{e^2 + 1}{4} \approx 2.097$$

$$\bar{y} = \frac{\frac{1}{2}\int_1^e (\ln x)^2\, dx}{1} = \frac{e - 2}{2} \approx 0.359$$

$$(\bar{x}, \bar{y}) = \left(\frac{e^2 + 1}{4}, \frac{e - 2}{2}\right) \approx (2.097, 0.359)$$

84. $y = x \sin x, \quad 0 \le x \le \pi$

(a) $$A = \int_0^\pi x \sin x\, dx$$

$$= -x \cos x + \int \cos x\, dx \text{ (Excercise 67)}$$

$$= -x \cos x + \sin x\Big]_0^\pi$$

$$= -\pi(-1) = \pi$$

(b) $$V = \int_0^\pi \pi(x \sin x)^2\, dx = \pi \int_0^\pi x^2 \sin^2 x\, dx$$

Let $u = x^2, du = 2x\, dx, dv = \sin^2 x\, dx = \dfrac{1 - \cos 2x}{2}\, dx, v = \dfrac{1}{2}x - \dfrac{\sin 2x}{4}$.

$$\int x^2 \sin^2 x\, dx = x^2\left(\frac{1}{2}x - \frac{\sin 2x}{4}\right) - \int\left(\frac{1}{2}x - \frac{\sin 2x}{4}\right)(2x\, dx)$$

$$= \frac{1}{2}x^3 - \frac{x^2 \sin 2x}{4} - \int\left(x^2 - \frac{x \sin 2x}{2}\right) dx$$

$$= \frac{1}{2}x^3 - \frac{x^2 \sin 2x}{4} - \frac{x^3}{3} + \int \frac{x \sin 2x}{2}\, dx$$

$$= \frac{1}{6}x^3 - \frac{1}{4}x^2 \sin 2x + \frac{1}{8}(\sin 2x - 2x \cos 2x) + C \qquad \text{(Integration by Parts)}$$

$$V = \pi \int_0^\pi x^2 \sin^2 x\, dx = \pi\left[\frac{1}{6}x^3 - \frac{1}{4}x^2 \sin 2x + \frac{1}{8}(\sin 2x - 2x \cos 2x)\right]_0^\pi = \frac{1}{6}\pi^4 - \frac{1}{4}\pi^2$$

(c) $V = \int_0^{\pi} 2\pi x(x \sin x)dx = 2\pi\Big[2 \cos x + 2x \sin x - x^2 \cos x\Big]_0^{\pi} = 2\pi(\pi^2 - 4) = 2\pi^3 - 8\pi$

(d) $m = \int_0^{\pi} x \sin(x)\, dx = \big[\sin x - x \cos x\big]_0^{\pi} = \pi$

$$M_x = \int_0^{\pi} \frac{1}{2}(x \sin x)^2\, dx = \frac{1}{2}\left(\frac{1}{6}\pi^3 - \frac{1}{4}\pi\right) \quad \text{(See part (a).)} = \frac{1}{12}\pi^3 - \frac{1}{8}\pi$$

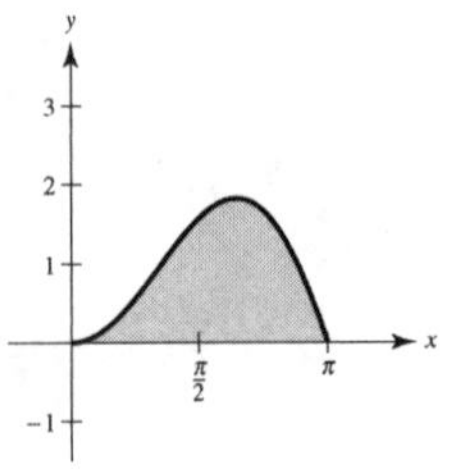

$$M_y = \int_0^{\pi} x(x \sin x)\, dx = \pi^2 - 4 \quad \text{(See part (b).)}$$

$$\bar{x} = \frac{M_y}{m} = \frac{\pi^2 - 4}{\pi} \approx 1.8684, \quad \bar{y} = \frac{M_x}{m} = \frac{(1/12)\pi^3 - (1/8)\pi}{\pi} = \frac{1}{2}\pi^2 - \frac{1}{8} \approx 0.6975$$

85. In Example 6, you showed that the centroid of an equivalent region was $(1, \pi/8)$. By symmetry, the centroid of this region is $(\pi/8, 1)$. You can also solve this problem directly.

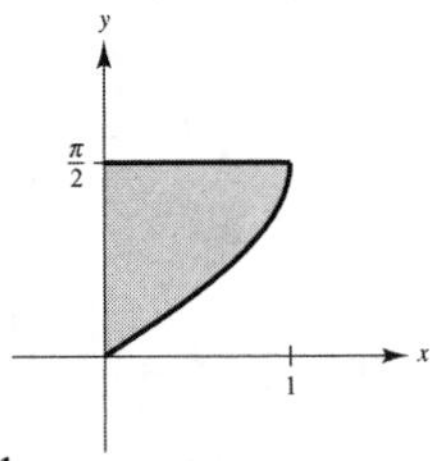

$$A = \int_0^1 \left(\frac{\pi}{2} - \arcsin x\right) dx = \left[\frac{\pi}{2}x - x \arcsin x - \sqrt{1 - x^2}\right]_0^1 \quad \text{(Example 3)}$$

$$= \left(\frac{\pi}{2} - \frac{\pi}{2} - 0\right) - (-1) = 1$$

$$\bar{x} = \frac{M_y}{A} = \int_0^1 x\left(\frac{\pi}{2} - \arcsin x\right) dx = \frac{\pi}{8}, \quad \bar{y} = \frac{M_x}{A} = \int_0^1 \frac{(\pi/2) + \arcsin x}{2}\left(\frac{\pi}{2} - \arcsin x\right) dx = 1$$

86. $f(x) = x^2, g(x) = 2^x$

$f(2) = g(2) = 4, f(4) = g(4) = 16$

$$m = \int_2^4 (x^2 - 2^x)\, dx = \left[\frac{x^3}{3} - \frac{1}{\ln 2}2^x\right]_2^4 = \left(\frac{64}{3} - \frac{16}{\ln 2}\right) - \left(\frac{8}{3} - \frac{4}{\ln 2}\right) = \frac{56}{3} - \frac{12}{\ln 2} \approx 1.3543$$

$$M_x = \int_2^4 \frac{1}{2}(x^2 + 2^x)(x^2 - 2^x)\, dx$$

$$= \frac{1}{2}\int_2^4 (x^4 - 2^{2x})\, dx$$

$$= \frac{1}{2}\left[\frac{x^5}{5} - \frac{2^{2x}}{2 \ln 2}\right]_2^4$$

$$= \frac{1}{2}\left[\left(\frac{1024}{5} - \frac{128}{\ln 2}\right) - \left(\frac{32}{5} - \frac{8}{\ln 2}\right)\right]$$

$$= \frac{496}{5} - \frac{60}{\ln 2} \approx 12.6383$$

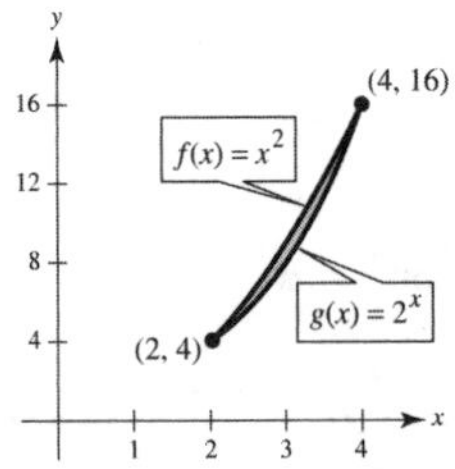

$$M_y = \int_2^4 x\left[x^2 - 2^x\right] dx = -\frac{56}{\ln 2} + \frac{12}{(\ln 2)^2} \approx 4.1855$$

$$(\bar{x}, \bar{y}) = \left(\frac{M_y}{m}, \frac{M_x}{m}\right) \approx (3.0905, 9.3318)$$

87. Average value $= \frac{1}{\pi}\int_0^{\pi} e^{-4t}(\cos 2t + 5\sin 2t)\,dt$

$$= \frac{1}{\pi}\left[e^{-4t}\left(\frac{-4\cos 2t + 2\sin 2t}{20}\right) + 5e^{-4t}\left(\frac{-4\sin 2t - 2\cos 2t}{20}\right)\right]_0^{\pi} \quad \text{(From Exercises 71 and 72)}$$

$$= \frac{7}{10\pi}\left(1 - e^{-4\pi}\right) \approx 0.223$$

88. (a) Average $= \int_1^2 (1.6t\ln t + 1)\,dt = \left[0.8t^2\ln t - 0.4t^2 + t\right]_1^2 = 3.2(\ln 2) - 0.2 \approx 2.018$

(b) Average $= \int_3^4 (1.6t\ln t + 1)\,dt = \left[0.8t^2\ln t - 0.4t^2 + t\right]_3^4 = 12.8(\ln 4) - 7.2(\ln 3) - 1.8 \approx 8.035$

89. $c(t) = 100{,}000 + 4000t,\ r = 5\%,\ t_1 = 10$

$$P = \int_0^{10}(100{,}000 + 4000t)e^{-0.05t}\,dt = 4000\int_0^{10}(25 + t)e^{-0.05t}\,dt$$

Let $u = 25 + t,\ dv = e^{-0.05t}dt,\ du = dt,\ v = -\frac{100}{5}e^{-0.05t}$.

$$P = 4000\left\{\left[(25 + t)\left(-\frac{100}{5}e^{-0.05t}\right)\right]_0^{10} + \frac{100}{5}\int_0^{10} e^{-0.05t}dt\right\} = 4000\left\{\left[(25 + t)\left(-\frac{100}{5}e^{-0.05t}\right)\right]_0^{10} - \left[\frac{10{,}000}{25}e^{-0.05t}\right]_0^{10}\right\} \approx \$931{,}265$$

90. $c(t) = 30{,}000 + 500t,\ r = 7\%,\ t_1 = 5$

$$P\int_0^5 (30{,}000 + 500t)e^{-0.07t}dt = 500\int_0^5 (60 + t)e^{-0.07t}\,dt$$

Let $u = 60 + t,\ dv = e^{-0.07t}dt,\ du = dt,\ v = -\frac{100}{7}e^{-0.07t}$.

$$P = 500\left\{\left[(60 + t)\left(-\frac{100}{7}e^{-0.07t}\right)\right]_0^5 + \frac{100}{7}\int_0^5 e^{-0.07t}\,dt\right\} = 500\left\{\left[(60 + t)\left(-\frac{100}{7}e^{-0.07t}\right)\right]_0^5 - \left[\frac{10{,}000}{49}e^{-0.07t}\right]_0^5\right\} \approx \$131{,}528.68$$

91. $$\int_{-\pi}^{\pi} x\sin nx\,dx = \left[-\frac{x}{n}\cos nx + \frac{1}{n^2}\sin nx\right]_{-\pi}^{\pi} = -\frac{\pi}{n}\cos \pi n - \frac{\pi}{n}\cos(-\pi n) = -\frac{2\pi}{n}\cos \pi n = \begin{cases} -(2\pi/n), & \text{if } n \text{ is even} \\ (2\pi/n), & \text{if } n \text{ is odd} \end{cases}$$

92. $$\int_{-\pi}^{\pi} x^2\cos nx\,dx = \left[\frac{x^2}{n}\sin nx + \frac{2x}{n^2}\cos nx - \frac{2}{n^3}\sin nx\right]_{-\pi}^{\pi}$$

$$= \frac{2\pi}{n^2}\cos n\pi + \frac{2\pi}{n^2}\cos(-n\pi)$$

$$= \frac{4\pi}{n^2}\cos n\pi$$

$$= \begin{cases} (4\pi/n^2), & \text{if } n \text{ is even} \\ -(4\pi/n^2), & \text{if } n \text{ is odd} \end{cases}$$

$$= \frac{(-1)^n 4\pi}{n^2}$$

93. Let $u = x$, $dv = \sin\left(\frac{n\pi}{2}x\right)dx$, $du = dx$, $v = -\frac{2}{n\pi}\cos\left(\frac{n\pi}{2}x\right)$.

$$I_1 = \int_0^1 x\sin\left(\frac{n\pi}{2}x\right)dx = \left[\frac{-2x}{n\pi}\cos\left(\frac{n\pi}{2}x\right)\right]_0^1 + \frac{2}{n\pi}\int_0^1 \cos\left(\frac{n\pi}{2}x\right)dx$$

$$= -\frac{2}{n\pi}\cos\left(\frac{n\pi}{2}\right) + \left[\left(\frac{2}{n\pi}\right)^2 \sin\left(\frac{n\pi}{2}x\right)\right]_0^1$$

$$= -\frac{2}{n\pi}\cos\left(\frac{n\pi}{2}\right) + \left(\frac{2}{n\pi}\right)^2 \sin\left(\frac{n\pi}{2}\right)$$

Let $u = (-x + 2)$, $dv = \sin\left(\frac{n\pi}{2}x\right)dx$, $du = -dx$, $v = -\frac{2}{n\pi}\cos\left(\frac{n\pi}{2}x\right)$.

$$I_2 = \int_1^2 (-x+2)\sin\left(\frac{n\pi}{2}x\right)dx = \left[\frac{-2(-x+2)}{n\pi}\cos\left(\frac{n\pi}{2}x\right)\right]_1^2 - \frac{2}{n\pi}\int_1^2 \cos\left(\frac{n\pi}{2}x\right)dx$$

$$= \frac{2}{n\pi}\cos\left(\frac{n\pi}{2}\right) - \left[\left(\frac{2}{n\pi}\right)^2 \sin\left(\frac{n\pi}{2}x\right)\right]_1^2$$

$$= \frac{2}{n\pi}\cos\left(\frac{n\pi}{2}\right) + \left(\frac{2}{n\pi}\right)^2 \sin\left(\frac{n\pi}{2}\right)$$

$$h(I_1 + I_2) = b_n = h\left[\left(\frac{2}{n\pi}\right)^2 \sin\left(\frac{n\pi}{2}\right) + \left(\frac{2}{n\pi}\right)^2 \sin\left(\frac{n\pi}{2}\right)\right] = \frac{8h}{(n\pi)^2}\sin\left(\frac{n\pi}{2}\right)$$

94. $f'(x) = xe^{-x}$

(a) $f(x) = \int xe^{-x}\,dx = -xe^{-x} - e^{-x} + C$

(Parts: $u = x$, $dv = e^{-x}\,dx$)

$f(0) = 0 = -1 + C \Rightarrow C = 1$

$f(x) = -xe^{-x} - e^{-x} + 1$

(b)

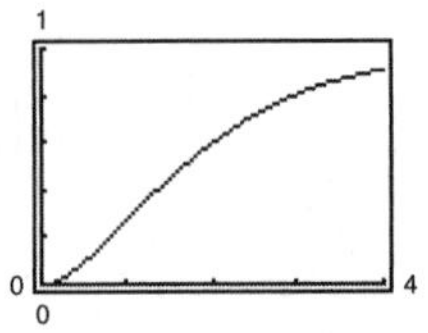

(c) Using $h = 0.05$ you obtain the points:

n	x_n	y_n
0	0	0
1	0.05	0
2	0.10	2.378×10^{-3}
3	0.15	0.0069
4	0.20	0.0134
⋮	⋮	⋮
80	4.0	0.9064

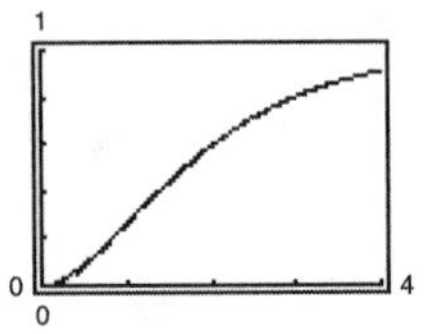

(d) Using $h = 0.1$ you obtain the points:

n	x_n	y_n
0	0	0
1	0.1	0
2	0.2	0.0090484
3	0.3	0.025423
4	0.4	0.047648
⋮	⋮	⋮
40	4.0	0.9039

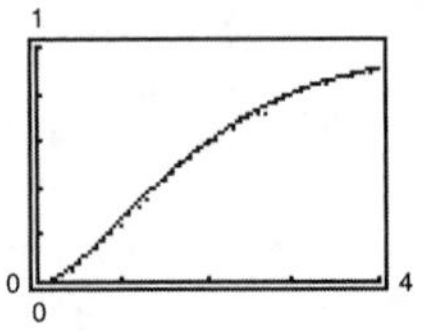

(e) The result in part (c) is better because h is smaller.

95. $f'(x) = 3x\sin(2x),\ f(0) = 0$

(a) $f(x) = \int 3x \sin 2x\, dx$

$= -\frac{3}{4}(2x\cos 2x - \sin 2x) + C$

(Parts: $u = 3x,\ dv = \sin 2x\, dx$)

$f(0) = 0 = -\frac{3}{4}(0) + C \Rightarrow C = 0$

$f(x) = -\frac{3}{4}(2x\cos 2x - \sin 2x)$

(b)

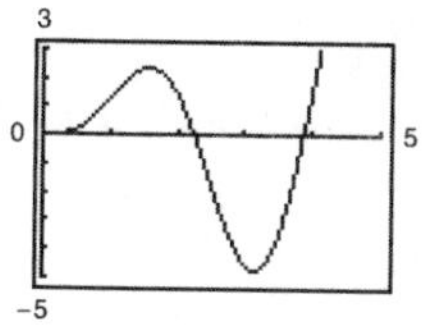

(c) Using $h = 0.05$, you obtain the points:

n	x_n	y_n
0	0	0
1	0.05	0
2	0.10	7.4875×10^{-4}
3	0.15	0.0037
4	0.20	0.0104
⋮	⋮	⋮
80	4.0	1.3181

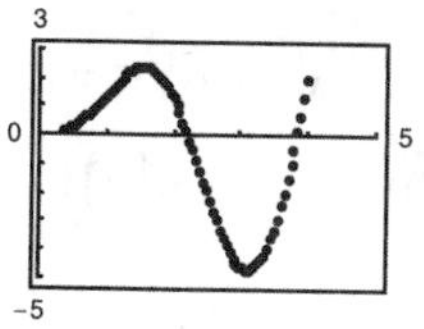

(d) Using $h = 0.1$, you obtain the points:

n	x_n	y_n
0	0	0
1	0.1	0
2	0.2	0.0060
3	0.3	0.0293
4	0.4	0.0801
⋮	⋮	⋮
40	4.0	1.0210

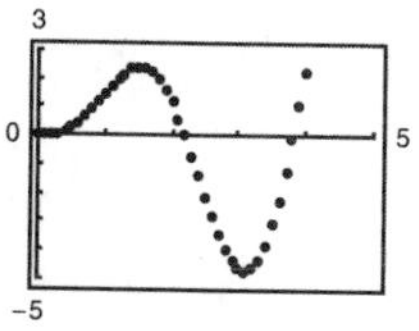

96. $f'(x) = \cos\sqrt{x},\ f(0) = 1$

(a) Let $w = \sqrt{x},\ w^2 = x,\ 2w\,dw = dx.$

$\int \cos\sqrt{x}\, dx = \int \cos w(2w\, dw)$

Now use parts: $u = 2w,\ dv = \cos w\, dw.$

$\int \cos\sqrt{x}\, dx = 2w\sin w + 2\cos w + C$

$= 2\sqrt{x}\sin\sqrt{x} + 2\cos\sqrt{x} + C$

$f(0) = 1 = 2 + C \Rightarrow C = -1$

$f(x) = 2\sqrt{x}\sin\sqrt{x} + 2\cos\sqrt{x} - 1$

(b)

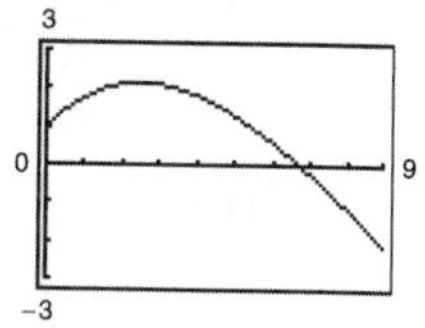

(c) Using $h = 0.05$, you obtain the points:

n	x_n	y_n
0	0	1
1	0.05	1.05
2	0.1	1.0988
3	0.15	1.1463
4	0.2	1.1926
⋮	⋮	⋮
80	4.0	1.8404

(d) Using $h = 0.1$, you obtain the points:

n	x_n	y_n
0	0	1
1	0.1	1.1
2	0.2	1.1950
3	0.3	1.2852
4	0.4	1.3706
⋮	⋮	⋮
80	4.0	1.8759

97. On $\left[0, \frac{\pi}{2}\right]$, $\sin x \le 1 \Rightarrow x \sin x \le x \Rightarrow \int_0^{\pi/2} x \sin x\, dx \le \int_0^{\pi/2} x\, dx.$

98. (a) $A = \int_0^{\pi} x \sin x\, dx = \left[\sin x - x \cos x\right]_0^{\pi} = \pi$

(b) $\int_{\pi}^{2\pi} x \sin x\, dx = \left[\sin x - x \cos x\right]_{\pi}^{2\pi} = -2\pi - \pi = -3\pi$

$A = 3\pi$

(c) $\int_{2\pi}^{3\pi} x \sin x\, dx = \left[\sin x - x \cos x\right]_{2\pi}^{3\pi} = 3\pi + 2\pi = 5\pi$

$A = 5\pi$

The area between $y = x \sin x$ and $y = 0$ on $\left[n\pi, (n + 1)\pi\right]$ is $(2n + 1)\pi$:

$$\int_{n\pi}^{(n+1)\pi} x \sin x\, dx = \left[\sin x - x \cos x\right]_{n\pi}^{(n+1)\pi} = \pm(n + 1)\pi \pm n\pi = \pm(2n + 1)\pi$$

$$A = \left|\pm(2n + 1)\pi\right| = (2n + 1)\pi$$

99. For any integrable function, $\int f(x)dx = C + \int f(x)dx$, but this cannot be used to imply that $C = 0$.

Section 8.3 Trigonometric Integrals

1. Let $u = \cos x$, $du = -\sin x\, dx$.

$$\int \cos^5 x \sin x\, dx = -\int \cos^5 x(-\sin x)\, dx = -\frac{\cos^6 x}{6} + C$$

2. $\int \cos^3 x \sin^4 x\, dx = \int \cos x\left(1 - \sin^2 x\right)\sin^4 x\, dx = \int\left(\sin^4 x - \sin^6 x\right)\cos x\, dx = \frac{\sin^5 x}{5} - \frac{\sin^7 x}{7} + C$

3. Let $u = \sin 2x$, $du = 2\cos 2x\, dx$.

$$\begin{aligned}\int \sin^7 2x \cos 2x\, dx &= \frac{1}{2}\int \sin^7 2x(2\cos 2x)\, dx \\ &= \frac{1}{2}\left(\frac{\sin^8 2x}{8}\right) + C \\ &= \frac{1}{16}\sin^8 2x + C\end{aligned}$$

4.
$$\begin{aligned}\int \sin^3 3x\, dx &= \int \sin^2 3x \sin 3x\, dx \\ &= \int\left(1 - \cos^2 3x\right)\sin 3x\, dx \\ &= \int \sin 3x\, dx - \int \cos^2 3x(\sin 3x\, dx) \\ &= -\frac{1}{3}\cos 3x + \frac{\cos^3 3x}{9} + C\end{aligned}$$

5.
$$\begin{aligned}\int \sin^3 x \cos^2 x\, dx &= \int\left(1 - \cos^2 x\right)\cos^2 x \sin x\, dx \\ &= \int\left(\cos^2 x - \cos^4 x\right)\sin x\, dx \\ &= -\int\left(\cos^2 x - \cos^4 x\right)(-\sin x)\, dx \\ &= -\frac{\cos^3 x}{3} + \frac{\cos^5 x}{5} + C\end{aligned}$$

6. Let $u = \sin\frac{x}{3}$, $du = \frac{1}{3}\cos\frac{x}{3}\, dx$.

$$\begin{aligned}\int \cos^3\frac{x}{3}\, dx &= \int\left(\cos\frac{x}{3}\right)\left(1 - \sin^2\frac{x}{3}\right)dx \\ &= 3\int\left(1 - \sin^2\frac{x}{3}\right)\left(\frac{1}{3}\cos\frac{x}{3}\right)dx \\ &= 3\left(\sin\frac{x}{3} - \frac{1}{3}\sin^3\frac{x}{3}\right) + C \\ &= 3\sin\frac{x}{3} - \sin^3\frac{x}{3} + C\end{aligned}$$

7. $\int \sin^3 2\theta\sqrt{\cos 2\theta}\,d\theta = \int (1 - \cos^2 2\theta)\sqrt{\cos 2\theta}\sin 2\theta\,d\theta$

$= \int \left[(\cos 2\theta)^{1/2} - (\cos 2\theta)^{5/2}\right]\sin 2\theta\,d\theta$

$= -\frac{1}{2}\int \left[(\cos 2\theta)^{1/2} - (\cos 2\theta)^{5/2}\right](-2\sin 2\theta)\,d\theta$

$= -\frac{1}{2}\left[\frac{2}{3}(\cos 2\theta)^{3/2} - \frac{2}{7}(\cos 2\theta)^{7/2}\right] + C$

$= -\frac{1}{3}(\cos 2\theta)^{3/2} + \frac{1}{7}(\cos 2\theta)^{7/2} + C$

8. $\int \frac{\cos^5 t}{\sqrt{\sin t}}\,dt = \int \cos t(1 - \sin^2 t)^2(\sin t)^{-1/2}\,dt$

$= \int (1 - 2\sin^2 t + \sin^4 t)(\sin t)^{-1/2}\cos t\,dt$

$= \int \left[(\sin t)^{-1/2} - 2(\sin t)^{3/2} + (\sin t)^{7/2}\right]\cos t\,dt$

$= 2\sqrt{\sin t} - \frac{4}{5}(\sin t)^{5/2} + \frac{2}{9}(\sin t)^{9/2} + C$

9. $\int \cos^2 3x\,dx = \int \frac{1 + \cos 6x}{2}\,dx = \frac{1}{2}\left(x + \frac{1}{6}\sin 6x\right) + C = \frac{1}{12}(6x + \sin 6x) + C$

10. $\int \sin^4 6\theta\,d\theta = \int \left(\frac{1 - \cos 12\theta}{2}\right)\left(\frac{1 - \cos 12\theta}{2}\right)d\theta$

$= \frac{1}{4}\int (1 - 2\cos 12\theta + \cos^2 12\theta)\,d\theta$

$= \frac{1}{4}\int \left(1 - 2\cos 12\theta + \frac{1 + \cos 24\theta}{2}\right)d\theta$

$= \frac{1}{4}\int \left(\frac{3}{2} - 2\cos 12\theta + \frac{1}{2}\cos 24\theta\right)d\theta$

$= \frac{1}{4}\left(\frac{3}{2}\theta - \frac{1}{6}\sin 12\theta + \frac{1}{48}\sin 24\theta\right) + C = \frac{3}{8}\theta - \frac{1}{24}\sin 12\theta + \frac{1}{192}\sin 24\theta + C$

11. Integration by parts:

$dv = \sin^2 x\,dx = \frac{1 - \cos 2x}{2} \Rightarrow v = \frac{x}{2} - \frac{\sin 2x}{4} = \frac{1}{4}(2x - \sin 2x)$

$u = x \Rightarrow du = dx$

$\int x\sin^2 x\,dx = \frac{1}{4}x(2x - \sin 2x) - \frac{1}{4}\int (2x - \sin 2x)\,dx$

$= \frac{1}{4}x(2x - \sin 2x) - \frac{1}{4}\left(x^2 + \frac{1}{2}\cos 2x\right) + C = \frac{1}{8}(2x^2 - 2x\sin 2x - \cos 2x) + C$

12. Use integration by parts twice.

$$dv = \sin^2 x\,dx = \frac{1 - \cos 2x}{2} \Rightarrow v = \frac{x}{2} - \frac{\sin 2x}{4} = \frac{1}{4}(2x - \sin 2x)$$

$$u = x^2 \Rightarrow du = 2x\,dx$$

$$dv = \sin 2x\,dx \Rightarrow v = -\frac{1}{2}\cos 2x$$

$$u = x \Rightarrow du = dx$$

$$\begin{aligned}\int x^2 \sin^2 x\,dx &= \frac{1}{4}x^2(2x - \sin 2x) - \frac{1}{2}\int\left(2x^2 - x\sin 2x\right)dx\\ &= \frac{1}{2}x^3 - \frac{1}{4}x^2\sin 2x - \frac{1}{3}x^3 + \frac{1}{2}\int x\sin 2x\,dx\\ &= \frac{1}{6}x^3 - \frac{1}{4}x^2\sin 2x + \frac{1}{2}\left(-\frac{1}{2}x\cos 2x + \frac{1}{2}\int\cos 2x\,dx\right)\\ &= \frac{1}{6}x^3 - \frac{1}{4}x^2\sin 2x - \frac{1}{4}x\cos 2x + \frac{1}{8}\sin 2x + C\\ &= \frac{1}{24}\left(4x^3 - 6x^2\sin 2x - 6x\cos 2x + 3\sin 2x\right) + C\end{aligned}$$

13. $\int_0^{\pi/2}\cos^7 x\,dx = \left(\frac{2}{3}\right)\left(\frac{4}{5}\right)\left(\frac{6}{7}\right) = \frac{16}{35},\ (n = 7)$

14. $\int_0^{\pi/2}\cos^9 x\,dx = \left(\frac{2}{3}\right)\left(\frac{4}{5}\right)\left(\frac{6}{7}\right)\left(\frac{8}{9}\right) = \frac{128}{315},\ (n = 9)$

15.
$$\begin{aligned}\int_0^{\pi/2}\cos^{10} x\,dx &= \left(\frac{1}{2}\right)\left(\frac{3}{4}\right)\left(\frac{5}{6}\right)\left(\frac{7}{8}\right)\left(\frac{9}{10}\right)\left(\frac{\pi}{2}\right)\\ &= \frac{63}{512}\pi,\ (n = 10)\end{aligned}$$

16. $\int_0^{\pi/2}\sin^5 x\,dx = \left(\frac{2}{3}\right)\left(\frac{4}{5}\right) = \frac{8}{15},\quad (n = 5)$

17. $\int_0^{\pi/2}\sin^6 x\,dx = \left(\frac{1}{2}\right)\left(\frac{3}{4}\right)\left(\frac{5}{6}\right)\frac{\pi}{2} = \frac{5\pi}{32},\ (n = 6)$

18. $\int_0^{\pi/2}\sin^8 x\,dx = \left(\frac{1}{2}\right)\left(\frac{3}{4}\right)\left(\frac{5}{6}\right)\left(\frac{7}{8}\right)\left(\frac{\pi}{2}\right) = \frac{35\pi}{256},\ (n = 8)$

19.
$$\begin{aligned}\int\sec 4x\,dx &= \frac{1}{4}\int\sec 4x(4\,dx)\\ &= \frac{1}{4}\ln\left|\sec 4x + \tan 4x\right| + C\end{aligned}$$

20.
$$\begin{aligned}\int\sec^4 2x\,dx &= \int\left(1 + \tan^2 2x\right)\sec^2 2x\,dx\\ &= \frac{1}{2}\tan 2x + \frac{\tan^3 2x}{6} + C\end{aligned}$$

21. $dv = \sec^2 \pi x\,dx \Rightarrow v = \frac{1}{\pi}\tan\pi x$

$u = \sec\pi x \Rightarrow du = \pi\sec\pi x\tan\pi x\,dx$

$$\int\sec^3\pi x\,dx = \frac{1}{\pi}\sec\pi x\tan\pi x - \int\sec\pi x\tan^2\pi x\,dx = \frac{1}{\pi}\sec\pi x\tan\pi x - \int\sec\pi x\left(\sec^2\pi x - 1\right)dx$$

$$2\int\sec^3\pi x\,dx = \frac{1}{\pi}\left(\sec\pi x\tan\pi x + \ln\left|\sec\pi x + \tan\pi x\right|\right) + C_1$$

$$\int\sec^3\pi x\,dx = \frac{1}{2\pi}\left(\sec\pi x\tan\pi x + \ln\left|\sec\pi x + \tan\pi x\right|\right) + C$$

22.
$$\begin{aligned}\int\tan^6 3x\,dx &= \int\left(\sec^2 3x - 1\right)\tan^4 3x\,dx\\ &= \int\tan^4 3x\sec^2 3x\,dx - \int\tan^4 3x\,dx\\ &= \int\tan^4 3x\sec^2 3x\,dx - \int\tan^2 3x\left(\sec^2 3x - 1\right)dx\\ &= \int\tan^4 3x\sec^2 3x\,dx - \int\tan^2 3x\sec^2 3x\,dx + \int\left(\sec^2 3x + 1\right)dx\\ &= \frac{\tan^5 3x}{15} - \frac{\tan^3 3x}{9} + \frac{\tan 3x}{3} + x + C\end{aligned}$$

23. $$\int \tan^5 \frac{x}{2}\,dx = \int\left(\sec^2 \frac{x}{2} - 1\right)\tan^3 \frac{x}{2}\,dx$$
$$= \int \tan^3 \frac{x}{2} \sec^2 \frac{x}{2}\,dx - \int \tan^3 \frac{x}{2}\,dx$$
$$= \frac{\tan^4 \frac{x}{2}}{2} - \int\left(\sec^2 \frac{x}{2} - 1\right)\tan \frac{x}{2}\,dx$$
$$= \frac{1}{2}\tan^4 \frac{x}{2} - \tan^2 \frac{x}{2} - 2\ln\left|\cos \frac{x}{2}\right| + C$$

24. $$\int \tan^3 \frac{\pi x}{2} \sec^2 \frac{\pi x}{2}\,dx = \frac{1}{2\pi}\tan^4 \frac{\pi x}{2} + C$$

25. Let $u = \sec 2t$, $du = 2\sec 2t \tan 2t$.

$$\int \tan^3 2t \cdot \sec^3 2t\,dt = \int\left(\sec^2 2t - 1\right)\sec^3 2t \cdot \tan 2t\,dt$$
$$= \int\left(\sec^4 2t - \sec^2 2t\right)\left(\sec 2t \tan 2t\right)dt = \frac{\sec^5 2t}{10} - \frac{\sec^3 2t}{6} + C$$

26. $$\int \tan^5 2x \sec^4 2x\,dx = \int \tan^5 2x\left(\tan^2 2x + 1\right)\sec^2 2x\,dx$$
$$= \int \tan^7 2x \sec^2 2x\,dx + \int \tan^5 2x \sec^2 2x\,dx$$
$$= \frac{1}{2}\left(\frac{\tan^8 2x}{8}\right) + \frac{1}{2}\left(\frac{\tan^6 2x}{6}\right) + C$$
$$= \frac{\tan^8 2x}{16} + \frac{\tan^6 2x}{12} + C$$

27. $$\int \sec^6 4x \tan 4x\,dx = \frac{1}{4}\int \sec^5 4x\left(4\sec 4x \tan 4x\right)dx$$
$$= \frac{\sec^6 4x}{24} + C$$

28. $$\int \sec^2 \frac{x}{2}\tan \frac{x}{2}\,dx = 2\int \sec \frac{x}{2}\left(\frac{1}{2}\sec \frac{x}{2}\tan \frac{x}{2}\right)dx$$
$$= \sec^2 \frac{x}{2} + C \quad \text{or}$$
$$\int \sec^2 \frac{x}{2}\tan \frac{x}{2}\,dx = 2\int \tan \frac{x}{2}\left(\frac{1}{2}\sec^2 \frac{x}{2}\right)dx = \tan^2 \frac{x}{2} + C$$

29. $$\int \sec^5 x \tan^3 x\,dx = \int \sec^4 x \tan^2 x\left(\sec x \tan x\right)dx$$
$$= \int \sec^4 x\left(\sec^2 x - 1\right)\left(\sec x \tan x\right)dx$$
$$= \int\left(\sec^6 x - \sec^4 x\right)\left(\sec x \tan x\right)dx$$
$$= \frac{\sec^7 x}{7} - \frac{\sec^5 x}{5} + C$$

30. $$\int \tan^3 3x\,dx = \int\left(\sec^2 3x - 1\right)\tan 3x\,dx$$
$$= \frac{1}{3}\int \tan 3x\left(3\sec^2 3x\right)dx + \frac{1}{3}\int \frac{-3\sin 3x}{\cos 3x}\,dx$$
$$= \frac{1}{6}\tan^2 3x + \frac{1}{3}\ln\left|\cos 3x\right| + C$$

31. $$\int \frac{\tan^2 x}{\sec x}\,dx = \int \frac{\left(\sec^2 x - 1\right)}{\sec x}\,dx$$
$$= \int\left(\sec x - \cos x\right)dx$$
$$= \ln\left|\sec x + \tan x\right| - \sin x + C$$

32. $$\int \frac{\tan^2 x}{\sec^5 x} = \int \frac{\sin^2 x}{\cos^2 x} \cdot \cos^5 x\,dx$$
$$= \int \sin^2 x \cdot \cos^3 x\,dx$$
$$= \int \sin^2 x\left(1 - \sin^2 x\right)\cos x\,dx$$
$$= \int\left(\sin^2 x - \sin^4 x\right)\cos x\,dx$$
$$= \frac{\sin^3 x}{3} - \frac{\sin^5 x}{5} + C$$

33. $$r = \int \sin^4(\pi\theta)\,d\theta = \frac{1}{4}\int\left[1 - \cos(2\pi\theta)\right]^2 d\theta$$
$$= \frac{1}{4}\int\left[1 - 2\cos(2\pi\theta) + \cos^2(2\pi\theta)\right]d\theta$$
$$= \frac{1}{4}\int\left[1 - 2\cos(2\pi\theta) + \frac{1 + \cos(4\pi\theta)}{2}\right]d\theta$$
$$= \frac{1}{4}\left[\theta - \frac{1}{\pi}\sin(2\pi\theta) + \frac{\theta}{2} + \frac{1}{8\pi}\sin(4\pi\theta)\right] + C$$
$$= \frac{1}{32\pi}\left[12\pi\theta - 8\sin(2\pi\theta) + \sin(4\pi\theta)\right] + C$$

34. $s = \int \sin^2 \frac{\alpha}{2} \cos^2 \frac{\alpha}{2}\,d\alpha$

$= \int \left(\frac{1 - \cos\alpha}{2}\right)\left(\frac{1 + \cos\alpha}{2}\right) d\alpha = \int \frac{1 - \cos^2\alpha}{4}\,d\alpha$

$= \frac{1}{4}\int \sin^2\alpha\,d\alpha = \frac{1}{8}\int (1 - \cos 2\alpha)\,d\alpha$

$= \frac{1}{8}\left(\theta - \frac{\sin 2\alpha}{2}\right) + C$

$= \frac{1}{16}(2\alpha - \sin 2\alpha) + C$

35. $y = \int \tan^3 3x \sec 3x\,dx$

$= \int (\sec^2 3x - 1)\sec 3x \tan 3x\,dx$

$= \frac{1}{3}\int \sec^2 3x(3\sec 3x \tan 3x)\,dx - \frac{1}{3}\int 3\sec 3x \tan 3x\,dx$

$= \frac{1}{9}\sec^3 3x - \frac{1}{3}\sec 3x + C$

36. $y = \int \sqrt{\tan x}\,\sec^4 x\,dx$

$= \int \tan^{1/2} x(\tan^2 x + 1)\sec^2 x\,dx$

$= \int (\tan^{5/2} x + \tan^{1/2} x)\sec^2 x\,dx$

$= \frac{2}{7}\tan^{7/2} x + \frac{2}{3}\tan^{3/2} x + C$

37. (a)

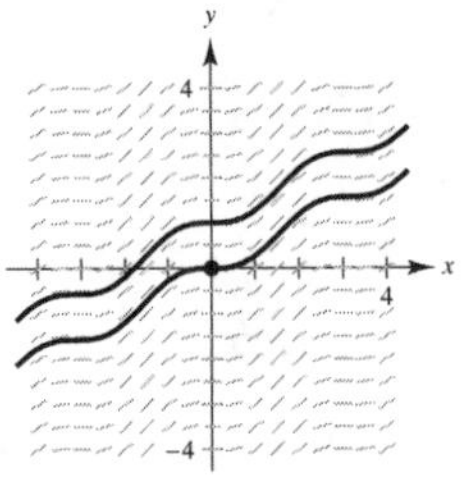

(b) $\frac{dy}{dx} = \sin^2 x, \quad (0, 0)$

$y = \int \sin^2 x\,dx = \int \frac{1 - \cos 2x}{2}\,dx$

$= \frac{1}{2}x - \frac{\sin 2x}{4} + C$

$(0, 0)$: $0 = C$, $y = \frac{1}{2}x - \frac{\sin 2x}{4}$

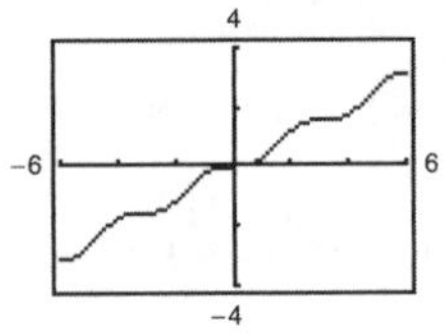

38. (a)

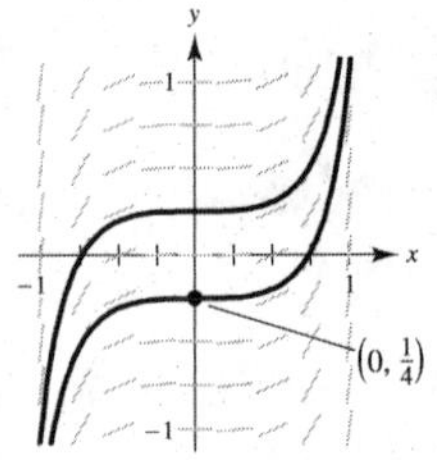

(b) $\frac{dy}{dx} = \sec^2 x \tan^2 x, \quad \left(0, -\frac{1}{4}\right)$

$y = \int \sec^2 x \tan^2 x\,dx \qquad u = \tan x, du = \sec^2 x\,dx$

$y = \frac{\tan^3 x}{3} + C$

$\left(0, -\frac{1}{4}\right)$: $-\frac{1}{4} = C \Rightarrow y = \frac{1}{3}\tan^3 x - \frac{1}{4}$

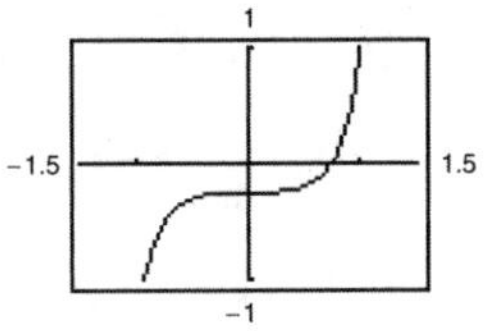

39. $\frac{dy}{dx} = \frac{3\sin x}{y}, \; y(0) = 2$

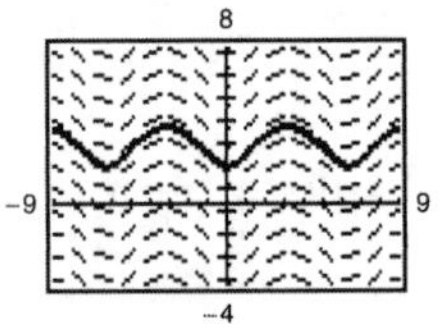

40. $\frac{dy}{dx} = 3\sqrt{y}\tan^2 x, \; y(0) = 3$

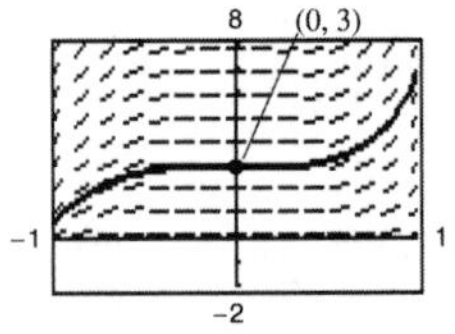

41. $$\begin{aligned}\int \cos 2x \cos 6x\,dx &= \frac{1}{2}\int\left[\cos\left((2-6)x\right) + \cos\left((2+6)x\right)\right]dx\\ &= \frac{1}{2}\int\left[\cos(-4x) + \cos 8x\right]dx\\ &= \frac{1}{2}\int\left(\cos 4x + \cos 8x\right)dx\\ &= \frac{1}{2}\left[\frac{\sin 4x}{4} + \frac{\sin 8x}{8}\right] + C\\ &= \frac{\sin 4x}{8} + \frac{\sin 8x}{16} + C\\ &= \frac{1}{16}(2\sin 4x + \sin 8x) + C\end{aligned}$$

42. $$\begin{aligned}\int \cos(5\theta)\cos(3\theta)\,d\theta &= \frac{1}{2}\int\left[\cos(5-3)\theta + \cos(5+3)\theta\right]d\theta\\ &= \frac{1}{2}\int\left(\cos 2\theta + \cos 8\theta\right)d\theta\\ &= \frac{1}{2}\left[\frac{\sin 2\theta}{2} + \frac{\sin 8\theta}{8}\right] + C\\ &= \frac{\sin 2\theta}{4} + \frac{\sin 8\theta}{16} + C\end{aligned}$$

43. $$\begin{aligned}\int \sin 2x \cos 4x\,dx &= \frac{1}{2}\int\left[\sin\left((2-4)x\right) + \sin\left((2+4)x\right)\right]dx\\ &= \frac{1}{2}\int\left(\sin(-2x) + \sin 6x\right)dx\\ &= \frac{1}{2}\int\left(-\sin 2x + \sin 6x\right)dx\\ &= \frac{1}{2}\left[\frac{\cos 2x}{2} - \frac{\cos 6x}{6}\right] + C\\ &= \frac{1}{4}\cos 2x - \frac{1}{12}\cos 6x + C\\ &= \frac{1}{12}(3\cos 2x - \cos 6x) + C\end{aligned}$$

44. $$\begin{aligned}\int \sin(-7x)\cos(6x)\,dx &= -\int \sin 7x \cos 6x\,dx\\ &= -\frac{1}{2}\int\left[\sin(7-6)x + \sin(7+6)x\right]dx\\ &= -\frac{1}{2}\int\left(\sin x + \sin 13x\right)dx\\ &= -\frac{1}{2}\left[-\cos x - \frac{\cos 13x}{13}\right] + C\\ &= \frac{1}{2}\cos x + \frac{1}{26}\cos 13x + C\end{aligned}$$

45. $$\begin{aligned}\int \sin\theta \sin 3\theta\,d\theta &= \tfrac{1}{2}\int\left(\cos 2\theta - \cos 4\theta\right)d\theta\\ &= \tfrac{1}{2}\left(\tfrac{1}{2}\sin 2\theta - \tfrac{1}{4}\sin 4\theta\right) + C\\ &= \tfrac{1}{8}(2\sin 2\theta - \sin 4\theta) + C\end{aligned}$$

46. $\int \sin 5x \sin 4x\,dx = \frac{1}{2}\int(\cos x - \cos 9x)\,dx$

$= \frac{1}{2}\left(\sin x - \frac{\sin 9x}{9}\right) + C$

$= \frac{\sin x}{2} - \frac{\sin 9x}{18} + C$

$= \frac{1}{18}(9\sin x - \sin 9x) + C$

47. $\int \cot^3 2x\,dx = \int(\csc^2 2x - 1)\cot 2x\,dx$

$= -\frac{1}{2}\int \cot 2x(-2\csc^2 2x)\,dx - \frac{1}{2}\int\frac{2\cos 2x}{\sin 2x}\,dx$

$= -\frac{1}{4}\cot^2 2x - \frac{1}{2}\ln|\sin 2x| + C$

$= \frac{1}{4}\left(\ln|\csc^2 2x| - \cot^2 2x\right) + C$

48. $\int \tan^5\frac{x}{4}\sec^4\frac{x}{4}\,dx$

$= \int \tan^5\frac{x}{4}\left(\tan^2\frac{x}{4} + 1\right)\sec^2\frac{x}{4}\,dx$

$= \int\left(\tan^7\frac{x}{4} + \tan^5\frac{x}{4}\right)\sec^2\frac{x}{4}\,dx$

$= \frac{\tan^8\frac{x}{4}}{2} + \frac{2\tan^6\frac{x}{4}}{3} + C$

$= \frac{1}{2}\tan^8\frac{x}{4} + \frac{2}{3}\tan^6\frac{x}{4} + C$

49. $\int \csc^4 3x\,dx = \int \csc^2 3x(1 + \cot^2 3x)\,dx$

$= \int \csc^2 3x\,dx + \int \cot^2 3x\csc^2 3x\,dx$

$= -\frac{1}{3}\cot 3x - \frac{1}{9}\cot^3 3x + C$

50. $\int \cot^3\frac{x}{2}\csc^4\frac{x}{2}\,dx = \int\cot^2\frac{x}{2}\csc^3\frac{x}{2}\left(\csc\frac{x}{2}\cot\frac{x}{2}\right)dx$

$= \int\left(\csc^2\frac{x}{2} - 1\right)\csc^3\frac{x}{2}\left(\csc\frac{x}{2}\cot\frac{x}{2}\right)dx$

$= \int\left(\csc^5\frac{x}{2} - \csc^3\frac{x}{2}\right)\left(\csc\frac{x}{2}\cot\frac{x}{2}\right)dx$

$= -\frac{1}{3}\csc^6\frac{x}{2} + \frac{1}{2}\csc^4\frac{x}{2} + C$

51. $\int\frac{\cot^2 t}{\csc t}\,dt = \int\frac{\csc^2 t - 1}{\csc t}\,dt$

$= \int(\csc t - \sin t)\,dt$

$= \ln|\csc t - \cot t| + \cos t + C$

52. $\int\frac{\cot^3 t}{\csc t}\,dt = \int\frac{\cos^3 t}{\sin^2 t}\,dt = \int\frac{(1 - \sin^2 t)\cos t}{\sin^2 t}\,dt$

$= \int\frac{\cos t}{\sin^2 t}\,dt - \int\cos t\,dt$

$= \frac{-1}{\sin t} - \sin t + C = -\csc t - \sin t + C$

53. $\int\frac{1}{\sec x\tan x}\,dx = \int\frac{\cos^2 x}{\sin x}\,dx = \int\frac{1 - \sin^2 x}{\sin x}\,dx$

$= \int(\csc x - \sin x)\,dx$

$= \ln|\csc x - \cot x| + \cos x + C$

54. $\int\frac{\sin^2 x - \cos^2 x}{\cos x}\,dx = \int\frac{1 - 2\cos^2 x}{\cos x}\,dx$

$= \int(\sec x - 2\cos x)\,dx$

$= \ln|\sec x + \tan x| - 2\sin x + C$

55. $\int(\tan^4 t - \sec^4 t)\,dt = \int(\tan^2 t + \sec^2 t)(\tan^2 t - \sec^2 t)\,dt, \qquad (\tan^2 t - \sec^2 t = -1)$

$= -\int(\tan^2 t + \sec^2 t)\,dt = -\int(2\sec^2 t - 1)\,dt = -2\tan t + t + C$

56. $\int\frac{1-\sec t}{\cos t - 1}\,dt = \int\frac{\cos t - 1}{(\cos t - 1)\cos t}\,dt = \int \sec t\,dt = \ln\left|\sec t + \tan t\right| + C$

57. $\int_{-\pi}^{\pi}\sin^2 x\,dx = 2\int_0^{\pi}\frac{1-\cos 2x}{2}\,dx$

$= \left[x - \frac{1}{2}\sin 2x\right]_0^{\pi} = \pi$

58. $\int_0^{\pi/3}\tan^2 x\,dx = \int_0^{\pi/3}(\sec^2 x - 1)\,dx$

$= \left[\tan x - x\right]_0^{\pi/3} = \sqrt{3} - \frac{\pi}{3}$

59. $\int_0^{\pi/4} 6\tan^3 x\,dx = 6\int_0^{\pi/4}(\sec^2 x - 1)\tan x\,dx$

$= 6\int_0^{\pi/4}\left[\tan x\sec^2 x - \tan x\right]dx$

$= 6\left[\frac{\tan^2 x}{2} + \ln\left|\cos x\right|\right]_0^{\pi/4}$

$= 6\left[\frac{1}{2} + \ln\left(\frac{\sqrt{2}}{2}\right)\right] = 6\left(\frac{1}{2} - \ln\sqrt{2}\right)$

$= 3(1 - \ln 2)$

60. $\int_0^{\pi/3}\sec^{3/2} x\tan x\,dx = \int_0^{\pi/3}\sec^{1/2} x(\sec x\tan x)\,dx$

$= \left[\frac{2}{3}\sec^{3/2} x\right]_0^{\pi/3}$

$= \frac{2}{3}\left(2\sqrt{2} - 1\right)$

61. Let $u = 1 + \sin t$, $du = \cos t\,dt$.

$\int_0^{\pi/2}\frac{\cos t}{1+\sin t}\,dt = \left[\ln\left|1+\sin t\right|\right]_0^{\pi/2} = \ln 2$

62. $\int_{\pi/6}^{\pi/3}\sin 6x\cos 4x\,dx = \frac{1}{2}\int_{\pi/6}^{\pi/3}(\sin 2x + \sin 10x)\,dx$

$= \left[-\frac{\cos 2x}{4} - \frac{\cos 10x}{20}\right]_{\pi/6}^{\pi/3}$

$= \left(\frac{1}{8} + \frac{1}{40}\right) - \left(-\frac{1}{8} - \frac{1}{40}\right) = \frac{3}{10}$

63. $\int_{-\pi/2}^{\pi/2} 3\cos^3 x\,dx = 3\int_{-\pi/2}^{\pi/2}(1 - \sin^2 x)\cos x\,dx$

$= 3\left[\sin x - \frac{\sin^3 x}{3}\right]_{-\pi/2}^{\pi/2}$

$= 3\left[\left(1 - \frac{1}{3}\right) - \left(-1 + \frac{1}{3}\right)\right] = 4$

64. $\int_{-\pi/2}^{\pi/2}(\sin^2 x + 1)\,dx = \int_{-\pi/2}^{\pi/2}\left(\frac{1-\cos 2x}{2} + 1\right)dx$

$= \int_{-\pi/2}^{\pi/2}\left(\frac{3}{2} - \frac{1}{2}\cos 2x\right)dx$

$= \left[\frac{3}{2}x - \frac{1}{4}\sin 2x\right]_{-\pi/2}^{\pi/2} = \frac{3\pi}{2}$

65. (a) Save one sine factor and convert the remaining factors to cosines. Then expand and integrate.

(b) Save one cosine factor and convert the remaining factors to sines. Then expand and integrate.

(c) Make repeated use of the power reducing formulas to convert the integrand to odd powers of the cosine. Then proceed as in part (b).

66. (a) Save a secant-squared factor and convert the remaining factors to tangents. Then expand and integrate.

(b) Save a secant-squared factor and convert the remaining factors to secants. Then expand and integrate.

(c) Convert a tangent-squared factor to a secant-squared factor, then expand and repeat if necessary.

(d) Use integration by parts.

67. (a) $\int \sin x \cos x\, dx = \frac{\sin^2 x}{2} + C$

(b) $-\int \cos x(-\sin x)\, dx = -\frac{\cos^2 x}{2} + C$

(c) $dv = \cos x\, dx \Rightarrow v = \sin x$

$u = \sin x \Rightarrow du = \cos x\, dx$

$$\int \sin x \cos x\, dx = \sin^2 x - \int \sin x \cos x\, dx$$

$$2\int \sin x \cos x\, dx = \sin^2 x$$

$$\int \sin x \cos x\, dx = \frac{\sin^2 x}{2} + C$$

(Answers will vary.)

(d) $\int \sin x \cos x\, dx = \int \frac{1}{2}\sin 2x\, dx = -\frac{1}{4}\cos 2x + C$

The answers all differ by a constant.

68. (a) f has a maximum at the points where f' changes from positive to negative: $x = -\pi, \pi$.

(b) f has a minimum at the points where f' changes from negative to positive: $x = 0$.

69. (a) Let $u = \tan 3x$, $du = 3\sec^2 3x\, dx$.

$$\int \sec^4 3x \tan^3 3x\, dx = \int \sec^2 3x \tan^3 3x \sec^2 3x\, dx = \frac{1}{3}\int\left(\tan^2 3x + 1\right)\tan^3 3x\left(3\sec^2 3x\right) dx$$

$$= \frac{1}{3}\int\left(\tan^5 3x + \tan^3 3x\right)\left(3\sec^2 3x\right) dx = \frac{\tan^6 3x}{18} + \frac{\tan^4 3x}{12} + C_1$$

Or let $u = \sec 3x$, $du = 3\sec 3x \tan 3x\, dx$.

$$\int \sec^4 3x \tan^3 3x\, dx = \int \sec^3 3x \tan^2 3x \sec 3x \tan 3x\, dx$$

$$= \frac{1}{3}\int \sec^3 3x\left(\sec^2 3x - 1\right)\left(3\sec 3x \tan 3x\right) dx = \frac{\sec^6 3x}{18} - \frac{\sec^4 3x}{12} + C$$

(b)

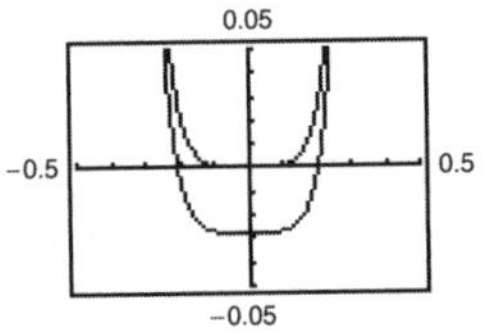

(c) $$\frac{\sec^6 3x}{18} - \frac{\sec^4 3x}{12} + C = \frac{\left(1 + \tan^2 3x\right)^3}{18} - \frac{\left(1 + \tan^2 3x\right)^2}{12} + C$$

$$= \frac{1}{18}\tan^6 3x + \frac{1}{6}\tan^4 3x + \frac{1}{6}\tan^2 3x + \frac{1}{18} - \frac{1}{12}\tan^4 3x - \frac{1}{6}\tan^2 3x - \frac{1}{12} + C$$

$$= \frac{\tan^6 3x}{18} + \frac{\tan^4 3x}{12} + \left(\frac{1}{18} - \frac{1}{12}\right) + C$$

$$= \frac{\tan^6 3x}{18} + \frac{\tan^4 3x}{12} + C_2$$

70. (a) Let $u = \tan x$, $du = \sec^2 x\, dx$.

$$\int \sec^2 x \tan x\, dx = \frac{1}{2}\tan^2 x + C_1$$

Or let $u = \sec x$, $du = \sec x \tan x\, dx$.

$$\int \sec x(\sec x \tan x)\, dx = \frac{1}{2}\sec^2 x + C$$

(b)

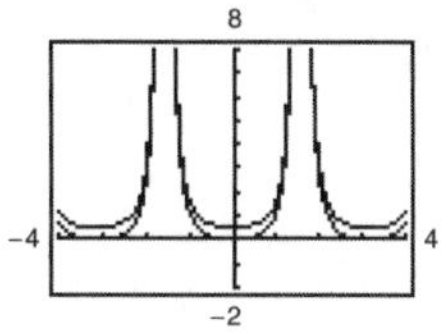

(c) $$\frac{1}{2}\sec^2 x + C = \frac{1}{2}(\tan^2 x + 1) + C = \frac{1}{2}\tan^2 x + \left(\frac{1}{2} + C\right) = \frac{1}{2}\tan^2 x + C_2$$

71. $$A = \int_0^{\pi/2}(\sin x - \sin^3 x)\, dx = \int_0^{\pi/2}\sin x\, dx - \int_0^{\pi/2}\sin^3 x\, dx = \left[-\cos x\right]_0^{\pi/2} - \frac{2}{3} \quad \text{(Wallis's Formula)} = 1 - \frac{2}{3} = \frac{1}{3}$$

72. $$A = \int_0^1 \sin^2(\pi x)\, dx = \int_0^1 \frac{1 - \cos(2\pi x)}{2}\, dx = \left[\frac{1}{2}x - \frac{\sin 2\pi x}{4\pi}\right]_0^1 = \frac{1}{2}$$

73. $$A = \int_{-\pi/4}^{\pi/4}\left[\cos^2 x - \sin^2 x\right] dx = \int_{-\pi/4}^{\pi/4}\cos 2x\, dx = \left[\frac{\sin 2x}{2}\right]_{-\pi/4}^{\pi/4} = \frac{1}{2} + \frac{1}{2} = 1$$

74. $$A = \int_{-\pi/2}^{\pi/4}\left[\cos^2 x - \sin x \cos x\right] dx = \int_{-\pi/2}^{\pi/4}\left[\frac{1 + \cos 2x}{2} - \sin x \cos x\right] dx = \left[\frac{1}{2}x + \frac{\sin 2x}{4} - \frac{\sin^2 x}{2}\right]_{-\pi/2}^{\pi/4} = \left(\frac{\pi}{8} + \frac{1}{4} - \frac{1}{4}\right) - \left(-\frac{\pi}{4} - \frac{1}{2}\right) = \frac{3\pi}{8} + \frac{1}{2}$$

75. Disks

$R(x) = \tan x$, $r(x) = 0$

$$V = 2\pi\int_0^{\pi/4}\tan^2 x\, dx = 2\pi\int_0^{\pi/4}(\sec^2 x - 1)\, dx = 2\pi\left[\tan x - x\right]_0^{\pi/4} = 2\pi\left(1 - \frac{\pi}{4}\right) \approx 1.348$$

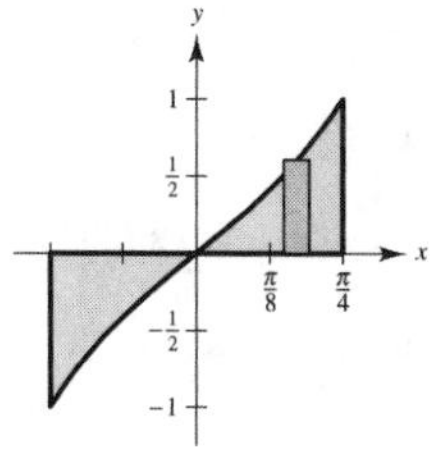

76. $$V = \pi\int_0^{\pi/2}\left[\cos^2\left(\frac{x}{2}\right) - \sin^2\left(\frac{x}{2}\right)\right] dx = \pi\int_0^{\pi/2}\cos x\, dx = \pi\left[\sin x\right]_0^{\pi/2} = \pi$$

77. (a) $$V = \pi\int_0^{\pi}\sin^2 x\, dx = \frac{\pi}{2}\int_0^{\pi}(1 - \cos 2x)\, dx = \frac{\pi}{2}\left[x - \frac{1}{2}\sin 2x\right]_0^{\pi} = \frac{\pi^2}{2}$$

(b) $$A = \int_0^{\pi}\sin x\, dx = \left[-\cos x\right]_0^{\pi} = 1 + 1 = 2$$

Let $u = x$, $dv = \sin x\, dx$, $du = dx$, $v = -\cos x$.

$$\bar{x} = \frac{1}{A}\int_0^{\pi} x \sin x\, dx = \frac{1}{2}\left[\left[-x\cos x\right]_0^{\pi} + \int_0^{\pi}\cos x\, dx\right] = \frac{1}{2}\left[-x\cos x + \sin x\right]_0^{\pi} = \frac{\pi}{2}$$

$$\bar{y} = \frac{1}{2A}\int_0^{\pi}\sin^2 x\, dx = \frac{1}{8}\int_0^{\pi}(1 - \cos 2x)\, dx = \frac{1}{8}\left[x - \frac{1}{2}\sin 2x\right]_0^{\pi} = \frac{\pi}{8}$$

$$(\bar{x}, \bar{y}) = \left(\frac{\pi}{2}, \frac{\pi}{8}\right)$$

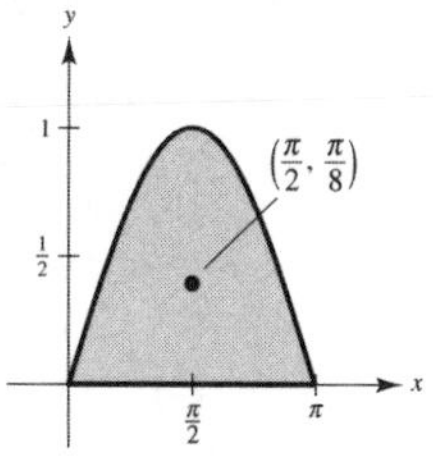

78. (a) $V = \pi\int_0^{\pi/2} \cos^2 x\,dx = \frac{\pi}{2}\int_0^{\pi/2}(1 + \cos 2x)\,dx = \frac{\pi}{2}\left[x + \frac{1}{2}\sin 2x\right]_0^{\pi/2} = \frac{\pi^2}{4}$

(b) $A = \int_0^{\pi/2} \cos x\,dx = \left[\sin x\right]_0^{\pi/2} = 1$

Let $u = x$, $dv = \cos x\,dx$, $du = dx$, $v = \sin x$.

$$\bar{x} = \int_0^{\pi/2} x\cos x\,dx = \left[x\sin x\right]_0^{\pi/2} - \int_0^{\pi/2}\sin x\,dx = \left[x\sin x + \cos x\right]_0^{\pi/2} = \frac{\pi}{2} - 1 = \frac{\pi - 2}{2}$$

$$\bar{y} = \frac{1}{2}\int_0^{\pi/2}\cos^2 x\,dx = \frac{1}{4}\int_0^{\pi/2}(1 + \cos 2x)\,dx = \frac{1}{4}\left[x + \frac{1}{2}\sin 2x\right]_0^{\pi/2} = \frac{\pi}{8}$$

$$(\bar{x}, \bar{y}) = \left(\frac{\pi - 2}{2}, \frac{\pi}{8}\right)$$

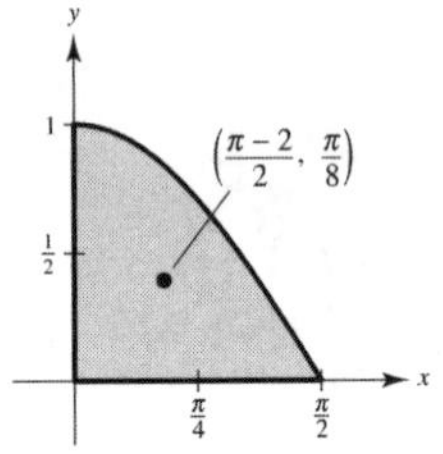

79. $dv = \sin x\,dx \Rightarrow v = -\cos x$

$u = \sin^{n-1} x \Rightarrow du = (n - 1)\sin^{n-2} x\cos x\,dx$

$$\int \sin^n x\,dx = -\sin^{n-1} x\cos x + (n - 1)\int \sin^{n-2} x\cos^2 x\,dx = -\sin^{n-1} x\cos x + (n - 1)\int \sin^{n-2} x\left(1 - \sin^2 x\right)dx$$

$$= -\sin^{n-1} x\cos x + (n - 1)\int \sin^{n-2} x\,dx - (n - 1)\int \sin^n x\,dx$$

Therefore, $n\int \sin^n x\,dx = -\sin^{n-1} x\cos x + (n - 1)\int \sin^{n-2} x\,dx$

$$\int \sin^n x\,dx = \frac{-\sin^{n-1} x\cos x}{n} + \frac{n - 1}{n}\int \sin^{n-2} x\,dx.$$

80. $dv = \cos x\,dx \Rightarrow v = \sin x$

$u = \cos^{n-1} x \Rightarrow du = -(n - 1)\cos^{n-2} x\sin x\,dx$

$$\int \cos^n x\,dx = \cos^{n-1} x\sin x + (n - 1)\int \cos^{n-2} x\sin^2 x\,dx$$

$$= \cos^{n-1} x\sin x + (n - 1)\int \cos^{n-2} x\left(1 - \cos^2 x\right)dx = \cos^{n-1} x\sin x + (n - 1)\int \cos^{n-2} x\,dx - (n - 1)\int \cos^n x\,dx$$

Therefore, $n\int \cos^n x\,dx = \cos^{n-1} x\sin x + (n - 1)\int \cos^{n-2} x\,dx$

$$\int \cos^n x\,dx = \frac{\cos^{n-1} x\sin x}{n} + \frac{n - 1}{n}\int \cos^{n-2} x\,dx.$$

81. Let $u = \sin^{n-1} x$, $du = (n-1)\sin^{n-2} x \cos x\, dx$, $dv = \cos^m x \sin x\, dx$, $v = \dfrac{-\cos^{m+1} x}{m+1}$.

$$\begin{aligned}
\int \cos^m x \sin^n x\, dx &= \frac{-\sin^{n-1} x \cos^{m+1} x}{m+1} + \frac{n-1}{m+1}\int \sin^{n-2} x \cos^{m+2} x\, dx \\
&= \frac{-\sin^{n-1} x \cos^{m+1} x}{m+1} + \frac{n-1}{m+1}\int \sin^{n-2} x \cos^m x\left(1 - \sin^2 x\right) dx \\
&= \frac{-\sin^{n-1} x \cos^{m+1} x}{m+1} + \frac{n-1}{m+1}\int \sin^{n-2} x \cos^m x\, dx - \frac{n-1}{m+1}\int \sin^n x \cos^m x\, dx \\
\frac{m+n}{m+1}\int \cos^m x \sin^n x\, dx &= \frac{-\sin^{n-1} x \cos^{m+1} x}{m+1} + \frac{n-1}{m+1}\int \sin^{n-2} x \cos^m x\, dx \\
\int \cos^m x \sin^n x\, dx &= \frac{-\cos^{m+1} x \sin^{n-1} x}{m+n} + \frac{n-1}{m+n}\int \cos^m x \sin^{n-2} x\, dx
\end{aligned}$$

82. Let $u = \sec^{n-2} x$, $du = (n-2)\sec^{n-2} x \tan x\, dx$, $dv = \sec^2 x\, dx$, $v = \tan x$.

$$\begin{aligned}
\int \sec^n x\, dx &= \sec^{n-2} x \tan x - \int (n-2)\sec^{n-2} x \tan^2 x\, dx \\
&= \sec^{n-2} x \tan x - (n-2)\int \sec^{n-2} x\left(\sec^2 x - 1\right) dx \\
&= \sec^{n-2} x \tan x - (n-2)\left[\int \sec^n x\, dx - \int \sec^{n-2} x\, dx\right] \\
(n-1)\int \sec^n x\, dx &= \sec^{n-2} x \tan x + (n-2)\int \sec^{n-2} x\, dx \\
\int \sec^n x\, dx &= \frac{1}{n-1}\sec^{n-2} x \tan x + \frac{n-2}{n-1}\int \sec^{n-2} x\, dx
\end{aligned}$$

83.
$$\begin{aligned}
\int \sin^5 x\, dx &= -\frac{\sin^4 x \cos x}{5} + \frac{4}{5}\int \sin^3 x\, dx \\
&= -\frac{\sin^4 x \cos x}{5} + \frac{4}{5}\left(-\frac{\sin^2 x \cos x}{3} + \frac{2}{3}\int \sin x\, dx\right) \\
&= -\frac{1}{5}\sin^4 x \cos x - \frac{4}{15}\sin^2 x \cos x - \frac{8}{15}\cos x + C \\
&= -\frac{\cos x}{15}\left(3\sin^4 x + 4\sin^2 x + 8\right) + C
\end{aligned}$$

84.
$$\begin{aligned}
\int \cos^4 x\, dx &= \frac{\cos^3 x \sin x}{4} + \frac{3}{4}\int \cos^2 x\, dx \\
&= \frac{\cos^3 x \sin x}{4} + \frac{3}{4}\left(\frac{\cos x \sin x}{2} + \frac{1}{2}\int dx\right) \\
&= \frac{1}{4}\cos^3 x \sin x + \frac{3}{8}\cos x \sin x + \frac{3}{8}x + C \\
&= \frac{1}{8}\left(2\cos^3 x \sin x + 3\cos x \sin x + 3x\right) + C
\end{aligned}$$

85.
$$\begin{aligned}
\int \sec^4 \frac{2\pi x}{5}\, dx &= \frac{5}{2\pi}\int \sec^4\left(\frac{2\pi x}{5}\right)\frac{2\pi}{5}\, dx \\
&= \frac{5}{2\pi}\left[\frac{1}{3}\sec^2\left(\frac{2\pi x}{5}\right)\tan\left(\frac{2\pi x}{5}\right) + \frac{2}{3}\int \sec^2\left(\frac{2\pi x}{5}\right)\frac{2\pi}{5}\, dx\right] \\
&= \frac{5}{6\pi}\left[\sec^2\left(\frac{2\pi x}{5}\right)\tan\left(\frac{2\pi x}{5}\right) + 2\tan\left(\frac{2\pi x}{5}\right)\right] + C \\
&= \frac{5}{6\pi}\tan\left(\frac{2\pi x}{5}\right)\left[\sec^2\left(\frac{2\pi x}{5}\right) + 2\right] + C
\end{aligned}$$

86. $$\int \sin^4 x \cos^2 x\,dx = -\frac{\cos^3 x \sin^3 x}{6} + \frac{1}{2}\int \cos^2 x \sin^2 x\,dx$$
$$= -\frac{\cos^3 x \sin^3 x}{6} + \frac{1}{2}\left(-\frac{\cos^3 x \sin x}{4} + \frac{1}{4}\int \cos^2 x\,dx\right)$$
$$= -\frac{1}{6}\cos^3 x \sin^3 x - \frac{1}{8}\cos^3 x \sin x + \frac{1}{8}\left(\frac{\cos x \sin x}{2} + \frac{x}{2}\right) + C$$
$$= -\frac{1}{48}\left(8\cos^3 x \sin^3 x + 6\cos^3 x \sin x - 3\cos x \sin x - 3x\right) + C$$

87. $f(t) = a_0 + a_1 \cos\dfrac{\pi t}{6} + b_1 \sin\dfrac{\pi t}{6}$

$$a_0 = \frac{1}{12}\int_0^{12} f(t)\,dt,\; a_1 = \frac{1}{6}\int_0^{12} f(t)\cos\frac{\pi t}{6}\,dt,\; b_1 = \frac{1}{6}\int_0^{12} f(t)\sin\frac{\pi t}{6}\,dt$$

(a) $$a_0 \approx \frac{1}{12}\cdot\frac{(12-0)}{3(12)}[33.5 + 4(35.4) + 2(44.7) + 4(55.6) + 2(67.4) + 4(76.2) + 2(80.4) + 4(79.0) + 2(72.0) + 4(61.0) + 2(49.3) + 4(38.6) + 33.5]$$
$$\approx 57.72$$
$a_1 \approx -23.36$

$b_1 \approx -2.75$ (Answers will vary.)

$$H(t) \approx 57.72 - 23.36\cos\left(\frac{\pi t}{6}\right) - 2.75\sin\left(\frac{\pi t}{6}\right)$$

(b) $$L(t) \approx 42.04 - 20.91\cos\left(\frac{\pi t}{6}\right) - 4.33\sin\left(\frac{\pi t}{6}\right)$$

(c) 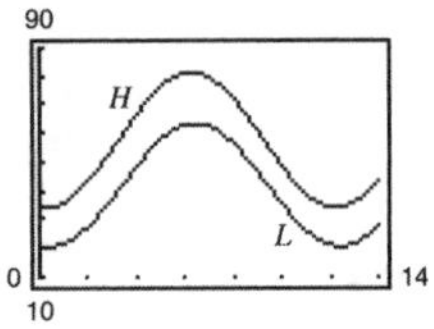

Temperature difference is greatest in the summer ($t \approx 4.9$ or end of May).

88. (a) n is odd and $n \geq 3$.

$$\int_0^{\pi/2} \cos^n x\,dx = \left[\frac{\cos^{n-1} x \sin x}{n}\right]_0^{\pi/2} + \frac{n-1}{n}\int_0^{\pi/2} \cos^{n-2} x\,dx$$

$$= \frac{n-1}{n}\left(\left[\frac{\cos^{n-3} x \sin x}{n-2}\right]_0^{\pi/2} + \frac{n-3}{n-2}\int_0^{\pi/2} \cos^{n-4} x\,dx\right)$$

$$= \frac{n-1}{n}\cdot\frac{n-3}{n-2}\left(\left[\frac{\cos^{n-5} x \sin x}{n-4}\right]_0^{\pi/2} + \frac{n-5}{n-4}\int_0^{\pi/2} \cos^{n-6} x\,dx\right)$$

$$= \frac{n-1}{n}\cdot\frac{n-3}{n-2}\cdot\frac{n-5}{n-4}\int_0^{\pi/2} \cos^{n-6} x\,dx$$

$$= \frac{n-1}{n}\cdot\frac{n-3}{n-2}\cdot\frac{n-5}{n-4}\cdots\int_0^{\pi/2} \cos x\,dx$$

$$= \left[\frac{n-1}{n}\cdot\frac{n-3}{n-2}\cdot\frac{n-5}{n-4}\cdots(\sin x)\right]_0^{\pi/2}$$

$$= \frac{n-1}{n}\cdot\frac{n-3}{n-2}\cdot\frac{n-5}{n-4}\cdots 1 \quad \text{(Reverse the order.)}$$

$$= (1)\left(\frac{2}{3}\right)\left(\frac{4}{5}\right)\left(\frac{6}{7}\right)\cdots\left(\frac{n-1}{n}\right) = \left(\frac{2}{3}\right)\left(\frac{4}{5}\right)\left(\frac{6}{7}\right)\cdots\left(\frac{n-1}{n}\right)$$

(b) n is even and $n \geq 2$.

$$\int_0^{\pi/2} \cos^n x\,dx = \frac{n-1}{n}\cdot\frac{n-3}{n-2}\cdot\frac{n-5}{n-4}\cdots\int_0^{\pi/2} \cos^2 x\,dx \quad \text{(From part (a))}$$

$$= \left[\frac{n-1}{n}\cdot\frac{n-3}{n-2}\cdot\frac{n-5}{n-4}\cdots\left(\frac{x}{2} + \frac{1}{4}\sin 2x\right)\right]_0^{\pi/2}$$

$$= \frac{n-1}{n}\cdot\frac{n-3}{n-2}\cdot\frac{n-5}{n-4}\cdots\frac{\pi}{4} \quad \text{(Reverse the order.)}$$

$$= \left(\frac{\pi}{2}\cdot\frac{1}{2}\right)\left(\frac{3}{4}\right)\left(\frac{5}{6}\right)\cdots\left(\frac{n-1}{n}\right) = \left(\frac{1}{2}\right)\left(\frac{3}{4}\right)\left(\frac{5}{6}\right)\cdots\left(\frac{n-1}{n}\right)\left(\frac{\pi}{2}\right)$$

89. $$\int_{-\pi}^{\pi} \cos(mx)\cos(nx)\,dx = \frac{1}{2}\left[\frac{\sin(m+n)x}{m+n} + \frac{\sin(m-n)x}{m-n}\right]_{-\pi}^{\pi} = 0, \quad (m \neq n)$$

$$\int_{-\pi}^{\pi} \sin(mx)\sin(nx)\,dx = \frac{1}{2}\int_{-\pi}^{\pi}\left[\cos(m-n)x - \cos(m+n)x\right]dx$$

$$= \frac{1}{2}\left[\frac{\sin(m-n)x}{m-n} - \frac{\sin(m+n)x}{m+n}\right]_{-\pi}^{\pi} = 0, \quad (m \neq n)$$

$$\int_{-\pi}^{\pi} \sin(mx)\cos(nx)\,dx = \frac{1}{2}\int_{-\pi}^{\pi}\left[\sin(m+n)x + \sin(m-n)x\right]dx$$

$$= -\frac{1}{2}\left[\frac{\cos(m+n)x}{m+n} + \frac{\cos(m-n)x}{m-n}\right]_{-\pi}^{\pi}, \quad (m \neq n)$$

$$= -\frac{1}{2}\left[\left(\frac{\cos(m+n)\pi}{m+n} + \frac{\cos(m-n)\pi}{m-n}\right) - \left(\frac{\cos(m+n)(-\pi)}{m+n} + \frac{\cos(m-n)(-\pi)}{m-n}\right)\right]$$

$$= 0, \text{ because } \cos(-\theta) = \cos\theta.$$

$$\int_{-\pi}^{\pi} \sin(mx)\cos(mx)\,dx = \frac{1}{m}\left[\frac{\sin^2(mx)}{2}\right]_{-\pi}^{\pi} = 0$$

90. $f(x) = \sum_{i=1}^{N} a_i \sin(ix)$

(a) $$f(x)\sin(nx) = \left[\sum_{i=1}^{N} a_i \sin(ix)\right]\sin(nx)$$

$$\int_{-\pi}^{\pi} f(x)\sin(nx)\,dx = \int_{-\pi}^{\pi}\left[\sum_{i=1}^{N} a_i \sin(ix)\right]\sin(nx)\,dx$$

$$= \int_{-\pi}^{\pi} a_n \sin^2(nx)\,dx \quad \text{(by Exercise 89)}$$

$$= \int_{-\pi}^{\pi} a_n \frac{1-\cos(2nx)}{2}\,dx = \left[\frac{a_n}{2}\left(x - \frac{\sin(2nx)}{2n}\right)\right]_{-\pi}^{\pi} = \frac{a_n}{2}(\pi + \pi) = a_n\pi$$

So, $a_n = \frac{1}{\pi}\int_{-\pi}^{\pi} f(x)\sin(nx)\,dx.$

(b) $f(x) = x$

$$a_1 = \frac{1}{\pi}\int_{-\pi}^{\pi} x\sin x\,dx = 2$$

$$a_2 = \frac{1}{\pi}\int_{-\pi}^{\pi} x\sin 2x\,dx = -1$$

$$a_3 = \frac{1}{\pi}\int_{-\pi}^{\pi} x\sin 3x\,dx = \frac{2}{3}$$

Section 8.4 Trigonometric Substitution

1. Use $x = 3\tan\theta$.

2. Use $x = 2\sin\theta$.

3. Use $x = 5\sin\theta$.

4. Use $x = 5\sec\theta$.

5. Let $x = 4\sin\theta$, $dx = 4\cos\theta\,d\theta$, $\sqrt{16-x^2} = 4\cos\theta$.

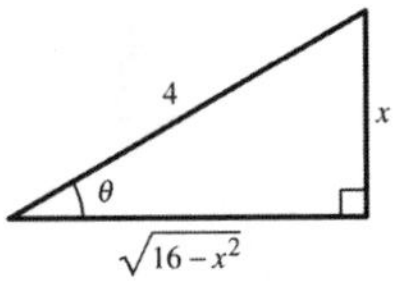

$$\int \frac{1}{(16-x^2)^{3/2}}\,dx = \int \frac{4\cos\theta}{(4\cos\theta)^3}\,d\theta = \frac{1}{16}\int \sec^2\theta\,d\theta = \frac{1}{16}\tan\theta + C = \frac{1}{16}\left(\frac{x}{\sqrt{16-x^2}}\right) + C$$

6. Same substitution as in Exercise 5

$$\int \frac{4}{x^2\sqrt{16-x^2}}\,dx = 4\int \frac{4\cos\theta}{(4\sin\theta)^2(4\cos\theta)}\,d\theta = \frac{1}{4}\int \csc^2\theta\,d\theta = -\frac{1}{4}\cot\theta + C = -\frac{1}{4}\frac{\sqrt{16-x^2}}{x} + C = \frac{-\sqrt{16-x^2}}{4x} + C$$

7. Same substitution as in Exercise 5

$$\begin{aligned}\int \frac{\sqrt{16 - x^2}}{x}\,dx &= \int \frac{4\cos\theta}{4\sin\theta}4\cos\theta\,d\theta\\ &= 4\int \frac{\cos^2\theta}{\sin\theta}\,d\theta\\ &= 4\int \frac{1 - \sin^2\theta}{\sin\theta}\,d\theta\\ &= 4\int(\csc\theta - \sin\theta)\,d\theta\\ &= -4\ln\left|\csc\theta + \cot\theta\right| + 4\cos\theta + C\\ &= -4\ln\left|\frac{4}{x} + \frac{\sqrt{16 - x^2}}{x}\right| + 4\frac{\sqrt{16 - x^2}}{4} + C\\ &= -4\ln\left|\frac{4 + \sqrt{16 - x^2}}{x}\right| + \sqrt{16 - x^2} + C\\ &= 4\ln\left|\frac{4 - \sqrt{16 - x^2}}{x}\right| + \sqrt{16 - x^2} + C\end{aligned}$$

8. Same substitution as in Exercise 5

$$\begin{aligned}\int \frac{x^3}{\sqrt{16 - x^2}}\,dx &= \int \frac{(4\sin\theta)^3}{4\cos\theta}4\cos\theta\,d\theta\\ &= 64\int \sin^3\theta\,d\theta\\ &= 64\int\left(1 - \cos^2\theta\right)\sin\theta\,d\theta\\ &= 64\left[-\cos\theta + \frac{\cos^3\theta}{3}\right] + C\\ &= 64\left[-\frac{\sqrt{16 - x^2}}{4} + \frac{\left(16 - x^2\right)^{3/2}}{64(3)}\right] + C\\ &= -16\sqrt{16 - x^2} + \frac{1}{3}\left(16 - x^2\right)^{3/2} + C\\ &= -\frac{1}{3}\sqrt{16 - x^2}\left[48 - \left(16 - x^2\right)\right] + C\\ &= -\frac{1}{3}\sqrt{16 - x^2}\left(32 + x^2\right) + C\end{aligned}$$

9. Let $x = 5\sec\theta$, $dx = 5\sec\theta\tan\theta\,d\theta$,

$\sqrt{x^2 - 25} = 5\tan\theta$.

$$\begin{aligned}\int \frac{1}{\sqrt{x^2 - 25}}\,dx &= \int \frac{5\sec\theta\tan\theta}{5\tan\theta}\,d\theta\\ &= \int \sec\theta\,d\theta\\ &= \ln\left|\sec\theta + \tan\theta\right| + C\\ &= \ln\left|\frac{x}{5} + \frac{\sqrt{x^2 - 25}}{5}\right| + C\\ &= \ln\left|x + \sqrt{x^2 - 25}\right| + C\end{aligned}$$

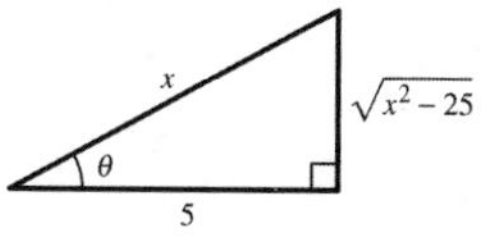

10. Same substitution as in Exercise 9

$$\int \frac{\sqrt{x^2 - 25}}{x}\,dx = \int \frac{5\tan\theta}{5\sec\theta} 5\sec\theta\tan\theta\,d\theta$$
$$= 5\int \tan^2\theta\,d\theta$$
$$= 5\int\left(\sec^2\theta - 1\right)d\theta$$
$$= 5\left(\tan\theta - \theta\right) + C$$
$$= 5\left(\frac{\sqrt{x^2 - 25}}{5} - \operatorname{arcsec}\frac{x}{5}\right) + C$$
$$= \sqrt{x^2 - 25} - 5\operatorname{arcsec}\frac{x}{5} + C$$

$\left[\textbf{Note: } \operatorname{arcsec}\left(\frac{x}{5}\right) = \arctan\left(\frac{\sqrt{x^2 - 25}}{5}\right)\right]$

11. Same substitution as in Exercise 9

$$\int x^3\sqrt{x^2 - 25}\,dx = \int (5\sec\theta)^3(5\tan\theta)(5\sec\theta\tan\theta)\,d\theta$$
$$= 3125\int \sec^4\theta\tan^2\theta\,d\theta$$
$$= 3125\int\left(1 + \tan^2\theta\right)\tan^2\theta\sec^2\theta\,d\theta$$
$$= 3125\int\left(\tan^2\theta + \tan^4\theta\right)\sec^2\theta\,d\theta$$
$$= 3125\left[\frac{\tan^3\theta}{3} + \frac{\tan^5\theta}{5}\right] + C$$
$$= 3125\left[\frac{\left(x^2 - 25\right)^{3/2}}{125(3)} + \frac{\left(x^2 - 25\right)^{5/2}}{5^5(5)}\right] + C$$
$$= \frac{1}{15}\left(x^2 - 25\right)^{3/2}\left[125 + 3\left(x^2 - 25\right)\right] + C$$
$$= \frac{1}{15}\left(x^2 - 25\right)^{3/2}\left(50 + 3x^2\right) + C$$

12. Same substitution as in Exercise 9

$$\int \frac{x^3}{\sqrt{x^2 - 25}}\,dx = \int \frac{(5\sec\theta)^3}{5\tan\theta} 5\sec\theta\tan\theta\,d\theta$$
$$= 125\int \sec^4\theta\,d\theta$$
$$= 125\int\left(\tan^2\theta + 1\right)\sec^2\theta\,d\theta$$
$$= 125\left(\frac{\tan^3\theta}{3} + \tan\theta\right) + C$$
$$= \frac{125}{3}\frac{\left(x^2 - 25\right)^{3/2}}{125} + 125\frac{\sqrt{x^2 - 25}}{5} + C$$
$$= \frac{1}{3}\left(x^2 - 25\right)^{3/2} + 25\left(x^2 - 25\right)^{1/2} + C$$
$$= \frac{1}{3}\sqrt{x^2 - 25}\left(x^2 - 25 + 75\right) + C$$
$$= \frac{1}{3}\sqrt{x^2 - 25}\left(50 + x^2\right) + C$$

13. Let $x = \tan\theta$, $dx = \sec^2\theta\, d\theta$, $\sqrt{1 + x^2} = \sec\theta$.

$$\int x\sqrt{1 + x^2}\, dx = \int \tan\theta(\sec\theta)\sec^2\theta\, d\theta = \frac{\sec^3\theta}{3} + C = \frac{1}{3}\left(1 + x^2\right)^{3/2} + C$$

Note: This integral could have been evaluated with the Power Rule.

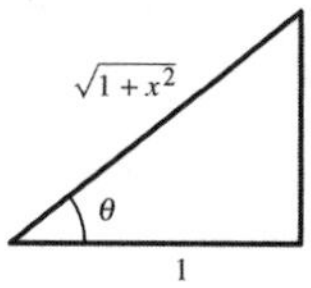

14. Same substitution as in Exercise 13

$$\int \frac{9x^3}{\sqrt{1 + x^2}}\, dx = 9\int \frac{\tan^3\theta}{\sec\theta}\sec^2\theta\, d\theta = 9\int\left(\sec^2\theta - 1\right)\sec\theta\tan\theta\, d\theta = 9\left[\frac{\sec^3\theta}{3} - \sec\theta\right] + C$$

$$= 3\sec\theta\left(\sec^2\theta - 3\right) + C = 3\sqrt{1 + x^2}\left[\left(1 + x^2\right) - 3\right] + C = 3\sqrt{1 + x^2}\left(x^2 - 2\right) + C$$

15. Same substitution as in Exercise 13

$$\int \frac{1}{\left(1 + x^2\right)^2}\, dx = \int \frac{1}{\left(\sqrt{1 + x^2}\right)^4}\, dx = \int \frac{\sec^2\theta\, d\theta}{\sec^4\theta}$$

$$= \int \cos^2\theta\, d\theta = \frac{1}{2}\int\left(1 + \cos 2\theta\right) d\theta$$

$$= \frac{1}{2}\left[\theta + \frac{\sin 2\theta}{2}\right]$$

$$= \frac{1}{2}\left[\theta + \sin\theta\cos\theta\right] + C$$

$$= \frac{1}{2}\left[\arctan x + \left(\frac{x}{\sqrt{1 + x^2}}\right)\left(\frac{1}{\sqrt{1 + x^2}}\right)\right] + C$$

$$= \frac{1}{2}\left(\arctan x + \frac{x}{1 + x^2}\right) + C$$

16. Same substitution as in Exercise 13

$$\int \frac{x^2}{\left(1 + x^2\right)^2}\, dx = \int \frac{x^2}{\left(\sqrt{1 + x^2}\right)^4}\, dx = \int \frac{\tan^2\theta\sec^2\theta\, d\theta}{\sec^4\theta} = \int \sin^2\theta\, d\theta$$

$$= \frac{1}{2}\int\left(1 - \cos 2\theta\right) d\theta = \frac{1}{2}\left[\theta - \frac{\sin 2\theta}{2}\right] = \frac{1}{2}\left[\theta - \sin\theta\cos\theta\right] + C$$

$$= \frac{1}{2}\left[\arctan x - \left(\frac{x}{\sqrt{1 + x^2}}\right)\left(\frac{1}{\sqrt{1 + x^2}}\right)\right] + C = \frac{1}{2}\left(\arctan x - \frac{x}{1 + x^2}\right) + C$$

17. Let $u = 4x$, $a = 3$, $du = 4\, dx$.

$$\int\sqrt{9 + 16x^2}\, dx = \tfrac{1}{4}\int\sqrt{(4x)^2 + 3^2}\,(4)\, dx$$

$$= \tfrac{1}{4}\cdot\tfrac{1}{2}\left[4x\sqrt{16x^2 + 9} + 9\ln\left|4x + \sqrt{16x^2 + 9}\right|\right] + C$$

$$= \tfrac{1}{2}x\sqrt{16x^2 + 9} + \tfrac{9}{8}\ln\left|4x + \sqrt{16x^2 + 9}\right| + C$$

18. Let $u = x, a = 2, du = dx$.

$$\int\sqrt{4 + x^2}\,dx = \int\sqrt{x^2 + 2^2}\,dx$$
$$= \frac{1}{2}\left[x\sqrt{x^2 + 4} + 4\ln\left|x + \sqrt{x^2 + 4}\right|\right] + C$$
$$= \frac{x}{2}\sqrt{x^2 + 4} + 2\ln\left|x + \sqrt{x^2 + 4}\right| + C$$

19. $\int\sqrt{25 - 4x^2}\,dx = \int 2\sqrt{\frac{25}{4} - x^2}\,dx, \quad a = \frac{5}{2}$

$$= 2\left(\frac{1}{2}\right)\left[\frac{25}{4}\arcsin\left(\frac{2x}{5}\right) + x\sqrt{\frac{25}{4} - x^2}\right] + C$$
$$= \frac{25}{4}\arcsin\left(\frac{2x}{5}\right) + \frac{x}{2}\sqrt{25 - 4x^2} + C$$

20. Let $u = \sqrt{5}x, a = 1, du = \sqrt{5}\,dx$.

$$\int\sqrt{5x^2 - 1}\,dx = \frac{1}{\sqrt{5}}\int\sqrt{\left(\sqrt{5}x\right)^2 - 1}\,\sqrt{5}\,dx$$
$$= \frac{1}{\sqrt{5}}\left(\frac{1}{2}\right)\left(\sqrt{5}x\sqrt{5x^2 - 1} - \ln\left|\sqrt{5}x + \sqrt{5x^2 - 1}\right|\right) + C$$
$$= \frac{x}{2}\sqrt{5x^2 - 1} - \frac{\sqrt{5}}{10}\ln\left|\sqrt{5}x + \sqrt{5x^2 - 1}\right| + C$$

21. $\int\frac{1}{\sqrt{16 - x^2}}\,dx = \arcsin\left(\frac{x}{4}\right) + C$

22. Let $x = 6\sin\theta, dx = 6\cos\theta\,d\theta, \sqrt{36 - x^2} = 6\cos\theta$.

$$\int\frac{x^2}{\sqrt{36 - x^2}}\,dx = \int\frac{36\sin^2\theta}{6\cos\theta}(6\cos\theta\,d\theta)$$
$$= 36\int\sin^2\theta\,d\theta$$
$$= 18\int(1 - \cos 2\theta)\,d\theta$$
$$= 18\left(\theta - \frac{\sin 2\theta}{2}\right) + C$$
$$= 18(\theta - \sin\theta\cos\theta) + C$$
$$= 18\left(\arcsin\frac{x}{6} - \frac{x}{6}\cdot\frac{\sqrt{36 - x^2}}{6}\right) + C$$
$$= 18\arcsin\frac{x}{6} - \frac{x\sqrt{36 - x^2}}{2} + C$$

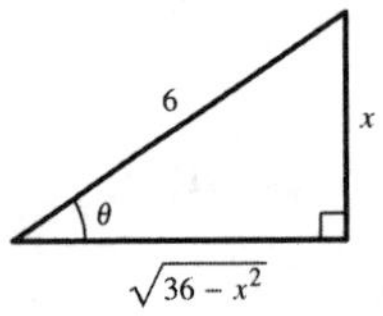

23. Let $x = 2\sin\theta$, $dx = 2\cos\theta\, d\theta$,

$\sqrt{4 - x^2} = 2\cos\theta$.

$$\begin{aligned}\int\sqrt{16 - 4x^2}\,dx &= 2\int\sqrt{4 - x^2}\,dx\\ &= 2\int 2\cos\theta(2\cos\theta\,d\theta)\\ &= 8\int\cos^2\theta\,d\theta\\ &= 4\int(1 + \cos 2\theta)\,d\theta\\ &= 4\left(\theta + \frac{1}{2}\sin 2\theta\right) + C\\ &= 4\theta + 4\sin\theta\cos\theta + C\\ &= 4\arcsin\left(\frac{x}{2}\right) + x\sqrt{4 - x^2} + C\end{aligned}$$

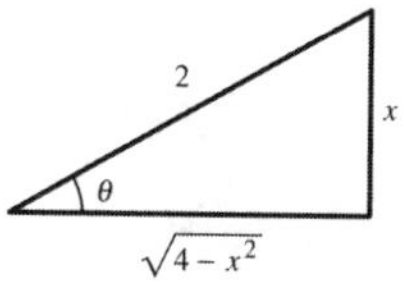

24. Let $x = 2\sec\theta$, $dx = 2\sec\theta\tan\theta\,d\theta$,

$\sqrt{x^2 - 4} = 2\tan\theta$.

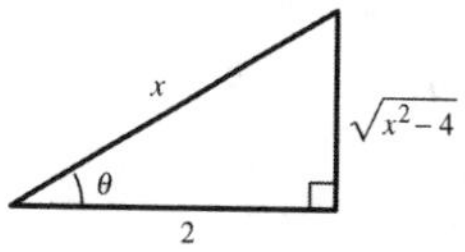

$$\begin{aligned}\int\frac{1}{\sqrt{x^2 - 4}}\,dx &= \int\frac{2\sec\theta\tan\theta}{2\tan\theta}\,d\theta\\ &= \int\sec\theta\,d\theta\\ &= \ln\left|\sec\theta + \tan\theta\right| + C\\ &= \ln\left|\frac{x}{2} + \frac{\sqrt{x^2 - 4}}{2}\right| + C\\ &= \ln\left|x + \sqrt{x^2 - 4}\right| + C\end{aligned}$$

25. Let $x = \sin\theta$, $dx = \cos\theta\,d\theta$, $\sqrt{1 - x^2} = \cos\theta$.

$$\begin{aligned}\int\frac{\sqrt{1 - x^2}}{x^4}\,dx &= \int\frac{\cos\theta(\cos\theta\,d\theta)}{\sin^4\theta}\\ &= \int\cot^2\theta\csc^2\theta\,d\theta\\ &= -\frac{1}{3}\cot^3\theta + C\\ &= -\frac{\left(1 - x^2\right)^{3/2}}{3x^3} + C\end{aligned}$$

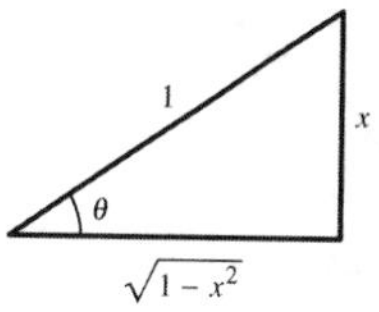

26. Let $5x = 2\tan\theta$, $5dx = 2\sec^2\theta\,d\theta$, $\sqrt{25x^2 + 4} = \sqrt{4\tan^2\theta + 4} = 2\sec\theta$.

$$\begin{aligned}\int\frac{\sqrt{25x^2 + 4}}{x^4}\,dx &= \int\frac{2\sec\theta}{\left(\frac{2}{5}\tan\theta\right)^4}\left(\frac{2}{5}\sec^2\theta\right)d\theta\\ &= \frac{125}{4}\int\frac{\cos\theta}{\sin^4\theta}\,d\theta\\ &= \frac{125}{4}\left(\frac{1}{(-3)\sin^3\theta}\right) + C\\ &= -\frac{125}{12}\csc^3\theta + C\\ &= -\frac{125}{12}\left(\frac{\sqrt{25x^2 + 4}}{5x}\right)^3\\ &= -\frac{\left(25x^2 + 4\right)^{3/2}}{12x^3} + C\end{aligned}$$

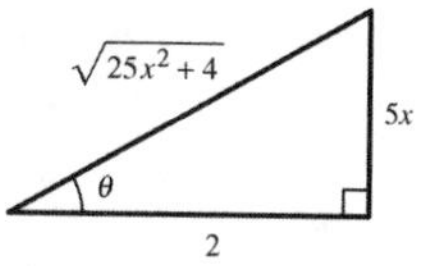

27. Let $2x = 3\tan\theta \Rightarrow x = \frac{3}{2}\tan\theta$, $dx = \frac{3}{2}\sec^2\theta\,d\theta$, $\sqrt{4x^2+9} = 3\sec\theta$.

$$\int \frac{1}{x\sqrt{4x^2+9}}\,dx = \int \frac{(3/2)\sec^2\theta\,d\theta}{(3/2)\tan\theta\,3\sec\theta}$$
$$= \frac{1}{3}\int \csc\theta\,d\theta$$
$$= -\frac{1}{3}\ln\left|\csc\theta + \cot\theta\right| + C$$
$$= -\frac{1}{3}\ln\left|\frac{\sqrt{4x^2+9}+3}{2x}\right| + C$$

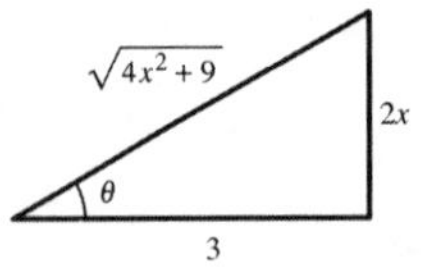

28. Let $3x = \tan\theta$, $3dx = \sec^2\theta\,d\theta$, $\sqrt{9x^2+1} = \sec\theta$.

$$\int \frac{1}{x\sqrt{9x^2+1}}\,dx = \int \frac{1}{\frac{1}{3}\tan\theta\sec\theta}\left(\frac{1}{3}\sec^2\theta\right)d\theta$$
$$= \int \frac{\sec\theta}{\tan\theta}\,d\theta$$
$$= \int \csc\theta\,d\theta$$
$$= \ln\left|\csc\theta - \cot\theta\right| + C$$
$$= \ln\left|\frac{\sqrt{9x^2+1}}{3x} - \frac{1}{3x}\right| + C$$
$$= \ln\left|\frac{\sqrt{9x^2+1}-1}{3x}\right| + C$$

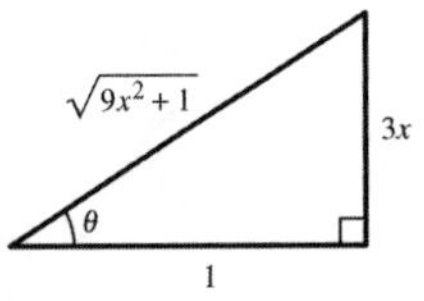

(**Note:** This equals $-\ln\left|\frac{\sqrt{9x^2+1}+1}{3x}\right| + C$.)

29. Let $u = x^2 + 3$, $du = 2x\,dx$.

$$\int \frac{-3x}{\left(x^2+3\right)^{3/2}}\,dx = -\frac{3}{2}\int \left(x^2+3\right)^{-3/2}(2x)\,dx$$
$$= -\frac{3}{2}\frac{\left(x^2+3\right)^{-1/2}}{(-1/2)} + C$$
$$= \frac{3}{\sqrt{x^2+3}} + C$$

30. Let $x = \sqrt{5}\tan\theta$, $dx = \sqrt{5}\sec^2\theta$, $x^2 + 5 = 5\sec^2\theta$.

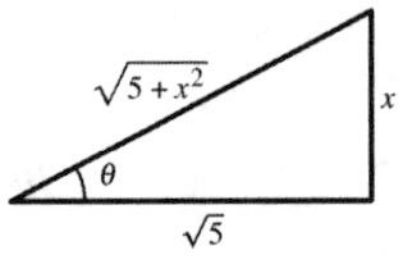

$$\int \frac{1}{\left(x^2+5\right)^{3/2}}\,dx = \int \frac{\sqrt{5}\sec^2\theta}{\left(\sqrt{5}\sec\theta\right)^3}\,d\theta$$
$$= \frac{1}{5}\int \cos\theta\,d\theta$$
$$= \frac{1}{5}\sin\theta + C = \frac{x}{5\sqrt{5+x^2}} + C$$

31. Let $e^x = \sin\theta$, $e^x\,dx = \cos\theta\,d\theta$, $\sqrt{1-e^{2x}} = \cos\theta$.

$$\begin{aligned}\int e^x\sqrt{1-e^{2x}}\,dx &= \int \cos^2\theta\,d\theta\\ &= \frac{1}{2}\int(1+\cos 2\theta)\,d\theta\\ &= \frac{1}{2}\left(\theta + \frac{\sin 2\theta}{2}\right)\\ &= \frac{1}{2}(\theta + \sin\theta\cos\theta) + C\\ &= \frac{1}{2}\left(\arcsin e^x + e^x\sqrt{1-e^{2x}}\right) + C\end{aligned}$$

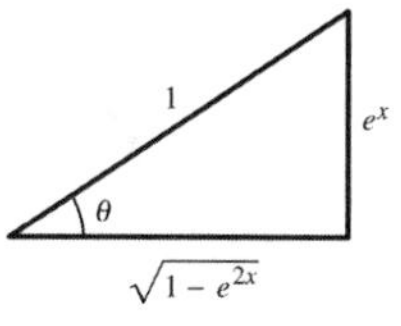

32. Let $\sqrt{x} = \sin\theta$, $x = \sin^2\theta$, $dx = 2\sin\theta\cos\theta\,d\theta$, $\sqrt{1-x} = \cos\theta$.

$$\begin{aligned}\int\frac{\sqrt{1-x}}{\sqrt{x}}\,dx &= \int\frac{\cos\theta(2\sin\theta\cos\theta\,d\theta)}{\sin\theta}\\ &= 2\int\cos^2\theta\,d\theta\\ &= \int(1+\cos 2\theta)\,d\theta\\ &= (\theta + \sin\theta\cos\theta) + C\\ &= \arcsin\sqrt{x} + \sqrt{x}\sqrt{1-x} + C\end{aligned}$$

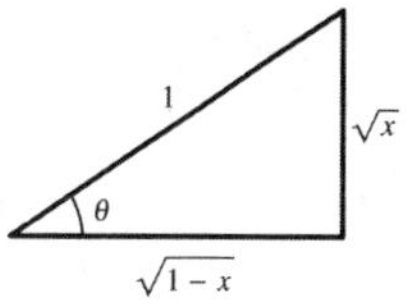

33. Let $x = \sqrt{2}\tan\theta$, $dx = \sqrt{2}\sec^2\theta\,d\theta$, $x^2 + 2 = 2\sec^2\theta$.

$$\begin{aligned}\int\frac{1}{4+4x^2+x^4}\,dx &= \int\frac{1}{(x^2+2)^2}\,dx = \int\frac{\sqrt{2}\sec^2\theta\,d\theta}{4\sec^4\theta}\\ &= \frac{\sqrt{2}}{4}\int\cos^2\theta\,d\theta\\ &= \frac{\sqrt{2}}{4}\left(\frac{1}{2}\right)\int(1+\cos 2\theta)\,d\theta\\ &= \frac{\sqrt{2}}{8}\left(\theta + \frac{1}{2}\sin 2\theta\right) + C\\ &= \frac{\sqrt{2}}{8}(\theta + \sin\theta\cos\theta) + C\\ &= \frac{\sqrt{2}}{8}\left(\arctan\frac{x}{\sqrt{2}} + \frac{x}{\sqrt{x^2+2}}\cdot\frac{\sqrt{2}}{\sqrt{x^2+2}}\right) = \frac{1}{4}\left(\frac{x}{x^2+2} + \frac{1}{\sqrt{2}}\arctan\frac{x}{\sqrt{2}}\right) + C\end{aligned}$$

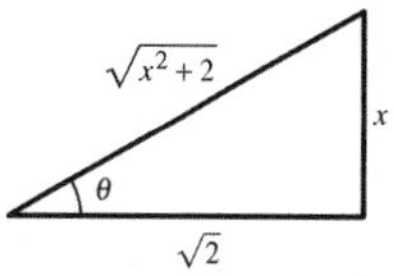

34. Let $x = \tan\theta$, $dx = \sec^2\theta\,d\theta$, $x^2 + 1 = \sec^2\theta$.

$$\begin{aligned}\int\frac{x^3+x+1}{x^4+2x^2+1}\,dx &= \frac{1}{4}\int\frac{4x^3+4x}{x^4+2x^2+1}\,dx + \int\frac{1}{(x^2+1)^2}\,dx\\ &= \frac{1}{4}\ln(x^4+2x^2+1) + \int\frac{\sec^2\theta\,d\theta}{\sec^4\theta}\\ &= \frac{1}{2}\ln(x^2+1) + \frac{1}{2}\int(1+\cos 2\theta)\,d\theta\\ &= \frac{1}{2}\ln(x^2+1) + \frac{1}{2}(\theta + \sin\theta\cos\theta) + C\\ &= \frac{1}{2}\left(\ln(x^2+1) + \arctan x + \frac{x}{x^2+1}\right) + C\end{aligned}$$

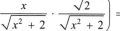

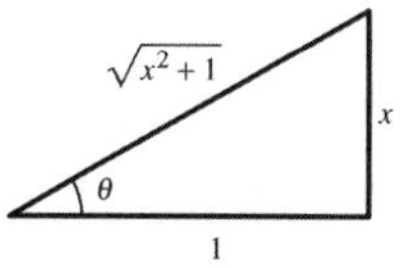

35. Use integration by parts. Because $x > \frac{1}{2}$,

$$u = \operatorname{arcsec} 2x \Rightarrow du = \frac{1}{x\sqrt{4x^2 - 1}}\,dx,\ dv = dx \Rightarrow v = x$$

$$\int \operatorname{arcsec} 2x\,dx = x \operatorname{arcsec} 2x - \int \frac{1}{\sqrt{4x^2 - 1}}\,dx$$

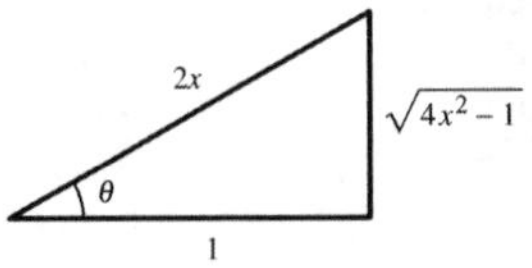

$$2x = \sec\theta,\ dx = \frac{1}{2}\sec\theta\tan\theta\,d\theta,\ \sqrt{4x^2 - 1} = \tan\theta$$

$$\int \operatorname{arcsec} 2x\,dx = x \operatorname{arcsec} 2x - \int \frac{(1/2)\sec\theta\tan\theta\,d\theta}{\tan\theta} = x \operatorname{arcsec} 2x - \frac{1}{2}\int \sec\theta\,d\theta$$

$$= x \operatorname{arcsec} 2x - \frac{1}{2}\ln\left|\sec\theta + \tan\theta\right| + C = x \operatorname{arcsec} 2x - \frac{1}{2}\ln\left|2x + \sqrt{4x^2 - 1}\right| + C.$$

36. $u = \arcsin x \Rightarrow du = \frac{1}{\sqrt{1 - x^2}}\,dx,\ dv = x\,dx \Rightarrow v = \frac{x^2}{2}$

$$\int x \arcsin x\,dx = \frac{x^2}{2}\arcsin x - \frac{1}{2}\int \frac{x^2}{\sqrt{1 - x^2}}\,dx$$

$$x = \sin\theta,\ dx = \cos\theta\,d\theta,\ \sqrt{1 - x^2} = \cos\theta$$

$$\int x \arcsin x\,dx = \frac{x^2}{2}\arcsin x = \frac{1}{2}\int \frac{\sin^2\theta}{\cos\theta}\cos\theta\,d\theta = \frac{x^2}{2}\arcsin x - \frac{1}{4}\int (1 - \cos 2\theta)\,d\theta$$

$$= \frac{x^2}{2}\arcsin x - \frac{1}{4}\left(\theta - \frac{1}{2}\sin 2\theta\right) + C = \frac{x^2}{2}\arcsin x - \frac{1}{4}(\theta - \sin\theta\cos\theta) + C$$

$$= \frac{x^2}{2}\arcsin x - \frac{1}{4}\left(\arcsin x - x\sqrt{1 - x^2}\right) + C = \frac{1}{4}\left[\left(2x^2 - 1\right)\arcsin x + x\sqrt{1 - x^2}\right] + C$$

37. $\int \frac{1}{\sqrt{4x - x^2}}\,dx = \int \frac{1}{\sqrt{4 - (x - 2)^2}}\,dx = \arcsin\left(\frac{x - 2}{2}\right) + C$

38. Let $x - 1 = \sin\theta,\ dx = \cos\theta\,d\theta,\ \sqrt{1 - (x - 1)^2} = \sqrt{2x - x^2} = \cos\theta.$

$$\int \frac{x^2}{\sqrt{2x - x^2}}\,dx = \int \frac{x^2}{\sqrt{1 - (x - 1)^2}}\,dx$$

$$= \int \frac{(1 + \sin\theta)^2(\cos\theta\,d\theta)}{\cos\theta}$$

$$= \int \left(1 + 2\sin\theta + \sin^2\theta\right)d\theta$$

$$= \int \left(\frac{3}{2} + 2\sin\theta - \frac{1}{2}\cos 2\theta\right)d\theta$$

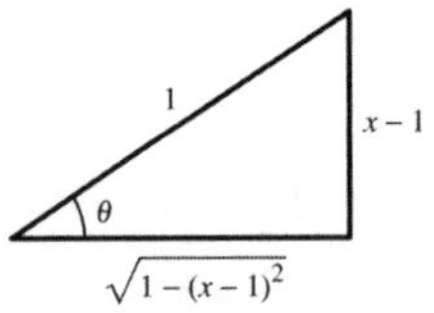

$$= \frac{3}{2}\theta - 2\cos\theta - \frac{1}{4}\sin 2\theta + C$$

$$= \frac{3}{2}\theta - 2\cos\theta - \frac{1}{2}\sin\theta\cos\theta + C$$

$$= \frac{3}{2}\arcsin(x - 1) - 2\sqrt{2x - x^2} - \frac{1}{2}(x - 1)\sqrt{2x - x^2} + C$$

$$= \frac{3}{2}\arcsin(x - 1) - \frac{1}{2}\sqrt{2x - x^2}(x + 3) + C$$

39. $x^2 + 6x + 12 = x^2 + 6x + 9 + 3 = (x + 3)^2 + \left(\sqrt{3}\right)^2$

Let $x + 3 = \sqrt{3}\tan\theta,\ dx = \sqrt{3}\sec^2\theta\,d\theta.$

$$\sqrt{x^2 + 6x + 12} = \sqrt{(x+3)^2 + \left(\sqrt{3}\right)^2} = \sqrt{3}\sec\theta$$

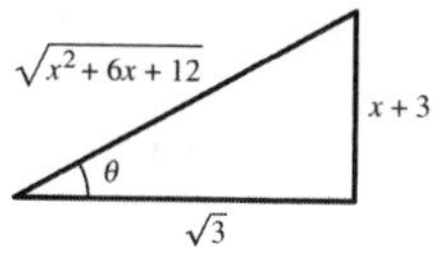

$$\begin{aligned}
\int \frac{x}{\sqrt{x^2 + 6x + 12}}\,dx &= \int \frac{\sqrt{3}\tan\theta - 3}{\sqrt{3}\sec\theta}\sqrt{3}\sec^2\theta\,d\theta \\
&= \int \sqrt{3}\sec\theta\tan\theta\,d\theta - 3\int \sec\theta\,d\theta \\
&= \sqrt{3}\sec\theta - 3\ln\left|\sec\theta + \tan\theta\right| + C \\
&= \sqrt{3}\left(\frac{\sqrt{x^2 + 6x + 12}}{\sqrt{3}}\right) - 3\ln\left|\frac{\sqrt{x^2 + 6x + 12}}{\sqrt{3}} + \frac{x+3}{\sqrt{3}}\right| + C \\
&= \sqrt{x^2 + 6x + 12} - 3\ln\left|\sqrt{x^2 + 6x + 12} + (x + 3)\right| + C
\end{aligned}$$

40. Let $x - 3 = 2\sec\theta,\ dx = 2\sec\theta\tan\theta\,d\theta,\ \sqrt{(x-3)^2 - 4} = 2\tan\theta.$

$$\begin{aligned}
\int \frac{x}{\sqrt{x^2 - 6x + 5}}\,dx &= \int \frac{x}{\sqrt{(x-3)^2 - 4}}\,dx \\
&= \int \frac{(2\sec\theta + 3)}{2\tan\theta}(2\sec\theta\tan\theta)\,d\theta \\
&= \int \left(2\sec^2\theta + 3\sec\theta\right)d\theta \\
&= 2\tan\theta + 3\ln\left|\sec\theta + \tan\theta\right| + C_1 \\
&= 2\left[\frac{\sqrt{(x-3)^3 - 4}}{2}\right] + 3\ln\left|\frac{x-3}{2} + \frac{\sqrt{(x-3)^2 - 4}}{2}\right| + C_1 \\
&= \sqrt{x^2 - 6x + 5} + 3\ln\left|(x - 3) + \sqrt{x^2 - 6x + 5}\right| + C
\end{aligned}$$

41. Let $t = \sin\theta,\ dt = \cos\theta\,d\theta,\ 1 - t^2 = \cos^2\theta.$

(a) $\displaystyle\int \frac{t^2}{\left(1 - t^2\right)^{3/2}}\,dt = \int \frac{\sin^2\theta\cos\theta\,d\theta}{\cos^3\theta} = \int \tan^2\theta\,d\theta = \int\left(\sec^2\theta - 1\right)d\theta = \tan\theta - \theta + C = \frac{t}{\sqrt{1 - t^2}} - \arcsin t + C$

So, $\displaystyle\int_0^{\sqrt{3}/2} \frac{t^2}{\left(1 - t^2\right)^{3/2}}\,dt = \left[\frac{t}{\sqrt{1 - t^2}} - \arcsin t\right]_0^{\sqrt{3}/2} = \frac{\sqrt{3}/2}{\sqrt{1/4}} - \arcsin\frac{\sqrt{3}}{2} = \sqrt{3} - \frac{\pi}{3} \approx 0.685.$

(b) When $t = 0,\ \theta = 0.$ When $t = \sqrt{3}/2,\ \theta = \pi/3.$ So,

$$\int_0^{\sqrt{3}/2} \frac{t^2}{\left(1 - t^2\right)^{3/2}}\,dt = \left[\tan\theta - \theta\right]_0^{\pi/3} = \sqrt{3} - \frac{\pi}{3} \approx 0.685.$$

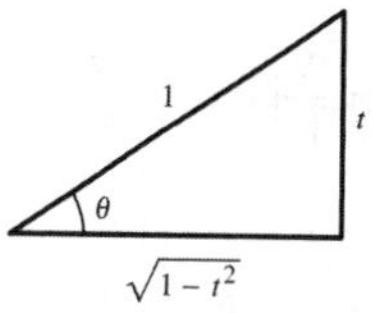

42. Same substitution as in Exercise 41

(a) $$\int \frac{1}{\left(1-t^2\right)^{5/2}}\,dt = \int \frac{\cos\theta\,d\theta}{\cos^5\theta} = \int \sec^4\theta\,d\theta = \int\left(\tan^2\theta + 1\right)\sec^2\theta\,d\theta$$

$$= \frac{1}{3}\tan^3\theta + \tan\theta + C = \frac{1}{3}\left(\frac{t}{\sqrt{1-t^2}}\right)^3 + \frac{t}{\sqrt{1-t^2}} + C$$

So, $$\int_0^{\sqrt{3}/2} \frac{1}{\left(1-t^2\right)^{5/2}}\,dt = \left[\frac{t^3}{3\left(1-t^2\right)^{3/2}} + \frac{t}{\sqrt{1-t^2}}\right]_0^{\sqrt{3}/2} = \frac{3\sqrt{3}/8}{3(1/4)^{3/2}} + \frac{\sqrt{3}/2}{\sqrt{1/4}} = \sqrt{3} + \sqrt{3} = 2\sqrt{3} \approx 3.464.$$

(b) When $t = 0, \theta = 0$. When $t = \sqrt{3}/2, \theta = \pi/3$. So,

$$\int_0^{\sqrt{3}/2} \frac{1}{\left(1-t^2\right)^{5/2}}\,dt = \left[\frac{1}{3}\tan^3\theta + \tan\theta\right]_0^{\pi/3} = \frac{1}{3}\left(\sqrt{3}\right)^3 + \sqrt{3} = 2\sqrt{3} \approx 3.464.$$

43. (a) Let $x = 3\tan\theta, dx = 3\sec^2\theta\,d\theta, \sqrt{x^2+9} = 3\sec\theta$.

$$\int \frac{x^3}{\sqrt{x^2+9}}\,dx = \int \frac{\left(27\tan^3\theta\right)\left(3\sec^2\theta\,d\theta\right)}{3\sec\theta}$$

$$= 27\int\left(\sec^2\theta - 1\right)\sec\theta\tan\theta\,d\theta$$

$$= 27\left[\frac{1}{3}\sec^3\theta - \sec\theta\right] + C = 9\left[\sec^3\theta - 3\sec\theta\right] + C$$

$$= 9\left[\left(\frac{\sqrt{x^2+9}}{3}\right)^3 - 3\left(\frac{\sqrt{x^2+9}}{3}\right)\right] + C = \frac{1}{3}\left(x^2+9\right)^{3/2} - 9\sqrt{x^2+9} + C$$

So, $$\int_0^3 \frac{x^3}{\sqrt{x^2+9}}\,dx = \left[\frac{1}{3}\left(x^2+9\right)^{3/2} - 9\sqrt{x^2+9}\right]_0^3$$

$$= \left(\frac{1}{3}\left(54\sqrt{2}\right) - 27\sqrt{2}\right) - (9 - 27) = 18 - 9\sqrt{2} = 9\left(2 - \sqrt{2}\right) \approx 5.272.$$

(b) When $x = 0, \theta = 0$. When $x = 3, \theta = \pi/4$. So,

$$\int_0^3 \frac{x^3}{\sqrt{x^2+9}}\,dx = 9\left[\sec^3\theta - 3\sec\theta\right]_0^{\pi/4} = 9\left(2\sqrt{2} - 3\sqrt{2}\right) - 9(1 - 3) = 9\left(2 - \sqrt{2}\right) \approx 5.272.$$

44. (a) Let $5x = 3\sin\theta, dx = \frac{3}{5}\cos\theta\,d\theta, \sqrt{9-25x^2} = 3\cos\theta$.

$$\int\sqrt{9-25x^2}\,dx = \int(3\cos\theta)\frac{3}{5}\cos\theta\,d\theta$$

$$= \frac{9}{5}\int\frac{1+\cos 2\theta}{2}\,d\theta$$

$$= \frac{9}{10}\left(\theta + \frac{1}{2}\sin 2\theta\right) + C$$

$$= \frac{9}{10}(\theta + \sin\theta\cos\theta) + C$$

$$= \frac{9}{10}\left(\arcsin\frac{5x}{3} + \frac{5x}{3}\cdot\frac{\sqrt{9-25x^2}}{3}\right) + C$$

So, $$\int_0^{3/5}\sqrt{9-25x^2}\,dx = \frac{9}{10}\left[\arcsin\frac{5x}{3} + \frac{5x\sqrt{9-25x^2}}{9}\right]_0^{3/5} = \frac{9}{10}\left[\frac{\pi}{2}\right] = \frac{9\pi}{20}.$$

(b) When $x = 0, \theta = 0$. When $x = \dfrac{3}{5}, \theta = \dfrac{\pi}{2}$.

So, $\displaystyle\int_0^{3/5} \sqrt{9 - 25x^2}\, dx = \left[\frac{9}{10}(\theta + \sin\theta\cos\theta)\right]_0^{\pi/2} = \frac{9}{10}\left(\frac{\pi}{2}\right) = \frac{9\pi}{20}.$

45. (a) Let $x = 3\sec\theta$, $dx = 3\sec\theta\tan\theta\, d\theta$, $\sqrt{x^2 - 9} = 3\tan\theta$.

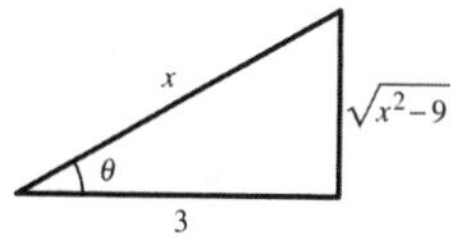

$$\begin{aligned}
\int \frac{x^2}{\sqrt{x^2 - 9}}\, dx &= \int \frac{9\sec^2\theta}{3\tan\theta} 3\sec\theta\tan\theta\, d\theta \\
&= 9\int \sec^3\theta\, d\theta \\
&= 9\left(\frac{1}{2}\sec\theta\tan\theta + \frac{1}{2}\int\sec\theta\, d\theta\right) \quad \text{(8.3 Exercise 102 or Example 5, Section 8.2)} \\
&= \frac{9}{2}\left(\sec\theta\tan\theta + \ln\left|\sec\theta + \tan\theta\right|\right) \\
&= \frac{9}{2}\left(\frac{x}{3} \cdot \frac{\sqrt{x^2 - 9}}{3} + \ln\left|\frac{x}{3} + \frac{\sqrt{x^2 - 9}}{3}\right|\right)
\end{aligned}$$

So,

$$\begin{aligned}
\int_4^6 \frac{x^2}{\sqrt{x^2 - 9}}\, dx &= \frac{9}{2}\left[\frac{x\sqrt{x^2 - 9}}{9} + \ln\left|\frac{x}{3} + \frac{\sqrt{x^2 - 9}}{3}\right|\right]_4^6 \\
&= \frac{9}{2}\left[\left(\frac{6\sqrt{27}}{9} + \ln\left|2 + \frac{\sqrt{27}}{3}\right|\right) - \left(\frac{4\sqrt{7}}{9} + \ln\left|\frac{4}{3} + \frac{\sqrt{7}}{3}\right|\right)\right] \\
&= 9\sqrt{3} - 2\sqrt{7} + \frac{9}{2}\left[\ln\left(\frac{6 + \sqrt{27}}{3}\right) - \ln\left(\frac{4 + \sqrt{7}}{3}\right)\right] \\
&= 9\sqrt{3} - 2\sqrt{7} + \frac{9}{2}\ln\left(\frac{6 + 3\sqrt{3}}{4 + \sqrt{7}}\right) \approx 12.644.
\end{aligned}$$

(b) When $x = 4, \theta = \operatorname{arcsec}\left(\dfrac{4}{3}\right)$. When $x = 6, \theta = \operatorname{arcsec}(2) = \dfrac{\pi}{3}$.

$$\begin{aligned}
\int_4^6 \frac{x^2}{\sqrt{x^2 - 9}}\, dx &= \frac{9}{2}\Big[\sec\theta\tan\theta + \ln\left|\sec\theta + \tan\theta\right|\Big]_{\operatorname{arcsec}(4/3)}^{\pi/3} \\
&= \frac{9}{2}\left(2 \cdot \sqrt{3} + \ln\left|2 + \sqrt{3}\right|\right) - \frac{9}{2}\left(\frac{4}{3}\left(\frac{\sqrt{7}}{3}\right) + \ln\left|\frac{4}{3} + \frac{\sqrt{7}}{3}\right|\right) \\
&= 9\sqrt{3} - 2\sqrt{7} + \frac{9}{2}\ln\left(\frac{6 + 3\sqrt{3}}{4 + \sqrt{7}}\right) \approx 12.644
\end{aligned}$$

46. (a) Let $x = 4\sec\theta$, $dx = 4\sec\theta\tan\theta\,d\theta$, $\sqrt{x^2-16} = 4\tan\theta$.

$$\begin{aligned}\int\frac{\sqrt{x^2-16}}{x^2}\,dx &= \int\frac{4\tan\theta}{16\sec^2\theta}(4\sec\theta\tan\theta)\,d\theta\\ &= \int\frac{\tan^2\theta}{\sec\theta}\,d\theta\\ &= \int\frac{\sin^2\theta}{\cos\theta}\,d\theta\\ &= \int\frac{1-\cos^2\theta}{\cos\theta}\,d\theta\\ &= \int\sec\theta\,d\theta - \int\cos\theta\,d\theta\\ &= \ln\left|\sec\theta+\tan\theta\right| - \sin\theta + C\\ &= \ln\left|\frac{x}{4}+\frac{\sqrt{x^2-16}}{4}\right| - \frac{\sqrt{x^2-16}}{x} + C\end{aligned}$$

So,

$$\begin{aligned}\int_4^8\frac{\sqrt{x^2-16}}{x^2}\,dx &= \left[\ln\left|\frac{x}{4}+\frac{\sqrt{x^2-16}}{4}\right| - \frac{\sqrt{x^2-16}}{x}\right]_4^8\\ &= \left[\ln\left(2+\frac{\sqrt{48}}{4}\right) - \frac{\sqrt{48}}{8}\right] - \left[\ln(1)\right]\\ &= \ln\left(2+\sqrt{3}\right) - \frac{\sqrt{3}}{2}.\end{aligned}$$

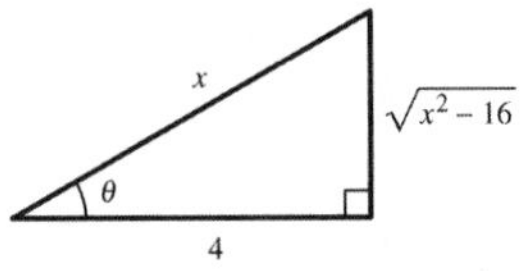

(b) When $x = 4$, $\theta = 0$, and when $x = 8$, $\theta = \dfrac{\pi}{3}$. So,

$$\begin{aligned}\int_4^8\frac{\sqrt{x^2-16}}{x^2}\,dx &= \left[\ln\left|\sec\theta+\tan\theta\right| - \sin\theta\right]_0^{\pi/3}\\ &= \ln\left|2+\sqrt{3}\right| - \frac{\sqrt{3}}{2}.\end{aligned}$$

47. (a) Let $u = a\sin\theta$, $\sqrt{a^2-u^2} = a\cos\theta$, where $-\pi/2 \le \theta \le \pi/2$.

(b) Let $u = a\tan\theta$, $\sqrt{a^2+u^2} = a\sec\theta$, where $-\pi/2 < \theta < \pi/2$.

(c) Let $u = a\sec\theta$, $\sqrt{u^2-a^2} = \tan\theta$ if $u > a$ and $\sqrt{u^2-a^2} = -\tan\theta$ if $u < -a$, where $0 \le \theta < \pi/2$ or $\pi/2 < \theta \le \pi$.

48. (a) Substitution: $u = x^2+1$, $du = 2x\,dx$

(b) Trigonometric substitution: $x = \sec\theta$

49. (a) $u = x^2 + 9, du = 2x\,dx$

$$\int \frac{x}{x^2+9}\,dx = \frac{1}{2}\int \frac{du}{u} = \frac{1}{2}\ln|u| + C = \frac{1}{2}\ln(x^2+9) + C$$

Let $x = 3\tan\theta, x^2 + 9 = 9\sec^2\theta, dx = 3\sec^2\theta\,d\theta.$

$$\int \frac{x}{x^2+9}\,dx = \int \frac{3\tan\theta}{9\sec^2\theta}3\sec^2\theta\,d\theta = \int \tan\theta\,d\theta$$
$$= -\ln|\cos\theta| + C_1$$
$$= -\ln\left|\frac{3}{\sqrt{x^2+9}}\right| + C_1$$
$$= -\ln 3 + \ln\sqrt{x^2+9} + C_1 = \frac{1}{2}\ln(x^2+9) + C_2$$

The answers are equivalent.

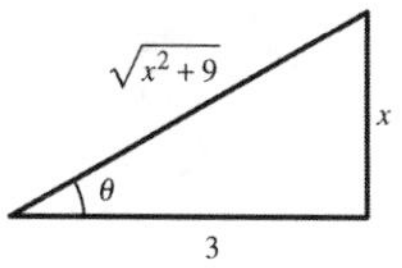

(b) $$\int \frac{x^2}{x^2+9}\,dx = \int \frac{x^2+9-9}{x^2+9}\,dx = \int\left(1 - \frac{9}{x^2+9}\right)dx = x - 3\arctan\left(\frac{x}{3}\right) + C$$

Let $x = 3\tan\theta, x^2 + 9 = 9\sec^2\theta, dx = 3\sec^2\theta\,d\theta.$

$$\int \frac{x^2}{x^2+9}\,dx = \int \frac{9\tan^2\theta}{9\sec^2\theta}3\sec^2\theta\,d\theta$$
$$= 3\int \tan^2\theta\,d\theta = 3\int(\sec^2\theta - 1)\,d\theta$$
$$= 3\tan\theta - 3\theta + C_1$$
$$= x - 3\arctan\left(\frac{x}{3}\right) + C_1$$

The answers are equivalent.

50. (a) The graph of f is increasing when $f' > 0 : 0 < x < \infty.$

The graph of f is decreasing when $f' < 0 : -\infty < x < 0.$

(b) The graph of f is concave upward when the graph of f' is increasing. There are no such intervals.

The graph of f is concave downward when the graph of f' is decreasing:

$-\infty < x < 0$ and $0 < x < \infty.$

51. True

$$\int \frac{dx}{\sqrt{1-x^2}} = \int \frac{\cos\theta\,d\theta}{\cos\theta} = \int d\theta$$

52. False

$$\int \frac{\sqrt{x^2-1}}{x}\,dx = \int \frac{\tan\theta}{\sec\theta}(\sec\theta\tan\theta\,d\theta) = \int \tan^2\theta\,d\theta$$

53. False

$$\int_0^{\sqrt{3}} \frac{dx}{\left(\sqrt{1+x^2}\right)^3} = \int_0^{\pi/3} \frac{\sec^2\theta\,d\theta}{\sec^3\theta} = \int_0^{\pi/3}\cos\theta\,d\theta$$

54. True

$$\int_{-1}^{1} x^2\sqrt{1-x^2}\,dx = 2\int_0^1 x^2\sqrt{1-x^2}\,dx$$
$$= 2\int_0^{\pi/2}(\sin^2\theta)(\cos\theta)(\cos\theta\,d\theta)$$
$$= 2\int_0^{\pi/2}\sin^2\theta\cos^2\theta\,d\theta$$

55. $A = 4\int_0^a \frac{b}{a}\sqrt{a^2 - x^2}\,dx$

$= \frac{4b}{a}\int_0^a \sqrt{a^2 - x^2}\,dx$

$= \left[\frac{4b}{a}\left(\frac{1}{2}\right)\left(a^2 \arcsin\frac{x}{a} + x\sqrt{a^2 - x^2}\right)\right]_0^a$

$= \frac{2b}{a}\left(a^2\left(\frac{\pi}{2}\right)\right) = \pi ab$

Note: See Theorem 8.2 for $\int\sqrt{a^2 - x^2}\,dx$.

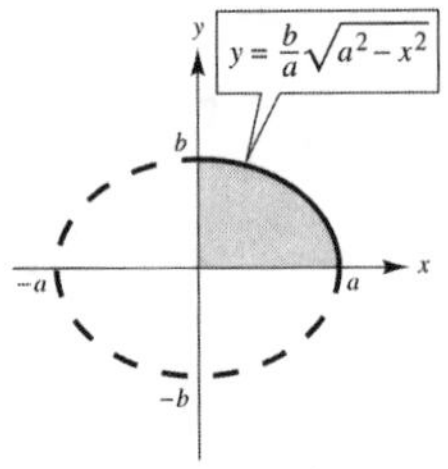

56. $x^2 + y^2 = a^2$

$x = \pm\sqrt{a^2 - y^2}$

$A = 2\int_h^a \sqrt{a^2 - y^2}\,dy$

$= \left[a^2 \arcsin\left(\frac{y}{a}\right) + y\sqrt{a^2 - y^2}\right]_h^a$ (Theorem 8.2)

$= \left(a^2\frac{\pi}{2}\right) - \left(a^2 \arcsin\left(\frac{h}{a}\right) + h\sqrt{a^2 - h^2}\right)$

$= \frac{a^2\pi}{2} - a^2 \arcsin\left(\frac{h}{a}\right) - h\sqrt{a^2 - h^2}$

57. (a) $x^2 + (y - k)^2 = 25$

Radius of circle $= 5$

$k^2 = 5^2 + 5^2 = 50$

$k = 5\sqrt{2}$

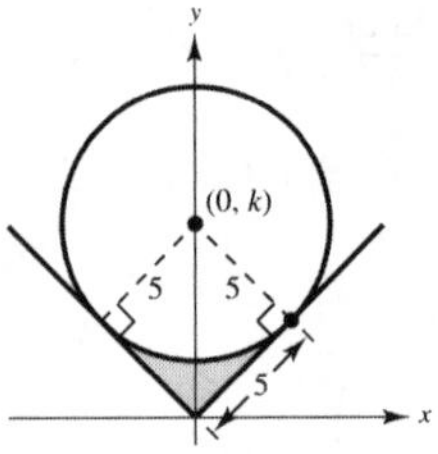

(b) Area $=$ square $- \frac{1}{4}$(circle)

$= 25 - \frac{1}{4}\pi(5)^2 = 25\left(1 - \frac{\pi}{4}\right)$

(c) Area $= r^2 - \frac{1}{4}\pi r^2 = r^2\left(1 - \frac{\pi}{4}\right)$

58. (a) Place the center of the circle at $(0, 1)$; $x^2 + (y - 1)^2 = 1$. The depth d satisfies $0 \le d \le 2$. The volume is

$$V = 3 \cdot 2\int_0^d \sqrt{1 - (y - 1)^2}\,dy = 6 \cdot \frac{1}{2}\left[\arcsin(y - 1) + (y - 1)\sqrt{1 - (y - 1)^2}\right]_0^d \quad \text{(Theorem 8.2 (1))}$$

$$= 3\left[\arcsin(d - 1) + (d - 1)\sqrt{1 - (d - 1)^2} - \arcsin(-1)\right]$$

$$= \frac{3\pi}{2} + 3\arcsin(d - 1) + 3(d - 1)\sqrt{2d - d^2}.$$

(b)

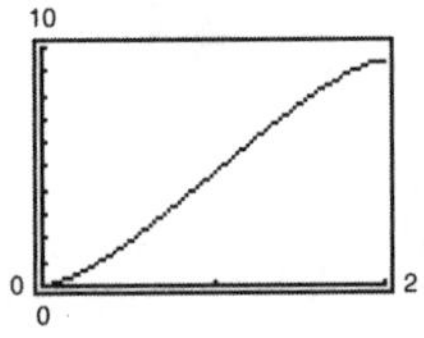

(c) The full tank holds $3\pi \approx 9.4248$ cubic meters. The horizontal lines

$y = \frac{3\pi}{4}, y = \frac{3\pi}{2}, y = \frac{9\pi}{4}$

intersect the curve at $d = 0.596, 1.0, 1.404$. The dipstick would have these markings on it.

(d) $V = 6\int_0^d \sqrt{1 - (y - 1)^2}\, dy$

$$\frac{dV}{dt} = \frac{dV}{dd} \cdot \frac{dd}{dt} = 6\sqrt{1 - (d - 1)^2} \cdot d'(t) = \frac{1}{4} \Rightarrow d'(t) = \frac{1}{24\sqrt{1 - (d - 1)^2}}$$

(e)

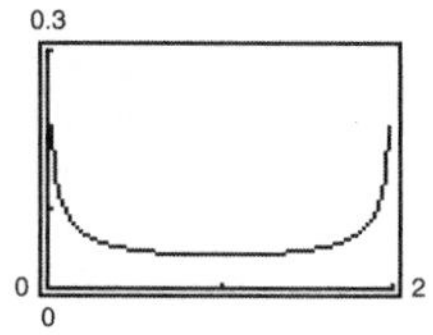

The minimum occurs at $d = 1$, which is the widest part of the tank.

59. Let $x - 3 = \sin\theta$, $dx = \cos\theta\, d\theta$, $\sqrt{1 - (x - 3)^2} = \cos\theta$.

Shell Method:

$$V = 4\pi\int_2^4 x\sqrt{1 - (x - 3)^2}\, dx$$

$$= 4\pi\int_{-\pi/2}^{\pi/2} (3 + \sin\theta)\cos^2\theta\, d\theta$$

$$= 4\pi\left[\frac{3}{2}\int_{-\pi/2}^{\pi/2} (1 + \cos 2\theta)\, d\theta + \int_{-\pi/2}^{\pi/2} \cos^2\theta \sin\theta\, d\theta\right]$$

$$= 4\pi\left[\frac{3}{2}\left(\theta + \frac{1}{2}\sin 2\theta\right) - \frac{1}{3}\cos^3\theta\right]_{-\pi/2}^{\pi/2} = 6\pi^2$$

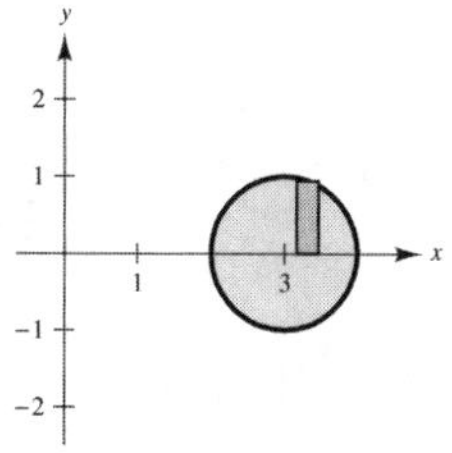

60. Let $x - h = r\sin\theta$, $dx = r\cos\theta\, d\theta$, $\sqrt{r^2 - (x - h)^2} = r\cos\theta$.

Shell Method:

$$V = 4\pi\int_{h-r}^{h+r} x\sqrt{r^2 - (x - h)^2}\, dx$$

$$= 4\pi\int_{-\pi/2}^{\pi/2} (h + r\sin\theta) r\cos\theta(r\cos\theta)\, d\theta = 4\pi r^2\int_{-\pi/2}^{\pi/2} (h + r\sin\theta)\cos^2\theta\, d\theta$$

$$= 4\pi r^2\left[\frac{h}{2}\int_{-\pi/2}^{\pi/2} (1 + \cos 2\theta)\, d\theta + r\int_{-\pi/2}^{\pi/2} \sin\theta\cos^2\theta\, d\theta\right]$$

$$= 2\pi r^2 h\left[\theta + \frac{1}{2}\sin 2\theta\right]_{-\pi/2}^{\pi/2} - \left[4\pi r^3\left(\frac{\cos^3\theta}{3}\right)\right]_{-\pi/2}^{\pi/2} = 2\pi^2 r^2 h$$

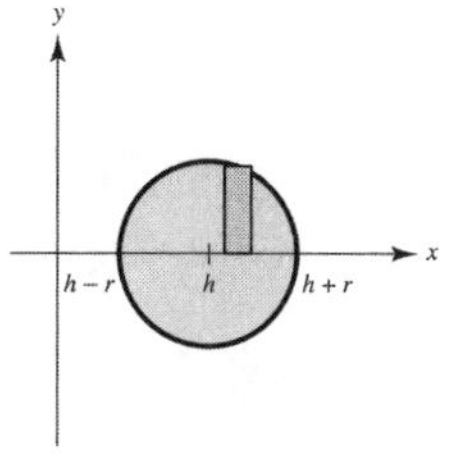

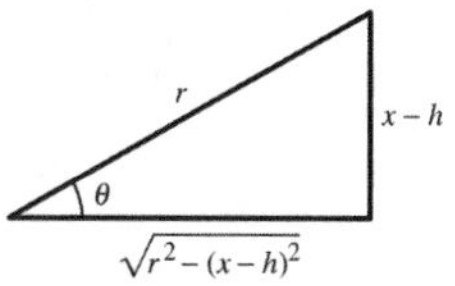

61. $y = \ln x,\ y' = \dfrac{1}{x},\ 1 + (y')^2 = 1 + \dfrac{1}{x^2} = \dfrac{x^2 + 1}{x^2}$

Let $x = \tan\theta,\ dx = \sec^2\theta\, d\theta,\ \sqrt{x^2 + 1} = \sec\theta.$

$$s = \int_1^5 \sqrt{\frac{x^2+1}{x^2}}\,dx = \int_1^5 \frac{\sqrt{x^2+1}}{x}\,dx$$

$$= \int_a^b \frac{\sec\theta}{\tan\theta}\sec^2\theta\,d\theta = \int_a^b \frac{\sec\theta}{\tan\theta}\left(1 + \tan^2\theta\right)d\theta$$

$$= \int_a^b (\csc\theta + \sec\theta\tan\theta)\,d\theta = \Big[-\ln\left|\csc\theta + \cot\theta\right| + \sec\theta\Big]_a^b$$

$$= \left[-\ln\left|\frac{\sqrt{x^2+1}}{x} + \frac{1}{x}\right| + \sqrt{x^2+1}\right]_1^5$$

$$= \left[-\ln\left(\frac{\sqrt{26}+1}{5}\right) + \sqrt{26}\right] - \left[-\ln\left(\sqrt{2}+1\right) + \sqrt{2}\right]$$

$$= \ln\left[\frac{5\left(\sqrt{2}+1\right)}{\sqrt{26}+1}\right] + \sqrt{26} - \sqrt{2} \approx 4.367 \text{ or } \ln\left[\frac{\sqrt{26}-1}{5\left(\sqrt{2}-1\right)}\right] + \sqrt{26} - \sqrt{2}$$

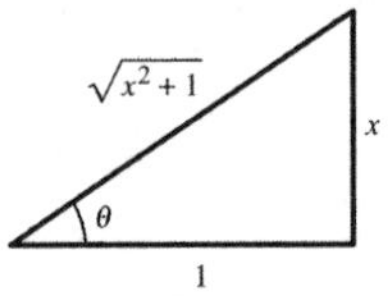

62. $y = \frac{1}{2}x^2,\ y' = x,\ 1 + (y')^2 = 1 + x^2$

$$s = \int_0^4 \sqrt{1 + x^2}\,dx = \frac{1}{2}\left[x\sqrt{x^2+1} + \ln\left|x + \sqrt{x^2+1}\right|\right]_0^4 \quad \text{(Theorem 8.2)}$$

$$= \frac{1}{2}\left[4\sqrt{17} + \ln\left(4 + \sqrt{17}\right)\right] \approx 9.2936$$

63. Length of one arch of sine curve: $y = \sin x,\ y' = \cos x$

$$L_1 = \int_0^\pi \sqrt{1 + \cos^2 x}\,dx$$

Length of one arch of cosine curve: $y = \cos x,\ y' = -\sin x$

$$L_2 = \int_{-\pi/2}^{\pi/2} \sqrt{1 + \sin^2 x}\,dx$$

$$= \int_{-\pi/2}^{\pi/2} \sqrt{1 + \cos^2\left(x - \frac{\pi}{2}\right)}\,dx, \quad u = x - \frac{\pi}{2},\ du = dx$$

$$= \int_{-\pi}^{0} \sqrt{1 + \cos^2 u}\,du$$

$$= \int_0^\pi \sqrt{1 + \cos^2 u}\,du = L_1$$

64. (a) Along line: $d_1 = \sqrt{a^2 + a^4} = a\sqrt{1 + a^2}$

Along parabola: $y = x^2,\ y' = 2x$

$$d_2 = \int_0^a \sqrt{1 + 4x^2}\,dx$$

$$= \frac{1}{4}\left[2x\sqrt{4x^2+1} + \ln\left|2x + \sqrt{4x^2+1}\right|\right]_0^a \quad \text{(Theorem 8.2)}$$

$$= \frac{1}{4}\left[2a\sqrt{4a^2+1} + \ln\left(2a + \sqrt{4a^2+1}\right)\right]$$

(b) For $a = 1$, $d_1 = \sqrt{2}$ and $d_2 = \dfrac{\sqrt{5}}{2} + \dfrac{1}{4}\ln\left(2 + \sqrt{5}\right) \approx 1.4789$.

For $a = 10$, $d_1 = 10\sqrt{101} \approx 100.4988$ and $d_2 \approx 101.0473$.

(c) As a increases, $d_2 - d_1 \to 0$.

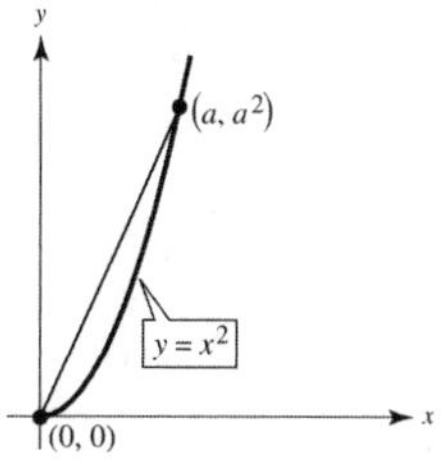

65. Let $x = 3\tan\theta$, $dx = 3\sec^2\theta\, d\theta$, $\sqrt{x^2 + 9} = 3\sec\theta$.

$$A = 2\int_0^4 \frac{3}{\sqrt{x^2+9}}\,dx = 6\int_0^4 \frac{dx}{\sqrt{x^2+9}} = 6\int_a^b \frac{3\sec^2\theta\, d\theta}{3\sec\theta}$$

$$= 6\int_a^b \sec\theta\, d\theta = \Big[6\ln\left|\sec\theta + \tan\theta\right|\Big]_a^b = \left[6\ln\left|\frac{\sqrt{x^2+9}+x}{3}\right|\right]_0^4 = 6\ln 3$$

$\bar{x} = 0$ (by symmetry)

$$\bar{y} = \frac{1}{2}\left(\frac{1}{A}\right)\int_{-4}^4 \left(\frac{3}{\sqrt{x^2+9}}\right)^2 dx = \frac{9}{12\ln 3}\int_{-4}^4 \frac{1}{x^2+9}\,dx = \frac{3}{4\ln 3}\left[\frac{1}{3}\arctan\frac{x}{3}\right]_{-4}^4 = \frac{2}{4\ln 3}\arctan\frac{4}{3} \approx 0.422$$

$$(\bar{x}, \bar{y}) = \left(0, \frac{1}{2\ln 3}\arctan\frac{4}{3}\right) \approx (0, 0.422)$$

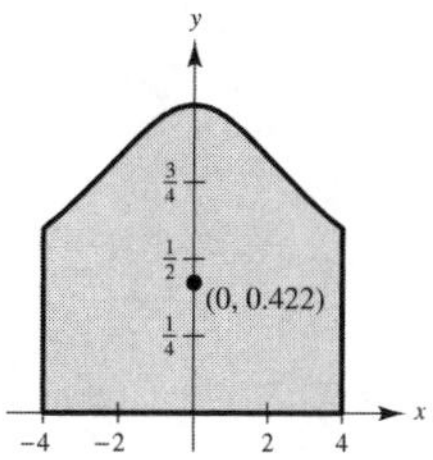

66. First find where the curves intersect.

$$y^2 = 16 - (x-4)^2 = \frac{1}{16}x^4$$

$$16^2 - 16(x-4)^2 = x^4$$

$$16^2 - 16x^2 + 128x - 16^2 = x^4$$

$$x^4 + 16x^2 - 128x = 0$$

$$x(x-4)(x^2 + 4x + 32) = 0$$

$$\Rightarrow x = 0, 4$$

$$A = \int_0^4 \frac{1}{4}x^2\,dx + \frac{1}{4}\pi(4)^2 = \left[\frac{1}{12}x^3\right]_0^4 + 4\pi = \frac{16}{3} + 4\pi$$

$$M_y = \int_0^4 x\left(\frac{1}{4}x^2\right)dx + \int_4^8 x\sqrt{16 - (x-4)^2}\,dx$$

$$= \left[\frac{x^4}{16}\right]_0^4 + \int_4^8 (x-4)\sqrt{16-(x-4)^2}\,dx + \int_4^8 4\sqrt{16-(x-4)^2}\,dx$$

$$= 16 + \left[\frac{-1}{3}\left(16 - (x-4)^2\right)^{3/2}\right]_4^8 + 2\left[16\arcsin\frac{x-4}{4} + (x-4)\sqrt{16-(x-4)^2}\right]_4^8$$

$$= 16 + \frac{1}{3}\left(16^{3/2}\right) + 2\left[16\left(\frac{\pi}{2}\right)\right] = 16 + \frac{64}{3} + 16\pi = \frac{112}{3} + 16\pi$$

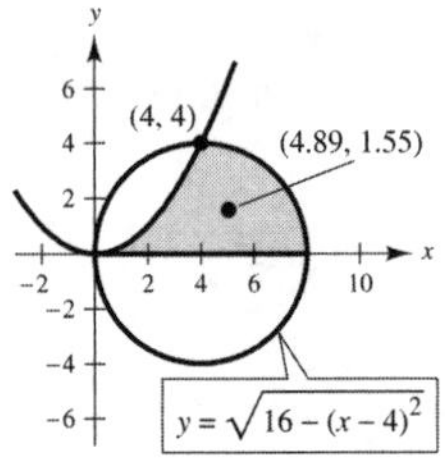

$$M_x = \int_0^4 \frac{1}{2}\left(\frac{1}{4}x^2\right)^2 dx + \int_4^8 \frac{1}{2}\left(16 - (x-4)^2\right)dx = \left[\frac{1}{32}\cdot\frac{x^5}{5}\right]_0^4 + \left[8x - \frac{(x-4)^3}{6}\right]_4^8 = \frac{32}{5} + \left(64 - \frac{64}{6}\right) - 32 = \frac{416}{15}$$

$$\bar{x} = \frac{M_y}{A} = \frac{112/3 + 16\pi}{16/3 + 4\pi} = \frac{112 + 48\pi}{16 + 12\pi} = \frac{28 + 12\pi}{4 + 3\pi} \approx 4.89$$

$$\bar{y} = \frac{M_x}{A} = \frac{416/15}{(16/3) + 4\pi} = \frac{104}{5(4 + 3\pi)} \approx 1.55$$

$$(\bar{x}, \bar{y}) \approx (4.89, 1.55)$$

67. $y = x^2, y' = 2x, 1 + (y') = 1 + 4x^2$

$$2x = \tan\theta,\ dx = \frac{1}{2}\sec^2\theta\,d\theta,\ \sqrt{1 + 4x^2} = \sec\theta$$

(For $\int \sec^5\theta\,d\theta$ and $\int \sec^3\theta\,d\theta$, see Exercise 82 in Section 8.3.)

$$S = 2\pi\int_0^{\sqrt{2}} x^2\sqrt{1 + 4x^2}\,dx = 2\pi\int_a^b \left(\frac{\tan\theta}{2}\right)^2(\sec\theta)\left(\frac{1}{2}\sec^2\theta\right)d\theta$$

$$= \frac{\pi}{4}\int_a^b \sec^3\theta\tan^2\theta\,d\theta = \frac{\pi}{4}\left[\int_a^b \sec^5\theta\,d\theta - \int_a^b \sec^3\theta\,d\theta\right]$$

$$= \frac{\pi}{4}\left\{\frac{1}{4}\left[\sec^3\theta\tan\theta + \frac{3}{2}\left(\sec\theta\tan\theta + \ln\left|\sec\theta + \tan\theta\right|\right)\right] - \frac{1}{2}\left(\sec\theta\tan\theta + \ln\left|\sec\theta + \tan\theta\right|\right)\right\}\Bigg]_a^b$$

$$= \frac{\pi}{4}\left[\frac{1}{4}\left[\left(1 + 4x^2\right)^{3/2}(2x)\right] - \frac{1}{8}\left[\left(1 + 4x^2\right)^{1/2}(2x) + \ln\left|\sqrt{1 + 4x^2} + 2x\right|\right]\right]_0^{\sqrt{2}}$$

$$= \frac{\pi}{4}\left[\frac{54\sqrt{2}}{4} - \frac{6\sqrt{2}}{8} - \frac{1}{8}\ln\left(3 + 2\sqrt{2}\right)\right]$$

$$= \frac{\pi}{4}\left(\frac{51\sqrt{2}}{4} - \frac{\ln\left(3 + 2\sqrt{2}\right)}{8}\right) = \frac{\pi}{32}\left[102\sqrt{2} - \ln\left(3 + 2\sqrt{2}\right)\right] \approx 13.989$$

68. Let $r = L\tan\theta$, $dr = L\sec^2\theta\,d\theta$, $r^2 + L^2 = L^2\sec^2\theta$.

$$\frac{1}{R}\int_0^R \frac{2mL}{\left(r^2+L^2\right)^{3/2}}\,dr = \frac{2mL}{R}\int_a^b \frac{L\sec^2\theta\,d\theta}{L^3\sec^3\theta} = \frac{2m}{RL}\int_a^b \cos\theta\,d\theta = \left[\frac{2m}{RL}\sin\theta\right]_a^b = \left[\frac{2m}{RL}\frac{r}{\sqrt{r^2+L^2}}\right]_0^R = \frac{2m}{L\sqrt{R^2+L^2}}$$

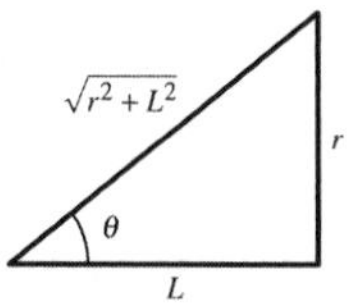

69. (a) Area of representative rectangle: $2\sqrt{1-y^2}\,\Delta y$

Force: $2(62.4)(3-y)\sqrt{1-y^2}\,\Delta y$

$$F = 124.8\int_{-1}^{1}(3-y)\sqrt{1-y^2}\,dy$$
$$= 124.8\left[3\int_{-1}^{1}\sqrt{1-y^2}\,dy - \int_{-1}^{1}y\sqrt{1-y^2}\,dy\right]$$
$$= 124.8\left[\frac{3}{2}\left(\arcsin y + y\sqrt{1-y^2}\right) + \frac{1}{2}\left(\frac{2}{3}\right)\left(1-y^2\right)^{3/2}\right]_{-1}^{1} = (62.4)3\left[\arcsin 1 - \arcsin(-1)\right] = 187.2\pi \text{ lb}$$

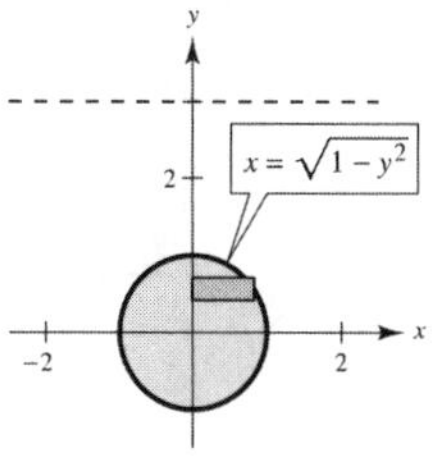

(b) $$F = 124.8\int_{-1}^{1}(d-y)\sqrt{1-y^2}\,dy = 124.8d\int_{-1}^{1}\sqrt{1-y^2}\,dy - 124.8\int_{-1}^{1}y\sqrt{1-y^2}\,dy$$
$$= 124.8\left(\frac{d}{2}\right)\left[\arcsin y + y\sqrt{1-y^2}\right]_{-1}^{1} - 124.8(0) = 62.4\pi d \text{ lb}$$

70. (a) $$F_{\text{inside}} = 48\int_{-1}^{0.8}(0.8-y)(2)\sqrt{1-y^2}\,dy$$
$$= 96\left[0.8\int_{-1}^{0.8}\sqrt{1-y^2}\,dy - \int_{-1}^{0.8}y\sqrt{1-y^2}\,dy\right]$$
$$= 96\left[\frac{0.8}{2}\left(\arcsin y + y\sqrt{1-y^2}\right) + \frac{1}{3}\left(1-y^2\right)^{3/2}\right]_{-1}^{0.8}$$
$$\approx 96(1.263) \approx 121.3 \text{ lb}$$

(b) $$F_{\text{outside}} = 64\int_{-1}^{0.4}(0.4-y)(2)\sqrt{1-y^2}\,dy$$
$$= 128\left[0.4\int_{-1}^{0.4}\sqrt{1-y^2}\,dy - \int_{-1}^{0.4}y\sqrt{1-y^2}\,dy\right]$$
$$= 128\left[\frac{0.4}{2}\left(\arcsin y + y\sqrt{1-y^2}\right) + \frac{1}{3}\left(1-y^2\right)^{3/2}\right]_{-1}^{0.4} \approx 92.98$$

71. Let $u = a\sin\theta$, $du = a\cos\theta\,d\theta$, $\sqrt{a^2 - u^2} = a\cos\theta$.

$$\int\sqrt{a^2-u^2}\,du = \int a^2\cos^2\theta\,d\theta = a^2\int\frac{1+\cos 2\theta}{2}\,d\theta$$
$$= \frac{a^2}{2}\left(\theta + \frac{1}{2}\sin 2\theta\right) + C = \frac{a^2}{2}(\theta + \sin\theta\cos\theta) + C$$
$$= \frac{a^2}{2}\left[\arcsin\frac{u}{a} + \left(\frac{u}{a}\right)\left(\frac{\sqrt{a^2+u^2}}{a}\right)\right] + C = \frac{1}{2}\left(a^2\arcsin\frac{u}{a} + u\sqrt{a^2-u^2}\right) + C$$

Let $u = a\sec\theta$, $du = a\sec\theta\tan\theta\,d\theta$, $\sqrt{u^2 - a^2} = a\tan\theta$.

$$\int\sqrt{u^2-a^2}\,du = \int a\tan\theta(a\sec\theta\tan\theta)\,d\theta = a^2\int\tan^2\theta\sec\theta\,d\theta$$
$$= a^2\int(\sec^2\theta - 1)\sec\theta\,d\theta = a^2\int(\sec^3\theta - \sec\theta)\,d\theta$$
$$= a^2\left[\frac{1}{2}\sec\theta\tan\theta + \frac{1}{2}\int\sec\theta\,d\theta\right] - a^2\int\sec\theta\,d\theta = a^2\left[\frac{1}{2}\sec\theta\tan\theta - \frac{1}{2}\ln|\sec\theta + \tan\theta|\right]$$
$$= \frac{a^2}{2}\left[\frac{u}{a}\cdot\frac{\sqrt{u^2-a^2}}{a} - \ln\left|\frac{u}{a} + \frac{\sqrt{u^2-a^2}}{a}\right|\right] + C_1 = \frac{1}{2}\left[u\sqrt{u^2-a^2} - a^2\ln\left|u + \sqrt{u^2-a^2}\right|\right] + C$$

Let $u = a\tan\theta$, $du = a\sec^2\theta\,d\theta$, $\sqrt{u^2 + a^2} = a\sec\theta$.

$$\int\sqrt{u^2+a^2}\,du = \int(a\sec\theta)(a\sec^2\theta)\,d\theta$$
$$= a^2\int\sec^3\theta\,d\theta = a^2\left[\frac{1}{2}\sec\theta\tan\theta + \frac{1}{2}\ln|\sec\theta + \tan\theta|\right] + C_1$$
$$= \frac{a^2}{2}\left[\frac{\sqrt{u^2+a^2}}{a}\cdot\frac{u}{a} + \ln\left|\frac{\sqrt{u^2+a^2}}{a} + \frac{u}{a}\right|\right] + C_1 = \frac{1}{2}\left[u\sqrt{u^2+a^2} + a^2\ln\left|u + \sqrt{u^2+a^2}\right|\right] + C$$

72. $y = \sin x$ on $[0, 2]$

$y' = \cos x$

$s_1 = 2\int_0^{\pi}\sqrt{1+\cos^2 x}\,dx \quad (\approx 3.820197789)$

Ellipse: $x^2 + 2y^2 = 2$

Upper half: $y = \sqrt{1 - \frac{1}{2}x^2}$, $-\sqrt{2} \le x \le \sqrt{2}$

$$y' = \frac{-x}{2\sqrt{1-(1/2)x^2}}$$

$$s_2 = 2\int_{-\sqrt{2}}^{\sqrt{2}}\sqrt{1 + \frac{x^2}{4(1-(1/2)x^2)}}\,dx = 2\int_{-\sqrt{2}}^{\sqrt{2}}\sqrt{1 + \frac{x^2}{4-2x^2}}\,dx$$

Let $x = \sqrt{2}\sin\theta$, $dx = \sqrt{2}\cos\theta\,d\theta$, $x^2 = 2\sin^2\theta$, $4 - 2x^2 = 4 - 4\sin^2\theta = 4\cos^2\theta$.

$$s_2 = 2\int_{-\pi/2}^{\pi/2}\sqrt{1 + \frac{2\sin^2\theta}{4\cos^2\theta}}\sqrt{2}\cos\theta\,d\theta$$
$$= 2\int_{-\pi/2}^{\pi/2}\frac{\sqrt{4\cos^2\theta + 2\sin^2\theta}}{2\cos\theta}\sqrt{2}\cos\theta\,d\theta$$
$$= 2\int_{-\pi/2}^{\pi/2}\frac{\sqrt{2+2\cos^2\theta}}{\sqrt{2}}\,d\theta = 2\int_{-\pi/2}^{\pi/2}\sqrt{1+\cos^2\theta}\,d\theta = 2\int_0^{\pi}\sqrt{1+\cos^2\theta}\,d\theta = s_1$$

73. Large circle: $x^2 + y^2 = 25$

$$y = \sqrt{25 - x^2}, \text{ upper half}$$

From the right triangle, the center of the small circle is $(0, 4)$.

$$x^2 + (y - 4)^2 = 9$$

$$y = 4 + \sqrt{9 - x^2}, \quad \text{upper half}$$

$$A = 2\int_0^3 \left[\left(4 + \sqrt{9 - x^2}\right) - \sqrt{25 - x^2}\right] dx$$

$$= 2\left[4x + \frac{1}{2}\left[9 \arcsin\left(\frac{x}{3}\right) + x\sqrt{9 - x^2}\right] - \frac{1}{2}\left[25 \arcsin\left(\frac{x}{5}\right) + x\sqrt{25 - x^2}\right]\right]_0^3$$

$$= 2\left[12 + \frac{9}{2}\arcsin(1) - \frac{25}{2}\arcsin\frac{3}{5} - 6\right]$$

$$= 12 + \frac{9\pi}{2} - 25 \arcsin\frac{3}{5} \approx 10.050$$

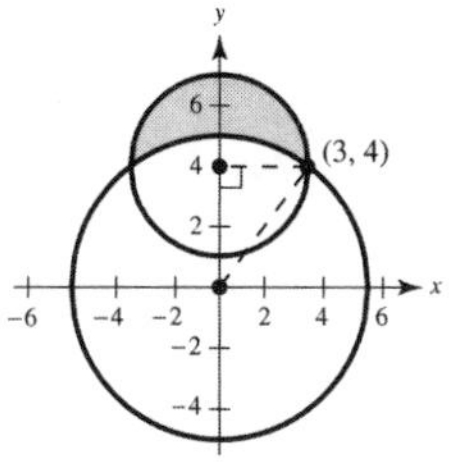

74. The left circle has equation $(x + 2)^2 + y^2 = 9$. The shaded area is four times the area in the first quadrant, under the curve $y = \sqrt{9 - (x + 2)^2}$.

$$A = 4\int_0^1 \sqrt{9 - (x + 2)^2}\, dx$$

Let $x + 2 = 3 \sin \theta$, $dx = 3 \cos \theta\, d\theta$, $\sqrt{9 - (x + 2)^2} = 3 \cos \theta$

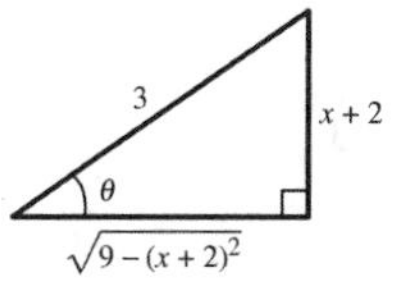

$$\int\sqrt{9 - (x + 2)^2}\, dx = \int 3 \cos \theta (3 \cos \theta)\, d\theta = 9\int\frac{1 + \cos 2\theta}{2}\, d\theta$$

$$= \frac{9}{2}\left(\theta + \frac{\sin 2\theta}{2}\right) + C = \frac{9}{2}(\theta + \sin\theta\cos\theta) + C$$

$$= \frac{9}{2}\left(\arcsin\left(\frac{x + 2}{3}\right) + \left(\frac{x + 2}{3}\right)\left(\frac{\sqrt{9 - (x + 2)^2}}{3}\right)\right) + C$$

$$A = 4 \cdot \frac{9}{2}\left[\arcsin\left(\frac{x + 2}{3}\right) + \left(\frac{x + 2}{3}\right)\left(\frac{\sqrt{9 - (x + 2)^2}}{3}\right)\right]_0^1 = 18\left[\left(\frac{\pi}{2} + 0\right) - \left(\arcsin\frac{2}{3} + \frac{2}{3}\frac{\sqrt{5}}{3}\right)\right] = 9\pi - 18 \arcsin\frac{2}{3} - 4\sqrt{5}$$

75. Let $I = \int_0^1 \frac{\ln(x+1)}{x^2+1}\,dx$

Let $x = \frac{1-u}{1+u}$, $dx = \frac{-2}{(1+u)^2}\,du$

$x + 1 = \frac{2}{1+u}$, $x^2 + 1 = \frac{2+2u^2}{(1+u)^2}$

$$I = \int_1^0 \frac{\ln\left(\frac{2}{1+u}\right)}{\left(\frac{2+2u^2}{(1+u)^2}\right)}\left(\frac{-2}{(1+u)^2}\right)du$$

$$= \int_1^0 \frac{-\ln\left(\frac{2}{1+u}\right)}{1+u^2}\,du = \int_0^1 \frac{\ln\left(\frac{2}{1+u}\right)}{1+u^2}\,du = \int_0^1 \frac{\ln 2}{1+u^2} - \int_0^1 \frac{\ln(1+u)}{1+u^2}\,du = (\ln 2)\left[\arctan u\right]_0^1 - I$$

$$\Rightarrow 2I = \ln 2\left(\frac{\pi}{4}\right)$$

$$I = \frac{\pi}{8}\ln 2 \approx 0.272198$$

Section 8.5 Partial Fractions

1. $\frac{4}{x^2 - 8x} = \frac{4}{x(x-8)} = \frac{A}{x} + \frac{B}{x-8}$

2. $\frac{2x^2+1}{(x-3)^3} = \frac{A}{x-3} + \frac{B}{(x-3)^2} + \frac{C}{(x-3)^3}$

3. $\frac{2x-3}{x^3+10x} = \frac{2x-3}{x(x^2+10)} = \frac{A}{x} + \frac{Bx+C}{x^2+10}$

4. $\frac{2x-1}{x(x^2+1)^2} = \frac{A}{x} + \frac{Bx+C}{x^2+1} + \frac{Dx+E}{(x^2+1)^2}$

5. $\frac{1}{x^2-9} = \frac{1}{(x-3)(x+3)} = \frac{A}{x+3} + \frac{B}{x-3}$

$1 = A(x-3) + B(x+3)$

When $x = 3$, $1 = 6B \Rightarrow B = \frac{1}{6}$.

When $x = -3$, $1 = -6A \Rightarrow A = -\frac{1}{6}$.

$$\int \frac{1}{x^2-9}\,dx = -\frac{1}{6}\int \frac{1}{x+3}\,dx + \frac{1}{6}\int \frac{1}{x-3}\,dx$$
$$= -\frac{1}{6}\ln|x+3| + \frac{1}{6}\ln|x-3| + C$$
$$= \frac{1}{6}\ln\left|\frac{x-3}{x+3}\right| + C$$

6. $\frac{2}{9x^2-1} = \frac{2}{(3x-1)(3x+1)} = \frac{A}{3x-1} + \frac{B}{3x+1}$

$2 = A(3x+1) + B(3x-1)$

When $x = \frac{1}{3}$, $2 = 2A \Rightarrow A = 1$.

When $x = -\frac{1}{3}$, $2 = -2B \Rightarrow B = -1$.

$$\int \frac{2}{9x^2-1}\,dx = \int \frac{1}{3x-1}\,dx + \int \frac{-1}{3x+1}\,dx$$
$$= \frac{1}{3}\ln|3x-1| - \frac{1}{3}\ln|3x+1| + C$$
$$= \frac{1}{3}\ln\left|\frac{3x-1}{3x+1}\right| + C$$

7. $\frac{5}{x^2+3x-4} = \frac{5}{(x+4)(x-1)} = \frac{A}{x+4} + \frac{B}{x-1}$

$5 = A(x-1) + B(x+4)$

When $x = 1$, $5 = 5B \Rightarrow B = 1$.

When $x = -4$, $5 = -5A \Rightarrow A = -1$.

$$\int \frac{5}{x^2+3x-4}\,dx = \int \frac{-1}{x+4}\,dx + \int \frac{1}{x-1}\,dx$$
$$= -\ln|x+4| + \ln|x-1| + C$$
$$= \ln\left|\frac{x-1}{x+4}\right| + C$$

8. $\dfrac{3-x}{3x^2-2x-1} = \dfrac{3-x}{(3x+1)(x-1)} = \dfrac{A}{3x+1} + \dfrac{B}{x-1}$

$3 - x = A(x-1) + B(3x+1)$

When $x = 1$, $2 = 4B \Rightarrow B = \dfrac{1}{2}$.

When $x = -\dfrac{1}{3}$, $\dfrac{10}{3} = -\dfrac{4}{3}A \Rightarrow A = -\dfrac{5}{2}$.

$$\int \frac{3-x}{3x^2-2x-1}\,dx = -\frac{5}{2}\int \frac{1}{3x+1}\,dx + \frac{1}{2}\int \frac{1}{x-1}\,dx$$

$$= \frac{-5}{6}\ln|3x+1| + \frac{1}{2}\ln|x-1| + C$$

9. $\dfrac{x^2+12x+12}{x(x+2)(x-2)} = \dfrac{A}{x} + \dfrac{B}{x+2} + \dfrac{C}{x-2}$

$x^2 + 12x + 12 = A(x+2)(x-2) + Bx(x-2) + Cx(x+2)$

When $x = 0, 12 = -4A \Rightarrow A = -3$.

When $x = -2, -8 = 8B \Rightarrow B = -1$.

When $x = 2, 40 = 8C \Rightarrow C = 5$.

$$\int \frac{x^2+12x+12}{x^3-4x}\,dx = 5\int \frac{1}{x-2}\,dx - \int \frac{1}{x+2}\,dx - 3\int \frac{1}{x}\,dx = 5\ln|x-2| - \ln|x+2| - 3\ln|x| + C$$

10. $\dfrac{x^3-x+3}{x^2+x-2} = x - 1 + \dfrac{2x+1}{(x+2)(x-1)} = x - 1 + \dfrac{A}{x+2} + \dfrac{B}{x-1}$

$2x + 1 = A(x-1) + B(x+2)$

When $x = -2, -3 = -3A \Rightarrow A = 1$.

When $x = 1, 3 = 3B \Rightarrow B = 1$.

$$\int \frac{x^3-x+3}{x^2+x-2}\,dx = \int\left(x - 1 + \frac{1}{x+2} + \frac{1}{x-1}\right)dx = \frac{x^2}{2} - x + \ln|x+2| + \ln|x-1| + C = \frac{x^2}{2} - x + \ln\left|x^2+x-2\right| + C$$

11. $\dfrac{2x^3-4x^2-15x+5}{x^2-2x-8} = 2x + \dfrac{x+5}{(x-4)(x+2)} = 2x + \dfrac{A}{x-4} + \dfrac{B}{x+2}$

$x + 5 = A(x+2) + B(x-4)$

When $x = 4, 9 = 6A \Rightarrow A = \dfrac{3}{2}$.

When $x = -2, 3 = -6B \Rightarrow B = -\dfrac{1}{2}$.

$$\int \frac{2x^3-4x^2-15x+5}{x^2-2x-8}\,dx = \int\left(2x + \frac{3/2}{x-4} - \frac{1/2}{x+2}\right)dx = x^2 + \frac{3}{2}\ln|x-4| - \frac{1}{2}\ln|x+2| + C$$

12. $\dfrac{x+2}{x^2+5x} = \dfrac{x+2}{x(x+5)} = \dfrac{A}{x} + \dfrac{B}{x+5}$

$$x+2 = A(x+5) + Bx$$

When $x = -5$, $-3 = -5B \Rightarrow B = \dfrac{3}{5}$.

When $x = 0$, $2 = 5A \Rightarrow A = \dfrac{2}{5}$.

$$\int \frac{x+2}{x^2+5x}\,dx = \frac{2}{5}\int \frac{1}{x}\,dx + \frac{3}{5}\int \frac{1}{x+5}\,dx$$

$$= \frac{2}{5}\ln|x| + \frac{3}{5}\ln|x+5| + C$$

13. $\dfrac{4x^2+2x-1}{x^2(x+1)} = \dfrac{A}{x} + \dfrac{B}{x^2} + \dfrac{C}{x+1}$

$$4x^2 + 2x - 1 = Ax(x+1) + B(x+1) + Cx^2$$

When $x = 0$, $B = -1$.

When $x = -1$, $C = 1$.

When $x = 1$, $A = 3$.

$$\int \frac{4x^2+2x-1}{x^3+x^2}\,dx = \int\left(\frac{3}{x} - \frac{1}{x^2} + \frac{1}{x+1}\right)dx$$

$$= 3\ln|x| + \frac{1}{x} + \ln|x+1| + C$$

$$= \frac{1}{x} + \ln|x^4 + x^3| + C$$

14. $\dfrac{5x-2}{(x-2)^2} = \dfrac{A}{x-2} + \dfrac{B}{(x-2)^2}$

$$5x - 2 = A(x-2) + B$$

When $x = 2$, $8 = B$.

When $x = 0$, $-2 = -2A + B = -2A + 8 \Rightarrow A = 5$.

$$\int \frac{5x-2}{(x-2)^2}\,dx = \int \frac{5}{x-2}\,dx + \int \frac{8}{(x-2)^2}\,dx$$

$$= 5\ln|x-2| - \frac{8}{x-2} + C$$

15. $\dfrac{x^2+3x-4}{x^3-4x^2+4x} = \dfrac{x^2+3x-4}{x(x-2)^2} = \dfrac{A}{x} + \dfrac{B}{(x-2)} + \dfrac{C}{(x-2)^2}$

$$x^2 + 3x - 4 = A(x-2)^2 + Bx(x-2) + Cx$$

When $x = 0$, $-4 = 4A \Rightarrow A = -1$.

When $x = 2$, $6 = 2C \Rightarrow C = 3$.

When $x = 1$, $0 = -1 - B + 3 \Rightarrow B = 2$.

$$\int \frac{x^2+3x-4}{x^3-4x^2+4x}\,dx = \int \frac{-1}{x}\,dx + \int \frac{2}{(x-2)}\,dx + \int \frac{3}{(x-2)^2}\,dx = -\ln|x| + 2\ln|x-2| - \frac{3}{(x-2)} + C$$

16. $\dfrac{8x}{x^3+x^2-x-1} = \dfrac{8x}{x^2(x+1)-(x+1)} = \dfrac{8x}{(x+1)(x-1)(x+1)}$

$$= \frac{A}{x-1} + \frac{B}{x+1} + \frac{C}{(x+1)^2}$$

$$8x = A(x+1)^2 + B(x-1)(x+1) + C(x-1)$$

When $x = 1$, $8 = 4A \Rightarrow A = 2$.

When $x = -1$, $-8 = -2C \Rightarrow C = 4$.

When $x = 0$, $0 = A - B - C = 2 - B - 4 \Rightarrow B = -2$.

$$\int \frac{8x}{x^3+x^2-x-1}\,dx = \int \frac{2}{x-1}\,dx + \int \frac{-2}{x+1}\,dx + \int \frac{4}{(x+1)^2}\,dx$$

$$= 2\ln|x-1| - 2\ln|x+1| - \frac{4}{x+1} + C$$

17. $\dfrac{x^2 - 1}{x(x^2 + 1)} = \dfrac{A}{x} + \dfrac{Bx + C}{x^2 + 1}$

$x^2 - 1 = A(x^2 + 1) + (Bx + C)x$

When $x = 0, A = -1.$

When $x = 1, 0 = -2 + B + C.$

When $x = -1, 0 = -2 + B - C.$

Solving these equations you have $A = -1, B = 2, C = 0.$

$$\int \frac{x^2 - 1}{x^3 + x}\,dx = -\int \frac{1}{x}\,dx + \int \frac{2x}{x^2 + 1}\,dx = -\ln|x| + \ln|x^2 + 1| + C = \ln\left|\frac{x^2 + 1}{x}\right| + C$$

18. $\dfrac{6x}{x^3 - 8} = \dfrac{6x}{(x - 2)(x^2 + 2x + 4)} = \dfrac{A}{x - 2} + \dfrac{Bx + C}{x^2 + 2x + 4}$

$6x = A(x^2 + 2x + 4) + (Bx + C)(x - 2)$

When $x = 2, 12 = 12A \Rightarrow A = 1.$

When $x = 0, 0 = 4 - 2C \Rightarrow C = 2.$

When $x = 1, 6 = 7 + (B + 2)(-1) \Rightarrow B = -1.$

$$\begin{aligned}\int \frac{6x}{x^3 - 8}\,dx &= \int \frac{1}{x - 2}\,dx + \int \frac{-x + 2}{x^2 + 2x + 4}\,dx \\ &= \int \frac{1}{x - 2}\,dx + \int \frac{-x - 1}{x^2 + 2x + 4}\,dx + \int \frac{3}{(x^2 + 2x + 1) + 3}\,dx \\ &= \ln|x - 2| - \frac{1}{2}\ln|x^2 + 2x + 4| + \frac{3}{\sqrt{3}}\arctan\left(\frac{x + 1}{\sqrt{3}}\right) + C \\ &= \ln|x - 2| - \frac{1}{2}\ln|x^2 + 2x + 4| + \sqrt{3}\arctan\left(\frac{\sqrt{3}(x + 1)}{3}\right) + C\end{aligned}$$

19. $\dfrac{x^2}{x^4 - 2x^2 - 8} = \dfrac{A}{x - 2} + \dfrac{B}{x + 2} + \dfrac{Cx + D}{x^2 + 2}$

$x^2 = A(x + 2)(x^2 + 2) + B(x - 2)(x^2 + 2) + (Cx + D)(x + 2)(x - 2)$

When $x = 2, 4 = 24A.$

When $x = -2, 4 = -24B.$

When $x = 0, 0 = 4A - 4B - 4D.$

When $x = 1, 1 = 9A - 3B - 3C - 3D.$

Solving these equations you have $A = \dfrac{1}{6}, B = -\dfrac{1}{6}, C = 0, D = \dfrac{1}{3}.$

$$\int \frac{x^2}{x^4 - 2x^2 - 8}\,dx = \frac{1}{6}\left(\int \frac{1}{x - 2}\,dx - \int \frac{1}{x + 2}\,dx + 2\int \frac{1}{x^2 + 2}\,dx\right) = \frac{1}{6}\left(\ln\left|\frac{x - 2}{x + 2}\right| + \sqrt{2}\arctan\frac{x}{\sqrt{2}}\right) + C$$

20. $\dfrac{x}{(2x-1)(2x+1)(4x^2+1)} = \dfrac{A}{2x-1} + \dfrac{B}{2x+1} + \dfrac{Cx+D}{4x^2+1}$

$$x = A(2x+1)(4x^2+1) + B(2x-1)(4x^2+1) + (Cx+D)(2x-1)(2x+1)$$

When $x = \frac{1}{2}, \frac{1}{2} = 4A.$

When $x = -\frac{1}{2}, -\frac{1}{2} = -4B.$

When $x = 0, 0 = A - B - D.$

When $x = 1, 1 = 15A + 5B + 3C + 3D.$

Solving these equations you have $A = \frac{1}{8}, B = \frac{1}{8}, C = -\frac{1}{2}, D = 0.$

$$\int \frac{x}{16x^4-1}\,dx = \frac{1}{8}\left(\int \frac{1}{2x-1}\,dx + \int \frac{1}{2x+1}\,dx - 4\int \frac{x}{4x^2+1}\,dx\right) = \frac{1}{16}\ln\left|\frac{4x^2-1}{4x^2+1}\right| + C$$

21. $\dfrac{x^2+5}{(x+1)(x^2-2x+3)} = \dfrac{A}{x+1} + \dfrac{Bx+C}{x^2-2x+3}$

$$\begin{aligned} x^2+5 &= A(x^2-2x+3) + (Bx+C)(x+1) \\ &= (A+B)x^2 + (-2A+B+C)x + (3A+C) \end{aligned}$$

When $x = -1, A = 1.$

By equating coefficients of like terms, you have $A + B = 1, -2A + B + C = 0, 3A + C = 5.$

Solving these equations you have $A = 1, B = 0, C = 2.$

$$\int \frac{x^2+5}{x^3-x^2+x+3}\,dx = \int \frac{1}{x+1}\,dx + 2\int \frac{1}{(x-1)^2+2}\,dx = \ln|x+1| + \sqrt{2}\arctan\left(\frac{x-1}{\sqrt{2}}\right) + C$$

22. $\dfrac{x^2+6x+4}{x^4+8x^2+16} = \dfrac{x^2+6x+4}{(x^2+4)^2} = \dfrac{Ax+B}{x^2+4} + \dfrac{Cx+D}{(x^2+4)^2}$

$$\begin{aligned} x^2+6x+4 &= (Ax+B)(x^2+4) + Cx + D \\ &= Ax^3 + Bx^2 + (4A+C)x + 4B + D \end{aligned}$$

By equating coefficients of like terms, you have

$A = 0, \quad B = 1, \quad 4A + C = 6, \quad 4B + D = 4.$

Solving these equations you have $A = 0, B = 1, C = 6, D = 0.$

$$\begin{aligned} \int \frac{x^2+6x+4}{x^4+8x^2+16}\,dx &= \int \frac{1}{x^2+4}\,dx + \int \frac{6x}{(x^2+4)^2}\,dx \\ &= \frac{1}{2}\arctan\frac{x}{2} - \frac{3}{x^2+4} + C \end{aligned}$$

23. $\dfrac{3}{4x^2 + 5x + 1} = \dfrac{3}{(4x + 1)(x + 1)} = \dfrac{A}{4x + 1} + \dfrac{B}{x + 1}$

$$3 = A(x + 1) + B(4x + 1)$$

When $x = -1, 3 = -3B \Rightarrow B = -1$.

When $-\dfrac{1}{4}, 3 = \dfrac{3}{4}A \Rightarrow A = 4$.

$$\begin{aligned}\int_0^2 \frac{3}{4x^2 + 5x + 1}\,dx &= \int_0^2 \frac{4}{4x + 1}\,dx + \int_0^2 \frac{-1}{x + 1}\,dx \\ &= \Big[\ln|4x + 1| - \ln|x + 1|\Big]_0^2 \\ &= \ln 9 - \ln 3 \\ &= 2\ln 3 - \ln 3 = \ln 3\end{aligned}$$

24. $\dfrac{x - 1}{x^2(x + 1)} = \dfrac{A}{x} + \dfrac{B}{x^2} + \dfrac{C}{x + 1}$

$$x - 1 = Ax(x + 1) + B(x + 1) + Cx^2$$

When $x = 0, B = -1$.

When $x = -1, C = -2$.

When $x = 1, 0 = 2A + 2B + C$.

Solving these equations you have $A = 2, B = -1, C = -2$.

$$\begin{aligned}\int_1^5 \frac{x - 1}{x^2(x + 1)}\,dx &= 2\int_1^5 \frac{1}{x}\,dx - \int_1^5 \frac{1}{x^2}\,dx - 2\int_1^5 \frac{1}{x + 1}\,dx \\ &= \left[2\ln|x| + \frac{1}{x} - 2\ln|x + 1|\right]_1^5 \\ &= \left[2\ln\left|\frac{x}{x + 1}\right| + \frac{1}{x}\right]_1^5 \\ &= 2\ln\frac{5}{3} - \frac{4}{5}\end{aligned}$$

25. $\dfrac{x + 1}{x(x^2 + 1)} = \dfrac{A}{x} + \dfrac{Bx + C}{x^2 + 1}$

$$x + 1 = A(x^2 + 1) + (Bx + C)x$$

When $x = 0, A = 1$.

When $x = 1, 2 = 2A + B + C$.

When $x = -1, 0 = 2A + B - C$.

Solving these equations we have $A = 1, B = -1, C = 1$.

$$\begin{aligned}\int_1^2 \frac{x + 1}{x(x^2 + 1)}\,dx &= \int_1^2 \frac{1}{x}\,dx - \int_1^2 \frac{x}{x^2 + 1}\,dx + \int_1^2 \frac{1}{x^2 + 1}\,dx \\ &= \left[\ln|x| - \frac{1}{2}\ln(x^2 + 1) + \arctan x\right]_1^2 \\ &= \frac{1}{2}\ln\frac{8}{5} - \frac{\pi}{4} + \arctan 2 \\ &\approx 0.557\end{aligned}$$

26.
$$\begin{aligned}\int_0^1 \frac{x^2 - x}{x^2 + x + 1}\,dx &= \int_0^1 dx - \int_0^1 \frac{2x + 1}{x^2 + x + 1}\,dx \\ &= \Big[x - \ln|x^2 + x + 1|\Big]_0^1 \\ &= 1 - \ln 3\end{aligned}$$

27. Let $u = \cos x$, $du = -\sin x\,dx$.

$$\frac{1}{u(u + 1)} = \frac{A}{u} + \frac{B}{u + 1}$$

$$1 = A(u + 1) + Bu$$

When $u = 0, A = 1$.

When $u = -1, B = -1$.

$$\begin{aligned}\int \frac{\sin x}{\cos x + \cos^2 x}\,dx &= -\int \frac{1}{u(u + 1)}\,du \\ &= \int \frac{1}{u + 1}\,du - \int \frac{1}{u}\,du \\ &= \ln|u + 1| - \ln|u| + C \\ &= \ln\left|\frac{u + 1}{u}\right| + C \\ &= \ln\left|\frac{\cos x + 1}{\cos x}\right| + C \\ &= \ln|1 + \sec x| + C\end{aligned}$$

28.
$$\begin{aligned}\int \frac{5\cos x}{\sin^2 x + 3\sin x - 4}\,dx &= 5\int \frac{1}{u^2 + 3u - 4}\,du \\ &= \ln\left|\frac{u - 1}{u + 4}\right| + C \\ &= \ln\left|\frac{-1 + \sin x}{4 + \sin x}\right| + C\end{aligned}$$

(From Exercise 7 with $u = \sin x$, $du = \cos x\,dx$)

29. Let $u = \tan x$, $du = \sec^2 x\,dx$.

$$\frac{1}{u^2 + 5u + 6} = \frac{1}{(u + 3)(u + 2)} = \frac{A}{u + 3} + \frac{B}{u + 2}$$

$$1 = A(u + 2) + B(u + 3)$$

When $u = -2, 1 = B$.

When $u = -3, 1 = -A \Rightarrow A = -1$.

$$\begin{aligned}\int \frac{\sec^2 x}{\tan^2 x + 5\tan x + 6}\,dx &= \int \frac{1}{u^2 + 5u + 6}\,du \\ &= \int \frac{-1}{u + 3}\,du + \int \frac{1}{u + 2}\,du \\ &= -\ln|u + 3| + \ln|u + 2| + C \\ &= \ln\left|\frac{\tan x + 2}{\tan x + 3}\right| + C\end{aligned}$$

30. $\dfrac{1}{u(u+1)} = \dfrac{A}{u} + \dfrac{B}{u+1}$, $u = \tan x$, $du = \sec^2 x\,dx$

$$1 = A(u+1) + Bu$$

When $u = 0$, $A = 1$.

When $u = -1$, $1 = -B \Rightarrow B = -1$.

$$\begin{aligned}\int \frac{\sec^2 x\,dx}{\tan x(\tan x + 1)} &= \int \frac{1}{u(u+1)}\,du\\ &= \int\left(\frac{1}{u} - \frac{1}{u+1}\right)du\\ &= \ln|u| - \ln|u+1| + C\\ &= \ln\left|\frac{u}{u+1}\right| + C\\ &= \ln\left|\frac{\tan x}{\tan x + 1}\right| + C\end{aligned}$$

31. Let $u = e^x$, $du = e^x\,dx$.

$$\frac{1}{(u-1)(u+4)} = \frac{A}{u-1} + \frac{B}{u+4}$$

$$1 = A(u+4) + B(u-1)$$

When $u = 1$, $A = \dfrac{1}{5}$.

When $u = -4$, $B = -\dfrac{1}{5}$.

$$\begin{aligned}\int \frac{e^x}{(e^x-1)(e^x+4)}\,dx &= \int \frac{1}{(u-1)(u+4)}\,du\\ &= \frac{1}{5}\left(\int \frac{1}{u-1}\,du - \int \frac{1}{u+4}\,du\right)\\ &= \frac{1}{5}\ln\left|\frac{u-1}{u+4}\right| + C\\ &= \frac{1}{5}\ln\left|\frac{e^x-1}{e^x+4}\right| + C\end{aligned}$$

32. Let $u = e^x$, $du = e^x\,dx$.

$$\frac{1}{(u^2+1)(u-1)} = \frac{A}{u-1} + \frac{Bu + C}{u^2+1}$$

$$1 = A(u^2+1) + (Bu + C)(u-1)$$

When $u = 1$, $A = \dfrac{1}{2}$.

When $u = 0$, $1 = A - C$.

When $u = -1$, $1 = 2A + 2B - 2C$.

Solving these equations you have $A = \dfrac{1}{2}$, $B = -\dfrac{1}{2}$, and $C = -\dfrac{1}{2}$.

$$\begin{aligned}\int \frac{e^x}{(e^{2x}+1)(e^x-1)}\,dx &= \int \frac{1}{(u^2+1)(u-1)}\,du\\ &= \frac{1}{2}\left(\int \frac{1}{u-1}\,du - \int \frac{u+1}{u^2+1}\,du\right)\\ &= \frac{1}{2}\left(\ln|u-1| - \frac{1}{2}\ln|u^2+1| - \arctan u\right) + C\\ &= \frac{1}{4}\left(2\ln|e^x-1| - \ln|e^{2x}+1| - 2\arctan e^x\right) + C\end{aligned}$$

33. Let $u = \sqrt{x}, u^2 = x, 2u\,du = dx$.

$$\int \frac{\sqrt{x}}{x-4}\,dx = \int \frac{u(2u)du}{u^2-4} = \int \left(\frac{2u^2-8}{u^2-4} + \frac{8}{u^2-4}\right)du = \int \left(2 + \frac{8}{u^2-4}\right)du$$

$$\frac{8}{u^2-4} = \frac{8}{(u-2)(u+2)} = \frac{A}{u-2} + \frac{B}{u+2}$$

$$8 = A(u+2) + B(u-2)$$

When $u = -2, 8 = -4B \Rightarrow B = -2$.

When $u = 2, 8 = 4A \Rightarrow A = 2$.

$$\int \left(2 + \frac{8}{u^2-4}\right)du = 2u + \int \left(\frac{2}{u-2} - \frac{2}{u+2}\right)du$$

$$= 2u + 2\ln|u-2| - 2\ln|u+2| + C$$

$$= 2\sqrt{x} + 2\ln\left|\frac{\sqrt{x}-2}{\sqrt{x}+2}\right| + C$$

34. Let $u = x^{1/6}, u^2 = x^{1/3}, u^3 = x^{1/2}, u^6 = x, 6u^5\,du = dx$.

$$\int \frac{1}{\sqrt{x}-\sqrt[3]{x}}\,dx = \int \frac{6u^5\,du}{u^3-u^2} = 6\int \frac{u^3\,du}{u-1}$$

$$= 6\int \left(u^2 + u + 1 + \frac{1}{u-1}\right)du \qquad \text{(long division)}$$

$$= 6\left(\frac{u^3}{3} + \frac{u^2}{2} + u + \ln|u-1|\right) + C$$

$$= 2\sqrt{x} + 3x^{1/3} + 6x^{1/6} + 6\ln|x^{1/6}-1| + C$$

35. $\dfrac{1}{x(a+bx)} = \dfrac{A}{x} + \dfrac{B}{a+bx}$

$$1 = A(a+bx) + Bx$$

When $x = 0, 1 = aA \Rightarrow A = 1/a$.

When $x = -a/b, 1 = -(a/b)B \Rightarrow B = -b/a$.

$$\int \frac{1}{x(a+bx)}\,dx = \frac{1}{a}\int \left(\frac{1}{x} - \frac{b}{a+bx}\right)dx$$

$$= \frac{1}{a}\left(\ln|x| - \ln|a+bx|\right) + C$$

$$= \frac{1}{a}\ln\left|\frac{x}{a+bx}\right| + C$$

36. $\dfrac{1}{a^2-x^2} = \dfrac{A}{a-x} + \dfrac{B}{a+x}$

$$1 = A(a+x) + B(a-x)$$

When $x = a, 1 = 2aA \Rightarrow A = 1/2a$.

When $x = -a, 1 = 2aB \Rightarrow B = 1/2a$.

$$\int \frac{1}{a^2-x^2}\,dx = \frac{1}{2a}\int \left(\frac{1}{a-x} + \frac{1}{a+x}\right)dx$$

$$= \frac{1}{2a}\left(-\ln|a-x| + \ln|a+x|\right) + C$$

$$= \frac{1}{2a}\ln\left|\frac{a+x}{a-x}\right| + C$$

37. $\dfrac{x}{(a+bx)^2} = \dfrac{A}{a+bx} + \dfrac{B}{(a+bx)^2}$

$x = A(a+bx) + B$

When $x = -a/b$, $B = -a/b$.

When $x = 0$, $0 = aA + B \Rightarrow A = 1/b$.

$$\begin{aligned}\int \frac{x}{(a+bx)^2}\,dx &= \int\left(\frac{1/b}{a+bx} + \frac{-a/b}{(a+bx)^2}\right)dx\\ &= \frac{1}{b}\int\frac{1}{a+bx}\,dx - \frac{a}{b}\int\frac{1}{(a+bx)^2}\,dx\\ &= \frac{1}{b^2}\ln|a+bx| + \frac{a}{b^2}\left(\frac{1}{a+bx}\right) + C\\ &= \frac{1}{b^2}\left(\frac{a}{a+bx} + \ln|a+bx|\right) + C\end{aligned}$$

38. $\dfrac{1}{x^2(a+bx)} = \dfrac{A}{x} + \dfrac{B}{x^2} + \dfrac{C}{a+bx}$

$1 = Ax(a+bx) + B(a+bx) + Cx^2$

When $x = 0$, $1 = Ba \Rightarrow B = 1/a$. When $x = -a/b$,

$1 = C(a^2/b^2) \Rightarrow C = b^2/a^2$. When $x = 1$,

$1 = (a+b)A + (a+b)B + C \Rightarrow A = -b/a^2$.

$$\begin{aligned}\int\frac{1}{x^2(a+bx)}\,dx &= \int\left(\frac{-b/a^2}{x} + \frac{1/a}{x^2} + \frac{b^2/a^2}{a+bx}\right)dx\\ &= -\frac{b}{a^2}\ln|x| - \frac{1}{ax} + \frac{b}{a^2}\ln|a+bx| + C\\ &= -\frac{1}{ax} + \frac{b}{a^2}\ln\left|\frac{a+bx}{x}\right| + C\\ &= -\frac{1}{ax} - \frac{b}{a^2}\ln\left|\frac{x}{a+bx}\right| + C\end{aligned}$$

39. Dividing x^3 by $x - 5$

40. (a) $\dfrac{N(x)}{D(x)} = \dfrac{A_1}{px+q} + \dfrac{A_2}{(px+q)^2} + \cdots + \dfrac{A_m}{(px+q)^m}$

(b) $\dfrac{N(x)}{D(x)} = \dfrac{A_1 + B_1x}{(ax^2+bx+c)} + \cdots + \dfrac{A_n + B_nx}{(ax^2+bx+c)^n}$

41. (a) Substitution: $u = x^2 + 2x - 8$

(b) Partial fractions

(c) Trigonometric substitution (tan) or inverse tangent rule

42. (a) Yes. Because $f' > 0$ on $(0, 5)$, f is increasing, and $f(3) > f(2)$. Therefore, $f(3) - f(2) > 0$.

(b) The area under the graph of f' is greater on the interval $[1, 2]$ because the graph is decreasing on $[1, 4]$.

43. $A = \displaystyle\int_0^1 \frac{12}{x^2+5x+6}\,dx$

$\dfrac{12}{x^2+5x+6} = \dfrac{12}{(x+2)(x+3)} = \dfrac{A}{x+2} + \dfrac{B}{x+3}$

$12 = A(x+3) + B(x+2)$

Let $x = -3$: $12 = B(-1) \Rightarrow B = -12$

Let $x = -2$: $12 = A(1) \Rightarrow A = 12$

$$\begin{aligned}A &= \int_0^1\left(\frac{12}{x+2} - \frac{12}{x+3}\right)dx\\ &= \Big[12\ln|x+2| - 12\ln|x+3|\Big]_0^1\\ &= 12(\ln 3 - \ln 4 - \ln 2 + \ln 3)\\ &= 12\ln\left(\frac{9}{8}\right) \approx 1.4134\end{aligned}$$

44. $A = 2\displaystyle\int_0^3\left(1 - \frac{7}{16-x^2}\right)dx = 2\int_0^3 dx - 14\int_0^3\frac{1}{16-x^2}\,dx$

$= \left[2x - \dfrac{14}{8}\ln\left|\dfrac{4+x}{4-x}\right|\right]_0^3$ (From Exercise 36)

$= 6 - \dfrac{7}{4}\ln 7 \approx 2.595$

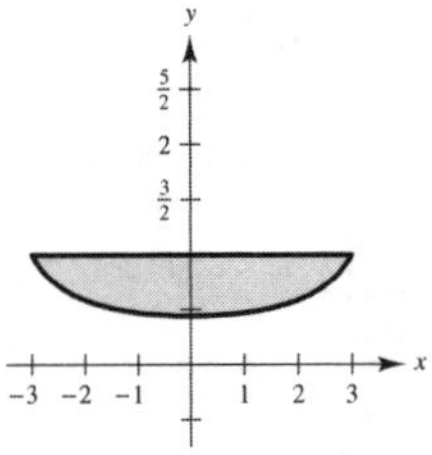

45. Average cost $= \dfrac{1}{80-75}\displaystyle\int_{75}^{80}\frac{124p}{(10+p)(100-p)}\,dp$

$$\begin{aligned}&= \frac{1}{5}\int_{75}^{80}\left(\frac{-124}{(10+p)11} + \frac{1240}{(100-p)11}\right)dp\\ &= \frac{1}{5}\left[\frac{-124}{11}\ln(10+p) - \frac{1240}{11}\ln(100-p)\right]_{75}^{80}\\ &\approx \frac{1}{5}(24.51) = 4.9\end{aligned}$$

Approximately \$490,000

46. (a)

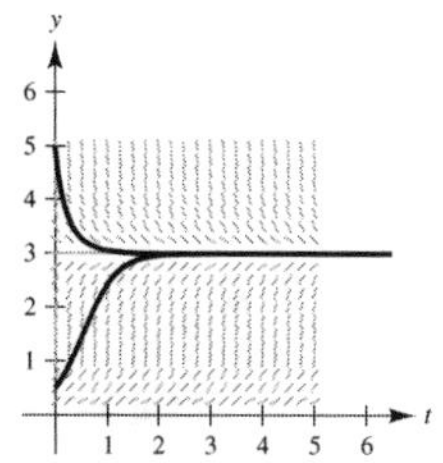

(b) The slope is negative because the function is decreasing.

(c) For $y > 0$, $\lim_{t\to\infty} y(t) = 3$.

(d) $$\frac{dy}{y(L-y)} = \frac{A}{y} + \frac{B}{L-y}$$

$$1 = A(L-y) + By \Rightarrow A = \frac{1}{L}, B = \frac{1}{L}$$

$$\int \frac{dy}{y(L-y)} = \int k\,dt$$

$$\frac{1}{L}\left[\int \frac{1}{y}\,dy + \int \frac{1}{L-y}\,dy\right] = \int k\,dt$$

$$\frac{1}{L}\left[\ln|y| - \ln|L-y|\right] = kt + C_1$$

$$\ln\left|\frac{y}{L-y}\right| = kLt + LC_1$$

$$C_2 e^{kLt} = \frac{y}{L-y}$$

When $t = 0$, $\dfrac{y_0}{L-y_0} = C_2 \Rightarrow \dfrac{y}{L-y} = \dfrac{y_0}{L-y_0}e^{kLt}$.

Solving for y, you obtain $y = \dfrac{y_0 L}{y_0 + (L-y_0)e^{-kLt}}$.

(e) $k = 1, L = 3$

(*i*) $y(0) = 5$: $\quad y = \dfrac{15}{5 - 2e^{-3t}}$

(*ii*) $y(0) = \dfrac{1}{2}$: $\quad y = \dfrac{3/2}{(1/2) + (5/2)e^{-3t}} = \dfrac{3}{1 + 5e^{-3t}}$

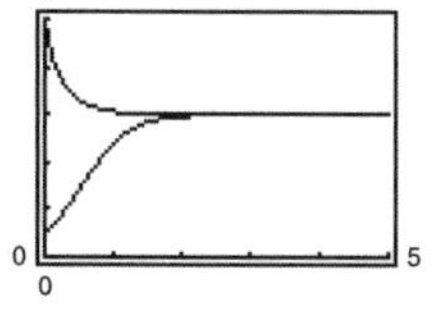

(f) $$\frac{dy}{dt} = ky(L-y)$$

$$\frac{d^2y}{dt^2} = k\left[y\left(\frac{-dy}{dt}\right) + (L-y)\frac{dy}{dt}\right] = 0$$

$$\Rightarrow \quad y\frac{dy}{dt} = (L-y)\frac{dy}{dt}$$

$$\Rightarrow \quad y = \frac{L}{2}$$

From the first derivative test, this is a maximum.

47. $V = \pi \int_0^3 \left(\frac{2x}{x^2+1}\right)^2 dx = 4\pi \int_0^3 \frac{x^2}{(x^2+1)^2} dx$

$$= 4\pi \int_0^3 \left(\frac{1}{x^2+1} - \frac{1}{(x^2+1)^2}\right) dx \qquad \text{(partial fractions)}$$

$$= 4\pi \left[\arctan x - \frac{1}{2}\left(\arctan x + \frac{x}{x^2+1}\right)\right]_0^3 \qquad \text{(trigonometric substitution)}$$

$$= 2\pi \left[\arctan x - \frac{x}{x^2+1}\right]_0^3 = 2\pi\left(\arctan 3 - \frac{3}{10}\right) \approx 5.963$$

$$A = \int_0^3 \frac{2x}{x^2+1} dx = \left[\ln(x^2+1)\right]_0^3 = \ln 10$$

$$\bar{x} = \frac{1}{A}\int_0^3 \frac{2x^2}{x^2+1} dx = \frac{1}{\ln 10}\int_0^3 \left(2 - \frac{2}{x^2+1}\right) dx = \frac{1}{\ln 10}\left[2x - 2\arctan x\right]_0^3 = \frac{2}{\ln 10}(3 - \arctan 3) \approx 1.521$$

$$\bar{y} = \frac{1}{A}\left(\frac{1}{2}\right)\int_0^3 \left(\frac{2x}{x^2+1}\right)^2 dx = \frac{2}{\ln 10}\int_0^3 \frac{x^2}{(x^2+1)^2} dx$$

$$= \frac{2}{\ln 10}\int_0^3 \left(\frac{1}{x^2+1} - \frac{1}{(x^2+1)^2}\right) dx \qquad \text{(partial fractions)}$$

$$= \frac{2}{\ln 10}\left[\arctan x - \frac{1}{2}\left(\arctan x + \frac{x}{x^2+1}\right)\right]_0^3 \qquad \text{(trigonometric substitution)}$$

$$= \frac{2}{\ln 10}\left[\frac{1}{2}\arctan x - \frac{x}{2(x^2+1)}\right]_0^3 = \frac{1}{\ln 10}\left[\arctan x - \frac{x}{x^2+1}\right]_0^3 = \frac{1}{\ln 10}\left(\arctan 3 - \frac{3}{10}\right) \approx 0.412$$

$(\bar{x}, \bar{y}) \approx (1.521, 0.412)$

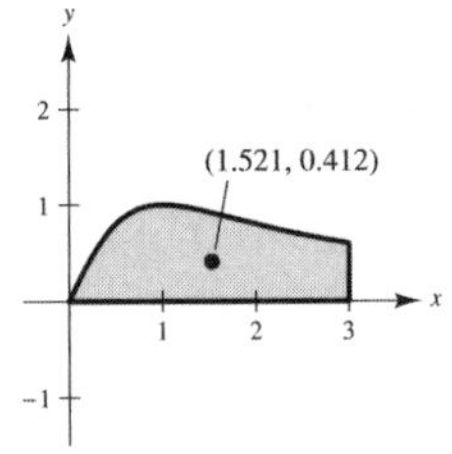

48. $y^2 = \frac{(2-x)^2}{(1+x)^2}, \quad [0, 1]$

$$V = \int_0^1 \pi \frac{(2-x)^2}{(1+x)^2} dx$$

$$= \pi\left[\int_0^1 \frac{4}{(1+x)^2} dx - \int_0^1 \frac{4x}{(1+x)^2} dx + \int_0^1 \frac{x^2}{(1+x)^2} dx\right]$$

$$= \pi\left[2 - (4\ln 2 - 2) + \frac{3}{2} - 2\ln 2\right]$$

$$= \pi\left(\frac{11}{2} - 6\ln 2\right) = \frac{\pi}{2}(11 - 12\ln 2)$$

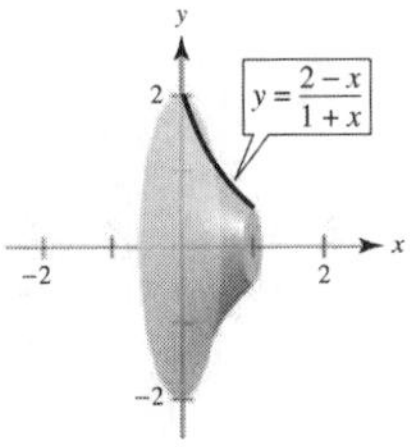

49. $$\frac{1}{(x+1)(n-x)} = \frac{A}{x+1} + \frac{B}{n-x}, A = B = \frac{1}{n+1}$$

$$\frac{1}{n+1}\int\left(\frac{1}{x+1} + \frac{1}{n-x}\right)dx = kt + C$$

$$\frac{1}{n+1}\ln\left|\frac{x+1}{n-x}\right| = kt + C$$

When $t = 0, x = 0, C = \frac{1}{n+1}\ln\frac{1}{n}$.

$$\frac{1}{n+1}\ln\left|\frac{x+1}{n-x}\right| = kt + \frac{1}{n+1}\ln\frac{1}{n}$$

$$\frac{1}{n+1}\left[\ln\left|\frac{x+1}{n-x}\right| - \ln\frac{1}{n}\right] = kt$$

$$\ln\frac{nx+n}{n-x} = (n+1)kt$$

$$\frac{nx+n}{n-x} = e^{(n+1)kt}$$

$$x = \frac{n\left[e^{(n+1)kt} - 1\right]}{n + e^{(n+1)kt}}$$ **Note:** $\lim_{t\to\infty} x = n$

50. (a) $$\frac{1}{(y_0 - x)(z_0 - x)} = \frac{A}{y_0 - x} + \frac{B}{z_0 - x},$$

$$A = \frac{1}{z_0 - y_0}, B = -\frac{1}{z_0 - y_0}, \quad (\text{Assume } y_0 \neq z_0.)$$

$$\frac{1}{z_0 - y_0}\int\left(\frac{1}{y_0 - x} - \frac{1}{z_0 - x}\right)dx = kt + C$$

$$\frac{1}{z_0 - y_0}\ln\left|\frac{z_0 - x}{y_0 - x}\right| = kt + C, \text{ when } t = 0, x = 0$$

$$C = \frac{1}{z_0 - y_0}\ln\frac{z_0}{y_0}$$

$$\frac{1}{z_0 - y_0}\left[\ln\left|\frac{z_0 - x}{y_0 - x}\right| - \ln\left(\frac{z_0}{y_0}\right)\right] = kt$$

$$\ln\left[\frac{y_0(z_0 - x)}{z_0(y_0 - x)}\right] = (z_0 - y_0)kt$$

$$\frac{y_0(z_0 - x)}{z_0(y_0 - x)} = e^{(z_0 - y_0)kt}$$

$$x = \frac{y_0 z_0\left[e^{(z_0 - y_0)kt} - 1\right]}{z_0 e^{(z_0 - y_0)kt} - y_0}$$

(b) (1) If $y_0 < z_0$, $\lim_{t\to\infty} x = y_0$.

(2) If $y_0 > z_0$, $\lim_{t\to\infty} x = z_0$.

(3) If $y_0 = z_0$, then the original equation is:

$$\int \frac{1}{(y_0 - x)^2}\,dx = \int k\,dt$$

$$(y_0 - x)^{-1} = kt + C_1$$

$$x = 0 \text{ when } t = 0 \Rightarrow \frac{1}{y_0} = C_1$$

$$\frac{1}{y_0 - x} = kt + \frac{1}{y_0} = \frac{kty_0 + 1}{y_0}$$

$$y_0 - x = \frac{y_0}{kty_0 + 1}$$

$$x = y_0 - \frac{y_0}{kty_0 + 1}$$

As $t \to \infty$, $x \to y_0 = x_0$.

51. $$\frac{x}{1 + x^4} = \frac{Ax + B}{x^2 + \sqrt{2}x + 1} + \frac{Cx + D}{x^2 - \sqrt{2}x + 1}$$

$$x = (Ax + B)(x^2 - \sqrt{2}x + 1) + (Cx + D)(x^2 + \sqrt{2}x + 1)$$

$$= (A + C)x^3 + (B + D - \sqrt{2}A + \sqrt{2}C)x^2 + (A + C - \sqrt{2}B + \sqrt{2}D)x + (B + D)$$

$0 = A + C \Rightarrow C = -A$

$0 = B + D - \sqrt{2}A + \sqrt{2}C$ $\qquad -2\sqrt{2}A = 0 \Rightarrow A = 0$ and $C = 0$

$1 = A + C - \sqrt{2}B + \sqrt{2}D$ $\qquad -2\sqrt{2}B = 1 \Rightarrow B = -\frac{\sqrt{2}}{4}$ and $D = \frac{\sqrt{2}}{4}$

$0 = B + D \Rightarrow D = -B$

So,

$$\int_0^1 \frac{x}{1 + x^4}\,dx = \int_0^1 \left(\frac{-\sqrt{2}/4}{x^2 + \sqrt{2}x + 1} + \frac{\sqrt{2}/4}{x^2 - \sqrt{2}x + 1}\right)dx$$

$$= \frac{\sqrt{2}}{4}\int_0^1 \left[\frac{-1}{\left[x + \left(\sqrt{2}/2\right)\right]^2 + (1/2)} + \frac{1}{\left[x - \left(\sqrt{2}/2\right)\right]^2 + (1/2)}\right]dx$$

$$= \frac{\sqrt{2}}{4}\cdot\frac{1}{1/\sqrt{2}}\left[-\arctan\left(\frac{x + \left(\sqrt{2}/2\right)}{1/\sqrt{2}}\right) + \arctan\left(\frac{x - \left(\sqrt{2}/2\right)}{1/\sqrt{2}}\right)\right]_0^1$$

$$= \frac{1}{2}\left[-\arctan\left(\sqrt{2}x + 1\right) + \arctan\left(\sqrt{2}x - 1\right)\right]_0^1$$

$$= \frac{1}{2}\left[\left(-\arctan\left(\sqrt{2} + 1\right) + \arctan\left(\sqrt{2} - 1\right)\right) - \left(-\arctan 1 + \arctan(-1)\right)\right]$$

$$= \frac{1}{2}\left[\arctan\left(\sqrt{2} - 1\right) - \arctan\left(\sqrt{2} + 1\right) + \frac{\pi}{4} + \frac{\pi}{4}\right].$$

Because $\arctan x - \arctan y = \arctan\left[(x - y)/(1 + xy)\right]$, you have:

$$\int_0^1 \frac{x}{1 + x^4}\,dx = \frac{1}{2}\left[\arctan\left(\frac{\left(\sqrt{2} - 1\right) - \left(\sqrt{2} + 1\right)}{1 + \left(\sqrt{2} - 1\right)\left(\sqrt{2} + 1\right)}\right) + \frac{\pi}{2}\right] = \frac{1}{2}\left[\arctan\left(\frac{-2}{2}\right) + \frac{\pi}{2}\right] = \frac{1}{2}\left(-\frac{\pi}{4} + \frac{\pi}{2}\right) = \frac{\pi}{8}$$

52. The partial fraction decomposition is:

$$\frac{x^4(1-x)^4}{1+x^2} = x^6 - 4x^5 + 5x^4 - 4x^2 + 4 - \frac{4}{1+x^2}$$

$$\int_0^1 \frac{x^4(1-x)^4}{1+x^2}\,dx = \left[\frac{x^7}{7} - \frac{2x^6}{3} + x^5 - \frac{4}{3}x^3 + 4x - 4\arctan x\right]_0^1$$

$$= \frac{1}{7} - \frac{2}{3} + 1 - \frac{4}{3} + 4 - 4\left(\frac{\pi}{4}\right)$$

$$= \frac{22}{7} - \pi$$

Note: You can easily verify this calculation with a graphing utility.

Section 8.6 Integration by Tables and Other Integration Techniques

1. By Formula 6: $(a = 5, b = 1)$

$$\int \frac{x^2}{5+x}\,dx = \left[-\frac{x}{2}(10 - x) + 25\ln|5 + x|\right] + C$$

2. By Formula 13: $(a = 4, b = 3)$

$$\int \frac{2}{x^2(4+3x)^2}\,dx = 2\left(\frac{-1}{16}\right)\left[\frac{4+6x}{x(4+3x)} + \frac{6}{4}\ln\left|\frac{x}{4+3x}\right|\right] + C$$

$$= -\frac{(2+3x)}{4x(3x+4)} - \frac{3}{16}\ln\left|\frac{x}{4+3x}\right| + C$$

3. By Formula 44: $\displaystyle\int \frac{1}{x^2\sqrt{1-x^2}}\,dx = -\frac{\sqrt{1-x^2}}{x} + C$

4. Let $u = x^2,\ du = 2x\,dx.$

$$\int \frac{\sqrt{64-x^4}}{x}\,du = \frac{1}{2}\int \frac{\sqrt{64-x^4}}{x^2}(2x\,dx)$$

$$= \frac{1}{2}\int \frac{\sqrt{64-u^2}}{u}\,du$$

By Formula 39: $(a = 8)$

$$\int \frac{\sqrt{64-u^2}}{u}\,dx = \frac{1}{2}\left[\sqrt{64-u^2} - 8\ln\left|\frac{8+\sqrt{64-u^2}}{u}\right|\right] + C$$

$$= \frac{1}{2}\sqrt{64-x^4} - 4\ln\left|\frac{8+\sqrt{64-x^4}}{x^2}\right| + C$$

5. By Formulas 51 and 49:

$$\begin{aligned}\int \cos^4 3x\,dx &= \frac{1}{3}\int \cos^4 3x\,(3)\,dx\\ &= \frac{1}{3}\left[\frac{\cos^3 3x \sin 3x}{4} + \frac{3}{4}\int \cos^2 3x\,dx\right]\\ &= \frac{1}{12}\cos^3 3x \sin 3x + \frac{1}{4}\cdot\frac{1}{3}\int \cos^2 3x\,(3)\,dx\\ &= \frac{1}{12}\cos^3 3x \sin 3x + \frac{1}{12}\cdot\frac{1}{2}(3x + \sin 3x \cos 3x) + C\\ &= \frac{1}{24}(2\cos^3 3x \sin 3x + 3x + \sin 3x \cos 3x) + C\end{aligned}$$

6. Let $u = \sqrt{x}$, $du = \dfrac{1}{2\sqrt{x}}\,dx$.

$$\begin{aligned}\int \frac{\sin^4\sqrt{x}}{\sqrt{x}}\,dx &= 2\int \sin^4 u\,du\\ &= 2\left[-\frac{\sin^3 u \cos u}{4} + \frac{3}{4}\int \sin^2 u\,du\right] && (\text{Formula 50, } n = 4)\\ &= 2\left[-\frac{\sin^3 u \cos u}{4} + \frac{3}{4}\cdot\frac{1}{2}(u - \sin u \cos u)\right] + C && (\text{Formula 48})\\ &= -\frac{1}{2}\sin^3 u \cos u + \frac{3}{4}u - \frac{3}{4}\sin u \cos u + C\\ &= -\frac{1}{2}\sin^3\sqrt{x}\,\cos\sqrt{x} + \frac{3}{4}\sqrt{x} - \frac{3}{4}\sin\sqrt{x}\cos\sqrt{x} + C\end{aligned}$$

7. By Formula 57: $\displaystyle\int \frac{1}{\sqrt{x}\left(1 - \cos\sqrt{x}\right)}\,dx = 2\int \frac{1}{1 - \cos\sqrt{x}}\left(\frac{1}{2\sqrt{x}}\right)dx = -2\left(\cot\sqrt{x} + \csc\sqrt{x}\right) + C$

$u = \sqrt{x}$, $du = \dfrac{1}{2\sqrt{x}}\,dx$

8. Let $u = 4x$, $du = 4\,dx$.

By Formula 72:

$$\begin{aligned}\int \frac{1}{1 + \cot 4x}\,dx &= \frac{1}{4}\int \frac{1}{1 + \cot 4x}(4\,dx)\\ &= \frac{1}{4}\cdot\frac{1}{2}\left(4x - \ln\left|\sin 4x + \cos 4x\right|\right) + C\\ &= \frac{1}{2}x - \frac{1}{8}\ln\left|\sin 4x + \cos 4x\right| + C\end{aligned}$$

9. By Formula 84:

$$\int \frac{1}{1 + e^{2x}}\,dx = 2x - \frac{1}{2}\ln\left(1 + e^{2x}\right) + C$$

10. By Formula 85: $(a = -4, b = 3)$

$$\begin{aligned}\int e^{-4x}\sin 3x\,dx &= \frac{e^{-4x}}{(-4)^2 + 3^2}(-4\sin 3x - 3\cos 3x) + C\\ &= \frac{e^{-4x}}{25}(-4\sin 3x - 3\cos 3x) + C\end{aligned}$$

11. By Formula 89: $(n = 7)$

$$\int x^7 \ln x\, dx = \frac{x^8}{64}\left[-1 + 8\ln x\right] + C = \frac{1}{64}x^8(8\ln x - 1) + C$$

12. By Formulas 90 and 91:
$$\begin{aligned}\int(\ln x)^3\, dx &= x(\ln x)^3 - 3\int(\ln x)^2\, dx\\ &= x(\ln x)^3 - 3x\left[2 - 2\ln x + (\ln x)^2\right] + C\\ &= x\left[(\ln x)^3 - 3(\ln x)^2 + 6\ln x - 6\right] + C\end{aligned}$$

13. (a) Let $u = 3x,\ x = \frac{u}{3},\ du = 3\,dx.$

$$\int x^2e^{3x}\, dx = \int\left(\frac{u}{3}\right)^2 e^u\frac{1}{3}du = \frac{1}{27}\int u^2e^u\, du$$

By Formulas 83 and 82:

$$\begin{aligned}\int x^2e^{3x}\, dx &= \frac{1}{27}\left[u^2e^u - 2\int ue^u\, du\right]\\ &= \frac{1}{27}\left[u^2e^u - 2\left((u-1)e^u\right)\right] + C\\ &= \frac{1}{27}e^{3x}\left(9x^2 - 6x + 2\right) + C\end{aligned}$$

(b) Integration by parts:

$$u = x^2,\ du = 2x\,dx,\ dv = e^{3x}\,dx,\ v = \frac{1}{3}e^{3x}$$

$$\int x^2e^{3x}\, dx = x^2\frac{1}{3}e^{3x} - \int\frac{2}{3}xe^{3x}\, dx$$

Parts again: $u = x,\ du = dx,\ dv = e^{3x},\ v = \frac{1}{3}e^{3x}$

$$\begin{aligned}\int x^2e^{3x}\, dx &= \frac{1}{3}x^2e^{3x} - \frac{2}{3}\left[\frac{x}{3}e^{3x} - \int\frac{1}{3}e^{3x}\, dx\right]\\ &= \frac{1}{3}x^2e^{3x} - \frac{2}{9}xe^{3x} + \frac{2}{27}e^{3x} + C\\ &= \frac{1}{27}e^{3x}\left[9x^2 - 6x + 2\right] + C\end{aligned}$$

14. (a) By Formula 89: $(n = 5)$

$$\int x^5 \ln x\, dx = \frac{x^6}{36}\ \left[-1 + 6\ln x\right] + C$$

(b) Integration by parts:

$$u = \ln x,\ \ du = \frac{1}{x}\,dx,\quad dv = x^5\,dx,\ \ v = \frac{x^6}{6}$$

$$\begin{aligned}\int x^5\ln x\, dx &= \frac{x^6}{6}\ln x - \int\frac{x^6}{6}\cdot\frac{1}{x}\, dx\\ &= \frac{x^6}{6}\ln x - \frac{x^6}{36} + C\end{aligned}$$

15. (a) By Formula 12: $(a = b = 1, u = x)$

$$\begin{aligned}\int\frac{1}{x^2(x+1)}\, dx &= \frac{-1}{1}\left(\frac{1}{x} + \frac{1}{1}\ln\left|\frac{x}{1+x}\right|\right) + C\\ &= \frac{-1}{x} - \ln\left|\frac{x}{1+x}\right| + C\\ &= \frac{-1}{x} + \ln\left|\frac{x+1}{x}\right| + C\end{aligned}$$

(b) Partial fractions:

$$\frac{1}{x^2(x+1)} = \frac{A}{x} + \frac{B}{x^2} + \frac{C}{x+1}$$

$$1 = Ax(x+1) + B(x+1) + Cx^2$$

$x = 0$: $1 = B$

$x = -1$: $1 = C$

$x = 1$: $1 = 2A + 2 + 1 \Rightarrow A = -1$

$$\begin{aligned}\int\frac{1}{x^2(x+1)}\, dx &= \int\left[\frac{-1}{x} + \frac{1}{x^2} + \frac{1}{x+1}\right] dx\\ &= -\ln|x| - \frac{1}{x} + \ln|x+1| + C\\ &= -\frac{1}{x} - \ln\left|\frac{x}{x+1}\right| + C\end{aligned}$$

16. (a) By Formula 24: $(a = 6)$

$$\int\frac{1}{x^2 - 36}\, dx = \frac{1}{12}\ln\left|\frac{x-6}{x+6}\right| + C$$

(b) Partial Fractions:

$$\frac{1}{x^2 - 36} = \frac{A}{x-6} + \frac{B}{x+6}$$

$$1 = A(x+6) + B(x-6)$$

When $x = -6$, $1 = -12B \Rightarrow B = -\frac{1}{12}.$

When $x = 6$, $1 = 12A \Rightarrow A = \frac{1}{12}.$

$$\begin{aligned}\int\frac{1}{x^2 - 36}\, dx &= \int\frac{1/12}{x-6}\, dx + \int\frac{-1/12}{x+6}\, dx\\ &= \frac{1}{12}\ln|x-6| - \frac{1}{12}\ln|x+6| + C\\ &= \frac{1}{12}\ln\left|\frac{x-6}{x+6}\right| + C\end{aligned}$$

17. By Formula 80:

$$\int x \operatorname{arccsc}(x^2 + 1)\,dx = \frac{1}{2}\int \operatorname{arccsc}(x^2 + 1)(2x)\,dx$$
$$= \frac{1}{2}\left[(x^2 + 1)\operatorname{arccsc}(x^2 + 1) + \ln\left|x^2 + 1 + \sqrt{(x^2 + 1)^2 - 1}\right|\right] + C$$
$$= \frac{1}{2}(x^2 + 1)\operatorname{arccsc}(x^2 + 1) + \frac{1}{2}\ln\left(x^2 + 1 + \sqrt{x^4 + 2x^2}\right) + C$$

18. By Formula 75: $u = 4x$

$$\int \arcsin 4x\,dx = \frac{1}{4}\int \arcsin 4x(4\,dx)$$
$$= \frac{1}{4}\left[4x \arcsin 4x + \sqrt{1 - (4x)^2}\right] + C$$
$$= x \arcsin 4x + \frac{1}{4}\sqrt{1 - 16x^2} + C$$

19. By Formula 35: $\displaystyle\int \frac{1}{x^2\sqrt{x^2 - 4}}\,dx = \frac{\sqrt{x^2 - 4}}{4x} + C$

20. By Formula 14: $(a = 8, b = 4, c = 1, b^2 < 4ac)$

$$\int \frac{1}{x^2 + 4x + 8}\,dx = \frac{2}{\sqrt{16}}\arctan\frac{2x + 4}{\sqrt{16}} + C$$
$$= \frac{1}{2}\arctan\left(\frac{x + 2}{2}\right) + C$$

21. By Formula 4: $(a = 2, b = -5)$

$$\int \frac{4x}{(2 - 5x)^2}\,dx = 4\left[\frac{1}{25}\left(\frac{2}{2 - 5x} + \ln|2 - 5x|\right)\right] + C$$
$$= \frac{4}{25}\left(\frac{2}{2 - 5x} + \ln|2 - 5x|\right) + C$$

22. By Formula 56: $u = \theta^4$, $du = 4\theta^3\,d\theta$

$$\int \frac{\theta^3}{1 + \sin\theta^4}\,d\theta = \frac{1}{4}\int \frac{1}{1 + \sin\theta^4}\,4\theta^3\,d\theta$$
$$= \frac{1}{4}\left(\tan\theta^4 - \sec\theta^4\right) + C$$

23. By Formula 76:

$$\int e^x \arccos e^x\,dx = e^x \arccos e^x - \sqrt{1 - e^{2x}} + C$$

$u = e^x$, $du = e^x\,dx$

24. By Formula 71:

$$\int \frac{e^x}{1 - \tan e^x}\,dx = \frac{1}{2}\left(e^x - \ln\left|\cos e^x - \sin e^x\right|\right) + C$$

$u = e^x$, $du = e^x\,dx$

25. By Formula 73:

$$\int \frac{x}{1 - \sec x^2}\,dx = \frac{1}{2}\int \frac{2x}{1 - \sec x^2}\,dx$$
$$= \frac{1}{2}\left(x^2 + \cot x^2 + \csc x^2\right) + C$$

26. By Formula 23: $\displaystyle\int \frac{1}{t\left[1 + (\ln t)^2\right]}\,dt = \int \frac{1}{1 + (\ln t)^2}\left(\frac{1}{t}\right)dt = \arctan(\ln t) + C$

$u = \ln t$, $du = \dfrac{1}{t}\,dt$

27. By Formula 14: $\displaystyle\int \frac{\cos\theta}{3 + 2\sin\theta + \sin^2\theta}\,d\theta = \frac{\sqrt{2}}{2}\arctan\left(\frac{1 + \sin\theta}{\sqrt{2}}\right) + C \quad \left(b^2 = 4 < 12 = 4ac\right)$

$u = \sin\theta$, $du = \cos\theta\,d\theta$

28. By Formula 27: $\displaystyle\int x^2\sqrt{2 + (3x)^2}\,dx = \frac{1}{27}\int (3x)^2\sqrt{\left(\sqrt{2}\right)^2 + (3x)^2}\,3\,dx$

$$= \frac{1}{8(27)}\left[3x\left(18x^2 + 2\right)\sqrt{2 + 9x^2} - 4\ln\left|3x + \sqrt{2 + 9x^2}\right|\right] + C$$

29. By Formula 35: $\displaystyle\int \frac{1}{x^2\sqrt{2 + 9x^2}}\,dx = 3\int \frac{3}{(3x)^2\sqrt{\left(\sqrt{2}\right)^2 + (3x)^2}}\,dx = -\frac{3\sqrt{2 + 9x^2}}{6x} + C = -\frac{\sqrt{2 + 9x^2}}{2x} + C$

30. By Formula 77: $\int \sqrt{x} \arctan\left(x^{3/2}\right) dx = \frac{2}{3}\int \arctan\left(x^{3/2}\right)\left(\frac{3}{2}\sqrt{x}\right) dx = \frac{2}{3}\left[x^{3/2} \arctan\left(x^{3/2}\right) - \ln \sqrt{1 + x^3}\right] + C$

31. By Formula 3: $\int \frac{\ln x}{x(3 + 2 \ln x)} dx = \frac{1}{4}\left(2 \ln|x| - 3 \ln\left|3 + 2 \ln|x|\right|\right) + C$

$u = \ln x, du = \frac{1}{x} dx$

32. By Formula 45: $\int \frac{e^x}{\left(1 - e^{2x}\right)^{3/2}} dx = \frac{e^x}{\sqrt{1 - e^{2x}}} + C$

$u = e^x, du = e^x\, dx$

33. By Formulas 1, 23, and 35: $\int \frac{x}{\left(x^2 - 6x + 10\right)^2} dx = \frac{1}{2}\int \frac{2x - 6 + 6}{\left(x^2 - 6x + 10\right)^2} dx$

$$= \frac{1}{2}\int \left(x^2 - 6x + 10\right)^{-2}(2x - 6)\, dx + 3\int \frac{1}{\left[(x - 3)^2 + 1\right]^2} dx$$

$$= -\frac{1}{2\left(x^2 - 6x + 10\right)} + \frac{3}{2}\left[\frac{x - 3}{x^2 - 6x + 10} + \arctan(x - 3)\right] + C$$

$$= \frac{3x - 10}{2\left(x^2 - 6x + 10\right)} + \frac{3}{2} \arctan(x - 3) + C$$

34. By Formula 41:

$$\int \sqrt{\frac{5 - x}{5 + x}}\, dx = \int \frac{\sqrt{5 - x}}{\sqrt{5 + x}} \cdot \frac{\sqrt{5 - x}}{\sqrt{5 - x}}\, dx$$

$$= \int \frac{5 - x}{\sqrt{25 - x^2}}\, dx$$

$$= \int \frac{5\, dx}{\sqrt{25 - x^2}} - \int \frac{x}{\sqrt{25 - x^2}}\, dx$$

$$= 5 \arcsin\left(\frac{x}{5}\right) + \sqrt{25 - x^2} + C$$

35. By Formula 31: $\int \frac{x}{\sqrt{x^4 - 6x^2 + 5}} dx = \frac{1}{2}\int \frac{2x}{\sqrt{\left(x^2 - 3\right)^2 - 4}} dx = \frac{1}{2} \ln\left|x^2 - 3 + \sqrt{x^4 - 6x^2 + 5}\right| + C$

$u = x^2 - 3, du = 2x\, dx$

36. By Formula 31: $\int \frac{\cos x}{\sqrt{\sin^2 x + 1}} dx = \ln\left|\sin x + \sqrt{\sin^2 x + 1}\right| + C$

$u = \sin x, du = \cos x\, dx$

37. By Formula 8:

$$\int \frac{e^{3x}}{\left(1 + e^x\right)^3} dx = \int \frac{\left(e^x\right)^2}{\left(1 + e^x\right)^3}\left(e^x\right) dx$$

$$= \frac{2}{1 + e^x} - \frac{1}{2\left(1 + e^x\right)^2} + \ln\left|1 + e^x\right| + C$$

$u = e^x, du = e^x\, dx$

38. By Formulas 64 and 68:

$$\int \cot^4 \theta \, d\theta = -\frac{\cot^3 \theta}{3} - \int \cot^2 \theta \, d\theta$$
$$= -\frac{\cot^3 \theta}{3} + \theta + \cot \theta + C$$

39. By Formula 81:

$$\int_0^1 xe^{x^2} \, dx = \left[\tfrac{1}{2}e^{x^2}\right]_0^1 = \tfrac{1}{2}(e - 1) \approx 0.8591$$

40. By Formula 21: $(a = 3, b = 2)$

$$\int_0^4 \frac{x}{\sqrt{3 + 2x}} \, dx = \left[\frac{-2(6 - 2x)}{12}\sqrt{3 + 2x}\right]_0^4$$
$$= \left[-\frac{1}{6}(6 - 2x)\sqrt{3 + 2x}\right]_0^4$$
$$= -\frac{1}{6}(-2)\sqrt{11} + \frac{1}{6}(6)\sqrt{3}$$
$$= \frac{\sqrt{11}}{3} + \sqrt{3}$$

41. By Formula 89: $(n = 4)$

$$\int_1^2 x^4 \ln x \, dx = \left[\frac{x^5}{25}(-1 + 5 \ln x)\right]_1^2$$
$$= \frac{32}{25}[-1 + 5 \ln 2] - \frac{1}{25}[-1 + 0]$$
$$= -\frac{31}{25} + \frac{32}{5} \ln 2 \approx 3.1961$$

42. By Formula 52: $u = 2x, du = 2dx$

$$\int_0^{\pi/2} x \sin 2x \, dx = \frac{1}{4}\int_0^{\pi/2} (2x) \sin 2x \, (2dx)$$
$$= \frac{1}{4}[\sin 2x - 2x \cos 2x]_0^{\pi/2}$$
$$= \frac{1}{4}[0 - \pi(-1)]$$
$$= \frac{\pi}{4}$$

43. By Formula 23, and letting $u = \sin x$:

$$\int_{-\pi/2}^{\pi/2} \frac{\cos x}{1 + \sin^2 x} \, dx = \left[\arctan(\sin x)\right]_{-\pi/2}^{\pi/2}$$
$$= \arctan(1) - \arctan(-1) = \frac{\pi}{2}$$

44. By Formula 7: $(a = 5, b = 2)$

$$\int_0^5 \frac{x^2}{(5 + 2x)^2} \, dx = \frac{1}{8}\left[2x - \frac{25}{5 + 2x} - 10 \ln|5 + 2x|\right]_0^5$$
$$= \frac{1}{8}\left[\left(10 - \frac{25}{15} - 10 \ln 15\right) - (-5 - 10 \ln 5)\right]$$
$$= \frac{5}{3} - \frac{1}{8}(10) \ln\left(\frac{15}{5}\right)$$
$$= \frac{5}{3} - \frac{5}{4} \ln 3$$

45. By Formulas 54 and 55:

$$\int t^3 \cos t \, dt = t^3 \sin t - 3\int t^2 \sin t \, dt$$
$$= t^3 \sin t - 3\left(-t^2 \cos t + 2\int t \cos t \, dt\right)$$
$$= t^3 \sin t + 3t^2 \cos t - 6\left(t \sin t - \int \sin t \, dt\right)$$
$$= t^3 \sin t + 3t^2 \cos t - 6t \sin t - 6 \cos t + C$$

So,

$$\int_0^{\pi/2} t^3 \cos t \, dt = \left[t^3 \sin t + 3t^2 \cos t - 6t \sin t - 6 \cos t\right]_0^{\pi/2}$$
$$= \left(\frac{\pi^3}{8} - 3\pi\right) + 6 = \frac{\pi^3}{8} + 6 - 3\pi \approx 0.4510.$$

46. By Formula 26: $(a = 4)$

$$\int_0^3 \sqrt{x^2 + 16}\,dx = \frac{1}{2}\left[x\sqrt{x^2 + 16} + 16 \ln\left|x + \sqrt{x^2 + 16}\right|\right]_0^3$$
$$= \frac{1}{2}\left[\left(3(5) + 16 \ln|3 + 5|\right) - \left(16 \ln 4\right)\right]$$
$$= \frac{15}{2} + 8 \ln 8 - 8 \ln 4$$
$$= \frac{15}{2} + 8 \ln 2$$

47. $$\frac{u^2}{(a + bu)^2} = \frac{1}{b^2} - \frac{(2a/b)u + (a^2/b^2)}{(a + bu)^2} = \frac{1}{b^2} + \frac{A}{a + bu} + \frac{B}{(a + bu)^2}$$

$$-\frac{2a}{b}u - \frac{a^2}{b^2} = A(a + bu) + B = (aA + B) + bAu$$

Equating the coefficients of like terms you have $aA + B = -a^2/b^2$ and $bA = -2a/b$. Solving these equations you have $A = -2a/b^2$ and $B = a^2/b^2$.

$$\int \frac{u^2}{(a + bu)^2}\,du = \frac{1}{b^2}\int du - \frac{2a}{b^2}\left(\frac{1}{b}\right)\int \frac{1}{a + bu} b\,du + \frac{a^2}{b^2}\left(\frac{1}{b}\right)\int \frac{1}{(a + bu)^2} b\,du = \frac{1}{b^2}u - \frac{2a}{b^3}\ln|a + bu| - \frac{a^2}{b^3}\left(\frac{1}{a + bu}\right) + C$$
$$= \frac{1}{b^3}\left(bu - \frac{a^2}{a + bu} - 2a \ln|a + bu|\right) + C$$

48. Integration by parts: $w = u^n$, $dw = nu^{n-1}\,du$, $dv = \dfrac{du}{\sqrt{a + bu}}$, $v = \dfrac{2}{b}\sqrt{a + bu}$

$$\int \frac{u^n}{\sqrt{a + bu}}\,du = \frac{2u^n}{b}\sqrt{a + bu} - \frac{2n}{b}\int u^{n-1}\sqrt{a + bu}\,du$$
$$= \frac{2u^n}{b}\sqrt{a + bu} - \frac{2n}{b}\int u^{n-1}\sqrt{a + bu} \cdot \frac{\sqrt{a + bu}}{\sqrt{a + bu}}\,du$$
$$= \frac{2u^n}{b}\sqrt{a + bu} - \frac{2n}{b}\int \frac{au^{n-1} + bu^n}{\sqrt{a + bu}}\,du$$
$$= \frac{2u^n}{b}\sqrt{a + bu} - \frac{2na}{b}\int \frac{u^{n-1}}{\sqrt{a + bu}}\,du - 2n\int \frac{u^n}{\sqrt{a + bu}}\,du$$

Therefore, $(2n + 1)\displaystyle\int \frac{u^n}{\sqrt{a + bu}}\,du = \frac{2}{b}\left[u^n\sqrt{a + bu} - na\int \frac{u^{n-1}}{\sqrt{a + bu}}\,du\right]$ and

$$\int \frac{u^n}{\sqrt{a + bu}} = \frac{2}{(2n + 1)b}\left[u^n\sqrt{a + bu} - na\int \frac{u^{n-1}}{\sqrt{a + bu}}\,du\right].$$

49. When you have $u^2 + a^2$:

$$u = a\tan\theta$$
$$du = a\sec^2\theta\,d\theta$$
$$u^2 + a^2 = a^2\sec^2\theta$$

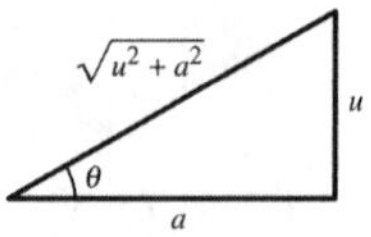

$$\int \frac{1}{\left(u^2 + a^2\right)^{3/2}}\,du = \int \frac{a\sec^2\theta\,d\theta}{a^3\sec^3\theta} = \frac{1}{a^2}\int \cos\theta\,d\theta = \frac{1}{a^2}\sin\theta + C = \frac{u}{a^2\sqrt{u^2 + a^2}} + C$$

When you have $u^2 - a^2$:

$$u = a\sec\theta$$
$$du = a\sec\theta\tan\theta\,d\theta$$
$$u^2 - a^2 = a^2\tan^2\theta$$

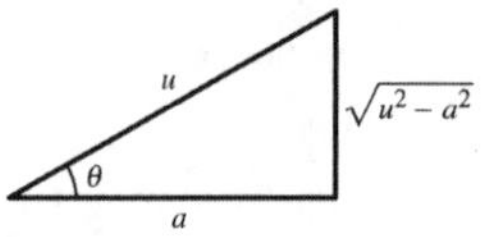

$$\int \frac{1}{\left(u^2 - a^2\right)^{3/2}}\,du = \int \frac{a\sec\theta\tan\theta\,d\theta}{a^3\tan^3\theta} = \frac{1}{a^2}\int \frac{\cos\theta}{\sin^2\theta}\,d\theta = \frac{1}{a^2}\int \csc\theta\cot\theta\,d\theta = -\frac{1}{a^2}\csc\theta + C = \frac{-u}{a^2\sqrt{u^2 - a^2}} + C$$

50. $\int u^n(\cos u)\,du = u^n\sin u - n\int u^{n-1}(\sin u)\,du$

$w = u^n,\ dv = \cos u\,du,\ dw = nu^{n-1}\,du,\ v = \sin u$

51.
$$\int(\arctan u)\,du = u\arctan u - \frac{1}{2}\int\frac{2u}{1 + u^2}\,du$$
$$= u\arctan u - \frac{1}{2}\ln\left(1 + u^2\right) + C$$
$$= u\arctan u - \ln\sqrt{1 + u^2} + C$$

$w = \arctan u,\ dv = du,\ dw = \dfrac{du}{1 + u^2},\ v = u$

52.
$$\int(\ln u)^n\,du = u(\ln u)^n - \int n(\ln u)^{n-1}\left(\frac{1}{u}\right)u\,du$$
$$= u(\ln u)^n - n\int(\ln u)^{n-1}\,du$$

$w = (\ln u)^n,\ dv = du,\ dw = n(\ln u)^{n-1}\left(\dfrac{1}{u}\right)du,$

$v = u$

53.
$$\int\frac{1}{2 - 3\sin\theta}\,d\theta = \int\left[\frac{\dfrac{2\,du}{1 + u^2}}{2 - 3\left(\dfrac{2u}{1 + u^2}\right)}\right],\ u = \tan\frac{\theta}{2}$$
$$= \int\frac{2}{2\left(1 + u^2\right) - 6u}\,du$$
$$= \int\frac{1}{u^2 - 3u + 1}\,du$$
$$= \int\frac{1}{\left(u - \dfrac{3}{2}\right)^2 - \dfrac{5}{4}}\,du$$
$$= \frac{1}{\sqrt{5}}\ln\left|\frac{\left(u - \dfrac{3}{2}\right) - \dfrac{\sqrt{5}}{2}}{\left(u - \dfrac{3}{2}\right) + \dfrac{\sqrt{5}}{2}}\right| + C$$
$$= \frac{1}{\sqrt{5}}\ln\left|\frac{2u - 3 - \sqrt{5}}{2u - 3 + \sqrt{5}}\right| + C$$
$$= \frac{1}{\sqrt{5}}\ln\left|\frac{2\tan\left(\dfrac{\theta}{2}\right) - 3 - \sqrt{5}}{2\tan\left(\dfrac{\theta}{2}\right) - 3 + \sqrt{5}}\right| + C$$

54.
$$\int\frac{\sin\theta}{1 + \cos^2\theta}\,d\theta = -\int\frac{-\sin\theta}{1 + (\cos\theta)^2}\,d\theta$$
$$= -\arctan(\cos\theta) + C$$

55. $$\int_0^{\pi/2} \frac{1}{1 + \sin\theta + \cos\theta}\,d\theta = \int_0^1 \left[\frac{\dfrac{2\,du}{1+u^2}}{1 + \dfrac{2u}{1+u^2} + \dfrac{1-u^2}{1+u^2}}\right]$$
$$= \int_0^1 \frac{1}{1+u}\,du$$
$$= \Big[\ln|1+u|\Big]_0^1$$
$$= \ln 2$$

$u = \tan\dfrac{\theta}{2}$

56. $$\int_0^{\pi/2} \frac{1}{3 - 2\cos\theta}\,d\theta = \int_0^1 \left[\frac{\dfrac{2u}{1+u^2}}{3 - \dfrac{2(1-u^2)}{1+u^2}}\right]$$
$$= 2\int_0^1 \frac{1}{5u^2 + 1}\,du$$
$$= \left[\frac{2}{\sqrt{5}}\arctan\left(\sqrt{5}\,u\right)\right]_0^1$$
$$= \frac{2}{\sqrt{5}}\arctan\sqrt{5}$$

$u = \tan\dfrac{\theta}{2}$

57. $$\int \frac{\sin\theta}{3 - 2\cos\theta}\,d\theta = \frac{1}{2}\int \frac{2\sin\theta}{3 - 2\cos\theta}\,d\theta$$
$$= \frac{1}{2}\ln|u| + C$$
$$= \frac{1}{2}\ln(3 - 2\cos\theta) + C$$

$u = 3 - 2\cos\theta,\ du = 2\sin\theta\,d\theta$

58. $$\int \frac{\cos\theta}{1+\cos\theta}\,d\theta = \int \frac{\cos\theta(1-\cos\theta)}{(1+\cos\theta)(1-\cos\theta)}\,d\theta$$
$$= \int \frac{\cos\theta - \cos^2\theta}{\sin^2\theta}\,d\theta$$
$$= \int\left(\csc\theta\cot\theta - \cot^2\theta\right)d\theta$$
$$= \int\left(\csc\theta\cot\theta - \left(\csc^2\theta - 1\right)\right)d\theta$$
$$= -\csc\theta + \cot\theta + \theta + C$$

59. $$\int \frac{\sin\sqrt{\theta}}{\sqrt{\theta}}\,d\theta = 2\int \sin\sqrt{\theta}\left(\frac{1}{2\sqrt{\theta}}\right)d\theta$$
$$= -2\cos\sqrt{\theta} + C$$

$u = \sqrt{\theta},\ du = \dfrac{1}{2\sqrt{\theta}}\,d\theta$

60. $$\int \frac{4}{\csc\theta - \cot\theta}\,d\theta = \int \frac{4}{\left(\dfrac{1}{\sin\theta}\right) - \left(\dfrac{\cos\theta}{\sin\theta}\right)}\,d\theta$$
$$= 4\int \frac{\sin\theta}{1 - \cos\theta}\,d\theta$$
$$= 4\ln|1 - \cos\theta| + C$$

$u = 1 - \cos\theta,\ du = \sin\theta\,d\theta$

61. By Formula 21: $(a = 3, b = 1)$

$$A = \int_0^6 \frac{x}{\sqrt{x+3}}\,dx = \left[\frac{-2(6-x)}{3}\sqrt{x+3}\right]_0^6$$
$$= 4\sqrt{3} \approx 6.928 \text{ square units}$$

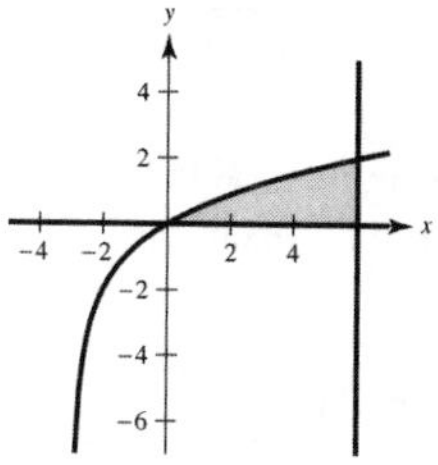

62. $$A = \int_0^2 \frac{x}{1 + e^{x^2}}\,dx$$
$$= \frac{1}{2}\int_0^2 \frac{2x\,dx}{1 + e^{x^2}}$$
$$= \frac{1}{2}\left[x^2 - \ln\left(1 + e^{x^2}\right)\right]_0^2$$
$$= \frac{1}{2}\left[4 - \ln\left(1 + e^4\right)\right] + \frac{1}{2}\ln 2$$
$$\approx 0.337 \text{ square units}$$

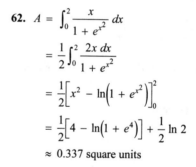

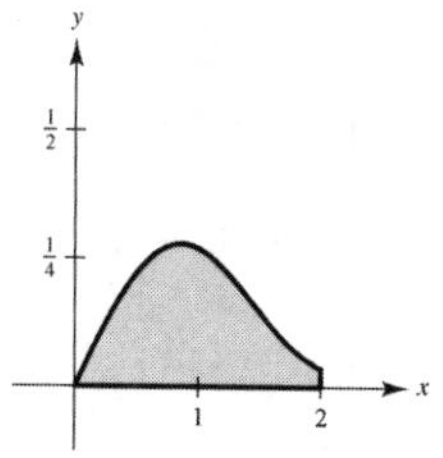

63. (a) $n = 1$: $u = \ln x,\ du = \frac{1}{x}\,dx,\ dv = x\,dx,\ v = \frac{x^2}{2}$

$$\int x \ln x\,dx = \frac{x^2}{2}\ln x - \int\left(\frac{x^2}{2}\right)\frac{1}{x}\,dx = \frac{x^2}{2}\ln x - \frac{x^2}{4} + C$$

$n = 2$: $u = \ln x,\ du = \frac{1}{x}\,dx,\ dv = x^2\,dx,\ v = \frac{x^3}{3}$

$$\int x^2 \ln x\,dx = \frac{x^3}{3}\ln x - \int\left(\frac{x^3}{3}\right)\frac{1}{x}\,dx = \frac{x^3}{3}\ln x - \frac{x^3}{9} + C$$

$n = 3$: $u = \ln x,\ du = \frac{1}{x}\,dx,\ dv = x^3\,dx,\ v = \frac{x^4}{4}$

$$\int x^3 \ln x\,dx = \frac{x^4}{4}\ln x - \int\left(\frac{x^4}{4}\right)\frac{1}{x}\,dx = \frac{x^4}{4}\ln x - \frac{x^4}{16} + C$$

(b) $\int x^n \ln x\,dx = \frac{x^{n+1}}{n+1}\ln x - \frac{x^{n+1}}{(n+1)^2} + C$

64. A reduction formula reduces an integral to the sum of a function and a simpler integral. For example, see Formulas 50, 54.

65. (a) Arctangent Formula, Formula 23,

$\int \frac{1}{u^2 + 1}\,du,\ u = e^x$

(b) Log Rule: $\int \frac{1}{u}\,du,\ u = e^x + 1$

(c) Substitution: $u = x^2,\ du = 2x\,dx$, then Formula 81

(d) Integration by parts

(e) Cannot be integrated.

(f) Formula 16 with $u = e^{2x}$

66. (a) The slope of f at $x = -1$ is approximately $0.5\ (f' > 0$ at $x = -1)$.

(b) $f' > 0$ on $(-\infty, 0)$, so f is increasing on $(-\infty, 0)$.

$f' < 0$ on $(0, \infty)$, so f is decreasing on $(0, \infty)$.

67. False. You might need to convert your integral using substitution or algebra.

68. True

69.
$$\begin{aligned} W &= \int_0^5 2000xe^{-x}\,dx \\ &= -2000\int_0^5 -xe^{-x}\,dx \\ &= 2000\int_0^5 (-x)e^{-x}(-1)\,dx \\ &= 2000\Big[(-x)e^{-x} - e^{-x}\Big]_0^5 \\ &= 2000\left(-\frac{6}{e^5} + 1\right) \\ &\approx 1919.145 \text{ ft-lb} \end{aligned}$$

70.
$$\begin{aligned} W &= \int_0^5 \frac{500x}{\sqrt{26 - x^2}}\,dx \\ &= -250\int_0^5 \left(26 - x^2\right)^{-1/2}(-2x)\,dx \\ &= \left[-500\sqrt{26 - x^2}\right]_0^5 \\ &= 500\left(\sqrt{26} - 1\right) \\ &\approx 2049.51 \text{ ft-lb} \end{aligned}$$

71.

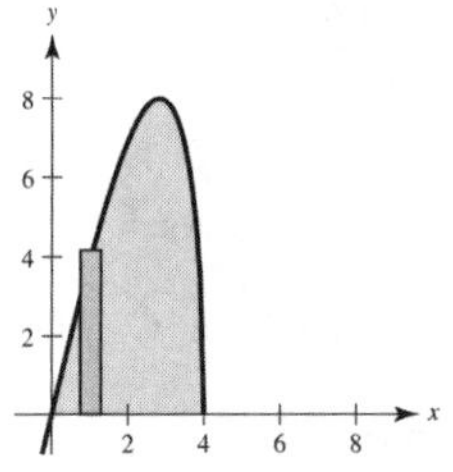

$$\begin{aligned} V &= 2\pi\int_0^4 x\left(x\sqrt{16 - x^2}\right)dx \\ &= 2\pi\int_0^4 x^2\sqrt{16 - x^2}\,dx \end{aligned}$$

By Formula 38: $(a = 4)$

$$\begin{aligned} V &= 2\pi\left[\frac{1}{8}\left(x\left(2x^2 - 16\right)\sqrt{16 - x^2} + 256\arcsin\left(\frac{x}{4}\right)\right)\right]_0^4 \\ &= 2\pi\left[32\left(\frac{\pi}{2}\right)\right] = 32\pi^2 \end{aligned}$$

72. (a) $V = 20(2)\int_0^3 \frac{2}{\sqrt{1+y^2}}\,dy$

$= \left[80 \ln\left|y + \sqrt{1+y^2}\right|\right]_0^3$

$= 80 \ln\left(3 + \sqrt{10}\right)$

$\approx 145.5 \text{ ft}^3$

$W = 148\left(80 \ln\left(3 + \sqrt{10}\right)\right)$

$= 11{,}840 \ln\left(3 + \sqrt{10}\right)$

$\approx 21{,}530.4$ lb

(b) By symmetry, $\bar{x} = 0$.

$M = \rho(2)\int_0^3 \frac{2}{\sqrt{1+y^2}}\,dy$

$= \left[4\rho \ln\left|y + \sqrt{1+y^2}\right|\right]_0^3$

$= 4\rho \ln\left(3 + \sqrt{10}\right)$

$M_x = 2\rho\int_0^3 \frac{2y}{\sqrt{1+y^2}}\,dy$

$= \left[4\rho\sqrt{1+y^2}\right]_0^3$

$= 4\rho\left(\sqrt{10} - 1\right)$

$\bar{y} = \frac{M_x}{M} = \frac{4\rho\left(\sqrt{10}-1\right)}{4\rho \ln\left(3+\sqrt{10}\right)} \approx 1.19$

Centroid: $(\bar{x}, \bar{y}) \approx (0, 1.19)$

73. $\frac{1}{2-0}\int_0^2 \frac{5000}{1+e^{4.8-1.9t}}\,dt = \frac{2500}{-1.9}\int_0^2 \frac{-1.9\,dt}{1+e^{4.8-1.9t}}$

$= -\frac{2500}{1.9}\left[(4.8 - 1.9t) - \ln\left(1 + e^{4.8-1.9t}\right)\right]_0^2$

$= -\frac{2500}{1.9}\left[\left(1 - \ln(1+e)\right) - \left(4.8 - \ln\left(1 + e^{4.8}\right)\right)\right]$

$= \frac{2500}{1.9}\left[3.8 + \ln\left(\frac{1+e}{1+e^{4.8}}\right)\right] \approx 401.4$

74. Let $I = \int_0^{\pi/2} \frac{dx}{1 + (\tan x)^{\sqrt{2}}}$.

For $x = \frac{\pi}{2} - u$, $dx = -du$, and

$$I = \int_{\pi/2}^0 \frac{-du}{1 + \left(\tan(\pi/2 - u)\right)^{\sqrt{2}}} = \int_0^{\pi/2} \frac{du}{1 + (\cot u)^{\sqrt{2}}} = \int_0^{\pi/2} \frac{(\tan u)^{\sqrt{2}}}{(\tan u)^{\sqrt{2}} + 1}\,du.$$

$$2I = \int_0^{\pi/2} \frac{dx}{1 + (\tan x)^{\sqrt{2}}} + \int_0^{\pi/2} \frac{(\tan x)^{\sqrt{2}}}{(\tan x)^{\sqrt{2}} + 1}\,dx = \int_0^{\pi/2} dx = \frac{\pi}{2}$$

So, $I = \frac{\pi}{4}$.

Section 8.7 Indeterminate Forms and L'Hôpital's Rule

1. $\lim_{x\to 0} \frac{\sin 4x}{\sin 3x} \approx 1.3333 \left(\text{exact: } \frac{4}{3}\right)$

x	−0.1	−0.01	−0.001	0.001	0.01	0.1
$f(x)$	1.3177	1.3332	1.3333	1.3333	1.3332	1.3177

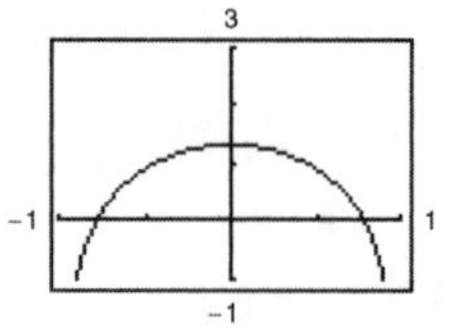

2. $\lim_{x\to 0} \frac{1 - e^x}{x} \approx -1$

x	−0.1	−0.01	−0.001	0.001	0.01	0.1
$f(x)$	−0.9516	−0.9950	−0.9995	−1.00005	−1.005	−1.0517

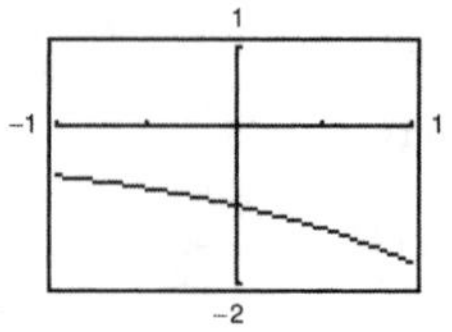

3. $\lim_{x\to\infty} x^5 e^{-x/100} \approx 0$

x	1	10	10^2	10^3	10^4	10^5
$f(x)$	0.9900	90,484	3.7×10^9	4.5×10^{10}	0	0

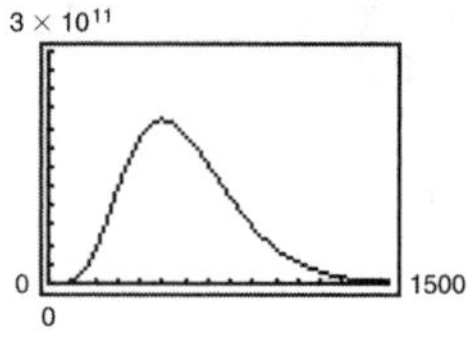

4. $\lim_{x\to\infty} \frac{6x}{\sqrt{3x^2 - 2x}} \approx 3.4641 \left(\text{exact: } \frac{6}{\sqrt{3}}\right)$

x	1	10	10^2	10^3	10^4	10^5
$f(x)$	6	3.5857	3.4757	3.4653	3.4642	3.4641

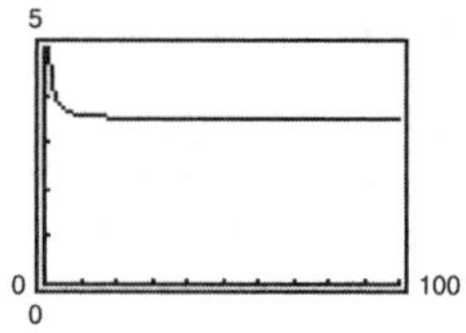

5. (a) $\lim_{x\to 4} \frac{3(x-4)}{x^2 - 16} = \lim_{x\to 4} \frac{3(x-4)}{(x-4)(x+4)} = \lim_{x\to 4} \frac{3}{x+4} = \frac{3}{8}$

(b) $\lim_{x\to 4} \frac{3(x-4)}{x^2 - 16} = \lim_{x\to 4} \frac{d/dx[3(x-4)]}{d/dx[x^2 - 16]} = \lim_{x\to 4} \frac{3}{2x} = \frac{3}{8}$

6. (a) $\lim_{x\to -4} \frac{2x^2 + 13x + 20}{x+4} = \lim_{x\to -4} \frac{(x+4)(2x+5)}{x+4} = \lim_{x\to -4}(2x+5) = -8 + 5 = -3$

(b) $\lim_{x\to -4} \frac{2x^2 + 13x + 20}{x+4} = \lim_{x\to -4} \frac{d/dx[2x^2 + 13x + 20]}{d/dx[x+4]} = \lim_{x\to -4} \frac{4x+13}{1} = -3$

7. (a) $\lim_{x\to 6} \frac{\sqrt{x+10} - 4}{x - 6} = \lim_{x\to 6} \frac{\sqrt{x+10} - 4}{x-6} \cdot \frac{\sqrt{x+10} + 4}{\sqrt{x+10} + 4} = \lim_{x\to 6} \frac{(x+10) - 16}{(x-6)(\sqrt{x+10} + 4)} = \lim_{x\to 6} \frac{1}{\sqrt{x+10} + 4} = \frac{1}{8}$

(b) $\lim_{x\to 6} \frac{\sqrt{x+10} - 4}{x - 6} = \lim_{x\to 6} \frac{d/dx[\sqrt{x+10} - 4]}{d/dx[x - 6]} = \lim_{x\to 6} \frac{\frac{1}{2}(x+10)^{-1/2}}{1} = 1/8$

8. (a) $\lim_{x\to 0} \frac{\sin 6x}{4x} = \lim_{x\to 0}\left(\frac{3}{2}\cdot\frac{\sin 6x}{6x}\right) = \frac{3}{2}(1) = \frac{3}{2}$

(b) $\lim_{x\to 0} \frac{\sin 6x}{4x} = \lim_{x\to 0} \frac{d/dx[\sin 6x]}{d/dx[4x]} = \lim_{x\to 0} \frac{6\cos 6x}{4} = \frac{3}{2}$

9. (a) $\lim_{x\to\infty} \frac{5x^2 - 3x + 1}{3x^2 - 5} = \lim_{x\to\infty} \frac{5 - (3/x) + (1/x^2)}{3 - (5/x^2)} = \frac{5}{3}$

(b) $\lim_{x\to\infty} \frac{5x^2 - 3x + 1}{3x^2 - 5} = \lim_{x\to\infty} \frac{(d/dx)[5x^2 - 3x + 1]}{(d/dx)[3x^2 - 5]} = \lim_{x\to\infty} \frac{10x - 3}{6x} = \lim_{x\to\infty} \frac{(d/dx)[10x - 3]}{(d/dx)[6x]} = \lim_{x\to\infty} \frac{10}{6} = \frac{5}{3}$

10. (a) $\lim_{x\to\infty} \frac{4x - 3}{5x^2 + 1} = \lim_{x\to\infty} \frac{4/x - 3/x^2}{5 + 1/x^2} = 0$

(b) $\lim_{x\to\infty} \frac{4x - 3}{5x^2 + 1} = \lim_{x\to\infty} \frac{(d/dx)[4x - 3]}{(d/dx)[5x^2 + 1]} = \lim_{x\to\infty} \frac{4}{10x} = 0$

11. $\lim_{x\to 3} \frac{x^2 - 2x - 3}{x - 3} = \lim_{x\to 3} \frac{2x - 2}{1} = 4$

12. $\lim_{x\to -2} \frac{x^2 - 3x - 10}{x + 2} = \lim_{x\to -2} \frac{2x - 3}{1} = -7$

13. $\lim_{x\to 0} \frac{\sqrt{25 - x^2} - 5}{x} = \lim_{x\to 0} \frac{\frac{1}{2}(25 - x^2)^{-1/2}(-2x)}{1}$

$= \lim_{x\to 0} \frac{-x}{\sqrt{25 - x^2}} = 0$

14. $\lim_{x\to 5^-} \frac{\sqrt{25 - x^2}}{x - 5} = \lim_{x\to 5^-} \frac{\frac{1}{2}(25 - x^2)^{-1/2}(-2x)}{1}$

$= \lim_{x\to 5^-} \frac{-x}{\sqrt{25 - x^2}} = -\infty$

15. $\lim_{x\to 0^+} \frac{e^x - (1 + x)}{x^3} = \lim_{x\to 0^+} \frac{e^x - 1}{3x^2} = \lim_{x\to 0^+} \frac{e^x}{6x} = \infty$

16. $\lim_{x\to 1} \frac{\ln x^3}{x^2 - 1} = \lim_{x\to 1} \frac{3\ln x}{x^2 - 1} = \lim_{x\to 1} \frac{3/x}{2x} = \frac{3}{2}$

17. $\lim_{x\to 1} \frac{x^{11} - 1}{x^4 - 1} = \lim_{x\to 1} \frac{11x^{10}}{4x^3} = \frac{11}{4}$

18. $\lim_{x\to 1} \frac{x^a - 1}{x^b - 1} = \lim_{x\to 1} \frac{ax^{a-1}}{bx^{b-1}} = \frac{a}{b}$

19. $\lim_{x\to 0} \frac{\sin 3x}{\sin 5x} = \lim_{x\to 0} \frac{3\cos 3x}{5\cos 5x} = \frac{3}{5}$

20. $\lim_{x\to 0} \frac{\sin ax}{\sin bx} = \lim_{x\to 0} \frac{a\cos ax}{b\cos bx} = \frac{a}{b}$

21. $\lim_{x\to 0} \frac{\arcsin x}{x} = \lim_{x\to 0} \frac{1/\sqrt{1 - x^2}}{1} = 1$

22. $\lim_{x\to 1} \frac{\arctan x - (\pi/4)}{x - 1} = \lim_{x\to 1} \frac{1/(1 + x^2)}{1} = \frac{1}{2}$

23. $\lim_{x\to\infty} \frac{5x^2 + 3x - 1}{4x^2 + 5} = \lim_{x\to\infty} \frac{10x + 3}{8x} = \lim_{x\to\infty} \frac{10}{8} = \frac{5}{4}$

24. $\lim_{x\to\infty} \frac{5x + 3}{x^3 - 6x + 2} = \lim_{x\to\infty} \frac{5}{3x^2 - 6} = 0$

25. $\lim_{x\to\infty} \frac{x^2 + 4x + 7}{x - 6} = \lim_{x\to\infty} \frac{2x + 4}{1} = \infty$

26. $\lim_{x\to\infty} \frac{x^3}{x + 1} = \lim_{x\to\infty} \frac{3x^2}{1} = \infty$

27. $\lim_{x\to\infty} \frac{x^3}{e^{x/2}} = \lim_{x\to\infty} \frac{3x^2}{(1/2)e^{x/2}}$

$= \lim_{x\to\infty} \frac{6x}{(1/4)e^{x/2}} = \lim_{x\to\infty} \frac{6}{(1/8)e^{x/2}} = 0$

28. $\lim_{x\to\infty} \frac{x^3}{e^{x^2}} = \lim_{x\to\infty} \frac{3x^2}{2xe^{x^2}}$

$= \lim_{x\to\infty} \frac{6x}{(4x^2 + 2)e^{x^2}}$

$= \lim_{x\to\infty} \frac{6}{4x(2x^2 + 3)e^{x^2}} = 0$

29. $\lim_{x\to\infty} \frac{x}{\sqrt{x^2 + 1}} = \lim_{x\to\infty} \frac{1}{\sqrt{1 + (1/x^2)}} = 1$

Note: L'Hôpital's Rule does not work on this limit. See Exercise 83.

30. $\lim_{x\to\infty} \frac{x^2}{\sqrt{x^2+1}} = \lim_{x\to\infty} \frac{x}{\sqrt{1+(1/x)^2}} = \infty$

31. $\lim_{x\to\infty} \frac{\cos x}{x} = 0$ by Squeeze Theorem

$\left(\frac{\cos x}{x} \le \frac{1}{x}, \text{ for } x > 0\right)$

32. $\lim_{x\to\infty} \frac{\sin x}{x-\pi} = 0$

Note: Use the Squeeze Theorem for $x > \pi$.

$$-\frac{1}{x-\pi} \le \frac{\sin x}{x-\pi} \le \frac{1}{x-\pi}$$

33. $\lim_{x\to\infty} \frac{\ln x}{x^2} = \lim_{x\to\infty} \frac{1/x}{2x} = \lim_{x\to\infty} \frac{1}{2x^2} = 0$

34. $\lim_{x\to\infty} \frac{\ln x^4}{x^3} = \lim_{x\to\infty} \frac{4\ln x}{x^3} = \lim_{x\to\infty} \frac{4/x}{3x^2} = \lim_{x\to\infty} \frac{4}{3x^3} = 0$

35. $\lim_{x\to\infty} \frac{e^x}{x^4} = \lim_{x\to\infty} \frac{e^x}{4x^3}$

$= \lim_{x\to\infty} \frac{e^x}{12x^2}$

$= \lim_{x\to\infty} \frac{e^x}{24x}$

$= \lim_{x\to\infty} \frac{e^x}{24} = \infty$

36. $\lim_{x\to\infty} \frac{e^{x/2}}{x} = \lim_{x\to\infty} \frac{(1/2)e^{x/2}}{1} = \infty$

37. $\lim_{x\to 0} \frac{\sin 5x}{\tan 9x} = \lim_{x\to 0} \frac{5\cos 5x}{9\sec^2 9x} = \frac{5}{9}$

38. $\lim_{x\to 1} \frac{\ln x}{\sin \pi x} = \lim_{x\to 1} \frac{1/x}{\pi \cos \pi x} = -\frac{1}{\pi}$

39. $\lim_{x\to 0} \frac{\arctan x}{\sin x} = \lim_{x\to 0} \frac{1/(1+x^2)}{\cos x} = 1$

40. $\lim_{x\to 0} \frac{x}{\arctan 2x} = \lim_{x\to 0} \frac{1}{2/(1+4x^2)} = 1/2$

41. $\lim_{x\to\infty} \frac{\int_1^x \ln\left(e^{4t-1}\right) dt}{x}$

$= \lim_{x\to\infty} \frac{\int_1^x (4t-1)\, dt}{x}$

$= \lim_{x\to\infty} \frac{4x-1}{1} = \infty$

42. $\lim_{x\to 1^+} \frac{\int_1^x \cos\theta\, d\theta}{x-1} = \lim_{x\to 1^+} \frac{\cos x}{1} = \cos(1)$

43. (a) $\lim_{x\to\infty} x \ln x$, not indeterminate

(b) $\lim_{x\to\infty} x \ln x = (\infty)(\infty) = \infty$

(c)

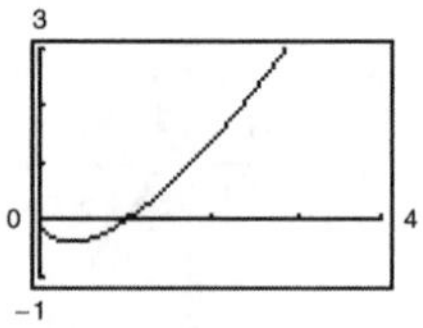

44. (a) $\lim_{x\to 0^+} x^3 \cot x = (0)(\infty)$

(b) $\lim_{x\to 0^+} x^3 \cot x = \lim_{x\to 0^+} \frac{x^3}{\tan x} = \lim_{x\to 0^+} \frac{3x^2}{\sec^2 x} = 0$

(c)

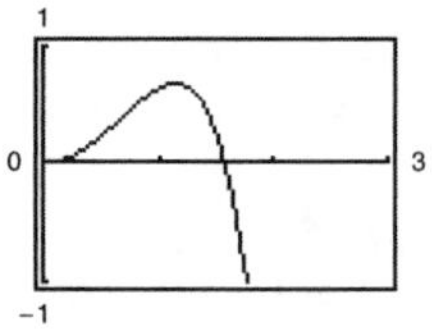

45. (a) $\lim_{x\to\infty} \left(x \sin \frac{1}{x}\right) = (\infty)(0)$

(b) $\lim_{x\to\infty} x \sin \frac{1}{x} = \lim_{x\to\infty} \frac{\sin(1/x)}{1/x}$

$= \lim_{x\to\infty} \frac{\left(-1/x^2\right)\cos(1/x)}{-1/x^2}$

$= \lim_{x\to\infty} \cos\left(\frac{1}{x}\right) = 1$

(c)

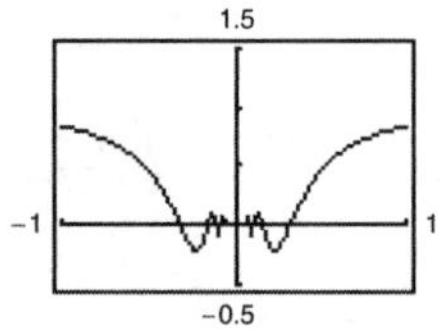

46. (a) $\lim_{x\to\infty}\left(x \tan \frac{1}{x}\right) = (\infty)(0)$

(b) $\lim_{x\to\infty} x \tan \frac{1}{x} = \lim_{x\to\infty} \frac{\tan(1/x)}{1/x}$

$= \lim_{x\to\infty} \frac{-(1/x^2)\sec^2(1/x)}{-(1/x^2)}$

$= \lim_{x\to\infty} \sec^2\left(\frac{1}{x}\right) = 1$

(c) 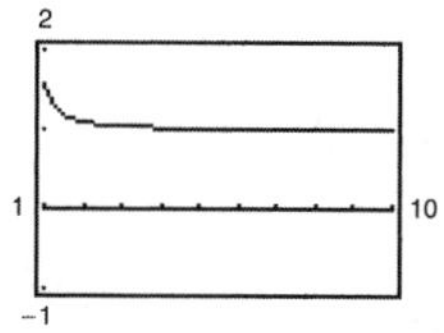

47. (a) $\lim_{x\to 0^+} x^{1/x} = 0^\infty = 0$, not indeterminate

(See Exercise 108).

(b) Let $y = x^{1/x}$

$\ln y = \ln x^{1/x} = \frac{1}{x}\ln x.$

Because $x \to 0^+, \frac{1}{x}\ln x \to (\infty)(-\infty) = -\infty$. So,

$\ln y \to -\infty \Rightarrow y \to 0^+.$

Therefore, $\lim_{x\to 0^+} x^{1/x} = 0.$

(c) 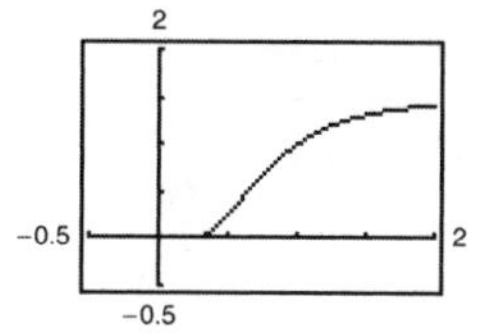

48. (a) $\lim_{x\to 0^+} (e^x + x)^{2/x} = 1^\infty$

(b) Let $y = \lim_{x\to 0^+} (e^x + x)^{2/x}$.

$\ln y = \lim_{x\to 0^+} \frac{2\ln(e^x + x)}{x}$

$= \lim_{x\to 0^+} \frac{2(e^x + 1)/(e^x + x)}{1} = 4$

So, $\ln y = 4 \Rightarrow y = e^4 \approx 54.598$. Therefore,

$\lim_{x\to 0^+} (e^x + x)^{2/x} = e^4.$

(c) 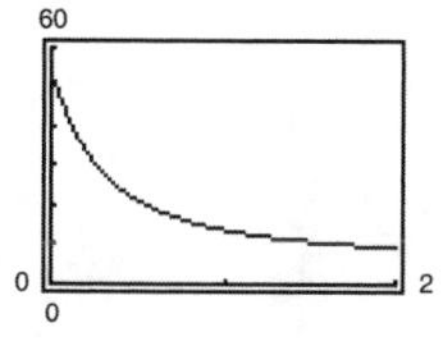

49. (a) $\lim_{x\to\infty} x^{1/x} = \infty^0$

(b) Let $y = \lim_{x\to\infty} x^{1/x}$.

$\ln y = \lim_{x\to\infty} \frac{\ln x}{x} = \lim_{x\to\infty}\left(\frac{1/x}{1}\right) = 0$

So, $\ln y = 0 \Rightarrow y = e^0 = 1$. Therefore,

$\lim_{x\to\infty} x^{1/x} = 1.$

(c) 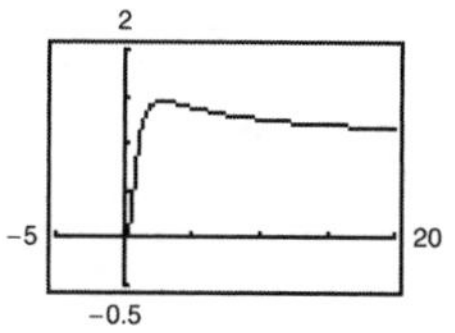

50. (a) $\lim_{x\to\infty}\left(1 + \frac{1}{x}\right)^x = 1^\infty$

(b) Let $y = \lim_{x\to\infty}\left(1 + \frac{1}{x}\right)^x$.

$\ln y = \lim_{x\to\infty}\left[x \ln\left(1 + \frac{1}{x}\right)\right] = \lim_{x\to\infty} \frac{\ln[1 + (1/x)]}{1/x}$

$= \lim_{x\to\infty} \frac{\left[\frac{(-1/x^2)}{1 + (1/x)}\right]}{(-1/x^2)} = \lim_{x\to\infty} \frac{1}{1 + (1/x)} = 1$

So, $\ln y = 1 \Rightarrow y = e^1 = e$. Therefore,

$\lim_{x\to\infty}\left(1 + \frac{1}{x}\right)^x = e.$

(c) 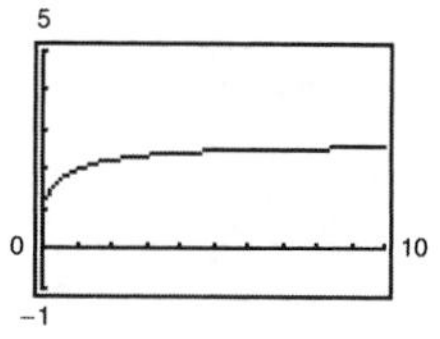

51. (a) $\lim_{x\to 0^+} (1 + x)^{1/x} = 1^\infty$

(b) Let $y = \lim_{x\to 0^+} (1 + x)^{1/x}$.

$$\ln y = \lim_{x\to 0^+} \frac{\ln(1 + x)}{x}$$

$$= \lim_{x\to 0^+} \left(\frac{1/(1 + x)}{1}\right) = 1$$

So, $\ln y = 1 \Rightarrow y = e^1 = e$.

Therefore, $\lim_{x\to 0^+} (1 + x)^{1/x} = e$.

(c)

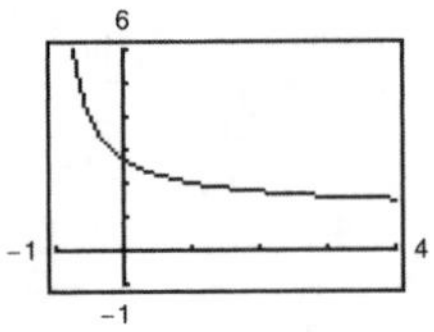

52. (a) $\lim_{x\to\infty} (1 + x)^{1/x} = \infty^0$

(b) Let $y = \lim_{x\to\infty} (1 + x)^{1/x}$.

$$\ln y = \lim_{x\to\infty} \frac{\ln(1 + x)}{x} = \lim_{x\to\infty} \left(\frac{1/(1 + x)}{1}\right) = 0$$

So, $\ln y = 0 \Rightarrow y = e^0 = 1$.

Therefore, $\lim_{x\to\infty} (1 + x)^{1/x} = 1$.

(c)

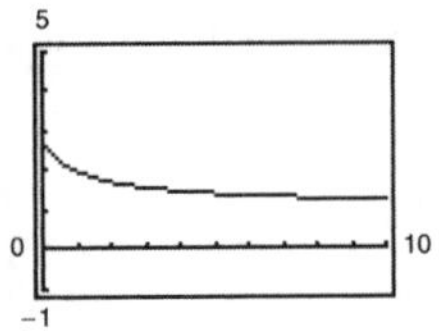

53. (a) $\lim_{x\to 0^+} \left[3(x)^{x/2}\right] = 0^0$

(b) Let $y = \lim_{x\to 0^+} 3(x)^{x/2}$.

$$\ln y = \lim_{x\to 0^+} \left[\ln 3 + \frac{x}{2} \ln x\right]$$

$$= \lim_{x\to 0^+} \left[\ln 3 + \frac{\ln x}{2/x}\right]$$

$$= \lim_{x\to 0^+} \ln 3 + \lim_{x\to 0^+} \frac{1/x}{-2/x^2}$$

$$= \lim_{x\to 0^+} \ln 3 - \lim_{x\to 0^+} \frac{x}{2}$$

$$= \ln 3$$

So, $\lim_{x\to 0^+} 3(x)^{x/2} = 3$.

(c)

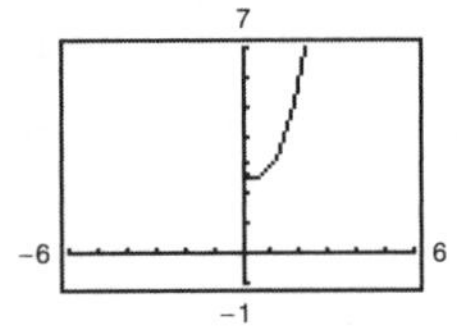

54. (a) $\lim_{x\to 4^+} \left[3(x - 4)\right]^{x-4} = 0^0$

(b) Let $y = \lim_{x\to 4^+} \left[3(x - 4)\right]^{x-4}$.

$$\ln y = \lim_{x\to 4^+} (x - 4) \ln\left[3(x - 4)\right]$$

$$= \lim_{x\to 4^+} \frac{\ln\left[3(x - 4)\right]}{1/(x - 4)}$$

$$= \lim_{x\to 4^+} \frac{1/(x - 4)}{-1/(x - 4)^2}$$

$$= \lim_{x\to 4^+} \left[-(x - 4)\right] = 0$$

So, $\lim_{x\to 4^+} \left[3(x - 4)\right]^{x-4} = 1$.

(c)

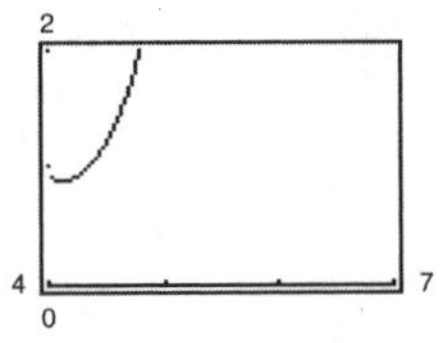

55. (a) $\lim_{x\to 1^+} (\ln x)^{x-1} = 0^0$

(b) Let $y = (\ln x)^{x-1}$.

$$\ln y = \ln\left[(\ln x)^{x-1}\right] = (x-1)\ln(\ln x) = \frac{\ln(\ln x)}{(x-1)^{-1}}$$

$$\lim_{x\to 1^+} \ln y = \lim_{x\to 1^+} \frac{\ln(\ln x)}{(x-1)^{-1}} = \lim_{x\to 1^+} \frac{1/(x \ln x)}{-(x-1)^{-2}} = \lim_{x\to 1^+} \frac{-(x-1)^2}{x \ln x} = \lim_{x\to 1^+} \frac{-2(x-1)}{1+\ln x} = 0$$

Because $\lim_{x\to 1^+} \ln y = 0$, $\lim_{x\to 1^+} y = 1$.

(c)

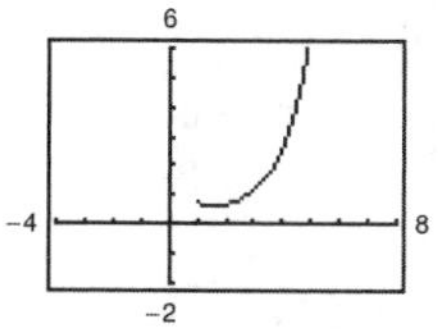

56. $\cos\left(\frac{\pi}{2} - x\right) = \sin x$

(a) $\lim_{x\to 0^+} \left[\cos\left(\frac{\pi}{2} - x\right)\right]^x = \lim_{x\to 0^+} [\sin x]^x = 0^0$

(b) Let $y = (\sin x)^x$

$$\ln y = x \ln(\sin x) = \frac{\ln(\sin x)}{1/x}$$

$$\lim_{x\to 0^+} \frac{\ln(\sin x)}{1/x} = \lim_{x\to 0^+} \frac{\cos x/\sin x}{-1/x^2} = \lim_{x\to 0^+} \frac{-x^2 \cos x}{\sin x} = \lim_{x\to 0^+} \frac{x}{\sin x}\left(\frac{-x \cos x}{1}\right) = 0$$

So, $\lim_{x\to 0^+} \left[\cos\left(\frac{\pi}{2} - x\right)\right]^x = 1$.

(c)

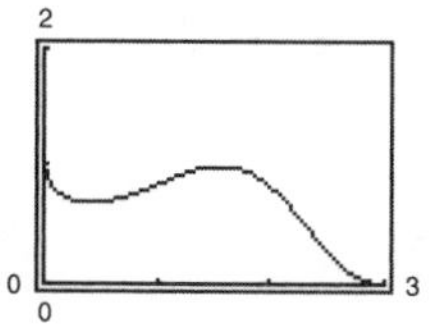

57. (a) $\lim_{x\to 2^+} \left(\frac{8}{x^2 - 4} - \frac{x}{x - 2}\right) = \infty - \infty$

(b) $$\lim_{x\to 2^+} \left(\frac{8}{x^2 - 4} - \frac{x}{x - 2}\right) = \lim_{x\to 2^+} \frac{8 - x(x+2)}{x^2 - 4} = \lim_{x\to 2^+} \frac{(2-x)(4+x)}{(x+2)(x-2)} = \lim_{x\to 2^+} \frac{-(x+4)}{x+2} = \frac{-3}{2}$$

(c)

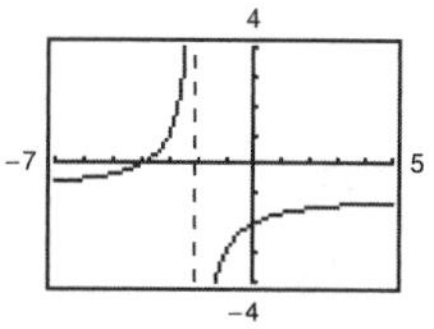

58. (a) $\lim_{x\to 2^+} \left(\frac{1}{x^2 - 4} - \frac{\sqrt{x-1}}{x^2 - 4}\right) = \infty - \infty$

(b) $$\lim_{x\to 2^+} \left(\frac{1}{x^2 - 4} - \frac{\sqrt{x-1}}{x^2 - 4}\right) = \lim_{x\to 2^+} \frac{1 - \sqrt{x-1}}{x^2 - 4} = \lim_{x\to 2^+} \frac{-1/\left(2\sqrt{x-1}\right)}{2x} = \lim_{x\to 2^+} \frac{-1}{4x\sqrt{x-1}} = \frac{-1}{8}$$

(c)

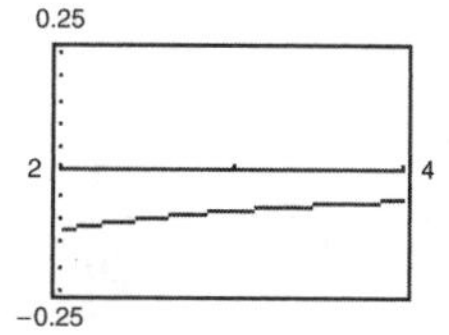

59. (a) $\lim_{x\to 1^+} \left(\frac{3}{\ln x} - \frac{2}{x - 1}\right) = \infty - \infty$

(b) $$\lim_{x\to 1^+} \left(\frac{3}{\ln x} - \frac{2}{x - 1}\right) = \lim_{x\to 1^+} \frac{3x - 3 - 2\ln x}{(x-1)\ln x} = \lim_{x\to 1^+} \frac{3 - (2/x)}{\left[(x-1)/x\right] + \ln x} = \infty$$

(c)

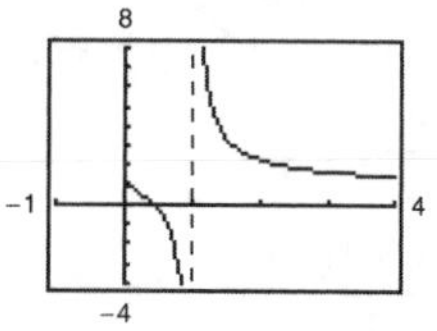

60. (a) $\lim_{x\to 0^+}\left(\frac{10}{x}-\frac{3}{x^2}\right) = \infty - \infty$

(b) $\lim_{x\to 0^+}\left(\frac{10}{x}-\frac{3}{x^2}\right) = \lim_{x\to 0^+}\left(\frac{10x-3}{x^2}\right) = -\infty$

(c)

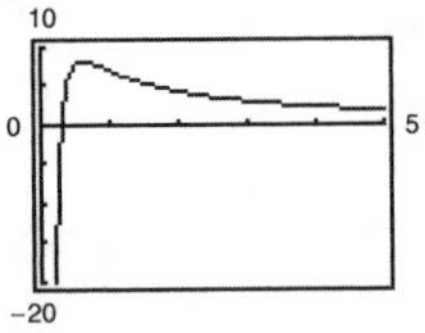

61. $\frac{0}{0}, \frac{\infty}{\infty}, 0\cdot\infty, 1^\infty, 0^0, \infty-\infty, \infty^0$

62. See Theorem 8.4.

63. (a) Let $f(x) = x^2 - 25$ and $g(x) = x - 5$.

(b) Let $f(x) = (x-5)^2$ and $g(x) = x^2 - 25$.

(c) Let $f(x) = x^2 - 25$ and $g(x) = (x-5)^3$.

(Answers will vary.)

64. Let $f(x) = x + 25$ and $g(x) = x$.

(Answers will vary.)

65. (a) Yes: $\frac{0}{0}$

(b) No: $\frac{0}{-1}$

(c) Yes: $\frac{\infty}{\infty}$

(d) Yes: $\frac{0}{0}$

(e) No: $\frac{-1}{0}$

(f) Yes: $\frac{0}{0}$

66. (a) From the graph, $\lim_{x\to 1^-} f(x) = \infty$.

(b) From the graph, $\lim_{x\to 1^+} f(x) = -\infty$.

(c) From the graph, $\lim_{x\to 1} f(x)$ does not exist.

67.

x	10	10^2	10^4	10^6	10^8	10^{10}
$\frac{(\ln x)^4}{x}$	2.811	4.498	0.720	0.036	0.001	0.000

68.

x	1	5	10	20	30	40	50	100
$\frac{e^x}{x^5}$	2.718	0.047	0.220	151.614	4.40×10^5	2.30×10^9	1.66×10^{13}	2.69×10^{33}

69. $\lim_{x\to\infty}\frac{x^2}{e^{5x}} = \lim_{x\to\infty}\frac{2x}{5e^{5x}} = \lim_{x\to\infty}\frac{2}{25e^{5x}} = 0$

70. $\lim_{x\to\infty}\frac{x^3}{e^{2x}} = \lim_{x\to\infty}\frac{3x^2}{2e^{2x}} = \lim_{x\to\infty}\frac{6x}{4e^{2x}} = \lim_{x\to\infty}\frac{6}{8e^{2x}} = 0$

71. $\lim_{x\to\infty}\frac{(\ln x)^3}{x} = \lim_{x\to\infty}\frac{3(\ln x)^2(1/x)}{1}$

$= \lim_{x\to\infty}\frac{3(\ln x)^2}{x}$

$= \lim_{x\to\infty}\frac{6(\ln x)(1/x)}{1}$

$= \lim_{x\to\infty}\frac{6(\ln x)}{x} = \lim_{x\to\infty}\frac{6}{x} = 0$

72. $\lim_{x\to\infty}\frac{(\ln x)^2}{x^3} = \lim_{x\to\infty}\frac{(2\ln x)/x}{3x^2}$

$= \lim_{x\to\infty}\frac{2\ln x}{3x^3}$

$= \lim_{x\to\infty}\frac{2/x}{9x^2} = \lim_{x\to\infty}\frac{2}{9x^3} = 0$

73. $\lim_{x\to\infty} \frac{(\ln x)^n}{x^m} = \lim_{x\to\infty} \frac{n(\ln x)^{n-1}/x}{mx^{m-1}}$

$= \lim_{x\to\infty} \frac{n(\ln x)^{n-1}}{mx^m}$

$= \lim_{x\to\infty} \frac{n(n-1)(\ln x)^{n-2}}{m^2x^m}$

$= \cdots = \lim_{x\to\infty} \frac{n!}{m^n x^m} = 0$

74. $\lim_{x\to\infty} \frac{x^m}{e^{nx}} = \lim_{x\to\infty} \frac{mx^{m-1}}{ne^{nx}}$

$= \lim_{x\to\infty} \frac{m(m-1)x^{m-2}}{n^2e^{nx}}$

$= \cdots = \lim_{x\to\infty} \frac{m!}{n^m e^{nx}} = 0$

75. $y = x^{1/x}, x > 0$

Horizontal asymptote: $y = 1$ (See Exercise 49.)

$\ln y = \frac{1}{x}\ln x$

$\left(\frac{1}{y}\right)\frac{dy}{dx} = \frac{1}{x}\left(\frac{1}{x}\right) + (\ln x)\left(-\frac{1}{x^2}\right)$

$\frac{dy}{dx} = x^{1/x}\left(\frac{1}{x^2}\right)(1 - \ln x) = x^{(1/x)-2}(1 - \ln x) = 0$

Critical number:	$x = e$	
Intervals:	$(0, e)$	(e, ∞)
Sign of dy/dx:	$+$	$-$
$y = f(x)$:	Increasing	Decreasing

Relative maximum: $(e, e^{1/e})$

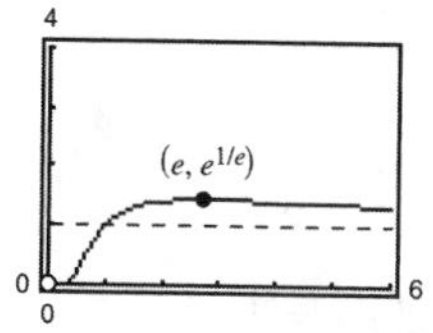

76. $y = x^x, x > 0$

$\lim_{x\to\infty} x^x = \infty$ and $\lim_{x\to 0^+} x^x = 1$

No horizontal asymptotes

$\ln y = x \ln x$

$\left(\frac{1}{y}\right)\frac{dy}{dx} = x\left(\frac{1}{x}\right) + \ln x$

$\frac{dy}{dx} = x^x(1 + \ln x) = 0$

Critical number:	$x = e^{-1}$	
Intervals:	$(0, e^{-1})$	$(e^{-1}, 0)$
Sign of dy/dx:	$-$	$+$
$y = f(x)$:	Decreasing	Increasing

Relative maximum: $\left(e^{-1}, (e^{-1})^{e^{-1}}\right) = \left(\frac{1}{e}, \left(\frac{1}{e}\right)^{1/e}\right)$

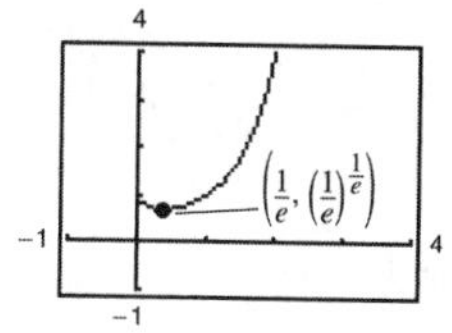

77. $y = 2xe^{-x}$

$\lim_{x\to\infty} \frac{2x}{e^x} = \lim_{x\to\infty} \frac{2}{e^x} = 0$

Horizontal asymptote: $y = 0$

$\frac{dy}{dx} = 2x(-e^{-x}) + 2e^{-x}$

$= 2e^{-x}(1 - x) = 0$

Critical number:	$x = 1$	
Intervals:	$(-\infty, 1)$	$(1, \infty)$
Sign of dy/dx:	$+$	$-$
$y = f(x)$:	Increasing	Decreasing

Relative maximum: $\left(1, \frac{2}{e}\right)$

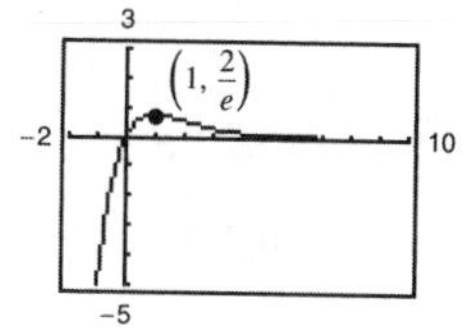

78. $y = \dfrac{\ln x}{x}$

Horizontal asymptote: $y = 0$ (See Example 2.)

$$\frac{dy}{dx} = \frac{x(1/x) - (\ln x)(1)}{x^2} = \frac{1 - \ln x}{x^2} = 0$$

Critical number: $x = e$

Intervals:	$(0, e)$	(e, ∞)
Sign of dy/dx:	$+$	$-$
$y = f(x)$:	Increasing	Decreasing

Relative maximum: $\left(e, \dfrac{1}{e}\right)$

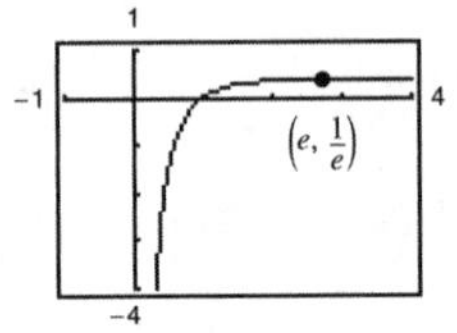

79. $\displaystyle\lim_{x\to 2} \frac{3x^2 + 4x + 1}{x^2 - x - 2} = \frac{21}{0}$

Limit is not of the form $\dfrac{0}{0}$ or $\dfrac{\infty}{\infty}$.

L'Hôpital's Rule does not apply.

80. $\displaystyle\lim_{x\to 0} \frac{e^{2x} - 1}{e^x} = \frac{0}{1} = 0$

Limit is not of the form $0/0$ or ∞/∞.

L'Hôpital's Rule does not apply.

81. $\displaystyle\lim_{x\to\infty} \frac{e^{-x}}{1 + e^{-x}} = \frac{0}{1 + 0} = 0$

Limit is not of the form $0/0$ or ∞/∞.

L'Hôpital's Rule does not apply.

82. $\displaystyle\lim_{x\to\infty} x \cos\frac{1}{x} = \infty(1) = \infty$

Limit is not of the form $0/0$ or ∞/∞.

L'Hôpital's Rule does not apply.

83. (a) Applying L'Hôpital's Rule twice results in the original limit, so L'Hôpital's Rule fails:

$$\lim_{x\to\infty} \frac{x}{\sqrt{x^2 + 1}} = \lim_{x\to\infty} \frac{1}{x/\sqrt{x^2 + 1}}$$
$$= \lim_{x\to\infty} \frac{\sqrt{x^2 + 1}}{x}$$
$$= \lim_{x\to\infty} \frac{x/\sqrt{x^2 + 1}}{1}$$
$$= \lim_{x\to\infty} \frac{x}{\sqrt{x^2 + 1}}$$

(b) $\displaystyle\lim_{x\to\infty} \frac{x}{\sqrt{x^2 + 1}} = \lim_{x\to\infty} \frac{x/x}{\sqrt{x^2 + 1}/x}$

$$= \lim_{x\to\infty} \frac{1}{\sqrt{1 + 1/x^2}} = \frac{1}{\sqrt{1 + 0}} = 1$$

(c)

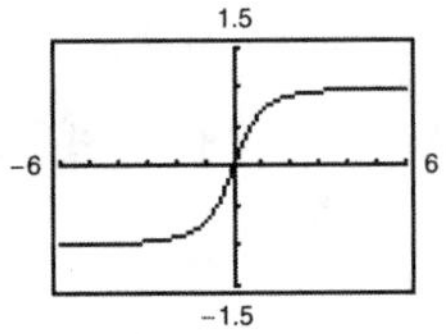

84. (a) Applying L'Hôpital's Rule twice results in the original limit, so L'Hôpital's Rule fails:

$\displaystyle\lim_{x\to\pi/2^-} \frac{\tan x}{\sec x}$ is indeterminant: $\dfrac{\infty}{\infty}$

$$\lim_{x\to\pi/2^-} \frac{\tan x}{\sec x} = \lim_{x\to\pi/2^-} \frac{\sec^2 x}{\sec x \tan x}$$
$$= \lim_{x\to\pi/2^-} \frac{\sec x}{\tan x} \quad \left(\frac{\infty}{\infty}\right)$$
$$= \lim_{x\to\pi/2^-} \frac{\sec x \tan x}{\sec^2 x}$$
$$= \lim_{x\to\pi/2^-} \frac{\tan x}{\sec x}$$

(b) $\displaystyle\lim_{x\to\pi/2^-} \frac{\tan x}{\sec x} = \lim_{x\to\pi/2^-} \frac{\sin x}{\cos x}(\cos x)$

$$= \lim_{x\to\pi/2^-} \sin x = 1$$

(c)

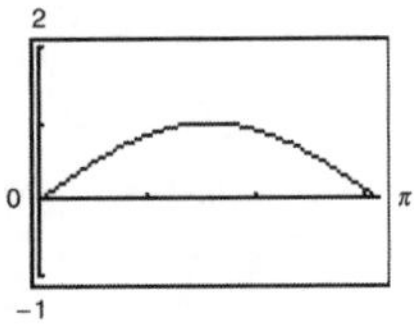

85. $f(x) = \sin(3x),\ g(x) = \sin(4x)$

$f'(x) = 3\cos(3x),\ g'(x) = 4\cos(4x)$

$$y_1 = \frac{f(x)}{g(x)} = \frac{\sin 3x}{\sin 4x},$$

$$y_2 = \frac{f'(x)}{g'(x)} = \frac{3\cos 3x}{4\cos 4x}$$

As $x \to 0$, $y_1 \to 0.75$ and $y_2 \to 0.75$

By L'Hôpital's Rule,

$$\lim_{x\to 0}\frac{\sin 3x}{\sin 4x} = \lim_{x\to 0}\frac{3\cos 3x}{4\cos 4x} = \frac{3}{4}$$

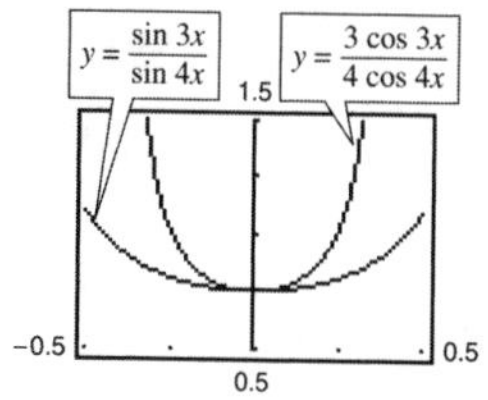

86. $f(x) = e^{3x} - 1,\ g(x) = x$

$f'(x) = 3e^{3x},\ g'(x) = 1$

$$y_1 = \frac{f(x)}{g(x)} = \frac{e^{3x} - 1}{x},$$

$$y_2 = \frac{f'(x)}{g'(x)} = 3e^{3x}$$

As $x \to 0$, $y_1 \to 3$ and $y_2 \to 3$

By L'Hôpital's Rule,

$$\lim_{x\to 0}\frac{e^{3x} - 1}{x} = \lim_{x\to 0}\frac{3e^{3x}}{1} = 3$$

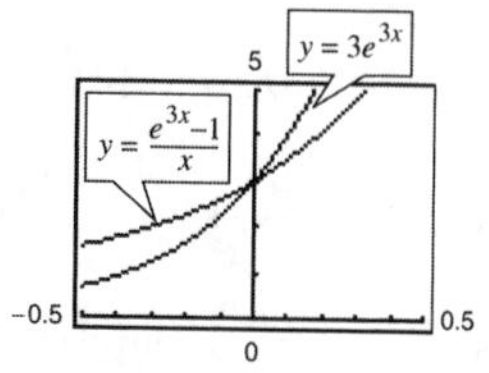

87. $$\lim_{k\to 0}\frac{32\left(1 - e^{-kt} + \dfrac{v_0ke^{-kt}}{32}\right)}{k} = \lim_{k\to 0}\frac{32(1 - e^{-kt})}{k} + \lim_{k\to 0}\left(v_0e^{-kt}\right) = \lim_{k\to 0}\frac{32(0 + te^{-kt})}{1} + \lim_{k\to 0}\left(\frac{v_0}{e^{kt}}\right) = 32t + v_0$$

88. $$A = P\left(1 + \frac{r}{n}\right)^{nt}$$

$$\ln A = \ln P + nt\ln\left(1 + \frac{r}{n}\right) = \ln P + \frac{\ln\left(1 + \dfrac{r}{n}\right)}{\dfrac{1}{nt}}$$

$$\lim_{n\to\infty}\left[\frac{\ln\left(1 + \dfrac{r}{n}\right)}{\dfrac{1}{nt}}\right] = \lim_{n\to\infty}\left[\frac{-\dfrac{r}{n^2}\left(\dfrac{1}{1 + (r/n)}\right)}{-\left(\dfrac{1}{n^2t}\right)}\right] = \lim_{n\to\infty}\left[rt\left(\frac{1}{1 + \dfrac{r}{n}}\right)\right] = rt$$

Because $\lim_{n\to\infty}\ln A = \ln P + rt$, you have $\lim_{n\to\infty} A = e^{(\ln P + rt)} = e^{\ln P}e^{rt} = Pe^{rt}$. Alternatively,

$$\lim_{n\to\infty} A = \lim_{n\to\infty} P\left(1 + \frac{r}{n}\right)^{nt} = \lim_{n\to\infty} P\left[\left(1 + \frac{r}{n}\right)^{n/r}\right]^{rt} = Pe^{rt}.$$

89. Let N be a fixed value for n. Then

$$\lim_{x\to\infty}\frac{x^{N-1}}{e^x} = \lim_{x\to\infty}\frac{(N - 1)x^{N-2}}{e^x} = \lim_{x\to\infty}\frac{(N - 1)(N - 2)x^{N-3}}{e^x} = \cdots = \lim_{x\to\infty}\left[\frac{(N - 1)!}{e^x}\right] = 0.$$ (See Exercise 74.)

90. (a) $m = \dfrac{dy}{dx} = \dfrac{y - \left(y + \sqrt{144 - x^2}\right)}{x - 0} = -\dfrac{\sqrt{144 - x^2}}{x}$

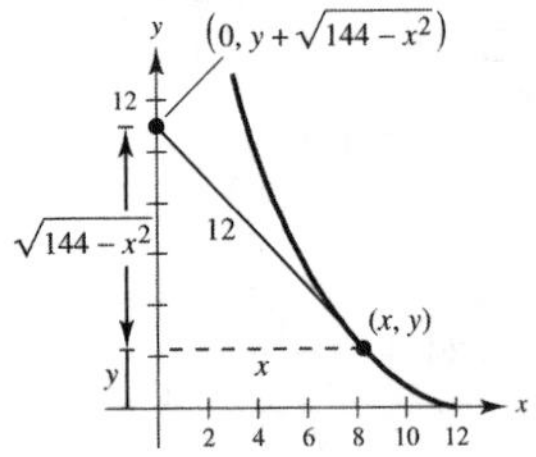

(b) $y = -\displaystyle\int \frac{\sqrt{144 - x^2}}{x}\,dx$

Let $x = 12 \sin \theta$, $dx = 12 \cos \theta\, d\theta$, $\sqrt{144 - x^2} = 12 \cos \theta$.

$$y = -\int \frac{12 \cos \theta}{12 \sin \theta} 12 \cos \theta\, d\theta = -12 \int \frac{1 - \sin^2 \theta}{\sin \theta}\, d\theta$$

$$= -12 \int (\csc \theta - \sin \theta)\, d\theta = -12 \ln\left|\csc \theta - \cot \theta\right| - 12 \cos \theta + C$$

$$= -12 \ln\left|\frac{12}{x} - \frac{\sqrt{144 - x^2}}{x}\right| - 12\left(\frac{\sqrt{144 - x^2}}{12}\right) + C = -12 \ln\left|\frac{12 - \sqrt{144 - x^2}}{x}\right| - \sqrt{144 - x^2} + C$$

When $x = 12$, $y = 0 \Rightarrow C = 0$. So, $y = -12 \ln\left(\dfrac{12 - \sqrt{144 - x^2}}{x}\right) - \sqrt{144 - x^2}$.

Note: $\dfrac{12 - \sqrt{144 - x^2}}{x} > 0$ for $0 < x \le 12$

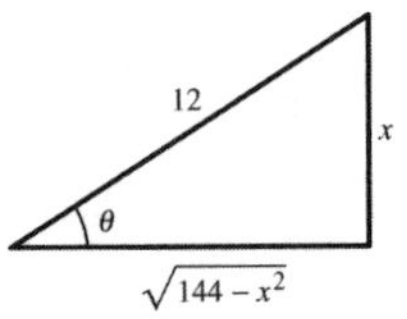

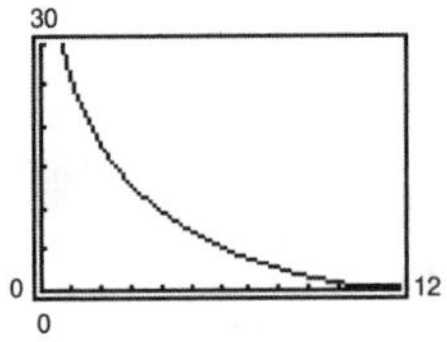

(c) Vertical asymptote: $x = 0$

(d) $y + \sqrt{144 - x^2} = 12 \Rightarrow y = 12 - \sqrt{144 - x^2}$

So,
$$12 - \sqrt{144 - x^2} = -12 \ln\left(\frac{12 - \sqrt{144 - x^2}}{x}\right) - \sqrt{144 - x^2}$$
$$-1 = \ln\left(\frac{12 - \sqrt{144 - x^2}}{x}\right)$$
$$xe^{-1} = 12 - \sqrt{144 - x^2}$$
$$\left(xe^{-1} - 12\right)^2 = \left(-\sqrt{144 - x^2}\right)^2$$
$$x^2e^{-2} - 24xe^{-1} + 144 = 144 - x^2$$
$$x^2\left(e^{-2} + 1\right) - 24xe^{-1} = 0$$
$$x\left[x\left(e^{-2} + 1\right) - 24e^{-1}\right] = 0$$
$$x = 0 \text{ or } x = \frac{24e^{-1}}{e^{-2} + 1} \approx 7.77665.$$

Therefore,
$$s = \int_{7.77665}^{12} \sqrt{1 + \left(-\frac{\sqrt{144 - x^2}}{x}\right)^2}\, dx = \int_{7.77665}^{12} \sqrt{\frac{x^2 + \left(144 - x^2\right)}{x^2}}\, dx$$
$$= \int_{7.77665}^{12} \frac{12}{x}\, dx = \Big[12 \ln|x|\Big]_{7.77665}^{12} = 12(\ln 12 - \ln 7.77665) \approx 5.2 \text{ meters.}$$

91. $f(x) = x^3, g(x) = x^2 + 1, [0, 1]$
$$\frac{f(b) - f(a)}{g(b) - g(a)} = \frac{f'(c)}{g'(c)}$$
$$\frac{f(1) - f(0)}{g(1) - g(0)} = \frac{3c^2}{2c}$$
$$\frac{1}{1} = \frac{3c}{2}$$
$$c = \frac{2}{3}$$

92. $f(x) = \frac{1}{x}, g(x) = x^2 - 4, [1, 2]$
$$\frac{f(2) - f(1)}{g(2) - g(1)} = \frac{f'(c)}{g'(c)}$$
$$\frac{-1/2}{3} = \frac{-1/c^2}{2c}$$
$$-\frac{1}{6} = -\frac{1}{2c^3}$$
$$2c^3 = 6$$
$$c = \sqrt[3]{3}$$

93. $f(x) = \sin x, g(x) = \cos x, \left[0, \frac{\pi}{2}\right]$
$$\frac{f(\pi/2) - f(0)}{g(\pi/2) - g(0)} = \frac{f'(c)}{g'(c)}$$
$$\frac{1}{-1} = \frac{\cos c}{-\sin c}$$
$$-1 = -\cot c$$
$$c = \frac{\pi}{4}$$

94. $f(x) = \ln x, g(x) = x^3, [1, 4]$
$$\frac{f(4) - f(1)}{g(4) - g(1)} = \frac{f'(c)}{g'(c)}$$
$$\frac{\ln 4}{63} = \frac{1/c}{3c^2} = \frac{1}{3c^3}$$
$$3c^3 \ln 4 = 63$$
$$c^3 = \frac{21}{\ln 4}$$
$$c = \sqrt[3]{\frac{21}{\ln 4}} \approx 2.474$$

95. False. L'Hôpital's Rule does not apply because $\lim_{x \to 0} \left(x^2 + x + 1\right) \neq 0.$
$$\lim_{x \to 0^+} \frac{x^2 + x + 1}{x} = \lim_{x \to 0^+}\left(x + 1 + \frac{1}{x}\right) = 1 + \infty = \infty$$

96. False. If $y = e^x/x^2$, then

$$y' = \frac{x^2e^x - 2xe^x}{x^4} = \frac{xe^x(x-2)}{x^4} = \frac{e^x(x-2)}{x^3}.$$

97. True

98. False. Let $f(x) = x$ and $g(x) = x + 1$. Then

$$\lim_{x\to\infty} \frac{x}{x+1} = 1, \text{ but } \lim_{x\to\infty}\left[x - (x+1)\right] = -1.$$

99. Area of triangle: $\frac{1}{2}(2x)(1 - \cos x) = x - x\cos x$

Shaded area: Area of rectangle − Area under curve

$$\begin{aligned} 2x(1-\cos x) - 2\int_0^x (1 - \cos t)\,dt &= 2x(1-\cos x) - 2\left[t - \sin t\right]_0^x \\ &= 2x(1-\cos x) - 2(x - \sin x) \\ &= 2\sin x - 2x\cos x \end{aligned}$$

$$\begin{aligned} \text{Ratio: } \lim_{x\to 0}\frac{x - x\cos x}{2\sin x - 2x\cos x} &= \lim_{x\to 0}\frac{1 + x\sin x - \cos x}{2\cos x + 2x\sin x - 2\cos x} \\ &= \lim_{x\to 0}\frac{1 + x\sin x - \cos x}{2x\sin x} \\ &= \lim_{x\to 0}\frac{x\cos x + \sin x + \sin x}{2x\cos x + 2\sin x} \\ &= \lim_{x\to 0}\frac{x\cos x + 2\sin x}{2x\cos x + 2\sin x}\cdot\frac{1/\cos x}{1/\cos x} = \lim_{x\to 0}\frac{x + 2\tan x}{2x + 2\tan x} = \lim_{x\to 0}\frac{1 + 2\sec^2 x}{2 + 2\sec^2 x} = \frac{3}{4} \end{aligned}$$

100. (a) $\sin\theta = BD$

$\cos\theta = DO \Rightarrow AD = 1 - \cos\theta$

Area $\Delta ABD = \frac{1}{2}bh = \frac{1}{2}(1 - \cos\theta)\sin\theta = \frac{1}{2}\sin\theta - \frac{1}{2}\sin\theta\cos\theta$

(b) Area of sector: $\frac{1}{2}\theta$

Shaded area: $\frac{1}{2}\theta - \text{Area }\Delta OBD = \frac{1}{2}\theta - \frac{1}{2}(\cos\theta)(\sin\theta) = \frac{1}{2}\theta - \frac{1}{2}\sin\theta\cos\theta$

(c) $$R = \frac{(1/2)\sin\theta - (1/2)\sin\theta\cos\theta}{(1/2)\theta - (1/2)\sin\theta\cos\theta} = \frac{\sin\theta - \sin\theta\cos\theta}{\theta - \sin\theta\cos\theta}$$

(d) $$\lim_{\theta\to 0} R = \lim_{\theta\to 0}\frac{\sin\theta - (1/2)\sin 2\theta}{\theta - (1/2)\sin 2\theta} = \lim_{\theta\to 0}\frac{\cos\theta - \cos 2\theta}{1 - \cos 2\theta} = \lim_{\theta\to 0}\frac{-\sin\theta + 2\sin 2\theta}{2\sin 2\theta} = \lim_{\theta\to 0}\frac{-\cos\theta + 4\cos 2\theta}{4\cos 2\theta} = \frac{3}{4}$$

101. $$\lim_{x\to 0}\frac{4x - 2\sin 2x}{2x^3} = \lim_{x\to 0}\frac{4 - 4\cos 2x}{6x^2} = \lim_{x\to 0}\frac{8\sin 2x}{12x} = \lim_{x\to 0}\frac{16\cos 2x}{12} = \frac{16}{12} = \frac{4}{3}$$

Let $c = \frac{4}{3}$.

102. Let $y = \left(e^x + x\right)^{1/x}$.

$$\ln y = \frac{1}{x}\ln\left(e^x + x\right) = \frac{\ln\left(e^x + x\right)}{x}$$

$$\lim_{x\to 0}\frac{\ln\left(e^x + x\right)}{x} = \lim_{x\to 0}\frac{e^x + 1}{e^x + x} = \frac{2}{1} = 2$$

So, $\lim_{x\to 0}\left(e^x + x\right)^{1/x} = e^2$.

Let $c = e^2 \approx 7.389$.

103. $$\lim_{x\to 0}\frac{a - \cos bx}{x^2} = 2$$

Near $x = 0$, $\cos bx \approx 1$ and $x^2 \approx 0 \Rightarrow a = 1$.

Using L'Hôpital's Rule,

$$\lim_{x\to 0}\frac{1 - \cos bx}{x^2} = \lim_{x\to 0}\frac{b\sin bx}{2x} = \lim_{x\to 0}\frac{b^2\cos bx}{2} = 2.$$

So, $b^2 = 4$ and $b = \pm 2$.

Answer: $a = 1, b = \pm 2$

104. $f(x) = \dfrac{x^k - 1}{k}$

$k = 1, \quad f(x) = x - 1$

$k = 0.1, \quad f(x) = \dfrac{x^{0.1} - 1}{0.1} = 10\left(x^{0.1} - 1\right)$

$k = 0.01, \quad f(x) = \dfrac{x^{0.01} - 1}{0.01} = 100\left(x^{0.01} - 1\right)$

$\lim\limits_{k\to 0^+} \dfrac{x^k - 1}{k} = \lim\limits_{k\to 0^+} \dfrac{x^k(\ln x)}{1} = \ln x$

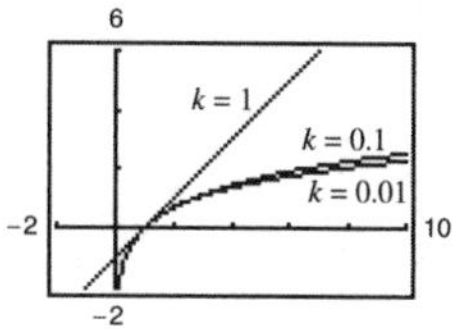

105. (a) $\lim\limits_{h\to 0} \dfrac{f(x+h) - f(x-h)}{2h} = \lim\limits_{h\to 0} \dfrac{f'(x+h)(1) - f'(x-h)(-1)}{2} = \lim\limits_{h\to 0}\left[\dfrac{f'(x+h) + f'(x-h)}{2}\right] = \dfrac{f'(x) + f'(x)}{2} = f'(x)$

(b)

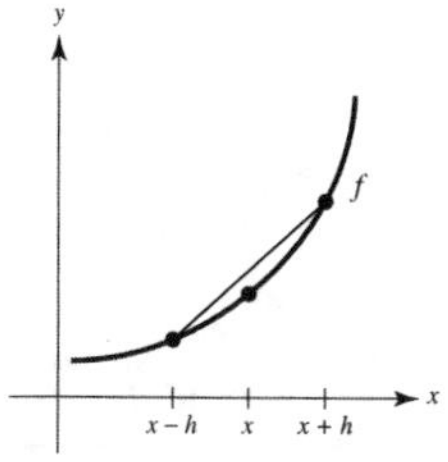

Graphically, the slope of the line joining $(x - h, f(x - h))$ and $(x + h, f(x + h))$ is approximately $f'(x)$.

So, $\lim\limits_{h\to 0} \dfrac{f(x+h) - f(x-h)}{2h} = f'(x)$.

106.
$$\lim_{h\to 0} \frac{f(x+h) - 2f(x) + f(x-h)}{h^2} = \lim_{h\to 0} \frac{f'(x+h)(1) + f'(x-h)(-1)}{2h}$$
$$= \lim_{h\to 0} \frac{f'(x+h) - f'(x-h)}{2h}$$
$$= \lim_{h\to 0} \frac{f''(x+h)(1) - f''(x-h)(-1)}{2}$$
$$= \lim_{h\to 0} \frac{f''(x+h) + f''(x-h)}{2}$$
$$= \frac{f''(x) + f''(x)}{2} = f''(x)$$

107. (a) $\lim\limits_{x\to 0^+} (-x \ln x)$ is the form $0 \cdot \infty$.

(b) $\lim\limits_{x\to 0^+} \dfrac{-\ln x}{1/x} = \lim\limits_{x\to 0^+} \dfrac{-1/x}{-1/x^2} = \lim\limits_{x\to 0^+} (x) = 0$

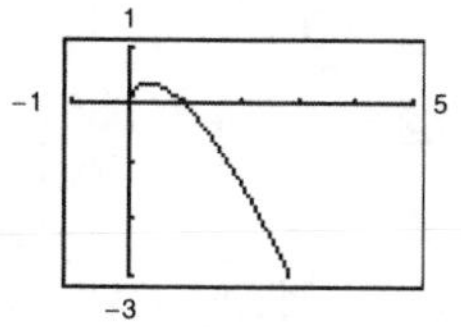

108. $\lim\limits_{x\to a} f(x)^{g(x)}$

$y = f(x)^{g(x)}$

$\ln y = g(x) \ln f(x)$

$\lim\limits_{x\to a} g(x) \ln f(x) = (\infty)(-\infty) = -\infty$

As $x \to a$, $\ln y \Rightarrow -\infty$, and therefore $y = 0$. So,

$\lim\limits_{x\to a} f(x)^{g(x)} = 0.$

109. $\lim_{x \to a} f(x)^{g(x)}$

$$y = f(x)^{g(x)}$$
$$\ln y = g(x) \ln f(x)$$
$$\lim_{x \to a} g(x) \ln f(x) = (-\infty)(-\infty) = \infty$$

As $x \to a$, $\ln y \Rightarrow \infty$, and therefore $y = \infty$. So, $\lim_{x \to a} f(x)^{g(x)} = \infty$.

110. $$f'(a)(b-a) - \int_a^b f''(t)(t-b)\,dt = f'(a)(b-a) - \left\{\left[f'(t)(t-b)\right]_a^b - \int_a^b f'(t)\,dt\right\}$$
$$= f'(a)(b-a) + f'(a)(a-b) + \left[f(t)\right]_a^b = f(b) - f(a)$$

$$dv = f''(t)\,dt \Rightarrow v = f'(t)$$
$$u = t - b \Rightarrow du = dt$$

111. (a) $\lim_{x \to 0^+} x^{(\ln 2)/(1+\ln x)}$ is of form 0^0.

Let $y = x^{(\ln 2)/(1+\ln x)}$

$$\ln y = \frac{\ln 2}{1 + \ln x} \ln x$$

$$\lim_{x \to 0^+} \ln y = \frac{\ln 2(1/x)}{1/x} = \ln 2.$$

So, $\lim_{x \to 0^+} x^{(\ln 2)/(1+\ln x)} = 2$.

(b) $\lim_{x \to \infty} x^{(\ln 2)/(1+\ln x)}$ is of form ∞^0.

Let $y = x^{(\ln 2)/(1+\ln x)}$

$$\ln y = \frac{\ln 2}{1 + \ln x} \ln x$$

$$\lim_{x \to \infty} \ln y = \frac{\ln 2(1/x)}{1/x} = \ln 2.$$

So, $\lim_{x \to \infty} x^{(\ln 2)/(1+\ln x)} = 2$.

(c) $\lim_{x \to 0}(x+1)^{(\ln 2)/(x)}$ is of form 1^∞.

Let $y = (x+1)^{(\ln 2)/(x)}$

$$\ln y = \frac{\ln 2}{x} \ln(x+1)$$

$$\lim_{x \to 0} \ln y = \lim_{x \to 0} \frac{(\ln 2)1/(x+1)}{1} = \ln 2.$$

So, $\lim_{x \to 0}(x+1)^{(\ln 2)/(x)} = 2$.

112. $$\lim_{x \to a} \frac{\sqrt{2a^3x - x^4} - a\sqrt[3]{a^2x}}{a - \sqrt[4]{ax^3}}$$
$$= \lim_{x \to a} \frac{\frac{1}{2}(2a^3x - x^4)^{-1/2}(2a^3 - 4x^3) - \frac{a}{3}(a^2x)^{-2/3}a^2}{-\frac{1}{4}(ax^3)^{-3/4}}$$
$$= \frac{\frac{1}{2}(a^4)^{-1/2}(-2a^3) - \frac{a^3}{3}(a^3)^{-2/3}}{-\frac{1}{4}(ax^3)^{-3/4}(3ax^2)}$$
$$= \frac{a + \frac{a}{3}}{\frac{1}{4}(a^{-3})(3a^3)}$$
$$= \frac{\frac{4}{3}a}{\frac{3}{4}} = \frac{16}{9}a$$

113. (a) $h(x) = \dfrac{x + \sin x}{x}$

$$\lim_{x \to \infty} h(x) = 1$$

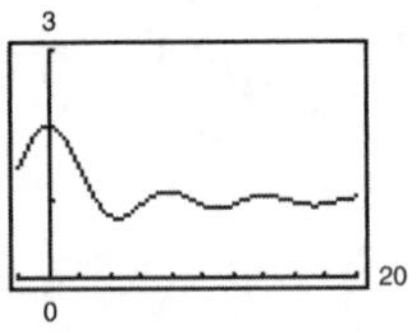

(b) $h(x) = \dfrac{x + \sin x}{x} = \dfrac{x}{x} + \dfrac{\sin x}{x} = 1 + \dfrac{\sin x}{x}, x > 0$

So, $\lim_{x \to \infty} h(x) = \lim_{x \to \infty}\left[1 + \dfrac{\sin x}{x}\right] = 1 + 0 = 1.$

(c) No. $h(x)$ is not an indeterminate form.

114. (a) $\lim_{x\to\infty} \dfrac{f(x)}{g(x)} = \lim_{x\to\infty} \dfrac{x + x\sin x}{x^2 - 4} = \lim_{x\to\infty} \dfrac{1 + \sin x}{x - 4/x} = 0$

$\left(\text{Because } |1 + \sin x| \le 1 \text{ and } x \to \infty.\right)$

(b) $\lim_{x\to\infty} f(x) = \lim_{x\to\infty} x(1 + \sin x) = \infty$

$\lim_{x\to\infty} g(x) = \lim_{x\to\infty} \left(x^2 - 4\right) = \infty$

(c) $\lim_{x\to\infty} \dfrac{f'(x)}{g'(x)} = \lim_{x\to\infty} \dfrac{1 + \sin x + x\cos x}{2x}$ undefined

(d) No. If $\lim_{x\to\infty} \dfrac{f'(x)}{g'(x)}$ does not exist, then you cannot assume anything about $\lim_{x\to\infty} \dfrac{f(x)}{g(x)}$.

115. Let $f(x) = \left[\dfrac{1}{x} \cdot \dfrac{a^x - 1}{a - 1}\right]^{1/x}$.

For $a > 1$ and $x > 0$,

$$\ln f(x) = \frac{1}{x}\left[\ln\frac{1}{x} + \ln\left(a^x - 1\right) - \ln(a - 1)\right] = -\frac{\ln x}{x} + \frac{\ln\left(a^x - 1\right)}{x} - \frac{\ln(a - 1)}{x}.$$

As $x \to \infty$, $\dfrac{\ln x}{x} \to 0$, $\dfrac{\ln(a - 1)}{x} \to 0$, and $\dfrac{\ln\left(a^x - 1\right)}{x} = \dfrac{\ln\left[\left(1 - a^{-x}\right)a^x\right]}{x} = \dfrac{\ln\left(1 - a^{-x}\right)}{x} + \ln a \to \ln a.$

So, $\ln f(x) \to \ln a$.

For $0 < a < 1$ and $x > 0$,

$$\ln f(x) = \frac{-\ln x}{x} + \frac{\ln\left(1 - a^x\right)}{x} - \frac{\ln(1 - a)}{x} \to 0 \text{ as } x \to \infty.$$

Combining these results, $\lim_{x\to\infty} f(x) = \begin{cases} a & \text{if } a > 1 \\ 1 & \text{if } 0 < a < 1 \end{cases}$.

Section 8.8 Improper Integrals

1. $\displaystyle\int_0^1 \frac{dx}{5x - 3}$ is improper because $5x - 3 = 0$ when $x = \dfrac{3}{5}$, and $0 \le \dfrac{3}{5} \le 1$.

2. $\displaystyle\int_1^2 \frac{dx}{x^3}$ is not improper because $f(x) = \dfrac{1}{x^3}$ is continuous on $[1, 2]$.

3. $\displaystyle\int_0^1 \frac{2x - 5}{x^2 - 5x + 6}\,dx = \int_0^1 \frac{2x - 5}{(x - 2)(x - 3)}\,dx$ is not improper because

$\dfrac{2x - 5}{(x - 2)(x - 3)}$ is continuous on $[0, 1]$.

4. $\displaystyle\int_1^\infty \ln\left(x^2\right)dx$ is improper because the upper limit of integration is ∞.

5. $\displaystyle\int_0^2 e^{-x}\,dx$ is not improper because $f(x) = e^{-x}$ is continuous on $[0, 2]$.

6. $\displaystyle\int_0^\infty \cos x\,dx$ is improper because the upper limit of integration is ∞.

7. $\displaystyle\int_{-\infty}^\infty \frac{\sin x}{4 + x^2}\,dx$ is improper because the limits of integration are $-\infty$ and ∞.

8. $\displaystyle\int_0^{\pi/4} \csc x\,dx$ is improper because $f(x) = \csc x$ is undefined at $x = 0$.

9. Infinite discontinuity at $x = 0$.

$$\int_0^4 \frac{1}{\sqrt{x}}\,dx = \lim_{b\to0^+}\int_b^4 \frac{1}{\sqrt{x}}\,dx$$
$$= \lim_{b\to0^+}\left[2\sqrt{x}\right]_b^4$$
$$= \lim_{b\to0^+}\left(4 - 2\sqrt{b}\right) = 4$$

Converges

10. Infinite discontinuity at $x = 3$.

$$\int_3^4 \frac{1}{(x-3)^{3/2}}\,dx = \lim_{b\to3^+}\int_b^4 (x-3)^{-3/2}\,dx$$
$$= \lim_{b\to3^+}\left[-2(x-3)^{-1/2}\right]_b^4$$
$$= \lim_{b\to3^+}\left[-2 + \frac{2}{\sqrt{b-3}}\right] = \infty$$

Diverges

11. Infinite discontinuity at $x = 1$.

$$\int_0^2 \frac{1}{(x-1)^2}\,dx = \int_0^1 \frac{1}{(x-1)^2}\,dx + \int_1^2 \frac{1}{(x-1)^2}\,dx$$
$$= \lim_{b\to1^-}\int_0^b \frac{1}{(x-1)^2}\,dx + \lim_{c\to1^+}\int_c^2 \frac{1}{(x-1)^2}\,dx$$
$$= \lim_{b\to1^-}\left[-\frac{1}{x-1}\right]_0^b + \lim_{c\to1^+}\left[-\frac{1}{x-1}\right]_c^2$$
$$= (\infty - 1) + (-1 + \infty)$$

Diverges

12. Infinite limit of integration.

$$\int_{-\infty}^0 e^{3x}\,dx = \lim_{b\to-\infty}\int_b^0 e^{3x}\,dx$$
$$= \lim_{b\to-\infty}\left[\tfrac{1}{3}e^{3x}\right]_b^0$$
$$= \lim_{b\to-\infty}\left[\tfrac{1}{3} - \tfrac{1}{3}e^{3b}\right] = \tfrac{1}{3}$$

Converges

13. $\displaystyle\int_{-1}^1 \frac{1}{x^2}\,dx \neq -2$

because the integrand is not defined at $x = 0$.

The integral diverges.

14. $\displaystyle\int_{-2}^2 \frac{-2}{(x-1)^3}\,dx \neq \frac{8}{9}$

because the integral is not defined at $x = 1$. The integral diverges.

15. $\displaystyle\int_0^\infty e^{-x}\,dx \neq 0$. You need to evaluate the limit.

$$\lim_{b\to\infty}\int_0^b e^{-x}\,dx = \lim_{b\to\infty}\left[-e^{-x}\right]_0^b$$
$$= \lim_{b\to\infty}\left[-e^{-b} + 1\right] = 1$$

16. $\displaystyle\int_0^\pi \sec x\,dx \neq 0$ because $\sec x$ is not defined at $x = \pi/2$.

The integral diverges.

17.
$$\int_1^\infty \frac{1}{x^3}\,dx = \lim_{b\to\infty}\int_1^b x^{-3}\,dx$$
$$= \lim_{b\to\infty}\left[\frac{x^{-2}}{-2}\right]_1^b$$
$$= \lim_{b\to\infty}\left[\frac{-1}{2b^2} + \frac{1}{2}\right] = \frac{1}{2}$$

18.
$$\int_1^\infty \frac{6}{x^4}\,dx = \lim_{b\to\infty} 6\int_1^b x^{-4}\,dx$$
$$= \lim_{b\to\infty} 6\left[\frac{x^{-3}}{-3}\right]_1^b$$
$$= \lim_{b\to\infty}\left[\frac{-2}{b^3} + 2\right] = 2$$

19.
$$\int_1^\infty \frac{3}{\sqrt[3]{x}}\,dx = \lim_{b\to\infty}\int_1^b 3x^{-1/3}\,dx$$
$$= \lim_{b\to\infty}\left[\frac{9}{2}x^{2/3}\right]_1^b = \infty$$

Diverges

20.
$$\int_1^\infty \frac{4}{\sqrt[4]{x}}\,dx = \lim_{b\to\infty}\int_1^b 4x^{-1/4}\,dx$$
$$= \lim_{b\to\infty}\left[\frac{16}{3}x^{3/4}\right]_1^b = \infty$$ Diverges

21.
$$\int_{-\infty}^0 xe^{-4x}\,dx = \lim_{b\to-\infty}\int_b^0 xe^{-4x}\,dx$$
$$= \lim_{b\to-\infty}\left[\left(\frac{-x}{4} - \frac{1}{16}\right)e^{-4x}\right]_b^0$$ (Integration by parts)
$$= \lim_{b\to-\infty}\left[-\frac{1}{16} + \frac{b}{4} + \frac{1}{16}e^{-4b}\right] = -\infty$$

Diverges

22. $\int_0^{\infty} xe^{-x/3}\,dx = \lim_{b\to\infty}\int_0^b xe^{-x/3}\,dx$

$= \lim_{b\to\infty}\left[(-3x - 9)e^{-x/3}\right]_0^b$

$= \lim_{b\to\infty}\left[(-3b - 9)e^{-b/3} + 9\right] = 9$

23. $\int_0^{\infty} x^2e^{-x}\,dx = \lim_{b\to\infty}\int_0^b x^2e^{-x}\,dx$

$= \lim_{b\to\infty}\left[-e^{-x}\left(x^2 + 2x + 2\right)\right]_0^b$

$= \lim_{b\to\infty}\left(-\frac{b^2 + 2b + 2}{e^b} + 2\right) = 2$

Because $\lim_{b\to\infty}\left(-\frac{b^2 + 2b + 2}{e^b}\right) = 0$ by L'Hôpital's Rule.

24. $\int_0^{\infty} e^{-x}\cos x\,dx = \lim_{b\to\infty}\frac{1}{2}\left[e^{-x}(-\cos x + \sin x)\right]_0^b$

$= \frac{1}{2}\left[0 - (-1)\right] = \frac{1}{2}$

25. $\int_4^{\infty}\frac{1}{x(\ln x)^3}\,dx = \lim_{b\to\infty}\int_4^b(\ln x)^{-3}\frac{1}{x}\,dx$

$= \lim_{b\to\infty}\left[-\frac{1}{2}(\ln x)^{-2}\right]_4^b$

$= \lim_{b\to\infty}\left[-\frac{1}{2}(\ln b)^{-2} + \frac{1}{2}(\ln 4)^{-2}\right]$

$= \frac{1}{2}\frac{1}{(2\ln 2)^2} = \frac{1}{2(\ln 4)^2}$

26. $\int_1^{\infty}\frac{\ln x}{x}\,dx = \lim_{b\to\infty}\int_1^b\frac{\ln x}{x}\,dx$

$= \lim_{b\to\infty}\left[\frac{(\ln x)^2}{2}\right]_1^b = \infty$

Diverges

27. $\int_{-\infty}^{\infty}\frac{4}{16 + x^2}\,dx = \int_{-\infty}^0\frac{4}{16 + x^2}\,dx + \int_0^{\infty}\frac{4}{16 + x^2}\,dx$

$= \lim_{b\to-\infty}\int_b^0\frac{4}{16 + x^2}\,dx + \lim_{c\to\infty}\int_0^c\frac{4}{16 + x^2}\,dx$

$= \lim_{b\to-\infty}\left[\arctan\left(\frac{x}{4}\right)\right]_b^0 + \lim_{c\to\infty}\left[\arctan\left(\frac{x}{4}\right)\right]_0^c$

$= \lim_{b\to-\infty}\left[0 - \arctan\left(\frac{b}{4}\right)\right] + \lim_{c\to\infty}\left[\arctan\left(\frac{c}{4}\right) - 0\right]$

$= -\left(-\frac{\pi}{2}\right) + \frac{\pi}{2} = \pi$

28. $\int_0^{\infty}\frac{x^3}{(x^2 + 1)^2}\,dx = \lim_{b\to\infty}\int_0^b\frac{x}{x^2 + 1}\,dx - \lim_{b\to\infty}\int_0^b\frac{x}{(x^2 + 1)^2}\,dx = \lim_{b\to\infty}\left[\frac{1}{2}\ln(x^2 + 1) + \frac{1}{2(x^2 + 1)}\right]_0^b = \infty - \frac{1}{2}$

Diverges

29. $\int_0^{\infty}\frac{1}{e^x + e^{-x}}\,dx = \lim_{b\to\infty}\int_0^b\frac{e^x}{1 + e^{2x}}\,dx$

$= \lim_{b\to\infty}\left[\arctan(e^x)\right]_0^b$

$= \frac{\pi}{2} - \frac{\pi}{4} = \frac{\pi}{4}$

30. $\int_0^{\infty}\frac{e^x}{1 + e^x}\,dx = \lim_{b\to\infty}\left[\ln(1 + e^x)\right]_0^b = \infty - \ln 2$

Diverges

31. $\int_0^{\infty}\cos\pi x\,dx = \lim_{b\to\infty}\left[\frac{1}{\pi}\sin\pi x\right]_0^b$

Diverges because $\sin\pi b$ does not approach a limit as $b\to\infty$.

32. $\int_0^{\infty}\sin\frac{x}{2}\,dx = \lim_{b\to\infty}\left[-2\cos\frac{x}{2}\right]_0^b$

Diverges because $\cos\frac{x}{2}$ does not approach a limit as $x\to\infty$.

33. $\int_0^1\frac{1}{x^2}\,dx = \lim_{b\to0^+}\left[\frac{-1}{x}\right]_b^1 = \lim_{b\to0^+}\left(-1 + \frac{1}{b}\right) = -1 + \infty$

Diverges

34. $\int_0^5 \frac{10}{x}\,dx = \lim_{b\to 0^+} \int_b^5 \frac{10}{x}\,dx$

$= \lim_{b\to 0^+} \left[10 \ln x\right]_b^5$

$= \lim_{b\to 0^+} \left(10 \ln 5 - 10 \ln b\right) = \infty$

Diverges

35. $\int_0^2 \frac{1}{\sqrt[3]{x-1}}\,dx = \int_0^1 \frac{1}{\sqrt[3]{x-1}}\,dx + \int_1^2 \frac{1}{\sqrt[3]{x-1}}\,dx = \lim_{b\to 1^-}\left[\frac{3}{2}(x-1)^{2/3}\right]_0^b + \lim_{c\to 1^+}\left[\frac{3}{2}(x-1)^{2/3}\right]_c^2 = \frac{-3}{2} + \frac{3}{2} = 0$

36. $\int_0^8 \frac{3}{\sqrt{8-x}}\,dx = \lim_{b\to 8^-} 3\int_0^b (8-x)^{-1/2}\,dx$

$= \lim_{b\to 8^-}\left[-6\sqrt{8-x}\right]_0^b$

$= \lim_{b\to 8^-}\left(-6\sqrt{8-b} + 6\sqrt{8}\right)$

$= 12\sqrt{2}$

37. $\int_0^1 x \ln x\,dx = \lim_{b\to 0^+}\left[\frac{x^2}{2}\ln|x| - \frac{x^2}{4}\right]_b^1$

$= \lim_{b\to 0^+}\left(\frac{-1}{4} - \frac{b^2 \ln b}{2} + \frac{b^2}{4}\right) = \frac{-1}{4}$

because $\lim_{b\to 0^+}\left(b^2 \ln b\right) = 0$ by L'Hôpital's Rule.

38. $\int_0^e \ln x^2\,dx = \lim_{b\to 0^+}\int_0^e 2\ln x\,dx$

$= \lim_{b\to 0^+}\left[2x\ln x - 2x\right]_b^e$

$= \lim_{b\to 0^+}\left[(2e - 2e) - (2b\ln b - 2b)\right]$

$= 0$

39. $\int_0^{\pi/2} \tan\theta\,d\theta = \lim_{b\to(\pi/2)^-}\left[\ln|\sec\theta|\right]_0^b = \infty$

Diverges

40. $\int_0^{\pi/2} \sec\theta\,d\theta = \lim_{b\to(\pi/2)^-}\left[\ln|\sec\theta + \tan\theta|\right]_0^b = \infty$

Diverges

41. $\int_2^4 \frac{2}{x\sqrt{x^2-4}}\,dx = \lim_{b\to 2^+}\int_b^4 \frac{2}{x\sqrt{x^2-4}}\,dx$

$= \lim_{b\to 2^+}\left[\operatorname{arcsec}\left|\frac{x}{2}\right|\right]_b^4$

$= \lim_{b\to 2^+}\left(\operatorname{arcsec} 2 - \operatorname{arcsec}\left(\frac{b}{2}\right)\right)$

$= \frac{\pi}{3} - 0 = \frac{\pi}{3}$

42. $\int_3^6 \frac{1}{\sqrt{36-x^2}}\,dx = \lim_{b\to 6^-}\int_3^b \frac{1}{\sqrt{36-x^2}}\,dx$

$= \lim_{b\to 6^-}\left[\arcsin\frac{x}{6}\right]_3^b$

$= \lim_{b\to 6^-}\left[\arcsin\frac{b}{6} - \arcsin\frac{1}{2}\right]$

$= \frac{\pi}{2} - \frac{\pi}{6} = \frac{\pi}{3}$

43. $\int_3^5 \frac{1}{\sqrt{x^2-9}}\,dx = \lim_{b\to 3^+}\left[\ln\left|x + \sqrt{x^2-9}\right|\right]_b^5$

$= \lim_{b\to 3^+}\left[\ln 9 - \ln\left(b + \sqrt{b^2-9}\right)\right]$

$= \ln 9 - \ln 3$

$= \ln\frac{9}{3} = \ln 3$

44. $\int_0^5 \frac{1}{25-x^2}\,dx = \lim_{b\to 5^-}\int_0^b \frac{1}{25-x^2}\,dx$

$= \lim_{b\to 5^-}\int_0^b \frac{1}{10}\left(\frac{1}{x+5} - \frac{1}{x-5}\right)dx$ (partial fractions)

$= \lim_{b\to 5^-}\left[\frac{1}{10}\ln\left|\frac{x+5}{x-5}\right|\right]_0^b$

$= \infty - 0$ Diverges

45. $$\int_3^{\infty} \frac{1}{x\sqrt{x^2-9}}\,dx = \lim_{b\to 3^+}\int_b^5 \frac{1}{x\sqrt{x^2-9}}\,dx + \lim_{c\to\infty}\int_5^{\infty}\frac{1}{x\sqrt{x^2-9}}\,dx$$
$$= \lim_{b\to 3^+}\left[\frac{1}{3}\operatorname{arcsec}\frac{x}{3}\right]_b^5 + \lim_{c\to\infty}\left[\frac{1}{3}\operatorname{arcsec}\left(\frac{x}{3}\right)\right]_5^{\infty}$$
$$= \lim_{b\to 3^+}\left[\frac{1}{3}\operatorname{arcsec}\left(\frac{5}{3}\right) - \frac{1}{3}\operatorname{arcsec}\left(\frac{b}{3}\right)\right] + \lim_{c\to\infty}\left[\frac{1}{3}\operatorname{arcsec}\left(\frac{c}{3}\right) - \frac{1}{3}\operatorname{arcsec}\left(\frac{5}{3}\right)\right] = -0 + \frac{1}{3}\left(\frac{\pi}{2}\right) = \frac{\pi}{6}$$

46. $$\int_4^{\infty}\frac{\sqrt{x^2-16}}{x^2}\,dx = \lim_{b\to\infty}\int_4^b \frac{\sqrt{x^2-16}}{x^2}\,dx$$
$$= \lim_{b\to\infty}\left[\frac{-\sqrt{x^2-16}}{x} + \ln\left|x + \sqrt{x^2-16}\right|\right]_4^b \qquad \text{(Formula 30)}$$
$$= \lim_{b\to\infty}\left[-\frac{\sqrt{b^2-16}}{b} + \ln\left|b + \sqrt{b^2-16}\right| - \ln 4\right] = \infty$$

Diverges

47. $$\int_0^{\infty}\frac{4}{\sqrt{x}(x+6)}\,dx = \int_0^1 \frac{4}{\sqrt{x}(x+6)}\,dx + \int_1^{\infty}\frac{4}{\sqrt{x}(x+6)}\,dx$$

Let $u = \sqrt{x}, u^2 = x, 2u\,du = dx$.

$$\int\frac{4}{\sqrt{x}(x+6)}\,dx = \int\frac{4(2u\,du)}{u(u^2+6)} = 8\int\frac{du}{u^2+6} = \frac{8}{\sqrt{6}}\arctan\left(\frac{u}{\sqrt{6}}\right) + C = \frac{8}{\sqrt{6}}\arctan\left(\frac{\sqrt{x}}{\sqrt{6}}\right) + C$$

So, $$\int_0^{\infty}\frac{4}{\sqrt{x}(x+6)}\,dx = \lim_{b\to 0^+}\left[\frac{8}{\sqrt{6}}\arctan\left(\frac{\sqrt{x}}{\sqrt{6}}\right)\right]_b^1 + \lim_{c\to\infty}\left[\frac{8}{\sqrt{6}}\arctan\left(\frac{\sqrt{x}}{\sqrt{6}}\right)\right]_1^c$$
$$= \left[\frac{8}{\sqrt{6}}\arctan\left(\frac{1}{\sqrt{6}}\right) - \frac{8}{\sqrt{6}}(0)\right] + \left[\frac{8}{\sqrt{6}}\left(\frac{\pi}{2}\right) - \frac{8}{\sqrt{6}}\arctan\left(\frac{1}{\sqrt{6}}\right)\right] = \frac{8\pi}{2\sqrt{6}} = \frac{2\pi\sqrt{6}}{3}.$$

48. $$\int\frac{1}{x\ln x}\,dx = \ln\left|\ln|x|\right| + C$$

So,

$$\int_1^{\infty}\frac{1}{x\ln x}\,dx = \int_1^e \frac{1}{x\ln x}\,dx + \int_e^{\infty}\frac{1}{x\ln x}\,dx = \lim_{b\to 1^+}\left[\ln(\ln x)\right]_1^e + \lim_{c\to\infty}\left[\ln(\ln x)\right]_e^{\infty}.$$

Diverges

49. If $p = 1$, $$\int_1^{\infty}\frac{1}{x}\,dx = \lim_{b\to\infty}\int_1^b \frac{1}{x}\,dx = \lim_{b\to\infty}\left[\ln x\right]_1^b$$
$$= \lim_{b\to\infty}(\ln b) = \infty.$$

Diverges. For $p \neq 1$,

$$\int_1^{\infty}\frac{1}{x^p}\,dx = \lim_{b\to\infty}\left[\frac{x^{1-p}}{1-p}\right]_1^b = \lim_{b\to\infty}\left(\frac{b^{1-p}}{1-p} - \frac{1}{1-p}\right).$$

This converges to $\dfrac{1}{p-1}$ if $1 - p < 0$ or $p > 1$.

50. If $p = 1$, $$\int_0^1 \frac{1}{x}\,dx = \lim_{a\to 0^+}\left[\ln x\right]_a^1 = \lim_{a\to 0^+} -\ln a = \infty.$$

Diverges. For $p \neq 1$,

$$\int_0^1 \frac{1}{x^p}\,dx = \lim_{a\to 0^+}\left[\frac{x^{1-p}}{1-p}\right]_a^1 = \lim_{a\to 0^+}\left(\frac{1}{1-p} - \frac{a^{1-p}}{1-p}\right).$$

This converges to $\dfrac{1}{1-p}$ if $1 - p > 0$ or $p < 1$.

51. For $n = 1$:

$$\int_0^\infty xe^{-x}\,dx = \lim_{b\to\infty}\int_0^b xe^{-x}\,dx$$
$$= \lim_{b\to\infty}\left[-e^{-x}x - e^{-x}\right]_0^b \qquad \left(\text{Parts: } u = x, dv = e^{-x}\,dx\right)$$
$$= \lim_{b\to\infty}\left(-e^{-b}b - e^{-b} + 1\right)$$
$$= \lim_{b\to\infty}\left(\frac{-b}{e^b} - \frac{1}{e^b} + 1\right) = 1 \quad \text{(L'Hôpital's Rule)}$$

Assume that $\int_0^\infty x^n e^{-x}\,dx$ converges. Then for $n + 1$ you have

$$\int x^{n+1}e^{-x}\,dx = -x^{n+1}e^{-x} + (n+1)\int x^n e^{-x}\,dx$$

by parts $\left(u = x^{n+1}, du = (n+1)x^n\,dx, dv = e^{-x}\,dx, v = -e^{-x}\right)$.

So,

$$\int_0^\infty x^{n+1}e^{-x}\,dx = \lim_{b\to\infty}\left[-x^{n+1}e^{-x}\right]_0^b + (n+1)\int_0^\infty x^n e^{-x}\,dx = 0 + (n+1)\int_0^\infty x^n e^{-x}\,dx, \text{ which converges.}$$

52. (a) $\int_1^\infty e^{-x}\,dx = \lim_{b\to\infty}\int_1^b e^{-x}\,dx = \lim_{b\to\infty}\left[-e^{-b}\right]_1^b = 1$

Because $e^{-x^2} \le e^{-x}$ on $[1, \infty)$

and

$\int_1^\infty e^{-x}\,dx$

converges, then so does

$\int_1^\infty e^{-x^2}\,dx.$

(b) $\int_1^\infty \frac{1}{x^5}\,dx$ converges (see Exercise 49).

Because $\frac{1}{x^5 + 1} < \frac{1}{x^5}$ on $[1, \infty)$, then $\int_1^\infty \frac{1}{x^5 + 1}\,dx$ also converges.

53. $\int_0^1 \frac{1}{x^5}\,dx$ diverges by Exercise 50. $(p = 5)$

54. $\int_0^1 \frac{1}{x^{1/5}}\,dx$ converges by Exercise 50. $\left(p = \frac{1}{5}\right)$

55. $\int_1^\infty \frac{1}{x^5}\,dx$ converges by Exercise 49. $(p = 5)$

56. $\int_0^\infty x^4 e^{-x}\,dx$ converges by Exercise 51. $(n = 4)$

57. Because $\frac{1}{x^2 + 5} \le \frac{1}{x^2}$ on $[1, \infty)$ and

$\int_1^\infty \frac{1}{x^2}\,dx$ converges by Exercise 49,

$\int_1^\infty \frac{1}{x^2 + 5}\,dx$ converges.

58. Because $\frac{1}{\sqrt{x - 1}} \ge \frac{1}{x}$ on $[2, \infty)$ and $\int_2^\infty \frac{1}{x}\,dx$ diverges

by Exercise 55, $\int_2^\infty \frac{1}{\sqrt{x - 1}}\,dx$ diverges.

59. Because $\frac{1}{\sqrt[3]{x(x - 1)}} \ge \frac{1}{\sqrt[3]{x^2}}$ on $[2, \infty)$ and

$\int_2^\infty \frac{1}{\sqrt[3]{x^2}}\,dx$ diverges by Exercise 49,

$\int_2^\infty \frac{1}{\sqrt[3]{x(x - 1)}}\,dx$ diverges.

60. Because $\frac{1}{\sqrt{x}(1 + x)} \le \frac{1}{x^{3/2}}$ on $[1, \infty)$ and

$\int_1^\infty \frac{1}{x^{3/2}}\,dx$ converges by Exercise 49,

$\int_1^\infty \frac{1}{\sqrt{x}(1 + x)}\,dx$ converges.

61. $\int_1^{\infty} \frac{2}{x^2}\,dx$ converges, and $\frac{1 - \sin x}{x^2} \le \frac{2}{x^2}$ on $[1, \infty)$, so $\int_1^{\infty} \frac{1 - \sin x}{x^2}\,dx$ converges.

62. $\int_0^{\infty} \frac{1}{e^x}\,dx = \int_0^{\infty} e^{-x}\,dx$ converges, and $\frac{1}{e^x} \ge \frac{1}{e^x + x}$ on $[0, \infty)$, so $\int_0^{\infty} \frac{1}{e^x + x}\,dx$ converges.

63. Answers will vary. *Sample answer:*

An integral with infinite integration limits or an integral with an infinite discontinuity at or between the integration limits

64. When the limit of the integral exists, the improper integral converges. When the limit does not exist, the improper integral diverges.

65. $\int_{-1}^{1} \frac{1}{x^3}\,dx = \int_{-1}^{0} \frac{1}{x^3}\,dx + \int_0^1 \frac{1}{x^3}\,dx$

These two integrals diverge by Exercise 50.

66. $\frac{10}{x^2 - 2x} = \frac{10}{x(x - 2)} \Rightarrow x = 0, 2.$

You must analyze three improper integrals, and each must converge in order for the original integral to converge.

$$\int_0^3 f(x)\,dx = \int_0^1 f(x)\,dx + \int_1^2 f(x)\,dx + \int_2^3 f(x)\,dx$$

67. $A = \int_{-\infty}^{1} e^x\,dx$

$$= \lim_{b\to-\infty} \int_b^1 e^x\,dx$$

$$= \lim_{b\to-\infty} \left[e^x\right]_b^1$$

$$= \lim_{b\to-\infty} \left(e - e^b\right) = e$$

68. $A = \int_0^1 -\ln x\,dx$

$$= -\lim_{b\to0^+} \int_b^1 \ln x\,dx$$

$$= -\lim_{b\to0^+} \left[x \ln x - x\right]_b^1$$

$$= -\lim_{b\to0^+} \left[(0 - 1) - b \ln b + b\right]$$

$$= 1$$

Note: $\lim_{b\to0^+} b \ln b = \lim_{b\to0^+} \frac{\ln b}{1/b} = \lim_{b\to0^+} \frac{1/b}{-1/b^2} = 0$

69. $A = \int_{-\infty}^{\infty} \frac{1}{x^2 + 1}\,dx$

$$= \lim_{b\to-\infty} \int_b^0 \frac{1}{x^2 + 1}\,dx + \lim_{b\to\infty} \int_0^b \frac{1}{x^2 + 1}\,dx$$

$$= \lim_{b\to-\infty} \left[\arctan(x)\right]_b^0 + \lim_{b\to\infty} \left[\arctan(x)\right]_0^b$$

$$= \lim_{b\to-\infty} \left[0 - \arctan(b)\right] + \lim_{b\to\infty} \left[\arctan(b) - 0\right]$$

$$= -\left(-\frac{\pi}{2}\right) + \frac{\pi}{2} = \pi$$

70. $A = \int_{-\infty}^{\infty} \frac{8}{x^2 + 4}\,dx$

$$= \lim_{b\to-\infty} \int_b^0 \frac{8}{x^2 + 4}\,dx + \lim_{b\to\infty} \int_0^b \frac{8}{x^2 + 4}\,dx$$

$$= \lim_{b\to-\infty} \left[4 \arctan\left(\frac{x}{2}\right)\right]_b^0 + \lim_{b\to\infty} \left[4 \arctan\left(\frac{x}{2}\right)\right]_0^b$$

$$= \lim_{b\to-\infty} \left[0 - 4 \arctan\left(\frac{b}{2}\right)\right] + \lim_{b\to\infty} \left[4 \arctan\left(\frac{b}{2}\right) - 0\right]$$

$$= -4\left(\frac{-\pi}{2}\right) + 4\left(\frac{\pi}{2}\right) = 4\pi$$

71. (a) $A = \int_0^{\infty} e^{-x}\,dx$

$$= \lim_{b\to\infty} \left[-e^{-x}\right]_0^b = 0 - (-1) = 1$$

(b) **Disk:**

$$V = \pi \int_0^{\infty} \left(e^{-x}\right)^2\,dx$$

$$= \lim_{b\to\infty} \pi \left[-\frac{1}{2} e^{-2x}\right]_0^b = \frac{\pi}{2}$$

(c) **Shell:**

$$V = 2\pi \int_0^{\infty} x e^{-x}\,dx$$

$$= \lim_{b\to\infty} 2\pi \left[-e^{-x}(x + 1)\right]_0^b = 2\pi$$

72. (a) $A = \int_1^{\infty} \frac{1}{x^2}\,dx = \left[-\frac{1}{x}\right]_1^{\infty} = 1$

(b) **Disk:**

$$V = \pi \int_1^{\infty} \frac{1}{x^4}\,dx = \lim_{b\to\infty} \left[-\frac{\pi}{3x^3}\right]_1^b = \frac{\pi}{3}$$

(c) **Shell:**

$$V = 2\pi \int_1^{\infty} x\left(\frac{1}{x^2}\right)dx = \lim_{b\to\infty} \left[2\pi(\ln x)\right]_1^b = \infty$$

Diverges

73. $x^{2/3} + y^{2/3} = 4$

$\frac{2}{3}x^{-1/3} + \frac{2}{3}y^{-1/3}y' = 0$

$y' = \frac{-y^{1/3}}{x^{1/3}}$

$\sqrt{1 + (y')^2} = \sqrt{1 + \frac{y^{2/3}}{x^{2/3}}} = \sqrt{\frac{x^{2/3} + y^{2/3}}{x^{2/3}}} = \sqrt{\frac{4}{x^{2/3}}} = \frac{2}{x^{1/3}}, \quad (x > 0)$

$s = 4\int_0^8 \frac{2}{x^{1/3}}\,dx = \lim_{b\to 0^+}\left[8 \cdot \frac{3}{2}x^{2/3}\right]_b^8 = 48$

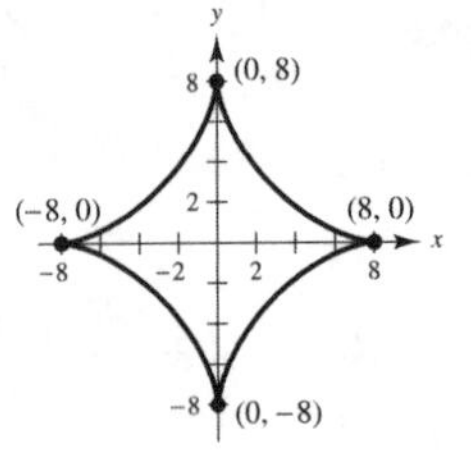

74. $y = \sqrt{16 - x^2}, \quad 0 \le x \le 4$

$y' = \frac{-x}{\sqrt{16 - x^2}}$

$s = \int_0^4 \sqrt{1 + \frac{x^2}{16 - x^2}}\,dx = \int_0^4 \frac{4}{\sqrt{16 - x^2}}\,dx = \lim_{t\to 4^-}\int_0^t \frac{4}{\sqrt{16 - x^2}}\,dx = \lim_{t\to 4^-}\left[4\arcsin\left(\frac{x}{4}\right)\right]_0^t = \lim_{t\to 4^-} 4\arcsin\left(\frac{t}{4}\right) = 2\pi$

75. $(x - 2)^2 + y^2 = 1$

$2(x - 2) + 2yy' = 0$

$y' = \frac{-(x - 2)}{y}$

$\sqrt{1 + (y')^2} = \sqrt{1 + \left[(x - 2)^2/y^2\right]} = \frac{1}{y}$ (Assume $y > 0$.)

$S = 4\pi\int_1^3 \frac{x}{y}\,dx = 4\pi\int_1^3 \frac{x}{\sqrt{1 - (x - 2)^2}}\,dx = 4\pi\int_1^3 \left[\frac{x - 2}{\sqrt{1 - (x - 2)^2}} + \frac{2}{\sqrt{1 - (x - 2)^2}}\right]dx$

$= \lim_{\substack{a\to 1^+ \\ b\to 3^-}} 4\pi\left[-\sqrt{1 - (x - 2)^2} + 2\arcsin(x - 2)\right]_a^b = 4\pi\left[0 + 2\arcsin(1) - 2\arcsin(-1)\right] = 8\pi^2$

76. $y = 2e^{-x}$

$y' = -2e^{-x}$

$S = 2\pi\int_0^\infty \left(2e^{-x}\right)\sqrt{1 + 4e^{-2x}}\,dx$

Let $u = e^{-x}$, $du = -e^{-x}\,dx$.

$\int e^{-x}\sqrt{1 + 4e^{-2x}}\,dx = -\int\sqrt{1 + 4u^2}\,du$

$= -\frac{1}{4}\left[2u\sqrt{4u^2 + 1} + \ln\left|2u + \sqrt{4u^2 + 1}\right|\right] + C$

$= -\frac{1}{4}\left[2e^{-x}\sqrt{4e^{-2x} + 1} + \ln\left|2e^{-x} + \sqrt{4e^{-2x} + 1}\right|\right] + C$

$S = 4\pi\lim_{b\to\infty}\int_0^b \left(e^{-x}\right)\sqrt{1 + 4e^{-2x}}\,dx$

$= -\pi\lim_{b\to\infty}\left[2e^{-x}\sqrt{4e^{-2x} + 1} + \ln\left|2e^{-x} + \sqrt{4e^{-2x} + 1}\right|\right]_0^b = \pi\left[2\sqrt{5} + \ln\left(2 + \sqrt{5}\right)\right] \approx 18.5849$

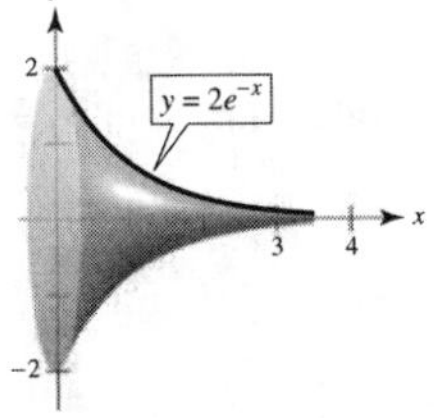

77. (a) $F(x) = \frac{K}{x^2}, 5 = \frac{K}{(4000)^2}, K = 80{,}000{,}000$

$$W = \int_{4000}^{\infty} \frac{80{,}000{,}000}{x^2}\,dx = \lim_{b\to\infty}\left[\frac{-80{,}000{,}000}{x}\right]_{4000}^{b} = 20{,}000 \text{ mi-ton}$$

(b) $\frac{W}{2} = 10{,}000 = \left[\frac{-80{,}000{,}000}{x}\right]_{4000}^{b} = \frac{-80{,}000{,}000}{b} + 20{,}000$

$$\frac{80{,}000{,}000}{b} = 10{,}000$$

$$b = 8000$$

Therefore, the rocket has traveled 4000 miles above the earth's surface.

78. (a) $F(x) = \frac{k}{x^2}, 10 = \frac{k}{4000^2}, k = 10(4000^2)$

$$W = \int_{4000}^{\infty} \frac{10(4000^2)}{x^2}\,dx = \lim_{b\to\infty}\left[\frac{-10(4000^2)}{x}\right]_{4000}^{b}$$

$$= \frac{10(4000^2)}{4000} = 40{,}000 \text{ mi-ton}$$

(b) $\frac{W}{2} = 20{,}000 = \left[\frac{-10(4000^2)}{x}\right]_{4000}^{b}$

$$= \frac{-10(4000^2)}{b} + 40{,}000$$

$$\frac{10(4000^2)}{b} = 20{,}000$$

$$b = 8000$$

Therefore, the rocket has traveled 4000 miles above the earth's surface.

79. (a) $\int_{-\infty}^{\infty} \frac{1}{7}e^{-t/7}\,dt = \int_0^{\infty} \frac{1}{7}e^{-t/7}\,dt = \lim_{b\to\infty}\left[-e^{-t/7}\right]_0^b = 1$

(b) $\int_0^4 \frac{1}{7}e^{-t/7}\,dt = \left[-e^{-t/7}\right]_0^4 = -e^{-4/7} + 1$

$\approx 0.4353 = 43.53\%$

(c) $\int_0^{\infty} t\left[\frac{1}{7}e^{-t/7}\right]dt = \lim_{b\to\infty}\left[-te^{-t/7} - 7e^{-t/7}\right]_0^b$

$= 0 + 7 = 7$

80. (a) $\int_{-\infty}^{\infty} \frac{2}{5}e^{-2t/5}\,dt = \int_0^{\infty} \frac{2}{5}e^{-2t/5}\,dt = \lim_{b\to\infty}\left[-e^{-2t/5}\right]_0^b = 1$

(b) $\int_0^4 \frac{2}{5}e^{-2t/5}\,dt = \left[-e^{-2t/5}\right]_0^4 = -e^{-8/5} + 1$

$\approx 0.7981 = 79.81\%$

(c) $\int_0^{\infty} t\left[\frac{2}{5}e^{-2t/5}\right]dt = \lim_{b\to\infty}\left[-te^{2t/5} - \frac{5}{2}e^{-2t/5}\right]_0^b = \frac{5}{2}$

81. (a) $C = 650{,}000 + \int_0^5 25{,}000\,e^{-0.06t}\,dt = 650{,}000 - \left[\frac{25{,}000}{0.06}e^{-0.06t}\right]_0^5 \approx \$757{,}992.41$

(b) $C = 650{,}000 + \int_0^{10} 25{,}000e^{-0.06t}\,dt \approx \$837{,}995.15$

(c) $C = 650{,}000 + \int_0^{\infty} 25{,}000e^{-0.06t}\,dt = 650{,}000 - \lim_{b\to\infty}\left[\frac{25{,}000}{0.06}e^{-0.06t}\right]_0^b \approx \$1{,}066{,}666.67$

82. (a) $C = 650{,}000 + \int_0^5 25{,}000(1 + 0.08t)e^{-0.06t}\,dt$

$$= 650{,}000 + 25{,}000\left[-\frac{1}{0.06}e^{-0.06t} - 0.08\left(\frac{t}{0.06}e^{-0.06t} + \frac{1}{(0.06)^2}e^{-0.06t}\right)\right]_0^5 \approx \$778{,}512.58$$

(b) $C = 650{,}000 + \int_0^{10} 25{,}000(1 + 0.08t)e^{-0.06t}\,dt$

$$= 650{,}000 + 25{,}000\left[-\frac{1}{0.06}e^{-0.06t} - 0.08\left(\frac{t}{0.06}e^{-0.06t} + \frac{1}{(0.06)^2}e^{-0.06t}\right)\right]_0^{10} \approx \$905{,}718.14$$

(c) $C = 650{,}000 + \int_0^{\infty} 25{,}000(1 + 0.08t)e^{-0.06t}\,dt$

$$= 650{,}000 + 25{,}000\lim_{b\to\infty}\left[-\frac{1}{0.06}e^{-0.06t} - 0.08\left(\frac{t}{0.06}e^{-0.06t} + \frac{1}{(0.06)^2}e^{-0.06t}\right)\right]_0^b \approx \$1{,}622{,}222.22$$

83. Let $K = \dfrac{2\pi NI\,r}{k}$. Then

$$P = K\int_c^{\infty} \frac{1}{\left(r^2 + x^2\right)^{3/2}}\,dx.$$

Let

$x = r\tan\theta,\ dx = r\sec^2\theta\,d\theta,\ \sqrt{r^2 + x^2} = r\sec\theta.$

$$\int \frac{1}{\left(r^2 + x^2\right)^{3/2}}\,dx = \int \frac{r\sec^2\theta\,d\theta}{r^3\sec^3\theta} = \frac{1}{r^2}\int \cos\theta\,d\theta$$
$$= \frac{1}{r^2}\sin\theta + C = \frac{1}{r^2}\frac{x}{\sqrt{r^2 + x^2}} + C$$

So,

$$P = K\frac{1}{r^2}\lim_{b\to\infty}\left[\frac{x}{\sqrt{r^2 + x^2}}\right]_c^b$$
$$= \frac{K}{r^2}\left[1 - \frac{c}{\sqrt{r^2 + c^2}}\right]$$
$$= \frac{K\left(\sqrt{r^2 + c^2} - c\right)}{r^2\sqrt{r^2 + c^2}}$$
$$= \frac{2\pi NI\left(\sqrt{r^2 + c^2} - c\right)}{kr\sqrt{r^2 + c^2}}.$$

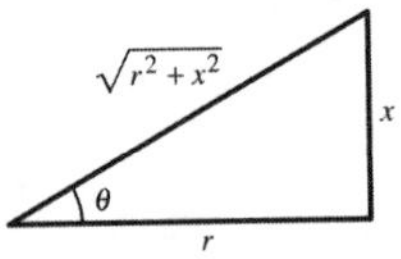

84.
$$F = \int_0^{\infty} \frac{GM\delta}{(a + x)^2}\,dx$$
$$= \lim_{b\to\infty}\left[\frac{-GM\delta}{a + x}\right]_0^b$$
$$= \frac{GM\delta}{a}$$

85. False. $f(x) = 1/(x + 1)$ is continuous on $[0, \infty)$, $\lim_{x\to\infty} 1/(x + 1) = 0$, but

$$\int_0^{\infty} \frac{1}{x + 1}\,dx = \lim_{b\to\infty}\left[\ln|x + 1|\right]_0^b = \infty.$$

Diverges

86. False. This is equivalent to Exercise 85.

87. True

88. True

89. (a)
$$\int_{-\infty}^{\infty} \sin x\,dx = \int_{-\infty}^{0} \sin x\,dx + \int_0^{\infty} \sin x\,dx$$
$$= \lim_{b\to-\infty}\int_b^0 \sin x\,dx + \lim_{c\to\infty}\int_0^c \sin x\,dx$$
$$= \lim_{b\to-\infty}\left[-\cos x\right]_b^0 + \lim_{c\to\infty}\left[-\cos x\right]_0^c$$

Because $\lim_{b\to-\infty}\left[-\cos b\right]$ diverges, as does $\lim_{c\to\infty}\left[-\cos c\right]$,

$\int_{-\infty}^{\infty} \sin x\,dx$ diverges.

(b)
$$\lim_{a\to\infty}\int_{-a}^{a} \sin x\,dx = \lim_{a\to\infty}\left[-\cos x\right]_{-a}^{a}$$
$$= \lim_{a\to\infty}\left[-\cos(a) + \cos(-a)\right] = 0$$

(c) The definition of $\int_{-\infty}^{\infty} f(x)\,dx$ is not

$\lim_{a\to\infty}\int_{-a}^{a} f(x)\,dx.$

90. (a) $b = 3$ (infinite discontinuity at 3)

(b) $b = 4$ (infinite discontinuity at 4)

(c) $b = 3$ (or $b = 4$) (infinite discontinuity at 3)

(d) $b = 0$ (infinite discontinuity at 0)

(e) $b = \pi/4$ (infinite discontinuity at $\pi/4$)

(f) $b = \pi/2$ (infinite discontinuity at $\pi/2$)

91. (a)
$$\int_1^{\infty} \frac{1}{x}\,dx = \lim_{b\to\infty}\left[\ln|x|\right]_1^b = \infty$$
$$\int_1^{\infty} \frac{1}{x^2}\,dx = \lim_{b\to\infty}\left[-\frac{1}{x}\right]_1^b = 1$$

$\int_1^{\infty} \frac{1}{x^n}\,dx$ will converge if $n > 1$ and will diverge if $n \le 1$.

(b) It would appear to converge.

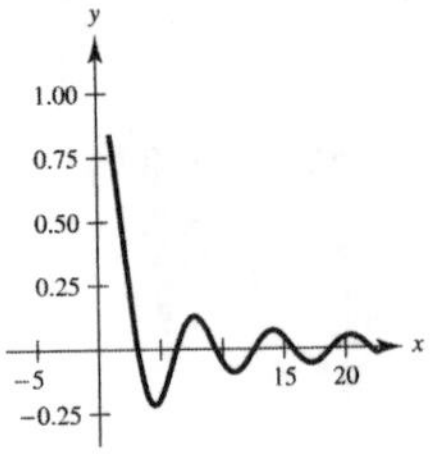

(c) Let $dv = \sin x\,dx \Rightarrow v = -\cos x$

$u = \dfrac{1}{x} \Rightarrow du = -\dfrac{1}{x^2}\,dx.$

$$\int_1^{\infty} \frac{\sin x}{x}\,dx = \lim_{b\to0}\left[-\frac{\cos x}{x}\right]_1^b - \int_1^{\infty} \frac{\cos x}{x^2}\,dx$$
$$= \cos 1 - \int_1^{\infty} \frac{\cos x}{x^2}\,dx$$

Converges

92. (a) Yes, the integrand is not defined at $x = \pi/2$.

(b)

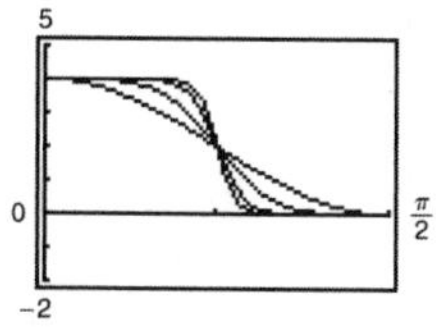

(c) As $n \to \infty$, the integral approaches $4(\pi/4) = \pi$.

(d) $I_n = \int_0^{\pi/2} \frac{4}{1 + (\tan x)^n}\,dx$

$I_2 \approx 3.14159$

$I_4 \approx 3.14159$

$I_8 \approx 3.14159$

$I_{12} \approx 3.14159$

93. (a) $f(x) = \frac{1}{2.85\sqrt{2\pi}} e^{-(x-70)^2/16.245}$

$\int_{50}^{90} f(x)\,dx \approx 1.0$

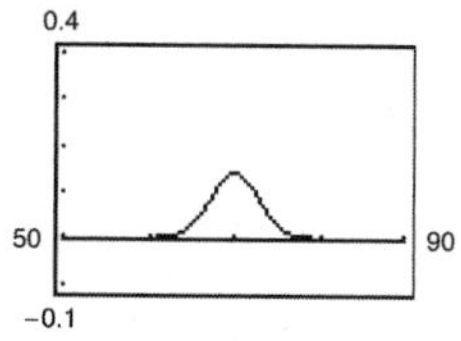

(b) $P(72 \le x < \infty) \approx 0.2414$

(c) $0.5 - P(70 \le x \le 72) \approx 0.5 - 0.2586 = 0.2414$

These are the same answers because of symmetry,

$P(70 \le x < \infty) = 0.5$

and

$$0.5 = P(70 \le x < \infty)$$
$$= P(70 \le x \le 72) + P(72 \le x < \infty).$$

94. (a) The area under the curve is greater on the interval $26 \le x \le 28$ than on the interval $22 \le x \le 24$. So, the probability is greater for choosing a car getting between 26 and 28 miles per gallon.

(b) The area under the curve is greater on the interval $x \ge 30$ than on the interval $20 \le x \le 22$. So, the probability is greater for choosing a car getting at least 30 miles per gallon.

95. $f(t) = 1$

$$F(s) = \int_0^\infty e^{-st}\,dt = \lim_{b\to\infty}\left[-\frac{1}{s}e^{-st}\right]_0^b = \frac{1}{s},\ s > 0$$

96. $f(t) = t$

$$F(s) = \int_0^\infty te^{-st}\,dt = \lim_{b\to\infty}\left[\frac{1}{s^2}(-st - 1)e^{-st}\right]_0^b$$
$$= \frac{1}{s^2},\ s > 0$$

97. $f(t) = t^2$

$$F(s) = \int_0^\infty t^2e^{-st}\,dt = \lim_{b\to\infty}\left[\frac{1}{s^3}(-s^2t^2 - 2st - 2)e^{-st}\right]_0^b$$
$$= \frac{2}{s^3},\ s > 0$$

98. $f(t) = e^{at}$

$$F(s) = \int_0^\infty e^{at}e^{-st}\,dt = \int_0^\infty e^{t(a-s)}\,dt$$
$$= \lim_{b\to\infty}\left[\frac{1}{a-s}e^{t(a-s)}\right]_0^b$$
$$= 0 - \frac{1}{a-s} = \frac{1}{s-a},\ s > a$$

99. $f(t) = \cos at$

$$F(s) = \int_0^\infty e^{-st}\cos at\,dt$$
$$= \lim_{b\to\infty}\left[\frac{e^{-st}}{s^2 + a^2}(-s\cos at + a\sin at)\right]_0^b$$
$$= 0 + \frac{s}{s^2 + a^2} = \frac{s}{s^2 + a^2},\ s > 0$$

100. $f(t) = \sin at$

$$F(s) = \int_0^\infty e^{-st}\sin at\,dt$$
$$= \lim_{b\to\infty}\left[\frac{e^{-st}}{s^2 + a^2}(-s\sin at - a\cos at)\right]_0^b$$
$$= 0 + \frac{a}{s^2 + a^2} = \frac{a}{s^2 + a^2},\ s > 0$$

101. $f(t) = \cosh at$

$$F(s) = \int_0^\infty e^{-st}\cosh at\,dt = \int_0^\infty e^{-st}\left(\frac{e^{at}+e^{-at}}{2}\right)dt = \frac{1}{2}\int_0^\infty\left[e^{t(-s+a)} + e^{t(-s-a)}\right]dt$$

$$= \lim_{b\to\infty}\frac{1}{2}\left[\frac{1}{(-s+a)}e^{t(-s+a)} + \frac{1}{(-s-a)}e^{t(-s-a)}\right]_0^b = 0 - \frac{1}{2}\left[\frac{1}{(-s+a)} + \frac{1}{(-s-a)}\right]$$

$$= \frac{-1}{2}\left[\frac{1}{(-s+a)} + \frac{1}{(-s-a)}\right] = \frac{s}{s^2-a^2},\ s > |a|$$

102. $f(t) = \sinh at$

$$F(s) = \int_0^\infty e^{-st}\sinh at\,dt = \int_0^\infty e^{-st}\left(\frac{e^{at}-e^{-at}}{2}\right)dt = \frac{1}{2}\int_0^\infty\left[e^{t(-s+a)} - e^{t(-s-a)}\right]dt$$

$$= \lim_{b\to\infty}\frac{1}{2}\left[\frac{1}{(-s+a)}e^{t(-s+a)} - \frac{1}{(-s-a)}e^{t(-s-a)}\right]_0^b = 0 - \frac{1}{2}\left[\frac{1}{(-s+a)} - \frac{1}{(-s-a)}\right]$$

$$= \frac{-1}{2}\left[\frac{1}{(-s+a)} - \frac{1}{(-s-a)}\right] = \frac{a}{s^2-a^2},\ s > |a|$$

103. $\Gamma(n) = \int_0^\infty x^{n-1}e^{-x}\,dx$

(a) $\Gamma(1) = \int_0^\infty e^{-x}\,dx = \lim_{b\to\infty}\left[-e^{-x}\right]_0^b = 1$

$\Gamma(2) = \int_0^\infty xe^{-x}\,dx = \lim_{b\to\infty}\left[-e^{-x}(x+1)\right]_0^b = 1$

$\Gamma(3) = \int_0^\infty x^2e^{-x}\,dx = \lim_{b\to\infty}\left[-x^2e^{-x} - 2xe^{-x} - 2e^{-x}\right]_0^b = 2$

(b) $\Gamma(n+1) = \int_0^\infty x^ne^{-x}\,dx = \lim_{b\to\infty}\left[-x^ne^{-x}\right]_0^b + \lim_{b\to\infty} n\int_0^b x^{n-1}e^{-x}\,dx = 0 + n\Gamma(n) \qquad \left(u = x^n,\ dv = e^{-x}\,dx\right)$

(c) $\Gamma(n) = (n-1)!$

104. For $n = 1$,

$$I_1 = \int_0^\infty \frac{x}{\left(x^2+1\right)^4}\,dx = \lim_{b\to\infty}\frac{1}{2}\int_0^b\left(x^2+1\right)^{-4}(2x\,dx) = \lim_{b\to\infty}\left[-\frac{1}{6}\cdot\frac{1}{\left(x^2+1\right)^3}\right]_0^b = \frac{1}{6}.$$

For $n > 1$,

$$I_n = \int_0^\infty \frac{x^{2n-1}}{\left(x^2+1\right)^{n+3}}\,dx = \lim_{b\to\infty}\left[\frac{-x^{2n-2}}{2(n+2)\left(x^2+1\right)^{n+2}}\right]_0^b + \frac{n-1}{n+2}\int_0^\infty\frac{x^{2n-3}}{\left(x^2+1\right)^{n+2}}\,dx = 0 + \frac{n-1}{n+2}\left(I_{n-1}\right)$$

$$\left(\text{Parts: } u = x^{2n-2},\ du = (2n-2)x^{2n-3}\,dx,\ dv = \frac{x}{\left(x^2+1\right)^{n+3}}\,dx,\ v = \frac{-1}{2(n+2)\left(x^2+1\right)^{n+2}}\right)$$

(a) $\displaystyle\int_0^\infty \frac{x}{\left(x^2+1\right)^4}\,dx = \lim_{b\to\infty}\left[-\frac{1}{6\left(x^2+1\right)^3}\right]_0^b = \frac{1}{6}$

(b) $\displaystyle\int_0^\infty \frac{x^3}{\left(x^2+1\right)^5}\,dx = \frac{1}{4}\int_0^\infty\frac{x}{\left(x^2+1\right)^4}\,dx = \frac{1}{4}\left(\frac{1}{6}\right) = \frac{1}{24}$

(c) $\displaystyle\int_0^\infty \frac{x^5}{\left(x^2+1\right)^6} = \frac{2}{5}\int_0^\infty\frac{x^3}{\left(x^2+1\right)^5}\,dx = \frac{2}{5}\left(\frac{1}{24}\right) = \frac{1}{60}$

105. $$\int_0^\infty \left(\frac{1}{\sqrt{x^2+1}} - \frac{c}{x+1}\right) dx = \lim_{b\to\infty} \int_0^b \left(\frac{1}{\sqrt{x^2+1}} - \frac{c}{x+1}\right) dx$$

$$= \lim_{b\to\infty}\left[\ln\left|x + \sqrt{x^2+1}\right| - c\ln|x+1|\right]_0^b$$

$$= \lim_{b\to\infty}\left[\ln\left(b + \sqrt{b^2+1}\right) - \ln(b+1)^c\right] = \lim_{b\to\infty}\ln\left[\frac{b+\sqrt{b^2+1}}{(b+1)^c}\right]$$

This limit exists for $c = 1$, and you have

$$\lim_{b\to\infty}\ln\left[\frac{b+\sqrt{b^2+1}}{(b+1)}\right] = \ln 2.$$

106. $$\int_1^\infty \left(\frac{cx}{x^2+2} - \frac{1}{3x}\right) dx = \lim_{b\to\infty}\int_1^b \left(\frac{cx}{x^2+2} - \frac{1}{3x}\right) dx$$

$$= \lim_{b\to\infty}\left[\frac{c}{2}\ln\left(x^2+2\right) - \frac{1}{3}\ln|x|\right]_1^b$$

$$= \lim_{b\to\infty}\ln\left[\frac{\left(x^2+2\right)^{c/2}}{x^{1/3}}\right]_1^b$$

$$= \lim_{b\to\infty}\left[\ln\frac{\left(b^2+2\right)^{c/2}}{b^{1/3}} - \ln 3^{c/2}\right]$$

This limit exists if $c = 1/3$, and you have

$$\lim_{b\to\infty}\left[\ln\frac{\left(b^2+2\right)^{1/6}}{b^{1/3}} - \ln 3^{1/6}\right] = -\ln 3^{1/6} = \frac{-\ln 3}{6}.$$

107. $$f(x) = \begin{cases} x\ln x, & 0 < x \le 2 \\ 0, & x = 0 \end{cases}$$

$$V = \pi\int_0^2 (x\ln x)^2\, dx$$

Let $u = \ln x, e^u = x, e^u\, du = dx.$

$$V = \pi\int_{-\infty}^{\ln 2} e^{2u}u^2\left(e^4\, du\right)$$

$$= \pi\int_{-\infty}^{\ln 2} e^{3u}u^2\, du$$

$$= \lim_{b\to-\infty}\left[\pi\left[\frac{u^2}{3} - \frac{2u}{9} + \frac{2}{27}\right]e^{3u}\right]_b^{\ln 2}$$

$$= 8\pi\left[\frac{(\ln 2)^2}{3} - \frac{2\ln 2}{9} + \frac{2}{27}\right] \approx 2.0155$$

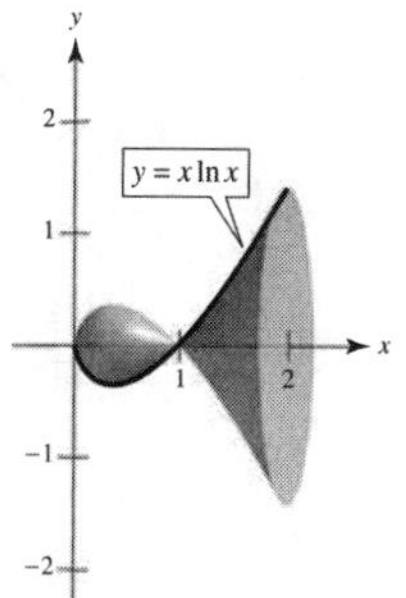

108. $V = \pi \int_0^1 (-\ln x)^2\, dx$

$= \lim_{b\to 0^+} \pi \int_b^1 (\ln x)^2\, dx$

$= \lim_{b\to 0^+} \pi x\left[(\ln x)^2 - 2\ln x + 2\right]_b^1$

$= \lim_{b\to 0^+} \pi\left[2 - b(\ln b)^2 - 2b\ln b - 2b\right]$

$= 2\pi$

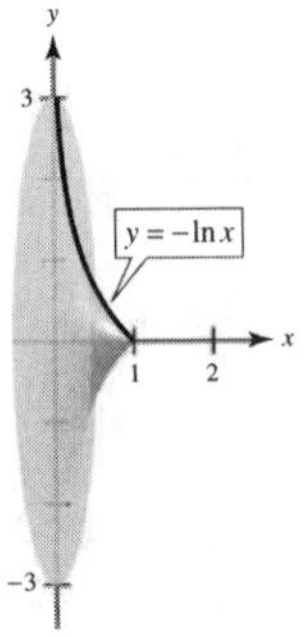

109. $u = \sqrt{x}, u^2 = x, 2u\, du = dx$

$$\int_0^1 \frac{\sin x}{\sqrt{x}}\, dx = \int_0^1 \frac{\sin(u^2)}{u}(2u\, du) = \int_0^1 2\sin(u^2)\, du$$

Trapezoidal Rule $(n = 5)$: 0.6278

110. $u = \sqrt{1-x}, 1 - x = u^2, 2u\, du = -dx$

$$\int_0^1 \frac{\cos x}{\sqrt{1-x}}\, dx = \int_1^0 \frac{\cos(1-u^2)}{u}(-2u\, du)$$

$$= \int_0^1 2\cos(1-u^2)\, du$$

Trapezoidal Rule $(n = 5)$: 1.4997

111. Assume $a < b$. The proof is similar if $a > b$.

$$\int_{-\infty}^a f(x)\, dx + \int_a^\infty f(x)\, dx = \lim_{c\to-\infty}\int_c^a f(x)\, dx + \lim_{d\to\infty}\int_a^d f(x)\, dx$$

$$= \lim_{c\to-\infty}\int_c^a f(x)\, dx + \lim_{d\to\infty}\left[\int_a^b f(x)\, dx + \int_b^d f(x)\, dx\right]$$

$$= \lim_{c\to-\infty}\int_c^a f(x)\, dx + \int_a^b f(x)\, dx + \lim_{d\to\infty}\int_b^d f(x)\, dx$$

$$= \lim_{c\to-\infty}\left[\int_c^a f(x)\, dx + \int_a^b f(x)\, dx\right] + \lim_{d\to\infty}\int_b^d f(x)\, dx$$

$$= \lim_{c\to-\infty}\int_c^b f(x)\, dx + \lim_{d\to\infty}\int_b^d f(x)\, dx$$

$$= \int_{-\infty}^b f(x)\, dx + \int_b^\infty f(x)\, dx$$

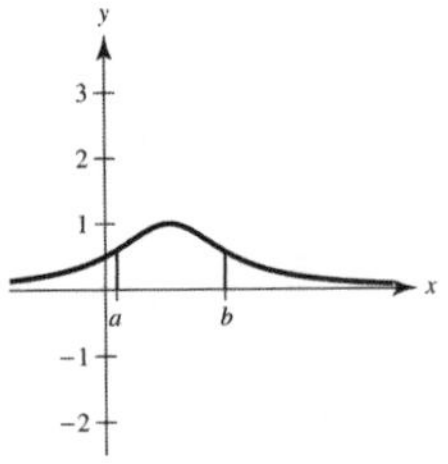

Review Exercises for Chapter 8

1. $\int x\sqrt{x^2 - 36}\, dx = \frac{1}{2}\int (x^2 - 36)^{1/2}(2x)\, dx$

$= \frac{1}{2}\left[\frac{(x^2-36)^{3/2}}{3/2}\right] + C$

$= \frac{1}{3}(x^2 - 36)^{3/2} + C$

2. $\int xe^{x^2-1}\, dx = \frac{1}{2}\int e^{x^2-1}(2x)\, dx = \frac{1}{2}e^{x^2-1} + C$

3. $\int \frac{x}{x^2 - 49}\, dx = \frac{1}{2}\int \frac{2x}{x^2 - 49}\, dx = \frac{1}{2}\ln\left|x^2 - 49\right| + C$

4. $\int \frac{x}{\sqrt[3]{4 - x^2}}\, dx = -\frac{1}{2}\int (4 - x^2)^{-1/3}(-2x)\, dx$

$= -\frac{1}{2}\frac{(4 - x^2)^{2/3}}{(2/3)} + C$

$= -\frac{3}{4}(4 - x^2)^{2/3} + C$

5. Let $u = \ln(2x)$, $du = \frac{1}{x}\,dx$.

$$\int_1^e \frac{\ln(2x)}{x}\,dx = \int_{\ln 2}^{1+\ln 2} u\,du$$
$$= \left[\frac{u^2}{2}\right]_{\ln 2}^{1+\ln 2}$$
$$= \frac{1}{2}\left[1 + 2\ln 2 + (\ln 2)^2 - (\ln 2)^2\right]$$
$$= \frac{1}{2} + \ln 2 \approx 1.1931$$

6. Let $u = 2x - 3$, $du = 2\,dx$, $x = \frac{1}{2}(u + 3)$.

$$\int_{3/2}^{2} 2x\sqrt{2x - 3}\,dx = \int_0^1 (u + 3)u^{1/2}\left(\frac{1}{2}\right)du$$
$$= \frac{1}{2}\int_0^1 \left(u^{3/2} + 3u^{1/2}\right)du$$
$$= \frac{1}{2}\left[\frac{2}{5}u^{5/2} + 2u^{3/2}\right]_0^1$$
$$= \frac{1}{2}\left(\frac{2}{5} + 2\right) = \frac{6}{5}$$

7. $$\int \frac{100}{\sqrt{100 - x^2}}\,dx = 100\arcsin\left(\frac{x}{10}\right) + C$$

8. $$\int \frac{2x}{x - 3}\,dx = \int\left(2 + \frac{6}{x - 3}\right)dx$$
$$= 2x + 6\ln|x - 3| + C$$

9. $$\int xe^{3x}\,dx = \frac{x}{3}e^{3x} - \int \frac{1}{3}e^{3x}\,dx$$
$$= \frac{x}{3}e^{3x} - \frac{1}{9}e^{3x} + C$$
$$= \frac{1}{9}e^{3x}(3x - 1) + C$$

$dv = e^{3x}\,dx \Rightarrow v = \frac{1}{3}e^{3x}$

$u = x \Rightarrow du = dx$

10. $$\int x^3e^x\,dx = x^3e^x - \int 3x^2e^x\,dx$$
$$= x^3e^x - \left(3x^2e^x - \int 6xe^x\,dx\right)$$
$$= x^3e^x - 3x^2e^x + \int 6xe^x\,dx$$
$$= x^3e^x - 3x^2e^x + \left(6xe^x - \int 6e^x\,dx\right)$$
$$= x^3e^x - 3x^2e^x + 6xe^x - 6e^x + C$$
$$= \left(x^3 - 3x^2 + 6x - 6\right)e^x + C$$

(1) $dv = e^x\,dx \Rightarrow v = e^x$

$u = x^3 \Rightarrow du = 3x^2\,dx$

(2) $dv = e^x\,dx \Rightarrow v = e^x$

$u = 3x^2 \Rightarrow du = 6x\,dx$

(3) $dv = e^x\,dx \Rightarrow v = e^x$

$u = 6x \Rightarrow du = 6\,dx$

11. $$\int e^{2x}\sin 3x\,dx = -\frac{1}{3}e^{2x}\cos 3x + \frac{2}{3}\int e^{2x}\cos 3x\,dx$$
$$= -\frac{1}{3}e^{2x}\cos 3x + \frac{2}{3}\left(\frac{1}{3}e^{2x}\sin 3x - \frac{2}{3}\int e^{2x}\sin 3x\,dx\right)$$
$$\frac{13}{9}\int e^{2x}\sin 3x\,dx = -\frac{1}{3}e^{2x}\cos 3x + \frac{2}{9}e^{2x}\sin 3x$$
$$\int e^{2x}\sin 3x\,dx = \frac{e^{2x}}{13}(2\sin 3x - 3\cos 3x) + C$$

(1) $dv = \sin 3x\,dx \Rightarrow v = -\frac{1}{3}\cos 3x$

$u = e^{2x} \Rightarrow du = 2e^{2x}\,dx$

(2) $dv = \cos 3x\,dx \Rightarrow v = \frac{1}{3}\sin 3x$

$u = e^{2x} \Rightarrow du = 2e^{2x}\,dx$

12. $\int x\sqrt{x-1}\,dx = \frac{2}{3}x(x-1)^{3/2} - \int \frac{2}{3}(x-1)^{3/2}\,dx$

$= \frac{2}{3}x(x-1)^{3/2} - \frac{4}{15}(x-1)^{5/2} + C$

$= \frac{2}{15}(x-1)^{3/2}\left(5x - 2(x-1)\right) + C$

$= \frac{2}{15}(x-1)^{3/2}(3x+2) + C$

$dv = (x-1)^{1/2}\,dx \Rightarrow v = \frac{2}{3}(x-1)^{3/2}$

$h = x \Rightarrow du = dx$

13. $\int x^2 \sin 2x\,dx = -\frac{1}{2}x^2\cos 2x + \int x\cos 2x\,dx$

$= -\frac{1}{2}x^2\cos 2x + \frac{1}{2}x\sin 2x - \frac{1}{2}\int \sin 2x\,dx$

$= -\frac{1}{2}x^2\cos 2x + \frac{x}{2}\sin 2x + \frac{1}{4}\cos 2x + C$

(1) $dv = \sin 2x\,dx \Rightarrow v = -\frac{1}{2}\cos 2x$

$u = x^2 \Rightarrow du = 2x\,dx$

(2) $dv = \cos 2x\,dx \Rightarrow v = \frac{1}{2}\sin 2x$

$u = x \Rightarrow du = dx$

14. $\int \ln\sqrt{x^2-4}\,dx = \frac{1}{2}\int \ln(x^2-4)\,dx$

$= \frac{1}{2}\left[x\ln(x^2-4) - \int \frac{2x^2}{x^2-4}\,dx\right]$

$= \frac{1}{2}x\ln(x^2-4) - \int\left(1 + \frac{4}{x^2-4}\right)dx$

$= \frac{1}{2}x\ln(x^2-4) - x - \ln\left|\frac{x-2}{x+2}\right| + C$

$dv = dx \Rightarrow v = x$

$u = \ln(x^2-4) \Rightarrow du = \frac{2x}{x^2-4}\,dx$

15. $\int x\arcsin 2x\,dx = \frac{x^2}{2}\arcsin 2x - \int \frac{x^2}{\sqrt{1-4x^2}}\,dx$

$= \frac{x^2}{2}\arcsin 2x - \frac{1}{4}\int \frac{(2x)^2}{\sqrt{1-(2x)^2}}\,dx$

$= \frac{x^2}{2}\arcsin 2x - \frac{1}{4}\left(\frac{1}{2}\right)\left[-(2x)\sqrt{1-4x^2} + \arcsin 2x\right] + C$ (by Formula 43 of Integration Tables)

$= \frac{1}{8}\left[(4x^2-1)\arcsin 2x + 2x\sqrt{1-4x^2}\right] + C$

$dv = x\,dx \Rightarrow v = \frac{x^2}{2}$

$u = \arcsin 2x \Rightarrow du = \frac{2}{\sqrt{1-4x^2}}\,dx$

16. $\int \arctan 2x\,dx = x\arctan 2x - \int \frac{2x}{1+4x^2}\,dx$

$= x\arctan 2x - \frac{1}{4}\ln(1+4x^2) + C$

$dv = dx \Rightarrow v = x$

$u = \arctan 2x \Rightarrow du = \frac{2}{1+4x^2}\,dx$

17. $$\int \cos^3(\pi x - 1)\,dx = \int\left[1 - \sin^2(\pi x - 1)\right]\cos(\pi x - 1)\,dx$$
$$= \frac{1}{\pi}\left[\sin(\pi x - 1) - \frac{1}{3}\sin^3(\pi x - 1)\right] + C$$
$$= \frac{1}{3\pi}\sin(\pi x - 1)\left[3 - \sin^2(\pi x - 1)\right] + C$$
$$= \frac{1}{3\pi}\sin(\pi x - 1)\left[3 - \left(1 - \cos^2(\pi x - 1)\right)\right] + C$$
$$= \frac{1}{3\pi}\sin(\pi x - 1)\left[2 + \cos^2(\pi x - 1)\right] + C$$

18. $$\int \sin^2\frac{\pi x}{2}\,dx = \int \frac{1}{2}(1 - \cos \pi x)\,dx = \frac{1}{2}\left(x - \frac{1}{\pi}\sin \pi x\right) + C = \frac{1}{2\pi}(\pi x - \sin \pi x) + C$$

19. $$\int \sec^4\left(\frac{x}{2}\right)dx = \int\left[\tan^2\left(\frac{x}{2}\right) + 1\right]\sec^2\left(\frac{x}{2}\right)dx$$
$$= \int \tan^2\left(\frac{x}{2}\right)\sec^2\left(\frac{x}{2}\right)dx + \int \sec^2\left(\frac{x}{2}\right)dx$$
$$= \frac{2}{3}\tan^3\left(\frac{x}{2}\right) + 2\tan\left(\frac{x}{2}\right) + C = \frac{2}{3}\left[\tan^3\left(\frac{x}{2}\right) + 3\tan\left(\frac{x}{2}\right)\right] + C$$

20. $$\int \tan\theta \sec^4\theta\,d\theta = \int\left(\tan^3\theta + \tan\theta\right)\sec^2\theta\,d\theta + \tfrac{1}{4}\tan^4\theta + \tfrac{1}{2}\tan^2\theta + C_1$$

or

$$\int \tan\theta\sec^4\theta\,d\theta = \int \sec^3\theta(\sec\theta\tan\theta)\,d\theta + \tfrac{1}{4}\sec^4\theta + C_2$$

21. $$\int \frac{1}{1 - \sin\theta}\,d\theta = \int \frac{1}{1 - \sin\theta}\cdot\frac{1 + \sin\theta}{1 + \sin\theta}\,d\theta = \int \frac{1 + \sin\theta}{\cos^2\theta}\,d\theta = \int\left(\sec^2\theta + \sec\theta\tan\theta\right)d\theta = \tan\theta + \sec\theta + C$$

22. $$\int \cos 2\theta(\sin\theta + \cos\theta)^2\,d\theta = \int\left(\cos^2\theta - \sin^2\theta\right)(\sin\theta + \cos\theta)^2\,d\theta$$
$$= \int(\sin\theta + \cos\theta)^3(\cos\theta - \sin\theta)\,d\theta = \tfrac{1}{4}(\sin\theta + \cos\theta)^4 + C$$

23. $A = \int_{\pi/4}^{3\pi/4} \sin^4 x\,dx$. Using the Table of Integrals,

$$\int \sin^4 x\,dx = -\frac{\sin^3 x\cos x}{4} + \frac{3}{4}\int \sin^2 x\,dx = \frac{-\sin^3 x\cos x}{4} + \frac{3}{4}\left[\frac{1}{2}(x - \sin x\cos x)\right] + C$$

$$\int_{\pi/4}^{3\pi/4} \sin^4 x\,dx = \left[\frac{-\sin^3 x\cos x}{4} + \frac{3}{8}x - \frac{3}{8}\sin x\cos x\right]_{\pi/4}^{3\pi/4} = \left(\frac{1}{16} + \frac{9\pi}{32} + \frac{3}{16}\right) - \left(\frac{-1}{16} + \frac{3\pi}{32} - \frac{3}{16}\right) = \frac{3\pi}{16} + \frac{1}{2} \approx 1.0890$$

24. $$A = \int_0^{\pi/4} \sin 3x\cos 2x\,dx$$
$$= \frac{1}{2}\int_0^{\pi/4}[\sin x + \sin 5x]\,dx$$
$$= \frac{1}{2}\left[-\cos x - \frac{1}{5}\cos 5x\right]_0^{\pi/4}$$
$$= \frac{1}{2}\left[-\frac{\sqrt{2}}{2} - \frac{1}{5}\left(-\frac{\sqrt{2}}{2}\right) + 1 + \frac{1}{5}\right]$$
$$= \frac{3}{5} - \frac{\sqrt{2}}{5} \approx 0.317$$

25. $\displaystyle\int \frac{-12}{x^2\sqrt{4-x^2}}\,dx = \int \frac{-24\cos\theta\,d\theta}{(4\sin^2\theta)(2\cos\theta)}$

$\displaystyle = -3\int \csc^2\theta\,d\theta$

$= 3\cot\theta + C$

$\displaystyle = \frac{3\sqrt{4-x^2}}{x} + C$

$x = 2\sin\theta,\ dx = 2\cos\theta\,d\theta,\ \sqrt{4-x^2} = 2\cos\theta$

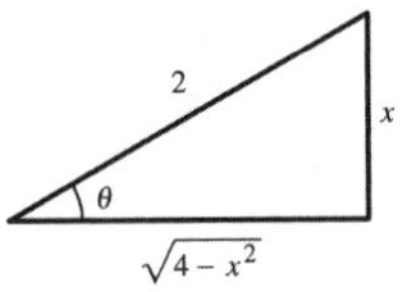

26. $\displaystyle\int \frac{\sqrt{x^2-9}}{x}\,dx = \int \frac{3\tan\theta}{3\sec\theta}(3\sec\theta\tan\theta\,d\theta)$

$\displaystyle = 3\int \tan^2\theta\,d\theta$

$\displaystyle = 3\int (\sec^2\theta - 1)\,d\theta$

$= 3(\tan\theta - \theta) + C$

$\displaystyle = \sqrt{x^2-9} - 3\,\text{arcsec}\left(\frac{x}{3}\right) + C$

$x = 3\sec\theta,\ dx = 3\sec\theta\tan\theta\,d\theta,\ \sqrt{x^2-9} = 3\tan\theta$

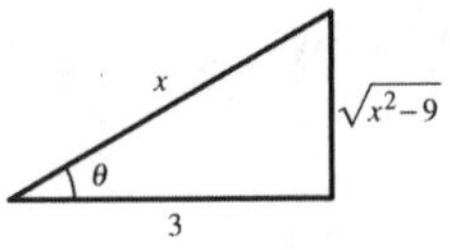

27. $x = 2\tan\theta$

$dx = 2\sec^2\theta\,d\theta$

$4 + x^2 = 4\sec^2\theta$

$\displaystyle\int \frac{x^3}{\sqrt{4+x^2}}\,dx = \int \frac{8\tan^3\theta}{2\sec\theta}2\sec^2\theta\,d\theta$

$\displaystyle = 8\int \tan^3\theta\sec\theta\,d\theta$

$\displaystyle = 8\int (\sec^2\theta - 1)\tan\theta\sec\theta\,d\theta$

$\displaystyle = 8\left[\frac{\sec^3\theta}{3} - \sec\theta\right] + C$

$\displaystyle = 8\left[\frac{(x^2+4)^{3/2}}{24} - \frac{\sqrt{x^2+4}}{2}\right] + C$

$\displaystyle = \sqrt{x^2+4}\left[\frac{1}{3}(x^2+4) - 4\right] + C$

$\displaystyle = \frac{1}{3}x^2\sqrt{x^2+4} - \frac{8}{3}\sqrt{x^2+4} + C$

$\displaystyle = \frac{1}{3}(x^2+4)^{1/2}(x^2-8) + C$

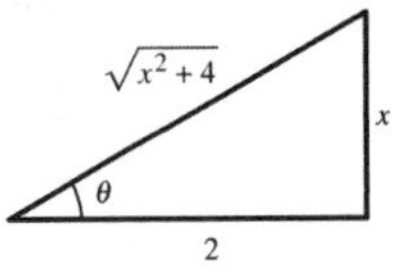

28. $\displaystyle\int \sqrt{25-9x^2}\,dx = \frac{1}{3}\int \sqrt{5^2-(3x)^2}\,(3)\,dx$

$\displaystyle = \frac{1}{3}\frac{1}{2}\left[25\arcsin\left(\frac{3x}{5}\right) + 3x\sqrt{25-9x^2}\right] + C = \frac{25}{6}\arcsin\left(\frac{3x}{5}\right) + \frac{x}{2}\sqrt{25-9x^2} + C$

(Theorem 8.2)

29. $x = 4\tan\theta,\ dx = 4\sec^2\theta\, d\theta,\ \sqrt{16 + x^2} = 4\sec\theta$

$$\int \frac{6x^3}{\sqrt{16+x^2}}\,dx = \int \frac{6(4\tan\theta)^3}{4\sec\theta}\,4\sec^2\theta\,d\theta$$
$$= 384\int \tan^3\theta\sec\theta\,d\theta$$
$$= 384\int (\sec^2\theta - 1)\sec\theta\tan\theta\,d\theta$$
$$= 384\left[\frac{\sec^3\theta}{3} - \sec\theta\right] + C$$
$$= \frac{384}{3}\cdot\frac{(16+x^2)^{3/2}}{64} - \frac{384\sqrt{16+x^2}}{4} + C$$
$$= 2\sqrt{x^2+16}(16 + x^2 - 48) + C$$
$$= 2\sqrt{x^2+16}(x^2 - 32) + C$$

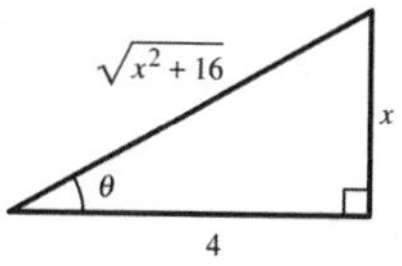

$$\int_0^1 \frac{6x^3}{\sqrt{16+x^2}}\,dx = \left[2\sqrt{x^2+16}\,(x^2-32)\right]_0^1$$
$$= 2\sqrt{17}(-31) - 2(4)(-32)$$
$$= 256 - 62\sqrt{17}$$

30. $x = 3\sec\theta,\ dx = 3\sec\theta\tan\theta\,d\theta,\ \sqrt{x^2 - 9} = 3\tan\theta$

$$\int x^3\sqrt{x^2-9}\,dx = \int 27\sec^3\theta\,(3\tan\theta)\,3\sec\theta\tan\theta\,d\theta$$
$$= 243\int \sec^4\theta\tan^2\theta\,d\theta$$
$$= 243\int (1+\tan^2\theta)\tan^2\theta\sec^2\theta\,d\theta$$
$$= 243\left[\frac{\tan^3\theta}{3} + \frac{\tan^5\theta}{5}\right] + C$$
$$= 243\left[\frac{(x^2-9)^{3/2}}{81} + \frac{(x^2-9)^{5/2}}{1215}\right] + C$$

$$\int_3^4 x^3\sqrt{x^2-9}\,dx = 243\left[\frac{(x^2-9)^{3/2}}{81} + \frac{(x^2-9)^{5/2}}{1215}\right]_3^4$$
$$= 243\left[\frac{7^{3/2}}{81} + \frac{7^{5/2}}{1215}\right]$$
$$= 243\left[\frac{7\sqrt{7}}{81} + \frac{49\sqrt{7}}{1215}\right]$$
$$= \frac{154}{5}\sqrt{7}$$

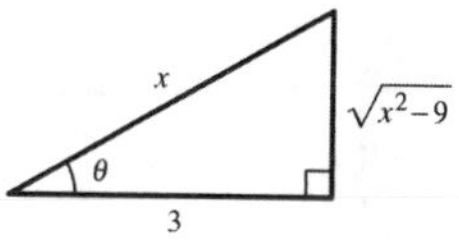

31. (a) Let $x = 2\tan\theta$, $dx = 2\sec^2\theta\, d\theta$.

$$\begin{aligned}\int \frac{x^3}{\sqrt{4+x^2}}\,dx &= \int \frac{8\tan^3\theta}{2\sec\theta}\,2\sec^2\theta\,d\theta\\ &= 8\int \tan^3\theta\sec\theta\,d\theta\\ &= 8\int \frac{\sin^3\theta}{\cos^4\theta}\,d\theta\\ &= 8\int \left(1-\cos^2\theta\right)\cos^{-4}\theta\sin\theta\,d\theta\\ &= 8\int \left(\cos^{-4}\theta - \cos^{-2}\theta\right)\sin\theta\,d\theta\\ &= 8\left[\frac{\cos^{-3}\theta}{3} - \frac{\cos^{-1}\theta}{-1}\right] + C\\ &= \frac{8}{3}\sec\theta\left(\sec^2\theta - 3\right) + C\\ &= \frac{8}{3}\left(\frac{\sqrt{4+x^2}}{2}\right)\left(\frac{4+x^2}{4} - 3\right) + C\\ &= \frac{1}{3}\sqrt{4+x^2}\left(x^2-8\right) + C\end{aligned}$$

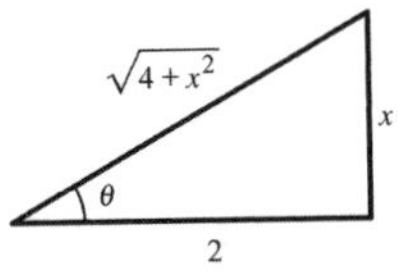

(b) $$\begin{aligned}\int \frac{x^3}{\sqrt{4+x^2}}\,dx &= \int \frac{x^2}{\sqrt{4+x^2}}\,x\,dx\\ &= \int \frac{(u^2-4)u\,du}{u}\\ &= \int \left(u^2-4\right)du\\ &= \frac{1}{3}u^3 - 4u + C\\ &= \frac{u}{3}\left(u^2-12\right) + C\\ &= \frac{\sqrt{4+x^2}}{3}\left(x^2-8\right) + C\end{aligned}$$

$u^2 = 4 + x^2$, $2u\,du = 2x\,dx$

(c) $$\begin{aligned}\int \frac{x^3}{\sqrt{4+x^2}}\,dx &= x^2\sqrt{4+x^2} - \int 2x\sqrt{4+x^2}\,dx\\ &= x^2\sqrt{4+x^2} - \frac{2}{3}\left(4+x^2\right)^{3/2} + C = \frac{\sqrt{4+x^2}}{3}\left(x^2-8\right) + C\end{aligned}$$

$dv = \dfrac{x}{\sqrt{4+x^2}}\,dx \Rightarrow v = \sqrt{4+x^2}$

$u = x^2 \Rightarrow du = 2x\,dx$

32. (a) $\int x\sqrt{4+x}\,dx = 64\int \tan^3\theta \sec^3\theta\,d\theta$

$= 64\int\left(\sec^4\theta - \sec^2\theta\right)\sec\theta\tan\theta\,d\theta$

$= \dfrac{64\sec^3\theta}{15}\left(3\sec^3\theta - 5\right) + C$

$= \dfrac{2(4+x)^{3/2}}{15}(3x-8) + C$

$x = 4\tan^2\theta,\ dx = 8\tan\theta\sec^2\theta\,d\theta,$

$\sqrt{4+x} = 2\sec\theta$

(b) $\int x\sqrt{4+x}\,dx = 2\int\left(u^4 - 4u^2\right)du$

$= \dfrac{2u^3}{15}\left(3u^2 - 20\right) + C$

$= \dfrac{2(4+x)^{3/2}}{15}(3x-8) + C$

$u^2 = 4 + x,\ dx = 2u\,du$

(c) $\int x\sqrt{4+x}\,dx = \int\left(u^{3/2} - 4u^{1/2}\right)du$

$= \dfrac{2u^{3/2}}{15}(3u - 20) + C$

$= \dfrac{2(4+x)^{3/2}}{15}(3x-8) + C$

$u = 4 + x,\ du = dx$

(d) $\int x\sqrt{4+x}\,dx = \dfrac{2x}{3}(4+x)^{3/2} - \dfrac{2}{3}\int(4+x)^{3/2}\,dx$

$= \dfrac{2x}{3}(4+x)^{3/2} - \dfrac{4}{15}(4+x)^{5/2} + C$

$= \dfrac{2(4+x)^{3/2}}{15}(3x-8) + C$

$dv = \sqrt{4+x}\,dx \Rightarrow\ v = \dfrac{2}{3}(4+x)^{3/2}$

$u = x \qquad\qquad \Rightarrow\ du = dx$

33. $\dfrac{x-39}{x^2 - x - 12} = \dfrac{x-39}{(x-4)(x+3)} = \dfrac{A}{x-4} + \dfrac{B}{x+3}$

$x - 39 = A(x+3) + B(x-4)$

When $x = -3$, $-42 = -7B \Rightarrow B = 6$.

When $x = 4$, $\ -35 = 7A \Rightarrow A = -5$.

$\int\dfrac{x-39}{x^2 - x - 12}\,dx = \int\dfrac{-5}{x-4}\,dx + \int\dfrac{6}{x+3}\,dx$

$= -5\ln|x-4| + 6\ln|x+3| + C$

34. $\dfrac{5x-2}{x^2 - x} = \dfrac{5x-2}{x(x-1)} = \dfrac{A}{x} + \dfrac{B}{x-1}$

$5x - 2 = A(x-1) + Bx$

When $x = 1$, $\ 3 = B$.

When $x = 0$, $\ -2 = -A \Rightarrow A = 2$.

$\int\dfrac{5x-2}{x^2 - x}\,dx = \int\left(\dfrac{2}{x} + \dfrac{3}{x-1}\right)dx$

$= 2\ln|x| + 3\ln|x-1| + C$

35. $\dfrac{x^2 + 2x}{(x - 1)(x^2 + 1)} = \dfrac{A}{x - 1} + \dfrac{Bx + C}{x^2 + 1}$

$x^2 + 2x = A(x^2 + 1) + (Bx + C)(x - 1)$

When $x = 1$, $3 = 2A \Rightarrow A = \dfrac{3}{2}$.

When $x = 0$, $0 = A - C \Rightarrow C = \dfrac{3}{2}$.

When $x = 2$, $8 = 5A + 2B + C \Rightarrow B = -\dfrac{1}{2}$.

$$\begin{aligned}\int \frac{x^2 + 2x}{x^3 - x^2 + x - 1}\,dx &= \frac{3}{2}\int \frac{1}{x - 1}\,dx - \frac{1}{2}\int \frac{x - 3}{x^2 + 1}\,dx\\ &= \frac{3}{2}\int \frac{1}{x - 1}\,dx - \frac{1}{4}\int \frac{2x}{x^2 + 1}\,dx + \frac{3}{2}\int \frac{1}{x^2 + 1}\,dx\\ &= \frac{3}{2}\ln\left|x - 1\right| - \frac{1}{4}\ln\left|x^2 + 1\right| + \frac{3}{2}\arctan x + C\\ &= \frac{1}{4}\left[6\ln\left|x - 1\right| - \ln\left(x^2 + 1\right) + 6\arctan x\right] + C\end{aligned}$$

36. $\dfrac{4x - 2}{3(x - 1)^2} = \dfrac{A}{x - 1} + \dfrac{B}{(x - 1)^2}$

$4x - 2 = 3A(x - 1) + 3B$

When $x = 1$, $2 = 3B \Rightarrow B = \dfrac{2}{3}$.

When $x = 2$, $6 = 3A + 3B \Rightarrow A = \dfrac{4}{3}$.

$$\int \frac{4x - 2}{3(x - 1)^2}\,dx = \frac{4}{3}\int \frac{1}{x - 1}\,dx + \frac{2}{3}\int \frac{1}{(x - 1)^2}\,dx = \frac{4}{3}\ln\left|x - 1\right| - \frac{2}{3(x - 1)} + C = \frac{2}{3}\left(2\ln\left|x - 1\right| - \frac{1}{x - 1}\right) + C$$

37. $\dfrac{x^2}{x^2 + 5x - 24} = 1 - \dfrac{5x - 24}{x^2 + 5x - 24} = 1 - \dfrac{5x - 24}{(x + 8)(x - 3)}$

$\dfrac{5x - 24}{(x + 8)(x - 3)} = \dfrac{A}{x + 8} + \dfrac{B}{x - 3}$

$5x - 24 = A(x - 3) + B(x + 8)$

When $x = 3$, $-9 = 11B \Rightarrow B = -\dfrac{9}{11}$.

When $x = -8$, $-64 = -11A \Rightarrow A = \dfrac{64}{11}$.

$$\begin{aligned}\int \frac{x^2}{x^2 + 5x - 24}\,dx &= \int\left[1 - \frac{64/11}{x + 8} + \frac{9/11}{x - 3}\right]dx\\ &= x - \frac{64}{11}\ln\left|x + 8\right| + \frac{9}{11}\ln\left|x - 3\right| + C\end{aligned}$$

38. $u = \tan\theta, du = \sec^2\theta\, d\theta$

$$\frac{1}{u(u-1)} = \frac{A}{u} + \frac{B}{u-1}$$

$$1 = A(u-1) + Bu$$

When $u = 0$, $1 = -A \Rightarrow A = -1$.

When $u = 1$, $1 = B$.

$$\int \frac{\sec^2\theta}{\tan\theta(\tan\theta - 1)}\, d\theta = \int \frac{1}{u(u-1)}\, du = \int \frac{1}{u-1}\, du - \int \frac{1}{u}\, du$$

$$= \ln|u-1| - \ln|u| + C = \ln\left|\frac{\tan\theta - 1}{\tan\theta}\right| + C = \ln|1 - \cot\theta| + C$$

39. Using Formula 4: $(a = 4, b = 5)$

$$\int \frac{x}{(4+5x)^2}\, dx = \frac{1}{25}\left(\frac{4}{4+5x} + \ln|4+5x|\right) + C$$

40. Using Formula 21: $(a = 4, b = 5)$

$$\int \frac{x}{\sqrt{4+5x}}\, dx = \frac{-2(8-5x)}{75}\sqrt{4+5x} + C$$

$$= \frac{10x - 16}{75}\sqrt{4+5x} + C$$

41. Let $u = x^2, du = 2x\, dx$.

$$\int_0^{\sqrt{\pi}/2} \frac{x}{1+\sin x^2}\, dx = \frac{1}{2}\int_0^{\pi/4} \frac{1}{1+\sin u}\, du$$

$$= \frac{1}{2}\Big[\tan u - \sec u\Big]_0^{\pi/4}$$

$$= \frac{1}{2}\Big[\big(1 - \sqrt{2}\big) - (0 - 1)\Big]$$

$$= 1 - \frac{\sqrt{2}}{2}$$

42. Let $u = x^2, du = 2x\, dx$.

$$\int_0^1 \frac{x}{1+e^{x^2}}\, dx = \frac{1}{2}\int_0^1 \frac{1}{1+e^u}\, du$$

$$= \frac{1}{2}\Big[u - \ln(1+e^u)\Big]_0^1$$

$$= \frac{1}{2}\Big[\big(1 - \ln(1+e)\big) + \ln 2\Big]$$

$$= \frac{1}{2}\left[1 + \ln\left(\frac{2}{1+e}\right)\right]$$

43. $$\int \frac{x}{x^2+4x+8}\, dx = \frac{1}{2}\left[\ln|x^2+4x+8| - 4\int \frac{1}{x^2+4x+8}\, dx\right] \qquad \text{(Formula 15)}$$

$$= \frac{1}{2}\Big[\ln|x^2+4x+8|\Big] - 2\left[\frac{2}{\sqrt{32-16}}\arctan\left(\frac{2x+4}{\sqrt{32-16}}\right)\right] + C \qquad \text{(Formula 14)}$$

$$= \frac{1}{2}\ln|x^2+4x+8| - \arctan\left(1 + \frac{x}{2}\right) + C$$

44. $$\int \frac{3}{2x\sqrt{9x^2-1}}\, dx = \frac{3}{2}\int \frac{1}{3x\sqrt{(3x)^2-1}}\, 3\, dx \qquad (u = 3x)$$

$$= \frac{3}{2}\,\text{arcsec}|3x| + C \qquad \text{(Formula 33)}$$

45. $$\int \frac{1}{\sin \pi x \cos \pi x}\, dx = \frac{1}{\pi}\int \frac{1}{\sin \pi x \cos \pi x}(\pi)\, dx \qquad (u = \pi x)$$

$$= \frac{1}{\pi}\ln|\tan \pi x| + C \qquad \text{(Formula 58)}$$

46. $\displaystyle\int \frac{1}{1+\tan \pi x}\,dx = \frac{1}{\pi}\int \frac{1}{1+\tan \pi x}(\pi)\,dx \qquad (u = \pi x)$

$\displaystyle = \frac{1}{\pi}\left[\frac{1}{2}\left(\pi x + \ln\left|\cos \pi x + \sin \pi x\right|\right)\right] + C \qquad \text{(Formula 71)}$

47. $dv = dx \quad \Rightarrow \quad v = x$

$u = (\ln x)^n \quad \Rightarrow \quad du = n(\ln x)^{n-1}\dfrac{1}{x}\,dx$

$\displaystyle\int (\ln x)^n\,dx = x(\ln x)^n - n\int (\ln x)^{n-1}\,dx$

48. $\displaystyle\int \tan^n x\,dx = \int \tan^{n-2} x(\sec^2 x - 1)\,dx$

$\displaystyle = \int \tan^{n-2} x \sec^2 x\,dx - \int \tan^{n-2} x\,dx$

$\displaystyle = \frac{1}{n-1}\tan^{n-1} x - \int \tan^{n-2} x\,dx$

49. $\displaystyle\int \theta \sin\theta \cos\theta\,d\theta = \frac{1}{2}\int \theta \sin 2\theta\,d\theta$

$\displaystyle = -\frac{1}{4}\theta\cos 2\theta + \frac{1}{4}\int \cos 2\theta\,d\theta = -\frac{1}{4}\theta\cos 2\theta + \frac{1}{8}\sin 2\theta + C = \frac{1}{8}(\sin 2\theta - 2\theta\cos 2\theta) + C$

$dv = \sin 2\theta\,d\theta \Rightarrow v = -\dfrac{1}{2}\cos 2\theta$

$u = \theta \quad \Rightarrow du = d\theta$

50. $\displaystyle\int \frac{\csc\sqrt{2x}}{\sqrt{x}}\,dx = \sqrt{2}\int \csc\sqrt{2x}\left(\frac{1}{\sqrt{2x}}\right)dx = -\sqrt{2}\ln\left|\csc\sqrt{2x} + \cot\sqrt{2x}\right| + C$

$u = \sqrt{2x},\ du = \dfrac{1}{\sqrt{2x}}\,dx$

51. $\displaystyle\int \frac{x^{1/4}}{1+x^{1/2}}\,dx = 4\int \frac{u(u^3)}{1+u^2}\,du$

$\displaystyle = 4\int\left(u^2 - 1 + \frac{1}{u^2+1}\right)du$

$\displaystyle = 4\left(\frac{1}{3}u^3 - u + \arctan u\right) + C$

$\displaystyle = \frac{4}{3}\left[x^{3/4} - 3x^{1/4} + 3\arctan\left(x^{1/4}\right)\right] + C$

$u = \sqrt[4]{x},\ x = u^4,\ dx = 4u^3\,du$

52. $\displaystyle\int \sqrt{1+\sqrt{x}}\,dx = \int u(4u^3 - 4u)\,du = \int (4u^4 - 4u^2)\,du = \frac{4u^5}{5} - \frac{4u^3}{3} + C = \frac{4}{15}\left(1+\sqrt{x}\right)^{3/2}\left(3\sqrt{x} - 2\right) + C$

$u = \sqrt{1+\sqrt{x}},\ x = u^4 - 2u^2 + 1,\ dx = (4u^3 - 4u)\,du$

53. $\int \sqrt{1 + \cos x}\, dx = \int \frac{\sqrt{1 + \cos x}}{1} \cdot \frac{\sqrt{1 - \cos x}}{\sqrt{1 - \cos x}}\, dx$

$$= \int \frac{\sin x}{\sqrt{1 - \cos x}}\, dx$$

$$= \int (1 - \cos x)^{-1/2}(\sin x)\, dx$$

$$= 2\sqrt{1 - \cos x} + C$$

$u = 1 - \cos x,\ du = \sin x\, dx$

54. $\frac{3x^3 + 4x}{(x^2 + 1)^2} = \frac{Ax + B}{x^2 + 1} + \frac{Cx + D}{(x^2 + 1)^2}$

$3x^3 + 4x = (Ax + B)(x^2 + 1) + Cx + D = Ax^3 + Bx^2 + (A + C)x + (B + D)$

$A = 3,\ B = 0,\ A + C = 4 \Rightarrow C = 1,$

$B + D = 0 \Rightarrow D = 0$

$$\int \frac{3x^3 + 4x}{(x^2 + 1)^2}\, dx = 3\int \frac{x}{x^2 + 1}\, dx + \int \frac{x}{(x^2 + 1)^2}\, dx = \frac{3}{2}\ln(x^2 + 1) - \frac{1}{2(x^2 + 1)} + C$$

55. $\int \cos x \ln(\sin x)\, dx = \sin x \ln(\sin x) - \int \cos x\, dx = \sin x \ln(\sin x) - \sin x + C$

$dv = \cos x\, dx \Rightarrow v = \sin x$

$u = \ln(\sin x) \Rightarrow du = \frac{\cos x}{\sin x}\, dx$

56. $\int (\sin\theta + \cos\theta)^2\, d\theta = \int (\sin^2\theta + 2\sin\theta\cos\theta + \cos^2\theta)\, d\theta$

$$= \int (1 + \sin 2\theta)\, d\theta = \theta - \frac{1}{2}\cos 2\theta + C = \frac{1}{2}(2\theta - \cos 2\theta) + C$$

57. $y = \int \frac{25}{x^2 - 25}\, dx = 25\left(\frac{1}{10}\right)\ln\left|\frac{x - 5}{x + 5}\right| + C$

$$= \frac{5}{2}\ln\left|\frac{x - 5}{x + 5}\right| + C$$

(Formula 24)

58. $y = \int \frac{\sqrt{4 - x^2}}{2x}\, dx = \int \frac{2\cos\theta(2\cos\theta)\, d\theta}{4\sin\theta}$

$$= \int (\csc\theta - \sin\theta)\, d\theta$$

$$= \left[-\ln\left|\csc\theta + \cos\theta\right| + \cos\theta\right] + C$$

$$= -\ln\left|\frac{2 + \sqrt{4 - x^2}}{x}\right| + \frac{\sqrt{4 - x^2}}{2} + C$$

$x = 2\sin\theta,\ dx = 2\cos\theta\, d\theta,\ \sqrt{4 - x^2} = 2\cos\theta$

59. $y = \int \ln(x^2 + x)\,dx = x\ln|x^2 + x| - \int \frac{2x^2 + x}{x^2 + x}\,dx$

$= x\ln|x^2 + x| - \int \frac{2x + 1}{x + 1}\,dx$

$= x\ln|x^2 + x| - \int 2\,dx + \int \frac{1}{x + 1}\,dx$

$= x\ln|x^2 + x| - 2x + \ln|x + 1| + C$

$dv = dx \quad \Rightarrow \quad v = x$

$u = \ln(x^2 + x) \Rightarrow du = \frac{2x + 1}{x^2 + x}\,dx$

60. $y = \int \sqrt{1 - \cos\theta}\,d\theta$

$= \int \frac{\sin\theta}{\sqrt{1 + \cos\theta}}\,d\theta$

$= -\int (1 + \cos\theta)^{-1/2}(-\sin\theta)\,d\theta$

$= -2\sqrt{1 + \cos\theta} + C$

$u = 1 + \cos\theta, du = -\sin\theta\,d\theta$

61. $\int_2^{\sqrt{5}} x(x^2 - 4)^{3/2}\,dx = \left[\frac{1}{5}(x^2 - 4)^{5/2}\right]_2^{\sqrt{5}} = \frac{1}{5}$

62. $\int_0^1 \frac{x}{(x - 2)(x - 4)}\,dx = \left[2\ln|x - 4| - \ln|x - 2|\right]_0^1$

$= 2\ln 3 - 2\ln 4 + \ln 2$

$= \ln\frac{9}{8} \approx 0.118$

63. $\int_1^4 \frac{\ln x}{x}\,dx = \left[\frac{1}{2}(\ln x)^2\right]_1^4 = \frac{1}{2}(\ln 4)^2 \approx 0.961$

64. $\int_0^2 xe^{3x}\,dx = \left[\frac{e^{3x}}{9}(3x - 1)\right]_0^2 = \frac{1}{9}(5e^6 + 1) \approx 224.238$

65. $\int_0^{\pi} x\sin x\,dx = \left[-x\cos x + \sin x\right]_0^{\pi} = \pi$

66. $\int_0^5 \frac{x}{\sqrt{4 + x}}\,dx = \left[\frac{2x - 16}{3}\sqrt{4 + x}\right]_0^5$

$= -2(3) + \frac{16}{3}(2) = \frac{14}{3}$

67. $A = \int_0^4 x\sqrt{4 - x}\,dx = \int_2^0 (4 - u^2)\,u(-2u)\,du$

$= \int_2^0 2(u^4 - 4u^2)\,du$

$= \left[2\left(\frac{u^5}{5} - \frac{4u^3}{3}\right)\right]_2^0 = \frac{128}{15}$

$u = \sqrt{4 - x}, x = 4 - u^2, dx = -2u\,du$

68. $A = \int_0^4 \frac{1}{25 - x^2}\,dx$

$= \left[-\frac{1}{10}\ln\left|\frac{x - 5}{x + 5}\right|\right]_0^4 = -\frac{1}{10}\ln\frac{1}{9} = \frac{1}{10}\ln 9 \approx 0.220$

69. By symmetry, $\bar{x} = 0, A = \frac{1}{2}\pi.$

$\bar{y} = \frac{2}{\pi}\left(\frac{1}{2}\right)\int_{-1}^{1}\left(\sqrt{1 - x^2}\right)^2\,dx = \frac{1}{\pi}\left[x - \frac{1}{3}x^3\right]_{-1}^{1} = \frac{4}{3\pi}$

$(\bar{x}, \bar{y}) = \left(0, \frac{4}{3\pi}\right)$

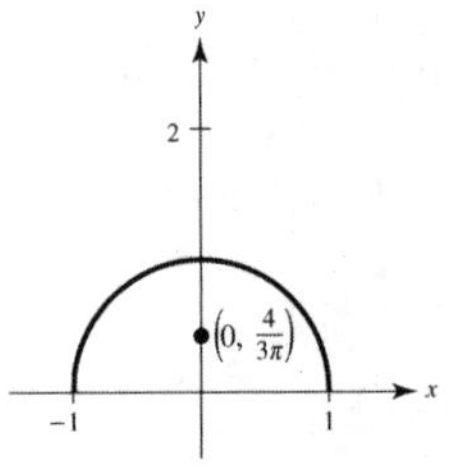

70. By symmetry, $\bar{y} = 0.$

$A = \pi + 4\pi = 5\pi$

$\bar{x} = \frac{1(\pi) + 4(4\pi)}{\pi + 4\pi}$

$= \frac{17\pi}{5\pi} = 3.4$

$(\bar{x}, \bar{y}) = (3.4, 0)$

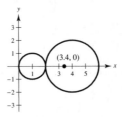

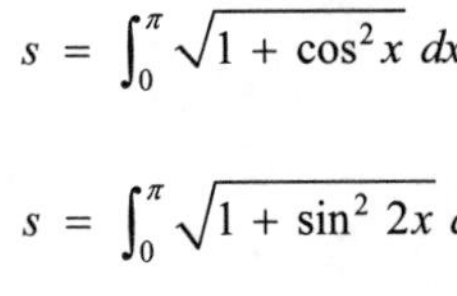

71. $s = \int_0^{\pi} \sqrt{1 + \cos^2 x}\,dx \approx 3.82$

72. $s = \int_0^{\pi} \sqrt{1 + \sin^2 2x}\,dx \approx 3.82$

73. $\lim_{x\to 1}\left[\dfrac{(\ln x)^2}{x-1}\right] = \lim_{x\to 1}\left[\dfrac{2(1/x)\ln x}{1}\right] = 0$

74. $\lim_{x\to 0}\dfrac{\sin \pi x}{\sin 5\pi x} = \lim_{x\to 0}\dfrac{\pi\cos\pi x}{5\pi\cos 5\pi x} = \dfrac{1}{5}$

75. $\lim_{x\to\infty}\dfrac{e^{2x}}{x^2} = \lim_{x\to\infty}\dfrac{2e^{2x}}{2x} = \lim_{x\to\infty}\dfrac{4e^{2x}}{2} = \infty$

76. $\lim_{x\to\infty} xe^{-x^2} = \lim_{x\to\infty}\dfrac{x}{e^{x^2}} = \lim_{x\to\infty}\dfrac{1}{2xe^{x^2}} = 0$

77. $y = \lim_{x\to\infty}(\ln x)^{2/x}$

$\ln y = \lim_{x\to\infty}\dfrac{2\ln(\ln x)}{x} = \lim_{x\to\infty}\left[\dfrac{2/(x\ln x)}{1}\right] = 0$

Because $\ln y = 0$, $y = 1$.

78. $y = \lim_{x\to 1^+}(x-1)^{\ln x}$

$\ln y = \lim_{x\to 1^+}\left[(\ln x)\ln(x-1)\right]$

$= \lim_{x\to 1^+}\left[\dfrac{\ln(x-1)}{\dfrac{1}{\ln x}}\right] = \lim_{x\to 1^+}\left[\dfrac{\dfrac{1}{x-1}}{\left(\dfrac{1}{x}\right)\dfrac{-1}{\ln^2 x}}\right] = \lim_{x\to 1^+}\left[\dfrac{-\ln^2 x}{\dfrac{x-1}{x}}\right] = \lim_{x\to 1^+}\left[\dfrac{-2\left(\dfrac{1}{x}\right)(\ln x)}{\dfrac{1}{x^2}}\right]$

$= \lim_{x\to 1^+} 2x(\ln x) = 0$

Because $\ln y = 0$, $y = 1$.

79. $\lim_{n\to\infty} 1000\left(1+\dfrac{0.09}{n}\right)^n = 1000\lim_{n\to\infty}\left(1+\dfrac{0.09}{n}\right)^n$

Let $y = \lim_{n\to\infty}\left(1+\dfrac{0.09}{n}\right)^n$.

$\ln y = \lim_{n\to\infty} n\ln\left(1+\dfrac{0.09}{n}\right) = \lim_{n\to\infty}\dfrac{\ln\left(1+\dfrac{0.09}{n}\right)}{\dfrac{1}{n}} = \lim_{n\to\infty}\left(\dfrac{\dfrac{-0.09/n^2}{1+(0.09/n)}}{-\dfrac{1}{n^2}}\right) = \lim_{n\to\infty}\dfrac{0.09}{1+\left(\dfrac{0.09}{n}\right)} = 0.09$

So, $\ln y = 0.09 \Rightarrow y = e^{0.09}$ and $\lim_{n\to\infty} 1000\left(1+\dfrac{0.09}{n}\right)^n = 1000e^{0.09} \approx 1094.17$.

80. $\lim_{x\to 1^+}\left(\dfrac{2}{\ln x} - \dfrac{2}{x-1}\right) = \lim_{x\to 1^+}\left[\dfrac{2x-2-2\ln x}{(\ln x)(x-1)}\right]$

$= \lim_{x\to 1^+}\left[\dfrac{2-(2/x)}{(x-1)(1/x)+\ln x}\right]$

$= \lim_{x\to 1^+}\dfrac{2x-2}{(x-1)+x\ln x}$

$= \lim_{x\to 1^+}\dfrac{2}{1+1+\ln x} = 1$

81. $\int_0^{16}\dfrac{1}{\sqrt[4]{x}}\,dx = \lim_{b\to 0^+}\left[\dfrac{4}{3}x^{3/4}\right]_b^{16} = \dfrac{32}{3}$

82. $\int_0^2\dfrac{7}{x-2}\,dx = \lim_{b\to 2^-}\left[7\ln|x-2|\right]_0^b$

$= -\infty$ Diverges

83. $\int_1^\infty x^2\ln x\,dx = \lim_{b\to\infty}\left[\dfrac{x^3}{9}(-1+3\ln x)\right]_1^b = \infty$

Diverges

84. $\int_0^\infty\dfrac{e^{-1/x}}{x^2}\,dx = \lim_{\substack{a\to 0^+\\ b\to\infty}}\left[e^{-1/x}\right]_a^b = 1 - 0 = 1$

85. Let $u = \ln x$, $du = \dfrac{1}{x}\,dx$, $dv = x^{-2}\,dx$, $v = -x^{-1}$.

$\int\dfrac{\ln x}{x^2}\,dx = \dfrac{-\ln x}{x} + \int\dfrac{1}{x^2}\,dx = \dfrac{-\ln x}{x} - \dfrac{1}{x} + C$

$\int_1^\infty\dfrac{\ln x}{x^2}\,dx = \lim_{b\to\infty}\left[\dfrac{-\ln x}{x} - \dfrac{1}{x}\right]_1^b$

$= \lim_{b\to\infty}\left(\dfrac{-\ln b}{b} - \dfrac{1}{b}\right) - (-1)$

$= 0 + 1 = 1$

86. $\int_1^\infty\dfrac{1}{\sqrt[4]{x}}\,dx = \lim_{b\to\infty}\int_1^b x^{-1/4}\,dx$

$= \lim_{b\to\infty}\left[\dfrac{4}{3}x^{3/4}\right]_1^b$

$= \lim_{b\to\infty}\left[\dfrac{4}{3}b^{3/4} - \dfrac{4}{3}\right] = \infty$

Diverges

87. $\int_2^\infty \frac{1}{x\sqrt{x^2-4}}\,dx = \int_2^3 \frac{1}{x\sqrt{x^2-4}}\,dx + \int_3^\infty \frac{1}{x\sqrt{x^2-4}}\,dx$

$$= \lim_{b\to 2^+}\left[\frac{1}{2}\operatorname{arcsec}\left(\frac{x}{2}\right)\right]_b^3 + \lim_{c\to\infty}\left[\frac{1}{2}\operatorname{arcsec}\left(\frac{x}{2}\right)\right]_3^c$$

$$= \frac{1}{2}\operatorname{arcsec}\left(\frac{3}{2}\right) - \frac{1}{2}(0) + \frac{1}{2}\left(\frac{\pi}{2}\right) - \frac{1}{2}\operatorname{arcsec}\left(\frac{3}{2}\right)$$

$$= \frac{\pi}{4}$$

88. Let $u = \sqrt{x}$, $du = \frac{1}{2\sqrt{x}}\,dx \Rightarrow dx = 2u\,du$.

$$\int \frac{2}{\sqrt{x}(x+4)}\,dx = \int \frac{2}{u(u^2+4)}2u\,du = \int \frac{4}{u^2+4}\,du = 2\arctan\left(\frac{u}{2}\right) + C = 2\arctan\left(\frac{\sqrt{x}}{2}\right) + C$$

$$\int_0^\infty \frac{2}{\sqrt{x}(x+4)}\,dx = \lim_{b\to 0^+}\left[2\arctan\left(\frac{\sqrt{x}}{2}\right)\right]_b^1 + \lim_{c\to\infty}\left[2\arctan\left(\frac{\sqrt{x}}{2}\right)\right]_1^c = \left(2\arctan\frac{1}{2} - 0\right) + 2\left(\frac{\pi}{2}\right) - 2\arctan\frac{1}{2} = \pi$$

89. $\int_0^{t_0} 500{,}000e^{-0.05t}\,dt = \left[\frac{500{,}000}{-0.05}e^{-0.05t}\right]_0^{t_0}$

$$= \frac{-500{,}000}{0.05}\left(e^{-0.05t_0} - 1\right)$$

$$= 10{,}000{,}000\left(1 - e^{-0.05t_0}\right)$$

(a) $t_0 = 20$: \$6,321,205.59

(b) $t_0 \to \infty$: \$10,000,000

90. $V = \pi\int_0^\infty \left(xe^{-x}\right)^2\,dx$

$$= \pi\int_0^\infty x^2e^{-2x}\,dx$$

$$= \lim_{b\to\infty}\left[-\frac{\pi e^{-2x}}{4}\left(2x^2 + 2x + 1\right)\right]_0^b = \frac{\pi}{4}$$

91. (a) $P(13 \le x < \infty) = \frac{1}{0.95\sqrt{2\pi}}\int_{13}^\infty e^{-(x-12.9)^2/2(0.95)^2}\,dx \approx 0.4581$

(b) $P(15 \le x < \infty) = \frac{1}{0.95\sqrt{2\pi}}\int_{15}^\infty e^{-(x-12.9)^2/2(0.95)^2}\,dx \approx 0.0135$

Problem Solving for Chapter 8

1. (a) $\int_{-1}^1 \left(1 - x^2\right)dx = \left[x - \frac{x^3}{3}\right]_{-1}^1 = 2\left(1 - \frac{1}{3}\right) = \frac{4}{3}$

$$\int_{-1}^1 \left(1 - x^2\right)^2 dx = \int_{-1}^1 \left(1 - 2x^2 + x^4\right)dx = \left[x - \frac{2x^3}{3} + \frac{x^5}{5}\right]_{-1}^1 = 2\left(1 - \frac{2}{3} + \frac{1}{5}\right) = \frac{16}{15}$$

(b) Let $x = \sin u$, $dx = \cos u\,du$, $1 - x^2 = 1 - \sin^2 u = \cos^2 u$.

$$\int_{-1}^1 \left(1 - x^2\right)^n dx = \int_{-\pi/2}^{\pi/2} \left(\cos^2 u\right)^n \cos u\,du$$

$$= \int_{-\pi/2}^{\pi/2} \cos^{2n+1} u\,du$$

$$= 2\left[\frac{2}{3}\cdot\frac{4}{5}\cdot\frac{6}{7}\cdots\frac{(2n)}{(2n+1)}\right] \quad \text{(Wallis's Formula)}$$

$$= 2\left[\frac{2^2\cdot 4^2\cdot 6^2\cdots(2n)^2}{2\cdot 3\cdot 4\cdot 5\cdots(2n)(2n+1)}\right]$$

$$= \frac{2\left(2^{2n}\right)(n!)^2}{(2n+1)!} = \frac{2^{2n+1}(n!)^2}{(2n+1)!}$$

2. (a) $\int_0^1 \ln x\, dx = \lim_{b\to 0^+} \left[x \ln - x\right]_b^1$

$= (-1) - \lim_{b\to 0^+} (b \ln b - b) = -1$

Note: $\lim_{b\to 0^+} b \ln b = \lim_{b\to 0^+} \dfrac{\ln b}{b^{-1}} = \lim_{b\to 0^+} \dfrac{1/b}{-1/b^2} = 0$

$\int_0^1 (\ln x)^2 dx = \lim_{b\to 0^+} \left[x(\ln x)^2 - 2x \ln x + 2x\right]_b^1$

$= 2 - \lim_{b\to 0^+} \left(b (\ln b)^2 - 2b \ln b + 2b\right) = 2$

(b) Note first that $\lim_{b\to 0^+} b(\ln b)^n = 0$ (Mathematical induction).

Also, $\int (\ln x)^{n+1}\, dx = x(\ln x)^{n+1} - (n + 1)\int (\ln x)^n\, dx.$

Assume $\int_0^1 (\ln x)^n\, dx = (-1)^n n!$.

Then, $\int_0^1 (\ln x)^{n+1}\, dx = \lim_{b\to 0^+} \left[x(\ln x)^{n+1}\right]_b^1 - (n + 1)\int_0^1 (\ln x)^n\, dx = 0 - (n + 1)(-1)^n n! = (-1)^{n+1}(n + 1)!$.

3.

$$\lim_{x\to\infty} \left(\frac{x + c}{x - c}\right)^x = 9$$

$$\lim_{x\to\infty} x \ln\left(\frac{x + c}{x - c}\right) = \ln 9$$

$$\lim_{x\to\infty} \frac{\ln(x + c) - \ln(x - c)}{1/x} = \ln 9$$

$$\lim_{x\to\infty} \frac{\dfrac{1}{x + c} - \dfrac{1}{x - c}}{-\dfrac{1}{x^2}} = \ln 9$$

$$\lim_{x\to\infty} \frac{-2c}{(x + c)(x - c)}(-x^2) = \ln 9$$

$$\lim_{x\to\infty} \left(\frac{2cx^2}{x^2 - c^2}\right) = \ln 9$$

$$2c = \ln 9$$

$$2c = 2 \ln 3$$

$$c = \ln 3$$

4.

$$\lim_{x\to\infty} \left(\frac{x - c}{x + c}\right)^x = \frac{1}{4}$$

$$\lim_{x\to\infty} x \ln\left(\frac{x - c}{x + c}\right) = \ln \frac{1}{4}$$

$$\lim_{x\to\infty} \frac{\ln(x - c) - \ln(x + c)}{1/x} = -\ln 4$$

$$\lim_{x\to\infty} \frac{\dfrac{1}{x - c} - \dfrac{1}{x + c}}{-\dfrac{1}{x^2}} = -\ln 4$$

$$\lim_{x\to\infty} \frac{2c}{(x - c)(x + c)}(-x^2) = -\ln 4$$

$$\lim_{x\to\infty} \frac{2cx^2}{x^2 - c^2} = \ln 4$$

$$2c = \ln 4$$

$$2x = 2 \ln 2$$

$$c = \ln 2$$

5. $\sin\theta = \dfrac{PB}{OP} = PB, \cos\theta = OB$

$AQ = \widehat{AP} = \theta$

$BR = OR + OB = OR + \cos\theta$

The triangles ΔAQR and ΔBPR are similar:

$$\frac{AR}{AQ} = \frac{BR}{BP} \Rightarrow \frac{OR + 1}{\theta} = \frac{OR + \cos\theta}{\sin\theta}$$

$$\sin\theta(OR) + \sin\theta = (OR)\theta + \theta\cos\theta$$

$$OR = \frac{\theta\cos\theta - \sin\theta}{\sin\theta - \theta}$$

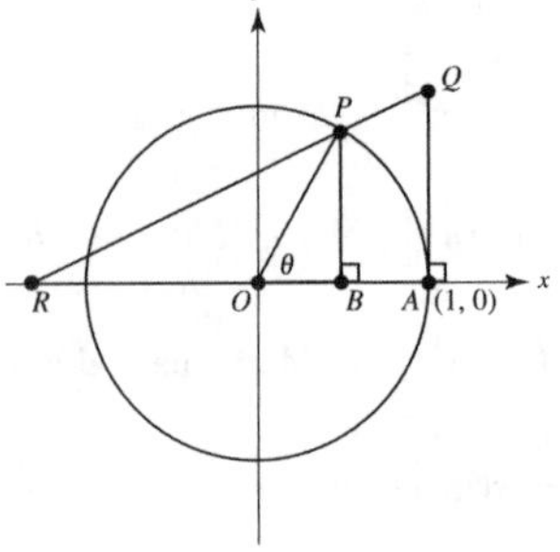

$$\begin{aligned}\lim_{\theta\to 0^+} OR &= \lim_{\theta\to 0^+} \frac{\theta\cos\theta - \sin\theta}{\sin\theta - \theta}\\ &= \lim_{\theta\to 0^+} \frac{-\theta\sin\theta + \cos\theta - \cos\theta}{\cos\theta - 1}\\ &= \lim_{\theta\to 0^+} \frac{-\theta\sin\theta}{\cos\theta - 1}\\ &= \lim_{\theta\to 0^+} \frac{-\sin\theta - \theta\cos\theta}{-\sin\theta}\\ &= \lim_{\theta\to 0^+} \frac{\cos\theta + \cos\theta - \theta\sin\theta}{\cos\theta}\\ &= 2\end{aligned}$$

6. $\sin\theta = BD, \cos\theta = OD$

$$\text{Area } \Delta DAB = \frac{1}{2}(DA)(BD) = \frac{1}{2}(1 - \cos\theta)\sin\theta$$

$$\text{Shaded area} = \frac{\theta}{2} - \frac{1}{2}(1)(BD) = \frac{\theta}{2} - \frac{1}{2}\sin\theta$$

$$R = \frac{\Delta DAB}{\text{Shaded area}} = \frac{1/2(1 - \cos\theta)\sin\theta}{1/2(\theta - \sin\theta)}$$

$$\begin{aligned}\lim_{\theta\to 0^+} R &= \lim_{\theta\to 0^+} \frac{(1 - \cos\theta)\sin\theta}{\theta - \sin\theta} = \lim_{\theta\to 0^+} \frac{(1 - \cos\theta)\cos\theta + \sin^2\theta}{1 - \cos\theta}\\ &= \lim_{\theta\to 0^+} \frac{(1 - \cos\theta)(-\sin\theta) + \cos\theta\sin\theta + 2\sin\theta\cos\theta}{\sin\theta}\\ &= \lim_{\theta\to 0^+} \frac{-\sin\theta - 4\cos\theta\sin\theta}{\sin\theta} = \lim_{\theta\to 0} \frac{4\cos\theta - 1}{1} = 3\end{aligned}$$

7. (a)

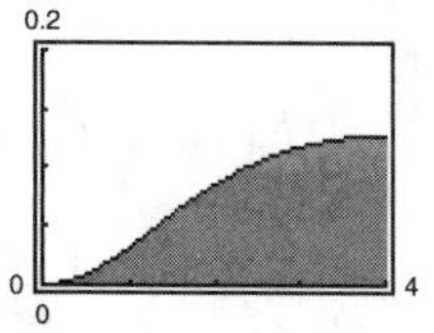

Area ≈ 0.2986

(b) Let $x = 3\tan\theta$, $dx = 3\sec^2\theta\, d\theta$, $x^2 + 9 = 9\sec^2\theta$.

$$\begin{aligned}\int \frac{x^2}{(x^2+9)^{3/2}}\,dx &= \int \frac{9\tan^2\theta}{(9\sec^2\theta)^{3/2}}(3\sec^2\theta\,d\theta)\\ &= \int \frac{\tan^2\theta}{\sec\theta}\,d\theta\\ &= \int \frac{\sin^2\theta}{\cos\theta}\,d\theta\\ &= \int \frac{1-\cos^2\theta}{\cos\theta}\,d\theta\\ &= \ln\left|\sec\theta + \tan\theta\right| - \sin\theta + C\end{aligned}$$

$$\begin{aligned}\text{Area} = \int_0^4 \frac{x^2}{(x^2+9)^{3/2}}\,dx &= \Big[\ln\left|\sec\theta + \tan\theta\right| - \sin\theta\Big]_0^{\tan^{-1}(4/3)}\\ &= \left[\ln\left(\frac{\sqrt{x^2+9}}{3} + \frac{x}{3}\right) - \frac{x}{\sqrt{x^2+9}}\right]_0^4\\ &= \ln\left(\frac{5}{3} + \frac{4}{3}\right) - \frac{4}{5} = \ln 3 - \frac{4}{5}\end{aligned}$$

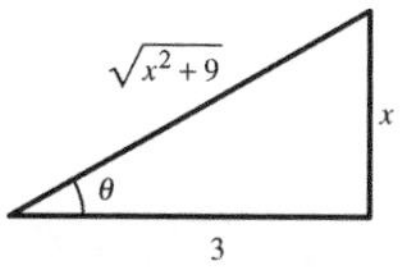

(c) $x = 3\sinh u$, $dx = 3\cosh u\,du$, $x^2 + 9 = 9\sinh^2 u + 9 = 9\cosh^2 u$

$$\begin{aligned}A = \int_0^4 \frac{x^2}{(x^2+9)^{3/2}}\,dx &= \int_0^{\sinh^{-1}(4/3)} \frac{9\sinh^2 u}{(9\cosh^2 u)^{3/2}}(3\cosh u\,du) = \int_0^{\sinh^{-1}(4/3)} \tanh^2 u\,du\\ &= \int_0^{\sinh^{-1}(4/3)} (1 - \operatorname{sech}^2 u)\,du = \Big[u - \tanh u\Big]_0^{\text{t}\sinh^{-1}(4/3)}\\ &= \sinh^{-1}\left(\frac{4}{3}\right) - \tanh\left(\sinh^{-1}\left(\frac{4}{3}\right)\right) = \ln\left(\frac{4}{3} + \sqrt{\frac{16}{9} + 1}\right) - \tanh\left[\ln\left(\frac{4}{3} + \sqrt{\frac{16}{9} + 1}\right)\right]\\ &= \ln\left(\frac{4}{3} + \frac{5}{3}\right) - \tanh\left(\ln\left(\frac{4}{3} + \frac{5}{3}\right)\right) = \ln 3 - \tanh(\ln 3)\\ &= \ln 3 - \frac{3 - (1/3)}{3 + (1/3)} = \ln 3 - \frac{4}{5}\end{aligned}$$

8. $u = \tan\dfrac{x}{2}, \cos x = \dfrac{1-u^2}{1+u^2},$

$$2 + \cos x = 2 + \frac{1-u^2}{1+u^2} = \frac{3+u^2}{1+u^2}$$

$$dx = \frac{2\,du}{1+u^2}$$

$$\int_0^{\pi/2} \frac{1}{2+\cos x}\,dx = \int_0^1 \left(\frac{1+u^2}{3+u^2}\right)\left(\frac{2}{1+u^2}\right) du$$

$$= \int_0^1 \frac{2}{3+u^2}\,du$$

$$= \left[2\frac{1}{\sqrt{3}}\arctan\left(\frac{u}{\sqrt{3}}\right)\right]_0^1$$

$$= \frac{2}{\sqrt{3}}\arctan\left(\frac{1}{\sqrt{3}}\right)$$

$$= \frac{2}{\sqrt{3}}\left(\frac{\pi}{6}\right) = \frac{\pi\sqrt{3}}{9} \approx 0.6046$$

9. $y = \ln(1-x^2), y' = \dfrac{-2x}{1-x^2}$

$$1 + (y')^2 = 1 + \frac{4x^2}{(1-x^2)^2}$$

$$= \frac{1 - 2x^2 + x^4 + 4x^2}{(1-x^2)^2}$$

$$= \left(\frac{1+x^2}{1-x^2}\right)^2$$

$$\text{Arc length} = \int_0^{1/2} \sqrt{1+(y')^2}\,dx$$

$$= \int_0^{1/2} \left(\frac{1+x^2}{1-x^2}\right) dx$$

$$= \int_0^{1/2} \left(-1 + \frac{2}{1-x^2}\right) dx$$

$$= \int_0^{1/2} \left(-1 + \frac{1}{x+1} + \frac{1}{1-x}\right) dx$$

$$= \left[-x + \ln(1+x) - \ln(1-x)\right]_0^{1/2}$$

$$= \left(-\frac{1}{2} + \ln\frac{3}{2} - \ln\frac{1}{2}\right)$$

$$= -\frac{1}{2} + \ln 3 - \ln 2 + \ln 2$$

$$= \ln 3 - \frac{1}{2} \approx 0.5986$$

10. Let $u = cx, du = c\,dx$.

$$\int_0^b e^{-c^2x^2}\,dx = \int_0^{cb} e^{-u^2}\frac{du}{c} = \frac{1}{c}\int_0^{cb} e^{-u^2}\,du$$

As $b \to \infty, cb \to \infty$. So, $\displaystyle\int_0^\infty e^{-c^2x^2}dx = \frac{1}{c}\int_0^\infty e^{-x^2}\,dx.$

$\bar{x} = 0$ by symmetry.

$$\bar{y} = \frac{M_x}{m} = \frac{2\displaystyle\int_0^\infty \frac{\left(e^{-c^2x^2}\right)^2}{2}\,dx}{2\displaystyle\int_0^\infty e^{-c^2x^2}\,dx}$$

$$= \left(\frac{1}{2}\right)\frac{\displaystyle\int_0^\infty e^{-2c^2x^2}\,dx}{\displaystyle\int_0^\infty e^{-c^2x^2}\,dx}$$

$$= \left(\frac{1}{2}\right)\frac{\dfrac{1}{\sqrt{2}c}\displaystyle\int_0^\infty e^{-x^2}\,dx}{\dfrac{1}{c}\displaystyle\int_0^\infty e^{-x^2}\,dx}$$

$$= \frac{1}{2\sqrt{2}} = \frac{\sqrt{2}}{4}$$

So, $(\bar{x}, \bar{y}) = \left(0, \dfrac{\sqrt{2}}{4}\right)$.

11. Using a graphing utility,

(a) $\lim_{x\to 0^+}\left(\cot x + \frac{1}{x}\right) = \infty.$

(b) $\lim_{x\to 0^+}\left(\cot x - \frac{1}{x}\right) = 0.$

(c) $\lim_{x\to 0^+}\left(\cot x + \frac{1}{x}\right)\left(\cot x - \frac{1}{x}\right) \approx -\frac{2}{3}.$

Analytically,

(a) $\lim_{x\to 0^+}\left(\cot x + \frac{1}{x}\right) = \infty + \infty = \infty.$

(b)
$$\begin{aligned}
\lim_{x\to 0^+}\left(\cot x - \frac{1}{x}\right) &= \lim_{x\to 0^+}\frac{x\cot x - 1}{x} = \lim_{x\to 0^+}\frac{x\cos x - \sin x}{x\sin x}\\
&= \lim_{x\to 0^+}\frac{\cos x - x\sin x - \cos x}{\sin x + x\cos x}\\
&= \lim_{x\to 0^+}\frac{-x\sin x}{\sin x + x\cos x}\\
&= \lim_{x\to 0^+}\frac{-\sin x - x\cos x}{\cos x + \cos x - x\sin x} = 0.
\end{aligned}$$

(c)
$$\begin{aligned}
\left(\cot x + \frac{1}{x}\right)\left(\cot x - \frac{1}{x}\right) &= \cot^2 x - \frac{1}{x^2}\\
&= \frac{x^2\cot^2 x - 1}{x^2}
\end{aligned}$$

$$\begin{aligned}
\lim_{x\to 0^+}\frac{x^2\cot^2 x - 1}{x^2} &= \lim_{x\to 0^+}\frac{2x\cot^2 x - 2x^2\cot x\csc^2 x}{2x}\\
&= \lim_{x\to 0^+}\frac{\cot^2 x - x\cot x\csc^2 x}{1}\\
&= \lim_{x\to 0^+}\frac{\cos^2 x\sin x - x\cos x}{\sin^3 x}\\
&= \lim_{x\to 0^+}\frac{\left(1 - \sin^2 x\right)\sin x - x\cos x}{\sin^3 x}\\
&= \lim_{x\to 0^+}\frac{\sin x - x\cos x}{\sin^3 x} - 1.
\end{aligned}$$

Now,
$$\begin{aligned}
\lim_{x\to 0^+}\frac{\sin x - x\cos x}{\sin^3 x} &= \lim_{x\to 0^+}\frac{\cos x - \cos x + x\sin x}{3\sin^2 x\cos x}\\
&= \lim_{x\to 0^+}\frac{x}{3\sin x\cdot\cos x}\\
&= \lim_{x\to 0^+}\left(\frac{x}{\sin x}\right)\frac{1}{3\cos x} = \frac{1}{3}.
\end{aligned}$$

So, $\lim_{x\to 0^+}\left(\cot x + \frac{1}{x}\right)\left(\cot x - \frac{1}{x}\right) = \frac{1}{3} - 1 = -\frac{2}{3}.$

The form $0\cdot\infty$ is indeterminant.

12. (a) Let $y = f^{-1}(x)$, $f(y) = x$, $dx = f'(y)\,dy$.

$$\int f^{-1}(x)\,dx = \int yf'(y)\,dy$$
$$= yf(y) - \int f(y)\,dy \qquad \begin{bmatrix} u = y, du = dy \\ dv = f'(y)\,dy, v = f(y) \end{bmatrix}$$
$$= xf^{-1}(x) - \int f(y)\,dy$$

(b) $f^{-1}(x) = \arcsin x = y$, $f(x) = \sin x$

$$\int \arcsin x\,dx = x\arcsin x - \int \sin y\,dy = x\arcsin x + \cos y + C = x\arcsin x + \sqrt{1 - x^2} + C$$

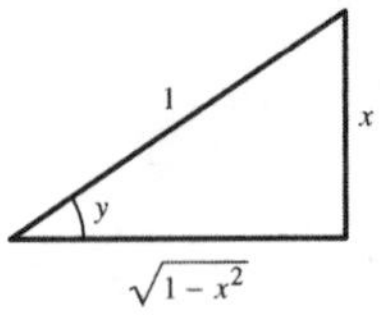

(c) $f(x) = e^x$, $f^{-1}(x) = \ln x = y \quad x = 1 \Leftrightarrow y = 0; x = e \Leftrightarrow y = 1$

$$\int_1^e \ln x\,dx = \left[x\ln x\right]_1^e - \int_0^1 e^y dy = e - \left[e^y\right]_0^1 = e - (e - 1) = 1$$

13. $x^4 + 1 = (x^2 + ax + b)(x^2 + cx + d)$

$$= x^4 + (a + c)x^3 + (ac + b + d)x^2 + (ad + bc)x + bd$$

$a = -c, b = d = 1, a = \sqrt{2}$

$$x^4 + 1 = \left(x^2 + \sqrt{2}x + 1\right)\left(x^2 - \sqrt{2}x + 1\right)$$

$$\int_0^1 \frac{1}{x^4 + 1}\,dx = \int_0^1 \frac{Ax + B}{x^2 + \sqrt{2}x + 1}\,dx + \int_0^1 \frac{Cx + D}{x^2 - \sqrt{2}x + 1}\,dx$$
$$= \int_0^1 \frac{\frac{1}{2} + \frac{\sqrt{2}}{4}x}{x^2 + \sqrt{2}x + 1}\,dx - \int_0^1 \frac{-\frac{1}{2} + \frac{\sqrt{2}}{4}x}{x^2 + \sqrt{2}x + 1}\,dx$$
$$= \frac{\sqrt{2}}{4}\left[\arctan\left(\sqrt{2}x + 1\right) + \arctan\left(\sqrt{2}x - 1\right)\right]_0^1 + \frac{\sqrt{2}}{8}\left[\ln\left(x^2 + \sqrt{2}x + 1\right) - \ln\left(x^2 - \sqrt{2}x + 1\right)\right]_0^1$$
$$= \frac{\sqrt{2}}{4}\left[\arctan\left(\sqrt{2} + 1\right) + \arctan\left(\sqrt{2} - 1\right)\right] + \frac{\sqrt{2}}{8}\left[\ln\left(2 + \sqrt{2}\right) - \ln\left(2 - \sqrt{2}\right)\right] - \frac{\sqrt{2}}{4}\left[\frac{\pi}{4} - \frac{\pi}{4}\right] - \frac{\sqrt{2}}{8}[0]$$
$$\approx 0.5554 + 0.3116$$
$$\approx 0.8670$$

14. $\dfrac{N(x)}{D(x)} = \dfrac{P_1}{x - c_1} + \dfrac{P_2}{x - c_2} + \cdots + \dfrac{P_n}{x - c_n}$

$N(x) = P_1(x - c_2)(x - c_3)\ldots(x - c_n) + P_2(x - c_1)(x - c_3)\ldots(x - c_n) + \cdots + P_n(x - c_1)(x - c_2)\ldots(x - c_{n-1})$

Let $x = c_1$: $N(c_1) = P_1(c_1 - c_2)(c_1 - c_3)\ldots(c_1 - c_n)$

$$P_1 = \frac{N(c_1)}{(c_1 - c_2)(c_1 - c_3)\ldots(c_1 - c_n)}$$

Let $x = c_2$: $N(c_2) = P_2(c_2 - c_1)(c_2 - c_3)\ldots(c_2 - c_n)$

$$P_2 = \frac{N(c_2)}{(c_2 - c_1)(c_2 - c_3)\ldots(c_2 - c_n)}$$

$\vdots \qquad \vdots$

Let $x = c_n$: $N(c_n) = P_n(c_n - c_1)(c_n - c_2)\ldots(c_n - c_{n-1})$

$$P_n = \frac{N(c_n)}{(c_n - c_1)(c_n - c_2)\ldots(c_n - c_{n-1})}$$

If $D(x) = (x - c_1)(x - c_2)(x - c_3)\ldots(x - c_n)$, then by the Product Rule

$D'(x) = (x - c_2)(x - c_3)\ldots(x - c_n) + (x - c_1)(x - c_3)\ldots(x - c_n) + \cdots + (x - c_1)(x - c_2)(x - c_3)\ldots(x - c_{n-1})$

and

$D'(c_1) = (c_1 - c_2)(c_1 - c_3)\ldots(c_1 - c_n)$

$D'(c_2) = (c_2 - c_1)(c_2 - c_3)\ldots(c_2 - c_n)$

$\vdots$

$D'(c_n) = (c_n - c_1)(c_n - c_2)\ldots(c_n - c_{n-1})$.

So, $P_k = N(c_k)/D'(c_k)$ for $k = 1, 2, \ldots, n$.

15. $\dfrac{x^3 - 3x^2 + 1}{x^4 - 13x^2 + 12x} = \dfrac{P_1}{x} + \dfrac{P_2}{x - 1} + \dfrac{P_3}{x + 4} + \dfrac{P_4}{x - 3} \Rightarrow c_1 = 0, c_2 = 1, c_3 = -4, c_4 = 3$

$N(x) = x^3 - 3x^2 + 1$

$D'(x) = 4x^3 - 26x + 12$

$$P_1 = \frac{N(0)}{D'(0)} = \frac{1}{12}$$

$$P_2 = \frac{N(1)}{D'(1)} = \frac{-1}{-10} = \frac{1}{10}$$

$$P_3 = \frac{N(-4)}{D'(-4)} = \frac{-111}{-140} = \frac{111}{140}$$

$$P_4 = \frac{N(3)}{D'(3)} = \frac{1}{42}$$

So, $\dfrac{x^3 - 3x^2 + 1}{x^4 - 13x^2 + 12x} = \dfrac{1/12}{x} + \dfrac{1/10}{x - 1} + \dfrac{111/140}{x + 4} + \dfrac{1/42}{x - 3}$.

16. (a) Let $x = \dfrac{\pi}{2} - u$, $dx = du$.

$$I = \int_0^{\pi/2} \frac{\sin x}{\cos x + \sin x}\,dx = \int_{\pi/2}^{0} \frac{\sin\left(\frac{\pi}{2} - u\right)}{\cos\left(\frac{\pi}{2} - u\right) + \sin\left(\frac{\pi}{2} - u\right)}(-du) = \int_0^{\pi/2} \frac{\cos u}{\sin u + \cos u}\,du$$

So,

$$2I = \int_0^{\pi/2} \frac{\sin x}{\cos x + \sin x}\,dx + \int_0^{\pi/2} \frac{\cos x}{\sin x + \cos x}\,dx = \int_0^{\pi/2} 1\,dx = \frac{\pi}{2} \Rightarrow I = \frac{\pi}{4}.$$

(b) $I = \int_{\pi/2}^{0} \frac{\sin^n\left(\frac{\pi}{2} - u\right)}{\cos^n\left(\frac{\pi}{2} - u\right) + \sin^n\left(\frac{\pi}{2} - u\right)}(-du) = \int_0^{\pi/2} \frac{\cos^n u}{\sin^n u + \cos^n u}\,du$

So, $2I = \int_0^{\pi/2} 1\,dx = \frac{\pi}{2} \Rightarrow I = \frac{\pi}{4}.$

17. Consider $\int \frac{1}{\ln x}\,dx.$

Let $u = \ln x$, $du = \frac{1}{x}\,dx$, $x = e^u$. Then $\int \frac{1}{\ln x}\,dx = \int \frac{1}{u}e^u\,du = \int \frac{e^u}{u}\,du.$

If $\int \frac{1}{\ln x}\,dx$ were elementary, then $\int \frac{e^u}{u}\,du$ would be too, which is false.

So, $\int \frac{1}{\ln x}\,dx$ is not elementary.

18. $s(t) = \int\left[-32t + 12{,}000 \ln \frac{50{,}000}{50{,}000 - 400t}\right]dt = -16t^2 + 12{,}000\int\left[\ln 50{,}000 - \ln(50{,}000 - 400t)\right]dt$

$= 16t^2 + 12{,}000t \ln 50{,}000 - 12{,}000\left[t \ln(50{,}000 - 400t) - \int \frac{-400t}{50{,}000 - 400t}\,dt\right]$

$= -16t^2 + 12{,}000t \ln \frac{50{,}000}{50{,}000 - 400t} + 12{,}000t\int\left[1 - \frac{50{,}000}{50{,}000 - 400t}\right]dt$

$= -16t^2 + 12{,}000t \ln \frac{50{,}000}{50{,}000 - 400t} + 12{,}000t + 1{,}500{,}000 \ln(50{,}000 - 400t) + C$

$s(0) = 1{,}500{,}000 \ln 50{,}000 + C = 0$

$C = -1{,}500{,}000 \ln 50{,}000$

$s(t) = -16t^2 + 12{,}000t\left[1 + \ln\frac{50{,}000}{50{,}000 - 400t}\right] + 1{,}500{,}000 \ln \frac{50{,}000 - 400t}{50{,}000}$

When $t = 100$, $s(100) \approx 557{,}168.626$ feet.

19. By parts,

$\int_a^b f(x)g''(x)\,dx = \left[f(x)g'(x)\right]_a^b - \int_a^b f'(x)g'(x)\,dx \qquad \left[u = f(x), dv = g''(x)\,dx\right]$

$= -\int_a^b f'(x)g'(x)\,dx$

$= \left[-f'(x)g(x)\right]_a^b + \int_a^b g(x)f''(x)\,dx \qquad \left[u = f'(x), dv = g'(x)\,dx\right]$

$= \int_a^b f''(x)g(x)\,dx.$

20. Let $u = (x - a)(x - b)$, $du = \left[(x - a) + (x - b)\right]dx$, $dv = f''(x)\,dx$, $v = f'(x)$.

$\int_a^b (x - a)(x - b)\,dx = \left[(x - a)(x - b)f'(x)\right]_a^b - \int_a^b \left[(x - a) + (x - b)\right]f'(x)\,dx$

$= -\int_a^b (2x - a - b)f'(x)\,dx \qquad \begin{pmatrix} u = 2x - a - b \\ dv = f'(x)\,dx \end{pmatrix}$

$= \left[-(2x - a - b)f(x)\right]_a^b + \int_a^b 2f(x)\,dx = 2\int_a^b f(x)\,dx$

21. $\int_2^{\infty}\left[\frac{1}{x^5}+\frac{1}{x^{10}}+\frac{1}{x^{15}}\right]dx < \int_2^{\infty}\frac{1}{x^5-1}\,dx < \int_2^{\infty}\left[\frac{1}{x^5}+\frac{1}{x^{10}}+\frac{2}{x^{15}}\right]dx$

$$\lim_{b\to\infty}\left[-\frac{1}{4x^4}-\frac{1}{9x^9}-\frac{1}{14x^{14}}\right]_2^b < \int_2^{\infty}\frac{1}{x^5-1}\,dx < \lim_{b\to\infty}\left[-\frac{1}{4x^4}-\frac{1}{9x^9}-\frac{1}{7x^{14}}\right]_2^b$$

$$0.015846 < \int_2^{\infty}\frac{2}{x^5-1}\,dx < 0.015851$$

22. $\frac{1}{2}V = \int_0^{\arcsin(c)}\pi(c-\sin x)^2\,dx + \int_{\arcsin(c)}^{\pi/2}\pi(\sin x - c)^2\,dx = \frac{2c^2\pi - 8c + \pi}{4}\pi = f(c)$

$f'(c) = \frac{4c\pi - 8}{4}\pi = 0 \Rightarrow c = \frac{2}{\pi}$

For $c = 0, \frac{1}{2}V = \frac{\pi^2}{4} \approx 2.4674.$

For $c = 1, \frac{1}{2}V = \frac{\pi}{4}(3\pi - 8) \approx 1.1190.$

For $c = \frac{2}{\pi}, \frac{1}{2}V = \frac{\pi^2 - 8}{4} \approx 0.4674.$

(a) Maximum: $c = 0$

(b) Minimum: $c = \frac{2}{\pi}$

CHAPTER 9
Infinite Series

Section 9.1 Sequences.....855

Section 9.2 Series and Convergence.....865

Section 9.3 The Integral Test and p-Series.....876

Section 9.4 Comparisons of Series.....886

Section 9.5 Alternating Series.....893

Section 9.6 The Ratio and Root Tests.....901

Section 9.7 Taylor Polynomials and Approximations.....914

Section 9.8 Power Series.....926

Section 9.9 Representation of Functions by Power Series.....939

Section 9.10 Taylor and Maclaurin Series.....948

Review Exercises.....966

Problem Solving.....979

CHAPTER 9
Infinite Series

Section 9.1 Sequences

1. $a_n = 3^n$

$a_1 = 3^1 = 3$

$a_2 = 3^2 = 9$

$a_3 = 3^3 = 27$

$a_4 = 3^4 = 81$

$a_5 = 3^5 = 243$

2. $a_n = \left(-\frac{2}{5}\right)^n$

$a_1 = \left(-\frac{2}{5}\right)^1 = -\frac{2}{5}$

$a_2 = \left(-\frac{2}{5}\right)^2 = \frac{4}{25}$

$a_3 = \left(-\frac{2}{5}\right)^3 = -\frac{8}{125}$

$a_4 = \left(-\frac{2}{5}\right)^4 = \frac{16}{625}$

$a_5 = \left(-\frac{2}{5}\right)^5 = -\frac{32}{3125}$

3. $a_n = \sin\dfrac{n\pi}{2}$

$a_1 = \sin\dfrac{\pi}{2} = 1$

$a_2 = \sin \pi = 0$

$a_3 = \sin\dfrac{3\pi}{2} = -1$

$a_4 = \sin 2\pi = 0$

$a_5 = \sin\dfrac{5\pi}{2} = 1$

4. $a_n = \dfrac{3n}{n + 4}$

$a_1 = \dfrac{3(1)}{1 + 4} = \dfrac{3}{5}$

$a_2 = \dfrac{3(2)}{2 + 4} = \dfrac{6}{6} = 1$

$a_3 = \dfrac{3(3)}{3 + 4} = \dfrac{9}{7}$

$a_4 = \dfrac{3(4)}{4 + 4} = \dfrac{12}{8} = \dfrac{3}{2}$

$a_5 = \dfrac{3(5)}{5 + 4} = \dfrac{15}{9} = \dfrac{5}{3}$

5. $a_n = (-1)^{n+1}\left(\dfrac{2}{n}\right)$

$a_1 = \dfrac{2}{1} = 2$

$a_2 = -\dfrac{2}{2} = -1$

$a_3 = \dfrac{2}{3}$

$a_4 = -\dfrac{2}{4} = -\dfrac{1}{2}$

$a_5 = \dfrac{2}{5}$

6. $a_n = 2 + \dfrac{2}{n} - \dfrac{1}{n^2}$

$a_1 = 2 + 2 - 1 = 3$

$a_2 = 2 + 1 - \dfrac{1}{4} = \dfrac{11}{4}$

$a_3 = 2 + \dfrac{2}{3} - \dfrac{1}{9} = \dfrac{23}{9}$

$a_4 = 2 + \dfrac{2}{4} - \dfrac{1}{16} = \dfrac{39}{16}$

$a_5 = 2 + \dfrac{2}{5} - \dfrac{1}{25} = \dfrac{59}{25}$

7. $a_1 = 3, a_{k+1} = 2(a_k - 1)$

$a_2 = 2(a_1 - 1)$

$= 2(3 - 1) = 4$

$a_3 = 2(a_2 - 1)$

$= 2(4 - 1) = 6$

$a_4 = 2(a_3 - 1)$

$= 2(6 - 1) = 10$

$a_5 = 2(a_4 - 1)$

$= 2(10 - 1) = 18$

8. $a_1 = 6, a_{k+1} = \frac{1}{3}a_k^2$

$a_2 = \frac{1}{3}a_1^2 = \frac{1}{3}(6^2) = 12$

$a_3 = \frac{1}{3}a_2^2 = \frac{1}{3}(12^2) = 48$

$a_4 = \frac{1}{3}a_3^2 = \frac{1}{3}(48^2) = 768$

$a_5 = \frac{1}{3}a_4^2 = \frac{1}{3}(768)^2 = 196,608$

9. $a_n = \dfrac{10}{n+1}, a_1 = \dfrac{10}{1+1} = 5, a_2 = \dfrac{10}{3}$

Matches (c).

10. $a_n = \dfrac{10n}{n+1}, a_1 = \dfrac{10}{2} = 5, a_2 = \dfrac{20}{3}$

Matches (a).

11. $a_n = (-1)^n, a_1 = -1, a_2 = 1, a_3 = -1, \ldots$

Matches (d).

12. $a_n = \dfrac{(-1)^n}{n}, a_1 = \dfrac{-1}{1} = -1, a_2 = \dfrac{1}{2}.$

Matches (b).

13. $a_n = 3n - 1$

$a_5 = 3(5) - 1 = 14$

$a_6 = 3(6) - 1 = 17$

Add 3 to preceding term.

14. $a_n = 3 + 5n$

$a_6 = 3 + 5(6) = 33$

$a_7 = 3 + 5(7) = 38$

Add 5 to preceding term.

15. $a_{n+1} = 2a_n, a_1 = 5$

$a_5 = 2(40) = 80$

$a_6 = 2(80) = 160$

Multiply the preceding term by 2.

16. $a_n = -\frac{1}{3}a_{n-1}, a_1 = 6$

$a_5 = -\frac{1}{3}\left(-\frac{2}{9}\right) = \frac{2}{27}$

$a_6 = -\frac{1}{3}\left(\frac{2}{27}\right) = -\frac{2}{81}$

Multiply the preceding term by $-\frac{1}{3}$.

17. $\dfrac{(n+1)!}{n!} = \dfrac{n!(n+1)}{n!} = n + 1$

18. $\dfrac{n!}{(n+2)!} = \dfrac{n!}{(n+2)(n+1)n!} = \dfrac{1}{(n+2)(n+1)}$

19. $\dfrac{(2n-1)!}{(2n+1)!} = \dfrac{(2n-1)!}{(2n-1)!(2n)(2n+1)} = \dfrac{1}{2n(2n+1)}$

20. $\dfrac{(2n+2)!}{(2n)!} = \dfrac{(2n)!(2n+1)(2n+2)}{(2n)!}$

$= (2n+1)(2n+2)$

21. $\displaystyle\lim_{n\to\infty} \frac{5n^2}{n^2+2} = 5$

22. $\displaystyle\lim_{n\to\infty}\left(6 + \frac{2}{n^2}\right) = 6 + 0 = 6$

23. $\displaystyle\lim_{n\to\infty} \frac{2n}{\sqrt{n^2+1}} = \lim_{n\to\infty} \frac{2}{\sqrt{1 + (1/n^2)}} = \frac{2}{1} = 2$

24. $\displaystyle\lim_{n\to\infty} \cos\left(\frac{2}{n}\right) = 1$

25.

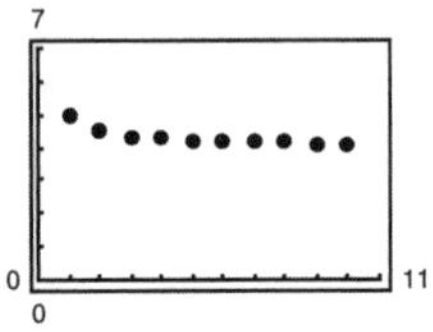

The graph seems to indicate that the sequence converges to 4. Analytically,

$$\lim_{n\to\infty} a_n = \lim_{n\to\infty} \frac{4n+1}{n} = \lim_{x\to\infty} \frac{4x+1}{x} = 4.$$

26.

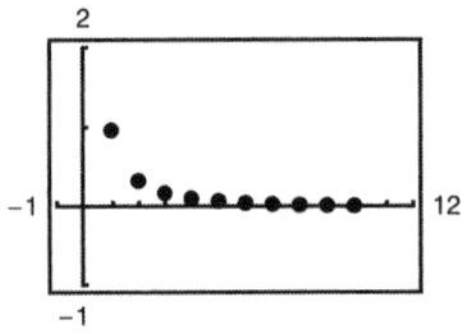

The graph seems to indicate that the sequence converges to 0. Analytically,

$$\lim_{n\to\infty} a_n = \lim_{n\to\infty} \frac{1}{n^{3/2}} = \lim_{x\to\infty} \frac{1}{x^{3/2}} = 0.$$

27.

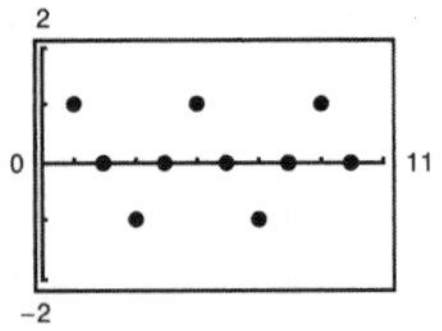

The graph seems to indicate that the sequence diverges. Analytically, the sequence is

$\{a_n\} = \{1, 0, -1, 0, 1, \ldots\}.$

So, $\displaystyle\lim_{n\to\infty} a_n$ does not exist.

28.

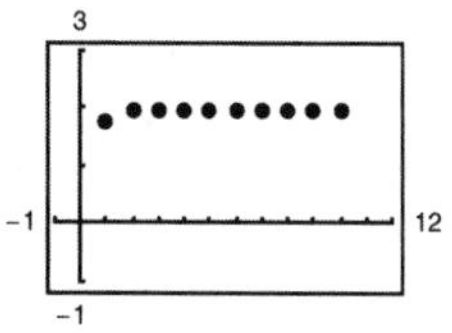

The graph seems to indicate that the sequence converges to 2. Analytically,

$$\lim_{n\to\infty} a_n = \lim_{n\to\infty}\left(2 - \frac{1}{4^n}\right) = 2 - 0 = 2.$$

29. $\lim_{n\to\infty} \frac{5}{n+2} = 0$, converges

30. $\lim_{n\to\infty}\left(8 + \frac{5}{n}\right) = 8 + 0 = 8$, converges

31. $\lim_{n\to\infty} (-1)^n\left(\frac{n}{n+1}\right)$

does not exist (oscillates between −1 and 1), diverges.

32. $\lim_{n\to\infty} \frac{1 + (-1)^n}{n^2} = 0$, converges

33. $$\lim_{n\to\infty} \frac{10n^2 + 3n + 7}{2n^2 - 6} = \lim_{n\to\infty} \frac{10 + 3/n + 7/n^2}{2 - 6/n^2} = \frac{10}{2} = 5, \text{ converges}$$

34. $\lim_{n\to\infty} \frac{\sqrt[3]{n}}{\sqrt[3]{n} + 1} = 1$, converges

35. $$\lim_{n\to\infty} \frac{\ln(n^3)}{2n} = \lim_{n\to\infty} \frac{3}{2}\frac{\ln(n)}{n} = \lim_{n\to\infty} \frac{3}{2}\left(\frac{1}{n}\right) = 0, \text{ converges}$$

(L'Hôpital's Rule)

36. $\lim_{n\to\infty} \frac{5^n}{3^n} = \lim_{n\to\infty}\left(\frac{5}{3}\right)^n = \infty$, diverges

37. $\lim_{n\to\infty} \frac{(n+1)!}{n!} = \lim_{n\to\infty} (n+1) = \infty$, diverges

38. $\lim_{n\to\infty} \frac{(n-2)!}{n!} = \lim_{n\to\infty} \frac{1}{n(n-1)} = 0$, converges

39. $\lim_{n\to\infty} \frac{n^p}{e^n} = 0$, converges

$(p > 0, n \ge 2)$

40. $a_n = n \sin \frac{1}{n}$

Let $f(x) = x \sin \frac{1}{x}$.

$$\lim_{x\to\infty} x \sin\frac{1}{x} = \lim_{x\to\infty} \frac{\sin(1/x)}{1/x} = \lim_{x\to\infty} \frac{(-1/x^2)\cos(1/x)}{-1/x^2} = \lim_{x\to\infty} \cos\frac{1}{x} = \cos 0 = 1 \text{ (L'Hôpital's Rule)}$$

or,

$\lim_{x\to\infty} \frac{\sin(1/x)}{1/x} = \lim_{y\to 0^+} \frac{\sin(y)}{y} = 1$. Therefore,

$\lim_{n\to\infty} n \sin\frac{1}{n} = 1$, converges.

41. $\lim_{n\to\infty} 2^{1/n} = 2^0 = 1$, converges

42. $\lim_{n\to\infty} -3^{-n} = \lim_{n\to\infty} \frac{-1}{3^n} = 0$, converges

43. $\lim_{n\to\infty} \frac{\sin n}{n} = \lim_{n\to\infty} (\sin n)\frac{1}{n} = 0$,

converges (because $(\sin n)$ is bounded)

44. $\lim_{n\to\infty} \frac{\cos \pi n}{n^2} = 0$, converges

45. $a_n = -4 + 6n$

46. $a_n = \frac{1}{n!}$

47. $a_n = n^2 - 3$

48. $a_n = \frac{(-1)^{n-1}}{n^2}$

49. $a_n = \frac{n+1}{n+2}$

50. $a_n = (2n)!, n = 1, 2, 3, \ldots$

51. $a_n = 1 + \frac{1}{n} = \frac{n+1}{n}$

52. $a_n = \frac{n}{(n+1)(n+2)}$

53. $a_n = 4 - \frac{1}{n} < 4 - \frac{1}{n+1} = a_{n+1},$

Monotonic; $|a_n| < 4$, bounded

54. Let $f(x) = \frac{3x}{x+2}$. Then $f'(x) = \frac{6}{(x+2)^2}$.

So, f is increasing which implies $\{a_n\}$ is increasing.

$|a_n| < 3$, bounded

55. $a_n = ne^{-n/2}$

$a_1 = 0.6065$

$a_2 = 0.7358$

$a_3 = 0.6694$

Not monotonic; $|a_n| \le 0.7358$, bounded

56. $a_n = \left(-\frac{2}{3}\right)^n$

$a_1 = -\frac{2}{3}$

$a_2 = \frac{4}{9}$

$a_3 = -\frac{8}{27}$

Not monotonic; $|a_n| \le \frac{2}{3}$, bounded

57. $a_n = \left(\frac{2}{3}\right)^n > \left(\frac{2}{3}\right)^{n+1} = a_{n+1}$

Monotonic; $|a_n| \le \frac{2}{3}$, bounded

58. $a_n = \left(\frac{3}{2}\right)^n < \left(\frac{3}{2}\right)^{n+1} = a_{n+1}$

Monotonic; $\lim_{n\to\infty} a_n = \infty$, not bounded

59. $a_n = \sin\left(\frac{n\pi}{6}\right)$

$a_1 = 0.500$

$a_2 = 0.8660$

$a_3 = 1.000$

$a_4 = 0.8660$

Not monotonic; $|a_n| \le 1$, bounded

60. $a_n = \frac{\cos n}{n}$

$a_1 = 0.5403$

$a_2 = -0.2081$

$a_3 = -0.3230$

$a_4 = -0.1634$

Not monotonic; $|a_n| \le 1$, bounded

61. (a) $a_n = 7 + \frac{1}{n}$

$\left|7 + \frac{1}{n}\right| \le 8 \Rightarrow \{a_n\}$, bounded

$a_n = 7 + \frac{1}{n} > 7 + \frac{1}{n+1} = a_{n+1} \Rightarrow \{a_n\}$, monotonic

Therefore, $\{a_n\}$ converges.

(b)

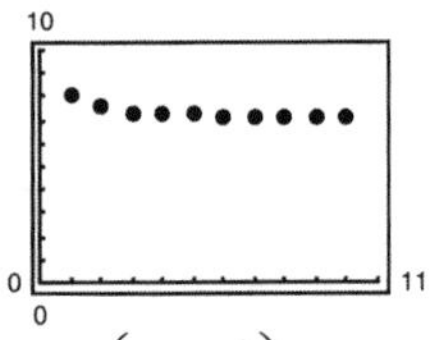

$\lim_{n\to\infty}\left(7 + \frac{1}{n}\right) = 7$

62. (a) $a_n = 5 - \frac{2}{n}$

$\left|5 - \frac{2}{n}\right| \le 5 \Rightarrow \{a_n\}$, bounded

$a_n = 5 - \frac{2}{n} < 5 - \frac{2}{n+1} = a_n + 1 \Rightarrow \{a_n\}$, monotonic

Therefore, $\{a_n\}$ converges.

(b)

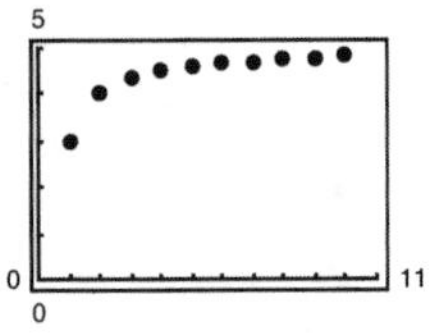

$\lim_{n\to\infty}\left(5 - \frac{2}{n}\right) = 5 - 0 = 5$

63. (a) $a_n = \frac{1}{3}\left(1 - \frac{1}{3^n}\right)$

$\left|\frac{1}{3}\left(1 - \frac{1}{3^n}\right)\right| < \frac{1}{3} \Rightarrow \{a_n\}$, bounded

$a_n = \frac{1}{3}\left(1 - \frac{1}{3^n}\right) < \frac{1}{3}\left(1 - \frac{1}{3^{n+1}}\right)$

$= a_{n+1} \Rightarrow \{a_n\}$, monotonic

Therefore, $\{a_n\}$ converges.

(b)

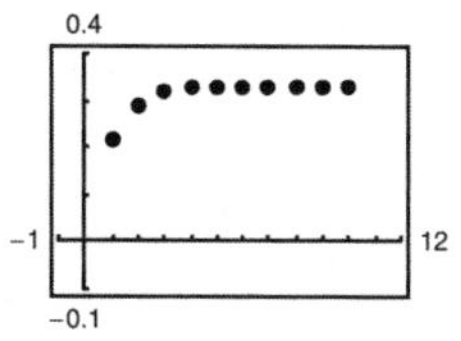

$\lim_{n\to\infty}\left[\frac{1}{3}\left(1 - \frac{1}{3^n}\right)\right] = \frac{1}{3}$

64. (a) $a_n = 2 + \frac{1}{5^n}$

$\left|2 + \frac{1}{5^n}\right| < 3 \Rightarrow \{a_n\}$, bounded

$a_n = 2 + \frac{1}{5^n} > 2 + \frac{1}{5^{n+1}} = a_{n+1} \Rightarrow \{a_n\}$, monotonic

Therefore, $\{a_n\}$ converges.

(b)

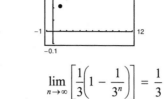

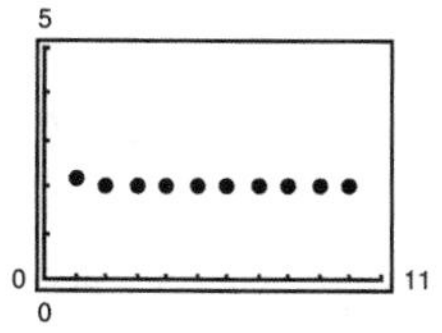

$\lim_{n\to\infty}\left(2 + \frac{1}{5^n}\right) = 2 + 0 = 2$

65. $\{a_n\}$ has a limit because it is a bounded, monotonic sequence. The limit is less than or equal to 4, and greater than or equal to 2.

$2 \le \lim_{n\to\infty} a_n \le 4$

66. The sequence $\{a_n\}$ could converge or diverge. If $\{a_n\}$ is increasing, then it converges to a limit less than or equal to 1. If $\{a_n\}$ is decreasing, then it could converge (example: $a_n = 1/n$) or diverge (example: $a_n = -n$).

67. $A_n = P\left(1 + \frac{r}{12}\right)^n$

(a) Because $P > 0$ and $\left(1 + \frac{r}{12}\right) > 1$, the sequence diverges. $\lim_{n\to\infty} A_n = \infty$

(b) $P = 10{,}000$, $r = 0.055$, $A_n = 10{,}000\left(1 + \frac{0.055}{12}\right)^n$

$A_0 = 10{,}000$
$A_1 = 10{,}045.83$
$A_2 = 10{,}091.88$
$A_3 = 10{,}138.13$
$A_4 = 10{,}184.60$
$A_5 = 10{,}231.28$
$A_6 = 10{,}278.17$
$A_7 = 10{,}325.28$
$A_8 = 10{,}372.60$
$A_9 = 10{,}420.14$
$A_{10} = 10{,}467.90$

68. (a) $A_n = 100(401)(1.0025^n - 1)$

$A_0 = 0$

$A_1 = 100.25$

$A_2 = 200.75$

$A_3 = 301.50$

$A_4 = 402.51$

$A_5 = 503.76$

$A_6 = 605.27$

(b) $A_{60} = 6480.83$

(c) $A_{240} = 32{,}912.28$

69. No, it is not possible. See the "Definition of the Limit of a sequence". The number L is unique.

70. (a) A sequence is a function whose domain is the set of positive integers.

(b) A sequence converges if it has a limit. See the definition.

(c) A sequence is monotonic if its terms are nondecreasing, or nonincreasing.

(d) A sequence is bounded if it is bounded below $(a_n \geq N$ for some $N)$ and bounded above $(a_n \leq M$ for some $M)$.

71. (a) $a_n = 10 - \frac{1}{n}$

(b) Impossible. The sequence converges by Theorem 9.5.

(c) $a_n = \frac{3n}{4n + 1}$

(d) Impossible. An unbounded sequence diverges.

72. The graph on the left represents a sequence with alternating signs because the terms alternate from being above the x-axis to being below the x-axis.

73. (a) $A_n = (0.8)^n\ 4{,}500{,}000{,}000$

(b) $A_1 = \$3{,}600{,}000{,}000$

$A_2 = \$2{,}880{,}000{,}000$

$A_3 = \$2{,}304{,}000{,}000$

$A_4 = \$1{,}843{,}200{,}000$

(c) $\lim_{n\to\infty} A_n = \lim_{n\to\infty} (0.8)^n(4.5) = 0$, converges

74. $P_n = 25{,}000(1.045)^n$

$P_1 = \$26{,}125.00$

$P_2 = \$27{,}300.63$

$P_3 = \$28{,}529.15$

$P_4 = \$29{,}812.97$

$P_5 = \$31{,}154.55$

75. $a_n = \sqrt[n]{n} = n^{1/n}$

$a_1 = 1^{1/1} = 1$

$a_2 = \sqrt{2} \approx 1.4142$

$a_3 = \sqrt[3]{3} \approx 1.4422$

$a_4 = \sqrt[4]{4} \approx 1.4142$

$a_5 = \sqrt[5]{5} \approx 1.3797$

$a_6 = \sqrt[6]{6} \approx 1.3480$

Let $y = \lim_{n\to\infty} n^{1/n}$.

$$\ln y = \lim_{n\to\infty}\left(\frac{1}{n}\ln n\right) = \lim_{n\to\infty}\frac{\ln n}{n} = \lim_{n\to\infty}\frac{1/n}{n} = 0$$

Because $\ln y = 0$, you have $y = e^0 = 1$. Therefore, $\lim_{n\to\infty} \sqrt[n]{n} = 1$.

76. $a_n = \left(1 + \frac{1}{n}\right)^n$

$a_1 = 2.0000$

$a_2 = 2.2500$

$a_3 \approx 2.3704$

$a_4 \approx 2.4414$

$a_5 \approx 2.4883$

$a_6 \approx 2.5216$

$$\lim_{n\to\infty}\left(1 + \frac{1}{n}\right)^n = e$$

77. Because

$\lim_{n\to\infty} s_n = L > 0,$

there exists for each $\varepsilon > 0$,

an integer N such that $|s_n - L| < \varepsilon$ for every $n > N$.

Let $\varepsilon = L > 0$ and you have,

$|s_n - L| < L, -L < s_n - L < L$, or $0 < s_n < 2L$ for each $n > N$.

78. (a) $a_n = 0.072n^2 + 0.02n + 5.8$ (b) For 2020, $n = 20$: $a_{20} = \$35$ trillion

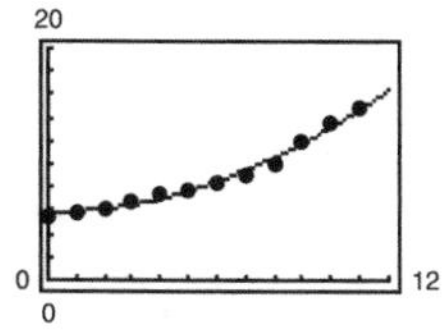

79. True

80. True

81. True

82. False. Let $a_n = (-1)^n$ and $b_n = (-1)^{n+1}$ then $\{a_n\}$ and $\{b_n\}$ diverge. But $\{a_n + b_n\} = \{(-1)^n + (-1)^{n+1}\}$ converges to 0.

83. $a_{n+2} = a_n + a_{n+1}$

(a) $a_1 = 1$ $\quad a_7 = 8 + 5 = 13$

$a_2 = 1$ $\quad a_8 = 13 + 8 = 21$

$a_3 = 1 + 1 = 2$ $\quad a_9 = 21 + 13 = 34$

$a_4 = 2 + 1 = 3$ $\quad a_{10} = 34 + 21 = 55$

$a_5 = 3 + 2 = 5$ $\quad a_{11} = 55 + 34 = 89$

$a_6 = 5 + 3 = 8$ $\quad a_{12} = 89 + 55 = 144$

(b) $b_n = \dfrac{a_{n+1}}{a_n},\ n \geq 1$

$b_1 = \frac{1}{1} = 1$ $\quad b_6 = \frac{13}{8} = 1.625$

$b_2 = \frac{2}{1} = 2$ $\quad b_7 = \frac{21}{13} \approx 1.6154$

$b_3 = \frac{3}{2} = 1.5$ $\quad b_8 = \frac{34}{21} \approx 1.6190$

$b_4 = \frac{5}{3} \approx 1.6667$ $\quad b_9 = \frac{55}{34} \approx 1.6176$

$b_5 = \frac{8}{5} = 1.6$ $\quad b_{10} = \frac{89}{55} \approx 1.6182$

(c) $$1 + \frac{1}{b_{n-1}} = 1 + \frac{1}{a_n/a_{n-1}}$$

$$= 1 + \frac{a_{n-1}}{a_n} = \frac{a_n + a_{n-1}}{a_n} = \frac{a_{n+1}}{a_n} = b_n$$

(d) If $\lim_{n\to\infty} b_n = \rho$, then $\lim_{n\to\infty}\left(1 + \dfrac{1}{b_{n-1}}\right) = \rho$.

Because $\lim_{n\to\infty} b_n = \lim_{n\to\infty} b_{n-1}$, you have

$1 + (1/\rho) = \rho$.

$$\rho + 1 = \rho^2$$
$$0 = \rho^2 - \rho - 1$$
$$\rho = \frac{1 \pm \sqrt{1 + 4}}{2} = \frac{1 \pm \sqrt{5}}{2}$$

Because a_n, and therefore b_n, is positive,

$$\rho = \frac{1 + \sqrt{5}}{2} \approx 1.6180.$$

84. Let $f(x) = \sin(\pi x)$

$\lim_{x\to\infty} \sin(\pi x)$ does not exist.

$a_n = f(n) = \sin(\pi n) = 0$ for all n

$\lim_{n\to\infty} a_n = 0$, coverges

85. (a) $a_1 = \sqrt{2} \approx 1.4142$

$a_2 = \sqrt{2 + \sqrt{2}} \approx 1.8478$

$a_3 = \sqrt{2 + \sqrt{2 + \sqrt{2}}} \approx 1.9616$

$a_4 = \sqrt{2 + \sqrt{2 + \sqrt{2 + \sqrt{2}}}} \approx 1.9904$

$a_5 = \sqrt{2 + \sqrt{2 + \sqrt{2 + \sqrt{2 + \sqrt{2}}}}} \approx 1.9976$

(b) $a_n = \sqrt{2 + a_{n-1}}, \quad n \geq 2,\ a_1 = \sqrt{2}$

(c) First use mathematical induction to show that $a_n \le 2$; clearly $a_1 \le 2$. So assume $a_k \le 2$. Then

$$\begin{aligned} a_k + 2 &\le 4 \\ \sqrt{a_k + 2} &\le 2 \\ a_{k+1} &\le 2. \end{aligned}$$

Now show that $\{a_n\}$ is an increasing sequence. Because $a_n \ge 0$ and $a_n \le 2$,

$$\begin{aligned} (a_n - 2)(a_n + 1) &\le 0 \\ a_n^2 - a_n - 2 &\le 0 \\ a_n^2 &\le a_n + 2 \\ a_n &\le \sqrt{a_n + 2} \\ a_n &\le a_{n+1}. \end{aligned}$$

Because $\{a_n\}$ is a bounding increasing sequence, it converges to some number L, by Theorem 9.5.

$$\lim_{n\to\infty} a_n = L \Rightarrow \sqrt{2 + L} = L \Rightarrow 2 + L = L^2 \Rightarrow L^2 - L - 2 = 0$$
$$\Rightarrow (L - 2)(L + 1) = 0 \Rightarrow L = 2 \quad (L \ne -1)$$

86. (a) Use mathematical induction to show that

$$a_n \le \frac{1 + \sqrt{1 + 4k}}{2}.$$

[Note that if $k = 2$, and $a_n \le 3$, and if $k = 6$, then $a_n \le 3$.] Clearly,

$$a_1 = \sqrt{k} \le \frac{\sqrt{1 + 4k}}{2} \le \frac{1 + \sqrt{1 + 4k}}{2}.$$

Before proceeding to the induction step, note that

$$\begin{aligned} 2 + 2\sqrt{1 + 4k} + 4k &= 2 + 2\sqrt{1 + 4k} + 4k \\ \frac{1 + \sqrt{1 + 4k}}{2} + k &= \frac{1 + 2\sqrt{1 + 4k} + 1 + 4k}{4} \\ \frac{1 + \sqrt{1 + 4k}}{2} + k &= \left[\frac{1 + \sqrt{1 + 4k}}{2}\right]^2 \\ \sqrt{\frac{1 + \sqrt{1 + 4k}}{2} + k} &= \frac{1 + \sqrt{1 + 4k}}{2}. \end{aligned}$$

So assume $a_n \le \dfrac{1 + \sqrt{1 + 4k}}{2}$. Then

$$\begin{aligned} a_n + k &\le \frac{1 + \sqrt{1 + 4k}}{2} + k \\ \sqrt{a_n + k} &\le \sqrt{\frac{1 + \sqrt{1 + 4k}}{2} + k} \\ a_{n+1} &\le \frac{1 + \sqrt{1 + 4k}}{2}. \end{aligned}$$

$\{a_n\}$ is increasing because

$$\begin{aligned} \left(a_n - \frac{1 + \sqrt{1 + 4k}}{2}\right)\left(a_n - \frac{1 - \sqrt{1 + 4k}}{2}\right) &\le 0 \\ a_n^2 - a_n - k &\le 0 \\ a_n^2 &\le a_n + k \\ a_n &\le \sqrt{a_n + k} \\ a_n &\le a_{n+1}. \end{aligned}$$

(b) Because $\{a_n\}$ is bounded and increasing, it has a limit L.

(c) $\lim_{n\to\infty} a_n = L$ implies that

$$L = \sqrt{k + L} \Rightarrow L^2 = k + L$$
$$\Rightarrow L^2 - L - k = 0$$
$$\Rightarrow L = \frac{1 \pm \sqrt{1 + 4k}}{2}.$$

Because $L > 0$, $L = \dfrac{1 + \sqrt{1 + 4k}}{2}.$

87. (a)

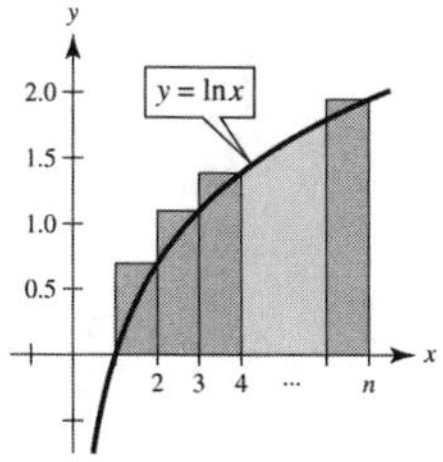

$$\int_1^n \ln x \, dx < \ln 2 + \ln 3 + \cdots + \ln n$$
$$= \ln(1 \cdot 2 \cdot 3 \cdots n) = \ln(n!)$$

(b)

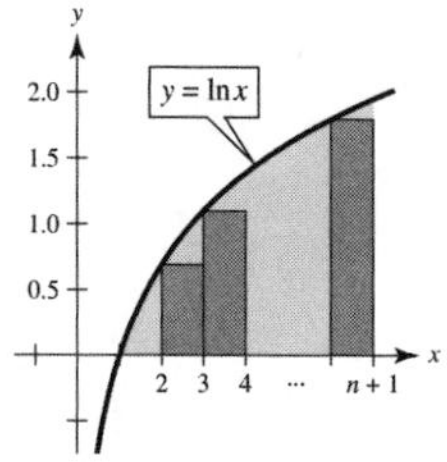

$$\int_1^{n+1} \ln x \, dx > \ln 2 + \ln 3 + \cdots + \ln n = \ln(n!)$$

(c) $\int \ln x \, dx = x \ln x - x + C$

$$\int_1^n \ln x \, dx = n \ln n - n + 1 = \ln n^n - n + 1$$

From part (a): $\ln n^n - n + 1 < \ln(n!)$

$$e^{\ln n^n - n + 1} < n!$$
$$\frac{n^n}{e^{n-1}} < n!$$

$$\int_1^{n+1} \ln x \, dx = (n + 1)\ln(n + 1) - (n + 1) + 1$$
$$= \ln(n + 1)^{n+1} - n$$

From part (b): $\ln(n + 1)^{n+1} - n > \ln(n!)$

$$e^{\ln(n+1)^{n+1} - n} > n!$$
$$\frac{(n + 1)^{n+1}}{e^n} > n!$$

(d) $$\frac{n^n}{e^{n-1}} < n! < \frac{(n + 1)^{n+1}}{e^n}$$

$$\frac{n}{e^{1-(1/n)}} < \sqrt[n]{n!} < \frac{(n + 1)^{(n+1)/n}}{e}$$

$$\frac{1}{e^{1-(1/n)}} < \frac{\sqrt[n]{n!}}{n} < \frac{(n + 1)^{1+(1/n)}}{ne}$$

$$\lim_{n\to\infty} \frac{1}{e^{1-(1/n)}} = \frac{1}{e}$$

$$\lim_{n\to\infty} \frac{(n + 1)^{1+(1/n)}}{ne} = \lim_{n\to\infty} \frac{(n + 1)}{n} \frac{(n + 1)^{1/n}}{e}$$
$$= (1)\frac{1}{e}$$
$$= \frac{1}{e}$$

By the Squeeze Theorem, $\lim_{n\to\infty} \dfrac{\sqrt[n]{n!}}{n} = \dfrac{1}{e}.$

(e) $n = 20$: $\dfrac{\sqrt[20]{20!}}{20} \approx 0.4152$

$n = 50$: $\dfrac{\sqrt[50]{50!}}{50} \approx 0.3897$

$n = 100$: $\dfrac{\sqrt[100]{100!}}{100} \approx 0.3799$

$\dfrac{1}{e} \approx 0.3679$

88. For a given $\varepsilon > 0$, you must find $M > 0$ such that

$$|a_n - L| = \left|\frac{1}{n^3}\right| < \varepsilon$$

whenever $n > M$. That is,

$$n^3 > \frac{1}{\varepsilon} \text{ or } n > \left(\frac{1}{\varepsilon}\right)^{1/3}.$$

So, let $\varepsilon > 0$ be given. Let M be an integer satisfying $M > (1/\varepsilon)^{1/3}$. For $n > M$, you have

$$n > \left(\frac{1}{\varepsilon}\right)^{1/3}$$

$$n^3 > \frac{1}{\varepsilon}$$

$$\varepsilon > \frac{1}{n^3} \Rightarrow \left|\frac{1}{n^3} - 0\right| < \varepsilon.$$

So, $\lim_{n\to\infty} \frac{1}{n^3} = 0.$

89. For a given $\varepsilon > 0$, you must find $M > 0$ such that $|a_n - L| = |r^n| \varepsilon$ whenever $n > M$. That is,

$n \ln|r| < \ln(\varepsilon)$ or

$$n > \frac{\ln(\varepsilon)}{\ln|r|} \text{ (because } \ln|r| < 0 \text{ for } |r| < 1).$$

So, let $\varepsilon > 0$ be given. Let M be an integer satisfying

$$M > \frac{\ln(\varepsilon)}{\ln|r|}.$$

For $n > M$, you have

$$n > \frac{\ln(\varepsilon)}{\ln|r|}$$

$$n \ln|r| < \ln(\varepsilon)$$

$$\ln|r|^n < \ln(\varepsilon)$$

$$|r|^n < \varepsilon$$

$$|r^n - 0| < \varepsilon.$$

So, $\lim_{n\to\infty} r^n = 0$ for $-1 < r < 1$.

90. Answers will vary. *Sample answer*:

$$\{a_n\} = \{(-1)^n\} = \{-1, 1, -1, 1, \ldots\} \text{ diverges}$$

$$\{a_{2n}\} = \{(-1)^{2n}\} = \{1, 1, 1, 1, \ldots\} \text{ converges}$$

91. If $\{a_n\}$ is bounded, monotonic and nonincreasing, then $a_1 \ge a_2 \ge a_3 \ge \cdots \ge a_n \ge \cdots$. Then $-a_1 \le -a_2 \le -a_3 \le \cdots \le -a_n \le \cdots$ is a bounded, monotonic, nondecreasing sequence which converges by the first half of the theorem. Because $\{-a_n\}$ converges, then so does $\{a_n\}$.

92. Define $a_n = \dfrac{x_{n+1} + x_{n-1}}{x_n}, \quad n \ge 1.$

$$x_{n+1}^2 - x_n x_{n+2} = 1 = x_n^2 - x_{n-1}x_{n+1} \Rightarrow$$

$$x_{n+1}(x_{n+1} + x_{n-1}) = x_n(x_n + x_{n+2})$$

$$\frac{x_{n+1} + x_{n-1}}{x_n} = \frac{x_{n+2} + x_n}{x_{n+1}}$$

$$a_n = a_{n+1}$$

Therefore, $a_1 = a_2 = \ldots = a$. So,

$$x_{n+1} = a_n x_n - x_{n-1} = a x_n - x_{n-1}.$$

93. $T_n = n! + 2^n$

Use mathematical induction to verify the formula.

$T_0 = 1 + 1 = 2$

$T_1 = 1 + 2 = 3$

$T_2 = 2 + 4 = 6$

Assume $T_k = k! + 2^k$. Then

$$\begin{aligned} T_{k+1} &= (k + 1 + 4)T_k - 4(k + 1)T_{k-1} + \big(4(k + 1) - 8\big)T_{k-2} \\ &= (k + 5)\left[k! + 2^k\right] - 4(k + 1)\left((k - 1)! + 2^{k-1}\right) + (4k - 4)\left((k - 2)! + 2^{k-2}\right) \\ &= \left[(k + 5)(k)(k - 1) - 4(k + 1)(k - 1) + 4(k - 1)\right](k - 2)! + \left[(k + 5)4 - 8(k + 1) + 4(k - 1)\right]2^{k-2} \\ &= \left[k^2 + 5k - 4k - 4 + 4\right](k - 1)! + 8 \cdot 2^{k-2} \\ &= (k + 1)! + 2^{k+1}. \end{aligned}$$

By mathematical induction, the formula is valid for all n.

Section 9.2 Series and Convergence

1. $S_1 = 1$

$S_2 = 1 + \frac{1}{4} = 1.2500$

$S_3 = 1 + \frac{1}{4} + \frac{1}{9} \approx 1.3611$

$S_4 = 1 + \frac{1}{4} + \frac{1}{9} + \frac{1}{16} \approx 1.4236$

$S_5 = 1 + \frac{1}{4} + \frac{1}{9} + \frac{1}{16} + \frac{1}{25} \approx 1.4636$

2. $S_1 = \frac{1}{6} \approx 0.1667$

$S_2 = \frac{1}{6} + \frac{1}{6} \approx 0.3333$

$S_3 = \frac{1}{6} + \frac{1}{6} + \frac{3}{20} \approx 0.4833$

$S_4 = \frac{1}{6} + \frac{1}{6} + \frac{3}{20} + \frac{2}{15} \approx 0.6167$

$S_5 = \frac{1}{6} + \frac{1}{6} + \frac{3}{20} + \frac{2}{15} + \frac{5}{42} \approx 0.7357$

3. $S_1 = 3$

$S_2 = 3 - \frac{9}{2} = -1.5$

$S_3 = 3 - \frac{9}{2} + \frac{27}{4} = 5.25$

$S_4 = 3 - \frac{9}{2} + \frac{27}{4} - \frac{81}{8} = -4.875$

$S_5 = 3 - \frac{9}{2} + \frac{27}{4} - \frac{81}{8} + \frac{243}{16} = 10.3125$

4. $S_1 = 1$

$S_2 = 1 + \frac{1}{2} = \frac{3}{2}$

$S_3 = 1 + \frac{1}{2} + \frac{1}{4} = \frac{7}{4}$

$S_4 = 1 + \frac{1}{2} + \frac{1}{4} + \frac{1}{6} = \frac{23}{12}$

$S_5 = 1 + \frac{1}{2} + \frac{1}{4} + \frac{1}{6} + \frac{1}{8} = \frac{49}{24}$

5. $S_1 = 3$

$S_2 = 3 + \frac{3}{2} = 4.5$

$S_3 = 3 + \frac{3}{2} + \frac{3}{4} = 5.250$

$S_4 = 3 + \frac{3}{2} + \frac{3}{4} + \frac{3}{8} = 5.625$

$S_5 = 3 + \frac{3}{2} + \frac{3}{4} + \frac{3}{8} + \frac{3}{16} = 5.8125$

6. $S_1 = 1$

$S_2 = 1 - \frac{1}{2} = 0.5$

$S_3 = 1 - \frac{1}{2} + \frac{1}{6} \approx 0.6667$

$S_4 = 1 - \frac{1}{2} + \frac{1}{6} - \frac{1}{24} \approx 0.6250$

$S_5 = 1 - \frac{1}{2} + \frac{1}{6} - \frac{1}{24} + \frac{1}{120} \approx 0.6333$

7. $\sum_{n=0}^{\infty} \left(\frac{7}{6}\right)^n$

Geometric series

$r = \frac{7}{6} > 1$

Diverges by Theorem 9.6

8. $\sum_{n=0}^{\infty} 4(-1.05)^n$

Geometric series

$|r| = |-1.05| = 1.05 > 1$

Diverges by Theorem 9.6

9. $\sum_{n=1}^{\infty} \frac{n}{n+1}$

$\lim_{n\to\infty} \frac{n}{n+1} = 1 \neq 0$

Diverges by Theorem 9.9

10. $\sum_{n=1}^{\infty} \frac{n}{2n+3}$

$\lim_{n\to\infty} \frac{n}{2n+3} = \frac{1}{2} \neq 0$

Diverges by Theorem 9.9

11. $\sum_{n=1}^{\infty} \frac{n^2}{n^2+1}$

$\lim_{n\to\infty} \frac{n^2}{n^2+1} = 1 \neq 0$

Diverges by Theorem 9.9

12. $\sum_{n=1}^{\infty} \frac{n}{\sqrt{n^2+1}}$

$\lim_{n\to\infty} \frac{n}{\sqrt{n^2+1}} = \lim_{n\to\infty} \frac{1}{\sqrt{1+(1/n^2)}} = 1 \neq 0$

Diverges by Theorem 9.9

13. $\sum_{n=1}^{\infty} \frac{2^n+1}{2^{n+1}}$

$\lim_{n\to\infty} \frac{2^n+1}{2^{n+1}} = \lim_{n\to\infty} \frac{1+2^{-n}}{2} = \frac{1}{2} \neq 0$

Diverges by Theorem 9.9

14. $\sum_{n=1}^{\infty} \frac{n!}{2^n}$

$\lim_{n\to\infty} \frac{n!}{2^n} = \infty$

Diverges by Theorem 9.9

15. $\sum_{n=0}^{\infty} \left(\frac{5}{6}\right)^n$

Geometric series with $r = \frac{5}{6} < 1$

Converges by Theorem 9.6

16. $\sum_{n=0}^{\infty} 2\left(-\frac{1}{2}\right)^n$

Geometric series with $|r| = \left|-\frac{1}{2}\right| < 1$

Converges by Theorem 9.6

17. $\sum_{n=0}^{\infty} (0.9)^n$

Geometric series with $r = 0.9 < 1$

Converges by Theorem 9.6

18. $\sum_{n=0}^{\infty} (-0.6)^n$

Geometric series with $|r| = |-0.6| < 1$

Converges by Theorem 9.6

19. $\sum_{n=1}^{\infty} \frac{1}{n(n+1)} = \sum_{n=1}^{\infty} \left(\frac{1}{n} - \frac{1}{n+1}\right) = \left(1 - \frac{1}{2}\right) + \left(\frac{1}{2} - \frac{1}{3}\right) + \left(\frac{1}{3} - \frac{1}{4}\right) + \left(\frac{1}{4} - \frac{1}{5}\right) + \cdots,\quad S_n = 1 - \frac{1}{n+1}$

$$\sum_{n=1}^{\infty} \frac{1}{n(n+1)} = \lim_{n\to\infty} S_n = \lim_{n\to\infty}\left(1 - \frac{1}{n+1}\right) = 1$$

20. $\sum_{n=1}^{\infty} \frac{1}{n(n+2)} = \sum_{n=1}^{\infty} \left(\frac{1}{2n} - \frac{1}{2(n+2)}\right) = \left(\frac{1}{2} - \frac{1}{6}\right) + \left(\frac{1}{4} - \frac{1}{8}\right) + \left(\frac{1}{6} - \frac{1}{10}\right) + \left(\frac{1}{8} - \frac{1}{12}\right) + \left(\frac{1}{10} - \frac{1}{14}\right) + \cdots$

$$\sum_{n=1}^{\infty} \frac{1}{n(n+2)} = \lim_{n\to\infty} S_n = \lim_{n\to\infty}\left[\frac{1}{2} + \frac{1}{4} - \frac{1}{2(n+1)} - \frac{1}{2(n+2)}\right] = \frac{1}{2} + \frac{1}{4} = \frac{3}{4}$$

21. (a) $\sum_{n=1}^{\infty} \frac{6}{n(n+3)} = 2\sum_{n=1}^{\infty}\left(\frac{1}{n} - \frac{1}{n+3}\right) = 2\left[\left(1 - \frac{1}{4}\right) + \left(\frac{1}{2} - \frac{1}{5}\right) + \left(\frac{1}{3} - \frac{1}{6}\right) + \left(\frac{1}{4} - \frac{1}{7}\right) + \cdots\right]$

$$\left(S_n = 2\left[1 + \frac{1}{2} + \frac{1}{3} - \left(\frac{1}{n+1} + \frac{1}{n+2} + \frac{1}{n+3}\right)\right]\right) = 2\left(1 + \frac{1}{2} + \frac{1}{3}\right) = \frac{11}{3} \approx 3.667$$

(b)

n	5	10	20	50	100
S_n	2.7976	3.1643	3.3936	3.5513	3.6078

(c)

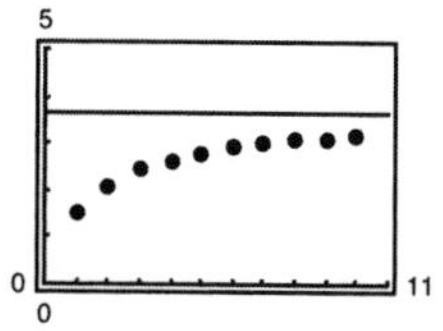

(d) The terms of the series decrease in magnitude slowly. So, the sequence of partial sums approaches the sum slowly.

22. (a) $\sum_{n=1}^{\infty} \frac{4}{n(n+4)} = \sum_{n=1}^{\infty}\left(\frac{1}{n} - \frac{1}{n+4}\right)$

$$= \left(1 - \frac{1}{5}\right) + \left(\frac{1}{2} - \frac{1}{6}\right) + \left(\frac{1}{3} - \frac{1}{7}\right) + \left(\frac{1}{4} - \frac{1}{8}\right) + \left(\frac{1}{5} - \frac{1}{9}\right) + \left(\frac{1}{6} - \frac{1}{10}\right) + \cdots$$

$$= 1 + \frac{1}{2} + \frac{1}{3} + \frac{1}{4} = \frac{25}{12} \approx 2.0833$$

(b)

n	5	10	20	50	100
S_n	1.5377	1.7607	1.9051	2.0071	2.0443

(c)

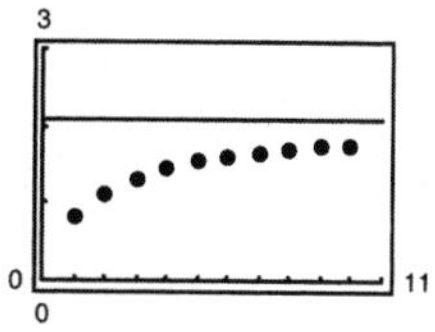

(d) The terms of the series decrease in magnitude slowly. So, the sequence of partial sums approaches the sum slowly.

23. (a) $\sum_{n=1}^{\infty} 2(0.9)^{n-1} = \sum_{n=0}^{\infty} 2(0.9)^n = \frac{2}{1 - 0.9} = 20$

(b)

n	5	10	20	50	100
S_n	8.1902	13.0264	17.5685	19.8969	19.9995

(c)

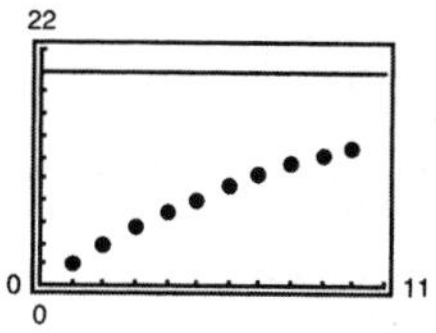

(d) The terms of the series decrease in magnitude slowly. So, the sequence of partial sums approaches the sum slowly.

24. (a) $\sum_{n=1}^{\infty} 10\left(-\frac{1}{4}\right)^{n-1} = \sum_{n=0}^{\infty} 10\left(-\frac{1}{4}\right)^n = \frac{10}{1 - (-1/4)} = 8$

(b)

n	5	10	20	50	100
S_n	8.0078	7.99999	8.0000	8.0000	8.0000

(c)

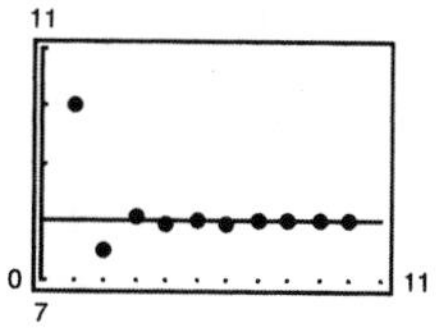

(d) The terms of the series decrease in magnitude rapidly. So, the sequence of partial sums approaches the sum rapidly.

25. $\sum_{n=0}^{\infty} 5\left(\frac{2}{3}\right)^n = \frac{5}{1 - (2/3)} = 15$

26. $\sum_{n=0}^{\infty} \left(-\frac{1}{5}\right)^n = \frac{1}{1 - (-1/5)} = \frac{5}{6}$

27. $\sum_{n=1}^{\infty} \frac{4}{n(n+2)} = 2\sum_{n=1}^{\infty}\left(\frac{1}{n} - \frac{1}{n+2}\right)$

$$S_n = 2\left[\left(1 - \frac{1}{3}\right) + \left(\frac{1}{2} - \frac{1}{4}\right) + \left(\frac{1}{3} - \frac{1}{5}\right) + \cdots + \left(\frac{1}{n-1} - \frac{1}{n+1}\right) + \left(\frac{1}{n} - \frac{1}{n+2}\right)\right] = 2\left(1 + \frac{1}{2} - \frac{1}{n+1} - \frac{1}{n+2}\right)$$

$$\sum_{n=1}^{\infty} \frac{4}{n(n+2)} = \lim_{n\to\infty} S_n = \lim_{n\to\infty} 2\left(1 + \frac{1}{2} - \frac{1}{n+1} - \frac{1}{n+2}\right) = 3$$

28. $\sum_{n=1}^{\infty} \frac{1}{(2n+1)(2n+3)} = \frac{1}{2}\sum_{n=1}^{\infty}\left(\frac{1}{2n+1} - \frac{1}{2n+3}\right)$

$$S_n = \frac{1}{2}\left[\left(\frac{1}{3} - \frac{1}{5}\right) + \left(\frac{1}{5} - \frac{1}{7}\right) + \left(\frac{1}{7} - \frac{1}{9}\right) + \cdots + \left(\frac{1}{2n+1} - \frac{1}{2n+3}\right)\right] = \frac{1}{2}\left(\frac{1}{3} - \frac{1}{2n+3}\right)$$

$$\sum_{n=1}^{\infty} \frac{1}{(2n+1)(2n+3)} = \lim_{n\to\infty} S_n = \lim_{n\to\infty} \frac{1}{2}\left(\frac{1}{3} - \frac{1}{2n+3}\right) = \frac{1}{6}$$

29. $\sum_{n=0}^{\infty} 8\left(\frac{3}{4}\right)^n = \frac{8}{1 - (3/4)} = 32$

30. $\sum_{n=0}^{\infty} 9\left(-\frac{1}{3}\right)^n = \frac{9}{1 - (-1/3)} = \frac{27}{4}$

31. $\sum_{n=0}^{\infty}\left(\frac{1}{2^n} - \frac{1}{3^n}\right) = \sum_{n=0}^{\infty}\left(\frac{1}{2}\right)^n - \sum_{n=0}^{\infty}\left(\frac{1}{3}\right)^n$

$$= \frac{1}{1 - (1/2)} - \frac{1}{1 - (1/3)} = 2 - \frac{3}{2} = \frac{1}{2}$$

32. $\sum_{n=0}^{\infty}\left[(0.3)^n + (0.8)^n\right] = \sum_{n=0}^{\infty}\left(\frac{3}{10}\right)^n + \sum_{n=0}^{\infty}\left(\frac{8}{10}\right)^n$

$$= \frac{1}{1 - (3/10)} + \frac{1}{1 - (8/10)}$$

$$= \frac{10}{7} + 5 = \frac{45}{7}$$

33. Note that $\sin(1) \approx 0.8415 < 1$. The series $\sum_{n=1}^{\infty}\left[\sin(1)\right]^n$ is geometric with $r = \sin(1) < 1$. So,

$$\sum_{n=1}^{\infty}\left[\sin(1)\right]^n = \sin(1)\sum_{n=0}^{\infty}\left[\sin(1)\right]^n = \frac{\sin(1)}{1 - \sin(1)} \approx 5.3080.$$

34. $S_n = \sum_{k=1}^{n} \frac{1}{9k^2 + 3k - 2} = \sum_{k=1}^{n} \frac{1}{(3k-1)(3k+2)}$

$$= \sum_{k=1}^{n}\left[\frac{1}{9k-3} - \frac{1}{9k+6}\right] = \frac{1}{3}\sum_{k=1}^{n}\left[\frac{1}{3k-1} - \frac{1}{3k+2}\right]$$

$$= \frac{1}{3}\left[\left(\frac{1}{2} - \frac{1}{5}\right) + \left(\frac{1}{5} - \frac{1}{8}\right) + \left(\frac{1}{8} - \frac{1}{11}\right) + \cdots + \left(\frac{1}{3n-1} - \frac{1}{3n+2}\right)\right] = \frac{1}{3}\left(\frac{1}{2} - \frac{1}{3n+2}\right)$$

$$\lim_{n\to\infty} S_n = \lim_{n\to\infty} \frac{1}{3}\left(\frac{1}{2} - \frac{1}{3n+2}\right) = \frac{1}{6}$$

35. (a) $0.\overline{4} = \sum_{n=0}^{\infty} \frac{4}{10}\left(\frac{1}{10}\right)^n$

(b) Geometric series with $a = \frac{4}{10}$ and $r = \frac{1}{10}$

$$S = \frac{a}{1-r} = \frac{4/10}{1-(1/10)} = \frac{4}{9}$$

36. (a) $0.\overline{36} = \sum_{n=0}^{\infty} \frac{36}{100}\left(\frac{1}{100}\right)^n$

(b) Geometric series with $a = \frac{36}{100}$ and $r = \frac{1}{100}$

$$S = \frac{a}{1-r} = \frac{36/100}{1-(1/100)} = \frac{36}{99} = \frac{4}{11}$$

37. (a) $0.\overline{81} = \sum_{n=0}^{\infty} \frac{81}{100}\left(\frac{1}{100}\right)^n$

(b) Geometric series with $a = \frac{81}{100}$ and $r = \frac{1}{100}$

$$S = \frac{a}{1-r} = \frac{81/100}{1-(1/100)} = \frac{81}{99} = \frac{9}{11}$$

38. (a) $0.\overline{01} = \sum_{n=1}^{\infty} \left(\frac{1}{100}\right)^n = \frac{1}{100}\sum_{n=0}^{\infty} \left(\frac{1}{100}\right)^n$

(b) $0.\overline{01} = \frac{1}{100} \cdot \frac{1}{1-(1/100)} = \frac{1}{100} \cdot \frac{100}{99} = \frac{1}{99}$

39. (a) $0.0\overline{75} = \sum_{n=0}^{\infty} \frac{3}{40}\left(\frac{1}{100}\right)^n$

(b) Geometric series with $a = \frac{3}{40}$ and $r = \frac{1}{100}$

$$S = \frac{a}{1-r} = \frac{3/40}{99/100} = \frac{5}{66}$$

40. (a) $0.2\overline{15} = \frac{1}{5} + \sum_{n=0}^{\infty} \frac{3}{200}\left(\frac{1}{100}\right)^n$

(b) Geometric series with $a = \frac{3}{200}$ and $r = \frac{1}{100}$

$$S = \frac{1}{5} + \frac{a}{1-r} = \frac{1}{5} + \frac{3/200}{99/100} = \frac{71}{330}$$

41. $\sum_{n=0}^{\infty} (1.075)^n$

Geometric series with $r = 1.075$

Diverges by Theorem 9.6

42. $\sum_{n=1}^{\infty} \frac{3^n}{1000}$

Geometric series with $r = 3 > 1$.

Diverges by Theorem 9.6

43. $\sum_{n=1}^{\infty} \frac{n+10}{10n+1}$

$$\lim_{n\to\infty} \frac{n+10}{10n+1} = \frac{1}{10} \neq 0$$

Diverges by Theorem 9.9

44. $\sum_{n=1}^{\infty} \frac{4n+1}{3n-1}$

$$\lim_{n\to\infty} \frac{4n+1}{3n-1} = \frac{4}{3} \neq 0$$

Diverges by Theorem 9.9

45. $\sum_{n=1}^{\infty} \left(\frac{1}{n} - \frac{1}{n+2}\right)$

$$S_n = \left(1 - \frac{1}{3}\right) + \left(\frac{1}{2} - \frac{1}{4}\right) + \left(\frac{1}{3} - \frac{1}{5}\right) + \cdots + \left(\frac{1}{n-1} - \frac{1}{n+1}\right) + \left(\frac{1}{n} - \frac{1}{n+2}\right) = 1 + \frac{1}{2} - \frac{1}{n+1} - \frac{1}{n+2}$$

$$\sum_{n=1}^{\infty} \left(\frac{1}{n} - \frac{1}{n+2}\right) = \lim_{n\to\infty} S_n = \lim_{n\to\infty} \left(1 + \frac{1}{2} - \frac{1}{n+1} - \frac{1}{n+2}\right) = \frac{3}{2}, \text{ converges}$$

46. $\sum_{n=1}^{\infty} \left(\frac{1}{n+1} - \frac{1}{n+2}\right)$

$$S_n = \left(\frac{1}{2} - \frac{1}{3}\right) + \left(\frac{1}{3} - \frac{1}{4}\right) + \cdots + \left(\frac{1}{n+1} - \frac{1}{n+2}\right) = \frac{1}{2} - \frac{1}{n+2}$$

$$\sum_{n=1}^{\infty} \left(\frac{1}{n+1} - \frac{1}{n+2}\right) = \lim_{n\to\infty} S_n = \lim_{n\to\infty} \left(\frac{1}{2} - \frac{1}{n+2}\right) = \frac{1}{2}, \text{ converges}$$

47. $\sum_{n=1}^{\infty} \frac{3^n}{n^3}$

$$\lim_{n\to\infty} \frac{3^n}{n^3} = \lim_{n\to\infty} \frac{(\ln 2)3^n}{3n^2}$$
$$= \lim_{n\to\infty} \frac{(\ln 2)^2 3^n}{6n} = \lim_{n\to\infty} \frac{(\ln n)^3 3^n}{6} = \infty$$

(by L'Hôpital's Rule); diverges by Theorem 9.9

48. $\sum_{n=0}^{\infty} \frac{3}{5^n} = 3\sum_{n=0}^{\infty}\left(\frac{1}{5}\right)^n$, convergent

Geometric series with $r = \frac{1}{5}$

49. Because $n > \ln(n)$, the terms $a_n = \frac{n}{\ln(n)}$ do not approach 0 as $n \to \infty$. So, the series $\sum_{n=2}^{\infty} \frac{n}{\ln(n)}$ diverges.

50. $S_n = \sum_{k=1}^{n} \ln\left(\frac{1}{k}\right) = \sum_{k=1}^{n} -\ln(k)$

$= 0 - \ln 2 - \ln 3 - \cdots - \ln(n)$

Because $\lim_{n\to\infty} S_n$ diverges, $\sum_{n=1}^{\infty} \ln\left(\frac{1}{n}\right)$ diverges.

51. For $k \neq 0$,

$$\lim_{n\to\infty}\left(1 + \frac{k}{n}\right)^n = \lim_{n\to\infty}\left[\left(1 + \frac{k}{n}\right)^{n/k}\right]^k$$
$$= e^k \neq 0.$$

For $k = 0, \lim_{n\to\infty}(1 + 0)^n = 1 \neq 0.$

So, $\sum_{n=1}^{\infty}\left[1 + \frac{k}{n}\right]^n$ diverges.

52. $\sum_{n=1}^{\infty} e^{-n} = \sum_{n=1}^{\infty}\left(\frac{1}{e}\right)^n$ converges because it is geometric with

$|r| = \frac{1}{e} < 1.$

53. $\lim_{n\to\infty} \arctan n = \frac{\pi}{2} \neq 0$

So, $\sum_{n=1}^{\infty} \arctan n$ diverges.

54. $S_n = \sum_{k=1}^{n} \ln\left(\frac{k+1}{k}\right)$

$= \ln\left(\frac{2}{1}\right) + \ln\left(\frac{3}{2}\right) + \cdots + \ln\left(\frac{n+1}{n}\right)$

$= (\ln 2 - \ln 1) + (\ln 3 - \ln 2) + \cdots + (\ln(n+1) - \ln n)$

$= \ln(n+1) - \ln(1) = \ln(n+1)$

Diverges

55. See definitions on page 595.

56. $\lim_{n\to\infty} a_n = 5$ means that the limit of the sequence $\{a_n\}$ is 5.

$\sum_{n=1}^{\infty} a_n = a_1 + a_2 + \cdots = 5$ means that the limit of the partial sums is 5.

57. The series given by

$$\sum_{n=0}^{\infty} ar^n = a + ar + ar^2 + \cdots + ar^n + \cdots, a \neq 0$$

is a geometric series with ratio r. When $0 < |r| < 1$, the series converges to $a/(1 - r)$. The series diverges if $|r| \geq 1$.

58. If $\lim_{n\to\infty} a_n \neq 0$, then $\sum_{n=1}^{\infty} a_n$ diverges.

59. (a) $\sum_{n=1}^{\infty} a_n = a_1 + a_2 + a_3 + \cdots$

(b) $\sum_{k=1}^{\infty} a_k = a_1 + a_2 + a_3 + \cdots$

These are the same. The third series is different, unless $a_1 = a_2 = \cdots = a$ is constant.

(c) $\sum_{n=1}^{\infty} a_k = a_k + a_k + \cdots$

60. (a) Yes, the new series will still diverge.

(b) Yes, the new series will converge.

61. $\sum_{n=1}^{\infty}(3x)^n = (3x)\sum_{n=0}^{\infty}(3x)^n$

Geometric series: converges for $|3x| < 1 \Rightarrow |x| < \frac{1}{3}$

$$f(x) = (3x)\sum_{n=0}^{\infty}(3x)^n = (3x)\frac{1}{1-3x} = \frac{3x}{1-3x}, \ |x| < \frac{1}{3}$$

62. $\sum_{n=0}^{\infty}\left(\frac{2}{x}\right)^n$

Geometric series: converges for

$\left|\frac{2}{x}\right| < 1 \Rightarrow |x| > 2 \Rightarrow x < -2 \text{ or } x > 2$

$f(x) = \sum_{n=0}^{\infty}\left(\frac{2}{x}\right)^n = \frac{1}{1-(2/x)} = \frac{x}{x-2}, x > 2 \text{ or } x < -2$

63. $\sum_{n=1}^{\infty}(x-1)^n = (x-1)\sum_{n=0}^{\infty}(x-1)^n$

Geometric series: converges for $|x-1| < 1 \Rightarrow 0 < x < 2$

$f(x) = (x-1)\sum_{n=0}^{\infty}(x-1)^n$

$= (x-1)\frac{1}{1-(x-1)} = \frac{x-1}{2-x}, \quad 0 < x < 2$

64. $\sum_{n=0}^{\infty} 5\left(\frac{x-2}{3}\right)^n$

Geometric series: converges for

$\left|\frac{x-2}{3}\right| < 1 \Rightarrow |x-2| < 3 \Rightarrow -1 < x < 5$

$f(x) = \sum_{n=0}^{\infty} 5\left(\frac{x-2}{3}\right)^n$

$= \frac{5}{1-\left(\frac{x-2}{3}\right)} = \frac{5}{(3-x+2)/3}$

$= \frac{15}{5-x}, -1 < x < 5$

65. $\sum_{n=0}^{\infty}(-1)^n x^n = \sum_{n=0}^{\infty}(-x)^n$

Geometric series: converges for

$|-x| < 1 \Rightarrow |x| < 1 \Rightarrow -1 < x < 1$

$f(x) = \sum_{n=0}^{\infty}(-x)^n = \frac{1}{1+x}, \quad -1 < x < 1$

66. $\sum_{n=0}^{\infty}(-1)^n x^{2n} = \sum_{n=0}^{\infty}\left(-x^2\right)^n$

Geometric series: converges for

$\left|-x^2\right| < 1 \Rightarrow -1 < x < 1$

$f(x) = \sum_{n=0}^{\infty}\left(-x^2\right)^n = \frac{1}{1-\left(-x^2\right)} = \frac{1}{1+x^2}, -1 < x < 1$

67. (a) x is the common ratio.

(b) $1 + x + x^2 + \cdots = \sum_{n=0}^{\infty} x^n = \frac{1}{1-x}, \quad |x| < 1$

(c) $y_1 = \frac{1}{1-x}$

$y_2 = S_3 = 1 + x + x^2$

$y_3 = S_5 = 1 + x + x^2 + x^3 + x^4$

Answers will vary.

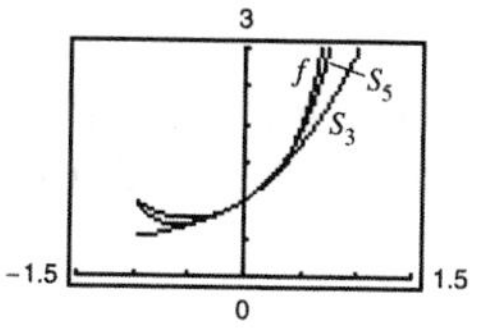

68. (a) $\left(-\frac{x}{2}\right)$ is the common ratio.

(b) $$1 - \frac{x}{2} + \frac{x^2}{4} - \frac{x^3}{8} + \cdots = \sum_{n=0}^{\infty}\left(-\frac{x}{2}\right)^n = \frac{1}{1-(-x/2)} = \frac{2}{2+x}, \quad |x| < 2$$

(c) $$y_1 = \frac{2}{2+x}$$

$$y_2 = S_3 = 1 - \frac{x}{2} + \frac{x^2}{4}$$

$$y_3 = S_5 = 1 - \frac{x}{2} + \frac{x^2}{4} - \frac{x^3}{8} + \frac{x^4}{16}$$

Answers will vary.

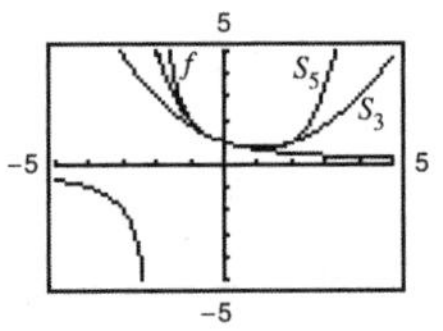

69. $$\frac{1}{n(n+1)} < 0.0001$$

$$10{,}000 < n^2 + n$$

$$0 < n^2 + n - 10{,}000$$

$$n = \frac{-1 \pm \sqrt{1^2 - 4(1)(-10{,}000)}}{2}$$

Choosing the positive value for n you have $n \approx 99.5012$. The first *term* that is less than 0.0001 is $n = 100$.

$$\left(\frac{1}{8}\right)^n < 0.0001$$

$$10{,}000 < 8^n$$

This inequality is true when $n = 5$. This series converges at a faster rate.

70. $$\frac{1}{2^n} < 0.0001$$

$$10{,}000 < 2^n$$

This inequality is true when $n = 14$.

$$(0.01)^n < 0.0001$$

$$10{,}000 < 10^n$$

This inequality is true when $n = 5$. This series converges at a faster rate.

71. $$\sum_{i=0}^{n-1} 8000(0.95)^i = \frac{8000\left[1 - 0.95^n\right]}{1 - 0.95} = 160{,}000\left[1 - 0.95^n\right], \quad n > 0$$

72. $$V(t) = 475{,}000(1 - 0.3)^n = 475{,}000(0.7)^n$$

$$V(5) = 475{,}000(0.7)^5 = \$79{,}833.25$$

73. $\sum_{i=0}^{\infty} 200(0.75)^i = 800$ million dollars

74. $\sum_{i=0}^{\infty} 200(0.60)^i = 500$ million dollars

75. $$D_1 = 16$$

$$D_2 = \underbrace{0.81(16)}_{\text{up}} + \underbrace{0.81(16)}_{\text{down}} = 32(0.81)$$

$$D_3 = 16(0.81)^2 + 16(0.81)^2 = 32(0.81)^2$$

$$\vdots$$

$$D = 16 + 32(0.81) + 32(0.81)^2 + \cdots$$

$$= -16 + \sum_{n=0}^{\infty} 32(0.81)^n = -16 + \frac{32}{1 - 0.81}$$

≈ 152.42 feet

76. The ball in Exercise 75 takes the following times for each fall.

$s_1 = -16t^2 + 16$	$s_1 = 0$ if $t = 1$
$s_2 = -16t^2 + 16(0.81)$	$s_2 = 0$ if $t = 0.9$
$s_3 = -16t^2 + 16(0.81)^2$	$s_3 = 0$ if $t = (0.9)^2$
$\vdots$	$\vdots$
$s_n = -16t^2 + 16(0.81)^{n-1}$	$s_n = 0$ if $t = (0.9)^{n-1}$

Beginning with s_2, the ball takes the same amount of time to bounce up as it takes to fall. The total elapsed time before the ball comes to rest is

$$t = 1 + 2\sum_{n=1}^{\infty}(0.9)^n = -1 + 2\sum_{n=0}^{\infty}(0.9)^n$$

$$= -1 + \frac{2}{1 - 0.9} = 19 \text{ seconds.}$$

77. $P(n) = \frac{1}{2}\left(\frac{1}{2}\right)^n$

$P(2) = \frac{1}{2}\left(\frac{1}{2}\right)^2 = \frac{1}{8}$

$\sum_{n=0}^{\infty} \frac{1}{2}\left(\frac{1}{2}\right)^n = \frac{1/2}{1-(1/2)} = 1$

78. $P(n) = \frac{1}{3}\left(\frac{2}{3}\right)^n$

$P(2) = \frac{1}{3}\left(\frac{2}{3}\right)^2 = \frac{4}{27}$

$\sum_{n=0}^{\infty} \frac{1}{3}\left(\frac{2}{3}\right)^n = \frac{1/3}{1-(2/3)} = 1$

79. (a) $\sum_{n=1}^{\infty} \left(\frac{1}{2}\right)^n = \sum_{n=0}^{\infty} \frac{1}{2}\left(\frac{1}{2}\right)^n = \frac{1}{2}\frac{1}{\left(1-(1/2)\right)} = 1$

(b) No, the series is not geometric.

(c) $\sum_{n=1}^{\infty} n\left(\frac{1}{2}\right)^n = 2$

80. Person 1: $\frac{1}{2} + \frac{1}{2^4} + \frac{1}{2^7} + \cdots = \frac{1}{2}\sum_{n=0}^{\infty}\left(\frac{1}{8}\right)^n = \frac{1}{2}\frac{1}{1-(1/8)} = \frac{4}{7}$

Person 2: $\frac{1}{2^2} + \frac{1}{2^5} + \frac{1}{2^8} + \cdots = \frac{1}{4}\sum_{n=0}^{\infty}\left(\frac{1}{8}\right)^n = \frac{1}{4}\frac{1}{1-(1/8)} = \frac{2}{7}$

Person 3: $\frac{1}{2^3} + \frac{1}{2^6} + \frac{1}{2^9} + \cdots = \frac{1}{8}\sum_{n=0}^{\infty}\left(\frac{1}{8}\right)^n = \frac{1}{8}\frac{1}{1-(1/8)} = \frac{1}{7}$

Sum: $\frac{4}{7} + \frac{2}{7} + \frac{1}{7} = 1$

81. (a) $64 + 32 + 16 + 8 + 4 + 2 = 126 \text{ in.}^2$

(b) $\sum_{n=0}^{\infty} 64\left(\frac{1}{2}\right)^n = \frac{64}{1-(1/2)} = 128 \text{ in.}^2$

Note: This is one-half of the area of the original square

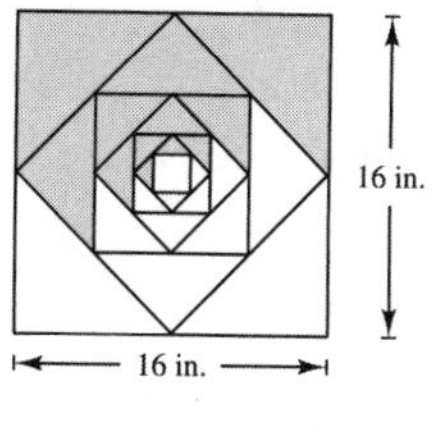

82. (a) $\sin\theta = \frac{|Y_{y_1}|}{z} \Rightarrow |Y_{y_1}| = z\sin\theta$

$\sin\theta = \frac{|x_1y_1|}{|Y_{y_1}|} \Rightarrow |x_1y_1| = |Y_{y_1}|\sin\theta = z\sin^2\theta$

$\sin\theta = \frac{|x_1y_2|}{|x_1y_1|} \Rightarrow |x_1y_2| = |x_1y_1|\sin\theta = z\sin^3\theta$

Total: $z\sin\theta + z\sin^2\theta + z\sin^3\theta + \cdots = z\frac{\sin\theta}{1-\sin\theta}$

(b) If $z = 1$ and $\theta = \frac{\pi}{6}$, then total $= \frac{1/2}{1-(1/2)} = 1$.

83. $\sum_{n=1}^{20} 100{,}000\left(\frac{1}{1.06}\right)^n = \frac{100{,}000}{1.06}\sum_{i=0}^{19}\left(\frac{1}{1.06}\right)^i = \frac{100{,}000}{1.06}\left[\frac{1 - 1.06^{-20}}{1 - 1.06^{-1}}\right] \quad \left(n = 20, r = 1.06^{-1}\right) \approx \$1{,}146{,}992.12$

The $2,000,000 sweepstakes has a present value of $1,146,992.12. After accruing interest over the 20-year period, it attains its full value.

84.
$$\sum_{n=0}^{12t-1} P\left(1 + \frac{r}{12}\right)^n = \frac{P\left[1 - \left(1 + \frac{r}{12}\right)^{12t}\right]}{1 - \left(1 + \frac{r}{12}\right)}$$
$$= P\left(-\frac{12}{r}\right)\left[\left(1 - \left(1 + \frac{r}{12}\right)^{12t}\right)\right]$$
$$= P\left(\frac{12}{r}\right)\left[\left(1 + \frac{r}{12}\right)^{12t} - 1\right]$$
$$\sum_{n=0}^{12t-1} P\left(e^{r/12}\right)^n = \frac{P\left(1 - \left(e^{r/12}\right)^{12t}\right)}{1 - e^{r/12}} = \frac{P\left(e^{rt} - 1\right)}{e^{r/12} - 1}$$

85. $w = \sum_{i=0}^{n-1} 0.01(2)^i = \frac{0.01\left(1 - 2^n\right)}{1 - 2} = 0.01\left(2^n - 1\right)$

(a) When $n = 29$: $w = \$5{,}368{,}709.11$

(b) When $n = 30$: $w = \$10{,}737{,}418.23$

(c) When $n = 31$: $w = \$21{,}474{,}836.47$

86. Surface area $= 4\pi(1)^2 + 9\left(4\pi\left(\frac{1}{3}\right)^2\right) + 9^2 \cdot 4\pi\left(\frac{1}{9}\right)^2 + \cdots = 4(\pi + \pi + \cdots) = \infty$

87. $P = 45, \quad r = 0.03, \quad t = 20$

(a) $A = 45\left(\frac{12}{0.03}\right)\left[\left(1 + \frac{0.03}{12}\right)^{12(20)} - 1\right] \approx \$14{,}773.59$

(b) $A = \frac{45\left(e^{0.03(20)} - 1\right)}{e^{0.03/12} - 1} \approx \$14{,}779.65$

88. $P = 75, \quad r = 0.055, \quad t = 25$

(a) $A = 75\left(\frac{12}{0.055}\right)\left[\left(1 + \frac{0.055}{12}\right)^{12(25)} - 1\right] \approx \$48{,}152.81$

(b) $A = \frac{75\left(e^{0.055(25)} - 1\right)}{e^{0.055/12} - 1} \approx \$48{,}245.07$

89. $P = 100, \quad r = 0.04, \quad t = 35$

(a) $A = 100\left(\frac{12}{0.04}\right)\left[\left(1 + \frac{0.04}{12}\right)^{12(35)} - 1\right] \approx \$91{,}373.09$

(b) $A = \frac{100\left(e^{0.04(35)} - 1\right)}{e^{0.04/12} - 1} \approx \$91{,}503.32$

90. $P = 30, \quad r = 0.06, \quad t = 50$

(a) $A = 30\left(\frac{12}{0.06}\right)\left[\left(1 + \frac{0.06}{12}\right)^{12(50)} - 1\right] \approx 113{,}615.73$

(b) $A = \frac{30\left(e^{0.06(50)} - 1\right)}{e^{0.06/12} - 1} \approx \$114{,}227.18$

91. False. $\lim_{n\to\infty} \frac{1}{n} = 0$, but $\sum_{n=1}^{\infty} \frac{1}{n}$ diverges.

92. True

93. False; $\sum_{n=1}^{\infty} ar^n = \left(\frac{a}{1 - r}\right) - a$

The formula requires that the geometric series begins with $n = 0$.

94. True

$$\lim_{n\to\infty} \frac{n}{1000(n + 1)} = \frac{1}{1000} \neq 0$$

95. True

$$0.74999\ldots = 0.74 + \frac{9}{10^3} + \frac{9}{10^4} + \cdots$$
$$= 0.74 + \frac{9}{10^3}\sum_{n=0}^{\infty}\left(\frac{1}{10}\right)^n$$
$$= 0.74 + \frac{9}{10^3} \cdot \frac{1}{1 - (1/10)}$$
$$= 0.74 + \frac{9}{10^3} \cdot \frac{10}{9}$$
$$= 0.74 + \frac{1}{100} = 0.75$$

96. True

97. Let $\sum a_n = \sum_{n=0}^{\infty} 1$ and $\sum b_n = \sum_{n=0}^{\infty} (-1)$.

Both are divergent series.

$$\sum(a_n + b_n) = \sum_{n=0}^{\infty}[1 + (-1)] = \sum_{n=0}^{\infty}[1 - 1] = 0$$

98. If $\Sigma(a_n + b_n)$ converged, then

$\Sigma(a_n + b_n) - \Sigma a_n = \Sigma b_n$ would converge, which is a contradiction. So, $\Sigma(a_n + b_n)$ diverges.

99. (a) $\dfrac{1}{a_{n+1}a_{n+2}} - \dfrac{1}{a_{n+2}a_{n+3}} = \dfrac{a_{n+3} - a_{n+1}}{a_{n+1}a_{n+2}a_{n+3}} = \dfrac{a_{n+2}}{a_{n+1}a_{n+2}a_{n+3}} = \dfrac{1}{a_{n+1}a_{n+3}}$

(b)
$$S_n = \sum_{k=0}^{n} \frac{1}{a_{k+1}a_{k+3}}$$
$$= \sum_{k=0}^{n}\left[\frac{1}{a_{k+1}a_{k+2}} - \frac{1}{a_{k+2}a_{k+3}}\right]$$
$$= \left[\frac{1}{a_1a_2} - \frac{1}{a_2a_3}\right] + \left[\frac{1}{a_2a_3} - \frac{1}{a_3a_4}\right] + \cdots + \left[\frac{1}{a_{n+1}a_{n+2}} - \frac{1}{a_{n+2}a_{n+3}}\right] = \frac{1}{a_1a_2} - \frac{1}{a_{n+2}a_{n+3}} = 1 - \frac{1}{a_{n+2}a_{n+3}}$$
$$\sum_{n=0}^{\infty} \frac{1}{a_{n+1}a_{n+3}} = \lim_{n\to\infty} S_n = \lim_{n\to\infty}\left[1 - \frac{1}{a_{n+2}a_{n+3}}\right] = 1$$

100. Let $\{S_n\}$ be the sequence of partial sums for the convergent series

$\sum_{n=1}^{\infty} a_n = L$. Then $\lim_{n\to\infty} S_n = L$ and because

$$R_n = \sum_{k=n+1}^{\infty} a_k = L - S_n,$$

you have

$$\lim_{n\to\infty} R_n = \lim_{n\to\infty}(L - S_n) = \lim_{n\to\infty} L - \lim_{n\to\infty} S_n = L - L = 0.$$

101. $\dfrac{1}{r} + \dfrac{1}{r^2} + \dfrac{1}{r^3} + \cdots = \displaystyle\sum_{n=0}^{\infty} \frac{1}{r}\left(\frac{1}{r}\right)^n = \frac{1/r}{1 - (1/r)} = \frac{1}{r - 1} \quad \left(\text{since} \left|\frac{1}{r}\right| < 1\right)$

This is a geometric series which converges if

$$\left|\frac{1}{r}\right| < 1 \Leftrightarrow |r| > 1.$$

102. The entire rectangle has area 2 because the height is 1 and the base is $1 + \frac{1}{2} + \frac{1}{4} + \cdots = 2$. The squares all lie inside the rectangle, and the sum of their areas is

$$1 + \frac{1}{2^2} + \frac{1}{3^2} + \frac{1}{4^2} + \cdots.$$

So, $\displaystyle\sum_{n=1}^{\infty} \frac{1}{n^2} < 2$.

103. The series is telescoping:

$$S_n = \sum_{k=1}^{n} \frac{6^k}{(3^{k+1} - 2^{k+1})(3^k - 2^k)}$$
$$= \sum_{k=1}^{n}\left[\frac{3^k}{3^k - 2^k} - \frac{3^{k+1}}{3^{k+1} - 2^{k+1}}\right]$$
$$= 3 - \frac{3^{n+1}}{3^{n+1} - 2^{n+1}}$$
$$\lim_{n\to\infty} S_n = 3 - 1 = 2$$

104. $f(1) = 0, f(2) = 1, f(3) = 2, f(4) = 4, \ldots$

In general: $f(n) = \begin{cases} n^2/4, & n \text{ even} \\ (n^2 - 1)/4, & n \text{ odd.} \end{cases}$

(See below for a proof of this.)

$x + y$ and $x - y$ are either both odd or both even. If both even, then

$$f(x + y) - f(x - y) = \frac{(x + y)^2}{4} - \frac{(x - y)^2}{4} = xy.$$

If both odd,

$$f(x + y) - f(x - y) = \frac{(x + y)^2 - 1}{4} - \frac{(x - y)^2 - 1}{4} = xy.$$

Proof by induction that the formula for $f(n)$ is correct. It is true for $n = 1$. Assume that the formula is valid for k. If k is even, then $f(k) = k^2/4$ and

$$f(k + 1) = f(k) + \frac{k}{2} = \frac{k^2}{4} + \frac{k}{2} = \frac{k^2 + 2k}{4} = \frac{(k + 1)^2 - 1}{4}.$$

The argument is similar if k is odd.

Section 9.3 The Integral Test and *p*-Series

1. $\displaystyle\sum_{n=1}^{\infty} \frac{1}{n + 3}$

Let

$f(x) = \dfrac{1}{x + 3},\ f'(x) = -\dfrac{1}{(x + 3)^2} < 0$ for $x \geq 1$.

f is positive, continuous, and decreasing for $x \geq 1$.

$$\int_1^{\infty} \frac{1}{x + 3}\,dx = \Big[\ln(x + 3)\Big]_1^{\infty} = \infty$$

So, the series diverges by Theorem 9.10.

2. $\displaystyle\sum_{n=1}^{\infty} \frac{2}{3n + 5}$

Let $f(x) = \dfrac{2}{3x + 5}$.

f is positive, continuous, and decreasing for $x \geq 1$.

$$\int_1^{\infty} \frac{2}{3x + 5}\,dx = \left[\frac{2}{3}\ln(3x + 5)\right]_1^{\infty} = \infty$$

So, the series diverges by Theorem 9.10.

3. $\displaystyle\sum_{n=1}^{\infty} \frac{1}{2^n}$

Let $f(x) = \dfrac{1}{2^x},\ f'(x) = -(\ln 2)2^{-x} < 0$ for $x \geq 1$.

f is positive, continuous, and decreasing for $x \geq 1$.

$$\int_1^{\infty} \frac{1}{2^x}\,dx = \left[\frac{-1}{(\ln 2)\,2^x}\right]_1^{\infty} = \frac{1}{2 \ln 2}$$

So, the series converges by Theorem 9.10.

4. $\displaystyle\sum_{n=1}^{\infty} 3^{-n}$

Let $f(x) = \dfrac{1}{3^x},\ f'(x) = -(\ln 3)3^{-x} < 0$ for $x \geq 1$.

f is positive, continuous, and decreasing for $x \geq 1$.

$$\int_1^{\infty} \frac{1}{3^x}\,dx = \left[\frac{-1}{(\ln 3)\,3^x}\right]_1^{\infty} = \frac{1}{3 \ln 3}$$

So, the series converges by Theorem 9.10.

5. $\displaystyle\sum_{n=1}^{\infty} e^{-n}$

Let $f(x) = e^{-x},\ f'(x) = -e^{-x} < 0$ for $x \geq 1$.

f is positive, continuous, and decreasing for $x \geq 1$.

$$\int_1^{\infty} e^{-x}\,dx = \Big[-e^{-x}\Big]_1^{\infty} = \frac{1}{e}$$

So, the series converges by Theorem 9.10.

6. $\displaystyle\sum_{n=1}^{\infty} ne^{-n/2}$

Let $f(x) = xe^{-x/2},\ f'(x) = \dfrac{2 - x}{2e^{x/2}} < 0$ for $x \geq 3$.

f is positive, continuous, and decreasing for $x \geq 3$

$$\int_3^{\infty} xe^{-x/2}\,dx = \Big[-2(x + 2)e^{-x/2}\Big]_3^{\infty} = 10e^{-3/2}$$

So, the series converges by Theorem 9.10.

7. $\sum_{n=1}^{\infty} \frac{1}{n^2 + 1}$

Let

$$f(x) = \frac{1}{x^2 + 1},\ f'(x) = -\frac{2x}{\left(x^2 + 1\right)^2} < 0 \text{ for } x \geq 1.$$

f is positive, continuous, and decreasing for $x \geq 1$.

$$\int_1^{\infty} \frac{1}{x^2 + 1}\,dx = \left[\arctan x\right]_1^{\infty} = \frac{\pi}{4}$$

So, the series converges by Theorem 9.10.

8. $\sum_{n=1}^{\infty} \frac{1}{2n + 1}$

Let

$$f(x) = \frac{1}{2x + 1},\ f'(x) = -\frac{2}{\left(2x + 1\right)^2} < 0 \text{ for } x \geq 1.$$

f is positive, continuous, and decreasing for $x \geq 1$.

$$\int_1^{\infty} \frac{1}{2x + 1}\,dx = \left[\ln \sqrt{2x + 1}\right]_1^{\infty} = \infty$$

So, the series diverges by Theorem 9.10.

9. $\sum_{n=1}^{\infty} \frac{\ln(n + 1)}{n + 1}$

Let $f(x) = \frac{\ln(x + 1)}{x + 1},\ f'(x) = \frac{1 - \ln(x + 1)}{(x + 1)^2} < 0$ for $x \geq 2$.

f is positive, continuous, and decreasing for $x \geq 2$.

$$\int_1^{\infty} \frac{\ln(x + 1)}{x + 1}\,dx = \left[\frac{\left[\ln(x + 1)\right]^2}{2}\right]_1^{\infty} = \infty$$

So, the series diverges by Theorem 9.10.

10. $\sum_{n=2}^{\infty} \frac{\ln n}{\sqrt{n}}$

Let $f(x) = \frac{\ln x}{\sqrt{x}},\ f'(x) = \frac{2 - \ln x}{2x^{3/2}}$.

f is positive, continuous, and decreasing for $x > e^2 \approx 7.4$.

$$\int_2^{\infty} \frac{\ln x}{\sqrt{x}}\,dx = \left[2\sqrt{x}(\ln x - 2)\right]_2^{\infty} = \infty$$

So, the series diverges by Theorem 9.10.

11. $\sum_{n=1}^{\infty} \frac{1}{\sqrt{n}\left(\sqrt{n} + 1\right)}$

Let $f(x) = \frac{1}{\sqrt{x}\left(\sqrt{x} + 1\right)}$,

$$f'(x) = -\frac{1 + 2\sqrt{x}}{2x^{3/2}\left(\sqrt{x} + 1\right)^2} < 0.$$

f is positive, continuous, and decreasing for $x \geq 1$.

$$\int_1^{\infty} \frac{1}{\sqrt{x}\left(\sqrt{x} + 1\right)}\,dx = \left[2\ln\left(\sqrt{x} + 1\right)\right]_1^{\infty} = \infty$$

So, the series diverges by Theorem 9.10.

12. $\sum_{n=1}^{\infty} \frac{n}{n^2 + 3}$

Let $f(x) = \frac{x}{x^2 + 3},\ f'(x) = \frac{3 - x^2}{\left(x^2 + 3\right)} < 0$ for $x \geq 2$.

f is positive, continuous, and decreasing for $x \geq 2$

$$\int_1^{\infty} \frac{x}{x^2 + 3}\,dx = \left[\ln\sqrt{x^2 + 3}\right]_1^{\infty} = \infty$$

So, the series diverges by Theorem 9.10.

13. $\sum_{n=1}^{\infty} \frac{\arctan n}{n^2 + 1}$

Let $f(x) = \frac{\arctan x}{x^2 + 1}$,

$$f'(x) = \frac{1 - 2x \arctan x}{\left(x^2 + 1\right)^2} < 0 \text{ for } x \geq 1.$$

f is positive, continuous, and decreasing for $x \geq 1$.

$$\int_1^{\infty} \frac{\arctan x}{x^2 + 1}\,dx = \left[\frac{(\arctan x)^2}{2}\right]_1^{\infty} = \frac{3\pi^2}{32}$$

So, the series converges by Theorem 9.10.

14. $\displaystyle\sum_{n=2}^{\infty} \frac{\ln n}{n^3}$

Let $f(x) = \dfrac{\ln x}{x^3}$, $f'(x) = \dfrac{1 - 3\ln x}{x^4}$.

f is positive, continuous, and decreasing for $x > 2$.

$$\int_2^{\infty} \frac{\ln x}{x^3}\,dx = \left[-\frac{(2\ln x + 1)}{4x^4}\right]_2^{\infty} = \frac{2\ln 2 + 1}{16}$$

So, the series converges by Theorem 9.10.

15. $\displaystyle\sum_{n=1}^{\infty} \frac{\ln n}{n^2}$

Let $f(x) = \dfrac{\ln x}{x^2}$, $f'(x) = \dfrac{1 - 2\ln x}{x^3}$.

f is positive, continuous, and decreasing for $x > e^{1/2} \approx 1.6$.

$$\int_1^{\infty} \frac{\ln x}{x^2}\,dx = \left[\frac{-(\ln x + 1)}{x}\right]_1^{\infty} = 1$$

So, the series converges by Theorem 9.10.

16. $\displaystyle\sum_{n=2}^{\infty} \frac{1}{n\sqrt{\ln n}}$

Let $f(x) = \dfrac{1}{x\sqrt{\ln x}}$, $f'(x) = -\dfrac{2\ln x + 1}{2x^2(\ln x)^{3/2}}$.

f is positive, continuous, and decreasing for $x \ge 2$.

$$\int_2^{\infty} \frac{1}{x\sqrt{\ln x}}\,dx = \left[2\sqrt{\ln x}\right]_2^{\infty} = \infty$$

So, the series diverges by Theorem 9.10.

17. $\displaystyle\sum_{n=1}^{\infty} \frac{1}{(2n + 3)^3}$

Let $f(x) = (2x + 3)^{-3}$, $f'(x) = \dfrac{-6}{(2x + 3)^4} < 0$

f is positive, continuous, and decreasing for $x \ge 1$.

$$\int_1^{\infty} (2x + 3)^{-3}\,dx = \left[\frac{-1}{4(2x + 3)^2}\right]_1^{\infty} = \frac{1}{100}$$

So, the series converges by Theorem 9.10.

18. $\displaystyle\sum_{n=1}^{\infty} \frac{n + 2}{n + 1}$

Let $f(x) = \dfrac{x + 2}{x + 1} = 1 + \dfrac{1}{x + 1}$, $f'(x) = \dfrac{-1}{(x + 1)^2} < 0$

f is positive, continuous, and decreasing for $x \ge 1$.

$$\int_1^{\infty} \frac{x + 2}{x + 1}\,dx = \left[x + \ln(x + 1)\right]_1^{\infty} = \infty$$

So, the series diverges by Theorem 9.10.

[**Note:** $\displaystyle\lim_{n\to\infty} \frac{n + 2}{n + 1} = 1 \ne 0$, so the series diverges.]

19. $\displaystyle\sum_{n=1}^{\infty} \frac{4n}{2n^2 + 1}$

Let $f(x) = \dfrac{4x}{2x^2 + 1}$, $f'(x) = \dfrac{-4(2x^2 - 1)}{(2x^2 + 1)^2} < 0$

for $x \ge 1$.

f is positive, continuous, and decreasing for $x \ge 1$.

$$\int_1^{\infty} \frac{4x}{2x^2 + 1}\,dx = \left[\ln(2x^2 + 1)\right]_1^{\infty} = \infty$$

So, the series diverges by Theorem 9.10.

20. $\displaystyle\sum_{n=1}^{\infty} \frac{1}{\sqrt{n + 2}}$

Let $f(x) = \dfrac{1}{\sqrt{x + 2}}$, $f'(x) = \dfrac{-1}{2(x + 2)^{3/2}} < 0$.

f is positive, continuous, and decreasing for $x \ge 1$.

$$\int_1^{\infty} \frac{1}{(x + 2)^{1/2}}\,dx = \left[2\sqrt{x + 2}\right]_1^{\infty} = \infty$$

So, the series diverges by Theorem 9.10.

21. $\displaystyle\sum_{n=1}^{\infty} \frac{n}{n^4 + 1}$

Let $f(x) = \dfrac{x}{x^4 + 1}$, $f'(x) = \dfrac{1 - 3x^4}{(x^4 + 1)^2} < 0$ for $x > 1$.

f is positive, continuous, and decreasing for $x > 1$.

$$\int_1^{\infty} \frac{x}{x^4 + 1}\,dx = \left[\frac{1}{2}\arctan(x^2)\right]_1^{\infty} = \frac{\pi}{8}$$

So, the series converges by Theorem 9.10.

22. $\sum_{n=1}^{\infty} \frac{n}{n^4 + 2n^2 + 1} = \sum_{n=1}^{\infty} \frac{n}{(n^2 + 1)^2}$

Let $f(x) = \frac{x}{(x^2 + 1)^2}, f'(x) = \frac{-(3x^2 - 1)}{(x^2 + 1)^3} < 0$ for $x \geq 1$.

f is positive, continuous, and decreasing for $x \geq 1$.

$$\int_1^{\infty} \frac{x}{(x^2 + 1)^2}\,dx = \left[\frac{-1}{2(x^2 + 1)}\right]_1^{\infty} = \frac{1}{4}$$

So, the series converges by Theorem 9.10.

23. $\sum_{n=1}^{\infty} \frac{n^{k-1}}{n^k + c}$

Let

$$f(x) = \frac{x^{k-1}}{x^k + c},\quad f'(x) = \frac{x^{k-2}\left[c(k - 1) - x^k\right]}{(x^k + c)^2} < 0$$

for $x > \sqrt[k]{c(k - 1)}$.

f is positive, continuous, and decreasing for $x > \sqrt[k]{c(k - 1)}$.

$$\int_1^{\infty} \frac{x^{k-1}}{x^k + c}\,dx = \left[\frac{1}{k}\ln(x^k + c)\right]_1^{\infty} = \infty$$

So, the series diverges by Theorem 9.10.

24. $\sum_{n=1}^{\infty} n^k e^{-n}$

Let $f(x) = \frac{x^k}{e^x},\ f'(x) = \frac{x^{k-1}(k - x)}{e^x} < 0$ for $x > k$.

f is positive, continuous, and decreasing for $x > k$.

Use integration by parts.

$$\int_1^{\infty} x^k e^{-x}\,dx = \left[-x^k e^{-x}\right]_1^{\infty} + k\int_1^{\infty} x^{k-1}e^{-x}\,dx$$

$$= \frac{1}{e} + \frac{k}{e} + \frac{k(k - 1)}{e} + \cdots + \frac{k!}{e}$$

So, the series converges by Theorem 9.10.

25. Let $f(x) = \frac{(-1)^x}{x}, f(n) = a_n$.

The function f is not positive for $x \geq 1$.

26. Let $f(x) = e^{-x}\cos x, f(n) = a_n$.

The function f is not positive for $x \geq 1$.

27. Let $f(x) = \frac{2 + \sin x}{x}, f(n) = a_n$.

The function f is not decreasing for $x \geq 1$.

28. Let $f(x) = \left(\frac{\sin x}{x}\right)^2, f(n) = a_n$.

The function f is not decreasing for $x \geq 1$.

29. $\sum_{n=1}^{\infty} \frac{1}{n^3}$

Let $f(x) = \frac{1}{x^3}$.

f is positive, continuous, and decreasing for $x \geq 1$.

$$\int_1^{\infty} \frac{1}{x^3}\,dx = \left[-\frac{1}{2x^2}\right]_1^{\infty} = \frac{1}{2}$$

Converges by Theorem 9.10

30. $\sum_{n=1}^{\infty} \frac{1}{n^{1/2}}$

Let $f(x) = \frac{1}{x^{1/2}} = \frac{1}{\sqrt{x}}$.

f is positive, continuous, and decreasing for $x \geq 1$.

$$\int_1^{\infty} \frac{1}{x^{1/2}}\,dx = \left[2x^{1/2}\right]_1^{\infty} = \infty$$

Diverges by Theorem 9.10

31. $\sum_{n=1}^{\infty} \frac{1}{n^{1/4}}$

Let $f(x) = \frac{1}{x^{1/4}}, f'(x) = \frac{-1}{4x^{5/4}} < 0$ for $x \geq 1$

f is positive, continuous, and decreasing for $x \geq 1$.

$$\int_1^{\infty} \frac{1}{x^{1/4}}\,dx = \left[\frac{4x^{3/4}}{3}\right]_1^{\infty} = \infty$$

Diverges by Theorem 9.10

32. $\sum_{n=1}^{\infty} \frac{1}{n^5}$

Let $f(x) = \frac{1}{x^5}$.

f is positive, continuous, and decreasing for $x \geq 1$.

$$\int_1^{\infty} \frac{1}{x^5}\,dx = \left[-\frac{1}{4x^4}\right]_1^{\infty} = \frac{1}{4}$$

Converges by Theorem 9.10

33. $\sum_{n=1}^{\infty} \frac{1}{\sqrt[5]{n}} = \sum_{n=1}^{\infty} \frac{1}{n^{1/5}}$

Divergent p-series with $p = \frac{1}{5} < 1$

34. $\sum_{n=1}^{\infty} \frac{3}{n^{5/3}}$

Convergent p-series with $p = \frac{5}{3} > 1$

35. $\sum_{n=1}^{\infty} \frac{1}{n^{3/2}}$

Convergent p-series with $p = \frac{3}{2} > 1$

36. $\sum_{n=1}^{\infty} \frac{1}{n^{2/3}}$

Divergent p-series with $p = \frac{2}{3} < 1$

37. $\sum_{n=1}^{\infty} \frac{1}{n^{1.04}}$

Convergent p-series with $p = 1.04 > 1$

38. $\sum_{n=1}^{\infty} \frac{1}{n^{\pi}}$

Convergent p-series with $p = \pi > 1$

39. (a)

n	5	10	20	50	100
S_n	3.7488	3.75	3.75	3.75	3.75

The partial sums approach the sum 3.75 very rapidly.

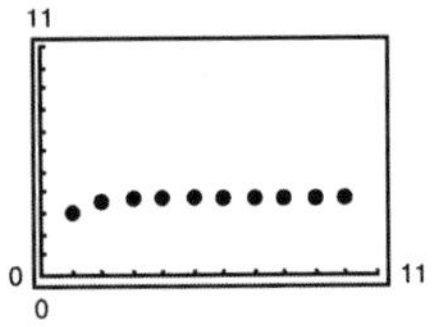

(b)

n	5	10	20	50	100
S_n	1.4636	1.5498	1.5962	1.6251	1.635

The partial sums approach the sum $\pi^2/6 \approx 1.6449$ slower than the series in part (a).

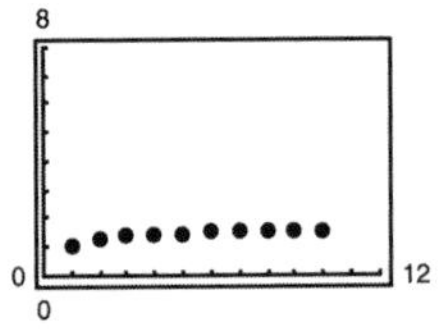

40. $\sum_{n=1}^{N} \frac{1}{n} = 1 + \frac{1}{2} + \frac{1}{3} + \frac{1}{4} + \cdots + \frac{1}{N} > M$

(a)

M	2	4	6	8
N	4	31	227	1674

(b) No. Because the terms are decreasing (approaching zero), more and more terms are required to increase the partial sum by 2.

41. Let f be positive, continuous, and decreasing for $x \geq 1$ and $a_n = f(n)$. Then,

$$\sum_{n=1}^{\infty} a_n \text{ and } \int_1^{\infty} f(x)\,dx$$

either both converge or both diverge (Theorem 9.10). See Example 1, page 620.

42. A series of the form $\sum_{n=1}^{\infty} \frac{1}{n^p}$ is a p-series, $p > 0$.

The p-series converges if $p > 1$ and diverges if $0 < p \leq 1$.

43. Your friend is not correct. The series

$$\sum_{n=10,000}^{\infty} \frac{1}{n} = \frac{1}{10,000} + \frac{1}{10,001} + \cdots$$

is the harmonic series, starting with the 10,000th term, and therefore diverges.

44. $\sum_{n=1}^{6} a_n \geq \int_1^7 f(x)\,dx \geq \sum_{n=2}^{7} a_n$

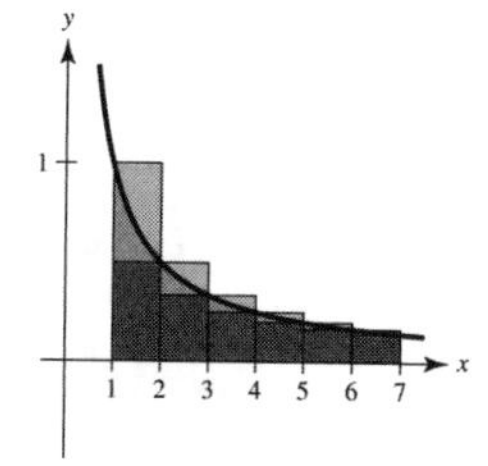

45. (a)

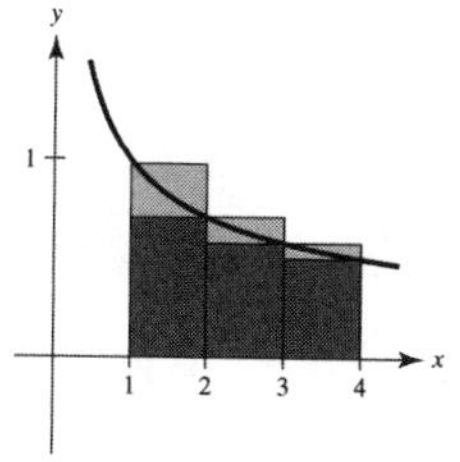

$$\sum_{n=1}^{\infty} \frac{1}{\sqrt{n}} > \int_1^{\infty} \frac{1}{\sqrt{x}}\,dx$$

The area under the rectangle is greater than the area under the curve.

Because $\int_1^{\infty} \frac{1}{\sqrt{x}}\,dx = \left[2\sqrt{x}\right]_1^{\infty} = \infty$, diverges,

$\sum_{n=1}^{\infty} \frac{1}{\sqrt{n}}$ diverges.

(b)

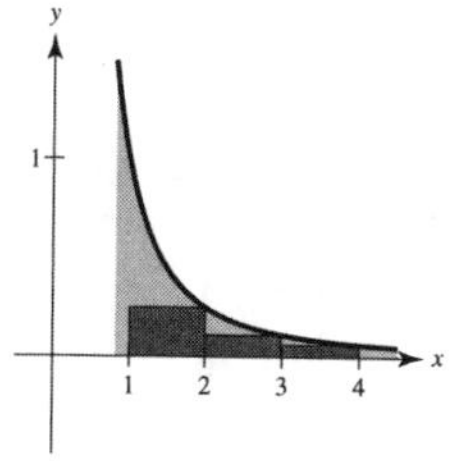

$$\sum_{n=2}^{\infty} \frac{1}{n^2} < \int_1^{\infty} \frac{1}{x^2}\,dx$$

The area under the rectangles is less than the area under the curve.

Because $\int_1^{\infty} \frac{1}{x^2}\,dx = \left[-\frac{1}{x}\right]_1^{\infty} = 1$, converges,

$\sum_{n=2}^{\infty} \frac{1}{n^2}$ converges $\left(\text{and so does } \sum_{n=1}^{\infty} \frac{1}{n^2}\right)$.

46. Answers will vary. *Sample answer*: The graph of the partial sums of the first series seems to be increasing without bound; therefore, the series diverges. The graph of the partial sums of the second series seems to be approaching a limit; therefore the series converges.

47. $\sum_{n=2}^{\infty} \frac{1}{n(\ln n)^p}$

If $p = 1$, then the series diverges by the Integral Test. If $p \neq 1$,

$$\int_2^{\infty} \frac{1}{x(\ln x)^p}\,dx = \int_2^{\infty} (\ln x)^{-p} \frac{1}{x}\,dx = \left[\frac{(\ln x)^{-p+1}}{-p+1}\right]_2^{\infty}.$$

Converges for $-p + 1 < 0$ or $p > 1$

48. $\sum_{n=2}^{\infty} \frac{\ln n}{n^p}$

If $p = 1$, then the series diverges by the Integral Test. If $p \neq 1$,

$$\int_2^{\infty} \frac{\ln x}{x^p}\,dx = \int_2^{\infty} x^{-p} \ln x\,dx = \left[\frac{x^{-p+1}}{(-p+1)^2}\left[-1 + (-p+1)\ln x\right]\right]_2^{\infty}.$$ (Use integration by parts.)

Converges for $-p + 1 < 0$ or $p > 1$

49. $\sum_{n=1}^{\infty} \frac{n}{(1+n^2)^p}$

If $p = 1$, $\sum_{n=1}^{\infty} \frac{n}{1+n^2}$ diverges (see Example 1). Let

$$f(x) = \frac{x}{(1+x^2)^p},\ p \neq 1$$

$$f'(x) = \frac{1-(2p-1)x^2}{(1+x^2)^{p+1}}.$$

For a fixed $p > 0$, $p \neq 1$, $f'(x)$ is eventually negative. f is positive, continuous, and eventually decreasing.

$$\int_1^{\infty} \frac{x}{(1+x^2)^p}\,dx = \left[\frac{1}{(x^2+1)^{p-1}(2-2p)}\right]_1^{\infty}$$

For $p > 1$, this integral converges. For $0 < p < 1$, it diverges.

50. $\sum_{n=1}^{\infty} n(1+n^2)^p$

Because $p > 0$, the series diverges for all values of p.

51. $\sum_{n=1}^{\infty} \left(\frac{3}{p}\right)^n$, Geometric series.

Converges for $\left|\frac{3}{p}\right| < 1 \Rightarrow |p| > 3 \Rightarrow p > 3$

52. $\sum_{n=3}^{\infty} \frac{1}{n \ln n[\ln(\ln n)]^p}$

If $p = 1$, then

$$\int_3^{\infty} \frac{1}{x \ln x[\ln(\ln x)]}\,dx = \left[\ln(\ln(\ln x))\right]_3^{\infty} = \infty,$$ so the series diverges by the Integral Test.

If $p \neq 1$,

$$\int_3^{\infty} \frac{1}{x \ln x[\ln(\ln x)]^p}\,dx = \left[\frac{[\ln(\ln x)]^{-p+1}}{-p+1}\right]_3^{\infty}.$$

This converges for $-p + 1 < 0 \Rightarrow p > 1$.

So, the series converges for $p > 1$, and diverges for $0 < p \leq 1$.

53.

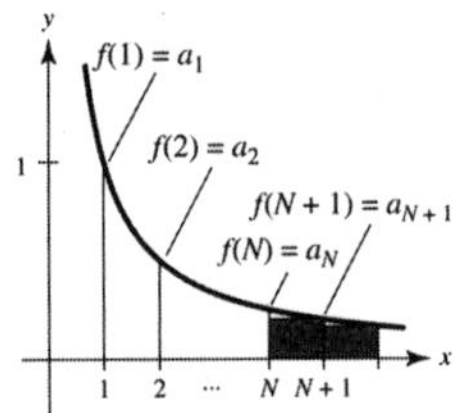

$$S_N = \sum_{n=1}^{N} a_n = a_1 + a_2 + \cdots + a_N$$

$$R_N = S - S_N = \sum_{n=N+1}^{\infty} a_n > 0$$

$$R_N = S - S_N = \sum_{n=N+1}^{\infty} a_n = a_{N+1} + a_{N+2} + \cdots$$

$$\leq \int_N^{\infty} f(x)\,dx$$

So, $0 \leq R_n \leq \int_N^{\infty} f(x)\,dx$

54. From Exercise 53, you have:

$$0 \leq S - S_N \leq \int_N^{\infty} f(x)\,dx$$

$$S_N \leq S \leq S_N + \int_N^{\infty} f(x)\,dx$$

$$\sum_{n=1}^{N} a_n \leq S \leq \sum_{n=1}^{N} a_n + \int_N^{\infty} f(x)\,dx$$

55. $S_5 = 1 + \frac{1}{2^2} + \frac{1}{3^2} + \frac{1}{4^2} + \frac{1}{5^2} \approx 1.4636$

$$0 \leq R_5 \leq \int_5^{\infty} \frac{1}{x^2}\,dx = \left[-\frac{1}{x}\right]_5^{\infty} = \frac{1}{5} = 0.2$$

$$1.4636 \leq \sum_{n=1}^{\infty} \frac{1}{n^2} \leq 1.4636 + 0.2 = 1.6636$$

56. $S_6 = 1 + \frac{1}{2^5} + \cdots + \frac{1}{6^5} \approx 1.0368$

$$0 \leq R_6 \leq \int_6^{\infty} \frac{1}{x^5}\,dx = \left[-\frac{1}{4x^4}\right]_6^{\infty} \approx 0.0002$$

$$1.0368 \leq \sum_{n=1}^{\infty} \frac{1}{n^5} \leq 1.0368 + 0.0002 = 1.0370$$

57. $S_{10} = \frac{1}{2} + \frac{1}{5} + \frac{1}{10} + \frac{1}{17} + \frac{1}{26} + \frac{1}{37} + \frac{1}{50} + \frac{1}{65} + \frac{1}{82} + \frac{1}{101} \approx 0.9818$

$0 \le R_{10} \le \int_{10}^{\infty} \frac{1}{x^2+1}\,dx = \left[\arctan x\right]_{10}^{\infty} = \frac{\pi}{2} - \arctan 10 \approx 0.0997$

$0.9818 \le \sum_{n=1}^{\infty} \frac{1}{n^2+1} \le 0.9818 + 0.0997 = 1.0815$

58. $S_{10} = \frac{1}{2(\ln 2)^3} + \frac{1}{3(\ln 3)^3} + \frac{1}{4(\ln 4)^3} + \cdots + \frac{1}{11(\ln 11)^3} \approx 1.9821$

$0 \le R_{10} \le \int_{10}^{\infty} \frac{1}{(x+1)\left[\ln(x+1)\right]^3}\,dx = \left[-\frac{1}{2\left[\ln(x+1)\right]^2}\right]_{10}^{\infty} = \frac{1}{2(\ln 11)^3} \approx 0.0870$

$1.9821 \le \sum_{n=1}^{\infty} \frac{1}{(n+1)\left[\ln(n+1)\right]^3} \le 1.9821 + 0.0870 = 2.0691$

59. $S_4 = \frac{1}{e} + \frac{2}{e^4} + \frac{3}{e^9} + \frac{4}{e^{16}} \approx 0.4049$

$0 \le R_4 \le \int_4^{\infty} xe^{-x^2}\,dx = \left[-\frac{1}{2}e^{-x^2}\right]_4^{\infty} = \frac{e^{-16}}{2} \approx 5.6 \times 10^{-8}$

$0.4049 \le \sum_{n=1}^{\infty} ne^{-n^2} \le 0.4049 + 5.6 \times 10^{-8}$

60. $S_4 = \frac{1}{e} + \frac{1}{e^2} + \frac{1}{e^3} + \frac{1}{e^4} \approx 0.5713$

$0 \le R_4 \le \int_4^{\infty} e^{-x}\,dx = \left[-e^{-x}\right]_4^{\infty} \approx 0.0183$

$0.5713 \le \sum_{n=0}^{\infty} e^{-n} \le 0.5713 + 0.0183 = 0.5896$

61. $0 \le R_N \le \int_N^{\infty} \frac{1}{x^4}\,dx = \left[-\frac{1}{3x^3}\right]_N^{\infty} = \frac{1}{3N^3} < 0.001$

$\frac{1}{N^3} < 0.003$

$N^3 > 333.33$

$N > 6.93$

$N \ge 7$

62. $0 \le R_N \le \int_N^{\infty} \frac{1}{x^{3/2}}\,dx = \left[-\frac{2}{x^{1/2}}\right]_N^{\infty} = \frac{2}{\sqrt{N}} < 0.001$

$N^{-1/2} < 0.0005$

$\sqrt{N} > 2000$

$N \ge 4{,}000{,}000$

63. $R_N \le \int_N^{\infty} e^{-x/2}\,dx = \left[-2e^{-x/2}\right]_N^{\infty} = \frac{2}{e^{N/2}} < 0.001$

$\frac{2}{e^{N/2}} < 0.001$

$e^{N/2} > 2000$

$\frac{N}{2} > \ln 2000$

$N > 2\ln 2000 \approx 15.2$

$N \ge 16$

64. $R_N \le \int_N^{\infty} \frac{1}{x^2+1}\,dx = \left[\arctan x\right]_N^{\infty}$

$= \frac{\pi}{2} - \arctan N < 0.001$

$-\arctan N < 0.001 - \frac{\pi}{2}$

$\arctan N > \frac{\pi}{2} - 0.001$

$N > \tan\left(\frac{\pi}{2} - 0.001\right)$

$N \ge 1000$

65. (a) $\sum_{n=2}^{\infty} \frac{1}{n^{1.1}}$. This is a convergent p-series with $p = 1.1 > 1$. $\sum_{n=2}^{\infty} \frac{1}{n \ln n}$ is a divergent series. Use the Integral Test.

$f(x) = \frac{1}{x \ln x}$ is positive, continuous, and decreasing for $x \ge 2$.

$\int_2^{\infty} \frac{1}{x \ln x}\,dx = \left[\ln|\ln x|\right]_2^{\infty} = \infty$

(b) $\sum_{n=2}^{6} \frac{1}{n^{1.1}} = \frac{1}{2^{1.1}} + \frac{1}{3^{1.1}} + \frac{1}{4^{1.1}} + \frac{1}{5^{1.1}} + \frac{1}{6^{1.1}} \approx 0.4665 + 0.2987 + 0.2176 + 0.1703 + 0.1393$

$\sum_{n=2}^{6} \frac{1}{n \ln n} = \frac{1}{2 \ln 2} + \frac{1}{3 \ln 3} + \frac{1}{4 \ln 4} + \frac{1}{5 \ln 5} + \frac{1}{6 \ln 6} \approx 0.7213 + 0.3034 + 0.1803 + 0.1243 + 0.0930$

For $n \geq 4$, the terms of the convergent series **seem** to be larger than those of the divergent series.

(c) $\frac{1}{n^{1.1}} < \frac{1}{n \ln n}$

$n \ln n < n^{1.1}$

$\ln n < n^{0.1}$

This inequality holds when $n \geq 3.5 \times 10^{15}$. Or, $n > e^{40}$. Then $\ln e^{40} = 40 < \left(e^{40}\right)^{0.1} = e^4 \approx 55$.

66. (a) $\int_{10}^{\infty} \frac{1}{x^p}\,dx = \left[\frac{x^{-p+1}}{-p+1}\right]_{10}^{\infty} = \frac{1}{(p-1)10^{p-1}},\ p > 1$

(b) $f(x) = \frac{1}{x^p}$

$R_{10}(p) = \sum_{n=11}^{\infty} \frac{1}{n^p}$

$\leq$ Area under the graph of f over the interval $[10, \infty)$

(c) The horizontal asymptote is $y = 0$. As n increases, the error decreases.

67. (a) Let $f(x) = 1/x$. f is positive, continuous, and decreasing on $[1, \infty)$.

$S_n - 1 \leq \int_1^n \frac{1}{x}\,dx$

$S_n - 1 \leq \ln n$

So, $S_n \leq 1 + \ln n$. Similarly,

$S_n \geq \int_1^{n+1} \frac{1}{x}\,dx = \ln(n+1)$.

So, $\ln(n+1) \leq S_n \leq 1 + \ln n$.

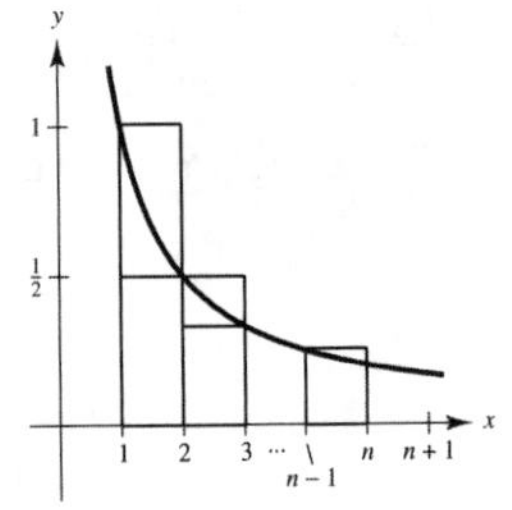

(b) Because $\ln(n+1) \leq S_n \leq 1 + \ln n$, you have $\ln(n+1) - \ln n \leq S_n - \ln n \leq 1$. Also, because $\ln x$ is an increasing function, $\ln(n+1) - \ln n > 0$ for $n \geq 1$. So, $0 \leq S_n - \ln n \leq 1$ and the sequence $\{a_n\}$ is bounded.

(c) $a_n - a_{n+1} = [S_n - \ln n] - [S_{n+1} - \ln(n+1)] = \int_n^{n+1} \frac{1}{x}\,dx - \frac{1}{n+1} \geq 0$

So, $a_n \geq a_{n+1}$ and the sequence is decreasing.

(d) Because the sequence is bounded and monotonic, it converges to a limit, γ.

(e) $a_{100} = S_{100} - \ln 100 \approx 0.5822$ (Actually $\gamma \approx 0.577216$.)

68. $\sum_{n=2}^{\infty}\ln\left(1-\frac{1}{n^2}\right)=\sum_{n=2}^{\infty}\ln\left(\frac{n^2-1}{n^2}\right)=\sum_{n=2}^{\infty}\ln\frac{(n+1)(n-1)}{n^2}=\sum_{n=2}^{\infty}\left[\ln(n+1)+\ln(n-1)-2\ln n\right]$

$$=\left(\cancel{\ln 3}+\cancel{\ln 1}-2\ln 2\right)+\left(\cancel{\ln 4}+\ln 2-\cancel{2\ln 3}\right)+\left(\cancel{\ln 5}+\cancel{\ln 3}-\cancel{2\ln 4}\right)+\left(\cancel{\ln 6}+\cancel{\ln 4}-\cancel{2\ln 5}\right)$$
$$+\left(\cancel{\ln 7}+\cancel{\ln 5}-\cancel{2\ln 6}\right)+\left(\cancel{\ln 8}+\cancel{\ln 6}-\cancel{2\ln 7}\right)+\left(\cancel{\ln 9}+\cancel{\ln 7}-\cancel{2\ln 8}\right)+\cdots=-\ln 2$$

69. $\sum_{n=2}^{\infty}x^{\ln n}$

(a) $x=1$: $\sum_{n=2}^{\infty}1^{\ln n}=\sum_{n=2}^{\infty}1$, diverges

(b) $x=\frac{1}{e}$: $\sum_{n=2}^{\infty}\left(\frac{1}{e}\right)^{\ln n}=\sum_{n=2}^{\infty}e^{-\ln n}=\sum_{n=2}^{\infty}\frac{1}{n}$, diverges

(c) Let x be given, $x>0$. Put $x=e^{-p}\Leftrightarrow \ln x=-p$.

$$\sum_{n=2}^{\infty}x^{\ln n}=\sum_{n=2}^{\infty}e^{-p\ln n}=\sum_{n=2}^{\infty}n^{-p}=\sum_{n=2}^{\infty}\frac{1}{n^p}$$

This series converges for $p>1\Rightarrow x<\frac{1}{e}$.

70. $\xi(x)=\sum_{n=1}^{\infty}n^{-x}=\sum_{n=1}^{\infty}\frac{1}{n^x}$

Converges for $x>1$ by Theorem 9.11

71. Let $f(x)=\frac{1}{3x-2}$, $f'(x)=\frac{-3}{(3x-2)^2}<0$ for $x\geq 1$

f is positive, continuous, and decreasing for $x\geq 1$.

$$\int_1^{\infty}\frac{1}{3x-2}\,dx=\left[\frac{1}{3}\ln|3x-2|\right]_1^{\infty}=\infty$$

So, the series $\sum_{n=1}^{\infty}\frac{1}{3n-2}$

diverges by Theorem 9.10.

72. $\sum_{n=2}^{\infty}\frac{1}{n\sqrt{n^2-1}}$

Let $f(x)=\frac{1}{x\sqrt{x^2-1}}$.

f is positive, continuous, and decreasing for $x\geq 2$.

$$\int_2^{\infty}\frac{1}{x\sqrt{x^2-1}}\,dx=\left[\operatorname{arcsec}x\right]_2^{\infty}=\frac{\pi}{2}-\frac{\pi}{3}$$

Converges by Theorem 9.10

73. $\sum_{n=1}^{\infty}\frac{1}{n\sqrt[4]{n}}=\sum_{n=1}^{\infty}\frac{1}{n^{5/4}}$

p-series with $p=\frac{5}{4}$

Converges by Theorem 9.11

74. $3\sum_{n=1}^{\infty}\frac{1}{n^{0.95}}$

p-series with $p=0.95$

Diverges by Theorem 9.11

75. $\sum_{n=0}^{\infty}\left(\frac{2}{3}\right)^n$

Geometric series with $r=\frac{2}{3}$

Converges by Theorem 9.6

76. $\sum_{n=0}^{\infty}(1.042)^n$ is geometric with $r=1.042>1$. Diverges by Theorem 9.6.

77. $\sum_{n=1}^{\infty}\frac{n}{\sqrt{n^2+1}}$

$$\lim_{n\to\infty}\frac{n}{\sqrt{n^2+1}}=\lim_{n\to\infty}\frac{1}{\sqrt{1+(1/n^2)}}=1\neq 0$$

Diverges by Theorem 9.9

78. $\sum_{n=1}^{\infty}\left(\frac{1}{n^2}-\frac{1}{n^3}\right)=\sum_{n=1}^{\infty}\frac{1}{n^2}-\sum_{n=1}^{\infty}\frac{1}{n^3}$

Because these are both convergent p-series, the difference is convergent.

79. $\sum_{n=1}^{\infty}\left(1+\frac{1}{n}\right)^n$

$$\lim_{n\to\infty}\left(1+\frac{1}{n}\right)^n=e\neq 0$$

Fails nth-Term Test

Diverges by Theorem 9.9

80. $\sum_{n=2}^{\infty} \ln(n)$

$\lim_{n\to\infty} \ln(n) = \infty$

Diverges by Theorem 9.9

81. $\sum_{n=2}^{\infty} \frac{1}{n(\ln n)^3}$

Let $f(x) = \frac{1}{x(\ln x)^3}$.

f is positive, continuous, and decreasing for $x \ge 2$.

$$\int_2^{\infty} \frac{1}{x(\ln x)^3}\,dx = \int_2^{\infty} (\ln x)^{-3}\frac{1}{x}\,dx$$

$$= \left[\frac{(\ln x)^{-2}}{-2}\right]_2^{\infty}$$

$$= \left[-\frac{1}{2(\ln x)^2}\right]_2^{\infty} = \frac{1}{2(\ln 2)^2}$$

Converges by Theorem 9.10. See Exercise 47.

82. $\sum_{n=2}^{\infty} \frac{\ln n}{n^3}$

Let $f(x) = \frac{\ln x}{x^3}$.

f is positive, continuous, and decreasing for $x \ge 2$ since

$f'(x) = \frac{1 - 3\ln x}{x^4} < 0$ for $x \ge 2$.

$$\int_2^{\infty} \frac{\ln x}{x^3}\,dx = \left[-\frac{\ln x}{2x^2}\right]_2^{\infty} + \frac{1}{2}\int_2^{\infty} \frac{1}{x^3}\,dx$$

$$= \frac{\ln 2}{8} + \left[-\frac{1}{4x^2}\right]_2^{\infty}$$

$$= \frac{\ln 2}{8} + \frac{1}{16} \text{ (Use integration by parts.)}$$

Converges by Theorem 9.10. See Exercise 34.

Section 9.4 Comparisons of Series

1. (a) $\sum_{n=1}^{\infty} \frac{6}{n^{3/2}} = \frac{6}{1} + \frac{6}{2^{3/2}} + \cdots;\ S_1 = 6$

$\sum_{n=1}^{\infty} \frac{6}{n^{3/2} + 3} = \frac{6}{4} + \frac{6}{2^{3/2} + 3} + \cdots;\ S_1 = \frac{3}{2}$

$\sum_{n=1}^{\infty} \frac{6}{n\sqrt{n^2 + 0.5}} = \frac{6}{1\sqrt{1.5}} + \frac{6}{2\sqrt{4.5}} + \cdots;\ S_1 = \frac{6}{\sqrt{1.5}} \approx 4.9$

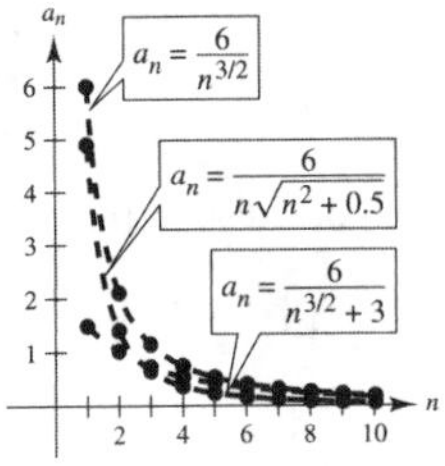

(b) The first series is a p-series. It converges $\left(p = \frac{3}{2} > 1\right)$.

(c) The magnitude of the terms of the other two series are less than the corresponding terms at the convergent p-series. So, the other two series converge.

(d) The smaller the magnitude of the terms, the smaller the magnitude of the terms of the sequence of partial sums.

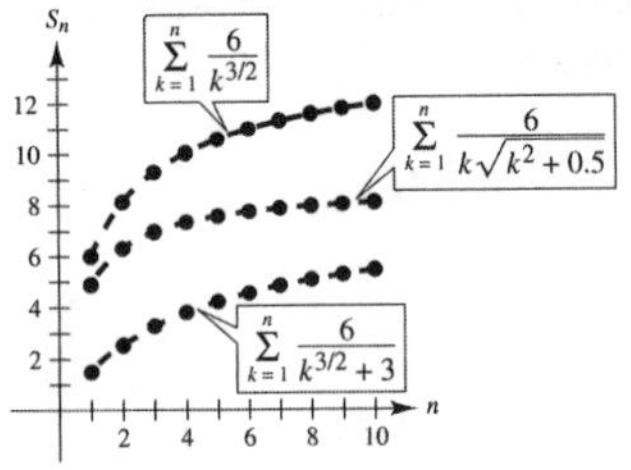

2. (a) $\sum_{n=1}^{\infty} \frac{2}{\sqrt{n}} = 2 + \frac{2}{\sqrt{2}} + \cdots \quad S_1 = 2$

$\sum_{n=1}^{\infty} \frac{2}{\sqrt{n} - 0.5} = \frac{2}{0.5} + \frac{2}{\sqrt{2} - 0.5} + \cdots \quad S_1 = 4$

$\sum_{n=1}^{\infty} \frac{4}{\sqrt{n} + 0.5} = \frac{4}{\sqrt{1.5}} + \frac{4}{\sqrt{2.5}} + \cdots \quad S_1 \approx 3.3$

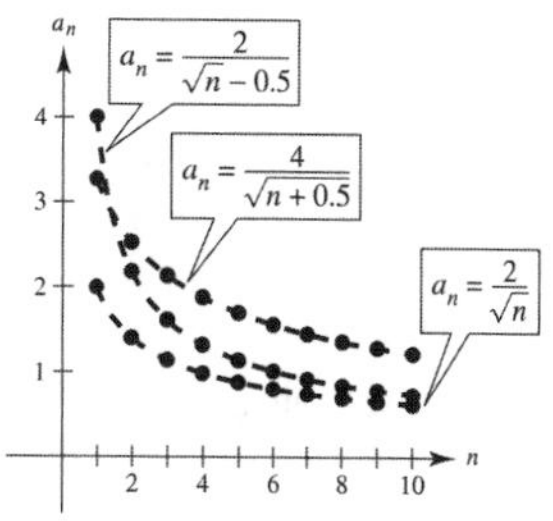

(b) The first series is a p-series. It diverges $\left(p = \frac{1}{2} < 1\right)$.

(c) The magnitude of the terms of the other two series are greater than the corresponding terms of the divergent p-series. So, the other two series diverge.

(d) The larger the magnitude of the terms, the larger the magnitude of the terms of the sequence of partial sums.

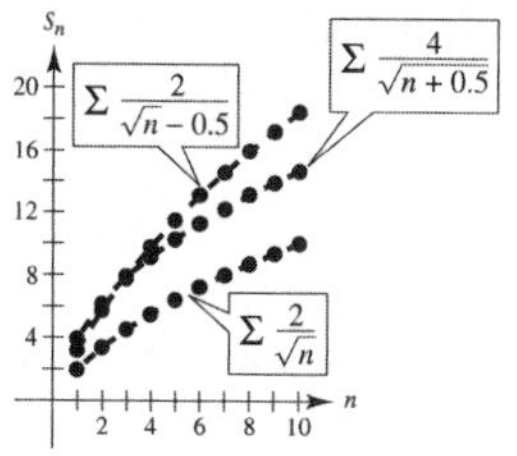

3. $\frac{1}{2n - 1} > \frac{1}{2n} > 0$ for $n \geq 1$

Therefore,

$$\sum_{n=1}^{\infty} \frac{1}{2n - 1}$$

diverges by comparison with the divergent p-series

$$\frac{1}{2}\sum_{n=1}^{\infty} \frac{1}{n}.$$

4. $\frac{1}{3n^2 + 2} < \frac{1}{3n^2}$

Therefore,

$$\sum_{n=1}^{\infty} \frac{1}{3n^2 + 2}$$

converges by comparison with the convergent p-series

$$\frac{1}{3}\sum_{n=1}^{\infty} \frac{1}{n^2}.$$

5. $\frac{1}{\sqrt{n} - 1} > \frac{1}{\sqrt{n}}$ for $n \geq 2$

Therefore,

$$\sum_{n=2}^{\infty} \frac{1}{\sqrt{n} - 1}$$

diverges by comparison with the divergent p-series

$$\sum_{n=2}^{\infty} \frac{1}{\sqrt{n}}.$$

6. $\frac{4^n}{5^n + 3} < \left(\frac{4}{5}\right)^n$

Therefore,

$$\sum_{n=0}^{\infty} \frac{4^n}{5^n + 3}$$

converges by comparison with the convergent geometric series

$$\sum_{n=0}^{\infty} \left(\frac{4}{5}\right)^n.$$

7. For $n \geq 3$, $\frac{\ln n}{n + 1} > \frac{1}{n + 1} > 0$.

Therefore,

$$\sum_{n=1}^{\infty} \frac{\ln n}{n + 1}$$

diverges by comparison with the divergent series

$$\sum_{n=1}^{\infty} \frac{1}{n + 1}.$$

Note: $\sum_{n=1}^{\infty} \frac{1}{n + 1}$ diverges by the Integral Test.

8. $\dfrac{1}{\sqrt{n^3+1}} < \dfrac{1}{n^{3/2}}$

Therefore,

$$\sum_{n=1}^{\infty} \frac{1}{\sqrt{n^3+1}}$$

converges by comparison with the convergent p-series

$$\sum_{n=1}^{\infty} \frac{1}{n^{3/2}}.$$

9. For $n > 3$, $\dfrac{1}{n^2} > \dfrac{1}{n!} > 0$.

Therefore,

$$\sum_{n=0}^{\infty} \frac{1}{n!}$$

converges by comparison with the convergent p-series

$$\sum_{n=1}^{\infty} \frac{1}{n^2}.$$

10. $\dfrac{1}{4\sqrt[3]{n}-1} > \dfrac{1}{4\sqrt[4]{n}}$

Therefore,

$$\sum_{n=1}^{\infty} \frac{1}{4\sqrt[3]{n}-1}$$

diverges by comparison with the divergent p-series

$$\frac{1}{4}\sum_{n=1}^{\infty} \frac{1}{\sqrt[4]{n}}.$$

11. $0 < \dfrac{1}{e^{n^2}} \le \dfrac{1}{e^n}$

Therefore,

$$\sum_{n=0}^{\infty} \frac{1}{e^{n^2}}$$

converges by comparison with the convergent geometric series

$$\sum_{n=0}^{\infty} \left(\frac{1}{e}\right)^n.$$

12. $\dfrac{3^n}{2^n-1} > \left(\dfrac{3}{2}\right)^n$ for $n \ge 1$

Therefore,

$$\sum_{n=1}^{\infty} \frac{3^n}{2^n-1}$$

diverges by comparison with the divergent geometric series

$$\sum_{n=1}^{\infty} \left(\frac{3}{2}\right)^n.$$

13. $\displaystyle\lim_{n\to\infty} \frac{n/(n^2+1)}{1/n} = \lim_{n\to\infty} \frac{n^2}{n^2+1} = 1$

Therefore,

$$\sum_{n=1}^{\infty} \frac{n}{n^2+1}$$

diverges by a limit comparison with the divergent p-series

$$\sum_{n=1}^{\infty} \frac{1}{n}.$$

14. $\displaystyle\lim_{n\to\infty} \frac{5/(4^n+1)}{1/4^n} = \lim_{n\to\infty} \frac{5\cdot 4^n}{4^n+1} = 5$

Therefore,

$$\sum_{n=1}^{\infty} \frac{5}{4^n+1}$$

converges by a limit comparison with the convergent geometric series

$$\sum_{n=1}^{\infty} \left(\frac{1}{4}\right)^n.$$

15. $\displaystyle\lim_{n\to\infty} \frac{1/\sqrt{n^2+1}}{1/n} = \lim_{n\to\infty} \frac{n}{\sqrt{n^2+1}} = 1$

Therefore,

$$\sum_{n=0}^{\infty} \frac{1}{\sqrt{n^2+1}}$$

diverges by a limit comparison with the divergent p-series

$$\sum_{n=1}^{\infty} \frac{1}{n}.$$

16. $\displaystyle\lim_{n\to\infty} \frac{(2^n+1)/(5^n+1)}{(2/5)^n} = \lim_{n\to\infty} \frac{2^n+1}{5^n+1}\cdot\frac{5^n}{2^n} = 1$

Therefore,

$$\sum_{n=1}^{\infty} \frac{2^n+1}{5^n+1}$$

converges by a limit comparison with the convergent geometric series

$$\sum_{n=1}^{\infty} \left(\frac{2}{5}\right)^n.$$

17. $\lim_{n\to\infty} \dfrac{\dfrac{2n^2 - 1}{3n^5 + 2n + 1}}{1/n^3} = \lim_{n\to\infty} \dfrac{2n^5 - n^3}{3n^5 + 2n + 1} = \dfrac{2}{3}$

Therefore,

$$\sum_{n=1}^{\infty} \frac{2n^2 - 1}{3n^5 + 2n + 1}$$

converges by a limit comparison with the convergent p-series

$$\sum_{n=1}^{\infty} \frac{1}{n^3}.$$

18. $\lim_{n\to\infty} \dfrac{1/n^2(n + 3)}{1/n^3} = \lim_{n\to\infty} \dfrac{n^3}{n^2(n + 3)} = 1$

Therefore,

$$\sum_{n=1}^{\infty} \frac{1}{n^2(n + 3)}$$

converges by a limit comparison with the convergent p-series

$$\sum_{n=1}^{\infty} \frac{1}{n^3}.$$

19. $\lim_{n\to\infty} \dfrac{1/\left(n\sqrt{n^2 + 1}\right)}{1/n^2} = \lim_{n\to\infty} \dfrac{n^2}{n\sqrt{n^2 + 1}} = 1$

Therefore,

$$\sum_{n=1}^{\infty} \frac{1}{n\sqrt{n^2 + 1}}$$

converges by a limit comparison with the convergent p-series

$$\sum_{n=1}^{\infty} \frac{1}{n^2}.$$

20. $\lim_{n\to\infty} \dfrac{n/\left[(n + 1)2^{n-1}\right]}{1/\left(2^{n-1}\right)} = \lim_{n\to\infty} \dfrac{n}{n + 1} = 1$

Therefore,

$$\sum_{n=1}^{\infty} \frac{n}{(n + 1)2^{n-1}}$$

converges by a limit comparison with the convergent geometric series

$$\sum_{n=1}^{\infty} \left(\frac{1}{2}\right)^{n-1}.$$

21. $\lim_{n\to\infty} \dfrac{\left(n^{k-1}\right)/\left(n^k + 1\right)}{1/n} = \lim_{n\to\infty} \dfrac{n^k}{n^k + 1} = 1$

Therefore,

$$\sum_{n=1}^{\infty} \frac{n^{k-1}}{n^k + 1}$$

diverges by a limit comparison with the divergent p-series

$$\sum_{n=1}^{\infty} \frac{1}{n}.$$

22. $\lim_{n\to\infty} \dfrac{\sin(1/n)}{1/n} = \lim_{n\to\infty} \dfrac{\left(-1/n^2\right)\cos(1/n)}{-1/n^2} = \lim_{n\to\infty} \cos\left(\dfrac{1}{n}\right) = 1$

Therefore,

$$\sum_{n=1}^{\infty} \sin\left(\frac{1}{n}\right)$$

diverges by a limit comparison with the divergent p-series

$$\sum_{n=1}^{\infty} \frac{1}{n}.$$

23. $\displaystyle\sum_{n=1}^{\infty} \frac{\sqrt[1/3]{n}}{n} = \sum_{n=1}^{\infty} \frac{1}{n^{2/3}}$

Diverges;

p-series with $p = \dfrac{2}{3}$

24. $\displaystyle\sum_{n=0}^{\infty} 5\left(-\frac{4}{3}\right)^n$

Diverges;

Geometric series with $|r| = \left|-\dfrac{4}{3}\right| = \dfrac{4}{3} > 1$

25. $\displaystyle\sum_{n=1}^{\infty} \frac{1}{5^n + 1}$

Converges;

Direct comparison with convergent geometric series

$$\sum_{n=1}^{\infty} \left(\frac{1}{5}\right)^n$$

26. $\displaystyle\sum_{n=3}^{\infty} \frac{1}{n^3 - 8}$

Converges; limit comparison with $\displaystyle\sum_{n=3}^{\infty} \frac{1}{n^3}$

27. $\displaystyle\sum_{n=1}^{\infty} \frac{2n}{3n - 2}$

Diverges; n^{th}-Term Test

$\lim_{n\to\infty} \dfrac{2n}{3n - 2} = \dfrac{2}{3} \neq 0$

28. $\sum_{n=1}^{\infty}\left(\frac{1}{n+1}-\frac{1}{n+2}\right)=\left(\frac{1}{2}-\frac{1}{3}\right)+\left(\frac{1}{3}-\frac{1}{4}\right)+\left(\frac{1}{4}-\frac{1}{5}\right)+\cdots=\frac{1}{2}$

Converges; telescoping series

29. $\sum_{n=1}^{\infty}\frac{n}{\left(n^2+1\right)^2}$

Converges; Integral Test

30. $\sum_{n=1}^{\infty}\frac{3}{n(n+3)}$

Converges; telescoping series

$\sum_{n=1}^{\infty}\left(\frac{1}{n}-\frac{1}{n+3}\right)$

31. $\lim_{n\to\infty}\frac{a_n}{1/n}=\lim_{n\to\infty}na_n$. By given conditions $\lim_{n\to\infty}na_n$ is finite and nonzero. Therefore,

$\sum_{n=1}^{\infty}a_n$

diverges by a limit comparison with the p-series

$\sum_{n=1}^{\infty}\frac{1}{n}.$

32. If $j<k-1$, then $k-j>1$. The p-series with $p=k-j$ converges and because

$\lim_{n\to\infty}\frac{P(n)/Q(n)}{1/n^{k-j}}=L>0$, the series $\sum_{n=1}^{\infty}\frac{P(n)}{Q(n)}$

converges by the Limit Comparison Test. Similarly, if $j\ge k-1$, then $k-j\le 1$ which implies that

$\sum_{n=1}^{\infty}\frac{P(n)}{Q(n)}$

diverges by the Limit Comparison Test.

33. $\frac{1}{2}+\frac{2}{5}+\frac{3}{10}+\frac{4}{17}+\frac{5}{26}+\cdots=\sum_{n=1}^{\infty}\frac{n}{n^2+1}$,

which diverges because the degree of the numerator is only one less than the degree of the denominator.

34. $\frac{1}{3}+\frac{1}{8}+\frac{1}{15}+\frac{1}{24}+\frac{1}{35}+\cdots=\sum_{n=2}^{\infty}\frac{1}{n^2-1}$,

which converges because the degree of the numerator is two less than the degree of the denominator.

35. $\sum_{n=1}^{\infty}\frac{1}{n^3+1}$

converges because the degree of the numerator is three less than the degree of the denominator.

36. $\sum_{n=1}^{\infty}\frac{n^2}{n^3+1}$

diverges because the degree of the numerator is only one less than the degree of the denominator.

37. $\lim_{n\to\infty}n\left(\frac{n^3}{5n^4+3}\right)=\lim_{n\to\infty}\frac{n^4}{5n^4+3}=\frac{1}{5}\neq 0$

Therefore, $\sum_{n=1}^{\infty}\frac{n^3}{5n^4+3}$ diverges.

38. $\lim_{n\to\infty}\frac{n}{\ln n}=\lim_{n\to\infty}\frac{1}{1/n}=\lim_{n\to\infty}n=\infty\neq 0$

Therefore, $\sum_{n=2}^{\infty}\frac{1}{\ln n}$ diverges.

39. $\frac{1}{200}+\frac{1}{400}+\frac{1}{600}+\cdots=\sum_{n=1}^{\infty}\frac{1}{200n}$

diverges, (harmonic)

40. $\frac{1}{200}+\frac{1}{210}+\frac{1}{220}+\cdots=\sum_{n=0}^{\infty}\frac{1}{200+10n}$

diverges

41. $\frac{1}{201}+\frac{1}{204}+\frac{1}{209}+\frac{1}{216}=\sum_{n=1}^{\infty}\frac{1}{200+n^2}$

converges

42. $\frac{1}{201}+\frac{1}{208}+\frac{1}{227}+\frac{1}{264}+\cdots=\sum_{n=1}^{\infty}\frac{1}{200+n^3}$

converges

43. Some series diverge or converge very slowly. You cannot decide convergence or divergence of a series by comparing the first few terms.

44. See Theorem 9.12, page 612. One example is

$\sum_{n=1}^{\infty}\frac{1}{n^2+1}$ converges because $\frac{1}{n^2+1}<\frac{1}{n^2}$ and

$\sum_{n=1}^{\infty}\frac{1}{n^2}$ converges (p-series).

45. See Theorem 9.13, page 614. One example is

$\sum_{n=2}^{\infty}\frac{1}{\sqrt{n-1}}$ diverges because $\lim_{n\to\infty}\frac{1/\sqrt{n-1}}{1/\sqrt{n}}=1$ and

$\sum_{n=2}^{\infty}\frac{1}{\sqrt{n}}$ diverges (p-series).

46. This is not correct. The beginning terms do not affect the convergence or divergence of a series. In fact,

$$\frac{1}{1000} + \frac{1}{1001} + \cdots = \sum_{n=1000}^{\infty} \frac{1}{n} \text{ diverges (harmonic)}$$

and $1 + \frac{1}{4} + \frac{1}{9} + \cdots = \sum_{n=1}^{\infty} \frac{1}{n^2}$ converges (p-series).

47. (a) $\sum_{n=1}^{\infty} \frac{1}{(2n-1)^2} = \sum_{n=1}^{\infty} \frac{1}{4n^2 - 4n + 1}$

converges because the degree of the numerator is two less than the degree of the denominator. (See Exercise 32.)

(b)

n	5	10	20	50	100
S_n	1.1839	1.2087	1.2212	1.2287	1.2312

(c) $\sum_{n=3}^{\infty} \frac{1}{(2n-1)^2} = \frac{\pi^2}{8} - S_2 \approx 0.1226$

(d) $\sum_{n=10}^{\infty} \frac{1}{(2n-1)^2} = \frac{\pi^2}{8} - S_9 \approx 0.0277$

48.

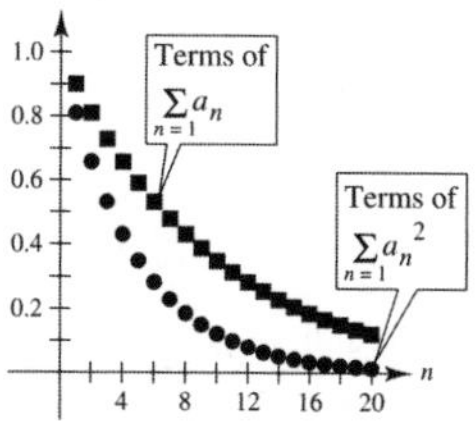

For $0 < a_n < 1$, $0 < a_n^2 < a_n < 1$.

So, the lower terms are those of Σa_n^2.

49. False. Let $a_n = \frac{1}{n^3}$ and $b_n = \frac{1}{n^2}$. $0 < a_n \le b_n$ and both

$\sum_{n=1}^{\infty} \frac{1}{n^3}$ and $\sum_{n=1}^{\infty} \frac{1}{n^2}$ converge.

50. True

51. True

52. False. Let $a_n = 1/n$, $b_n = 1/n$, $c_n = 1/n^2$. Then,

$a_n \le b_n + c_n$, but $\sum_{n=1}^{\infty} c_n$ converges.

53. True

54. False. $\sum_{n=1}^{\infty} a_n$ could converge or diverge.

For example, let $\sum_{n=1}^{\infty} b_n = \sum_{n=1}^{\infty} \frac{1}{\sqrt{n}}$, which diverges.

$0 < \frac{1}{n} < \frac{1}{\sqrt{n}}$ and $\sum_{n=1}^{\infty} \frac{1}{n}$ diverges, but

$0 < \frac{1}{n^2} < \frac{1}{\sqrt{n}}$ and $\sum_{n=1}^{\infty} \frac{1}{n^2}$ converges.

55. Because $\sum_{n=1}^{\infty} b_n$ converges, $\lim_{n\to\infty} b_n = 0$. There exists N such that $b_n < 1$ for $n > N$. So, $a_n b_n < a_n$ for $n > N$ and $\sum_{n=1}^{\infty} a_n b_n$ converges by comparison to the convergent series $\sum_{i=1}^{\infty} a_n$.

56. Because $\sum_{n=1}^{\infty} a_n$ converges, then

$\sum_{n=1}^{\infty} a_n a_n = \sum_{n=1}^{\infty} a_n^2$ converges by Exercise 55.

57. $\sum \frac{1}{n^2}$ and $\sum \frac{1}{n^3}$ both converge, and therefore, so does

$$\sum \left(\frac{1}{n^2}\right)\left(\frac{1}{n^3}\right) = \sum \frac{1}{n^5}.$$

58. $\sum \frac{1}{n^2}$ converge, and therefore, so does

$$\sum\left(\frac{1}{n^2}\right)^2 = \sum\frac{1}{n^4}.$$

59. Suppose $\lim_{n\to\infty} \frac{a_n}{b_n} = 0$ and Σb_n converges.

From the definition of limit of a sequence, there exists $M > 0$ such that

$$\left|\frac{a_n}{b_n} - 0\right| < 1$$

whenever $n > M$. So, $a_n < b_n$ for $n > M$. From the Comparison Test, Σa_n converges.

60. Suppose $\lim_{n\to\infty} \frac{a_n}{b_n} = \infty$ and Σb_n diverges. From the definition of limit of a sequence, there exists $M > 0$ such that

$$\frac{a_n}{b_n} > 1$$

for $n > M$. So, $a_n > b_n$ for $n > M$. By the Comparison Test, Σa_n diverges.

61. (a) Let $\sum a_n = \sum \frac{1}{(n+1)^3}$, and $\sum b_n = \sum \frac{1}{n^2}$, converges.

$$\lim_{n\to\infty} \frac{a_n}{b_n} = \lim_{n\to\infty} \frac{1/\left[(n+1)^3\right]}{1/(n^2)} = \lim_{n\to\infty} \frac{n^2}{(n+1)^3} = 0$$

By Exercise 59, $\sum_{n=1}^{\infty} \frac{1}{(n+1)^3}$ converges.

(b) Let $\sum a_n = \sum \frac{1}{\sqrt{n}\pi^n}$, and $\sum b_n = \sum \frac{1}{\pi^n}$, converges.

$$\lim_{n\to\infty} \frac{a_n}{b_n} = \lim_{n\to\infty} \frac{1/\left(\sqrt{n}\pi^n\right)}{1/(\pi^n)} = \lim_{n\to\infty} \frac{1}{\sqrt{n}} = 0$$

By Exercise 59, $\sum_{n=1}^{\infty} \frac{1}{\sqrt{n}\pi^n}$ converges.

62. (a) Let $\sum a_n = \sum \frac{\ln n}{n}$, and $\sum b_n = \sum \frac{1}{n}$, diverges.

$$\lim_{n\to\infty} \frac{a_n}{b_n} = \lim_{n\to\infty} \frac{(\ln n)/n}{1/n} = \lim_{n\to\infty} \ln n = \infty$$

By Exercise 60, $\sum_{n=1}^{\infty} \frac{\ln n}{n}$ diverges.

(b) Let $\sum a_n = \sum \frac{1}{\ln n}$, and $\sum b_n = \sum \frac{1}{n}$, diverges.

$$\lim_{n\to\infty} \frac{a_n}{b_n} = \lim_{n\to\infty} \frac{n}{\ln n} = \infty$$

By Exercise 60, $\sum \frac{1}{\ln n}$ diverges.

63. Because $\lim_{n\to\infty} a_n = 0$, the terms of $\Sigma \sin(a_n)$ are positive for sufficiently large n. Because

$$\lim_{n\to\infty} \frac{\sin(a_n)}{a_n} = 1 \text{ and } \sum a_n$$

converges, so does $\Sigma \sin(a_n)$.

64.
$$\sum_{n=1}^{\infty} \frac{1}{1+2+\cdots+n} = \sum_{n=1}^{\infty} \frac{1}{\left[n(n+1)\right]/2} = \sum_{n=1}^{\infty} \frac{2}{n(n+1)}$$

Because $\Sigma 1/n^2$ converges, and

$$\lim_{n\to\infty} \frac{2/\left[n(n+1)\right]}{1/(n^2)} = \lim_{n\to\infty} \frac{2n^2}{n(n+1)} = 2,$$

$\sum \frac{1}{1+2+\cdots+n}$ converges.

65. First note that $f(x) = \ln x - x^{1/4} = 0$ when $x \approx 5503.66$. That is,

$\ln n < n^{1/4}$ for $n > 5504$

which implies that

$\frac{\ln n}{n^{3/2}} < \frac{1}{n^{5/4}}$ for $n > 5504$.

Because $\sum_{n=1}^{\infty} \frac{1}{n^{5/4}}$ is a convergent p-series,

$$\sum_{n=1}^{\infty} \frac{\ln n}{n^{3/2}}$$

converges by direct comparison.

66. The series diverges. For $n > 1$,

$$n < 2^n$$
$$n^{1/n} < 2$$
$$\frac{1}{n^{1/n}} > \frac{1}{2}$$
$$\frac{1}{n^{(n+1)/n}} > \frac{1}{2n}$$

Because $\sum \frac{1}{2n}$ diverges, so does $\sum \frac{1}{n^{(n+1)/n}}$.

67. Consider two cases:

If $a_n \geq \frac{1}{2^{n+1}}$, then $a_n^{1/(n+1)} \geq \left(\frac{1}{2^{n+1}}\right)^{1/(n+1)} = \frac{1}{2}$, and

$$a_n^{n/(n+1)} = \frac{a_n}{a_n^{1/(n+1)}} \leq 2a_n.$$

If $a_n \leq \frac{1}{2^{n+1}}$, then $a_n^{n/(n+1)} \leq \left(\frac{1}{2^{n+1}}\right)^{n/(n+1)} = \frac{1}{2^n}$, and

combining, $a_n^{n/(n+1)} \leq 2a_n + \frac{1}{2^n}$.

Because $\sum_{n=1}^{\infty}\left(2a_n + \frac{1}{2^n}\right)$ converges, so does $\sum_{n=1}^{\infty} a_n^{n/(n+1)}$ by the Comparison Test.

Section 9.5 Alternating Series

1. $\sum_{n=1}^{\infty} \frac{(-1)^{n-1}}{2n-1} = \frac{\pi}{4} \approx 0.7854$

(a)

n	1	2	3	4	5	6	7	8	9	10
S_n	1	0.6667	0.8667	0.7238	0.8349	0.7440	0.8209	0.7543	0.8131	0.7605

(b)

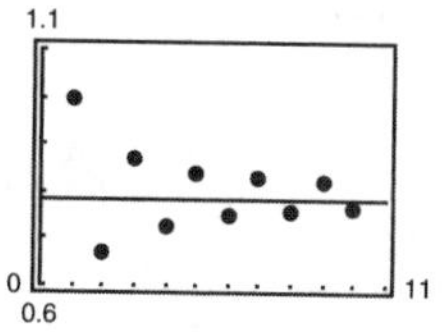

(c) The points alternate sides of the horizontal line $y = \frac{\pi}{4}$ that represents the sum of the series.

The distance between successive points and the line decreases.

(d) The distance in part (c) is always less than the magnitude of the next term of the series.

2. $\sum_{n=1}^{\infty} \frac{(-1)^{n-1}}{(n-1)!} = \frac{1}{e} \approx 0.3679$

(a)

n	1	2	3	4	5	6	7	8	9	10
S_n	1	0	0.5	0.3333	0.375	0.3667	0.3681	0.3679	0.3679	0.3679

(b)

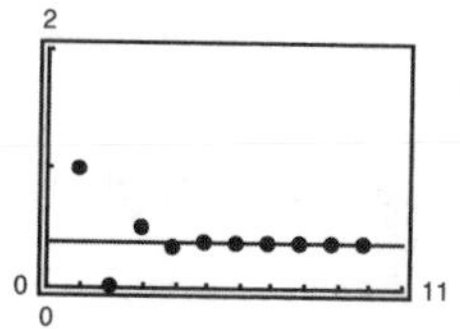

(c) The points alternate sides of the horizontal line $y = \frac{1}{e}$ that represents the sum of the series.

The distance between successive points and the line decreases.

(d) The distance in part (c) is always less than the magnitude of the next series.

3. $\sum_{n=1}^{\infty} \frac{(-1)^{n-1}}{n^2} = \frac{\pi^2}{12} \approx 0.8225$

(a)

n	1	2	3	4	5	6	7	8	9	10
S_n	1	0.75	0.8611	0.7986	0.8386	0.8108	0.8312	0.8156	0.8280	0.8180

(b)

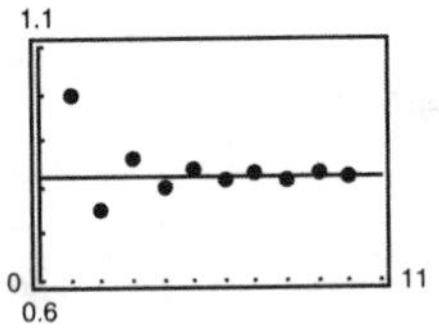

(c) The points alternate sides of the horizontal line $y = \frac{\pi^2}{12}$ that represents the sum of the series.

The distance between successive points and the line decreases.

(d) The distance in part (c) is always less than the magnitude of the next term in the series.

4. $\sum_{n=1}^{\infty} \frac{(-1)^{n-1}}{(2n-1)!} = \sin(1) \approx 0.8415$

(a)

n	1	2	3	4	5	6	7	8	9	10
S_n	1	0.8333	0.8417	0.8415	0.8415	0.8415	0.8415	0.8415	0.8415	0.8415

(b)

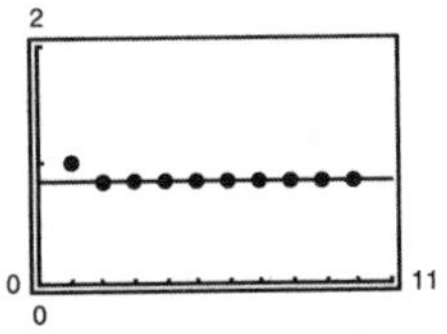

(c) The points alternate sides of the horizontal line $y = \sin(1)$ that represents the sum of the series.

The distance between successive points and the line decreases.

(d) The distance in part (c) is always less than the magnitude of the next series.

5. $\sum_{n=1}^{\infty} \frac{(-1)^{n+1}}{n+1}$

$a_{n+1} = \frac{1}{n+2} < \frac{1}{n+1} = a_n$

$\lim_{n\to\infty} a_n = \lim_{n\to\infty} \frac{1}{n+1} = 0$

Converges by Theorem 9.14

6. $\sum_{n=1}^{\infty} \frac{(-1)^{n+1} n}{3n+2}$

$\lim_{n\to\infty} \frac{n}{3n+2} = \frac{1}{3}$

Diverges by nth-Term test

7. $\sum_{n=1}^{\infty} \frac{(-1)^n}{3^n}$

$a_{n+1} = \frac{1}{3^{n+1}} < \frac{1}{3^n} = a_n$

$\lim_{n\to 0} \frac{1}{3^n} = 0$

Converges by Theorem 9.14

(Note: $\sum_{n=1}^{\infty} \left(\frac{-1}{3}\right)^n$ is a convergent geometric series)

8. $\sum_{n=1}^{\infty} \frac{(-1)^n}{e^n}$

$$a_{n+1} = \frac{1}{e^{n+1}} < \frac{1}{e^n} = a_n$$

$$\lim_{n\to\infty} \frac{1}{e^n} = 0$$

Converges by Theorem 9.14

(**Note:** $\sum_{n=1}^{\infty} \left(\frac{-1}{e}\right)^n$ is a convergent geometric series)

9. $\sum_{n=1}^{\infty} \frac{(-1)^n (5n - 1)}{4n + 1}$

$$\lim_{n\to\infty} \frac{5n - 1}{4n + 1} = \frac{5}{4}$$

Diverges by nth-Term test

10. $\sum_{n=1}^{\infty} \frac{(-1)^{n+1} n}{n^2 + 5}$

Let $f(x) = \frac{x}{x^2 + 5}$, $f'(x) = \frac{-(x^2 - 5)}{(x^2 + 5)^2} < 0$ for $x \geq 3$

So, $a_{n+1} < a_n$ for $n \geq 3$

$$\lim_{n\to\infty} \frac{n}{n^2 + 5} = 0$$

Converges by Theorem 9.14

11. $\sum_{n=1}^{\infty} \frac{(-1)^n n}{\ln(n + 1)}$

$$\lim_{n\to\infty} \frac{n}{\ln(n + 1)} = \infty$$

Diverges by nth-Term test

12. $\sum_{n=1}^{\infty} \frac{(-1)^n}{\ln(n + 1)}$

$$a_{n+1} = \frac{1}{\ln(n + 2)} < \frac{1}{\ln(n + 1)} = a_n$$

$$\lim_{n\to\infty} \frac{1}{\ln(n + 1)} = 0$$

Converges by Theorem 9.14

13. $\sum_{n=1}^{\infty} \frac{(-1)^n}{\sqrt{n}}$

$$a_{n+1} = \frac{1}{\sqrt{n + 1}} < \frac{1}{\sqrt{n}} = a_n$$

$$\lim_{n\to\infty} \frac{1}{\sqrt{n}} = 0$$

Converges by Theorem 9.14

14. $\sum_{n=1}^{\infty} \frac{(-1)^{n+1} n^2}{n^2 + 4}$

$$\lim_{n\to\infty} \frac{n^2}{n^2 + 4} = 1$$

Diverges by nth-Term test

15. $\sum_{n=1}^{\infty} \frac{(-1)^{n+1}(n + 1)}{\ln(n + 1)}$

$$\lim_{n\to\infty} \frac{n + 1}{\ln(n + 1)} = \lim_{n\to\infty} \frac{1}{1/(n + 1)} = \lim_{n\to\infty} (n + 1) = \infty$$

Diverges by the nth-Term Test

16. $\sum_{n=1}^{\infty} \frac{(-1)^{n+1}\ln(n+1)}{n+1}$

$a_{n+1} = \frac{\ln[(n+1)+1]}{(n+1)+1} < \frac{\ln(n+1)}{n+1}$ for $n \ge 2$

$\lim_{n\to\infty} \frac{\ln(n+1)}{n+1} = \lim_{n\to\infty} \frac{1/(n+1)}{1} = 0$

Converges by Theorem 9.14

17. $\sum_{n=1}^{\infty} \sin\left[\frac{(2n-1)\pi}{2}\right] = \sum_{n=1}^{\infty} (-1)^{n+1}$

Diverges by the nth-Term Test

18. $\sum_{n=1}^{\infty} \frac{1}{n}\cos n\pi = \sum_{n=1}^{\infty} \frac{(-1)^n}{n}$

$a_{n+1} = \frac{1}{n+1} < \frac{1}{n} = a_n$

$\lim_{n\to\infty} \frac{1}{n} = 0$

Converges by Theorem 9.14

19. $\sum_{n=0}^{\infty} \frac{(-1)^n}{n!}$

$a_{n+1} = \frac{1}{(n+1)!} < \frac{1}{n!} = a_n$

$\lim_{n\to\infty} \frac{1}{n!} = 0$

Converges by Theorem 9.14

20. $\sum_{n=0}^{\infty} \frac{(-1)^n}{(2n+1)!}$

$a_{n+1} = \frac{1}{(2n+3)!} < \frac{1}{(2n+1)!} = a_n$

$\lim_{n\to\infty} \frac{1}{(2n+1)!} = 0$

Converges by Theorem 9.14

21. $\sum_{n=1}^{\infty} \frac{(-1)^{n+1}\sqrt{n}}{n+2}$

$a_{n+1} = \frac{\sqrt{n+1}}{(n+1)+2} < \frac{\sqrt{n}}{n+2}$ for $n \ge 2$

$\lim_{n\to\infty} \frac{\sqrt{n}}{n+2} = 0$

Converges by Theorem 9.14

22. $\sum_{n=1}^{\infty} \frac{(-1)^{n+1}\sqrt{n}}{\sqrt[3]{n}}$

$\lim_{n\to\infty} \frac{n^{1/2}}{n^{1/3}} = \lim_{n\to\infty} n^{1/6} = \infty$

Diverges by the nth-Term Test

23. $\sum_{n=1}^{\infty} \frac{(-1)^{n+1} n!}{1\cdot 3\cdot 5\cdots(2n-1)}$

$a_{n+1} = \frac{(n+1)!}{1\cdot 3\cdot 5\cdots(2n-1)(2n+1)} = \frac{n!}{1\cdot 3\cdot 5\cdots(2n-1)}\cdot\frac{n+1}{2n+1} = a_n\left(\frac{n+1}{2n+1}\right) < a_n$

$\lim_{n\to\infty} a_n = \lim_{n\to\infty} \frac{n!}{1\cdot 3\cdot 5\cdots(2n-1)} = \lim_{n\to\infty} \frac{1\cdot 2\cdot 3\cdots n}{1\cdot 3\cdot 5\cdots(2n-1)} = \lim_{n\to\infty} 2\left[\frac{3}{3}\cdot\frac{4}{5}\cdot\frac{5}{7}\cdots\frac{n}{2n-3}\right]\cdot\frac{1}{2n-1} = 0$

Converges by Theorem 9.14

24. $\sum_{n=1}^{\infty} (-1)^{n+1} \frac{1\cdot 3\cdot 5\cdots(2n-1)}{1\cdot 4\cdot 7\cdots(3n-2)}$

$a_{n+1} = \frac{1\cdot 3\cdot 5\cdots(2n-1)(2n+1)}{1\cdot 4\cdot 7\cdots(3n-2)(3n+1)} = a_n\left(\frac{2n+1}{3n+1}\right) < a_n$

$\lim_{n\to\infty} a_n = \lim_{n\to\infty} 3\left[\frac{5}{4}\cdot\frac{7}{7}\cdot\frac{9}{10}\cdots\frac{2n-1}{3n-5}\right]\frac{1}{3n-2} = 0$

Converges by Theorem 9.14

25. $\sum_{n=1}^{\infty} \frac{(-1)^{n+1}(2)}{e^n - e^{-n}} = \sum_{n=1}^{\infty} \frac{(-1)^{n+1}(2e^n)}{e^{2n}-1}$

Let $f(x) = \frac{2e^x}{e^{2x}-1}$. Then

$f'(x) = \frac{-2e^x(e^{2x}+1)}{(e^{2x}-1)^2} < 0.$

So, $f(x)$ is decreasing. Therefore, $a_{n+1} < a_n$, and

$\lim_{n\to\infty} \frac{2e^n}{e^{2n}-1} = \lim_{n\to\infty} \frac{2e^n}{2e^{2n}} = \lim_{n\to\infty} \frac{1}{e^n} = 0.$

The series converges by Theorem 9.14.

26. $\sum_{n=1}^{\infty} \frac{2(-1)^{n+1}}{e^n + e^{-n}} = \sum_{n=1}^{\infty} \frac{(-1)^{n+1}(2e^n)}{e^{2n} + 1}$

Let $f(x) = \frac{2e^x}{e^{2x} + 1}$. Then

$$f'(x) = \frac{2e^{2x}(1 - e^{2x})}{(e^{2x} + 1)^2} < 0 \text{ for } x > 0.$$

So, $f(x)$ is decreasing for $x > 0$ which implies $a_{n+1} < a_n$.

$$\lim_{n\to\infty} \frac{2e^n}{e^{2n} + 1} = \lim_{n\to\infty} \frac{2e^n}{2e^{2n}} = \lim_{n\to\infty} \frac{1}{e^n} = 0$$

The series converges by Theorem 9.14.

27. $S_6 = \sum_{n=0}^{5} \frac{(-1)^n 5}{n!} = \frac{11}{6}$

$$|R_6| = |S - S_6| \le a_7 = \frac{5}{720} = \frac{1}{144}$$

$$\frac{11}{6} - \frac{1}{144} \le S \le \frac{11}{6} + \frac{1}{144}$$

$$1.8264 \le S \le 1.8403$$

28. $S_6 = \sum_{n=1}^{6} \frac{4(-1)^{n+1}}{\ln(n + 1)} \approx 2.7067$

$$|R_6| = |S - S_6| \le a_7 = \frac{4}{\ln 8} \approx 1.9236$$

$$0.7831 \le S \le 4.6303$$

29. $S_6 = \sum_{n=1}^{6} \frac{(-1)^{n+1} 2}{n^3} \approx 1.7996$

$$|R_6| = |S - S_6| \le a_7 = \frac{2}{7^3} \approx 0.0058$$

$$1.7796 - 0.0058 \le S \le 1.7796 + 0.0058$$

$$1.7938 \le S \le 1.8054$$

30. $S_6 = \sum_{n=1}^{6} \frac{(-1)^{n+1} n}{3^n} \approx 0.1852$

$$|R_6| = |S - S_6| \le a_7 = \frac{7}{3^7} \approx 0.0032$$

$$0.1852 - 0.0032 \le S \le 0.1852 + 0.0032$$

$$0.1820 \le S \le 0.1884$$

31. $\sum_{n=1}^{\infty} \frac{(-1)^{n+1}}{n^3}$

By Theorem 9.15,

$$|R_N| \le a_{N+1} = \frac{1}{(N + 1)^3} < 0.001$$

$$\Rightarrow (N + 1)^3 > 1000 \Rightarrow N + 1 > 10.$$

Use 10 terms.

32. $\sum_{n=1}^{\infty} \frac{(-1)^{n+1}}{n^2}$

By Theorem 9.15,

$$|R_N| \le a_{N+1} = \frac{1}{(N + 1)^2} < 0.001$$

$$\Rightarrow (N + 1)^2 > 1000.$$

By trial and error, this inequality is valid when $N = 31(32^2 = 1024)$.

Use 31 terms.

33. $\sum_{n=1}^{\infty} \frac{(-1)^{n+1}}{2n^3 - 1}$

By Theorem 9.15,

$$|R_N| \le a_{N+1} = \frac{1}{2(N + 1)^3 - 1} < 0.001$$

$$\Rightarrow 2(N + 1)^3 - 1 > 1000.$$

By trial and error, this inequality is valid when $N = 7[2(8^3) - 1 = 1024]$.

Use 7 terms.

34. $\sum_{n=1}^{\infty} \frac{(-1)^{n+1}}{n^5}$

By Theorem 9.15,

$$|R_N| \le a_{N+1} = \frac{1}{(N + 1)^5} < 0.001$$

$$\Rightarrow (N + 1)^5 > 1000.$$

By trial and error, this inequality is valid when $N = 3(4^5 = 1024)$.

Use 3 terms.

35. $\sum_{n=0}^{\infty} \frac{(-1)^n}{n!}$

By Theorem 9.15,

$$|R_N| \le a_{N+1} = \frac{1}{(N + 1)!} < 0.001$$

$$\Rightarrow (N + 1)! > 1000.$$

By trial and error, this inequality is valid when $N = 6(7! = 5040)$. Use 7 terms since the sum begins with $n = 0$.

36. $\sum_{n=0}^{\infty} \frac{(-1)^n}{(2n)!}$

By Theorem 9.15,

$$|R_N| \le a_{N+1} = \frac{1}{(2(N+1))!} = \frac{1}{(2N+2)!} < 0.001$$

$$\Rightarrow (2N+2)! > 1000.$$

By trial and error, this inequality is valid when $N = 3(8! = 40{,}320)$. Use 4 terms since the sum begins with $n = 0$.

37. $\sum_{n=1}^{\infty} \frac{(-1)^n}{2^n}$

$\sum_{n=1}^{\infty} \frac{1}{2^n}$ is a convergent geometric series.

Therefore, $\sum_{n=1}^{\infty} \frac{(-1)^n}{2^n}$ converges absolutely.

38. $\sum_{n=1}^{\infty} \frac{(-1)^{n+1}}{n^2}$

$\sum_{n=1}^{\infty} \frac{1}{n^2}$ is a convergent p-series.

Therefore, $\sum_{n=1}^{\infty} \frac{(-1)^{n+1}}{n^2}$ converges absolutely.

39. $\sum_{n=1}^{\infty} \frac{(-1)^n}{n!}$

$\frac{1}{n!} < \frac{1}{n^2}$ for $n \ge 4$

and $\sum_{n=1}^{\infty} \frac{1}{n^2}$ is a convergent p-series.

So, $\sum_{n=1}^{\infty} \frac{1}{n!}$ converges, and

$\sum_{n=1}^{\infty} \frac{(-1)^n}{n!}$ converges absolutely.

40. $\sum_{n=1}^{\infty} \frac{(-1)^{n+1}}{n+3}$

The series converges by the Alternating Series Test. But, the series

$$\sum_{n=1}^{\infty} \frac{1}{n+3}$$

diverges by comparison to $\sum_{n=1}^{\infty} \frac{1}{n}$.

Therefore, $\sum_{n=1}^{\infty} \frac{(-1)^{n+1}}{n+3}$ converges conditionally.

41. $\sum_{n=1}^{\infty} \frac{(-1)^{n+1}}{\sqrt{n}}$

The given series converges by the Alternating Series Test, but does not converge absolutely because

$$\sum_{n=1}^{\infty} \frac{1}{\sqrt{n}}$$

is a divergent p-series. Therefore, the series converges conditionally.

42. $\sum_{n=1}^{\infty} \frac{(-1)^{n+1}}{n\sqrt{n}}$

$\sum_{n=1}^{\infty} \frac{1}{n\sqrt{n}} = \sum_{n=1}^{\infty} \frac{1}{n^{3/2}}$ which is a convergent p-series.

Therefore, the given series converges absolutely.

43. $\sum_{n=1}^{\infty} \frac{(-1)^{n+1} n^2}{(n+1)^2}$

$$\lim_{n \to \infty} \frac{n^2}{(n+1)^2} = 1$$

Therefore, the series diverges by the nth-Term Test.

44. $\sum_{n=1}^{\infty} \frac{(-1)^{n+1}(2n+3)}{n+10}$

$$\lim_{n \to \infty} \frac{2n+3}{n+10} = 2$$

Therefore, the series diverges by the nth-Term Test.

45. $\sum_{n=2}^{\infty} \frac{(-1)^n}{n \ln n}$

The series converges by the Alternating Series Test.

Let $f(x) = \frac{1}{x \ln x}$.

$$\int_2^{\infty} \frac{1}{x \ln x}\,dx = \Big[\ln(\ln x)\Big]_2^{\infty} = \infty$$

By the Integral Test, $\sum_{n=2}^{\infty} \frac{1}{n \ln n}$ diverges.

So, the series $\sum_{n=2}^{\infty} \frac{(-1)^n}{n \ln n}$ converges conditionally.

46. $\sum_{n=0}^{\infty} \frac{(-1)^n}{e^{n^2}}$

$\sum_{n=0}^{\infty} \frac{1}{e^{n^2}}$ converges by a comparison to the convergent geometric series $\sum_{n=0}^{\infty} \left(\frac{1}{e}\right)^n$. Therefore, the given series converges absolutely.

47. $\sum_{n=2}^{\infty} \frac{(-1)^n n}{n^3 - 5}$

$\sum_{n=2}^{\infty} \frac{n}{n^3 - 5}$ converges by a limit comparison to the p-series $\sum_{n=2}^{\infty} \frac{1}{n^2}$. Therefore, the given series converges absolutely.

48. $\sum_{n=1}^{\infty} \frac{(-1)^{n+1}}{n^{4/3}}$

$\sum_{n=1}^{\infty} \frac{1}{n^{4/3}}$ is a convergent p-series. Therefore, the given series converges absolutely.

49. $\sum_{n=0}^{\infty} \frac{(-1)^n}{(2n + 1)!}$

$$\sum_{n=0}^{\infty} \frac{1}{(2n + 1)!}$$

is convergent by comparison to the convergent geometric series

$$\sum_{n=0}^{\infty} \left(\frac{1}{2}\right)^n$$

because

$$\frac{1}{(2n + 1)!} < \frac{1}{2^n} \text{ for } n > 0.$$

Therefore, the given series converges absolutely.

50. $\sum_{n=0}^{\infty} \frac{(-1)^n}{\sqrt{n + 4}}$

The given series converges by the Alternating Series Test, but

$$\sum_{n=0}^{\infty} \frac{1}{\sqrt{n + 4}}$$

diverges by a limit comparison to the divergent p-series

$$\sum_{n=1}^{\infty} \frac{1}{\sqrt{n}}.$$

Therefore, the given series converges conditionally.

51. $\sum_{n=0}^{\infty} \frac{\cos n\pi}{n + 1} = \sum_{n=0}^{\infty} \frac{(-1)^n}{n + 1}$

The given series converges by the Alternating Series Test, but

$$\sum_{n=0}^{\infty} \frac{|\cos n\pi|}{n + 1} = \sum_{n=0}^{\infty} \frac{1}{n + 1}$$

diverges by a limit comparison to the divergent harmonic series,

$$\sum_{n=1}^{\infty} \frac{1}{n}.$$

$\lim_{n \to \infty} \frac{|\cos n\pi|/(n + 1)}{1/n} = 1$, therefore, the series converges conditionally.

52. $\sum_{n=1}^{\infty} (-1)^{n+1} \arctan n$

$$\lim_{n \to \infty} \arctan n = \frac{\pi}{2} \neq 0$$

Therefore, the series diverges by the nth-Term Test.

53. $\sum_{n=1}^{\infty} \frac{\cos n\pi}{n^2} = \sum_{n=1}^{\infty} \frac{(-1)^n}{n^2}$

$\sum_{n=1}^{\infty} \frac{1}{n^2}$ is a convergent p-series.

Therefore, the given series converges absolutely.

54. $\sum_{n=1}^{\infty} \frac{\sin[(2n - 1)\pi/2]}{n} = \sum_{n=1}^{\infty} \frac{(-1)^{n+1}}{n}$

The given series converges by the Alternating Series Test, but

$$\sum_{n=1}^{\infty} \left| \frac{\sin[(2n - 1)\pi/2]}{n} \right| = \sum_{n=1}^{\infty} \frac{1}{n}$$

is a divergent p-series. Therefore, the series converges conditionally.

55. An alternating series is a series whose terms alternate in sign.

56. See Theorem 9.14.

57. $|S - S_N| = |R_N| \le a_{N+1}$ (Theorem 9.15)

58. $\sum a_n$ is absolutely convergent if $\sum |a_n|$ converges.

$\sum a_n$ is conditionally convergent if $\sum |a_n|$ diverges, but $\sum a_n$ converges.

59. (a) False. For example, let $a_n = \frac{(-1)^n}{n}$.

Then $\sum a_n = \sum \frac{(-1)^n}{n}$ converges

and $\sum(-a_n) = \sum \frac{(-1)^{n+1}}{n}$ converges.

But, $\sum |a_n| = \sum \frac{1}{n}$ diverges.

(b) True. For if $\sum |a_n|$ converged, then so would $\sum a_n$ by Theorem 9.16.

60. (b). The partial sums alternate above and below the horizontal line representing the sum.

61. True. $S_{100} = -1 + \frac{1}{2} - \frac{1}{3} + \cdots + \frac{1}{100}$

Because the next term $-\frac{1}{101}$ is negative, S_{100} is an overestimate of the sum.

62. False. Let

$$\sum a_n = \sum b_n = \sum \frac{(-1)^n}{\sqrt{n}}.$$

Then both converge by the Alternating Series Test. But,

$\sum a_n b_n = \sum \frac{1}{n}$, which diverges.

63. $\sum_{n=1}^{\infty} (-1)^n \frac{1}{n^p}$

If $p = 0$, then $\sum_{n=1}^{\infty} (-1)^n$ diverges.

If $p < 0$, then $\sum_{n=1}^{\infty} (-1)^n n^{-p}$ diverges.

If $p > 0$, then $\lim_{n\to\infty} \frac{1}{n^p} = 0$ and

$$a_{n+1} = \frac{1}{(n+1)^p} < \frac{1}{n^p} = a_n.$$

Therefore, the series converges for $p > 0$.

64. $\sum_{n=1}^{\infty} (-1)^n \frac{1}{n+p}$

Assume that $n + p \neq 0$ so that $a_n = 1/(n+p)$ are defined for all n. For all p,

$$\lim_{n\to\infty} a_n = \lim_{n\to\infty} \frac{1}{n+p} = 0$$

$$a_{n+1} = \frac{1}{n+1+p} < \frac{1}{n+p} < a_n.$$

Therefore, the series converges for all p.

65. Because

$$\sum_{n=1}^{\infty} |a_n|$$

converges you have $\lim_{n\to\infty} |a_n| = 0$. So, there must exist an $N > 0$ such that $|a_N| < 1$ for all $n > N$ and it follows that $a_n^2 \le |a_n|$ for all $n > N$. So, by the Comparison Test,

$$\sum_{n=1}^{\infty} a_n^2$$

converges. Let $a_n = 1/n$ to see that the converse is false.

66. $\sum_{n=1}^{\infty} \frac{(-1)^{n-1}}{n}$ converges, but $\sum_{n=1}^{\infty} \frac{1}{n}$ diverges.

67. $\sum_{n=1}^{\infty} \frac{1}{n^2}$ converges, and so does $\sum_{n=1}^{\infty} \frac{1}{n^4}$.

68. (a) $\sum_{n=1}^{\infty} \frac{x^n}{n}$

converges absolutely (by comparison) for $-1 < x < 1$, because

$$\left|\frac{x^n}{n}\right| < |x^n| \text{ and } \sum x^n$$

is a convergent geometric series for $-1 < x < 1$.

(b) When $x = -1$, you have the convergent alternating series

$$\sum_{n=1}^{\infty} \frac{(-1)^n}{n}.$$

When $x = 1$, you have the divergent harmonic series $1/n$. Therefore,

$\sum_{n=1}^{\infty} \frac{x^n}{n}$ converges conditionally for $x = -1$.

69. (a) No, the series does not satisfy $a_{n+1} \le a_n$ for all n. For example, $\frac{1}{9} < \frac{1}{8}$.

(b) Yes, the series converges.

$$S_{2n} = \frac{1}{2} - \frac{1}{3} + \cdots + \frac{1}{2^n} - \frac{1}{3^n}$$
$$= \left(\frac{1}{2} + \cdots + \frac{1}{2^n}\right) - \left(\frac{1}{3} + \cdots + \frac{1}{3^n}\right)$$
$$= \left(1 + \frac{1}{2} + \cdots + \frac{1}{2^n}\right) - \left(1 + \frac{1}{3} + \cdots + \frac{1}{3^n}\right)$$

As $n \to \infty$,

$$S_{2n} \to \frac{1}{1 - (1/2)} - \frac{1}{1 - (1/3)} = 2 - \frac{3}{2} = \frac{1}{2}.$$

70. (a) No, the series does not satisfy $a_{n+1} \le a_n$:

$$\sum_{n=1}^{\infty}(-1)^{n+1}a_n = 1 - \frac{1}{8} + \frac{1}{\sqrt{3}} - \frac{1}{64} + \cdots \text{ and}$$

$$\frac{1}{8} < \frac{1}{\sqrt{3}}.$$

(b) No, the series diverges because $\sum \frac{1}{\sqrt{n}}$ diverges.

71. $\sum_{n=1}^{\infty} \frac{10}{n^{3/2}} = 10\sum_{n=1}^{\infty} \frac{1}{n^{3/2}}$,

convergent p-series

72. $\sum_{n=1}^{\infty} \frac{3}{n^2 + 5}$

converges by limit comparison to convergent p-series

$\sum \frac{1}{n^2}$.

73. Diverges by nth-Term Test

$\lim_{n\to\infty} a_n = \infty$

74. Converges by limit comparison to convergent geometric series $\sum \frac{1}{2^n}$.

75. Convergent geometric series

$\left(r = \frac{7}{8} < 1\right)$

76. Diverges by nth-Term Test

$\lim_{n\to\infty} a_n = \frac{3}{2}$

77. Convergent geometric series $\left(r = 1/\sqrt{e}\right)$ or Integral Test

78. Converges (conditionally) by Alternating Series Test

79. Converges (absolutely) by Alternating Series Test

80. Diverges by comparison to Divergent Harmonic Series:

$\frac{\ln n}{n} > \frac{1}{n}$ for $n \ge 3$

81. The first term of the series is zero, not one. You cannot regroup series terms arbitrarily.

82. $s = 1 - \frac{1}{2} + \frac{1}{3} - \frac{1}{4} + \cdots$

$$S = 1 + \frac{1}{3} - \frac{1}{2} + \frac{1}{5} + \frac{1}{7} - \frac{1}{4} + \frac{1}{9} + \frac{1}{11} - \frac{1}{6} + \cdots$$

(i) $s_{4n} = 1 - \frac{1}{2} + \frac{1}{3} - \frac{1}{4} + \frac{1}{5} - \frac{1}{6} + \cdots + \frac{1}{4n-1} - \frac{1}{4n}$

$\frac{1}{2}s_{2n} = \frac{1}{2} - \frac{1}{4} + \frac{1}{6} - \frac{1}{8} + \frac{1}{10} - \cdots + \frac{1}{4n-2} - \frac{1}{4n}$

Adding: $s_{4n} + \frac{1}{2}s_{2n} = 1 + \frac{1}{3} - \frac{1}{2} + \frac{1}{5} + \frac{1}{7} - \frac{1}{4} + \cdots + \frac{1}{4n-3} + \frac{1}{4n-1} - \frac{1}{2n} = S_{3n}$

(ii) $\lim_{n\to\infty} s_n = s$ (In fact, $s = \ln 2$.)

$s \ne 0$ because $s > \frac{1}{2}$.

$S = \lim_{n\to\infty} S_{3n} = s_{4n} + \frac{1}{2}s_{2n} = s + \frac{1}{2}s = \frac{3}{2}s$

So, $S \ne s$.

Section 9.6 The Ratio and Root Tests

1. $\frac{(n+1)!}{(n-2)!} = \frac{(n+1)(n)(n-1)(n-2)!}{(n-2)!} = (n+1)(n)(n-1)$

2. $\frac{(2k-2)!}{(2k)!} = \frac{(2k-2)!}{(2k)(2k-1)(2k-2)!} = \frac{1}{(2k)(2k-1)}$

3. Use the Principle of Mathematical Induction. When $k = 1$, the formula is valid because $1 = \frac{(2(1))!}{2^1 \cdot 1!}$. Assume that

$$1 \cdot 3 \cdot 5 \cdots (2n-1) = \frac{(2n)!}{2^n n!}$$

and show that

$$1 \cdot 3 \cdot 5 \cdots (2n-1)(2n+1) = \frac{(2n+2)!}{2^{n+1}(n+1)!}.$$

To do this, note that:

$$\begin{aligned}
1 \cdot 3 \cdot 5 \cdots (2n-1)(2n+1) &= \left[1 \cdot 3 \cdot 5 \cdots (2n-1)\right](2n+1) \\
&= \frac{(2n)!}{2^n n!} \cdot (2n+1) \quad \text{(Induction hypothesis)} \\
&= \frac{(2n)!(2n+1)}{2^n n!} \cdot \frac{(2n+2)}{2(n+1)} \\
&= \frac{(2n)!(2n+1)(2n+2)}{2^{n+1} n!(n+1)} \\
&= \frac{(2n+2)!}{2^{n+1}(n+1)!}
\end{aligned}$$

The formula is valid for all $n \geq 1$.

4. Use the Principle of Mathematical Induction. When $k = 3$, the formula is valid because $\frac{1}{1} = \frac{2^3 3!(3)(5)}{6!} = 1$. Assume that

$$\frac{1}{1 \cdot 3 \cdot 5 \cdots (2n-5)} = \frac{2^n n!(2n-3)(2n-1)}{(2n)!}$$

and show that

$$\frac{1}{1 \cdot 3 \cdot 5 \cdots (2n-5)(2n-3)} = \frac{2^{n+1}(n+1)!(2n-1)(2n+1)}{(2n+2)!}.$$

To do this, note that:

$$\begin{aligned}
\frac{1}{1 \cdot 3 \cdot 5 \cdots (2n-5)(2n-3)} &= \frac{1}{1 \cdot 3 \cdot 5 \cdots (2n-5)} \cdot \frac{1}{(2n-3)} \\
&= \frac{2^n n!\cancel{(2n-3)}(2n-1)}{(2n)!} \cdot \frac{1}{\cancel{(2n-3)}} \\
&= \frac{2^n n!(2n-1)}{(2n)!} \cdot \frac{(2n+1)(2n+2)}{(2n+1)(2n+2)} \\
&= \frac{2^n(2)(n+1)n!(2n-1)(2n+1)}{(2n)!(2n+1)(2n+2)} \\
&= \frac{2^{n+1}(n+1)!(2n-1)(2n+1)}{(2n+2)!}
\end{aligned}$$

The formula is valid for all $n \geq 3$.

5. $\sum_{n=1}^{\infty} n\left(\frac{3}{4}\right)^n = 1\left(\frac{3}{4}\right) + 2\left(\frac{9}{16}\right) + \cdots$

$S_1 = \frac{3}{4}, S_2 \approx 1.875$

Matches (d).

6. $\sum_{n=1}^{\infty} \left(\frac{3}{4}\right)^n \left(\frac{1}{n!}\right) = \frac{3}{4} + \frac{9}{16}\left(\frac{1}{2}\right) + \cdots$

$S_1 = \frac{3}{4}, S_2 = 1.03$

Matches (c).

7. $\sum_{n=1}^{\infty} \frac{(-3)^{n+1}}{n!} = 9 - \frac{3^3}{2} + \cdots$

$S_1 = 9$

Matches (f).

8. $\sum_{n=1}^{\infty} \frac{(-1)^{n-1}4}{(2n)!} = \frac{4}{2} - \frac{4}{24} + \cdots$

$S_1 = 2$

Matches (b).

9. $\sum_{n=1}^{\infty} \left(\frac{4n}{5n-3}\right)^n = \frac{4}{2} + \left(\frac{8}{7}\right)^2 + \cdots$

$S_1 = 2, S_2 = 3.31$

Matches (a).

10. $\sum_{n=0}^{\infty} 4e^{-n} = 4 + \frac{4}{e} + \cdots$

$S_1 = 4$

Matches (e).

11. (a) Ratio Test: $\lim_{n\to\infty}\left|\frac{a_{n+1}}{a_n}\right| = \lim_{n\to\infty} \frac{(n+1)^3(1/2)^{n+1}}{n^3(1/2)^n}$

$= \lim_{n\to\infty}\left(\frac{n+1}{n}\right)^3 \frac{1}{2} = \frac{1}{2} < 1$, converges

(b)

n	5	10	15	20	25
S_n	13.7813	24.2363	25.8468	25.9897	25.9994

(c)

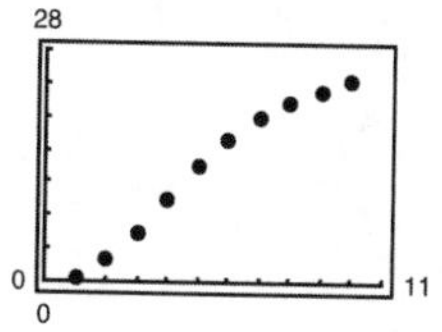

(d) The sum is approximately 26.

(e) The more rapidly the terms of the series approach 0, the more rapidly the sequence of partial sums approaches the sum of the series.

12. (a) Ratio Test: $\lim_{n\to\infty}\left|\frac{a_{n+1}}{a_n}\right| = \lim_{n\to\infty} \frac{\dfrac{(n+1)^2+1}{(n+1)!}}{\dfrac{n^2+1}{n!}} = \lim_{n\to\infty}\left(\frac{n^2+2n+2}{n^2+1}\right)\left(\frac{1}{n+1}\right) = 0 < 1$, converges

(b)

n	5	10	15	20	25
S_n	7.0917	7.1548	7.1548	7.1548	7.1548

(c)

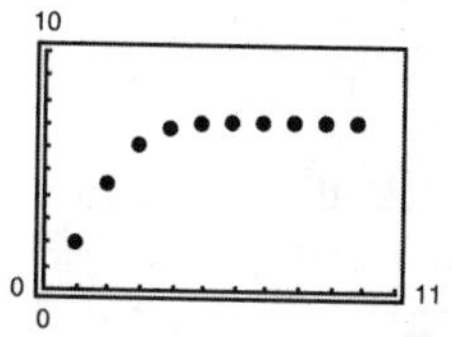

(d) The sum is approximately 7.15485

(e) The more rapidly the terms of the series approach 0, the more rapidly the sequence of the partial sums approaches the sum of the series.

13. $\sum_{n=1}^{\infty} \frac{1}{5^n}$

$$\lim_{n\to\infty}\left|\frac{a_{n+1}}{a_n}\right| = \lim_{n\to\infty}\left|\frac{1/5^{(n+1)}}{1/5^n}\right| = \lim_{n\to\infty}\frac{5^n}{5^{n+1}} = \frac{1}{5} < 1$$

Therefore, the series converges by the Ratio Test.

14. $\sum_{n=1}^{\infty} \frac{1}{n!}$

$$\lim_{n\to\infty}\left|\frac{a_{n+1}}{a_n}\right| = \lim_{n\to\infty}\frac{1/(n+1)!}{1/n!}$$

$$= \lim_{n\to\infty}\frac{n!}{(n+1)!} = \lim_{n\to\infty}\frac{1}{n+1} = 0$$

Therefore, the series converges by the Ratio Test.

15. $\sum_{n=0}^{\infty} \frac{n!}{3^n}$

$$\lim_{n\to\infty}\left|\frac{a_{n+1}}{a_n}\right| = \lim_{n\to\infty}\left|\frac{(n+1)!}{3^{n+1}}\cdot\frac{3^n}{n!}\right| = \lim_{n\to\infty}\frac{n+1}{3} = \infty$$

Therefore, by the Ratio Test, the series diverges.

16. $\sum_{n=0}^{\infty} \frac{2^n}{n!}$

$$\lim_{n\to\infty}\left|\frac{a_{n+1}}{a_n}\right| = \lim_{n\to\infty}\left|\frac{2^{(n+1)}/(n+1)!}{2^n/n!}\right|$$

$$= \lim_{n\to\infty}\frac{2}{n+1} = 0 < 1$$

Therefore, the series converges by the Ratio Test.

17. $\sum_{n=1}^{\infty} n\left(\frac{6}{5}\right)^n$

$$\lim_{n\to\infty}\left|\frac{a_{n+1}}{a_n}\right| = \lim_{n\to\infty}\frac{(n+1)(6/5)^{n+1}}{n(6/5)^n}$$

$$= \lim_{n\to\infty}\frac{n+1}{n}\left(\frac{6}{5}\right) = \frac{6}{5} > 1$$

Therefore, the series diverges by the Ratio Test.

18. $\sum_{n=1}^{\infty} n\left(\frac{7}{8}\right)^n$

$$\lim_{n\to\infty}\left|\frac{a_{n+1}}{a_n}\right| = \lim_{n\to\infty}\left|\frac{(n+1)(7/8)^{n+1}}{n(7/8)^n}\right|$$

$$= \lim_{n\to\infty}\frac{n+1}{n}\left(\frac{7}{8}\right) = \frac{7}{8} < 1$$

Therefore, the series converges by the Ratio Test.

19. $\sum_{n=1}^{\infty} \frac{n}{4^n}$

$$\lim_{n\to\infty}\left|\frac{a_{n+1}}{a_n}\right| = \lim_{n\to\infty}\frac{(n+1)/4^{n+1}}{n/4^n} = \lim_{n\to\infty}\frac{n+1}{4n} = 1/4 < 1$$

Therefore, the series converges by the Ratio Test.

20. $\sum_{n=1}^{\infty} \frac{5^n}{n^4}$

$$\lim_{n\to\infty}\left|\frac{a_{n+1}}{a_n}\right| = \lim_{n\to\infty}\left|\frac{5^{(n+1)}/(n+1)^4}{5^n/n^4}\right|$$

$$= \lim_{n\to\infty} 5\left(\frac{n+1}{n}\right)^4 = 5 > 1$$

Therefore, the series diverges by the Ratio Test.

21. $\sum_{n=1}^{\infty} \frac{n^3}{3^n}$

$$\lim_{n\to\infty}\left|\frac{a_{n+1}}{a_n}\right| = \lim_{n\to\infty}\left|\frac{(n+1)^3/3^{(n+1)}}{n^3/3^n}\right|$$

$$= \lim_{n\to\infty}\left(\frac{n+1}{n}\right)^3\frac{1}{3} = \frac{1}{3} < 1$$

Therefore, the series converges by the Ratio Test.

22. $\sum_{n=1}^{\infty} \frac{(-1)^{n+1}(n+2)}{n(n+1)}$

$$a_{n+1} = \frac{n+3}{(n+1)(n+2)} \le \frac{n+2}{n(n+1)} = a_n$$

$$\lim_{n\to\infty}\frac{n+2}{n(n+1)} = 0$$

Therefore, by Theorem 9.14, the series converges.

Note: The Ratio Test is inconclusive because

$$\lim_{n\to\infty}\left|\frac{a_{n+1}}{a_n}\right| = 1.$$

The series converges conditionally.

23. $\sum_{n=0}^{\infty} \frac{(-1)^n 2^n}{n!}$

$$\lim_{n\to\infty}\left|\frac{a_{n+1}}{a_n}\right| = \lim_{n\to\infty}\left|\frac{2^{n+1}}{(n+1)!}\cdot\frac{n!}{2^n}\right|$$

$$= \lim_{n\to\infty}\frac{2}{n+1} = 0$$

Therefore, by the Ratio Test, the series converges.

24. $\sum_{n=1}^{\infty} \frac{(-1)^{n-1}(3/2)^n}{n^2}$

$$\lim_{n\to\infty}\left|\frac{a_{n+1}}{a_n}\right| = \lim_{n\to\infty}\left|\frac{(3/2)^{n+1}}{n^2+2n+1}\cdot\frac{n^2}{(3/2)^n}\right|$$

$$= \lim_{n\to\infty}\frac{3n^2}{2(n^2+2n+1)} = \frac{3}{2} > 1$$

Therefore, by the Ratio Test, the series diverges.

25. $\sum_{n=1}^{\infty} \frac{n!}{n3^n}$

$$\lim_{n\to\infty}\left|\frac{a_{n+1}}{a_n}\right| = \lim_{n\to\infty}\left|\frac{(n+1)!}{(n+1)3^{n+1}}\cdot\frac{n3^n}{n!}\right| = \lim_{n\to\infty}\frac{n}{3} = \infty$$

Therefore, by the Ratio Test, the series diverges.

26. $\sum_{n=1}^{\infty} \frac{(2n)!}{n^5}$

$$\lim_{n\to\infty}\left|\frac{a_{n+1}}{a_n}\right| = \lim_{n\to\infty}\left|\frac{(2n+2)!}{(n+1)^5}\cdot\frac{n^5}{(2n)!}\right|$$

$$= \lim_{n\to\infty}\frac{(2n+2)(2n+1)n^5}{(n+1)^5} = \infty$$

Therefore, by the Ratio Test, the series diverges.

27. $\sum_{n=0}^{\infty} \frac{e^n}{n!}$

$$\lim_{n\to\infty}\left|\frac{a_{n+1}}{a_n}\right| = \lim_{n\to\infty}\frac{e^{n+1}/(n+1)!}{e^n/n!}$$

$$= \lim_{n\to\infty} e\left(\frac{n!}{(n+1)!}\right) = \lim_{n\to\infty}\frac{e}{n+1} = 0$$

Therefore, the series converges by the Ratio Test.

28. $\sum_{n=1}^{\infty} \frac{n!}{n^n}$

$$\lim_{n\to\infty}\left|\frac{a_{n+1}}{a_n}\right| = \lim_{n\to\infty}\frac{(n+1)!/(n+1)^{n+1}}{n!/n^n}$$

$$= \lim_{n\to\infty}\left(\frac{n}{n+1}\right)^n = \frac{1}{e}$$

Therefore, the series converges by the Ratio Test.

29. $\sum_{n=0}^{\infty} \frac{6^n}{(n+1)^n}$

$$\lim_{n\to\infty}\left|\frac{a_{n+1}}{a_n}\right| = \lim_{n\to\infty}\frac{6^{n+1}/(n+2)^{n+1}}{6^n/(n+1)^n} = \lim_{n\to\infty}\frac{6}{n+2}\left(\frac{n+1}{n+2}\right)^n = 0\left(\frac{1}{e}\right) = 0.$$

To find $\lim_{n\to\infty}\left(\frac{n+1}{n+2}\right)^n$: Let $y = \left(\frac{n+1}{n+2}\right)^n$

$$\ln y = n\ln\left(\frac{n+1}{n+2}\right) = \frac{\ln(n+1) - \ln(n+2)}{1/n}$$

$$\lim_{n\to\infty}[\ln y] = \lim_{n\to\infty}\left[\frac{1/(n+1) - 1/(n+2)}{-1/n^2}\right] = \lim_{n\to\infty}\left[\frac{-n^2[(n+2)-(n+1)]}{(n+1)(n+2)}\right] = -1$$

by L'Hôpital's Rule. So, $y \to \frac{1}{e}$.

Therefore, the series converges by the Ratio Test.

30. $\sum_{n=0}^{\infty} \frac{(n!)^2}{(3n)!}$

$$\lim_{n\to\infty}\left|\frac{a_{n+1}}{a_n}\right| = \lim_{n\to\infty}\left|\frac{[(n+1)!]^2}{(3n+3)!}\cdot\frac{(3n)!}{(n!)^2}\right| = \lim_{n\to\infty}\frac{(n+1)^2}{(3n+3)(3n+2)(3n+1)} = 0$$

Therefore, by the Ratio Test, the series converges.

31. $\displaystyle\sum_{n=0}^{\infty} \frac{5^n}{2^n + 1}$

$$\lim_{n\to\infty}\left|\frac{a_{n+1}}{a_n}\right| = \lim_{n\to\infty}\frac{5^{n+1}/(2^{n+1}+1)}{5^n/(2^n+1)} = \lim_{n\to\infty}\frac{5(2^n+1)}{(2^{n+1}+1)} = \lim_{n\to\infty}\frac{5(1+1/2^n)}{2+1/2^n} = \frac{5}{2} > 1$$

Therefore, the series diverges by the Ratio Test.

32. $\displaystyle\sum_{n=0}^{\infty} \frac{(-1)^n 2^{4n}}{(2n+1)!}$

$$\lim_{n\to\infty}\left|\frac{a_{n+1}}{a_n}\right| = \lim_{n\to\infty}\left|\frac{2^{4n+4}}{(2n+3)!}\cdot\frac{(2n+1)!}{2^{4n}}\right| = \lim_{n\to\infty}\frac{2^4}{(2n+3)(2n+2)} = 0$$

Therefore, by the Ratio Test, the series converges.

33. $\displaystyle\sum_{n=0}^{\infty} \frac{(-1)^{n+1} n!}{1\cdot 3\cdot 5\cdots(2n+1)}$

$$\lim_{n\to\infty}\left|\frac{a_{n+1}}{a_n}\right| = \lim_{n\to\infty}\left|\frac{(n+1)!}{1\cdot 3\cdot 5\cdots(2n+1)(2n+3)}\cdot\frac{1\cdot 3\cdot 5\cdots(2n+1)}{n!}\right| = \lim_{n\to\infty}\frac{n+1}{2n+3} = \frac{1}{2}$$

Therefore, by the Ratio Test, the series converges.

Note: The first few terms of this series are $-1 + \frac{1}{1\cdot 3} - \frac{2!}{1\cdot 3\cdot 5} + \frac{3!}{1\cdot 3\cdot 5\cdot 7} - \cdots$.

34. $\displaystyle\sum_{n=1}^{\infty} \frac{(-1)^n 2\cdot 4\cdot 6\cdots 2n}{2\cdot 5\cdot 8\cdots(3n-1)}$

$$\lim_{n\to\infty}\left|\frac{a_{n+1}}{a_n}\right| = \lim_{n\to\infty}\left|\frac{2\cdot 4\cdots 2n(2n+2)}{2\cdot 5\cdots(3n-1)(3n+2)}\cdot\frac{2\cdot 5\cdots(3n-1)}{2\cdot 4\cdots 2n}\right| = \lim_{n\to\infty}\frac{2n+2}{3n+2} = \frac{2}{3}$$

Therefore, by the Ratio Test, the series converges.

Note: The first few terms of this series are $-\frac{2}{2} + \frac{2\cdot 4}{2\cdot 5} - \frac{2\cdot 4\cdot 6}{2\cdot 5\cdot 8} + \cdots$

35. $\displaystyle\sum_{n=1}^{\infty} \frac{1}{5^n}$

$$\lim_{n\to\infty}\sqrt[n]{|a_n|} = \lim_{n\to\infty}\left[\frac{1}{5^n}\right]^{1/n} = \frac{1}{5} < 1$$

Therefore, by the Root Test, the series converges.

36. $\displaystyle\sum_{n=1}^{\infty} \frac{1}{n^n}$

$$\lim_{n\to\infty}\sqrt[n]{|a_n|} = \lim_{n\to\infty}\left[\frac{1}{n^n}\right]^{1/n} = \lim_{n\to\infty}\frac{1}{n} = 0$$

Therefore, by the Root Test, the series converges.

37. $\displaystyle\sum_{n=1}^{\infty} \left(\frac{n}{2n+1}\right)^n$

$$\lim_{n\to\infty}\sqrt[n]{|a_n|} = \lim_{n\to\infty}\sqrt[n]{\left(\frac{n}{2n+1}\right)^n} = \lim_{n\to\infty}\frac{n}{2n+1} = \frac{1}{2}$$

Therefore, by the Root Test, the series converges.

38. $\displaystyle\sum_{n=1}^{\infty} \left(\frac{2n}{n+1}\right)^n$

$$\lim_{n\to\infty}\sqrt[n]{|a_n|} = \lim_{n\to\infty}\sqrt[n]{\left(\frac{2n}{n+1}\right)^n} = \lim_{n\to\infty}\frac{2n}{n+1} = 2$$

Therefore, by the Root Test, the series diverges.

39. $\displaystyle\sum_{n=1}^{\infty} \left(\frac{3n+2}{n+3}\right)^n$

$$\lim_{n\to\infty}\sqrt[n]{|a_n|} = \lim_{n\to\infty}\sqrt[n]{\left(\frac{3n+2}{n+3}\right)^n}$$
$$= \lim_{n\to\infty}\frac{3n+2}{n+3} = 3 > 1$$

Therefore, the series diverges by the Root Test.

40. $\sum_{n=1}^{\infty}\left(\frac{n-2}{5n+1}\right)^n$

$$\lim_{n\to\infty}\sqrt[n]{|a_n|} = \lim_{n\to\infty}\sqrt[n]{\left|\frac{n-2}{5n+1}\right|^n}$$

$$= \lim_{n\to\infty}\left|\frac{n-2}{5n+1}\right| = \frac{1}{5} < 1$$

Therefore, the series converges by the Root Test.

41. $\sum_{n=2}^{\infty}\frac{(-1)^n}{(\ln n)^n}$

$$\lim_{n\to\infty}\sqrt[n]{|a_n|} = \lim_{n\to\infty}\sqrt[n]{\left|\frac{(-1)^n}{(\ln n)^n}\right|} = \lim_{n\to\infty}\frac{1}{|\ln n|} = 0$$

Therefore, by the Root Test, the series converges.

42. $\sum_{n=1}^{\infty}\left(\frac{-3n}{2n+1}\right)^{3n}$

$$\lim_{n\to\infty}\sqrt[n]{|a_n|} = \lim_{n\to\infty}\sqrt[n]{\left|\left(\frac{-3n}{2n+1}\right)^{3n}\right|}$$

$$= \lim_{n\to\infty}\left(\frac{3n}{2n+1}\right)^3 = \left(\frac{3}{2}\right)^3 = \frac{27}{8}$$

Therefore, by the Root Test, the series diverges.

43. $\sum_{n=1}^{\infty}\left(2\sqrt[n]{n}+1\right)^n$

$$\lim_{n\to\infty}\sqrt[n]{|a_n|} = \lim_{n\to\infty}\sqrt[n]{\left(2\sqrt[n]{n}+1\right)^n} = \lim_{n\to\infty}\left(2\sqrt[n]{n}+1\right)$$

To find $\lim_{n\to\infty}\sqrt[n]{n}$, let $y = \lim_{n\to\infty}\sqrt[n]{n}$. Then

$$\ln y = \lim_{n\to\infty}\left(\ln\sqrt[n]{n}\right)$$

$$= \lim_{n\to\infty}\frac{1}{n}\ln n = \lim_{n\to\infty}\frac{\ln n}{n} = \lim_{n\to\infty}\frac{1/n}{1} = 0.$$

So, $\ln y = 0$, so $y = e^0 = 1$ and

$\lim_{n\to\infty}\left(2\sqrt[n]{n}+1\right) = 2(1) + 1 = 3.$

Therefore, by the Root Test, the series diverges.

44. $\sum_{n=0}^{\infty}e^{-3n} = \sum_{n=0}^{\infty}\frac{1}{e^{3n}}$

$$\lim_{n\to\infty}\sqrt[n]{|a_n|} = \lim_{n\to\infty}\sqrt[n]{\frac{1}{e^{3n}}} = \lim_{n\to\infty}\left(\frac{1}{3^n}\right)^{1/n} = \frac{1}{3}$$

Therefore, the series converges by the Root Test.

45. $\sum_{n=1}^{\infty}\frac{n}{3^n}$

$$\lim_{n\to\infty}\sqrt[n]{|a_n|} = \lim_{n\to\infty}\left(\frac{n}{3^n}\right)^{1/n} = \lim_{n\to\infty}\frac{n^{1/n}}{3} = \frac{1}{3}$$

Therefore, the series converges by the Root Test.

Note: You can use L'Hôpital's Rule to show $\lim_{n\to\infty}n^{1/n} = 1$:

Let $y = n^{1/n}$, $\ln y = \frac{1}{n}\ln n = \frac{\ln n}{n}$

$$\lim_{n\to\infty}\frac{\ln n}{n} = \lim_{n\to\infty}\frac{1/n}{1} = 0 \Rightarrow y \to 1$$

46. $\sum_{n=1}^{\infty}\left(\frac{n}{500}\right)^n$

$$\lim_{n\to\infty}\sqrt[n]{|a_n|} = \lim_{n\to\infty}\sqrt[n]{\left(\frac{n}{500}\right)^n} = \lim_{n\to\infty}\left(\frac{n}{500}\right) = \infty$$

Therefore, by the Root Test, the series diverges.

47. $\sum_{n=1}^{\infty}\left(\frac{1}{n}-\frac{1}{n^2}\right)^n$

$$\lim_{n\to\infty}\sqrt[n]{|a_n|} = \lim_{n\to\infty}\sqrt[n]{\left(\frac{1}{n}-\frac{1}{n^2}\right)^n}$$

$$= \lim_{n\to\infty}\left(\frac{1}{n}-\frac{1}{n^2}\right) = 0 - 0 = 0 < 1$$

Therefore, by the Root Test, the series converges.

48. $\sum_{n=1}^{\infty}\left(\frac{\ln n}{n}\right)^n$

$$\lim_{n\to\infty}\sqrt[n]{|a_n|} = \lim_{n\to\infty}\sqrt[n]{\left(\frac{\ln n}{n}\right)^n} = \lim_{n\to\infty}\frac{\ln n}{n} = 0 < 1$$

Therefore, by the Root Test, the series converges.

49. $\sum_{n=2}^{\infty}\frac{n}{(\ln n)^n}$

$$\lim_{n\to\infty}\sqrt[n]{|a_n|} = \lim_{n\to\infty}\sqrt[n]{\frac{n}{(\ln n)^n}} = \lim_{n\to\infty}\frac{n^{1/n}}{\ln n} = 0$$

Therefore, by the Root Test, the series converges.

50. $\sum_{n=1}^{\infty}\frac{(n!)^n}{(n^n)^2} = \sum_{n=1}^{\infty}\frac{(n!)^n}{(n^2)^n}$

$$\lim_{n\to\infty}\sqrt[n]{|a_n|} = \lim_{n\to\infty}\sqrt[n]{\frac{(n!)}{(n^2)^n}} = \lim_{n\to\infty}\frac{n!}{n^2} = \infty$$

Therefore, by the Root Test, the series diverges.

51. $\sum_{n=1}^{\infty} \frac{(-1)^{n+1}5}{n}$

$$a_{n+1} = \frac{5}{n+1} < \frac{5}{n} = a_n$$

$$\lim_{n\to\infty} \frac{5}{n} = 0$$

Therefore, by the Alternating Series Test, the series converges (conditional convergence).

52. $\sum_{n=1}^{\infty} \frac{100}{n} = 100\sum_{n=1}^{\infty} \frac{1}{n}$

This is the divergent harmonic series.

53. $\sum_{n=1}^{\infty} \frac{3}{n\sqrt{n}} = 3\sum_{n=1}^{\infty} \frac{1}{n^{3/2}}$

This is a convergent p-series.

54. $\sum_{n=1}^{\infty} \left(\frac{2\pi}{3}\right)^n$

Because $|r| = \frac{2\pi}{3} > 1$, this is a divergent Geometric Series.

55. $\sum_{n=1}^{\infty} \frac{5n}{2n-1}$

$$\lim_{n\to\infty} \frac{5n}{2n-1} = \frac{5}{2}$$

Therefore, the series diverges by the nth-Term Test

56. $\sum_{n=1}^{\infty} \frac{n}{2n^2+1}$

$$\lim_{n\to\infty} \frac{n/(2n^2+1)}{1/n} = \lim_{n\to\infty} \frac{n^2}{2n^2+1} = \frac{1}{2} > 0$$

This series diverges by limit comparison to the divergent harmonic series

$$\sum_{n=1}^{\infty} \frac{1}{n}.$$

57. $\sum_{n=1}^{\infty} \frac{(-1)^n 3^{n-2}}{2^n} = \sum_{n=1}^{\infty} \frac{(-1)^n 3^n 3^{-2}}{2^n} = \sum_{n=1}^{\infty} \frac{1}{9}\left(-\frac{3}{2}\right)^n$

Because $|r| = \frac{3}{2} > 1$, this is a divergent geometric series.

58. $\sum_{n=1}^{\infty} \frac{10}{3\sqrt{n^3}}$

$$\lim_{n\to\infty} \frac{10/3n^{3/2}}{1/n^{3/2}} = \frac{10}{3}$$

Therefore, the series converges by a Limit Comparison Test with the p-series

$$\sum_{n=1}^{\infty} \frac{1}{n^{3/2}}.$$

59. $\sum_{n=1}^{\infty} \frac{10n+3}{n2^n}$

$$\lim_{n\to\infty} \frac{(10n+3)/n2^n}{1/2^n} = \lim_{n\to\infty} \frac{10n+3}{n} = 10$$

Therefore, the series converges by a Limit Comparison Test with the geometric series

$$\sum_{n=0}^{\infty} \left(\frac{1}{2}\right)^n.$$

60. $\sum_{n=1}^{\infty} \frac{2^n}{4n^2-1}$

$$\lim_{n\to\infty} \frac{2^n}{4n^2-1} = \lim_{n\to\infty} \frac{(\ln 2)2^n}{8n} = \lim_{n\to\infty} \frac{(\ln 2)^2 2^n}{8} = \infty$$

Therefore, the series diverges by the nth-Term Test.

61. $\left|\frac{\cos n}{3^n}\right| \le \frac{1}{3^n}$

Therefore the series $\sum_{n=1}^{\infty} \left|\frac{\cos n}{3^n}\right|$ converges by Direct comparison with the convergent geometric series $\sum_{n=1}^{\infty} \frac{1}{3^n}$. So, $\sum \frac{\cos n}{3^n}$ converges.

62. $\sum_{n=2}^{\infty} \frac{(-1)^n}{n \ln n}$

$$a_{n+1} = \frac{1}{(n+1)\ln(n+1)} \le \frac{1}{n\ln(n)} = a_n$$

$$\lim_{n\to\infty} \frac{1}{n\ln(n)} = 0$$

Therefore, by the Alternating Series Test, the series converges.

63. $\sum_{n=1}^{\infty} \frac{n!}{n7^n}$

$$\lim_{n\to\infty}\left|\frac{a_{n+1}}{a_n}\right| = \lim_{n\to\infty}\left|\frac{(n+1)!/(n+1)7^{n+1}}{n!/n7^n}\right| = \lim_{n\to\infty}\frac{(n+1)!\,n}{(n+1)\,n!}7 = \lim_{n\to\infty} 7n = \infty$$

Therefore, the series diverges by the Ratio Test.

64. $\sum_{n=1}^{\infty} \frac{\ln(n)}{n^2}$

$$\frac{\ln(n)}{n^2} \le \frac{1}{n^{3/2}}$$

Therefore, the series converges by comparison with the p-series

$$\sum_{n=1}^{\infty}\frac{1}{n^{3/2}}.$$

65. $\sum_{n=1}^{\infty} \frac{(-1)^n 3^{n-1}}{n!}$

$$\lim_{n\to\infty}\left|\frac{a_{n+1}}{a_n}\right| = \lim_{n\to\infty}\left|\frac{3^n}{(n+1)!}\cdot\frac{n!}{3^{n-1}}\right| = \lim_{n\to\infty}\frac{3}{n+1} = 0$$

Therefore, by the Ratio Test, the series converges. (Absolutely)

66. $\sum_{n=1}^{\infty} \frac{(-1)^n 3^n}{n2^n}$

$$\lim_{n\to\infty}\left|\frac{a_{n+1}}{a_n}\right| = \lim_{n\to\infty}\left|\frac{3^{n+1}}{(n+1)2^{n+1}}\cdot\frac{n2^n}{3^n}\right| = \lim_{n\to\infty}\frac{3n}{2(n+1)} = \frac{3}{2}$$

Therefore, by the Ratio Test, the series diverges.

67. $\sum_{n=1}^{\infty} \frac{(-3)^n}{3\cdot 5\cdot 7\cdots(2n+1)}$

$$\lim_{n\to\infty}\left|\frac{a_{n+1}}{a_n}\right| = \lim_{n\to\infty}\left|\frac{(-3)^{n+1}}{3\cdot 5\cdot 7\cdots(2n+1)(2n+3)}\cdot\frac{3\cdot 5\cdot 7\cdots(2n+1)}{(-3)^n}\right| = \lim_{n\to\infty}\frac{3}{2n+3} = 0$$

Therefore, by the Ratio Test, the series converges.

68. $\sum_{n=1}^{\infty} \frac{3\cdot 5\cdot 7\cdots(2n+1)}{18^n(2n-1)n!}$

$$\lim_{n\to\infty}\left|\frac{a_{n+1}}{a_n}\right| = \lim_{n\to\infty}\left|\frac{3\cdot 5\cdot 7\cdots(2n+1)(2n+3)}{18^{n+1}(2n+1)(2n-1)n!}\cdot\frac{18^n(2n-1)n!}{3\cdot 5\cdot 7\cdots(2n+1)}\right| = \lim_{n\to\infty}\frac{(2n+3)(2n-1)}{18(2n+1)(2n-1)} = \frac{1}{18}$$

Therefore, by the Ratio Test, the series converge.

69. (a) and (c) are the same.

$$\sum_{n=1}^{\infty}\frac{n5^n}{n!} = \sum_{n=0}^{\infty}\frac{(n+1)5^{n+1}}{(n+1)!} = 5 + \frac{(2)(5)^2}{2!} + \frac{(3)(5)^3}{3!} + \frac{(4)(5)^4}{4!} + \cdots$$

70. (b) and (c) are the same.

$$\sum_{n=0}^{\infty}(n+1)\left(\tfrac{3}{4}\right)^n = \sum_{n=1}^{\infty}n\left(\tfrac{3}{4}\right)^{n-1} = 1 + 2\left(\tfrac{3}{4}\right) + 3\left(\tfrac{3}{4}\right)^2 + 4\left(\tfrac{3}{4}\right)^3 + \cdots$$

71. (a) and (b) are the same.

$$\sum_{n=0}^{\infty}\frac{(-1)^n}{(2n+1)!} = \sum_{n=1}^{\infty}\frac{(-1)^{n-1}}{(2n-1)!} = 1 - \frac{1}{3!} + \frac{1}{5!} - \cdots$$

72. (a) and (b) are the same.

$$\sum_{n=2}^{\infty}\frac{(-1)^n}{(n-1)2^{n-1}} = \sum_{n=1}^{\infty}\frac{(-1)^{n+1}}{n2^n}$$
$$= \frac{1}{2} - \frac{1}{2\cdot 2^2} + \frac{1}{3\cdot 2^3} - \cdots$$

73. Replace n with $n+1$.

$$\sum_{n=1}^{\infty}\frac{n}{7^n} = \sum_{n=0}^{\infty}\frac{n+1}{7^{n+1}}$$

74. Replace n with $n+2$.

$$\sum_{n=2}^{\infty}\frac{9^n}{(n-2)!} = \sum_{n=0}^{\infty}\frac{9^{n+2}}{n!}$$

75. (a) Because

$$\frac{3^{10}}{2^{10}10!} \approx 1.59\times 10^{-5},$$

use 9 terms.

(b) $\displaystyle\sum_{k=1}^{9}\frac{(-3)^k}{2^k k!} \approx -0.7769$

76. (a) Use 10 terms, $k = 9$, see Exercise 3.

(b) $$\sum_{k=0}^{\infty}\frac{(-3)^k}{1\cdot 3\cdot 5\ldots(2k+1)} = \sum_{k=0}^{\infty}\frac{(-3)^k 2^k k!}{(2k)!(2k+1)}$$
$$= \sum_{k=0}^{\infty}\frac{(-6)^k k!}{(2k+1)!} \approx 0.40967$$

77. $$\lim_{n\to\infty}\left|\frac{a_{n+1}}{a_n}\right| = \lim_{n\to\infty}\left|\frac{(4n-1)/(3n+2)a_n}{a_n}\right|$$
$$= \lim_{n\to\infty}\frac{4n-1}{3n+2} = \frac{4}{3} > 1$$

The series diverges by the Ratio Test.

78. $$\lim_{n\to\infty}\left|\frac{a_{n+1}}{a_n}\right| = \lim_{n\to\infty}\left|\frac{(2n+1)/(5n-4)a_n}{a_n}\right|$$
$$= \lim_{n\to\infty}\frac{2n+1}{5n-4} = \frac{2}{5} < 1$$

The series converges by the Ratio Test.

79. $$\lim_{n\to\infty}\left|\frac{a_{n+1}}{a_n}\right| = \lim_{n\to\infty}\left|\frac{(\sin n+1)/(\sqrt{n})a_n}{a_n}\right|$$
$$= \lim_{n\to\infty}\frac{\sin n+1}{\sqrt{n}} = 0 < 1$$

The series converges by the Ratio Test.

80. $$\lim_{n\to\infty}\left|\frac{a_{n+1}}{a_n}\right| = \lim_{n\to\infty}\left|\frac{(\cos n+1)/(n)a_n}{a_n}\right|$$
$$= \lim_{n\to\infty}\frac{\cos n+1}{n} = 0 < 1$$

The series converges by the Ratio Test.

81. $$\lim_{n\to\infty}\left|\frac{a_{n+1}}{a_n}\right| = \lim_{n\to\infty}\left|\frac{(1+(1)/(n))a_n}{a_n}\right| = \lim_{n\to\infty}\left(1+\frac{1}{n}\right) = 1$$

The Ratio Test is inconclusive.

But, $\lim_{n\to\infty} a_n \neq 0$, so the series diverges.

82. The series diverges because $\lim_{n\to\infty} a_n \neq 0$.

$$a_1 = \tfrac{1}{4}$$
$$a_2 = \left(\tfrac{1}{4}\right)^{1/2} = \tfrac{1}{2}$$
$$a_3 = \left(\tfrac{1}{2}\right)^{1/3} \approx 0.7937$$

In general, $a_{n+1} > a_n > 0$.

83. $$\lim_{n\to\infty}\left|\frac{a_{n+1}}{a_n}\right| = \lim_{n\to\infty}\left|\frac{\dfrac{1\cdot 2\cdots n(n+1)}{1\cdot 3\cdots(2n-1)(2n+1)}}{\dfrac{1\cdot 2\cdots n}{1\cdot 3\cdots(2n-1)}}\right|$$
$$= \lim_{n\to\infty}\frac{n+1}{2n+1} = \frac{1}{2} < 1$$

The series converges by the Ratio Test.

84. $\displaystyle\sum_{n=0}^{\infty}\frac{n+1}{3^n}$

$$\lim_{n\to\infty}\sqrt[n]{|a_n|} = \lim_{n\to\infty}\sqrt[n]{\frac{n+1}{3^n}} = \lim_{n\to\infty}\frac{\sqrt[n]{n+1}}{3}$$

Let $\quad y = \lim_{n\to\infty}\sqrt[n]{n+1}$

$$\ln y = \lim_{n\to\infty}\left(\ln\sqrt[n]{n+1}\right)$$
$$= \lim_{n\to\infty}\frac{1}{n}\ln(n+1)$$
$$= \lim_{n\to\infty}\frac{\ln(n+1)}{n} = \frac{1}{n+1} = 0.$$

Because $\ln y = 0$, $y = e^0 = 1$, so

$$\lim_{n\to\infty}\frac{\sqrt{n+1}}{3} = \frac{1}{3}.$$

Therefore, by the Root Test, the series converges.

85. $\displaystyle\sum_{n=3}^{\infty} \frac{1}{(\ln n)^n}$

$$\lim_{n\to\infty} \sqrt[n]{|a_n|} = \lim_{n\to\infty} \sqrt[n]{\frac{1}{(\ln n)^n}} = \lim_{n\to\infty} \frac{1}{\ln n} = 0$$

Therefore, by the Root Test, the series converges.

86. $$\lim_{n\to\infty}\left|\frac{a_{n+1}}{a_n}\right| = \lim_{n\to\infty}\left|\frac{\dfrac{1\cdot3\cdot5\cdots(2n-1)(2n+1)}{1\cdot2\cdot3\cdots(2n-1)(2n)(2n+1)}}{\dfrac{1\cdot3\cdot5\cdots(2n-1)}{1\cdot2\cdot3\cdots(2n-1)}}\right|$$

$$= \lim_{n\to\infty}\frac{2n+1}{(2n)(2n+1)} = 0 < 1$$

The series converges by the Ratio Test.

87. $\displaystyle\sum_{n=0}^{\infty} 2\left(\frac{x}{3}\right)^n$

$$\lim_{n\to\infty}\left|\frac{a_{n+1}}{a_n}\right| = \lim_{n\to\infty}\left|\frac{2(x/3)^{n+1}}{2(x/3)^n}\right| = \lim_{n\to\infty}\left|\frac{x}{3}\right| = \left|\frac{x}{3}\right|$$

For the series to converge, $\left|\dfrac{x}{3}\right| < 1 \Rightarrow -3 < x < 3$.

For $x = 3$, $\displaystyle\sum_{n=0}^{\infty} 2(1)^n$ diverges.

For $x = -3$, $\displaystyle\sum_{n=0}^{\infty} 2(-1)^n$ diverges.

88. $\displaystyle\sum_{n=0}^{\infty}\left(\frac{x-3}{5}\right)^n$, Geometric series

For the series to converge,

$$\left|\frac{x-3}{5}\right| < 1 \Rightarrow |x-3| < 5$$

$$\Rightarrow -2 < x < 8.$$

For $x = 8$, $\displaystyle\sum_{n=0}^{\infty} 1^n$ diverges.

For $x = -2$, $\displaystyle\sum_{n=0}^{\infty} (-1)^n$ diverges.

(Note: You could also use the Ratio Test.)

89. $\displaystyle\sum_{n=1}^{\infty} \frac{(-1)^n(x+1)^n}{n}$

$$\lim_{n\to\infty}\left|\frac{a_{n+1}}{a_n}\right| = \lim_{n\to\infty}\left|\frac{(x+1)^{n+1}/(n+1)}{x^n/n}\right|$$

$$= \lim_{n\to\infty}\left|\frac{n}{n+1}(x+1)\right| = |x+1|$$

For the series to converge,

$$|x+1| < 1 \Rightarrow -1 < x+1 < 1$$

$$\Rightarrow -2 < x < 0.$$

For $x = 0$, $\displaystyle\sum_{n=1}^{\infty}\frac{(-1)^n}{n}$ converges.

For $x = -2$, $\displaystyle\sum_{n=1}^{\infty}\frac{(-1)^n(-1)^n}{n} = \sum_{n=1}^{\infty}\frac{1}{n}$ diverges.

90. $\displaystyle\sum_{n=0}^{\infty} 3(x-4)^n$, Geometric series

For the series to converge,

$$|x-4| < 1 \Rightarrow -1 < x-4 < 1 \Rightarrow 3 < x < 5.$$

For $x = 1$, $\displaystyle\sum_{n=0}^{\infty} 3(-3)^n$ diverges.

For $x = -1$, $\displaystyle\sum_{n=0}^{\infty} 3(-5)^n$ diverges.

91. $\displaystyle\sum_{n=0}^{\infty} n!\left(\frac{x}{2}\right)^n$

$$\lim_{n\to\infty}\left|\frac{a_{n+1}}{a_n}\right| = \lim_{n\to\infty}\frac{(n+1)!\left|\dfrac{x}{2}\right|^{n+1}}{n!\left|\dfrac{x}{2}\right|^n}$$

$$= \lim_{n\to\infty}(n+1)\left|\frac{x}{2}\right| = \infty$$

The series converges only at $x = 0$.

92. $\displaystyle\sum_{n=0}^{\infty} \frac{(x+1)^n}{n!}$

$$\lim_{n\to\infty}\left|\frac{a_{n+1}}{a_n}\right| = \lim_{n\to\infty}\frac{|x+1|^{n+1}}{(n+1)!}\bigg/\frac{|x+1|^n}{n!} = \lim_{n\to\infty}\frac{|x+1|}{n+1} = 0$$

The series converges for all x.

93. See Theorem 9.17, page 627.

94. See Theorem 9.18, page 630.

95. No. Let $a_n = \dfrac{1}{n + 10{,}000}$.

The series $\displaystyle\sum_{n=1}^{\infty} \frac{1}{n + 10{,}000}$ diverges.

96. (a) Converges (Ratio Test)
(b) Inconclusive (See Ratio Test)
(c) Diverges (Ratio Test)
(d) Diverges (Root Test)
(e) Inconclusive (See Root Test)
(f) Diverges (Root Test, $e > 1$)

97. The series converges absolutely. See Theorem 9.17.

98. For $0 < a_n < 1$, $a_n < \sqrt{a_n}$.

Thus, the series $\displaystyle\sum_{n=1}^{\infty} a_n$ is the lower series, indicated by the round dots.

99. Assume that

$\displaystyle\lim_{n\to\infty} \left|a_{n+1}/a_n\right| = L > 1$ or that $\displaystyle\lim_{n\to\infty} \left|a_{n+1}/a_n\right| = \infty$.

Then there exists $N > 0$ such that $\left|a_{n+1}/a_n\right| > 1$ for all $n > N$. Therefore,

$\left|a_{n+1}\right| > \left|a_n\right|,\ n > N \Rightarrow \displaystyle\lim_{n\to\infty} a_n \neq 0 \Rightarrow \sum a_n$ diverges.

100. First, let

$$\lim_{n\to\infty} \sqrt[n]{\left|a_n\right|} = r < 1$$

and choose R such that $0 \le r < R < 1$. There must exist some $N > 0$ such that $\sqrt[n]{\left|a_n\right|} < R$ for all $n > N$. So, for $n > N \left|a_n\right| < R^n$ and because the geometric series

$$\sum_{n=0}^{\infty} R^n$$

converges, you can apply the Comparison Test to conclude that

$$\sum_{n=1}^{\infty} \left|a_n\right|$$

converges which in turn implies that $\displaystyle\sum_{n=1}^{\infty} a_n$ converges.

Second, let

$$\lim_{n\to\infty} \sqrt[n]{\left|a_n\right|} = r > R > 1.$$

Then there must exist some $M > 0$ such that $\sqrt[n]{\left|a_n\right|} > R$ for infinitely many $n > M$. So, for infinitely many $n > M$, you have $\left|a_n\right| > R^n > 1$ which implies that $\displaystyle\lim_{n\to\infty} a_n \neq 0$ which in turn implies that

$$\sum_{n=1}^{\infty} a_n \text{ diverges.}$$

101. $\displaystyle\sum_{n=1}^{\infty} \frac{1}{n^{3/2}}$

$$\lim_{n\to\infty} \left|\frac{a_{n+1}}{a_n}\right| = \lim_{n\to\infty} \left|\frac{1}{(n+1)^{3/2}} \cdot \frac{n^{3/2}}{1}\right| = \lim_{n\to\infty} \left(\frac{n}{n+1}\right)^{3/2} = 1$$

102. $\displaystyle\sum_{n=1}^{\infty} \frac{1}{n^{1/2}}$

$$\lim_{n\to\infty} \left|\frac{a_{n+1}}{a_n}\right| = \lim_{n\to\infty} \left|\frac{1}{(n+1)^{1/2}} \cdot \frac{n^{1/2}}{1}\right| = \lim_{n\to\infty} \left(\frac{n}{n+1}\right)^{1/2} = 1$$

103. $\displaystyle\sum_{n=1}^{\infty} \frac{1}{n^4}$

$$\lim_{n\to\infty} \left|\frac{a_{n+1}}{a_n}\right| = \lim_{n\to\infty} \left|\frac{1}{(n+1)^4} \cdot \frac{n^4}{1}\right| = \lim_{n\to\infty} \left(\frac{n}{n+1}\right)^4 = 1$$

104. $\displaystyle\sum_{n=1}^{\infty} \frac{1}{n^p}$

$$\lim_{n\to\infty} \left|\frac{a_{n+1}}{a_n}\right| = \lim_{n\to\infty} \left|\frac{1}{(n+1)^p} \cdot \frac{n^p}{1}\right| = \lim_{n\to\infty} \left(\frac{n}{n+1}\right)^p = 1$$

105. $\sum_{n=1}^{\infty} \frac{1}{n^p}$, p-series

$$\lim_{n\to\infty} \sqrt[n]{|a_n|} = \lim_{n\to\infty} \sqrt[n]{\frac{1}{n^p}} = \lim_{n\to\infty} \frac{1}{n^{p/n}} = 1$$

So, the Root Test is inconclusive.

Note: $\lim_{n\to\infty} n^{p/n} = 1$ because if $y = n^{p/n}$, then

$\ln y = \frac{p}{n} \ln n$ and $\frac{p}{n} \ln n \to 0$ as $n \to \infty$.

So $y \to 1$ as $n \to \infty$.

106. Ratio Test:

$$\lim_{n\to\infty} \left|\frac{a_{n+1}}{a_n}\right| = \lim_{n\to\infty} \frac{n(\ln n)^p}{(n+1)(\ln(n+1))^p} = 1, \text{ inconclusive.}$$

Root Test:

$$\lim_{n\to\infty} \sqrt[n]{|a_n|} = \lim_{n\to\infty} \sqrt[n]{\frac{1}{n(\ln n)^p}} = \lim_{n\to\infty} \frac{1}{n^{1/n}(\ln n)^{p/n}}$$

$\lim_{n\to\infty} n^{1/n} = 1$. Furthermore, let $y = (\ln n)^{p/n} \Rightarrow$

$\ln y = \frac{p}{n} \ln(\ln n)$.

$$\lim_{n\to\infty} \ln y = \lim_{n\to\infty} \frac{p \ln(\ln n)}{n} = \lim_{n\to\infty} \frac{p}{\ln(n)(1/n)} = 0 \Rightarrow \lim_{n\to\infty} (\ln n)^{p/n} = 1.$$

So, $\lim_{n\to\infty} \frac{1}{n^{1/n}(\ln n)^{p/n}} = 1$, inconclusive.

107. $\sum_{n=1}^{\infty} \frac{(n!)^2}{(xn)!}$, x positive integer

(a) $x = 1$: $\sum \frac{(n!)^2}{n!} = \sum n!$, diverges

(b) $x = 2$: $\sum \frac{(n!)^2}{(2n)!}$ converges by the Ratio Test:

$$\lim_{n\to\infty} \frac{[(n+1)!]^2}{(2n+2)!} \Big/ \frac{(n!)^2}{(2n)!} = \lim_{n\to\infty} \frac{(n+1)^2}{(2n+2)(2n+1)} = \frac{1}{4} < 1$$

(c) $x = 3$: $\sum \frac{(n!)^2}{(3n)!}$ converges by the Ratio Test:

$$\lim_{n\to\infty} \frac{[(n+1)!]^2}{(3n+3)!} \Big/ \frac{(n!)^2}{(3n)!} = \lim_{n\to\infty} \frac{(n+1)^2}{(3n+3)(3n+2)(3n+1)} = 0 < 1$$

(d) Use the Ratio Test:

$$\lim_{n\to\infty} \frac{[(n+1)!]^2}{[x(n+1)]!} \Big/ \frac{(n!)^2}{(xn)!} = \lim_{n\to\infty} (n+1)^2 \frac{(xn)!}{(xn+x)!}$$

The cases $x = 1, 2, 3$ were solved above. For $x > 3$, the limit is 0. So, the series converges for all integers $x \geq 2$.

108. For $n = 1, 2, 3, \ldots, -|a_n| \le a_n \le |a_n| \Rightarrow -\sum_{n=1}^{k}|a_n| \le \sum_{n=1}^{k} a_n \le \sum_{n=1}^{k}|a_n|$.

Taking limits as $k \to \infty$, $-\sum_{n=1}^{\infty}|a_n| \le \sum_{n=1}^{\infty} a_n \le \sum_{n=1}^{\infty}|a_n| \Rightarrow \left|\sum_{n=1}^{\infty} a_n\right| \le \sum_{n=1}^{\infty}|a_n|$.

109. First prove Abel's Summation Theorem:

If the partial sums of $\sum a_n$ are bounded and if $\{b_n\}$ decreases to zero, then $\sum a_n b_n$ converges.

Let $S_k = \sum_{i=1}^{k} a_i$. Let M be a bound for $\{|S_k|\}$.

$$\begin{aligned} a_1b_1 + a_2b_2 + \cdots + a_nb_n &= S_1b_1 + (S_2 - S_1)b_2 + \cdots + (S_n - S_{n-1})b_n \\ &= S_1(b_1 - b_2) + S_2(b_2 - b_3) + \cdots + S_{n-1}(b_{n-1} - b_n) + S_nb_n \\ &= \sum_{i=1}^{n-1} S_i(b_i - b_{i+1}) + S_nb_n \end{aligned}$$

The series $\sum_{i=1}^{\infty} S_i(b_i - b_{i+1})$ is absolutely convergent because $|S_i(b_i - b_{i+1})| \le M(b_i - b_{i+1})$ and $\sum_{i=1}^{\infty}(b_i - b_{i+1})$ converges to b_1.

Also, $\lim_{n\to\infty} S_nb_n = 0$ because $\{S_n\}$ bounded and $b_n \to 0$. Thus, $\sum_{n=1}^{\infty} a_nb_n = \lim_{n\to\infty}\sum_{i=1}^{n} a_ib_i$ converges.

Now let $b_n = \dfrac{1}{n}$ to finish the problem.

110. Using the Ratio Test,

$$\lim_{n\to\infty}\left|\frac{a_{n+1}}{a_n}\right| = \lim_{n\to\infty}\left[\frac{n!}{(n+1)^n}\left(\frac{19}{7}\right)^n \Big/ \frac{(n-1)!}{n^{n-1}}\left(\frac{19}{7}\right)^{n-1}\right] = \lim_{n\to\infty}\left[\frac{n \cdot n^{n-1}}{(n+1)^n}\left(\frac{19}{7}\right)\right] = \lim_{n\to\infty}\left[\frac{1}{(1 + (1/n))^n}\left(\frac{19}{7}\right)\right] = \frac{19}{7}\cdot\frac{1}{e} < 1$$

So, the series converges.

Section 9.7 Taylor Polynomials and Approximations

1. $y = -\frac{1}{2}x^2 + 1$

Parabola

Matches (d)

2. $y = \frac{1}{8}x^4 - \frac{1}{2}x^2 + 1$

y-axis symmetry

Three relative extrema

Matches (c)

3. $y = e^{-1/2}\left[(x + 1) + 1\right]$

Linear

Matches (a)

4. $y = e^{-1/2}\left[\frac{1}{3}(x - 1)^3 - (x - 1) + 1\right]$

Cubic

Matches (b)

5. $f(x) = \dfrac{\sqrt{x}}{4}$, $C = 4$, $f(4) = \dfrac{1}{2}$

$$\begin{aligned} f'(x) &= \frac{1}{8\sqrt{x}}, \quad f'(4) = \frac{1}{16} \\ P_1(x) &= f(4) + f'(4)(x - 4) \\ &= \frac{1}{2} + \frac{1}{16}(x - 4) \\ &= \frac{1}{16}x + \frac{1}{4} \end{aligned}$$

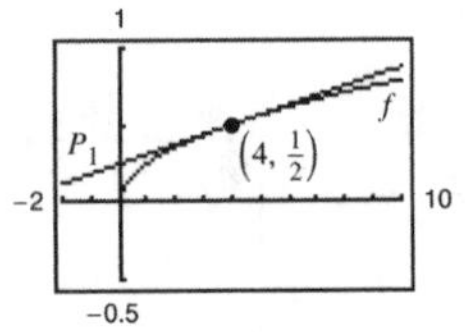

P_1 is the first-degree Taylor polynomial for f at 4.

6. $f(x) = \dfrac{6}{\sqrt[3]{x}} = 6x^{-1/3}$ $\qquad f(8) = 3$

$f'(x) = -2x^{-4/3}$ $\qquad f'(8) = -\dfrac{1}{8}$

$P_1(x) = f(8) + f'(8)(x - 8)$

$= 3 - \dfrac{1}{8}(x - 8) = -\dfrac{1}{8}x + 4$

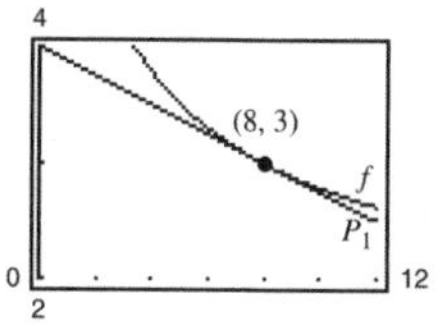

P_1 is the first degree Taylor polynomial for f at 8.

7. $f(x) = \sec x$ $\qquad f\left(\dfrac{\pi}{4}\right) = \sqrt{2}$

$f'(x) = \sec x \tan x$ $\qquad f'\left(\dfrac{\pi}{4}\right) = \sqrt{2}$

$P_1(x) = f\left(\dfrac{\pi}{4}\right) + f'\left(\dfrac{\pi}{4}\right)\left(x - \dfrac{\pi}{4}\right)$

$P_1(x) = \sqrt{2} + \sqrt{2}\left(x - \dfrac{\pi}{4}\right)$

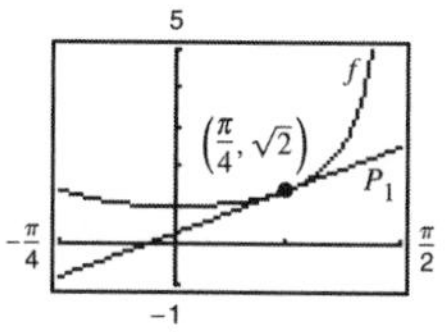

P_1 is called the first degree Taylor polynomial for f at $\dfrac{\pi}{4}$.

8. $f(x) = \tan x$ $\qquad f\left(\dfrac{\pi}{4}\right) = 1$

$f'(x) = \sec^2 x$ $\qquad f'\left(\dfrac{\pi}{4}\right) = 2$

$P_1 = f\left(\dfrac{\pi}{4}\right) + f'\left(\dfrac{\pi}{4}\right)\left(x - \dfrac{\pi}{4}\right) = 1 + 2\left(x - \dfrac{\pi}{4}\right)$

$P_1(x) = 2x + 1 - \dfrac{\pi}{2}$

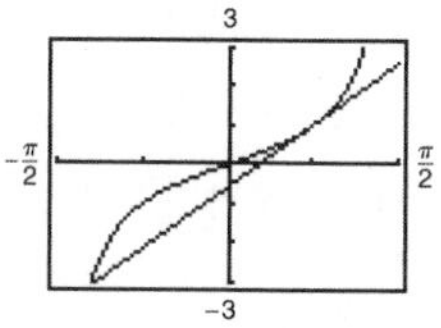

P_1 is called the first degree Taylor polynomial for f at $\dfrac{\pi}{4}$.

9. $f(x) = \dfrac{4}{\sqrt{x}} = 4x^{-1/2}$ $\qquad f(1) = 4$

$f'(x) = -2x^{-3/2}$ $\qquad f'(1) = -2$

$f''(x) = 3x^{-5/2}$ $\qquad f''(1) = 3$

$P_2 = f(1) + f'(1)(x - 1) + \dfrac{f''(1)}{2}(x - 1)^2$

$= 4 - 2(x - 1) + \dfrac{3}{2}(x - 1)^2$

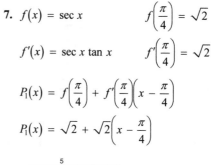

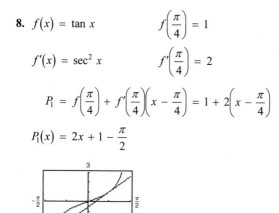

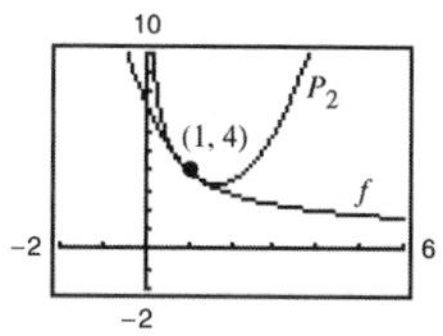

x	0	0.8	0.9	1.0	1.1	1.2	2
$f(x)$	Error	4.4721	4.2164	4.0	3.8139	3.6515	2.8284
$P_2(x)$	7.5	4.46	4.215	4.0	3.815	3.66	3.5

10. $f(x) = \sec x \qquad f\left(\dfrac{\pi}{4}\right) = \sqrt{2}$

$f'(x) = \sec x \tan x \qquad f'\left(\dfrac{\pi}{4}\right) = \sqrt{2}$

$f''(x) = \sec^3 x + \sec x \tan^2 x \qquad f''\left(\dfrac{\pi}{4}\right) = 3\sqrt{2}$

$$P_2(x) = f\left(\frac{\pi}{4}\right) + f'\left(\frac{\pi}{4}\right)\left(x - \frac{\pi}{4}\right) + \frac{f''(\pi/4)}{2}\left(x - \frac{\pi}{4}\right)^2$$

$$P_2(x) = \sqrt{2} + \sqrt{2}\left(x - \frac{\pi}{4}\right) + \frac{3}{2}\sqrt{2}\left(x - \frac{\pi}{4}\right)^2$$

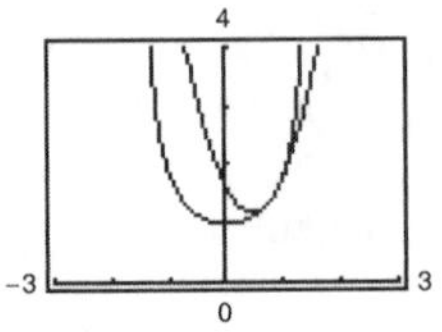

x	−2.15	0.585	0.685	$\pi/4$	0.885	0.985	1.785
$f(x)$	−1.8270	1.1995	1.2913	1.4142	1.5791	1.8088	−4.7043
$P_2(x)$	15.5414	1.2160	1.2936	1.4142	1.5761	1.7810	4.9475

11. $f(x) = \cos x$

$P_2(x) = 1 - \frac{1}{2}x^2$

$P_4(x) = 1 - \frac{1}{2}x^2 + \frac{1}{24}x^4$

$P_6(x) = 1 - \frac{1}{2}x^2 + \frac{1}{24}x^4 - \frac{1}{720}x^6$

(a)

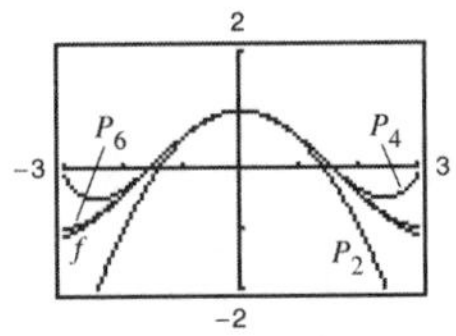

(b) $f'(x) = -\sin x \qquad P_2'(x) = -x$

$f''(x) = -\cos x \qquad P_2''(x) = -1$

$f''(0) = P_2''(0) = -1$

$f'''(x) = \sin x \qquad P_4'''(x) = x$

$f^{(4)}(x) = \cos x \qquad P_4^{(4)}(x) = 1$

$f^{(4)}(0) = 1 = P_4^{(4)}(0)$

$f^{(5)}(x) = -\sin x \qquad P_6^{(5)}(x) = -x$

$f^{(6)}(x) = -\cos x \qquad P^{(6)}(x) = -1$

$f^{(6)}(0) = -1 = P_6^{(6)}(0)$

(c) In general, $f^{(n)}(0) = P_n^{(n)}(0)$ for all n.

12. $f(x) = x^2e^x,\ f(0) = 0$

(a) $f'(x) = (x^2 + 2x)e^x \qquad f'(0) = 0$

$f''(x) = (x^2 + 4x + 2)e^x \qquad f''(0) = 2$

$f'''(x) = (x^2 + 6x + 6)e^x \qquad f'''(0) = 6$

$f^{(4)}(x) = (x^2 + 8x + 12)e^x \qquad f^{(4)}(0) = 12$

$$P_2(x) = \frac{2x^2}{2!} = x^2$$

$$P_3(x) = x^2 + \frac{6x^3}{3!} = x^2 + x^3$$

$$P_4(x) = x^2 + x^3 + \frac{12x^4}{4!} = x^2 + x^3 + \frac{x^4}{2}$$

(b)

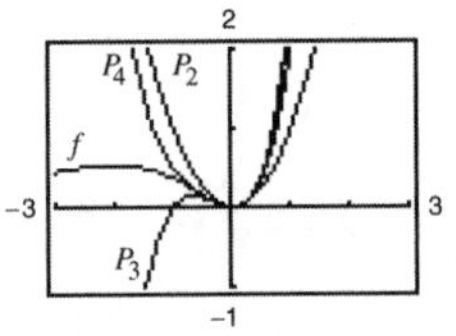

(c) $f''(0) = 2 = P_2''(0)$

$f'''(0) = 6 = P_3'''(0)$

$f^{(4)}(0) = 12 = P_4^{(4)}(0)$

(d) $f^{(n)}(0) = P_n^{(n)}(0)$

13. $f(x) = e^{4x}$ $\quad f(0) = 1$

$f'(x) = 4e^{4x}$ $\quad f'(0) = 4$

$f''(x) = 16e^{4x}$ $\quad f''(0) = 16$

$f'''(x) = 64e^{4x}$ $\quad f'''(0) = 64$

$f^{(4)}(x) = 256e^{4x}$ $\quad f^{(4)}(0) = 256$

$$P_4(x) = f(0) + f'(0)x + \frac{f''(0)}{2!}x^2 + \frac{f'''(0)}{3!}x^3 + \frac{f^{(4)}(0)}{4!}x^4$$
$$= 1 + 4x + 8x^2 + \frac{32}{3}x^3 + \frac{32}{3}x^4$$

14. $f(x) = e^{-x}$ $\quad f(0) = 1$

$f'(x) = -e^{-x}$ $\quad f'(0) = -1$

$f''(x) = e^{-x}$ $\quad f''(0) = 1$

$f'''(x) = -e^{-x}$ $\quad f'''(0) = -1$

$f^{(4)}(x) = e^{-x}$ $\quad f^{(4)}(0) = 1$

$f^{(5)}(x) = -e^{-x}$ $\quad f^{(5)}(0) = -1$

$$P_5(x) = f(0) + f'(0)x + \frac{f'(0)}{2!}x^2 + \frac{f'''(0)}{3!}x^3 + \frac{f^{(4)}(0)}{4!}x^4$$
$$+ \frac{f^{(5)}(0)}{5!}x^5 = 1 - x + \frac{x^2}{2} - \frac{x^3}{6} + \frac{x^4}{24} - \frac{x^5}{120}$$

15. $f(x) = e^{-x/2}$ $\quad f(0) = 1$

$f'(x) = -\frac{1}{2}e^{-x/2}$ $\quad f'(0) = -\frac{1}{2}$

$f''(x) = \frac{1}{4}e^{-x/2}$ $\quad f''(0) = \frac{1}{4}$

$f'''(x) = -\frac{1}{8}e^{-x/2}$ $\quad f'''(0) = -\frac{1}{8}$

$f^{(4)}(x) = \frac{1}{16}e^{-x/2}$ $\quad f^{(4)}(0) = \frac{1}{16}$

$$P_4(x) = f(0) + f'(0)x + \frac{f''(0)}{2!}x^2 + \frac{f'''(0)}{3!}x^3 + \frac{f^{(4)}(0)}{4!}x^4$$
$$= 1 - \frac{1}{2}x + \frac{1}{8}x^2 - \frac{1}{48}x^3 + \frac{1}{384}x^4$$

16.
$$f(x) = e^{x/3} \qquad f(0) = 1$$
$$f'(x) = \frac{1}{3}e^{x/3} \qquad f'(0) = \frac{1}{3}$$
$$f''(x) = \frac{1}{9}e^{x/3} \qquad f''(0) = \frac{1}{9}$$
$$f'''(x) = \frac{1}{27}e^{x/3} \qquad f'''(0) = \frac{1}{27}$$
$$f^{(4)}(x) = \frac{1}{81}e^{x/3} \qquad f^{(4)}(x) = \frac{1}{81}$$

$$P_4(x) = f(0) + f'(0)x + \frac{f''(0)}{2!}x^2 + \frac{f'''(0)}{3!}x^3 + \frac{f^{(4)}(0)}{4!}x^4$$
$$= 1 + \frac{1}{3}x + \frac{1/9}{2!}x^2 + \frac{1/27}{3!}x^3 + \frac{1/81}{4!}x^4$$
$$= 1 + \frac{1}{3}x + \frac{1}{18}x^2 + \frac{1}{162}x^3 + \frac{1}{1944}x^4$$

17.
$$f(x) = \sin x \qquad f(0) = 0$$
$$f'(x) = \cos x \qquad f'(0) = 1$$
$$f''(x) = -\sin x \qquad f''(0) = 0$$
$$f'''(x) = -\cos x \qquad f'''(0) = -1$$
$$f^{(4)}(x) = \sin x \qquad f^{(4)}(0) = 0$$
$$f^{(5)}(x) = \cos x \qquad f^{(5)}(0) = 1$$

$$P_5(x) = 0 + (1)x + \frac{0}{2!}x^2 + \frac{-1}{3!}x^3 + \frac{0}{4!}x^4 + \frac{1}{5!}x^5$$
$$= x - \frac{1}{6}x^3 + \frac{1}{120}x^5$$

18.
$$f(x) = \cos \pi x \qquad f(0) = 1$$
$$f'(x) = -\pi \sin \pi x \qquad f'(0) = 0$$
$$f''(x) = -\pi^2 \cos \pi x \qquad f''(0) = -\pi^2$$
$$f'''(x) = \pi^3 \sin \pi x \qquad f'''(0) = 0$$
$$f^{(4)}(x) = \pi^4 \cos \pi x \qquad f^{(4)}(0) = \pi^4$$

$$P_4(x) = f(0) + f'(0)x + \frac{f''(0)}{2!}x^2 + \frac{f'''(0)}{3!}x^3 + \frac{f^{(4)}(0)}{4!}x^4$$
$$= 1 - \frac{\pi^2}{2}x^2 + \frac{\pi^4}{24}x^4$$

19.
$$f(x) = xe^x \qquad f(0) = 0$$
$$f'(x) = xe^x + e^x \qquad f'(0) = 1$$
$$f''(x) = xe^x + 2e^x \qquad f''(0) = 2$$
$$f'''(x) = xe^x + 3e^x \qquad f'''(0) = 3$$
$$f^{(4)}(x) = xe^x + 4e^x \qquad f^{(4)}(0) = 4$$

$$P_4(x) = 0 + x + \frac{2}{2!}x^2 + \frac{3}{3!}x^3 + \frac{4}{4!}x^4$$
$$= x + x^2 + \frac{1}{2}x^3 + \frac{1}{6}x^4$$

20. $f(x) = x^2e^{-x}$ $\quad f(0) = 0$

$f'(x) = 2xe^{-x} - x^2e^{-x}$ $\quad f'(0) = 0$

$f''(x) = 2e^{-x} - 4xe^{-x} + x^2e^{-x}$ $\quad f''(0) = 2$

$f'''(x) = -6e^{-x} + 6xe^{-x} - x^2e^{-x}$ $\quad f'''(0) = -6$

$f^{(4)}(x) = 12e^{-x} - 8xe^{-x} + x^2e^{-x}$ $\quad f^{(4)}(0) = 12$

$$P_4(x) = 0 + 0x + \frac{2}{2!}x^2 + \frac{-6}{3!}x^3 + \frac{12}{4!}x^4$$
$$= x^2 - x^3 + \frac{1}{2}x^4$$

21. $f(x) = \dfrac{1}{x+1} = (x+1)^{-1}$ $\quad f(0) = 1$

$f'(x) = -(x+1)^{-2}$ $\quad f'(0) = -1$

$f''(x) = 2(x+1)^{-3}$ $\quad f''(0) = 2$

$f'''(x) = -6(x+1)^{-4}$ $\quad f'''(0) = -6$

$f^{(4)}(x) = 24(x+1)^{-5}$ $\quad f^{(4)}(0) = 24$

$f^{(5)}(x) = -120(x+1)^{-6}$ $\quad f^{(5)}(0) = -120$

$$P_5(x) = 1 - x + \frac{2x^2}{2!} - \frac{6x^3}{3!} + \frac{24x^4}{4!} - \frac{120x^5}{5!}$$
$$= 1 - x + x^2 - x^3 + x^4 - x^5$$

22. $f(x) = \dfrac{x}{x+1} = \dfrac{x+1-1}{x+1}$ $\quad f(0) = 0$

$= 1 - (x+1)^{-1}$

$f'(x) = (x+1)^{-2}$ $\quad f'(0) = 1$

$f''(x) = -2(x+1)^{-3}$ $\quad f''(0) = -2$

$f'''(x) = 6(x+1)^{-4}$ $\quad f'''(0) = 6$

$f^{(4)}(x) = -24(x+1)^{-5}$ $\quad f^{(4)}(0) = -24$

$$P_4(x) = 0 + 1(x) - \frac{2}{2}x^2 + \frac{6}{6}x^3 - \frac{24}{24}x^4$$
$$= x - x^2 + x^3 - x^4$$

23. $f(x) = \sec x$ $\quad f(0) = 1$

$f'(x) = \sec x \tan x$ $\quad f'(0) = 0$

$f''(x) = \sec^3 x + \sec x \tan^2 x$ $\quad f''(0) = 1$

$$P_2(x) = 1 + 0x + \frac{1}{2!}x^2 = 1 + \frac{1}{2}x^2$$

24. $f(x) = \tan x$ $\quad f(0) = 0$

$f'(x) = \sec^2 x$ $\quad f'(0) = 1$

$f''(x) = 2\sec^2 x \tan x$ $\quad f''(0) = 0$

$f'''(x) = 4\sec^2 x \tan^2 x + 2\sec^4 x$ $\quad f'''(0) = 2$

$$P_3(x) = 0 + 1(x) + 0 + \tfrac{2}{6}x^3 = x + \tfrac{1}{3}x^3$$

25. $f(x) = \dfrac{2}{x} = 2x^{-1}$ $\quad f(1) = 2$

$f'(x) = -2x^{-2}$ $\quad f'(1) = -2$

$f''(x) = 4x^{-3}$ $\quad f''(1) = 4$

$f'''(x) = -12x^{-4}$ $\quad f'''(1) = -12$

$$P_3(x) = 2 - 2(x-1) + \frac{4}{2!}(x-1)^2 - \frac{12}{3!}(x-1)^3$$
$$= 2 - 2(x-1) + 2(x-1)^2 - 2(x-1)^3$$

26. $f(x) = \dfrac{1}{x^2} = x^{-2}$ $\quad f(2) = 1/4$

$f'(x) = -2x^{-3}$ $\quad f'(2) = -1/4$

$f''(x) = 6x^{-4}$ $\quad f''(2) = 3/8$

$f'''(x) = -24x^{-5}$ $\quad f'''(2) = -3/4$

$f^{(4)}(x) = 120x^{-6}$ $\quad f^{(4)}(2) = 15/8$

$$P_4(x) = \frac{1}{4} - \frac{1}{4}(x-2) + \frac{3/8}{2!}(x-2)^2 - \frac{3/4}{3!}(x-2)^3 + \frac{15/8}{4!}(x-2)^4$$
$$= \frac{1}{4} - \frac{1}{4}(x-2) + \frac{3}{16}(x-2)^2 - \frac{1}{8}(x-2)^3 + \frac{5}{64}(x-2)^4$$

27. $f(x) = \sqrt{x} = x^{1/2}$ $\quad f(4) = 2$

$f'(x) = \frac{1}{2}x^{-1/2}$ $\quad f'(4) = \frac{1}{4}$

$f''(x) = -\frac{1}{4}x^{-3/2}$ $\quad f''(4) = -\frac{1}{32}$

$f'''(x) = \frac{3}{8}x^{-5/2}$ $\quad f'''(4) = \frac{3}{256}$

$$P_3(x) = 2 + \frac{1}{4}(x-4) - \frac{1/32}{2!}(x-4)^2 + \frac{3/256}{3!}(x-4)^3$$
$$= 2 + \frac{1}{4}(x-4) - \frac{1}{64}(x-4)^2 + \frac{1}{512}(x-4)^3$$

28. $f(x) = x^{1/3}$ $\quad f(8) = 2$

$f'(x) = \frac{1}{3}x^{-2/3}$ $\quad f'(8) = \frac{1}{12}$

$f''(x) = -\frac{2}{9}x^{-5/3}$ $\quad f''(8) = -\frac{1}{144}$

$f'''(x) = \frac{10}{27}x^{-8/3}$ $\quad f'''(8) = \frac{10}{27} \cdot \frac{1}{2^8} = \frac{5}{3456}$

$$P_3(x) = 2 + \frac{1}{12}(x-8) - \frac{1}{288}(x-8)^2 + \frac{5}{20{,}736}(x-8)^3$$

29. $f(x) = \ln x$ $\quad f(2) = \ln 2$

$f'(x) = \frac{1}{x} = x^{-1}$ $\quad f'(2) = 1/2$

$f''(x) = -x^{-2}$ $\quad f''(2) = -1/4$

$f'''(x) = 2x^{-3}$ $\quad f'''(2) = 1/4$

$f^{(4)}(x) = -6x^{-4}$ $\quad f^{(4)}(2) = -3/8$

$$P_4(x) = \ln 2 + \frac{1}{2}(x-2) - \frac{1/4}{2!}(x-2)^2 + \frac{1/4}{3!}(x-2)^3 - \frac{3/8}{4!}(x-2)^4$$
$$= \ln 2 + \frac{1}{2}(x-2) - \frac{1}{8}(x-2)^2 + \frac{1}{24}(x-2)^3 - \frac{1}{64}(x-2)^4$$

30. $f(x) = x^2 \cos x$ $\quad f(\pi) = -\pi^2$

$f'(x) = \cos x - x^2 \sin x$ $\quad f'(\pi) = -2\pi$

$f''(x) = 2\cos x - 4x \sin x - x^2 \cos x$ $\quad f''(\pi) = -2 + \pi^2$

$$P_2(x) = -\pi^2 - 2\pi(x-\pi) + \frac{(\pi^2 - 2)}{2}(x-\pi)^2$$

31. (a) $P_3(x) = \pi x + \frac{\pi^3}{3}x^3$

(b) $Q_3(x) = 1 + 2\pi\left(x - \frac{1}{4}\right) + 2\pi^2\left(x - \frac{1}{4}\right)^2 + \frac{8}{3}\pi^3\left(x - \frac{1}{4}\right)^3$

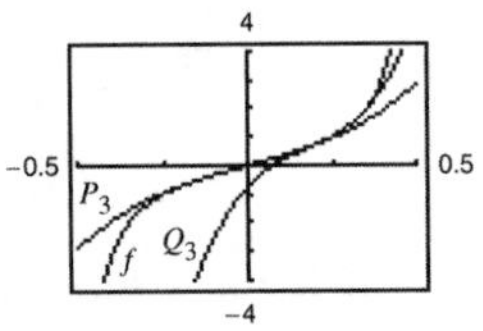

32. (a) $P_4(x) = 1 + 0x + \dfrac{-2}{2!}x^2 + \dfrac{0}{3!}x^3 + \dfrac{24}{4!}x^4 = 1 - x^2 + x^4$

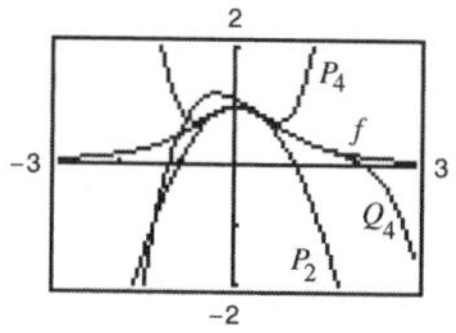

(b) $Q_4(x) = \dfrac{1}{2} + \left(-\dfrac{1}{2}\right)(x-1) + \dfrac{1/2}{2!}(x-1)^2 + \dfrac{0}{3!}(x-1)^3 + \dfrac{-3}{4!}(x-1)^4 = \dfrac{1}{2} - \dfrac{1}{2}(x-1) + \dfrac{1}{4}(x-1)^2 - \dfrac{1}{8}(x-1)^4$

33. $f(x) = \sin x$

$P_1(x) = x$

$P_3(x) = x - \frac{1}{6}x^3$

$P_5(x) = x - \frac{1}{6}x^3 + \frac{1}{120}x^5$

(a)

x	0.00	0.25	0.50	0.75	1.00
$\sin x$	0.0000	0.2474	0.4794	0.6816	0.8415
$P_1(x)$	0.0000	0.2500	0.5000	0.7500	1.0000
$P_3(x)$	0.0000	0.2474	0.4792	0.6797	0.8333
$P_5(x)$	0.0000	0.2474	0.4794	0.6817	0.8417

(b)

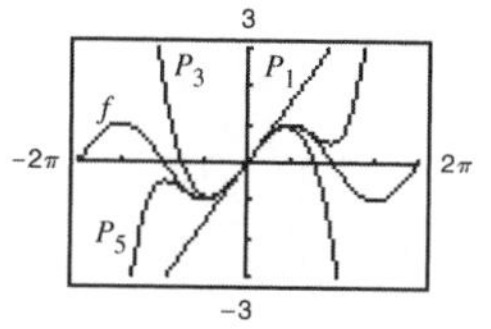

(c) As the distance increases, the accuracy decreases.

34. (a) $f(x) = e^x \qquad f(1) = e$

$f'(x) = e^x \qquad f'(1) = e$

$f''(x) = f'''(x) = f^{(4)}(x) = e^x$ and $f''(1) = f'''(1) = f^{(4)}(1) = e$

$P_1(x) = e + e(x-1)$

$P_2(x) = e + e(x-1) + \dfrac{e}{2}(x-1)^2$

$P_4(x) = e + e(x-1) + \dfrac{e}{2}(x-1)^2 + \dfrac{e}{6}(x-1)^3 + \dfrac{e}{24}(x-1)^4$

x	1.00	1.25	1.50	1.75	2.00
e^x	e	3.4903	4.4817	5.7546	7.3891
$P_1(x)$	e	3.3979	4.0774	4.7570	5.4366
$P_2(x)$	e	3.4828	4.4172	5.5215	6.7957
$P_4(x)$	e	3.4903	4.4809	5.7485	7.3620

(b)

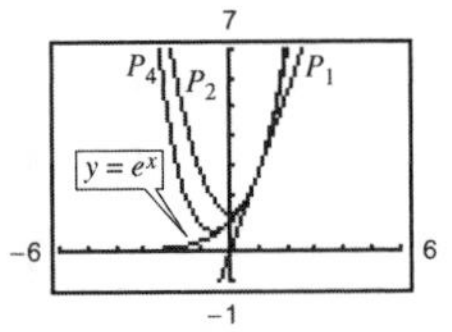

(c) As the degree increases, the accuracy increases. As the distance from x to 1 increases, the accuracy decreases.

35. $f(x) = \arcsin x$

(a) $P_3(x) = x + \dfrac{x^3}{6}$

(b)

x	–0.75	–0.50	–0.25	0	0.25	0.50	0.75
$f(x)$	–0.848	–0.524	–0.253	0	0.253	0.524	0.848
$P_3(x)$	–0.820	–0.521	–0.253	0	0.253	0.521	0.820

(c)

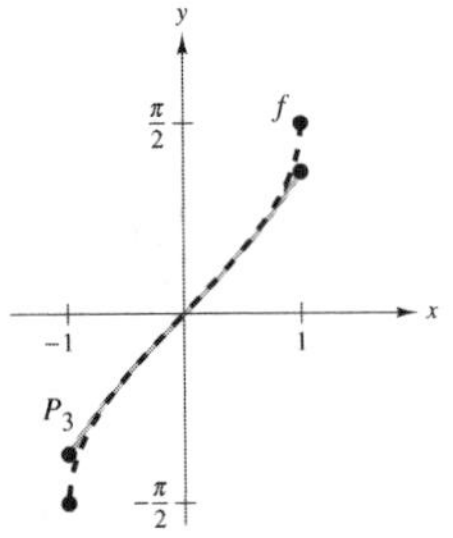

36. (a) $f(x) = \arctan x$

$P_3(x) = x - \dfrac{x^3}{3}$

(b)

x	–0.75	–0.50	–0.25	0	0.25	0.50	0.75
$f(x)$	–0.6435	–0.4636	–0.2450	0	0.2450	0.4636	0.6435
$P_3(x)$	–0.6094	–0.4583	–0.2448	0	0.2448	0.4583	0.6094

(c)

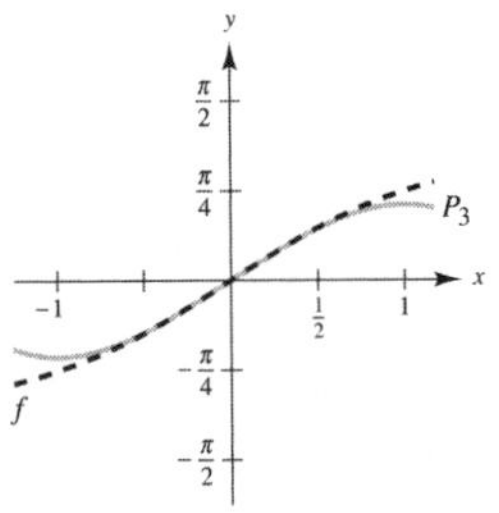

37. $f(x) = \cos x$

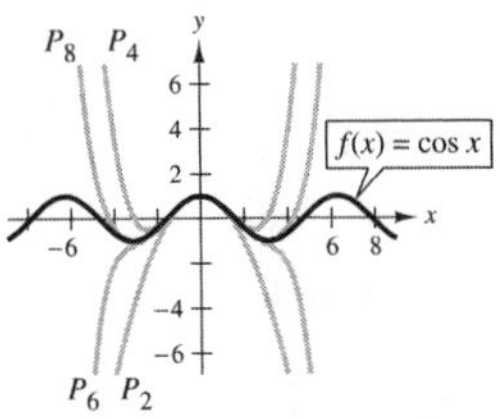

38. $f(x) = \arctan x$

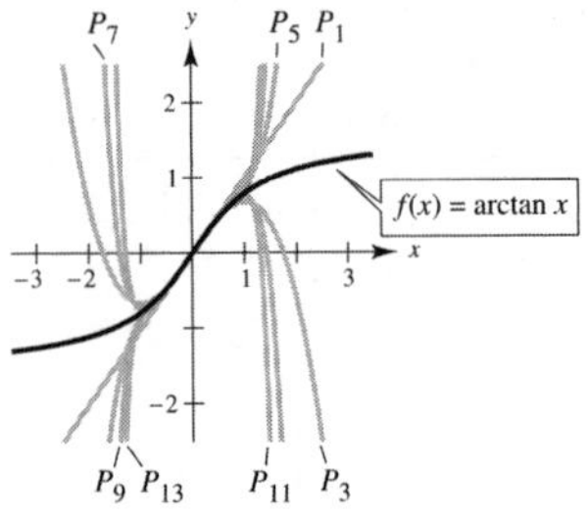

39. $f(x) = \ln(x^2 + 1)$

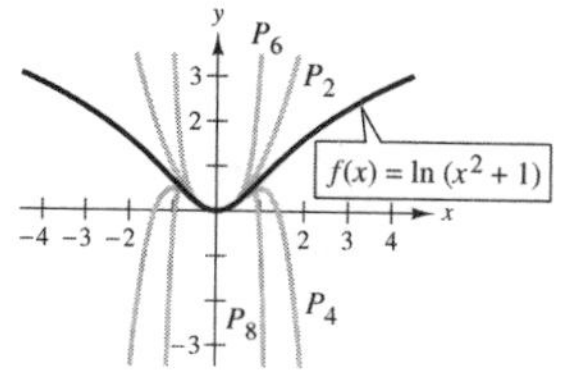

40. $f(x) = 4xe^{-x^2/4}$

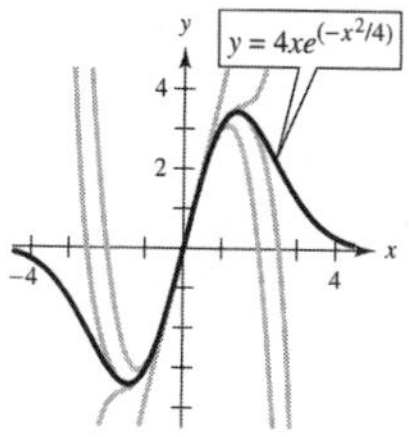

41. $f(x) = e^{3x} \approx 1 + 3x + \frac{9}{2}x^2 + \frac{9}{2}x^3 + \frac{27}{8}x^4$

$f\left(\frac{1}{2}\right) \approx 4.3984$

42. $f(x) = x^2e^{-x} \approx x^2 - x^3 + \frac{1}{2}x^4$

$f\left(\frac{1}{5}\right) \approx 0.0328$

43. $f(x) = \ln x \approx \ln(2) + \frac{1}{2}(x - 2) - \frac{1}{8}(x - 2)^2 + \frac{1}{24}(x - 2)^3 - \frac{1}{64}(x - 2)^4$

$f(2.1) \approx 0.7419$

44. $f(x) = x^2 \cos x \approx -\pi^2 - 2\pi(x - \pi) + \left(\frac{\pi^2 - 2}{2}\right)(x - \pi)^2$

$f\left(\frac{7\pi}{8}\right) \approx -6.7954$

45. $f(x) = \cos x; f^{(5)}(x) = -\sin x \Rightarrow$ Max on [0, 0.3] is 1.

$R_4(x) \le \frac{1}{5!}(0.3)^5 = 2.025 \times 10^{-5}$

Note: you could use $R_5(x)$: $f^{(6)}(x) = -\cos x$, max on [0, 0.3] is 1.

$R_5(x) \le \frac{1}{6!}(0.3)^6 = 1.0125 \times 10^{-6}$

Exact error: $0.000001 = 1.0 \times 10^{-6}$

46. $f(x) = e^x; f^{(6)}(x) = e^x \Rightarrow$ Max on [0, 1] is e^1.

$R_5(x) \le \frac{e^1}{6!}(1)^6 \approx 0.00378 = 3.78 \times 10^{-3}$

47. $f(x) = \arcsin x; f^{(4)}(x) = \frac{x(6x^2 + 9)}{(1 - x^2)^{7/2}} \Rightarrow$ Max on

[0, 0.4] is $f^{(4)}(0.4) \approx 7.3340$.

$R_3(x) \le \frac{7.3340}{4!}(0.4)^4 \approx 0.00782 = 7.82 \times 10^{-3}$. The exact error is 8.5×10^{-4}. [Note: You could use R_4.]

48. $f(x) = \arctan x; f^{(4)}(x) = \frac{24x(x^2 + 1)}{(1 - x^2)^4}$

$\Rightarrow$ Max on [0, 0.4] is $f^{(4)}(0.4) \approx 22.3672$.

$R_3(x) \le \frac{22.3672}{4!}(0.4)^4 \approx 0.0239$

49. $g(x) = \sin x$

$\left|g^{(n+1)}(x)\right| \le 1$ for all x.

$R_n(x) \le \frac{1}{(n + 1)!}(0.3)^{n+1} < 0.001$

By trial and error, $n = 3$.

50. $f(x) = \cos x$

$\left|f^{(n+1)}(x)\right| \le 1$ for all x and all n.

$$\left|R_n(x)\right| = \left|\frac{f^{(n+1)}(z)x^{n+1}}{(n+1)!}\right| \le \frac{(0.1)^{n+1}}{(n+1)!} < 0.001$$

By trial and error, $n = 2$.

51. $f(x) = e^x$

$f^{(n+1)}(x) = e^x$

Max on $[0, 0.6]$ is $e^{0.6} \approx 1.8221$.

$$R_n \le \frac{1.8221}{(n+1)!}(0.6)^{n+1} < 0.001$$

By trial and error, $n = 5$.

52. $f(x) = \ln x,\ f'(x) = x^{-1},\ f''(x) = -x^{-2}, \ldots$

$$f^{(n+1)}(x) = (-1)^n \frac{n!}{x^{n+1}}$$

The maximum value of $\left|f^{(n+1)}(x)\right|$ on $[1, 1.25]$ is $n!$

$$\left|R_n\right| \le \frac{n!}{(n+1)!}(0.25)^{n+1} < 0.001$$

$$\frac{(0.25)^{n+1}}{n+1} < 0.001$$

By trial and error, $n = 3$

53. $f(x) = \ln(x+1)$

$$f^{(n+1)}(x) = \frac{(-1)^n n!}{(x+1)^{n+1}} \Rightarrow \text{Max on } [0, 0.5] \text{ is } n!.$$

$$R_n \le \frac{n!}{(n+1)!}(0.5)^{n+1} = \frac{(0.5)^{n+1}}{n+1} < 0.0001$$

By trial and error, $n = 9$. (See Example 9.) Using 9 terms, $\ln(1.5) \approx 0.4055$.

54. $f(x) = e^{-\pi x},\ f(1.3)$

$f'(x) = (-\pi)e^{-\pi x}$

$$f^{(n+1)}(x) = (-\pi)^{n+1}e^{-\pi x} \le \left|(-\pi)^{n+1}\right| \text{ on } [0, 1.3]$$

$$\left|R_n\right| \le \frac{(\pi)^{n+1}}{(n+1)!}(1.3)^{n+1} < 0.0001$$

By trial and error, $n = 16$. Using 16 terms, $e^{-\pi(1.3)} \approx 0.01684$.

55. $f(x) = e^x \approx 1 + x + \dfrac{x^2}{2} + \dfrac{x^3}{6},\ x < 0$

$$R_3(x) = \frac{e^z}{4!}x^4 < 0.001$$

$$e^z x^4 < 0.024$$

$$\left|xe^{z/4}\right| < 0.3936$$

$$|x| < \frac{0.3936}{e^{z/4}} < 0.3936,\ z < 0$$

$$-0.3936 < x < 0$$

56. $f(x) = \sin x \approx x - \dfrac{x^3}{3!}$

$$\left|R_3(x)\right| = \left|\frac{\sin z}{4!}x^4\right| \le \frac{\left|x^4\right|}{4!} < 0.001$$

$$x^4 < 0.024$$

$$|x| < 0.3936$$

$$-0.3936 < x < 0.3936$$

57. $f(x) = \cos x \approx 1 - \dfrac{x^2}{2!} + \dfrac{x^4}{4!}$, fifth degree polynomial

$\left|f^{(n+1)}(x)\right| \le 1$ for all x and all n.

$$\left|R_5(x)\right| \le \frac{1}{6!}|x|^6 < 0.001$$

$$|x|^6 < 0.72$$

$$|x| < 0.9467$$

$$-0.9467 < x < 0.9467$$

Note: Use a graphing utility to graph $y = \cos x - \left(1 - x^2/2 + x^4/24\right)$ in the viewing window $[-0.9467, 0.9467] \times [-0.001, 0.001]$ to verify the answer.

58. $f(x) = e^{-2x} \approx 1 - 2x + 2x^2 - \dfrac{4}{3}x^3$

$f'(x) = -2e^{-2x},\ f''(x) = 4e^{-2x},$

$f'''(x) = -8e^{-2x},\ f^{(4)}(x) = 16e^{-2x}$

$$R_3(x) = \frac{f^4(z)}{4!}(x-0)^4 = \frac{16e^{-2z}}{24}x^4 = \frac{2}{3}e^{-2z}x^4 < 0.001$$

$$e^{-2z}x^4 < 0.0015$$

$$x < \left(\frac{0.0015}{e^{-2z}}\right)^{1/4} \approx 0.1970e^{2z} < 0.1970, \text{ for } z < 0.$$

So, $0 < x < 0.1970$.

In fact, by graphing $f(x) = e^{-2x}$ and $y = 1 - 2x + 2x^2 - \dfrac{4}{3}x^3$, you can verify that $\left|f(x) - y\right| < 0.001$ on $(-0.19294, 0.20068)$.

59. The graph of the approximating polynomial P and the elementary function f both pass through the point $(c, f(c))$ and the slopes of P and f agree at $(c, f(c))$. Depending on the degree of P, the nth derivatives of P and f agree at $(c, f(c))$.

60. $f(c) = P_2(c)$, $f'(c) = P_2'(c)$, and $f''(c) = P_2''(c)$

61. See definition on page 638.

62. See Theorem 9.19, page 642.

63. As the degree of the polynomial increases, the graph of the Taylor polynomial becomes a better and better approximation of the function within the interval of convergence. Therefore, the accuracy is increased.

64.

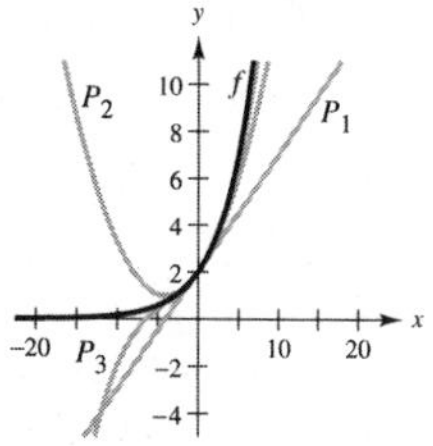

65. (a) $f(x) = e^x$

$$P_4(x) = 1 + x + \frac{1}{2}x^2 + \frac{1}{6}x^3 + \frac{1}{24}x^4$$

$$g(x) = xe^x$$

$$Q_5(x) = x + x^2 + \frac{1}{2}x^3 + \frac{1}{6}x^4 + \frac{1}{24}x^5$$

$$Q_5(x) = x\,P_4(x)$$

(b) $f(x) = \sin x$

$$P_5(x) = x - \frac{x^3}{3!} + \frac{x^5}{5!}$$

$$g(x) = x \sin x$$

$$Q_6(x) = x\,P_5(x) = x^2 - \frac{x^4}{3!} + \frac{x^6}{5!}$$

(c) $g(x) = \dfrac{\sin x}{x} = \dfrac{1}{x}P_5(x) = 1 - \dfrac{x^2}{3!} + \dfrac{x^4}{5!}$

66. (a) $P_5(x) = x - \dfrac{x^3}{3!} + \dfrac{x^5}{5!}$ for $f(x) = \sin x$

$$P_5'(x) = 1 - \frac{x^2}{2!} + \frac{x^4}{4!}$$

This is the Maclaurin polynomial of degree 4 for $g(x) = \cos x$.

(b) $Q_6(x) = 1 - \dfrac{x^2}{2} + \dfrac{x^4}{4!} - \dfrac{x^6}{6!}$ for $\cos x$

$$Q_6'(x) = -x + \frac{x^3}{3!} - \frac{x^5}{5!} = -P_5(x)$$

(c) $R(x) = 1 + x + \dfrac{x^2}{2!} + \dfrac{x^3}{3!} + \dfrac{x^4}{4!}$

$$R'(x) = 1 + x + \frac{x^2}{2!} + \frac{x^3}{3!}$$

The first four terms are the same!

67. (a) $Q_2(x) = -1 + \dfrac{\pi^2(x+2)^2}{32}$

(b) $R_2(x) = -1 + \dfrac{\pi^2(x-6)^2}{32}$

(c) No. The polynomial will be linear. Horizontal translations of the result in part (a) are possible only at $x = -2 + 8n$ (where n is an integer) because the period of f is 8.

68. Let f be an odd function and P_n be the nth Maclaurin polynomial for f. Because f is odd, f' is even:

$$\begin{aligned} f'(-x) &= \lim_{h\to 0}\frac{f(-x+h) - f(-x)}{h} \\ &= \lim_{h\to 0}\frac{-f(x-h) + f(x)}{h} \\ &= \lim_{h\to 0}\frac{f(x+(-h)) - f(x)}{-h} = f'(x). \end{aligned}$$

Similarly, f'' is odd, f''' is even, etc. Therefore, $f, f'', f^{(4)}$, etc. are all odd functions, which implies that $f(0) = f''(0) = \cdots = 0$. So, in the formula

$$P_n(x) = f(0) + f'(0)x + \frac{f''(0)x^2}{2!} + \cdots \text{all the}$$

coefficients of the even power of x are zero.

69. Let f be an even function and P_n be the nth Maclaurin polynomial for f. Because f is even, f' is odd, f'' is even, f''' is odd, etc. All of the odd derivatives of f are odd and so, all of the odd powers of x will have coefficients of zero. P_n will only have terms with even powers of x.

70. Let

$$P_n(x) = a_0 + a_1(x - c) + a_2(x - c)^2 + \cdots + a_n(x - c)^n$$

where $a_i = \dfrac{f^{(i)}(c)}{i!}$.

$$P_n(c) = a_0 = f(c)$$

For

$$1 \le k \le n, \quad P_n^{(k)}(c) = a_n k! = \left(\frac{f^{(k)}(c)}{k!}\right)k! = f^{(k)}(c).$$

71. As you move away from $x = c$, the Taylor Polynomial becomes less and less accurate.

Section 9.8 Power Series

1. Centered at 0

2. Centered at 0

3. Centered at 2

4. Centered at π

5. $\displaystyle\sum_{n=0}^{\infty} (-1)^n \frac{x^n}{n+1}$

$$L = \lim_{n\to\infty}\left|\frac{u_{n+1}}{u_n}\right| = \lim_{n\to\infty}\left|\frac{(-1)^{n+1}x^{n+1}}{n+2}\cdot\frac{n+1}{(-1)^n x^n}\right|$$

$$= \lim_{n\to\infty}\left|\frac{n+1}{n+2}\right||x| = |x|$$

$$|x| < 1 \Rightarrow R = 1$$

6. $\displaystyle\sum_{n=0}^{\infty} (3x)^n$

$$L = \lim_{n\to\infty}\left|\frac{u_{n+1}}{u_n}\right| = \lim_{n\to\infty}\left|\frac{(3x)^{n+1}}{(3x)^n}\right|$$

$$= \lim_{n\to\infty}|3x| = 3|x|$$

$$3|x| < 1 \Rightarrow |x| < \frac{1}{3} \Rightarrow R = \frac{1}{3}$$

7. $\displaystyle\sum_{n=1}^{\infty} \frac{(4x)^n}{n^2}$

$$L = \lim_{n\to\infty}\left|\frac{u_{n+1}}{u_n}\right|$$

$$= \lim_{n\to\infty}\left|\frac{(4x)^{n+1}/(n+1)^2}{(4x)^n/n^2}\right| = \lim_{n\to\infty}\left|\frac{n^2}{(n+1)^2}(4x)\right| = 4|x|$$

$$4|x| < 1 \Rightarrow R = \frac{1}{4}$$

8. $\displaystyle\sum_{n=0}^{\infty} \frac{(-1)^n x^n}{5^n}$

$$L = \lim_{n\to\infty}\left|\frac{u_{n+1}}{u_n}\right| = \lim_{n\to\infty}\left|\frac{(-1)^{n+1}x^{n+1}/5^{n+1}}{(-1)^n x^n/5^n}\right| = \lim_{n\to\infty}\frac{|x|}{5} = \frac{|x|}{5}$$

$$\frac{|x|}{5} < 1 \Rightarrow R = 5$$

9. $\displaystyle\sum_{n=0}^{\infty} \frac{x^{2n}}{(2n)!}$

$$L = \lim_{n\to\infty}\left|\frac{u_{n+1}}{u_n}\right|$$

$$= \lim_{n\to\infty}\left|\frac{x^{(2n+2)}/(2n+2)!}{x^{2n}/(2n)!}\right|$$

$$= \lim_{n\to\infty}\left|\frac{x^2}{(2n+2)(2n+1)}\right| = 0$$

So, the series converges for all $x \Rightarrow R = \infty$.

10. $\displaystyle\sum_{n=0}^{\infty} \frac{(2n)!x^{2n}}{n!}$

$$L = \lim_{n\to\infty}\left|\frac{u_n + 1}{u_n}\right| = \lim_{n\to\infty}\left|\frac{(2n+2)!x^{2n+2}/(n+1)!}{(2n)!x^{2n}/n!}\right|$$

$$= \lim_{n\to\infty}\left|\frac{(2n+2)(2n+1)x^2}{(n+1)}\right| = \infty$$

The series only converges at $x = 0$. $R = 0$.

11. $\displaystyle\sum_{n=0}^{\infty} \left(\frac{x}{4}\right)^n$

Because the series is geometric, it converges only if $\left|\dfrac{x}{4}\right| < 1$, or $-4 < x < 4$.

12. $\sum_{n=0}^{\infty} (2x)^n$

Because the series is geometric, it converges only if $|2x| < 1 \Rightarrow |x| < \frac{1}{2}$ or $-\frac{1}{2} < x < \frac{1}{2}$.

13. $\sum_{n=1}^{\infty} \frac{(-1)^n x^n}{n}$

$$\lim_{n\to\infty}\left|\frac{u_{n+1}}{u_n}\right| = \lim_{n\to\infty}\left|\frac{(-1)^{n+1}x^{n+1}}{n+1}\cdot\frac{n}{(-1)^n x^n}\right|$$

$$= \lim_{n\to\infty}\left|\frac{nx}{n+1}\right| = |x|$$

Interval: $-1 < x < 1$

When $x = 1$, the alternating series $\sum_{n=1}^{\infty}\frac{(-1)^n}{n}$ converges.

When $x = -1$, the p-series $\sum_{n=1}^{\infty}\frac{1}{n}$ diverges.

Therefore, the interval of convergence is $(-1, 1]$.

14. $\sum_{n=0}^{\infty}(-1)^{n+1}(n+1)x^n$

$$\lim_{n\to\infty}\left|\frac{u_n+1}{u_n}\right| = \lim_{n\to\infty}\left|\frac{(-1)^{n+2}(n+2)x^{n+1}}{(-1)^n(n+1)x^n}\right|$$

$$= \lim_{n\to\infty}\left|\frac{(n+2)x}{n+1}\right| = |x|$$

Interval: $-1 < x < 1$

When $x = 1$, the series $\sum_{n=0}^{\infty}(-1)^{n+1}(n+1)$ diverges.

When $x = -1$, the series $\sum_{n=0}^{\infty}-(n+1)$ diverges.

Therefore, the interval of convergence is $(-1, 1)$.

15. $\sum_{n=0}^{\infty}\frac{x^{5n}}{n!}$

$$\lim_{n\to\infty}\left|\frac{u_{n+1}}{u_n}\right| = \lim_{n\to\infty}\left|\frac{x^{5(n+1)}/(n+1)!}{5^n/n!}\right| = \lim_{n\to\infty}\left|\frac{x^5}{n+1}\right| = 0$$

The series converges for all x. The interval of convergence is $(-\infty, \infty)$.

16. $\sum_{n=0}^{\infty}\frac{(3x)^n}{(2n)!}$

$$\lim_{n\to\infty}\left|\frac{u_{n+1}}{u_n}\right| = \lim_{n\to\infty}\left|\frac{(3x)^{n+1}}{(2n+1)!}\cdot\frac{(2n)!}{(3x)^n}\right|$$

$$= \lim_{n\to\infty}\left|\frac{3x}{(2n+2)(2n+1)}\right| = 0$$

Therefore, the interval of convergence is $(-\infty, \infty)$.

17. $\sum_{n=0}^{\infty}(2n)!\left(\frac{x}{3}\right)^n$

$$\lim_{n\to\infty}\left|\frac{u_{n+1}}{u_n}\right| = \lim_{n\to\infty}\left|\frac{(2n+2)!(x/3)^{n+1}}{(2n)!(x/3)^n}\right|$$

$$= \left|\frac{(2n+2)(2n+1)x}{3}\right| = \infty$$

The series converges only for $x = 0$.

18. $\sum_{n=0}^{\infty}\frac{(-1)^n x^n}{(n+1)(n+2)}$

$$\lim_{n\to\infty}\left|\frac{u_{n+1}}{u_n}\right| = \lim_{n\to\infty}\left|\frac{(-1)^{n+1}x^{n+1}}{(n+2)(n+3)}\cdot\frac{(n+1)(n+2)}{(-1)^n x^n}\right| = \lim_{n\to\infty}\left|\frac{(n+1)x}{n+3}\right| = |x|$$

Interval: $-1 < x < 1$

When $x = 1$, the alternating series $\sum_{n=0}^{\infty}\frac{(-1)^n}{(n+1)(n+2)}$ converges.

When $x = -1$, the series $\sum_{n=0}^{\infty}\frac{1}{(n+1)(n+2)}$ converges by limit comparison to $\sum_{n=1}^{\infty}\frac{1}{n^2}$.

Therefore, the interval of convergence is $[-1, 1]$.

19. $\displaystyle\sum_{n=1}^{\infty} \frac{(-1)^{n+1} x^n}{6^n}$

Because the series is geometric, it converges only if

$$\left|\frac{x}{6}\right| < 1 \Rightarrow |x| < 6 \text{ or } -6 < x < 6.$$

20. $\displaystyle\sum_{n=0}^{\infty} \frac{(-1)^n n!(x-5)^n}{3^n}$

$$\lim_{n\to\infty}\left|\frac{u_{n+1}}{u_n}\right| = \lim_{n\to\infty}\left|\frac{(-1)^{n+1}(n+1)!(x-5)^{n+1}/3^{n+1}}{(-1)^n n!(x-5)^n/3^n}\right| = \lim_{n\to\infty}\left|\frac{(n+1)(x-5)}{3}\right| = \infty$$

The series converges only for $x = 5$.

21. $\displaystyle\sum_{n=1}^{\infty} \frac{(-1)^{n+1}(x-4)^n}{n9^n}$

$$\lim_{n\to\infty}\left|\frac{u_{n+1}}{u_n}\right| = \lim_{n\to\infty}\left|\frac{(-1)^{n+2}(x-4)^{n+1}/((n+1)9^{n+1})}{(-1)^n(x-4)^n/(n9^n)}\right|$$

$$= \lim_{n\to\infty}\left|\frac{n}{n+1}\frac{(x-4)}{9}\right| = \frac{1}{9}|x-4|$$

$R = 9$

Interval: $-5 < x < 13$

When $x = 13$, $\displaystyle\sum_{n=1}^{\infty}\frac{(-1)^{n+1}9^n}{n9^n} = \sum_{n=1}^{\infty}\frac{(-1)^{n+1}}{n}$ converges.

When $x = -5$, $\displaystyle\sum_{n=1}^{\infty}\frac{(-1)^{n+1}(-9)^n}{n9^n} = \sum_{n=1}^{\infty}\frac{-1}{n}$ diverges.

Therefore, the interval of convergence is $(-5, 13]$.

22. $\displaystyle\sum_{n=0}^{\infty} \frac{(x-3)^{n+1}}{(n+1)4^{n+1}}$

$$\lim_{n\to\infty}\left|\frac{u_{n+1}}{u_n}\right| = \lim_{n\to\infty}\left|\frac{(x-3)^{n+2}/[(n+2)4^{n+2}]}{(x-3)^{n+1}/[(n+1)4^{n+1}]}\right|$$

$$= \lim_{n\to\infty}\left|\frac{(x-3)(n+1)}{4(n+2)}\right| = \left|\frac{x-3}{4}\right|$$

$R = 4$

Interval: $-1 < x < 7$

When $x = 7$, $\displaystyle\sum_{n=0}^{\infty}\frac{4^{n+1}}{(n+1)4^{n+1}} = \sum_{n=0}^{\infty}\frac{1}{n+1}$ diverges.

When $x = -1$, $\displaystyle\sum_{n=0}^{\infty}\frac{(-4)^{n+1}}{(n+1)4^{n+1}} = \sum_{n=0}^{\infty}\frac{(-1)^{n+1}}{(n+1)}$ converges.

Therefore, the interval of convergence is $[-1, 7)$.

23. $\displaystyle\sum_{n=0}^{\infty} \frac{(-1)^{n+1}(x-1)^{n+1}}{n+1}$

$$\lim_{n\to\infty}\left|\frac{u_{n+1}}{u_n}\right| = \lim_{n\to\infty}\left|\frac{(-1)^{n+2}(x-1)^{n+2}}{n+2}\cdot\frac{n+1}{(-1)^{n+1}(x-1)^{n+1}}\right| = \lim_{n\to\infty}\left|\frac{(n+1)(x-1)}{n+2}\right| = |x-1|$$

$R = 1$

Center: $x = 1$

Interval: $-1 < x - 1 < 1$ or $0 < x < 2$

When $x = 0$, the series $\displaystyle\sum_{n=0}^{\infty}\frac{1}{n+1}$ diverges by the integral test.

When $x = 2$, the alternating series $\displaystyle\sum_{n=0}^{\infty}\frac{(-1)^{n+1}}{n+1}$ converges.

Therefore, the interval of convergence is $(0, 2]$.

24. $\sum_{n=1}^{\infty} \frac{(-1)^{n+1}(x-2)^n}{n2^n}$

$$\lim_{n\to\infty}\left|\frac{a_{n+1}}{a_n}\right| = \lim_{n\to\infty}\left|\frac{(-1)^{n+2}(x-2)^{n+1}}{(n+1)2^{n+1}} \bigg/ \frac{(-1)^{n+1}(x-2)^n}{n2^n}\right| = \lim_{n\to\infty}\left|\frac{x-2}{2}\frac{n}{n+1}\right| = \left|\frac{x-2}{2}\right|$$

$$\left|\frac{x-2}{2}\right| < 1 \Rightarrow -2 < x - 2 < 2 \Rightarrow 0 < x < 4$$

when $x = 0$,

$$\sum_{n=1}^{\infty}\frac{(-1)^{n+1}(-2)^n}{n2^n} = \sum_{n=1}^{\infty}\frac{(-1)(2^n)}{n2^n} = \sum_{n=1}^{\infty}\frac{(-1)}{n} \text{ diverges.}$$

when $x = 4$,

$$\sum_{n=1}^{\infty}\frac{(-1)^{n+1}2n}{n2^n} = \sum_{n=1}^{\infty}\frac{(-1)^{n+1}}{n} \text{ converges.}$$

Therefore the interval of convergence is $(0, 4]$.

25. $\sum_{n=1}^{\infty}\left(\frac{x-3}{3}\right)^{n-1}$ is geometric. It converges if

$$\left|\frac{x-3}{3}\right| < 1 \Rightarrow |x-3| < 3 \Rightarrow 0 < x < 6.$$

Therefore, the interval of convergence is $(0, 6)$.

26. $\sum_{n=0}^{\infty}\frac{(-1)^n x^{2n+1}}{2n+1}$

$$\lim_{n\to\infty}\left|\frac{u_{n+1}}{u_n}\right| = \lim_{n\to\infty}\left|\frac{(-1)^{n+1}x^{2n+3}}{(2n+3)}\cdot\frac{(2n+1)}{(-1)^n x^{2n+1}}\right|$$

$$= \lim_{n\to\infty}\left|\frac{(2n+1)}{(2n+3)}x^2\right| = |x^2|$$

$R = 1$

Interval: $-1 < x < 1$

When $x = 1$, $\sum_{n=0}^{\infty}\frac{(-1)^n}{2n+1}$ converges.

When $x = -1$, $\sum_{n=0}^{\infty}\frac{(-1)^{n+1}}{2n+1}$ converges.

Therefore, the interval of convergence is $[-1, 1]$.

27. $\sum_{n=1}^{\infty}\frac{n}{n+1}(-2x)^{n-1}$

$$\lim_{n\to\infty}\left|\frac{u_{n+1}}{u_n}\right| = \lim_{n\to\infty}\left|\frac{(n+1)(-2x)^n}{n+2}\cdot\frac{n+1}{n(-2x)^{n-1}}\right|$$

$$= \lim_{n\to\infty}\left|\frac{(-2x)(n+1)^2}{n(n+2)}\right| = 2|x|$$

$R = \frac{1}{2}$

Interval: $-\frac{1}{2} < x < \frac{1}{2}$

When $x = -\frac{1}{2}$, the series $\sum_{n=1}^{\infty}\frac{n}{n+1}$ diverges by the nth Term Test.

When $x = \frac{1}{2}$, the alternating series

$$\sum_{n=1}^{\infty}\frac{(-1)^{n-1}n}{n+1} \text{ diverges.}$$

Therefore, the interval of convergence is $\left(-\frac{1}{2}, \frac{1}{2}\right)$.

28. $\sum_{n=0}^{\infty}\frac{(-1)^n x^{2n}}{n!}$

$$\lim_{n\to\infty}\left|\frac{u_{n+1}}{u_n}\right| = \lim_{n\to\infty}\left|\frac{(-1)^{n+1}x^{2n+2}}{(n+1)!}\cdot\frac{n!}{(-1)^n x^{2n}}\right|$$

$$= \lim_{n\to\infty}\left|\frac{x^2}{n+1}\right| = 0$$

Therefore, the interval of convergence is $(-\infty, \infty)$.

29. $\sum_{n=0}^{\infty} \frac{x^{3n+1}}{(3n+1)!}$

$$\lim_{n\to\infty}\left|\frac{u_{n+1}}{u_n}\right| = \lim_{n\to\infty}\left|\frac{x^{3n+4}/(3n+4)!}{x^{3n+1}/(3n+1)!}\right|$$

$$= \lim_{n\to\infty}\left|\frac{x^3}{(3n+4)(3n+3)(3n+2)}\right| = 0$$

Therefore, the interval of convergence is $(-\infty, \infty)$.

30. $\sum_{n=1}^{\infty} \frac{n!x^n}{(2n)!}$

$$\lim_{n\to\infty}\left|\frac{u_{n+1}}{u_n}\right| = \lim_{n\to\infty}\left|\frac{(n+1)!x^{n+1}}{(2n+2)!}\cdot\frac{(2n)!}{n!x^n}\right|$$

$$= \lim_{n\to\infty}\left|\frac{(n+1)x}{(2n+2)(2n+1)}\right| = 0$$

Therefore, the interval of convergence is $(-\infty, \infty)$.

31. $\sum_{n=1}^{\infty} \frac{2\cdot 3\cdot 4\cdots(n+1)x^n}{n!} = \sum_{n=1}^{\infty}(n+1)x^n$

$$\lim_{n\to\infty}\left|\frac{a_{n+1}}{a_n}\right| = \lim_{n\to\infty}\left|\frac{(n+2)x^{n+1}}{(n+1)x^n}\right| = \lim_{n\to\infty}\left|\frac{n+2}{n+1}x\right| = |x|$$

Converges if $|x| < 1 \Rightarrow -1 < x < 1$.

At $x = \pm 1$, diverges.

Therefore the interval of convergence is $(-1, 1)$.

32. $\sum_{n=1}^{\infty} \frac{2\cdot 4\cdot 6\cdots(2n)}{3\cdot 5\cdot 7\cdots(2n+1)}\left(x^{2n+1}\right)$

$$\lim_{n\to\infty}\left|\frac{u_{n+1}}{u_n}\right| = \lim_{n\to\infty}\left|\frac{2\cdot 4\cdots(2n)(2n+2)x^{2n+3}}{3\cdot 5\cdot 7\cdots(2n+1)(2n+3)}\cdot\frac{3\cdot 5\cdots(2n+1)}{2\cdot 4\cdots(2n)x^{2n+1}}\right| = \lim_{n\to\infty}\left|\frac{(2n+2)x^2}{(2n+3)}\right| = \left|x^2\right|$$

$R = 1$

When $x = \pm 1$, the series diverges by comparing it to

$$\sum_{n=1}^{\infty}\frac{1}{2n+1}$$

which diverges.

Therefore, the interval of convergence is $(-1, 1)$.

33. $\sum_{n=1}^{\infty} \frac{(-1)^{n+1}3\cdot 7\cdot 11\cdots(4n-1)(x-3)^n}{4^n}$

$$\lim_{n\to\infty}\left|\frac{u_{n+1}}{u_n}\right| = \lim_{n\to\infty}\left|\frac{(-1)^{n+2}\cdot 3\cdot 7\cdot 11\cdots(4n-1)(4n+3)(x-3)^{n+1}}{4^{n+1}}\cdot\frac{4^n}{(-1)^{n+1}\cdot 3\cdot 7\cdot 11\cdots(4n-1)(x-3)^n}\right|$$

$$= \lim_{n\to\infty}\left|\frac{(4n+3)(x-3)}{4}\right| = \infty$$

$R = 0$

Center: $x = 3$

Therefore, the series converges only for $x = 3$.

34. $\sum_{n=1}^{\infty} \frac{n!(x+1)^n}{1 \cdot 3 \cdot 5 \cdots (2n-1)}$

$$\lim_{n\to\infty}\left|\frac{a_{n+1}}{a_n}\right| = \lim_{n\to\infty}\left|\frac{(n+1)!(x+1)^{n+1}}{1 \cdot 3 \cdot 5 \cdots (2n-1)(2n+1)} \Big/ \frac{(n)!(x+1)^n}{1 \cdot 3 \cdot 5 \cdots (2n-1)}\right| = \lim_{n\to\infty}\left|\frac{(n+1)(x+1)}{2n+1}\right| = \frac{1}{2}\left|x+1\right|$$

Converges if $\frac{1}{2}\left|x+1\right| < 1 \Rightarrow -2 < x+1 < 2 \Rightarrow -3 < x < 1.$

At $x = 1$, $a_n = \frac{n!2^n}{1 \cdot 3 \cdot 5 \cdots (2n-1)} = \frac{2 \cdot 4 \cdot 6 \cdots 2n}{1 \cdot 3 \cdot 5 \cdots (2n-1)} > 1,$ diverges.

At $x = -3$, $a_n = \frac{n!(-2)^n}{1 \cdot 3 \cdots (2n-1)} = (-1)^n \frac{2 \cdot 4 \cdots 2n}{1 \cdot 3 \cdots (2n-1)},$ diverges.

Therefore, the interval of convergence is $(-3, 1)$.

35. $\sum_{n=1}^{\infty} \frac{(x-c)^{n-1}}{c^{n-1}}$

$$\lim_{n\to\infty}\left|\frac{u_{n+1}}{u_n}\right| = \lim_{n\to\infty}\left|\frac{(x-c)^n}{c^n} \cdot \frac{c^{n-1}}{(x-c)^{n-1}}\right| = \frac{1}{c}\left|x-c\right|$$

$R = c$

Center: $x = c$

Interval: $-c < x - c < c$ or $0 < x < 2c$

When $x = 0$, the series $\sum_{n=1}^{\infty} (-1)^{n-1}$ diverges.

When $x = 2c$, the series $\sum_{n=1}^{\infty} 1$ diverges.

Therefore, the interval of convergence is $(0, 2c)$.

36. $\sum_{n=0}^{\infty} \frac{(n!)^k x^n}{(kn)!}$, k is a positive integer.

$$\lim_{n\to\infty}\left|\frac{a_{n+1}}{a_n}\right| = \lim_{n\to\infty}\left|\frac{\left[(n+1)!\right]^k x^{n+1}}{\left[k(n+1)\right]!} \Big/ \frac{(n!)^k x^n}{(kn)!}\right| = \lim_{n\to\infty}\left|\frac{(n+1)^k x}{(k+nk)(k-1+nk)\cdots(1+nk)}\right| = \frac{|x|}{k^k}$$

Converges if $\frac{|x|}{k^k} < 1 \Rightarrow R = k^k.$

37. $\sum_{n=0}^{\infty} \left(\frac{x}{k}\right)^n$

Because the series is geometric, it converges only if $|x/k| < 1$ or $-k < x < k$.

Therefore, the interval of convergence is $(-k, k)$.

38. $\displaystyle\sum_{n=1}^{\infty}\frac{(-1)^{n+1}(x-c)^n}{nc^n}$

$$\lim_{n\to\infty}\left|\frac{u_{n+1}}{u_n}\right| = \lim_{n\to\infty}\left|\frac{(-1)^{n+2}(x-c)^{n+1}}{(n+1)c^{n+1}}\cdot\frac{nc^n}{(-1)^{n+1}(x-c)^n}\right| = \lim_{n\to\infty}\left|\frac{n(x-c)}{c(n+1)}\right| = \frac{1}{c}|x-c|$$

$R = c$

Center: $x = c$

Interval: $-c < x - c < c$ or $0 < x < 2c$

When $x = 0$, the p-series $\displaystyle\sum_{n=1}^{\infty}\frac{-1}{n}$ diverges.

When $x = 2c$, the alternating series $\displaystyle\sum_{n=1}^{\infty}\frac{(-1)^{n+1}}{n}$ converges.

Therefore, the interval of convergence is $(0, 2c]$.

39. $\displaystyle\sum_{n=1}^{\infty}\frac{k(k+1)\cdots(k+n-1)x^n}{n!}$

$$\lim_{n\to\infty}\left|\frac{u_{n+1}}{u_n}\right| = \lim_{n\to\infty}\left|\frac{k(k+1)\cdots(k+n-1)(k+n)x^{n+1}}{(n+1)!}\cdot\frac{n!}{k(k+1)\cdots(k+n-1)x^n}\right| = \lim_{n\to\infty}\left|\frac{(k+n)x}{n+1}\right| = |x|$$

$R = 1$

When $x = \pm 1$, the series diverges and the interval of convergence is $(-1, 1)$.

$$\left[\frac{k(k+1)\cdots(k+n-1)}{1\cdot 2\cdots n} \geq 1\right]$$

40. $\displaystyle\sum_{n=1}^{\infty}\frac{n!(x-c)^n}{1\cdot 3\cdot 5\cdots(2n-1)}$

$$\lim_{n\to\infty}\left|\frac{u_{n+1}}{u_n}\right| = \lim_{n\to\infty}\left|\frac{(n+1)!(x-c)^{n+1}}{1\cdot 3\cdot 5\cdots(2n-1)(2n+1)}\cdot\frac{1\cdot 3\cdot 5\cdots(2n-1)}{n!(x-c)}\right| = \lim_{n\to\infty}\left|\frac{(n+1)(x-c)}{2n+1}\right| = \frac{1}{2}|x-c|$$

$R = 2$

Interval: $-2 < x - c < 2$ or $c - 2 < x < c + 2$

The series diverges at the endpoints. Therefore, the interval of convergence is $(c-2, c+2)$.

$$\left[\frac{n!(c+2-c)^n}{1\cdot 3\cdot 5\cdots(2n-1)} = \frac{n!2^2}{1\cdot 3\cdot 5\cdots(2n-1)} = \frac{2\cdot 4\cdot 6\cdots(2n)}{1\cdot 3\cdot 5\cdots(2n-1)} > 1\right]$$

41. $\displaystyle\sum_{n=0}^{\infty}\frac{x^n}{n!} = 1 + \frac{x}{1} + \frac{x^2}{2} + \cdots = \sum_{n=1}^{\infty}\frac{x^{n-1}}{(n-1)!}$

42. $\displaystyle\sum_{n=0}^{\infty}(-1)^{n+1}(n+1)x^n = \sum_{n=1}^{\infty}(-1)^n(n)x^{n-1}$

43. $\displaystyle\sum_{n=0}^{\infty}\frac{x^{2n+1}}{(2n+1)!} = \sum_{n=1}^{\infty}\frac{x^{2n-1}}{(2n-1)!}$

Replace n with $n - 1$.

44. $\displaystyle\sum_{n=0}^{\infty}\frac{(-1)^n x^{2n+1}}{2n+1} = \sum_{n=1}^{\infty}\frac{(-1)^{n-1}x^{2n-1}}{2n-1}$

Replace n with $n - 1$.

45. (a) $f(x) = \displaystyle\sum_{n=0}^{\infty}\left(\frac{x}{3}\right)^n, (-3, 3)$ (Geometric)

(b) $f'(x) = \displaystyle\sum_{n=1}^{\infty}\frac{n}{3}\left(\frac{x}{3}\right)^{n-1}, (-3, 3)$

(c) $f''(x) = \displaystyle\sum_{n=2}^{\infty}\frac{n(n-1)}{9}\left(\frac{x}{3}\right)^{n-2}, (-3, 3)$

(d) $\displaystyle\int f(x)\,dx = \sum_{n=0}^{\infty}\frac{3}{n+1}\left(\frac{x}{3}\right)^{n+1}, [-3, 3)$

$$\left[\sum\frac{3}{n+1}\left(\frac{-3}{3}\right)^{n+1} = \sum\frac{(-1)^{n+1}3}{n+1}, \text{ converges}\right]$$

46. (a) $f(x) = \sum_{n=1}^{\infty} \frac{(-1)^{n+1}(x-5)^n}{n5^n}$, $(0, 10]$

(b) $f'(x) = \sum_{n=1}^{\infty} \frac{(-1)^{n+1}(x-5)^{n-1}}{5^n}$, $(0, 10)$

(c) $f''(x) = \sum_{n=2}^{\infty} \frac{(-1)^{n+1}(n-1)(x-5)^{n-2}}{5^n}$, $(0, 10)$

(d) $\int f(x)\,dx = \sum_{n=1}^{\infty} \frac{(-1)^{n+1}(x-5)^{n+1}}{n(n+1)5^n}$, $[0, 10]$

47. (a) $f(x) = \sum_{n=0}^{\infty} \frac{(-1)^{n+1}(x-1)^{n+1}}{n+1}$, $(0, 2]$

(b) $f'(x) = \sum_{n=0}^{\infty} (-1)^{n+1}(x-1)^n$, $(0, 2)$

(c) $f''(x) = \sum_{n=1}^{\infty} (-1)^{n+1}n(x-1)^{n-1}$, $(0, 2)$

(d) $\int f(x)\,dx = \sum_{n=1}^{\infty} \frac{(-1)^{n+1}(x-1)^{n+2}}{(n+1)(n+2)}$, $[0, 2]$

48. (a) $f(x) = \sum_{n=1}^{\infty} \frac{(-1)^{n+1}(x-2)^n}{n}$, $(1, 3]$

(b) $f'(x) = \sum_{n=1}^{\infty} (-1)^{n+1}(x-2)^{n-1}$, $(1, 3)$

(c) $f''(x) = \sum_{n=2}^{\infty} (-1)^{n+1}(n-1)(x-2)^{n-2}$, $(1, 3)$

(d) $\int f(x)\,dx = \sum_{n=1}^{\infty} \frac{(-1)^{n+1}(x-2)^{n+1}}{n(n+1)}$, $[1, 3]$

49. A series of the form

$$\sum_{n=0}^{\infty} a_n(x-c)^n = a_0 + a_1(x-c) + a_2(x-c)^2 + \cdots + a_n(x-c)^n + \cdots$$

is called a power series centered at c, where c is constant.

50. The set of all values of x for which the power series converges is the interval of convergence. If the power series converges for all x, then the radius of convergence is $R = \infty$. If the power series converges at only c, then $R = 0$. Otherwise, according to Theorem 8.20, there exists a real number $R > 0$ (radius of convergence) such that the series converges absolutely for $|x - c| < R$ and diverges for $|x - c| > R$.

51. The interval of convergence of a power series is the set of all values of x for which the power series converges.

52. A single point, an interval, or the entire real line.

53. You differentiate and integrate the power series term by term. The radius of convergence remains the same. However, the interval of convergence might change.

54. Answers will vary.

$\sum_{n=1}^{\infty} \frac{x^n}{n}$ converges for $-1 \le x < 1$. At $x = -1$, the convergence is conditional because $\sum \frac{1}{n}$ diverges.

$\sum_{n=1}^{\infty} \frac{x^n}{n^2}$ converges for $-1 \le x \le 1$. At $x = \pm 1$, the convergence is absolute.

55. Many answers possible.

(a) $\sum_{n=1}^{\infty} \left(\frac{x}{2}\right)^n$ Geometric: $\left|\frac{x}{2}\right| < 1 \Rightarrow |x| < 2$

(b) $\sum_{n=1}^{\infty} \frac{(-1)^n x^n}{n}$ converges for $-1 < x \le 1$

(c) $\sum_{n=1}^{\infty} (2x+1)^n$ Geometric:

$|2x + 1| < 1 \Rightarrow -1 < x < 0$

(d) $\sum_{n=1}^{\infty} \frac{(x-2)^n}{n4^n}$ converges for $-2 \le x < 6$

56. (a) $g(1) = \sum_{n=0}^{\infty} \left(\frac{1}{3}\right)^n = 1 + \frac{1}{3} + \frac{1}{9} + \ldots$

$S_1 = 1, S_2 = \frac{4}{3}, \ldots$

Matches (iii).

(b) $g(2) = \sum_{n=0}^{\infty} \left(\frac{2}{3}\right)^n = 1 + \frac{2}{3} + \frac{4}{9} + \ldots$

$S_1 = 1, S_2 = \frac{5}{3}, \ldots$

Matches (i).

(c) $g(3) = \sum_{n=0}^{\infty} \left(\frac{3}{3}\right)^n = 1 + 1 + 1 + \ldots$

$S_1 = 1, S_2 = 2, \ldots$

Matches (ii).

(d) $g(-2) = \sum_{n=0}^{\infty} \left(\frac{-2}{3}\right)^n = 1 - \frac{2}{3} + \frac{4}{9} - \ldots$ (alternating)

$S_1 = 1, S_2 = \frac{1}{3}, S_3 = \frac{7}{9}, \ldots$

Matches (iv).

57. (a) $f(x) = \sum_{n=0}^{\infty} \frac{x^{2n+1}}{(2n+1)!}$

$$\lim_{n\to\infty}\left|\frac{u_{n+1}}{u_n}\right| = \lim_{n\to\infty}\left|\frac{x^{2n+3}}{(2n+3)!}\cdot\frac{(2n+1)!}{x^{2n+1}}\right|$$

$$= \lim_{n\to\infty}\left|\frac{x^2}{(2n+2)(2n+3)}\right| = 0$$

Therefore, the interval of convergence is $(-\infty\ \infty)$.

$$g(x) = \sum_{n=0}^{\infty}\frac{(-1)^n x^{2n}}{(2n)!}$$

$$\lim_{n\to\infty}\left|\frac{u_{n+1}}{u_n}\right| = \lim_{n\to\infty}\left|\frac{(-1)^{n+1}x^{2n+2}}{(2n+2)!}\cdot\frac{(2n)!}{(-1)^n x^{2n}}\right|$$

$$= \lim_{n\to\infty}\frac{1}{2n+2} = 0$$

Therefore, the interval of convergence is $(-\infty\ \infty)$.

(b) $f'(x) = \sum_{n=0}^{\infty}\frac{(-1)^n x^{2n}}{(2n)!} = g(x)$

(c) $g'(x) = \sum_{n=1}^{\infty}\frac{(-1)^n x^{2n-1}}{(2n-1)!} = \sum_{n=0}^{\infty}\frac{(-1)^{n+1}x^{2n+1}}{(2n+1)!}$

$$= -\sum_{n=0}^{\infty}\frac{(-1)^n x^{2n+1}}{(2n+1)!} = -f(x)$$

(d) $f(x) = \sin x$ and

$g(x) = \cos x$

58. (a) $f(x) = \sum_{n=0}^{\infty}\frac{x^n}{n!}$

$$\lim_{n\to\infty}\left|\frac{u_{n+1}}{u_n}\right| = \lim_{n\to\infty}\left|\frac{x^{n+1}}{(n+1)!}\cdot\frac{n!}{x^n}\right|$$

$$= \lim_{n\to\infty}\left|\frac{x}{n+1}\right| = 0$$

The series converges for all x. Therefore, the interval of convergence is $(-\infty, \infty)$.

(b) $f'(x) = \sum_{n=1}^{\infty}\frac{nx^{n-1}}{n!} = \sum_{n=1}^{\infty}\frac{x^{n-1}}{(n-1)!} = \sum_{n=0}^{\infty}\frac{x^n}{n!} = f(x)$

(c) $f(x) = \sum_{n=1}^{\infty}\frac{x^n}{n!} = 1 + x + \frac{x^2}{2!} + \frac{x^3}{3!} + \frac{x^4}{4!} + \cdots$

$f(0) = 1$

(d) $f(x) = e^x$

59. $y = \sum_{n=0}^{\infty}\frac{(-1)^n x^{2n+1}}{(2n+1)!} = \sum_{n=1}^{\infty}\frac{(-1)^{n-1}x^{2n-1}}{(2n-1)!}$

$$y' = \sum_{n=0}^{\infty}\frac{(-1)^n(2n+1)x^{2n}}{(2n+1)!} = \sum_{n=0}^{\infty}\frac{(-1)^n x^{2n}}{(2n)!}$$

$$y'' = \sum_{n=1}^{\infty}\frac{(-1)^n(2n)x^{2n-1}}{(2n)!} = \sum_{n=1}^{\infty}\frac{(-1)x^{2n-1}}{(2n-1)!}$$

$$y'' + y = \sum_{n=1}^{\infty}\frac{(-1)^n x^{2n-1}}{(2n-1)!} + \sum_{n=1}^{\infty}\frac{(-1)^{n-1}x^{2n-1}}{(2n-1)!} = 0$$

60. $y = \sum_{n=0}^{\infty}\frac{(-1)^n x^{2n}}{(2n)!} = \sum_{n=1}^{\infty}\frac{(-1)^{n-1}x^{2n-2}}{(2n-2)!}$

$$y' = \sum_{n=1}^{\infty}\frac{(-1)^n(2n)x^{2n-1}}{(2n)!} = \sum_{n=1}^{\infty}\frac{(-1)^n x^{2n-1}}{(2n-1)!}$$

$$y'' = \sum_{n=1}^{\infty}\frac{(-1)^n(2n-1)x^{2n-2}}{(2n-1)!} = \sum_{n=1}^{\infty}\frac{(-1)^n x^{2n-2}}{(2n-2)!}$$

$$y'' + y = \sum_{n=1}^{\infty}\frac{(-1)^n x^{2n-2}}{(2n-2)!} + \sum_{n=1}^{\infty}\frac{(-1)^{n-1}x^{2n-2}}{(2n-2)!} = 0$$

61. $y = \sum_{n=0}^{\infty}\frac{x^{2n+1}}{(2n+1)!} = \sum_{n=1}^{\infty}\frac{x^{2n-1}}{(2n-1)!}$

$$y' = \sum_{n=0}^{\infty}\frac{(2n+1)x^{2n}}{(2n+1)!} = \sum_{n=0}^{\infty}\frac{x^{2n}}{(2n)!}$$

$$y'' = \sum_{n=1}^{\infty}\frac{(2n)x^{2n-1}}{(2n)!} = \sum_{n=1}^{\infty}\frac{x^{2n-1}}{(2n-1)!} = y$$

$$y'' - y = 0$$

62. $y = \sum_{n=0}^{\infty}\frac{x^{2n}}{(2n)!} = \sum_{n=1}^{\infty}\frac{x^{2n-2}}{(2n-2)!}$

$$y' = \sum_{n=1}^{\infty}\frac{(2n-2)x^{2n-1}}{(2n-2)!} = \sum_{n=1}^{\infty}\frac{x^{2n-1}}{(2n-1)!}$$

$$y'' = \sum_{n=1}^{\infty}\frac{(2n-1)x^{2n-2}}{(2n-1)!} = \sum_{n=1}^{\infty}\frac{x^{2n-2}}{(2n-2)!} = y$$

$$y'' - y = 0$$

63. $y = \sum_{n=0}^{\infty} \frac{x^{2n}}{2^n n!}$ $\qquad y' = \sum_{n=1}^{\infty} \frac{2nx^{2n-1}}{2^n n!}$ $\qquad y'' = \sum_{n=1}^{\infty} \frac{2n(2n-1)x^{2n-2}}{2^n n!}$

$$\begin{aligned} y'' - xy' - y &= \sum_{n=1}^{\infty} \frac{2n(2n-1)x^{2n-2}}{2^n n!} - \sum_{n=1}^{\infty} \frac{2nx^{2n}}{2^n n!} - \sum_{n=0}^{\infty} \frac{x^{2n}}{2^n n!} \\ &= \sum_{n=1}^{\infty} \frac{2n(2n-1)x^{2n-2}}{2^n n!} - \sum_{n=0}^{\infty} \frac{(2n+1)x^{2n}}{2^n n!} \\ &= \sum_{n=0}^{\infty} \left[\frac{(2n+2)(2n+1)x^{2n}}{2^{n+1}(n+1)!} - \frac{(2n+1)x^{2n}}{2^n n!} \cdot \frac{2(n+1)}{2(n+1)} \right] \\ &= \sum_{n=0}^{\infty} \frac{2(n+1)x^{2n}\left[(2n+1) - (2n+1)\right]}{2^{n+1}(n+1)!} = 0 \end{aligned}$$

64.

$$y = 1 + \sum_{n=1}^{\infty} \frac{(-1)^n x^{4n}}{2^{2n} n! \cdot 3 \cdot 7 \cdot 11 \cdots (4n-1)}$$

$$y' = \sum_{n=1}^{\infty} \frac{(-1)^n 4nx^{4n-1}}{2^{2n} n! \cdot 3 \cdot 7 \cdot 11 \cdots (4n-1)}$$

$$y'' = \sum_{n=1}^{\infty} \frac{(-1)^n 4n(4n-1)x^{4n-2}}{2^{2n} n! \cdot 3 \cdot 7 \cdot 11 \cdots (4n-1)} = -x^2 + \sum_{n=2}^{\infty} \frac{(-1)^n 4nx^{4n-2}}{2^{2n} n! \cdot 3 \cdot 7 \cdot 11 \cdots (4n-5)}$$

$$\begin{aligned} y'' + x^2 y &= -x^2 + \sum_{n=2}^{\infty} \frac{(-1)^n 4nx^{4n-2}}{2^{2n} n! \cdot 3 \cdot 7 \cdot 11 \cdots (4n-5)} + \sum_{n=1}^{\infty} \frac{(-1)^n x^{4n+2}}{2^{2n} n! \cdot 3 \cdot 7 \cdot 11 \cdots (4n-1)} + x^2 \\ &= \sum_{n=1}^{\infty} \frac{(-1)^{n+1} 4(n+1)x^{4n+2}}{2^{2n+2}(n+1)! \cdot 3 \cdot 7 \cdot 11 \cdots (4n-1)} - \sum_{n=1}^{\infty} \frac{(-1)^{n+1} x^{4n+2}}{2^{2n} n! \cdot 3 \cdot 7 \cdot 11 \cdots (4n-1)} \frac{2^2(n+1)}{2^2(n+1)} = 0 \end{aligned}$$

65. $J_0(x) = \sum_{k=0}^{\infty} \frac{(-1)^k x^{2k}}{2^{2k}(k!)^2}$

(a) $\lim_{k\to\infty} \left| \frac{u_{k+1}}{u_k} \right| = \lim_{k\to\infty} \left| \frac{(-1)^{k+1} x^{2k+2}}{2^{2k+2}\left[(k+1)!\right]^2} \cdot \frac{2^{2k}(k!)^2}{(-1)^k x^{2k}} \right| = \lim_{k\to\infty} \left| \frac{(-1)x^2}{2^2(k+1)^2} \right| = 0$

Therefore, the interval of convergence is $-\infty < x < \infty$.

(b)

$$J_0 = \sum_{k=0}^{\infty} (-1)^k \frac{x^{2k}}{4^k (k!)^2}$$

$$J_0' = \sum_{k=1}^{\infty} (-1)^k \frac{2kx^{2k-1}}{4^k (k!)^2} = \sum_{k=0}^{\infty} (-1)^{k+1} \frac{(2k+2)x^{2k+1}}{4^{k+1}\left[(k+1)!\right]^2}$$

$$J_0'' = \sum_{k=1}^{\infty} (-1)^k \frac{2k(2k-1)x^{2k-2}}{4^k (k!)^2} = \sum_{k=0}^{\infty} (-1)^{k+1} \frac{(2k+2)(2k+1)x^{2k}}{4^{k+1}\left[(k+1)!\right]^2}$$

$$\begin{aligned} x^2 J_0'' + xJ_0' + x^2 J_0 &= \sum_{k=0}^{\infty} (-1)^{k+1} \frac{2(2k+1)x^{2k+2}}{4^{k+1}(k+1)!k!} + \sum_{k=0}^{\infty} (-1)^{k+1} \frac{2x^{2k+2}}{4^{k+1}(k+1)!k!} + \sum_{k=0}^{\infty} (-1)^k \frac{x^{2k+2}}{4^k (k!)^2} \\ &= \sum_{k=0}^{\infty} \frac{(-1)^k x^{2k+2}}{4^k (k!)^2} \left[(-1)\frac{2(2k+1)}{4(k+1)} + (-1)\frac{2}{4(k+1)} + 1 \right] \\ &= \sum_{k=0}^{\infty} \frac{(-1)^k x^{2k+2}}{4^k (k!)^2} \left[\frac{-4k-2}{4k+4} - \frac{2}{4k+4} + \frac{4k+4}{4k+4} \right] = 0 \end{aligned}$$

(c) $P_6(x) = 1 - \dfrac{x^2}{4} + \dfrac{x^4}{64} - \dfrac{x^6}{2304}$

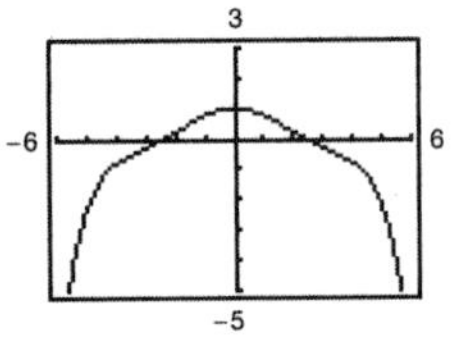

(d) $\int_0^1 J_0 dx = \int_0^1 \sum_{k=0}^{\infty} \dfrac{(-1)^k x^{2k}}{4^k (k!)^2} dx = \left[\sum_{k=0}^{\infty} \dfrac{(-1)^k x^{2k+1}}{4^k (k!)^2 (2k+1)}\right]_0^1 = \sum_{k=0}^{\infty} \dfrac{(-1)^k}{4^k (k!)^2 (2k+1)} = 1 - \dfrac{1}{12} + \dfrac{1}{320} \approx 0.92$

(integral is approximately 0.9197304101)

66. $J_1(x) = x\sum_{k=0}^{\infty} \dfrac{(-1)^k x^{2k}}{2^{2k+1} k!(k+1)!} = \sum_{k=0}^{\infty} \dfrac{(-1)^k x^{2k+1}}{2^{2k+1} k!(k+1)!}$

(a) $\lim_{k\to\infty} \left|\dfrac{u_{k+1}}{u_k}\right| = \lim_{k\to\infty} \left|\dfrac{(-1)^{k+1} x^{2k+3}}{2^{2k+3}(k+1)!(k+2)!} \cdot \dfrac{2^{2k+1} k!(k+1)!}{(-1)^k x^{2k+1}}\right| = \lim_{k\to\infty} \left|\dfrac{(-1)x^2}{2^2(k+2)(k+1)}\right| = 0$

Therefore, the interval of convergence is $-\infty < x < \infty$.

(b) $J_1(x) = \sum_{k=0}^{\infty} \dfrac{(-1)^k x^{2k+1}}{2^{2k+1} k!(k+1!)}$

$J_1'(x) = \sum_{k=0}^{\infty} \dfrac{(-1)^k (2k+1) x^{2k}}{2^{2k+1} k!(k+1)!}$

$J_1''(x) = \sum_{k=1}^{\infty} \dfrac{(-1)^k (2k+1)(2k) x^{2k-1}}{2^{2k+1} k!(k+1)!}$

$$\begin{aligned}
x^2 J_1'' + xJ_1' + (x^2 - 1)J_1 &= \sum_{k=1}^{\infty} \frac{(-1)^k (2k+1)(2k) x^{2k+1}}{2^{2k+1} k!(k+1)!} + \sum_{k=0}^{\infty} \frac{(-1)^k (2k+1) x^{2k+1}}{2^{2k+1} k!(k+1)!} \\
&\quad + \sum_{k=0}^{\infty} \frac{(-1)^k x^{2k+3}}{2^{2k+1} k!(k+1)!} - \sum_{k=0}^{\infty} \frac{(-1)^k x^{2k+1}}{2^{2k+1} k!(k+1)!} \\
&= \left[\sum_{k=1}^{\infty} \frac{(-1)^k (2k+1)(2k) x^{2k+1}}{2^{2k+1} k!(k+1)!} + \frac{x}{2} + \sum_{k=1}^{\infty} \frac{(-1)^k (2k+1) x^{2k+1}}{2^{2k+1} k!(k+1)!} - \frac{x}{2} - \sum_{k=1}^{\infty} \frac{(-1)^k x^{2k+1}}{2^{2k+1} k!(k+1)!}\right] \\
&\quad + \sum_{k=0}^{\infty} \frac{(-1)^k x^{2k+3}}{2^{2k+1} k!(k+1)!} \\
&= \sum_{k=1}^{\infty} \frac{(-1)^k x^{2k+1}\left[(2k+1)(2k) + (2k+1) - 1\right]}{2^{2k+1} k!(k+1)!} + \sum_{k=0}^{\infty} \frac{(-1)^k x^{2k+3}}{2^{2k+1} k!(k+1)!} \\
&= \sum_{k=1}^{\infty} \frac{(-1)^k x^{2k+1} 4k(k+1)}{2^{2k+1} k!(k+1)!} + \sum_{k=0}^{\infty} \frac{(-1)^k x^{2k+3}}{2^{2k+1} k!(k+1)!} \\
&= \sum_{k=1}^{\infty} \frac{(-1)^k x^{2k+1}}{2^{2k-1}(k-1)!k!} + \sum_{k=0}^{\infty} \frac{(-1)^k x^{2k+3}}{2^{2k+1} k!(k+1)!} \\
&= \sum_{k=0}^{\infty} \frac{(-1)^{k+1} x^{2k+3}}{2^{2k+1} k!(k+1)!} + \sum_{k=0}^{\infty} \frac{(-1)^k x^{2k+3}}{2^{2k+1} k!(k+1)!} \\
&= \sum_{n=0}^{\infty} \frac{(-1)^k x^{2k+3}\left[(-1) + 1\right]}{2^{2k+1} k!(k+1)!} = 0
\end{aligned}$$

(c) $P_7(x) = \dfrac{x}{2} - \dfrac{1}{16}x^3 + \dfrac{1}{384}x^5 - \dfrac{1}{18{,}432}x^7$

(d) $J_0'(x) = \displaystyle\sum_{k=0}^{\infty} \frac{(-1)^{k+1}2(k+1)x^{2k+1}}{2^{2k+2}(k+1)!(k+1)!} = \sum_{k=0}^{\infty} \frac{(-1)^{k+1}x^{2k+1}}{2^{2k+1}k!(k+1)!}$

$-J_1(x) = -\displaystyle\sum_{k=0}^{\infty} \frac{(-1)^{k}x^{2k+1}}{2^{2k+1}k!(k+1)!} = \sum_{k=0}^{\infty} \frac{(-1)^{k+1}x^{2k+1}}{2^{2k+1}k!(k+1)!}$ **Note:** $J_0'(x) = -J_1(x)$

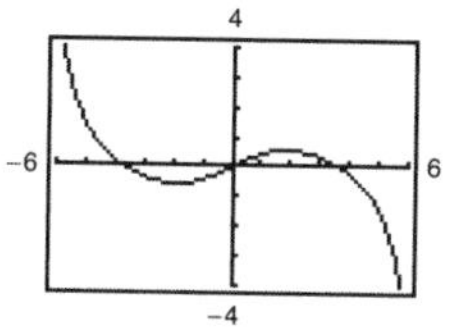

67. $\displaystyle\sum_{n=0}^{\infty} \left(\frac{x}{4}\right)^n, \quad (-4, 4)$

(a) $\displaystyle\sum_{n=0}^{\infty} \left(\frac{(5/2)}{4}\right)^n = \sum_{n=0}^{\infty} \left(\frac{5}{8}\right)^n = \frac{1}{1 - 5/8} = \frac{8}{3}$

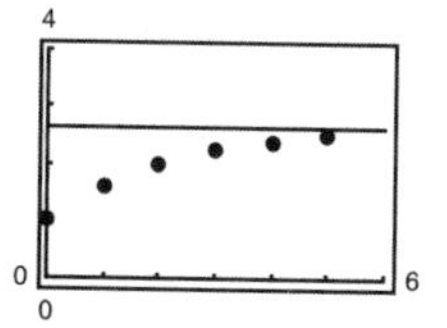

(b) $\displaystyle\sum_{n=0}^{\infty} \left(\frac{(-5/2)}{4}\right)^n = \sum_{n=0}^{\infty} \left(-\frac{5}{8}\right)^n = \frac{1}{1 + 5/8} = \frac{8}{13}$

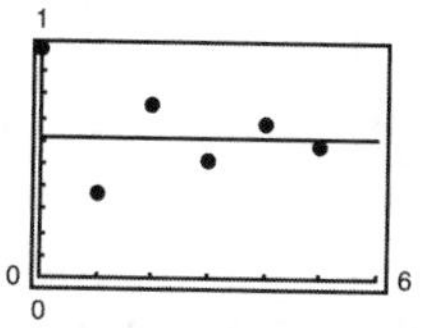

(c) The alternating series converges more rapidly. The partial sums of the series of positive terms approaches the sum from below. The partial sums of the alternating series alternate sides of the horizontal line representing the sum.

(d)

M	10	100	1000	10,000
N	5	14	24	35

68. $\displaystyle\sum_{n=0}^{\infty} (3x)^n$ converges on $\left(-\dfrac{1}{3}, \dfrac{1}{3}\right)$.

(a) $x = \dfrac{1}{6}$: $\displaystyle\sum_{n=0}^{\infty} \left(3\left(\frac{1}{6}\right)\right)^n = \sum_{n=0}^{\infty} \left(\frac{1}{2}\right)^n = \frac{1}{1 - (1/2)} = 2$

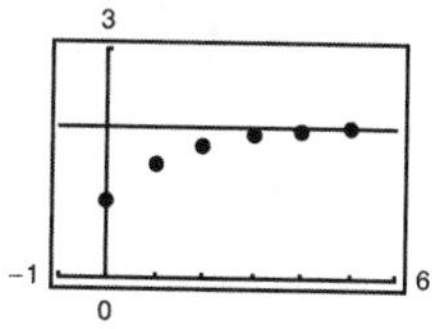

(b) $x = -\dfrac{1}{6}$: $\displaystyle\sum_{n=0}^{\infty} \left(3\left(-\frac{1}{6}\right)\right)^n = \sum_{n=0}^{\infty} \left(-\frac{1}{2}\right)^n = \frac{1}{1 + (1/2)} = \frac{2}{3}$

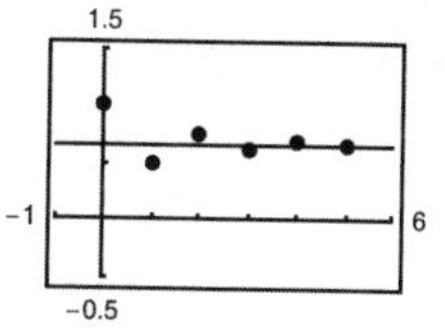

(c) The alternating series converges more rapidly. The partial sums in (a) approach the sum 2 from below. The partial sums in (b) alternate sides of the horizontal line $y = \dfrac{2}{3}$.

(d) $\displaystyle\sum_{n=0}^{N} \left(3 \cdot \frac{2}{3}\right)^n = \sum_{n=0}^{N} 2^n > M$

M	10	100	1000	10,000
N	3	6	9	13

69. $f(x) = \sum_{n=0}^{\infty} (-1)^n \dfrac{x^{2n}}{(2n)!} = \cos x$

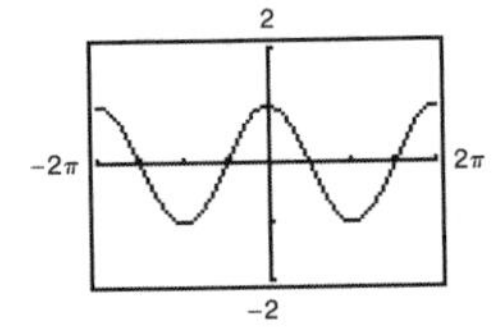

70. $f(x) = \sum_{n=0}^{\infty} (-1)^n \dfrac{x^{2n+1}}{(2n+1)!} = \sin x$

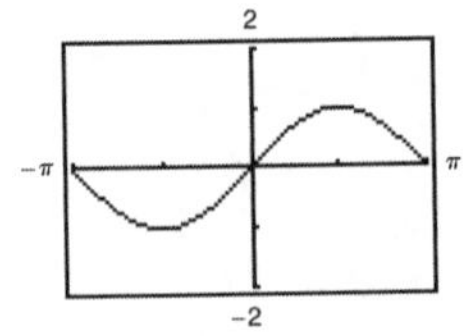

71. $f(x) = \sum_{n=0}^{\infty} (-1)^n x^n = \sum_{n=0}^{\infty} (-x)^n$ Geometric

$= \dfrac{1}{1-(-x)} = \dfrac{1}{1+x}$ for $-1 < x < 1$

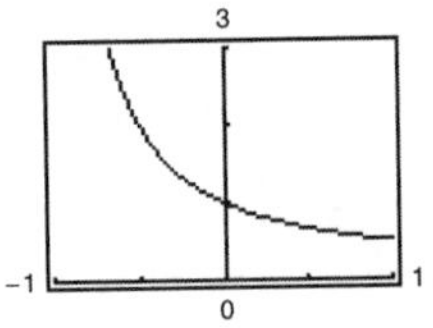

72. $f(x) = \sum_{n=0}^{\infty} (-1)^n \dfrac{x^{2n+1}}{2n+1} = \arctan x, -1 \le x \le 1$

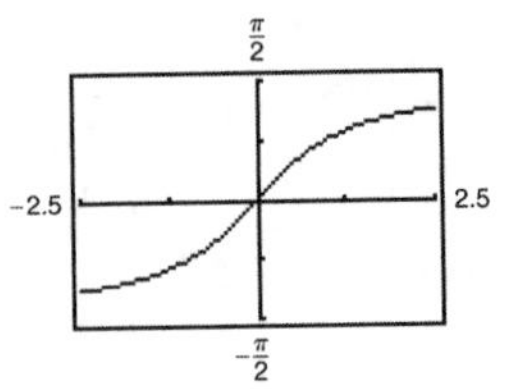

73. False;

$$\sum_{n=1}^{\infty} \frac{(-1)^n x^n}{n2^n}$$

converges for $x = 2$ but diverges for $x = -2$.

74. False; it is not possible. See Theorem 9.20.

75. True; the radius of convergence is $R = 1$ for both series.

76. True

$$\int_0^1 f(x)\,dx = \int_0^1 \left(\sum_{n=0}^{\infty} a_n x^n\right) dx$$

$$= \left[\sum_{n=0}^{\infty} \frac{a_n x^{n+1}}{n+1}\right]_0^1 = \sum_{n=0}^{\infty} \frac{a_n}{n+1}$$

77. $\lim_{n\to\infty} \left|\dfrac{a_{n+1}}{a_n}\right| = \lim_{n\to\infty} \left|\dfrac{(n+1+p)!}{(n+1)!(n+1+q)!} x^{n+1} \Big/ \dfrac{(n+p)!}{n!(n+q)!} x^n\right| = \lim_{n\to\infty} \left|\dfrac{(n+1+p)x}{(n+1)(n+1+q)}\right| = 0$

So, the series converges for all x: $R = \infty$.

78. (a) $g(x) = 1 + 2x + x^2 + 2x^3 + x^4 + \cdots$

$S_{2n} = 1 + 2x + x^2 + 2x^3 + x^4 + \cdots + x^{2n} + 2x^{2n+1} = (1 + x^2 + x^4 + \cdots + x^{2n}) + 2x(1 + x^2 + x^4 + \cdots + x^{2n})$

$\lim_{n\to\infty} S_{2n} = \sum_{n=0}^{\infty} x^{2n} + 2x\sum_{n=0}^{\infty} x^{2n}$

Each series is geometric, $R = 1$, and the interval of convergence is $(-1, 1)$.

(b) For $|x| < 1$, $g(x) = \dfrac{1}{1-x^2} + 2x\dfrac{1}{1-x^2} = \dfrac{1+2x}{1-x^2}$.

79. (a) $f(x) = \sum_{n=0}^{\infty} c_n x^n, c_{n+3} = c_n$

$= c_0 + c_1 x + c_2 x^2 + c_0 x^3 + c_1 x^4 + c_2 x^5 + c_0 x^6 + \cdots$

$S_{3n} = c_0(1 + x^3 + \cdots + x^{3n}) + c_1 x(1 + x^3 + \cdots + x^{3n}) + c_2 x^2(1 + x^3 + \cdots + x^{3n})$

$\lim_{n\to\infty} S_{3n} = c_0\sum_{n=0}^{\infty} x^{3n} + c_1 x\sum_{n=0}^{\infty} x^{3n} + c_2 x^2\sum_{n=0}^{\infty} x^{3n}$

Each series is geometric, $R = 1$, and the interval of convergence is $(-1, 1)$.

(b) For $|x| < 1$, $f(x) = c_0\dfrac{1}{1-x^3} + c_1 x\dfrac{1}{1-x^3} + c_2 x^2\dfrac{1}{1-x^3} = \dfrac{c_0 + c_1 x + c_2 x^2}{1-x^3}$.

80. For the series $\sum c_n x^n$,

$$\lim_{n\to\infty}\left|\frac{a_{n+1}}{a_n}\right| = \lim_{n\to\infty}\left|\frac{c_{n+1}x^{n+1}}{c_n x^n}\right| = \lim_{n\to\infty}\left|\frac{c_{n+1}}{c_n}x\right| < 1 \Rightarrow |x| < \left|\frac{c_n}{c_{n+1}}\right| = R$$

For the series $\sum c_n x^{2n}$,

$$\lim_{n\to\infty}\left|\frac{b_{n+1}}{b_n}\right| = \lim_{n\to\infty}\left|\frac{c_{n+1}x^{2n+2}}{c_n x^{2n}}\right| = \lim_{n\to\infty}\left|\frac{c_{n+1}}{c_n}x^2\right| < 1 \Rightarrow \left|x^2\right| < \left|\frac{c_n}{c_{n+1}}\right| = R \Rightarrow |x| < \sqrt{R}.$$

81. At $x = x_0 + R$, $\sum_{n=0}^{\infty} c_n(x - x_0)^n = \sum_{n=0}^{\infty} c_n R^n$, diverges.

At $x = x_0 - R$, $\sum_{n=0}^{\infty} c_n(x - x_0)^n = \sum_{n=0}^{\infty} c_n(-R)^n$, converges.

Furthermore, at $x = x_0 - R$,

$$\sum_{n=0}^{\infty}\left|c_n(x - x_0)^n\right| = \sum_{n=0}^{\infty} C_n R^n, \text{ diverges.}$$

So, the series converges conditionally at $x_0 - R$.

Section 9.9 Representation of Functions by Power Series

1. (a) $$\frac{1}{4 - x} = \frac{1/4}{1 - (x/4)}$$

$$= \frac{a}{1 - r} = \sum_{n=0}^{\infty}\left(\frac{1}{4}\right)\left(\frac{x}{4}\right)^n = \sum_{n=0}^{\infty}\frac{x^n}{4^{n+1}}$$

This series converges on $(-4, 4)$.

(b) $$\begin{array}{r l}
 & \frac{1}{4} + \frac{x}{16} + \frac{x^2}{64} + \cdots \\
4 - x \,\big) & 1 \\
 & 1 - \frac{x}{4} \\
\hline
 & \frac{x}{4} \\
 & \frac{x}{4} - \frac{x^2}{16} \\
\hline
 & \frac{x^2}{16} \\
 & \frac{x^2}{16} - \frac{x^3}{64} \\
\hline
 & \vdots
\end{array}$$

2. (a) $$\frac{1}{2 + x} = \frac{1/2}{1 - (-x/2)} = \frac{a}{1 - r}$$

$$= \sum_{n=0}^{\infty}\frac{1}{2}\left(-\frac{x}{2}\right)^n = \sum_{n=0}^{\infty}\frac{(-1)^n x^n}{2^{n+1}}$$

This series converges on $(-2, 2)$.

(b) $$\begin{array}{r l}
 & \frac{1}{2} - \frac{x}{4} + \frac{x^2}{8} - \frac{x^3}{16} + \cdots \\
2 + x \,\big) & 1 \\
 & 1 + \frac{x}{2} \\
\hline
 & -\frac{x}{2} \\
 & -\frac{x}{2} - \frac{x^2}{4} \\
\hline
 & \frac{x^2}{4} \\
 & \frac{x^2}{4} + \frac{x^3}{8} \\
\hline
 & -\frac{x^3}{8} \\
 & -\frac{x^3}{8} - \frac{x^4}{16} \\
\hline
 & \vdots
\end{array}$$

3. (a) $\dfrac{4}{3+x} = \dfrac{4/3}{1-(-x/3)} = \dfrac{a}{1-r}$

$$= \sum_{n=0}^{\infty} \frac{4}{3}\left(\frac{-x}{3}\right)^n = \sum_{n=0}^{\infty} \frac{4(-1)^n x^n}{3^{n+1}}$$

The series converges on $(-3, 3)$.

(b)
$$\begin{array}{r l}
 & \frac{4}{3} - \frac{4}{9}x + \frac{4x^2}{27} - \cdots \\
3 + x \,\big) & 4 \\
 & 4 + \frac{4}{3}x \\ \hline
 & -\frac{4}{3}x \\
 & -\frac{4}{3}x - \frac{4x^2}{9} \\ \hline
 & \frac{4x^2}{9} \\
 & \frac{4x^2}{9} + \frac{4x^3}{27} \\ \hline
 & -\frac{4x^3}{27} \\
 & \vdots
\end{array}$$

4. (a) $\dfrac{2}{5-x} = \dfrac{2/5}{1-(x/5)} = \dfrac{a}{1-r}$

$$= \sum_{n=0}^{\infty} \frac{2}{5}\left(\frac{x}{5}\right)^n = \sum_{n=0}^{\infty} \frac{2x^n}{5^{n+1}}$$

This series converges on $(-5, 5)$.

(b)
$$\begin{array}{r l}
 & \frac{2}{5} + \frac{2x}{25} + \frac{2x^2}{125} + \cdots \\
5 - x \,\big) & 2 \\
 & 2 - \frac{2}{5}x \\ \hline
 & \frac{2}{5}x \\
 & \frac{2x}{5} - \frac{2x^2}{25} \\ \hline
 & \frac{2x^2}{25} \\
 & \vdots
\end{array}$$

5. $\dfrac{1}{3-x} = \dfrac{1}{2-(x-1)} = \dfrac{1/2}{1-\left(\dfrac{x-1}{2}\right)} = \dfrac{a}{1-r}$

$$= \sum_{n=0}^{\infty} \frac{1}{2}\left(\frac{x-1}{2}\right)^n = \sum_{n=0}^{\infty} \frac{(x-1)^n}{2^{n+1}}$$

Interval of convergence: $\left|\dfrac{x-1}{2}\right| < 1 \Rightarrow |x-1| < 2 \Rightarrow (-1, 3)$

6. $\dfrac{2}{6-x} = \dfrac{2}{8-(x+2)} = \dfrac{1/4}{1-\left(\dfrac{x+2}{8}\right)} = \dfrac{a}{1-r}$

$$= \sum_{n=0}^{\infty} \frac{1}{4}\left(\frac{x+2}{8}\right)^n = \sum_{n=0}^{\infty} \frac{(x+2)^n}{2^{3n+2}}$$

Interval of convergence: $\left|\dfrac{x+2}{8}\right| < 1 \Rightarrow |x+2| < 8 \Rightarrow (-10, 6)$

7. $\dfrac{1}{1-3x} = \dfrac{a}{1-r} = \displaystyle\sum_{n=0}^{\infty} (3x)^n$

Interval of convergence: $|3x| < 1 \Rightarrow \left(\dfrac{1}{3}, \dfrac{1}{3}\right)$

8. $\dfrac{1}{1-5x} = \dfrac{a}{1-r} = \displaystyle\sum_{n=0}^{\infty} (5x)^n$

Interval of convergence: $|5x| < 1 \Rightarrow \left(\dfrac{1}{5}, \dfrac{1}{5}\right)$

9. $\dfrac{5}{2x-3} = \dfrac{5}{-9+2(x+3)} = \dfrac{-5/9}{1-\dfrac{2}{9}(x+3)} = \dfrac{a}{1-r}$

$$= -\frac{5}{9}\sum_{n=0}^{\infty}\left(\frac{2}{9}(x+3)\right)^n, \quad \left|\frac{2}{9}(x+3)\right| < 1$$

$$= -5\sum_{n=0}^{\infty} \frac{2^n}{9^{n+1}}(x+3)^n$$

Interval of convergence: $\left|\dfrac{2}{9}(x+3)\right| < 1 \Rightarrow \left(-\dfrac{15}{2}, \dfrac{3}{2}\right)$

10. $\dfrac{3}{2x-1} = \dfrac{3}{3+2(x-2)} = \dfrac{1}{1+(2/3)(x-2)} = \dfrac{a}{1-r}$

$= \displaystyle\sum_{n=0}^{\infty}\left[-\frac{2}{3}(x-2)\right]^n$

$= \displaystyle\sum_{n=0}^{\infty}\frac{(-2)^n(x-2)^n}{3^n}$

Interval of convergence: $|x-2| < \dfrac{3}{2} \Rightarrow \left(\dfrac{1}{2}, \dfrac{7}{2}\right)$

11. $\dfrac{3}{3x+4} = \dfrac{3/4}{1+\frac{3}{4}x} = \dfrac{3/4}{1-\left(-\frac{3}{4}x\right)} = \dfrac{a}{1-r}$

$= \displaystyle\sum_{n=0}^{\infty}\frac{3}{4}\left(-\frac{3}{4}x\right)^n = \sum_{n=0}^{\infty}\frac{(-1)^n 3^{n+1} x^n}{4^{n+1}}$

Interval of convergence:

$\left|-\dfrac{3}{4}x\right| < 1 \Rightarrow |3x| < 4 \Rightarrow |x| < \dfrac{4}{3} \Rightarrow \left(-\dfrac{4}{3}, \dfrac{4}{3}\right)$

12. $\dfrac{4}{3x+2} = \dfrac{4}{11+3(x-3)} = \dfrac{4/11}{1-(-3/11)(x-3)} = \dfrac{a}{1-r}$

$= \dfrac{4}{11}\displaystyle\sum_{n=0}^{\infty}\left[\frac{-3(x-3)}{11}\right]^n$

$= 4\displaystyle\sum_{n=0}^{\infty}\frac{(-3)^n(x-3)^n}{11^{n+1}}$

Interval of convergence: $\left|-\dfrac{3}{11}(x-3)\right| < 1 \Rightarrow \left(3, \dfrac{20}{3}\right)$

13. $\dfrac{4x}{x^2+2x-3} = \dfrac{3}{x+3} + \dfrac{1}{x-1}$

$= \dfrac{1}{1-(-x/3)} + \dfrac{-1}{1-x}$

$= \displaystyle\sum_{n=0}^{\infty}\left(-\frac{x}{3}\right)^n - \sum_{n=0}^{\infty}x^n = \sum_{n=0}^{\infty}\left[\frac{1}{(-3)^n} - 1\right]x^n$

Interval of convergence: $\left|\dfrac{x}{3}\right| < 1$ and $|x| < 1 \Rightarrow (-1, 1)$

14. $\dfrac{3x-8}{3x^2+5x-2} = \dfrac{2}{x+2} - \dfrac{3}{3x-1}$

$= \dfrac{1}{1-(-x/2)} + \dfrac{3}{1-3x}$

$= \displaystyle\sum_{n=0}^{\infty}\left(-\frac{x}{2}\right)^n + 3\sum_{n=0}^{\infty}(3x)^n$

$= \displaystyle\sum_{n=0}^{\infty}\left[\left(-\frac{1}{2}\right)^n + 3^{n+1}\right]x^n$

Interval of convergence: $\left|\dfrac{x}{2}\right| < 1$ and

$|3x| < 1 \Rightarrow \left(-\dfrac{1}{3}, \dfrac{1}{3}\right)$

15. $\dfrac{2}{1-x^2} = \dfrac{1}{1-x} + \dfrac{1}{1+x}$

$= \displaystyle\sum_{n=0}^{\infty}\left(1+(-1)^n\right)x^n = 2\sum_{n=0}^{\infty}x^{2n}$

Interval of convergence: $|x^2| < 1$ or $(-1, 1)$ because

$\displaystyle\lim_{n\to\infty}\left|\frac{u_{n+1}}{u_n}\right| = \lim_{n\to\infty}\left|\frac{2x^{2n+2}}{2x^{2n}}\right| = |x^2|$

16. $\dfrac{5}{5+x^2} = \dfrac{1}{1-\left(\dfrac{-x^2}{5}\right)} = \dfrac{a}{1-r} = \displaystyle\sum_{n=0}^{\infty}\left(-\frac{x^2}{5}\right)^n = \sum_{n=0}^{\infty}\left(\frac{-1}{5}\right)^n x^{2n}$

Interval of convergence: $\left|\dfrac{x^2}{5}\right| < 1 \Rightarrow -5 < x^2 < 5 \Rightarrow \left(-\sqrt{5}, \sqrt{5}\right)$

17. $\dfrac{1}{1+x} = \displaystyle\sum_{n=0}^{\infty}(-1)^n x^n$

$\dfrac{1}{1-x} = \displaystyle\sum_{n=0}^{\infty}(-1)^n(-x)^n = \sum_{n=0}^{\infty}(-1)^{2n}x^n = \sum_{n=0}^{\infty}x^n$

$h(x) = \dfrac{-2}{x^2-1} = \dfrac{1}{1+x} + \dfrac{1}{1-x} = \displaystyle\sum_{n=0}^{\infty}(-1)^n x^n + \sum_{n=0}^{\infty}x^n = \sum_{n=0}^{\infty}\left[(-1)^n + 1\right]x^n$

$= 2 + 0x + 2x^2 + 0x^3 + 2x^4 + 0x^5 + 2x^6 + \cdots = 2\displaystyle\sum_{n=0}^{\infty}x^{2n}$, $(-1, 1)$ (See Exercise 15.)

18. $h(x) = \dfrac{x}{x^2 - 1} = \dfrac{1}{2(1 + x)} - \dfrac{1}{2(1 - x)} = \dfrac{1}{2}\sum_{n=0}^{\infty} (-1)^n x^n - \dfrac{1}{2}\sum_{n=0}^{\infty} x^n$ (See Exercise 17.)

$$= \frac{1}{2}\sum_{n=0}^{\infty}\left[(-1)^n - 1\right]x^n = \frac{1}{2}\left[0 - 2x + 0x^2 - 2x^3 + 0x^4 - 2x^5 + \cdots\right]$$

$$= \frac{1}{2}\sum_{n=0}^{\infty}(-2)x^{2n+1} = -\sum_{n=0}^{\infty} x^{2n+1}, (-1, 1)$$

19. By taking the first derivative, you have $\dfrac{d}{dx}\left[\dfrac{1}{x + 1}\right] = \dfrac{-1}{(x + 1)^2}$. Therefore,

$$\frac{-1}{(x + 1)^2} = \frac{d}{dx}\left[\sum_{n=0}^{\infty}(-1)^n x^n\right] = \sum_{n=1}^{\infty}(-1)^n nx^{n-1} = \sum_{n=0}^{\infty}(-1)^{n+1}(n + 1)x^n, (-1, 1).$$

20. By taking the second derivative, you have $\dfrac{d^2}{dx^2}\left[\dfrac{1}{x + 1}\right] = \dfrac{2}{(x + 1)^3}$. Therefore,

$$\frac{2}{(x + 1)^3} = \frac{d^2}{dx^2}\left[\sum_{n=0}^{\infty}(-1)^n x^n\right] = \frac{d}{dx}\left[\sum_{n=1}^{\infty}(-1)^n nx^{n-1}\right] = \sum_{n=2}^{\infty}(-1)^n n(n - 1)x^{n-2} = \sum_{n=0}^{\infty}(-1)^n(n + 2)(n + 1)x^n, (-1, 1).$$

21. By integrating, you have $\displaystyle\int \frac{1}{x + 1}\,dx = \ln(x + 1)$. Therefore,

$$\ln(x + 1) = \int\left[\sum_{n=0}^{\infty}(-1)^n x^n\right]dx = C + \sum_{n=0}^{\infty}\frac{(-1)^n x^{n+1}}{n + 1}, -1 < x \le 1.$$

To solve for C, let $x = 0$ and conclude that $C = 0$. Therefore,

$$\ln(x + 1) = \sum_{n=0}^{\infty}\frac{(-1)^n x^{n+1}}{n + 1}, (-1, 1].$$

22. By integrating, you have

$\displaystyle\int \frac{1}{1 + x}\,dx = \ln(1 + x) + C_1$ and $\displaystyle\int \frac{1}{1 - x}\,dx = -\ln(1 - x) + C_2$.

$f(x) = \ln(1 - x^2) = \ln(1 + x) - \left[-\ln(1 - x)\right]$. Therefore,

$$\ln(1 - x^2) = \int\frac{1}{1 + x}\,dx - \int\frac{1}{1 - x}\,dx$$

$$= \int\left[\sum_{n=0}^{\infty}(-1)^n x^n\right]dx - \int\left[\sum_{n=0}^{\infty}x^n\right]dx = \left[C_1 + \sum_{n=0}^{\infty}\frac{(-1)^n x^{n+1}}{n + 1}\right] - \left[C_2 + \sum_{n=0}^{\infty}\frac{x^{n+1}}{n + 1}\right]$$

$$= C + \sum_{n=0}^{\infty}\frac{\left[(-1)^n - 1\right]x^{n+1}}{n + 1} = C + \sum_{n=0}^{\infty}\frac{-2x^{2n+2}}{2n + 2} = C + \sum_{n=0}^{\infty}\frac{(-1)x^{2n+2}}{n + 1}$$

To solve for C, let $x = 0$ and conclude that $C = 0$. Therefore,

$$\ln(1 - x^2) = -\sum_{n=0}^{\infty}\frac{x^{2n+2}}{n + 1}, (-1, 1).$$

23. $\dfrac{1}{x^2 + 1} = \displaystyle\sum_{n=0}^{\infty}(-1)^n(x^2)^n = \sum_{n=0}^{\infty}(-1)^n x^{2n}, (-1, 1)$

24. $\frac{2x}{x^2+1} = 2x\sum_{n=0}^{\infty}(-1)^n x^{2n}$ (See Exercise 23.)

$$= \sum_{n=0}^{\infty}(-1)^n 2x^{2n+1}$$

Because $\frac{d}{dx}\left(\ln\left(x^2+1\right)\right) = \frac{2x}{x^2+1}$, you have

$$\ln\left(x^2+1\right) = \int\left[\sum_{n=0}^{\infty}(-1)^n 2x^{2n+1}\right]dx = C + \sum_{n=0}^{\infty}\frac{(-1)^n x^{2n+2}}{n+1}, -1 \le x \le 1.$$

To solve for C, let $x = 0$ and conclude that $C = 0$. Therefore,

$$\ln\left(x^2+1\right) = \sum_{n=0}^{\infty}\frac{(-1)^n x^{2n+2}}{n+1}, [-1, 1].$$

25. Because, $\frac{1}{x+1} = \sum_{n=0}^{\infty}(-1)^n x^n$, you have $\frac{1}{4x^2+1} = \sum_{n=0}^{\infty}(-1)^n\left(4x^2\right)^n = \sum_{n=0}^{\infty}(-1)^n 4^n x^{2n} = \sum_{n=0}^{\infty}(-1)^n(2x)^{2n}, \left(-\frac{1}{2}, \frac{1}{2}\right).$

26. Because $\int\frac{1}{4x^2+1}\,dx = \frac{1}{2}\arctan(2x)$, you can use the result of Exercise 25 to obtain

$$\arctan(2x) = 2\int\frac{1}{4x^2+1}\,dx = 2\int\left[\sum_{n=0}^{\infty}(-1)^n 4^n x^{2n}\right]dx = C + 2\sum_{n=0}^{\infty}\frac{(-1)^n 4^n x^{2n+1}}{2n+1}, -\frac{1}{2} < x < \frac{1}{2}.$$

To solve for C, let $x = 0$ and conclude that $C = 0$. Therefore,

$$\arctan(2x) = 2\sum_{n=0}^{\infty}\frac{(-1)^n 4^n x^{2n+1}}{2n+1}, \left(-\frac{1}{2}, \frac{1}{2}\right].$$

27. $x - \frac{x^2}{2} \le \ln(x+1) \le x - \frac{x^2}{2} + \frac{x^3}{3}$

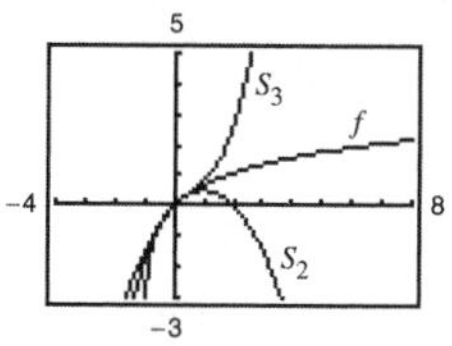

x	0.0	0.2	0.4	0.6	0.8	1.0
$S_2 = x - \frac{x^2}{2}$	0.000	0.180	0.320	0.420	0.480	0.500
$\ln(x+1)$	0.000	0.182	0.336	0.470	0.588	0.693
$S_3 = x - \frac{x^2}{2} + \frac{x^3}{3}$	0.000	0.183	0.341	0.492	0.651	0.833

28. $x - \frac{x^2}{2} + \frac{x^3}{3} - \frac{x^4}{4} \le \ln(x+1) \le x - \frac{x^2}{2} + \frac{x^3}{3} - \frac{x^4}{4} + \frac{x^5}{5}$

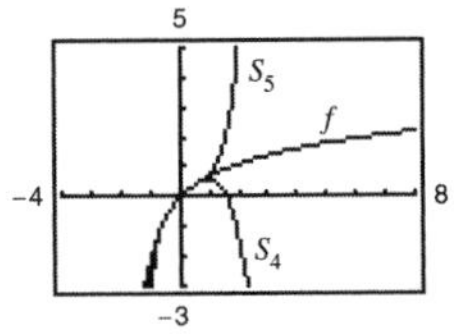

x	0.0	0.2	0.4	0.6	0.8	1.0
$S_4 = x - \frac{x^2}{2} + \frac{x^3}{3} - \frac{x^4}{4}$	0.0	0.18227	0.33493	0.45960	0.54827	0.58333
$\ln(x+1)$	0.0	0.18232	0.33647	0.47000	0.58779	0.69315
$S_5 = x - \frac{x^2}{2} + \frac{x^3}{3} - \frac{x^4}{4} + \frac{x^5}{5}$	0.0	0.18233	0.33698	0.47515	0.61380	0.78333

29. $\sum_{n=1}^{\infty} \frac{(-1)^{n+1}(x-1)^n}{n} = \frac{(x-1)}{1} - \frac{(x-1)^2}{2} + \frac{(x-1)^3}{3} - \cdots$

(a)

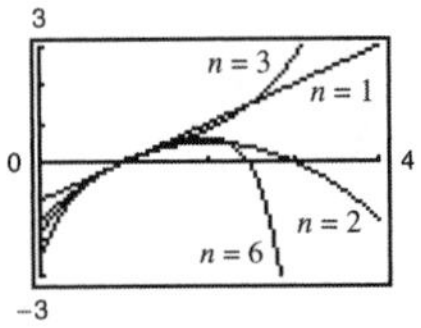

(b) From Example 4,

$$\sum_{n=1}^{\infty} \frac{(-1)^{n+1}(x-1)^n}{n} = \sum_{n=0}^{\infty} \frac{(-1)^n(x-1)^{n+1}}{n+1} = \ln x,\ 0 < x \le 2,\ R = 1.$$

(c) $x = 0.5$:

$$\sum_{n=1}^{\infty} \frac{(-1)^{n+1}(-1/2)^n}{n} = \sum_{n=1}^{\infty} \frac{-(1/2)^n}{n} \approx -0.693147$$

(d) This is an approximation of $\ln\left(\frac{1}{2}\right)$. The error is approximately 0. [The error is less than the first omitted term, $1/(51 \cdot 2^{51}) \approx 8.7 \times 10^{-18}$.]

30. $\sum_{n=0}^{\infty} \frac{(-1)^n x^{2n+1}}{(2n+1)!} = x - \frac{x^3}{3!} + \frac{x^5}{5!} - \cdots$

(a)

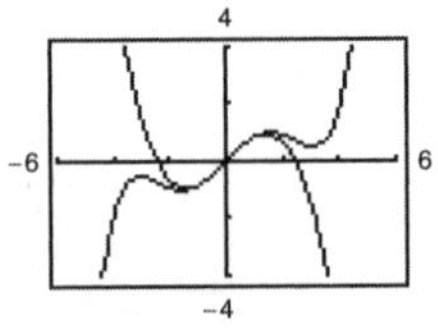

(b) $\sum_{n=0}^{\infty} \frac{(-1)^n x^{2n+1}}{(2n+1)!} = \sin x,\ R = \infty$

(c) $\sum_{n=0}^{\infty} \frac{(-1)^n (1/2)^{2n+1}}{(2n+1)!} \approx 0.4794255386$

(d) This is an approximation of $\sin\left(\frac{1}{2}\right)$. The error is approximately 0.

In Exercises 31–34, $\arctan x = \sum_{n=0}^{\infty} (-1)^n \frac{x^{2n+1}}{2n+1}$.

31. $\arctan \frac{1}{4} = \sum_{n=0}^{\infty} (-1)^n \frac{(1/4)^{2n+1}}{2n+1} = \sum_{n=0}^{\infty} \frac{(-1)^n}{(2n+1)4^{2n+1}} = \frac{1}{4} - \frac{1}{192} + \frac{1}{5120} + \cdots$

Because $\frac{1}{5120} < 0.001$, you can approximate the series by its first two terms: $\arctan \frac{1}{4} \approx \frac{1}{4} - \frac{1}{192} \approx 0.245$.

32.

$$\arctan x^2 = \sum_{n=0}^{\infty} (-1)^n \frac{x^{4n+2}}{2n+1}$$

$$\int \arctan x^2 \, dx = \sum_{n=0}^{\infty} (-1)^n \frac{x^{4n+3}}{(4n+3)(2n+1)} + C, C = 0$$

$$\int_0^{3/4} \arctan x^2 \, dx = \sum_{n=0}^{\infty} (-1)^n \frac{(3/4)^{4n+3}}{(4n+3)(2n+1)} = \sum_{n=0}^{\infty} (-1)^n \frac{3^{4n+3}}{(4n+3)(2n+1)4^{4n+3}}$$

$$= \frac{27}{192} - \frac{2187}{344{,}064} + \frac{177{,}147}{230{,}686{,}720}$$

Because $177{,}147/230{,}686{,}720 < 0.001$, you can approximate the series by its first two terms: 0.134.

33.

$$\frac{\arctan x^2}{x} = \frac{1}{x}\sum_{n=0}^{\infty} (-1)^n \frac{(x^2)^{2n+1}}{2n+1} = \sum_{n=0}^{\infty} (-1)^2 \frac{x^{4n+1}}{2n+1}$$

$$\int \frac{\arctan x^2}{x} \, dx = \sum_{n=0}^{\infty} (-1)^n \frac{x^{4n+2}}{(4n+2)(2n+1)} + C \text{ (Note: } C = 0)$$

$$\int_0^{1/2} \frac{\arctan x^2}{x} \, dx = \sum_{n=0}^{\infty} (-1)^n \frac{1}{(4n+2)(2n+1)2^{4n+2}} = \frac{1}{8} - \frac{1}{1152} + \cdots$$

Because $\frac{1}{1152} < 0.001$, you can approximate the series by its first term: $\int_0^{1/2} \frac{\arctan x^2}{x} \, dx \approx 0.125$.

34.

$$x^2 \arctan x = \sum_{n=0}^{\infty} (-1)^n \frac{x^{2n+3}}{2n+1}$$

$$\int x^2 \arctan x \, dx = \sum_{n=0}^{\infty} (-1)^n \frac{x^{2n+4}}{(2n+4)(2n+1)}$$

$$\int_0^{1/2} x^2 \arctan x \, dx = \sum_{n=0}^{\infty} (-1)^n \frac{1}{(2n+4)(2n+1)2^{2n+4}} = \frac{1}{64} - \frac{1}{1152} + \cdots$$

Because $\frac{1}{1152} < 0.001$, you can approximate the series by its first term: $\int_0^{1/2} x^2 \arctan x \, dx \approx 0.016$.

In Exercises 35–38, use $\frac{1}{1-x} = \sum_{n=0}^{\infty} x^n, |x| < 1$.

35. $\frac{1}{(1-x)^2} = \frac{d}{dx}\left[\frac{1}{1-x}\right] = \frac{d}{dx}\left[\sum_{n=0}^{\infty} x^n\right] = \sum_{n=1}^{\infty} nx^{n-1}, |x| < 1$

36. $\frac{x}{(1-x)^2} = x\sum_{n=1}^{\infty} nx^{n-1} = \sum_{n=1}^{\infty} nx^n, |x| < 1$

37.

$$\frac{1+x}{(1-x)^2} = \frac{1}{(1-x)^2} + \frac{x}{(1-x)^2}$$

$$= \sum_{n=1}^{\infty} n(x^{n-1} + x^n), \quad |x| < 1$$

$$= \sum_{n=0}^{\infty} (2n+1)x^n, \quad |x| < 1$$

38. $\frac{x(1+x)}{(1-x)^2} = x\sum_{n=0}^{\infty} (2n+1)x^n = \sum_{n=0}^{\infty} (2n+1)x^{n+1}, |x| < 1$

(See Exercise 37.)

39. $P(n) = \left(\frac{1}{2}\right)^n$

$$E(n) = \sum_{n=1}^{\infty} nP(n) = \sum_{n=1}^{\infty} n\left(\frac{1}{2}\right)^n = \frac{1}{2}\sum_{n=1}^{\infty} n\left(\frac{1}{2}\right)^{n-1}$$
$$= \frac{1}{2}\frac{1}{\left[1 - (1/2)\right]^2} = 2$$

Because the probability of obtaining a head on a single toss is $\frac{1}{2}$, it is expected that, on average, a head will be obtained in two tosses.

40. (a) $\frac{1}{3}\sum_{n=1}^{\infty} n\left(\frac{2}{3}\right)^n = \frac{2}{9}\sum_{n=1}^{\infty} n\left(\frac{2}{3}\right)^{n-1} = \frac{2}{9}\frac{1}{\left[1 - (2/3)\right]^2} = 2$

(b) $\frac{1}{10}\sum_{n=1}^{\infty} n\left(\frac{9}{10}\right)^n = \frac{9}{100}\sum_{n=1}^{\infty} n\left(\frac{9}{10}\right)^{n-1}$

$$= \frac{9}{100}\cdot\frac{1}{\left[1 - (9/10)\right]^2} = 9$$

41. Because $\frac{1}{1 + x} = \frac{1}{1 - (-x)}$, substitute $(-x)$ into the geometric series.

42. Because $\frac{1}{1 - x^2} = \frac{1}{1 - \left(x^2\right)}$, substitute $\left(x^2\right)$ into the geometric series.

43. Because $\frac{1}{1 + x} = 5\left(\frac{1}{1 - (-x)}\right)$, substitute $(-x)$ into the geometric series and then multiply the series by 5.

44. Because $\ln(1 - x) = -\int \frac{1}{1 - x}\,dx$, integrate the series and then multiply by (-1).

45. Let $\arctan x + \arctan y = \theta$. Then,

$$\tan(\arctan x + \arctan y) = \tan\theta$$
$$\frac{\tan(\arctan x) + \tan(\arctan y)}{1 - \tan(\arctan x)\tan(\arctan y)} = \tan\theta$$
$$\frac{x + y}{1 - xy} = \tan\theta$$
$$\arctan\left(\frac{x + y}{1 - xy}\right) = \theta.$$

Therefore,

$$\arctan x + \arctan y = \arctan\left(\frac{x + y}{1 - xy}\right) \text{ for } xy \neq 1.$$

46. (a) From Exercise 45, you have

$$\arctan\frac{120}{119} - \arctan\frac{1}{239} = \arctan\frac{120}{119} + \arctan\left(-\frac{1}{239}\right) = \arctan\left[\frac{(120/119) + (-1/239)}{1 - (120/119)(-1/239)}\right]$$
$$= \arctan\left(\frac{28{,}561}{28{,}561}\right) = \arctan 1 = \frac{\pi}{4}$$

(b) $$2\arctan\frac{1}{5} = \arctan\frac{1}{5} + \arctan\frac{1}{5} = \arctan\left[\frac{2(1/5)}{1 - (1/5)^2}\right] = \arctan\frac{10}{24} = \arctan\frac{5}{12}$$

$$4\arctan\frac{1}{5} = 2\arctan\frac{1}{5} + 2\arctan\frac{1}{5} = \arctan\frac{5}{12} + \arctan\frac{5}{12} = \arctan\left[\frac{2(5/12)}{1 - (5/12)^2}\right] = \arctan\frac{120}{119}$$

$$4\arctan\frac{1}{5} - \arctan\frac{1}{239} = \arctan\frac{120}{119} - \arctan\frac{1}{239} = \frac{\pi}{4} \text{ (see part (a).)}$$

47. (a) $$2\arctan\frac{1}{2} = \arctan\frac{1}{2} + \arctan\frac{1}{2} = \arctan\left[\frac{\frac{1}{2} + \frac{1}{2}}{1 - (1/2)^2}\right] = \arctan\frac{4}{3}$$

$$2\arctan\frac{1}{2} - \arctan\frac{1}{7} = \arctan\frac{4}{3} + \arctan\left(-\frac{1}{7}\right) = \arctan\left[\frac{(4/3) - (1/7)}{1 + (4/3)(1/7)}\right] = \arctan\frac{25}{25} = \arctan 1 = \frac{\pi}{4}$$

(b) $$\pi = 8\arctan\frac{1}{2} - 4\arctan\frac{1}{7} \approx 8\left[\frac{1}{2} - \frac{(0.5)^3}{3} + \frac{(0.5)^5}{5} - \frac{(0.5)^7}{7}\right] - 4\left[\frac{1}{7} - \frac{(1/7)^3}{3} + \frac{(1/7)^5}{5} - \frac{(1/7)^7}{7}\right] \approx 3.14$$

48. (a) $\arctan\frac{1}{2} + \arctan\frac{1}{3} = \arctan\left[\frac{(1/2) + (1/3)}{1 - (1/2)(1/3)}\right] = \arctan\left(\frac{5/6}{5/6}\right) = \frac{\pi}{4}$

(b) $\pi = 4\left[\arctan\frac{1}{2} + \arctan\frac{1}{3}\right]$

$$= 4\left[\frac{1}{2} - \frac{(1/2)^3}{3} + \frac{(1/2)^5}{5} - \frac{(1/2)^7}{7}\right] + 4\left[\frac{1}{3} - \frac{(1/3)^3}{3} + \frac{(1/3)^5}{5} - \frac{(1/3)^7}{7}\right] \approx 4(0.4635) + 4(0.3217) \approx 3.14$$

49. From Exercise 21, you have

$$\ln(x + 1) = \sum_{n=0}^{\infty}\frac{(-1)^n x^{n+1}}{n + 1} = \sum_{n=1}^{\infty}\frac{(-1)^{n-1}x^n}{n}$$
$$= \sum_{n=1}^{\infty}\frac{(-1)^{n+1}x^n}{n}.$$

So, $\sum_{n=1}^{\infty}(-1)^{n+1}\frac{1}{2^n n} = \sum_{n=1}^{\infty}\frac{(-1)^{n+1}(1/2)^n}{n}$

$$= \ln\left(\frac{1}{2} + 1\right) = \ln\frac{3}{2} \approx 0.4055.$$

50. From Exercise 49, you have

$$\sum_{n=1}^{\infty}(-1)^{n+1}\frac{1}{3^n n} = \sum_{n=1}^{\infty}\frac{(-1)^{n+1}(1/3)^n}{n}$$
$$= \ln\left(\frac{1}{3} + 1\right) = \ln\frac{4}{3} \approx 0.2877.$$

51. From Exercise 49, you have

$$\sum_{n=1}^{\infty}(-1)^{n+1}\frac{2^n}{5^n n} = \sum_{n=1}^{\infty}\frac{(-1)^{n+1}(2/5)^n}{n}$$
$$= \ln\left(\frac{2}{5} + 1\right) = \ln\frac{7}{5} \approx 0.3365.$$

52. From Example 5, you have $\arctan x = \sum_{n=0}^{\infty}(-1)^n\frac{x^{2n+1}}{2n + 1}$.

$$\sum_{n=0}^{\infty}(-1)^n\frac{1}{2n + 1} = \sum_{n=0}^{\infty}(-1)^n\frac{(1)^{2n+1}}{2n + 1}$$
$$= \arctan 1 = \frac{\pi}{4} \approx 0.7854$$

53. From Exercise 52, you have

$$\sum_{n=0}^{\infty}(-1)^n\frac{1}{2^{2n+1}(2n + 1)} = \sum_{n=0}^{\infty}(-1)^n\frac{(1/2)^{2n+1}}{2n + 1}$$
$$= \arctan\frac{1}{2} \approx 0.4636.$$

54. From Exercise 52, you have

$$\sum_{n=1}^{\infty}(-1)^{n+1}\frac{1}{3^{2n-1}(2n - 1)} = \sum_{n=0}^{\infty}(-1)^n\frac{1}{3^{2n+1}(2n + 1)}$$
$$= \sum_{n=0}^{\infty}(-1)^n\frac{(1/3)^{2n+1}}{2n + 1}$$
$$= \arctan\frac{1}{3} \approx 0.3218.$$

55. The series in Exercise 52 converges to its sum at a slower rate because its terms approach 0 at a much slower rate.

56. Because $\frac{d}{dx}\left[\sum_{n=0}^{\infty}a_n x^n\right] = \sum_{n=1}^{\infty}na_n x^{n-1}$, the radius of convergence is the same, 3.

57. Because the first series is the derivative of the second series, the second series converges for $|x + 1| < 4$ (and perhaps at the endpoints, $x = 3$ and $x = -5$.)

58.

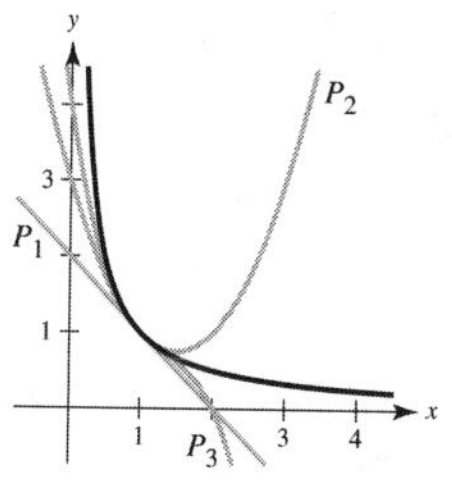

59. $\displaystyle\sum_{n=0}^{\infty} \frac{(-1)^n}{3^n(2n+1)}$

From Example 5 you have $\arctan x = \displaystyle\sum_{n=0}^{\infty} (-1)^n \frac{x^{2n+1}}{2n+1}$.

$$\sum_{n=0}^{\infty} \frac{(-1)^n}{3^n(2n+1)} = \sum_{n=0}^{\infty} \frac{(-1)^n}{\left(\sqrt{3}\right)^{2n}(2n+1)} \frac{\sqrt{3}}{\sqrt{3}}$$

$$= \sqrt{3}\sum_{n=0}^{\infty} \frac{(-1)^n\left(1/\sqrt{3}\right)^{2n+1}}{2n+1}$$

$$= \sqrt{3}\arctan\left(\frac{1}{\sqrt{3}}\right)$$

$$= \sqrt{3}\left(\frac{\pi}{6}\right) \approx 0.9068997$$

60. $\displaystyle\sum_{n=0}^{\infty} \frac{(-1)^n \pi^{2n+1}}{3^{2n+1}(2n+1)!} = \sum_{n=0}^{\infty} (-1)^n \frac{(\pi/3)^{2n+1}}{(2n+1)!}$

$$= \sin\left(\frac{\pi}{3}\right) = \frac{\sqrt{3}}{2} \approx 0.866025$$

61. Using a graphing utility, you obtain the following partial sums for the left hand side. Note that $1/\pi = 0.3183098862$.

$n = 0: S_0 \approx 0.3183098784$

$n = 1: S_1 = 0.3183098862$

62. You can verify that the statement is incorrect by calculating the constant terms of each side:

$$\sum_{n=0}^{\infty} x^n + \sum_{n=0}^{\infty}\left(\frac{x}{5}\right)^n = (1+1) + \left(x + \frac{x}{5}\right) + \cdots$$

$$\sum_{n=0}^{\infty}\left(1 + \frac{1}{5}\right)x^n = \left(1 + \frac{1}{5}\right) + \left(1 + \frac{1}{5}\right)x + \cdots$$

The formula should be

$$\sum_{n=0}^{\infty} x^n + \sum_{n=0}^{\infty}\left(\frac{x}{5}\right)^n = \sum_{n=0}^{\infty}\left[1 + \left(\frac{1}{5}\right)^n\right]x^n.$$

Section 9.10 Taylor and Maclaurin Series

1. For $c = 0$, you have:

$$f(x) = e^{2x}$$

$$f^{(n)}(x) = 2^n e^{2x} \Rightarrow f^{(n)}(0) = 2^n$$

$$e^{2x} = 1 + 2x + \frac{4x^2}{2!} + \frac{8x^3}{3!} + \frac{16x^4}{4!} + \cdots = \sum_{n=0}^{\infty} \frac{(2x)^n}{n!}.$$

2. For $c = 0$, you have:

$$f(x) = e^{-4x}$$

$$f^{(n)}(x) = (-4)^n e^{-4x} \Rightarrow f^{(n)}(0) = (-4)^n$$

$$e^{-4x} = 1 - 4x + \frac{16x^2}{2!} - \frac{64x^3}{3!} + \cdots = \sum_{n=0}^{\infty} \frac{(-1)^n (4x)^n}{n!}.$$

3. For $c = \pi/4$, you have:

$$f(x) = \cos(x) \qquad f\left(\frac{\pi}{4}\right) = \frac{\sqrt{2}}{2}$$

$$f'(x) = -\sin(x) \qquad f'\left(\frac{\pi}{4}\right) = -\frac{\sqrt{2}}{2}$$

$$f''(x) = -\cos(x) \qquad f''\left(\frac{\pi}{4}\right) = -\frac{\sqrt{2}}{2}$$

$$f'''(x) = \sin(x) \qquad f'''\left(\frac{\pi}{4}\right) = \frac{\sqrt{2}}{2}$$

$$f^{(4)}(x) = \cos(x) \qquad f^{(4)}\left(\frac{\pi}{4}\right) = \frac{\sqrt{2}}{2}$$

and so on. Therefore, you have:

$$\begin{aligned}
\cos x &= \sum_{n=0}^{\infty} \frac{f^{(n)}(\pi/4)\left[x - (\pi/4)\right]^n}{n!} \\
&= \frac{\sqrt{2}}{2}\left[1 - \left(x - \frac{\pi}{4}\right) - \frac{\left[x - (\pi/4)\right]^2}{2!} + \frac{\left[x - (\pi/4)\right]^3}{3!} + \frac{\left[x - (\pi/4)\right]^4}{4!} - \cdots\right] \\
&= \frac{\sqrt{2}}{2}\sum_{n=0}^{\infty} \frac{(-1)^{n(n+1)/2}\left[x - (\pi/4)\right]^n}{n!}.
\end{aligned}$$

[**Note:** $(-1)^{n(n+1)/2} = 1, -1, -1, 1, 1, -1, -1, 1, \ldots$]

4. For $c = \pi/4$, you have:

$$f(x) = \sin x \qquad f\left(\frac{\pi}{4}\right) = \frac{\sqrt{2}}{2}$$

$$f'(x) = \cos x \qquad f'\left(\frac{\pi}{4}\right) = \frac{\sqrt{2}}{2}$$

$$f''(x) = -\sin x \qquad f''\left(\frac{\pi}{4}\right) = -\frac{\sqrt{2}}{2}$$

$$f'''(x) = -\cos x \qquad f'''\left(\frac{\pi}{4}\right) = -\frac{\sqrt{2}}{2}$$

$$f^{(4)}(x) = \sin x \qquad f^{(4)}\left(\frac{\pi}{4}\right) = \frac{\sqrt{2}}{2}$$

and so on. Therefore you have:

$$\begin{aligned}
\sin x &= \sum_{n=0}^{\infty} \frac{f^{(n)}(\pi/4)\left[x - (\pi/4)\right]^n}{n!} \\
&= \frac{\sqrt{2}}{2}\left[1 + \left(x - \frac{\pi}{4}\right) - \frac{\left[x - (\pi/4)\right]^2}{2!} - \frac{\left[x - (\pi/4)\right]^3}{3!} + \frac{\left[x - (\pi/4)\right]^4}{4!} + \cdots\right] \\
&= \frac{\sqrt{2}}{2}\left\{\sum_{n=0}^{\infty} \frac{(-1)^{n(n-1)/2}\left[x - (\pi/4)\right]^{n+1}}{(n+1)!} + 1\right\}.
\end{aligned}$$

5. For $c = 1$, you have

$$f(x) = \frac{1}{x} = x^{-1} \qquad f(1) = 1$$
$$f'(x) = -x^{-2} \qquad f'(1) = -1$$
$$f''(x) = 2x^{-3} \qquad f''(1) = 2$$
$$f'''(x) = -6x^{-4} \qquad f'''(1) = -6$$

and so on. Therefore, you have

$$\frac{1}{x} = \sum_{n=0}^{\infty} \frac{f^{(n)}(1)(x-1)^n}{n!}$$
$$= 1 - (x-1) + \frac{2(x-1)^2}{2!} - \frac{6(x-1)^3}{3!} + \cdots$$
$$= 1 - (x-1) + (x-1)^2 - (x-1)^3 + \cdots$$
$$= \sum_{n=0}^{\infty} (-1)^n (x-1)^n$$

6. For $c = 2$, you have

$$f(x) = \frac{1}{1-x} = (1-x)^{-1} \qquad f(2) = -1$$
$$f'(x) = (1-x)^{-2} \qquad f'(2) = 1$$
$$f''(x) = 2(1-x)^{-3} \qquad f''(2) = -2$$
$$f'''(x) = 6(1-x)^{-4} \qquad f'''(2) = 6$$

and so on. Therefore you have

$$\frac{1}{1-x} = \sum_{n=0}^{\infty} \frac{f^{(n)}(2)(x-2)^n}{n!}$$
$$= -1 + (x-2) - (x-2)^2 + (x-2)^3 - \cdots$$
$$= \sum_{n=0}^{\infty} (-1)^{n+1}(x-2)^n$$

7. For $c = 1$, you have,

$$f(x) = \ln x \qquad f(1) = 0$$
$$f'(x) = \frac{1}{x} \qquad f'(1) = 1$$
$$f''(x) = -\frac{1}{x^2} \qquad f''(1) = -1$$
$$f'''(x) = \frac{2}{x^3} \qquad f'''(1) = 2$$
$$f^{(4)}(x) = -\frac{6}{x^4} \qquad f^{(4)}(1) = -6$$
$$f^{(5)}(x) = \frac{24}{x^5} \qquad f^{(5)}(1) = 24$$

and so on. Therefore, you have:

$$\ln x = \sum_{n=0}^{\infty} \frac{f^{(n)}(1)(x-1)^n}{n!}$$
$$= 0 + (x-1) - \frac{(x-1)^2}{2!} + \frac{2(x-1)^3}{3!} - \frac{6(x-1)^4}{4!} + \frac{24(x-1)^5}{5!} - \cdots$$
$$= (x-1) - \frac{(x-1)^2}{2} + \frac{(x-1)^3}{3} - \frac{(x-1)^4}{4} + \frac{(x-1)^5}{5} - \cdots$$
$$= \sum_{n=0}^{\infty} (-1)^n \frac{(x-1)^{n+1}}{n+1}.$$

8. For $c = 1$, you have:

$$f(x) = e^x$$
$$f^{(n)}(x) = e^x \Rightarrow f^{(n)}(1) = e$$
$$e^x = \sum_{n=0}^{\infty} \frac{f^{(n)}(1)(x-1)^n}{n!} = e\left[1 + (x-1) + \frac{(x-1)^2}{2!} + \frac{(x-1)^3}{3!} + \frac{(x-1)^4}{4!} + \cdots\right] = e\sum_{n=0}^{\infty} \frac{(x-1)^n}{n!}.$$

9. For $c = 0$, you have

$$\begin{aligned} f(x) &= \sin 3x & f(0) &= 0 \\ f'(x) &= 3\cos 3x & f'(0) &= 3 \\ f''(x) &= -9\sin 3x & f''(0) &= 0 \\ f'''(x) &= -27\cos 3x & f'''(0) &= -27 \\ f^{(4)}(x) &= 81\sin 3x & f^{(4)}(0) &= 0 \end{aligned}$$

and so on. Therefore you have

$$\sin 3x = \sum_{n=0}^{\infty} \frac{f^{(n)}(0)x^n}{n!} = 0 + 3x + 0 - \frac{27x^3}{3!} + 0 + \cdots = \sum_{n=0}^{\infty} \frac{(-1)^n (3x)^{2n+1}}{(2n+1)!}$$

10. For $c = 0$, you have.

$$\begin{aligned} f(x) &= \ln(x^2 + 1) & f(0) &= 0 \\ f'(x) &= \frac{2x}{x^2 + 1} & f'(0) &= 0 \\ f''(x) &= \frac{2 - 2x^2}{(x^2 + 1)^2} & f''(0) &= 2 \\ f'''(x) &= \frac{4x(x^2 - 3)}{(x^2 + 1)^3} & f'''(0) &= 0 \\ f^{(4)}(x) &= \frac{12(-x^4 + 6x^2 - 1)}{(x^2 + 1)^4} & f^{(4)}(0) &= -12 \\ f^{(5)}(x) &= \frac{48x(x^4 - 10x^2 + 5)}{(x^2 + 1)^5} & f^{(5)}(0) &= 0 \\ f^{(6)}(x) &= \frac{-240(5x^6 - 15x^4 + 15x^2 - 1)}{(x^2 + 1)^6} & f^{(6)}(0) &= 240 \end{aligned}$$

and so on. Therefore, you have:

$$\begin{aligned} \ln(x^2 + 1) &= \sum_{n=0}^{\infty} \frac{f^{(n)}(0)x^n}{n!} = 0 + 0x + \frac{2x^2}{2!} + \frac{0x^3}{3!} - \frac{12x^4}{4!} + \frac{0x^5}{5!} + \frac{240x^6}{6!} + \cdots \\ &= x^2 - \frac{x^4}{2} + \frac{x^6}{3} - \cdots = \sum_{n=0}^{\infty} \frac{(-1)^n x^{2n+2}}{n + 1}. \end{aligned}$$

11. For $c = 0$, you have:

$$\begin{aligned} f(x) &= \sec(x) & f(0) &= 1 \\ f'(x) &= \sec(x)\tan(x) & f'(0) &= 0 \\ f''(x) &= \sec^3(x) + \sec(x)\tan^2(x) & f''(0) &= 1 \\ f'''(x) &= 5\sec^3(x)\tan(x) + \sec(x)\tan^3(x) & f'''(0) &= 0 \\ f^{(4)}(x) &= 5\sec^5(x) + 18\sec^3(x)\tan^2(x) + \sec(x)\tan^4(x) & f^{(4)}(0) &= 5 \end{aligned}$$

$$\sec(x) = \sum_{n=0}^{\infty} \frac{f^{(n)}(0)x^n}{n!} = 1 + \frac{x^2}{2!} + \frac{5x^4}{4!} + \cdots.$$

12. For $c = 0$, you have;

$$f(x) = \tan(x) \qquad f(0) = 0$$
$$f'(x) = \sec^2(x) \qquad f'(0) = 1$$
$$f''(x) = 2\sec^2(x)\tan(x) \qquad f''(0) = 0$$
$$f'''(x) = 2\left[\sec^4(x) + 2\sec^2(x)\tan^2(x)\right] \qquad f'''(0) = 2$$
$$f^{(4)}(x) = 8\left[\sec^4(x)\tan(x) + \sec^2(x)\tan^3(x)\right] \qquad f^{(4)}(0) = 0$$
$$f^{(5)}(x) = 8\left[2\sec^6(x) + 11\sec^4(x)\tan^2(x) + 2\sec^2(x)\tan^4(x)\right] \qquad f^{(5)}(0) = 16$$
$$\tan(x) = \sum_{n=0}^{\infty}\frac{f^{(n)}(0)x^n}{n!} = x + \frac{2x^3}{3!} + \frac{16x^5}{5!} + \cdots = x + \frac{x^3}{3} + \frac{2}{15}x^5 + \cdots.$$

13. The Maclaurin series for $f(x) = \cos x$ is $\displaystyle\sum_{n=0}^{\infty}\frac{(-1)^n x^{2n}}{(2n)!}$.

Because $f^{(n+1)}(x) = \pm\sin x$ or $\pm\cos x$, you have $\left|f^{(n+1)}(z)\right| \le 1$ for all z. So by Taylor's Theorem,

$$0 \le \left|R_n(x)\right| = \left|\frac{f^{(n+1)}(z)}{(n+1)!}x^{n+1}\right| \le \frac{|x|^{n+1}}{(n+1)!}.$$

Because $\displaystyle\lim_{n\to\infty}\frac{|x|^{n+1}}{(n+1)!} = 0$, it follows that $R_n(x) \to 0$ as $n \to \infty$. So, the Maclaurin series for $\cos x$ converges to $\cos x$ for all x.

14. The Maclaurin series for $f(x) = e^{-2x}$ is $\displaystyle\sum_{n=0}^{\infty}\frac{(-2x)^n}{n!}$.

$f^{(n+1)}(x) = (-2)^{n+1}e^{-2x}$. So, by Taylor's Theorem,

$$0 \le \left|Rn(x)\right| = \left|\frac{f^{(n+1)}(z)}{(n+1)!}x^{n+1}\right| = \left|\frac{(-2)^{n+1}e^{-2z}}{(n+1)!}x^{n+1}\right|.$$

Because $\displaystyle\lim_{n\to\infty}\left|\frac{(-2)^{n+1}x^{n+1}}{(n+1)!}\right| = \lim_{n\to\infty}\left|\frac{(2x)^{n+1}}{(n+1)!}\right| = 0$, it follows that $R_n(x) \to 0$ as $n \to \infty$.

So, the Maclaurin Series for e^{-2x} converges to e^{-2x} for all x.

15. The Maclaurin series for $f(x) = \sinh x$ is $\displaystyle\sum_{n=0}^{\infty}\frac{x^{2n+1}}{(2n+1)!}$.

$f^{(n+1)}(x) = \sinh x$ (or $\cosh x$). For fixed x,

$$0 \le \left|R_n(x)\right| = \left|\frac{f^{(n+1)}(z)}{(n+1)!}x^{n+1}\right| = \left|\frac{\sinh(z)}{(n+1)!}x^{n+1}\right| \to 0 \text{ as } n \to \infty.$$

(The argument is the same if $f^{(n+1)}(x) = \cosh x$). So, the Maclaurin series for $\sinh x$ converges to $\sinh x$ for all x.

16. The Maclaurin series for $f(x) = \cosh x$ is $\displaystyle\sum_{n=0}^{\infty} \frac{x^{2n}}{(2n)!}$.

$f^{(n+1)}(x) = \sinh x$ (or $\cosh x$). For fixed x,

$$0 \le |R_n(x)| = \left|\frac{f^{(n+1)}(z)}{(n+1)!}x^{n+1}\right| = \left|\frac{\sinh(z)}{(n+1)!}x^{n+1}\right| \to 0 \text{ as } n \to \infty.$$

(The argument is the same if $f^{(n+1)}(x) = \cosh x$). So, the Maclaurin series for $\cosh x$ converges to $\cosh x$ for all x.

17. Because $(1+x)^{-k} = 1 - kx + \dfrac{k(k+1)x^2}{2!} - \dfrac{k(k+1)(k+2)x^3}{3!} + \cdots$, you have

$$(1+x)^{-2} = 1 - 2x + \frac{2(3)x^2}{2!} - \frac{2(3)(4)x^3}{3!} + \frac{2(3)(4)(5)x^4}{4!} - \cdots = 1 - 2x + 3x^2 - 4x^3 + 5x^4 - \cdots$$

$$= \sum_{n=0}^{\infty} (-1)^n (n+1)x^n.$$

18. Because $(1+x)^{-k} = 1 - kx + \dfrac{k(k+1)x^2}{2!} - \dfrac{k(k+1)(k+2)x^3}{3!} + \cdots$

you have

$$(1+x)^{-4} = 1 - 4x + \frac{4(5)}{2!}x^2 - \frac{4(5)(6)}{3!}x^3 + \frac{4(5)(6)(7)}{4!}x^4 = 1 - 4x + 10x^2 - 20x^3 + 35x^4 - \cdots = \sum_{n=0}^{\infty} (-1)^n \frac{(n+3)!}{3!n!}x^n$$

19. Because $(1+x)^{-k} = 1 - kx + \dfrac{k(k+1)x^2}{2!} - \dfrac{k(k+1)(k+2)x^3}{3!} + \cdots$, you have

$$[1 + (-x)]^{-1/2} = 1 + \left(\frac{1}{2}\right)x + \frac{(1/2)(3/2)x^2}{2!} + \frac{(1/2)(3/2)(5/2)x^3}{3!} + \cdots$$

$$= 1 + \frac{x}{2} + \frac{(1)(3)x^2}{2^2\,2!} + \frac{(1)(3)(5)x^3}{2^3\,3!} + \cdots$$

$$= 1 + \sum_{n=1}^{\infty} \frac{1 \cdot 3 \cdot 5 \cdots (2n-1)x^n}{2^n n!}.$$

20. Because $(1+x)^{-k} = 1 - kx + \dfrac{k(k+1)x^2}{2!} - \dfrac{k(k+1)(k+2)x^3}{3!} + \cdots$ you have

$$[1 + (-x^2)]^{-1/2} = 1 - \frac{1}{2}x^2 + \frac{(1/2)(3/2)}{2!}x^4 - \frac{(1/2)(3/2)(5/2)x^6}{3!} + \cdots$$

$$= 1 - \frac{1}{2}x^2 + \frac{(1)(3)}{2^2\,2!}x^4 - \frac{(1)(3)(5)}{2^3\,3!}x^6 + \cdots$$

$$= 1 + \sum_{n=1}^{\infty} \frac{1 \cdot 3 \cdot 5 \cdots (2n-1)}{2^n n!}x^{2n}$$

21. $\dfrac{1}{\sqrt{4+x^2}} = \left(\dfrac{1}{2}\right)\left[1 + \left(\dfrac{x}{2}\right)^2\right]^{-1/2}$ and because $(1+x)^{-1/2} = 1 + \displaystyle\sum_{n=1}^{\infty} \frac{(-1)^n 1 \cdot 3 \cdot 5 \cdots (2n-1)x^n}{2^n n!}$, you have

$$\frac{1}{\sqrt{4+x^2}} = \frac{1}{2}\left[1 + \sum_{n=1}^{\infty} \frac{(-1)^n 1 \cdot 3 \cdot 5 \cdots (2n-1)(x/2)^{2n}}{2^n n!}\right] = \frac{1}{2} + \sum_{n=1}^{\infty} \frac{(-1)^n 1 \cdot 3 \cdot 5 \cdots (2n-1)x^{2n}}{2^{3n+1} n!}.$$

22. $$\frac{1}{(2+x)^3} = \frac{1}{8}\left(1+\frac{x}{2}\right)^{-3}, \qquad k = -3$$

$$\frac{1}{(2+x)^3} = \frac{1}{8}\left\{1 - 3\left(\frac{x}{2}\right) + \frac{3(4)}{2!}\left(\frac{x}{2}\right)^2 - \frac{3(4)(5)}{3!}\left(\frac{x}{2}\right)^3 + \cdots\right\} = \frac{1}{8}\left[1 + \sum_{n=1}^{\infty}(-1)^n\frac{(n+2)!}{2^{n+1}n!}x^n\right]$$

23. $$\sqrt{1+x} = (1+x)^{1/2}, \qquad k = 1/2$$

$$\sqrt{1+x} = 1 + \frac{1}{2}x + \frac{1/2(-1/2)}{2!}x^2 + \frac{1/2(-1/2)(-3/2)}{3!}x^3 + \cdots = 1 + \frac{1}{2}x + \sum_{n=2}^{\infty}(-1)^{n+1}\frac{1\cdot3\cdot5\cdots(2n-3)}{2^n n!}x^n$$

24. $$(1+x)^{1/4} = 1 + \frac{1}{4}x + \frac{(1/4)(-3/4)}{2!}x^2 + \frac{(1/4)(-3/4)(-7/4)}{3!}x^3 + \cdots$$

$$= 1 + \frac{1}{4}x - \frac{3}{4^2 2!}x^2 + \frac{3\cdot7}{4^3 3!}x^3 - \frac{3\cdot7\cdot11}{4^4 4!}x^4 + \cdots$$

$$= 1 + \frac{1}{4}x + \sum_{n=2}^{\infty}\frac{(-1)^{n+1}3\cdot7\cdot11\cdots(4n-5)}{4^n n!}x^n$$

25. Because $(1+x)^{1/2} = 1 + \frac{x}{2} + \sum_{n=2}^{\infty}\frac{(-1)^{n+1}1\cdot3\cdot5\cdots(2n-3)x^n}{2^n n!}$

you have $(1+x^2)^{1/2} = 1 + \frac{x^2}{2} + \sum_{n=2}^{\infty}\frac{(-1)^{n+1}1\cdot3\cdot5\cdots(2n-3)x^{2n}}{2^n n!}$.

26. Because $(1+x)^{1/2} = 1 + \frac{x}{2} + \sum_{n=2}^{\infty}\frac{(-1)^{n+1}1\cdot3\cdot5\cdots(2n-3)x^n}{2^n n!}$

you have $(1+x^3)^{1/2} = 1 + \frac{x^3}{2} + \sum_{n=2}^{\infty}\frac{(-1)^{n+1}1\cdot3\cdot5\cdots(2n-3)x^{3n}}{2^n n!}$.

27. $$e^x = \sum_{n=0}^{\infty}\frac{x^n}{n!} = 1 + x + \frac{x^2}{2!} + \frac{x^3}{3!} + \frac{x^4}{4!} + \frac{x^5}{5!} + \cdots$$

$$e^{x^2/2} = \sum_{n=0}^{\infty}\frac{(x^2/2)^n}{n!} = \sum_{n=0}^{\infty}\frac{x^{2n}}{2^n n!} = 1 + \frac{x^2}{2} + \frac{x^4}{2^2 2!} + \frac{x^6}{2^3 3!} + \frac{x^8}{2^4 4!} + \cdots$$

28. $$e^x = \sum_{n=0}^{\infty}\frac{x^n}{n!} = 1 + x + \frac{x^2}{2!} + \frac{x^3}{3!} + \frac{x^4}{4!} + \frac{x^5}{5!} + \cdots$$

$$e^{-3x} = \sum_{n=0}^{\infty}\frac{(-3x)^n}{n!} = \sum_{n=0}^{\infty}\frac{(-1)^n 3^n x^n}{n!} = 1 - 3x + \frac{9x^2}{2!} - \frac{27x^3}{3!} + \frac{81x^4}{4!} - \frac{243x^5}{5!} + \cdots$$

29. $$\ln x = \sum_{n=1}^{\infty}(-1)^{n-1}\frac{(x-1)^n}{n}, \quad 0 < x \le 2$$

$$\ln(x+1) = \sum_{n=1}^{\infty}\frac{(-1)^{n-1}x^n}{n}, \quad -1 < x \le 1$$

30. $$\ln x = \sum_{n=1}^{\infty}\frac{(-1)^{n-1}(x-1)^n}{n}, \quad 0 < x \le 2$$

$$\ln(x^2+1) = \sum_{n=1}^{\infty}\frac{(-1)^{n-1}x^{2n}}{n}, \quad -1 < x \le 1$$

31. $$\sin x = \sum_{n=0}^{\infty}\frac{(-1)^n x^{2n+1}}{(2n+1)!}$$

$$\sin 3x = \sum_{n=0}^{\infty}\frac{(-1)^n(3x)^{2n+1}}{(2n+1)!}$$

32. $$\sin x = \sum_{n=0}^{\infty}\frac{(-1)^n x^{2n+1}}{(2n+1)!}$$

$$\sin \pi x = \sum_{n=0}^{\infty}\frac{(-1)^n(\pi x)^{2n+1}}{(2n+1)!}$$

33. $$\cos x = \sum_{n=0}^{\infty} \frac{(-1)^n x^{2n}}{(2n)!} = 1 - \frac{x^2}{2!} + \frac{x^4}{4!} - \frac{x^6}{6!} + \cdots$$

$$\cos 4x = \sum_{n=0}^{\infty} \frac{(-1)^n (4x)^{2n}}{(2n)!} = \sum_{n=0}^{\infty} \frac{(-1)^n 4^{2n} x^{2n}}{(2n)!}$$

$$= 1 - \frac{16x^2}{2!} + \frac{256x^4}{4!} - \cdots$$

34. $$\cos x = \sum_{n=0}^{\infty} \frac{(-1)^n x^{2n}}{(2n)!}$$

$$\cos \pi x = \sum_{n=0}^{\infty} \frac{(-1)^n (\pi x)^{2n}}{(2n)!}$$

35. $$\cos x = \sum_{n=0}^{\infty} \frac{(-1)^n x^{2n}}{(2n)!} = 1 - \frac{x^2}{2!} + \frac{x^4}{4!} - \cdots$$

$$\cos x^{3/2} = \sum_{n=0}^{\infty} \frac{(-1)^n (x^{3/2})^{2n}}{(2n)!}$$

$$= \sum_{n=0}^{\infty} \frac{(-1)^n x^{3n}}{(2n)!}$$

$$= 1 - \frac{x^3}{2!} + \frac{x^6}{4!} - \cdots$$

36. $$\sin x = \sum_{n=0}^{\infty} \frac{(-1)^n x^{2n+1}}{(2n+1)!}$$

$$2 \sin x^3 = 2\sum_{n=0}^{\infty} \frac{(-1)^n (x^3)^{2n+1}}{(2n+1)!}$$

$$= 2\left(x^3 - \frac{x^9}{3!} + \frac{x^{15}}{5!} - \cdots\right)$$

$$= 2x^3 - \frac{2x^9}{3!} + \frac{2x^{15}}{5!} - \cdots$$

37. $$e^x = 1 + x + \frac{x^2}{2!} + \frac{x^3}{3!} + \frac{x^4}{4!} + \frac{x^5}{5!} + \cdots$$

$$e^{-x} = 1 - x + \frac{x^2}{2!} - \frac{x^3}{3!} + \frac{x^4}{4!} - \frac{x^5}{5!} + \cdots$$

$$e^x - e^{-x} = 2x + \frac{2x^3}{3!} + \frac{2x^5}{5!} + \frac{2x^7}{7!} + \cdots$$

$$\sinh(x) = \frac{1}{2}(e^x - e^{-x})$$

$$= x + \frac{x^3}{3!} + \frac{x^5}{5!} + \frac{x^7}{7!} + \cdots = \sum_{n=0}^{\infty} \frac{x^{2n+1}}{(2n+1)!}$$

38. $$e^x = 1 + x + \frac{x^2}{2!} + \frac{x^3}{3!} + \cdots$$

$$e^{-x} = 1 - x + \frac{x^2}{2!} - \frac{x^3}{3!} + \cdots$$

$$e^x + e^{-x} = 2 + \frac{2x^2}{2!} + \frac{2x^4}{4!} + \cdots$$

$$2 \cos h(x) = e^x + e^{-x} = \sum_{n=0}^{\infty} 2\frac{x^{2n}}{(2n)!}$$

39. $$\cos^2(x) = \frac{1}{2}[1 + \cos(2x)]$$

$$= \frac{1}{2}\left[1 + 1 - \frac{(2x)^2}{2!} + \frac{(2x)^4}{4!} - \frac{(2x)^6}{6!} - \cdots\right] = \frac{1}{2}\left[1 + \sum_{n=0}^{\infty} \frac{(-1)^n (2x)^{2n}}{(2n)!}\right]$$

40. The formula for the binomial series gives $(1 + x)^{-1/2} = 1 + \sum_{n=1}^{\infty} \frac{(-1)^n 1 \cdot 3 \cdot 5 \cdots (2n-1)x^n}{2^n n!}$, which implies that

$$(1 + x^2)^{-1/2} = 1 + \sum_{n=1}^{\infty} \frac{(-1)^n 1 \cdot 3 \cdot 5 \cdots (2n-1)x^{2n}}{2^n n!}$$

$$\ln\left(x + \sqrt{x^2 + 1}\right) = \int \frac{1}{\sqrt{x^2 + 1}}\, dx$$

$$= x + \sum_{n=1}^{\infty} \frac{(-1)^n 1 \cdot 3 \cdot 5 \cdots (2n-1)x^{2n+1}}{2^n (2n+1) n!}$$

$$= x - \frac{x^3}{2 \cdot 3} + \frac{1 \cdot 3x^5}{2 \cdot 4 \cdot 5} - \frac{1 \cdot 3 \cdot 5x^7}{2 \cdot 4 \cdot 6 \cdot 7} + \cdots.$$

41. $$x \sin x = x\left(x - \frac{x^3}{3!} + \frac{x^5}{5!} - \cdots\right) = x^2 - \frac{x^4}{3!} + \frac{x^6}{5!} - \cdots = \sum_{n=0}^{\infty} \frac{(-1)^n x^{2n+2}}{(2n+1)!}$$

42. $x\cos x = x\left(1 - \dfrac{x^2}{2!} + \dfrac{x^4}{4!} - \cdots\right) = x - \dfrac{x^3}{2!} + \dfrac{x^5}{4!} - \cdots = \displaystyle\sum_{n=0}^{\infty} \frac{(-1)^n x^{2n+1}}{(2n)!}$

43. $\dfrac{\sin x}{x} = \dfrac{x - (x^3/3!) + (x^5/5!) - \cdots}{x} = 1 - \dfrac{x^2}{2!} + \dfrac{x^4}{4!} - \cdots = \displaystyle\sum_{n=0}^{\infty} \frac{(-1)^n x^{2n}}{(2n+1)!}, x \neq 0$

$= 1, x = 0$

44. $\dfrac{\arcsin x}{x} = \displaystyle\sum_{n=0}^{\infty} \frac{(2n)!x^{2n+1}}{(2^n n!)^2(2n+1)} \cdot \frac{1}{x} = \sum_{n=0}^{\infty} \frac{(2n)!x^{2n}}{(2^n n!)^2(2n+1)}, x \neq 0$

$= 1, x = 0$

45. $e^{ix} = 1 + ix + \dfrac{(ix)^2}{2!} + \dfrac{(ix)^3}{3!} + \dfrac{(ix)^4}{4!} + \cdots = 1 + ix - \dfrac{x^2}{2!} - \dfrac{ix^3}{3!} + \dfrac{x^4}{4!} + \dfrac{ix^5}{5!} - \dfrac{x^6}{6!} - \cdots$

$e^{-ix} = 1 - ix + \dfrac{(-ix)^2}{2!} + \dfrac{(-ix)^3}{3!} + \dfrac{(-ix)^4}{4!} + \cdots = 1 - ix - \dfrac{x^2}{2!} + \dfrac{ix^3}{3!} + \dfrac{x^4}{4!} - \dfrac{ix^5}{5!} - \dfrac{x^6}{6!} + \cdots$

$e^{ix} - e^{-ix} = 2ix - \dfrac{2ix^3}{3!} + \dfrac{2ix^5}{5!} - \dfrac{2ix^7}{7!} + \cdots$

$\dfrac{e^{ix} - e^{-ix}}{2i} = x - \dfrac{x^3}{3!} + \dfrac{x^5}{5!} - \dfrac{x^7}{7!} + \cdots = \displaystyle\sum_{n=0}^{\infty} \frac{(-1)^n x^{2n+1}}{(2n+1)!} = \sin(x)$

46. $e^{ix} + e^{-ix} = 2 - \dfrac{2x^2}{2!} + \dfrac{2x^4}{4!} - \dfrac{2x^6}{6!} + \cdots$ (See Exercise 45.)

$\dfrac{e^{ix} + e^{-ix}}{2} = 1 - \dfrac{x^2}{2!} + \dfrac{x^4}{4!} - \dfrac{x^6}{6!} + \cdots = \displaystyle\sum_{n=0}^{\infty} \frac{(-1)^n x^{2n}}{(2n)!} = \cos(x)$

47. $f(x) = e^x \sin x$

$= \left(1 + x + \dfrac{x^2}{2} + \dfrac{x^3}{6} + \dfrac{x^4}{24} + \cdots\right)\left(x - \dfrac{x^3}{6} + \dfrac{x^5}{120} - \cdots\right)$

$= x + x^2 + \left(\dfrac{x^3}{2} - \dfrac{x^3}{6}\right) + \left(\dfrac{x^4}{6} - \dfrac{x^4}{6}\right) + \left(\dfrac{x^5}{120} - \dfrac{x^5}{12} + \dfrac{x^5}{24}\right) + \cdots$

$= x + x^2 + \dfrac{x^3}{3} - \dfrac{x^5}{30} + \cdots$

$P_5(x) = x + x^2 + \dfrac{x^3}{3} - \dfrac{x^5}{30}$

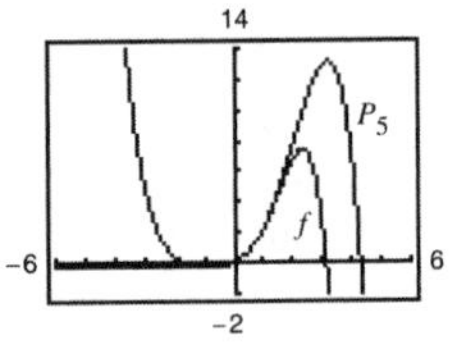

48. $g(x) = e^x \cos x$

$= \left(1 + x + \dfrac{x^2}{2} + \dfrac{x^4}{6} + \dfrac{x^4}{24} + \cdots\right)\left(1 - \dfrac{x^2}{2} + \dfrac{x^4}{24} - \cdots\right)$

$= 1 + x + \left(\dfrac{x^2}{2} - \dfrac{x^2}{2}\right) + \left(\dfrac{x^3}{6} - \dfrac{x^3}{2}\right) + \left(\dfrac{x^4}{24} - \dfrac{x^4}{4} + \dfrac{x^4}{24}\right) + \cdots$

$= 1 + x - \dfrac{x^3}{3} - \dfrac{x^4}{6} + \cdots$

$P_4(x) = 1 + x - \dfrac{x^3}{3} - \dfrac{x^4}{6}$

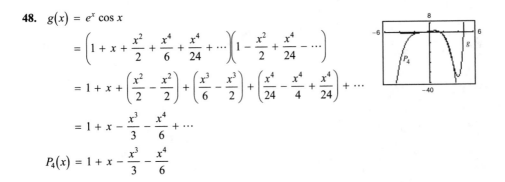

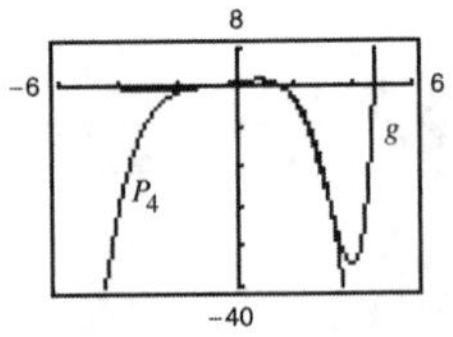

49. $h(x) = \cos x \ln(1+x)$

$$= \left(1 - \frac{x^2}{2} + \frac{x^4}{24} + \cdots\right)\left(x - \frac{x^2}{2} + \frac{x^3}{3} - \frac{x^4}{4} + \frac{x^5}{5} - \cdots\right)$$

$$= x - \frac{x^2}{2} + \left(\frac{x^3}{3} - \frac{x^3}{2}\right) + \left(\frac{x^4}{4} - \frac{x^4}{4}\right) + \left(\frac{x^5}{5} - \frac{x^5}{6} + \frac{x^5}{24}\right) + \cdots$$

$$= x - \frac{x^2}{2} - \frac{x^3}{6} + \frac{3x^5}{40} + \cdots$$

$$P_5(x) = x - \frac{x^2}{2} - \frac{x^3}{6} + \frac{3x^5}{40}$$

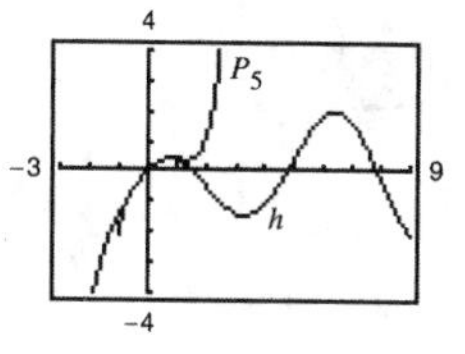

50. $f(x) = e^x \ln(1+x)$

$$= \left(1 + x + \frac{x^2}{2} + \frac{x^3}{6} + \frac{x^4}{24} + \cdots\right)\left(x - \frac{x^2}{2} + \frac{x^3}{3} - \frac{x^4}{4} + \frac{x^5}{5} - \cdots\right)$$

$$= x + \left(x^2 - \frac{x^2}{2}\right) + \left(\frac{x^3}{3} - \frac{x^3}{2} + \frac{x^3}{2}\right) + \left(-\frac{x^4}{4} + \frac{x^4}{3} - \frac{x^4}{4} + \frac{x^4}{6}\right) + \left(\frac{x^5}{5} - \frac{x^5}{4} + \frac{x^5}{6} - \frac{x^5}{12} + \frac{x^5}{24}\right) + \cdots$$

$$= x + \frac{x^2}{2} + \frac{x^3}{3} + \frac{3x^5}{40} + \cdots$$

$$P_5(x) = x + \frac{x^2}{2} + \frac{x^3}{3} + \frac{3x^5}{40}$$

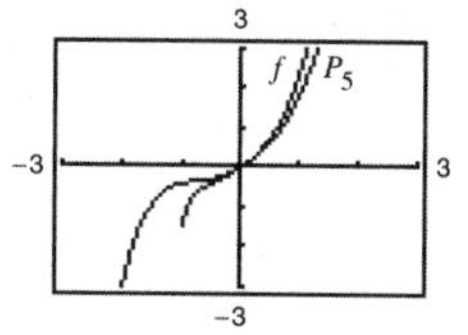

51. $g(x) = \dfrac{\sin x}{1+x}$. Divide the series for $\sin x$ by $(1+x)$.

$$\begin{array}{rl}
 & x - x^2 + \frac{5x^3}{6} - \frac{5x^4}{6} + \\
1 + x \,\big) & x + 0x^2 - \frac{x^3}{6} + 0x^4 + \frac{x^5}{120} + \cdots \\
 & \underline{x + x^2} \\
 & -x^2 - \frac{x^3}{6} \\
 & \underline{-x^2 - x^3} \\
 & \frac{5x^3}{6} + 0x^4 \\
 & \underline{\frac{5x^3}{6} + \frac{5x^4}{6}} \\
 & -\frac{5x^4}{6} + \frac{x^5}{120} \\
 & \underline{-\frac{5x^4}{6} - \frac{5x^5}{6}} \\
 & \vdots
\end{array}$$

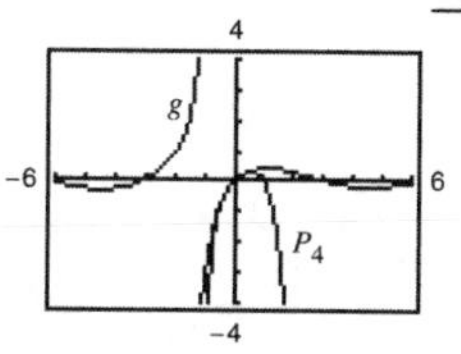

$$g(x) = x - x^2 + \frac{5x^3}{6} - \frac{5x^4}{6} + \cdots$$

$$P_4(x) = x - x^2 + \frac{5x^3}{6} - \frac{5x^4}{6}$$

52. $f(x) = \dfrac{e^x}{1+x}$. Divide the series for e^x by $(1+x)$.

$$\begin{array}{rl}
 & 1 + \frac{x^2}{2} - \frac{x^3}{3} + \frac{3x^4}{8} + \cdots \\
1 + x \,\big) & 1 + x + \frac{x^2}{2} + \frac{x^3}{6} + \frac{x^4}{24} + \frac{x^5}{120} + \cdots \\
 & \underline{1 + x} \\
 & 0 + \frac{x^2}{2} + \frac{x^3}{6} \\
 & \underline{\frac{x^2}{2} + \frac{x^3}{2}} \\
 & -\frac{x^3}{3} + \frac{x^4}{24} \\
 & \underline{-\frac{x^3}{3} - \frac{x^4}{3}} \\
 & \frac{3x^4}{8} + \frac{x^5}{120} \\
 & \underline{\frac{3x^4}{8} + \frac{3x^5}{8}} \\
 & \vdots
\end{array}$$

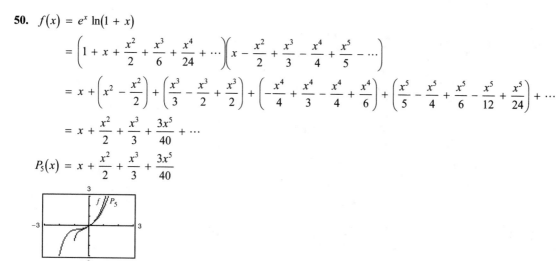

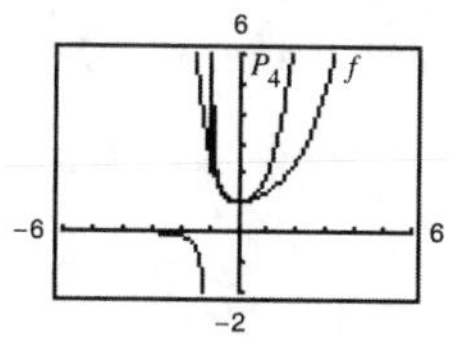

$$f(x) = 1 + \frac{x^2}{2} - \frac{x^3}{3} + \frac{3x^4}{8} - \cdots$$

$$P_4(x) = 1 + \frac{x^2}{2} - \frac{x^3}{3} + \frac{3x^4}{8}$$

53. $$\int_0^x \left(e^{-t^2} - 1\right) dt = \int_0^x \left[\left(\sum_{n=0}^{\infty} \frac{(-1)^n t^{2n}}{n!}\right) - 1\right] dt$$
$$= \int_0^x \left[\sum_{n=0}^{\infty} \frac{(-1)^{n+1} t^{2n+2}}{(n+1)!}\right] dt$$
$$= \left[\sum_{n=0}^{\infty} \frac{(-1)^{n+1} t^{2n+3}}{(2n+3)(n+1)!}\right]_0^x$$
$$= \sum_{n=0}^{\infty} \frac{(-1)^{n+1} x^{2n+3}}{(2n+3)(n+1)!}$$

54. $$\int_0^x \sqrt{1+t^3}\, dt = \int_0^x \left[1 + \frac{t^3}{2} + \sum_{n=2}^{\infty} \frac{(-1)^{n-1} 1 \cdot 3 \cdot 5 \cdots (2n-3) t^{3n}}{2^n n!}\right] dt$$
$$= \left[t + \frac{t^4}{8} + \sum_{n=2}^{\infty} \frac{(-1)^{n-1} 1 \cdot 3 \cdot 5 \cdots (2n-3) t^{3n+1}}{(3n+1)2^n n!}\right]_0^x$$
$$= x + \frac{x^4}{8} + \sum_{n=2}^{\infty} \frac{(-1)^{n-1} 1 \cdot 3 \cdot 5 \cdots (2n-3) x^{3n+1}}{(3n+1)2^n n!}$$

55. Because $\ln x = \sum_{n=0}^{\infty} \frac{(-1)^n (x-1)^{n+1}}{n+1} = (x-1) - \frac{(x-1)^2}{2} + \frac{(x-1)^3}{3} - \frac{(x-1)^4}{4} + \cdots, \quad (0 < x \le 2)$

you have $\ln 2 = 1 - \frac{1}{2} + \frac{1}{3} - \frac{1}{4} + \cdots = \sum_{n=1}^{\infty} (-1)^{n+1} \frac{1}{n} \approx 0.6931.$ (10,001 terms)

56. Because $\sin(x) = \sum_{n=0}^{\infty} \frac{(-1)^n x^{2n+1}}{(2n+1)!} = x - \frac{x^3}{3!} + \frac{x^5}{5!} - \frac{x^7}{7!} + \cdots$, you have

$\sin(1) = \sum_{n=0}^{\infty} \frac{(-1)^n}{(2n+1)!} = 1 - \frac{1}{3!} + \frac{1}{5!} - \frac{1}{7!} + \cdots \approx 0.8415.$ (4 terms)

57. Because $e^x = \sum_{n=0}^{\infty} \frac{x^n}{n!} = 1 + x + \frac{x^2}{2!} + \frac{x^3}{3!} + \cdots,$

you have $e^2 = 1 + 2 + \frac{2^2}{2!} + \frac{2^3}{3!} + \cdots = \sum_{n=0}^{\infty} \frac{2^n}{n!} \approx 7.3891.$ (12 terms)

58. Because $e^x = \sum_{n=0}^{\infty} \frac{x^n}{n!} = 1 + x + \frac{x^2}{2!} + \frac{x^3}{3!} + \frac{x^4}{4!} + \frac{x^5}{5!} + \cdots$, you have $e^{-1} = 1 - 1 + \frac{1}{2!} - \frac{1}{3!} + \frac{1}{4!} - \frac{1}{5!} + \cdots$

and $\frac{e-1}{e} = 1 - e^{-1} = 1 - \frac{1}{2!} + \frac{1}{3!} - \frac{1}{4!} + \frac{1}{5!} - \frac{1}{7!} + \cdots = \sum_{n=1}^{\infty} \frac{(-1)^{n-1}}{n!} \approx 0.6321.$ (6 terms)

59. Because

$$\cos x = \sum_{n=0}^{\infty} \frac{(-1)^n x^{2n}}{(2n)!} = 1 - \frac{x^2}{2!} + \frac{x^4}{4!} - \frac{x^6}{6!} + \frac{x^8}{8!} - \cdots$$

$$1 - \cos x = \frac{x^2}{2!} - \frac{x^4}{4!} + \frac{x^6}{6!} - \frac{x^8}{8!} + \cdots = \sum_{n=0}^{\infty} \frac{(-1)^n x^{2n+2}}{(2n+2)!}$$

$$\frac{1 - \cos}{x} = \frac{x}{2!} - \frac{x^3}{4!} + \frac{x^5}{6!} - \frac{x^7}{8!} + \cdots = \sum_{n=0}^{\infty} \frac{(-1)^n x^{2n+1}}{(2n+2)!}$$

you have $\lim_{x\to 0} \frac{1 - \cos x}{x} = \lim_{x\to 0} \sum_{n=0}^{\infty} \frac{(-1)x^{2n+1}}{(2n+2)!} = 0.$

60. Because

$$\sin x = \sum_{n=0}^{\infty} \frac{(-1)^n x^{2n+1}}{(2n+1)!} = x - \frac{x^3}{3!} + \frac{x^5}{5!} - \frac{x^7}{7!} + \cdots$$

$$\frac{\sin x}{x} = 1 - \frac{x^2}{3!} + \frac{x^4}{5!} - \frac{x^6}{7!} + \cdots = \sum_{n=0}^{\infty} \frac{(-1)^n x^{2n}}{(2n+1)!}$$

you have $\lim_{x\to 0} \frac{\sin x}{x} = \lim_{x\to 0} \sum_{n=0}^{\infty} \frac{(-1)^n x^{2n}}{(2n+1)!} = 1.$

61. Because $e^x = 1 + x + \frac{x^2}{2!} + \frac{x^3}{3!} + \cdots$

$$e^x - 1 = x + \frac{x^2}{2!} + \frac{x^3}{3!} + \cdots \sum_{n=0}^{\infty} \frac{x^{n+1}}{(n+1)!}$$

and $\frac{e^x - 1}{x} = 1 + \frac{x}{2!} + \frac{x^2}{3!} + \cdots \sum_{n=0}^{\infty} \frac{x^n}{(n+1)!}$

you have $\lim_{x\to 0} \frac{e^x - 1}{x} = \lim_{x\to 0} \sum \frac{x^n}{(n+1)!} = 1.$

62. Because $\ln(x + 1) = x - \frac{x^2}{2} + \frac{x^3}{3} - \cdots$

(See Exercise 29.)

$$\frac{\ln(x+1)}{x} = 1 - \frac{x}{2} + \frac{x^2}{3} - \cdots \sum_{n=0}^{\infty} \frac{(-1)^n x^n}{n+1}$$

you have $\lim_{x\to 0} \frac{\ln(x+1)}{x} = \lim_{x\to 0} \sum_{n=0}^{\infty} \frac{(-1)^n x^n}{n+1} = 1.$

63.
$$\int_0^1 e^{-x^3}\,dx = \int_0^1 \left[\sum_{n=0}^{\infty} \frac{(-x^3)^n}{n!}\right] dx$$
$$= \int_0^1 \left[\sum_{n=0}^{\infty} \frac{(-1)^n x^{3n}}{n!}\right] dx$$
$$= \left[\sum_{n=0}^{\infty} \frac{(-1)^n x^{3n+1}}{(3n+1)n!}\right]_0^1$$
$$= \sum_{n=0}^{\infty} \frac{(-1)^n}{(3n+1)n!}$$
$$= 1 - \frac{1}{4} + \frac{1}{14} - \cdots + (-1)^n \frac{1}{(3n+1)n!} + \cdots$$

Because $\frac{1}{[3(6) + 1]6!} < 0.0001$, you need 6 terms.

$$\int_0^1 e^{-x^2}\,dx \approx \sum_{n=0}^{5} \frac{(-1)^n}{(3n+1)n!} \approx 0.8075$$

64. $\int_0^{1/4} x \ln(x + 1)\,dx = \int_0^{1/4} \left(x^2 - \frac{x^3}{2} + \frac{x^4}{3} - \frac{x^5}{4} + \cdots\right) dx = \left[\frac{x^3}{3} - \frac{x^4}{4 \cdot 2} + \frac{x^5}{5 \cdot 3} - \frac{x^6}{6 \cdot 4} + \cdots\right]_0^{1/4}$

Because $\frac{(1/4)^5}{15} < 0.0001$, $\int_0^{1/4} x \ln(x + 1)\,dx \approx \frac{(1/4)^3}{3} - \frac{(1/4)^4}{8} \approx 0.00472.$

65. $\int_0^1 \frac{\sin x}{x}\,dx = \int_0^1 \left[\sum_{n=0}^{\infty} \frac{(-1)^n x^{2n}}{(2n+1)!}\right] dx = \left[\sum_{n=0}^{\infty} \frac{(-1)^n x^{2n+1}}{(2n+1)(2n+1)!}\right]_0^1 = \sum_{n=0}^{\infty} \frac{(-1)^n}{(2n+1)(2n+1)!}$

Because $1/(7 \cdot 7!) < 0.0001$, you need three terms:

$$\int_0^1 \frac{\sin x}{x}\,dx = 1 - \frac{1}{3 \cdot 3!} + \frac{1}{5 \cdot 5!} - \cdots \approx 0.9461. \quad \text{(using three nonzero terms)}$$

Note: You are using $\lim_{x \to 0^+} \frac{\sin x}{x} = 1$.

66. $\int_0^1 \cos x^2\,dx = \int_0^1 \left[\sum_{n=0}^{\infty} \frac{(-1)^n x^{4n}}{(2n)!}\right] dx$

$$= \left[\sum_{n=0}^{\infty} \frac{(-1)^n x^{4n+1}}{(4n+1)(2n)!}\right]_0^1$$

$$= \sum_{n=0}^{\infty} \frac{(-1)^n}{(4n+1)(2n)!}$$

$$\int_0^1 \cos x^2\,dx \approx \sum_{n=0}^{3} \frac{(-1)^n}{(4n+1)(2n)!} \approx 0.904523$$

Because $\frac{1}{[4(4)+1][2(4)]!} < 0.0001$, you need 4 terms.

67. $\int_0^{1/2} \frac{\arctan x}{x}\,dx = \int_0^{1/2} \left(1 - \frac{x^2}{3} + \frac{x^4}{5} - \frac{x^6}{7} + \cdots\right) dx$

$$= \left[x - \frac{x^3}{3^2} + \frac{x^5}{5^2} - \frac{x^7}{7^2} + \cdots\right]_0^{1/2}$$

Because $1/(9^2 2^9) < 0.0001$, you have

$$\int_0^{1/2} \frac{\arctan x}{x}\,dx \approx \left(\frac{1}{2} - \frac{1}{3^2 2^3} + \frac{1}{5^2 2^5} - \frac{1}{7^2 2^7} + \frac{1}{9^2 2^9}\right)$$

$$\approx 0.4872.$$

Note: You are using $\lim_{x \to 0^+} \frac{\arctan x}{x} = 1$.

68. $\int_0^{1/2} \arctan(x^2)\,dx = \int_0^1 \left[\sum_{n=0}^{\infty} \frac{(-1)^n x^{4n+2}}{2n+1}\right] dx$

$$= \left[\sum_{n=0}^{\infty} \frac{(-1)^n x^{4n+3}}{(4n+3)(2n+1)}\right]_0^{1/2}$$

$$= \sum_{n=0}^{\infty} \frac{(-1)^n}{(4n+3)(2n+1)2^{4n+3}}$$

Because $\frac{1}{(4n+3)(2n+1)2^{4n+3}} < 0.0001$

when $n = 2$, you need 2 terms.

$$\int_0^{1/2} \arctan(x^2)\,dx \approx \frac{1}{3(1) \cdot 2^3} - \frac{1}{7(3)2^7} \approx 0.041295$$

69. $\int_{0.1}^{0.3} \sqrt{1+x^3}\,dx = \int_{0.1}^{0.3} \left(1 + \frac{x^3}{2} - \frac{x^6}{8} + \frac{x^9}{16} - \frac{5x^{12}}{128} + \cdots\right) dx = \left[x + \frac{x^4}{8} - \frac{x^7}{56} + \frac{x^{10}}{160} - \frac{5x^{13}}{1664} + \cdots\right]_{0.1}^{0.3}$

Because $\frac{1}{56}(0.3^7 - 0.1^7) < 0.0001$, you need two terms.

$$\int_{0.1}^{0.3} \sqrt{1+x^3}\,dx = \left[(0.3 - 0.1) + \frac{1}{8}(0.3^4 - 0.1^4)\right] \approx 0.201.$$

70. $\sqrt{1+x^2} = \left(1+x^2\right)^{1/2} = 1 + \frac{1}{2}x^2 + \frac{\frac{1}{2}\left(-\frac{1}{2}\right)x^4}{2!} + \frac{\frac{1}{2}\left(-\frac{1}{2}\right)\left(-\frac{3}{2}\right)x^6}{3!} + \cdots = 1 + \frac{1}{2}x^2 - \frac{1}{8}x^4 + \frac{1}{16}x^6 - \cdots$

$$\int_0^{0.2} \sqrt{1+x^2}\,dx = \int_0^{0.2}\left[1 + \frac{1}{2}x^2 - \frac{1}{8}x^4 + \frac{1}{16}x^6 - \cdots\right]dx = \left[x + \frac{x^3}{6} - \frac{x^5}{40} + \frac{x^7}{112} - \cdots\right]_0^{0.2}$$

Because $\frac{(0.2)^5}{40} < 0.0001$, you need 2 terms.

$$\int_0^{0.2} \sqrt{1+x^2}\,dx \approx 0.2 + \frac{(0.2)^3}{6} \approx 0.201333$$

71. $\int_0^{\pi/2} \sqrt{x}\cos x\,dx = \int_0^{\pi/2}\left[\sum_{n=0}^{\infty}\frac{(-1)^n x^{(4n+1)/2}}{(2n)!}\right]dx = \left[\sum_{n=0}^{\infty}\frac{(-1)^n x^{(4n+3)/2}}{\left(\frac{4n+3}{2}\right)(2n)!}\right]_0^{\pi/2} = \left[\sum_{n=0}^{\infty}\frac{(-1)^n 2x^{(4n+3)/2}}{(4n+3)(2n)!}\right]_0^{\pi/2}$

Because $2(\pi/2)^{23/2}/(23\cdot 10!) < 0.0001$, you need five terms.

$$\int_0^1 \sqrt{x}\cos x\,dx = 2\left[\frac{(\pi/2)^{3/2}}{3} - \frac{(\pi/2)^{7/2}}{14} + \frac{(\pi/2)^{11/2}}{264} - \frac{(\pi/2)^{15/2}}{10{,}800} + \frac{(\pi/2)^{19/2}}{766{,}080}\right] \approx 0.7040.$$

72. $\int_{0.5}^1 \cos\sqrt{x}\,dx = \int_{0.5}^1\left(1 - \frac{x}{2!} + \frac{x^2}{4!} - \frac{x^3}{6!} + \frac{x^4}{8!} - \cdots\right)dx = \left[x - \frac{x^2}{2(2!)} + \frac{x^3}{3(4!)} - \frac{x^4}{4(6!)} + \frac{x^5}{5(8!)} - \cdots\right]_{0.5}^1$

Because $\frac{1}{201{,}600}\left(1 - 0.5^5\right) < 0.0001$, you have

$$\int_{0.5}^1 \cos\sqrt{x}\,dx \approx \left[(1-0.5) - \frac{1}{4}\left(1-0.5^2\right) + \frac{1}{72}\left(1-0.5^3\right) - \frac{1}{2880}\left(1-0.5^4\right) + \frac{1}{201{,}600}\left(1-0.5^5\right)\right] \approx 0.3243.$$

73. From Exercise 27, you have

$$\frac{1}{\sqrt{2\pi}}\int_0^1 e^{-x^2/2}\,dx = \frac{1}{\sqrt{2\pi}}\int_0^1 \sum_{n=0}^{\infty}\frac{(-1)^n x^{2n}}{2^n n!}\,dx = \frac{1}{\sqrt{2\pi}}\left[\sum_{n=0}^{\infty}\frac{(-1)^n x^{2n+1}}{2^n n!(2n+1)}\right]_0^1 = \frac{1}{\sqrt{2\pi}}\sum_{n=0}^{\infty}\frac{(-1)^n}{2^n n!(2n+1)}$$

$$\approx \frac{1}{\sqrt{2\pi}}\left(1 - \frac{1}{2\cdot 1\cdot 3} + \frac{1}{2^2\cdot 2!\cdot 5} - \frac{1}{2^3\cdot 3!\cdot 7}\right) \approx 0.3412.$$

74. From Exercise 27, you have

$$\frac{1}{\sqrt{2\pi}}\int_1^2 e^{-x^2/2}\,dx = \frac{1}{\sqrt{2\pi}}\int_1^2 \sum_{n=0}^{\infty}\frac{(-1)^n x^{2n}}{2^n n!}\,dx = \frac{1}{\sqrt{2\pi}}\left[\sum_{n=0}^{\infty}\frac{(-1)^n x^{2n+1}}{2^n n!(2n+1)}\right]_1^2$$

$$= \frac{1}{\sqrt{2\pi}}\sum_{n=0}^{\infty}\frac{(-1)^n\left(2^{n+1}-1\right)}{2^n n!(2n+1)}$$

$$\approx \frac{1}{\sqrt{2\pi}}\left(1 - \frac{7}{2\cdot 1\cdot 3} + \frac{31}{2^2\cdot 2!\cdot 5} - \frac{127}{2^3\cdot 3!\cdot 7} + \frac{511}{2^4\cdot 4!\cdot 9} - \frac{2047}{2^5\cdot 5!\cdot 11}\right.$$

$$\left. + \frac{8191}{2^6\cdot 6!\cdot 13} - \frac{32{,}767}{2^7\cdot 7!\cdot 15} + \frac{131{,}071}{2^8\cdot 8!\cdot 17} - \frac{524{,}287}{2^9\cdot 9!\cdot 19}\right) \approx 0.1359.$$

75. $f(x) = x \cos 2x = \sum_{n=0}^{\infty} \frac{(-1)^n 4^n x^{2n+1}}{(2n)!}$

$P_5(x) = x - 2x^3 + \frac{2x^5}{3}$

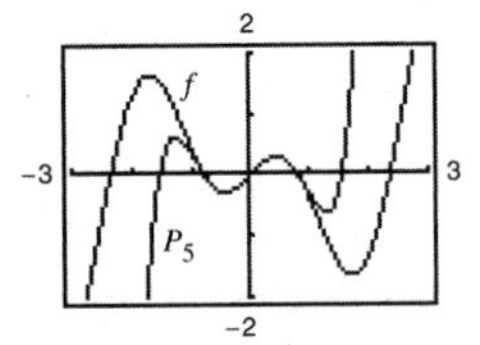

The polynomial is a reasonable approximation on the interval $\left[-\frac{3}{4}, \frac{3}{4}\right]$.

76. $f(x) = \sin \frac{x}{2} \ln(1 + x)$

$P_5(x) = \frac{x^2}{2} - \frac{x^3}{4} + \frac{7x^4}{48} - \frac{11x^5}{96}$

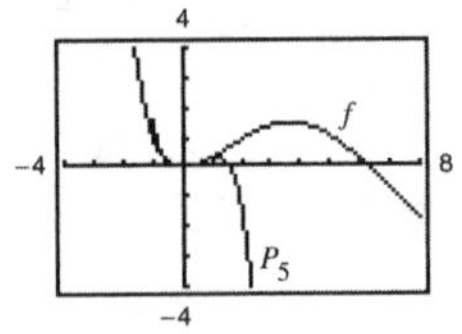

The polynomial is a reasonable approximation on the interval $(-0.60, 0.73)$.

77. $f(x) = \sqrt{x} \ln x, c = 1$

$P_5(x) = (x - 1) - \frac{(x-1)^3}{24} + \frac{(x-1)^4}{24} - \frac{71(x-1)^5}{1920}$

The polynomial is a reasonable approximation on the interval $\left[\frac{1}{4}, 2\right]$.

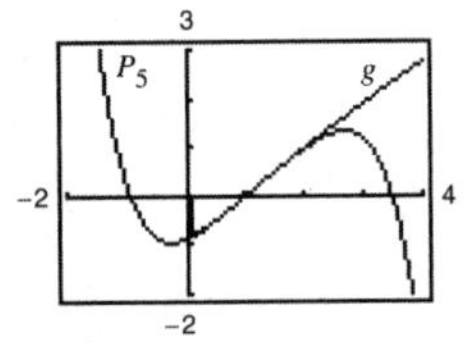

78. $f(x) = \sqrt[3]{x} \cdot \arctan x, c = 1$

$P_5(x) \approx 0.7854 + 0.7618(x - 1) - 0.3412\left[\frac{(x-1)^2}{2!}\right] - 0.0424\left[\frac{(x-1)^3}{3!}\right] + 1.3025\left[\frac{(x-1)^4}{4!}\right] - 5.5913\left[\frac{(x-1)^5}{5!}\right]$

The polynomial is a reasonable approximation on the interval $(0.48, 1.75)$.

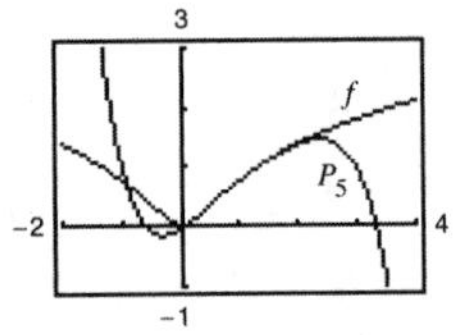

79. See Guidelines, page 668.

80. The binomial series is $(1 + x)^k = 1 + kx + \frac{k(k-1)}{2!}x^2 + \frac{k(k-1)(k-2)}{3!}x^3 + \cdots$. The radius of convergence is $R = 1$.

81. (a) Replace x with $(-x)$.

(b) Replace x with $3x$.

(c) Multiply series by x.

(d) Replace x with $2x$, then replace x with $-2x$, and add the two together.

82. (a) $y = x^2 - \dfrac{x^4}{3!} \Rightarrow$ even polynomial, degree 4

Matches (iii).

$$y = x\left(x - \frac{x^3}{3!}\right)$$

The second factor is the third-degree Taylor polynomial for $f(x) = \sin x$ at $c = 0$.

(b) $y = x - \dfrac{x^3}{2!} + \dfrac{x^5}{4!} \Rightarrow$ odd polynomial, degree 5

Matches (iv).

$$y = x\left(1 - \frac{x^2}{2!} + \frac{x^4}{4!}\right)$$

The second factor is the fourth-degree Taylor polynomial for $f(x) = \cos x$ at $c = 0$.

(c) $y = x + x^2 + \dfrac{x^3}{2!} \Rightarrow$ odd polynomial, degree 3

Matches (i).

$$y = x\left(1 + x + \frac{x^2}{2!}\right)$$

The second factor is the third-degree Taylor polynomial for $f(x) = e^x$ at $c = 0$.

(d) $y = x^2 - x^3 + x^4 \Rightarrow$ even polynomial, degree 4

Matches (ii).

$$y = x^2\left(1 - x + x^2\right)$$

The second factor is the second-degree Taylor polynomial for $f(x) = \dfrac{1}{1 + x}$ at $c = 0$.

83. $$y = \left(\tan\theta - \frac{g}{kv_0\cos\theta}\right)x - \frac{g}{k^2}\ln\left(1 - \frac{kx}{v_0\cos\theta}\right)$$

$$= (\tan\theta)x - \frac{gx}{kv_0\cos\theta} - \frac{g}{k^2}\left[-\frac{kx}{v_0\cos\theta} - \frac{1}{2}\left(\frac{kx}{v_0\cos\theta}\right)^2 - \frac{1}{3}\left(\frac{kx}{v_0\cos\theta}\right)^3 - \frac{1}{4}\left(\frac{kx}{v_0\cos\theta}\right)^4 - \cdots\right]$$

$$= (\tan\theta)x - \frac{gx}{kv_0\cos\theta} + \frac{gx}{kv_0\cos\theta} + \frac{gx^2}{2v_0^2\cos^2\theta} + \frac{gkx^3}{3v_0^3\cos^3\theta} + \frac{gk^2x^4}{4v_0^4\cos^4\theta} + \cdots$$

$$= (\tan\theta)x + \frac{gx^2}{2v_0^2\cos^2\theta} + \frac{kgx^3}{3v_0^3\cos^3\theta} + \frac{k^2gx^4}{4v_0^4\cos^4\theta} + \cdots$$

84. $\theta = 60°,\ v_0 = 64,\ k = \dfrac{1}{16},\ g = -32$

$$y = \sqrt{3}x - \frac{32x^2}{2(64)^2(1/2)^2} - \frac{(1/16)(32)x^3}{3(64)^3(1/2)^3} - \frac{(1/16)^2(32)x^4}{4(64)^4(1/2)^4} - \cdots$$

$$= \sqrt{3}x - 32\left[\frac{2^2x^2}{2(64)^2} + \frac{2^3x^3}{3(64)^3 16} + \frac{2^4x^4}{4(64)^4(16)^2} + \cdots\right]$$

$$= \sqrt{3}x - 32\sum_{n=2}^{\infty}\frac{2^n x^n}{n(64)^n(16)^{n-2}} = \sqrt{3}x - 32\sum_{n=2}^{\infty}\frac{x^n}{n(32)^n(16)^{n-2}}$$

85. $f(x) = \begin{cases} e^{-1/x^2}, & x \neq 0 \\ 0, & x = 0 \end{cases}$

(a)

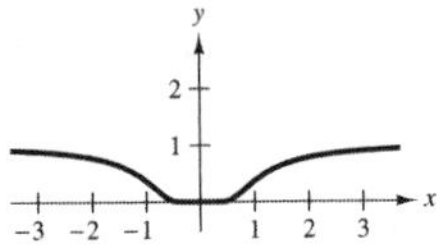

(b) $f'(0) = \lim_{x \to 0} \frac{f(x) - f(0)}{x - 0} = \lim_{x \to 0} \frac{e^{-1/x^2} - 0}{x}$

Let $y = \lim_{x \to 0} \frac{e^{-1/x^2}}{x}$. Then

$$\ln y = \lim_{x \to 0} \ln\left(\frac{e^{-1/x^2}}{x}\right) = \lim_{x \to 0^+}\left[-\frac{1}{x^2} - \ln x\right] = \lim_{x \to 0^+}\left[\frac{-1 - x^2 \ln x}{x^2}\right] = -\infty.$$

So, $y = e^{-\infty} = 0$ and you have $f'(0) = 0$.

(c) $\sum_{n=0}^{\infty} \frac{f^{(n)}(0)}{n!} x^n = f(0) + \frac{f'(0)x}{1!} + \frac{f''(0)x^2}{2!} + \cdots = 0 \neq f(x)$ This series converges to f at $x = 0$ only.

86. (a) $f(x) = \frac{\ln(x^2 + 1)}{x^2}$.

From Exercise 10, you obtain:

$$P = \frac{1}{x^2} \sum_{n=0}^{\infty} \frac{(-1)^n x^{2n+2}}{n+1} = \sum_{n=0}^{\infty} \frac{(-1)^n x^{2n}}{n+1}$$

$$P_8 = 1 - \frac{x^2}{2} + \frac{x^4}{3} - \frac{x^6}{4} + \frac{x^8}{5}.$$

(b)

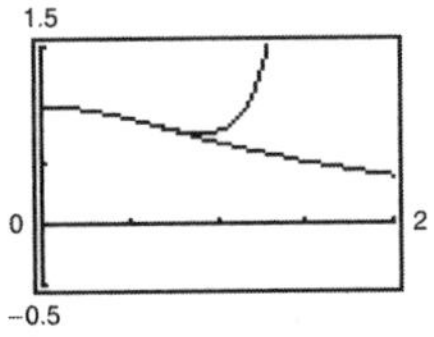

(c) $F(x) = \int_0^x \frac{\ln(t^2 + 1)}{t^2}\, dt$

$G(x) = \int_0^x P_8(t)\, dt$

x	0.25	0.50	0.75	1.00	1.50	2.00
$F(x)$	0.2475	0.4810	0.6920	0.8776	1.1798	1.4096
$G(x)$	0.2475	0.4810	0.6924	0.8865	1.6878	9.6063

(d) The curves are nearly identical for $0 < x < 1$. Hence, the integrals nearly agree on that interval.

87. By the Ratio Test: $\lim_{n \to \infty} \left| \frac{x^{n+1}}{(n+1)!} \cdot \frac{n!}{x^n} \right| = \lim_{n \to \infty} \frac{|x|}{n+1} = 0$ which shows that $\sum_{n=0}^{\infty} \frac{x^n}{n!}$ converges for all x.

88. $\ln\left(\dfrac{1+x}{1-x}\right) = \ln(1+x) - \ln(1-x)$

$$= \left(x - \frac{x^2}{2} + \frac{x^3}{3} - \cdots\right) - \left(-x - \frac{x^2}{2} - \frac{x^3}{3} - \cdots\right) = 2x + 2\frac{x^3}{3} + 2\frac{x^5}{5} + \cdots) = 2x\sum_{n=0}^{\infty}\frac{x^{2n}}{2n+1},\ R = 1$$

$$\ln 3 = \ln\left(\frac{1+1/2}{1-1/2}\right) \approx 2\left(\frac{1}{2}\right)\left[1 + \frac{(1/2)^2}{3} + \frac{(1/2)^4}{5} + \frac{(1/2)^6}{7}\right] = 1 + \frac{1}{12} + \frac{1}{80} + \frac{1}{448} \approx 1.098065$$

$(\ln 3 \approx 1.098612)$

89. $\dbinom{5}{3} = \dfrac{5\cdot 4\cdot 3}{3!} = \dfrac{60}{6} = 10$

90. $\dbinom{-2}{2} = \dfrac{(-2)(-3)}{2!} = 3$

91. $\dbinom{0.5}{4} = \dfrac{(0.5)(-0.5)(-1.5)(-2.5)}{4!} = -0.0390625 = -\dfrac{5}{128}$

92. $\dbinom{-1/3}{5} = \dfrac{(-1/3)(-4/3)(-7/3)(-10/3)(-13/3)}{5!}$

$$= \frac{-91}{729} \approx -0.12483$$

93. $(1+x)^k = \displaystyle\sum_{n=0}^{\infty}\binom{k}{n}x^n$

Example: $(1+x)^2 = \displaystyle\sum_{n=0}^{\infty}\binom{2}{n}x^n = 1 + 2x + x^2$

94. Assume $e = p/q$ is rational. Let $N > q$ and form the following.

$$e - \left[1 + 1 + \frac{1}{2!} + \cdots + \frac{1}{N!}\right] = \frac{1}{(N+1)!} + \frac{1}{(N+2)!} + \cdots$$

Set $a = N!\left[e - \left(1 + 1 + \cdots + \dfrac{1}{N!}\right)\right]$, a positive integer. But,

$$a = N!\left[\frac{1}{(N+1)!} + \frac{1}{(N+2)!} + \cdots\right] = \frac{1}{N+1} + \frac{1}{(N+1)(N+2)} + \cdots < \frac{1}{N+1} + \frac{1}{(N+1)^2} + \cdots$$

$$= \frac{1}{N+1}\left[1 + \frac{1}{N+1} + \frac{1}{(N+1)^2} + \cdots\right] = \frac{1}{N+1}\left[\frac{1}{1-\left(\frac{1}{N+1}\right)}\right] = \frac{1}{N},\ \text{a contradiction.}$$

95. $g(x) = \dfrac{x}{1-x-x^2} = a_0 + a_1x + a_2x^2 + \cdots$

$x = (1 - x - x^2)(a_0 + a_1x + a_2x^2 + \cdots)$

$x = a_0 + (a_1 - a_0)x + (a_2 - a_1 - a_0)x^2 + (a_3 - a_2 - a_1)x^3 + \cdots$

Equating coefficients,

$a_0 = 0$

$a_1 - a_0 = 1 \Rightarrow a_1 = 1$

$a_2 - a_1 - a_0 = 0 \Rightarrow a_2 = 1$

$a_3 - a_2 - a_1 = 0 \Rightarrow a_3 = 2$

$a_4 = a_3 + a_2 = 3$, etc.

In general, $a_n = a_{n-1} + a_{n-2}$. The coefficients are the Fibonacci numbers.

96. Assume the interval is $[-1, 1]$. Let $x \in [-1, 1]$,

$$f(1) = f(x) + (1 - x)f'(x) + \tfrac{1}{2}(1 - x)^2 f''(c),\ c \in (x, 1)$$

$$f(-1) = f(x) + (-1 - x)f'(x) + \tfrac{1}{2}(-1 - x)^2 f''(d),\ d \in (-1, x).$$

So, $f(1) - f(-1) = 2f'(x) + \tfrac{1}{2}(1 - x)^2 f''(c) - \tfrac{1}{2}(1 + x)^2 f''(d)$

$$2f'(x) = f(1) - f(-1) - \tfrac{1}{2}(1 - x)^2 f''(c) + \tfrac{1}{2}(1 + x)^2 f''(d).$$

Because $|f(x)| \le 1$ and $|f''(x)| \le 1$,

$$2|f'(x)| \le |f(1)| + |f(-1)| + \tfrac{1}{2}(1 - x)^2|f''(c)| + \tfrac{1}{2}(1 + x)^2|f''(d)| \le 1 + 1 + \tfrac{1}{2}(1 - x^2) + \tfrac{1}{2}(1 + x)^2 = 3 + x^2 \le 4.$$

So, $|f'(x)| \le 2$.

Note: Let $f(x) = \tfrac{1}{2}(x + 1)^2 - 1$. Then $|f'(x)| \le 1, |f''(x)| = 1$ and $f'(1) = 2$.

Review Exercises for Chapter 9

1. $a_n = 5^n$

$a_1 = 5^1 = 5$

$a_2 = 5^2 = 25$

$a_3 = 5^3 = 125$

$a_4 = 5^4 = 625$

$a_5 = 5^5 = 3125$

2. $a_n = \dfrac{3^n}{n!}$

$a_1 = \dfrac{3^1}{1!} = 3$

$a_2 = \dfrac{3^2}{2!} = \dfrac{9}{2}$

$a_3 = \dfrac{3^3}{3!} = \dfrac{9}{2}$

$a_4 = \dfrac{3^4}{4!} = \dfrac{27}{8}$

$a_5 = \dfrac{3^5}{5!} = \dfrac{81}{40}$

3. $a_n = \left(-\tfrac{1}{4}\right)^n$

$a_1 = \left(-\tfrac{1}{4}\right)^1 = -\tfrac{1}{4}$

$a_2 = \left(-\tfrac{1}{4}\right)^2 = \tfrac{1}{16}$

$a_3 = \left(-\tfrac{1}{4}\right)^3 = -\tfrac{1}{64}$

$a_4 = \left(-\tfrac{1}{4}\right)^4 = \tfrac{1}{256}$

$a_5 = \left(-\tfrac{1}{4}\right)^5 = -\tfrac{1}{1024}$

4. $a_n = \dfrac{2n}{n + 5}$

$a_1 = \dfrac{2(1)}{1 + 5} = \dfrac{1}{3}$

$a_2 = \dfrac{2(2)}{2 + 5} = \dfrac{4}{7}$

$a_3 = \dfrac{2(3)}{3 + 5} = \dfrac{3}{4}$

$a_4 = \dfrac{2(4)}{4 + 5} = \dfrac{8}{9}$

$a_5 = \dfrac{2(5)}{5 + 5} = 1$

5. $a_n = 4 + \dfrac{2}{n}$: 6, 5, 4.67, …

Matches (a).

6. $a_n = 4 - \dfrac{n}{2}$: 3.5, 3, …

Matches (c).

7. $a_n = 10(0.3)^{n-1}$: 10, 3, …

Matches (d).

8. $a_n = 6\left(-\tfrac{2}{3}\right)^{n-1}$: 6, −4, …

Matches (b).

9. $a_n = \dfrac{5n + 2}{n}$

The sequence seems to converge to 5.

$$\lim_{n\to\infty} a_n = \lim_{n\to\infty} \frac{5n+2}{n} = \lim_{n\to\infty}\left(5 + \frac{2}{n}\right) = 5$$

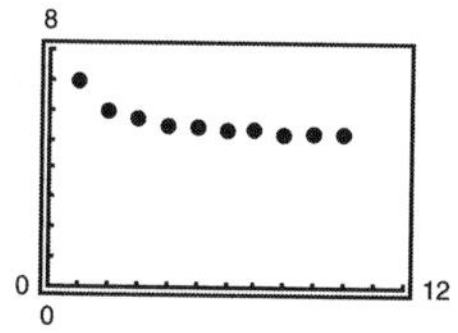

10. $a_n = \sin\dfrac{n\pi}{2}$

The sequence seems to diverge (oscillates).

$\sin\dfrac{n\pi}{2}$: 1, 0, −1, 0, 1, 0, ...

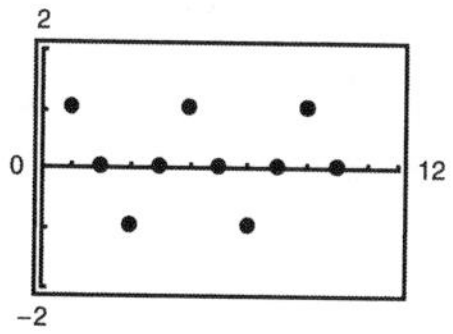

11. $\displaystyle\lim_{n\to\infty}\left[\left(\frac{2}{5}\right)^n + 5\right] = 0 + 5 = 5$

Converges

12. $\displaystyle\lim_{n\to\infty}\left[3 - \frac{2}{n^2 - 1}\right] = 3 - 0 = 3$

Converges

13. $\displaystyle\lim_{n\to\infty}\frac{n^3+1}{n^2} = \infty$

Diverges

14. $\displaystyle\lim_{n\to\infty}\frac{1}{\sqrt{n}} = 0$

Converges

15. $\displaystyle\lim_{n\to\infty}\frac{n}{n^2+1} = 0$

Converges

16. $\displaystyle\lim_{n\to\infty}\frac{n}{\ln(n)} = \lim_{n\to\infty}\frac{1}{1/n} = \infty$

Diverges

17. $$\lim_{n\to\infty}\left(\sqrt{n+1} - \sqrt{n}\right) = \lim_{n\to\infty}\left(\sqrt{n+1} - \sqrt{n}\right)\frac{\sqrt{n+1}+\sqrt{n}}{\sqrt{n+1}+\sqrt{n}} = \lim_{n\to\infty}\frac{1}{\sqrt{n+1}+\sqrt{n}} = 0$$

Converges

18. $\displaystyle\lim_{n\to\infty}\frac{\sin\sqrt{n}}{\sqrt{n}} = 0$

Converges

19. $a_n = 5n - 2$

20. $a_n = n^2 - 6$

21. $a_n = \dfrac{1}{n! + 1}$

22. $a_n = \dfrac{n}{n^2 + 1}$

23. (a) $A_n = 8000\left(1 + \dfrac{0.05}{4}\right)^n, \quad n = 1, 2, 3, \ldots$

$A_1 = 8000\left(1 + \dfrac{0.05}{4}\right)^1 = \$8100.00$

$A_2 = \$8201.25$

$A_3 = \$8303.77$

$A_4 = \$8407.56$

$A_5 = \$8512.66$

$A_6 = \$8619.07$

$A_7 = \$8726.80$

$A_8 = \$8835.89$

(b) $A_{40} = \$13{,}148.96$

24. (a) $V_n = 175{,}000(0.70)^n, \quad n = 1, 2, \ldots$

(b) $V_5 = 175{,}000(0.70)^5 \approx \$29{,}412.25$

25. $S_1 = 3$

$S_2 = 3 + \dfrac{3}{2} = \dfrac{9}{2} = 4.5$

$S_3 = 3 + \dfrac{3}{2} + 1 = \dfrac{11}{2} = 5.5$

$S_4 = 3 + \dfrac{3}{2} + 1 + \dfrac{3}{4} = \dfrac{25}{4} = 6.25$

$S_5 = 3 + \dfrac{3}{2} + 1 + \dfrac{3}{4} + \dfrac{3}{5} = \dfrac{137}{20} = 6.85$

26. $S_1 = -\frac{1}{2} = -0.5$

$S_2 = -\frac{1}{2} + \frac{1}{4} = -\frac{1}{4} = -0.25$

$S_3 = -\frac{1}{2} + \frac{1}{4} - \frac{1}{8} = -\frac{3}{8} = -0.375$

$S_4 = -\frac{1}{2} + \frac{1}{4} - \frac{1}{8} + \frac{1}{16} = -\frac{5}{16} = -0.3125$

$S_5 = -\frac{1}{2} + \frac{1}{4} - \frac{1}{8} + \frac{1}{16} - \frac{1}{32} = -\frac{11}{32} = -0.34375$

27. (a)

n	5	10	15	20	25
S_n	13.2	113.3	873.8	6648.5	50,500.3

The series diverges $\left(\text{geometric } r = \frac{3}{2} > 1\right)$.

(b)

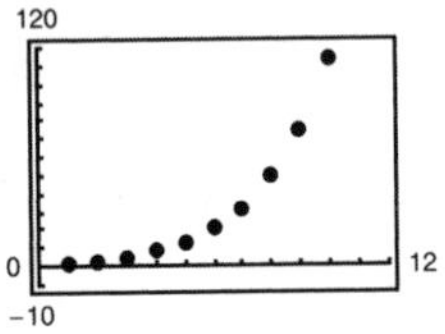

28. (a)

n	5	10	15	20	25
S_n	0.3917	0.3228	0.3627	0.3344	0.3564

The series converges by the Alternating Series Test.

(b)

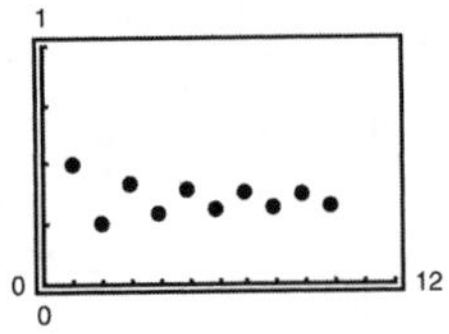

29. (a)

n	5	10	15	20	25
S_n	0.4597	0.4597	0.4597	0.4597	0.4597

The series converges by the Alternating Series Test.

(b)

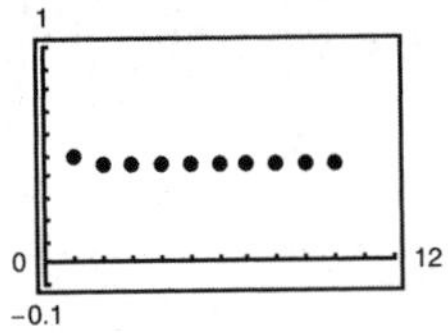

30. (a)

n	5	10	15	20	25
S_n	0.8333	0.9091	0.9375	0.9524	0.9615

The series converges, by the Limit Comparison Test with $\sum \frac{1}{n^2}$.

(b)

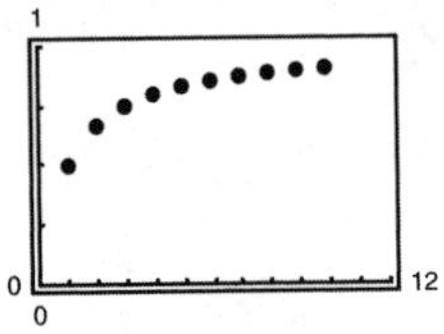

31. $\sum_{n=0}^{\infty}\left(\frac{2}{5}\right)^n = \frac{1}{1-(2/5)} = \frac{5}{3}$ (Geometric series)

32. $\sum_{n=0}^{\infty}\frac{3^{n+2}}{7^n} = 9\sum_{n=0}^{\infty}\left(\frac{3}{7}\right)^n = 9\left(\frac{1}{1-(3/7)}\right)$

$= 9\cdot\frac{7}{4} = \frac{63}{4}$ (Geometric series)

33. $\sum_{n=1}^{\infty}\left[(0.6)^n + (0.8)^n\right] = \sum_{n=0}^{\infty}0.6(0.6)^n + \sum_{n=0}^{\infty}0.8(0.8)^n = (0.6)\frac{1}{1-0.6} + (0.8)\frac{1}{1-0.8} = \frac{6}{10}\cdot\frac{10}{4} + \frac{8}{10}\cdot\frac{10}{2} = \frac{11}{2} = 5.5$

34. $\sum_{n=0}^{\infty}\left[\left(\frac{2}{3}\right)^n - \frac{1}{(n+1)(n+2)}\right] = \sum_{n=0}^{\infty}\left(\frac{2}{3}\right)^n - \sum_{n=0}^{\infty}\left(\frac{1}{n+1} - \frac{1}{n+2}\right)$

$= \frac{1}{1-(2/3)} - \left[\left(1-\frac{1}{2}\right) + \left(\frac{1}{2}-\frac{1}{3}\right) + \left(\frac{1}{3}-\frac{1}{4}\right) + \cdots\right] = 3 - 1 = 2$

35. (a) $0.\overline{09} = 0.09 + 0.0009 + 0.000009 + \cdots = 0.09(1 + 0.01 + 0.0001 + \cdots) = \sum_{n=0}^{\infty}(0.09)(0.01)^n$

(b) $0.\overline{09} = \frac{0.09}{1-0.01} = \frac{1}{11}$

36. (a) $0.\overline{64} = 0.64 + 0.0064 + 0.000064 + \cdots = 0.64(1 + 0.01 + 0.0001 + \cdots) = 0.64\sum_{n=0}^{\infty}(0.01)^n$

(b) $0.\overline{64} = \frac{0.64}{1-0.01} = \frac{64}{99}$

37. Diverges. Geometric series with $a = 1$ and $|r| = 1.67 > 1$.

38. Converges. Geometric series with $a = 1$ and $|r| = |0.36| < 1$.

39. Diverges. nth-Term Test. $\lim_{n\to\infty} a_n \neq 0$.

40. Diverges. nth-Term Test, $\lim_{n\to\infty} a_n = \frac{2}{3}$.

41. $D_1 = 8$

$D_2 = 0.7(8) + 0.7(8) = 16(0.7)$

$\vdots$

$D = 8 + 16(0.7) + 16(0.7)^2 + \cdots + 16(0.7)^n + \cdots$

$= -8 + \sum_{n=0}^{\infty}16(0.7)^n = -8 + \frac{16}{1-0.7} = 45\frac{1}{3}$ meters

42. (See Exercise 84 in Section 9.2)

$A = P\left(\frac{12}{r}\right)\left[\left(1+\frac{r}{12}\right)^{12t} - 1\right]$

$= 125\left(\frac{12}{0.035}\right)\left[\left(1+\frac{0.035}{12}\right)^{12(10)} - 1\right] \approx \$17{,}929.06$

43. $\sum_{n=1}^{\infty}\frac{2}{6n+1}$

Let $f(x) = \frac{2}{6x+1}$, $f'(x) = \frac{-12}{(6x+1)^2} < 0$ for $x \geq 1$

f is positive, continuous, and decreasing for $x \geq 1$.

$\int_1^{\infty}\frac{2}{6x+1}\,dx = \left[\frac{1}{3}\ln(6x+1)\right]_1^{\infty} = \infty$, diverges.

So, the series diverges by Theorem 9.10.

44. $\sum_{n=1}^{\infty}\frac{1}{\sqrt[4]{n^3}} = \sum_{n=1}^{\infty}\frac{1}{n^{3/4}}$

Divergent p-series, $p = \frac{3}{4} < 1$

45. $\sum_{n=1}^{\infty}\frac{1}{n^{5/2}}$ is a p-series with $p = \frac{5}{2} > 1$.

So, the series converges.

46. $\sum_{n=1}^{\infty} \frac{1}{5^n}$

Let $f(x) = \frac{1}{5^x}$, $f'(x) = -(\ln 5)5^{-x} < 0$ for $x \geq 1$.

f is positive, continuous, and decreasing for $x \geq 1$.

$$\int_1^{\infty} \frac{1}{5^x}\,dx = \left[\frac{-1}{(\ln 5)5^x}\right]_1^{\infty} = \frac{1}{5 \ln 5}$$

So, the series converges by Theorem 9.10.

47. $\sum_{n=1}^{\infty} \left(\frac{1}{n^2} - \frac{1}{n}\right) = \sum_{n=1}^{\infty} \frac{1}{n^2} - \sum_{n=1}^{\infty} \frac{1}{n}$

Because the second series is a divergent p-series while the first series is a convergent p-series, the difference diverges.

48. $\sum_{n=1}^{\infty} \frac{\ln n}{n^4}$

Let $f(x) = \frac{\ln x}{x^4}$, $f'(x) = \frac{1}{x^5} - \frac{4 \ln x}{x^3} < 0$.

f is positive, continuous, and decreasing for $x > 1$.

$$\int_1^{\infty} x^{-4} \ln x\,dx = \lim_{b\to\infty}\left[\frac{-\ln x}{3x^3} - \frac{1}{9x^3}\right]_1^b = 0 + \frac{1}{9} = \frac{1}{9}$$

So, the series converges by Theorem 9.10.

49. $\sum_{n=2}^{\infty} \frac{1}{\sqrt[3]{n} - 1}$

$$\frac{1}{\sqrt[3]{n} - 1} > \frac{1}{\sqrt[3]{n}}$$

Therefore, the series diverges by comparison with the divergent p-series

$$\sum_{n=2}^{\infty} \frac{1}{\sqrt[3]{n}} = \sum_{n=2}^{\infty} \frac{1}{n^{1/3}}.$$

50. $\sum_{n=1}^{\infty} \frac{n}{\sqrt{n^3 + 3n}}$

$$\lim_{n\to\infty} \frac{n/\sqrt{n^3 + 3n}}{1/\sqrt{n}} = \lim_{n\to\infty} \frac{n^{3/2}}{\sqrt{n^3 + 3n}} = 1$$

By a limit comparison test with the divergent p-series $\sum_{n=1}^{\infty} \frac{1}{n^{1/2}}$, the series diverges.

51. $\sum_{n=1}^{\infty} \frac{1}{\sqrt{n^3 + 2n}}$

$$\lim_{n\to\infty} \frac{1/\sqrt{n^3 + 2n}}{1/(n^{3/2})} = \lim_{n\to\infty} \frac{n^{3/2}}{\sqrt{n^3 + 2n}} = 1$$

By a limit comparison test with the convergent p-series $\sum_{n=1}^{\infty} \frac{1}{n^{3/2}}$, the series converges.

52. $\sum_{n=1}^{\infty} \frac{n+1}{n(n+2)}$

$$\lim_{n\to\infty} \frac{(n+1)/n(n+2)}{1/n} = \lim_{n\to\infty} \frac{n+1}{n+2} = 1$$

By a limit comparison test with $\sum_{n=1}^{\infty} \frac{1}{n}$, the series diverges.

53. $\sum_{n=1}^{\infty} \frac{1 \cdot 3 \cdot 5 \cdots (2n-1)}{2 \cdot 4 \cdot 6 \cdots (2n)}$

$$a_n = \frac{1 \cdot 3 \cdot 5 \cdots (2n-1)}{2 \cdot 4 \cdot 6 \cdots (2n)} = \left(\frac{3}{2} \cdot \frac{5}{4} \cdots \frac{2n-1}{2n-2}\right)\frac{1}{2n} > \frac{1}{2n}$$

Because $\sum_{n=1}^{\infty} \frac{1}{2n} = \frac{1}{2}\sum_{n=1}^{\infty} \frac{1}{n}$ diverges (harmonic series), so does the original series.

54. Because $\sum_{n=1}^{\infty} \frac{1}{3^n}$ converges, $\sum_{n=1}^{\infty} \frac{1}{3^n - 5}$ converges by the Limit Comparison Test.

55. $\sum_{n=1}^{\infty} \frac{(-1)^n}{n^5}$ converges by the Alternating Series Test.

$$\lim_{n\to\infty} \frac{1}{n^5} = 0 \text{ and } a_{n+1} = \frac{1}{(n+1)^5} < \frac{1}{n^5} = a_n.$$

56. $\sum_{n=1}^{\infty} \frac{(-1)^n (n+1)}{n^2 + 1}$ converges by the Alternating Series Test. $\lim_{n\to\infty} \frac{n+1}{n^2+1} = 0$ and if

$$f(x) = \frac{x+1}{x^2+1},\ f'(x) = \frac{-(x^2 + 2x - 1)}{(x^2+1)^2} < 0 \Rightarrow \text{terms}$$

are decreasing. So, $a_{n+1} < a_n$.

57. $\sum_{n=2}^{\infty} \frac{(-1)^n - n}{n^2 - 3}$ converges by the Alternating Series Test.

$\lim_{n\to\infty} \frac{n}{n^2 - 3} = 0$ and if

$f(x) = \frac{n}{n^2 - 3}, f'(x) = \frac{-(n^2 + 3)}{(n^2 - 3)^2} < 0 \Rightarrow$ terms are

decreasing. So, $a_{n+1} < a_n$.

58. $\sum_{n=1}^{\infty} \frac{(-1)^n \sqrt{n}}{n + 1}$

$$a_{n+1} = \frac{\sqrt{n+1}}{n+2} \le \frac{\sqrt{n}}{n+1} = a_n$$

$$\lim_{n\to\infty} \frac{\sqrt{n}}{n+1} = 0$$

By the Alternating Series Test, the series converges.

59. Diverges by the nth-Term Test.

$$\lim_{n\to\infty} \frac{n}{n-3} = 1 \ne 0$$

60. Converges by the Alternating Series Test.

$$a_{n+1} = \frac{3\ln(n+1)}{n+1} < \frac{3\ln n}{n} = a_n, \lim_{n\to\infty} \frac{3\ln n}{n} = 0$$

61. $\lim_{n\to\infty} \sqrt[n]{\left(\frac{3n-1}{2n+5}\right)^n} = \lim_{n\to\infty} \left(\frac{3n-1}{2n+5}\right) = \frac{3}{2} > 1$

Diverges by Root Test.

62. $\lim_{n\to\infty} \sqrt[n]{\left(\frac{4n}{7n-1}\right)^n} = \lim_{n\to\infty} \left(\frac{4n}{7n-1}\right) = \frac{4}{7} < 1$

Converges by Root Test.

63. $\sum_{n=1}^{\infty} \frac{n}{e^{n^2}}$

$$\begin{aligned}\lim_{n\to\infty} \left|\frac{a_{n+1}}{a_n}\right| &= \lim_{n\to\infty} \left|\frac{n+1}{e^{(n+1)^2}} \cdot \frac{e^{n^2}}{n}\right| \\ &= \lim_{n\to\infty} \left|\frac{e^{n^2}(n+1)}{e^{n^2+2n+1}n}\right| \\ &= \lim_{n\to\infty} \left(\frac{1}{e^{2n+1}}\right)\left(\frac{n+1}{n}\right) \\ &= (0)(1) = 0 < 1\end{aligned}$$

By the Ratio Test, the series converges.

64. $\sum_{n=1}^{\infty} \frac{n!}{e^n}$

$$\begin{aligned}\lim_{n\to\infty} \left|\frac{a_{n+1}}{a_n}\right| &= \lim_{n\to\infty} \left|\frac{(n+1)!}{e^{n+1}} \cdot \frac{e^n}{n!}\right| \\ &= \lim_{n\to\infty} \frac{n+1}{e} = \infty\end{aligned}$$

By the Ratio Test, the series diverges.

65. $\sum_{n=1}^{\infty} \frac{2^n}{n^3}$

$$\lim_{n\to\infty} \left|\frac{a_{n+1}}{a_n}\right| = \lim_{n\to\infty} \left|\frac{2^{n+1}}{(n+1)^3} \cdot \frac{n^3}{2^n}\right| = \lim_{n\to\infty} \frac{2n^3}{(n+1)^3} = 2$$

By the Ratio Test, the series diverges.

66. $\sum_{n=1}^{\infty} \frac{1 \cdot 3 \cdot 5 \cdots (2n-1)}{2 \cdot 5 \cdot 8 \cdots (3n-1)}$

$$\lim_{n\to\infty} \left|\frac{a_{n+1}}{a_n}\right| = \lim_{n\to\infty} \left|\frac{1 \cdot 3 \cdots (2n-1)(2n+1)}{2 \cdot 5 \cdots (3n-1)(3n+2)} \cdot \frac{2 \cdot 5 \cdots (3n-1)}{1 \cdot 3 \cdots (2n-1)}\right| = \lim_{n\to\infty} \frac{2n+1}{3n+2} = \frac{2}{3}$$

By the Ratio Test, the series converges.

67. (a) Ratio Test: $\lim_{n\to\infty} \left|\frac{a_{n+1}}{a_n}\right| = \lim_{n\to\infty} \frac{(n+1)(3/5)^{n+1}}{n(3/5)^n} = \lim_{n\to\infty} \left(\frac{n+1}{n}\right)\left(\frac{3}{5}\right) = \frac{3}{5} < 1$, converges

(b)

n	5	10	15	20	25
S_n	2.8752	3.6366	3.7377	3.7488	3.7499

(c)

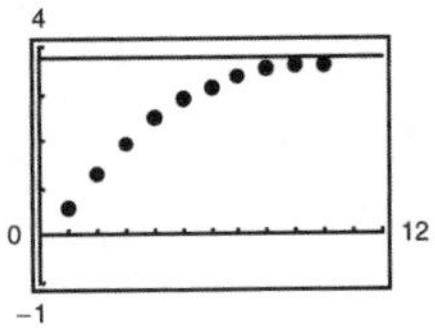

(d) The sum is approximately 3.75.

68. (a) The series converges by the Alternating Series Test.

(b)

n	5	10	15	20	25
S_n	0.0871	0.0669	0.0734	0.0702	0.0721

(c)

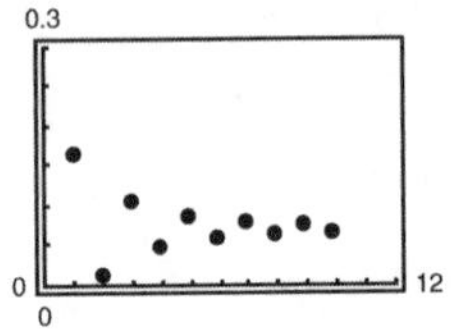

(d) The sum is approximately 0.0714.

69.
$$f(x) = e^{-2x}, \quad f(0) = 1$$
$$f'(x) = -2e^{-2x}, \quad f'(0) = -2$$
$$f''(x) = 4e^{-2x}, \quad f''(0) = 4$$
$$f'''(x) = -8e^{-2x}, \quad f'''(0) = -8$$
$$P_3(x) = f(0) + f'(0)x + \frac{f''(0)}{2!}x^2 + \frac{f'''(0)}{3!}x^3$$
$$= 1 - 2x + 2x^2 - \frac{4}{3}x^3$$

70.
$$f(x) = \cos \pi x, \quad f(0) = 1$$
$$f'(x) = -\pi \sin \pi x, \quad f'(0) = 0$$
$$f''(x) = -\pi^2 \cos \pi x, \quad f''(0) = -\pi^2$$
$$f'''(x) = \pi^3 \sin \pi x, \quad f'''(0) = 0$$
$$f^{(4)}(x) = \pi^4 \cos \pi x, \quad f^{(4)}(0) = \pi^4$$
$$P_4(x) = f(0) + f'(0)x + \frac{f''(0)}{2!}x^2 + \frac{f'''(0)}{3!}x^3 + \frac{f^{(4)}(0)}{4!}x^4$$
$$= 1 - \frac{\pi^2x^2}{2} + \frac{\pi^4x^4}{24}$$

71.
$$f(x) = e^{-3x} \quad f(0) = 1$$
$$f'(x) = -3e^{-3x} \quad f'(0) = -3$$
$$f''(x) = 9e^{-3x} \quad f''(0) = 9$$
$$f'''(x) = -27e^{-3x} \quad f'''(0) = -27$$
$$P_3(x) = f(0) + f'(0)x + \frac{f''(0)}{2!}x^2 + \frac{f'''(0)}{3!}x^3$$
$$= 1 - 3x + \frac{9}{2}x^2 - \frac{9}{2}x^3$$

72. $f(x) = \tan x \qquad f\left(-\frac{\pi}{4}\right) = -1$

$f'(x) = \sec^2 x \qquad f'\left(-\frac{\pi}{4}\right) = 2$

$f''(x) = 2\sec^2 x \tan x \qquad f''\left(-\frac{\pi}{4}\right) = -4$

$f'''(x) = 4\sec^2 x \tan^2 x + 2\sec^4 x \qquad f'''\left(-\frac{\pi}{4}\right) = 16$

$$P_3(x) = f\left(-\frac{\pi}{4}\right) + f'\left(-\frac{\pi}{4}\right)\left(x + \frac{\pi}{4}\right) + \frac{f''\left(-\frac{\pi}{4}\right)}{2!}\left(x + \frac{\pi}{4}\right)^2 + \frac{f'''\left(-\frac{\pi}{4}\right)}{3!}\left(x + \frac{\pi}{4}\right)^3$$

$$= -1 + 2\left(x + \frac{\pi}{4}\right) - 2\left(x + \frac{\pi}{4}\right)^2 + \frac{8}{3}\left(x + \frac{\pi}{4}\right)^3$$

73. $f(x) = \cos x$

$\left|f^{(n+1)}(x)\right| \le 1$ for all x and all n.

$$\left|R_n(x)\right| = \left|\frac{f^{(n+1)}(z)\,x^{n+1}}{(n+1)!}\right| \le \frac{(0.75)^{n+1}}{(n+1)!} < 0.001$$

By trial and error, $n = 5$. (3 terms)

74. $f(x) = e^x$, $f^{(n+1)} = e^x$

Maximum on $[-0.25, 0]$ is $e^0 = 1$.

$$|R_n| \le \frac{f^{(n+1)}(z)\,x^{n+1}}{(n+1)!} \le \frac{(-0.25)^{n+1}}{(n+1)!} < 0.001$$

By trial and error, $n = 3$.

75. $\displaystyle\sum_{n=0}^{\infty}\left(\frac{x}{10}\right)^n$

Geometric series which converges only if $|x/10| < 1$ or $-10 < x < 10$.

76. $\displaystyle\sum_{n=0}^{\infty}(5x)^n$

Geometric series which converges only if $|5x| < 1 \Rightarrow |x| < \frac{1}{5}$ or $-\frac{1}{5} < x < \frac{1}{5}$.

77. $\displaystyle\sum_{n=0}^{\infty}\frac{(-1)^n(x-2)^n}{(n+1)^2}$

$$\lim_{n\to\infty}\left|\frac{u_{n+1}}{u_n}\right| = \lim_{n\to\infty}\left|\frac{(-1)^{n+1}(x-2)^{n+1}}{(n+2)^2}\cdot\frac{(n+1)^2}{(-1)^n(x-2)^n}\right| = |x-2|$$

$R = 1$

Center: 2

Because the series converges when $x = 1$ and when $x = 3$, the interval of convergence is $[1, 3]$.

78. $\displaystyle\sum_{n=1}^{\infty}\frac{3^n(x-2)^n}{n}$

$$\lim_{n\to\infty}\left|\frac{u_{n+1}}{u_n}\right| = \lim_{n\to\infty}\left|\frac{3^{n+1}(x-2)^{n+1}}{n+1}\cdot\frac{n}{3^n(x-2)^n}\right| = 3|x-2|$$

$R = \frac{1}{3}$

Center: 2

Because the series converges at $\frac{5}{3}$ and diverges at $\frac{7}{3}$, the interval of convergence is $\left[\frac{5}{3}, \frac{7}{3}\right)$.

79. $\displaystyle\sum_{n=0}^{\infty}n!(x-2)^n$

$$\lim_{n\to\infty}\left|\frac{u_{n+1}}{u_n}\right| = \lim_{n\to\infty}\left|\frac{(n+1)!(x-2)^{n+1}}{n!(x-2)^n}\right| = \infty$$

which implies that the series converges only at the center $x = 2$.

80. $\sum_{n=0}^{\infty} \frac{(x-2)^n}{2^n} = \sum_{n=0}^{\infty} \left(\frac{x-2}{2}\right)^n$

Geometric series which converges only if

$\left|\frac{x-2}{2}\right| < 1$ or $0 < x < 4$.

81. (a) $f(x) = \sum_{n=0}^{\infty} \left(\frac{x}{5}\right)^n, (-5, 5)$ (Geometric)

(b) $f'(x) = \sum_{n=1}^{\infty} \frac{n}{5}\left(\frac{x}{5}\right)^{n-1}, (-5, 5)$

(c) $f''(x) = \sum_{n=2}^{\infty} \frac{n(n-1)}{25}\left(\frac{x}{5}\right)^{n-2}, (-5, 5)$

(d) $\int f(x)\, dx = \sum_{n=0}^{\infty} \frac{5}{n+1}\left(\frac{x}{5}\right)^{n+1}, (-5, 5)$

$\left[\sum_{n=0}^{\infty} \frac{5}{n+1}\left(\frac{-5}{5}\right)^{n+1} = \sum_{n=0}^{\infty} \frac{(-1)^{n+1} 5}{n+1}, \text{ converges}\right]$

82. (a) $f(x) = \sum_{n=1}^{\infty} \frac{(-1)^{n+1}(x-4)^n}{n}, (3, 5)$

$\left[\sum_{n=1}^{\infty} (-1)^{n+1} \frac{(3-4)^n}{n} = \sum_{n=1}^{\infty} \frac{(-1)^{2n+1}}{n}, \text{ diverges}\right]$

$\left[\sum_{n=1}^{\infty} (-1)^{n+1} \frac{(5-4)^n}{n} = \sum \frac{(-1)^{n+1}}{n}, \text{ converges}\right]$

(b) $f'(x) = \sum_{n=1}^{\infty} (-1)^{n+1}(x-4)^{n-1}, (3, 5)$

(c) $f''(x) = \sum_{n=2}^{\infty} (-1)^{n+1}(n-1)(x-4)^{n-2}, (3, 5)$

(d) $\int f(x)\, dx = \sum_{n=1}^{\infty} (-1)^{n+1} \frac{(x-4)^{n+1}}{n(n+1)}, [3, 5]$

$\left[\sum_{n=1}^{\infty} (-1)^{n+1} \frac{(3-4)^{n+1}}{n(n+1)} \text{ and } \sum_{n=1}^{\infty} (-1)^{n+1} \frac{5-4}{n(n+1)}, \text{ both converge}\right]$

83.

$$y = \sum_{n=0}^{\infty} (-1)^n \frac{x^{2n}}{4^n (n!)^2}$$

$$y' = \sum_{n=1}^{\infty} \frac{(-1)^n (2n) x^{2n-1}}{4^n (n!)^2} = \sum_{n=0}^{\infty} \frac{(-1)^{n+1}(2n+2)x^{2n+1}}{4^{n+1}[(n+1)!]^2}$$

$$y'' = \sum_{n=0}^{\infty} \frac{(-1)^{n+1}(2n+2)(2n+1)x^{2n}}{4^{n+1}[(n+1)!]^2}$$

$$x^2y'' + xy' + x^2y = \sum_{n=0}^{\infty} \frac{(-1)^{n+1}(2n+2)(2n+1)x^{2n+2}}{4^{n+1}[(n+1)!]^2} + \sum_{n=0}^{\infty} \frac{(-1)^{n+1}(2n+2)x^{2n+2}}{4^{n+1}[(n+1)!]^2} + \sum_{n=0}^{\infty} (-1)^n \frac{x^{2n+2}}{4^n (n!)^2}$$

$$= \sum_{n=0}^{\infty} \left[(-1)^{n+1} \frac{(2n+2)(2n+1)}{4^{n+1}[(n+1)!]^2} + \frac{(-1)^{n+1}(2n+2)}{4^{n+1}[(n+1)!]^2} + \frac{(-1)^n}{4^n (n!)^2} \right] x^{2n+2}$$

$$= \sum_{n=0}^{\infty} \left[\frac{(-1)^{n+1}(2n+2)(2n+1+1)}{4^{n+1}[(n+1)!]^2} + (-1)^n \frac{1}{4^n (n!)^2} \right] x^{2n+2}$$

$$= \sum_{n=0}^{\infty} \left[\frac{(-1)^{n+1} 4(n+1)^2}{4^{n+1}[(n+1)!]^2} + (-1)^n \frac{1}{4^n (n!)^2} \right] x^{2n+2}$$

$$= \sum_{n=0}^{\infty} \left[\frac{(-1)^{n+1} 1}{4^n (n!)^2} + (-1)^n \frac{1}{4^n (n!)^2} \right] x^{2n+2} = 0$$

84.

$$y = \sum_{n=0}^{\infty} \frac{(-3)^n x^{2n}}{2^n n!}$$

$$y' = \sum_{n=1}^{\infty} \frac{(-3)^n (2n) x^{2n-1}}{2^n n!} = \sum_{n=0}^{\infty} \frac{(-3)^{n+1}(2n+2)x^{2n+1}}{2^{n+1}(n+1)!}$$

$$y'' = \sum_{n=0}^{\infty} \frac{(-3)^{n+1}(2n+2)(2n+1)x^{2n}}{2^{n+1}(n+1)!}$$

$$y'' + 3xy' + 3y = \sum_{n=0}^{\infty} \frac{(-3)^{n+1}(2n+2)(2n+1)x^{2n}}{2^{n+1}(n+1)!} + \sum_{n=0}^{\infty} \frac{(-1)^{n+1}3^{n+2}(2n+2)x^{2n+2}}{2^{n+1}(n+1)!} + \sum_{n=0}^{\infty} \frac{(-1)^n 3^{n+1} x^{2n}}{2^n n!}$$

$$= \sum_{n=0}^{\infty} \frac{(-1)^{n+1}3^{n+1}(2n+2)x^{2n}}{2^n n!} + \sum_{n=0}^{\infty} \frac{(-1)^{n+1}3^{n+2}x^{2n+2}}{2^n n!} + \sum_{n=0}^{\infty} \frac{(-1)^n 3^{n+1}x^{2n}}{2^n n!}$$

$$= \sum_{n=0}^{\infty} \frac{(-1)^n 3^{n+1} x^{2n}}{2^n n!}\left[-(2n+1)+1\right] + \sum_{n=0}^{\infty} \frac{(-1)^{n+1}3^{n+2}x^{2n+2}}{2^n n!}$$

$$= \sum_{n=0}^{\infty} \frac{(-1)^n 3^{n+1} x^{2n}}{2^n n!}(-2n) + \sum_{n=0}^{\infty} \frac{(-1)^{n+1}3^{n+2}x^{2n+2}}{2^n n!}$$

$$= \sum_{n=1}^{\infty} \frac{(-1)^{n+1} 3^{n+1} x^{2n}}{2^n n!}(2n) + \sum_{n=1}^{\infty} \frac{(-1)^n 3^{n+1}x^{2n}}{2^{n-1}(n-1)!} \cdot \frac{2n}{2n}$$

$$= \sum_{n=1}^{\infty} \frac{(-1)^n 3^{n+1} x^{2n}}{2^n n!}\left[-2n + 2n\right] = 0$$

85. $\dfrac{2}{3-x} = \dfrac{2/3}{1-(x/3)} = \dfrac{a}{1-r}$

$$\sum_{n=0}^{\infty} \frac{2}{3}\left(\frac{x}{3}\right)^n = \sum_{n=0}^{\infty} \frac{2x^n}{3^{n+1}}$$

86. $\dfrac{3}{2+x} = \dfrac{3/2}{1+(x/2)} = \dfrac{3/2}{1-(-x/2)} = \dfrac{a}{1-r}$

$$\sum_{n=0}^{\infty} \frac{3}{2}\left(-\frac{x}{2}\right)^n = \sum_{n=0}^{\infty} \frac{(-1)^n 3x^n}{2^{n+1}}$$

87. $\dfrac{6}{4-x} = \dfrac{6}{3-(x-1)} = \dfrac{2}{1-\left(\dfrac{x-1}{3}\right)} = \dfrac{a}{1-r}$

$= \displaystyle\sum_{n=0}^{\infty} 2\left(\frac{x-1}{3}\right)^n = 2\sum_{n=0}^{\infty}\frac{(x-1)^n}{3^n}$

Interval of convergence:

$\left|\dfrac{x-1}{3}\right| < 1 \Rightarrow |x-1| < 3 \Rightarrow (-2, 4)$

88. $\dfrac{1}{3-2x} = \dfrac{1/3}{1-\left(\dfrac{2}{3}x\right)} = \dfrac{a}{1-r}$

$= \displaystyle\sum_{n=0}^{\infty}\frac{1}{3}\left(\frac{2x}{3}\right)^n = \frac{1}{3}\sum_{n=0}^{\infty}\left(\frac{2x}{3}\right)^n$

Interval of convergence:

$\left|\dfrac{2x}{3}\right| < 1 \Rightarrow |2x| < 3 \Rightarrow |x| < \dfrac{3}{2} \Rightarrow \left(-\dfrac{3}{2}, \dfrac{3}{2}\right)$

89. $\ln x = \displaystyle\sum_{n=1}^{\infty}(-1)^{n+1}\frac{(x-1)^n}{n}, \quad 0 < x \le 2$

$\ln\left(\dfrac{5}{4}\right) = \displaystyle\sum_{n=1}^{\infty}(-1)^{n+1}\left(\frac{(5/4)-1}{n}\right)^n = \sum_{n=1}^{\infty}(-1)^{n+1}\frac{1}{4^n n} \approx 0.2231$

90. $\ln x = \displaystyle\sum_{n=1}^{\infty}(-1)^{n+1}\frac{(x-1)^n}{n}, \quad 0 < x \le 2$

$\ln\left(\dfrac{6}{5}\right) = \displaystyle\sum_{n=1}^{\infty}(-1)^{n+1}\left(\frac{(6/5)-1}{n}\right)^n$

$= \displaystyle\sum_{n=1}^{\infty}(-1)^{n+1}\frac{1}{5^n n} \approx 0.1823$

91. $e^x = \displaystyle\sum_{n=0}^{\infty}\frac{x^n}{n!}, \quad -\infty < x < \infty$

$e^{1/2} = \displaystyle\sum_{n=0}^{\infty}\frac{(1/2)^n}{n!} = \sum_{n=0}^{\infty}\frac{1}{2^n n!} \approx 1.6487$

92. $e^x = \displaystyle\sum_{n=0}^{\infty}\frac{x^n}{n!}, \quad -\infty < x < \infty$

$e^{2/3} = \displaystyle\sum_{n=0}^{\infty}\frac{(2/3)^n}{n!} = \sum_{n=0}^{\infty}\frac{2^n}{3^n n!} \approx 1.9477$

93. $\cos x = \displaystyle\sum_{n=0}^{\infty}(-1)^n\frac{x^{2n}}{(2n)!}, \quad -\infty < x < \infty$

$\cos\left(\dfrac{2}{3}\right) = \displaystyle\sum_{n=0}^{\infty}(-1)^n\frac{2^{2n}}{3^{2n}(2n)!} = 0.7859$

94. $\sin x = \displaystyle\sum_{n=0}^{\infty}(-1)^n\frac{x^{2n+1}}{(2n+1)!}, \quad -\infty < x < \infty$

$\sin\left(\dfrac{1}{3}\right) = \displaystyle\sum_{n=0}^{\infty}(-1)^n\frac{1}{3^{2n+1}(2n+1)!} \approx 0.3272$

95. $f(x) = \sin x$

$f'(x) = \cos x$

$f''(x) = -\sin x$

$f'''(x) = -\cos x, \cdots$

$\sin(x) = \displaystyle\sum_{n=0}^{\infty}\frac{f^{(n)}(x)[x-(3\pi/4)]^n}{n!}$

$= \dfrac{\sqrt{2}}{2} - \dfrac{\sqrt{2}}{2}\left(x - \dfrac{3\pi}{4}\right) - \dfrac{\sqrt{2}}{2\cdot 2!}\left(x - \dfrac{3\pi}{4}\right)^2 + \cdots = \dfrac{\sqrt{2}}{2}\displaystyle\sum_{n=0}^{\infty}\frac{(-1)^{n(n+1)/2}[x-(3\pi/4)]^n}{n!}$

96. $f(x) = \cos x$

$f'(x) = -\sin x$

$f''(x) = -\cos x$

$f'''(x) = \sin x$

$\cos x = \displaystyle\sum_{n=0}^{\infty}\frac{f^{(n)}(-\pi/4)[x+(\pi/4)]^n}{n!} = \frac{\sqrt{2}}{2} + \frac{\sqrt{2}}{2}\left(x + \frac{\pi}{4}\right) - \frac{\sqrt{2}}{2\cdot 2!}\left(x + \frac{\pi}{4}\right)^2 - \frac{\sqrt{2}}{2\cdot 3!}\left(x + \frac{\pi}{4}\right)^3 + \frac{\sqrt{2}}{2\cdot 4!}\left(x + \frac{\pi}{4}\right)^4 + \cdots$

$= \dfrac{\sqrt{2}}{2}\left[1 + \left(x + \dfrac{\pi}{4}\right) + \displaystyle\sum_{n=1}^{\infty}\frac{(-1)^{[n(n+1)]/2}[x+(\pi/4)]^{n+1}}{(n+1)!}\right]$

97. $3^x = \left(e^{\ln(3)}\right)^x = e^{x\ln(3)}$ and because $e^x = \sum_{n=0}^{\infty} \frac{x^n}{n!}$, you have

$$3^x = \sum_{n=0}^{\infty} \frac{(x\ln 3)^n}{n!} = 1 + x\ln 3 + \frac{x^2[\ln 3]^2}{2!} + \frac{x^3[\ln 3]^3}{3!} + \frac{x^4[\ln 3]^4}{4!} + \cdots.$$

98.

$$f(x) = \csc(x)$$
$$f'(x) = -\csc(x)\cot(x)$$
$$f''(x) = \csc^3(x) + \csc(x)\cot^2(x)$$
$$f'''(x) = -5\csc^3(x)\cot(x) - \csc(x)\cot^3(x)$$
$$f^{(4)}(x) = 5\csc^5(x) + 15\csc^3(x)\cot^2(x) + \csc(x)\cot^4(x)$$
$$\csc(x) = \sum_{n=0}^{\infty} \frac{f^{(n)}(\pi/2)\left[x - (\pi/2)\right]^n}{n!} = 1 + \frac{1}{2!}\left(x - \frac{\pi}{2}\right)^2 + \frac{5}{4!}\left(x - \frac{\pi}{2}\right)^4 + \cdots$$

99.

$$f(x) = \frac{1}{x}$$
$$f'(x) = -\frac{1}{x^2}$$
$$f''(x) = \frac{2}{x^3}$$
$$f'''(x) = -\frac{6}{x^4}, \ldots$$
$$\frac{1}{x} = \sum_{n=0}^{\infty} \frac{f^{(n)}(-1)(x+1)^n}{n!} = \sum_{n=0}^{\infty} \frac{-n!(x+1)^n}{n!} = -\sum_{n=0}^{\infty} (x+1)^n, -2 < x < 0$$

100.

$$f(x) = \sqrt{x} = x^{1/2}$$
$$f'(x) = \frac{1}{2}x^{-1/2}$$
$$f''(x) = -\left(\frac{1}{2}\right)\left(\frac{1}{2}\right)x^{-3/2}$$
$$f'''(x) = \left(\frac{1}{2}\right)\left(\frac{1}{2}\right)\left(\frac{3}{2}\right)x^{-5/2}$$
$$f^{(4)}(x) = -\left(\frac{1}{2}\right)\left(\frac{1}{2}\right)\left(\frac{3}{2}\right)\left(\frac{5}{2}\right)x^{-7/2}, \ldots$$
$$\sqrt{x} = \sum_{n=0}^{\infty} \frac{f^{(n)}(4)(x-4)^n}{n!} = 2 + \frac{(x-4)}{2^2} - \frac{(x-4)^2}{2^5 2!} + \frac{1\cdot 3(x-4)^3}{2^8 3!} - \frac{1\cdot 3\cdot 5(x-4)^4}{2^{11}4!} + \cdots$$
$$= 2 + \frac{(x-4)}{2^2} + \sum_{n=2}^{\infty} \frac{(-1)^{n+1} 1\cdot 3\cdot 5\cdots(2n-3)(x-4)^n}{2^{3n-1}n!}$$

101.

$$(1+x)^k = 1 + kx + \frac{k(k-1)x^2}{2!} + \frac{k(k-1)(k-2)x^3}{3!} + \cdots$$
$$(1+x)^{1/5} = 1 + \frac{x}{5} + \frac{(1/5)(-4/5)x^2}{2!} + \frac{1/5(-4/5)(-9/5)x^3}{3!} + \cdots$$
$$= 1 + \frac{1}{5}x - \frac{1\cdot 4x^2}{5^2 2!} + \frac{1\cdot 4\cdot 9x^3}{5^3 3!} - \cdots = 1 + \frac{x}{5} + \sum_{n=2}^{\infty} \frac{(-1)^{n+1} 4\cdot 9\cdot 14\cdots(5n-6)x^n}{5^n n!} = 1 + \frac{x}{5} - \frac{2}{25}x^2 + \frac{6}{125}x^3 - \cdots$$

102.
$$\begin{aligned} h(x) &= (1+x)^{-3} \\ h'(x) &= -3(1+x)^{-4} \\ h''(x) &= 12(1+x)^{-5} \\ h'''(x) &= -60(1+x)^{-6} \\ h^{(4)}(x) &= 360(1+x)^{-7} \\ h^{(5)}(x) &= -2520(1+x)^{-8} \end{aligned}$$

$$\frac{1}{(1+x)^3} = 1 - 3x + \frac{12x^2}{2!} - \frac{60x^3}{3!} + \frac{360x^4}{4!} - \frac{2520x^5}{5!} + \cdots = \sum_{n=0}^{\infty} \frac{(-1)^n (n+2)!x^n}{2n!} = \sum_{n=0}^{\infty} \frac{(-1)^n (n+2)(n+1)x^n}{2}$$

103. (a)
$$\begin{aligned} f(x) &= e^{2x} & f(0) &= 1 \\ f'(x) &= 2e^{2x} & f'(0) &= 2 \\ f''(x) &= 4e^{2x} & f''(0) &= 4 \\ f'''(x) &= 8e^{2x} & f'''(0) &= 8 \end{aligned}$$

$$P(x) = 1 + 2x + \frac{4x^2}{2!} + \frac{8x^3}{3!} = 1 + 2x + 2x^2 + \frac{4}{3}x^3$$

(b) $$e^x = \sum_{n=0}^{\infty} \frac{x^n}{n!},\ e^{2x} = \sum_{n=0}^{\infty} \frac{(2x)^n}{n!}$$

$$P(x) = 1 + 2x + 2x^2 + \frac{4}{3}x^3$$

(c) $$e^x \cdot e^x = \left(1 + x + \frac{x^2}{2!} + \cdots\right)\left(1 + x + \frac{x^2}{2!} + \cdots\right)$$

$$P(x) = 1 + 2x + 2x^2 + \frac{4}{3}x^3$$

104. (a)
$$\begin{aligned} f(x) &= \sin 2x & f(0) &= 0 \\ f'(x) &= 2\cos 2x & f'(0) &= 2 \\ f''(x) &= -4\sin 2x & f''(0) &= 0 \\ f'''(x) &= -8\cos 2x & f'''(0) &= -8 \\ f^{(4)}(x) &= 16\sin 2x & f^{(4)}(0) &= 0 \\ f^{(5)}(x) &= 32\cos 2x & f^{(5)}(0) &= 32 \\ f^{(6)}(x) &= -64\sin 2x & f^{(6)}(0) &= 0 \\ f^{(7)}(x) &= -128\cos 2x & f^{(7)}(0) &= -128 \end{aligned}$$

$$\sin 2x = 0 + 2x + \frac{0x^2}{2!} - \frac{8x^3}{3!} + \frac{0x^4}{4!} + \frac{32x^5}{5!} + \frac{0x^6}{6!} - \frac{128x^7}{7!} + \cdots = 2x - \frac{4}{3}x^3 + \frac{4}{15}x^5 - \frac{8}{315}x^7 + \cdots$$

(b) $$\sin x = \sum_{n=0}^{\infty} \frac{(-1)^n x^{2n+1}}{(2n+1)!}$$

$$\begin{aligned} \sin 2x &= \sum_{n=0}^{\infty} \frac{(-1)^n (2x)^{2n+1}}{(2n+1)!} = 2x - \frac{(2x)^3}{3!} + \frac{(2x)^5}{5!} - \frac{(2x)^7}{7!} + \cdots \\ &= 2x - \frac{8x^3}{6} + \frac{32x^5}{120} - \frac{128x^7}{5040} + \cdots = 2x - \frac{4}{3}x^3 + \frac{4}{15}x^5 - \frac{8}{315}x^7 + \cdots \end{aligned}$$

(c) $\sin 2x = 2\sin x\cos x$

$$= 2\left(x - \frac{x^3}{6} + \frac{x^5}{120} - \frac{x^7}{5040} + \cdots\right)\left(1 - \frac{x^2}{2} + \frac{x^4}{24} - \frac{x^6}{720} + \cdots\right)$$

$$= 2\left[x + \left(-\frac{x^3}{2} - \frac{x^3}{6}\right) + \left(\frac{x^5}{24} + \frac{x^5}{12} + \frac{x^5}{120}\right) + \left(-\frac{x^7}{720} - \frac{x^7}{144} - \frac{x^7}{240} - \frac{x^7}{5040}\right) + \cdots\right]$$

$$= 2\left(x - \frac{2x^3}{3} + \frac{2x^5}{15} - \frac{4x^7}{315} + \cdots\right) = 2x - \frac{4}{3}x^3 + \frac{4}{15}x^5 - \frac{8}{315}x^7 + \cdots$$

105. $$e^x = \sum_{n=0}^{\infty}\frac{x^n}{n!} = 1 + x + \frac{x^2}{2!} + \frac{x^3}{3!} + \ldots$$

$$e^{6x} = \sum_{n=0}^{\infty}\frac{(6x)^n}{n!} = 1 + 6x + \frac{(6x)^2}{2!} + \frac{(6x)^3}{3!} + \ldots$$

$$= 1 + 6x + 18x^2 + 36x^3 + \ldots$$

106. $$\ln x = \sum_{n=1}^{\infty}(-1)^{n-1}\frac{(x-1)^n}{n}, 0 < x \le 2$$

$$\ln(x-1) = \sum_{n=1}^{\infty}(-1)^{n-1}\frac{(x-1-1)^n}{n}$$

$$= \sum_{n=1}^{\infty}(-1)^{n-1}\frac{(x-2)^n}{n}, 1 < x \le 3$$

107. $$\sin x = \sum_{n=0}^{\infty}\frac{(-1)^n x^{2n+1}}{(2n+1)!}$$

$$\sin 2x = \sum_{n=0}^{\infty}\frac{(-1)^n (2x)^{2n+1}}{(2n+1)!}$$

$$= 2x - \frac{4}{3}x^3 + \frac{4}{15}x^5 - \cdots$$

108. $$\cos x = \sum_{n=0}^{\infty}\frac{(-1)^n x^{2n}}{(2n)!}$$

$$\cos 3x = \sum_{n=0}^{\infty}(-1)^n\frac{(3x)^{2n}}{(2n)!}$$

$$= 1 - \frac{9}{2}x^2 + \frac{27x^4}{8} - \cdots$$

109. $$\arctan x = x - \frac{x^3}{3} + \frac{x^5}{5} - \frac{x^7}{7} + \frac{x^9}{9} - \cdots$$

$$\frac{\arctan x}{\sqrt{x}} = \sqrt{x} - \frac{x^{5/2}}{3} + \frac{x^{9/2}}{5} - \frac{x^{13/2}}{7} + \frac{x^{17/2}}{9} - \cdots$$

$$\lim_{x\to 0^+}\frac{\arctan x}{\sqrt{x}} = 0$$

By L'Hôpital's Rule,

$$\lim_{x\to 0^+}\frac{\arctan x}{\sqrt{x}} = \lim_{x\to 0^+}\frac{\left(\dfrac{1}{1+x^2}\right)}{\left(\dfrac{1}{2\sqrt{x}}\right)} = \lim_{x\to 0^+}\frac{2\sqrt{x}}{1+x^2} = 0.$$

110. $$\arcsin x = x + \frac{x^3}{2\cdot 3} + \frac{1\cdot 3x^5}{2\cdot 4\cdot 5} + \frac{1\cdot 3\cdot 5x^7}{2\cdot 4\cdot 6\cdot 7} + \cdots$$

$$\frac{\arcsin x}{x} = 1 + \frac{x^2}{2\cdot 3} + \frac{1\cdot 3x^4}{2\cdot 4\cdot 5} + \frac{1\cdot 3\cdot 5x^6}{2\cdot 4\cdot 6\cdot 7} + \cdots$$

$$\lim_{x\to 0}\frac{\arcsin x}{x} = 1$$

By L'Hôpital's Rule,

$$\lim_{x\to 0}\frac{\arcsin x}{x} = \lim_{x\to 0}\frac{\left(\dfrac{1}{\sqrt{1-x^2}}\right)}{1} = 1.$$

Problem Solving for Chapter 9

1. (a) $$1\left(\frac{1}{3}\right) + 2\left(\frac{1}{9}\right) + 4\left(\frac{1}{27}\right) + \cdots = \sum_{n=0}^{\infty}\frac{1}{3}\left(\frac{2}{3}\right)^n = \frac{1/3}{1-(2/3)} = 1$$

(b) $0, \frac{1}{3}, \frac{2}{3}, 1$, etc.

(c) $$\lim_{n\to\infty} C_n = 1 - \sum_{n=0}^{\infty}\frac{1}{3}\left(\frac{2}{3}\right)^n = 1 - 1 = 0$$

2. (a) Let $\varepsilon > 0$ be given. $\lim_{n\to\infty} a_{2n} = L$ means there exists M_1 such that $|a_{2n} - L| < \varepsilon$ for $n > M_1$. $\lim_{n\to\infty} a_{2n+1} = L$ means there exists M_2 such that $|a_{2n+1} - L| < \varepsilon$ for $n > M_2$. Let $M = \max\{2M_1, 2M_2 + 1\}$. Then for $n > M$, and $n = 2m$ even, you have $2m > M > 2M_1 \Rightarrow m > M_1 \Rightarrow |a_{2m} - L| < \varepsilon$. And for $n > M, n = 2_{m+1}$ odd, you have $2_{m+1} > M > 2M_2 + 1 \Rightarrow m > M_2 \Rightarrow |a_{2n+1} - L| < \varepsilon$. So, $\lim_{n\to\infty} a_n = L$.

(b) $a_1 = 1, a_{n+1} = 1 + \dfrac{1}{1 + a_n}$.

$$a_2 = 1 + \frac{1}{1 + a_1} = 1 + \frac{1}{1 + 1} = \frac{3}{2} = 1.5$$

$$a_3 = 1 + \frac{1}{1 + a_2} = 1 + \frac{1}{1 + (3/2)} = \frac{7}{5} = 1.4$$

$$a_4 = \frac{17}{12} = 1.41\overline{6}$$

$$a_5 = \frac{41}{29} \approx 1.4140$$

$$a_6 = \frac{99}{70} \approx 1.41429$$

$$a_7 = \frac{239}{169} \approx 1.414201$$

$$a_8 = \frac{577}{408} \approx 1.414216$$

Using mathematical induction, you can show that the odd terms are increasing and the even terms are decreasing. Both sequences are bounded in $[1, 2]$. So, both sequences converge.

Let $\lim_{n\to\infty} a_{2n} = L$. Then $\lim_{n\to\infty} a_{2n+2} = L$, and

$$a_{n+2} = 1 + \frac{1}{1 + a_{n+1}} = 1 + \frac{1}{1 + \left[1 + \dfrac{1}{1 + a_n}\right]} = 1 + \frac{1}{1 + \left[\dfrac{2 + a_n}{1 + a_n}\right]} = 1 + \frac{1}{\left(\dfrac{3 + 2a_n}{1 + a_n}\right)} = 1 + \frac{1 + a_n}{3 + 2a_n} = \frac{4 + 3a_n}{3 + 2a_n}$$

$$\Rightarrow a_{2n+2} = \frac{4 + 3a_{2n}}{3 + 2a_{2n}}$$

So, $L = \dfrac{4 + 3L}{3 + 2L} \Rightarrow 2L^2 = 4 \Rightarrow L = \sqrt{2}$. Similarly, $\lim_{n\to\infty} a_{2n+1} = \sqrt{2}$. So by part (a), $\lim_{n\to\infty} a_n = L = \sqrt{2}$

3. Let $S = \displaystyle\sum_{n=1}^{\infty} \frac{1}{(2n - 1)^2} = \frac{1}{1^2} + \frac{1}{3^2} + \frac{1}{5^2} + \cdots$.

Then

$$\frac{\pi^2}{6} = \frac{1}{1^2} + \frac{1}{2^2} + \frac{1}{3^2} + \frac{1}{4^2} + \cdots$$

$$= S + \frac{1}{2^2} + \frac{1}{4^2} + \cdots$$

$$= S + \frac{1}{2^2}\left[1 + \frac{1}{2^2} + \frac{1}{3^2} + \cdots\right] = S + \frac{1}{2^2}\left(\frac{\pi^2}{6}\right).$$

So, $S = \dfrac{\pi^2}{6} - \dfrac{1}{4}\dfrac{\pi^2}{6} = \dfrac{\pi^2}{6}\left(\dfrac{3}{4}\right) = \dfrac{\pi^2}{8}$.

4. If there are n rows, then $a_n = \dfrac{n(n+1)}{2}$.

For one circle, $a_1 = 1$ and $r_1 = \dfrac{1}{3}\left(\dfrac{\sqrt{3}}{2}\right) = \dfrac{\sqrt{3}}{6} = \dfrac{1}{2\sqrt{3}}$.

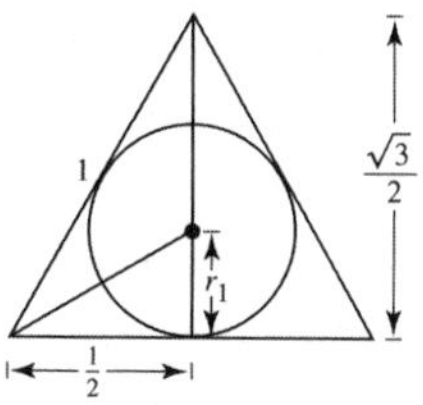

For three circles, $a_2 = 3$ and $1 = 2\sqrt{3}r_2 + 2r_2$

$$r_2 = \frac{1}{2 + 2\sqrt{3}}.$$

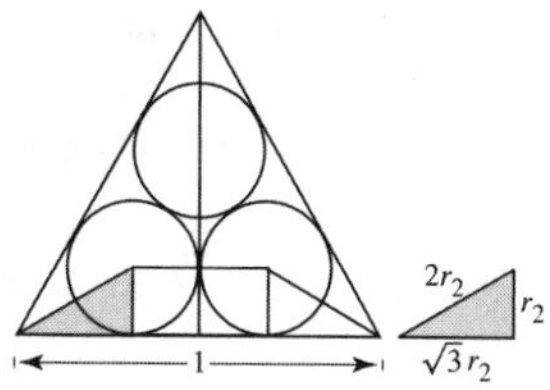

For six circles, $a_3 = 6$ and $1 = 2\sqrt{3}r_3 + 4r_3$

$$r_3 = \frac{1}{2\sqrt{3} + 4}.$$

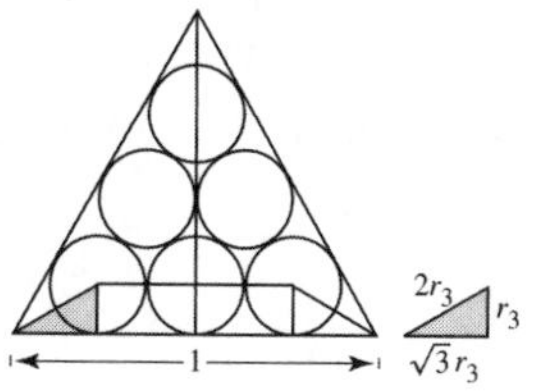

Continuing this pattern, $r_n = \dfrac{1}{2\sqrt{3} + 2(n-1)}$.

$$\text{Total Area} = \left(\pi r_n^2\right)a_n = \pi\left(\frac{1}{2\sqrt{3} + 2(n-1)}\right)^2 \frac{n(n+1)}{2}$$

$$A_n = \frac{\pi}{2}\frac{n(n+1)}{\left[2\sqrt{3} + 2(n-1)\right]^2}$$

$$\lim_{n\to\infty} A_n = \frac{\pi}{2}\cdot\frac{1}{4} = \frac{\pi}{8}$$

5. (a) Position the three blocks as indicated in the figure. The bottom block extends $1/6$ over the edge of the table, the middle block extends $1/4$ over the edge of the bottom block, and the top block extends $1/2$ over the edge of the middle block.

The centers of gravity are located at

bottom block: $\dfrac{1}{6} - \dfrac{1}{2} = -\dfrac{1}{3}$

middle block: $\dfrac{1}{6} + \dfrac{1}{4} - \dfrac{1}{2} = -\dfrac{1}{12}$

top block: $\dfrac{1}{6} + \dfrac{1}{4} + \dfrac{1}{2} - \dfrac{1}{2} = \dfrac{5}{12}$.

The center of gravity of the top 2 blocks is

$\left(-\dfrac{1}{12} + \dfrac{5}{12}\right)\Big/2 = \dfrac{1}{6}$, which lies over the bottom block. The center of gravity of the 3 blocks is

$\left(-\dfrac{1}{3} - \dfrac{1}{12} + \dfrac{5}{12}\right)\Big/3 = 0$ which lies over the table.

So, the far edge of the top block lies

$\dfrac{1}{6} + \dfrac{1}{4} + \dfrac{1}{2} = \dfrac{11}{12}$ beyond the edge of the table.

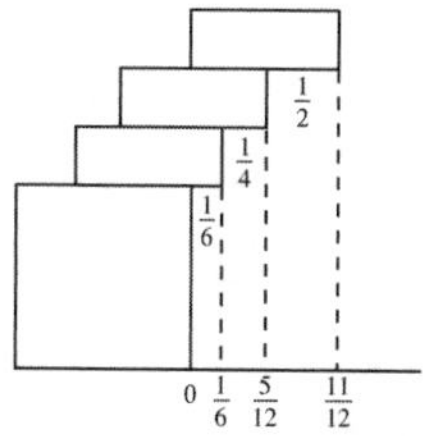

(b) Yes. If there are n blocks, then the edge of the top block lies $\displaystyle\sum_{i=1}^{n}\frac{1}{2i}$ from the edge of the table. Using 4 blocks,

$$\sum_{i=1}^{4}\frac{1}{2i} = \frac{1}{2} + \frac{1}{4} + \frac{1}{6} + \frac{1}{8} = \frac{25}{24}$$

which shows that the top block extends beyond the table.

(c) The blocks can extend any distance beyond the table because the series diverges:

$$\sum_{i=1}^{\infty}\frac{1}{2i} = \frac{1}{2}\sum_{i=1}^{\infty}\frac{1}{i} = \infty.$$

6. (a) $\sum a_n x^n = 1 + 2x + 3x^2 + x^3 + 2x^4 + 3x^5 + \cdots$

$$= \left(1 + x^3 + x^6 + \cdots\right) + 2\left(x + x^4 + x^7 + \cdots\right) + 3\left(x^2 + x^5 + x^8 + \cdots\right)$$

$$= \left(1 + x^3 + x^6 + \cdots\right)\left(1 + 2x + 3x^2\right) = \left(1 + 2x + 3x^2\right)\frac{1}{1 - x^3}$$

$R = 1$ because each series in the second line has $R = 1$.

(b) $\sum a_n x^n = \left(a_0 + a_1 x + \cdots + a_{p-1}x^{p-1}\right) + \left(a_0 x^p + a_1 x^{p+1} + \cdots\right) + \cdots$

$= a_0\left(1 + x^p + \cdots\right) + a_1 x\left(1 + x^p + \cdots\right) + \cdots + a_{p-1}x^{p-1}\left(1 + x^p + \cdots\right)$

$= \left(a_0 + a_1 x + \cdots + a_{p-1}x^{p-1}\right)\left(1 + x^p + \cdots\right) = \left(a_0 + a_1 x + \cdots + a_{p-1}x^{p-1}\right)\dfrac{1}{1 - x^p}$

$R = 1$

(Assume all $a_n > 0$.)

7. (a) $e^x = 1 + x + \dfrac{x^2}{2!} + \cdots = \displaystyle\sum_{n=0}^{\infty} \frac{x^n}{n!}$

$xe^x = \displaystyle\sum_{n=0}^{\infty} \frac{x^{n+1}}{n!}$

$\displaystyle\int xe^x\,dx = xe^x - e^x + C = \sum_{n=0}^{\infty} \frac{x^{n+2}}{(n+2)n!}$

Letting $x = 0$, you have $C = 1$. Letting $x = 1$,

$e - e + 1 = \displaystyle\sum_{n=0}^{\infty} \frac{1}{(n+2)n!} = \frac{1}{2} + \sum_{n=1}^{\infty} \frac{1}{(n+2)n!}.$

So, $\displaystyle\sum_{n=1}^{\infty} \frac{1}{(n+2)n!} = \frac{1}{2}.$

(b) Differentiating, $xe^x + e^x = \displaystyle\sum_{n=0}^{\infty} \frac{(n+1)x^n}{n!}.$

Letting $x = 1$, $2e = \displaystyle\sum_{n=0}^{\infty} \frac{n+1}{n!} \approx 5.4366.$

8. Let $a_1 = \displaystyle\int_0^{\pi} \frac{\sin x}{x}\,dx$, $a_2 = -\displaystyle\int_{\pi}^{2\pi} \frac{\sin x}{x}\,dx$,

$a_3 = \displaystyle\int_{2\pi}^{3\pi} \frac{\sin x}{x}\,dx$, etc. Then,

$\displaystyle\int_0^{\infty} \frac{\sin x}{x}\,dx = a_1 - a_2 + a_3 - a_4 + \cdots.$

Because $\lim_{n\to\infty} a_n = 0$ and $a_{n+1} < a_n$, this series converges.

9. $a - \dfrac{b}{2} + \dfrac{a}{3} - \dfrac{b}{4} + \cdots = \displaystyle\sum_{n=1}^{\infty} \frac{(-1)^{n+1}(a+b) + (a-b)}{2n}$

If $a = b$, $\displaystyle\sum_{n=1}^{\infty} \frac{(-1)^{n+1}(2a)}{2n} = a\sum_{n=1}^{\infty} \frac{(-1)^{n+1}}{n}$ converges conditionally.

If $a \neq b$, $\displaystyle\sum_{n=1}^{\infty} \frac{(-1)^{n+1}(a+b)}{2n} + \sum_{n=1}^{\infty} \frac{a-b}{2n}$ diverges.

No values of a and b give absolute convergence. $a = b$ implies conditional convergence.

10. (a) $a_1 = 3.0$

$a_2 \approx 1.73205$

$a_3 \approx 2.17533$

$a_4 \approx 2.27493$

$a_5 \approx 2.29672$

$a_6 \approx 2.30146$

$\lim_{n\to\infty} a_n = \dfrac{1 + \sqrt{13}}{2}$ [See part (b) for proof.]

(b) Use mathematical induction to show the sequence is increasing. Clearly,

$a_2 = \sqrt{a + a_1} = \sqrt{a\sqrt{a}} > \sqrt{a} = a_1$. Now assume $a_n > a_{n-1}$. Then

$a_n + a > a_{n-1} + a$

$\sqrt{a_n + a} > \sqrt{a_{n-1} + a}$

$a_{n+1} > a_n.$

Use mathematical induction to show that the sequence is bounded above by a. Clearly,

$a_1 = \sqrt{a} < a$. Now assume $a_n < a$. Then $a > a_n$ and $a - 1 > 1$ implies

$a(a - 1) > a_n(1)$

$a^2 - a > a_n$

$a^2 > a_n + a$

$a > \sqrt{a_n + a} = a_{n+1}.$

So, the sequence converges to some number L. To find L, assume $a_{n+1} \approx a_n \approx L$:

$L = \sqrt{a + L} \Rightarrow L^2 = a + L \Rightarrow L^2 - L - a = 0$

$L = \dfrac{1 \pm \sqrt{1 + 4a}}{2}.$

So, $L = \dfrac{1 + \sqrt{1 + 4a}}{2}.$

11. Let $b_n = a_n r^n$.

$(b_n)^{1/n} = (a_n r^n)^{1/n} = a_n^{1/n} \cdot r \to Lr$ as $n \to \infty$.

$Lr < \dfrac{1}{r}r = 1.$

By the Root Test, $\sum b_n$ converges $\Rightarrow \sum a_n r^n$ converges.

12. (a) $\displaystyle\sum_{n=1}^{\infty}\frac{1}{2^{n+(-1)^n}} = \frac{1}{2^{1-1}} + \frac{1}{2^{2+1}} + \frac{1}{2^{3-1}} + \frac{1}{2^{4+1}} + \frac{1}{2^{5-1}} + \cdots$

$$S_1 = \frac{1}{2^0} = 1$$
$$S_1 = 1 + \frac{1}{8} = \frac{9}{8}$$
$$S_3 = \frac{9}{8} + \frac{1}{4} = \frac{11}{8}$$
$$S_4 = \frac{11}{8} + \frac{1}{32} = \frac{45}{32}$$
$$S_5 = \frac{45}{32} + \frac{1}{16} = \frac{47}{32}$$

(b) $\displaystyle\frac{a_{n+1}}{a_n} = \frac{2^{n+(-1)^n}}{2^{(n+1)+(-1)^{n+1}}} = \frac{2(-1)^n}{2^{1+(-1)^{n+1}}}$

This sequence is $\frac{1}{8}, 2, \frac{1}{8}, 2, \ldots$ which diverges.

(c) $\displaystyle\sqrt[n]{\frac{1}{2^{n+(-1)^n}}} = \left(\frac{1}{2^n \cdot 2^{(-1)^n}}\right)^{1/n} = \frac{1}{2\cdot\sqrt[n]{2^{(-1)^n}}} \to \frac{1}{2} < 1$

converges because $\left\{2^{(-1)^n}\right\} = \frac{1}{2}, 2, \frac{1}{2}, 2, \ldots$

and $\sqrt[n]{1/2} \to 1$ and $\sqrt[n]{2} \to 1$.

13. (a) $\displaystyle\frac{1}{0.99} = \frac{1}{1-0.01} = \sum_{n=0}^{\infty}(0.01)^n$

$$= 1 + 0.01 + (0.01)^2 + \cdots$$
$$= 1.010101\cdots$$

(b) $\displaystyle\frac{1}{0.98} = \frac{1}{1-0.02} = \sum_{n=0}^{\infty}(0.02)^n$

$$= 1 + 0.02 + (0.02)^2 + \cdots$$
$$= 1 + 0.02 + 0.0004 + \cdots$$
$$= 1.0204081632\cdots$$

14. $S_6 = 130 + 70 + 40 = 240$

$S_7 = 240 + 130 + 70 = 440$

$S_8 = 440 + 240 + 130 = 810$

$S_9 = 810 + 440 + 240 = 1490$

$S_{10} = 1490 + 810 + 440 = 2740$

15. (a) $\displaystyle\text{Height} = 2\left[1 + \frac{1}{\sqrt{2}} + \frac{1}{\sqrt{3}} + \cdots\right]$

$$= 2\sum_{n=1}^{\infty}\frac{1}{n^{1/2}} = \infty\left(p\text{-series}, p = \frac{1}{2} < 1\right)$$

(b) $\displaystyle S = 4\pi\left[1 + \frac{1}{2} + \frac{1}{3} + \cdots\right] = 4\pi\sum_{n=1}^{\infty}\frac{1}{n} = \infty$

(c) $\displaystyle W = \frac{4}{3}\pi\left[1 + \frac{1}{2^{3/2}} + \frac{1}{3^{3/2}} + \cdots\right]$

$$= \frac{4}{3}\pi\sum_{n=1}^{\infty}\frac{1}{n^{3/2}} \text{ converges.}$$

16. (a) $\displaystyle\sum_{n=1}^{\infty}\frac{1}{2n} = \frac{1}{2}\sum_{n=1}^{\infty}\frac{1}{n}$ diverges (harmonic)

(b) Let $f(x) = \sin x$. By the Mean Value Theorem,

$$|f(x) - f(y)| = f'(c)|x - y|$$
$$= \cos(c)|x - y| \le |x - y|,$$

where c is between x and y. So,

$$0 \le \left|\sin\left(\frac{1}{2n}\right) - \sin\left(\frac{1}{2n+1}\right)\right|$$
$$\le \left|\frac{1}{2n} - \frac{1}{2n+1}\right| = \frac{1}{2n(2n+1)}$$

Because $\displaystyle\sum_{n=1}^{\infty}\frac{1}{2n(2n+1)}$ converges, the Comparison Theorem tells us that

$$\sum_{n=1}^{\infty}\left[\sin\left(\frac{1}{2n}\right) - \sin\left(\frac{1}{2n+1}\right)\right] \text{ converges.}$$

CHAPTER 10
Conics, Parametric Equations, and Polar Coordinates

Section 10.1 Conics and Calculus 985

Section 10.2 Plane Curves and Parametric Equations 1003

Section 10.3 Parametric Equations and Calculus 1013

Section 10.4 Polar Coordinates and Polar Graphs 1028

Section 10.5 Area and Arc Length in Polar Coordinates 1045

Section 10.6 Polar Equations of Conics and Kepler's Laws 1058

Review Exercises 1066

Problem Solving 1084

CHAPTER 10
Conics, Parametric Equations, and Polar Coordinates

Section 10.1 Conics and Calculus

1. $y^2 = 4x$ Parabola

Vertex: $(0, 0)$

$p = 1 > 0$

Opens to the right

Matches (a).

2. $(x + 4)^2 = -2(y - 2)$ Parabola

Vertex: $(-4, 2)$

Opens downward

Matches (e).

3. $\dfrac{y^2}{16} - \dfrac{x^2}{1} = 1$ Hyperbola

Vertices: $(0, \pm 4)$

Matches (c).

4. $\dfrac{(x - 2)^2}{16} + \dfrac{(y + 1)^2}{4} = 1$ Ellipse

Center: $(2, -1)$

Matches (b).

5. $\dfrac{x^2}{4} + \dfrac{y^2}{9} = 1$ Ellipse

Center: $(0, 0)$

Vertices: $(0, \pm 3)$

Matches (f).

6. $\dfrac{(x - 2)^2}{9} - \dfrac{y^2}{4} = 1$ Hyperbola

Vertices: $(5, 0), (-1, 0)$

Matches (d).

7 $y^2 = -8x = 4(-2)x$

Vertex: $(0, 0)$

Focus: $(-2, 0)$

Directrix: $x = 2$

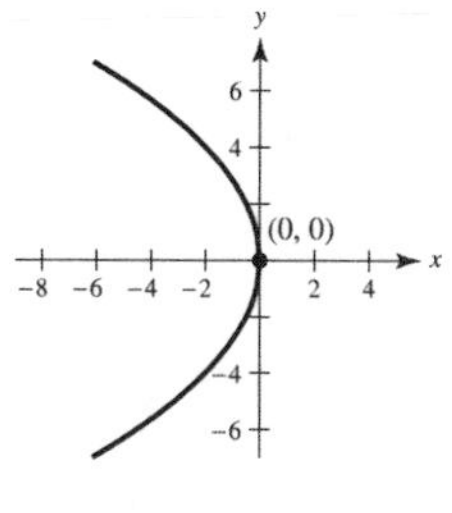

8. $x^2 + 6y = 0$

$$x^2 = -6y = 4\left(-\frac{3}{2}\right)y$$

Vertex: $(0, 0)$

Focus: $\left(0, -\dfrac{3}{2}\right)$

Directrix: $y = \dfrac{3}{2}$

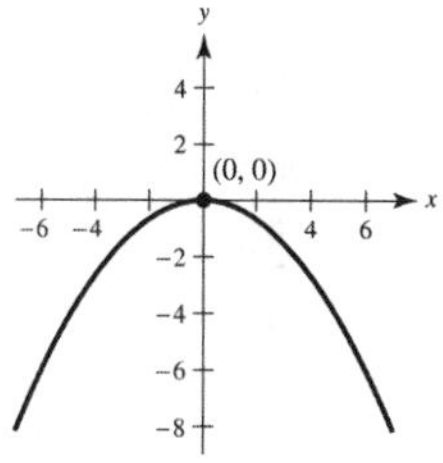

9. $(x + 5) + (y - 3)^2 = 0$

$$(y - 3)^2 = -(x + 5) = 4\left(-\tfrac{1}{4}\right)(x + 5)$$

Vertex: $(-5, 3)$

Focus: $\left(-\tfrac{21}{4}, 3\right)$

Directrix: $x = -\tfrac{19}{4}$

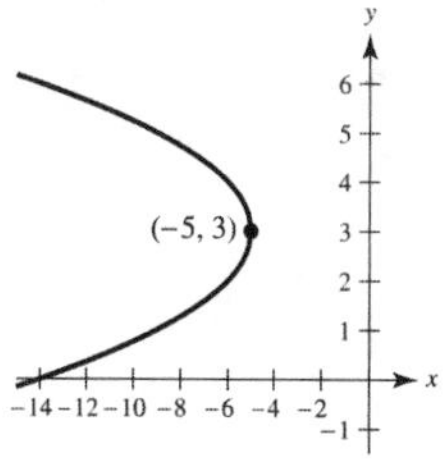

10. $(x - 6)^2 + 8(y + 7) = 0$

$$(x - 6)^2 = -8(y + 7) = 4(-2)(y + 7)$$

Vertex: $(6, -7)$

Focus: $(6, -9)$

Directrix: $y = -5$

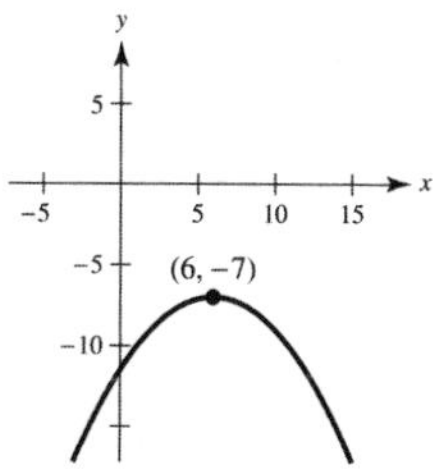

11. $y^2 - 4y - 4x = 0$

$$y^2 - 4y + 4 = 4x + 4$$
$$(y - 2)^2 = 4(1)(x + 1)$$

Vertex: $(-1, 2)$

Focus: $(0, 2)$

Directrix: $x = -2$

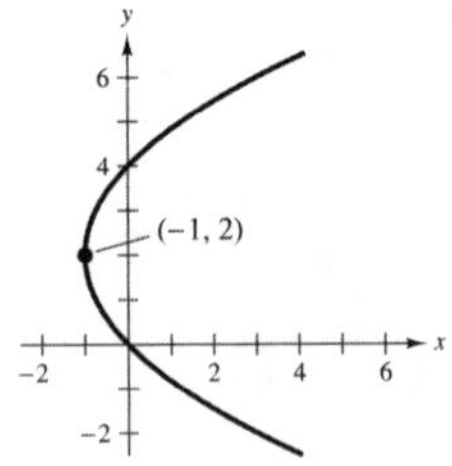

12. $y^2 + 6y + 8x + 25 = 0$

$$y^2 + 6y + 9 = -8x - 25 + 9$$
$$(y + 3)^2 = 4(-2)(x + 2)$$

Vertex: $(-2, -3)$

Focus: $(-4, -3)$

Directrix: $x = 0$

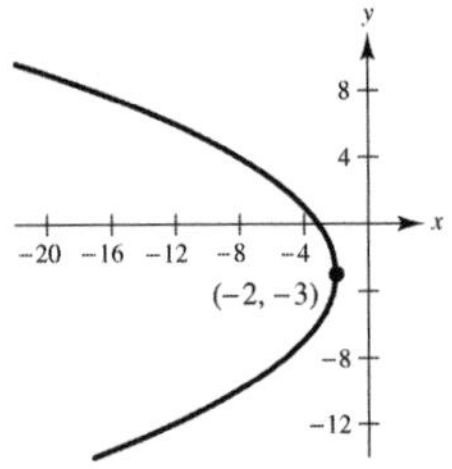

13. $x^2 + 4x + 4y - 4 = 0$

$$x^2 + 4x + 4 = -4y + 4 + 4$$
$$(x + 2)^2 = 4(-1)(y - 2)$$

Vertex: $(-2, 2)$

Focus: $(-2, 1)$

Directrix: $y = 3$

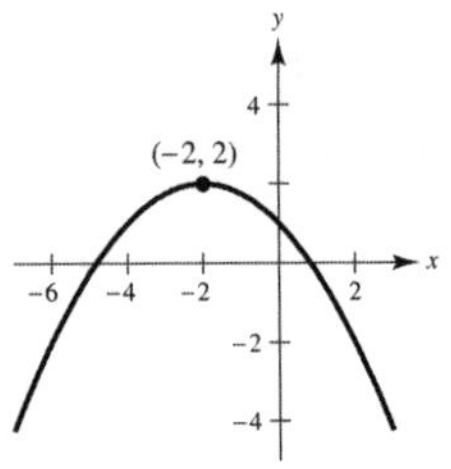

14. $y^2 + 4y + 8x - 12 = 0$

$$y^2 + 4y + 4 = -8x + 12 + 4$$
$$(y + 2)^2 = 4(-2)(x - 2)$$

Vertex: $(2, -2)$

Focus: $(0, -2)$

Directrix: $x = 4$

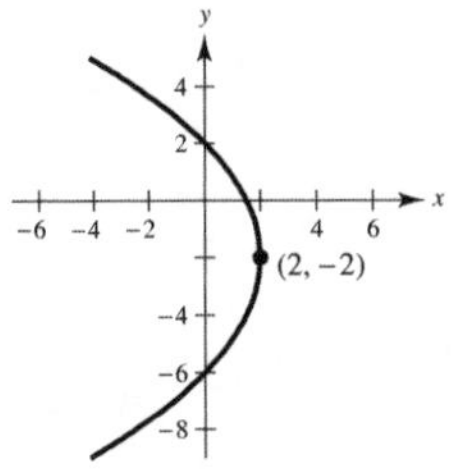

15.
$$(y - 4)^2 = 4(-2)(x - 5)$$
$$y^2 - 8y + 16 = -8x + 40$$
$$y^2 - 8y + 8x - 24 = 0$$

16.
$$(x + 2)^2 = 4(-2)(y - 1)$$
$$x^2 + 4x + 8y - 4 = 0$$

17.
$$(x - 0)^2 = 4(8)(y - 5)$$
$$x^2 = 4(8)(y - 5)$$
$$x^2 - 32y + 160 = 0$$

18. Vertex: $(0, 2)$

$$(y - 2)^2 = 4(2)(x - 0)$$
$$y^2 - 8x - 4y + 4 = 0$$

19. Vertex: $(0, 4)$, vertical axis

$$(x - 0)^2 = 4p(y - 4)$$

$(-2, 0)$ on parabola: $(-2)^2 = 4p(-4)$

$$4 = -16p$$
$$p = -\tfrac{1}{4}$$
$$x^2 = 4\left(-\tfrac{1}{4}\right)(y - 4)$$
$$x^2 = -(y - 4)$$
$$x^2 + y - 4 = 0$$

20. Vertex: $(2, 4)$, vertical axis

$$(x-2)^2 = 4p(y-4)$$

$(0, 0)$ on parabola: $(-2)^2 = 4p(0-4)$

$$4 = -16p$$

$$p = -\tfrac{1}{4}$$

$$(x-2)^2 = 4\left(-\tfrac{1}{4}\right)(y-4)$$

$$x^2 - 4x + 4 = -y + 4$$

$$x^2 - 4x + y = 0$$

21. Because the axis of the parabola is vertical, the form of the equation is $y = ax^2 + bx + c$. Now, substituting the values of the given coordinates into this equation, you obtain

$3 = c, 4 = 9a + 3b + c, 11 = 16a + 4b + c.$

Solving this system, you have $a = \frac{5}{3}, b = -\frac{14}{3}, c = 3.$

So,

$y = \frac{5}{3}x^2 - \frac{14}{3}x + 3$ or $5x^2 - 14x - 3y + 9 = 0.$

22. From Example 2: $4p = 8$ or $p = 2$

Vertex: $(4, 0)$

$$(x-4)^2 = 8(y-0)$$

$$x^2 - 8x - 8y + 16 = 0$$

23. $16x^2 + y^2 = 16$

$$x^2 + \frac{y^2}{16} = 1$$

$a^2 = 16, b^2 = 1, c^2 = 16 - 1 = 15$

Center: $(0, 0)$

Foci: $\left(0, \pm\sqrt{15}\right)$

Vertices: $(0, \pm 4)$

$$e = \frac{c}{a} = \frac{\sqrt{15}}{4}$$

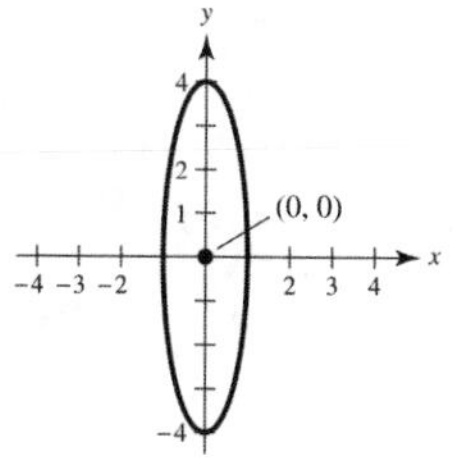

24. $3x^2 + 7y^2 = 63$

$$\frac{x^2}{21} + \frac{y^2}{9} = 1$$

$a^2 = 21, b^2 = 9, c^2 = 21 - 9 = 12$

Center: $(0, 0)$

Foci: $\left(\pm 2\sqrt{3}, 0\right)$

Vertices: $\left(\pm\sqrt{21}, 0\right)$

$$e = \frac{c}{a} = \frac{2\sqrt{3}}{\sqrt{21}} = \frac{2\sqrt{7}}{7}$$

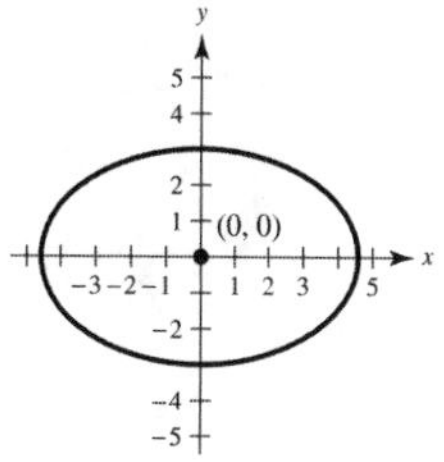

25. $$\frac{(x-3)^2}{16} + \frac{(y-1)^2}{25} = 1$$

$a^2 = 25, b^2 = 16, c^2 = 25 - 16 = 9$

Center: $(3, 1)$

Foci: $(3, 1+3) = (3, 4), (3, 1-3) = (3, -2)$

Vertices: $(3, 6), (3, -4)$

$$e = \frac{c}{a} = \frac{3}{5}$$

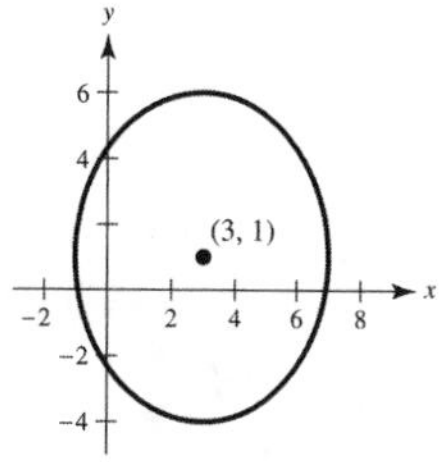

26. $$(x+4)^2 + \frac{(y+6)^2}{1/4} = 1$$

$$a^2 = 1, b^2 = \frac{1}{4}, c^2 = 1 - \frac{1}{4} = \frac{3}{4}$$

Center: $(-4, -6)$

Foci: $\left(-4 \pm \frac{\sqrt{3}}{2}, -6\right)$

Vertices: $(-5, -6), (-3, -6)$

$$e = \frac{c}{a} = \frac{\sqrt{3}}{2}$$

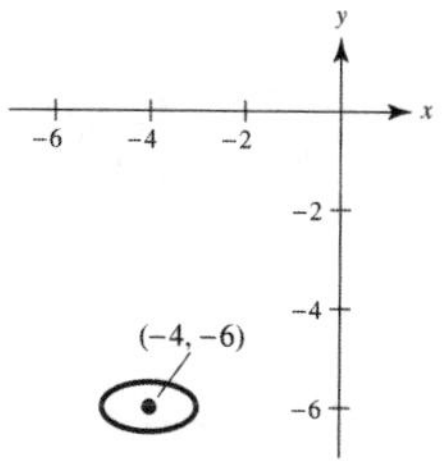

27. $$9x^2 + 4y^2 + 36x - 24y + 36 = 0$$
$$9(x^2 + 4x + 4) + 4(y^2 - 6y + 9) = -36 + 36 + 36$$
$$= 36$$
$$\frac{(x+2)^2}{4} + \frac{(y-3)^2}{9} = 1$$

$a^2 = 9, b^2 = 4, c^2 = 5$

Center: $(-2, 3)$

Foci: $\left(-2, 3 \pm \sqrt{5}\right)$

Vertices: $(-2, 6), (-2, 0)$

$$e = \frac{\sqrt{5}}{3}$$

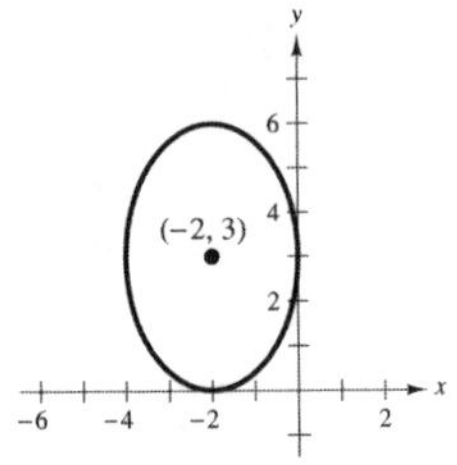

28. $$16x^2 + 25y^2 - 64x + 150y + 279 = 0$$
$$16(x^2 - 4x + 4) + 25(y^2 + 6y + 0) = -279 + 64 + 225$$
$$= 10$$
$$\frac{(x-2)^2}{(5/8)} + \frac{(y+3)^2}{(2/5)} = 1$$

$$a^2 = \frac{5}{8}, b^2 = \frac{2}{5}, c^2 = a^2 - b^2 = \frac{9}{40}$$

Center: $(2, -3)$

Foci: $\left(2 \pm \frac{3\sqrt{10}}{20}, -3\right)$

Vertices: $\left(2 \pm \frac{\sqrt{10}}{4}, -3\right)$

$$e = \frac{c}{a} = \frac{3}{5}$$

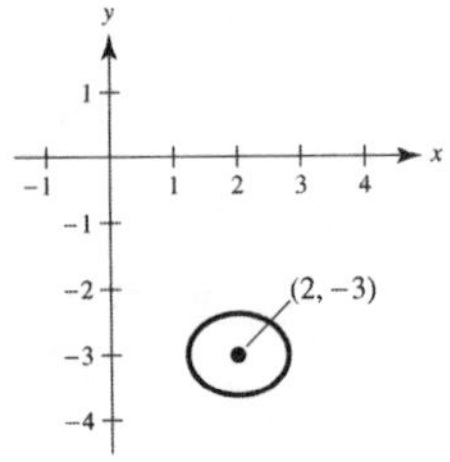

29. Center: $(0, 0)$

Focus: $(5, 0)$

Vertex: $(6, 0)$

Horizontal major axis

$a = 6, c = 5 \Rightarrow b = \sqrt{a^2 - c^2} = \sqrt{11}$

$$\frac{x^2}{36} + \frac{y^2}{11} = 1$$

30. Vertices: $(0, 3), (8, 3)$

Eccentricity: $\frac{3}{4}$

Horizontal major axis

Center: $(4, 3)$

$$a = 4, e = \frac{c}{a} \Rightarrow c = 4\left(\frac{3}{4}\right) = 3$$
$$\Rightarrow b = \sqrt{16 - 9} = \sqrt{7}$$

$$\frac{(x-4)^2}{16} + \frac{(y-3)^2}{7} = 1$$

31. Vertices: $(3, 1), (3, 9)$

Minor axis length: 6

Vertical major axis

Center: $(3, 5)$

$a = 4, b = 3$

$$\frac{(x-3)^2}{9} + \frac{(y-5)^2}{16} = 1$$

32. Foci: $(0, \pm 9)$

Major axis length: 22

Vertical major axis

Center: $(0, 0)$

$c = 9, a = 11 \Rightarrow b = \sqrt{40}$

$$\frac{x^2}{40} + \frac{y^2}{121} = 1$$

33. Center: $(0, 0)$

Horizontal major axis

Points on ellipse: $(3, 1), (4, 0)$

Because the major axis is horizontal,

$$\left(\frac{x^2}{a^2}\right) + \left(\frac{y^2}{b^2}\right) = 1.$$

Substituting the values of the coordinates of the given points into this equation, you have

$$\left(\frac{9}{a^2}\right) + \left(\frac{1}{b^2}\right) = 1, \text{ and } \frac{16}{a^2} = 1.$$

The solution to this system is $a^2 = 16, b^2 = \frac{16}{7}$.

So,

$$\frac{x^2}{16} + \frac{y^2}{16/7} = 1, \frac{x^2}{16} + \frac{7y^2}{16} = 1.$$

34. Center: $(1, 2)$

Vertical major axis

Points on ellipse: $(1, 6), (3, 2)$

From the sketch, you can see that

$h = 1, k = 2, a = 4, b = 2$

$$\frac{(x-1)^2}{4} + \frac{(y-2)^2}{16} = 1.$$

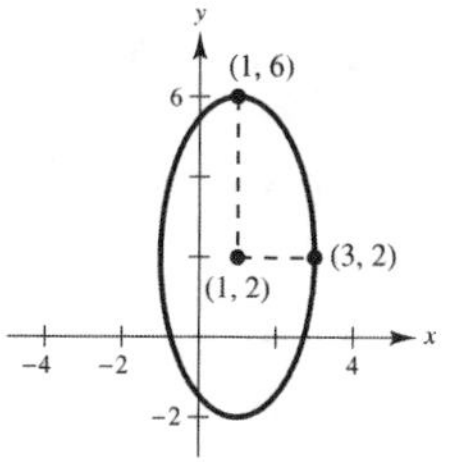

35. $\dfrac{x^2}{25} - \dfrac{y^2}{16} = 1$

$a = 5, b = 4, c = \sqrt{25 + 16} = \sqrt{41}$

Center: $(0, 0)$

Vertices: $(\pm 5, 0)$

Foci: $(\pm\sqrt{41}, 0)$

Asymptotes: $y = \pm\dfrac{b}{a}x = \pm\dfrac{4}{5}x$

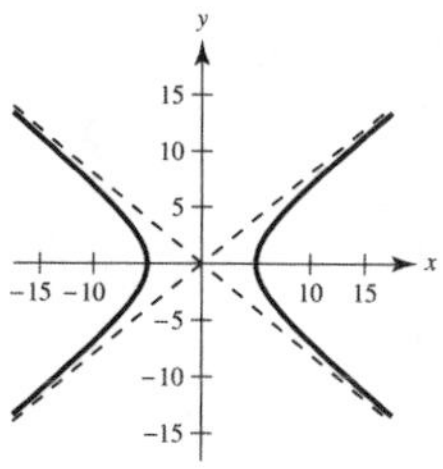

36. $\dfrac{(y+3)^2}{225} - \dfrac{(x-5)^2}{64} = 1$

$a = 15, b = 8, c = \sqrt{225 + 64} = 17$

Center: $(5, -3)$

Vertices: $(5, 12), (5, -18)$

Foci: $(5, 14), (5, -20)$

Asymptotes: $y = k \pm \dfrac{a}{b}(x - h) = -3 \pm \dfrac{15}{8}(x - 5)$

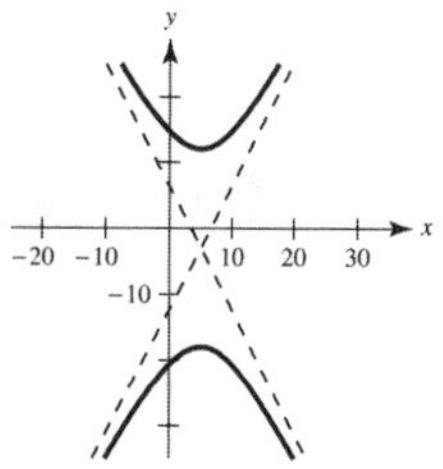

37.

$$9x^2 - y^2 - 36x - 6y + 18 = 0$$
$$9(x^2 - 4x + 4) - (y^2 + 6y + 9) = -18 + 36 - 9$$
$$\frac{(x-2)^2}{1} - \frac{(y+3)^2}{9} = 1$$

$a = 1, b = 3, c = \sqrt{10}$

Center: $(2, -3)$

Vertices: $(1, -3), (3, -3)$

Foci: $(2 \pm \sqrt{10}, -3)$

Asymptotes: $y = -3 \pm 3(x - 2)$

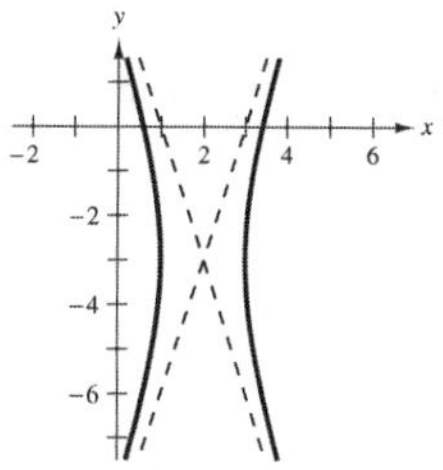

38. $y^2 - 16x^2 + 64x - 208 = 0$

$$y^2 - 16(x^2 - 4x + 4) = 208 - 64 = 144$$

$$\frac{y^2}{144} - \frac{(x-2)^2}{9} = 1$$

$a = 12, b = 3, c = \sqrt{144 + 9} = \sqrt{153}$

Center: $(2, 0)$

Vertices: $(2, 12), (2, -12)$

Foci: $(2, \pm\sqrt{153})$

Asymptotes: $y = \pm\frac{12}{3}(x - 2) = \pm 4(x - 2)$

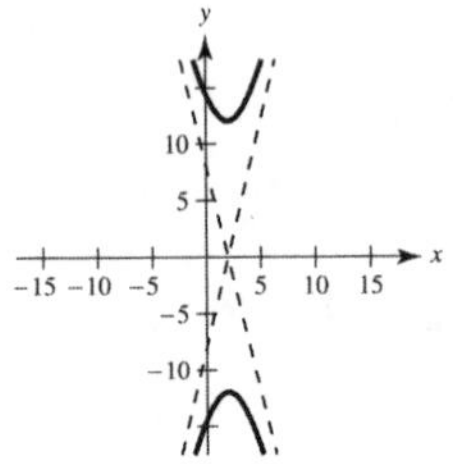

39. $x^2 - 9y^2 + 2x - 54y - 80 = 0$

$$(x^2 + 2x + 1) - 9(y^2 + 6y + 9) = 80 + 1 - 81 = 0$$

$$(x+1)^2 - 9(y+3)^2 = 0$$

$$y + 3 = \pm\frac{1}{3}(x + 1)$$

$$y = -3 \pm \frac{1}{3}(x + 1)$$

Degenerate hyperbola is two lines intersecting at $(-1, -3)$.

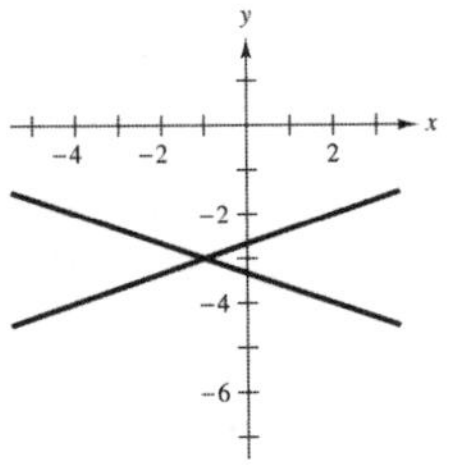

40. $9(x^2 + 6x + 9) - 4(y^2 - 2y + 1) = -78 + 81 - 4 = -1$

$$9(x+3)^2 - 4(y-1)^2 = -1$$

$$\frac{(y-1)^2}{1/4} - \frac{(x+3)^2}{1/9} = 1$$

$a = \frac{1}{2}, b = \frac{1}{3}, c = \frac{\sqrt{13}}{6}$

Center: $(-3, 1)$

Vertices: $\left(-3, \frac{1}{2}\right), \left(-3, \frac{3}{2}\right)$

Foci: $\left(-3, 1 \pm \frac{1}{6}\sqrt{13}\right)$

Asymptotes: $y = 1 \pm \frac{3}{2}(x + 3)$

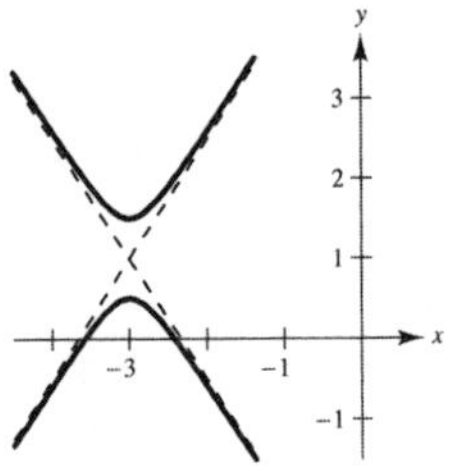

41. Vertices: $(\pm 1, 0)$

Asymptotes: $y = \pm 5x$

Horizontal transverse axis

Center: $(0, 0)$

$a = 1, \frac{b}{a} = 5 \Rightarrow b = 5$

$$\frac{x^2}{1} - \frac{y^2}{25} = 1$$

42. Vertices: $(0, \pm 4)$

Asymptotes: $y = \pm 2x$

Vertical transverse axis

$a = 4, \frac{a}{b} = 2 \Rightarrow b = 2$

$$\frac{y^2}{16} - \frac{x^2}{4} = 1$$

43. Vertices: $(2, \pm 3)$

Point on graph: $(0, 5)$

Vertical transverse axis

Center: $(2, 0)$

$a = 3$

So, the equation is of the form

$$\frac{y^2}{9} - \frac{(x-2)^2}{b^2} = 1.$$

Substituting the coordinates of the point $(0, 5)$, you have

$$\frac{25}{9} - \frac{4}{b^2} = 1 \quad \text{or} \quad b^2 = \frac{9}{4}.$$

So, the equation is $\dfrac{y^2}{9} - \dfrac{(x-2)^2}{9/4} = 1.$

44. Vertices: $(2, \pm 3)$

Foci: $(2, \pm 5)$

Vertical transverse axis

Center: $(2, 0)$

$a = 3, c = 5, b^2 = c^2 - a^2 = 16$

So, $\dfrac{y^2}{9} - \dfrac{(x-2)^2}{16} = 1.$

45. Center: $(0, 0)$

Vertex: $(0, 2)$

Focus: $(0, 4)$

Vertical transverse axis

$a = 2, c = 4, b^2 = c^2 - a^2 = 12$

So, $\dfrac{y^2}{4} - \dfrac{x^2}{12} = 1.$

46. Center: $(0, 0)$

Vertex: $(6, 0)$

Focus: $(10, 0)$

Horizontal transverse axis

$a = 6, c = 10, b^2 = c^2 - a^2 = 100 - 36 = 64$

$$\frac{x^2}{36} - \frac{y^2}{64} = 1$$

47. Vertices: $(0, 2), (6, 2)$

Asymptotes: $y = \frac{2}{3}x, y = 4 - \frac{2}{3}x$

Horizontal transverse axis

Center: $(3, 2)$

$a = 3$

Slopes of asymptotes: $\pm\frac{b}{a} = \pm\frac{2}{3}$

So, $b = 2$. Therefore,

$$\frac{(x-3)^2}{9} - \frac{(y-2)^2}{4} = 1.$$

48. Focus: $(20, 0)$

Asymptotes: $y = \pm\frac{3}{4}x$

Horizontal transverse axis

Center: $(0, 0)$

$c = 20$

$$\frac{b}{a} = \frac{3}{4} \Rightarrow b = \frac{3}{4}a$$

$$c^2 = 400 = a^2 + b^2 = a^2 + \frac{9}{16}a^2 = \frac{25}{16}a^2$$

$$\Rightarrow a^2 = 256 \quad \text{and} \quad b^2 = 144$$

$$\frac{x^2}{256} - \frac{y^2}{144} = 1$$

49. (a) $\dfrac{x^2}{9} - y^2 = 1, \dfrac{2x}{9} - 2yy' = 0, \dfrac{x}{9y} = y'$

At $x = 6$: $y = \pm\sqrt{3}, y' = \dfrac{\pm 6}{9\sqrt{3}} = \dfrac{\pm 2\sqrt{3}}{9}$

At $(6, \sqrt{3})$: $y - \sqrt{3} = \dfrac{2\sqrt{3}}{9}(x - 6)$

or $2x - 3\sqrt{3}y - 3 = 0$

At $(6, -\sqrt{3})$: $y + \sqrt{3} = \dfrac{-2\sqrt{3}}{9}(x - 6)$

or $2x + 3\sqrt{3}y - 3 = 0$

(b) From part (a) you know that the slopes of the normal lines must be $\mp 9/(2\sqrt{3})$.

At $(6, \sqrt{3})$: $y - \sqrt{3} = -\dfrac{9}{2\sqrt{3}}(x - 6)$

or $9x + 2\sqrt{3}y - 60 = 0$

At $(6, -\sqrt{3})$: $y + \sqrt{3} = \dfrac{9}{2\sqrt{3}}(x - 6)$

or $9x - 2\sqrt{3}y - 60 = 0$

50. (a) $\frac{y^2}{4} - \frac{x^2}{2} = 1,\ y^2 - 2x^2 = 4,\ 2yy' - 4x = 0,$

$y' = \frac{4x}{2y} = \frac{2x}{y}$

At $x = 4$: $y = \pm 6,\ y' = \frac{\pm 2(4)}{6} = \pm\frac{4}{3}$

At $(4, 6)$: $y - 6 = -\frac{4}{3}(x - 4)$ or $4x + 3y - 34 = 0$

At $(4, -6)$: $y + 6 = -\frac{4}{3}(x - 4)$ or $4x + 3y + 2 = 0$

(b) From part (a) you know that the slopes of the normal lines must be $\mp 3/4$.

At $(4, 6)$: $y - 6 = -\frac{3}{4}(x - 4)$ or $3x + 4y - 36 = 0$

At $(4, -6)$: $y + 6 = \frac{3}{4}(x - 4)$ or $3x - 4y - 36 = 0$

51.
$$x^2 + 4y^2 - 6x + 16y + 21 = 0$$
$$(x^2 - 6x + 9) + 4(y^2 + 4y + 4) = -21 + 9 + 16$$
$$(x - 3)^2 + 4(y + 2)^2 = 4$$

Ellipse

52. $4x^2 - y^2 - 4x - 3 = 0$
$$4\left(x^2 - x + \tfrac{1}{4}\right) - y^2 = 3 + 1$$
$$4\left(x - \tfrac{1}{2}\right)^2 - y^2 = 4$$

Hyperbola

53. $25x^2 - 10x - 200y - 119 = 0$
$$25\left(x^2 - \tfrac{2}{5}x + \tfrac{1}{25}\right) = 200y + 119 + 1$$
$$25\left(x - \tfrac{1}{5}\right)^2 = 200(y + 1)$$

Parabola

54.
$$y^2 - 4y = x + 5$$
$$y^2 - 4y + 4 = x + 5 + 4$$
$$(y - 2)^2 = x + 9$$

Parabola

55.
$$9x^2 + 9y^2 - 36x + 6y + 34 = 0$$
$$9(x^2 - 4x + 4) + 9\left(y^2 + \tfrac{2}{3}y + \tfrac{1}{9}\right) = -34 + 36 + 1$$
$$9(x - 2)^2 + 9\left(y + \tfrac{1}{3}\right)^2 = 3$$

Circle (Ellipse)

56.
$$2x(x - y) = y(3 - y - 2x)$$
$$2x^2 - 2xy = 3y - y^2 - 2xy$$
$$2x^2 + y^2 - 3y = 0$$
$$2x^2 + \left(y - \tfrac{3}{2}\right)^2 = \tfrac{9}{4}$$

Ellipse

57.
$$3(x - 1)^2 = 6 + 2(y + 1)^2$$
$$3(x - 1)^2 - 2(y + 1)^2 = 6$$
$$\frac{(x - 1)^2}{2} - \frac{(y + 1)^2}{3} = 1$$

Hyperbola

58.
$$9(x + 3)^2 = 36 - 4(y - 2)^2$$
$$9(x + 3)^2 + 4(y - 2)^2 = 36$$
$$\frac{(x + 3)^2}{4} + \frac{(y - 2)^2}{9} = 1$$

Ellipse

59. (a) A parabola is the set of all points (x, y) that are equidistant from a fixed line (directrix) and a fixed point (focus) not on the line.

(b) For directrix $y = k - p$: $(x - h)^2 = 4p(y - k)$

For directrix $x = h - p$: $(y - k)^2 = 4p(x - h)$

(c) If P is a point on a parabola, then the tangent line to the parabola at P makes equal angles with the line passing through P and the focus, and with the line passing through P parallel to the axis of the parabola.

60. (a) An ellipse is the set of all points (x, y), the sum of whose distance from two distinct fixed points (foci) is constant.

(b) $\frac{(x - h)^2}{a^2} + \frac{(y - k)^2}{b^2} = 1$ or $\frac{(x - h)^2}{b^2} + \frac{(y - k)^2}{a^2} = 1$

61. (a) A hyperbola is the set of all points (x, y) for which the absolute value of the difference between the distances from two distinct fixed points (foci) is constant.

(b) Transverse axis is horizontal:

$$\frac{(x - h)^2}{a^2} - \frac{(y - k)^2}{b^2} = 1$$

Transverse axis is vertical:

$$\frac{(y - k)^2}{a^2} - \frac{(x - h)^2}{b^2} = 1$$

(c) Transverse axis is horizontal:

$y = k + (b/a)(x - h)$ and $y = k - (b/a)(x - h)$

Transverse axis is vertical:

$y = k + (a/b)(x - h)$ and $y = k - (a/b)(x - h)$

62. $e = \frac{c}{a},\ c = \sqrt{a^2 - b^2},\ 0 < e < 1$

For $e \approx 0$, the ellipse is nearly circular.

For $e \approx 1$, the ellipse is elongated.

63. $9x^2 + 4y^2 - 36x - 24y - 36 = 0$

(a) $9(x^2 - 4x + 4) + 4(y^2 - 6y + 9) = 36 + 36 + 36$

$9(x - 2)^2 + 4(y - 3)^2 = 108$

$\dfrac{(x - 2)^2}{12} + \dfrac{(y - 3)^2}{27} = 1$

Ellipse

(b) $9x^2 - 4y^2 - 36x - 24y - 36 = 0$

$9(x^2 - 4x + 4) - 4(y^2 + 6y + 9) = 36 + 36 - 36$

$\dfrac{(x - 2)^2}{4} - \dfrac{(y + 3)^2}{9} = 1$

Hyperbola

(c) $4x^2 + 4y^2 - 36x - 24y - 36 = 0$

$4\left(x^2 - 9x + \dfrac{81}{4}\right) + 4(y^2 - 6y + 9) = 36 + 81 + 36$

$\left(x - \dfrac{9}{2}\right)^2 + (y - 3)^2 = \dfrac{153}{4}$

Circle

(d) *Sample answer:* Eliminate the y^2-term

64. (a) A circle is formed when a plane intersects the top or bottom half of a double-napped cone and is perpendicular to the axis of the cone.

(b) An ellipse is formed when a plane intersects only the top or bottom half of a double-napped cone but is not parallel or perpendicular to the axis of the cone, is not parallel to the side of the cone, and does not intersect the vertex.

(c) A parabola is formed when a plane intersects the top or bottom half of a double-napped cone, is parallel to the side of the cone, and does not intersect the vertex.

(d) A hyperbola is formed when a plane intersects both halves of a double-napped cone, is parallel to the axis of the cone, and does not intersect the vertex.

65. Assume that the vertex is at the origin.

$x^2 = 4py$

$(3)^2 = 4p(1)$

$\frac{9}{4} = p$

The pipe is located $\frac{9}{4}$ meters from the vertex.

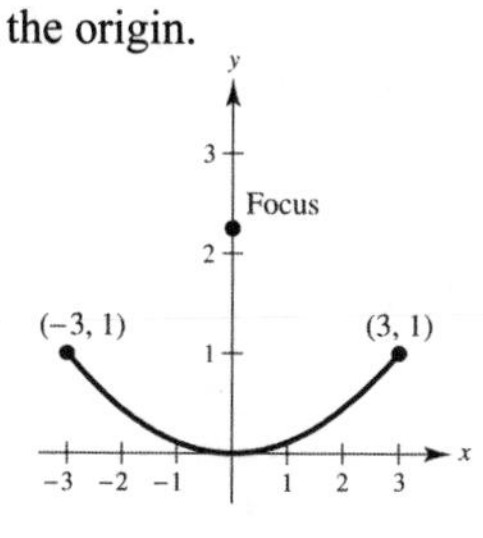

66. Assume that the vertex is at the origin.

(a) $x^2 = 4py$

$8^2 = 4p\left(\dfrac{3}{100}\right)$

$\dfrac{1600}{3} = p$

$x^2 = 4\left(\dfrac{1600}{3}\right)y = \dfrac{6400}{3}y$

(b) The deflection is 1 cm when

$y = \dfrac{2}{100} \Rightarrow x = \pm\sqrt{\dfrac{128}{3}} \approx \pm 6.53$ meters.

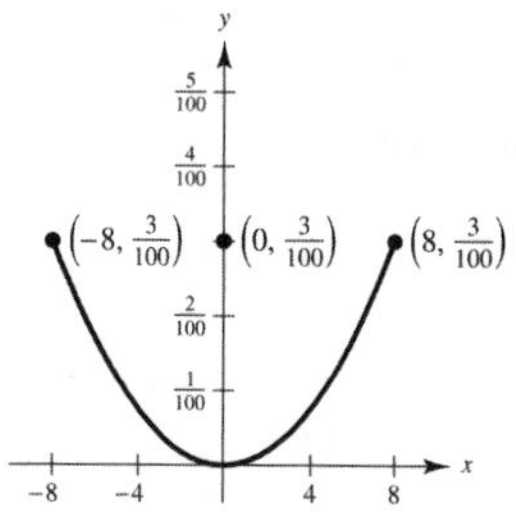

67. (a) Without loss of generality, place the coordinate system so that the equation of the parabola is $x^2 = 4py$ and, so,

$y' = \left(\dfrac{1}{2p}\right)x.$

So, for distinct tangent lines, the slopes are unequal and the lines intersect.

(b) $x^2 - 4x - 4y = 0$

$2x - 4 - 4\dfrac{dy}{dx} = 0$

$\dfrac{dy}{dx} = \dfrac{1}{2}x - 1$

At $(0, 0)$, the slope is -1: $y = -x$. At $(6, 3)$, the slope is 2: $y = 2x - 9$. Solving for x,

$-x = 2x - 9$

$-3x = -9$

$x = 3$

$y = -3.$

Point of intersection: $(3, -3)$

68. (a) Consider the parabola $x^2 = 4py$. Let m_0 be the slope of the one tangent line at (x_1, y_1) and so, $-\frac{1}{m_0}$ is the slope of the second at (x_2, y_2). Differentiating, $2x = 4py'$ or $y' = \frac{x}{2p}$, and you have:

$$m_0 = \frac{1}{2p}x_1 \quad \text{or} \quad x_1 = 2pm_0$$

$$\frac{-1}{m_0} = \frac{1}{2p}x_2 \quad \text{or} \quad x_2 = \frac{-2p}{m_0}.$$

Substituting these values of x into the equation $x^2 = 4py$, we have the coordinates of the points of tangency $(2pm_0, pm_0^2)$ and $(-2p/m_0, p/m_0^2)$ and the equations of the tangent lines are

$$(y - pm_0^2) = m_0(x - 2pm_0) \text{ and } \left(y - \frac{p}{m_0^2}\right) = \frac{-1}{m_0}\left(x + \frac{2p}{m_0}\right).$$

The point of intersection of these lines is $\left(\frac{p(m_0^2 - 1)}{m_0}, -p\right)$

and is on the directrix, $y = -p$.

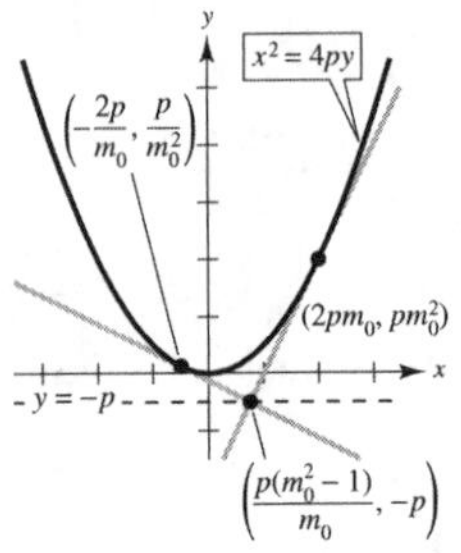

(b) $x^2 - 4x - 4y + 8 = 0$

$$(x - 2)^2 = 4(y - 1)$$

Vertex: $(2, 1)$

$$2x - 4 - 4\frac{dy}{dx} = 0$$

$$\frac{dy}{dx} = \frac{1}{2}x - 1$$

At $(-2, 5)$, $\frac{dy}{dx} = -2$. At $\left(3, \frac{5}{4}\right)$, $\frac{dy}{dx} = \frac{1}{2}$.

Tangent line at $(-2, 5)$: $y - 5 = -2(x + 2) \Rightarrow 2x + y - 1 = 0$.

Tangent line at $\left(3, \frac{5}{4}\right)$: $y - \frac{5}{4} = \frac{1}{2}(x - 3) \Rightarrow 2x - 4y - 1 = 0$.

Because $m_1m_2 = (-2)\left(\frac{1}{2}\right) = -1$, the lines are perpendicular.

Point of intersection: $-2x + 1 = \frac{1}{2}x - \frac{1}{4}$

$$-\frac{5}{2}x = -\frac{5}{4}$$

$$x = \frac{1}{2}$$

$$y = 0$$

Directrix: $y = 0$ and the point of intersection $\left(\frac{1}{2}, 0\right)$ lies on this line.

69. $x^2 = 4py$, $p = \frac{1}{4}, \frac{1}{2}, 1, \frac{3}{2}, 2$

As p increases, the graph becomes wider.

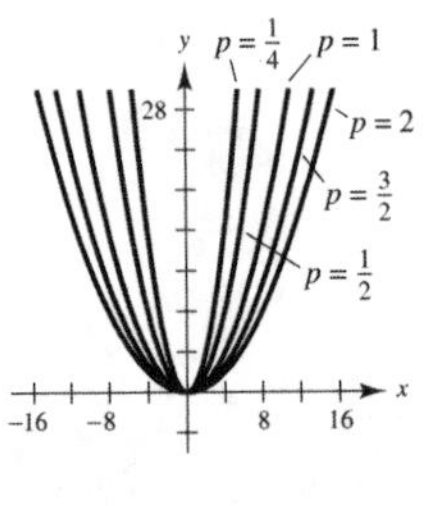

70. (a) Assume that $y = ax^2$.

$$20 = a(60)^2 \Rightarrow a = \frac{2}{360} = \frac{1}{180} \Rightarrow y = \frac{1}{180}x^2$$

(b) $f(x) = \frac{1}{180}x^2,\ f'(x) = \frac{1}{90}x$

$$S = 2\int_0^{60} \sqrt{1 + \left(\frac{1}{90}x\right)^2}\,dx = \frac{2}{90}\int_0^{60} \sqrt{90^2 + x^2}\,dx$$

$$= \frac{2}{90}\frac{1}{2}\left[x\sqrt{90^2 + x^2} + 90^2 \ln\left|x + \sqrt{90^2 + x^2}\right|\right]_0^{60} \quad \text{(Formula 26)}$$

$$= \frac{1}{90}\left[60\sqrt{11{,}700} + 90^2 \ln\left(60 + \sqrt{11{,}700}\right) - 90^2 \ln 90\right]$$

$$= \frac{1}{90}\left[1800\sqrt{13} + 90^2 \ln\left(60 + 30\sqrt{13}\right) - 90^2 \ln 90\right]$$

$$= 20\sqrt{13} + 90 \ln\left(\frac{60 + 30\sqrt{13}}{90}\right)$$

$$= 10\left[2\sqrt{13} + 9 \ln\left(\frac{2 + \sqrt{13}}{3}\right)\right] \approx 128.4 \text{ m}$$

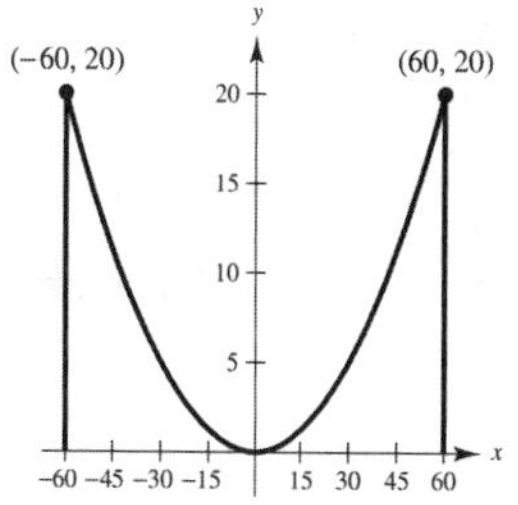

71. Parabola

Vertex: $(0, 4)$

$$x^2 = 4p(y - 4)$$
$$4^2 = 4p(0 - 4)$$
$$p = -1$$
$$x^2 = -4(y - 4)$$
$$y = 4 - \frac{x^2}{4}$$

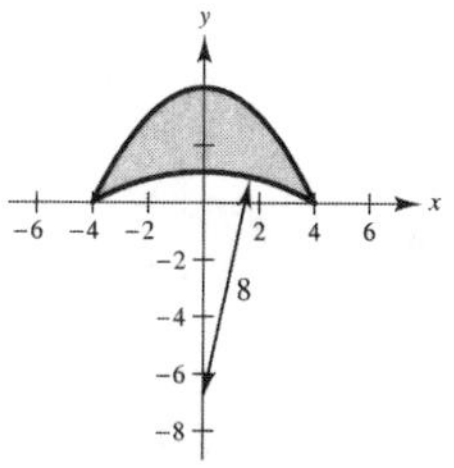

Circle

Center: $(0, k)$

Radius: 8

$$x^2 + (y - k)^2 = 64$$
$$4^2 + (0 - k)^2 = 64$$
$$k^2 = 48$$
$$k = -4\sqrt{3} \quad \text{(Center is on the negative } y\text{-axis.)}$$
$$x^2 + \left(y + 4\sqrt{3}\right)^2 = 64$$
$$y = -4\sqrt{3} \pm \sqrt{64 - x^2}$$

Because the y-value is positive when $x = 0$, we have $y = -4\sqrt{3} + \sqrt{64 - x^2}$.

$$A = 2\int_0^4 \left[\left(4 - \frac{x^2}{4}\right) - \left(-4\sqrt{3} + \sqrt{64 - x^2}\right)\right] dx$$

$$= 2\left[4x - \frac{x^3}{12} + 4\sqrt{3}x - \frac{1}{2}\left(x\sqrt{64 - x^2} + 64 \arcsin\frac{x}{8}\right)\right]_0^4$$

$$= 2\left(16 - \frac{64}{12} + 16\sqrt{3} - 2\sqrt{48} - 32 \arcsin\frac{1}{2}\right) = \frac{16\left(4 + 3\sqrt{3} - 2\pi\right)}{3} \approx 15.536 \text{ square feet}$$

72. $x^2 = 20y$

$y = \dfrac{x^2}{20}$

$y' = \dfrac{x}{10}$

$$S = 2\pi \int_0^r x\sqrt{1 + \left(\frac{x}{10}\right)^2}\,dx = 2\pi \int_0^r \frac{x\sqrt{100 + x^2}}{10}\,dx = \left[\frac{\pi}{10} \cdot \frac{2}{3}\left(100 + x^2\right)^{3/2}\right]_0^r = \frac{\pi}{15}\left[\left(100 + r^2\right)^{3/2} - 1000\right]$$

73. $e = \dfrac{c}{a}$

$0.0167 = \dfrac{c}{149{,}598{,}000}$

$c \approx 2{,}498{,}286.6$

Least distance: $a - c = 147{,}099{,}713.4$ km

Greatest distance: $a + c = 152{,}096{,}286.6$ km

74. $e = \dfrac{c}{a}$

$A + P = 2a$

$a = \dfrac{A + P}{2}$

$c = a - P = \dfrac{A + P}{2} - P = \dfrac{A - P}{2}$

$e = \dfrac{c}{a} = \dfrac{(A - P)/2}{(A + P)/2} = \dfrac{A - P}{A + P}$

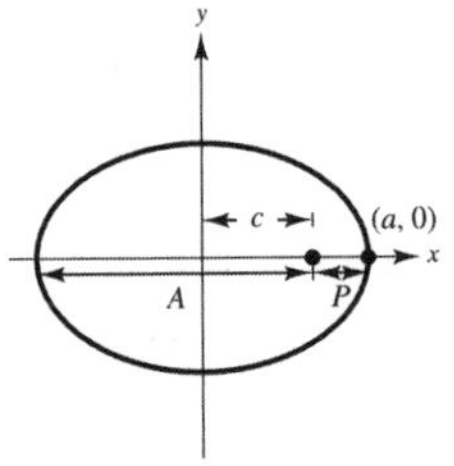

75. $e = \dfrac{A - P}{A + P}$

$= \dfrac{(123{,}000 + 4000) - (119 + 4000)}{(123{,}000 + 4000) + (119 + 4000)}$

$= \dfrac{122{,}881}{131{,}119} \approx 0.9372$

76. $e = \dfrac{A - P}{A + P}$ (Exercise 74)

$= \dfrac{(1865 + 4000) - (96 + 4000)}{(1865 + 4000) + (96 + 4000)} = \dfrac{1769}{9961} \approx 0.1776$

77. $e = \dfrac{A - P}{A + P} = \dfrac{35.29 - 0.59}{35.29 + 0.59} = 0.9671$

78. $\dfrac{x^2}{10^2} + \dfrac{y^2}{5^2} = 1$

$\dfrac{2x}{10^2} + \dfrac{2yy'}{5^2} = 0$

$y' = \dfrac{-5^2 x}{10^2 y} = \dfrac{-x}{4y}$

At $(-8, 3)$: $y' = \dfrac{8}{12} = \dfrac{2}{3}$

The equation of the tangent line is $y - 3 = \dfrac{2}{3}(x + 8)$. It will cross the y-axis when $x = 0$ and $y = \dfrac{2}{3}(8) + 3 = \dfrac{25}{3}$.

79. (a) $A = 4\int_0^2 \frac{1}{2}\sqrt{4 - x^2}\,dx = \left[x\sqrt{4 - x^2} + 4\arcsin\left(\frac{x}{2}\right)\right]_0^2 = 2\pi \quad \left[\text{or, } A = \pi ab = \pi(2)(1) = 2\pi\right]$

(b) **Disk:** $V = 2\pi\int_0^2 \frac{1}{4}\left(4 - x^2\right)dx = \frac{1}{2}\pi\left[4x - \frac{1}{3}x^3\right]_0^2 = \frac{8\pi}{3}$

$$y = \frac{1}{2}\sqrt{4 - x^2}$$

$$y' = \frac{-x}{2\sqrt{4 - x^2}}$$

$$\sqrt{1 + (y')^2} = \sqrt{1 + \frac{x^2}{16 - 4x^2}} = \sqrt{\frac{16 - 3x^2}{4y}}$$

$$S = 2(2\pi)\int_0^2 y\left(\frac{\sqrt{16 - 3x^2}}{4y}\right)dx = \pi\int_0^2 \sqrt{16 - 3x^2}\,dx$$

$$= \frac{\pi}{2\sqrt{3}}\left[\sqrt{3}x\sqrt{16 - 3x^2} + 16\arcsin\left(\frac{\sqrt{3}x}{4}\right)\right]_0^2 = \frac{2\pi}{9}\left(9 + 4\sqrt{3}\pi\right) \approx 21.48$$

(c) **Shell:** $V = 2\pi\int_0^2 x\sqrt{4 - x^2}\,dx = -\pi\int_0^2 -2x\left(4 - x^2\right)^{1/2}dx = -\frac{2\pi}{3}\left[\left(4 - x^2\right)^{3/2}\right]_0^2 = \frac{16\pi}{3}$

$$x = 2\sqrt{1 - y^2}$$

$$x' = \frac{-2y}{\sqrt{1 - y^2}}$$

$$\sqrt{1 + (x')^2} = \sqrt{1 + \frac{4y^2}{1 - y^2}} = \frac{\sqrt{1 + 3y^2}}{\sqrt{1 - y^2}}$$

$$S = 2(2\pi)\int_0^1 2\sqrt{1 - y^2}\frac{\sqrt{1 + 3y^2}}{\sqrt{1 - y^2}}\,dy = 8\pi\int_0^1 \sqrt{1 + 3y^2}\,dy$$

$$= \frac{8\pi}{2\sqrt{3}}\left[\sqrt{3}y\sqrt{1 + 3y^2} + \ln\left|\sqrt{3}y + \sqrt{1 + 3y^2}\right|\right]_0^1$$

$$= \frac{4\pi}{3}\left|6 + \sqrt{3}\ln\left(2 + \sqrt{3}\right)\right| \approx 34.69$$

80. (a) $A = 4\int_0^4 \frac{3}{4}\sqrt{16 - x^2}\,dx = \frac{3}{2}\left[x\sqrt{16 - x^2} + 16\arcsin\frac{x}{4}\right]_0^4 = 12\pi$

(b) **Disk:** $V = 2\pi\int_0^4 \frac{9}{16}\left(16 - x^2\right)dx = \frac{9\pi}{8}\left[\left(16x - \frac{1}{3}x^3\right)\right]_0^4 = 48\pi$

$$y = \frac{3}{4}\sqrt{16 - x^2}$$

$$y' = \frac{-3x}{4\sqrt{16 - x^2}}$$

$$\sqrt{1 + (y')^2} = \sqrt{1 + \frac{9x^2}{16\left(16 - x^2\right)}}$$

$$S = 2(2\pi)\int_0^4 \frac{3}{4}\sqrt{16 - x^2}\sqrt{\frac{16\left(16 - x^2\right) + 9x^2}{16\left(16 - x^2\right)}}\,dx = 4\pi\int_0^4 \frac{3}{4}\sqrt{16 - x^2}\frac{\sqrt{256 - 7x^2}}{4\sqrt{16 - x^2}}\,dx = \frac{3\pi}{4}\int_0^4 \sqrt{256 - 7x^2}\,dx$$

$$= \frac{3\pi}{8\sqrt{7}}\left[\sqrt{7}x\sqrt{256 - 7x^2} + 256\arcsin\frac{\sqrt{7}x}{16}\right]_0^4 = \frac{3\pi}{8\sqrt{7}}\left(48\sqrt{7} + 256\arcsin\frac{\sqrt{7}}{4}\right) \approx 138.93$$

(c) **Shell:** $V = 4\pi\int_0^4 x\left(\frac{3}{4}\sqrt{16 - x^2}\right)dx = 3\pi\left[\left(-\frac{1}{2}\right)\left(\frac{2}{3}\right)\left(16 - x^2\right)^{3/2}\right]_0^4 = 64\pi$

$$x = \frac{4}{3}\sqrt{9 - y^2}$$

$$x' = \frac{-4y}{3\sqrt{9 - y^2}}$$

$$\sqrt{1 + (x')^2} = \sqrt{1 + \frac{16y^2}{9(9 - y^2)}}$$

$$S = 2(2\pi)\int_0^3 \frac{4}{3}\sqrt{9 - y^2}\sqrt{\frac{9(9 - y^2) + 16y^2}{9(9 - y^2)}}\,dy$$

$$= 4\pi\int_0^3 \frac{4}{9}\sqrt{81 + 7y^2}\,dy$$

$$= \frac{16}{9}\left(\frac{\pi}{2\sqrt{7}}\right)\left[\sqrt{7}y\sqrt{81 + 7y^2} + 81\ln\left|\sqrt{7}y + \sqrt{81 + 7y^2}\right|\right]_0^3$$

$$= \frac{8\pi}{9\sqrt{7}}3\sqrt{7}(12) + 81\ln\left(3\sqrt{7} + 12\right) - 81\ln 9 \approx 168.53$$

81. From Example 5,

$$C = 4a\int_0^{\pi/2}\sqrt{1 - e^2\sin^2\theta}\,d\theta$$

For $\frac{x^2}{25} + \frac{y^2}{49} = 1$, you have

$$a = 7, b = 5, c = \sqrt{49 - 25} = 2\sqrt{6}, e = \frac{c}{a} = \frac{2\sqrt{6}}{7}.$$

$$C = 4(7)\int_0^{\pi/2}\sqrt{1 - \frac{24}{49}\sin^2\theta}\,d\theta \approx 28(1.3558) \approx 37.96$$

82. (a) $e = \frac{c}{a} = \frac{\sqrt{a^2 + b^2}}{a} \Rightarrow (ea)^2 - a^2 = b^2$. So,

$$\frac{(x - h)^2}{a^2} + \frac{(y - k)^2}{b^2} = 1$$

$$\frac{(x - h)^2}{a^2} + \frac{(y - k)^2}{a^2(1 - e^2)} = 1.$$

(b) $\frac{(x - 2)^2}{4} + \frac{(y - 3)^2}{4(1 - e^2)} = 1$

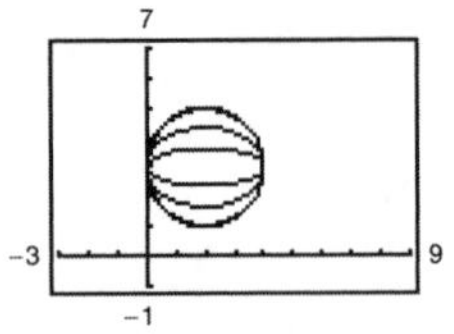

(c) As e approaches 0, the ellipse approaches a circle.

83. Area circle $= \pi r^2 = 100\pi$

Area ellipse $= \pi ab = \pi a(10)$

$2(100\pi) = 10\pi a \Rightarrow a = 20$

So, the length of the major axis is $2a = 40$.

84. (1) $\dfrac{x^2}{a^2} + \dfrac{y^2}{b^2} = 1$

$\dfrac{2x}{a^2} + \dfrac{2yy'}{b^2} = 0$

$y' = -\dfrac{xb^2}{ya^2}$

At P, $y' = -\dfrac{b^2}{a^2} \cdot \dfrac{x_0}{y_0} = m$.

(2) Slope of line through $(-c, 0)$ and (x_0, y_0): $m_1 = \dfrac{y_0}{x_0 + c}$

Slope of line through $(c, 0)$ and (x_0, y_0): $m_2 = \dfrac{y_0}{x_0 - c}$

(3) $\tan\alpha = \dfrac{m_2 - m}{1 + m_2 m} = \dfrac{\dfrac{y_0}{x_0 - c} - \left(-\dfrac{b^2x_0}{a^2y_0}\right)}{1 + \left(\dfrac{y_0}{x_0 - c}\right)\left(-\dfrac{b^2x_0}{a^2y_0}\right)} = \dfrac{a^2y_0^2 + b^2x_0(x_0 - c)}{a^2y_0(x_0 - c) - b^2x_0y_0}$

$= \dfrac{a^2y_0^2 + b^2x_0^2 - b^2x_0c}{x_0y_0(a^2 - b^2) - a^2y_0c} = \dfrac{a^2b^2 - b^2x_0c}{x_0y_0c^2 - a^2y_0c} = \dfrac{b^2(a^2 - x_0c)}{y_0c(x_0c - a^2)} = -\dfrac{b^2}{y_0c}$

$\alpha = \arctan\left(-\dfrac{b^2}{y_0c}\right) = -\arctan\left(\dfrac{b^2}{y_0c}\right)$

$\tan\beta = \dfrac{m_1 - m}{1 + m_1 m} = \dfrac{\dfrac{y_0}{x_0 + c} - \left(-\dfrac{b^2x_0}{a^2y_0}\right)}{1 + \left(\dfrac{y_0}{x_0 + c}\right)\left(-\dfrac{b^2x_0}{a^2y_0}\right)} = \dfrac{a^2y_0^2 + b^2x_0(x_0 + c)}{a^2y_0(x_0 + c) - b^2x_0y_0}$

$= \dfrac{a^2y_0^2 + b^2x_0^2 + b^2x_0c}{a^2x_0y_0 + a^2cy_0 - b^2x_0y_0} = \dfrac{a^2b^2 + b^2x_0c}{x_0y_0(a^2 - b^2) + a^2cy_0} = \dfrac{b^2(a^2 + x_0c)}{y_0c(x_0c + a^2)} = \dfrac{b^2}{y_0c}$

$\beta = \arctan\left(\dfrac{b^2}{y_0c}\right)$

Because $|\alpha| = |\beta|$, the tangent line to an ellipse at a point P makes equal angles with the line through P and the foci.

85. The transverse axis is horizontal since $(2, 2)$ and $(10, 2)$ are the foci (see definition of hyperbola).

Center: $(6, 2)$

$c = 4, 2a = 6, b^2 = c^2 - a^2 = 7$

So, the equation is $\dfrac{(x - 6)^2}{9} - \dfrac{(y - 2)^2}{7} = 1$.

86. Center: $(0, 0)$

Horizontal transverse axis

Foci: $(\pm c, 0)$

Vertices: $(\pm a, 0)$

The difference of the distances from any point on the hyperbola is constant. At a vertex, this constant difference is $(a + c) - (c - a) = 2a.$

Now, for any point (x, y) on the hyperbola, the difference of the distances between (x, y) and the two foci must also be $2a$.

$$\sqrt{(x-c)^2 + (y-0)^2} - \sqrt{(x+c)^2 + (y-0)^2} = 2a$$
$$\sqrt{(x-c)^2 + y^2} = 2a + \sqrt{(x+c)^2 + y^2}$$
$$(x-c)^2 + y^2 = 4a^2 + 4a\sqrt{(x+c)^2 + y^2} + (x+c)^2 + y^2$$
$$-4xc - 4a^2 = 4a\sqrt{(x+c)^2 + y^2}$$
$$-(xc + a^2) = a\sqrt{(x+c)^2 + y^2}$$
$$x^2c^2 + 2a^2cx + a^4 = a^2\left[x^2 + 2cx + c^2 + y^2\right]$$
$$x^2(c^2 - a^2) - a^2y^2 = a^2(c^2 - a^2)$$
$$\frac{x^2}{a^2} - \frac{y^2}{c^2 - a^2} = 1$$

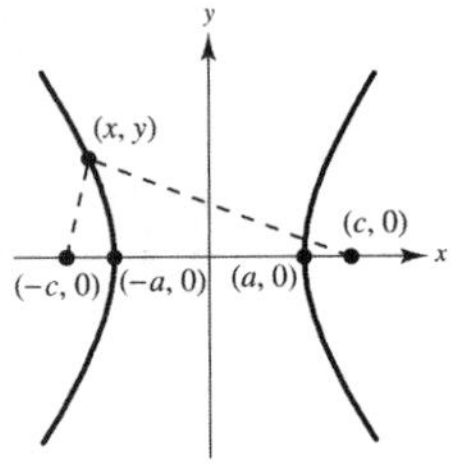

Because $a^2 + b^2 = c^2$, we have $\left(\dfrac{x^2}{a^2}\right) - \left(\dfrac{y^2}{b^2}\right) = 1.$

87. $c = 150, 2a = 0.001(186{,}000), a = 93,$

$b = \sqrt{150^2 - 93^2} = \sqrt{13{,}851}$

$$\frac{x^2}{93^2} - \frac{y^2}{13{,}851} = 1$$

When $y = 75$, you have

$$x^2 = 93^2\left(1 + \frac{75^2}{13{,}851}\right)$$

$x \approx 110.3$ mi.

88. The point (x, y) lies on the line between $(0, 10)$ and $(10, 0)$. So, $y = 10 - x$. The point also lies on the hyperbola $(x^2/36) - (y^2/64) = 1$. Using substitution, you have:

$$\frac{x^2}{36} - \frac{(10-x)^2}{64} = 1$$
$$16x^2 - 9(10-x)^2 = 576$$
$$7x^2 + 180x - 1476 = 0$$
$$x = \frac{-180 \pm \sqrt{180^2 - 4(7)(-1476)}}{2(7)}$$
$$= \frac{-180 \pm 192\sqrt{2}}{14} = \frac{-90 \pm 96\sqrt{2}}{7}$$

Choosing the positive value for x we have:

$$x = \frac{-90 + 96\sqrt{2}}{7} \approx 6.538 \text{ and}$$
$$y = \frac{160 - 96\sqrt{2}}{7} \approx 3.462$$

89. $$\frac{x^2}{a^2} - \frac{y^2}{b^2} = 1$$

$$\frac{2x}{a^2} - \frac{2yy'}{b^2} = 0 \text{ or } y' = \frac{b^2x}{a^2y}$$

$$y - y_0 = \frac{b^2x_0}{a^2y_0}(x - x_0)$$

$$a^2y_0y - a^2y_0^2 = b^2x_0x - b^2x_0^2$$

$$b^2x_0^2 - a^2y_0^2 = b^2x_0x - a^2y_0y$$

$$a^2b^2 = b^2x_0x - a^2y_0y$$

$$\frac{x_0x}{a^2} - \frac{y_0y}{b^2} = 1$$

90. $$Ax^2 + Cy^2 + Dx + Ey + F = 0 \quad \text{(Assume } A \neq 0 \text{ and } C \neq 0\text{; see (b) below)}$$

$$A\left(x^2 + \frac{D}{A}x\right) + C\left(y^2 + \frac{E}{C}y\right) = -F$$

$$A\left(x^2 + \frac{D}{A}x + \frac{D^2}{4A^2}\right) + C\left(y^2 + \frac{E}{C}y + \frac{E^2}{4C^2}\right) = -F + \frac{D^2}{4A} + \frac{E^2}{4C} = R$$

$$\frac{\left[x + \left(\frac{D}{2A}\right)\right]^2}{C} + \frac{\left[y + \left(\frac{E}{2C}\right)\right]^2}{A} = \frac{R}{AC}$$

(a) If $A = C$, you have

$$\left(x + \frac{D}{2A}\right)^2 + \left(y + \frac{E}{2C}\right)^2 = \frac{R}{A}$$

which is the standard equation of a circle.

(b) If $C = 0$, you have

$$A\left(x + \frac{D}{2A}\right)^2 = -F - Ey + \frac{D^2}{4A}.$$

If $A = 0$, you have

$$C\left(y + \frac{E}{2C}\right)^2 = -F - Dx + \frac{E^2}{4C}.$$

These are the equations of parabolas.

(c) If $AC < 0$, you have

$$\frac{\left[x + \left(\frac{D}{2A}\right)\right]^2}{\left|\frac{R}{A}\right|} + \frac{\left[y + \left(\frac{E}{2C}\right)\right]^2}{\left|\frac{R}{C}\right|} = 1$$

which is the equation of an ellipse.

(d) If $AC < 0$, you have

$$\frac{\left[x + \left(\frac{D}{2A}\right)\right]^2}{\left|\frac{R}{A}\right|} - \frac{\left[y + \left(\frac{E}{2C}\right)\right]^2}{\left|\frac{R}{C}\right|} = \pm 1$$

which is the equation of a hyperbola.

91. False. The parabola is equidistant from the directrix and focus and therefore cannot intersect the directrix.

92. True

93. True

94. False. $y^2 - x^2 + 2x + 2y = 0$ yields two intersecting lines: $y + 1 = \pm(x - 1)$

95. True

96. True

97. Let $\frac{x^2}{a^2} + \frac{y^2}{b^2} = 1$ be the equation of the ellipse with $a > b > 0$. Let $(\pm c, 0)$ be the foci, $c^2 = a^2 - b^2$. Let (u, v) be a point on the tangent line at $P(x, y)$, as indicated in the figure.

$$x^2b^2 + y^2a^2 = a^2b^2$$
$$2xb^2 + 2yy'a^2 = 0$$
$$y' = -\frac{b^2x}{a^2y} \quad \text{Slope at } P(x, y)$$

Now, $\frac{y - v}{x - u} = -\frac{b^2x}{a^2y}$

$$y^2a^2 - a^2vy = -b^2x^2 + b^2xu$$
$$y^2a^2 + x^2b^2 = a^2vy + b^2ux$$
$$a^2b^2 = a^2vy + b^2ux$$

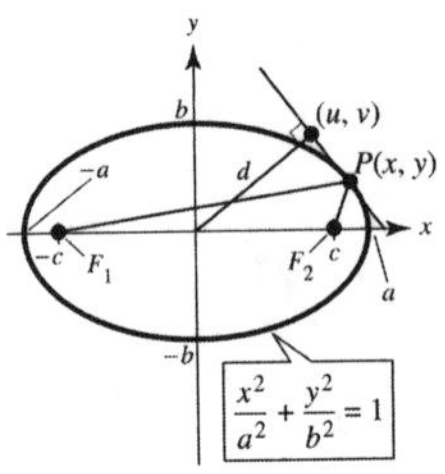

Because there is a right angle at (u, v),

$$\frac{v}{u} = \frac{a^2y}{b^2x}$$
$$vb^2x = a^2uy.$$

You have two equations:

$$a^2vy + b^2ux = a^2b^2$$
$$a^2uy - b^2vx = 0.$$

Multiplying the first by v and the second by u, and adding,

$$a^2v^2y + a^2u^2y = a^2b^2v$$
$$y\left[u^2 + v^2\right] = b^2v$$
$$yd^2 = b^2v$$
$$v = \frac{yd^2}{b^2}.$$

Similarly, $u = \frac{xd^2}{a^2}$.

From the figure, $u = d\cos\theta$ and $v = d\sin\theta$. So, $\cos\theta = \frac{xd}{a^2}$ and $\sin\theta = \frac{yd}{b^2}$.

$$\cos^2\theta + \sin^2\theta = \frac{x^2d^2}{a^4} + \frac{y^2d^2}{b^4} = 1$$
$$x^2b^4d^2 + y^2a^4d^2 = a^4b^4$$
$$d^2 = \frac{a^4b^4}{x^2b^4 + y^2a^4}$$

Let $r_1 = PF_1$ and $r_2 = PF_2$, $r_1 + r_2 = 2a$.

$$r_1r_2 = \frac{1}{2}\left[(r_1 + r_2)^2 - r_1^2 - r_2^2\right] = \frac{1}{2}\left[4a^2 - (x + c)^2 - y^2 - (x - c)^2 - y^2\right] = 2a^2 - x^2 - y^2 - c^2 = a^2 + b^2 - x^2 - y^2$$

Finally, $d^2r_1r_2 = \frac{a^4b^4}{x^2b^4 + y^2a^4} \cdot \left[a^2 + b^2 - x^2 - y^2\right]$

$$= \frac{a^4b^4}{b^2\left(b^2x^2\right) + a^2\left(a^2y^2\right)} \cdot \left[a^2 + b^2 - x^2 - y^2\right]$$
$$= \frac{a^4b^4}{b^2\left(a^2b^2 - a^2y^2\right) + a^2\left(a^2b^2 - b^2x^2\right)} \cdot \left[a^2 + b^2 - x^2 - y^2\right]$$
$$= \frac{a^4b^4}{a^2b^2\left[a^2 + b^2 - x^2 - y^2\right]} \cdot \left[a^2 + b^2 - x^2 - y^2\right] = a^2b^2, \text{ a constant!}$$

98. Consider circle $x^2 + y^2 = 2$ and hyperbola $y = \frac{9}{x}$.

Let $\left(u, \sqrt{2 - u^2}\right)$ and $\left(v, \frac{9}{v}\right)$ be points on the circle and hyperbola, respectively. We need to minimize the distance between these 2 points:

$$(\text{Distance})^2 = f(u, v) = (u - v)^2 + \left(\sqrt{2 + u^2} - \frac{9}{v}\right)^2.$$

The tangent lines at $(1, 1)$ and $(3, 3)$ are both perpendicular to $y = x$, and so they are parallel.

The minimum value is $(3 - 1)^2 + (3 - 1)^2 = 8$.

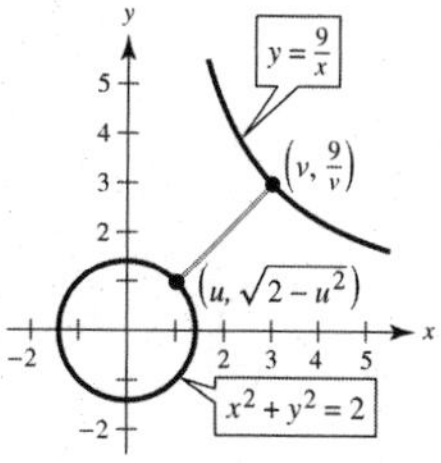

Section 10.2 Plane Curves and Parametric Equations

1. $x = 2t - 3$

$y = 3t + 1$

$t = \frac{x + 3}{2}$

$y = 3\left(\frac{x + 3}{2}\right) + 1 = \frac{3}{2}x + \frac{11}{2}$

$3x - 2y + 11 = 0$

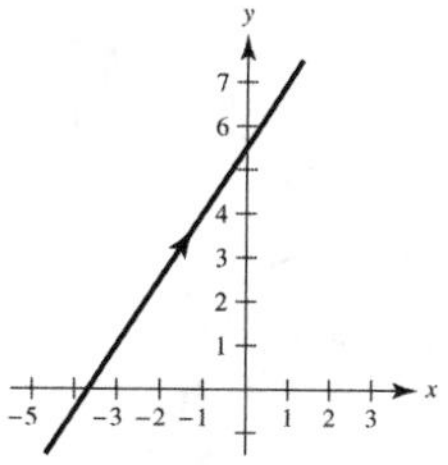

2. $x = 5 - 4t$

$y = 2 + 5t$

$t = \frac{5 - x}{4}$

$y = 2 + 5\left(\frac{5 - x}{4}\right) = -\frac{5}{4}x + \frac{33}{4}$

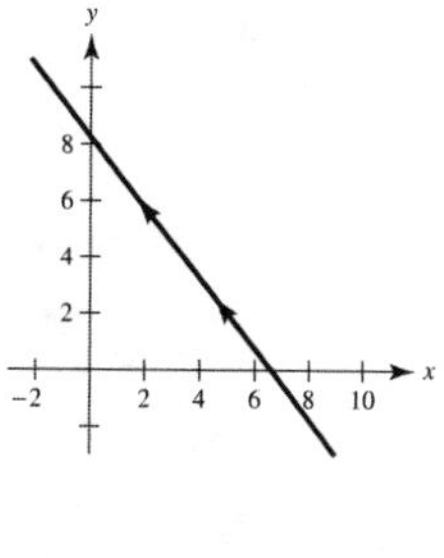

3. $x = t + 1$

$y = t^2$

$y = (x - 1)^2$

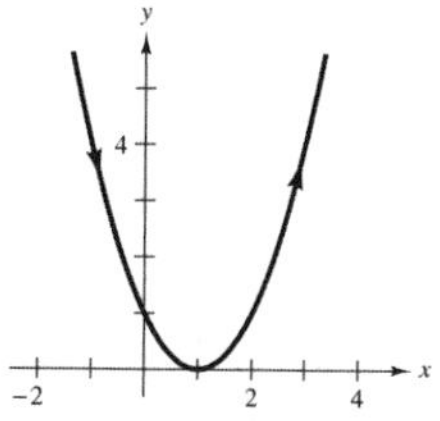

4. $x = 2t^2$

$y = t^4 + 1$

$y = \left(\frac{x}{2}\right)^2 + 1 = \frac{x^2}{4} + 1, x \geq 0$

For $t < 0$, the orientation is right to left.

For $t > 0$, the orientation is left to right.

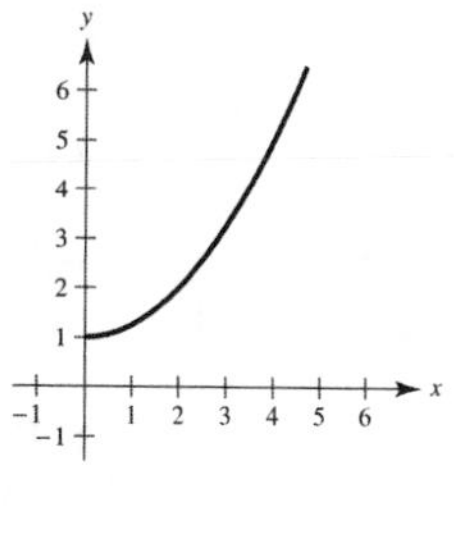

5. $x = t^3$

$y = \frac{1}{2}t^2$

$y = t^3$ implies $t = x^{1/3}$

$y = \frac{1}{2}x^{2/3}$

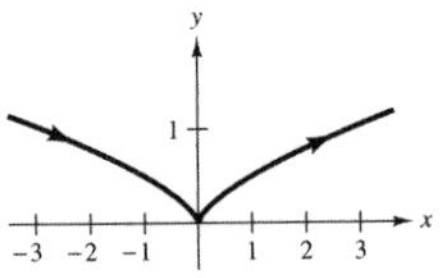

6. $x = t^2 + t,\ y = t^2 - t$

Subtracting the second equation from the first, you have

$x - y = 2t \quad \text{or} \quad t = \frac{x - y}{2}.$

$y = \frac{(x - y)^2}{4} - \frac{x - y}{2}$

t	-2	-1	0	1	2
x	2	0	0	2	6
y	6	2	0	0	2

Because the discriminant is

$B^2 - 4AC = (-2)^2 - 4(1)(1) = 0,$

the graph is a rotated parabola.

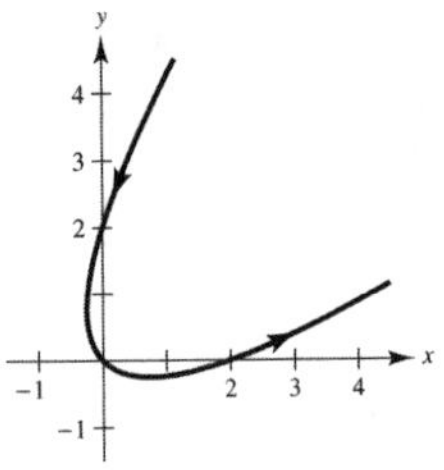

7. $x = \sqrt{t}$

$y = t - 5$

$x^2 = t$

$y = x^2 - 5,\ x \geq 0$

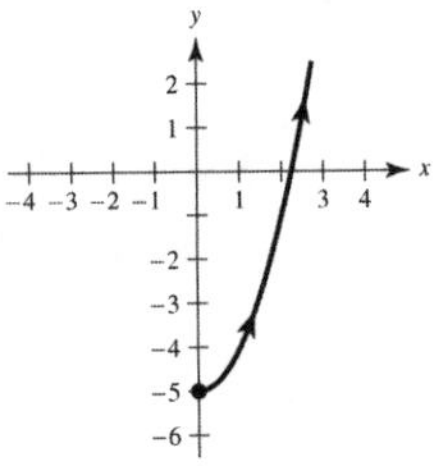

8. $x = \sqrt[4]{t}$

$y = 8 - t$

$x^4 = t$

$y = 8 - x^4,\ x \geq 0$

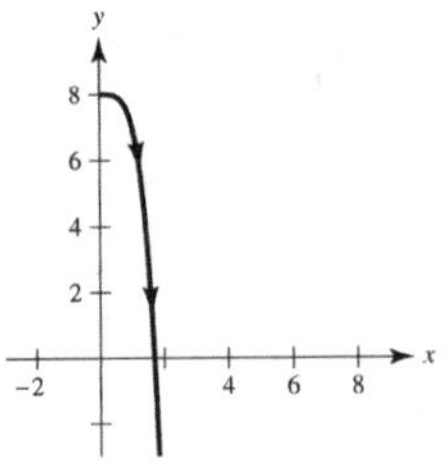

9. $x = t - 3$

$y = \frac{t}{t - 3}$

$t = x + 3$

$y = \frac{x + 3}{(x + 3) - 3} = 1 + \frac{3}{x} = \frac{x + 3}{x}$

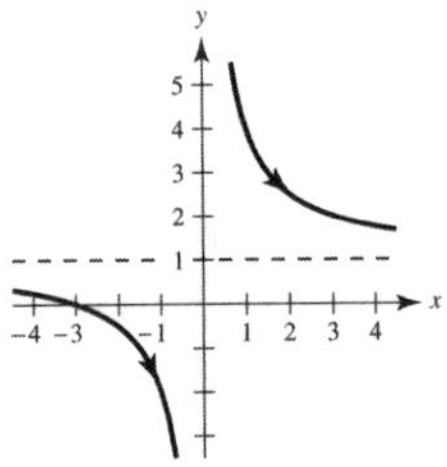

10. $x = 1 + \frac{1}{t}$

$y = t - 1$

$x = 1 + \frac{1}{t}$ implies $t = \frac{1}{x - 1}$

$y = \frac{1}{x - 1} - 1$

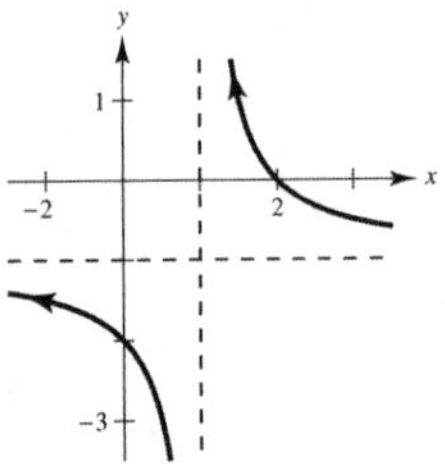

11. $x = 2t$

$y = |t - 2|$

$y = \left|\frac{x}{2} - 2\right| = \frac{|x - 4|}{2}$

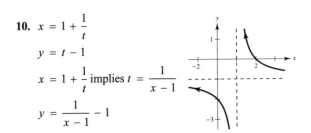

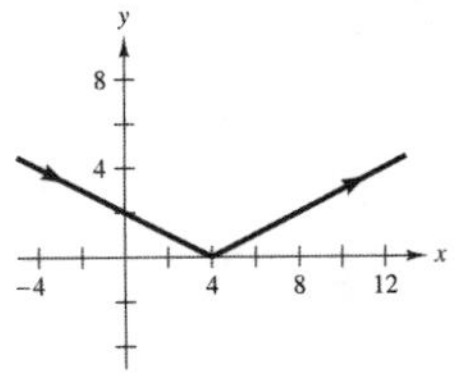

12. $x = |t - 1|$

$y = t + 2$

$x = |(y - 2) - 1| = |y - 3|$

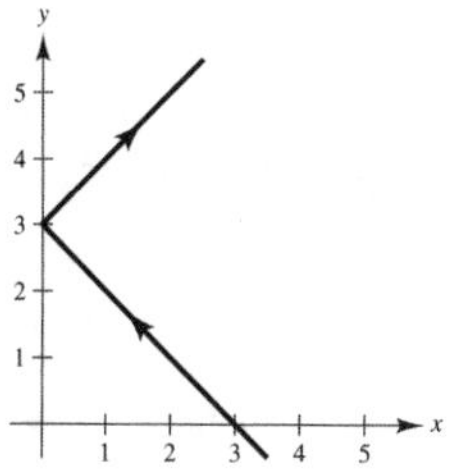

13. $x = e^t, x > 0$

$y = e^{3t} + 1$

$y = x^3 + 1, x > 0$

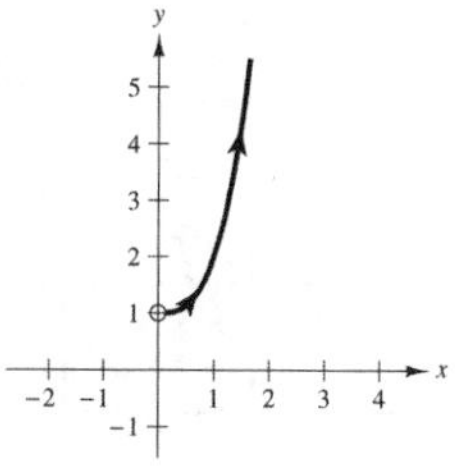

14. $x = e^{-t}, x > 0$

$y = e^{2t} - 1$

$y = x^{-2} - 1 = \dfrac{1}{x^2} - 1, x > 0$

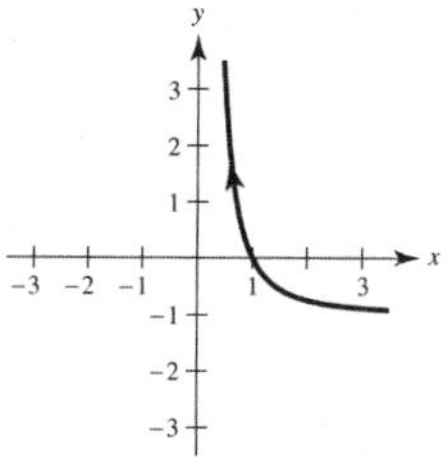

15. $x = \sec\theta$

$y = \cos\theta$

$0 \le \theta < \dfrac{\pi}{2}, \dfrac{\pi}{2} < \theta \le \pi$

$xy = 1$

$y = \dfrac{1}{x}$

$|x| \ge 1, |y| \le 1$

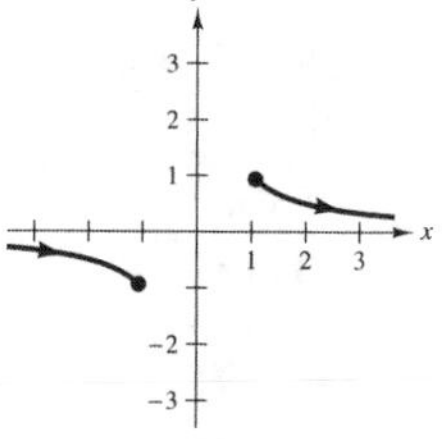

16. $x = \tan^2\theta$

$y = \sec^2\theta$

$\sec^2\theta = \tan^2\theta + 1$

$y = x + 1$

$x \ge 0$

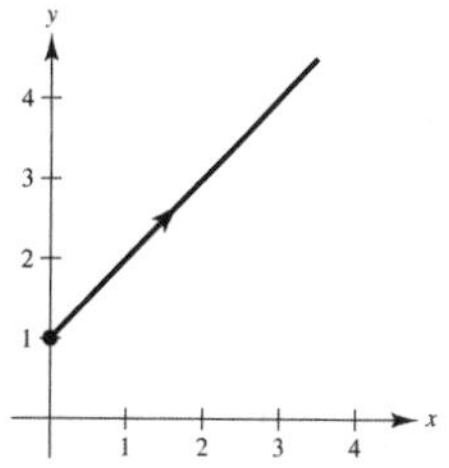

17. $x = 8\cos\theta$

$y = 8\sin\theta$

$x^2 + y^2 = 64\cos^2\theta + 64\sin^2\theta = 64(1) = 64$

Circle

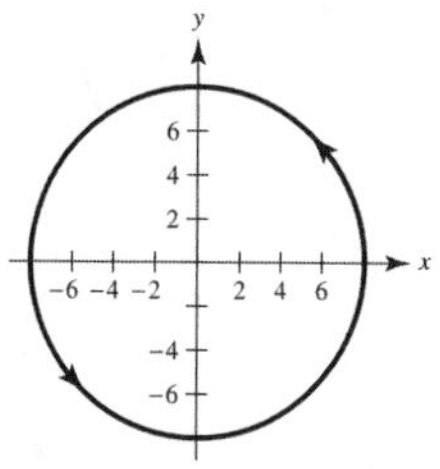

18. $x = 3\cos\theta$

$y = 7\sin\theta$

$\left(\dfrac{x}{3}\right)^2 + \left(\dfrac{y}{7}\right)^2 = \cos^2\theta + \sin^2\theta = 1$

$\dfrac{x^2}{9} + \dfrac{y^2}{49} = 1$

Ellipse

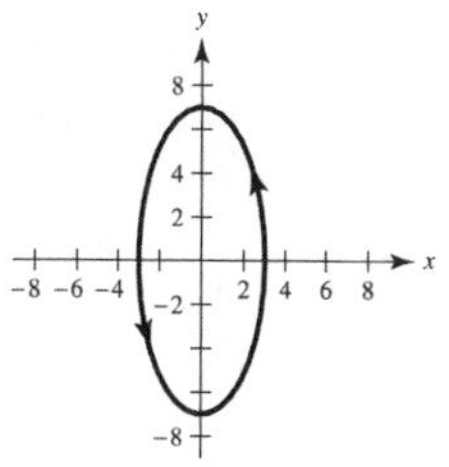

19. $x = 6 \sin 2\theta$

$y = 4 \cos 2\theta$

$\left(\dfrac{x}{6}\right)^2 + \left(\dfrac{y}{4}\right)^2 = \sin^2 2\theta + \cos^2 2\theta = 1$

$\dfrac{x^2}{36} + \dfrac{y^2}{16} = 1$

Ellipse

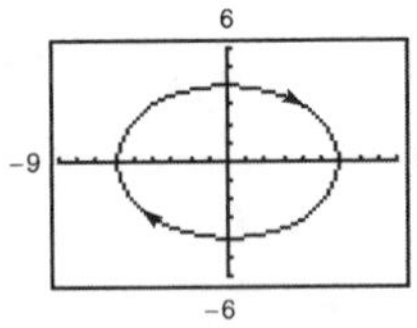

20. $x = \cos \theta$

$y = 2 \sin 2\theta$

$y = 4 \sin \theta \cos \theta$

$1 - x^2 = \sin^2 \theta$

$y = \pm 4x\sqrt{1 - x^2}$

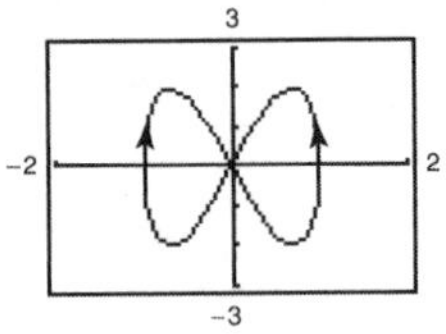

21. $x = 4 + 2 \cos \theta$

$y = -1 + \sin \theta$

$\dfrac{(x - 4)^2}{4} = \cos^2 \theta$

$\dfrac{(y + 1)^2}{1} = \sin^2 \theta$

$\dfrac{(x - 4)^2}{4} + \dfrac{(y + 1)^2}{1} = 1$

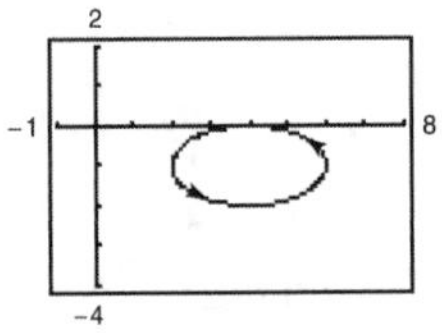

22. $x = -2 + 3 \cos \theta$

$y = -5 + 3 \sin \theta$

$(x + 2)^2 + (y + 5)^2 = 9 \cos^2 \theta + 9 \sin^2 \theta = 9$

$(x + 2)^2 + (y + 5)^2 = 9$

Circle

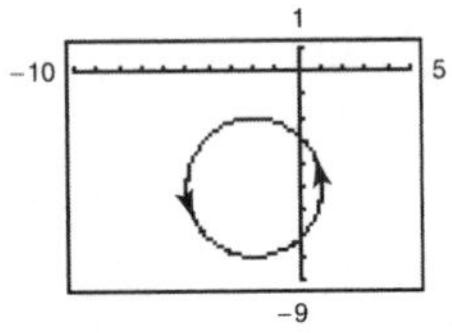

23. $x = -3 + 4 \cos \theta$

$y = 2 + 5 \sin \theta$

$x + 3 = 4 \cos \theta$

$y - 2 = 5 \sin \theta$

$\left(\dfrac{x + 3}{4}\right)^2 + \left(\dfrac{y - 2}{5}\right)^2 = \cos^2 \theta + \sin^2 \theta = 1$

$\dfrac{(x + 3)^2}{16} + \dfrac{(y - 2)^2}{25} = 1$

Ellipse

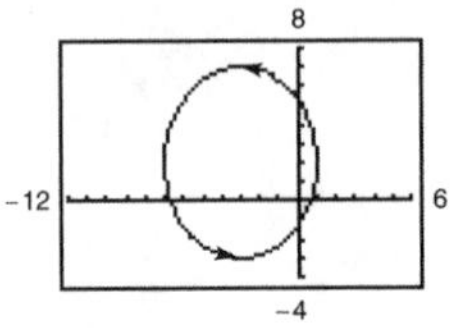

24. $x = \sec \theta$

$y = \tan \theta$

$x^2 = \sec^2 \theta$

$y^2 = \tan^2 \theta$

$x^2 - y^2 = 1$

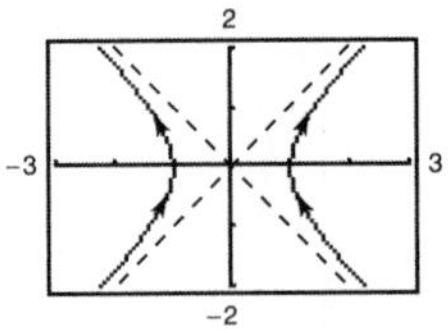

25. $x = 4 \sec \theta$

$y = 3 \tan \theta$

$\dfrac{x^2}{16} = \sec^2 \theta$

$\dfrac{y^2}{9} = \tan^2 \theta$

$\dfrac{x^2}{16} - \dfrac{y^2}{9} = 1$

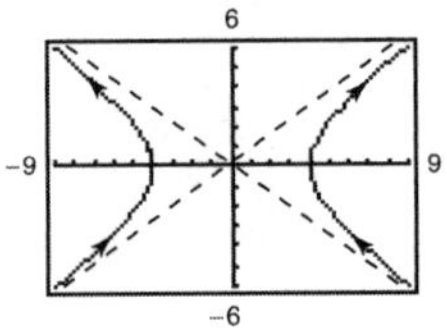

26. $x = \cos^3 \theta$

$y = \sin^3 \theta$

$x^{2/3} = \cos^2 \theta$

$y^{2/3} = \sin^2 \theta$

$x^{2/3} + y^{2/3} = 1$

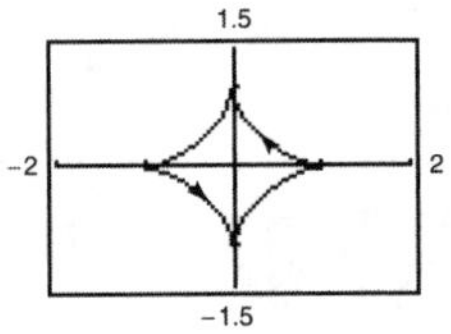

27. $x = t^3$

$y = 3 \ln t$

$y = 3 \ln \sqrt[3]{x} = \ln x$

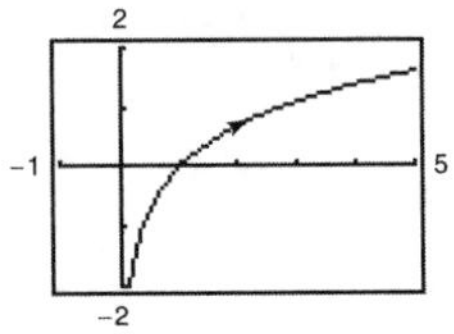

28. $x = \ln 2t$

$y = t^2$

$t = \dfrac{e^x}{2}$

$y = \dfrac{e^{2x}}{r} = \dfrac{1}{4}e^{2x}$

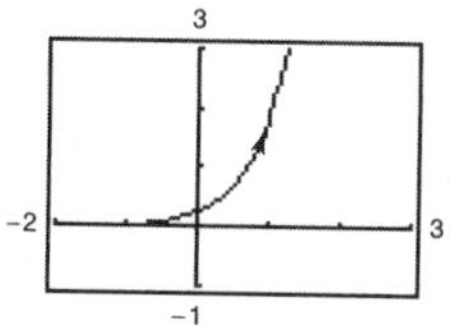

29. $x = e^{-t}$

$y = e^{3t}$

$e^t = \dfrac{1}{x}$

$e^t = \sqrt[3]{y}$

$\sqrt[3]{y} = \dfrac{1}{x}$

$y = \dfrac{1}{x^3}$

$x > 0$

$y > 0$

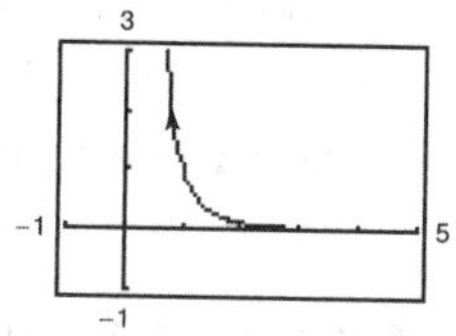

30. $x = e^{2t}$

$y = e^t$

$y^2 = x$

$y > 0$

$y = \sqrt{x}, x > 0$

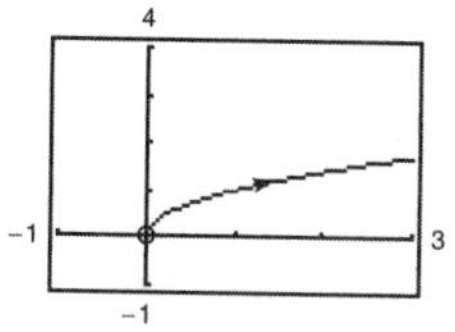

31. By eliminating the parameters in (a) – (d), you get $y = 2x + 1$. They differ from each other in orientation and in restricted domains. These curves are all smooth except for (b).

(a) $x = t, y = 2t + 1$

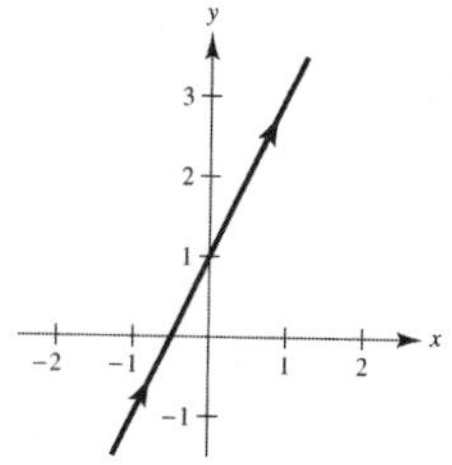

(b) $x = \cos\theta$ $\quad y = 2\cos\theta + 1$

$-1 \le x \le 1$ $\quad -1 \le y \le 3$

$\dfrac{dx}{d\theta} = \dfrac{dy}{d\theta} = 0$ when $\theta = 0, \pm\pi, \pm 2\pi, \ldots$.

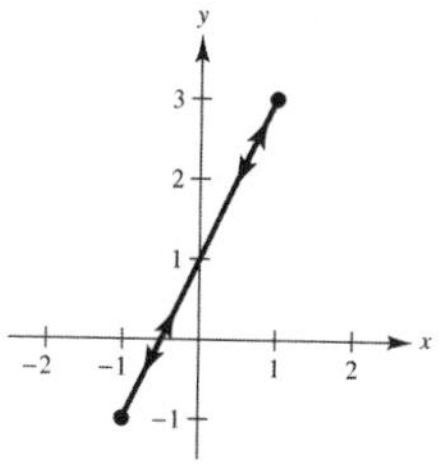

(c) $x = e^{-t}$ $\quad y = 2e^{-t} + 1$

$x > 0$ $\quad y > 1$

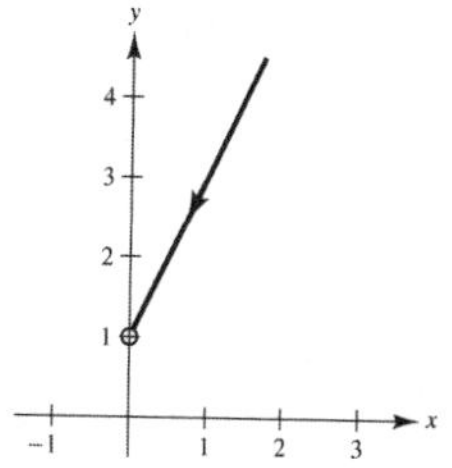

(d) $x = e^t$ $\quad y = 2e^t + 1$

$x > 0$ $\quad y > 1$

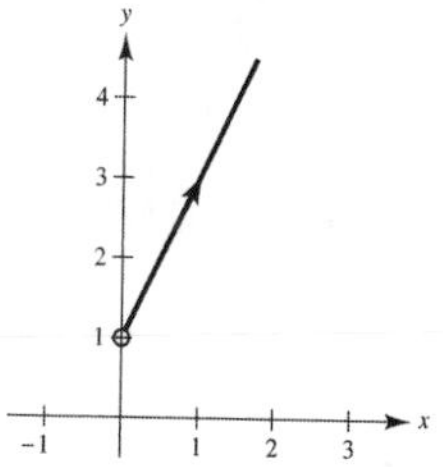

32. By eliminating the parameters in (a) – (d), you get $x^2 + y^2 = 4$. They differ from each other in orientation and in restricted domains. These curves are all smooth.

(a) $x = 2\cos\theta,\ y = 2\sin\theta$

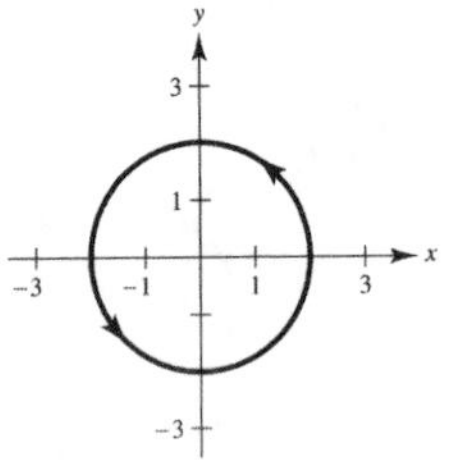

(b) $x = \dfrac{\sqrt{4t^2 - 1}}{|t|} = \sqrt{4 - \dfrac{1}{t^2}}$ $\quad y = \dfrac{1}{t}$

$x \ge 0,\ x \ne 2$ $\quad y \ne 0$

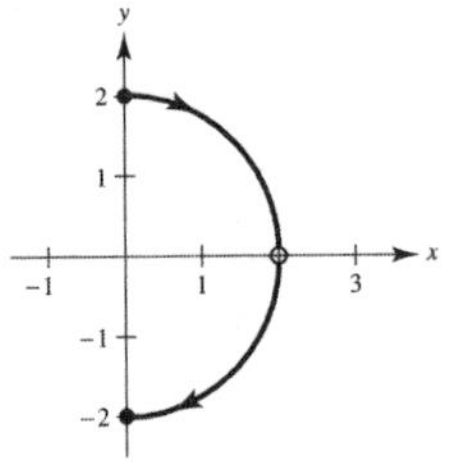

(c) $x = \sqrt{t}$ $\quad y = \sqrt{4 - t}$

$x \ge 0$ $\quad y \ge 0$

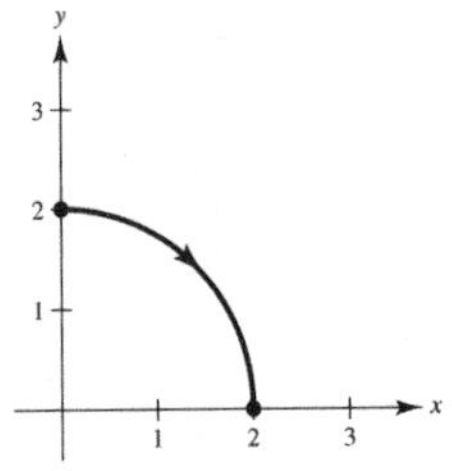

(d) $x = -\sqrt{4 - e^{2t}}$ $\quad y = e^t$

$-2 < x \le 0$ $\quad y > 0$

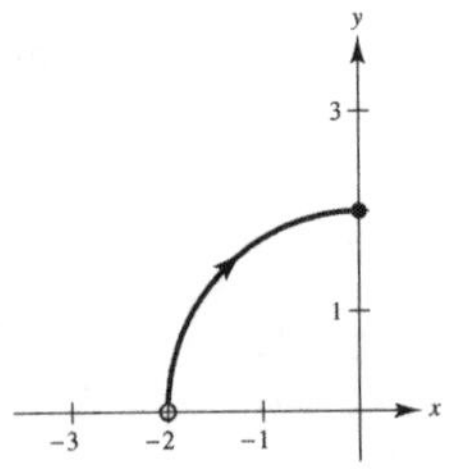

33. The curves are identical on $0 < \theta < \pi$. They are both smooth. They represent $y = 2(1 - x^2)$ for $-1 \le x \le 1$. The orientation is from right to left in part (a) and in part (b).

34. The orientations are reversed. The graphs are the same. They are both smooth.

35. (a)

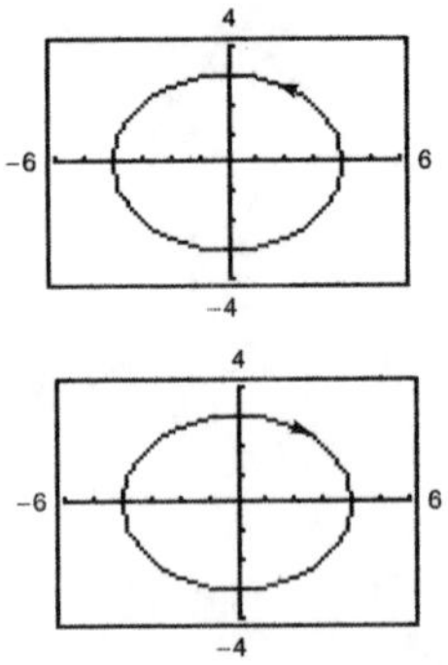

(b) The orientation of the second curve is reversed.

(c) The orientation will be reversed.

(d) Answers will vary. For example,

$x = 2\sec t \qquad x = 2\sec(-t)$

$y = 5\sin t \qquad y = 5\sin(-t)$

have the same graphs, but their orientations are reversed.

36. The set of points (x, y) corresponding to the rectangular equation of a set of parametric equations does not show the orientation of the curve nor any restriction on the domain of the original parametric equations.

37.

$$x = x_1 + t(x_2 - x_1)$$
$$y = y_1 + t(y_2 - y_1)$$
$$\frac{x - x_1}{x_2 - x_1} = t$$
$$y = y_1 + \left(\frac{x - x_1}{x_2 - x_1}\right)(y_2 - y_1)$$
$$y - y_1 = \frac{y_2 - y_1}{x_2 - x_1}(x - x_1)$$
$$y - y_1 = m(x - x_1)$$

38.

$$x = h + r\cos\theta$$
$$y = k + r\sin\theta$$
$$\cos\theta = \frac{x - h}{r}$$
$$\sin\theta = \frac{y - k}{r}$$
$$\cos^2\theta + \sin^2\theta = \frac{(x - h)^2}{r^2} + \frac{(y - k)^2}{r^2} = 1$$
$$(x - h)^2 + (y - k)^2 = r^2$$

39.
$$x = h + a\cos\theta$$
$$y = k + b\sin\theta$$
$$\frac{x-h}{a} = \cos\theta$$
$$\frac{y-k}{b} = \sin\theta$$
$$\frac{(x-h)^2}{a^2} + \frac{(y-k)^2}{b^2} = 1$$

40.
$$x = h + a\sec\theta$$
$$y = k + b\tan\theta$$
$$\frac{x-h}{a} = \sec\theta$$
$$\frac{y-k}{b} = \tan\theta$$
$$\frac{(x-h)^2}{a^2} - \frac{(y-k)^2}{b^2} = 1$$

41. From Exercise 37 you have

$x = 4t$

$y = -7t$

Solution not unique

42. From Exercise 37 you have

$x = 1 + 4t$

$y = 4 - 6t.$

Solution not unique

43. From Exercise 38 you have

$x = 3 + 2\cos\theta$

$y = 1 + 2\sin\theta$

Solution not unique

44. From Exercise 38 you have

$x = -6 + 4\cos\theta$

$y = 2 + 4\sin\theta$

45. From Exercise 39 you have

$a = 10, c = 8 \Rightarrow b = 6$

$x = 10\cos\theta$

$y = 6\sin\theta$

Center: $(0, 0)$

Solution not unique

46. From Exercise 39 you have

$a = 5, c = 3 \Rightarrow b = 4$

$x = 4 + 5\cos$

$y = 2 + 4\sin\theta.$

Center: $(4, 2)$

Solution not unique

47. From Exercise 40 you have

$a = 4, c = 5 \Rightarrow b = 3$

$x = 4\sec\theta$

$y = 3\tan\theta.$

Center: $(0, 0)$

Solution not unique

48. From Exercise 40 you have

$a = 1, c = 2 \Rightarrow b = \sqrt{3}$

$x = \sqrt{3}\tan\theta$

$y = \sec\theta.$

Center: $(0, 0)$

Solution not unique

The transverse axis is vertical, so, x and y are interchanged.

49. $y = 6x - 5$

Examples:

$x = t, y = 6t - 5$

$x = t + 1, y = 6t + 1$

50. $y = \dfrac{4}{x-1}$

Examples:

$x = t, y = \dfrac{4}{t-1}$

$x = t + 1, y = \dfrac{4}{t}$

51. $y = x^3$

Example

$x = t,\quad y = t^3$

$x = \sqrt[3]{t},\quad y = t$

$x = \tan t,\quad y = \tan^3 t$

52. $y = x^2$

Example

$x = t,\quad y = t^2$

$x = t^3,\quad y = t^6$

53. $y = 2x - 5$

At $(3, 1)$, $t = 0$: $x = 3 - t$

$y = 2(3 - t) - 5 = -2t + 1$

or, $x = t + 3$

$y = 2t + 1$

54. $y = 4x + 1$

At $(-2, -7), t = -1$: $\quad x = -1 + t$

$y = 4(-1 + t) + 1 = 4t - 3$

55. $y = x^2$

$t = 4$ at $(4, 16)$: $\quad x = t$

$y = t^2$

56. $y = 4 - x^2$

$t = 1$ at $(1, 3)$: $\quad x = t$

$y = 4 - t^2$

57. $x = 2(\theta - \sin\theta)$

$y = 2(1 - \cos\theta)$

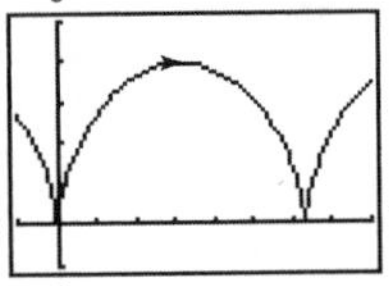

Not smooth at $\theta = 2n\pi$

58. $x = \theta + \sin\theta$

$y = 1 - \cos\theta$

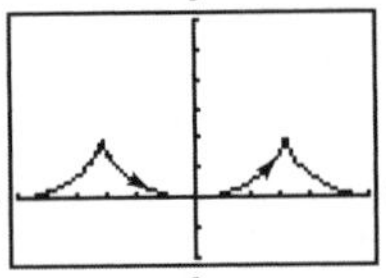

Not smooth at $x = (2n - 1)\pi$

59. $x = \theta - \frac{3}{2}\sin\theta$

$y = 1 - \frac{3}{2}\cos\theta$

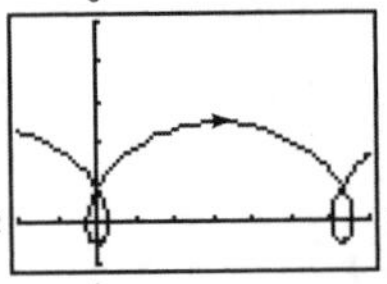

Smooth everywhere

60. $x = 2\theta - 4\sin\theta$

$y = 2 - 4\cos\theta$

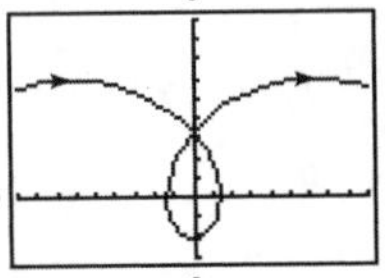

Smooth everywhere

61. $x = 3\cos^3\theta$

$y = 3\sin^3\theta$

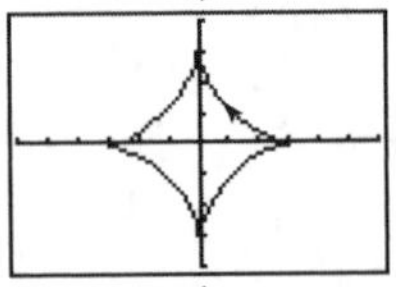

Not smooth at $(x, y) = (\pm 3, 0)$ and $(0, \pm 3)$, or $\theta = \frac{1}{2}n\pi$.

62. $x = 2\theta - \sin\theta$

$y = 2 - \cos\theta$

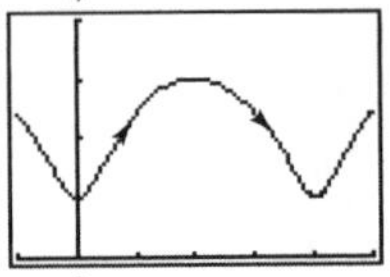

Smooth everywhere

63. $x = 2\cot\theta$

$y = 2\sin^2\theta$

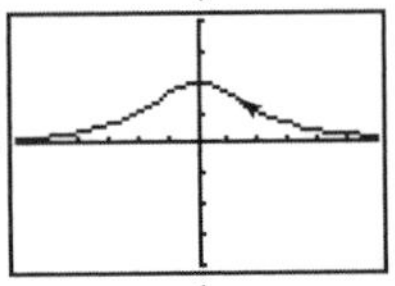

Smooth everywhere

64. $x = \dfrac{3t}{1 + t^3}$

$y = \dfrac{3t^2}{1 + t^3}$

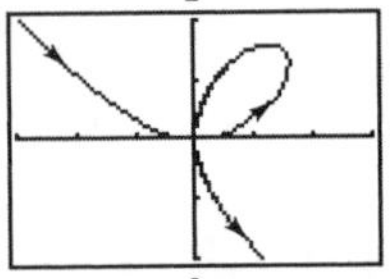

Smooth everywhere

65. If f and g are continuous functions of t on an interval I, then the equations $x = f(t)$ and $y = g(t)$ are called parametric equations and t is the parameter. The set of points (x, y) obtained as t varies over I is the graph. Taken together, the parametric equations and the graph are called a plane curve C.

66. Each point (x, y) in the plane is determined by the plane curve $x = f(t)$, $y = g(t)$. For each t, plot (x, y). As t increases, the curve is traced out in a specific direction called the orientation of the curve.

67. A curve C represented by $x = f(t)$ and $y = g(t)$ on an interval I is called smooth when f' and g' are continuous on I and not simultaneously 0, except possibly at the endpoints of I.

68. The graph matches (a) because $x = t \Rightarrow y = t^2 = x^2$. For (b), you have $y = t \Rightarrow x = t^2 = y^2$, which is not the correct parabola.

69. Matches (d) because $(4, 0)$ is on the graph.

70. Matches (a) because $(0, 2)$ is on the graph.

71. Matches (b) because $(1, 0)$ is on the graph.

72. Matches (c) because the graph is undefined when $\theta = 0$.

73. When the circle has rolled θ radians, you know that the center is at $(a\theta, a)$.

$$\sin\theta = \sin(180° - \theta) = \frac{|AC|}{b} = \frac{|BD|}{b} \text{ or } |BD| = b\sin\theta$$

$$\cos\theta = -\cos(180° - \theta) = \frac{|AP|}{-b} \text{ or } |AP| = -b\cos\theta$$

So, $x = a\theta - b\sin\theta$ and $y = a - b\cos\theta$.

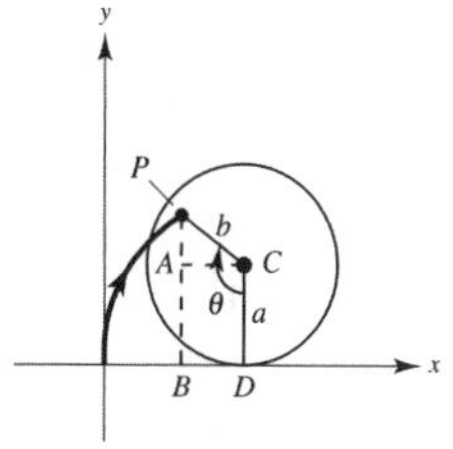

74. Let the circle of radius 1 be centered at C. A is the point of tangency on the line OC. $OA = 2$, $AC = 1$, $OC = 3$. $P = (x, y)$ is the point on the curve being traced out as the angle θ changes $\widehat{AB} = \widehat{AP}$, $\widehat{AB} = 2\theta$ and $\widehat{AP} = \alpha \Rightarrow \alpha = 2\theta$. Form the right triangle ΔCDP. The angle $OCE = (\pi/2) - \theta$ and

$$\angle DCP = \alpha - \left(\frac{\pi}{2} - \theta\right) = \alpha + \theta - \left(\frac{\pi}{2}\right) = 3\theta - \left(\frac{\pi}{2}\right).$$

$$x = OE + Ex = 3\sin\left(\frac{\pi}{2} - \theta\right) + \sin\left(3\theta - \frac{\pi}{2}\right) = 3\cos\theta - \cos 3\theta$$

$$y = EC - CD = 3\sin\theta - \cos\left(3\theta - \frac{\pi}{2}\right) = 3\sin\theta - \sin 3\theta$$

So, $x = 3\cos\theta - \cos 3\theta$, $y = 3\sin\theta - \sin 3\theta$.

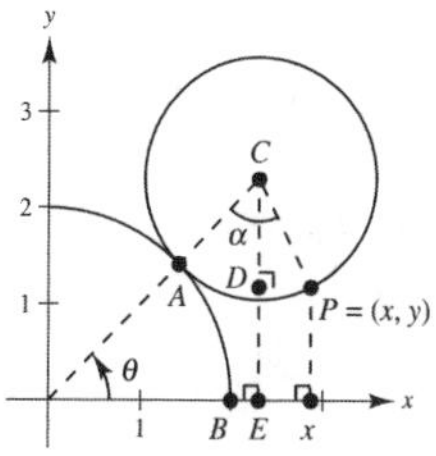

75. False

$x = t^2 \Rightarrow x \geq 0$

$y = t^2 \Rightarrow y \geq 0$

The graph of the parametric equations is only a portion of the line $y = x$ when $x \geq 0$.

76. False. Let $x = t^2$ and $y = t$. Then $x = y^2$ and y is not a function of x.

77. True. $y = \cos x$

78. $x = 8\cos t$, $y = 8\sin t$

(a) $\left(\dfrac{x}{8}\right)^2 + \left(\dfrac{y}{8}\right)^2 = \cos^2 t + \sin^2 t = 1$

$x^2 + y^2 = 64$ Circle radius 8,

Center: $(0, 0)$ Oriented counterclockwise

(b) Circle of radius 8, but Center: $(3, 6)$

(c) The orientation is reversed.

79. (a) $100 \text{ mi/hr} = \dfrac{(100)(5280)}{3600} = \dfrac{440}{3}$ ft/sec

$$x = (v_0\cos\theta)t = \left(\frac{440}{3}\cos\theta\right)t$$

$$y = h + (v_0\sin\theta)t - 16t^2 = 3 + \left(\frac{440}{3}\sin\theta\right)t - 16t^2$$

(b) 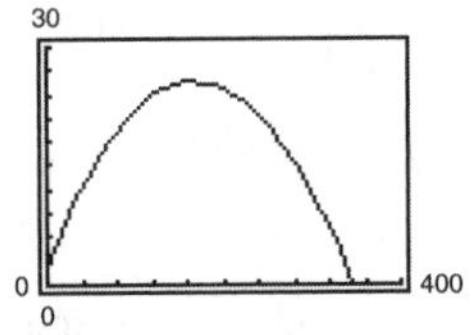

It is not a home run when $x = 400$, $y < 10$.

(c) 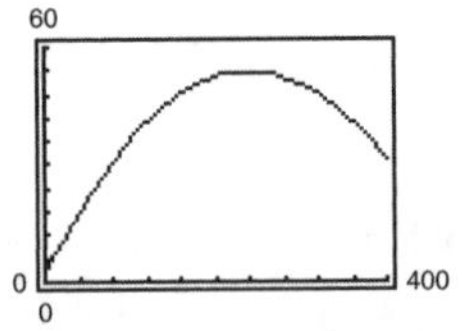

Yes, it's a home run when $x = 400$, $y > 10$.

(d) You need to find the angle θ (and time t) such that

$$x = \left(\frac{440}{3}\cos\theta\right)t = 400$$

$$y = 3 + \left(\frac{440}{3}\sin\theta\right)t - 16t^2 = 10.$$

From the first equation $t = 1200/440\cos\theta$. Substituting into the second equation,

$$10 = 3 + \left(\frac{440}{3}\sin\theta\right)\left(\frac{1200}{440\cos\theta}\right) - 16\left(\frac{1200}{440\cos\theta}\right)^2$$

$$7 = 400\tan\theta - 16\left(\frac{120}{44}\right)^2\sec^2\theta = 400\tan\theta - 16\left(\frac{120}{44}\right)^2\left(\tan^2\theta + 1\right).$$

You now solve the quadratic for $\tan\theta$:

$$16\left(\frac{120}{44}\right)^2\tan^2\theta - 400\tan\theta + 7 + 16\left(\frac{120}{44}\right)^2 = 0.$$

$\tan\theta \approx 0.35185 \Rightarrow \theta \approx 19.4°$

80. (a) $x = \left(v_0\cos\theta\right)t$

$y = h + \left(v_0\sin\theta\right)t - 16t^2$

$$t = \frac{x}{v_0\cos\theta} \Rightarrow y = h + \left(v_0\sin\theta\right)\frac{x}{v_0\cos\theta} - 16\left(\frac{x}{v_0\cos\theta}\right)^2$$

$$y = h + (\tan\theta)x - \frac{16\sec^2\theta}{v_0^2}x^2$$

(b) $y = 5 + x - 0.005x^2 = h + (\tan\theta)x - \dfrac{16\sec^2\theta}{v_0^2}x^2$

$h = 5$, $\tan\theta = 1 \Rightarrow \theta = \dfrac{\pi}{4}$, and

$$0.005 = \frac{16\sec^2(\pi/4)}{v_0^2} = \frac{16}{v_0^2}(2)$$

$$v_0^2 = \frac{32}{0.005} = 6400 \Rightarrow v_0 = 80.$$

So, $x = \left(80\cos(45°)\right)t$

$y = 5 + \left(80\sin(45°)\right)t - 16t^2.$

(c)

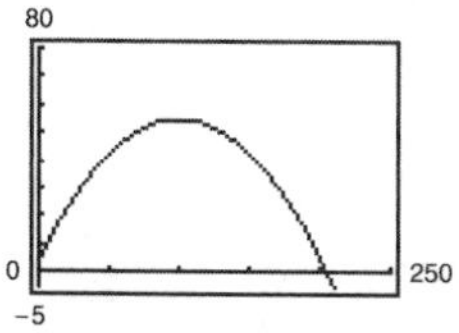

(d) Maximum height: $y = 55$ (at $x = 100$)

Range: 204.88

Section 10.3 Parametric Equations and Calculus

1. $\frac{dy}{dx} = \frac{dy/dt}{dx/dt} = \frac{-6}{2t} = -\frac{3}{t}$

2. $\frac{dy}{dx} = \frac{dy/dt}{dx/dt} = \frac{-1}{(1/3)t^{-2/3}} = -3t^{2/3}$

3. $\frac{dy}{dx} = \frac{dy/d\theta}{dx/d\theta} = \frac{-2\cos\theta\sin\theta}{2\sin\theta\cos\theta} = -1$

$\left[\text{Note: } x + y = 1 \Rightarrow y = 1 - x \text{ and } \frac{dy}{d\theta} = -1\right]$

4. $\frac{dy}{dx} = \frac{dy/d\theta}{dx/d\theta} = \frac{(-1/2)e^{-\theta/2}}{2e^{\theta}} = -\frac{1}{4}e^{-3\theta/2} = \frac{-1}{4e^{3\theta/2}}$

5. $x = 4t,\ y = 3t - 2$

$\frac{dy}{dx} = \frac{dy/dt}{dx/dt} = \frac{3}{4}$

$\frac{d^2y}{dx^2} = 0$

At $t = 3$, slope is $\frac{3}{4}$. (Line)

Neither concave upward nor downward

6. $x = \sqrt{t},\ y = 3t - 1$

$\frac{dy}{dx} = \frac{3}{1/(2\sqrt{t})} = 6\sqrt{t} = 6$ when $t = 1$.

$\frac{d^2y}{dx^2} = \frac{3/\sqrt{t}}{1/(2\sqrt{t})} = 6$

Concave upward

7. $x = t + 1,\ y = t^2 + 3t$

$\frac{dy}{dx} = \frac{2t + 3}{1} = 1$ when $t = -1$.

$\frac{d^2y}{dx^2} = 2$

Concave upward

8. $x = t^2 + 5t + 4,\ y = 4t$

$\frac{dy}{dx} = \frac{dy/dt}{dx/dt} = \frac{4}{2t + 5}$

$\frac{d^2y}{dx^2} = \frac{\frac{d}{dt}\left[\frac{4}{2t+5}\right]}{dx/dt} = \frac{\frac{-8}{(2t+5)^2}}{2t + 5} = \frac{-8}{(2t + 5)^3}$

At $t = 0$, $\frac{dy}{dx} = \frac{4}{5}$.

At $t = 0$, $\frac{d^2y}{dx^2} = -\frac{8}{125}$

Concave downward

9. $x = 4\cos\theta,\ y = 4\sin\theta$

$\frac{dy}{dx} = \frac{dy/d\theta}{dx/d\theta} = \frac{4\cos\theta}{-4\sin\theta} = \frac{-\cos\theta}{\sin\theta} = -\cot\theta$

$\frac{d^2y}{dx^2} = \frac{\frac{d}{d\theta}[-\cot\theta]}{dx/d\theta} = \frac{\csc^2\theta}{-4\sin\theta} = \frac{-1}{4\sin^3\theta} = -\frac{1}{4}\csc^3\theta$

At $\theta = \frac{\pi}{4}$, $\frac{dy}{dx} = -1$.

$\frac{d^2y}{dx^2} = \frac{-1}{4\left(\sqrt{2}/2\right)^3} = \frac{-\sqrt{2}}{2}$

Concave downward

10. $x = \cos\theta,\ y = 3\sin\theta$

$\frac{dy}{dx} = \frac{3\cos\theta}{-\sin\theta} = -3\cot\theta \cdot \frac{dy}{dx}$ is undefined when $\theta = 0$.

$\frac{d^2y}{dx^2} = \frac{3\csc^2\theta}{-\sin\theta} = \frac{-3}{\sin^3\theta} \cdot \frac{d^2y}{dx^2}$ is undefined when $\theta = 0$.

Neither concave upward nor downward

11. $x = 2 + \sec\theta,\ y = 1 + 2\tan\theta$

$$\frac{dy}{dx} = \frac{2\sec^2\theta}{\sec\theta\tan\theta} = \frac{2\sec\theta}{\tan\theta} = 2\csc\theta = 4 \text{ when } \theta = \frac{\pi}{6}.$$

$$\frac{d^2y}{dx^2} = \frac{d\left[\frac{dy}{dx}\right]}{\frac{dx}{d\theta}} = \frac{-2\csc\theta\cot\theta}{\sec\theta\tan\theta} = -2\cot^3\theta = -6\sqrt{3} \text{ when } \theta = \frac{\pi}{6}.$$

Concave downward

12. $x = \sqrt{t},\ y = \sqrt{t-1}$

$$\frac{dy}{dx} = \frac{1/\left(2\sqrt{t-1}\right)}{1/\left(2\sqrt{t}\right)} = \frac{\sqrt{t}}{\sqrt{t-1}} = \sqrt{2} \text{ when } t = 2.$$

$$\frac{d^2y}{dx^2} = \frac{\left[\sqrt{t-1}/\left(2\sqrt{t}\right) - \sqrt{t}\left(1/2\sqrt{t-1}\right)\right]/(t-1)}{1/\left(2\sqrt{t}\right)} = \frac{-1}{(t-1)^{3/2}} = -1 \text{ when } t = 2.$$

Concave downward

13. $x = \cos^3\theta,\ y = \sin^3\theta$

$$\frac{dy}{dx} = \frac{3\sin^2\theta\cos\theta}{-3\cos^2\theta\sin\theta} = -\tan\theta = -1 \text{ when } \theta = \frac{\pi}{4}.$$

$$\frac{d^2y}{dx^2} = \frac{-\sec^2\theta}{-3\cos^2\theta\sin\theta} = \frac{1}{3\cos^4\theta\sin\theta} = \frac{\sec^4\theta\csc\theta}{3} = \frac{4\sqrt{2}}{3} \text{ when } \theta = \frac{\pi}{4}.$$

Concave upward

14. $x = \theta - \sin\theta,\ y = 1 - \cos\theta$

$$\frac{dy}{dx} = \frac{\sin\theta}{1-\cos\theta} = 0 \text{ when } \theta = \pi.$$

$$\frac{d^2y}{dx^2} = \frac{\dfrac{\left[(1-\cos\theta)\cos\theta - \sin^2\theta\right]}{(1-\cos\theta)^2}}{(1-\cos\theta)} = \frac{-1}{(1-\cos\theta)^2} = -\frac{1}{4} \text{ when } \theta = \pi.$$

Concave downward

15. $x = 2\cot\theta,\ y = 2\sin^2\theta$

$$\frac{dy}{dx} = \frac{4\sin\theta\cos\theta}{-2\csc^2\theta} = -2\sin^3\theta\cos\theta$$

At $\left(-\frac{2}{\sqrt{3}}, \frac{3}{2}\right)$, $\theta = \frac{2\pi}{3}$, and $\frac{dy}{dx} = \frac{3\sqrt{3}}{8}$.

Tangent line: $y - \frac{3}{2} = \frac{3\sqrt{3}}{8}\left(x + \frac{2}{\sqrt{3}}\right)$

$3\sqrt{3}x - 8y + 18 = 0$

At $(0, 2)$, $\theta = \frac{\pi}{2}$, and $\frac{dy}{dx} = 0$.

Tangent line: $y - 2 = 0$

At $\left(2\sqrt{3}, \frac{1}{2}\right)$, $\theta = \frac{\pi}{6}$, and $\frac{dy}{dx} = -\frac{\sqrt{3}}{8}$.

Tangent line: $y - \frac{1}{2} = -\frac{\sqrt{3}}{8}\left(x - 2\sqrt{3}\right)$

$\sqrt{3}x + 8y - 10 = 0$

16. $x = 2 - 3\cos\theta,\ y = 3 + 2\sin\theta$

$$\frac{dy}{dx} = \frac{2\cos\theta}{3\sin\theta} = \frac{2}{3}\cot\theta$$

At $(-1, 3)$, $\theta = 0$, and $\frac{dy}{dx}$ is undefined.

Tangent line: $x = -1$

At $(2, 5)$, $\theta = \frac{\pi}{2}$, and $\frac{dy}{dx} = 0$.

Tangent line: $y = 5$

At $\left(\frac{4 + 3\sqrt{3}}{2}, 2\right)$, $\theta = \frac{7\pi}{6}$, and $\frac{dy}{dx} = \frac{2\sqrt{3}}{3}$.

Tangent line:

$$y - 2 = \frac{2\sqrt{3}}{3}\left(x - \frac{4 + 3\sqrt{3}}{2}\right)$$

$2\sqrt{3}x - 3y - 4\sqrt{3} - 3 = 0$

17. $x = t^2 - 4$

$y = t^2 - 2t$

$$\frac{dy}{dx} = \frac{dy/dt}{dx/dt} = \frac{2t - 2}{2t}$$

At $(0, 0)$, $t = 2$, $\dfrac{dy}{dx} = \dfrac{1}{2}$.

Tangent line: $y = \dfrac{1}{2}x$

$2y - x = 0$

At $(-3, -1)$, $t = 1$, $\dfrac{dy}{dx} = 0$.

Tangent line: $y = -1$

$y + 1 = 0$

At $(-3, 3)$, $t = -1$, $\dfrac{dy}{dx} = 2$.

Tangent line: $y - 3 = 2(x + 3)$

$2x - y + 9 = 0$

18. $x = t^4 + 2$

$y = t^3 + t$

$$\frac{dy}{dx} = \frac{dy/dt}{dx/dt} = \frac{3t^2 + 1}{4t^3}$$

At $(2, 0)$, $t = 0$, $\dfrac{dy}{dx}$ undefined.

Tangent line: $x = 2$ (vertical tangent)

At $(3, -2)$, $t = -1$, $\dfrac{dy}{dx} = -1$.

Tangent line: $y + 2 = -(x - 3)$

$y = -x + 1$

At $(18, 10)$, $t = 2$, $\dfrac{dy}{dx} = \dfrac{13}{32}$.

Tangent line: $y - 10 = \dfrac{13}{32}(x - 18)$

$y = \dfrac{13}{32}x + \dfrac{43}{16}$

19. $x = 6t$, $y = t^2 + 4$, $t = 1$

(a), (d)

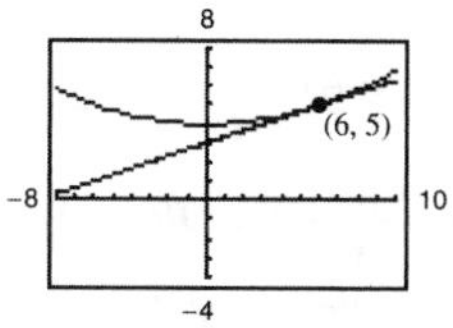

(b) At $t = 1$, $(x, y) = (6, 5)$, and

$$\frac{dx}{dt} = 6, \frac{dy}{dt} = 2, \frac{dy}{dx} = \frac{1}{3}.$$

(c) $y - 5 = \dfrac{1}{3}(x - 6)$

$y = \dfrac{1}{3}x + 3$

20. $x = t - 2$, $y = \dfrac{1}{t} + 3$, $t = 1$

(a), (d)

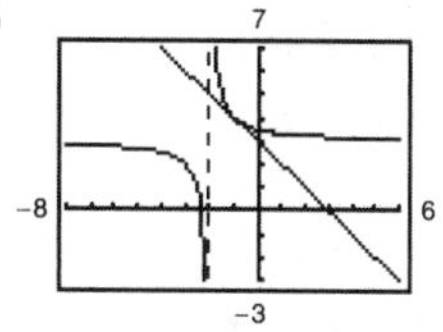

(b) At $t = 1$, $(x, y) = (-1, 4)$, and

$$\frac{dx}{dt} = 1, \frac{dy}{dt} = -1, \frac{dy}{dx} = -1.$$

(c) $y - 4 = -(x + 1)$

$y = -x + 3$

21. $x = t^2 - t + 2$, $y = t^3 - 3t$, $t = -1$

(a), (d)

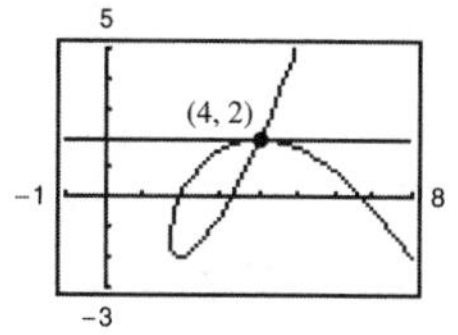

(b) At $t = -1$, $(x, y) = (4, 2)$, and

$$\frac{dx}{dt} = -3, \frac{dy}{dt} = 0, \frac{dy}{dx} = 0.$$

(c) $\dfrac{dy}{dx} = 0$. At $(4, 2)$, $y - 2 = 0(x - 4)$

$y = 2.$

22. $x = 3t - t^2$, $y = 2t^{3/2}$, $t = \dfrac{1}{4}$

(a), (d)

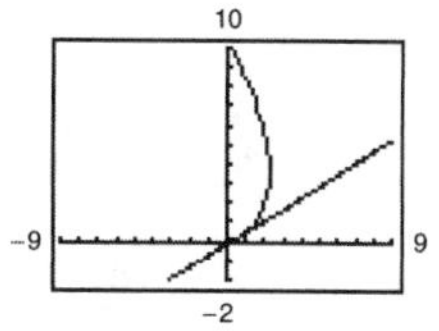

(b) At $t = \dfrac{1}{4}$, $(x, y) = \left(\dfrac{11}{16}, \dfrac{1}{4}\right)$, and

$$\frac{dx}{dt} = \frac{5}{2}, \frac{dy}{dt} = \frac{3}{2}, \frac{dy}{dx} = \frac{3/2}{5/2} = \frac{3}{5}.$$

(c) $\dfrac{dy}{dx} = \dfrac{3}{5}$. At $\left(\dfrac{11}{16}, \dfrac{1}{4}\right)$, $y - \dfrac{1}{4} = \dfrac{3}{5}\left(x - \dfrac{11}{16}\right)$

$y = \dfrac{3}{5}x - \dfrac{13}{80}.$

23. $x = 2\sin 2t,\ y = 3\sin t$ crosses itself at the origin, $(x, y) = (0, 0)$.

At this point, $t = 0$ or $t = \pi$.

$$\frac{dy}{dx} = \frac{3\cos t}{4\cos 2t}$$

At $t = 0$: $\dfrac{dy}{dx} = \dfrac{3}{4}$ and $y = \dfrac{3}{4}x$. Tangent Line

At $t = \pi$, $\dfrac{dy}{dx} = -\dfrac{3}{4}$ and $y = -\dfrac{3}{4}x$. Tangent Line

24. $x = 2 - \pi\cos t,\ y = 2t - \pi\sin t$ crosses itself at a point on the x-axis: $(2, 0)$. The corresponding t-values are $t = \pm\pi/2$.

$$\frac{dy}{dt} = 2 - \pi\cos t,\ \frac{dx}{dt} = \pi\sin t,\ \frac{dy}{dx} = \frac{2 - \pi\cos t}{\pi\sin t}$$

At $t = \dfrac{\pi}{2}$: $\dfrac{dy}{dx} = \dfrac{2}{\pi}$.

Tangent line: $y - 0 = \dfrac{2}{\pi}(x - 2)$

$$y = \frac{2}{\pi}x - \frac{4}{\pi}$$

At $t = -\dfrac{\pi}{2}$: $\dfrac{dy}{dx} = -\dfrac{2}{\pi}$.

Tangent line: $y - 0 = -\dfrac{2}{\pi}(x - 2)$

$$y = -\frac{2}{\pi}x + \frac{4}{\pi}$$

25. $x = t^2 - t,\ y = t^3 - 3t - 1$ crosses itself at the point $(x, y) = (2, 1)$.

At this point, $t = -1$ or $t = 2$.

$$\frac{dy}{dx} = \frac{3t^2 - 3}{2t - 1}$$

At $t = -1$, $\dfrac{dy}{dx} = 0$ and $y = 1$. Tangent Line

At $t = 2$, $\dfrac{dy}{dt} = \dfrac{9}{3} = 3$ and $y - 1 = 3(x - 2)$ or $y = 3x - 5$.

Tangent Line

26. $x = t^3 - 6t,\ y = t^2$ crosses itself at $(0, 6)$. The corresponding t-values are $t = \pm\sqrt{6}$.

$$\frac{dy}{dx} = \frac{2t}{3t^2 - 6}$$

At $t = \sqrt{6}$, $\dfrac{dy}{dx} = \dfrac{2\sqrt{6}}{12} = \dfrac{\sqrt{6}}{6}$.

Tangent line: $y - 6 = \dfrac{\sqrt{6}}{6}(x - 0)$

$$y = \frac{\sqrt{6}}{6}x + 6$$

At $t = -\sqrt{6}$, $\dfrac{dy}{dx} = -\dfrac{2\sqrt{6}}{12} = -\dfrac{\sqrt{6}}{6}$.

Tangent line: $y = -\dfrac{\sqrt{6}}{6}x + 6$

27. $x = \cos\theta + \theta\sin\theta,\ y = \sin\theta - \theta\cos\theta$

Horizontal tangents: $\dfrac{dy}{d\theta} = \theta\sin\theta = 0$ when $\theta = \pm\pi, \pm 2\pi, \pm 3\pi, \ldots$

Points: $(-1, [2n - 1]\pi), (1, 2n\pi)$ where n is an integer.

Points shown: $(1, 0), (-1, \pi), (1, -2\pi)$

Vertical tangents: $\dfrac{dx}{d\theta} = \theta\cos\theta = 0$ when $\theta = \pm\dfrac{\pi}{2}, \pm\dfrac{3\pi}{2}, \pm\dfrac{5\pi}{2}, \ldots$

Note: $\theta = 0$ corresponds to the cusp at $(x, y) = (1, 0)$.

$$\frac{dy}{dx} = \frac{\theta\sin\theta}{\theta\cos\theta} = \tan\theta = 0 \text{ at } \theta = 0$$

Points: $\left(\dfrac{(-1)^{n+1}(2n - 1)\pi}{2}, (-1)^{n+1}\right)$

Points shown: $\left(\dfrac{\pi}{2}, 1\right), \left(-\dfrac{3\pi}{2}, -1\right), \left(\dfrac{5\pi}{2}, 1\right)$

28. $x = 2\theta,\ y = 2(1 - \cos\theta)$

Horizontal tangents: $\dfrac{dy}{d\theta} = 2\sin\theta = 0$ when $\theta = 0, \pm\pi, \pm 2\pi, \ldots$

Points: $(4n\pi, 0), (2[2n - 1]\pi, 4)$ where n is an integer

Points shown: $(0, 0), (2\pi, 4), (4\pi, 0)$

Vertical tangents: $\dfrac{dx}{d\theta} = 2 \neq 0$; none

29. $x = 4 - t, y = t^2$

Horizontal tangents: $\frac{dy}{dt} = 2t = 0$ when $t = 0$.

Point: $(4, 0)$

Vertical tangents: $\frac{dx}{dt} = -1 \neq 0$ None

30. $x = t + 1, y = t^2 + 3t$

Horizontal tangents: $\frac{dy}{dt} = 2t + 3 = 0$ when $t = -\frac{3}{2}$

Point: $\left(-\frac{1}{2}, -\frac{9}{4}\right)$

Vertical tangents: $\frac{dx}{dt} = 1 \neq 0$; none

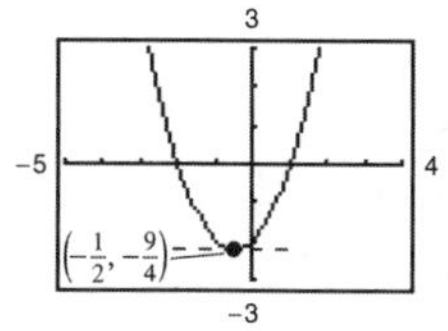

31. $x = t + 4, y = t^3 - 3t$

Horizontal tangents:

$\frac{dy}{dt} = 3t^2 - 3 = 3(t - 1)(t + 1) = 0 \Rightarrow t = \pm 1$

Points: $(5, -2), (3, 2)$

Vertical tangents: $\frac{dx}{dt} = 1 \neq 0$ None

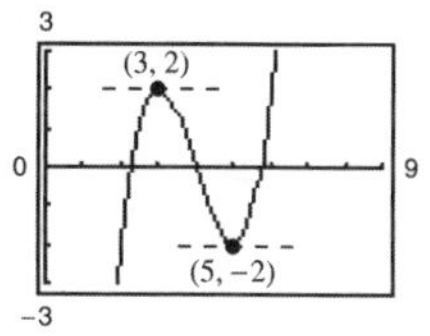

32. $x = t^2 - t + 2, y = t^3 - 3t$

Horizontal tangents: $\frac{dy}{dt} = 3t^2 - 3 = 0$ when $t = \pm 1$.

Points: $(2, -2), (4, 2)$

Vertical tangents: $\frac{dx}{dt} = 2t - 1 = 0$ when $t = \frac{1}{2}$.

Point: $\left(\frac{7}{4}, -\frac{11}{8}\right)$

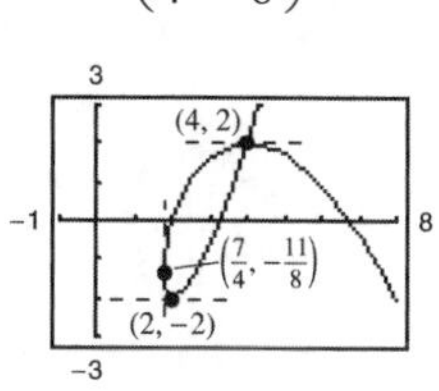

33. $x = 3 \cos \theta, y = 3 \sin \theta$

Horizontal tangents: $\frac{dy}{d\theta} = 3 \cos \theta = 0$ when

$\theta = \frac{\pi}{2}, \frac{3\pi}{2}$.

Points: $(0, 3), (0, -3)$

Vertical tangents: $\frac{dx}{d\theta} = -3 \sin \theta = 0$ when $\theta = 0, \pi$.

Points: $(3, 0), (-3, 0)$

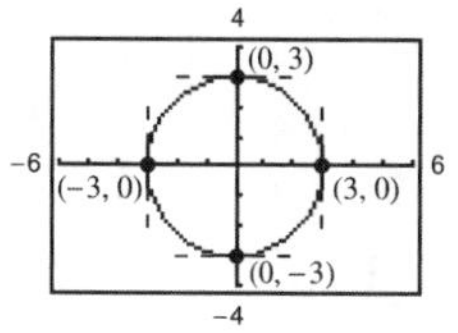

34. $x = \cos \theta, y = 2 \sin 2\theta$

Horizontal tangents: $\frac{dy}{d\theta} = 4 \cos 2\theta = 0$ when

$\theta = \frac{\pi}{4}, \frac{3\pi}{4}, \frac{5\pi}{4}, \frac{7\pi}{4}$.

Points: $\left(\frac{\sqrt{2}}{2}, 2\right), \left(-\frac{\sqrt{2}}{2}, -2\right), \left(-\frac{\sqrt{2}}{2}, 2\right), \left(\frac{\sqrt{2}}{2}, -2\right)$

Vertical tangents: $\frac{dx}{d\theta} = -\sin \theta = 0$ when $\theta = 0, \pi$.

Points: $(1, 0), (-1, 0)$

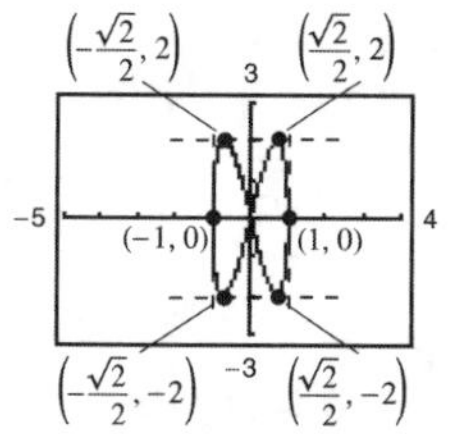

35. $x = 5 + 3 \cos \theta, y = -2 + \sin \theta$

Horizontal tangents: $\frac{dy}{dt} = \cos \theta = 0 \Rightarrow \theta = \frac{\pi}{2}, \frac{3\pi}{2}$

Points: $(5, -1), (5, -3)$

Vertical tangents: $\frac{dx}{dt} = -3 \sin \theta = 0 \Rightarrow \theta = 0, \pi$

Points: $(8, -2), (2, -2)$

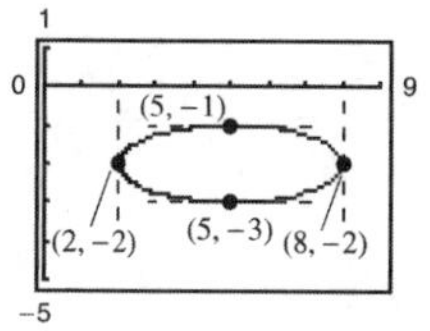

36. $x = 4\cos^2\theta,\ y = 2\sin\theta$

Horizontal tangents: $\dfrac{dy}{d\theta} = 2\cos\theta = 0$ when

$\theta = \dfrac{\pi}{2}, \dfrac{3\pi}{2}.$

Because $dx/d\theta = 0$ at $\pi/2$ and $3\pi/2$, exclude them.

Vertical tangents: $\dfrac{dx}{d\theta} = -8\cos\theta\sin\theta = 0$ when

$\theta = 0, \pi.$

Point: $(4, 0)$

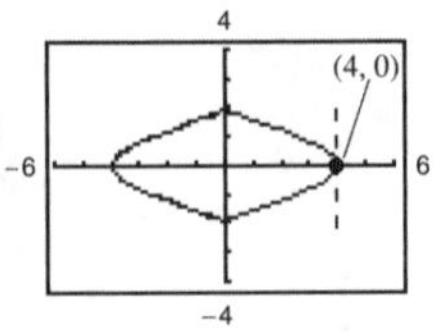

37. $x = \sec\theta,\ y = \tan\theta$

Horizontal tangents: $\dfrac{dy}{d\theta} = \sec^2\theta \neq 0$; None

Vertical tangents: $\dfrac{dx}{d\theta} = \sec\theta\tan\theta = 0$ when

$x = 0, \pi.$

Points: $(1, 0), (-1, 0)$

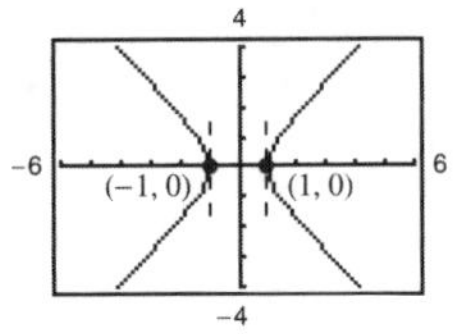

38. $x = \cos^2\theta,\ y = \cos\theta$

Horizontal tangents: $\dfrac{dy}{d\theta} = -\sin\theta = 0$ when $x = 0, \pi.$

Since $dx/d\theta = 0$ at these values, exclude them.

Vertical tangents: $\dfrac{dx}{d\theta} = -2\cos\theta\sin\theta = 0$ when

$\theta = \dfrac{\pi}{2}, \dfrac{3\pi}{2}.$

(Exclude $0, \pi$.)

Point: $(0, 0)$

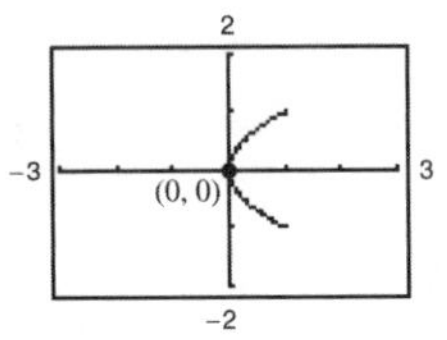

39. $x = 3t^2,\ y = t^3 - t$

$$\frac{dy}{dx} = \frac{dy/dt}{dx/dt} = \frac{3t^2 - 1}{6t} = \frac{t}{2} - \frac{1}{6t}$$

$$\frac{d^2y}{dx^2} = \frac{\dfrac{d}{dt}\left[\dfrac{t}{2} - \dfrac{1}{6t}\right]}{dx/dt} = \frac{\dfrac{1}{2} + \dfrac{1}{6t^2}}{6t} = \frac{6t^2 + 2}{36t^3}$$

Concave upward for $t > 0$

Concave downward for $t < 0$

40. $x = 2 + t^2,\ y = t^2 + t^3$

$$\frac{dy}{dx} = \frac{2t + 3t^2}{2t} = 1 + \frac{3}{2}t$$

$$\frac{d^2y}{dx^2} = \frac{3/2}{2t} = \frac{3}{4t}$$

Concave upward for $t > 0$

Concave downward for $t < 0$

41. $x = 2t + \ln t,\ y = 2t - \ln t,\ t > 0$

$$\frac{dy}{dx} = \frac{2 - (1/t)}{2 + (1/t)} = \frac{2t - 1}{2t + 1}$$

$$\frac{d^2y}{dx^2} = \left[\frac{(2t + 1)2 - (2t - 1)2}{(2t + 1)^2}\right] \Big/ \left(2 + \frac{1}{t}\right)$$

$$= \frac{4}{(2t + 1)^2} \cdot \frac{t}{2t + 1} = \frac{4t}{(2t + 1)^3}$$

Because $t > 0,\ \dfrac{d^2y}{dx^2} > 0$

Concave upward for $t > 0$

42. $x = t^2,\ y = \ln t,\ t > 0$

$$\frac{dy}{dx} = \frac{1/t}{2t} = \frac{1}{2t^2}$$

$$\frac{d^2y}{dx^2} = -\frac{1/t^3}{2t} = -\frac{1}{2t^4}$$

Because $t > 0,\ \dfrac{d^2y}{dx^2} < 0$

Concave downward for $t > 0$

43. $x = \sin t,\ y = \cos t,\ 0 < t < \pi$

$$\frac{dy}{dx} = -\frac{\sin t}{\cos t} = -\tan t$$

$$\frac{d^2y}{dx^2} = -\frac{\sec^2 t}{\cos t} = -\frac{1}{\cos^3 t}$$

Concave upward on $\pi/2 < t < \pi$

Concave downward on $0 < t < \pi/2$

44. $x = 4\cos t,\ y = 2\sin t,\ 0 < t < 2\pi$

$$\frac{dy}{dx} = \frac{dy/dt}{dx/dt} = \frac{2\cos t}{-4\sin t} = -\frac{1}{2}\cot t$$

$$\frac{d^2y}{dx^2} = \frac{\frac{d}{dt}\left[-\frac{1}{2}\cot t\right]}{dx/dt} = \frac{\frac{1}{2}\csc^2 t}{-4\sin t} = \frac{-1}{8\sin^3 t}$$

Concave upward on $\pi < t < 2\pi$

Concave downward on $0 < t < \pi$

45. $x = 3t + 5,\ y = 7 - 2t,\ -1 \le t \le 3$

$$\frac{dx}{dt} = 3,\ \frac{dy}{dt} = -2$$

$$s = \int_a^b \sqrt{\left(\frac{dx}{dt}\right)^2 + \left(\frac{dy}{dt}\right)^2}\,dt = \int_{-1}^{3} \sqrt{9 + 4}\,dt$$

$$\left[\sqrt{13}\,t\right]_{-1}^{3} = 4\sqrt{13} \approx 14.422$$

46. $x = 6t^2,\ y = 2t^3,\ 1 \le t \le 4$

$$\frac{dx}{dt} = 12t,\ \frac{dy}{dt} = 6t^2$$

$$s = \int_a^b \sqrt{\left(\frac{dx}{dt}\right)^2 + \left(\frac{dy}{dt}\right)^2}\,dt$$

$$= \int_1^4 \sqrt{144t^2 + 36t^4}\,dt$$

$$= \int_1^4 6t\sqrt{4 + t^2}\,dt$$

$$= \left[2\left(4 + t^2\right)^{3/2}\right]_1^4$$

$$= 2\left(20^{3/2} - 5^{3/2}\right)$$

$$= 70\sqrt{5} \approx 156.525$$

47. $x = e^{-t}\cos t,\ y = e^{-t}\sin t,\ 0 \le t \le \frac{\pi}{2}$

$$\frac{dx}{dt} = -e^{-t}\left(\sin t + \cos t\right),\ \frac{dy}{dt} = e^{-t}\left(\cos t - \sin t\right)$$

$$s = \int_0^{\pi/2} \sqrt{\left(\frac{dx}{dt}\right)^2 + \left(\frac{dy}{dt}\right)^2}\,dt$$

$$= \int_0^{\pi/2} \sqrt{2e^{-2t}}\,dt = -\sqrt{2}\int_0^{\pi/2} e^{-t}(-1)\,dt$$

$$= \left[-\sqrt{2}e^{-t}\right]_0^{\pi/2}$$

$$= \sqrt{2}\left(1 - e^{-\pi/2}\right) \approx 1.12$$

48. $x = \arcsin t,\ y = \ln\sqrt{1 - t^2},\ 0 \le t \le \frac{1}{2}$

$$\frac{dx}{dt} = \frac{1}{\sqrt{1 - t^2}},\ \frac{dy}{dt} = \frac{1}{2}\left(\frac{-2t}{1 - t^2}\right) = \frac{t}{1 - t^2}$$

$$s = \int_0^{1/2} \sqrt{\left(\frac{dx}{dt}\right)^2 + \left(\frac{dy}{dt}\right)^2}\,dt$$

$$= \int_0^{1/2} \sqrt{\frac{1}{\left(1 - t^2\right)^2}}\,dt = \int_0^{1/2} \frac{1}{1 - t^2}\,dt$$

$$= \left[-\frac{1}{2}\ln\left|\frac{t - 1}{t + 1}\right|\right]_0^{1/2}$$

$$= -\frac{1}{2}\ln\left(\frac{1}{3}\right) = \frac{1}{2}\ln(3) \approx 0.549$$

49. $x = \sqrt{t},\ y = 3t - 1,\ \frac{dx}{dt} = \frac{1}{2\sqrt{t}},\ \frac{dy}{dt} = 3$

$$s = \int_0^1 \sqrt{\frac{1}{4t} + 9}\,dt = \frac{1}{2}\int_0^1 \frac{\sqrt{1 + 36t}}{\sqrt{t}}\,dt$$

$$= \frac{1}{6}\int_0^6 \sqrt{1 + u^2}\,du$$

$$= \frac{1}{12}\left[\ln\left(\sqrt{1 + u^2} + u\right) + u\sqrt{1 + u^2}\right]_0^6$$

$$= \frac{1}{12}\left[\ln\left(\sqrt{37} + 6\right) + 6\sqrt{37}\right] \approx 3.249$$

$$u = 6\sqrt{t},\ du = \frac{3}{\sqrt{t}}\,dt$$

50. $x = t,\ y = \frac{t^5}{10} + \frac{1}{6t^3},\ \frac{dx}{dt} = 1,\ \frac{dy}{dt} = \frac{t^4}{2} - \frac{1}{2t^4}$

$$S = \int_1^2 \sqrt{1 + \left(\frac{t^4}{2} - \frac{1}{2t^4}\right)^2}\,dt$$

$$= \int_1^2 \sqrt{\left(\frac{t^4}{2} + \frac{1}{2t^4}\right)^2}\,dt$$

$$= \int_1^2 \left(\frac{t^4}{2} + \frac{1}{2t^4}\right)dt = \left[\frac{t^5}{10} - \frac{1}{6t^3}\right]_1^2 = \frac{779}{240}$$

51. $x = a\cos^3\theta,\ y = a\sin^3\theta,\ \frac{dx}{d\theta} = -3a\cos^2\theta\sin\theta,$

$$\frac{dy}{d\theta} = 3a\sin^2\theta\cos\theta$$

$$s = 4\int_0^{\pi/2} \sqrt{9a^2\cos^4\theta\sin^2\theta + 9a^2\sin^4\theta\cos^2\theta}\,d\theta$$

$$= 12a\int_0^{\pi/2} \sin\theta\cos\theta\sqrt{\cos^2\theta + \sin^2\theta}\,d\theta$$

$$= 6a\int_0^{\pi/2} \sin 2\theta\,d\theta = \left[-3a\cos 2\theta\right]_0^{\pi/2} = 6a$$

52. $x = a\cos\theta,\ y = a\sin\theta,\ \dfrac{dx}{d\theta} = -a\sin\theta,$

$\dfrac{dy}{d\theta} = a\cos\theta$

$$S = 4\int_0^{\pi/2} \sqrt{a^2\sin^2\theta + a^2\cos^2\theta}\,d\theta$$

$$= 4a\int_0^{\pi/2} d\theta = \left[4a\theta\right]_0^{\pi/2} = 2\pi a$$

53. $x = a(\theta - \sin\theta),\ y = a(1 - \cos\theta),$

$\dfrac{dx}{d\theta} = a(1 - \cos\theta),\ \dfrac{dy}{d\theta} = a\sin\theta$

$$s = 2\int_0^{\pi} \sqrt{a^2(1 - \cos\theta)^2 + a^2\sin^2\theta}\,d\theta$$

$$= 2\sqrt{2}a\int_0^{\pi} \sqrt{1 - \cos\theta}\,d\theta$$

$$= 2\sqrt{2}a\int_0^{\pi} \frac{\sin\theta}{\sqrt{1 + \cos\theta}}\,d\theta$$

$$= \left[-4\sqrt{2}a\sqrt{1 + \cos\theta}\right]_0^{\pi} = 8a$$

54. $x = \cos\theta + \theta\sin\theta,\ y = \sin\theta - \theta\cos\theta,$

$\dfrac{dx}{d\theta} = \theta\cos\theta$

$\dfrac{dy}{d\theta} = \theta\sin\theta$

$$S = \int_0^{2\pi} \sqrt{\theta^2\cos^2\theta + \theta^2\sin^2\theta}\,d\theta$$

$$= \int_0^{2\pi} \theta\,d\theta = \left[\frac{\theta^2}{2}\right]_0^{2\pi} = 2\pi^2$$

55. $x = (90\cos 30°)t,\ y = (90\sin 30°)t - 16t^2$

(a)

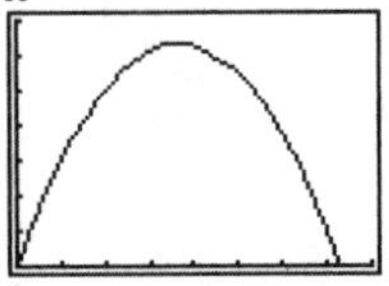

(b) Range: 219.2 ft, $\left(t = \dfrac{45}{16}\right)$

(c) $\dfrac{dx}{dt} = 90\cos 30°,\ \dfrac{dy}{dt} = 90\sin 30° - 32t$

$y = 0$ for $t = \dfrac{45}{16}$.

$$s = \int_0^{45/16} \sqrt{(90\cos 30°)^2 + (90\sin 30° - 32t)^2}\,dt$$

≈ 230.8 ft

56. $y = 0 \Rightarrow (90\sin\theta)t = 16t^2 \Rightarrow t = 0, \dfrac{90}{16}\sin\theta$

$$x = (90\cos\theta)t = (90\cos\theta)\frac{90}{16}\sin\theta$$

$$= \frac{90^2}{16}\sin\theta\cos\theta = \frac{90^2}{32}\sin 2\theta$$

$$x'(\theta) = \frac{90^2}{32}2\cos 2\theta = 0 \Rightarrow \theta = \frac{\pi}{4}$$

By the First Derivative Test, $\theta = \dfrac{\pi}{4}(45°)$ maximizes the range ($x = 253.125$ feet).

To maximize the arc length, you have

$\dfrac{dx}{dt} = 90\cos\theta,\ \dfrac{dy}{dt} = 90\sin\theta - 32t.$

$$s = \int_0^{(90/16)\sin\theta} \sqrt{(90\cos\theta)^2 + (90\sin\theta - 32t)^2}\,dt$$

$$= \frac{2025}{8}\sin\theta + \frac{2025}{16}\cos^2\theta\ln\left[\frac{1 + \sin\theta}{1 - \sin\theta}\right]$$

Using a graphing utility, we see that s is a maximum of approximately 303.67 feet at $\theta \approx 0.9855(56.5°)$.

57. $x = \dfrac{4t}{1 + t^3},\ y = \dfrac{4t^2}{1 + t^3}$

(a) $x^3 + y^3 = 4xy$

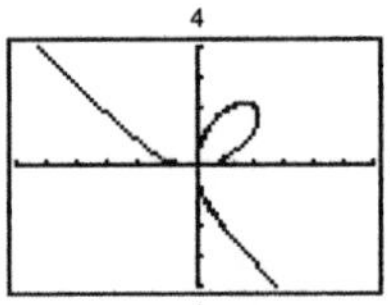

(b) $$\frac{dy}{dt} = \frac{(1 + t^3)(8t) - 4t^2(3t^2)}{(1 + t^3)^2}$$

$$= \frac{4t(2 - t^3)}{(1 + t^3)^2} = 0 \text{ when } t = 0 \text{ or } t = \sqrt[3]{2}.$$

Points: $(0, 0), \left(\dfrac{4\sqrt[3]{2}}{3}, \dfrac{4\sqrt[3]{4}}{3}\right) \approx (1.6799, 2.1165)$

(c) $$s = 2\int_0^1 \sqrt{\left[\frac{4(1 - 2t^3)}{(1 + t^3)^2}\right]^2 + \left[\frac{4t(2 - t^3)}{(1 + t^3)^2}\right]^2}\,dt$$

$$= 2\int_0^1 \sqrt{\frac{16}{(1 + t^3)^4}\left[t^8 + 4t^6 - 4t^5 - 4t^3 + 4t^2 + 1\right]}\,dt$$

$$= 8\int_0^1 \frac{\sqrt{t^8 + 4t^6 - 4t^5 - 4t^3 + 4t^2 + 1}}{(1 + t^3)^2}\,dt \approx 6.557$$

58. $x = 4\cot\theta = \dfrac{4}{\tan\theta}, y = 4\sin^2\theta, -\dfrac{\pi}{2} \le \theta \le \dfrac{\pi}{2}$

(a)

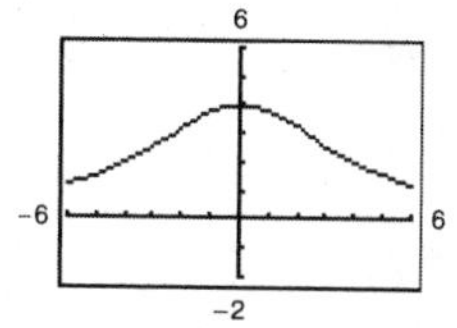

(b) $\dfrac{dy}{d\theta} = 8\sin\theta \cdot \cos\theta$

$\dfrac{dx}{d\theta} = -4\csc^2\theta$

$\dfrac{dy}{d\theta} = 0 \text{ for } \theta = 0, \pm\dfrac{\pi}{2}$

Horizontal tangent at $(x, y) = (0, 4)\left(\theta = \pm\dfrac{\pi}{2}\right)$

(Function is not defined at $\theta = 0$)

(c) Arc length over $\dfrac{\pi}{4} \le t \le \dfrac{\pi}{2}$: 4.5183

59. (a) $x = t - \sin t$

$y = 1 - \cos t$

$0 \le t \le 2\pi$

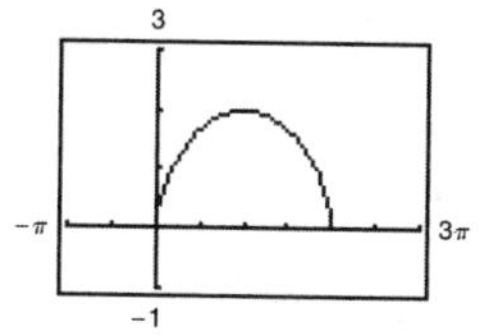

$x = 2t - \sin(2t)$

$y = 1 - \cos(2t)$

$0 \le t \le \pi$

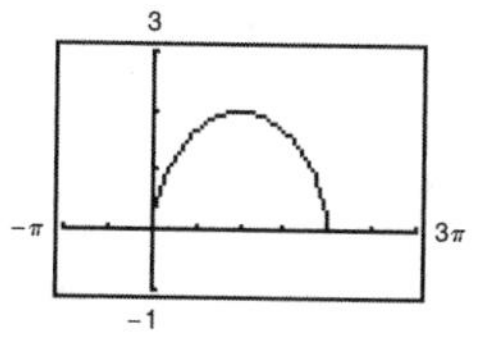

(b) The average speed of the particle on the second path is twice the average speed of a particle on the first path.

(c) $x = \frac{1}{2}t - \sin\left(\frac{1}{2}t\right)$

$y = 1 - \cos\left(\frac{1}{2}t\right)$

The time required for the particle to traverse the same path is $t = 4\pi$.

60. (a) First particle: $x = 3\cos t, y = 4\sin t, 0 \le t \le 2\pi$

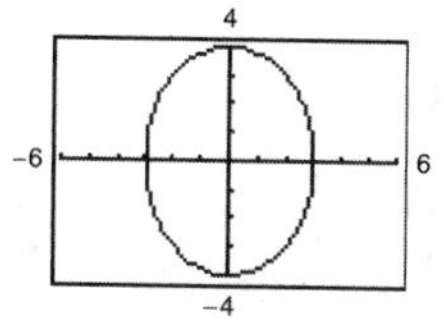

Second particle: $x = 4\sin t, y = 3\cos t,$
$0 \le t \le 2\pi$

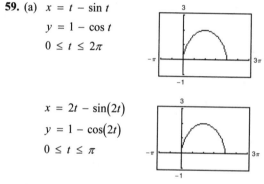

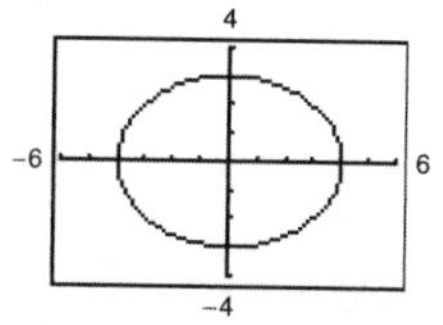

(b) There are 4 points of intersection.

(c) Suppose at time t that

$3\cos t = 4\sin t$ and $4\sin t = 3\cos t$

$\tan t = \frac{3}{4}$ and $\tan t = \frac{3}{4}$.

Yes, the particles are at the same place at the same time for $\tan t = \frac{3}{4}$. $t \approx 0.6435, 3.7851$. The intersection points are $(2.4, 2.4)$ and $(-2.4, -2.4)$

(d) The curves intersect twice, but not at the same time.

61. $x = 3t, \dfrac{dx}{dt} = 3$

$y = t + 2, \dfrac{dy}{dt} = 1$

$$S = 2\pi\int_0^4 (t + 2)\sqrt{3^2 + 1^2}\,dt$$

$$= 2\pi\sqrt{10}\left[\frac{t^2}{2} + 2t\right]_0^4$$

$$= 2\pi\sqrt{10}[8 + 8] = 32\sqrt{10}\pi \approx 317.9068$$

62. $x = \dfrac{1}{4}t^2 \quad \dfrac{dx}{dt} = \dfrac{t}{2}$

$y = t + 3, \dfrac{dy}{dt} = 1$

$$S = 2\pi\int_0^3 (t + 3)\sqrt{\left(\frac{t}{2}\right)^2 + 1}\,dt$$

$$= 2\pi\int_0^3 (t + 3)\sqrt{\frac{t^2}{4} + 1}\,dt$$

$$\approx 114.1999$$

63. $x = \cos^2\theta, \dfrac{dx}{d\theta} = -2\cos\theta\sin\theta$

$y = \cos\theta, \dfrac{dy}{d\theta} = -\sin\theta$

$$S = 2\pi\int_0^{\pi/2}\cos\theta\sqrt{4\cos^2\theta\sin^2\theta + \sin^2\theta}\,d\theta$$
$$= 2\pi\int_0^{\pi/2}\cos\theta\sin\theta\sqrt{4\cos^2\theta + 1}\,d\theta$$
$$= \frac{\left(5\sqrt{5} - 1\right)\pi}{6}$$
$$\approx 5.3304$$

64. $x = \theta + \sin\theta, \dfrac{dx}{d\theta} = 1 + \cos\theta$

$y = \theta + \cos\theta, \dfrac{dy}{d\theta} = 1 - \sin\theta$

$$S = 2\pi\int_0^{\pi/2}(\theta + \cos\theta)\sqrt{(1 + \cos\theta)^2 + (1 - \sin\theta)^2}\,d\theta$$
$$= 2\pi\int_0^{\pi/2}(\theta + \cos\theta)\sqrt{3 + 2\cos\theta - 2\sin\theta}\,d\theta$$
$$\approx 23.2433$$

65. $x = 2t, \dfrac{dx}{dt} = 2$

$y = 3t, \dfrac{dy}{dt} = 3$

(a) $S = 2\pi\int_0^3 3t\sqrt{4 + 9}\,dt$

$$= 6\sqrt{13}\pi\left[\frac{t^2}{2}\right]_0^3 = 6\sqrt{13}\pi\left(\frac{9}{2}\right) = 27\sqrt{13}\pi$$

(b) $S = 2\pi\int_0^3 2t\sqrt{4 + 9}\,dt$

$$= 4\sqrt{13}\pi\left[\frac{t^2}{2}\right]_0^3 = 4\sqrt{13}\pi\left(\frac{9}{2}\right) = 18\sqrt{13}\pi$$

66. $x = t, y = 4 - 2t, \dfrac{dx}{dt} = 1, \dfrac{dy}{dt} = -2$

(a) $S = 2\pi\int_0^2(4 - 2t)\sqrt{1 + 4}\,dt$

$$= \left[2\sqrt{5}\pi\left(4t - t^2\right)\right]_0^2 = 8\pi\sqrt{5}$$

(b) $S = 2\pi\int_0^2 t\sqrt{1 + 4}\,dt = \left[\sqrt{5}\pi t^2\right]_0^2 = 4\pi\sqrt{5}$

67. $x = 5\cos\theta\ \dfrac{dx}{d\theta} = -5\sin\theta$

$y = 5\sin\theta\ \dfrac{dy}{d\theta} = 5\cos\theta$

$$S = 2\pi\int_0^{\pi/2}5\cos\theta\sqrt{25\sin^2\theta + 25\cos^2\theta}\,d\theta$$
$$= 10\pi\int_0^{\pi/2}5\cos\theta\,d\theta$$
$$= 50\pi\left[\sin\theta\right]_0^{\pi/2} = 50\pi$$

[**Note:** This is the surface area of a hemisphere of radius 5]

68. $x = \frac{1}{3}t^3, y = t + 1, 1 \le t \le 2$, y-axis

$\dfrac{dx}{dt} = t^2, \dfrac{dy}{dt} = 1$

$$S = 2\pi\int_1^2\frac{1}{3}t^3\sqrt{t^4 + 1}\,dt = \frac{\pi}{9}\left[\left(x^4 + 1\right)^{3/2}\right]_1^2$$
$$= \frac{\pi}{9}\left(17^{3/2} - 2^{3/2}\right) \approx 23.48$$

69. $x = a\cos^3\theta, y = a\sin^3\theta, \dfrac{dx}{d\theta} = -3a\cos^2\theta\sin\theta, \dfrac{dy}{d\theta} = 3a\sin^2\theta\cos\theta$

$$S = 4\pi\int_0^{\pi/2}a\sin^3\theta\sqrt{9a^2\cos^4\theta\sin^2\theta + 9a^2\sin^4\theta\cos^2\theta}\,d\theta$$
$$= 12a^2\pi\int_0^{\pi/2}\sin^4\theta\cos\theta\,d\theta = \frac{12\pi a^2}{5}\left[\sin^5\theta\right]_0^{\pi/2} = \frac{12}{5}\pi a^2$$

70. $x = a\cos\theta,\ y = b\sin\theta,\ \dfrac{dx}{d\theta} = -a\sin\theta,\ \dfrac{dy}{d\theta} = b\cos\theta$

(a) $$S = 4\pi\int_0^{\pi/2} b\sin\theta\sqrt{a^2\sin^2\theta + b^2\cos^2\theta}\,d\theta$$

$$= 4\pi\int_0^{\pi/2} ab\sin\theta\sqrt{1 - \left(\frac{a^2 - b^2}{a^2}\right)\cos^2\theta}\,d\theta = \frac{-4ab\pi}{e}\int_0^{\pi/2}(-e\sin\theta)\sqrt{1 - e^2\cos^2\theta}\,d\theta$$

$$= \frac{-2ab\pi}{e}\left[e\cos\theta\sqrt{1 - e^2\cos^2\theta} + \arcsin(e\cos\theta)\right]_0^{\pi/2} = \frac{2ab\pi}{e}\left[e\sqrt{1 - e^2} + \arcsin(e)\right]$$

$$= 2\pi b^2 + \left(\frac{2\pi a^2 b}{\sqrt{a^2 - b^2}}\right)\arcsin\left(\frac{\sqrt{a^2 - b^2}}{a}\right) = 2\pi b^2 + 2\pi\left(\frac{ab}{e}\right)\arcsin(e)$$

$$\left(e = \frac{\sqrt{a^2 - b^2}}{a} = \frac{c}{a}\text{: eccentricity}\right)$$

(b) $$S = 4\pi\int_0^{\pi/2} a\cos\theta\sqrt{a^2\sin^2\theta + b^2\cos^2\theta}\,d\theta$$

$$= 4\pi\int_0^{\pi/2} a\cos\theta\sqrt{b^2 + c^2\sin^2\theta}\,d\theta = \frac{4a\pi}{c}\int_0^{\pi/2} c\cos\theta\sqrt{b^2 + c^2\sin^2\theta}\,d\theta$$

$$= \frac{2a\pi}{c}\left[c\sin\theta\sqrt{b^2 + c^2\sin^2\theta} + b^2\ln\left|c\sin\theta + \sqrt{b^2 + c^2\sin^2\theta}\right|\right]_0^{\pi/2}$$

$$= \frac{2a\pi}{c}\left[c\sqrt{b^2 + c^2} + b^2\ln\left|c + \sqrt{b^2 + c^2}\right| - b^2\ln b\right]$$

$$= 2\pi a^2 + \frac{2\pi ab^2}{\sqrt{a^2 - b^2}}\ln\left|\frac{a + \sqrt{a^2 - b^2}}{b}\right| = 2\pi a^2 + \left(\frac{\pi b^2}{e}\right)\ln\left|\frac{1 + e}{1 - e}\right|$$

71. $\dfrac{dy}{dx} = \dfrac{dy/dt}{dx/dt}$

See Theorem 10.7.

72. $x = t,\ y = 3 \Rightarrow \dfrac{dy}{dx} = 0$

73. $x = t,\ y = 6t - 5 \Rightarrow \dfrac{dy}{dx} = \dfrac{6}{1} = 6$

74. $s = \displaystyle\int_a^b\sqrt{\left(\frac{dx}{dt}\right)^2 + \left(\frac{dy}{dt}\right)^2}\,dt$

See Theorem 10.8.

75. (a) $S = 2\pi\displaystyle\int_a^b g(t)\sqrt{\left(\frac{dx}{dt}\right)^2 + \left(\frac{dy}{dt}\right)^2}\,dt$

(b) $S = 2\pi\displaystyle\int_a^b f(t)\sqrt{\left(\frac{dx}{dt}\right)^2 + \left(\frac{dy}{dt}\right)^2}\,dt$

76. (i) (a) $\dfrac{dx}{dt} < 0$ and $\dfrac{dy}{dx} < 0$ from the graph.

So, $\dfrac{dy}{dt} > 0$ because $\dfrac{dy}{dx} = \dfrac{dy/dt}{dx/dt}$.

(b) $\dfrac{dy}{dt} > 0$ and $\dfrac{dy}{dx} < 0$ from the graph.

So, $\dfrac{dx}{dt} < 0$ because $\dfrac{dy}{dx} = \dfrac{dy/dt}{dx/dt}$.

(ii) (a) $\dfrac{dx}{dt} < 0$ and $\dfrac{dy}{dx} > 0$ from the graph.

So, $\dfrac{dy}{dt} < 0$ because $\dfrac{dy}{dx} = \dfrac{dy/dt}{dx/dt}$.

(b) $\dfrac{dy}{dt} > 0$ and $\dfrac{dy}{dx} > 0$ from the graph.

So, $\dfrac{dx}{dt} > 0$ because $\dfrac{dy}{dx} = \dfrac{dy/dt}{dx/dt}$.

77. Let y be a continuous function of x on $a \le x \le b$. Suppose that $x = f(t)$, $y = g(t)$, and $f(t_1) = a$, $f(t_2) = b$. Then using integration by substitution, $dx = f'(t)\,dt$ and

$$\int_a^b y\,dx = \int_{t_1}^{t_2} g(t)f'(t)\,dt.$$

78. $x = r\cos\phi,\ y = r\sin\phi$

$$S = 2\pi\int_0^{\theta} r\sin\phi\sqrt{r^2\sin^2\phi + r^2\cos^2\phi}\,d\phi$$
$$= 2\pi r^2\int_0^{\theta}\sin\phi\,d\phi$$
$$= \left[-2\pi r^2\cos\phi\right]_0^{\theta}$$
$$= 2\pi r^2(1 - \cos\theta)$$

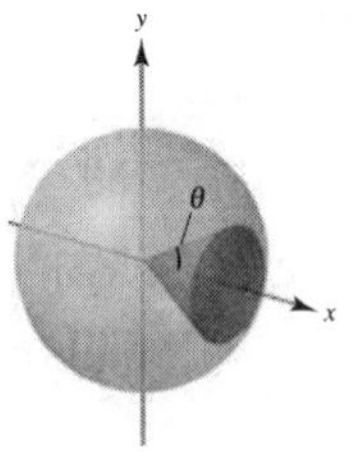

79. $x = 2\sin^2\theta$

$y = 2\sin^2\theta\tan\theta$

$\dfrac{dx}{d\theta} = 4\sin\theta\cos\theta$

$$A = \int_0^{\pi/2} 2\sin^2\theta\tan\theta(4\sin\theta\cos\theta)\,d\theta$$
$$= 8\int_0^{\pi/2}\sin^4\theta\,d\theta$$
$$= 8\left[\frac{-\sin^3\theta\cos\theta}{4} - \frac{3}{8}\sin\theta\cos\theta + \frac{3}{8}\theta\right]_0^{\pi/2} = \frac{3\pi}{2}$$

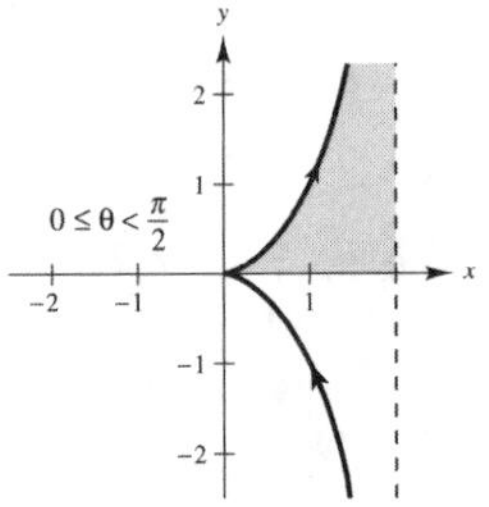

80. $x = 2\cot\theta,\ y = 2\sin^2\theta,\ \dfrac{dx}{d\theta} = -2\csc^2\theta$

$$A = 2\int_{\pi/2}^{0}\left(2\sin^2\theta\right)\left(-2\csc^2\theta\right)d\theta$$
$$= -8\int_{\pi/2}^{0}d\theta = \left[-8\theta\right]_{\pi/2}^{0} = 4\pi$$

81. πab is area of ellipse (d).

82. $\frac{3}{8}\pi a^2$ is area of asteroid (b).

83. $6\pi a^2$ is area of cardioid (f).

84. $2\pi a^2$ is area of deltoid (c).

85. $\frac{8}{3}ab$ is area of hourglass (a).

86. $2\pi ab$ is area of teardrop (e).

87. $x = \sqrt{t},\ y = 4 - t,\ 0 < t < 4$

$$A = \int_0^2 y\,dx = \int_0^4 (4 - t)\frac{1}{2\sqrt{t}}\,dt = \frac{1}{2}\int_0^4\left(4t^{-1/2} - t^{1/2}\right)dt = \left[\frac{1}{2}\left(8\sqrt{t} - \frac{2}{3}t\sqrt{t}\right)\right]_0^4 = \frac{16}{3}$$

$$\bar{x} = \frac{1}{A}\int_0^2 yx\,dx = \frac{3}{16}\int_0^4 (4 - t)\sqrt{t}\left(\frac{1}{2\sqrt{t}}\right)dt = \frac{3}{32}\int_0^4 (4 - t)\,dt = \left[\frac{3}{32}\left(4t - \frac{t^2}{2}\right)\right]_0^4 = \frac{3}{4}$$

$$\bar{y} = \frac{1}{A}\int_0^2 \frac{y^2}{2}\,dx = \frac{3}{32}\int_0^4 (4 - t)^2\frac{1}{2\sqrt{t}}\,dt = \frac{3}{64}\int_0^4\left(16t^{-1/2} - 8t^{1/2} + t^{3/2}\right)dt = \frac{3}{64}\left[32\sqrt{t} - \frac{16}{3}t\sqrt{t} + \frac{2}{5}t^2\sqrt{t}\right]_0^4 = \frac{8}{5}$$

$$(\bar{x}, \bar{y}) = \left(\frac{3}{4}, \frac{8}{5}\right)$$

88. $x = \sqrt{4-t},\ y = \sqrt{t},\ \dfrac{dx}{dt} = -\dfrac{1}{2\sqrt{4-t}},\ 0 \le t \le 4$

$$A = \int_4^0 \sqrt{t}\left(-\frac{1}{2\sqrt{4-t}}\right)dt = \int_0^2 \sqrt{4-u^2}\,du = \frac{1}{2}\left[u\sqrt{4-u^2} + 4\arcsin\frac{u}{2}\right]_0^2 = \pi$$

Let $u = \sqrt{4-t}$, then $du = -1\big/\left(2\sqrt{4-t}\right)dt$ and $\sqrt{t} = \sqrt{4-u^2}$.

$$\bar{x} = \frac{1}{\pi}\int_4^0 \sqrt{4-t}\sqrt{t}\left(-\frac{1}{2\sqrt{4-t}}\right)dt = -\frac{1}{2\pi}\int_4^0 \sqrt{t}\,dt = \left[-\frac{1}{2\pi}\frac{2}{3}t^{3/2}\right]_4^0 = \frac{8}{3\pi}$$

$$\bar{y} = \frac{1}{2\pi}\int_4^0 \left(\sqrt{t}\right)^2\left(-\frac{1}{2\sqrt{4-t}}\right)dt = -\frac{1}{4\pi}\int_4^0 \frac{t}{\sqrt{4-t}}\,dt = -\frac{1}{4\pi}\left[\frac{-2(8+t)}{3}\sqrt{4-t}\right]_4^0 = \frac{8}{3\pi}$$

$$(\bar{x}, \bar{y}) = \left(\frac{8}{3\pi}, \frac{8}{3\pi}\right)$$

89. $x = 6\cos\theta,\ y = 6\sin\theta,\ \dfrac{dx}{d\theta} = -6\sin\theta\,d\theta$

$$\begin{aligned} V &= 2\pi\int_{\pi/2}^0 (6\sin\theta)^2(-6\sin\theta)\,d\theta \\ &= -432\pi\int_{\pi/2}^0 \sin^3\theta\,d\theta \\ &= -432\pi\int_{\pi/2}^0 \left(1-\cos^2\theta\right)\sin\theta\,d\theta \\ &= -432\pi\left[-\cos\theta + \frac{\cos^3\theta}{3}\right]_{\pi/2}^0 \\ &= -432\pi\left(-1+\frac{1}{3}\right) = 288\pi \end{aligned}$$

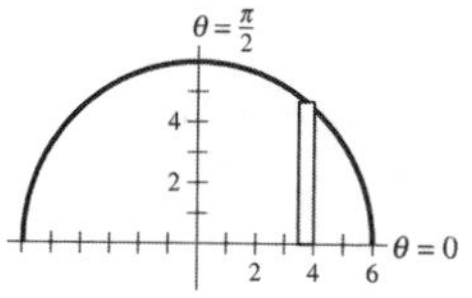

Note: Volume of sphere is $\frac{4}{3}\pi\left(6^3\right) = 288\pi$.

90. $x = \cos\theta,\ y = 3\sin\theta,\ \dfrac{dx}{d\theta} = -\sin\theta$

$$\begin{aligned} V &= 2\pi\int_{\pi/2}^0 (3\sin\theta)^2(-\sin\theta)\,d\theta \\ &= -18\pi\int_{\pi/2}^0 \sin^3\theta\,d\theta \\ &= -18\pi\left[-\cos\theta + \frac{\cos^3\theta}{3}\right]_{\pi/2}^0 = 12\pi \end{aligned}$$

91. $x = a(\theta - \sin\theta),\ y = a(1-\cos\theta)$

(a) $\dfrac{dy}{d\theta} = a\sin\theta,\ \dfrac{dx}{d\theta} = a(1-\cos\theta)$

$$\frac{dy}{dx} = \frac{a\sin\theta}{a(1-\cos\theta)} = \frac{\sin\theta}{1-\cos\theta}$$

$$\frac{d^2y}{dx^2} = \left[\frac{(1-\cos\theta)\cos\theta - \sin\theta(\sin\theta)}{(1-\cos\theta)^2}\right]\Bigg/\left[a(1-\cos\theta)\right] = \frac{\cos\theta - 1}{a(1-\cos\theta)^3} = \frac{-1}{a(\cos\theta-1)^2}$$

(b) At $\theta = \dfrac{\pi}{6}$, $x = a\left(\dfrac{\pi}{6} - \dfrac{1}{2}\right)$, $y = a\left(1 - \dfrac{\sqrt{3}}{2}\right)$, $\dfrac{dy}{dx} = \dfrac{1/2}{1-\sqrt{3}/2} = 2+\sqrt{3}$.

Tangent line: $y - a\left(1-\dfrac{\sqrt{3}}{2}\right) = \left(2+\sqrt{3}\right)\left(x - a\left(\dfrac{\pi}{6}-\dfrac{1}{2}\right)\right)$

(c) $\dfrac{dy}{dx} = \dfrac{\sin\theta}{1-\cos\theta} = 0 \Rightarrow \sin\theta = 0,\ 1-\cos\theta \ne 0$

Points of horizontal tangency: $(x, y) = \left(a(2n+1)\pi, 2a\right)$

(d) Concave downward on all open θ-intervals:

$\ldots, (-2\pi, 0), (0, 2\pi), (2\pi, 4\pi), \ldots$

(e) $s = \int_0^{2\pi} \sqrt{a^2 \sin^2 \theta + a^2(1 - \cos \theta)^2}\, d\theta$

$= a\int_0^{2\pi} \sqrt{2 - 2\cos\theta}\, d\theta = a\int_0^{2\pi} \sqrt{4\sin^2 \frac{\theta}{2}}\, d\theta = 2a\int_0^{2\pi} \sin\frac{\theta}{2}\, d\theta = \left[-4a\cos\left(\frac{\theta}{2}\right)\right]_0^{2\pi} = 8a$

92. $x = t^2\sqrt{3},\ y = 3t - \frac{1}{3}t^3$

(a)

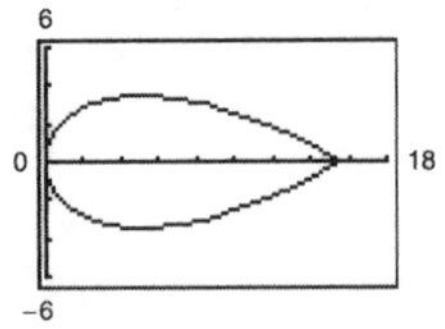

(b) $\frac{dx}{dt} = 2\sqrt{3}t,\ \frac{dy}{dt} = 3 - t^2,\ \frac{dy}{dx} = \frac{3 - t^2}{2\sqrt{3}t}$

$\frac{d^2y}{dx^2} = \left[\frac{2\sqrt{3}(t)(-2t) - (3 - t^2)2\sqrt{3}}{12t^2}\right] \Big/ \left[2\sqrt{3}t\right] = \frac{-2\sqrt{3}t^2 - 6\sqrt{3}}{(12t^2)(2\sqrt{3}t)} = -\frac{t^2 + 3}{12t^3}$

(c) $(x, y) = \left(\sqrt{3}, \frac{8}{3}\right)$ at $t = 1$. $\frac{dy}{dx} = \frac{2}{2\sqrt{3}} = \frac{\sqrt{3}}{3}$

$y - \frac{8}{3} = \frac{\sqrt{3}}{3}\left(x - \sqrt{3}\right)$

$y = \frac{\sqrt{3}}{3}x + \frac{5}{3}$

(d) $s = \int_{-3}^{3} \sqrt{12t^2 + (3 - t)^2}\, dt = \int_{-3}^{3} \sqrt{t^4 - 6t^2 + 9 + 12t^2}\, dt = \int_{-3}^{3} \sqrt{(t^2 + 3)^2}\, dt = \int_{-3}^{3} (t^2 + 3)\, dt = 36$

(e) $S = 2\pi \int_0^3 \left(3t - \frac{1}{3}t^3\right)\left(t^2 + 3\right) dt = 81\pi$

93. $x = t + u = r\cos\theta + r\theta\sin\theta$

$= r(\cos\theta + \theta\sin\theta)$

$y = v - w = r\sin\theta - r\theta\cos\theta$

$= r(\sin\theta - \theta\cos\theta)$

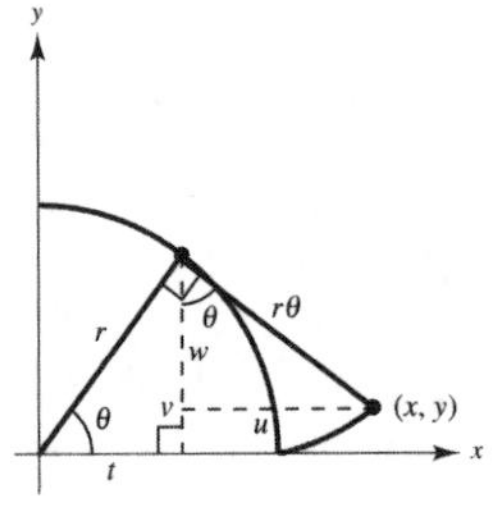

94. Focus on the region above the x-axis. From Exercise 99, the equation of the involute from $(1, 0)$ to $(-1, \pi)$ is

$x = \cos\theta + \theta\sin\theta$

$y = \sin\theta - \theta\cos\theta$

$0 \le \theta \le \pi.$

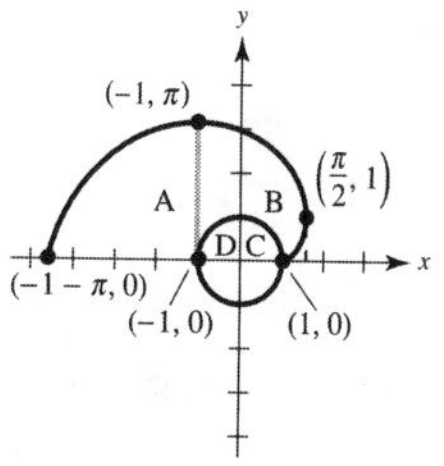

At $(-1, \pi)$, the string is fully extended and has length π.

$$\text{So, the area of region A is } \frac{1}{4}\pi(\pi^2) = \frac{1}{4}\pi^3.$$

You now need to find the area of region B.

$$\frac{dx}{d\theta} = -\sin\theta + \sin\theta + \theta\cos\theta = \theta\cos\theta = 0 \Rightarrow \theta = \frac{\pi}{2}.\ (\theta = 0 \text{ is cusp.})$$

So, the far right point on the involute is $(\pi/2, 1)$.

The area of the region $B + C + D$ is given by

$$\int_{\theta=\pi}^{\theta=\pi/2} y\,dx - \int_{\theta=0}^{\theta=\pi/2} y\,dx = \int_{\theta=\pi}^{\theta=0} y\,dx$$

where $y = \sin\theta - \theta\cos\theta$ and $dx = \theta\cos\theta\,d\theta$.

So, you can calculate

$$\int_{\pi}^{0}[\sin\theta - \theta\cos\theta]\theta\cos\theta\,d\theta = \frac{\pi}{6}(\pi^2 + 3).$$

Because the area of $C + D$ is $\pi/2$, you have

$$\text{Total area covered} = 2\left[\frac{1}{4}\pi^3 + \frac{\pi}{6}(\pi^2 + 3) - \frac{\pi}{2}\right] = \frac{5}{6}\pi^3.$$

95. (a)

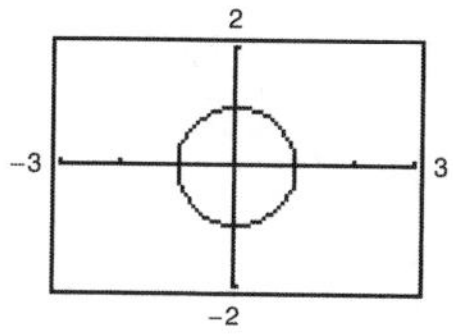

(b) $x = \dfrac{1 - t^2}{1 + t^2}, y = \dfrac{2t}{1 + t^2}, \quad -20 \le t \le 20$

The graph (for $-\infty < t < \infty$) is the circle $x^2 + y^2 = 1$, except the point $(-1, 0)$.

Verify:

$$x^2 + y^2 = \left(\frac{1 - t^2}{1 + t^2}\right)^2 + \left(\frac{2t}{1 + t^2}\right)^2$$

$$= \frac{1 - 2t^2 + t^4 + 4t^2}{(1 + t^2)^2} = \frac{(1 + t^2)^2}{(1 + t^2)^2} = 1$$

(c) As t increases from -20 to 0, the speed increases, and as t increases from 0 to 20, the speed decreases.

96. (a) $y = -12\ln\left(\dfrac{12 - \sqrt{144 - x^2}}{x}\right) - \sqrt{144 - x^2}$

$0 < x \le 12$

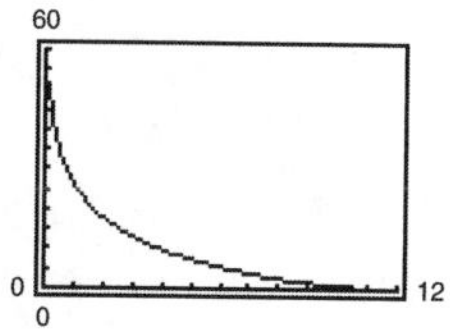

(b) $x = 12 \operatorname{sech} \dfrac{t}{12},\ y = t - 12 \tanh \dfrac{t}{12},\ 0 \le t$

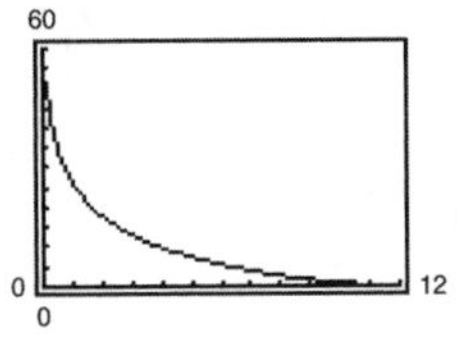

Same as the graph in (a), but has the advantage of showing the position of the object and any given time t.

(c) $\dfrac{dy}{dx} = \dfrac{1 - \operatorname{sech}^2(t/12)}{-\operatorname{sech}(t/12)\tanh(t/12)} = -\sinh \dfrac{t}{12}$

Tangent line: $y - \left(t_0 - 12 \tanh \dfrac{t_0}{12}\right) = -\sinh \dfrac{t_0}{12}\left(x - 12 \operatorname{sech} \dfrac{t_0}{12}\right)$

$$y = t_0 - \left(\sinh \frac{t_0}{12}\right)x$$

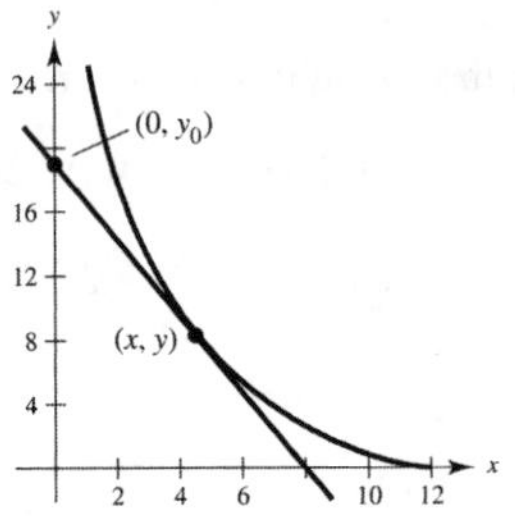

y-intercept: $(0, t_0)$

Distance between $(0, t_0)$ and (x, y): $d = \sqrt{\left(12 \operatorname{sech} \dfrac{t_0}{12}\right)^2 + \left(-12 \tanh \dfrac{t_0}{12}\right)^2} = 12$

$$d = 12 \text{ for any } t \ge 0.$$

97. False. $\dfrac{d^2y}{dx^2} = \dfrac{\dfrac{d}{dt}\left[\dfrac{g'(t)}{f'(t)}\right]}{f'(t)} = \dfrac{f'(t)g''(t) - g'(t)f''(t)}{\left[f'(t)\right]^3}$

98. False. Both dx/dt and dy/dt are zero when $t = 0$. By eliminating the parameter, you have $y = x^{2/3}$ which does not have a horizontal tangent at the origin.

Section 10.4 Polar Coordinates and Polar Graphs

1. $\left(8, \dfrac{\pi}{2}\right)$

$x = 8 \cos \dfrac{\pi}{2} = 0$

$y = 8 \sin \dfrac{\pi}{2} = 8$

$(x, y) = (0, 8)$

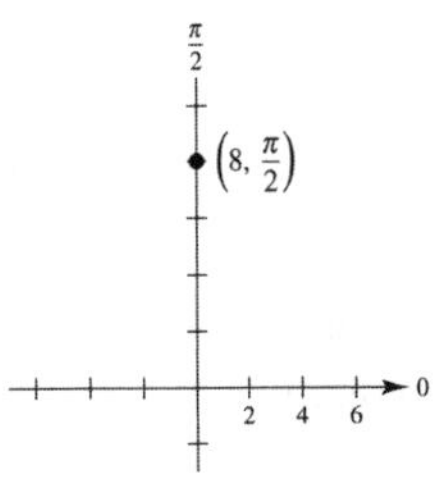

2. $\left(-2, \dfrac{5\pi}{3}\right)$

$x = -2 \cos \dfrac{5\pi}{3} = -2\left(\dfrac{1}{2}\right) = -1$

$y = -2 \sin \dfrac{5\pi}{3} = -2\left(\dfrac{-\sqrt{3}}{2}\right) = \sqrt{3}$

$(x, y) = \left(-1, \sqrt{3}\right)$

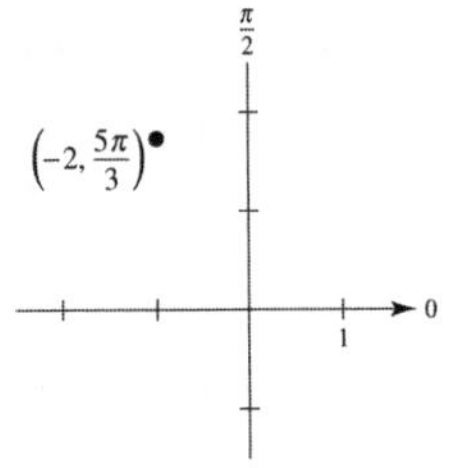

3. $\left(-4, -\dfrac{3\pi}{4}\right)$

$x = -4 \cos\left(\dfrac{-3\pi}{4}\right) = -4\left(-\dfrac{\sqrt{2}}{2}\right) = 2\sqrt{2}$

$y = -4 \sin\left(\dfrac{-3\pi}{4}\right) = -4\left(-\dfrac{\sqrt{2}}{2}\right) = 2\sqrt{2}$

$(x, y) = \left(2\sqrt{2}, 2\sqrt{2}\right)$

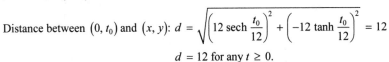

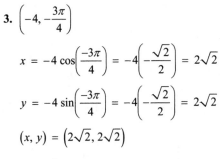

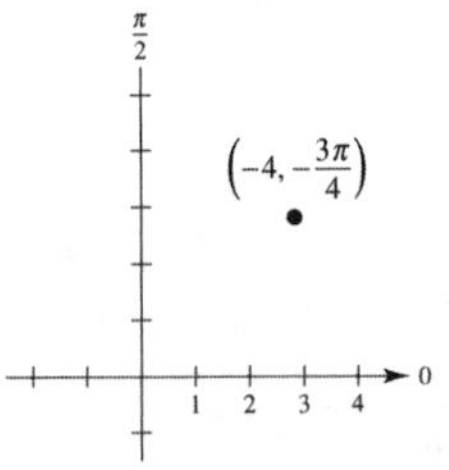

4. $\left(0, -\frac{7\pi}{6}\right)$

$x = 0\cos\left(-\frac{7\pi}{6}\right) = 0$

$y = 0\sin\left(-\frac{7\pi}{6}\right) = 0$

$(x, y) = (0, 0)$

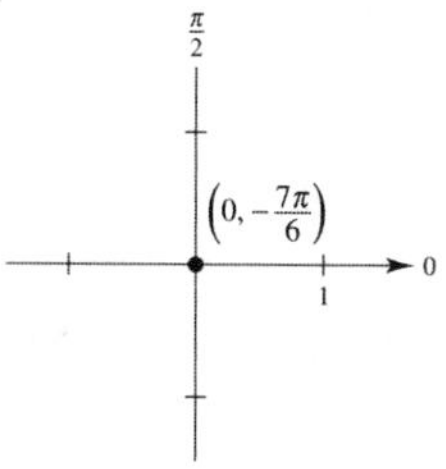

5. $(r, \theta) = \left(7, \frac{5\pi}{4}\right)$

$x = 7\cos\frac{5\pi}{4} = 7\left(\frac{-\sqrt{2}}{2}\right) = -\frac{7\sqrt{2}}{2}$

$y = 7\sin\frac{5\pi}{4} = 7\left(-\frac{\sqrt{2}}{2}\right) = -\frac{7\sqrt{2}}{2}$

$(x, y) = \left(-\frac{7\sqrt{2}}{2}, -\frac{7\sqrt{2}}{2}\right)$

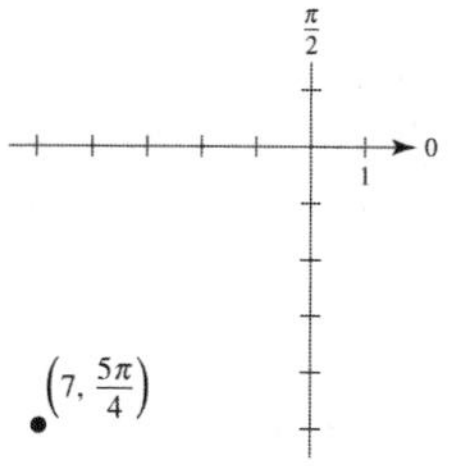

6. $(r, \theta) = \left(-2, \frac{11\pi}{6}\right)$

$x = -2\cos\left(\frac{11\pi}{6}\right) = -\sqrt{3}$

$y = -2\sin\left(\frac{11\pi}{6}\right) = 1$

$(x, y) = \left(-\sqrt{3}, 1\right)$

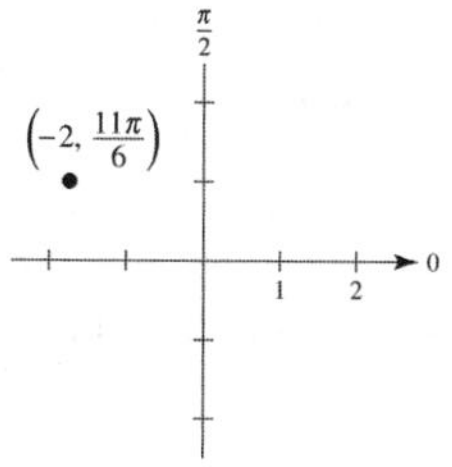

7. $\left(\sqrt{2}, 2.36\right)$

$x = \sqrt{2}\cos(2.36) \approx -1.004$

$y = \sqrt{2}\sin(2.36) \approx 0.996$

$(x, y) = (-1.004, 0.996)$

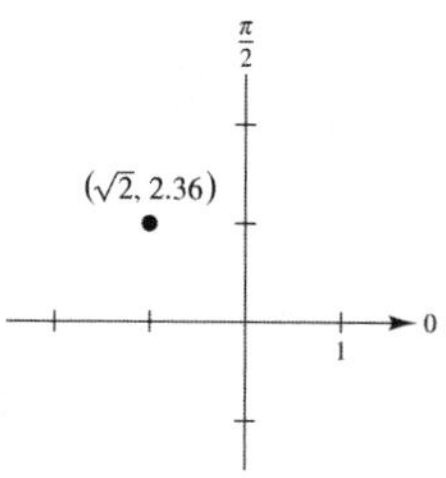

8. $(-3, -1.57)$

$x = -3\cos(-1.57) \approx -0.0024$

$y = -3\sin(-1.57) \approx 3$

$(x, y) = (-0.0024, 3)$

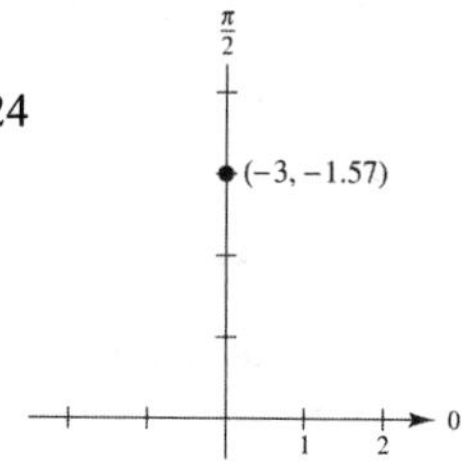

9. $(r, \theta) = (-4.5, 3.5)$

$x = -4.5\cos 3.5 \approx 4.2141$

$y = -4.5\sin 3.5 \approx 1.5785$

$(x, y) = (4.2141, 1.5785)$

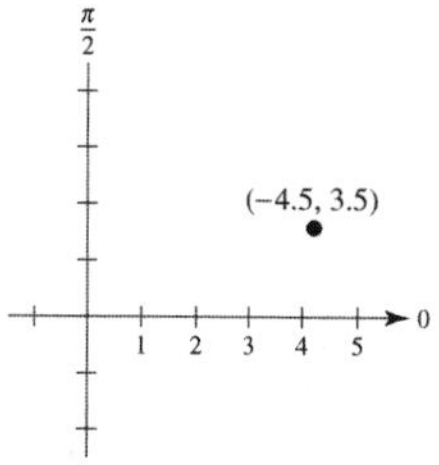

10. $(r, \theta) = (9.25, 1.2)$

$x = 9.25\cos 1.2 \approx 3.3518$

$y = 9.25\sin 1.2 \approx 8.6214$

$(x, y) = (3.3518, 8.6214)$

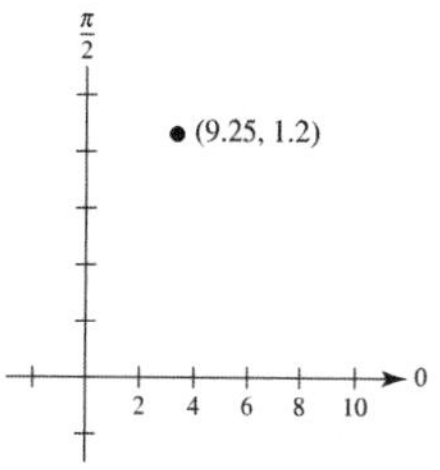

11. $(x, y) = (2, 2)$

$r = \sqrt{2^2 + 2^2} = 2\sqrt{2}$

$\tan\theta = \frac{2}{2} = 1$

$\theta = \frac{\pi}{4}, \frac{5\pi}{4}$

$\left(2\sqrt{2}, \frac{\pi}{4}\right), \left(-2\sqrt{2}, \frac{5\pi}{4}\right)$

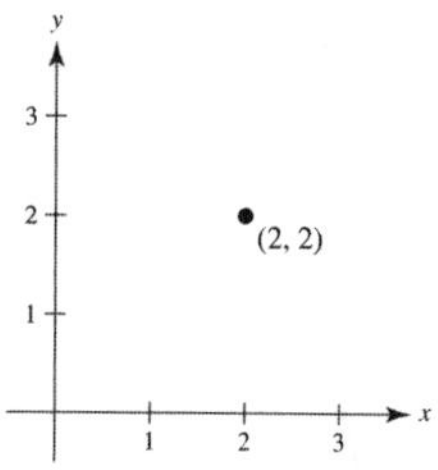

12. $(x, y) = (0, -6)$

$r = \pm 6$

$\tan\theta$ undefined

$\theta = \frac{\pi}{2}, \frac{3\pi}{2}$

$\left(6, \frac{3\pi}{2}\right), \left(-6, \frac{\pi}{2}\right)$

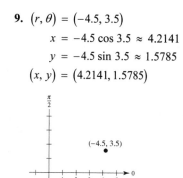

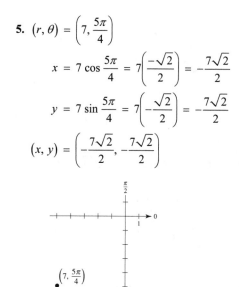

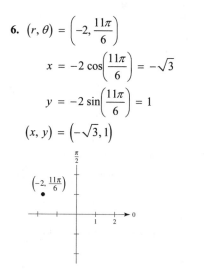

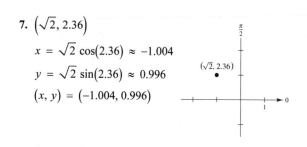

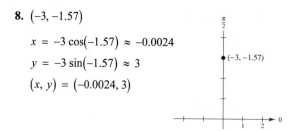

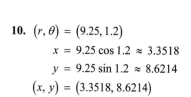

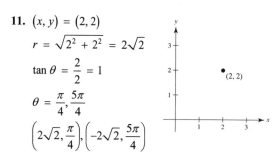

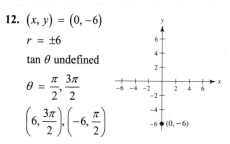

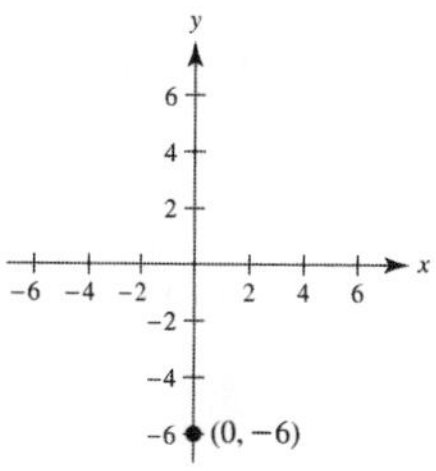

13. $(x, y) = (-3, 4)$

$r = \pm\sqrt{9 + 16} = \pm 5$

$\tan\theta = -\frac{4}{3}$

$\theta \approx 2.214, 5.356, (5, 2.214),$

$(-5, 5.356)$

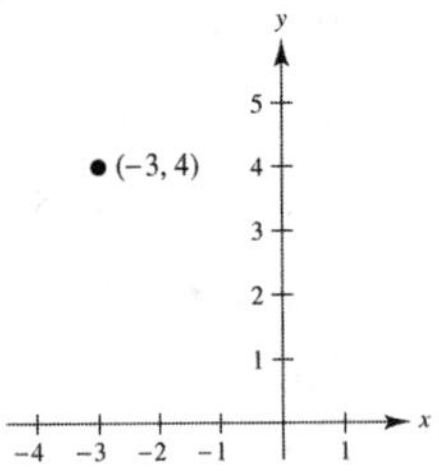

14. $(x, y) = (4, -2)$

$r = \pm\sqrt{16 + 4} = \pm 2\sqrt{5}$

$\tan\theta = -\frac{2}{4} = -\frac{1}{2}$

$\theta \approx -0.464$

$(2\sqrt{5}, -0.464), (-2\sqrt{5}, 2.678)$

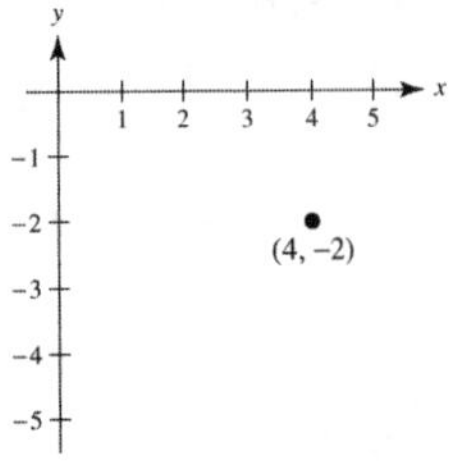

15. $(x, y) = (-1, -\sqrt{3})$

$r = \sqrt{4} = 2$

$\tan\theta = \frac{-\sqrt{3}}{-1} = \sqrt{3}$

$\theta = \frac{\pi}{3}, \frac{4\pi}{3}$

$\left(2, \frac{4\pi}{3}\right), \left(-2, \frac{\pi}{3}\right)$

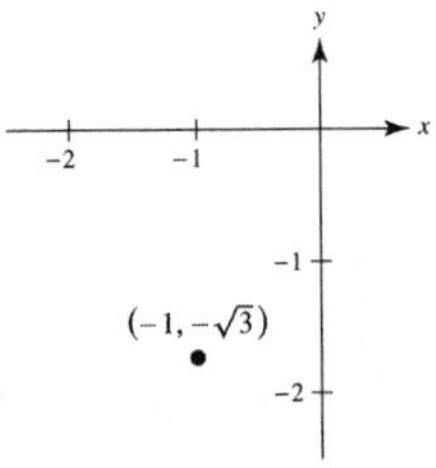

16. $(x, y) = (3, -\sqrt{3})$

$r = \sqrt{9 + 3} = 2\sqrt{3}$

$\tan\theta = -\frac{\sqrt{3}}{3}$

$(r, \theta) = \left(2\sqrt{3}, \frac{11\pi}{6}\right) = \left(-2\sqrt{3}, \frac{5\pi}{6}\right)$

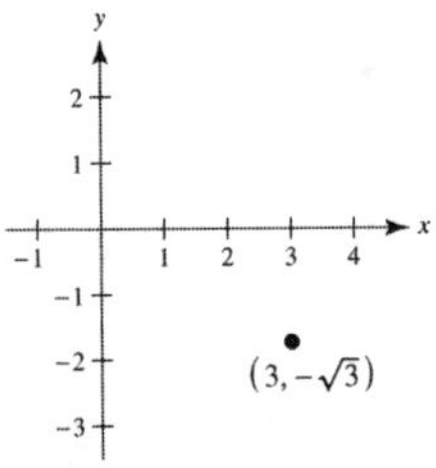

17. $(x, y) = (3, -2)$

$r = \sqrt{3^2 + (-2)^2} = \sqrt{13} \approx 3.6056$

$\tan\theta = -\frac{2}{3} \Rightarrow \theta \approx 5.6952$

$(r, \theta) \approx (3.6056, 5.6952) = (-3.6056, 2.5536)$

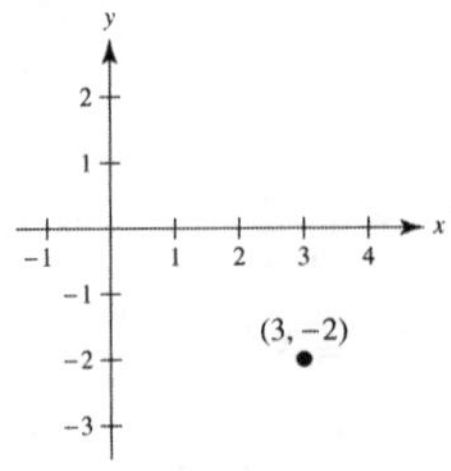

18. $(x, y) = (3\sqrt{2}, 3\sqrt{2})$

$r = \sqrt{(3\sqrt{2})^2 + (3\sqrt{2})^2} = \sqrt{18 + 18} = 6$

$\tan\theta = \frac{3\sqrt{2}}{3\sqrt{2}} = 1 \Rightarrow \theta = \frac{\pi}{4}$

$(r, \theta) = \left(6, \frac{\pi}{4}\right) = \left(-6, \frac{5\pi}{4}\right)$

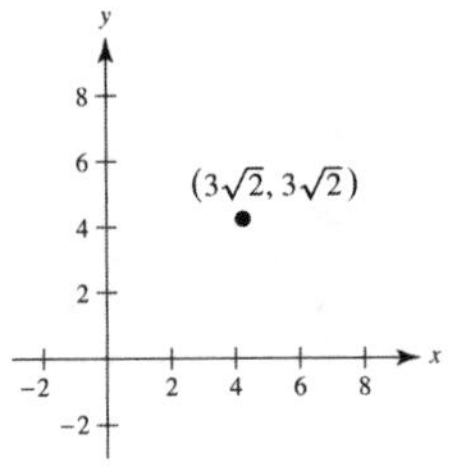

19. $(x, y) = \left(\frac{7}{4}, \frac{5}{2}\right)$

$r = \sqrt{\left(\frac{7}{4}\right)^2 + \left(\frac{5}{2}\right)^2} = \sqrt{\frac{149}{16}} = \frac{\sqrt{149}}{4} \approx 3.0516$

$\tan\theta = \frac{5/2}{7/4} = \frac{10}{7} \Rightarrow \theta \approx 0.9601$

$(r, \theta) \approx (3.0516, 0.9601) \approx (-3.0516, 4.1017)$

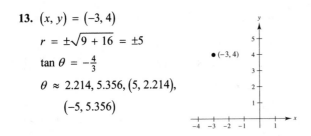

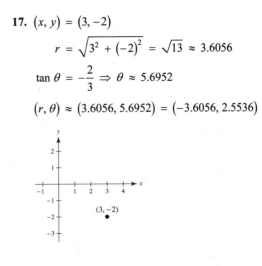

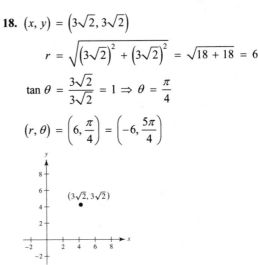

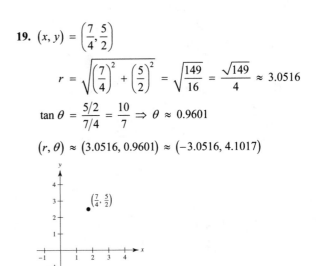

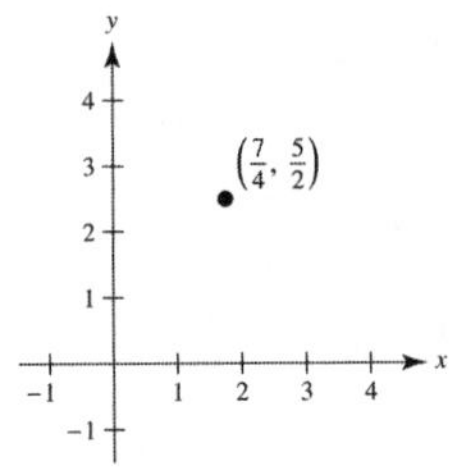

20. $(x, y) = (0, -5)$

$$r = 5, \theta = \frac{3\pi}{2}$$

$$(r, \theta) = \left(5, \frac{3\pi}{2}\right) = \left(-5, \frac{\pi}{2}\right)$$

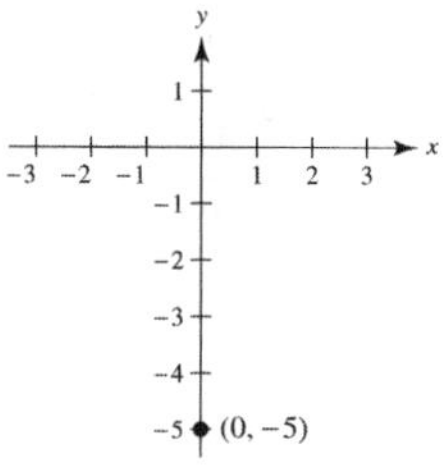

21. (a) $(x, y) = (4, 3.5)$

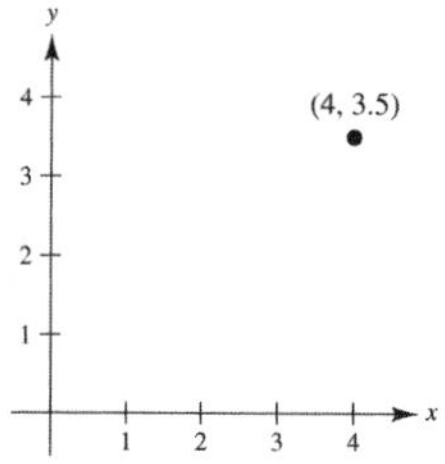

(b) $(r, \theta) = (4, 3.5)$

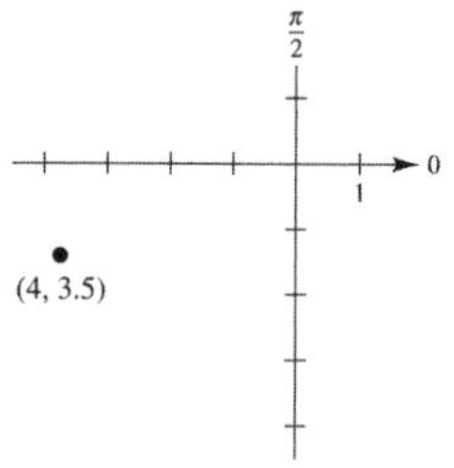

22. (a) Moving horizontally, the x-coordinate changes.
Moving vertically, the y-coordinate changes.
(b) Both r and θ values change.
(c) In polar mode, horizontal (or vertical) changes result in changes in both r and θ.

23. $x^2 + y^2 = 9$

$$r^2 = 9$$

$$r = 3$$

Circle

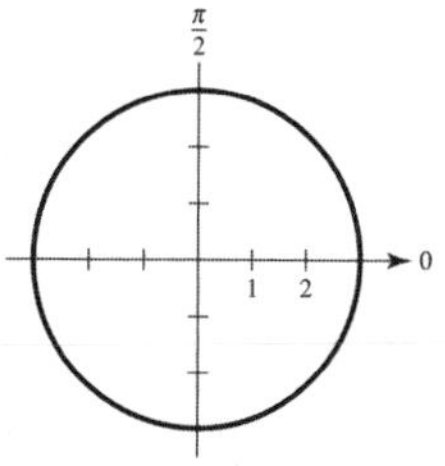

24. $x^2 - y^2 = 9$

$$(r\cos\theta)^2 - (r\sin\theta)^2 = 9$$

$$r^2(\cos^2\theta - \sin^2\theta) = 9$$

$$r^2\cos 2\theta = 9$$

$$r = \frac{3}{\sqrt{\cos 2\theta}}$$

Hyperbola

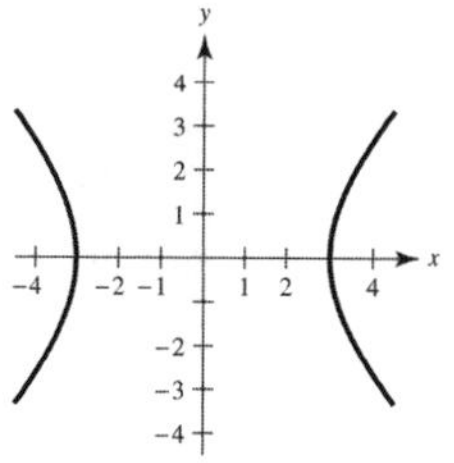

25. $x^2 + y^2 = a^2$

$$r = a$$

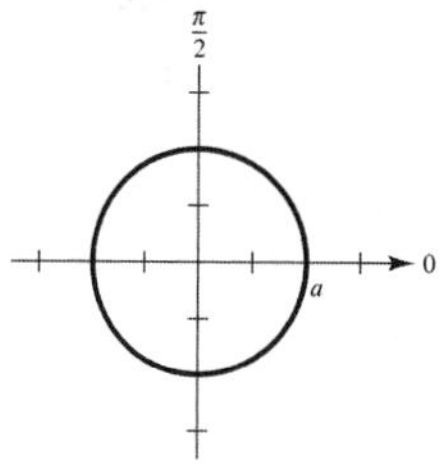

26. $x^2 + y^2 - 2ax = 0$

$$r^2 - 2ar\cos\theta = 0$$

$$r(r - 2a\cos\theta) = 0$$

$$r = 2a\cos\theta$$

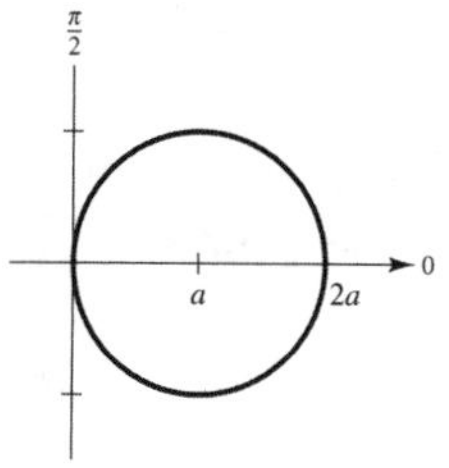

27. $y = 8$

$$r\sin\theta = 8$$

$$r = 8\csc\theta$$

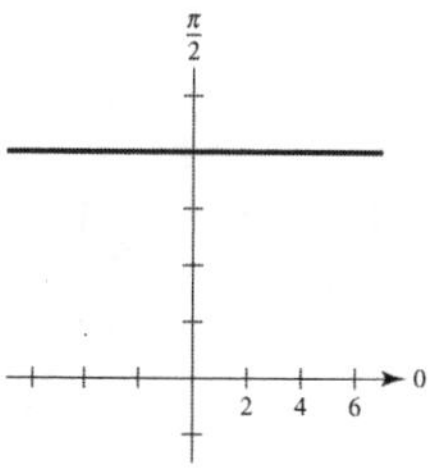

28. $x = 12$

$$r\cos\theta = 12$$

$$r = 12\sec\theta$$

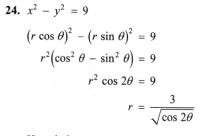

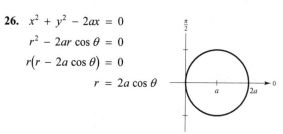

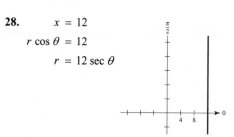

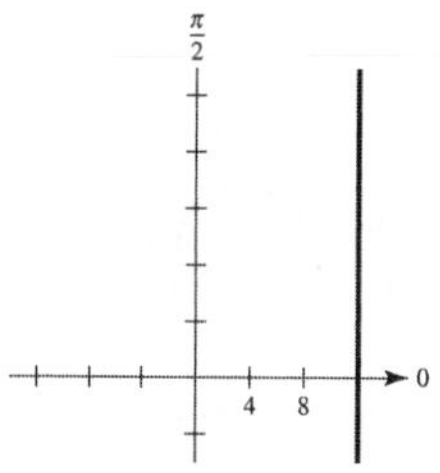

29.
$$3x - y + 2 = 0$$
$$3r\cos\theta - r\sin\theta + 2 = 0$$
$$r(3\cos\theta - \sin\theta) = -2$$
$$r = \frac{-2}{3\cos\theta - \sin\theta}$$

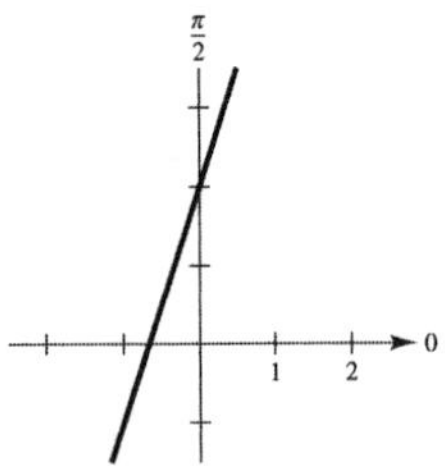

30.
$$xy = 4$$
$$(r\cos\theta)(r\sin\theta) = 4$$
$$r^2 = 4\sec\theta\csc\theta$$
$$= 8\csc 2\theta$$

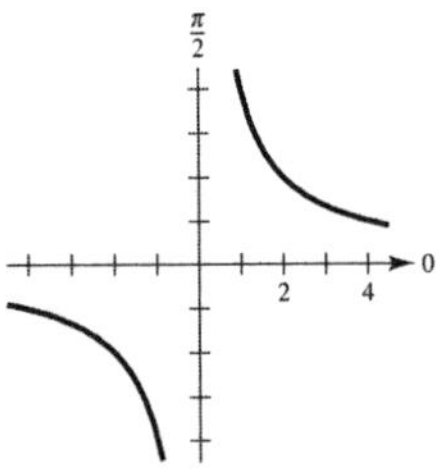

31.
$$y^2 = 9x$$
$$r^2\sin^2\theta = 9r\cos\theta$$
$$r = \frac{9\cos\theta}{\sin^2\theta}$$
$$r = 9\csc^2\theta\cos\theta$$

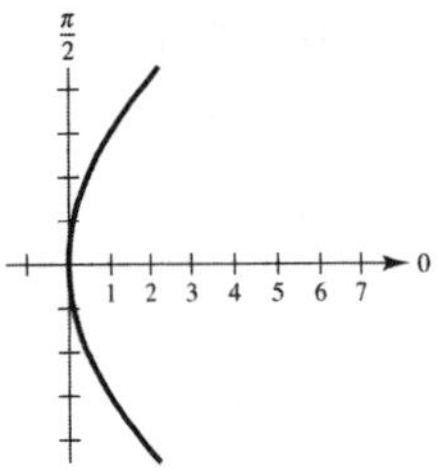

32.
$$(x^2 + y^2)^2 - 9(x^2 - y^2) = 0$$
$$(r^2)^2 - 9(r^2\cos^2\theta - r^2\sin^2\theta) = 0$$
$$r^2[r^2 - 9(\cos 2\theta)] = 0$$
$$r^2 = 9\cos 2\theta$$

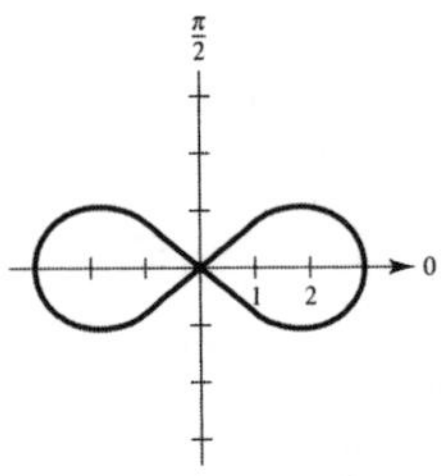

33. $r = 4$
$$r^2 = 16$$
$$x^2 + y^2 = 16$$
Circle

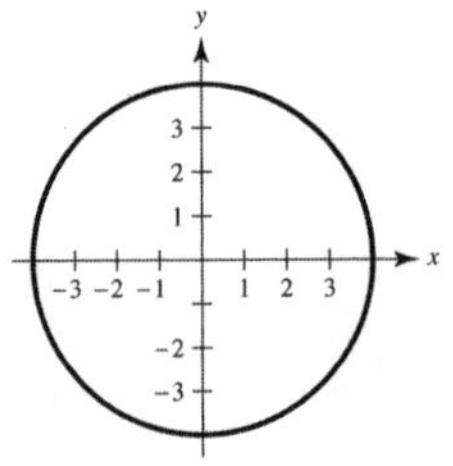

34. $r = -5$
$$r^2 = 25$$
$$x^2 + y^2 = 25$$
Circle

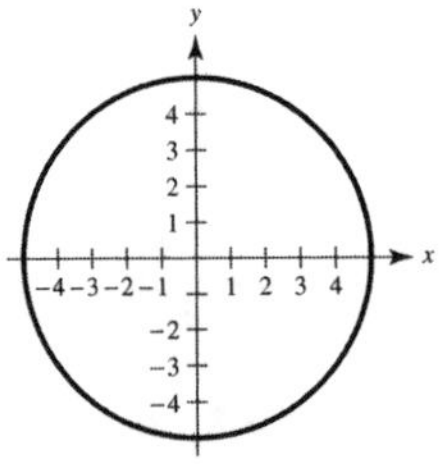

35. $r = 3\sin\theta$
$$r^2 = 3r\sin\theta$$
$$x^2 + y^2 = 3y$$
$$x^2 + \left(y^2 - 3y + \frac{9}{4}\right) = \frac{9}{4}$$
$$x^2 + \left(y - \frac{3}{2}\right)^2 = \frac{9}{4}$$
Circle

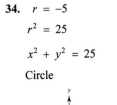

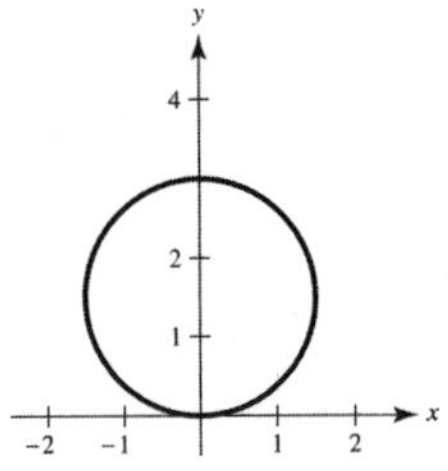

36.
$$\begin{aligned} r &= 5\cos\theta \\ r^2 &= 5r\cos\theta \\ x^2 + y^2 &= 5x \\ x^2 - 5x + \tfrac{25}{4} + y^2 &= \tfrac{25}{4} \\ \left(x - \tfrac{5}{2}\right)^2 + y^2 &= \left(\tfrac{5}{2}\right)^2 \end{aligned}$$

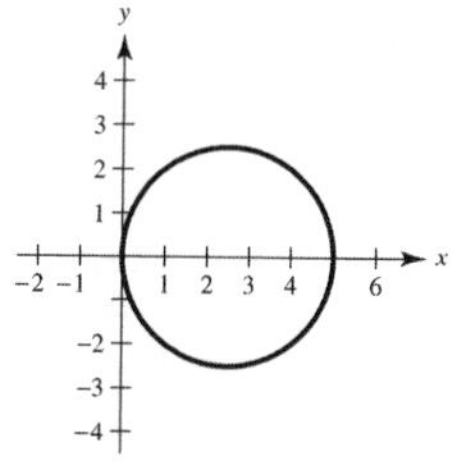

37.
$$\begin{aligned} r &= \theta \\ \tan r &= \tan\theta \\ \tan\sqrt{x^2 + y^2} &= \frac{y}{x} \\ \sqrt{x^2 + y^2} &= \arctan\frac{y}{x} \end{aligned}$$

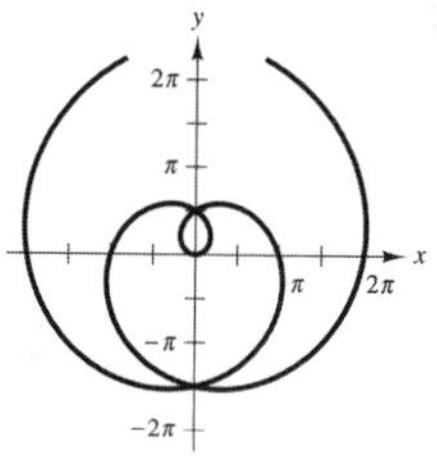

38.
$$\begin{aligned} \theta &= \frac{5\pi}{6} \\ \tan\theta &= \tan\frac{5\pi}{6} \\ \frac{y}{x} &= -\frac{\sqrt{3}}{3} \\ y &= -\frac{\sqrt{3}}{3}x \end{aligned}$$

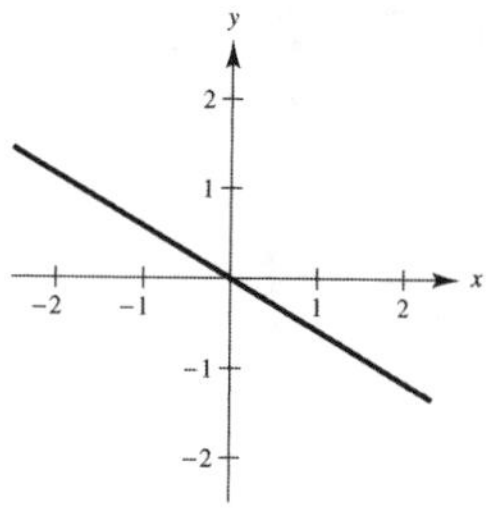

39.
$$\begin{aligned} r &= 3\sec\theta \\ r\cos\theta &= 3 \\ x &= 3 \\ x - 3 &= 0 \end{aligned}$$

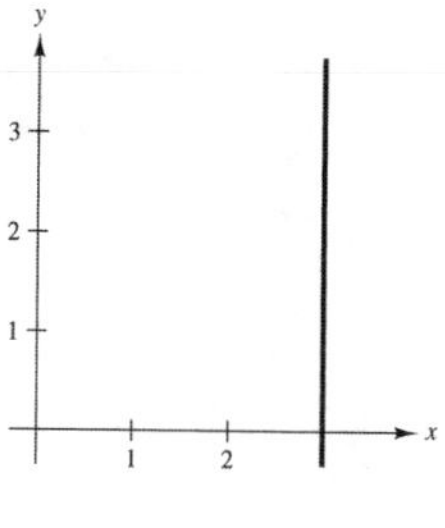

40.
$$\begin{aligned} r &= 2\csc\theta \\ r\sin\theta &= 2 \\ y &= 2 \\ y - 2 &= 0 \end{aligned}$$

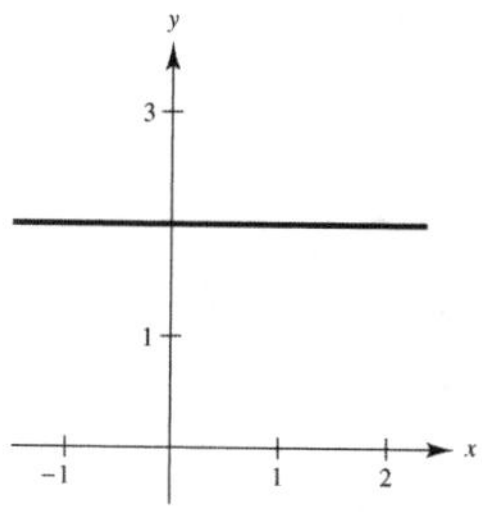

41.
$$\begin{aligned} r &= \sec\theta\tan\theta \\ r\cos\theta &= \tan\theta \\ x &= \frac{y}{x} \\ y &= x^2 \end{aligned}$$

Parabola

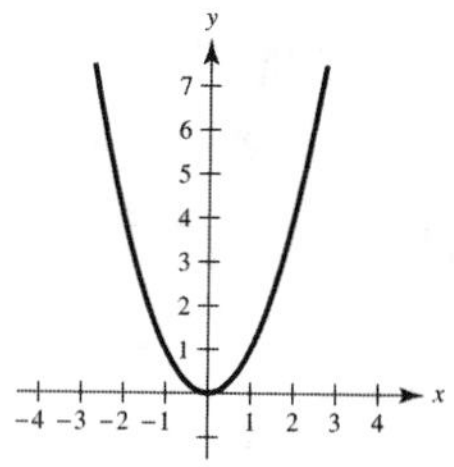

42.
$$\begin{aligned} r &= \cot\theta\csc\theta \\ r\sin\theta &= \cot\theta \\ y &= \frac{x}{y} \\ x &= y^2 \end{aligned}$$

Parabola

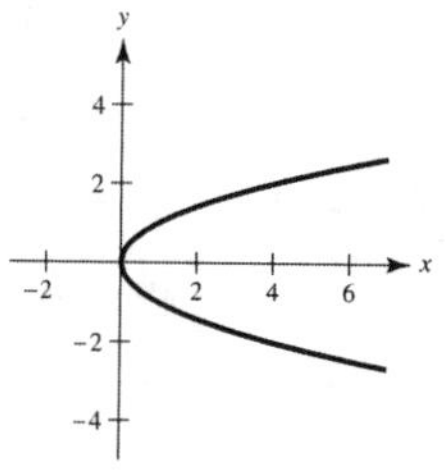

43. $r = 2 - 5\cos\theta$

$0 \le \theta < 2\pi$

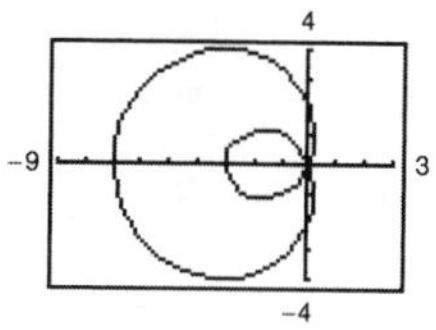

44. $r = 3(1 - 4\cos\theta)$

$0 \le \theta < 2\pi$

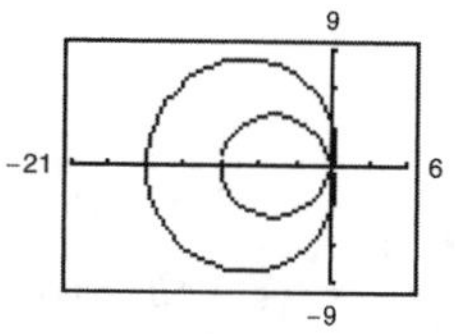

45. $r = 2 + \sin\theta$

$0 \le \theta < 2\pi$

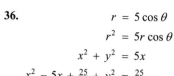

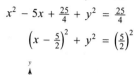

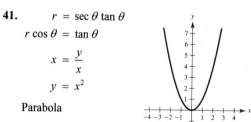

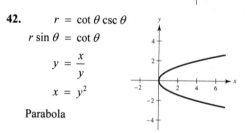

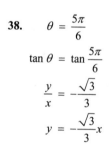

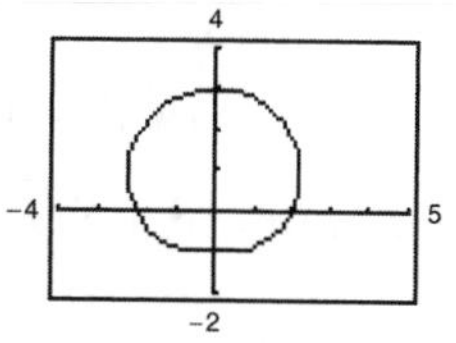

46. $r = 4 + 3\cos\theta$

$0 \le \theta < 2\pi$

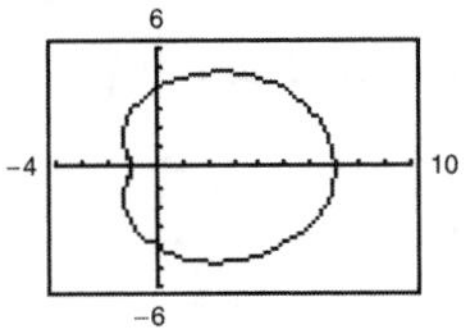

47. $r = \dfrac{2}{1 + \cos\theta}$

Traced out once on $-\pi < \theta < \pi$

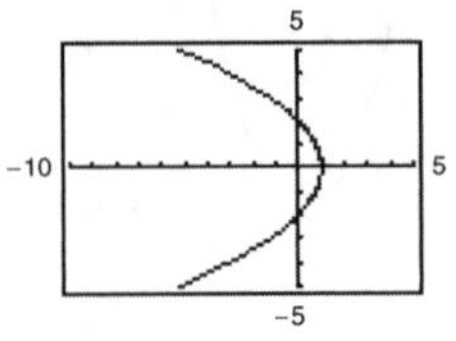

48. $r = \dfrac{2}{4 - 3\sin\theta}$

Traced out once on $0 \le \theta \le 2\pi$

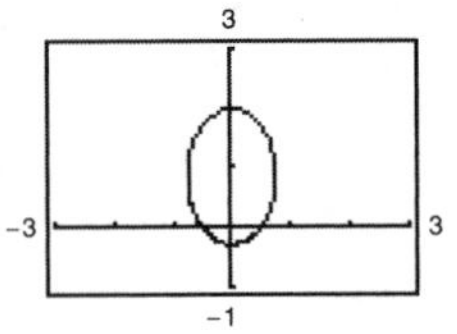

49. $r = 2\cos\left(\dfrac{3\theta}{2}\right)$

$0 \le \theta < 4\pi$

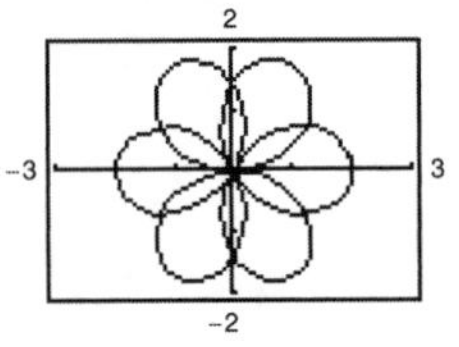

50. $r = 3\sin\left(\dfrac{5\theta}{2}\right)$

$0 \le \theta < 4\pi$

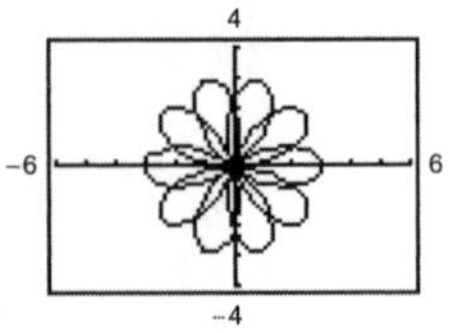

51. $r^2 = 4\sin 2\theta$

$r_1 = 2\sqrt{\sin 2\theta}$

$r_2 = -2\sqrt{\sin 2\theta}$

$0 \le \theta < \dfrac{\pi}{2}$

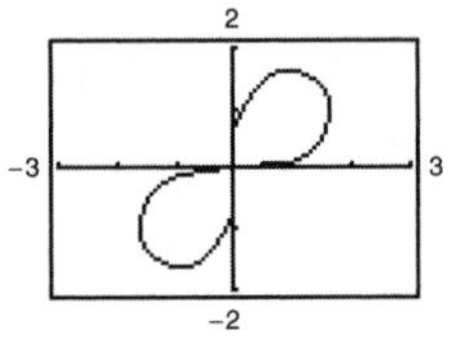

52. $r^2 = \dfrac{1}{\theta}$.

Graph as $r_1 = \dfrac{1}{\sqrt{\theta}}, r_2 = -\dfrac{1}{\sqrt{\theta}}$.

It is traced out once on $[0, \infty)$.

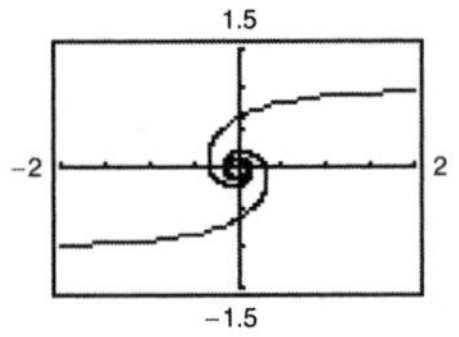

53.

$$\begin{aligned}
r &= 2(h\cos\theta + k\sin\theta)\\
r^2 &= 2r(h\cos\theta + k\sin\theta)\\
r^2 &= 2\left[h(r\cos\theta) + k(r\sin\theta)\right]\\
x^2 + y^2 &= 2(hx + ky)\\
x^2 + y^2 - 2hx - 2ky &= 0\\
\left(x^2 - 2hx + h^2\right) + \left(y^2 - 2ky + k^2\right) &= 0 + h^2 + k^2\\
(x - h)^2 + (y - k)^2 &= h^2 + k^2
\end{aligned}$$

Radius: $\sqrt{h^2 + k^2}$

Center: (h, k)

54. (a) The rectangular coordinates of (r_1, θ_1) are $(r_1 \cos \theta_1, r_1 \sin \theta_1)$. The rectangular coordinates of (r_2, θ_2) are $(r_2 \cos \theta_2, r_2 \sin \theta_2)$.

$$\begin{aligned} d^2 &= (x_2 - x_1)^2 + (y_2 - y_1)^2 \\ &= (r_2 \cos \theta_2 - r_1 \cos \theta_1)^2 + (r_2 \sin \theta_2 - r_1 \sin \theta_1)^2 \\ &= r_2^2 \cos^2 \theta_2 - 2r_1r_2 \cos \theta_1 \cos \theta_2 + r_1^2 \cos^2 \theta_1 + r_2^2 \sin^2 \theta_2^2 - 2r_1r_2 \sin \theta_1 \sin \theta_2 + r_1^2 \sin^2 \theta_1 \\ &= r_2^2(\cos^2 \theta_2 + \sin^2 \theta_2) + r_1^2(\cos^2 \theta_1 + \sin^2 \theta_1) - 2\, r_1r_2(\cos \theta_1 \cos \theta_2 + \sin \theta_1 \sin \theta_2) \\ &= r_1^2 + r_2^2 - 2r_1r_2 \cos(\theta_1 - \theta_2) \\ d &= \sqrt{r_1^2 + r_2^2 - 2r_1r_2 \cos(\theta_1 - \theta_2)} \end{aligned}$$

(b) If $\theta_1 = \theta_2$, the points lie on the same line passing through the origin. In this case,

$$\begin{aligned} d &= \sqrt{r_1^2 + r_2^2 - 2r_1r_2 \cos(0)} \\ &= \sqrt{(r_1 - r_2)^2} = |r_1 - r_2|. \end{aligned}$$

(c) If $\theta_1 - \theta_2 = 90°$, then $\cos(\theta_1 - \theta_2) = 0$ and $d = \sqrt{r_1^2 + r_2^2}$, the Pythagorean Theorem!

(d) Many answers are possible. For example, consider the two points $(r_1, \theta_1) = (1, 0)$ and $(r_2, \theta_2) = \left(2, \frac{\pi}{2}\right)$.

$$d = \sqrt{1 + 2^2 - 2(1)(2) \cos\left(0 - \frac{\pi}{2}\right)} = \sqrt{5}$$

Using $(r_1, \theta_1) = (-1, \pi)$ and $(r_2, \theta_2) = \left[2, \left(\frac{5\pi}{2}\right)\right]$, $d = \sqrt{(-1)^2 + (2)^2 - 2(-1)(2) \cos\left(\pi - \frac{5\pi}{2}\right)} = \sqrt{5}.$

You always obtain the same distance.

55. $\left(1, \frac{5\pi}{6}\right), \left(4, \frac{\pi}{3}\right)$

$$\begin{aligned} d &= \sqrt{1^2 + 4^2 - 2(1)(4) \cos\left(\frac{5\pi}{6} - \frac{\pi}{3}\right)} \\ &= \sqrt{17 - 8 \cos \frac{\pi}{2}} = \sqrt{17} \end{aligned}$$

56. $\left(8, \frac{7\pi}{4}\right), (5, \pi)$

$$\begin{aligned} d &= \sqrt{8^2 + 5^2 - 2(8)(5) \cos\left(\frac{7\pi}{4} - \pi\right)} \\ &= \sqrt{89 - 80 \cos \frac{3\pi}{4}} \\ &= \sqrt{89 - 80\left(-\frac{\sqrt{2}}{2}\right)} \\ &= \sqrt{89 + 40\sqrt{2}} \approx 12.0652 \end{aligned}$$

57. $(2, 0.5), (7, 1.2)$

$$\begin{aligned} d &= \sqrt{2^2 + 7^2 - 2(2)(7) \cos(0.5 - 1.2)} \\ &= \sqrt{53 - 28 \cos(-0.7)} \approx 5.6 \end{aligned}$$

58. $(4, 2.5), (12, 1)$

$$\begin{aligned} d &= \sqrt{4^2 + 12^2 - 2(4)(12) \cos(2.5 - 1)} \\ &= \sqrt{160 - 96 \cos 1.5} \approx 12.3 \end{aligned}$$

59. $r = 2 + 3 \sin \theta$

$$\begin{aligned} \frac{dy}{dx} &= \frac{3 \cos \theta \sin \theta + \cos \theta(2 + 3 \sin \theta)}{3 \cos \theta \cos \theta - \sin \theta(2 + 3 \sin \theta)} \\ &= \frac{2 \cos \theta(3 \sin \theta + 1)}{3 \cos 2\theta - 2 \sin \theta} = \frac{2 \cos \theta(3 \sin \theta + 1)}{6 \cos^2 \theta - 2 \sin \theta - 3} \end{aligned}$$

At $\left(5, \frac{\pi}{2}\right)$, $\frac{dy}{dx} = 0$.

At $(2, \pi)$, $\frac{dy}{dx} = -\frac{2}{3}$.

At $\left(-1, \frac{3\pi}{2}\right)$, $\frac{dy}{dx} = 0$.

60. $r = 2(1 - \sin\theta)$

$$\frac{dy}{dx} = \frac{-2\cos\theta\sin\theta + 2\cos\theta(1 - \sin\theta)}{-2\cos\theta\cos\theta - 2\sin\theta(1 - \sin\theta)}$$

At $(2, 0)$, $\dfrac{dy}{dx} = -1$.

At $\left(3, \dfrac{7\pi}{6}\right)$, $\dfrac{dy}{dx}$ is undefined.

At $\left(4, \dfrac{3\pi}{2}\right)$, $\dfrac{dy}{dx} = 0$.

61. (a), (b) $r = 3(1 - \cos\theta)$

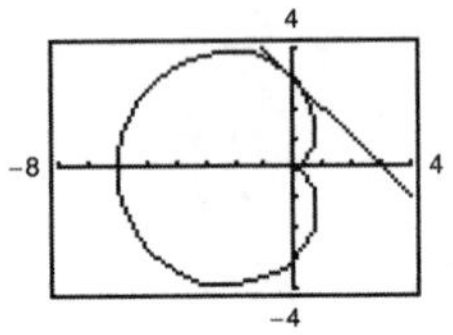

$(r, \theta) = \left(3, \dfrac{\pi}{2}\right) \Rightarrow (x, y) = (0, 3)$

Tangent line: $y - 3 = -1(x - 0)$

$y = -x + 3$

(c) At $\theta = \dfrac{\pi}{2}$, $\dfrac{dy}{dx} = -1.0$.

62. (a), (b) $r = 3 - 2\cos\theta$

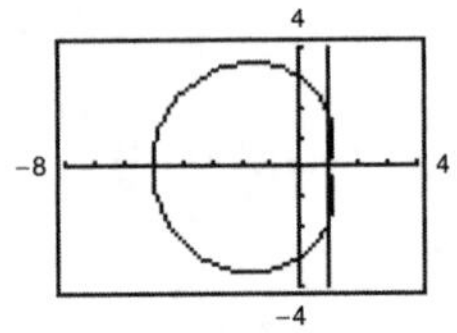

$(r, \theta) = (1, 0) \Rightarrow (x, y) = (1, 0)$

Tangent line: $x = 1$

(c) At $\theta = 0$, $\dfrac{dy}{dx}$ does not exist (vertical tangent).

63. (a), (b) $r = 3\sin\theta$

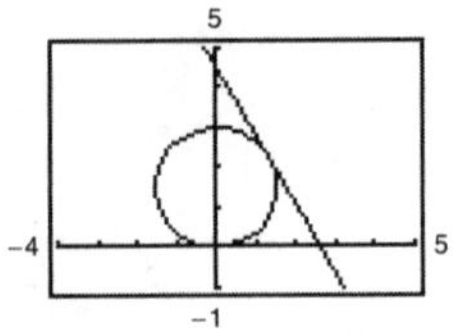

$(r, \theta) = \left(\dfrac{3\sqrt{3}}{2}, \dfrac{\pi}{3}\right) \Rightarrow (x, y) = \left(\dfrac{3\sqrt{3}}{4}, \dfrac{9}{4}\right)$

Tangent line: $y - \dfrac{9}{4} = -\sqrt{3}\left(x - \dfrac{3\sqrt{3}}{4}\right)$

$y = -\sqrt{3}x + \dfrac{9}{2}$

(c) At $\theta = \dfrac{\pi}{3}$, $\dfrac{dy}{dx} = -\sqrt{3} \approx -1.732$.

64. (a), (b) $r = 4$

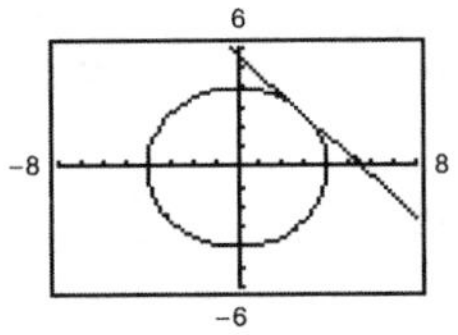

$(r, \theta) = \left(4, \dfrac{\pi}{4}\right) \Rightarrow (x, y) = \left(2\sqrt{2}, 2\sqrt{2}\right)$

Tangent line: $y - 2\sqrt{2} = -1\left(x - 2\sqrt{2}\right)$

$y = -x + 4\sqrt{2}$

(c) At $\theta = \dfrac{\pi}{4}$, $\dfrac{dy}{dx} = -1$.

65. $r = 1 - \sin\theta$

$$\frac{dy}{d\theta} = (1 - \sin\theta)\cos\theta - \cos\theta\sin\theta$$
$$= \cos\theta(1 - 2\sin\theta) = 0$$

$\cos\theta = 0$ or $\sin\theta = \dfrac{1}{2} \Rightarrow \theta = \dfrac{\pi}{2}, \dfrac{3\pi}{2}, \dfrac{\pi}{6}, \dfrac{5\pi}{6}$

Horizontal tangents: $\left(2, \dfrac{3\pi}{2}\right), \left(\dfrac{1}{2}, \dfrac{\pi}{6}\right), \left(\dfrac{1}{2}, \dfrac{5\pi}{6}\right)$

$$\frac{dx}{d\theta} = (-1 + \sin\theta)\sin\theta - \cos\theta\cos\theta$$
$$= -\sin\theta + \sin^2\theta + \sin^2\theta - 1$$
$$= 2\sin^2\theta - \sin\theta - 1$$
$$= (2\sin\theta + 1)(\sin\theta - 1) = 0$$

$\sin\theta = 1$ or $\sin\theta = -\dfrac{1}{2} \Rightarrow \theta = \dfrac{\pi}{2}, \dfrac{7\pi}{6}, \dfrac{11\pi}{6}$

Vertical tangents: $\left(\dfrac{3}{2}, \dfrac{7\pi}{6}\right), \left(\dfrac{3}{2}, \dfrac{11\pi}{6}\right)$

66. $r = a \sin\theta$

$$\frac{dy}{d\theta} = a\sin\theta\cos\theta + a\cos\theta\sin\theta$$
$$= 2a\sin\theta\cos\theta = 0$$

$$\theta = 0, \frac{\pi}{2}, \pi, \frac{3\pi}{2}$$

$$\frac{dx}{d\theta} = -a\sin^2\theta + a\cos^2\theta = a\left(1 - 2\sin^2\theta\right) = 0$$

$$\sin\theta = \pm\frac{1}{\sqrt{2}}, \theta = \frac{\pi}{4}, \frac{3\pi}{4}, \frac{5\pi}{4}, \frac{7\pi}{4}$$

Horizontal tangents: $(0, 0), \left(a, \frac{\pi}{2}\right)$

Vertical tangents: $\left(\frac{a\sqrt{2}}{2}, \frac{\pi}{4}\right), \left(\frac{a\sqrt{2}}{2}, \frac{3\pi}{4}\right)$

67. $r = 2\csc\theta + 3$

$$\frac{dy}{d\theta} = \left(2\csc\theta + 3\right)\cos\theta + \left(-2\csc\theta\cot\theta\right)\sin\theta$$
$$= 3\cos\theta = 0$$

$$\theta = \frac{\pi}{2}, \frac{3\pi}{2}$$

Horizontal tangents: $\left(5, \frac{\pi}{2}\right), \left(1, \frac{3\pi}{2}\right)$

68. $r = a\sin\theta\cos^2\theta$

$$\frac{dy}{d\theta} = a\sin\theta\cos^3\theta + \left[-2a\sin^2\theta\cos\theta + a\cos^3\theta\right]\sin\theta$$
$$= 2a\left[\sin\theta\cos^3\theta - \sin^3\theta\cos\theta\right]$$
$$= 2a\sin\theta\cos\theta\left(\cos^2\theta - \sin^2\theta\right) = 0$$

$$\theta = 0, \tan^2\theta = 1, \theta = \frac{\pi}{4}, \frac{3\pi}{4}$$

Horizontal tangents: $\left(\frac{\sqrt{2}a}{4}, \frac{\pi}{4}\right), \left(\frac{\sqrt{2}a}{4}, \frac{3\pi}{4}\right), (0, 0)$

69. $r = 5\sin\theta$

$$r^2 = 5r\sin\theta$$
$$x^2 + y^2 = 5y$$
$$x^2 + \left(y^2 - 5y + \tfrac{25}{4}\right) = \tfrac{25}{4}$$
$$x^2 + \left(y - \tfrac{5}{2}\right)^2 = \tfrac{25}{4}$$

Circle: center: $\left(0, \frac{5}{2}\right)$, radius: $\frac{5}{2}$

Tangent at pole: $\theta = 0$

Note: $f(\theta) = r = 5\sin\theta$

$f(0) = 0, f'(0) \neq 0$

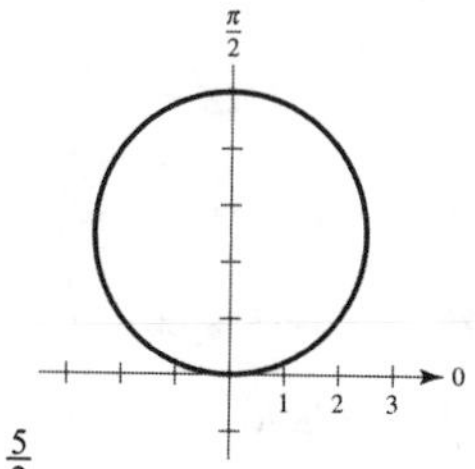

70. $r = 5\cos\theta$

$$r^2 = 5r\cos\theta$$
$$x^2 + y^2 = 5x$$
$$\left(x^2 - 5x + \frac{25}{4}\right) + y^2 = \frac{25}{4}$$
$$\left(x - \frac{5}{2}\right)^2 + y^2 = \frac{25}{4}$$

Circle: center: $\left(\frac{5}{2}, 0\right)$, radius: $\frac{5}{2}$

Tangent at pole: $\theta = \frac{\pi}{2}$.

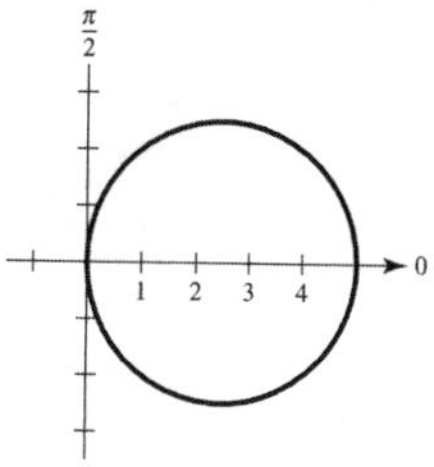

71. $r = 2\left(1 - \sin\theta\right)$

Cardioid

Symmetric to y-axis, $\theta = \frac{\pi}{2}$

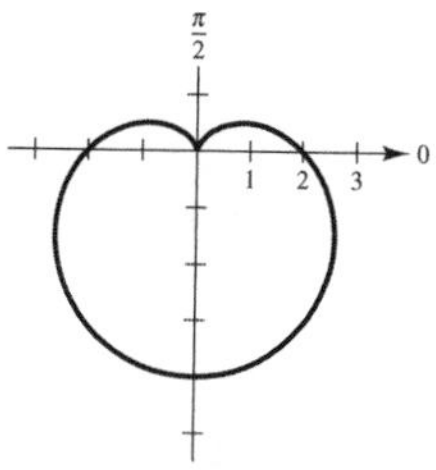

72. $r = 3\left(1 - \cos\theta\right)$

Cardioid

Symmetric to polar axis since r is a function of $\cos\theta$.

θ	0	$\frac{\pi}{3}$	$\frac{\pi}{2}$	$\frac{2\pi}{3}$	π
r	0	$\frac{3}{2}$	3	$\frac{9}{2}$	6

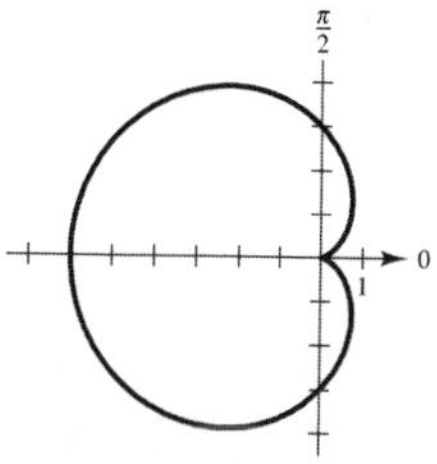

73. $r = 4\cos 3\theta$

Rose curve with three petals.

Tangents at pole: $(r = 0, r' \neq 0)$:

$\theta = \dfrac{\pi}{6}, \dfrac{\pi}{2}, \dfrac{5\pi}{6}$

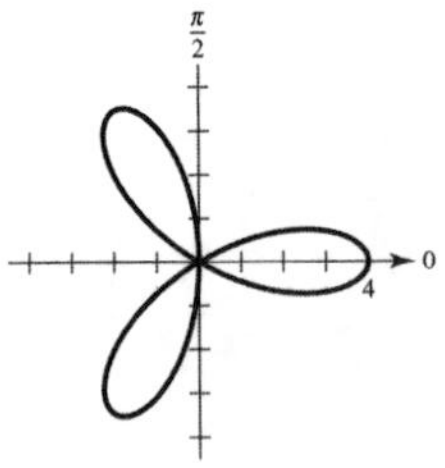

74. $r = -\sin(5\theta)$

Rose curve with five petals

Symmetric to $\theta = \dfrac{\pi}{2}$

Relative extrema occur when

$\dfrac{dr}{d\theta} = -5\cos(5\theta) = 0$ at $\theta = \dfrac{\pi}{10}, \dfrac{3\pi}{10}, \dfrac{5\pi}{10}, \dfrac{7\pi}{10}, \dfrac{9\pi}{10}.$

Tangents at the pole: $\theta = 0, \dfrac{\pi}{5}, \dfrac{2\pi}{5}, \dfrac{3\pi}{5}, \dfrac{4\pi}{5}$

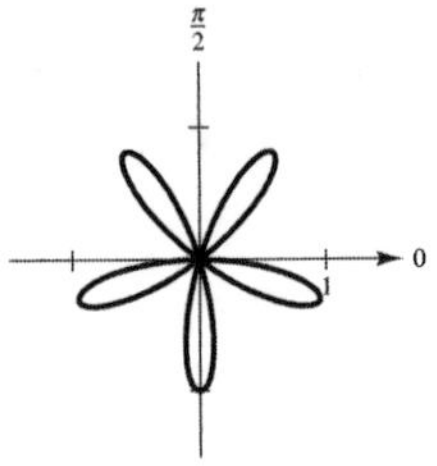

75. $r = 3\sin 2\theta$

Rose curve with four petals

Symmetric to the polar axis, $\theta = \dfrac{\pi}{2}$, and pole

Relative extrema: $\left(\pm 3, \dfrac{\pi}{4}\right), \left(\pm 3, \dfrac{5\pi}{4}\right)$

Tangents at the pole: $\theta = 0, \dfrac{\pi}{2}$

$\left(\theta = \pi, \dfrac{3\pi}{2} \text{ give the same tangents.}\right)$

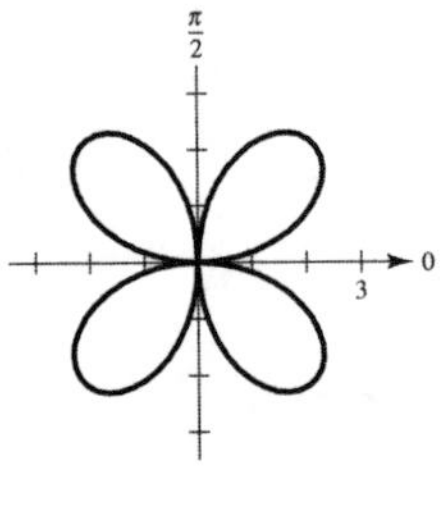

76. $r = 3\cos 2\theta$

Rose curve with four petals

Symmetric to the polar axis, $\theta = \dfrac{\pi}{2}$, and pole

Relative extrema: $(3, 0), \left(-3, \dfrac{\pi}{2}\right), (3, \pi), \left(-3, \dfrac{3\pi}{2}\right)$

Tangents at the pole: $\theta = \dfrac{\pi}{4}, \dfrac{3\pi}{4}$

$\theta = \dfrac{5\pi}{4}$ and $\dfrac{7\pi}{4}$ given the same tangents.

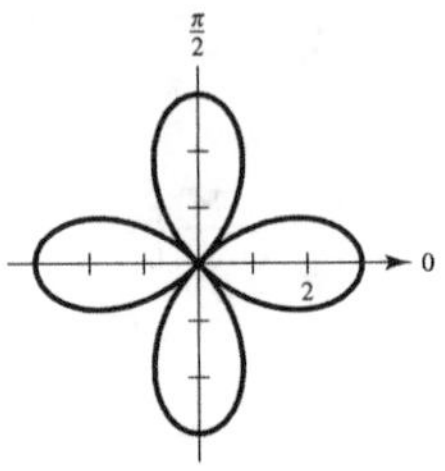

77. $r = 8$

Circle radius 8

$x^2 + y^2 = 64$

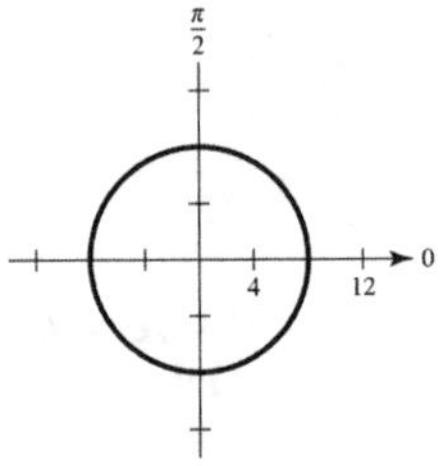

78. $r = 1$

Circle radius 1

$x^2 + y^2 = 1$

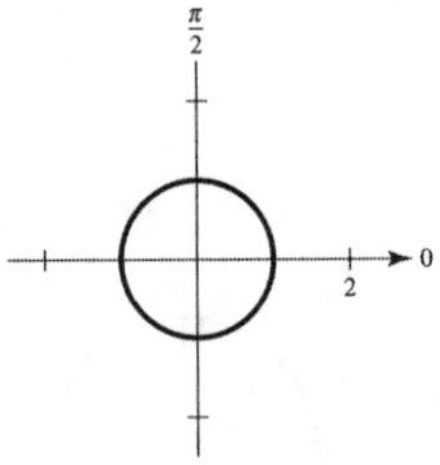

79. $r = 4(1 + \cos\theta)$

Cardioid

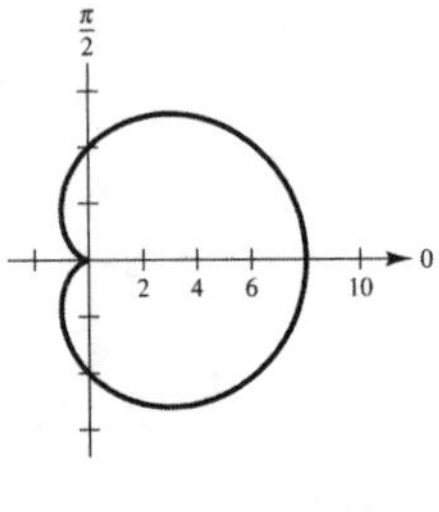

80. $r = 1 + \sin\theta$

Cardioid

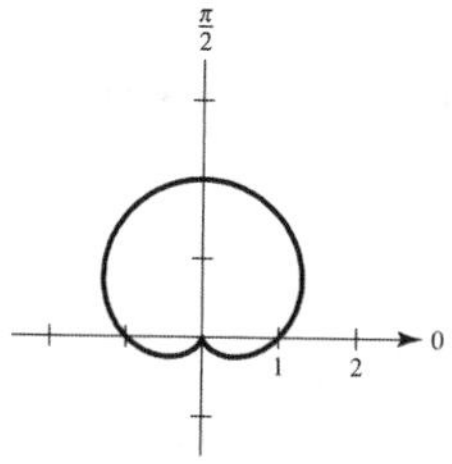

81. $r = 3 - 2\cos\theta$

Limaçon

Symmetric to polar axis

θ	0	$\frac{\pi}{3}$	$\frac{\pi}{2}$	$\frac{2\pi}{3}$	π
r	1	2	3	4	5

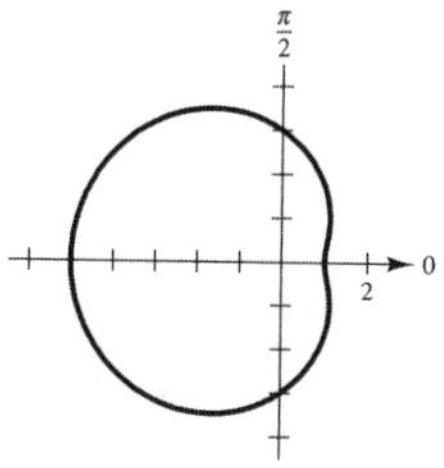

82. $r = 5 - 4\sin\theta$

Limaçon

Symmetric to $\theta = \frac{\pi}{2}$

θ	$-\frac{\pi}{2}$	$-\frac{\pi}{6}$	0	$\frac{\pi}{6}$	$\frac{\pi}{2}$
r	9	7	5	3	1

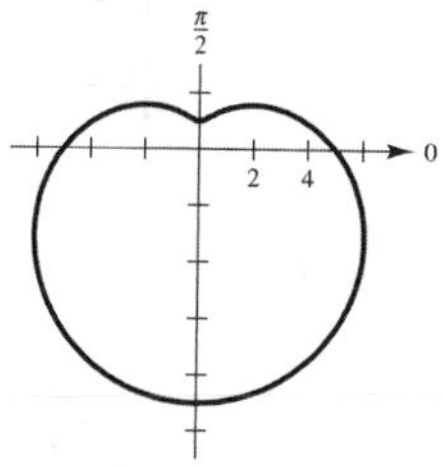

83.

$$r = 3\csc\theta$$
$$r\sin\theta = 3$$
$$y = 3$$

Horizontal line

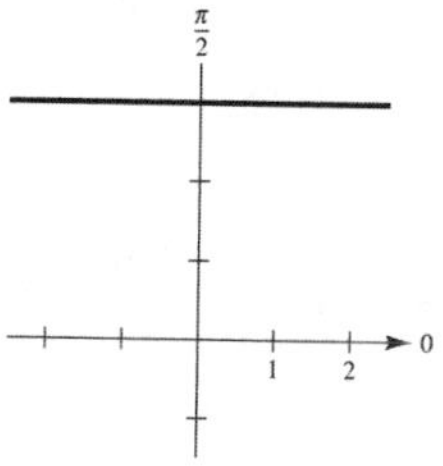

84.

$$r = \frac{6}{2\sin\theta - 3\cos\theta}$$
$$2r\sin\theta - 3r\cos\theta = 6$$
$$2y - 3x = 6$$

Line

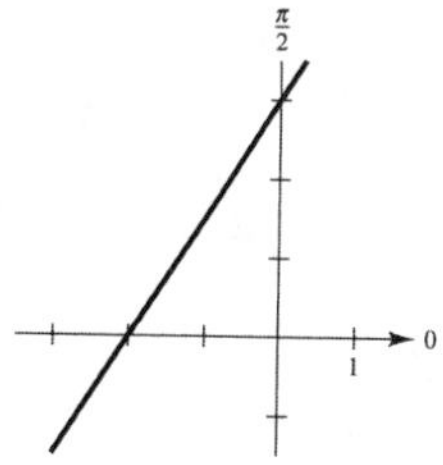

85. $r = 2\theta$

Spiral of Archimedes

Symmetric to $\theta = \frac{\pi}{2}$

θ	0	$\frac{\pi}{4}$	$\frac{\pi}{2}$	$\frac{3\pi}{4}$	π	$\frac{5\pi}{4}$	$\frac{3\pi}{2}$
r	0	$\frac{\pi}{2}$	π	$\frac{3\pi}{2}$	2π	$\frac{5\pi}{2}$	3π

Tangent at the pole: $\theta = 0$

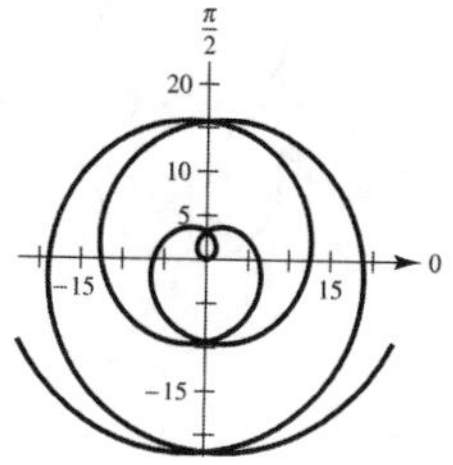

86. $r = \frac{1}{\theta}$

Hyperbolic spiral

θ	$\frac{\pi}{4}$	$\frac{\pi}{2}$	$\frac{3\pi}{4}$	π	$\frac{5\pi}{4}$	$\frac{3\pi}{2}$
r	$\frac{4}{\pi}$	$\frac{2}{\pi}$	$\frac{4}{3\pi}$	$\frac{1}{\pi}$	$\frac{4}{5\pi}$	$\frac{2}{3\pi}$

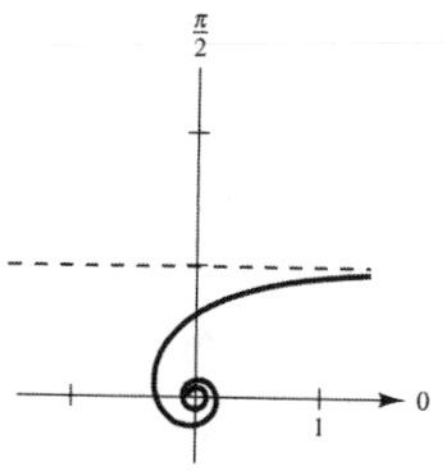

87. $r^2 = 4\cos(2\theta)$

$r = 2\sqrt{\cos 2\theta}, \quad 0 \le \theta \le 2\pi$

Lemniscate

Symmetric to the polar axis, $\theta = \dfrac{\pi}{2}$, and pole

Relative extrema: $(\pm 2, 0)$

θ	0	$\frac{\pi}{6}$	$\frac{\pi}{4}$
r	± 2	$\pm\sqrt{2}$	0

Tangents at the pole: $\theta = \dfrac{\pi}{4}, \dfrac{3\pi}{4}$

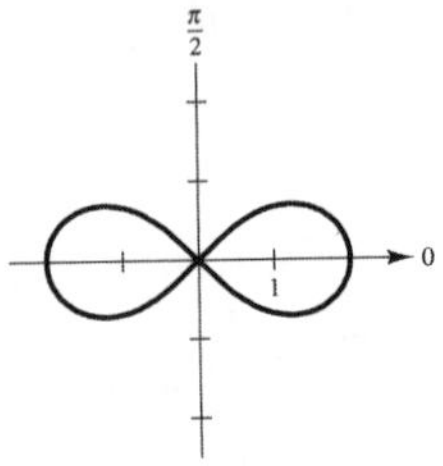

88. $r^2 = 4\sin\theta$

Lemniscate

Symmetric to the polar axis,

$\theta = \dfrac{\pi}{2}$, and pole

Relative extrema: $\left(\pm 2, \dfrac{\pi}{2}\right)$

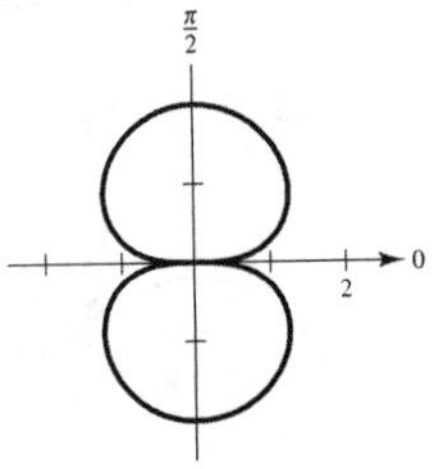

θ	0	$\frac{\pi}{6}$	$\frac{\pi}{2}$	$\frac{5\pi}{6}$	π
r	0	$\pm\sqrt{2}$	± 2	$\pm\sqrt{2}$	0

Tangent at the pole: $\theta = 0$

89. Because

$$r = 2 - \sec\theta = 2 - \frac{1}{\cos\theta},$$

the graph has polar axis symmetry and the tangents at the pole are

$$\theta = \frac{\pi}{3}, -\frac{\pi}{3}.$$

Furthermore,

$$r \Rightarrow -\infty \text{ as } \theta \Rightarrow \frac{\pi}{2^-}$$

$$r \Rightarrow \infty \text{ as } \theta \Rightarrow -\frac{\pi}{2^+}.$$

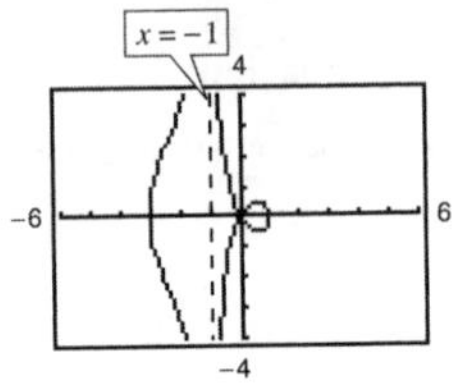

Also,

$$r = 2 - \frac{1}{\cos\theta}$$

$$= 2 - \frac{r}{r\cos\theta} = 2 - \frac{r}{x}$$

$$rx = 2x - r$$

$$r = \frac{2x}{1 + x}.$$

So, $r \Rightarrow \pm\infty$ as $x \Rightarrow -1$.

90. Because

$$r = 2 + \csc\theta = 2 + \frac{1}{\sin\theta},$$

the graphs has symmetry with respect to $\theta = \pi/2$. Furthermore,

$$r \Rightarrow \infty \text{ as } \theta \Rightarrow 0^+$$

$$r \Rightarrow \infty \text{ as } \theta \Rightarrow \pi^-.$$

Also, $r = 2 + \dfrac{1}{\sin\theta} = 2 + \dfrac{r}{\sin\theta} = 2 + \dfrac{r}{y}$

$$ry = 2y + r$$

$$r = \frac{2y}{y - 1}.$$

So, $r \Rightarrow \pm\infty$ as $y \Rightarrow 1$.

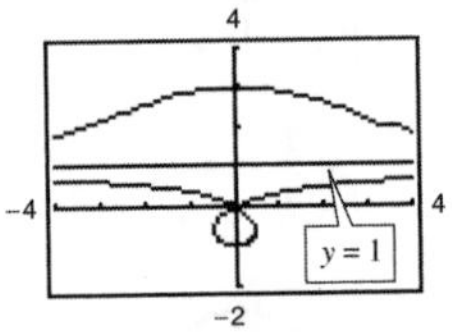

91. $r = \dfrac{2}{\theta}$

Hyperbolic spiral

$$r \Rightarrow \infty \text{ as } \theta \Rightarrow 0$$

$$r = \frac{2}{\theta} \Rightarrow \theta = \frac{2}{r} = \frac{2\sin\theta}{r\sin\theta} = \frac{2\sin\theta}{y}$$

$$y = \frac{2\sin\theta}{\theta}$$

$$\lim_{\theta\to 0}\frac{2\sin\theta}{\theta} = \lim_{\theta\to 0}\frac{2\cos\theta}{1}$$

$$= 2$$

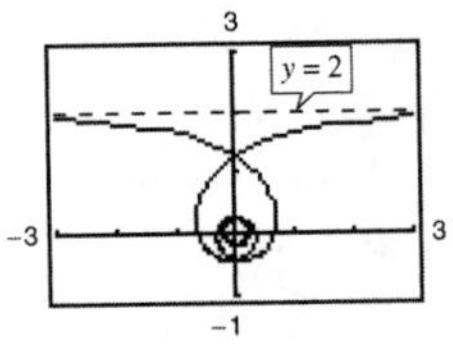

92. $r = 2\cos 2\theta \sec\theta$

Strophoid

$r \Rightarrow -\infty$ as $\theta \Rightarrow \dfrac{\pi^-}{2}$

$r \Rightarrow \infty$ as $\theta \Rightarrow \dfrac{-\pi^+}{2}$

$$r = 2\cos 2\theta \sec\theta = 2(2\cos^2\theta - 1)\sec\theta$$

$$r\cos\theta = 4\cos^2\theta - 2$$

$$x = 4\cos^2\theta - 2$$

$$\lim_{\theta\to\pm\pi/2}(4\cos^2\theta - 2) = -2$$

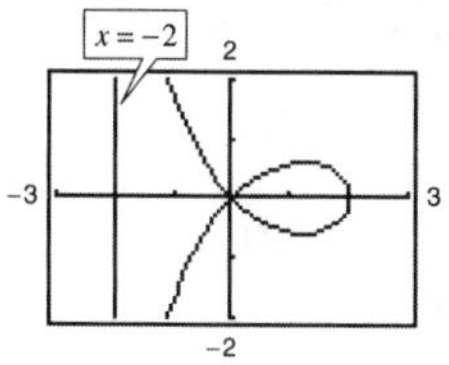

93. The rectangular coordinate system consists of all points of the form (x, y) where x is the directed distance from the y-axis to the point, and y is the directed distance from the x-axis to the point.

Every point has a unique representation.

The polar coordinate system uses (r, θ) to designate the location of a point.

r is the directed distance to the origin and θ is the angle the point makes with the positive x-axis, measured counterclockwise.

Points do not have a unique polar representation.

94. $x = r\cos\theta,\ y = r\sin\theta$

$x^2 + y^2 = r^2,\ \tan\theta = \dfrac{y}{x}$

95. Slope of tangent line to graph of $r = f(\theta)$ at (r, θ) is

$$\frac{dy}{dx} = \frac{f(\theta)\cos\theta + f'(\theta)\sin\theta}{-f(\theta)\sin\theta + f'(\theta)\cos\theta}.$$

If $f(\alpha) = 0$ and $f'(\alpha) \neq 0$, then $\theta = \alpha$ is tangent at the pole.

96. (a) The graph is a circle, where $a = 2$ is measured along the y-axis. So, the equation of the polar graph is $r = 2\sin\theta$.

(b) The graph is a rose curve with $n = 3$ petals and $a = 3$. So, the equation of the polar graph is $r = 3\sin 3\theta$.

(c) The graph is a rose curve with $2n = 4$ petals and $a = 4$. So, the equation of the polar graph is $r = 4\cos 2\theta$.

(d) The graph is a lemniscate with $a = 3$, which is measured along the x-axis. So, the equation of the polar graph is $r^2 = 9\cos 2\theta$.

97. $r = 4\sin\theta$

(a) $0 \le \theta \le \dfrac{\pi}{2}$

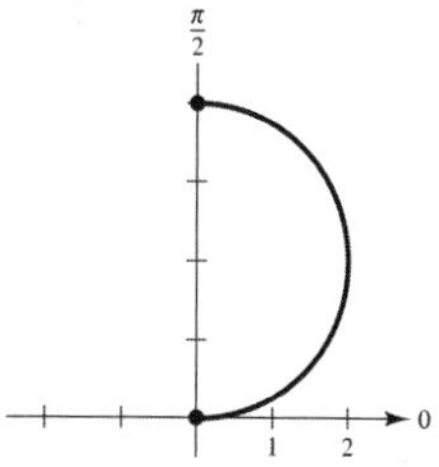

(b) $\dfrac{\pi}{2} \le \theta \le \pi$

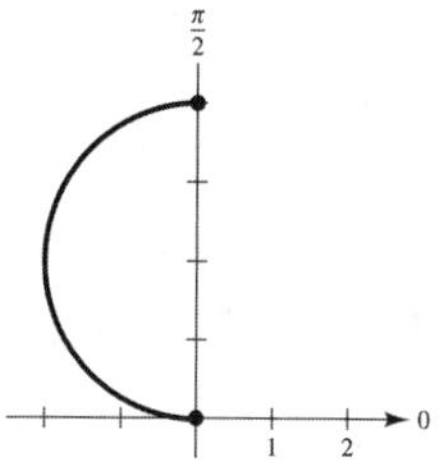

(c) $-\dfrac{\pi}{2} \le \theta \le \dfrac{\pi}{2}$

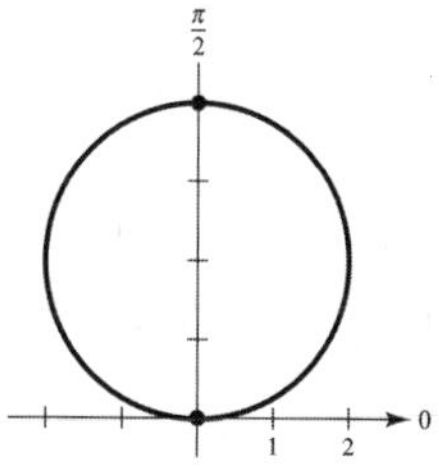

98. $r = 6[1 + \cos(\theta - \phi)]$

(a) $\phi = 0, r = 6[1 + \cos\theta]$

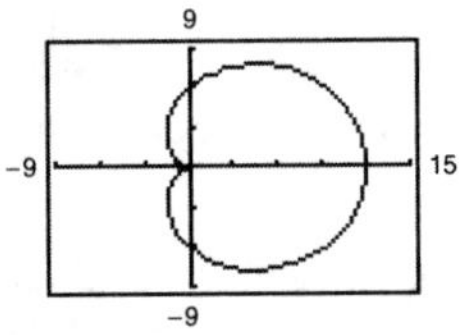

(b) $\theta = \dfrac{\pi}{4}, r = 6\left[1 + \cos\left(\theta - \dfrac{\pi}{4}\right)\right]$

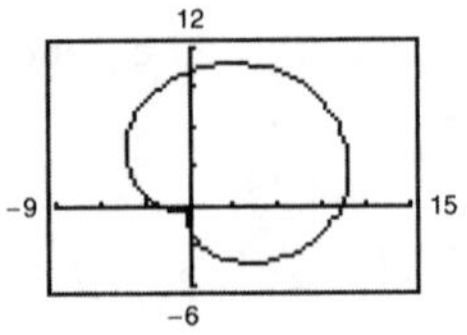

The graph of $r = 6[1 + \cos\theta]$ is rotated through the angle $\pi/4$.

(c) $\theta = \dfrac{\pi}{2}$

$$r = 6\left[1 + \cos\left(\theta - \frac{\pi}{2}\right)\right]$$

$$= 6\left[1 + \cos\theta\cos\frac{\pi}{2} + \sin\theta\sin\frac{\pi}{2}\right] = 6[1 + \sin\theta]$$

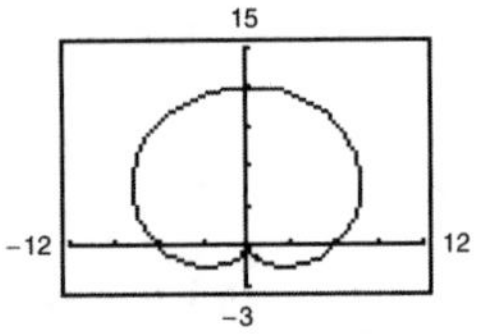

The graph of $r = 6[1 + \cos\theta]$ is rotated through the angle $\pi/2$.

99. Let the curve $r = f(\theta)$ be rotated by ϕ to form the curve $r = g(\theta)$. If (r_1, θ_1) is a point on $r = f(\theta)$, then $(r_1, \theta_1 + \phi)$ is on $r = g(\theta)$. That is,

$g(\theta_1 + \phi) = r_1 = f(\theta_1)$.

Letting $\theta = \theta_1 + \phi$, or $\theta_1 = \theta - \phi$, you see that

$g(\theta) = g(\theta_1 + \phi) = f(\theta_1) = f(\theta - \phi)$.

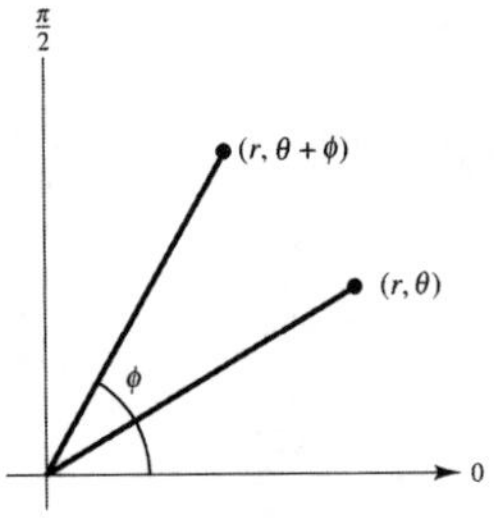

100. (a) $\sin\left(\theta - \dfrac{\pi}{2}\right) = \sin\theta\cos\left(\dfrac{\pi}{2}\right) - \cos\theta\sin\left(\dfrac{\pi}{2}\right)$

$$= -\cos\theta$$

$$r = f\left[\sin\left(\theta - \frac{\pi}{2}\right)\right]$$

$$= f(-\cos\theta)$$

(b) $\sin(\theta - \pi) = \sin\theta\cos\pi - \cos\theta\sin\pi$

$$= -\sin\theta$$

$$r = f[\sin(\theta - \pi)]$$

$$= f(-\sin\theta)$$

(c) $\sin\left(\theta - \dfrac{3\pi}{2}\right) = \sin\theta\cos\left(\dfrac{3\pi}{2}\right) - \cos\theta\sin\left(\dfrac{3\pi}{2}\right)$

$$= \cos\theta$$

$$r = f\left[\sin\left(\theta - \frac{3\pi}{2}\right)\right] = f(\cos\theta)$$

101. $r = 2 - \sin\theta$

(a) $r = 2 - \sin\left(\theta - \dfrac{\pi}{4}\right) = 2 - \dfrac{\sqrt{2}}{2}(\sin\theta - \cos\theta)$

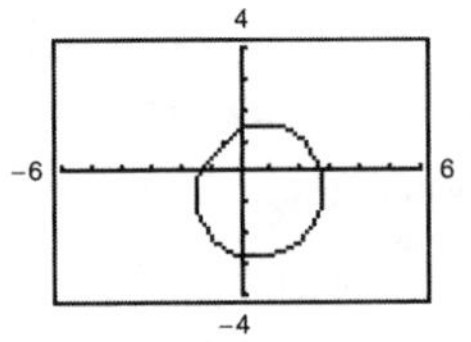

(b) $r = 2 - \sin\left(\theta - \dfrac{\pi}{2}\right) = 2 - (-\cos\theta) = 2 + \cos\theta$

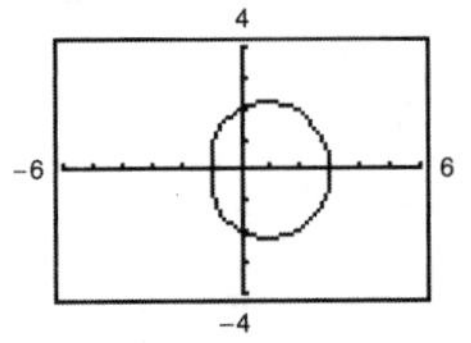

(c) $r = 2 - \sin(\theta - \pi) = 2 - (-\sin\theta) = 2 + \sin\theta$

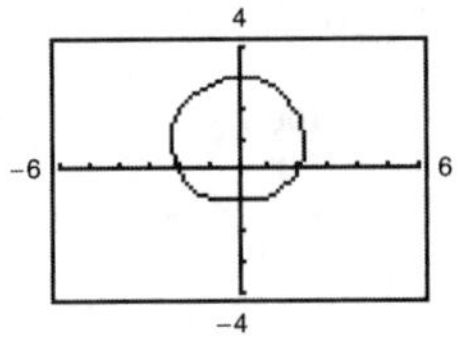

(d) $r = 2 - \sin\left(\theta - \dfrac{3\pi}{2}\right) = 2 - \cos\theta$

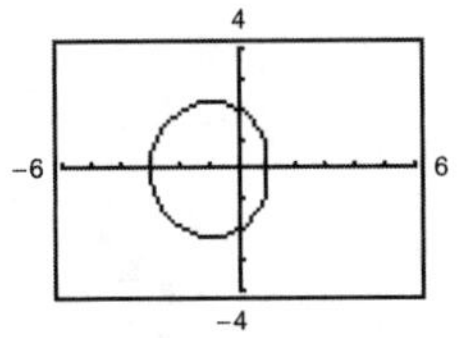

102. $r = 2\sin 2\theta = 4\sin\theta\cos\theta$

(a) $r = 4\sin\left(\theta - \dfrac{\pi}{6}\right)\cos\left(\theta - \dfrac{\pi}{6}\right)$

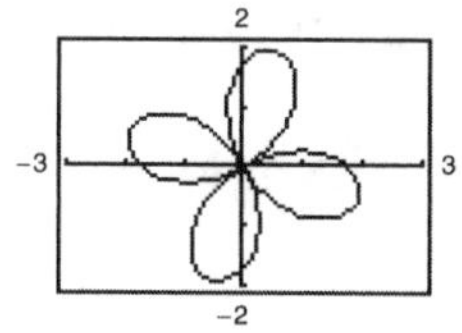

(b) $r = 4\sin\left(\theta - \dfrac{\pi}{2}\right)\cos\left(\theta - \dfrac{\pi}{2}\right) = -4\sin\theta\cos\theta$

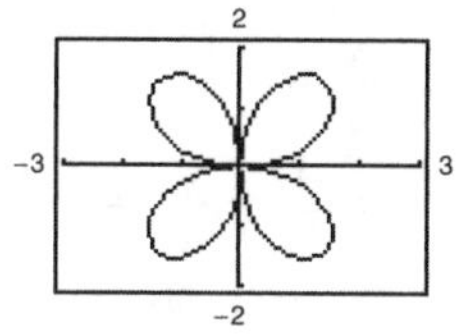

(c) $r = 4\sin\left(\theta - \dfrac{2\pi}{3}\right)\cos\left(\theta - \dfrac{2\pi}{3}\right)$

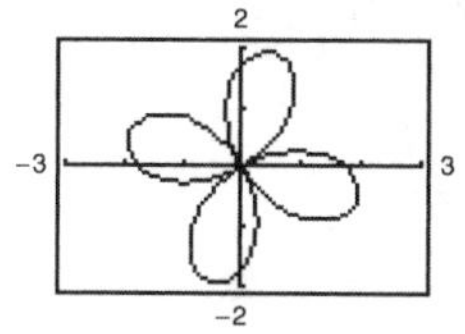

(d) $r = 4\sin(\theta - \pi)\cos(\theta - \pi) = 4\sin\theta\cos\theta$

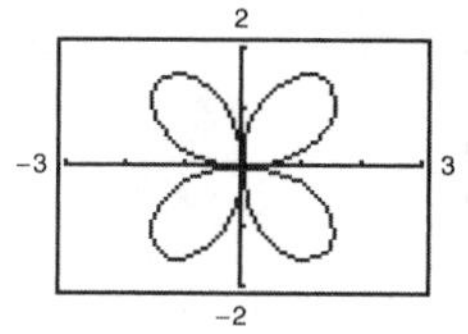

103. (a) $r = 1 - \sin\theta$

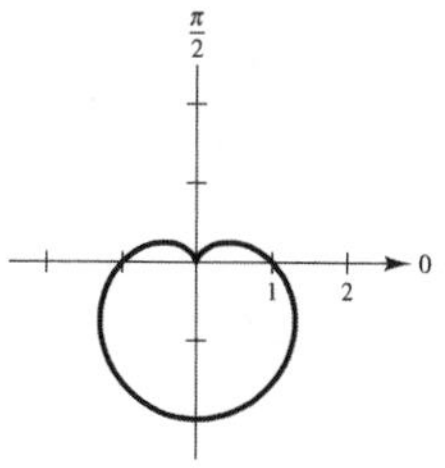

(b) $r = 1 - \sin\left(\theta - \dfrac{\pi}{4}\right)$

Rotate the graph of
$r = 1 - \sin\theta$
through the angle $\pi/4$.

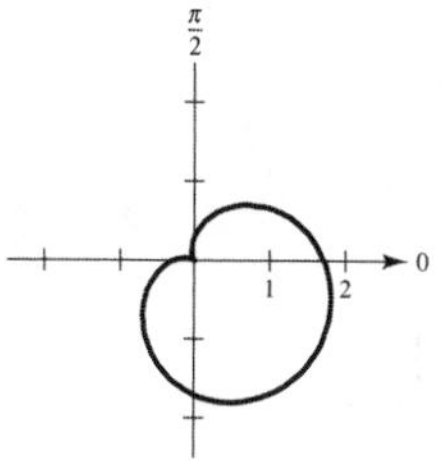

104. By Theorem 9.11, the slope of the tangent line through A and P is

$$\frac{f\cos\theta + f'\sin\theta}{-f\sin\theta + f'\cos\theta}.$$

This is equal to

$$\tan(\theta + \psi) = \frac{\tan\theta + \tan\psi}{1 - \tan\theta\tan\psi} = \frac{\sin\theta + \cos\theta\tan\psi}{\cos\theta - \sin\theta\tan\psi}.$$

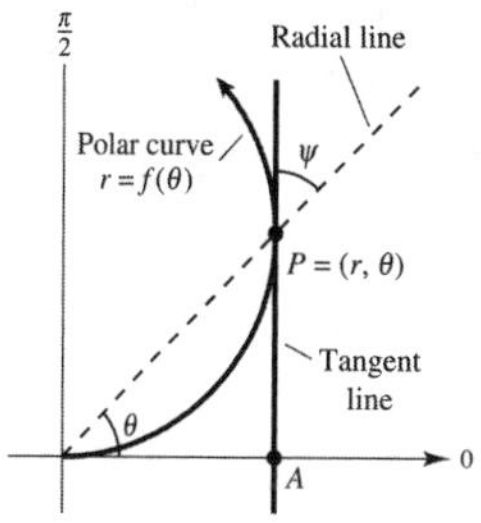

Equating the expressions and cross-multiplying, you obtain

$$\begin{aligned}
(f\cos\theta + f'\sin\theta)(\cos\theta - \sin\theta\tan\psi) &= (\sin\theta + \cos\theta\tan\psi)(-f\sin\theta + f'\cos\theta)\\
f\cos^2\theta - f\cos\theta\sin\theta\tan\psi + f'\sin\theta\cos\theta - f'\sin^2\theta\tan\psi &= -f\sin^2\theta - f\sin\theta\cos\theta\tan\psi + f'\sin\theta\cos\theta\\
&\quad + f'\cos^2\theta\tan\psi\\
f(\cos^2\theta + \sin^2\theta) &= f'\tan\psi(\cos^2\theta + \sin^2\theta)\\
\tan\psi &= \frac{f}{f'} = \frac{r}{dr/d\theta}.
\end{aligned}$$

105. $\tan\psi = \dfrac{r}{dr/d\theta} = \dfrac{2(1 - \cos\theta)}{2\sin\theta}$

At $\theta = \pi$, $\tan\psi$ is undefined $\Rightarrow \psi = \dfrac{\pi}{2}$.

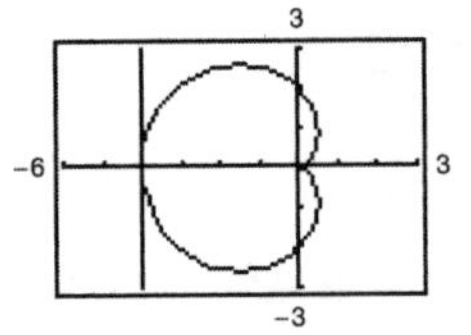

106. $\tan\psi = \dfrac{r}{dr/d\theta} = \dfrac{3(1 - \cos\theta)}{3\sin\theta}$

At $\theta = \dfrac{3\pi}{4}$, $\tan\psi = \dfrac{1 + \left(\sqrt{2}/2\right)}{\sqrt{2}/2} = \dfrac{2 + \sqrt{2}}{\sqrt{2}}$.

$\psi = \arctan\left(\dfrac{2 + \sqrt{2}}{\sqrt{2}}\right) \approx 1.178(\approx 67.5°)$

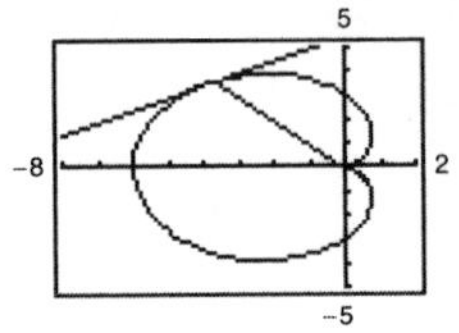

107. $r = 2\cos 3\theta$

$\tan\psi = \dfrac{r}{dr/d\theta} = \dfrac{2\cos 3\theta}{-6\sin 3\theta} = -\dfrac{1}{3}\cot 3\theta$

At $\theta = \dfrac{\pi}{4}$, $\tan\psi = -\dfrac{1}{3}\cot\left(\dfrac{3\pi}{4}\right) = \dfrac{1}{3}$.

$\psi = \arctan\left(\dfrac{1}{3}\right) \approx 18.4°$

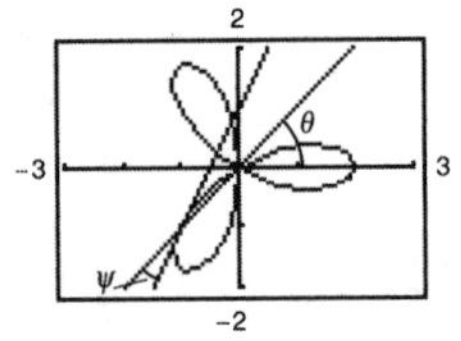

108. $\tan\psi = \dfrac{r}{dr/d\theta} = \dfrac{4\sin 2\theta}{8\cos 2\theta}$

At $\theta = \dfrac{\pi}{6}$, $\tan\psi = \dfrac{\sin(\pi/3)}{2\cos(\pi/3)} = \dfrac{\sqrt{3}}{2}$.

$\psi = \arctan\left(\dfrac{\sqrt{3}}{2}\right) \approx 0.7137(\approx 40.89°)$

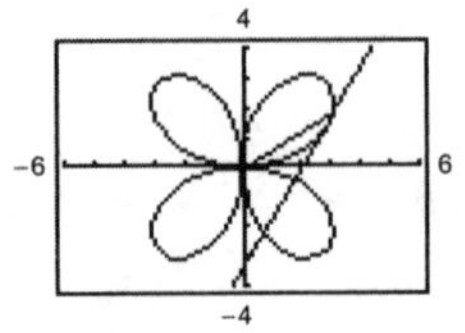

109. $r = \dfrac{6}{1 - \cos\theta} = 6(1 - \cos\theta)^{-1} \Rightarrow \dfrac{dr}{d\theta} = \dfrac{6\sin\theta}{(1 - \cos\theta)^2}$

$\tan\psi = \dfrac{r}{\dfrac{dr}{d\theta}} = \dfrac{\dfrac{6}{(1 - \cos\theta)}}{\dfrac{-6\sin\theta}{(1 - \cos\theta)^2}} = \dfrac{1 - \cos\theta}{-\sin\theta}$

At $\theta = \dfrac{2\pi}{3}$, $\tan\psi = \dfrac{1 - \left(-\dfrac{1}{2}\right)}{-\dfrac{\sqrt{3}}{2}} = -\sqrt{3}$.

$\psi = \dfrac{\pi}{3}, (60°)$

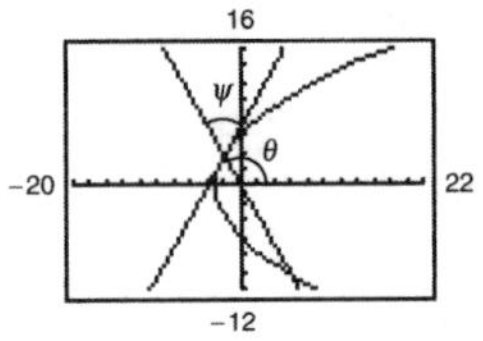

110. $\tan\psi = \dfrac{r}{dr/d\theta} = \dfrac{5}{0}$ undefined $\Rightarrow \psi = \dfrac{\pi}{2}$

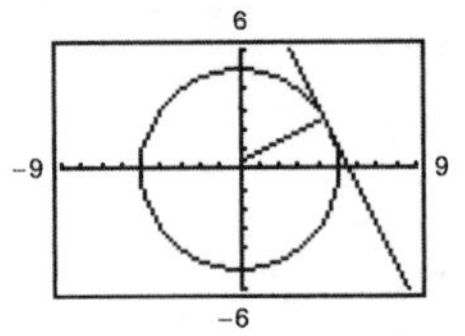

111. True

112. True

113. True

114. True

Section 10.5 Area and Arc Length in Polar Coordinates

1. $A = \frac{1}{2}\int_{\alpha}^{\beta}\left[f(\theta)\right]^2\,d\theta$

$= \frac{1}{2}\int_0^{\pi/2}\left[4\sin\theta\right]^2\,d\theta = 8\int_0^{\pi/2}\sin^2\theta\,d\theta$

2. $A = \frac{1}{2}\int_{\alpha}^{\beta}\left[f(\theta)\right]^2\,d\theta = \frac{1}{2}\int_{3\pi/4}^{5\pi/4}\left(\cos 2\theta\right)^2\,d\theta$

3. $A = \frac{1}{2}\int_{\alpha}^{\beta}\left[f(\theta)\right]^2\,d\theta = \frac{1}{2}\int_{\pi/2}^{3\pi/2}\left[3 - 2\sin\theta\right]^2\,d\theta$

4. $A = \frac{1}{2}\int_{\alpha}^{\beta}\left[f(\theta)\right]^2\,d\theta = \frac{1}{2}\int_0^{\pi/2}\left[1 - \cos 2\theta\right]^2\,d\theta$

5. $A = \frac{1}{2}\int_0^{\pi}\left[6\sin\theta\right]^2\,d\theta$

$= 18\int_0^{\pi}\frac{1 - \cos 2\theta}{2}\,d\theta = 9\left[\theta - \frac{\sin 2\theta}{2}\right]_0^{\pi} = 9\pi$

Note: $r = 6\sin\theta$ is circle of radius 3, $0 \le \theta \le \pi$.

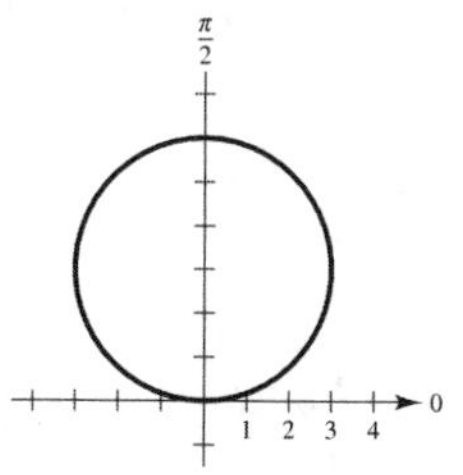

6. $A = \frac{1}{2}\int_0^{\pi}\left[3\cos\theta\right]^2\,d\theta$

$= \frac{9}{2}\int_0^{\pi}\frac{1 + \cos 2\theta}{2}\,d\theta$

$= \frac{9}{4}\left[\theta + \frac{\sin 2\theta}{2}\right]_0^{\pi} = \frac{9}{4}\pi$

Note: $r = 3\cos\theta$ is circle of radius $\frac{3}{2}$, $0 \le \theta \le \pi$.

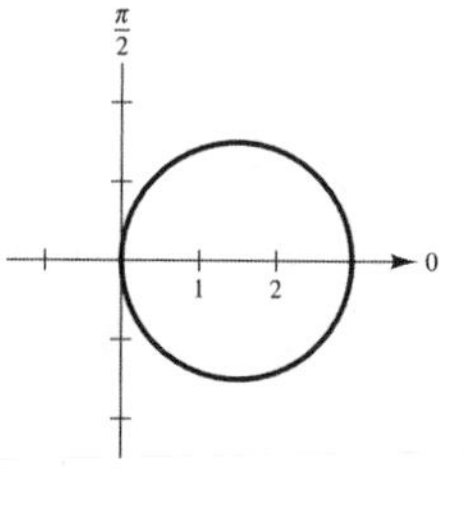

7. $A = 2\left[\frac{1}{2}\int_0^{\pi/6}\left(2\cos 3\theta\right)^2\,d\theta\right] = 2\left[\theta + \frac{1}{6}\sin 6\theta\right]_0^{\pi/6} = \frac{\pi}{3}$

8. $A = \frac{1}{2}\int_0^{\pi/3}\left[4\sin 3\theta\right]^2\,d\theta$

$= 8\int_0^{\pi/3}\sin^2 3\theta\,d\theta$

$= 8\int_0^{\pi/3}\frac{1 - \cos 6\theta}{2}\,d\theta$

$= 4\left[\theta - \frac{\sin 6\theta}{6}\right]_0^{\pi/3}$

$= 4\left[\frac{\pi}{3}\right] = \frac{4\pi}{3}$

9. $A = \frac{1}{2}\int_0^{\pi/2}\left[\sin 2\theta\right]^2\,d\theta$

$= \frac{1}{2}\int_0^{\pi/2}\frac{1 - \cos 4\theta}{2}\,d\theta$

$= \frac{1}{4}\left[\theta - \frac{\sin 4\theta}{4}\right]_0^{\pi/2}$

$= \frac{1}{4}\left[\frac{\pi}{2}\right] = \frac{\pi}{8}$

10. $A = 2\left[\frac{1}{2}\int_0^{\pi/10}\left(\cos 5\theta\right)^2\,d\theta\right]$

$= \frac{1}{2}\left[\theta + \frac{1}{10}\sin(10\theta)\right]_0^{\pi/10} = \frac{\pi}{20}$

11. $A = 2\left[\frac{1}{2}\int_{-\pi/2}^{\pi/2}\left(1 - \sin\theta\right)^2\,d\theta\right]$

$= \left[\frac{3}{2}\theta + 2\cos\theta - \frac{1}{4}\sin 2\theta\right]_{-\pi/2}^{\pi/2} = \frac{3\pi}{2}$

12. $A = 2\left[\frac{1}{2}\int_0^{\pi/2}\left(1 - \sin\theta\right)^2\,d\theta\right]$

$= \left[\frac{3}{2}\theta + 2\cos\theta - \frac{1}{4}\sin 2\theta\right]_0^{\pi/2} = \frac{3\pi - 8}{4}$

13. $A = \frac{1}{2}\int_0^{2\pi}\left[5 + 2\sin\theta\right]^2\,d\theta$

$= \frac{1}{2}\int_0^{2\pi}\left[25 + 20\sin\theta + 4\sin^2\theta\right]d\theta$

$= \frac{1}{2}\int_0^{2\pi}\left[25 + 20\sin\theta + 2\left(1 - \cos 2\theta\right)\right]d\theta$

$= \frac{1}{2}\left[27\theta - 20\cos\theta - \sin 2\theta\right]_0^{2\pi}$

$= \frac{1}{2}\left[27(2\pi)\right] = 27\pi$

14. $A = \frac{1}{2}\int_0^{2\pi} [4 - 4\cos\theta]^2\, d\theta$

$= 8\int_0^{2\pi} (1 - \cos\theta)^2\, d\theta$

$= 8\int_0^{2\pi} \left(1 - 2\cos\theta + \cos^2\theta\right) d\theta$

$= 8\int_0^{2\pi} \left(1 - 2\cos\theta + \frac{1 + \cos 2\theta}{2}\right) d\theta$

$= 8\left[\frac{3}{2}\theta - 2\sin\theta + \frac{1}{4}\sin 2\theta\right]_0^{2\pi}$

$= 8\left[\frac{3}{2}(2\pi)\right] = 24\pi$

15. On the interval $-\frac{\pi}{4} \le \theta \le 0$, $r = 2\sqrt{\cos 2\theta}$ traces out one-half of one leaf of the lemniscate. So,

$A = 4\frac{1}{2}\int_{-\pi/4}^{0} 4\cos 2\theta\, d\theta$

$= 8\left[\frac{\sin 2\theta}{2}\right]_{-\pi/4}^{0} = 8\left[\frac{1}{2}\right] = 4.$

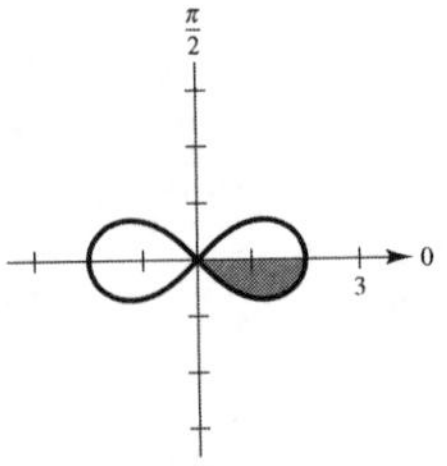

16. On the interval $0 \le \theta \le \pi/2$, $r = \sqrt{6\sin 2\theta}$ traces out half of the lemniscate. So

$A = 2\cdot\frac{1}{2}\int_0^{\pi/2} 6\sin 2\theta\, d\theta$

$= 6\left[\frac{-\cos 2\theta}{2}\right]_0^{\pi/2} = 6\left[\frac{1}{2} + \frac{1}{2}\right] = 6.$

17. $A = \left[2\frac{1}{2}\int_{2\pi/3}^{\pi} (1 + 2\cos\theta)^2\, d\theta\right]$

$= \left[3\theta + 4\sin\theta + \sin 2\theta\right]_{2\pi/3}^{\pi} = \frac{2\pi - 3\sqrt{3}}{2}$

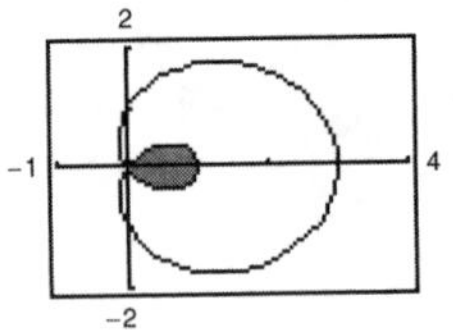

18. Half of the inner loop of $r = 2 - 4\cos\theta$ is traced out on the interval $0 \le \theta \le \frac{\pi}{3}$. So

$A = 2\cdot\frac{1}{2}\int_0^{\pi/3} (2 - 4\cos\theta)^2\, d\theta$

$= \int_0^{\pi/3} \left[4 - 16\cos\theta + 16\cos^2\theta\right] d\theta$

$= \int_0^{\pi/3} \left[4 - 16\cos\theta + 8[1 + \cos 2\theta]\right] d\theta$

$= \left[12\theta - 16\sin\theta + 4\sin 2\theta\right]_0^{\pi/3}$

$= 12(\pi/3) - 16\left(\sqrt{3}/2\right) + 4\left(\sqrt{3}/2\right)$

$= 4\pi - 6\sqrt{3}.$

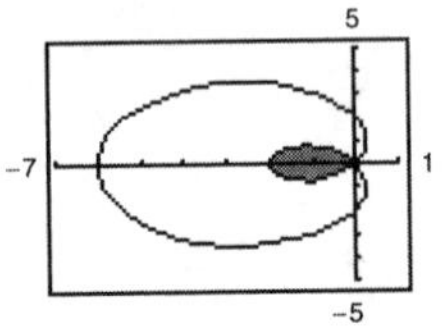

19. The inner loop of $r = 1 + 2\sin\theta$ is traced out on the interval $\frac{7\pi}{6} \le \theta \le \frac{11\pi}{6}$. So,

$A = \frac{1}{2}\int_{7\pi/6}^{11\pi/6} [1 + 2\sin\theta]^2\, d\theta$

$= \frac{1}{2}\int_{7\pi/6}^{11\pi/6} \left[1 + 4\sin\theta + 4\sin^2\theta\right] d\theta$

$= \frac{1}{2}\int_{7\pi/6}^{11\pi/6} \left[1 + 4\sin\theta + 2(1 - \cos 2\theta)\right] d\theta$

$= \frac{1}{2}\left[3\theta - 4\cos\theta - \sin 2\theta\right]_{7\pi/6}^{11\pi/6}$

$= \frac{1}{2}\left[\left(\frac{11\pi}{2} - 2\sqrt{3} + \frac{\sqrt{3}}{2}\right) - \left(\frac{7\pi}{2} + 2\sqrt{3} - \frac{\sqrt{3}}{2}\right)\right]$

$= \frac{1}{2}\left[2\pi - 3\sqrt{3}\right].$

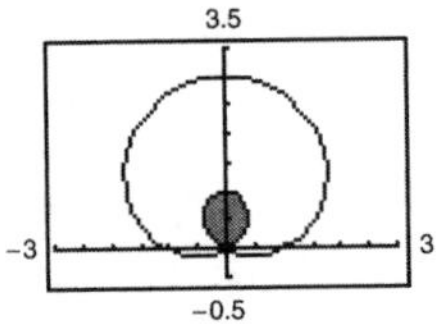

20. $$A = 2\left[\frac{1}{2}\int_{\arcsin(2/3)}^{\pi/2} (4 - 6\sin\theta)^2\,d\theta\right]$$

$$= \int_{\arcsin(2/3)}^{\pi/2} \left[16 - 48\sin\theta + 36\sin^2\theta\right]d\theta$$

$$= \int_{\arcsin(2/3)}^{\pi/2} \left[16 - 48\sin\theta + 36\left(\frac{1 - \cos 2\theta}{2}\right)\right]d\theta$$

$$= \left[34\theta + 48\cos\theta - 9\sin 2\theta\right]_{\arcsin(2/3)}^{\pi/2} \approx 1.7635$$

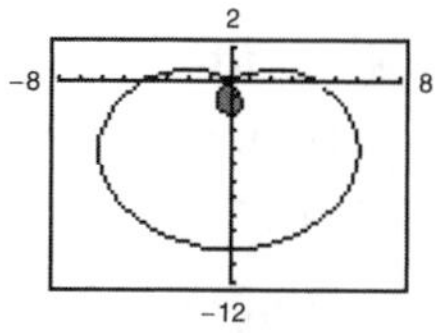

21. The area inside the outer loop is

$$2\left[\frac{1}{2}\int_0^{2\pi/3} (1 + 2\cos\theta)^2\,d\theta\right] = \left[3\theta + 4\sin\theta + \sin 2\theta\right]_0^{2\pi/3}$$

$$= \frac{4\pi + 3\sqrt{3}}{2}.$$

From the result of Exercise 17, the area between the loops is

$$A = \left(\frac{4\pi + 3\sqrt{3}}{2}\right) - \left(\frac{2\pi - 3\sqrt{3}}{2}\right) = \pi + 3\sqrt{3}.$$

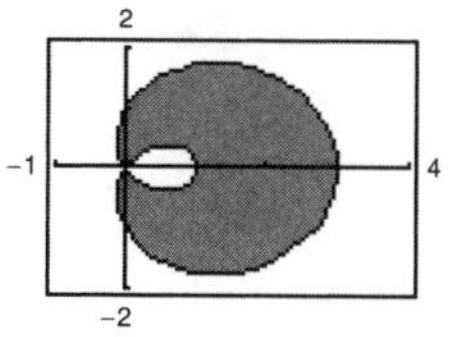

22. Four times the area in Exercise 21, $A = 4\left(\pi + 3\sqrt{3}\right)$. More specifically, you see that the area inside the outer loop is

$$2\left[\frac{1}{2}\int_{-\pi/6}^{\pi/2} \left(2(1 + 2\sin\theta)\right)^2\,d\theta\right] = \int_{-\pi/6}^{\pi/2} \left(4 + 16\sin\theta + 16\sin^2\theta\right)d\theta = 8\pi + 6\sqrt{3}.$$

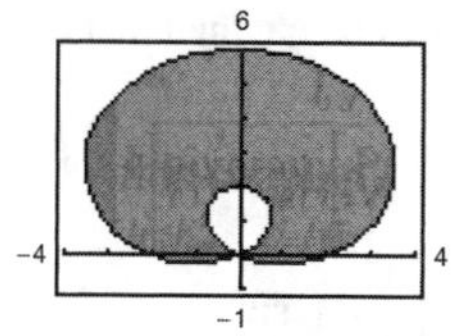

The area inside the inner loop is

$$2\frac{1}{2}\left[\int_{7\pi/6}^{3\pi/2} \left(2(1 + 2\sin\theta)\right)^2\,d\theta\right] = 4\pi - 6\sqrt{3}.$$

So, the area between the loops is $\left(8\pi + 6\sqrt{3}\right) - \left(4\pi - 6\sqrt{3}\right) = 4\pi + 12\sqrt{3}.$

23. The area inside the outer loop is

$$A = 2\cdot\frac{1}{2}\int_{5\pi/6}^{3\pi/2} \left[3 - 6\sin\theta\right]^2\,d\theta$$

$$= \int_{5\pi/6}^{3\pi/2} \left[9 - 36\sin\theta + 36\sin^2\theta\right]d\theta$$

$$= \int_{5\pi/6}^{3\pi/2} \left[9 - 36\sin\theta + 18(1 - \cos 2\theta)\right]d\theta$$

$$= \left[27\theta + 36\cos\theta - 9\sin 2\theta\right]_{5\pi/6}^{3\pi/2} = \left[\frac{81\pi}{2} - \left(\frac{45\pi}{2} - 18\sqrt{3} + \frac{9\sqrt{3}}{2}\right)\right] = 18\pi + \frac{27\sqrt{3}}{2}.$$

The area inside the inner loop is

$$A = 2\cdot\frac{1}{2}\int_{\pi/6}^{\pi/2} \left[3 - 6\sin\theta\right]^2\,d\theta$$

$$= \left[27\theta + 36\cos\theta - 9\sin 2\theta\right]_{\pi/6}^{\pi/2} = \left[\frac{27\pi}{2} - \left(\frac{9\pi}{2} + 18\sqrt{3} - \frac{9\sqrt{3}}{2}\right)\right] = 9\pi - \frac{27\sqrt{3}}{2}.$$

Finally, the area between the loops is

$$\left[18\pi + \frac{27\sqrt{3}}{2}\right] - \left[9\pi - \frac{27\sqrt{3}}{2}\right] = 9\pi + 27\sqrt{3}.$$

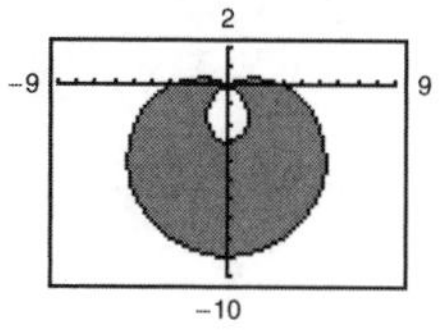

24. The area inside the outer loop is

$$A = 2 \cdot \frac{1}{2}\int_0^{2\pi/3}\left[\frac{1}{2} + \cos\theta\right]^2 d\theta$$
$$= \int_0^{2\pi/3}\left[\frac{1}{4} + \cos\theta + \frac{1 + \cos 2\theta}{2}\right] d\theta$$
$$= \left[\frac{3}{4}\theta + \sin\theta + \frac{\sin 2\theta}{4}\right]_0^{2\pi/3}$$
$$= \frac{3}{4}\left(\frac{2\pi}{3}\right) + \frac{\sqrt{3}}{2} - \frac{\sqrt{3}}{8}$$
$$= \frac{\pi}{2} + \frac{3\sqrt{3}}{8}.$$

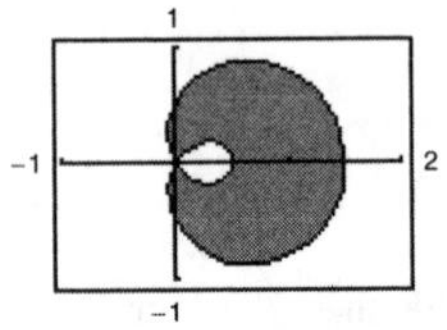

The area inside the inner loop is

$$A = 2 \cdot \frac{1}{2}\int_{2\pi/3}^{\pi}\left[\frac{1}{2} + \cos\theta\right]^2 d\theta$$
$$= \left[\frac{3}{4}\theta + \sin\theta + \frac{\sin 2\theta}{4}\right]_{2\pi/3}^{\pi}$$
$$= \frac{3}{4}\pi - \left(\frac{\pi}{2} + \frac{\sqrt{3}}{2} - \frac{\sqrt{3}}{8}\right) = \frac{\pi}{4} - \frac{3\sqrt{3}}{8}$$

Finally, the area between the loops is

$$\left[\frac{\pi}{2} + \frac{3\sqrt{3}}{8}\right] - \left[\frac{\pi}{4} - \frac{3\sqrt{3}}{8}\right] = \frac{\pi}{4} + \frac{3\sqrt{3}}{4}.$$

25. $r = 1 + \cos\theta$

$r = 1 - \cos\theta$

Solving simultaneously,

$$1 + \cos\theta = 1 - \cos\theta$$
$$2\cos\theta = 0$$
$$\theta = \frac{\pi}{2}, \frac{3\pi}{2}.$$

Replacing r by $-r$ and θ by $\theta + \pi$ in the first equation and solving, $-1 + \cos\theta = 1 - \cos\theta$, $\cos\theta = 1$, $\theta = 0$. Both curves pass through the pole, $(0, \pi)$, and $(0, 0)$, respectively.

Points of intersection: $\left(1, \frac{\pi}{2}\right), \left(1, \frac{3\pi}{2}\right), (0, 0)$

26. $r = 3(1 + \sin\theta)$

$r = 3(1 - \sin\theta)$

Solving simultaneously,

$$3(1 + \sin\theta) = 3(1 - \sin\theta)$$
$$2\sin\theta = 0$$
$$\theta = 0, \pi.$$

Replacing r by $-r$ and θ by $\theta + \pi$ in the first equation and solving, $-3(1 - \sin\theta) = 3(1 - \sin\theta)$, $\sin\theta = 1$, $\theta = \pi/2$. Both curves pass through the pole, $(0, 3\pi/2)$, and $(0, \pi/2)$, respectively.

Points of intersection: $(3, 0), (3, \pi), (0, 0)$

27. $r = 1 + \cos\theta$

$r = 1 - \sin\theta$

Solving simultaneously,

$$1 + \cos\theta = 1 - \sin\theta$$
$$\cos\theta = -\sin\theta$$
$$\tan\theta = -1$$
$$\theta = \frac{3\pi}{4}, \frac{7\pi}{4}.$$

Replacing r by $-r$ and θ by $\theta + \pi$ in the first equation and solving, $-1 + \cos\theta = 1 - \sin\theta$, $\sin\theta + \cos\theta = 2$, which has no solution. Both curves pass through the pole, $(0, \pi)$, and $(0, \pi/2)$, respectively.

Points of intersection:
$\left(\frac{2 - \sqrt{2}}{2}, \frac{3\pi}{4}\right), \left(\frac{2 + \sqrt{2}}{2}, \frac{7\pi}{4}\right), (0, 0)$

28. $r = 2 - 3\cos\theta$

$r = \cos\theta$

Solving simultaneously,

$$2 - 3\cos\theta = \cos\theta$$
$$\cos\theta = \frac{1}{2}$$
$$\theta = \frac{\pi}{3}, \frac{5\pi}{3}.$$

Both curves pass through the pole, $(0, \arccos 2/3)$, and $(0, \pi/2)$, respectively.

Points of intersection: $\left(\frac{1}{2}, \frac{\pi}{3}\right), \left(\frac{1}{2}, \frac{5\pi}{3}\right), (0, 0)$

29. $r = 4 - 5\sin\theta$

$r = 3\sin\theta$

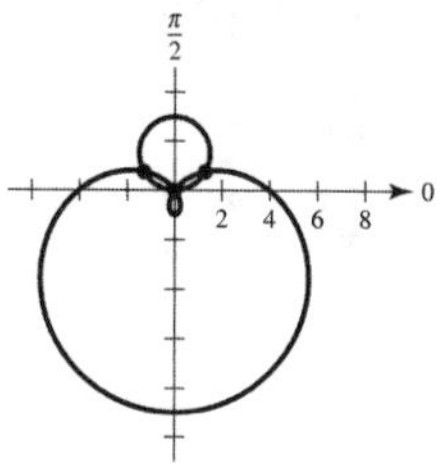

Solving simultaneously,

$$4 - 5\sin\theta = 3\sin\theta$$

$$\sin\theta = \frac{1}{2}$$

$$\theta = \frac{\pi}{6}, \frac{5\pi}{6}.$$

Both curves pass through the pole, $(0, \arcsin 4/5)$, and $(0, 0)$, respectively.

Points of intersection: $\left(\frac{3}{2}, \frac{\pi}{6}\right), \left(\frac{3}{2}, \frac{5\pi}{6}\right), (0, 0)$

30. $r = 3 + \sin\theta$

$r = 2\csc\theta$

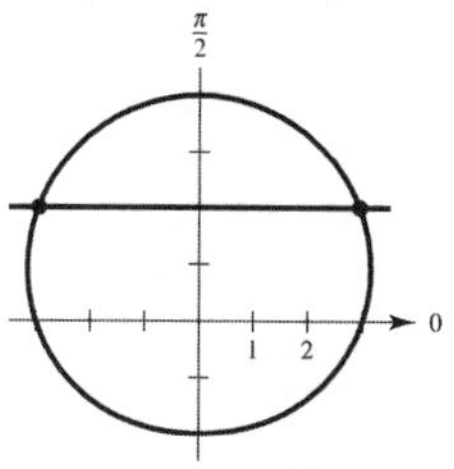

The graph of $r = 3 + \sin\theta$ is a limaçon symmetric to $\theta = \pi/2$, and the graph of $r = 2\csc\theta$ is the horizontal line $y = 2$. So, there are two points of intersection. Solving simultaneously,

$$3 + \sin\theta = 2\csc\theta$$

$$\sin^2\theta + 3\sin\theta - 2 = 0$$

$$\sin\theta = \frac{-3 \pm \sqrt{17}}{2}$$

$$\theta = \arcsin\left(\frac{\sqrt{17} - 3}{2}\right) \approx 0.596.$$

Points of intersection:

$\left(\frac{\sqrt{17} + 3}{2}, \arcsin\left(\frac{\sqrt{17} - 3}{2}\right)\right)$,

$\left(\frac{\sqrt{17} + 3}{2}, \pi - \arcsin\left(\frac{\sqrt{17} - 3}{2}\right)\right)$,

$(3.56, 0.596), (3.56, 2.545)$

31. $r = \frac{\theta}{2}$

$r = 2$

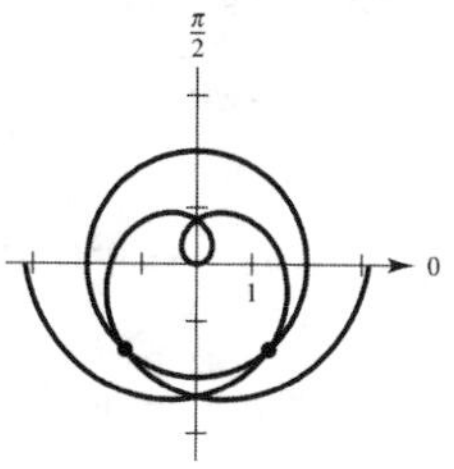

Solving simultaneously, you have

$\theta/2 = 2, \theta = 4.$

Points of intersection:

$(2, 4), (-2, -4)$

32. $\theta = \frac{\pi}{4}$

$r = 2$

Line of slope 1 passing through the pole and a circle of radius 2 centered at the pole.

Points of intersection: $\left(2, \frac{\pi}{4}\right), \left(-2, \frac{\pi}{4}\right)$

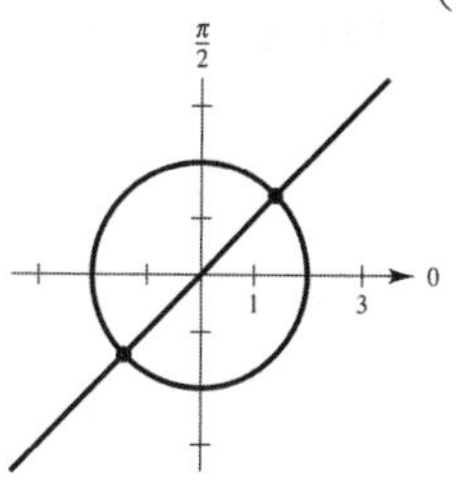

33. $r = \cos\theta$

$r = 2 - 3\sin\theta$

Points of intersection:

$(0, 0), (0.935, 0.363), (0.535, -1.006)$

The graphs reach the pole at different times (θ values).

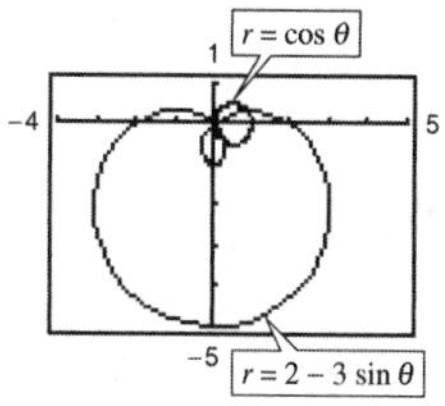

34. $r = 4\sin\theta$

$r = 2(1 + \sin\theta)$

Points of intersection: $(0, 0), \left(4, \frac{\pi}{2}\right)$

The graphs reach the pole at different times (θ values).

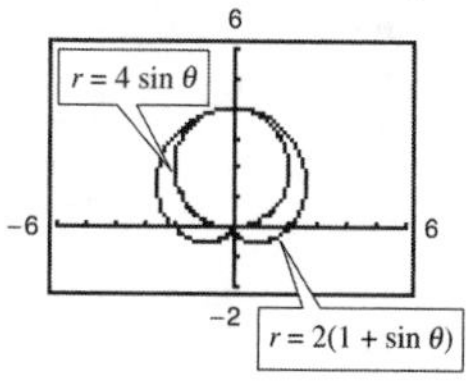

35. The points of intersection for one petal are $(2, \pi/12)$ and $(2, 5\pi/12)$. The area within one petal is

$$A = \frac{1}{2}\int_0^{\pi/12} (4 \sin 2\theta)^2\, d\theta + \frac{1}{2}\int_{\pi/12}^{5\pi/12} (2)^2\, d\theta + \frac{1}{2}\int_{5\pi/12}^{\pi/2} (4 \sin 2\theta)^2\, d\theta$$

$$= 16\int_0^{\pi/12} \sin^2(2\theta)\, d\theta + 2\int_{\pi/12}^{5\pi/12} d\theta \text{ (by symmetry of the petal)}$$

$$= 8\left[\theta - \frac{1}{4}\sin 4\theta\right]_0^{\pi/12} + \left[2\theta\right]_{\pi/12}^{5\pi/12} = \frac{4\pi}{3} - \sqrt{3}.$$

$$\text{Total area} = 4\left(\frac{4\pi}{3} - \sqrt{3}\right) = \frac{16\pi}{3} - 4\sqrt{3} = \frac{4}{3}\left(4\pi - 3\sqrt{3}\right)$$

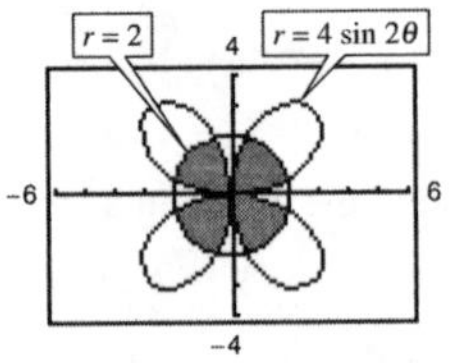

36. The common interior is 4 times the area in the first quadrant.

$$A = 4\frac{1}{2}\int_0^{\pi/2} \left[2(1 - \cos\theta)\right]^2\, d\theta$$

$$= 8\int_0^{\pi/2}\left(1 - 2\cos\theta + \frac{1 + \cos 2\theta}{2}\right) d\theta$$

$$= 8\left[\frac{3\theta}{2} - 2\sin\theta + \frac{\sin 2\theta}{4}\right]_0^{\pi/2}$$

$$= 8\left[\frac{3}{2}\left(\frac{\pi}{2}\right) - 2\right] = 6\pi - 16$$

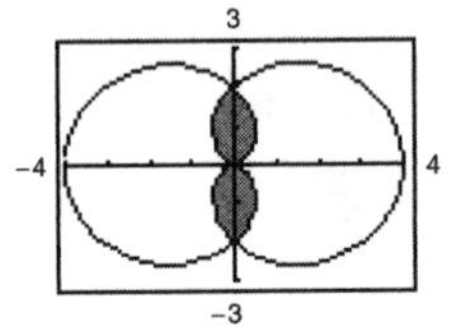

37. $A = 4\left[\frac{1}{2}\int_0^{\pi/2} (3 - 2\sin\theta)^2\, d\theta\right]$

$$= 2\left[11\theta + 12\cos\theta - \sin(2\theta)\right]_0^{\pi/2} = 11\pi - 24$$

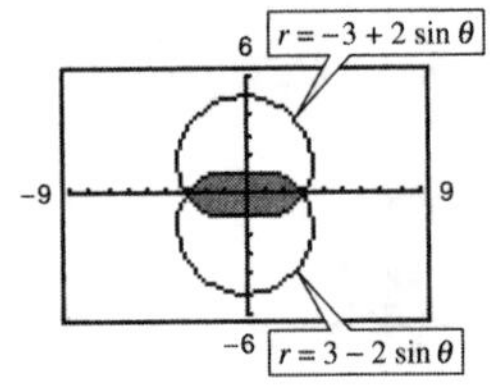

38. $r = 5 - 3\sin\theta$ and $r = 5 - 3\cos\theta$ intersect at $\theta = \pi/4$ and $\pi = 5\pi/4$.

$$A = 2\left[\frac{1}{2}\int_{\pi/4}^{5\pi/4} (5 - 3\sin\theta)^2\, d\theta\right]$$

$$= \left[\frac{59}{2}\theta + 30\cos\theta - \frac{9}{4}\sin 2\theta\right]_{\pi/4}^{5\pi/4}$$

$$= \left(\frac{59}{2}\left(\frac{5\pi}{4}\right) - 30\frac{\sqrt{2}}{2} - \frac{9}{4}\right) - \left(\frac{59}{2}\left(\frac{\pi}{4}\right) + 30\frac{\sqrt{2}}{2} - \frac{9}{4}\right)$$

$$= \frac{59\pi}{2} - 30\sqrt{2} \approx 50.251$$

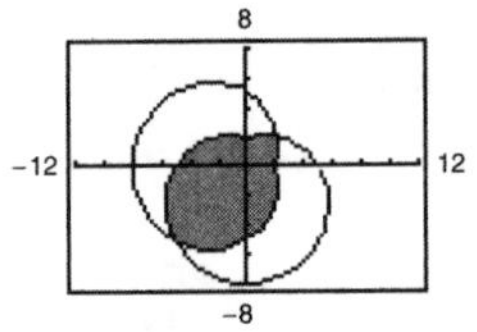

39. $A = 2\left[\frac{1}{2}\int_0^{\pi/6} (4\sin\theta)^2\, d\theta + \frac{1}{2}\int_{\pi/6}^{\pi/2} (2)^2\, d\theta\right]$

$$= 16\left[\frac{1}{2}\theta - \frac{1}{4}\sin(2\theta)\right]_0^{\pi/6} + \left[4\theta\right]_{\pi/6}^{\pi/2}$$

$$= \frac{8\pi}{3} - 2\sqrt{3} = \frac{2}{3}\left(4\pi - 3\sqrt{3}\right)$$

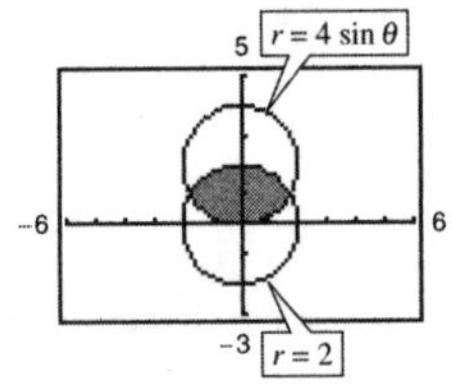

40. The common interior is given by

$$A = 2\frac{1}{2}\int_{\pi/4}^{\pi/2}[2\cos\theta]^2\,d\theta$$
$$= 4\int_{\pi/4}^{\pi/2}\frac{1+\cos 2\theta}{2}\,d\theta$$
$$= 2\left[\theta + \frac{\sin 2\theta}{2}\right]_{\pi/4}^{\pi/2}$$
$$= 2\left[\frac{\pi}{2} - \frac{\pi}{4} - \frac{1}{2}\right]$$
$$= \frac{\pi}{2} - 1$$

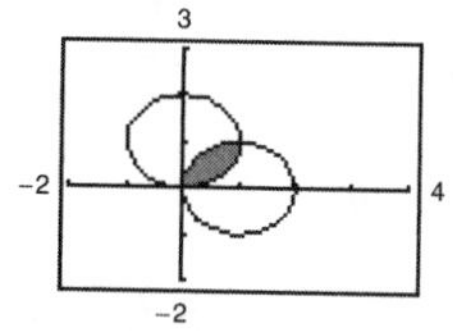

41. $r = 2\cos\theta = 1 \Rightarrow \theta = \pi/3$

$$A = 2\cdot\frac{1}{2}\int_0^{\pi/3}\left([2\cos\theta]^2 - 1\right)d\theta$$
$$= \int_0^{\pi/3}\left[2(1+\cos 2\theta) - 1\right]d\theta$$
$$= \left[\theta + \sin 2\theta\right]_0^{\pi/3}$$
$$= \frac{\pi}{3} + \frac{\sqrt{3}}{2}$$

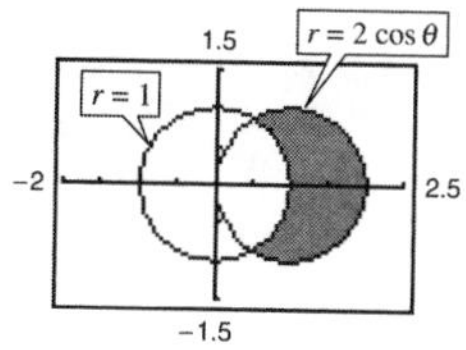

42. $3\sin\theta = 1 + \sin\theta \Rightarrow \sin\theta = 1/2 \Rightarrow \theta = \pi/6$

$$A = 2\cdot\frac{1}{2}\int_{\pi/6}^{\pi/2}\left([3\sin\theta]^2 - [1+\sin\theta]^2\right)d\theta$$
$$= \int_{\pi/6}^{\pi/2}\left[9\sin^2\theta - 1 - 2\sin\theta - \sin^2\theta\right]d\theta$$
$$= \int_{\pi/6}^{\pi/2}\left[4(1-\cos 2\theta) - 1 - 2\sin\theta\right]d\theta$$
$$= \left[3\theta - 2\sin 2\theta + 2\cos\theta\right]_{\pi/6}^{\pi/2}$$
$$= 3\frac{\pi}{2} - 3\frac{\pi}{6} + 2\frac{\sqrt{3}}{2} - 2\frac{\sqrt{3}}{2}$$
$$= \pi$$

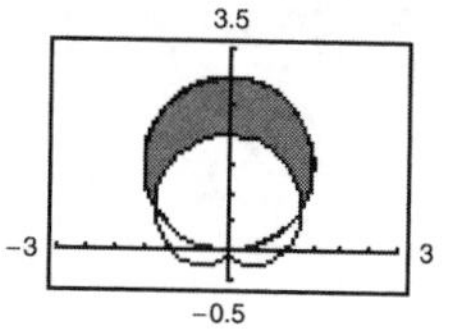

43. $$A = 2\left[\frac{1}{2}\int_0^{\pi}\left[a(1+\cos\theta)\right]^2\,d\theta\right] - \frac{a^2\pi}{4}$$
$$= a^2\left[\frac{3}{2}\theta + 2\sin\theta + \frac{\sin 2\theta}{4}\right]_0^{\pi} - \frac{a^2\pi}{4}$$
$$= \frac{3a^2\pi}{2} - \frac{a^2\pi}{4} = \frac{5a^2\pi}{4}$$

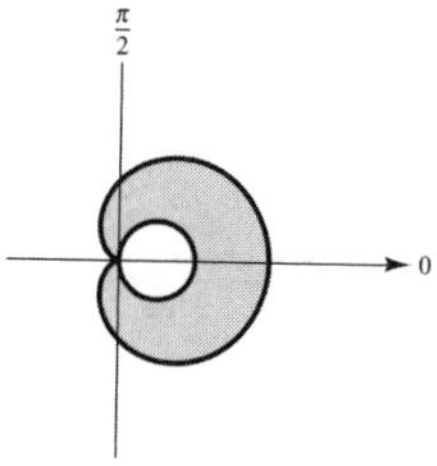

44. Area = Area of $r = 2a\cos\theta$ − Area of sector − twice area between $r = 2a\cos\theta$ and the lines $\theta = \frac{\pi}{3}, \theta = \frac{\pi}{2}$.

$$A = \pi a^2 - \left(\frac{\pi}{3}\right)a^2 - 2\left[\frac{1}{2}\int_{\pi/3}^{\pi/2}(2a\cos\theta)^2\,d\theta\right]$$
$$= \frac{2\pi a^2}{3} - 2a^2\int_{\pi/3}^{\pi/2}(1+\cos 2\theta)\,d\theta$$
$$= \frac{2\pi a^2}{3} - 2a^2\left[\theta + \frac{\sin 2\theta}{2}\right]_{\pi/3}^{\pi/2}$$
$$= \frac{2\pi a^2}{3} - 2a^2\left[\frac{\pi}{2} - \frac{\pi}{3} - \frac{\sqrt{3}}{4}\right] = \frac{2\pi a^2 + 3\sqrt{3}a^2}{6}$$

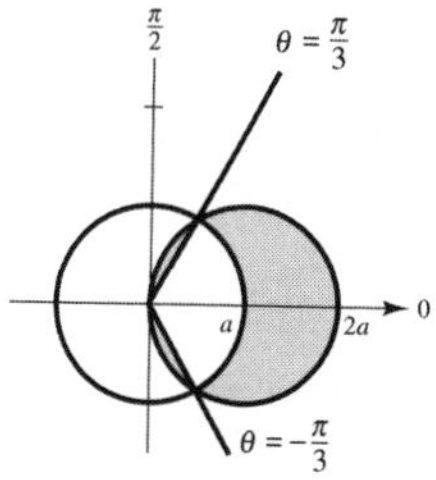

45. $$A = \frac{\pi a^2}{8} + \frac{1}{2}\int_{\pi/2}^{\pi}\left[a(1+\cos\theta)\right]^2\,d\theta$$
$$= \frac{\pi a^2}{8} + \frac{a^2}{2}\int_{\pi/2}^{\pi}\left(\frac{3}{2} + 2\cos\theta + \frac{\cos 2\theta}{2}\right)d\theta$$
$$= \frac{\pi a^2}{8} + \frac{a^2}{2}\left[\frac{3}{2}\theta + 2\sin\theta + \frac{\sin 2\theta}{4}\right]_{\pi/2}^{\pi}$$
$$= \frac{\pi a^2}{8} + \frac{a^2}{2}\left[\frac{3\pi}{2} - \frac{3\pi}{4} - 2\right] = \frac{a^2}{2}[\pi - 2]$$

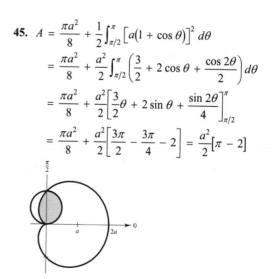

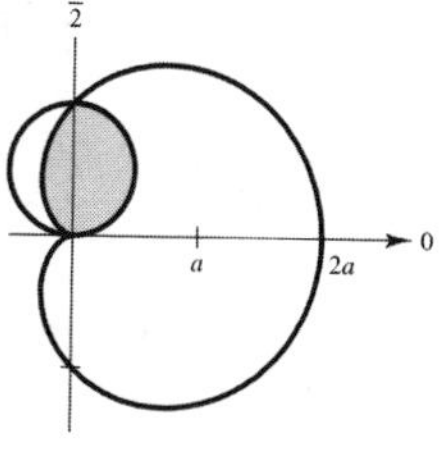

46. $r = a\cos\theta, r = a\sin\theta$

$\tan\theta = 1, \theta = \pi/4$

$$A = 2\left[\frac{1}{2}\int_0^{\pi/4}(a\sin\theta)^2\,d\theta\right]$$
$$= a^2\int_0^{\pi/4}\frac{1-\cos 2\theta}{2}\,d\theta$$
$$= \frac{1}{2}a^2\left[\theta - \frac{\sin 2\theta}{2}\right]_0^{\pi/4}$$
$$= \frac{1}{2}a^2\left[\frac{\pi}{4} - \frac{1}{2}\right]$$
$$= \frac{1}{8}a^2\pi - \frac{1}{4}a^2$$

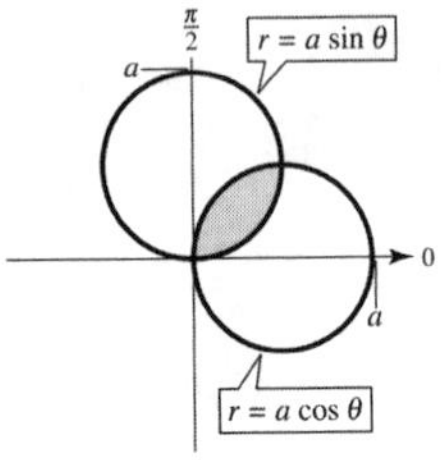

47. (a) $r = a\cos^2\theta$

$r^3 = ar^2\cos^2\theta$

$\left(x^2 + y^2\right)^{3/2} = ax^2$

(b)

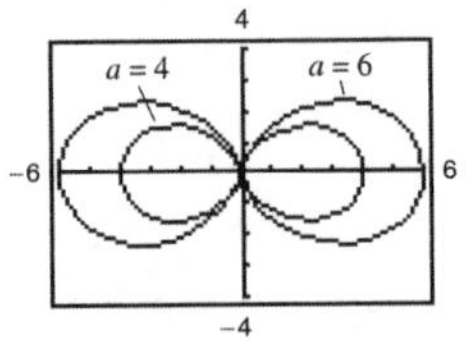

(c) $$A = 4\left(\frac{1}{2}\right)\int_0^{\pi/2}\left[\left(6\cos^2\theta\right)^2 - \left(4\cos^2\theta\right)^2\right]d\theta$$
$$= 40\int_0^{\pi/2}\cos^4\theta\,d\theta$$
$$= 10\int_0^{\pi/2}(1+\cos 2\theta)^2\,d\theta$$
$$= 10\int_0^{\pi/2}\left(1 + 2\cos 2\theta + \frac{1-\cos 4\theta}{2}\right)d\theta$$
$$= 10\left[\frac{3}{2}\theta + \sin 2\theta + \frac{1}{8}\sin 4\theta\right]_0^{\pi/2} = \frac{15\pi}{2}$$

48. By symmetry, $A_1 = A_2$ and $A_3 = A_4$.

$$A_1 = A_2 = \frac{1}{2}\int_{-\pi/3}^{\pi/6}\left[(2a\cos\theta)^2 - (a)^2\right]d\theta + \frac{1}{2}\int_{\pi/6}^{\pi/4}\left[(2a\cos\theta)^2 - (2a\sin\theta)^2\right]d\theta$$
$$= \frac{a^2}{2}\int_{-\pi/3}^{\pi/6}\left(4\cos^2\theta - 1\right)d\theta + 2a^2\int_{\pi/6}^{\pi/4}\cos 2\theta\,d\theta$$
$$= \frac{a^2}{2}[\theta + \sin 2\theta]_{-\pi/3}^{\pi/6} + a^2[\sin 2\theta]_{\pi/6}^{\pi/4} = \frac{a^2}{2}\left(\frac{\pi}{2} + \sqrt{3}\right) + a^2\left(1 - \frac{\sqrt{3}}{2}\right) = a^2\left(\frac{\pi}{4} + 1\right)$$

$$A_3 = A_4 = \frac{1}{2}\left(\frac{\pi}{2}\right)a^2 = \frac{\pi a^2}{4}$$

$$A_5 = \frac{1}{2}\left(\frac{5\pi}{6}\right)a^2 - 2\left(\frac{1}{2}\right)\int_{5\pi/6}^{\pi}(2a\sin\theta)^2\,d\theta$$
$$= \frac{5\pi a^2}{12} - 2a^2\int_{5\pi/6}^{\pi}(1-\cos 2\theta)\,d\theta$$
$$= \frac{5\pi a^2}{12} - a^2[2\theta - \sin 2\theta]_{5\pi/6}^{\pi} = \frac{5\pi a^2}{12} - a^2\left(\frac{\pi}{3} - \frac{\sqrt{3}}{2}\right) = a^2\left(\frac{\pi}{12} + \frac{\sqrt{3}}{2}\right)$$

$$A_6 = 2\left(\frac{1}{2}\right)\int_0^{\pi/6}(2a\sin\theta)^2\,d\theta + 2\left(\frac{1}{2}\right)\int_{\pi/6}^{\pi/4}a^2\,d\theta$$
$$= 2a^2\int_0^{\pi/6}(1-\cos 2\theta)\,d\theta + \left[a^2\theta\right]_{\pi/6}^{\pi/4}$$
$$= a^2[2\theta - \sin 2\theta]_0^{\pi/6} + \frac{\pi a^2}{12} = a^2\left(\frac{\pi}{3} - \frac{\sqrt{3}}{2}\right) + \frac{\pi a^2}{12} = a^2\left(\frac{5\pi}{12} - \frac{\sqrt{3}}{2}\right)$$

$$A_7 = 2\left(\frac{1}{2}\right)\int_{\pi/6}^{\pi/4}\left[(2a\sin\theta)^2 - (a)^2\right]d\theta$$
$$= a^2\int_{\pi/6}^{\pi/4}\left(4\sin^2\theta - 1\right)d\theta = a^2[\theta - \sin 2\theta]_{\pi/6}^{\pi/4} = a^2\left(\frac{\pi}{12} - 1 + \frac{\sqrt{3}}{2}\right)$$

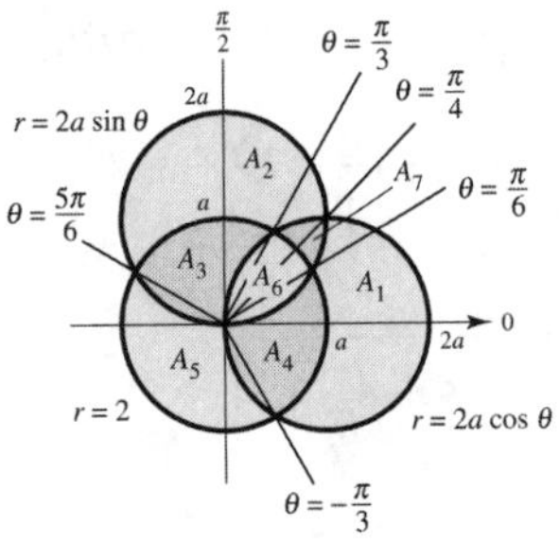

[Note: $A_1 + A_6 + A_7 + A_4 = \pi a^2$ = area of circle of radius a **]**

49. $r = a\cos(n\theta)$

For $n = 1$:

$r = a\cos\theta$

$$A = \pi\left(\frac{a}{2}\right)^2 = \frac{\pi a^2}{4}$$

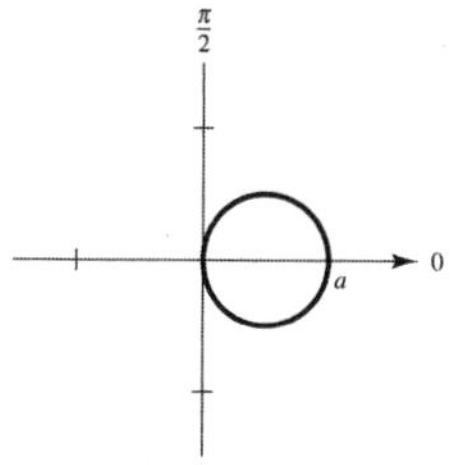

For $n = 2$:

$r = a\cos 2\theta$

$$A = 8\left(\frac{1}{2}\right)\int_0^{\pi/4}(a\cos 2\theta)^2\,d\theta = \frac{\pi a^2}{2}$$

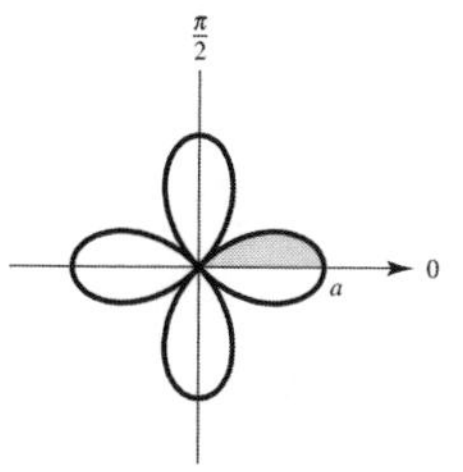

For $n = 3$:

$r = a\cos 3\theta$

$$A = 6\left(\frac{1}{2}\right)\int_0^{\pi/6}(a\cos 3\theta)^2\,d\theta = \frac{\pi a^2}{4}$$

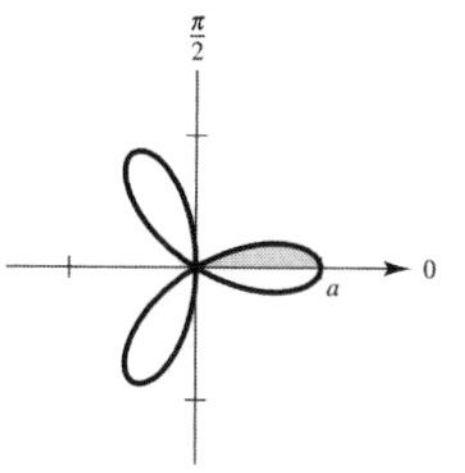

For $n = 4$:

$r = a\cos 4\theta$

$$A = 16\left(\frac{1}{2}\right)\int_0^{\pi/8}(a\cos 4\theta)^2\,d\theta = \frac{\pi a^2}{2}$$

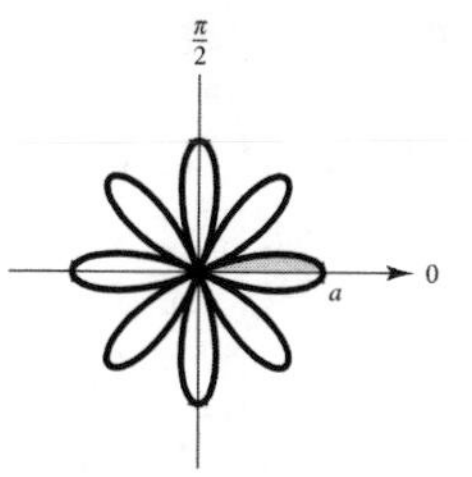

In general, the area of the region enclosed by $r = a\cos(n\theta)$ for $n = 1, 2, 3, \ldots$ is $(\pi a^2)/4$ if n is odd and is $(\pi a^2)/2$ if n is even.

50. $r = \sec\theta - 2\cos\theta, -\frac{\pi}{2} < \theta < \frac{\pi}{2}$

$$r\cos\theta = 1 - 2\cos^2\theta$$

$$x = 1 - 2\left(\frac{r^2\cos^2\theta}{r^2}\right) = 1 - 2\left(\frac{x^2}{x^2 + y^2}\right)$$

$$(x^2 + y^2)x = x^2 + y^2 - 2x^2$$

$$y^2(x - 1) = -x^2 - x^3$$

$$y^2 = \frac{x^2(1 + x)}{1 - x}$$

$$A = 2\left(\frac{1}{2}\right)\int_0^{\pi/4}(\sec\theta - 2\cos\theta)^2\,d\theta$$

$$= \int_0^{\pi/4}\left(\sec^2\theta - 4 + 4\cos^2\theta\right)d\theta$$

$$= \int_0^{\pi/4}\left(\sec^2\theta - 4 + 2(1 + \cos 2\theta)\right)d\theta$$

$$= \left[\tan\theta - 2\theta + \sin 2\theta\right]_0^{\pi/4} = 2 - \frac{\pi}{2}$$

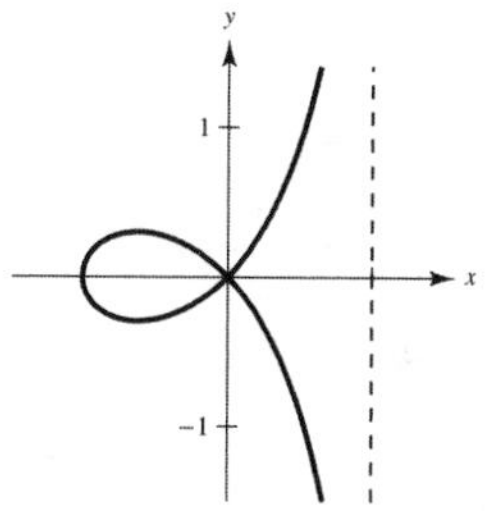

51. $r = 8, r^1 = 0$

$$s = \int_0^{2\pi}\sqrt{8^2 + 0^2}\,d\theta = 8\theta\Big]_0^{2\pi} = 16\pi$$

(circumference of circle of radius 8)

52. $r = a$

$r' = 0$

$$s = \int_0^{2\pi}\sqrt{a^2 + 0^2}\,d\theta = \left[a\theta\right]_0^{2\pi} = 2\pi a$$

(circumference of circle of radius a)

53. $r = 4\sin\theta$

$r' = 4\cos\theta$

$$s = \int_0^{\pi}\sqrt{(4\sin\theta)^2 + (4\cos\theta)^2}\,d\theta$$

$$= \int_0^{\pi}4\,d\theta = \left[4\theta\right]_0^{\pi} = 4\pi$$

(circumference of circle of radius 2)

54. $r = 2a\cos\theta$

$r' = -2a\sin\theta$

$$s = \int_{-\pi/2}^{\pi/2}\sqrt{(2a\cos\theta)^2 + (-2a\sin\theta)^2}\,d\theta$$

$$= \int_{-\pi/2}^{\pi/2}2a\,d\theta = \left[2a\theta\right]_{-\pi/2}^{\pi/2} = 2\pi a$$

55. $r = 1 + \sin\theta$

$r' = \cos\theta$

$$s = 2\int_{\pi/2}^{3\pi/2}\sqrt{(1+\sin\theta)^2 + (\cos\theta)^2}\,d\theta$$
$$= 2\sqrt{2}\int_{\pi/2}^{3\pi/2}\sqrt{1+\sin\theta}\,d\theta$$
$$= 2\sqrt{2}\int_{\pi/2}^{3\pi/2}\frac{-\cos\theta}{\sqrt{1-\sin\theta}}\,d\theta$$
$$= \left[4\sqrt{2}\sqrt{1-\sin\theta}\right]_{\pi/2}^{3\pi/2}$$
$$= 4\sqrt{2}\left(\sqrt{2} - 0\right) = 8$$

56. $r = 8(1 + \cos\theta),\ 0 \le \theta \le 2\pi$

$r' = -8\sin\theta$

$$s = 2\int_0^{\pi}\sqrt{[8(1+\cos\theta)]^2 + (-8\sin\theta]^2}\,d\theta$$
$$= 16\int_0^{\pi}\sqrt{1 + 2\cos\theta + \cos^2\theta + \sin^2\theta}\,d\theta$$
$$= 16\sqrt{2}\int_0^{\pi}\sqrt{1+\cos\theta}\,d\theta$$
$$= 16\sqrt{2}\int_0^{\pi}\sqrt{1+\cos\theta}\cdot\left(\frac{\sqrt{1-\cos\theta}}{\sqrt{1-\cos\theta}}\right)d\theta$$
$$= 16\sqrt{2}\int_0^{\pi}\frac{\sin\theta}{\sqrt{1-\cos\theta}}\,d\theta$$
$$= \left[32\sqrt{2}\sqrt{1-\cos\theta}\right]_0^{\pi}$$
$$= 64$$

57. $r = 2\theta,\ 0 \le \theta \le \dfrac{\pi}{2}$

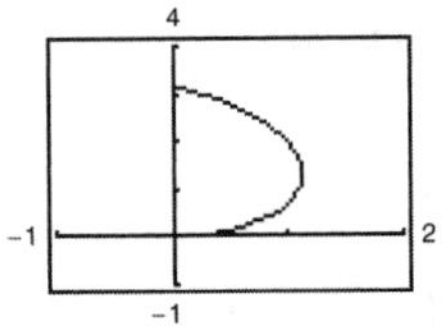

Length ≈ 4.16

58. $r = \sec\theta,\ 0 \le \theta \le \dfrac{\pi}{3}$

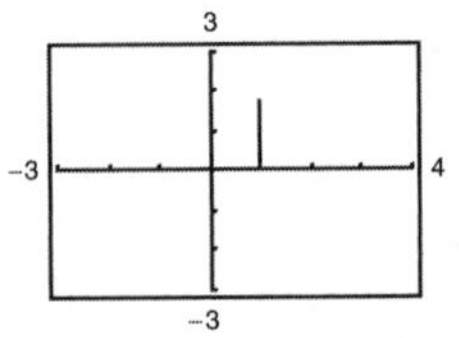

Length $\approx 1.73\left(\text{exact } \sqrt{3}\right)$

59. $r = \dfrac{1}{\theta},\ \pi \le \theta \le 2\pi$

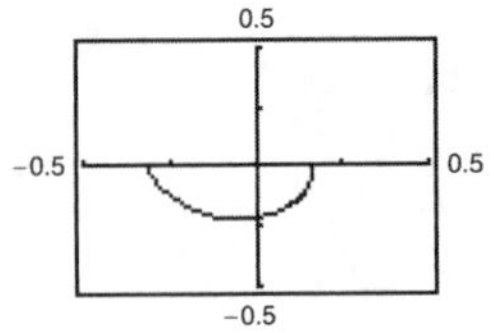

Length ≈ 0.71

60. $r = e^{\theta},\ 0 \le \theta \le \pi$

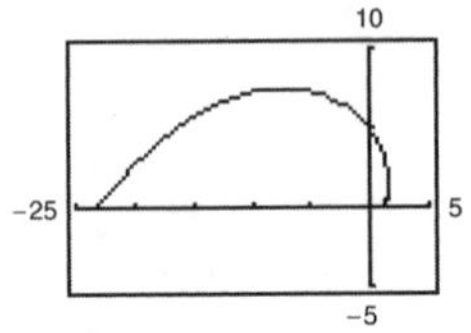

Length ≈ 31.31

61. $r = \sin(3\cos\theta),\ 0 \le \theta \le \pi$

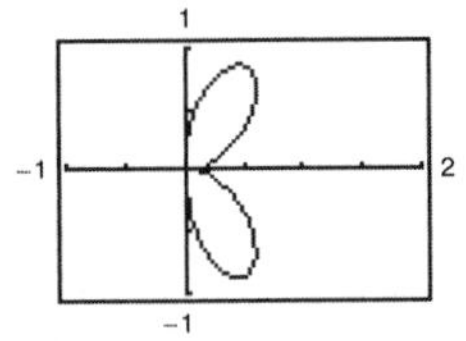

Length ≈ 4.39

62. $r = 2\sin(2\cos\theta),\ 0 \le \theta \le \pi$

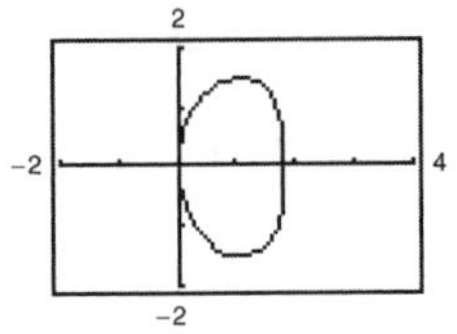

Length ≈ 7.78

63. $r = 6\cos\theta$

$r' = -6\sin\theta$

$$S = 2\pi\int_0^{\pi/2} 6\cos\theta\sin\theta\sqrt{36\cos^2\theta + 36\sin^2\theta}\,d\theta$$
$$= 72\pi\int_0^{\pi/2}\sin\theta\cos\theta\,d\theta$$
$$= \left[36\pi\sin^2\theta\right]_0^{\pi/2}$$
$$= 36\pi$$

64. $r = a\cos\theta$

$r' = -a\sin\theta$

$$S = 2\pi\int_0^{\pi/2} a\cos\theta(\cos\theta)\sqrt{a^2\cos\theta + a^2\sin^2\theta}\,d\theta$$

$$= 2\pi a^2\int_0^{\pi/2}\cos^2\theta\,d\theta = \pi a^2\int_0^{\pi/2}(1+\cos 2\theta)\,d\theta$$

$$= \left[\pi a^2\left(\theta + \frac{\sin 2\theta}{2}\right)\right]_0^{\pi/2} = \frac{\pi^2a^2}{2}$$

65. $r = e^{a\theta}$

$r' = ae^{a\theta}$

$$S = 2\pi\int_0^{\pi/2} e^{a\theta}\cos\theta\sqrt{\left(e^{a\theta}\right)^2 + \left(ae^{a\theta}\right)^2}\,d\theta$$

$$= 2\pi\sqrt{1+a^2}\int_0^{\pi/2} e^{2a\theta}\cos\theta\,d\theta$$

$$= 2\pi\sqrt{1+a^2}\left[\frac{e^{2a\theta}}{4a^2+1}(2a\cos\theta + \sin\theta)\right]_0^{\pi/2}$$

$$= \frac{2\pi\sqrt{1+a^2}}{4a^2+1}\left(e^{\pi a} - 2a\right)$$

66. $r = a(1+\cos\theta)$

$r' = -a\sin\theta$

$$S = 2\pi\int_0^{\pi} a(1+\cos\theta)\sin\theta\sqrt{a^2(1+\cos\theta)^2 + a^2\sin^2\theta}\,d\theta = 2\pi a^2\int_0^{\pi}\sin\theta(1+\cos\theta)\sqrt{2+2\cos\theta}\,d\theta$$

$$= -2\sqrt{2}\pi a^2\int_0^{\pi}(1+\cos\theta)^{3/2}(-\sin\theta)\,d\theta = -\frac{4\sqrt{2}\pi a^2}{5}\left[(1+\cos\theta)^{5/2}\right]_0^{\pi} = \frac{32\pi a^2}{5}$$

67. $r = 4\cos 2\theta$

$r' = -8\sin 2\theta$

$$S = 2\pi\int_0^{\pi/4} 4\cos 2\theta\sin\theta\sqrt{16\cos^2 2\theta + 64\sin^2\theta\, 2\theta}\,d\theta = 32\pi\int_0^{\pi/4}\cos 2\theta\sin\theta\sqrt{\cos^2 2\theta + 4\sin^2 2\theta}\,d\theta \approx 21.87$$

68. $r = \theta$

$r' = 1$

$$S = 2\pi\int_0^{\pi}\theta\sin\theta\sqrt{\theta^2+1}\,d\theta \approx 42.32$$

69. You will only find simultaneous points of intersection. There may be intersection points that do not occur with the same coordinates in the two graphs.

70. (a) $S = 2\pi\int_\alpha^\beta f(\theta)\sin\theta\sqrt{f(\theta)^2 + f'(\theta)^2}\,d\theta$

(b) $S = 2\pi\int_\alpha^\beta f(\theta)\cos\theta\sqrt{f(\theta)^2 + f'(\theta)^2}\,d\theta$

71. (a) $r = 10\cos\theta,\ 0 \le \theta < \pi$

Circle of radius 5

Area $= 25\pi$

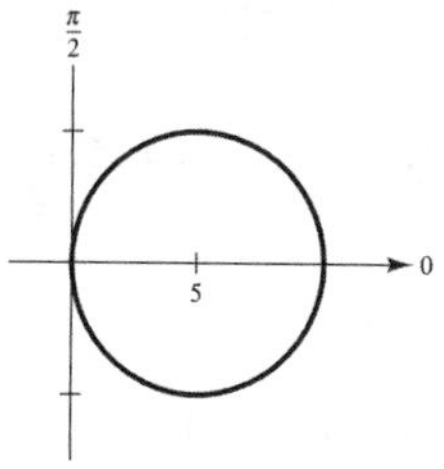

(b) $r = 5\sin\theta,\ 0 \le \theta < \pi$

Circle radius $5/2$

Area $= \frac{25}{4}\pi$

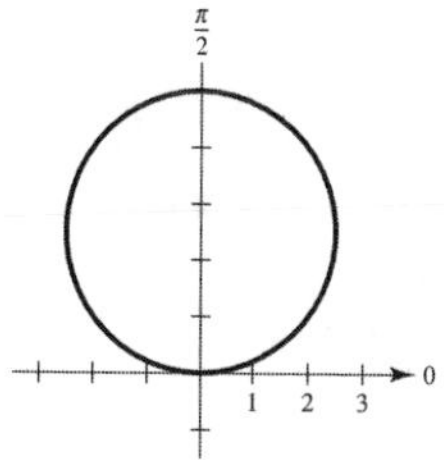

72. Graph (b) has a larger arc length because it has more leaves.

73. Revolve $r = 2$ about the line $r = 5 \sec \theta$.

$f(\theta) = 2, f'(\theta) = 0$

$$S = 2\pi \int_0^{2\pi} (5 - 2\cos\theta)\sqrt{2^2 + 0^2}\, d\theta$$

$$= 4\pi \int_0^{2\pi} (5 - 2\cos\theta)\, d\theta$$

$$= 4\pi\left[5\theta - 2\sin\theta\right]_0^{2\pi}$$

$$= 40\pi^2$$

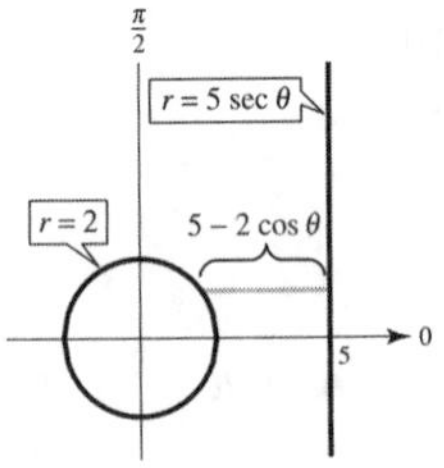

74. Revolve $r = a$ about the line $r = b \sec \theta$ where $b > a > 0$.

$f(\theta) = a$

$f'(\theta) = 0$

$$S = 2\pi \int_0^{2\pi} \left[b - a\cos\theta\right]\sqrt{a^2 + 0^2}\, d\theta$$

$$= 2\pi a\left[b\theta - a\sin\theta\right]_0^{2\pi}$$

$$= 2\pi a(2\pi b) = 4\pi^2 ab$$

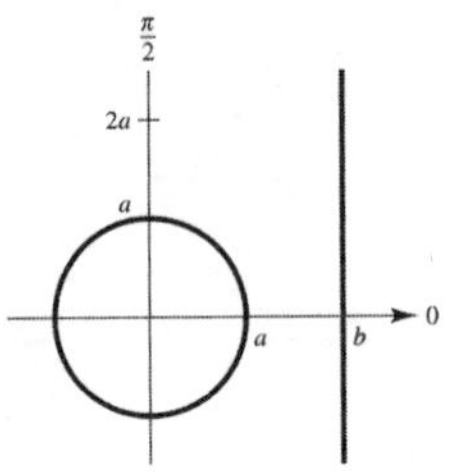

75. $r = 8\cos\theta, 0 \le \theta \le \pi$

(a) $A = \dfrac{1}{2}\displaystyle\int_0^{\pi} r^2\, d\theta = \dfrac{1}{2}\int_0^{\pi} 64\cos^2\theta\, d\theta = 32\int_0^{\pi}\dfrac{1 + \cos 2\theta}{2}\, d\theta = 16\left[\theta + \dfrac{\sin 2\theta}{2}\right]_0^{\pi} = 16\pi$

$\left(\text{Area circle} = \pi r^2 = \pi 4^2 = 16\pi\right)$

(b)

θ	0.2	0.4	0.6	0.8	1.0	1.2	1.4
A	6.32	12.14	17.06	20.80	23.27	24.60	25.08

(c), (d) For $\dfrac{1}{4}$ of area $(4\pi \approx 12.57)$: 0.42

For $\dfrac{1}{2}$ of area $(8\pi \approx 25.13)$: $1.57\left(\dfrac{\pi}{2}\right)$

For $\dfrac{3}{4}$ of area $(12\pi \approx 37.70)$: 2.73

(e) No, it does not depend on the radius.

76. $r = 3\sin\theta, 0 \le \theta \le \pi$

(a) $A = \frac{1}{2}\displaystyle\int_0^{\pi} r^2\, d\theta = \frac{9}{2}\int_0^{\pi}\sin^2\theta\, d\theta = \frac{9}{4}\int_0^{\pi}(1 - \cos 2\theta)\, d\theta = \frac{9}{4}\left[\theta - \frac{1}{2}\sin 2\theta\right]_0^{\pi} = \frac{9}{4}\pi$

$\left[\textbf{Note:}\text{ radius of circle is } \frac{3}{2} \Rightarrow A = \pi\left(\frac{3}{2}\right)^2 = \frac{9}{4}\pi\right]$

(b)

θ	0.2	0.4	0.6	0.8	1.0	1.2	1.4
A	0.0119	0.0930	0.3015	0.6755	1.2270	1.9401	2.7731

(c), (d) For $\dfrac{1}{8}$ of area $\left(\dfrac{1}{8}\dfrac{9}{4}\pi \approx 0.8836\right)$: $\theta \approx 0.88$

For $\dfrac{1}{4}$ of area $\left(\dfrac{1}{4}\dfrac{9}{4}\pi \approx 1.7671\right)$: $\theta \approx 1.15$

For $\dfrac{1}{2}$ of area $\left(\dfrac{1}{2}\dfrac{9}{4}\pi \approx 3.5343\right)$: $\theta = \dfrac{\pi}{2} \approx 1.57$

77.

$$\begin{aligned} r &= a\sin\theta + b\cos\theta \\ r^2 &= ar\sin\theta + br\cos\theta \\ x^2 + y^2 &= ay + bx \\ x^2 + y^2 - bx - ay &= 0 \text{ represents a circle.} \end{aligned}$$

78. $r = \sin\theta + \cos\theta$, Circle

$$A = \frac{1}{2}\int_\alpha^\beta r^2\,d\theta$$

$$= \frac{1}{2}\int_0^\pi (\sin\theta + \cos\theta)^2\,d\theta$$

$$= \frac{1}{2}\int_0^\pi (1 + 2\sin\theta\cos\theta)\,d\theta = \frac{1}{2}\Big[\theta + \sin^2\theta\Big]_0^\pi = \frac{\pi}{2}$$

Converting to rectangular form:

$$r^2 = r\sin\theta + r\cos\theta$$

$$x^2 + y^2 = y + x$$

$$\left(x^2 - x + \frac{1}{4}\right) + \left(y^2 - y + \frac{1}{4}\right) = \frac{1}{2}$$

$$\left(x - \frac{1}{2}\right)^2 + \left(y - \frac{1}{2}\right)^2 = \frac{1}{2}$$

Circle of radius $\frac{1}{\sqrt{2}}$ and center $\left(\frac{1}{2}, \frac{1}{2}\right)$

$$\text{Area} = \pi\left(\frac{1}{\sqrt{2}}\right)^2 = \frac{\pi}{2}$$

79. (a) $r = \theta, \theta \ge 0$

As a increases, the spiral opens more rapidly. If $\theta < 0$, the spiral is reflected about the y-axis.

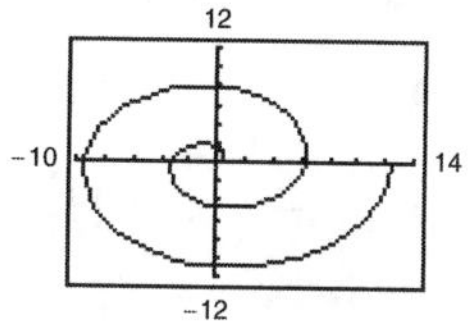

(b) $r = a\theta, \theta \ge 0$, crosses the polar axis for $\theta = n\pi$, n and integer. To see this

$r = a\theta \Rightarrow r\sin\theta = y = a\theta\sin\theta = 0$

for $\theta = n\pi$. The points are

$(r, \theta) = (an\pi, n\pi), n = 1, 2, 3, \ldots$

(c) $f(\theta) = \theta, f'(\theta) = 1$

$$s = \int_0^{2\pi} \sqrt{\theta^2 + 1}\,d\theta$$

$$= \frac{1}{2}\left[\ln\left(\sqrt{x^2+1} + x\right) + x\sqrt{x^2+1}\right]_0^{2\pi}$$

$$= \frac{1}{2}\ln\left(\sqrt{4\pi^2+1} + 2\pi\right) + \pi\sqrt{4\pi^2+1} \approx 21.2563$$

(d) $A = \frac{1}{2}\int_\alpha^\beta r^2\,dr = \frac{1}{2}\int_0^{2\pi}\theta^2\,d\theta = \left[\frac{\theta^3}{6}\right]_0^{2\pi} = \frac{4}{3}\pi^3$

80. $r = e^{\theta/6}$

$$A = \frac{1}{2}\int_0^{2\pi}\left(e^{\theta/6}\right)^2 d\theta - \frac{1}{2}\int_{-2\pi}^0 \left(e^{\theta/6}\right)^2 d\theta$$

$$= \frac{1}{2}\int_0^{2\pi} e^{\theta/3}\,d\theta - \frac{1}{2}\int_{-2\pi}^0 e^{\theta/3}\,d\theta$$

$$= \left[\frac{3}{2}e^{\theta/3}\right]_0^{2\pi} - \left[\frac{3}{2}e^{\theta/3}\right]_{-2\pi}^0$$

$$= \frac{3}{2}e^{2\pi/3} - \frac{3}{2} - \frac{3}{2} + \frac{3}{2}e^{-2\pi/3} = \frac{3}{2}\left[e^{2\pi/3} + e^{-2\pi/3} - 2\right]$$

$$\approx 9.3655$$

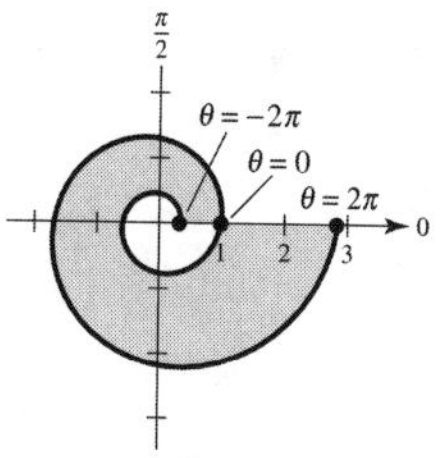

81. The smaller circle has equation $r = a\cos\theta$. The area of the shaded lune is:

$$A = 2\left(\frac{1}{2}\right)\int_0^{\pi/4}\left[(a\cos\theta)^2 - 1\right] d\theta$$

$$= \int_0^{\pi/4}\left[\frac{a^2}{2}(1 + \cos 2\theta) - 1\right] d\theta$$

$$= \left[\frac{a^2}{2}\left(\theta + \frac{\sin 2\theta}{2}\right) - \theta\right]_0^{\pi/4}$$

$$= \frac{a^2}{2}\left(\frac{\pi}{4} + \frac{1}{2}\right) - \frac{\pi}{4}$$

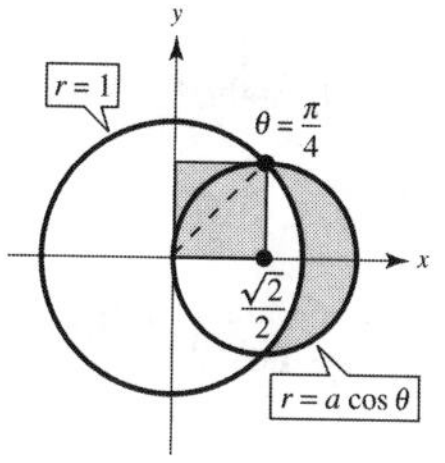

This equals the area of the square, $\left(\frac{\sqrt{2}}{2}\right)^2 = \frac{1}{2}$.

$$\frac{a^2}{2}\left(\frac{\pi}{4} + \frac{1}{2}\right) - \frac{\pi}{4} = \frac{1}{2}$$

$$\pi a^2 + 2a^2 - 2\pi - 4 = 0$$

$$a^2 = \frac{4 + 2\pi}{2 + \pi} = 2$$

$$a = \sqrt{2}$$

Smaller circle: $r = \sqrt{2}\cos\theta$

82. $x = \dfrac{3t}{1+t^3},\ y = \dfrac{3t^2}{1+t^3}$

(a) $x^3 + y^3 = \dfrac{27(t^3 + t^6)}{(1+t^3)^3} = \dfrac{27t^3}{(1+t^3)^2}$

$3xy = \dfrac{27t^3}{(1+t^3)^2}$

So, $x^3 + y^3 = 3xy$.

$(r\cos\theta)^3 + (r\sin\theta)^3 = 3(r\cos\theta)(r\sin\theta)$

$r = \dfrac{3\cos\theta\sin\theta}{\cos^3\theta + \sin^3\theta}$

(b)

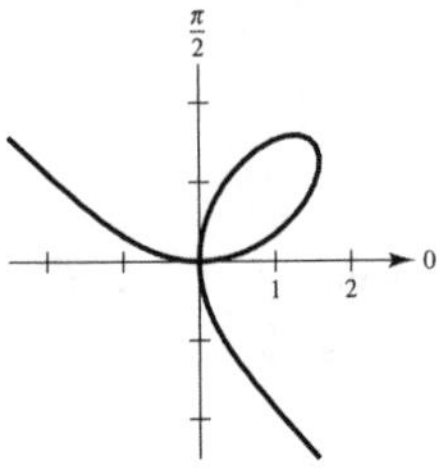

(c) $A = \dfrac{1}{2}\int_0^{\pi/2} r^2\,d\theta = \dfrac{3}{2}$

83. False. $f(\theta) = 1$ and $g(\theta) = -1$ have the same graphs.

84. False. $f(\theta) = 0$ and $g(\theta) = \sin 2\theta$ have only one point of intersection.

85. In parametric form,

$$s = \int_a^b \sqrt{\left(\frac{dx}{dt}\right)^2 + \left(\frac{dy}{dt}\right)^2}\,dt.$$

Using θ instead of t, you have $x = r\cos\theta = f(\theta)\cos\theta$ and $y = r\sin\theta = f(\theta)\sin\theta$. So,

$\dfrac{dx}{d\theta} = f'(\theta)\cos\theta - f(\theta)\sin\theta$ and

$\dfrac{dy}{d\theta} = f'(\theta)\sin\theta + f(\theta)\cos\theta.$

It follows that

$$\left(\frac{dx}{d\theta}\right)^2 + \left(\frac{dy}{d\theta}\right)^2 = [f(\theta)]^2 + [f'(\theta)]^2.$$

So, $s = \displaystyle\int_\alpha^\beta \sqrt{[f(\theta)]^2 + [f'(\theta)]^2}\,d\theta.$

Section 10.6 Polar Equations of Conics and Kepler's Laws

1. $r = \dfrac{2e}{1 + e\cos\theta}$

(a) $e = 1,\ r = \dfrac{2}{1+\cos\theta}$, parabola

(b) $e = 0.5$,

$r = \dfrac{1}{1 + 0.5\cos\theta} = \dfrac{2}{2+\cos\theta}$, ellipse

(c) $e = 1.5$,

$r = \dfrac{3}{1 + 1.5\cos\theta} = \dfrac{6}{2 + 3\cos\theta}$, hyperbola

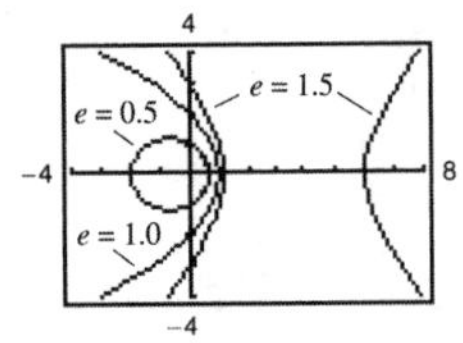

2. $r = \dfrac{2e}{1 - e\cos\theta}$

(a) $e = 1,\ r = \dfrac{2}{1-\cos\theta}$, parabola

(b) $e = 0.5$,

$r = \dfrac{1}{1 - 0.5\cos\theta} = \dfrac{2}{2-\cos\theta}$, ellipse

(c) $e = 1.5$,

$r = \dfrac{3}{1 - 1.5\cos\theta} = \dfrac{6}{2 - 3\cos\theta}$, hyperbola

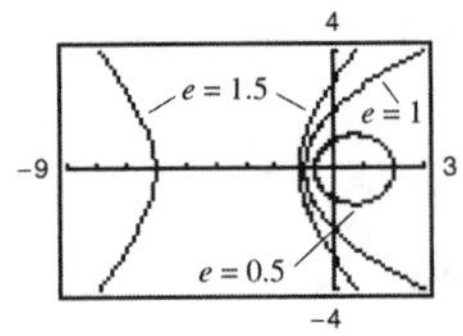

3. $r = \dfrac{2e}{1 - e \sin \theta}$

(a) $e = 1, r = \dfrac{2}{1 - \sin \theta}$, parabola

(b) $e = 0.5$,

$$r = \frac{1}{1 - 0.5 \sin \theta} = \frac{2}{2 - \sin \theta}, \text{ ellipse}$$

(c) $e = 1.5$,

$$r = \frac{3}{1 - 1.5 \sin \theta} = \frac{6}{2 - 3 \sin \theta}, \text{ hyperbola}$$

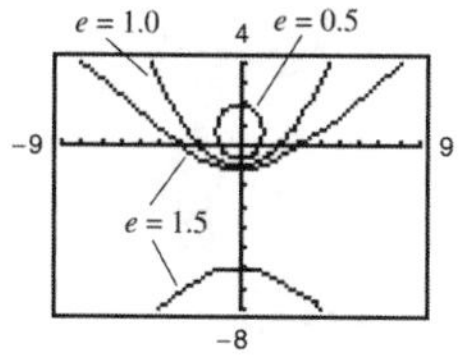

4. $r = \dfrac{2e}{1 + e \sin \theta}$

(a) $e = 1, r = \dfrac{2}{1 + \sin \theta}$, parabola

(b) $e = 0.5$,

$$r = \frac{1}{1 + 0.5 \sin \theta} = \frac{2}{2 + \sin \theta}, \text{ ellipse}$$

(c) $e = 1.5$,

$$r = \frac{3}{1 + 1.5 \sin \theta} = \frac{6}{2 + 3 \sin \theta}, \text{ hyperbola}$$

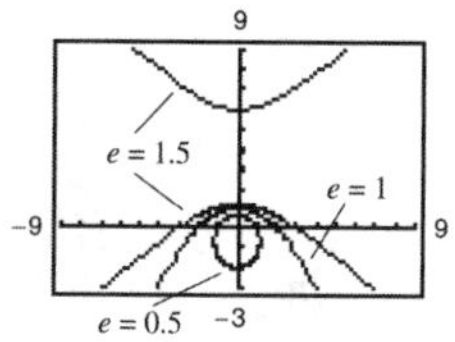

5. $r = \dfrac{4}{1 + e \sin \theta}$

(a) The conic is an ellipse. As $e \to 1^-$, the ellipse becomes more elliptical, and as $e \to 0^+$, it becomes more circular.

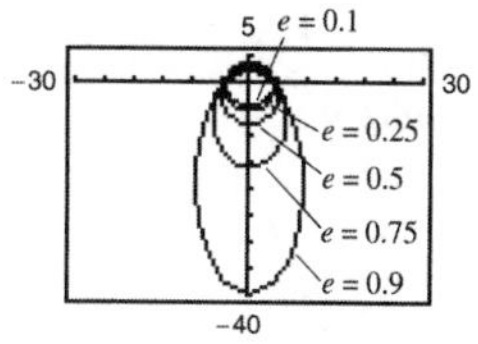

(b) The conic is a parabola.

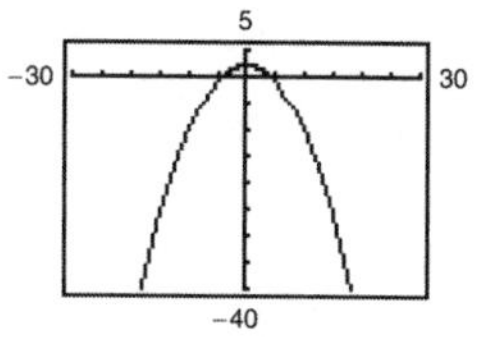

(c) The conic is a hyperbola. As $e \to 1^+$, the hyperbola opens more slowly, and as $e \to \infty$, it opens more rapidly.

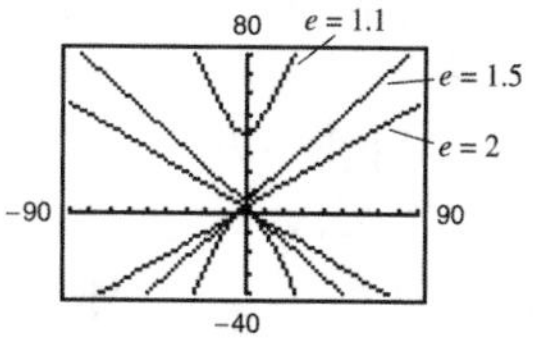

6. $r = \dfrac{4}{1 - 0.4 \cos \theta}$

(a) Because $e = 0.4 < 1$, the conic is an ellipse with vertical directrix to the left of the pole.

(b) $r = \dfrac{4}{1 + 0.4 \cos \theta}$

The ellipse is shifted to the left. The vertical directrix is to the right of the pole.

$$r = \frac{4}{1 - 0.4 \sin \theta}.$$

The ellipse has a horizontal directrix below the pole.

(c)

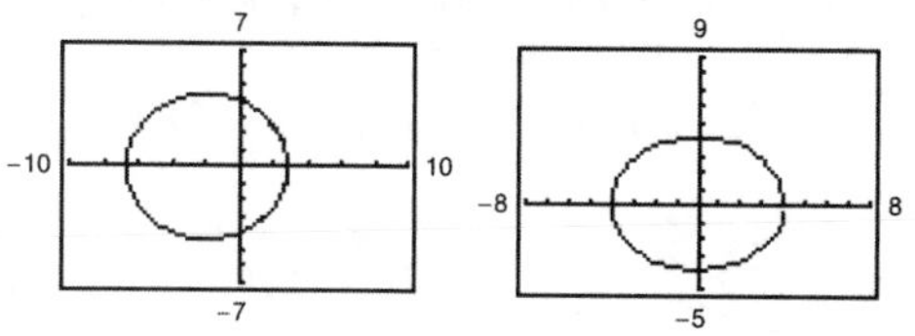

7. Parabola; Matches (c)

8. Ellipse; Matches (f)

9. Hyperbola; Matches (a)

10. Parabola; Matches (e)

11. Ellipse; Matches (b)

12. Hyperbola; Matches (d)

13. $r = \dfrac{1}{1 - \cos\theta}$

Parabola because $e = 1, d = 1$

Distance from pole to directrix: $|d| = 1$

Directrix: $x = -d = -1$

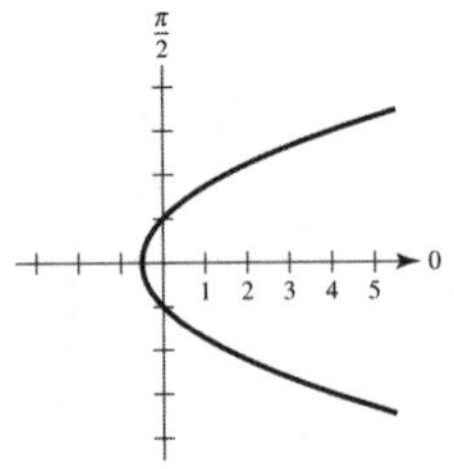

14. $r = \dfrac{6}{3 - 2\cos\theta} = \dfrac{2}{1 - \frac{2}{3}\cos\theta} = \dfrac{\left(\frac{2}{3}\right)3}{1 - \left(\frac{2}{3}\right)\cos\theta}$

Ellipse because $e = \dfrac{2}{3} < 1, d = 3$

Distance from directrix to pole: $|d| = 3$

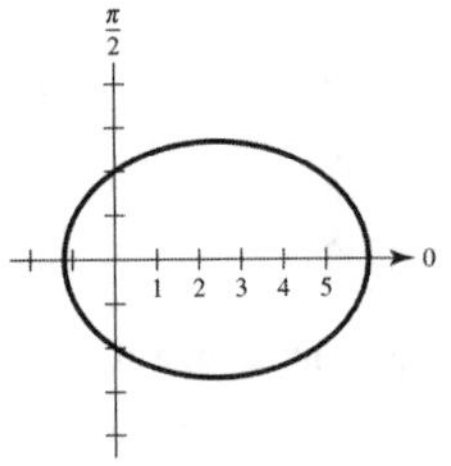

15. $r = \dfrac{3}{2 + 6\sin\theta} = \dfrac{3/2}{1 + 3\sin\theta}$

Hyperbola because $e = 3 > 0; d = 1/2$

Directrix: $y = 1/2$

Distance from pole to directrix: $|d| = 1/2$

Vertices: $(r, \theta) = (3/8, \pi/2), (-3/4, 3\pi/2)$

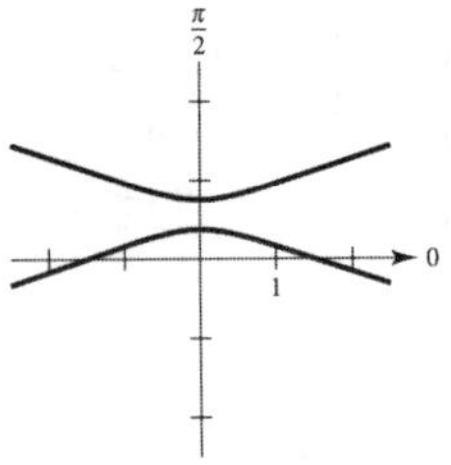

16. $r = \dfrac{4}{1 + \cos\theta}$

Parabola because $e = 1, d = 4$

Distance from pole to directrix: $|d| = 4$

Directrix: $x = 4$

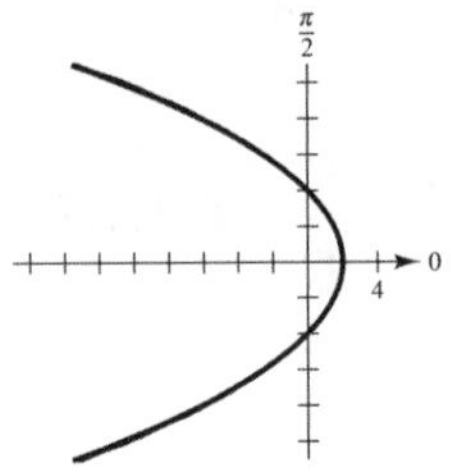

17. $r = \dfrac{5}{-1 + 2\cos\theta} = \dfrac{-5}{1 - 2\cos\theta}$

Hyperbola because $e = 2 > 1; d = -5/2$

Directrix: $x = 5/2$

Distance from pole to directrix: $|d| = 5/2$

Vertices: $(r, \theta) = (5, 0), (-5/3, \pi)$

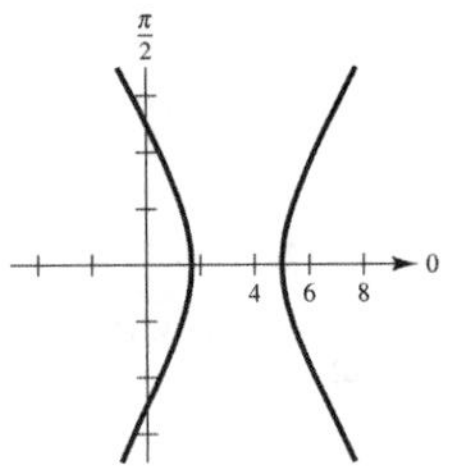

18. $r = \dfrac{10}{5 + 4\sin\theta} = \dfrac{2}{1 + \left(\frac{4}{5}\right)\sin\theta} = \dfrac{\left(\frac{4}{5}\right)\left(\frac{5}{2}\right)}{1 + \left(\frac{4}{5}\right)\sin\theta}$

Ellipse because $e = \dfrac{4}{5} < 1, d = \dfrac{5}{2}$

Distance from pole to directrix: $|d| = \dfrac{5}{2}$

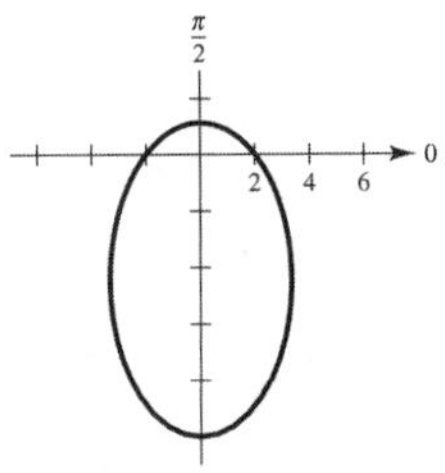

19. $r = \dfrac{6}{2 + \cos\theta} = \dfrac{3}{1 + (1/2)\cos\theta}$

Ellipse because $e = \dfrac{1}{2}; d = 6$

Directrix: $x = 6$

Distance from pole to directrix: $|d| = 6$

Vertices: $(r, \theta) = (2, 0), (6, \pi)$

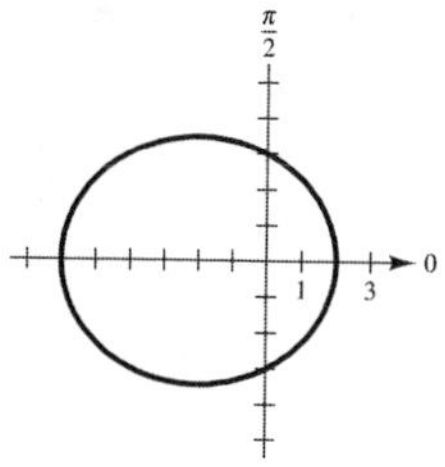

20. $r = \dfrac{-6}{3 + 7\sin\theta} = \dfrac{-2}{1 + (7/3)\sin\theta}$

Hyperbola because $e = 7/3 > 1; d = -6/7$

Directrix: $y = -6/7$

Distance from pole to directrix: $|d| = 6/7$

Vertices: $(r, \theta) = (-3/5, \pi/2), (3/2, 3\pi/2)$

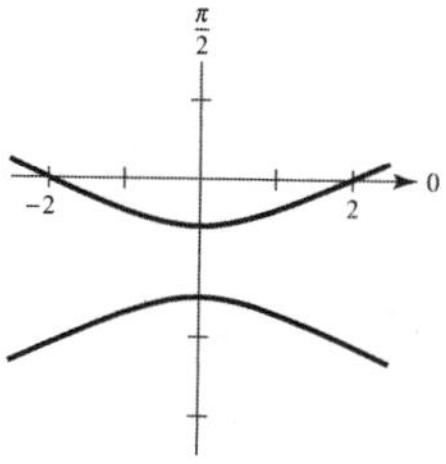

21. $r = \dfrac{300}{-12 + 6\sin\theta} = \dfrac{-25}{1 - \frac{1}{2}\sin\theta} = \dfrac{\frac{1}{2}(-50)}{1 - \frac{1}{2}\sin\theta}$

Ellipse because $e = \dfrac{1}{2}, d = -50$

Distance from pole to directrix: $|d| = 50$

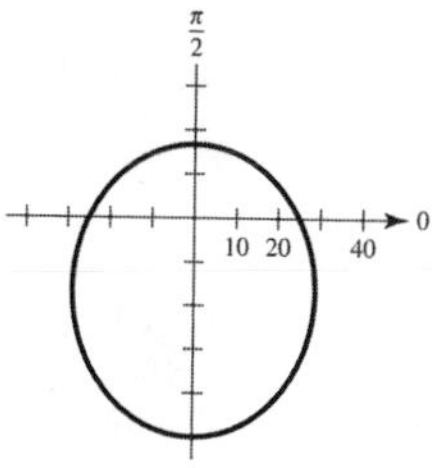

22. $r = \dfrac{1}{1 + \sin\theta}$

Parabola because $e = 1, d = 1$

Distance from pole to directrix: $|d| = 1$

Directrix: $y = 1$

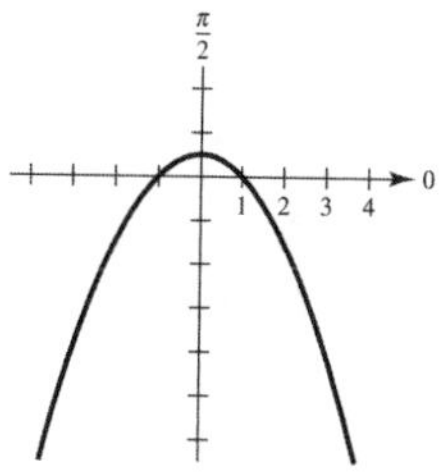

23. $r = \dfrac{3}{-4 + 2\sin\theta} = \dfrac{-\frac{3}{4}}{1 - \frac{1}{2}\sin\theta}$

$e = \dfrac{1}{2}$, Ellipse

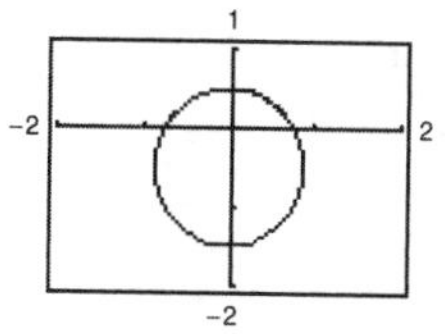

24. $r = \dfrac{-15}{2 + 8\sin\theta} = \dfrac{-\frac{15}{2}}{1 + 4\sin\theta}$

$e = 4$, Hyperbola

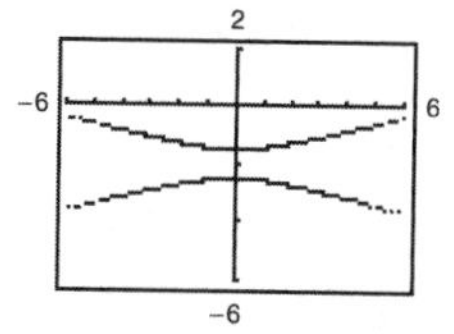

25. $r = \dfrac{-10}{1 - \cos\theta}$

$e = 1$, Parabola

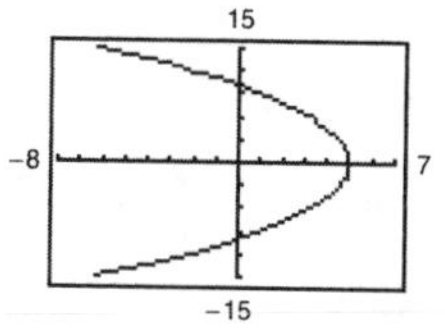

26. $r = \dfrac{6}{6 + 7\cos\theta} = \dfrac{1}{1 + \left(\frac{7}{6}\right)\cos\theta}$

$e = \dfrac{7}{6}$, Hyperbola

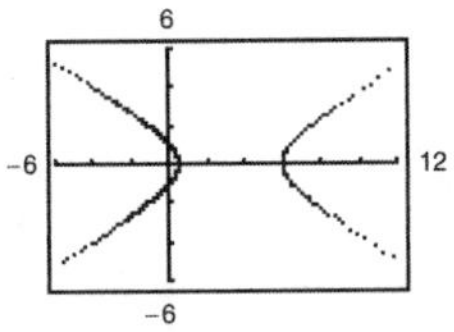

27. $r = \dfrac{4}{1 + \cos\left(\theta - \dfrac{\pi}{3}\right)}$

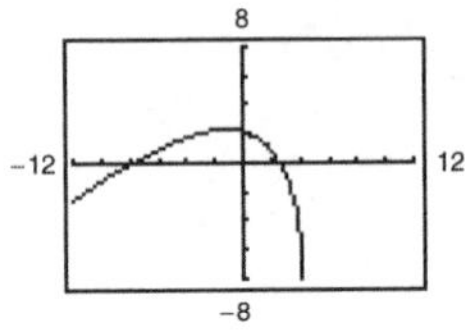

Rotate the graph of $r = \dfrac{4}{1 + \cos\theta}$

$\dfrac{\pi}{3}$ radian counterclockwise.

28. $r = \dfrac{10}{5 + 4\sin\left(\theta - \dfrac{\pi}{4}\right)}$

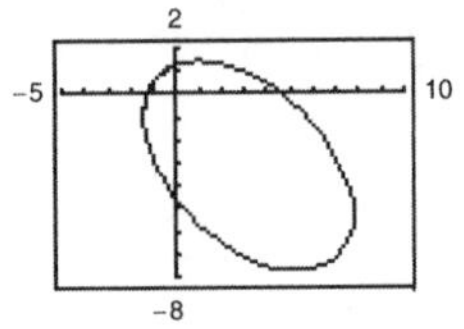

Rotate the graph of $r = \dfrac{10}{5 + 4\sin\theta}$

$\dfrac{\pi}{4}$ radian counterclockwise.

29. $r = \dfrac{6}{2 + \cos\left(\theta + \dfrac{\pi}{6}\right)}$

Rotate the graph of $r = \dfrac{6}{2 + \cos\theta}$

$\dfrac{\pi}{6}$ radian clockwise.

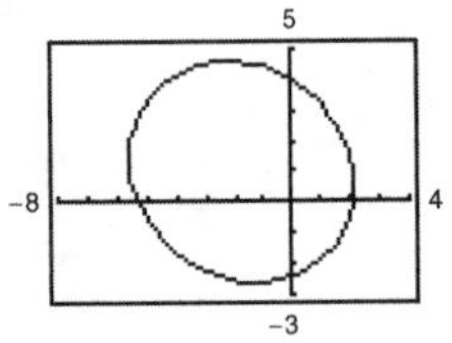

30. $r = \dfrac{-6}{3 + 7\sin(\theta + (2\pi/3))}$

Rotate graph of $r = \dfrac{-6}{3 + 7\sin\theta}$

$\dfrac{2\pi}{3}$ radians clockwise.

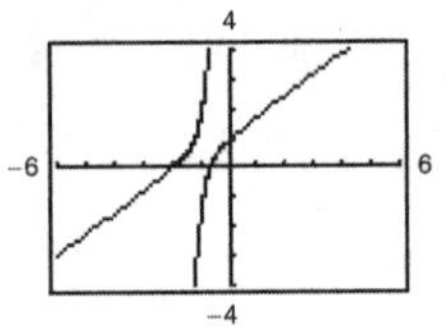

31. Change θ to $\theta + \dfrac{\pi}{6}$

$$r = \frac{8}{8 + 5\cos\left(\theta + \dfrac{\pi}{6}\right)}$$

32. Change θ to $\theta - \dfrac{\pi}{4}$

$$r = \frac{9}{1 + \sin\left(\theta - \dfrac{\pi}{4}\right)}$$

33. Parabola

$e = 1$

$x = -3 \Rightarrow d = 3$

$$r = \frac{ed}{1 - e\cos\theta} = \frac{3}{1 - \cos\theta}$$

34. Parabola

$e = 1,\ y = 4 \Rightarrow d = 4$

$$r = \frac{ed}{1 + e\sin\theta} = \frac{4}{1 + \sin\theta}$$

35. Ellipse

$e = \dfrac{1}{2},\ y = 1,\ d = 1$

$$r = \frac{ed}{1 + e\sin\theta} = \frac{1/2}{1 + (1/2)\sin\theta} = \frac{1}{2 + \sin\theta}$$

36. Ellipse

$e = \dfrac{3}{4},\ y = -2,\ d = 2$

$$r = \frac{ed}{1 - e\sin\theta} = \frac{2(3/4)}{1 - (3/4)\sin\theta} = \frac{6}{4 - 3\sin\theta}$$

37. Hyperbola

$e = 2, x = 1, d = 1$

$$r = \frac{ed}{1 + e\cos\theta} = \frac{2}{1 + 2\cos\theta}$$

38. Hyperbola

$e = \frac{3}{2}, x = -1, d = 1$

$$r = \frac{ed}{1 - e\cos\theta} = \frac{3/2}{1 - (3/2)\cos\theta} = \frac{3}{2 - 3\cos\theta}$$

39. Parabola

Vertex: $\left(1, -\frac{\pi}{2}\right)$

$e = 1, d = 2, r = \frac{2}{1 - \sin\theta}$

40. Parabola

Vertex: $(5, \pi)$

$e = 1, d = 10$

$$r = \frac{ed}{1 - e\cos\theta} = \frac{10}{1 - \cos\theta}$$

41. Ellipse

Vertices: $(2, 0), (8, \pi)$

$e = \frac{3}{5}, d = \frac{16}{3}$

$$r = \frac{ed}{1 + e\cos\theta} = \frac{16/5}{1 + (3/5)\cos\theta} = \frac{16}{5 + 3\cos\theta}$$

42. Ellipse

Vertices: $\left(2, \frac{\pi}{2}\right), \left(4, \frac{3\pi}{2}\right)$

$e = \frac{1}{3}, d = 8$

$$r = \frac{ed}{1 + e\sin\theta} = \frac{8/3}{1 + (1/3)\sin\theta} = \frac{8}{3 + \sin\theta}$$

43. Hyperbola

Vertices: $\left(1, \frac{3\pi}{2}\right), \left(9, \frac{3\pi}{2}\right)$

$e = \frac{5}{4}, d = \frac{9}{5}$

$$r = \frac{ed}{1 - e\sin\theta} = \frac{9/4}{1 - (5/4)\sin\theta} = \frac{9}{4 - 5\sin\theta}$$

44. Hyperbola

Vertices: $(2, 0), (10, 0)$

$e = \frac{3}{2}, d = \frac{10}{3}$

$$r = \frac{ed}{1 + e\cos\theta} = \frac{5}{1 + (3/2)\cos\theta} = \frac{10}{2 + 3\cos\theta}$$

45. Ellipse, $e = \frac{1}{2}$,

Directrix: $r = 4\sec\theta \Rightarrow x = r\cos\theta = 4$

$$r = \frac{ed}{1 + e\cos\theta} = \frac{\left(\frac{1}{2}\right)4}{1 + \frac{1}{2}\cos\theta} = \frac{4}{2 + \cos\theta}$$

46. Hyperbola, $e = 2$

Directrix: $r = -8\csc\theta \Rightarrow y = r\sin\theta = -8$

$$r = \frac{ed}{1 - e\sin\theta} = \frac{2(-8)}{1 - 2\sin\theta} = \frac{-16}{1 - 2\sin\theta}$$

47. Ellipse if $0 < e < 1$, parabola if $e = 1$, hyperbola if $e > 1$.

48. (a) Hyperbola $(e = 2 > 1)$

(b) Ellipse $\left(e = \frac{1}{10} < 1\right)$

(c) Parabola $(e = 1)$

(d) Rotated hyperbola $(e = 3)$

49. If the foci are fixed and $e \to 0$, then $d \to \infty$. To see this, compare the ellipses

$$r = \frac{1/2}{1 + (1/2)\cos\theta}, e = 1/2, d = 1$$

$$r = \frac{5/16}{1 + (1/4)\cos\theta}, e = 1/4, d = 5/4.$$

50. (a) The conic is an ellipse, so $0 < e < 1$.

(b) The conic is a parabola, so $e = 1$.

(c) The conic is a hyperbola, so $e > 1$.

(d) The conic is an ellipse, so $0 < e < 1$.

51.
$$\frac{x^2}{a^2} + \frac{y^2}{b^2} = 1$$
$$x^2b^2 + y^2a^2 = a^2b^2$$
$$b^2r^2\cos^2\theta + a^2r^2\sin^2\theta = a^2b^2$$
$$r^2\left[b^2\cos^2\theta + a^2\left(1 - \cos^2\theta\right)\right] = a^2b^2$$
$$r^2\left[a^2 + \cos^2\theta\left(b^2 - a^2\right)\right] = a^2b^2$$
$$r^2 = \frac{a^2b^2}{a^2 + \left(b^2 - a^2\right)\cos^2\theta} = \frac{a^2b^2}{a^2 - c^2\cos^2\theta}$$
$$= \frac{b^2}{1 - (c/a)^2\cos^2\theta} = \frac{b^2}{1 - e^2\cos^2\theta}$$

52.
$$\frac{x^2}{a^2} - \frac{y^2}{b^2} = 1$$
$$x^2b^2 - y^2a^2 = a^2b^2$$
$$b^2r^2\cos^2\theta - a^2r^2\sin^2\theta = a^2b^2$$
$$r^2\left[b^2\cos^2\theta - a^2\left(1 - \cos^2\theta\right)\right] = a^2b^2$$
$$r^2\left[-a^2 + \cos^2\theta\left(a^2 + b^2\right)\right] = a^2b^2$$
$$r^2 = \frac{a^2b^2}{-a^2 + c^2\cos^2\theta} = \frac{b^2}{-1 + \left(c^2/a^2\right)\cos^2\theta}$$
$$= \frac{-b^2}{1 - e^2\cos^2\theta}$$

53. $a = 5, c = 4, e = \frac{4}{5}, b = 3$
$$r^2 = \frac{9}{1 - (16/25)\cos^2\theta}$$

54. $a = 4, c = 5, b = 3, e = \frac{5}{4}$
$$r^2 = \frac{-9}{1 - (25/16)\cos^2\theta}$$

55. $a = 3, b = 4, c = 5, e = \frac{5}{3}$
$$r^2 = \frac{-16}{1 - (25/9)\cos^2\theta}$$

56. $a = 2, b = 1, c = \sqrt{3}, e = \frac{\sqrt{3}}{2}$
$$r^2 = \frac{1}{1 - (3/4)\cos^2\theta}$$

57. $A = 2\left[\frac{1}{2}\int_0^{\pi}\left(\frac{3}{2 - \cos\theta}\right)^2 d\theta\right]$
$$= 9\int_0^{\pi}\frac{1}{(2 - \cos\theta)^2}\,d\theta \approx 10.88$$

58. $A = \frac{1}{2}\int_0^{2\pi}\left(\frac{9}{4 + \cos\theta}\right)^2 d\theta \approx 17.52$

59. $A = 2\left[\frac{1}{2}\int_{-\pi/2}^{\pi/2}\left(\frac{2}{3 - 2\sin\theta}\right)^2 d\theta\right]$
$$= 4\int_{-\pi/2}^{\pi/2}\frac{1}{(3 - 2\sin\theta)^2}\,d\theta \approx 3.37$$

60. $A = \frac{1}{2}\int_0^{2\pi}\left[\frac{3}{6 + 5\sin\theta}\right]^2 d\theta \approx 4.65$

61. Vertices: $(123{,}000 + 4000, 0) = (127{,}000, 0)$
$(119 + 4000, \pi) = (4119, \pi)$

$$a = \frac{127{,}000 + 4119}{2} = 65{,}559.5$$
$$c = 65{,}559.5 - 4119 = 61{,}440.5$$
$$e = \frac{c}{a} = \frac{122{,}881}{131{,}119} \approx 0.93717$$
$$r = \frac{ed}{1 - e\cos\theta}$$
$$\theta = 0: r = \frac{ed}{1 - e}, \theta = \pi: r = \frac{ed}{1 + e}$$
$$2a = 2(65{,}559.5) = \frac{ed}{1 - e} + \frac{ed}{1 + e}$$
$$131{,}119 = d\left(\frac{e}{1 - e} + \frac{e}{1 + e}\right) = d\left(\frac{2e}{1 - e^2}\right)$$
$$d = \frac{131{,}119\left(1 - e^2\right)}{2e} \approx 8514.1397$$
$$r = \frac{7979.21}{1 - 0.93717\cos\theta} = \frac{1{,}046{,}226{,}000}{131{,}119 - 122{,}881\cos\theta}$$

When $\theta = 60° = \frac{\pi}{3}$, $r \approx 15{,}015$.

Distance between earth and the satellite is $r - 4000 \approx 11{,}015$ miles.

62. (a) $r = \dfrac{ed}{1 - e\cos\theta}$

When $\theta = 0$, $r = c + a = ea + a = a(1 + e)$.

So,

$$a(1 + e) = \frac{ed}{1 - e}$$
$$a(1 + e)(1 - e) = ed$$
$$a(1 - e^2) = ed.$$

So, $r = \dfrac{(1 - e^2)a}{1 - e\cos\theta}$.

(b) The perihelion distance is
$a - c = a - ea = a(1 - e)$.

When $\theta = \pi$, $r = \dfrac{(1 - e^2)a}{1 + e} = a(1 - e)$.

The aphelion distance is
$a + c = a + ea = a(1 + e)$.

When $\theta = 0$, $r = \dfrac{(1 - e^2)a}{1 - e} = a(1 + e)$.

63. $a = 1.496 \times 10^8$, $e = 0.0167$

$$r = \frac{(1 - e^2)a}{1 - e\cos\theta} = \frac{149{,}558{,}278.1}{1 - 0.0167\cos\theta}$$

Perihelion distance: $a(1 - e) \approx 147{,}101{,}680$ km

Aphelion distance: $a(1 + e) \approx 152{,}098{,}320$ km

64. $a = 1.427 \times 10^9$, $e = 0.0542$

$$r = \frac{(1 - e^2)a}{1 - e\cos\theta} = \frac{1{,}422{,}807{,}988}{1 - 0.0542\cos\theta}$$

Perihelion distance: $a(1 - e) \approx 1{,}349{,}656{,}600$ km

Aphelion distance: $a(1 + e) \approx 1{,}504{,}343{,}400$ km

65. $a = 4.498 \times 10^9$, $e = 0.0086$

$$r = \frac{(1 - e^2)a}{1 - e\cos\theta} = \frac{4{,}497{,}667{,}328}{1 - 0.0086\cos\theta}$$

Perihelion distance: $a(1 - e) \approx 4{,}459{,}317{,}200$ km

Aphelion distance: $a(1 + e) \approx 4{,}536{,}682{,}800$ km

66. $a = 5.791 \times 10^7$, $e = 0.2056$

$$r = \frac{(1 - e^2)a}{1 - e\cos\theta} \approx \frac{55{,}462{,}065.54}{1 - 0.2056\cos\theta}$$

Perihelion distance $\approx a(1 - e) \approx 46{,}003{,}704$ km

Aphelion distance $\approx a(1 + e) \approx 69{,}816{,}296$ km

67. $r = \dfrac{4.498 \times 10^9}{1 - 0.0086\cos\theta}$

(a) $A = \dfrac{1}{2}\displaystyle\int_0^{\pi/9} r^2\, d\theta \approx 3.591 \times 10^{18}\ \text{km}^2$

$$165\left[\frac{\frac{1}{2}\int_0^{\pi/2} r^2\, d\theta}{\frac{1}{2}\int_0^{2\pi} r^2\, d\theta}\right] \approx 9.322 \text{ yrs}$$

(b) $\dfrac{1}{2}\displaystyle\int_\pi^{\alpha} r^2\, d\theta = 3.591 \times 10^{18}$

By trial and error, $\alpha \approx \pi + 0.361$

$0.361 > \pi/9 \approx 0.349$ because the rays in part (a) are longer than those in part (b)

(c) For part (a),

$$s = \int_0^{\pi/9} \sqrt{r^2 + (dr/d\theta)^2} \approx 1.583 \times 10^9 \text{ km}$$

$$\text{Average per year} = \frac{1.583 \times 10^9}{9.322} \approx 1.698 \times 10^8 \text{ km/yr}$$

For part (b),

$$s = \int_{\pi}^{\pi+0.361} \sqrt{r^2 + (dr/d\theta)^2}\, d\theta \approx 1.610 \times 10^9 \text{ km}$$

$$\text{Average per year} = \frac{1.610 \times 10^9}{9.322} \approx 1.727 \times 10^8 \text{ km/yr}$$

68. $a = \frac{1}{2}(500) = 250$ au, $e \approx 0.995$

(a) $e = \frac{c}{a} \Rightarrow c \approx 248.75$

$b^2 = a^2 - c^2 \Rightarrow b \approx 24.969 \Rightarrow$ minor axis $= 2b \approx 49.9$ au

(b) $r = \frac{(1 - e^2)a}{1 - e\cos\theta} = \frac{2.49375}{1 - 0.995\cos\theta}$

(c) Perihelion distance: $a(1 - e) \approx 1.25$ au

Aphelion distance: $a(1 + e) \approx 498.75$ au

69. $r_1 = a + c, r_0 = a - c, r_1 - r_0 = 2c, r_1 + r_0 = 2a$

$$e = \frac{c}{a} = \frac{r_1 - r_0}{r_1 + r_0}$$

$$\frac{1 + e}{1 - e} = \frac{1 + \frac{c}{a}}{1 - \frac{c}{a}} = \frac{a + c}{a - c} = \frac{r_1}{r_0}$$

70. For a hyperbola,

$r_0 = c - a$ and $r_1 = c + a$.

So $r_1 + r_0 = 2c$ and $r_1 - r_0 = 2a$.

$$e = \frac{c}{a} = \frac{r_1 + r_0}{r_1 - r_0}$$

$$\frac{e + 1}{e - 1} = \frac{\frac{c}{a} + 1}{\frac{c}{a} - 1} = \frac{c + a}{c - a} = \frac{r_1}{r_0}$$

Review Exercises for Chapter 10

1. $4x^2 + y^2 = 4$

Ellipse

Vertex: $(1, 0)$.

Matches (e)

2. $4x^2 - y^2 = 4$

Hyperbola

Vertex: $(1, 0)$

Matches (c)

3. $y^2 = -4x$

Parabola opening to left.

Matches (b)

4. $y^2 - 4x^2 = 4$

Hyperbola

Vertex: $(0, 2)$

Matches (d)

5. $x^2 + 4y^2 = 4$

Ellipse

Vertex: $(0, 1)$

Matches (a)

6. $x^2 = 4y$

Parabola opening upward.

Matches (f)

7. $16x^2 + 16y^2 - 16x + 24y - 3 = 0$

$$\left(x^2 - x + \tfrac{1}{4}\right) + \left(y^2 + \tfrac{3}{2}y + \tfrac{9}{16}\right) = \tfrac{3}{16} + \tfrac{1}{4} + \tfrac{9}{16}$$

$$\left(x - \tfrac{1}{2}\right)^2 + \left(y + \tfrac{3}{4}\right)^2 = 1$$

Circle

Center: $\left(\frac{1}{2}, -\frac{3}{4}\right)$

Radius: 1

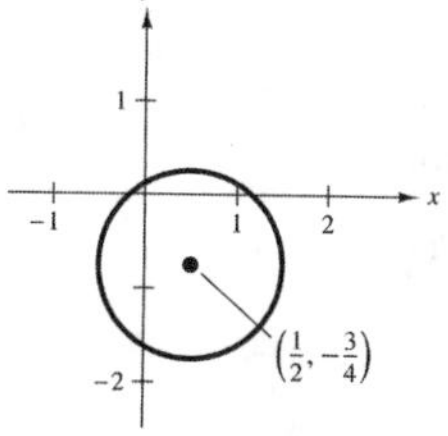

8. $y^2 - 12y - 8x + 20 = 0$

$$y^2 - 12y + 36 = 8x - 20 + 36$$

$$(y - 6)^2 = 4(2)(x + 2)$$

Parabola

Vertex: $(-2, 6)$

Directrix: $x = -2 - 2 = -4$

Focus: $(-2 + 2, 6) = (0, 6)$

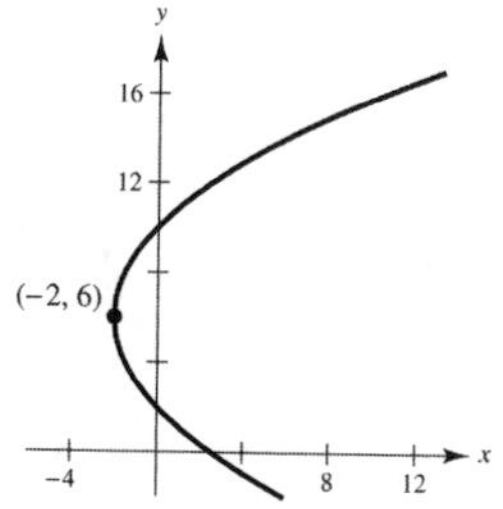

9. $3x^2 - 2y^2 + 24x + 12y + 24 = 0$

$$3\left(x^2 + 8x + 16\right) - 2\left(y^2 - 6y + 9\right) = -24 + 48 - 18$$

$$\frac{(x + 4)^2}{2} - \frac{(y - 3)^2}{3} = 1$$

Hyperbola

Center: $(-4, 3)$

Vertices: $\left(-4 \pm \sqrt{2}, 3\right)$

Foci: $\left(-4 \pm \sqrt{5}, 3\right)$

Eccentricity: $\dfrac{\sqrt{10}}{2}$

Asymptotes:

$$y = 3 \pm \sqrt{\frac{3}{2}}(x + 4)$$

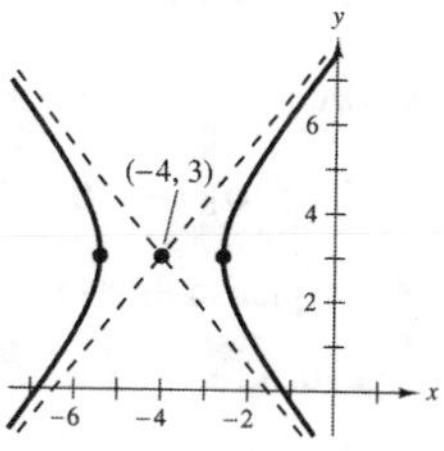

10. $5x^2 + y^2 - 20x + 19 = 0$

$$5\left(x^2 - 4x + 4\right) + y^2 = -19 + 20$$

$$5(x - 2)^2 + y^2 = 1$$

$$\frac{(x - 2)^2}{(1/5)} + y^2 = 1$$

Ellipse

Center: $(2, 0)$

Vertices: $(2, \pm 1)$

Foci: $\left(2, \pm\dfrac{4}{5}\right)$

Eccentricity: $\dfrac{4}{5}$

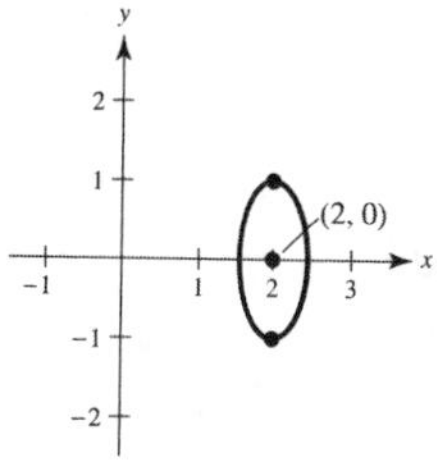

11. $3x^2 + 2y^2 - 12x + 12y + 29 = 0$

$$3\left(x^2 - 4x + 4\right) + 2\left(y^2 + 6y + 9\right) = -29 + 12 + 18$$

$$\frac{(x - 2)^2}{1/3} + \frac{(y + 3)^2}{1/2} = 1$$

Ellipse

Center: $(2, -3)$

Vertices: $\left(2, -3 \pm \dfrac{\sqrt{2}}{2}\right)$

Foci: $\left(2, -\dfrac{17}{6}\right), \left(2, -\dfrac{19}{6}\right)$

Eccentricity: $\dfrac{\sqrt{3}}{3}$

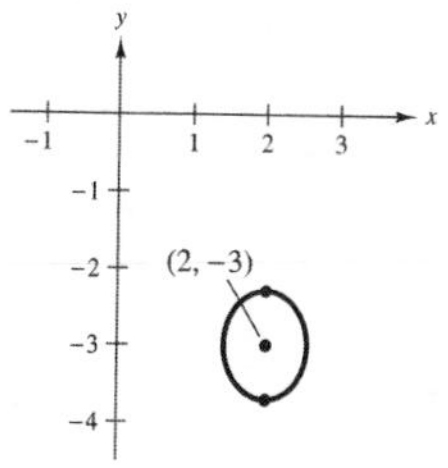

12. $12x^2 - 12y^2 - 12x + 24y - 45 = 0$

$$12\left(x^2 - x + \frac{1}{4}\right) - 12\left(y^2 - 2y + 1\right) = 45 + 3 - 12$$

$$12\left(x - \frac{1}{2}\right)^2 - 12(y - 1)^2 = 36$$

$$\frac{(x - 1/2)^2}{3} - \frac{(y - 1)^2}{3} = 1$$

Hyperbola

Center: $\left(\frac{1}{2}, 1\right)$

Vertices: $\left(\frac{1}{2} \pm \sqrt{3}, 1\right)$

Foci: $\left(\frac{1}{2} \pm \sqrt{6}, 1\right)$

Eccentricity: $\sqrt{2}$

Asymptotes: $y = 1 \pm \left(x - \frac{1}{2}\right)$

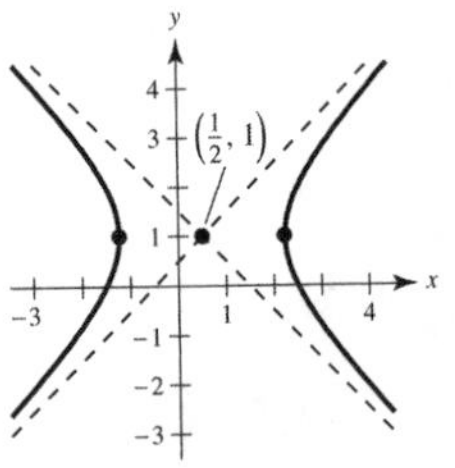

13. $x^2 - 6x - 8y + 1 = 0$

$$x^2 - 6x + 9 = 8y - 1 + 9$$

$$(x - 3)^2 = 8y + 8$$

$$(x - 3)^2 = 4(2)(y + 1)$$

Parabola

Vertex: $(3, -1)$

Directrix: $y = -2 - 1 = -3$

Focus: $(3, 1)$

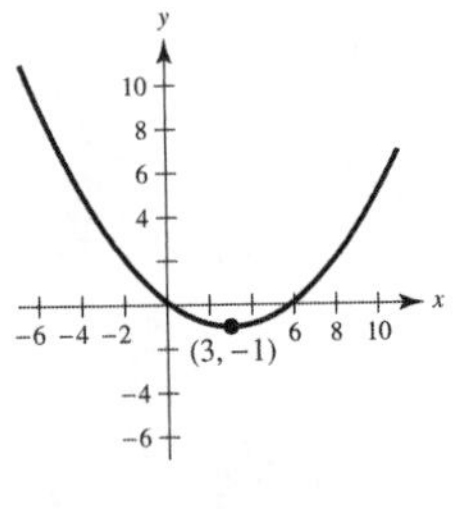

14. $9x^2 + 25y^2 + 18x - 100y - 116 = 0$

$$9\left(x^2 + 2x + 1\right) + 25\left(y^2 - 4y + 4\right) = 116 + 9 + 100$$

$$9(x + 1)^2 + 25(y - 2)^2 = 225$$

$$\frac{(x + 1)^2}{25} + \frac{(y - 2)^2}{9} = 1$$

Ellipse

Center: $(-1, 2)$

Vertices: $(4, 2)$

Foci: $(-5, 2), (3, 2)$

Eccentricity: $\frac{4}{5}$

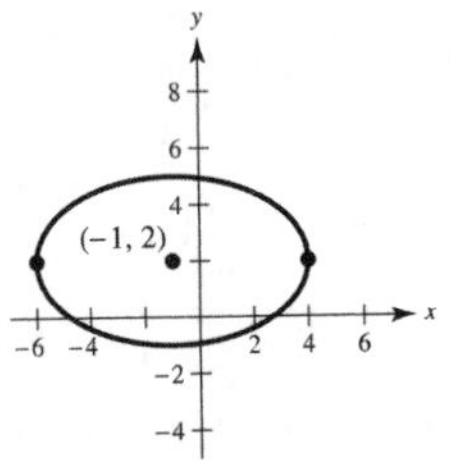

15. Vertex: $(0, 2)$

Directrix: $x = -3$

Parabola opens to the right.

$p = 3$

$(y - 2)^2 = 4(3)(x - 0)$

$y^2 - 4y - 12x + 4 = 0$

16. Vertex: $(2, 6)$

Focus: $(2, 4)$

Parabola opens downward, $p = -2$

$$(x - 2)^2 = 4(-2)(y - 6)$$

$$x^2 - 4x + 4 = -8y + 48$$

$$x^2 - 4x + 8y - 44 = 0$$

17. Center: $(0, 0)$

Vertices: $(7, 0), (-7, 0)$

Foci: $(5, 0), (-5, 0)$

Horizontal major axis

$a = 7, c = 5, b = \sqrt{49 - 25} = \sqrt{24} = 2\sqrt{6}$

$$\frac{x^2}{49} + \frac{y^2}{24} = 1$$

18. Center: $(0, 0)$

Solution points: $(1, 2), (2, 0)$

Substituting the values of the coordinates of the given points into

$$\left(\frac{x^2}{b^2}\right) + \left(\frac{y^2}{a^2}\right) = 1,$$

you obtain the system

$$\left(\frac{1}{b^2}\right) + \left(\frac{4}{a^2}\right) = 1, \frac{4}{b^2} = 1.$$

Solving the system, you have

$$a^2 = \frac{16}{3} \text{ and } b^2 = 4, \left(\frac{x^2}{4}\right) + \left(\frac{3y^2}{16}\right) = 1.$$

19. Vertices: $(3, 1), (3, 7)$

Center: $(3, 4)$

$$\text{Eccentricity} = \frac{2}{3} = \frac{c}{a} \Rightarrow a = 3, c = 2$$

Vertical major axis

$$b = \sqrt{9 - 4} = \sqrt{5}$$

$$\frac{(x-3)^2}{5} + \frac{(y-4)^2}{9} = 1$$

20. Foci: $(0, \pm 7) \Rightarrow c = 7$

Center: $(0, 0)$

Vertical major axis: $20 = 2a \Rightarrow a = 10$

$$b = \sqrt{100 - 49} = \sqrt{51}$$

$$\frac{x^2}{51} + \frac{y^2}{100} = 1$$

21. Vertices: $(0, \pm 8) \Rightarrow a = 8$

Center: $(0, 0)$

Vertical transverse axis

Asymptotes:

$$y = \pm 2x \Rightarrow \frac{a}{b} = 2 \Rightarrow \frac{8}{b} = 2 \Rightarrow b = 4$$

$$\frac{y^2}{64} - \frac{x^2}{16} = 1$$

22. Vertices: $(\pm 2, 0) \Rightarrow a = 2$

Center: $(0, 0)$

Horizontal transverse axis

Asymptotes:

$$y = \pm 32x \Rightarrow \frac{b}{a} = 32 \Rightarrow \frac{b}{2} = 32 \Rightarrow b = 64$$

$$\frac{x^2}{4} - \frac{y^2}{4096} = 1$$

23. Vertices: $(\pm 7, -1)$

Center: $(0, -1)$

Horizontal transverse axis

Foci: $(\pm 9, -1)$

$$a = 7, c = 9, b = \sqrt{81 - 49} = \sqrt{32} = 4\sqrt{2}$$

$$\frac{x^2}{49} - \frac{(y+1)^2}{32} = 1$$

24. Center: $(0, 0)$

Vertices: $(0, \pm 3) \Rightarrow a = 3$

Foci: $(0, \pm 6) \Rightarrow c = 6$

Vertical transverse axis

$$b = \sqrt{36 - 9} = \sqrt{27} = 3\sqrt{3}$$

$$\frac{y^2}{9} - \frac{x^2}{27} = 1$$

25. $y = \frac{1}{200}x^2$

(a) $x^2 = 200y$

$x^2 = 4(50)y$

Focus: $(0, 50)$

(b)

$$y = \frac{1}{200}x^2$$

$$y' = \frac{1}{100}x$$

$$\sqrt{1 + (y')^2} = \sqrt{1 + \frac{x^2}{10{,}000}}$$

$$S = 2\pi \int_0^{100} x\sqrt{1 + \frac{x^2}{10{,}000}}\, dx \approx 38{,}294.49$$

26. $\dfrac{x^2}{25} + \dfrac{y^2}{9} = 1$

(a) $y \pm 3\sqrt{1 - \dfrac{x^2}{25}} = \pm\dfrac{3}{5}\sqrt{25 - x^2}$

$$A = 4\int_0^5 \frac{3}{5}\sqrt{25 - x^2}\,dx = \frac{12}{5}\int_0^5 \sqrt{25 - x^2}\,dx = \left[\frac{12}{5} \cdot \frac{1}{2}\left(x\sqrt{25 - x^2} + 25 \arcsin \frac{x}{5}\right)\right]_0^5$$

$$= 15\pi$$

$\left[\text{or, } A = \pi ab = \pi(5)(3) = 15\pi\right]$

(b) **Disk:** $V = \pi \displaystyle\int_{-5}^{5} \left[\frac{3}{5}\sqrt{25 - x^2}\right]^2 dx$

$$= \frac{9}{25}\pi \int_{-5}^{5} \left(25 - x^2\right)dx$$

$$= \frac{18\pi}{25}\int_0^5 \left(25 - x^2\right)dx$$

$$= \frac{18\pi}{25}\left[25x - \frac{x^3}{3}\right]_0^5$$

$$= \frac{18\pi}{25}\left[125 - \frac{125}{3}\right] = 60\pi$$

27. $x = 1 + 8t,\ y = 3 - 4t$

$t = \dfrac{x - 1}{8} \Rightarrow y = 3 - 4\left(\dfrac{x - 1}{8}\right) = \dfrac{7}{2} - \dfrac{x}{2}$

$x + 2y - 7 = 0$, Line

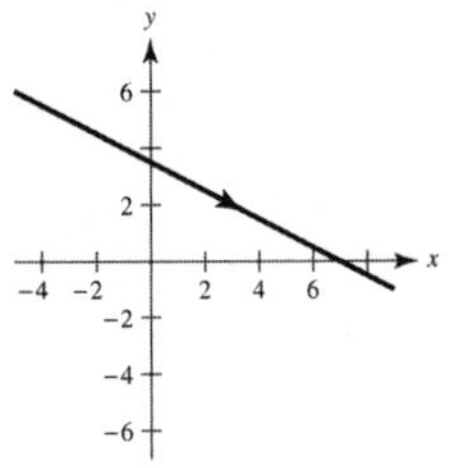

28. $x = t - 6,\ y = t^2$

$t = x + 6 \Rightarrow y = \left(x + 6\right)^2$, Parabola

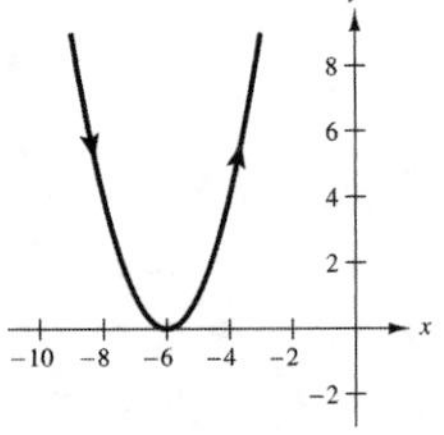

29. $x = e^t - 1,\ y = e^{3t}$

$e^t = x + 1 \Rightarrow y = \left(x + 1\right)^3, x > -1$

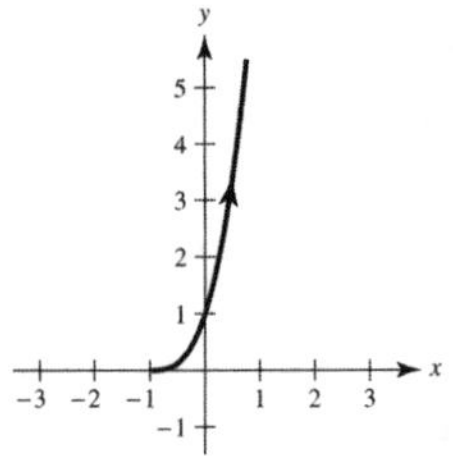

30. $x = e^{4t},\ y = t + 4$

$t = y - 4 \Rightarrow x = e^{4y-16}$

or, $4t = \ln x \Rightarrow y = \dfrac{\ln x}{4} + 4, x > 0$

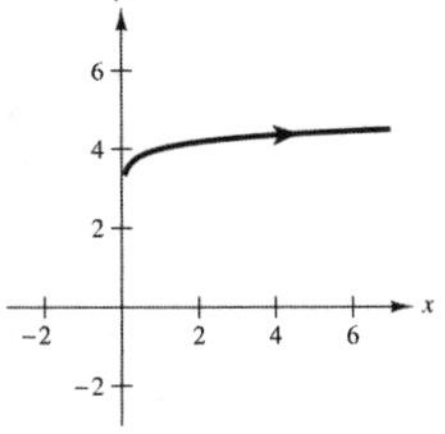

31. $x = 6\cos\theta,\ y = 6\sin\theta$

$$\left(\frac{x}{6}\right)^2 + \left(\frac{y}{6}\right)^2 = 1$$

$$x^2 + y^2 = 36$$

Circle

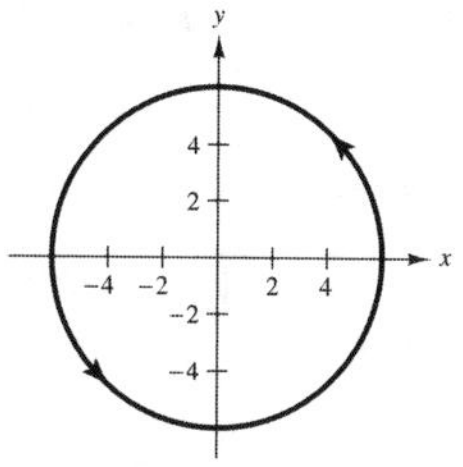

32. $x = 2 + 5\cos t,\ y = 3 + 2\sin t$

$$\left(\frac{x-2}{5}\right)^2 + \left(\frac{y-3}{2}\right)^2 = \cos^2 t + \sin^2 t = 1$$

$$\frac{(x-2)^2}{25} + \frac{(y-3)^2}{4} = 1 \quad \text{Ellipse}$$

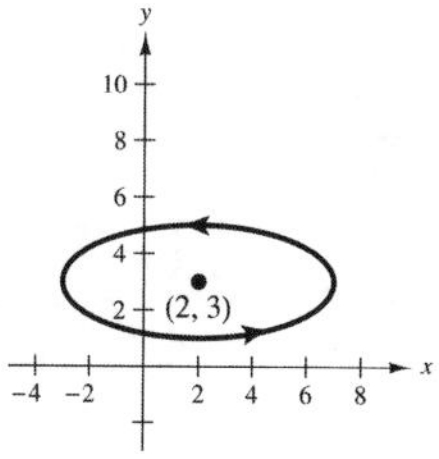

33. $x = 2 + \sec\theta,\ y = 3 + \tan\theta$

$$(x-2)^2 = \sec^2\theta = 1 + \tan^2\theta = 1 + (y-3)^2$$

$$(x-2)^2 - (y-3)^2 = 1$$

Hyperbola

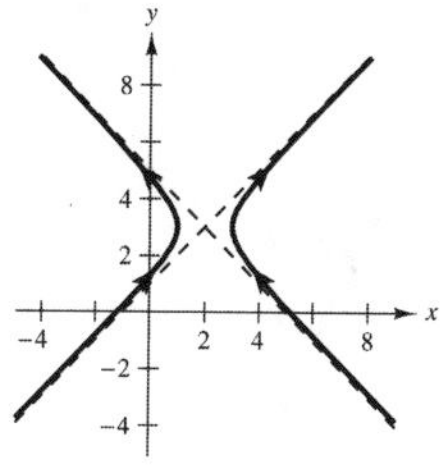

34. $x = 5\sin^3\theta,\ y = 5\cos^3\theta$

$$\left(\frac{x}{5}\right)^{2/3} + \left(\frac{y}{5}\right)^{2/3} = 1$$

$$x^{2/3} + y^{2/3} = 5^{2/3}$$

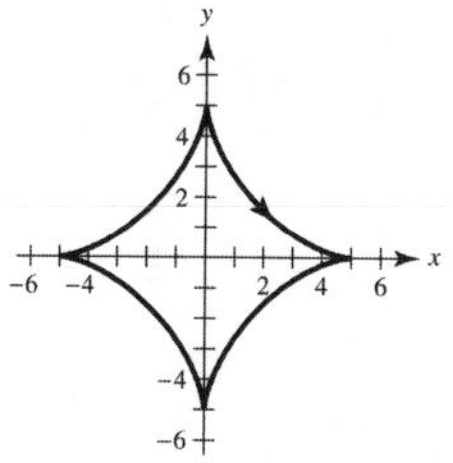

35. $y = 4x + 3$

Examples: $x = t,\ y = 4t + 3$

$x = t + 1,\ y = 4(t + 1) + 3 = 4t + 7$

36. $y = x^2 - 2$

Examples: $x = t,\ y = t^2 - 2$

$x = 2t,\ y = 4t^2 - 2$

37. $x = \cos 3\theta + 5\cos\theta$

$y = \sin 3\theta + 5\sin\theta$

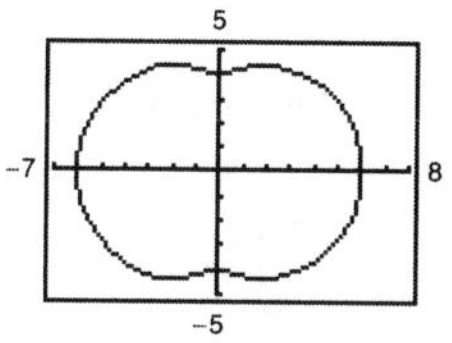

38. (a) $x = 2\cot\theta,\ y = 4\sin\theta\cos\theta,\ 0 < \theta < \pi$

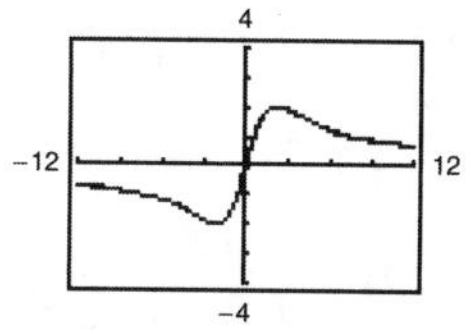

(b) $\left(4 + x^2\right)y = \left(4 + 4\cot^2\theta\right)4\sin\theta\cos\theta$

$= 16\csc^2\theta \cdot \sin\theta \cdot \cos\theta$

$= 16\dfrac{\cos\theta}{\sin\theta} = 8(2\cot\theta) = 8x$

39. $x = 2 + 5t,\ y = 1 - 4t$

$$\frac{dy}{dx} = \frac{dy/dt}{dx/dt} = \frac{-4}{5}$$

$$\frac{d^2y}{dx^2} = 0$$

At $t = 3$, the slope is $-\frac{4}{5}$. (Line)

Neither concave upward nor downward

40. $x = t - 6,\ y = t^2$

$$\frac{dy}{dx} = \frac{dy/dt}{dx/dt} = \frac{2t}{1} = 2t$$

$$\frac{d^2y}{dx^2} = \frac{\frac{d}{dt}\left[\frac{dy}{dx}\right]}{dx/dt} = \frac{2}{1} = 2 > 0$$

At $t = 5$, the slope is $2(5) = 10$ and $\dfrac{d^2y}{dx^2} = 2$.

Concave upward everywhere.

41. $x = \frac{1}{t},\ y = 2t + 3$

$$\frac{dy}{dx} = \frac{dy/dt}{dx/dt} = \frac{2}{\left(-1/t^2\right)} = -2t^2$$

$$\frac{d^2y}{dx^2} = \frac{\frac{d}{dt}\left[-2t^2\right]}{dx/dt} = \frac{-4t}{\left(-1/t^2\right)} = 4t^3$$

At $t = -1$, the slope is $\frac{dy}{dx} = -2$ and $\frac{d^2y}{dx^2} = -4$.

Concave downward

42. $x = \frac{1}{t},\ y = t^2$

$$\frac{dy}{dx} = \frac{dy/dt}{dx/dt} = \frac{2t}{\left(-1/t^2\right)} = -2t^3$$

$$\frac{d^2y}{dx^2} = \frac{\frac{d}{dt}\left[-2t^3\right]}{dx/dt} = \frac{-6t^2}{\left(-1/t^2\right)} = 6t^4$$

At $t = -2$, the slope is $\frac{dy}{dx} = 16$ and $\frac{d^2y}{dx^2} = 96$.

Concave upward

43. $x = 5 + \cos\theta,\ y = 3 + 4\sin\theta$

$$\frac{dy}{dx} = \frac{dy/d\theta}{dx/d\theta} = \frac{4\cos\theta}{-\sin\theta} = -4\cot\theta$$

$$\frac{d^2y}{dx^2} = \frac{\frac{d}{d\theta}\left[-4\cot\theta\right]}{dx/d\theta} = \frac{4\csc^2\theta}{-\sin\theta} = -4\csc^3\theta$$

At $\theta = \frac{\pi}{6}$, the slope is $\frac{dy}{dx} = -4\sqrt{3}$ and $\frac{d^2y}{dx^2} = -32$.

Concave downward

44. $x = 10\cos\theta,\ y = 10\sin\theta$

$$\frac{dy}{dx} = \frac{dy/dt}{dx/dt} = \frac{10\cos\theta}{-10\sin\theta} = -\cot\theta$$

$$\frac{d^2y}{dx^2} = \frac{\frac{d}{d\theta}\left[-\cot\theta\right]}{dx/d\theta} = \frac{\csc^2\theta}{-10\sin\theta} = -\frac{1}{10}\csc^3\theta$$

At $\theta = \frac{\pi}{4}$, the slope is $\frac{dy}{dx} = -1$ and $\frac{d^2y}{dx^2} = \frac{-\sqrt{2}}{5}$.

Concave downward

45. $x = \cos^3\theta,\ y = 4\sin^3\theta$

$$\frac{dy}{dx} = \frac{dy/d\theta}{dx/d\theta} = \frac{12\sin^2\theta\cos\theta}{3\cos^2\theta(-\sin\theta)} = -\frac{4\sin\theta}{\cos\theta} = -4\tan\theta$$

$$\frac{d^2y}{dx^2} = \frac{\frac{d}{d\theta}\left[-4\tan\theta\right]}{dx/d\theta} = \frac{-4\sec^2\theta}{3\cos^2\theta(-\sin\theta)} = \frac{4}{3\cos^4\theta\sin\theta} = \frac{4}{3}\sec^4\theta\csc\theta$$

At $\theta = \frac{\pi}{3}$, the slope is $\frac{dy}{dx} = -4\sqrt{3}$ and $\frac{d^2y}{dx^2} = \frac{4}{3\left(\frac{1}{16}\right)\left(\frac{\sqrt{3}}{2}\right)} = \frac{128}{3\sqrt{3}} = \frac{128\sqrt{3}}{9}$.

Concave upward

46. $x = e^t,\ y = e^{-t}$

$$\frac{dy}{dx} = \frac{dy/dt}{dx/dt} = \frac{-e^{-t}}{e^t} = -e^{-2t}$$

$$\frac{d^2y}{dx^2} = \frac{\frac{d}{dt}\left(-e^{-2t}\right)}{dx/dt} = \frac{2e^{-2t}}{e^t} = \frac{2}{e^{3t}}$$

At $t = 1$, the slope is $\frac{dy}{dx} = -\frac{1}{e^2}$ and $\frac{d^2y}{dx^2} = \frac{2}{e^3}$.

Concave upward

47. $x = \cot\theta,\ y = \sin 2\theta,\ \theta = \frac{\pi}{6}$

(a), (d)

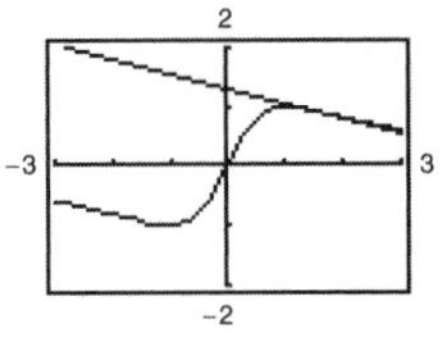

(b) At $\theta = \frac{\pi}{6}$, $\frac{dx}{d\theta} = -4$, $\frac{dy}{d\theta} = 1$, and $\frac{dy}{dx} = -\frac{1}{4}$.

(c) At $\theta = \frac{\pi}{6}$, $(x, y) = \left(\sqrt{3}, \frac{\sqrt{3}}{2}\right)$.

$$y - \frac{\sqrt{3}}{2} = -\frac{1}{4}\left(x - \sqrt{3}\right)$$

$$y = -\frac{1}{4}x + \frac{3\sqrt{3}}{4}$$

48. $x = \frac{1}{4}\tan\theta,\ y = 6\sin\theta,\ \theta = \frac{\pi}{3}$

(a), (d)

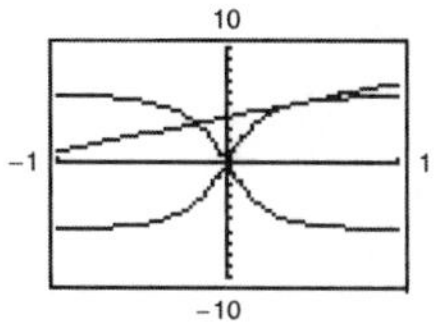

(b) At $\theta = \frac{\pi}{3},\ \frac{dx}{d\theta} = 1,\ \frac{dy}{d\theta} = 3$, and $\frac{dy}{dx} = 3$.

(c) At $\theta = \frac{\pi}{3},\ (x, y) = \left(\frac{\sqrt{3}}{4}, 3\sqrt{3}\right)$.

$$y - 3\sqrt{3} = 3\left(x - \frac{\sqrt{3}}{4}\right)$$

$$y = 3x + \frac{9\sqrt{3}}{4}$$

49. $x = 5 - t,\ y = 2t^2$

$\frac{dx}{dt} = -1,\ \frac{dy}{dt} = 4t$

Horizontal tangent at $t = 0$: $(5, 0)$

No vertical tangents

50. $x = t + 2,\ y = t^3 - 2t$

$\frac{dx}{dt} = 1,\ \frac{dy}{dt} = 3t^2 - 2$

$\frac{dy}{dt} = 0$ for $t = \pm\sqrt{\frac{2}{3}} = \frac{\pm\sqrt{6}}{3}$

Horizontal tangents:

$$t = \frac{\sqrt{2}}{3}: (x, y) = \left(\frac{\sqrt{6}}{3} + 2, \frac{2\sqrt{6}}{9} - \frac{2}{3}\sqrt{6}\right)$$

$$\approx (2.8165, -1.0887)$$

$$t = -\frac{\sqrt{6}}{3}: (x, y) = \left(-\frac{\sqrt{6}}{3} + 2, \frac{2}{3}\sqrt{6} - \frac{2\sqrt{6}}{9}\right)$$

$$\approx (1.1835, 1.0887)$$

No vertical tangents

51. $x = 2 + 2\sin\theta,\ y = 1 + \cos\theta$

$\frac{dx}{d\theta} = 2\cos\theta,\ \frac{dy}{d\theta} = -\sin\theta$

$\frac{dy}{d\theta} = 0$ for $\theta = 0, \pi, 2\pi, \ldots$

Horizontal tangents: $(x, y) = (2, 2), (2, 0)$

$\frac{dx}{d\theta} = 0$ for $\theta = \frac{\pi}{2}, \frac{3\pi}{2}, \ldots$

Vertical tangents: $(x, y) = (4, 1), (0, 1)$

52. $x = 2 - 2\cos\theta,\ y = 2\sin 2\theta$

$\frac{dx}{d\theta} = 2\sin\theta,\ \frac{dy}{d\theta} = 4\cos 2\theta$

$\frac{dy}{d\theta} = 0$ for $\theta = \frac{\pi}{4}, \frac{3\pi}{4}, \frac{5\pi}{4}, \frac{7\pi}{4}, \ldots$

Horizontal tangents: $(x, y) = \left(2 \pm \sqrt{2}, 2\right),$
$\left(2 \pm \sqrt{2}, -2\right)$

$\frac{dx}{d\theta} = 0$ for $\theta = 0, \pi, 2\pi, \ldots$

Vertical tangents: $(x, y) = (0, 0), (4, 0)$

53. $x = t^2 + 1,\ y = 4t^3 + 3,\ 0 \le t \le 2$

$\frac{dx}{dt} = 2t,\ \frac{dy}{dt} = 12t^2$

$$\begin{aligned} s &= \int_0^2 \sqrt{(2t)^2 + \left(12t^2\right)^2}\,dt \\ &= \int_0^2 \sqrt{4t^2 + 144t^4}\,dt \\ &= \int_0^2 2t\sqrt{1 + 36t^2}\,dt \\ &= \frac{1}{36}\left[\frac{2}{3}\left(1 + 36t^2\right)^{3/2}\right]_0^2 \\ &= \frac{1}{54}\left[145^{3/2} - 1\right] \approx 32.3154 \end{aligned}$$

54. $x = 6\cos\theta,\ y = 6\sin\theta,\ 0 \le \theta \le \pi$

$\frac{dx}{d\theta} = -6\sin\theta,\ \frac{dy}{d\theta} = 6\cos\theta$

$$\begin{aligned} s &= \int_0^\pi \sqrt{(-6\sin\theta)^2 + (6\cos\theta)^2}\,d\theta \\ &= \int_0^\pi \sqrt{36\sin^2\theta + 36\cos^2\theta}\,d\theta \\ &= 6\int_0^\pi \sqrt{\sin^2\theta + \cos^2\theta}\,d\theta \\ &= 6\int_0^\pi d\theta \\ &= 6\left[\theta\right]_0^\pi \\ &= 6\pi \text{ (one-half circumference of circle)} \end{aligned}$$

55. $x = t, y = 3t, 0 \le t \le 2$

$$\frac{dx}{dt} = 1, \frac{dy}{dt} = 3, \sqrt{\left(\frac{dx}{dt}\right)^2 + \left(\frac{dy}{dt}\right)^2} = \sqrt{1+9} = \sqrt{10}$$

(a) $S = 2\pi\int_0^2 3t\sqrt{10}\,dt = 6\sqrt{10}\,\pi\left[\frac{t^2}{2}\right]_0^2 = 12\sqrt{10}\,\pi \approx 119.215$

(b) $S = 2\pi\int_0^2 \sqrt{10}\,dt = 2\pi\left[\sqrt{10}t\right]_0^2 = 4\pi\sqrt{10} \approx 39.738$

56. $x = 2\cos\theta, y = 2\sin\theta, 0 \le \theta \le \frac{\pi}{2}$

$$\frac{dx}{dt} = -2\sin\theta, \frac{dy}{dt} = 2\cos\theta, \sqrt{\left(\frac{dx}{d\theta}\right)^2 + \left(\frac{dy}{d\theta}\right)^2} = 2$$

(a) $S = 2\pi\int_0^{\pi/2} 2\sin\theta(2)d\theta = 8\pi\left[-\cos\theta\right]_0^{\pi/2} = 8\pi$

(b) $S = 2\pi\int_0^{\pi/2} 2\cos\theta(2)d\theta = 8\pi\left[\sin\theta\right]_0^{\pi/2} = 8\pi$

[**Note:** The surface is a hemisphere: $\frac{1}{2}\left(4\pi\left(2^2\right)\right) = 8\pi$]

57. $x = 3\sin\theta, y = 2\cos\theta$

$$\begin{aligned} A &= \int_a^b y\,dx = \int_{-\pi/2}^{\pi/2} 2\cos\theta(3\cos\theta)\,d\theta \\ &= 6\int_{-\pi/2}^{\pi/2}\frac{1+\cos 2\theta}{2}\,d\theta \\ &= 3\left[\theta + \frac{\sin 2\theta}{2}\right]_{-\pi/2}^{\pi/2} \\ &= 3\left[\frac{\pi}{2} + \frac{\pi}{2}\right] = 3\pi \end{aligned}$$

58.
$$\begin{aligned} A &= \int_a^b y\,dx = \int_\pi^0 \sin\theta(-2\sin\theta)\,d\theta \\ &= -\int_\pi^0 \frac{1-\cos 2\theta}{2}\,d\theta \\ &= -\left[\theta - \frac{\sin 2\theta}{2}\right]_\pi^0 = \pi \end{aligned}$$

59. $(r, \theta) = \left(5, \frac{3\pi}{2}\right)$

$x = r\cos\theta = 5\cos\frac{3\pi}{2} = 0$

$y = r\sin\theta = 5\sin\frac{3\pi}{2} = -5$

$(x, y) = (0, -5)$

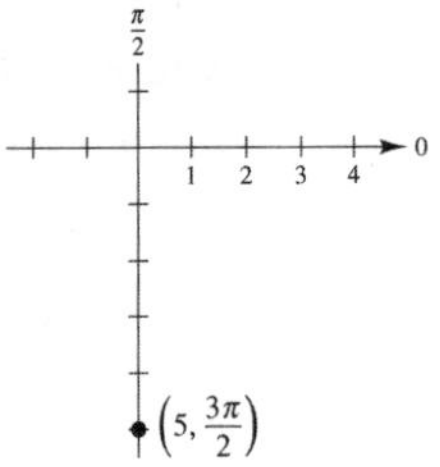

60. $(r, \theta) = \left(-6, \frac{7\pi}{6}\right)$

$x = r\cos\theta = -6\cos\frac{7\pi}{6} = (-6)\left(-\frac{\sqrt{3}}{2}\right) = 3\sqrt{3}$

$y = r\sin\theta = -6\sin\frac{7\pi}{6} = 3$

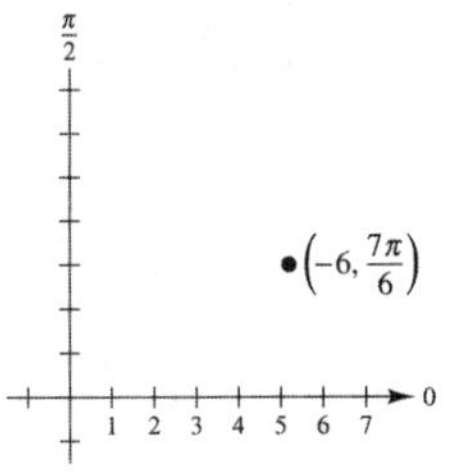

61. $(r, \theta) = \left(\sqrt{3}, 1.56\right)$

$(x, y) = \left(\sqrt{3}\cos(1.56), \sqrt{3}\sin(1.56)\right)$

$\approx (0.0187, 1.7319)$

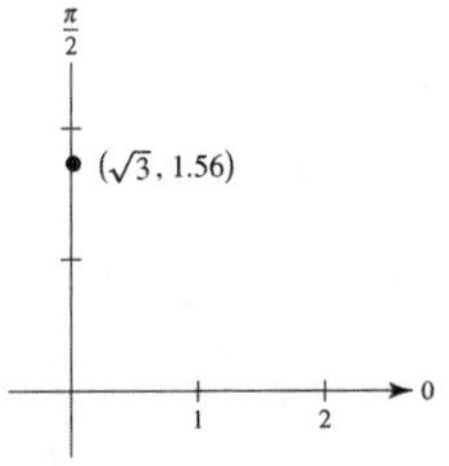

62. $(r, \theta) = (-2, -2.45)$

$(x, y) = (-2\cos(-2.45), -\sin(-2.45))$

$\approx (1.5405, 1.2755)$

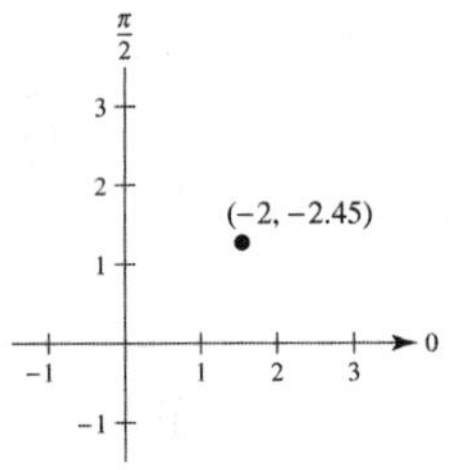

63. $(x, y) = (4, -4)$

$$r = \sqrt{4^2 + (-4)^2} = 4\sqrt{2}$$

$$\theta = \frac{7\pi}{4}$$

$$(r, \theta) = \left(4\sqrt{2}, \frac{7\pi}{4}\right), \left(-4\sqrt{2}, \frac{3\pi}{4}\right)$$

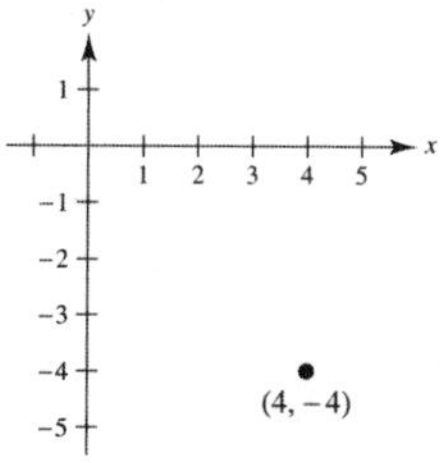

64. $(x, y) = (0, -7)$

$$r = \sqrt{0^2 + (-7)^2} = 7$$

$$\tan\theta \text{ undefined} \Rightarrow \theta = \frac{3\pi}{2}$$

$$(r, \theta) = \left(7, \frac{3\pi}{2}\right), \left(-7, \frac{\pi}{2}\right)$$

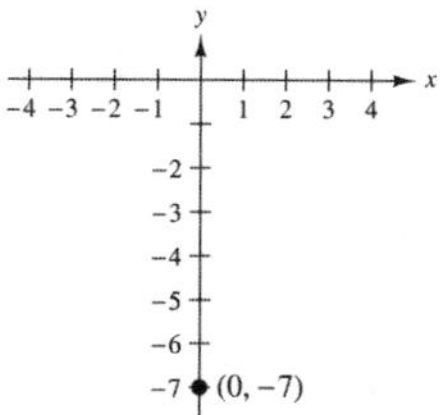

65. $(x, y) = (-1, 3)$

$$r = \sqrt{(-1)^2 + 3^2} = \sqrt{10}$$

$$\theta = \arctan(-3) \approx 1.89(108.43°)$$

$$(r, \theta) = \left(\sqrt{10}, 1.89\right), \left(-\sqrt{10}, 5.03\right)$$

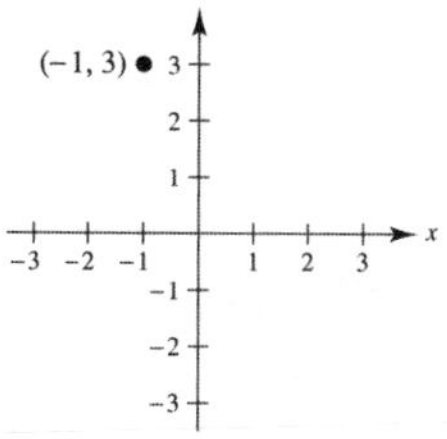

66. $(x, y) = \left(-\sqrt{3}, -\sqrt{3}\right)$

$$r = \sqrt{3 + 3} = \sqrt{6}$$

$$\tan\theta = 1 \Rightarrow \theta = \frac{\pi}{4}, \frac{5\pi}{4}$$

$$(r, \theta) = \left(\sqrt{6}, \frac{5\pi}{4}\right), \left(-\sqrt{6}, \frac{\pi}{4}\right)$$

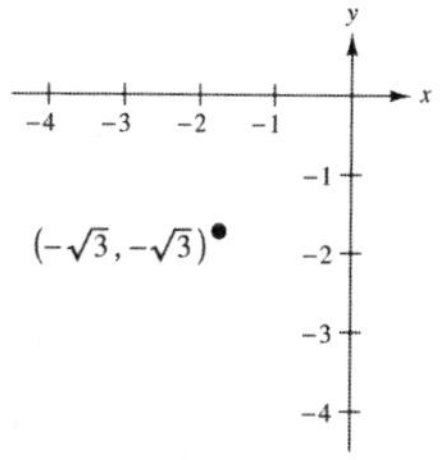

67. $x^2 + y^2 = 25$

$$r^2 = 25$$

$$r = 5$$

Circle

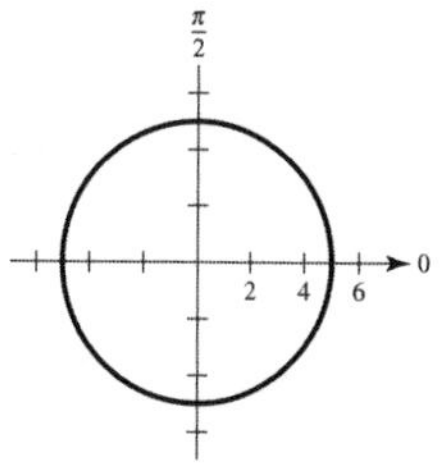

68.

$$x^2 - y^2 = 4$$

$$r^2\cos^2\theta - r^2\sin^2\theta = 4$$

$$r^2\cos 2\theta = 4$$

$$r^2 = \frac{4}{\cos 2\theta}$$

$$r = \frac{2}{\sqrt{\cos 2\theta}}$$

Hyperbola

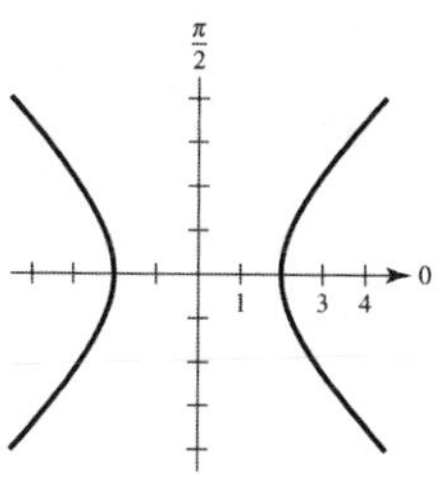

69.
$$y = 9$$
$$r \sin \theta = 9$$
$$r = \frac{9}{\sin \theta} = 9 \csc \theta$$

Horizontal line

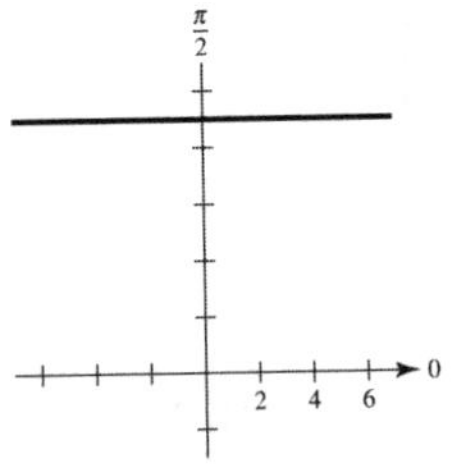

70.
$$x = 6$$
$$r \cos \theta = 6$$
$$r = \frac{6}{\cos \theta} = 6 \sec \theta$$

Vertical line

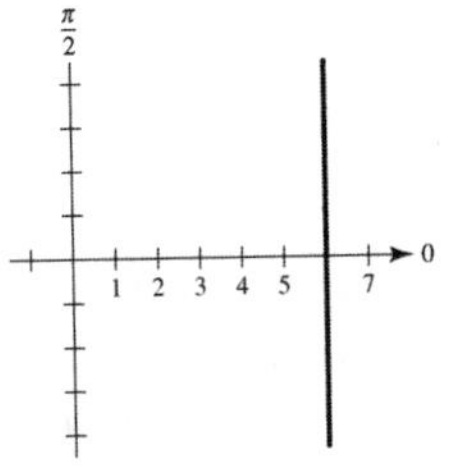

71.
$$x^2 = 4y$$
$$r^2 \cos^2 \theta = 4r \sin \theta$$
$$r = \frac{4 \sin \theta}{\cos^2 \theta} = 4 \tan \theta \sec \theta$$

Parabola

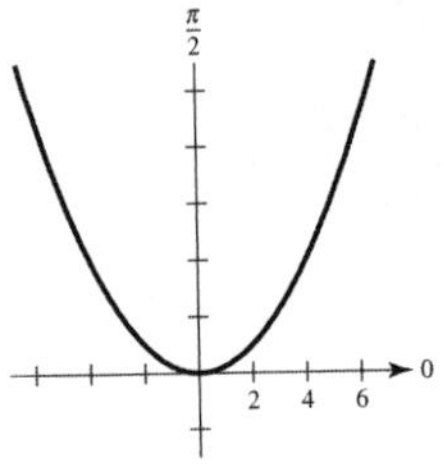

72. $x^2 + y^2 - 4x = 0$
$$r^2 - 4r \cos \theta = 0$$
$$r = 4 \cos \theta$$

Circle

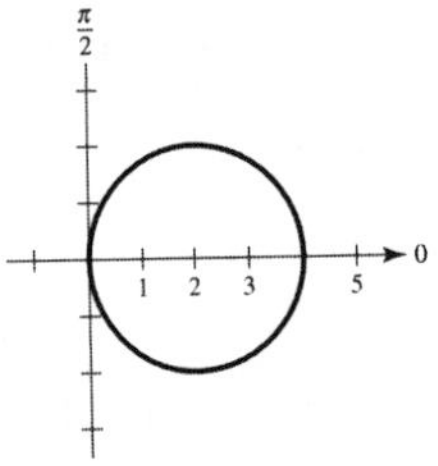

73.
$$r = 3 \cos \theta$$
$$r^2 = 3r \cos \theta$$
$$x^2 + y^2 = 3x$$
$$x^2 - 3x + \tfrac{9}{4} + y^2 = \tfrac{9}{4}$$
$$\left(x - \tfrac{3}{2}\right)^2 + y^2 = \tfrac{9}{4}$$

Circle

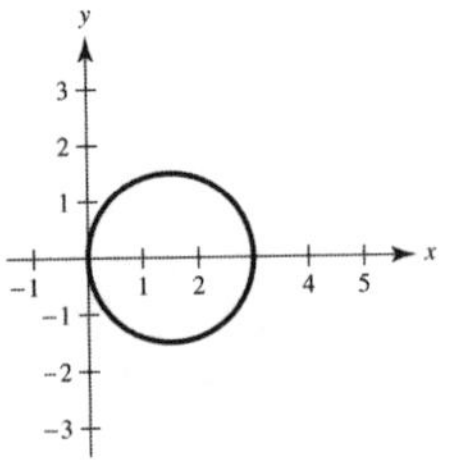

74.
$$r = 10$$
$$r^2 = 100$$
$$x^2 + y^2 = 100$$

Circle

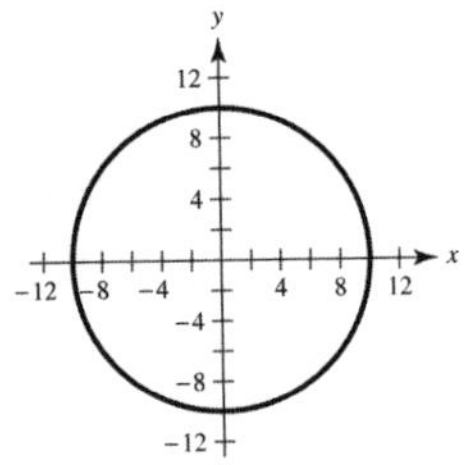

75.
$$r = 6 \sin \theta$$
$$r^2 = 6r \sin \theta$$
$$x^2 + y^2 = 6y$$
$$x^2 + y^2 - 6y + 9 = 9$$
$$x^2 + (y - 3)^2 = 9$$

Circle

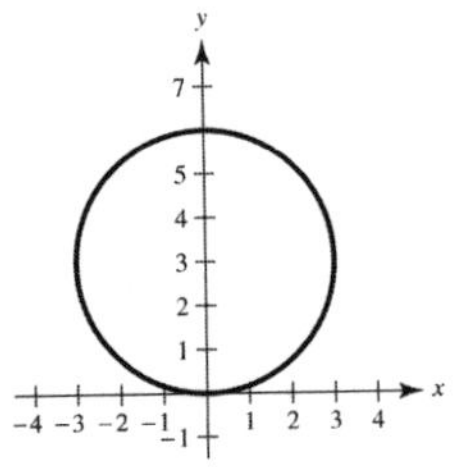

76. $r = 3 \csc \theta$

$r \sin \theta = 3$

$y = 3$

Horizontal line

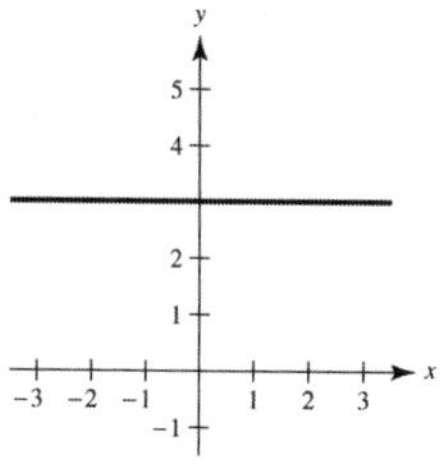

77. $r = -2 \sec \theta \tan \theta$

$r \cos \theta = -2 \tan \theta$

$x = -2\left(\dfrac{y}{x}\right)$

$x^2 = -2y$

$y = -\dfrac{1}{2}x^2$

Parabola

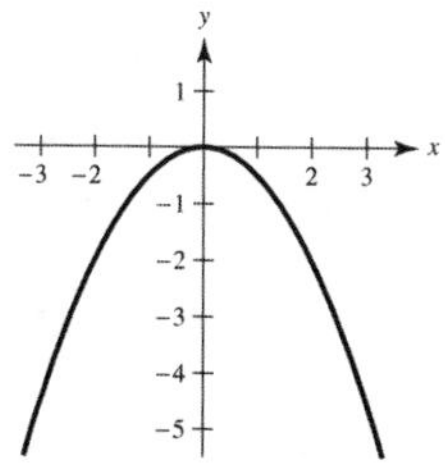

78. $\theta = \dfrac{3\pi}{4}$

$\tan \theta = -1$

$\dfrac{y}{x} = -1$

$y = -x$

Line

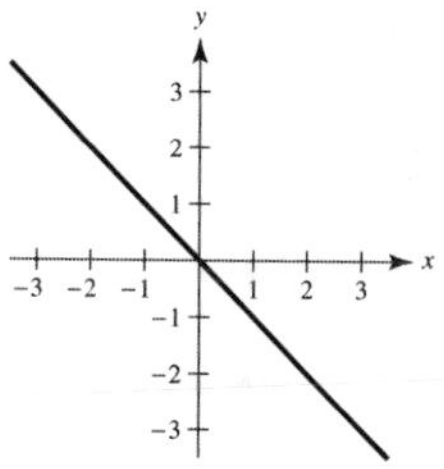

79. $r = \dfrac{3}{\cos \theta - (\pi/4)}$

Graph of $r = 3 \sec \theta$ rotated through an angle of $\pi/4$

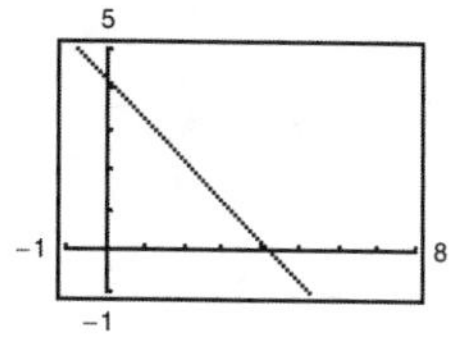

80. $r = 2 \sin \theta \cos^2 \theta$

Bifolium

Symmetric to $\theta = \pi/2$

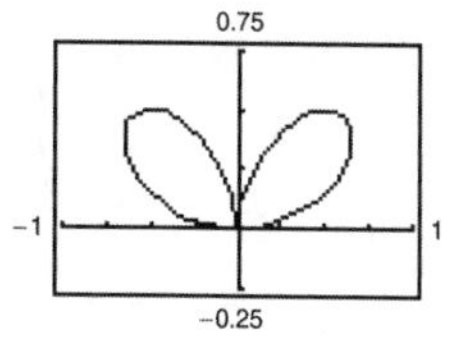

81. $r = 4 \cos 2\theta \sec \theta$

Strophoid

Symmetric to the polar axis

$r \Rightarrow \infty$ as $\theta \Rightarrow \dfrac{\pi^-}{2}$

$r \Rightarrow \infty$ as $\theta \Rightarrow \dfrac{-\pi^+}{2}$

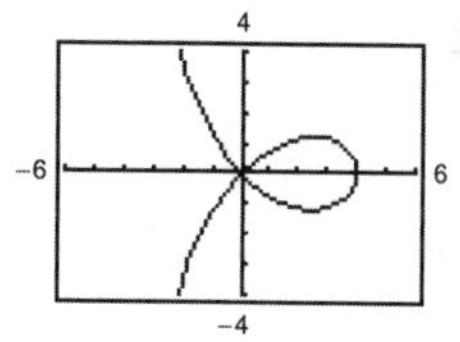

82. $r = 4(\sec \theta - \cos \theta)$

Semicubical parabola

Symmetric to the polar axis

$r \Rightarrow \infty$ as $\theta \Rightarrow \dfrac{\pi^-}{2}$

$r \Rightarrow \infty$ as $\theta \Rightarrow \dfrac{-\pi^+}{2}$

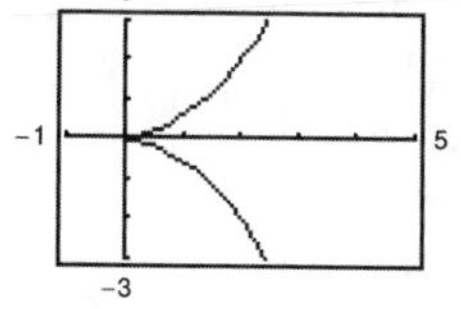

83. $r = 1 - \cos\theta$, Cardioid

$$\frac{dy}{dx} = \frac{(1 - \cos\theta)\cos\theta + (\sin\theta)\sin\theta}{-(1 - \cos\theta)\sin\theta + (\sin\theta)\cos\theta}$$

Horizontal tangents:

$$\cos\theta - \cos^2\theta + \sin^2\theta = 0$$
$$\cos\theta - \cos^2\theta + (1 - \cos^2\theta) = 0$$
$$2\cos^2\theta - \cos\theta - 1 = 0$$
$$(2\cos\theta + 1)(\cos\theta - 1) = 0$$
$$\cos\theta = -\frac{1}{2} \Rightarrow \theta = \frac{2\pi}{3}, \frac{4\pi}{3}$$
$$\cos\theta = 1 \Rightarrow \theta = 0$$

Vertical tangents:

$$-\sin\theta + 2\cos\theta\sin\theta = 0$$
$$\sin\theta(2\cos\theta - 1) = 0$$
$$\sin\theta = 0 \Rightarrow \theta = 0, \pi$$
$$\cos\theta = \frac{1}{2} \Rightarrow \theta = \frac{\pi}{3}, \frac{5\pi}{3}$$

Horizontal tangents: $\left(\frac{3}{2}, \frac{2\pi}{3}\right), \left(\frac{3}{2}, \frac{4\pi}{3}\right)$

Vertical tangents: $\left(\frac{1}{2}, \frac{\pi}{3}\right), \left(\frac{1}{2}, \frac{5\pi}{3}\right), (2\pi)$

(There is a cusp at the pole.)

84. $r = 3\tan\theta$

$$\frac{dy}{dx} = \frac{3\tan\theta\cos\theta + 3\sec^2\theta\sin\theta}{-3\tan\theta\sin\theta + 3\sec^2\theta\cos\theta}$$

Horizontal tangents:

$$3\tan\theta\cos\theta + 3\sec^2\theta\sin\theta = 0$$
$$\sin\theta + \sec^2\theta\sin\theta = 0$$
$$\sin\theta(1 + \sec^2\theta) = 0$$
$$\sin\theta = 0 \Rightarrow \theta = 0, \pi$$
$$1 + \sec^2\theta = 0 \text{ is undefined.}$$

Vertical tangents:

$$-3\tan\theta\sin\theta + 3\sec^2\theta\cos\theta = 0$$
$$-\frac{\sin^2\theta}{\cos\theta} + \frac{1}{\cos\theta} = 0$$
$$\frac{1}{\cos\theta}(1 - \sin^2\theta) = 0$$
$$\frac{1}{\cos\theta}(\cos^2\theta) = 0$$
$$\cos\theta = 0$$
$$\theta = \frac{\pi}{2}, \frac{3\pi}{2}$$

r is undefined at these points.

Horizontal tangent at the pole; no vertical tangents

85. $r = 4\sin 3\theta$, Rose curve with three petals

Tangents at the pole: $\sin 3\theta = 0$

$$\theta = 0, \frac{\pi}{3}, \frac{2\pi}{3}$$

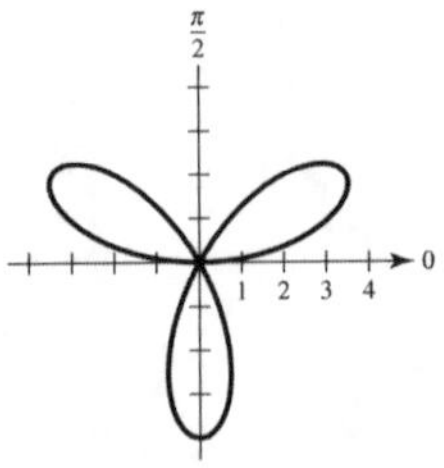

86. $r = 3\cos 4\theta$, Rose curve with eight petals

Tangents at the pole:
$\cos 4\theta = 0$

$$\theta = \frac{\pi}{8}, \frac{3\pi}{8}, \frac{5\pi}{8}, \frac{7\pi}{8}, \frac{9\pi}{8}, \frac{11\pi}{8}, \frac{13\pi}{8}, \frac{15\pi}{8}$$

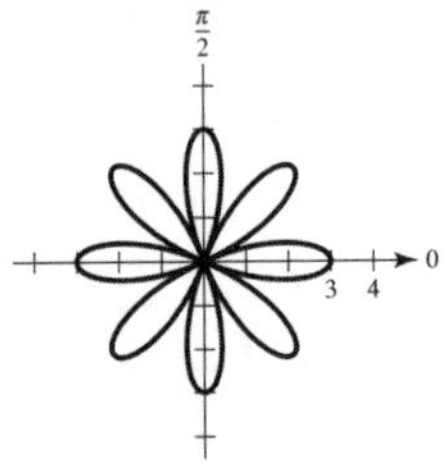

87. $r = 6$, Circle radius 6

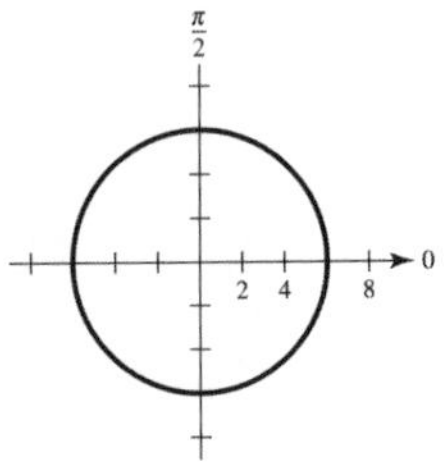

88. $\theta = \frac{\pi}{10}$, Line

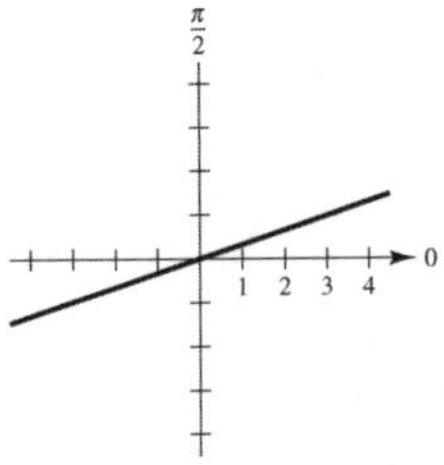

89. $r = -\sec\theta = \dfrac{-1}{\cos\theta}$

$r\cos\theta = -1,\ x = -1$

Vertical line

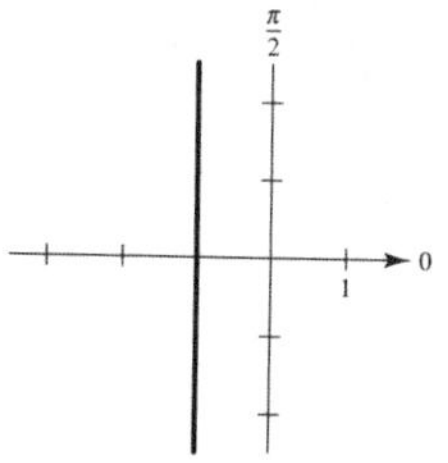

90. $r = 5\csc\theta \Rightarrow r\sin\theta = y = 5$

Horizontal line

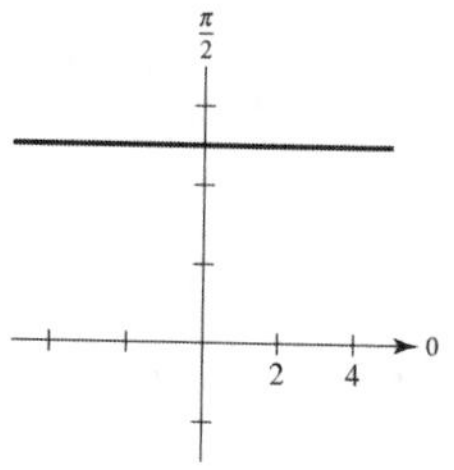

91. $r^2 = 4\sin^2 2\theta$

$r = \pm 2\sin(2\theta)$

Rose curve with four petals

Symmetric to the polar axis, $\theta = \dfrac{\pi}{2}$, and pole

Relative extrema: $\left(\pm 2, \dfrac{\pi}{4}\right), \left(\pm 2, \dfrac{3\pi}{4}\right)$

Tangents at the pole: $\theta = 0, \dfrac{\pi}{2}$

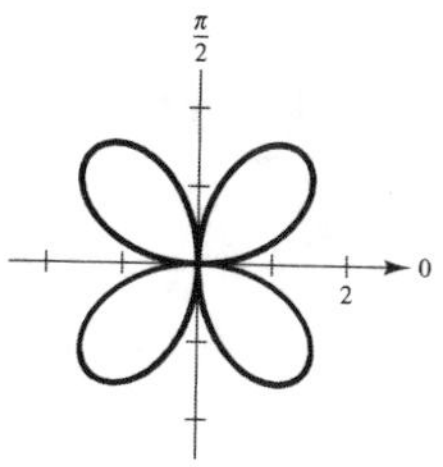

92. $r = 3 - 4\cos\theta$

Limaçon

Symmetric to polar axis

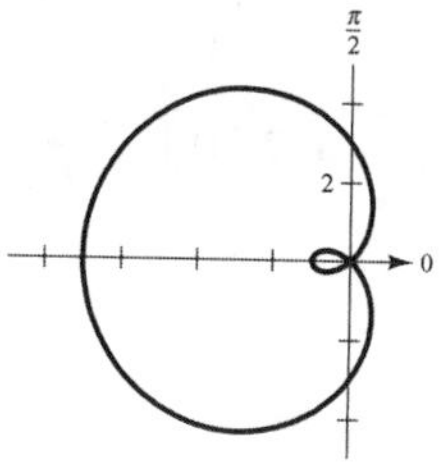

θ	0	$\dfrac{\pi}{3}$	$\dfrac{\pi}{2}$	$\dfrac{2\pi}{3}$	π
r	-1	1	3	5	7

93. $r = 4 - 3\cos\theta$

Limaçon

Symmetric to polar axis

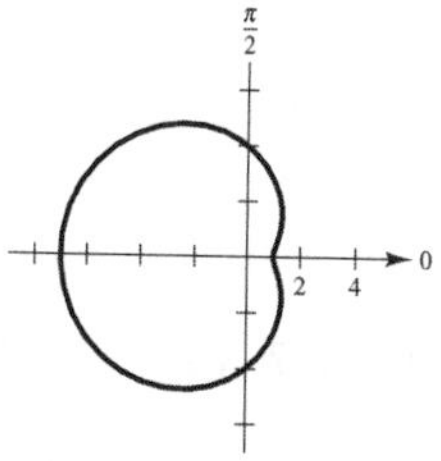

θ	0	$\dfrac{\pi}{3}$	$\dfrac{\pi}{2}$	$\dfrac{2\pi}{3}$	π
r	1	$\dfrac{5}{2}$	4	$\dfrac{11}{2}$	7

94. $r = 4\theta$

Spiral

Symmetric to $\theta = \dfrac{\pi}{2}$

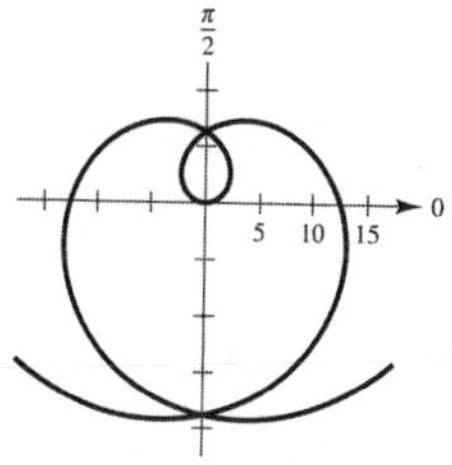

θ	0	$\dfrac{\pi}{4}$	$\dfrac{\pi}{2}$	$\dfrac{3\pi}{4}$	π	$\dfrac{3\pi}{2}$	2π
r	0	π	2π	3π	4π	6π	8π

95. $r = -3\cos 2\theta$

Rose curve with four petals

Symmetric to polar axis, $\theta = \dfrac{\pi}{2}$, and pole

Relative extrema: $(-3, 0), \left(3, \dfrac{\pi}{2}\right), (-3, \pi), \left(3, \dfrac{3\pi}{2}\right)$

Tangents at the pole: $\theta = \dfrac{\pi}{4}, \dfrac{3\pi}{4}$

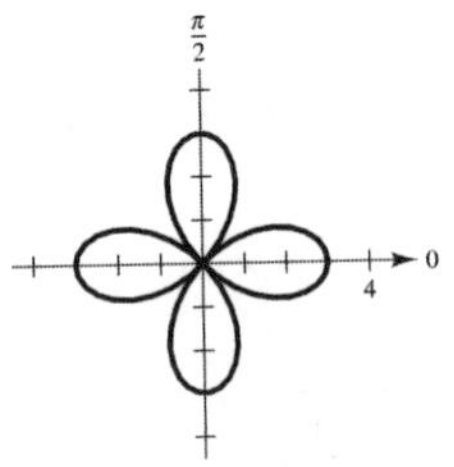

96. $r = \cos 5\theta$

Rose curve with five petals

Symmetric to polar axis

Relative extrema:

$(1, 0), \left(-1, \dfrac{\pi}{5}\right), \left(1, \dfrac{2\pi}{5}\right), \left(-1, \dfrac{3\pi}{5}\right), \left(1, \dfrac{4\pi}{5}\right)$

Tangents at the pole: $\theta = \dfrac{\pi}{10}, \dfrac{3\pi}{10}, \dfrac{\pi}{2}, \dfrac{7\pi}{10}, \dfrac{9\pi}{10}$

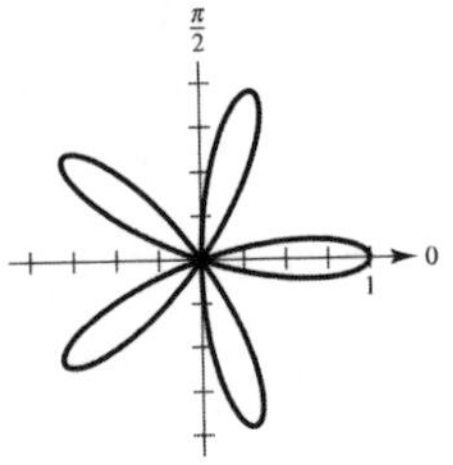

97. $A = 2 \cdot \dfrac{1}{2}\displaystyle\int_0^{\pi/10} [3\cos 5\theta]^2\, d\theta$

$= \displaystyle\int_0^{\pi/10} 9\left(\dfrac{1 + \cos(10\theta)}{2}\right) d\theta$

$= \dfrac{9}{2}\left[\theta + \dfrac{\sin(10\theta)}{2}\right]_0^{\pi/10} = \dfrac{9}{2}\left[\dfrac{\pi}{10}\right] = \dfrac{9\pi}{20}$

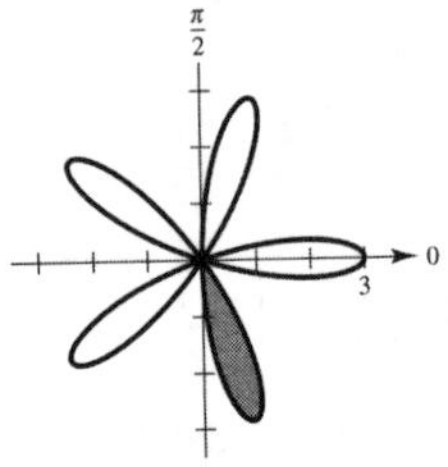

98. $A = 2 \cdot \dfrac{1}{2}\displaystyle\int_0^{\pi/12} [2\sin 6\theta]^2\, d\theta$

$= \displaystyle\int_0^{\pi/12} 4\left(\dfrac{1 - \cos 12\theta}{2}\right) d\theta$

$= 2\left[\theta - \dfrac{\sin 12\theta}{12}\right]_0^{\pi/12} = 2\left[\dfrac{\pi}{12}\right] = \dfrac{\pi}{6}$

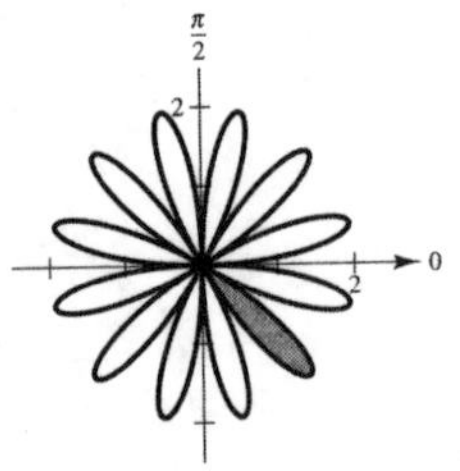

99. $r = 2 + \cos\theta$

$A = 2\left[\dfrac{1}{2}\displaystyle\int_0^{\pi} (2 + \cos\theta)^2\, d\theta\right] \approx 14.14, \left(\dfrac{9\pi}{2}\right)$

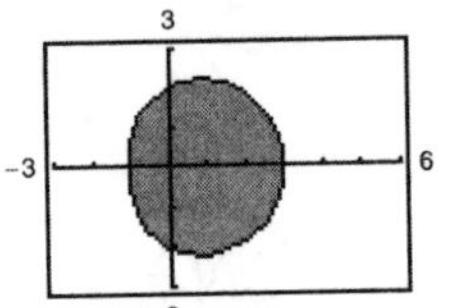

100. $r = 5(1 - \sin\theta)$

$A = 2\left[\dfrac{1}{2}\displaystyle\int_{\pi/2}^{3\pi/2} [5(1 - \sin\theta)]^2\right] d\theta \approx 117.81, \left(\dfrac{75\pi}{2}\right)$

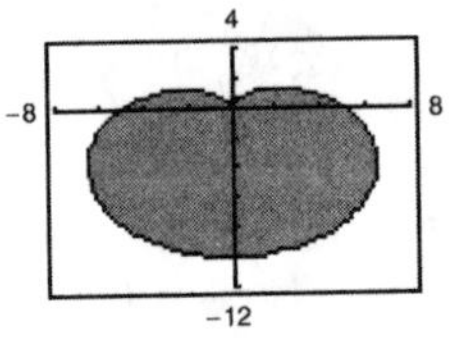

101. $r^2 = 4\sin 2\theta$

$A = 2\left[\dfrac{1}{2}\displaystyle\int_0^{\pi/2} 4\sin 2\theta\, d\theta\right] = 4$

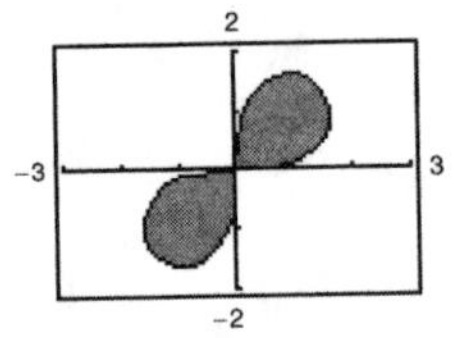

102. $r = 4\cos\theta, r = 2$

$$A = 2\left[\frac{1}{2}\int_0^{\pi/3} 4\,d\theta + \frac{1}{2}\int_{\pi/3}^{\pi/2}(4\cos\theta)^2\,d\theta\right] \approx 4.91$$

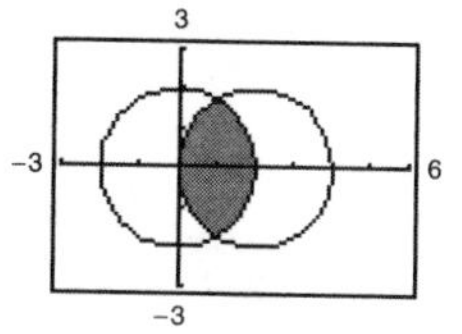

103. $r = 3 - 6\cos\theta$

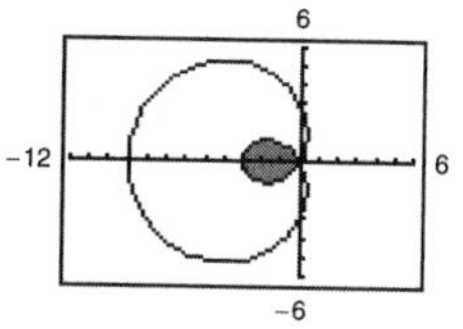

$$A = 2\left[\frac{1}{2}\int_0^{\pi/3}(3 - 6\cos\theta)^2\,d\theta\right]$$

$$= \int_0^{\pi/3}\left[9 - 36\cos\theta + 36\cos^2\theta\right]d\theta$$

$$= 9\int_0^{\pi/3}\left[1 - 4\cos\theta + 2(1 + \cos 2\theta)\right]d\theta$$

$$= 9\left[3\theta - 4\sin\theta + \sin 2\theta\right]_0^{\pi/3}$$

$$= 9\left[\pi - 2\sqrt{3} + \frac{\sqrt{3}}{2}\right] = \frac{18\pi - 27\sqrt{3}}{2}$$

104. $r = 2 + 4\sin\theta$

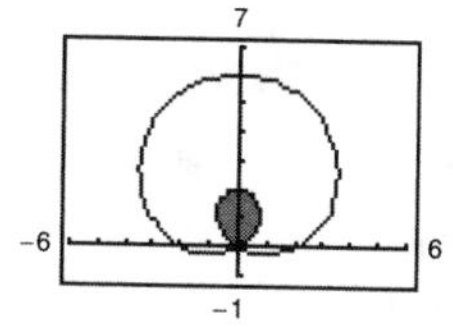

$$A = 2\left[\frac{1}{2}\int_{7\pi/6}^{3\pi/2}(2 + 4\sin\theta)^2\,d\theta\right]$$

$$= \int_{7\pi/6}^{3\pi/2}\left[4 + 16\sin\theta + 16\sin^2\theta\right]d\theta$$

$$= 4\int_{7\pi/6}^{3\pi/2}\left[1 + 4\sin\theta + 2(1 - \cos 2\theta)\right]d\theta$$

$$= 4\left[3\theta - 4\cos\theta - \sin 2\theta\right]_{7\pi/6}^{3\pi/2}$$

$$= 4\left[\left(\frac{9\pi}{2}\right) - \left(\frac{7\pi}{2} + 2\sqrt{3} - \frac{\sqrt{3}}{2}\right)\right]$$

$$= 4\pi - 6\sqrt{3}$$

105. $r = 3 - 6\cos\theta$

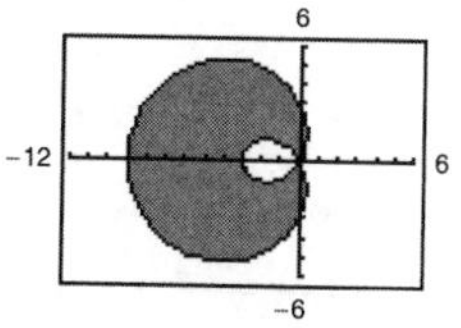

$$A = 2\left[\frac{1}{2}\int_{\pi/3}^{\pi}(3 - 6\cos\theta)^2\,d\theta - \frac{1}{2}\int_0^{\pi/3}(3 - 6\cos\theta)^2\,d\theta\right]$$

From Exercise 103 you have:

$$A = 9\left[3\theta - 4\sin\theta + \sin 2\theta\right]_{\pi/3}^{\pi} - 9\left[3\theta - 4\sin\theta + \sin 2\theta\right]_0^{\pi/3}$$

$$= 9\left[3\pi - \left(\pi - 2\sqrt{3} + \frac{\sqrt{3}}{2}\right)\right] - 9\left[\pi - 2\sqrt{3} + \frac{\sqrt{3}}{2}\right]$$

$$= 9\pi + 27\sqrt{3}$$

106. $r = 2 + 4 \sin \theta$

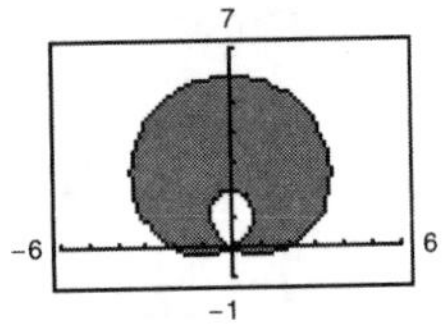

$$A = 2\left[\frac{1}{2}\int_{\pi/2}^{7\pi/6} (2 + 4 \sin \theta)^2 \, d\theta\right] - 2\left[\frac{1}{2}\int_{7\pi/2}^{3\pi/2} (2 + 4 \sin \theta)^2 \, d\theta\right]$$

From Exercise 104 you have:

$$= 4\left[3\theta - 4 \cos \theta - \sin 2\theta\right]_{\pi/2}^{7\pi/6} - 4\left[3\theta - 4 \cos \theta - \sin 2\theta\right]_{7\pi/6}^{3\pi/2}$$

$$= 4\left[\frac{7\pi}{2} + 2\sqrt{3} - \frac{\sqrt{3}}{2} - \frac{3\pi}{2}\right] - 4\left[\frac{9\pi}{2} - \left(\frac{7\pi}{2} + 2\sqrt{3} - \frac{\sqrt{3}}{2}\right)\right]$$

$$= 4\pi + 12\sqrt{3}$$

107. $r = 1 - \cos \theta, r = 1 + \sin \theta$

$1 - \cos \theta = 1 + \sin \theta$

$\tan \theta = -1 \Rightarrow \theta = \frac{3\pi}{4}, \frac{7\pi}{4}$

The graphs also intersect at the pole.

Points of intersection:

$\left(1 + \frac{\sqrt{2}}{2}, \frac{3\pi}{4}\right), \left(1 - \frac{\sqrt{2}}{2}, \frac{7\pi}{4}\right), (0, 0)$

108. $r = 1 + \sin \theta, r = 3 \sin \theta$

$1 + \sin \theta = 3 \sin \theta$

$1 = 2 \sin \theta$

$\frac{1}{2} = \sin \theta$

$\theta = \arcsin \frac{1}{2}, \frac{2\pi}{3} + \arcsin \frac{1}{2}$

The graphs also intersect at the pole.

Points of intersection:

$\left(\frac{3}{2}, \arcsin \frac{1}{2}\right), \left(\frac{3}{2}, \frac{2\pi}{3} + \arcsin \frac{1}{2}\right), (0, 0)$

109. $r = 5 \cos \theta, \frac{\pi}{2} \le \theta \le \pi$

$\frac{dr}{d\theta} = -5 \sin \theta$

$$s = \int_{\pi/2}^{\pi} \sqrt{\left(25 \cos^2 \theta\right) + \left(25 \sin^2 \theta\right)} \, d\theta$$

$$= \int_{\pi/2}^{\pi} 5 \, d\theta = \left[5\theta\right]_{\pi/2}^{\pi} = \frac{5\pi}{2} \quad \text{(Semicircle)}$$

110. $r = 3(1 - \cos \theta), 0 \le \theta \le \pi$

$\frac{dr}{d\theta} = 3 \sin \theta$

$$s = \int_0^{\pi} \sqrt{9(1 - \cos \theta)^2 + 9\sin^2 \theta} \, d\theta$$

$$= 3 \int_0^{\pi} \sqrt{1 - 2 \cos \theta + \cos^2 \theta + \sin^2 \theta} \, d\theta$$

$$= 3 \int_0^{\pi} \sqrt{2 - 2 \cos \theta} \, d\theta$$

$$= 3 \int_0^{\pi} \sqrt{4 \sin^2 \theta} \, d\theta$$

$$= 6 \int_0^{\pi} \sin \theta \, d\theta = \left[-6 \cos \theta\right]_0^{\pi} = 12$$

111. $f(\theta) = 1 + 4 \cos \theta$

$f'(\theta) = -4 \sin \theta$

$$\sqrt{f(\theta)^2 + f'(\theta)^2} = \sqrt{(1 + 4 \cos \theta)^2 + (-4 \sin \theta)^2}$$

$$= \sqrt{17 + 8 \cos \theta}$$

$$S = 2\pi \int_0^{\pi/2} (1 + 4 \cos \theta) \sin \theta \sqrt{17 + 8 \cos \theta} \, d\theta$$

$$= \frac{34\pi\sqrt{17}}{5} \approx 88.08$$

112. $f(\theta) = 2 \sin \theta$

$f'(\theta) = 2 \cos \theta$

$$\sqrt{f(\theta)^2 + f'(\theta)^2} = \sqrt{4 \sin^2 \theta + 4 \cos^2 \theta} = 2$$

$$S = 2\pi \int_0^{\pi/2} 2 \sin \theta \cos \theta(2) \, d\theta = 4\pi$$

113. $r = \dfrac{6}{1 - \sin\theta}$

$e = 1,$

Parabola

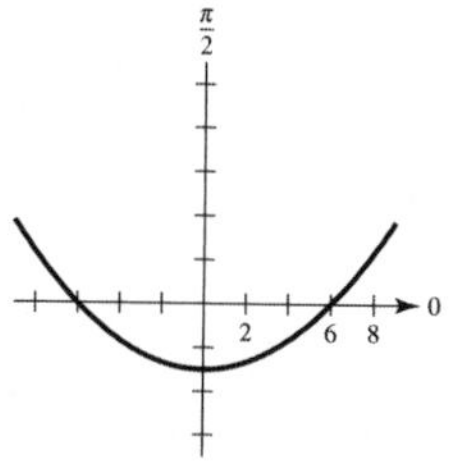

114. $r = \dfrac{2}{1 + \cos\theta}, e = 1$

Parabola

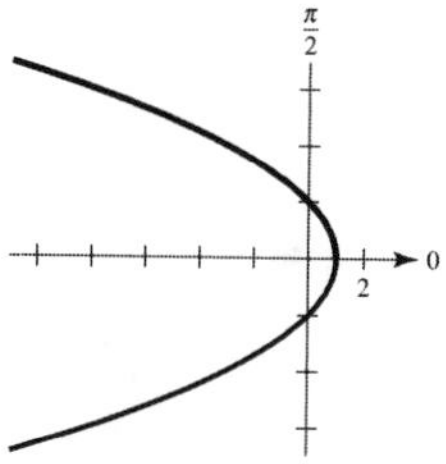

115. $r = \dfrac{6}{3 + 2\cos\theta} = \dfrac{2}{1 + (2/3)\cos\theta}, e = \dfrac{2}{3}$

Ellipse

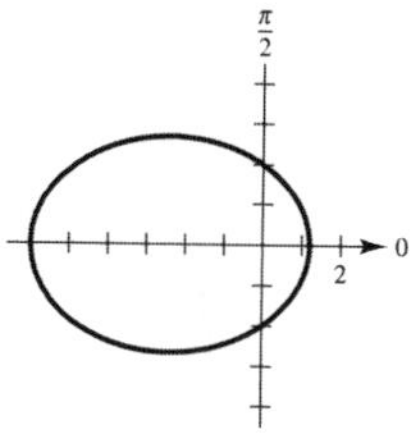

116. $r = \dfrac{4}{5 - 3\sin\theta} = \dfrac{4/5}{1 - (3/5)\sin\theta}, e = \dfrac{3}{5}$

Ellipse

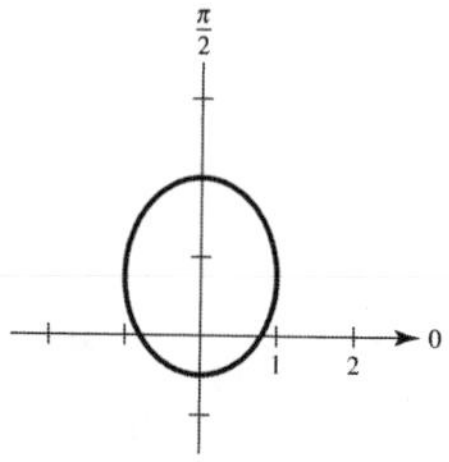

117. $r = \dfrac{4}{2 - 3\sin\theta} = \dfrac{2}{1 - (3/2)\sin\theta}, e = \dfrac{3}{2}$

Hyperbola

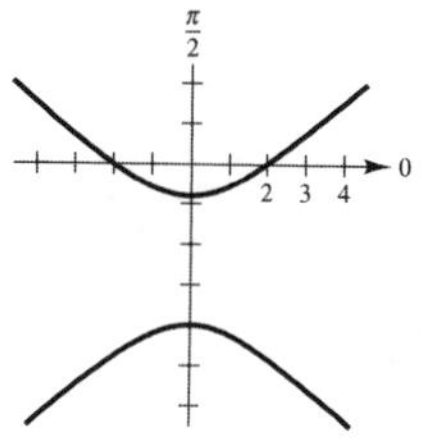

118. $r = \dfrac{8}{2 - 5\cos\theta} = \dfrac{4}{1 - (5/2)\cos\theta}, e = \dfrac{5}{2}$

Hyperbola

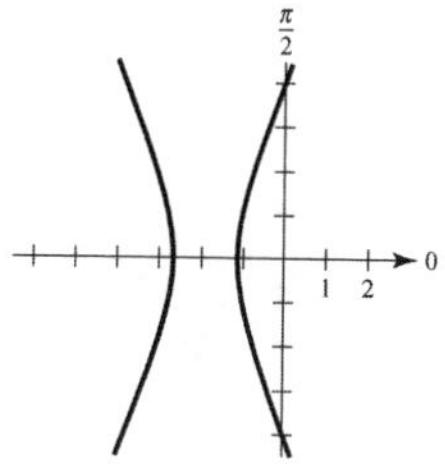

119. Parabola

$e = 1$

$x = 4 \Rightarrow d = 4$

$r = \dfrac{ed}{1 + e\cos\theta} = \dfrac{4}{1 + \cos\theta}$

120. Ellipse, $e = \dfrac{3}{4}, y = -2$

$d = 2$

$r = \dfrac{ed}{1 - e\sin\theta} = \dfrac{\left(\frac{3}{4}\right)2}{1 - \frac{3}{4}\sin\theta} = \dfrac{6}{4 - 3\sin\theta}$

121. Hyperbola, $e = 3, y = 3$

$d = 3$

$r = \dfrac{ed}{1 + e\sin\theta} = \dfrac{3(3)}{1 + 3\sin\theta} = \dfrac{9}{1 + 3\sin\theta}$

122. Parabola

Vertex: $\left(2, \dfrac{\pi}{2}\right)$

Focus: $(0, 0)$

$e = 1, d = 4$

$r = \dfrac{4}{1 + \sin\theta}$

123. Ellipse

Vertices: $(5, 0), (1, \pi)$

Focus: $(0, 0)$

$$a = 3, c = 2, e = \frac{2}{3}, d = \frac{5}{2}$$

$$r = \frac{\left(\frac{2}{3}\right)\left(\frac{5}{2}\right)}{1 - \left(\frac{2}{3}\right)\cos\theta} = \frac{5}{3 - 2\cos\theta}$$

124. Hyperbola

Vertices: $(1, 0), (7, 0)$

Focus: $(0, 0)$

$$a = 3, c = 4, e = \frac{4}{3}, d = \frac{7}{4}$$

$$r = \frac{\left(\frac{4}{3}\right)\left(\frac{7}{4}\right)}{1 + \left(\frac{4}{3}\right)\cos\theta} = \frac{7}{3 + 4\cos\theta}$$

Problem Solving for Chapter 10

1. (a)

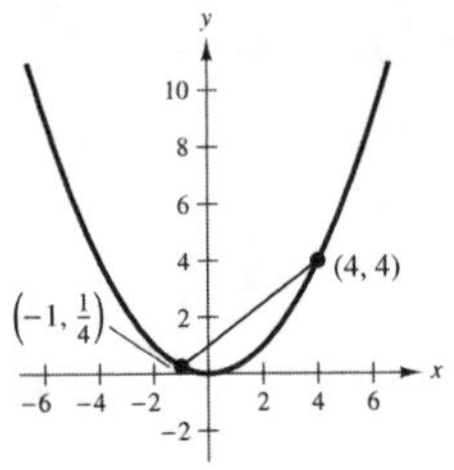

(b) $x^2 = 4y$

$2x = 4y'$

$y' = \frac{1}{2}x$

$y - 4 = 2(x - 4) \Rightarrow y = 2x - 4$ Tangent line at $(4, 4)$

$y - \frac{1}{4} = -\frac{1}{2}(x + 1) \Rightarrow y = -\frac{1}{2}x - \frac{1}{4}$ Tangent line at $\left(-1, \frac{1}{4}\right)$

Tangent lines have slopes of 2 and $-\frac{1}{2} \Rightarrow$ perpendicular.

(c) Intersection:

$2x - 4 = -\frac{1}{2}x - \frac{1}{4}$

$8x - 16 = -2x - 1$

$10x = 15$

$x = \frac{3}{2} \Rightarrow \left(\frac{3}{2}, -1\right)$

Point of intersection, $\left(\frac{3}{2}, -1\right)$, is on directrix $y = -1$.

2. Assume $p > 0$.

Let $y = mx + p$ be the equation of the focal chord.

First find x-coordinates of focal chord endpoints:

$$x^2 = 4py = 4p(mx + p)$$

$$x^2 - 4pmx - 4p^2 = 0$$

$$x = \frac{4pm \pm \sqrt{16p^2m^2 + 16p^2}}{2} = 2pm \pm 2p\sqrt{m^2 + 1}$$

$$x^2 = 4py, 2x = 4py' \Rightarrow y' = \frac{x}{2p}.$$

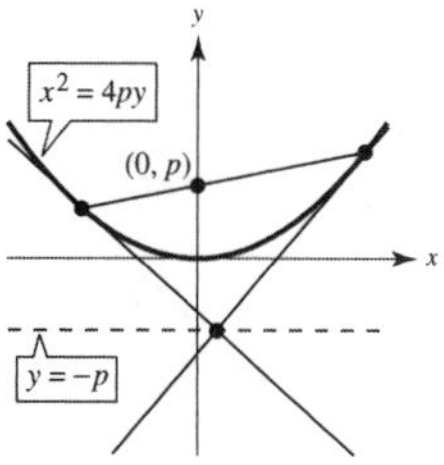

(a) The slopes of the tangent lines at the endpoints are perpendicular because

$$\frac{1}{2p}\left[2pm + 2p\sqrt{m^2 + 1}\right]\frac{1}{2p}\left[2pm - 2p\sqrt{m^2 + 1}\right] = \frac{1}{4p^2}\left[4p^2m^2 - 4p^2(m^2 + 1)\right] = \frac{1}{4p^2}\left[-4p^2\right] = -1$$

(b) Finally, you show that the tangent lines intersect at a point on the directrix $y = -p$.

Let $b = 2pm + 2p\sqrt{m^2 + 1}$ and $c = 2pm - 2p\sqrt{m^2 + 1}$.

$$b^2 = 8p^2m^2 + 4p^2 + 8p^2m\sqrt{m^2 + 1}$$

$$c^2 = 8p^2m^2 + 4p^2 - 8p^2m\sqrt{m^2 + 1}$$

$$\frac{b^2}{4p} = 2pm^2 + p + 2pm\sqrt{m^2 + 1}$$

$$\frac{c^2}{4p} = 2pm^2 + p - 2pm\sqrt{m^2 + 1}$$

Tangent line at $x = b$: $y - \frac{b^2}{4p} = \frac{b}{2p}(x - b) \Rightarrow y = \frac{bx}{2p} - \frac{b^2}{4p}$

Tangent line at $x = c$: $y - \frac{c^2}{4p} = \frac{c}{2p}(x - c) \Rightarrow y = \frac{cx}{2p} - \frac{c^2}{4p}$

Intersection of tangent lines:

$$\frac{bx}{2p} - \frac{b^2}{4p} = \frac{cx}{2p} - \frac{c^2}{4p}$$

$$2bx - b^2 = 2cx - c^2$$

$$2x(b - c) = b^2 - c^2$$

$$2x\left(4p\sqrt{m^2 + 1}\right) = 16p^2m\sqrt{m^2 + 1}$$

$$x = 2pm$$

Finally, the corresponding y-value is $y - p$, which shows that the intersection point lies on the directrix.

3. Consider $x^2 = 4py$ with focus $F = (0, p)$.

Let $P(a, b)$ be point on parabola.

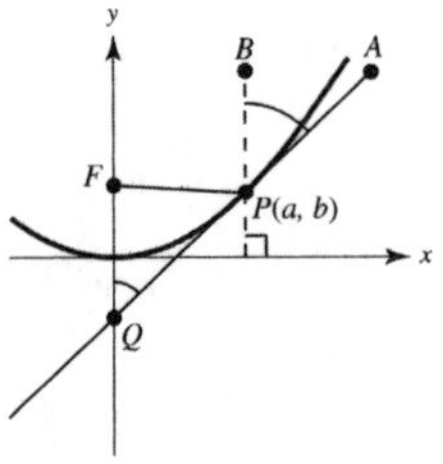

$$2x = 4py' \Rightarrow y' = \frac{x}{2p}$$

$$y - b = \frac{a}{2p}(x - a) \quad \text{Tangent line at } P$$

For $x = 0$, $y = b + \dfrac{a}{2p}(-a) = b - \dfrac{a^2}{2p} = b - \dfrac{4pb}{2p} = -b$.

So, $Q = (0, -b)$.

ΔFQP is isosceles because

$|FQ| = p + b$

$|FP| = \sqrt{(a-0)^2 + (b-p)^2} = \sqrt{a^2 + b^2 - 2bp + p^2} = \sqrt{4pb + b^2 - 2bp + p^2} = \sqrt{(b+p)^2} = b + p.$

So, $\measuredangle FQP = \measuredangle BPA = \measuredangle FPQ$.

4. (a) The first plane makes an angle of 70° with the positive x-axis, and is 150 miles from P:

$x_1 = \cos 70°(150 - 375t)$

$y_1 = \sin 70°(150 - 375t)$

Similarly for the second plane,

$x_2 = \cos 135°(190 - 450t)$

$= \cos 45°(-190 + 450t)$

$y_2 = \sin 135°(190 - 450t)$

$= \sin 45°(190 - 450t)$.

(b) $d = \sqrt{(x_2 - x_1)^2 + (y_2 - y_1)^2}$

$= \left[\left[\cos 45°(-190 + 450t) - \cos 70°(150 - 375t)\right]^2 + \left[\sin 45°(190 - 450t) - \sin 70°(150 - 375t)\right]^2\right]^{1/2}$

(c)

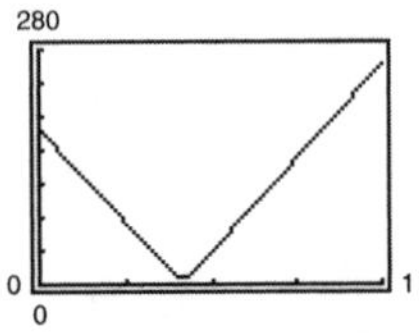

The minimum distance is 7.59 miles when $t = 0.4145$; Yes.

5. (a) $y^2 = \dfrac{t^2(1 - t^2)^2}{(1 + t^2)^2}$, $x^2 = \dfrac{(1 - t^2)^2}{(1 + t^2)^2}$

$$\frac{1 - x}{1 + x} = \frac{1 - \left(\dfrac{1 - t^2}{1 + t^2}\right)}{1 + \left(\dfrac{1 - t^2}{1 + t^2}\right)} = \frac{2t^2}{2} = t^2$$

So, $y^2 = x^2\left(\dfrac{1 - x}{1 + x}\right)$.

(b)

$$\begin{aligned} r^2 \sin^2 \theta &= r^2 \cos^2 \theta\left(\frac{1 - r\cos\theta}{1 + r\cos\theta}\right) \\ \sin^2 \theta(1 + r\cos\theta) &= \cos^2\theta(1 - r\cos\theta) \\ r\cos\theta\sin^2\theta + \sin^2\theta &= \cos^2\theta - r\cos^3\theta \\ r\cos\theta(\sin^2\theta + \cos^2\theta) &= \cos^2\theta - \sin^2\theta \\ r\cos\theta &= \cos 2\theta \\ r &= \cos 2\theta \cdot \sec\theta \end{aligned}$$

(c)

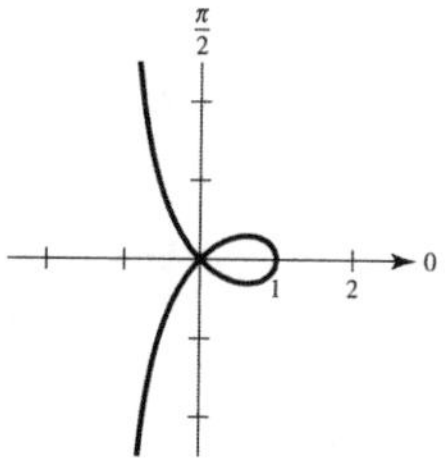

(d) $r(\theta) = 0$ for $\theta = \dfrac{\pi}{4}, \dfrac{3\pi}{4}$.

So, $y = x$ and $y = -x$ are tangent lines to curve at the origin.

(e) $y'(t) = \dfrac{(1 + t^2)(1 - 3t^2) - (t - t^3)(2t)}{(1 + t^2)^2} = \dfrac{1 - 4t^2 - t^4}{(1 + t^2)^2} = 0$

$t^4 + 4t^2 - 1 = 0 \Rightarrow t^2 = -2 \pm \sqrt{5} \Rightarrow x = \dfrac{1 - (-2 \pm \sqrt{5})}{1 + (-2 \pm \sqrt{5})} = \dfrac{3 \mp \sqrt{5}}{-1 \pm \sqrt{5}} = \dfrac{3 - \sqrt{5}}{-1 + \sqrt{5}} = \dfrac{\sqrt{5} - 1}{2}$

$\left(\dfrac{\sqrt{5} - 1}{2}, \pm \dfrac{\sqrt{5} - 1}{2}\sqrt{-2 + \sqrt{5}}\right)$

6. $y = a(1 - \cos\theta) \Rightarrow \cos\theta = \dfrac{a - y}{a}$

$$\begin{aligned} \theta &= \arccos\left(\frac{a - y}{a}\right) \\ x &= a(\theta - \sin\theta) \\ &= a\left(\arccos\left(\frac{a - y}{a}\right) - \sin\left(\arccos\left(\frac{a - y}{a}\right)\right)\right) \\ &= a\left(\arccos\left(\frac{a - y}{a}\right) - \frac{\sqrt{2ay - y^2}}{a}\right) \\ x &= a \cdot \arccos\left(\frac{a - y}{a}\right) - \sqrt{2ay - y^2}, \ 0 \le y \le 2a \end{aligned}$$

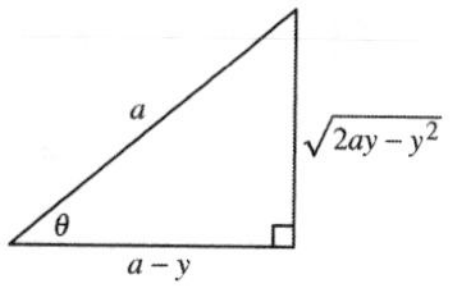

7. (a)

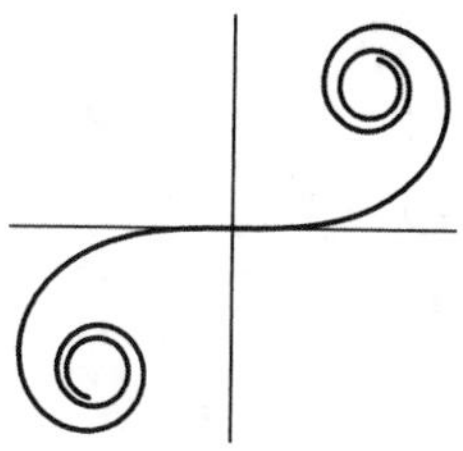

Generated by Mathematica

(b) $(-x, -y) = \left(-\int_0^t \cos\dfrac{\pi u^2}{2}\,du, -\int_0^t \sin\dfrac{\pi u^2}{2}\,du\right)$ is on the curve whenever (x, y) is on the curve.

(c) $x'(t) = \cos\dfrac{\pi t^2}{2}, y'(t) = \sin\dfrac{\pi t^2}{2}$,

$x'(t)^2 + y'(t)^2 = 1$

So, $s = \int_0^a dt = a$.

On $[-\pi, \pi]$, $s = 2\pi$.

8. (a) $A = 4\int_0^a \frac{b}{a}\sqrt{a^2 - x^2}\,dx = \frac{4b}{a}\left(\frac{1}{2}\right)\left[x\sqrt{a^2 - x^2} + a^2 \arcsin\left(\frac{x}{a}\right)\right]_0^a = \pi ab$

(b) **Disk:** $V = 2\pi\int_0^b \frac{a^2}{b^2}(b^2 - y^2)\,dy = \frac{2\pi a^2}{b^2}\int_0^b (b^2 - y^2)\,dy = \frac{2\pi a^2}{b^2}\left[b^2 y - \frac{1}{3}y^3\right]_0^b = \frac{4}{3}\pi a^2 b$

$$S = 4\pi\int_0^b \frac{a}{b}\sqrt{b^2 - y^2}\left(\frac{\sqrt{b^4 + (a^2 - b^2)y^2}}{b\sqrt{b^2 - y^2}}\right)dy$$

$$= \frac{4\pi a}{b^2}\int_0^b \sqrt{b^4 + c^2y^2}\,dy = \frac{2\pi a}{b^2 c}\left[cy\sqrt{b^4 + c^2y^2} + b^4 \ln\left|cy + \sqrt{b^4 + c^2y^2}\right|\right]_0^b$$

$$= \frac{2\pi a}{b^2 c}\left[b^2 c\sqrt{b^2 + c^2} + b^4 \ln\left|cb + b\sqrt{b^2 + c^2}\right| - b^4 \ln(b^2)\right]$$

$$= 2\pi a^2 + \frac{\pi ab^2}{c}\ln\left(\frac{c + a}{e}\right)^2 = 2\pi a^2 + \left(\frac{\pi b^2}{e}\right)\ln\left(\frac{1 + e}{1 - e}\right)$$

(c) **Disk:** $V = 2\pi\int_0^a \frac{b^2}{a^2}(a^2 - x^2)\,dx = \frac{2\pi b^2}{a^2}\int_0^a (a^2 - x^2)\,dx = \frac{2\pi b^2}{a^2}\left[a^2 x - \frac{1}{3}x^3\right]_0^a = \frac{4}{3}\pi ab^2$

$$S = 2(2\pi)\int_0^a \frac{b}{a}\sqrt{a^2 - x^2}\left(\frac{\sqrt{a^4 - (a^2 - b^2)x^2}}{a\sqrt{a^2 - x^2}}\right)dx$$

$$= \frac{4\pi b}{a^2}\int_0^a \sqrt{a^4 - c^2x^2}\,dx = \frac{2\pi b}{a^2 c}\left[cx\sqrt{a^4 - c^2x^2} + a^4 \arcsin\left(\frac{cx}{a^2}\right)\right]_0^a$$

$$= \frac{a\pi b}{a^2 c}\left[a^2 c\sqrt{a^2 - c^2} + a^4 \arcsin\left(\frac{c}{a}\right)\right] = 2\pi b^2 + 2\pi\left(\frac{ab}{e}\right)\arcsin(e)$$

9. $r = \dfrac{ab}{a\sin\theta + b\cos\theta},\quad 0 \le \theta \le \dfrac{\pi}{2}$

$$r(a\sin\theta + b\cos\theta) = ab$$

$$ay + bx = ab$$

$$\frac{y}{b} + \frac{x}{a} = 1$$

Line segment

Area $= \frac{1}{2}ab$

10. (a) Area $= \int_0^\alpha \frac{1}{2}r^2\,d\theta$

$= \frac{1}{2}\int_0^\alpha \sec^2\theta\,d\theta$

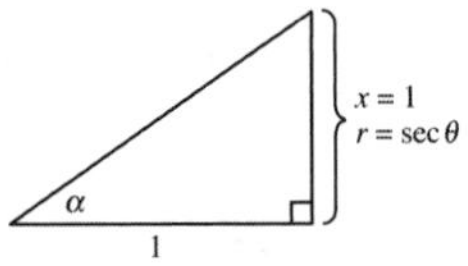

(b) $\tan\alpha = \frac{h}{1} \Rightarrow$ Area $= \frac{1}{2}(1)\tan\alpha$

$\Rightarrow \tan\alpha = \int_0^\alpha \sec^2\theta\,d\theta$

(c) Differentiating, $\frac{d}{d\alpha}(\tan\alpha) = \sec^2\alpha$.

11. Let (r, θ) be on the graph.

$$\sqrt{r^2 + 1 + 2r\cos\theta}\sqrt{r^2 + 1 - 2r\cos\theta} = 1$$
$$(r^2 + 1)^2 - 4r^2\cos^2\theta = 1$$
$$r^4 + 2r^2 + 1 - 4r^2\cos^2\theta = 1$$
$$r^2(r^2 - 4\cos^2\theta + 2) = 0$$
$$r^2 = 4\cos^2\theta - 2$$
$$r^2 = 2(2\cos^2\theta - 1)$$
$$r^2 = 2\cos 2\theta$$

12. For $t = \dfrac{\pi}{2}, \dfrac{3}{2}, \dfrac{5\pi}{2}, \dfrac{7\pi}{2}, \ldots$

$$y = \frac{2}{\pi}, \frac{-2}{3\pi}, \frac{2}{5\pi}, \frac{-2}{7\pi}, \ldots$$

So, the curve has length greater that

$$S = \frac{2}{\pi} + \frac{2}{3\pi} + \frac{2}{5\pi} + \frac{2}{7\pi} + \cdots$$
$$= \frac{2}{\pi}\left(1 + \frac{1}{3} + \frac{1}{5} + \frac{1}{7} + \cdots\right)$$
$$> \frac{2}{\pi}\left(\frac{1}{2} + \frac{1}{4} + \frac{1}{6} + \frac{1}{8} + \cdots\right)$$
$$= \infty. \text{ (Harmonic series)}$$

13. If a dog is located at (r, θ) in the first quadrant, then its neighbor is at $\left(r, \theta + \dfrac{\pi}{2}\right)$:

$(x_1, y_1) = (r\cos\theta, r\sin\theta)$ and

$(x_2, y_2) = (-r\sin\theta, r\cos\theta)$.

The slope joining these points is

$$\frac{r\cos\theta - r\sin\theta}{-r\sin\theta - r\cos\theta} = \frac{\sin\theta - \cos\theta}{\sin\theta + \cos\theta}$$
$$= \text{slope of tangent line at } (r, \theta).$$

$$\frac{dy}{dx} = \frac{\frac{dy}{dr}}{\frac{dx}{dr}} = \frac{\frac{dr}{d\theta}\sin\theta + r\cos\theta}{\frac{dr}{d\theta}\cos\theta - r\sin\theta} = \frac{\sin\theta - \cos\theta}{\sin\theta + \cos\theta}$$
$$\Rightarrow \frac{dr}{d\theta} = -r$$
$$\frac{dr}{r} = -d\theta$$
$$\ln r = -\theta + C_1$$
$$r = e^{-\theta + C_1}$$
$$r = Ce^{-\theta}$$

$$r\left(\frac{\pi}{4}\right) = \frac{d}{\sqrt{2}} \Rightarrow r = Ce^{-\pi/4} = \frac{d}{\sqrt{2}} \Rightarrow C = \frac{d}{\sqrt{2}}e^{\pi/4}$$

Finally, $r = \dfrac{d}{\sqrt{2}}e^{((\pi/4) - \theta)}, \theta \ge \dfrac{\pi}{4}$.

14. $\dfrac{x^2}{a^2} - \dfrac{y^2}{b^2} = 1,\ a^2 + b^2 = c^2,\ MF_2 - MF_1 = 2a$

$y' = \dfrac{b^2x}{a^2y}$

Tangent line at $M(x_0, y_0)$:

$$y - y_0 = \frac{b^2x_0}{a^2y_0}(x - x_0)$$

$$\frac{yy_0 - y_0^2}{b^2} = \frac{x_0x - x_0^2}{a^2}$$

$$\frac{x_0x}{a^2} - \frac{y_0y}{b^2} = \frac{x_0^2}{a^2} + \frac{y_0^2}{b^2}$$

$$\frac{x_0x}{a^2} - \frac{y_0y}{b^2} = 1$$

At $x = 0,\ y = -\dfrac{b^2}{y_0} \Rightarrow Q = \left(0, -\dfrac{b^2}{y_0}\right)$.

$QF_2 = QF_1 = \sqrt{c^2 + \dfrac{b^4}{y_0^2}} = d$

$MQ = \sqrt{x_0^2 + \left(y_0 + \dfrac{b^2}{y_0}\right)^2} = f$

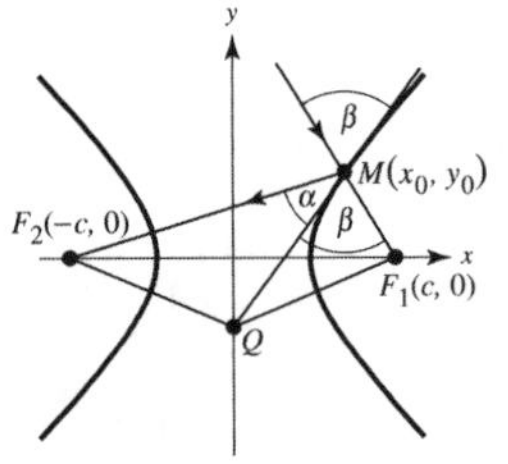

By the Law of Cosines,

$$(F_2Q)^2 = (MF_2)^2 + (MQ)^2 - 2(MF_2)(MQ)\cos\alpha$$

$$d^2 = (MF_2)^2 + f^2 - 2f(MF_2)\cos\alpha$$

$$(F_1Q)^2 = (MF_1)^2 + f^2 - 2f(MF_1)\cos\beta$$

$$d^2 = (MF_1)^2 + f^2 - 2f(MF_1)\cos\beta.$$

$$\cos\alpha = \frac{(MF_2)^2 f^2 - d^2}{2f(MF_2)},\ \cos\beta = \frac{(MF_1)^2 + f^2 - d^2}{2f(MF_1)}$$

$MF_2 = MF_1 + 2a$. Let $z = MF_1$.

Slopes: $MF_1 : \dfrac{y_0}{x_0 - c}$; $QF_1 : \dfrac{-b^2}{y_0c}$; $QF_2 : \dfrac{b^2}{y_0c}$

To show $\alpha = \beta$, consider

$$\left[(MF_2)^2 + f^2 - d^2\right]\left[2f(MF_1)\right] = \left[(MF_1)^2 + f^2 - d^2\right]\left[2f(MF_2)\right]$$

$$\Leftrightarrow \quad \left[(z + 2a)^2 + f^2 - d^2\right][z] = \left[z^2 + f^2 - d^2\right][z + 2a]$$

$$\Leftrightarrow \quad z^2 + 2az = f^2 - d^2$$

$$\Leftrightarrow \quad (x_0 - c)^2 + y_0^2 + 2az = \left(x_0^2 + \left(y_0 + \frac{b^2}{y_0}\right)^2\right) - \left(c^2 + \frac{b^4}{y_0^2}\right)$$

$$\Leftrightarrow \quad az - x_0c + a^2 = 0$$

$$\Leftrightarrow \quad a\sqrt{(x_0 - c)^2 + y_0^2} = x_0c - a^2$$

$$\Leftrightarrow \quad x_0^2b^2 - a^2y_0^2 = a^2b^2$$

$$\Leftrightarrow \quad \frac{x_0^2}{a^2} - \frac{y_0^2}{b^2} = 1.$$

So, $\alpha = \beta$ and the reflective property is verified.

15. (a) In ΔOCB, $\cos\theta = \dfrac{2a}{OB} \Rightarrow OB = 2a \cdot \sec\theta$.

In ΔOAC, $\cos\theta = \dfrac{OA}{2a} \Rightarrow OA = 2a \cdot \cos\theta$.

$$\begin{aligned} r = OP = AB = OB - OA &= 2a(\sec\theta - \cos\theta) \\ &= 2a\left(\frac{1}{\cos\theta} - \cos\theta\right) \\ &= 2a \cdot \frac{\sin^2\theta}{\cos\theta} \\ &= 2a \cdot \tan\theta \sin\theta \end{aligned}$$

(b) $x = r\cos\theta = (2a\tan\theta\sin\theta)\cos\theta = 2a\sin^2\theta$

$y = r\sin\theta = (2a\tan\theta\sin\theta)\sin\theta = 2a\tan\theta \cdot \sin^2\theta, -\dfrac{\pi}{2} < \theta < \dfrac{\pi}{2}$

Let $t = \tan\theta, -\infty < t < \infty$.

Then $\sin^2\theta = \dfrac{t^2}{1+t^2}$ and $x = 2a\dfrac{t^2}{1+t^2}, y = 2a\dfrac{t^3}{1+t^2}$.

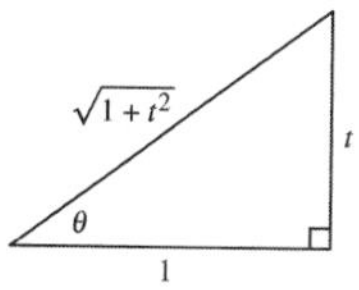

(c)
$$\begin{aligned} r &= 2a\tan\theta\sin\theta \\ r\cos\theta &= 2a\sin^2\theta \\ r^3\cos\theta &= 2a\,r^2\sin^2\theta \\ (x^2+y^2)x &= 2ay^2 \\ y^2 &= \frac{x^3}{(2a-x)} \end{aligned}$$

16. The curve is produced over the interval $0 \le \theta \le 10\pi$.

17.

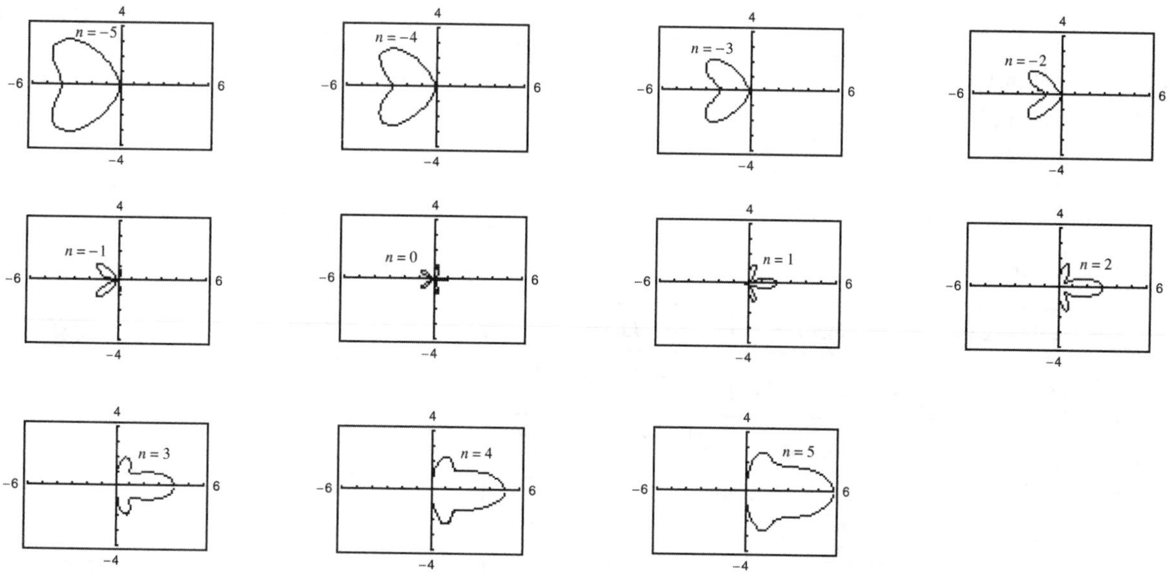

$n = 1, 2, 3, 4, 5$ produce "bells"; $n = -1, -2, -3, -4, -5$ produce "hearts".

Appendix C.1

1. $0.7 = \frac{7}{10}$

   Rational

2. $-3678 = \dfrac{-3678}{1}$

   Rational

3. $\dfrac{3\pi}{2}$

   Irrational (because π is irrational)

4. $3\sqrt{2} - 1$

   Irrational (because $\sqrt{2}$ is irrational)

5. $4.345\overline{1451}$

   Rational

6. $\frac{22}{7}$

   Rational

7. $\sqrt[3]{64} = 4$

   Rational

8. $0.81\overline{778177}$

   Rational

9. $4\frac{5}{8} = \frac{37}{8}$

   Rational

10. $\left(\sqrt{2}\right)^3 = 2\sqrt{2}$

    Irrational

11. Let $x = 0.36\overline{36}$.

$$\begin{aligned} 100x &= 36.36\overline{36} \\ -x &= -0.36\overline{36} \\ \hline 99x &= 36 \end{aligned}$$

$$x = \frac{36}{99} = \frac{4}{11}$$

12. Let $x = 0.318\overline{18}$.

$$\begin{aligned} 1000x &= 318.18\overline{18} \\ -10x &= -3.18\overline{18} \\ \hline 990x &= 315 \end{aligned}$$

$$x = \frac{315}{990} = \frac{7}{22}$$

13. Let $x = 0.297\overline{297}$.

$$\begin{aligned} 1000x &= 0.297\overline{297} \\ -x &= -297.297\overline{297} \\ \hline 999x &= 297 \end{aligned}$$

$$x = \frac{297}{999} = \frac{11}{37}$$

14. Let $x = 0.9900\overline{9900}$.

$$\begin{aligned} 10{,}000x &= 9900.9900\overline{9900} \\ -x &= -0.9900\overline{9900} \\ \hline 9999x &= 9900 \end{aligned}$$

$$x = \frac{9900}{9999} = \frac{100}{101}$$

15. Given $a < b$:

   (a) $a + 2 < b + 2$; True

   (b) $5b < 5a$; False

   (c) $5 - a > 5 - b$; True

   (d) $\dfrac{1}{a} < \dfrac{1}{b}$; False

   (e) $(a - b)(b - a) > 0$; False

   (f) $a^2 < b^2$; False

16.

Interval Notation	Set Notation	Graph
$[-2, 0)$	$\{x: -2 \le x < 0\}$	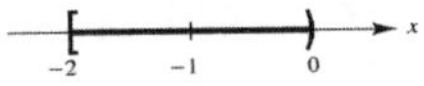
$(-\infty, -4]$	$\{x: x \le -4\}$	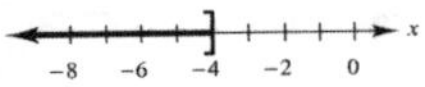
$\left[3, \frac{11}{2}\right]$	$\left\{x: 3 \le x \le \frac{11}{2}\right\}$	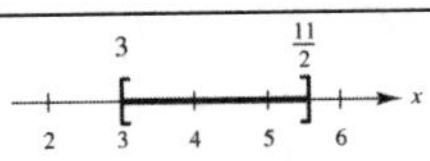
$(-1, 7)$	$\{x: -1 < x < 7\}$	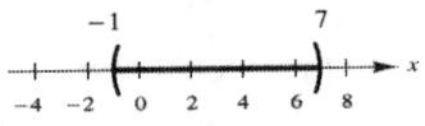

17. x is greater than -3 and less than 3.

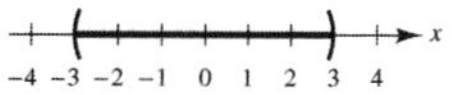

The interval is bounded.

18. x is greater than, or equal to, 4.

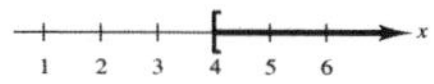

The interval is unbounded.

19. x is less than, or equal to, 5.

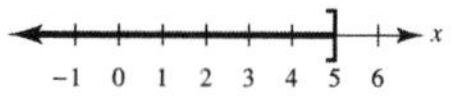

The interval is unbounded.

20. x is greater than or equal to 0, and less than 8.

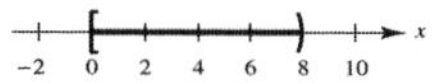

The interval is bounded.

21. $y \geq 4, [4, \infty)$

22. $q \geq 0, [0, \infty)$

23. $0.03 < r \leq 0.07, (0.03, 0.07]$

24. $T > 90°, (90°, \infty)$

25.
$$2x - 1 \geq 0$$
$$2x \geq 1$$
$$x \geq \tfrac{1}{2}$$

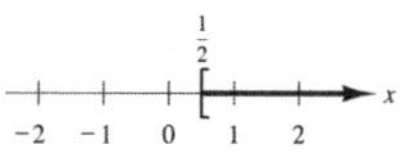

26.
$$3x + 1 \geq 2x + 2$$
$$3x \geq 2x + 1$$
$$x \geq 1$$

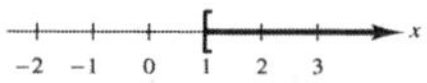

27.
$$-4 < 2x - 3 < 4$$
$$-1 < 2x < 7$$
$$-\tfrac{1}{2} < x < \tfrac{7}{2}$$

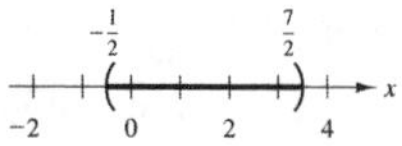

28.
$$0 \leq x + 3 < 5$$
$$-3 \leq x < 2$$

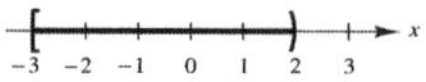

29.
$$\frac{x}{2} + \frac{x}{3} > 5$$
$$3x + 2x > 30$$
$$5x > 30$$
$$x > 6$$

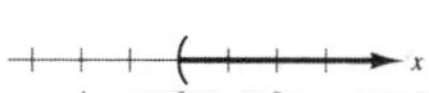

30.
$$x > \frac{1}{x}$$
$$x - \frac{1}{x} > 0$$
$$\frac{x^2 - 1}{x} > 0$$
$$\frac{(1 + x)(x - 1)}{x} > 0$$

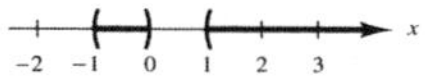

Test intervals: $(-\infty, -1), (-1, 0), (0, 1), (1, \infty)$

Solution: $-1 < x < 0$ or $x > 1$

31. $|x| < 1 \Rightarrow -1 < x < 1$

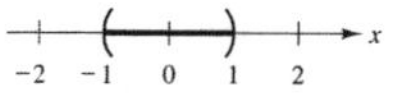

32.
$$\frac{x}{2} - \frac{x}{3} > 5$$
$$3x - 2x > 30$$
$$x > 30$$

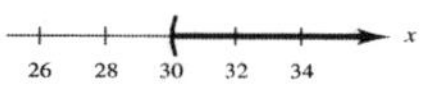

33.
$$\left|\frac{x - 3}{2}\right| \geq 5$$
$$x - 3 \geq 10 \quad \text{or} \quad x - 3 \leq -10$$
$$x \geq 13 \qquad\qquad x \leq -7$$

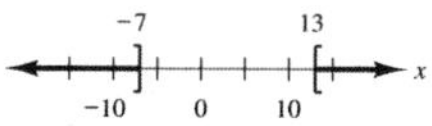

34. $\left|\frac{x}{2}\right| > 3 \Rightarrow x > 6$ or $x < -6$

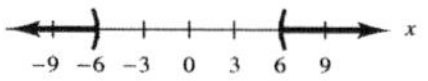

35.
$$|x - a| < b$$
$$-b < x - a < b$$
$$a - b < x < a + b$$

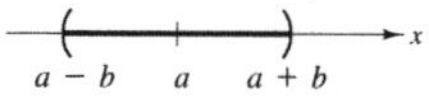

36.
$$|x + 2| < 5$$
$$-5 < x + 2 < 5$$
$$-7 < x < 3$$

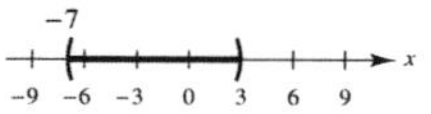

37.
$$|2x + 1| < 5$$
$$-5 < 2x + 1 < 5$$
$$-6 < 2x < 4$$
$$-3 < x < 2$$

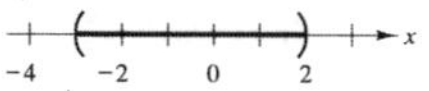

38.
$$|3x + 1| \geq 4$$
$$3x + 1 \geq 4 \quad \text{or} \quad 3x + 1 \leq -4$$
$$3x \geq 3 \qquad\qquad 3x \leq -5$$
$$x \geq 1 \qquad\qquad x \leq -\tfrac{5}{3}$$

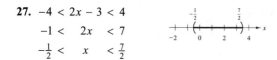

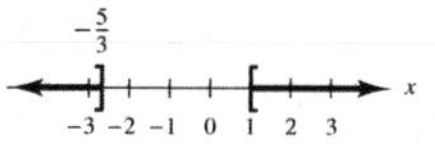

39. $\left|1 - \frac{2x}{3}\right| < 1$

$$-1 < 1 - \frac{2x}{3} < 1$$
$$-2 < -\frac{2x}{3} < 0$$
$$3 > x > 0$$

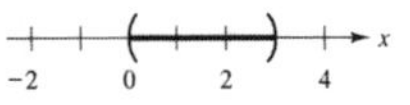

40. $|9 - 2x| < 1$

$$-1 < 9 - 2x < 1$$
$$-10 < -2x < -8$$
$$5 > x > 4$$

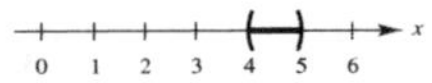

41. $x^2 \le 3 - 2x$

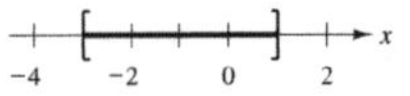

$$x^2 + 2x - 3 \le 0$$
$$(x + 3)(x - 1) \le 0$$

Test intervals: $(-\infty, -3), (-3, 1), (1, \infty)$

Solution: $-3 \le x \le 1$

42. $x^4 - x \le 0$

$$x(x^3 - 1) \le 0$$
$$x = 0$$
$$x = 1$$

Test intervals: $(-\infty, 0), (0, 1), (1, \infty)$

Solution: $0 \le x \le 1$

43. $x^2 + x - 1 \le 5$

$$x^2 + x - 6 \le 0$$
$$(x + 3)(x - 2) \le 0$$
$$x = -3$$
$$x = 2$$

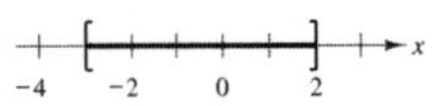

Test intervals: $(-\infty, -3), (-3, 2), (2, \infty)$

Solution: $-3 \le x \le 2$

44. $2x^2 + 1 < 9x - 3$

$$2x^2 - 9x + 4 < 0$$
$$(2x - 1)(x - 4) < 0$$
$$x = \frac{1}{2}$$
$$x = 4$$

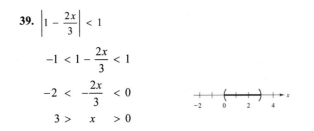

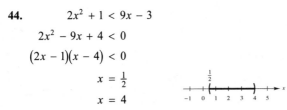

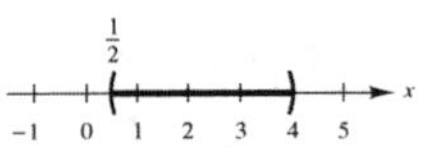

Test intervals: $\left(-\infty, \frac{1}{2}\right), \left(\frac{1}{2}, 4\right), (4, \infty)$

Solution: $\frac{1}{2} < x < 4$

45. $a = -1, b = 3$

Directed distance from a to b: 4

Directed distance from b to a: -4

Distance between a and b: 4

46. $a = -\frac{5}{2}, b = \frac{13}{4}$

Directed distance from a to b: $\frac{23}{4}$

Directed distance from b to a: $-\frac{23}{4}$

Distance between a and b: $\frac{23}{4}$

47. (a) $a = 126, b = 75$

Directed distance from a to b: -51

Directed distance from b to a: 51

Distance between a and b: 51

(b) $a = -126, b = -75$

Directed distance from a to b: 51

Directed distance from b to a: -51

Distance between a and b: 51

48. (a) $a = 9.34, b = -5.65$

Directed distance from a to b: -14.99

Directed distance from b to a: 14.99

Distance between a and b: 14.99

(b) $a = \frac{16}{5}, b = \frac{112}{75}$

Directed distance from a to b: $-\frac{128}{75}$

Directed distance from b to a: $\frac{128}{75}$

Distance between a and b: $\frac{128}{75}$

49. $a = -2, b = 2$

Midpoint: 0

Distance between midpoint and each endpoint: 2

$$|x - 0| \le 2$$
$$|x| \le 2$$

50. $a = -3, b = 3$

Midpoint: 0

Distance between midpoint and each endpoint: 3

$$|x - 0| \ge 3$$
$$|x| \ge 3$$

51. $a = 0, b = 4$

Midpoint: 2

Distance between midpoint and each endpoint: 2

$$|x - 2| > 2$$

52. $a = 20, b = 24$

Midpoint: 22

Distance between midpoint and each endpoint: 2

$|x - 22| \ge 2$

53. (a) All numbers that are at most 10 units from 12

$|x - 12| \le 10$

(b) All numbers that are at least 10 units from 12

$|x - 12| \ge 10$

54. (a) y is at most 2 units from a: $|y - a| \le 2$

(b) y is less than δ units from c: $|y - c| < \delta$

55. $a = -1, b = 3$

Midpoint: $\dfrac{-1 + 3}{2} = 1$

56. $a = -5, b = -\dfrac{3}{2}$

Midpoint: $\dfrac{-5 + (-3/2)}{2} = -\dfrac{13}{4}$

57. (a) $[7, 21]$

Midpoint: 14

(b) $[8.6, 11.4]$

Midpoint: 10

58. (a) $[-6.85, 9.35]$

Midpoint: 1.25

(b) $[-4.6, -1.3]$

Midpoint: −2.95

59. $R = 115.95x, C = 95x + 750, R > C$

$115.95x > 95x + 750$

$20.95x > 750$

$x > 35.7995$

$x \ge 36$ units

60. $C = 0.32m + 2300, C < 10{,}000$

$0.32m + 2300 < 10{,}000$

$0.32m < 7700$

$m < 24{,}062.5$ miles

61. $\left|\dfrac{x - 50}{5}\right| \ge 1.645$

$\dfrac{x - 50}{5} \le -1.645$ or $\dfrac{x - 50}{5} \ge 1.645$

$x - 50 \le -8.225 \qquad x - 50 \ge 8.225$

$x \le 41.775 \qquad x \ge 58.225$

$x \le 41 \qquad x \ge 59$

62. $|p - 2{,}250{,}000| < 125{,}000$

$-125{,}000 < p - 2{,}250{,}000 < 125{,}000$

$2{,}125{,}000 < p < 2{,}375{,}000$

High = 2,375,000 barrels

Low = 2,125,000 barrels

63. (a) $\pi \approx 3.1415926535$

$\frac{355}{113} \approx 3.141592920$

$\frac{355}{113} > \pi$

(b) $\pi \approx 3.1415926535$

$\frac{22}{7} \approx 3.142857143$

$\frac{22}{7} > \pi$

64. (a) $\frac{224}{151} \approx 1.483443709$

$\frac{144}{97} \approx 1.484536082$

$\frac{144}{97} > \frac{224}{151}$

(b) $\frac{73}{81} \approx 0.901234568$

$\frac{6427}{7132} \approx 0.901149748$

$\frac{73}{81} > \frac{6427}{7132}$

65. Speed of light: 2.998×10^8 meters per second

Distance traveled in one year = rate × time

$d = (2.998 \times 10^8) \times (365 \times 24 \times 60 \times 60)$

days × hours × minutes × seconds

$= (2.998 \times 10^8) \times (3.1536 \times 10^7) \approx 9.45 \times 10^{15}$

This is best estimated by (b).

66. The significant digits of a number are the digits of the number beginning with the first nonzero digit to the left of the decimal point (or the first digit to the right of the decimal point if there isn't a nonzero digit to the left of the decimal point) and ending with the last digit to the right. The following examples all have three significant digits.

100, 307, 0.123, 0.012, 0.001, 1.23, 12.3, 0.120, 0.300

67. False; 2 is a nonzero integer and the reciprocal of 2 is $\frac{1}{2}$.

68. True; if $x(x \neq 0)$ is rational, then $x = p/q$ where p and q are nonzero integers. The reciprocal of x is $1/x = q/p$ which is also the ratio of two integers.

69. True

70. False; $|0| = 0$ which is not positive.

71. True; if $x < 0$, then $|x| = -x = \sqrt{x^2}$.

72. True; because a and b are **distinct**, $a \neq b$ and one of the numbers must be larger than the other one.

73. If $a \geq 0$ and $b \geq 0$, then $|ab| = ab = |a|\,|b|$.

If $a < 0$ and $b < 0$, then $|ab| = ab = (-a)(-b) = |a|\,|b|$.

If $a \geq 0$ and $b < 0$, then $|ab| = -ab = a(-b) = |a|\,|b|$.

If $a < 0$ and $b \geq 0$, then $|ab| = -ab = (-a)b = |a|\,|b|$.

74. $|a - b| = |(-1)(b - a)| = |-1|\,|b - a| = (1)|b - a| = |b - a|$

75. $\left|\dfrac{a}{b}\right| = \left|a\left(\dfrac{1}{b}\right)\right| = |a|\left|\dfrac{1}{b}\right| = |a| \cdot \dfrac{1}{|b|} = \dfrac{|a|}{|b|}, b \neq 0$

76. If $a \geq 0$, then $|a| = a = \sqrt{a^2}$.

If $a < 0$, then $|a| = -a = \sqrt{(-a)^2} = \sqrt{a^2}$.

77. $n = 1$, $\quad |a| = |a|$

$n = 2$, $\quad |a^2| = |a \cdot a| = |a|\,|a| = |a|^2$

$n = 3$, $\quad |a^3| = |a^2 \cdot a| = |a^2|\,|a| = |a|^2|a| = |a|^3$

$\vdots$

$|a^n| = |a^{n-1}a| = |a^{n-1}|\,|a| = |a|^{n-1}|a| = |a|^n$

78. If $a \geq 0$, then $a = |a|$. So, $-|a| \leq a \leq |a|$.

If $a < 0$, then $a = -|a|$. So, $-|a| \leq a \leq |a|$.

79. $|a| \leq k \Leftrightarrow \sqrt{a^2} \leq k \Leftrightarrow a^2 \leq k^2 \Leftrightarrow a^2 - k^2 \leq 0 \Leftrightarrow (a + k)(a - k) \leq 0 \Leftrightarrow -k \leq a \leq k,\ \ k > 0$

80. $k \leq |a| \Leftrightarrow k \leq \sqrt{a^2} \Leftrightarrow k^2 \leq a^2 \Leftrightarrow 0 \leq a^2 - k^2 \Leftrightarrow 0 \leq (a + k)(a - k) \Leftrightarrow k \leq a$ or $a \leq -k,\ \ k > 0$

81. $\left.\begin{array}{l}|7 - 12| = |-5| = 5\\ |7| - |12| = 7 - 12 = -5\end{array}\right\} |7 - 12| > |7| - |12|$

$\left.\begin{array}{l}|12 - 7| = |5| = 5\\ |12| - |7| = 12 - 7 = 5\end{array}\right\} |12 - 7| = |12| - |7|$

You know that $|a||b| \geq ab$. So, $-2|a||b| \leq -2ab$. Because $a^2 = |a|^2$ and $b^2 = |b|^2$, you have

$$\begin{aligned}|a|^2 + |b|^2 - 2|a||b| &\leq a^2 + b^2 - 2ab\\ 0 \leq (|a| - |b|)^2 &\leq (a - b)^2\\ \sqrt{(|a| - |b|)^2} &\leq \sqrt{(a - b)^2}\\ ||a| - |b|| &\leq |a - b|.\end{aligned}$$

Because $|a| - |b| \leq ||a| - |b||$, you have $|a| - |b| \leq |a - b|$. So, $|a - b| \geq |a| - |b|$.

82. $$\begin{aligned}\frac{1}{2}(a + b + |a - b|) &= \frac{1}{2}(a + b) + \frac{1}{2}|a - b|\\ &= \frac{a + b}{2} + \frac{1}{2}|a - b|\\ &= \text{Midpoint} + \frac{1}{2}\text{ the distance between } a \text{ and } b\\ &= \max(a, b)\end{aligned}$$

$$\begin{aligned}\min(a, b) &= \text{Midpoint} - \frac{1}{2}\text{ the distance between } a \text{ and } b\\ &= \frac{a + b}{2} - \frac{1}{2}|a - b|\\ &= \frac{1}{2}(a + b - |a - b|)\end{aligned}$$

Appendix C.2

1. (a)

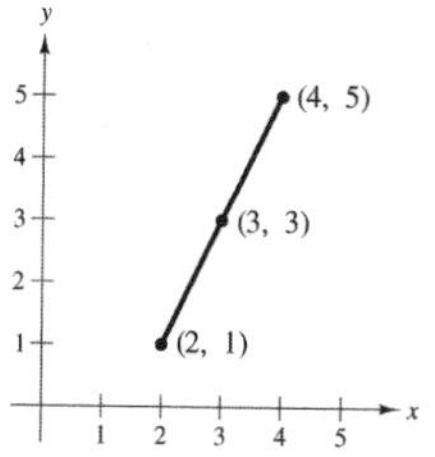

(b) $d = \sqrt{(4 - 2)^2 + (5 - 1)^2} = \sqrt{4 + 16} = \sqrt{20} = 2\sqrt{5}$

(c) Midpoint: $\left(\frac{4 + 2}{2}, \frac{5 + 1}{2}\right) = (3, 3)$

2. (a)

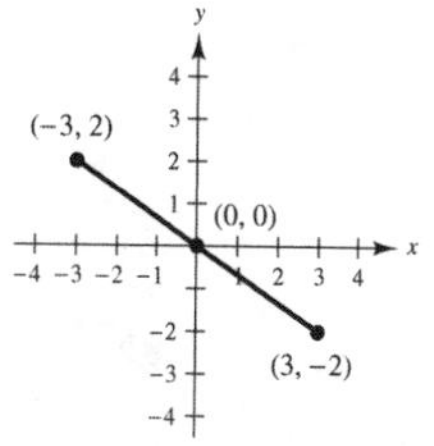

(b) $d = \sqrt{(3 + 3)^2 + (-2 - 2)^2} = \sqrt{36 + 16} = \sqrt{52} = 2\sqrt{13}$

(c) Midpoint: $\left(\frac{-3 + 3}{2}, \frac{2 + (-2)}{2}\right) = (0, 0)$

3. (a)

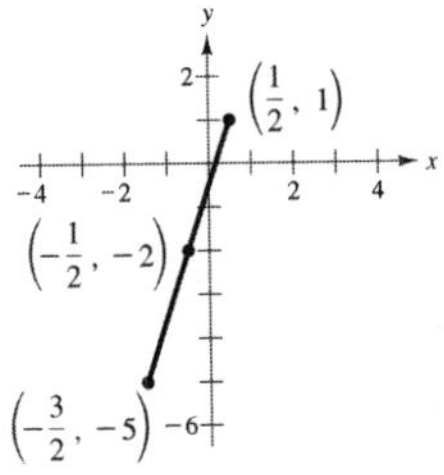

(b) $d = \sqrt{\left(\frac{1}{2} + \frac{3}{2}\right)^2 + (1 + 5)^2}$

$= \sqrt{4 + 36} = \sqrt{40} = 2\sqrt{10}$

(c) Midpoint: $\left(\frac{(-3/2) + (1/2)}{2}, \frac{-5 + 1}{2}\right) = \left(-\frac{1}{2}, -2\right)$

4. (a)

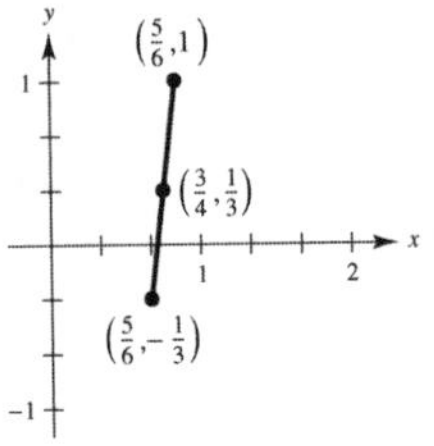

(b) $d = \sqrt{\left(\frac{5}{6} - \frac{4}{6}\right)^2 + \left(\frac{3}{3} + \frac{1}{3}\right)^2}$

$= \sqrt{\frac{1}{36} + \frac{64}{36}} = \frac{\sqrt{65}}{6}$

(c) Midpoint: $\left(\frac{(2/3) + (5/6)}{2}, \frac{(-1/3) + 1}{2}\right) = \left(\frac{3}{4}, \frac{1}{3}\right)$

5. (a)

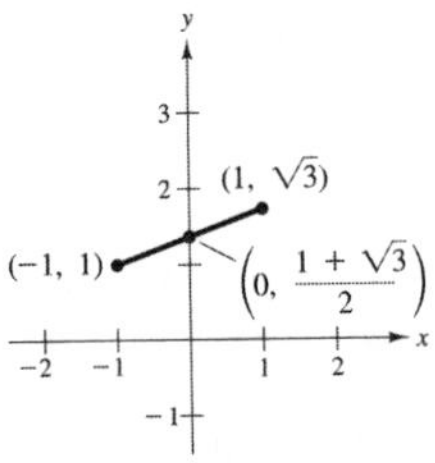

(b) $d = \sqrt{(-1 - 1)^2 + \left(1 - \sqrt{3}\right)^2}$

$= \sqrt{4 + 1 - 2\sqrt{3} + 3} = \sqrt{8 - 2\sqrt{3}}$

(c) Midpoint: $\left(\frac{-1 + 1}{2}, \frac{1 + \sqrt{3}}{2}\right) = \left(0, \frac{1 + \sqrt{3}}{2}\right)$

6. (a)

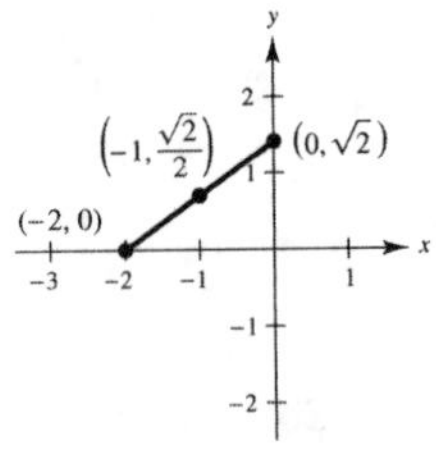

(b) $d = \sqrt{(-2 + 0)^2 + \left(0 - \sqrt{2}\right)^2}$

$= \sqrt{4 + 2} = \sqrt{6}$

(c) Midpoint: $\left(\frac{-2 + 0}{2}, \frac{0 + \sqrt{2}}{2}\right) = \left(-1, \frac{\sqrt{2}}{2}\right)$

7. $x = -2 \Rightarrow$ quadrants II, III

$y > 0 \Rightarrow$ quadrants I, II

Therefore, quadrant II

8. $y < -2 \Rightarrow$ quadrant III or IV

9. $xy > 0 \Rightarrow$ quadrants I or III

10. $(x, -y)$ in quadrant II $\Rightarrow$ (x, y) in quadrant III

11. $d_1 = \sqrt{9 + 36} = \sqrt{45}$

$d_2 = \sqrt{4 + 1} = \sqrt{5}$

$d_3 = \sqrt{25 + 25} = \sqrt{50}$

$(d_1)^2 + (d_2)^2 = (d_3)^2$

Right triangle

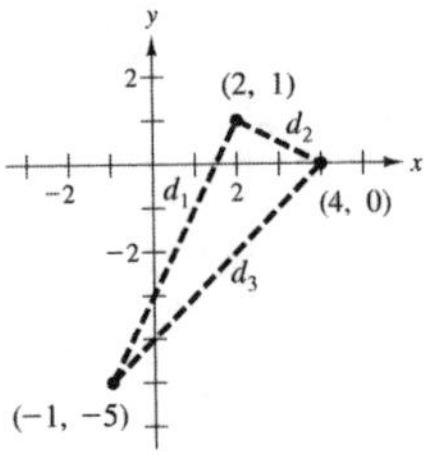

12. $d_1 = \sqrt{9 + 49} = \sqrt{58}$

$d_2 = \sqrt{25 + 4} = \sqrt{29}$

$d_3 = \sqrt{4 + 25} = \sqrt{29}$

$d_2 = d_3$

Isosceles triangle

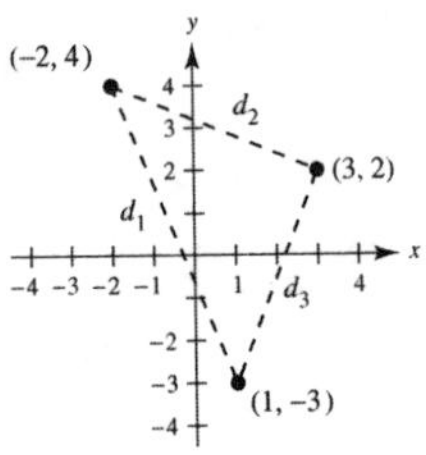

13. $d_1 = d_2 = d_3 = d_4 = \sqrt{5}$

Rhombus

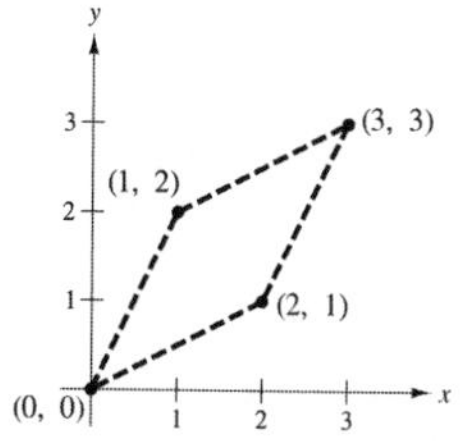

14. $d_1 = \sqrt{9 + 36} = \sqrt{45} = d_3$

$d_2 = \sqrt{1 + 9} = \sqrt{10} = d_4$

Parallelogram

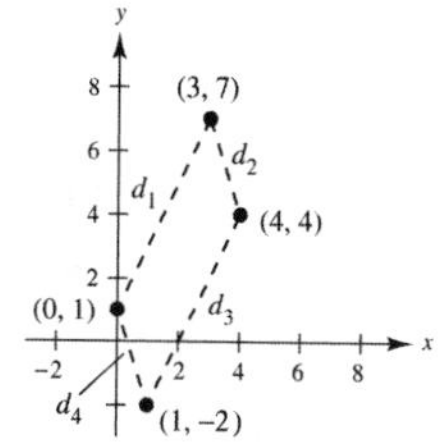

15.

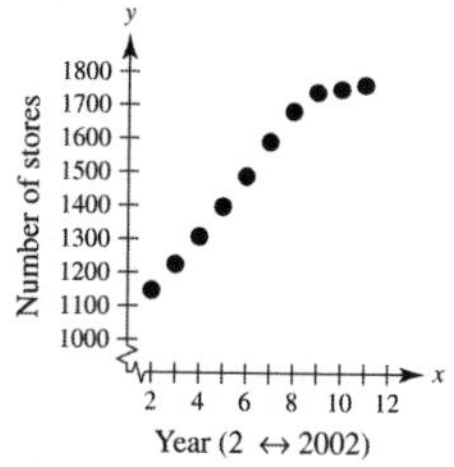

16.

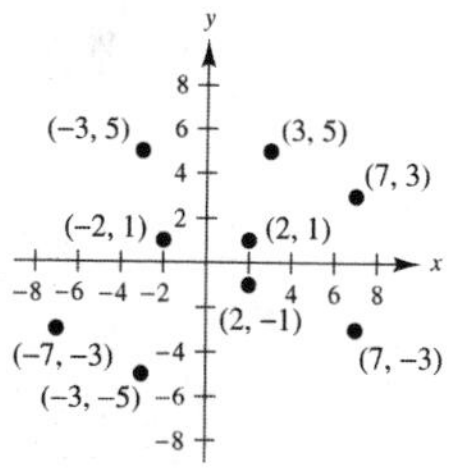

The new point $(-x, y)$ is located in a position symmetrical about the y-axis. Similarly, changing (x, y) to $(x, -y)$ moves the point to a position symmetrical about the x-axis.

17. $d_1 = \sqrt{4 + 16} = \sqrt{20} = 2\sqrt{5}$

$d_2 = \sqrt{1 + 4} = \sqrt{5}$

$d_3 = \sqrt{9 + 36} = 3\sqrt{5}$

$d_1 + d_2 = d_3$

Collinear

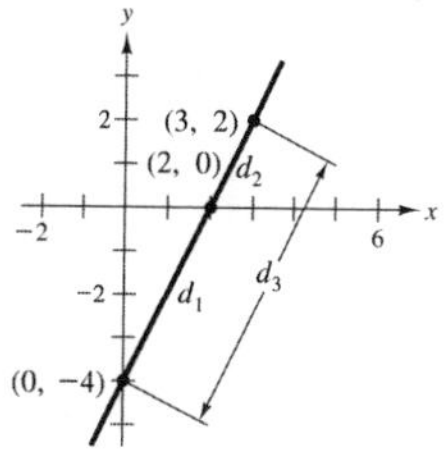

18. $d_1 = \sqrt{49 + 100} = \sqrt{149} \approx 12.2066$

$d_2 = \sqrt{25 + 49} = \sqrt{74} \approx 8.6023$

$d_3 = \sqrt{144 + 289} = \sqrt{433} \approx 20.8087$

$d_1 + d_2 \neq d_3$

Not collinear

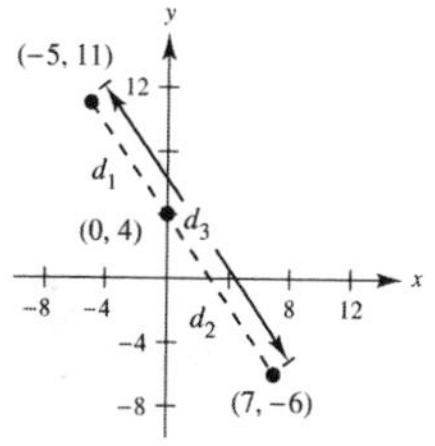

19. $d_1 = \sqrt{1 + 1} = \sqrt{2}$

$d_2 = \sqrt{9 + 4} = \sqrt{13}$

$d_3 = \sqrt{16 + 9} = 5$

$d_1 + d_2 \neq d_3$

Not collinear

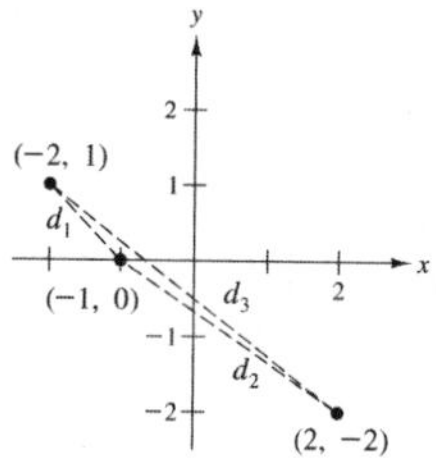

20. $d_1 = \sqrt{16+4} = \sqrt{20} = 2\sqrt{5}$

$d_2 = \sqrt{4+4} = \sqrt{8} = 2\sqrt{2}$

$d_3 = \sqrt{36+16} = \sqrt{52} = 2\sqrt{13}$

$d_1 + d_2 \neq d_3$

Not collinear

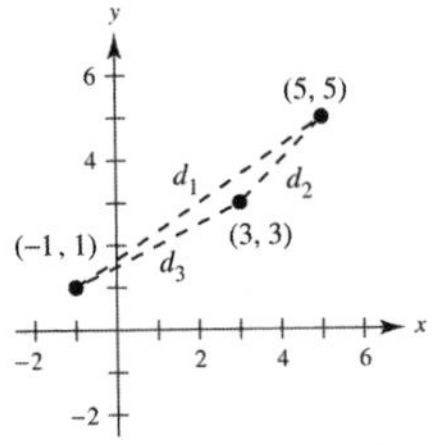

21. $5 = \sqrt{(x-0)^2 + (-4-0)^2}$

$5 = \sqrt{x^2 + 16}$

$25 = x^2 + 16$

$9 = x^2$

$x = \pm 3$

22. $5 = \sqrt{(x-2)^2 + (2+1)^2}$

$5 = \sqrt{(x-2)^2 + 9}$

$25 = (x-2)^2 + 9$

$16 = (x-2)^2$

$\pm 4 = x - 2$

$x = 2 \pm 4 = -2, 6$

23. $8 = \sqrt{(3-0)^2 + (y-0)^2}$

$8 = \sqrt{9 + y^2}$

$64 = 9 + y^2$

$55 = y^2$

$y = \pm\sqrt{55}$

24. $8 = \sqrt{(5-5)^2 + (y-1)^2}$

$8 = \sqrt{(y-1)^2}$

$8 = |y-1|$

$y - 1 = 8$ or $y - 1 = -8$

$y = 9$ $\qquad y = -7$

25. The midpoint of the given line segment is $\left(\frac{x_1+x_2}{2}, \frac{y_1+y_2}{2}\right)$.

The midpoint between (x_1, y_1) and $\left(\frac{x_1+x_2}{2}, \frac{y_1+y_2}{2}\right)$ is $\left(\frac{x_1+(x_1+x_2)/2}{2}, \frac{y_1+(y_1+y_2)/2}{2}\right) = \left(\frac{3x_1+x_2}{4}, \frac{3y_1+y_2}{4}\right)$.

The midpoint between $\left(\frac{x_1+x_2}{2}, \frac{y_1+y_2}{2}\right)$ and (x_2, y_2) is $\left(\frac{(x_1+x_2)/2+x_2}{2}, \frac{(y_1+y_2)/2+y_2}{2}\right) = \left(\frac{x_1+3x_2}{4}, \frac{y_1+3y_2}{4}\right)$.

Thus, the three points are $\left(\frac{3x_1+x_2}{4}, \frac{3y_1+y_2}{4}\right)$, $\left(\frac{x_1+x_2}{2}, \frac{y_1+y_2}{2}\right)$, $\left(\frac{x_1+3x_2}{4}, \frac{y_1+3y_2}{4}\right)$.

26. (a) $\left(\frac{3(1)+4}{4}, \frac{3(-2)+(-1)}{4}\right) = \left(\frac{7}{4}, -\frac{7}{4}\right)$

$\left(\frac{1+4}{2}, \frac{-2+(-1)}{2}\right) = \left(\frac{5}{2}, -\frac{3}{2}\right)$

$\left(\frac{1+3(4)}{4}, \frac{-2+3(-1)}{4}\right) = \left(\frac{13}{4}, -\frac{5}{4}\right)$

(b) $\left(\frac{3(-2)+0}{4}, \frac{3(-3)+0}{4}\right) = \left(-\frac{3}{2}, -\frac{9}{4}\right)$

$\left(\frac{-2+0}{2}, \frac{-3+0}{2}\right) = \left(-1, -\frac{3}{2}\right)$

$\left(\frac{-2+3(0)}{4}, \frac{-3+3(0)}{4}\right) = \left(-\frac{1}{2}, -\frac{3}{4}\right)$

27. Center: $(0, 0)$

Radius: 1

Matches graph (c)

28. Center: $(1, 3)$

Radius: 2

Matches graph (b)

29. Center: $(1, 0)$

Radius: 0

Matches graph (a)

30. Center: $\left(-\frac{1}{2}, \frac{3}{4}\right)$

Radius: $\frac{1}{2}$

Matches graph (d)

31. $(x - 0)^2 + (y - 0)^2 = (3)^2$

$$x^2 + y^2 - 9 = 0$$

32. $(x - 0)^2 + (y - 0)^2 = (5)^2$

$$x^2 + y^2 - 25 = 0$$

33. $(x - 2)^2 + (y + 1)^2 = (4)^2$

$$x^2 + y^2 - 4x + 2y - 11 = 0$$

34.

$$(x + 4)^2 + (y - 3)^2 = \left(\tfrac{5}{8}\right)^2$$

$$64(x + 4)^2 + 64(y - 3)^2 = 25$$

$$64x^2 + 64y^2 + 512x - 384y + 1575 = 0$$

35. Radius $= \sqrt{(-1 - 0)^2 + (2 - 0)^2} = \sqrt{5}$

$$(x + 1)^2 + (y - 2)^2 = 5$$

$$x^2 + 2x + 1 + y^2 - 4y + 4 = 5$$

$$x^2 + y^2 + 2x - 4y = 0$$

36. Radius $= \sqrt{[3 - (-1)]^2 + (-2 - 1)^2} = 5$

$$(x - 3)^2 + (y + 2)^2 = 25$$

$$x^2 - 6x + 9 + y^2 + 4y + 4 = 25$$

$$x^2 + y^2 - 6x + 4y - 12 = 0$$

37. Center $=$ Midpoint $= (3, 2)$

Radius $= \sqrt{10}$

$$(x - 3)^2 + (y - 2)^2 = \left(\sqrt{10}\right)^2$$

$$x^2 - 6x + 9 + y^2 - 4y + 4 = 10$$

$$x^2 + y^2 - 6x - 4y + 3 = 0$$

38. Center $=$ Midpoint $= (0, 0)$

Radius $= \sqrt{2}$

$$(x - 0)^2 + (y - 0)^2 = \left(\sqrt{2}\right)^2$$

$$x^2 + y^2 - 2 = 0$$

39. Place the center of Earth at the origin. Then you have

$$x^2 + y^2 = (22{,}000 + 4000)^2$$

$$x^2 + y^2 = 26{,}000^2.$$

40. Let d be the diameter of the water pipe and z be the distance between the water pipe and the corner of the wall. If you let y equal the hypotenuse of the triangle whose one vertex is located at the center of the air duct, then $y = z + d + (D/2)$. The hypotenuse of the triangle whose one vertex is located at the center of the water pipe is $z + (d/2)$. Using the Pythagorean Theorem, you can find z as follows.

$$\left(z + \frac{d}{2}\right)^2 = \left(\frac{d}{2}\right)^2 + \left(\frac{d}{2}\right)^2$$

$$\left(z + \frac{d}{2}\right)^2 = \frac{d^2}{2}$$

$$z + \frac{d}{2} = \frac{d}{\sqrt{2}}$$

$$z = \frac{d}{\sqrt{2}} - \frac{d}{2}$$

Now solve for d, using the fact that these are similar triangles.

$$\frac{\frac{d}{2}}{z + \frac{d}{2}} = \frac{\frac{D}{2}}{y}$$

$$\frac{\frac{d}{2}}{\frac{d}{\sqrt{2}} - \frac{d}{2} + \frac{d}{2}} = \frac{\frac{D}{2}}{z + d + \frac{D}{2}}$$

$$\frac{\frac{d}{2}}{\frac{d}{\sqrt{2}}} = \frac{\frac{D}{2}}{\frac{d}{\sqrt{2}} - \frac{d}{2} + d + \frac{D}{2}}$$

$$\frac{d}{2}\left(\frac{d}{\sqrt{2}} + \frac{d}{2} + \frac{D}{2}\right) = \frac{d}{\sqrt{2}} \cdot \frac{D}{2}$$

$$d\left(\frac{1}{\sqrt{2}} + \frac{1}{2}\right) + \frac{D}{2} = \frac{D}{\sqrt{2}}$$

$$d\left(\frac{2 + \sqrt{2}}{2\sqrt{2}}\right) = \frac{D}{\sqrt{2}} - \frac{D}{2}$$

$$d\left(\frac{2 + \sqrt{2}}{2\sqrt{2}}\right) = D\left(\frac{2 - \sqrt{2}}{2\sqrt{2}}\right)$$

$$d = D\left(\frac{2 - \sqrt{2}}{2 + \sqrt{2}}\right)$$

The diameter of the largest water pipe which can be run in the right angle corner behind the air duct is

$$D\left(\frac{2 - \sqrt{2}}{2 + \sqrt{2}}\right).$$

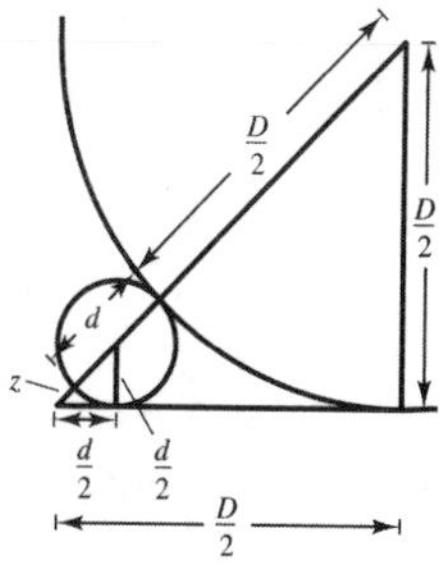

41. $x^2 + y^2 - 2x + 6y + 6 = 0$

$$(x^2 - 2x + 1) + (y^2 + 6y + 9) = -6 + 1 + 9$$

$$(x - 1)^2 + (y + 3)^2 = 4$$

Center: $(1, -3)$

Radius: 2

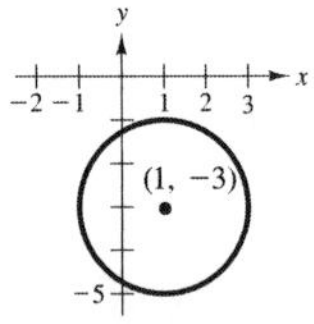

42. $x^2 + y^2 - 2x + 6y - 15 = 0$

$$(x^2 - 2x + 1) + (y^2 + 6y + 9) = 15 + 1 + 9$$

$$(x - 1)^2 + (y + 3)^2 = 25$$

Center: $(1, -3)$

Radius: 5

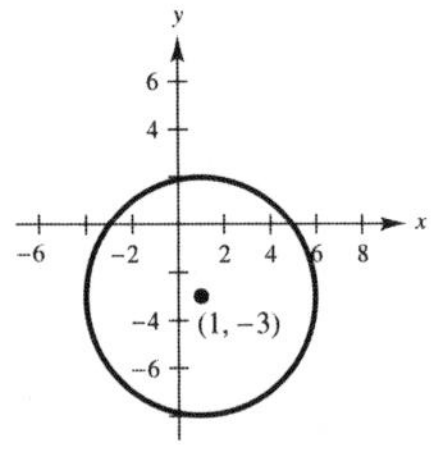

43. $x^2 + y^2 - 2x + 6y + 10 = 0$

$$(x^2 - 2x + 1) + (y^2 + 6y + 9) = -10 + 1 + 9$$

$$(x - 1)^2 + (y + 3)^2 = 0$$

Only a point $(1, -3)$

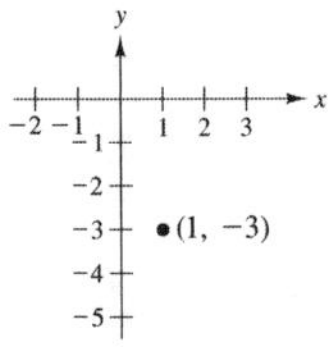

44. $3x^2 + 3y^2 - 6y - 1 = 0$

$$3x^2 + 3(y^2 - 2y + 1) = 1 + 3$$

$$x^2 + (y - 1)^2 = \frac{4}{3}$$

Center: $(0, 1)$

Radius: $\dfrac{2\sqrt{3}}{3}$

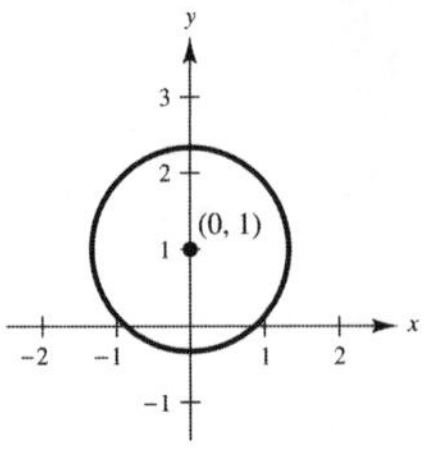

45. $2x^2 + 2y^2 - 2x - 2y - 3 = 0$

$$2\left(x^2 - x + \tfrac{1}{4}\right) + 2\left(y^2 - y + \tfrac{1}{4}\right) = 3 + \tfrac{1}{2} + \tfrac{1}{2}$$

$$\left(x - \tfrac{1}{2}\right)^2 + \left(y - \tfrac{1}{2}\right)^2 = 2$$

Center: $\left(\tfrac{1}{2}, \tfrac{1}{2}\right)$

Radius: $\sqrt{2}$

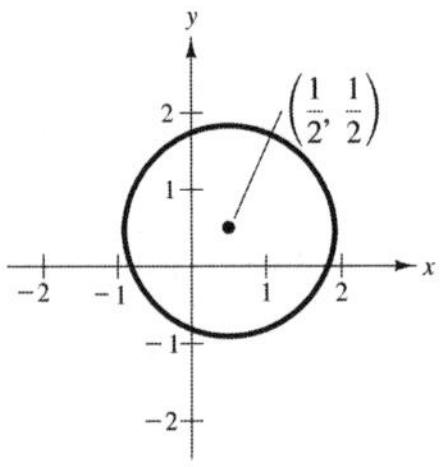

46. $4x^2 + 4y^2 - 4x + 2y - 1 = 0$

$$4\left(x^2 - x + \frac{1}{4}\right) + 4\left(y^2 + \frac{y}{2} + \frac{1}{16}\right) = 1 + 1 + \frac{1}{4}$$

$$4\left(x - \frac{1}{2}\right)^2 + 4\left(y + \frac{1}{4}\right)^2 = \frac{9}{4}$$

$$\left(x - \frac{1}{2}\right)^2 + \left(y + \frac{1}{4}\right)^2 = \frac{9}{16}$$

Center: $\left(\dfrac{1}{2}, -\dfrac{1}{4}\right)$

Radius: $\dfrac{3}{4}$

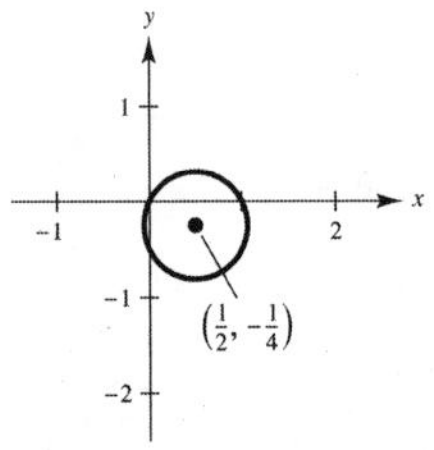

47. $$16x^2 + 16y^2 + 16x + 40y - 7 = 0$$

$$16\left(x^2 + x + \frac{1}{4}\right) + 16\left(y^2 + \frac{5y}{2} + \frac{25}{16}\right) = 7 + 4 + 25$$

$$16\left(x + \frac{1}{2}\right)^2 + 16\left(y + \frac{5}{4}\right)^2 = 36$$

$$\left(x + \frac{1}{2}\right)^2 + \left(y + \frac{5}{4}\right)^2 = \frac{9}{4}$$

Center: $\left(-\frac{1}{2}, -\frac{5}{4}\right)$

Radius: $\frac{3}{2}$

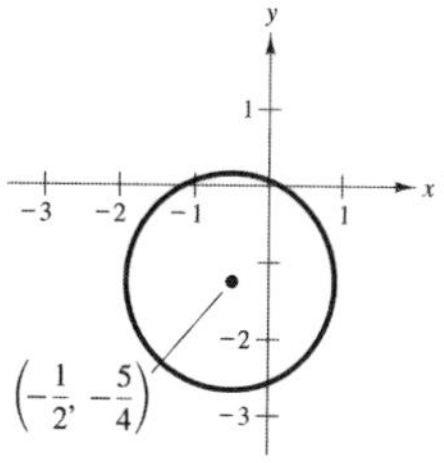

48. $$x^2 + y^2 - 4x + 2y + 3 = 0$$

$$\left(x^2 - 4x + 4\right) + \left(y^2 + 2y + 1\right) = -3 + 4 + 1$$

$$(x - 2)^2 + (y + 1)^2 = 2$$

Center: $(2, -1)$

Radius: $\sqrt{2}$

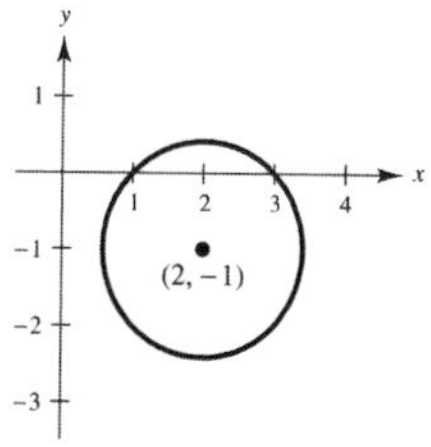

49. $$4x^2 + 4y^2 - 4x + 24y - 63 = 0$$

$$x^2 + y^2 - x + 6y = \frac{63}{4}$$

$$\left(x^2 - x + \frac{1}{4}\right) + \left(y^2 + 6y + 9\right) = \frac{63}{4} + \frac{1}{4} + 9$$

$$\left(x - \frac{1}{2}\right)^2 + (y + 3)^2 = 25$$

$$(y + 3)^2 = 25 - \left(x - \frac{1}{2}\right)^2$$

$$y + 3 = \pm\sqrt{25 - \left(x - \frac{1}{2}\right)^2}$$

$$y = -3 \pm \sqrt{25 - \left(x - \frac{1}{2}\right)^2}$$

$$= \frac{-6 \pm \sqrt{99 + 4x - 4x^2}}{2}$$

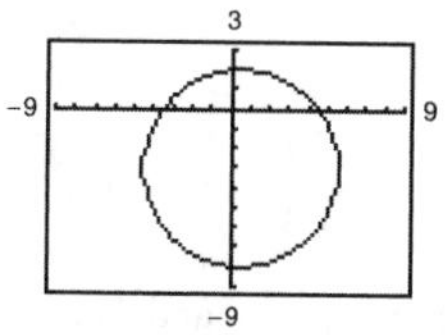

50. $$x^2 + y^2 - 8x - 6y - 11 = 0$$

$$\left(x^2 - 8x + 16\right) + \left(y^2 - 6y + 9\right) = 11 + 16 + 9$$

$$(x - 4)^2 + (y - 3)^2 = 36$$

$$(y - 3)^2 = 36 - (x - 4)^2$$

$$y - 3 = \pm\sqrt{36 - (x - 4)^2}$$

$$y = 3 \pm \sqrt{20 + 8x - x^2}$$

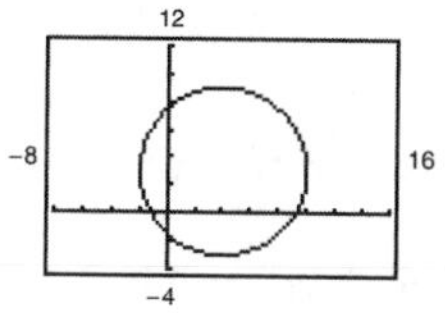

51. $$x^2 + y^2 - 4x + 2y + 1 \le 0$$
$$\left(x^2 - 4x + 4\right) + \left(y^2 + 2y + 1\right) \le -1 + 4 + 1$$
$$(x - 2)^2 + (y + 1)^2 \le 4$$

Center: $(2, -1)$

Radius: 2

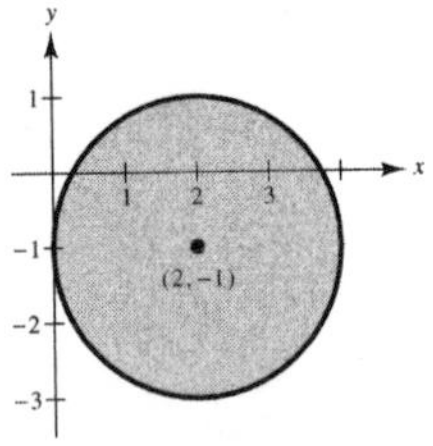

52. $(x - 1)^2 + \left(y - \frac{1}{2}\right)^2 > 1$

Center: $\left(1, \frac{1}{2}\right)$

Radius: 1

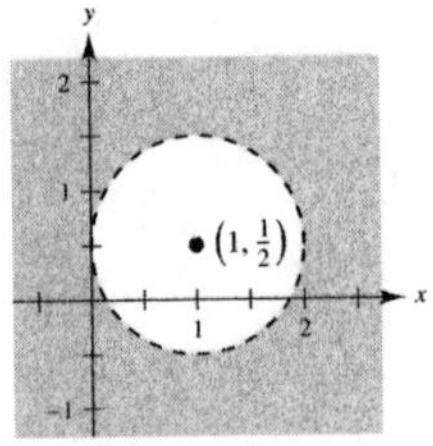

53. The distance between (x_1, y_1) and $\left(\frac{2x_1 + x_2}{3}, \frac{2y_1 + y_2}{3}\right)$ is

$$d = \sqrt{\left(x_1 - \frac{2x_1 + x_2}{3}\right)^2 + \left(y_1 - \frac{2y_1 + y_2}{3}\right)^2}$$
$$= \sqrt{\left(\frac{x_1 - x_2}{3}\right)^2 + \left(\frac{y_1 - y_2}{3}\right)^2}$$
$$= \sqrt{\frac{1}{9}\left[(x_1 - x_2)^2 + (y_1 - y_2)^2\right]} = \frac{1}{3}\sqrt{(x_1 - x_2)^2 + (y_1 - y_2)^2}$$

which is $\frac{1}{3}$ of the distance between (x_1, y_1) and (x_2, y_2).

$$\left(\frac{\left(\frac{2x_1 + x_2}{3}\right) + x_2}{2}, \frac{\left(\frac{2y_1 + y_2}{3}\right) + y_2}{2}\right) = \left(\frac{x_1 + 2x_2}{3}, \frac{y_1 + 2y_2}{3}\right)$$

is the second point of the trisection.

54. (a) $\left(\frac{2(1) + 4}{3}, \frac{2(-2) + 1}{3}\right) = (2, -1)$

$\left(\frac{1 + 2(4)}{3}, \frac{-2 + 2(1)}{3}\right) = (3, 0)$

(b) $\left(\frac{2(-2) + 0}{3}, \frac{2(-3) + 0}{3}\right) = \left(-\frac{4}{3}, -2\right)$

$\left(\frac{-2 + 2(0)}{3}, \frac{-3 + 2(0)}{3}\right) = \left(-\frac{2}{3}, -1\right)$

55. True; if $ab < 0$ then either a is positive and b is negative (Quadrant IV) or a is negative and b is positive (Quadrant II).

56. False

$$d = \sqrt{\left[(a + b) - (a - b)\right]^2 + (a - a)^2}$$
$$= \sqrt{(2b)^2 + 0^2} = \sqrt{4b^2} = 2|b|$$

57. True

58. True; if $ab = 0$ then $a = 0$ (y-axis) or $b = 0$ (x-axis).

59. Let one vertex be at $(0, 0)$ and another at $(a, 0)$.

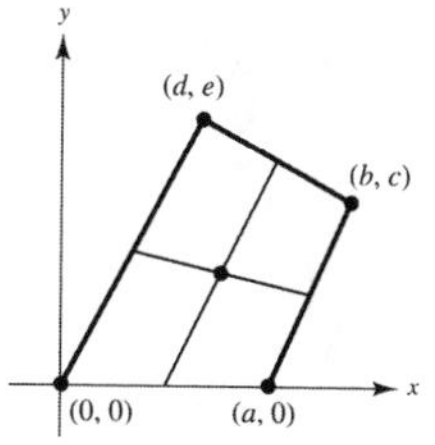

Midpoint of $(0, 0)$ and (d, e) is $\left(\frac{d}{2}, \frac{e}{2}\right)$.

Midpoint of (b, c) and $(a, 0)$ is $\left(\frac{a+b}{2}, \frac{c}{2}\right)$.

Midpoint of $(0, 0)$ and $(a, 0)$ is $\left(\frac{a}{2}, 0\right)$.

Midpoint of (b, c) and (d, e) is $\left(\frac{b+d}{2}, \frac{c+e}{2}\right)$.

Midpoint of line segment joining $\left(\frac{d}{2}, \frac{e}{2}\right)$ and $\left(\frac{a+b}{2}, \frac{c}{2}\right)$ is $\left(\frac{a+b+d}{4}, \frac{c+e}{4}\right)$.

Midpoint of line segment joining $\left(\frac{a}{2}, 0\right)$ and $\left(\frac{b+d}{2}, \frac{c+e}{2}\right)$ is $\left(\frac{a+b+d}{4}, \frac{c+e}{4}\right)$.

Therefore the line segments intersect at their midpoints.

60. Let the circle of radius r be centered at the origin. Let (a, b) and $(r, 0)$ be the endpoints of the chord. The midpoint M of the chord is $((a+r)/2, b/2)$. We will show that OM is perpendicular to MR by verifying that $d_1^2 + d_2^2 = d_3^2$.

$$d_1^2 = \left(\frac{a+r}{2} - 0\right)^2 + \left(\frac{b}{2} - 0\right)^2 = \left(\frac{a+r}{2}\right)^2 + \left(\frac{b}{2}\right)^2$$

$$d_2^2 = \left(\frac{a+r}{2} - r\right)^2 + \left(\frac{b}{2} - 0\right)^2 = \left(\frac{a-r}{2}\right)^2 + \left(\frac{b}{2}\right)^2$$

$$\begin{aligned} d_1^2 + d_2^2 &= \left(\frac{a^2 + 2ar + r^2}{4} + \frac{b^2}{4}\right) + \left(\frac{a^2 - 2ar + r^2}{4} + \frac{b^2}{4}\right) \\ &= \frac{a^2}{2} + \frac{r^2}{2} + \frac{b^2}{2} \\ &= \frac{1}{2}\left(a^2 + b^2\right) + \frac{1}{2}r^2 \\ &= \frac{1}{2}r^2 + \frac{1}{2}r^2 = r^2 = d_3^2 \end{aligned}$$

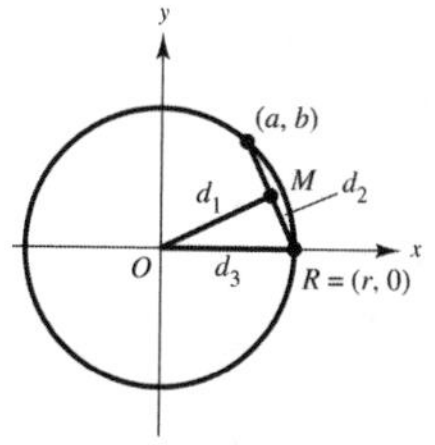

61. Let (a, b) be a point on the semicircle of radius r, centered at the origin. We will show that the angle at (a, b) is a right angle by verifying that $d_1^2 + d_2^2 = d_3^2$.

$$\begin{aligned} d_1^2 &= (a + r)^2 + (b - 0)^2 \\ d_2^2 &= (a - r)^2 + (b - 0)^2 \\ d_1^2 + d_2^2 &= \left(a^2 + 2ar + r^2 + b^2\right) + \left(a^2 - 2ar + r^2 + b^2\right) \\ &= 2a^2 + 2b^2 + 2r^2 \\ &= 2\left(a^2 + b^2\right) + 2r^2 \\ &= 2r^2 + 2r^2 \\ &= 4r^2 = (2r)^2 = d_3^2 \end{aligned}$$

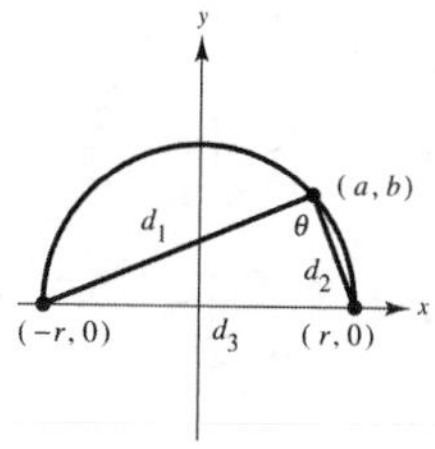

62. To show that $\left(\dfrac{x_1 + x_2}{2}, \dfrac{y_1 + y_2}{2}\right)$ is the midpoint of the line segment joining (x_1, y_1) and (x_2, y_2), we must show that $d_1 = d_2$ and $d_1 + d_2 = d_3$ (see graph).

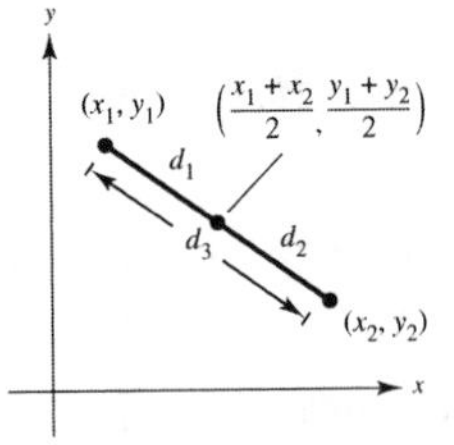

$$d_1 = \sqrt{\left(\frac{x_1 + x_2}{2} - x_1\right)^2 + \left(\frac{y_1 + y_2}{2} - y_1\right)^2}$$

$$= \sqrt{\left(\frac{x_2 - x_1}{2}\right)^2 + \left(\frac{y_2 - y_1}{2}\right)^2} = \frac{1}{2}\sqrt{(x_2 - x_1)^2 + (y_2 - y_1)^2}$$

$$d_2 = \sqrt{\left(x_2 - \frac{x_1 + x_2}{2}\right)^2 + \left(y_2 - \frac{y_1 - y_2}{2}\right)^2}$$

$$= \sqrt{\left(\frac{x_2 - x_1}{2}\right)^2 + \left(\frac{y_2 - y_1}{2}\right)^2} = \frac{1}{2}\sqrt{(x_2 - x_1)^2 + (y_2 - y_1)^2}$$

$$d_3 = \sqrt{(x_2 - x_1)^2 + (y_2 - y_1)^2}$$

Therefore, $d_1 = d_2$ and $d_1 + d_2 = d_3$.

Appendix C.3

1. (a) $396°, -324°$

(b) $240°, -480°$

2. (a) $660°, -60°$

(b) $300°, -60°$

3. (a) $\dfrac{19\pi}{9}, -\dfrac{17\pi}{9}$

(b) $\dfrac{10\pi}{3}, -\dfrac{2\pi}{3}$

4. (a) $\dfrac{7\pi}{4}, -\dfrac{\pi}{4}$

(b) $\dfrac{26\pi}{9}, -\dfrac{10\pi}{9}$

5. (a) $30\left(\dfrac{\pi}{180}\right) = \dfrac{\pi}{6} \approx 0.524$

(b) $150\left(\dfrac{\pi}{180}\right) = \dfrac{5\pi}{6} \approx 2.618$

(c) $315\left(\dfrac{\pi}{180}\right) = \dfrac{7\pi}{4} \approx 5.498$

(d) $120\left(\dfrac{\pi}{180}\right) = \dfrac{2\pi}{3} \approx 2.094$

6. (a) $-20\left(\dfrac{\pi}{180}\right) = -\dfrac{\pi}{9} \approx -0.349$

(b) $-240\left(\dfrac{\pi}{180}\right) = -\dfrac{4\pi}{3} \approx -4.189$

(c) $-270\left(\dfrac{\pi}{180}\right) = -\dfrac{3\pi}{2} \approx -4.712$

(d) $144\left(\dfrac{\pi}{180}\right) = -\dfrac{4\pi}{5} \approx 2.513$

7. (a) $\dfrac{3\pi}{2}\left(\dfrac{180}{\pi}\right) = 270°$

(b) $\dfrac{7\pi}{6}\left(\dfrac{180}{\pi}\right) = 210°$

(c) $-\dfrac{7\pi}{12}\left(\dfrac{180}{\pi}\right) = -105°$

(d) $-2.637\left(\dfrac{180}{\pi}\right) \approx -151.1°$

8. (a) $\dfrac{7\pi}{3}\left(\dfrac{180}{\pi}\right) = 420°$

(b) $-\dfrac{11\pi}{30}\left(\dfrac{180}{\pi}\right) = -66°$

(c) $\dfrac{11\pi}{6}\left(\dfrac{180}{\pi}\right) = 330°$

(d) $0.438\left(\dfrac{180}{\pi}\right) \approx 25.1°$

9.

r	8 ft	15 in.	85 cm	24 in.	$\dfrac{12{,}963}{\pi}$ mi
s	12 ft	24 in.	63.75π cm	96 in.	8642 mi
θ	1.5	1.6	$\dfrac{3\pi}{4}$	4	$\dfrac{2\pi}{3}$

10. (a) $50 \text{ mi/h} = \dfrac{50(5280)}{60} = 4400 \text{ ft/min}$

Circumference of tire: $C = 2.5\pi$ feet

Revolutions per minute: $\dfrac{4400}{2.5\pi} \approx 560.2$

(b) $\theta = \dfrac{4400}{2.5\pi}(2\pi) = 3520$ radians

Angular speed: $\dfrac{\theta}{t} = \dfrac{3520 \text{ radians}}{1 \text{ minute}} = 3520 \text{ rad/min}$

11. (a) $x = 3, y = 4, r = 5$

$\sin\theta = \frac{4}{5}$ $\quad\csc\theta = \frac{5}{4}$

$\cos\theta = \frac{3}{5}$ $\quad\sec\theta = \frac{5}{3}$

$\tan\theta = \frac{4}{3}$ $\quad\cot\theta = \frac{3}{4}$

(b) $x = -12, y = -5, r = 13$

$\sin\theta = -\frac{5}{13}$ $\quad\csc\theta = -\frac{13}{5}$

$\cos\theta = -\frac{12}{13}$ $\quad\sec\theta = -\frac{13}{12}$

$\tan\theta = \frac{5}{12}$ $\quad\cot\theta = \frac{12}{5}$

12. (a) $x = 8, y = -15, r = 17$

$\sin\theta = -\dfrac{15}{17}$ $\quad\csc\theta = -\dfrac{17}{15}$

$\cos\theta = \dfrac{8}{17}$ $\quad\sec\theta = \dfrac{17}{8}$

$\tan\theta = -\dfrac{15}{8}$ $\quad\cot\theta = -\dfrac{8}{15}$

(b) $x = 1, y = -1, r = \sqrt{2}$

$\sin\theta = -\dfrac{\sqrt{2}}{2}$ $\quad\csc\theta = -\sqrt{2}$

$\cos\theta = \dfrac{\sqrt{2}}{2}$ $\quad\sec\theta = \sqrt{2}$

$\tan\theta = -1$ $\quad\cot\theta = -1$

13. (a) $\sin\theta < 0 \Rightarrow \theta$ is in Quadrant III or IV.

$\cos\theta < 0 \Rightarrow \theta$ is in Quadrant II or III.

$\sin\theta < 0$ **and** $\cos\theta < 0 \Rightarrow \theta$ is in Quadrant III.

(b) $\sec\theta > 0 \Rightarrow \theta$ is in Quadrant I or IV.

$\cot\theta < 0 \Rightarrow \theta$ is in Quadrant II or IV.

$\sec\theta > 0$ **and** $\cot\theta < 0 \Rightarrow \theta$ is in Quadrant IV.

14. (a) $\sin\theta > 0 \Rightarrow \theta$ is in Quadrant I or II.

$\cos\theta < 0 \Rightarrow \theta$ is in Quadrant II or III.

$\sin\theta > 0$ **and** $\cos\theta < 0 \Rightarrow \theta$ is in Quadrant II.

(b) $\csc\theta < 0 \Rightarrow \theta$ is in Quadrant III or IV.

$\tan\theta > 0 \Rightarrow \theta$ is in Quadrant I or III.

$\csc\theta < 0$ **and** $\tan\theta > 0 \Rightarrow \theta$ is in Quadrant III.

15. $x^2 + 1^2 = 2^2 \Rightarrow x = \sqrt{3}$

$\cos\theta = \dfrac{x}{2} = \dfrac{\sqrt{3}}{2}$

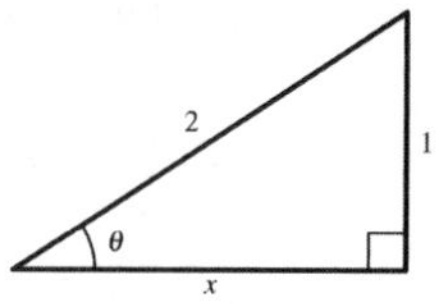

16. $x^2 + 1^2 = 3^2 \Rightarrow x = \sqrt{8} = 2\sqrt{2}$

$\tan\theta = \dfrac{1}{x} = \dfrac{1}{2\sqrt{2}} = \dfrac{\sqrt{2}}{4}$

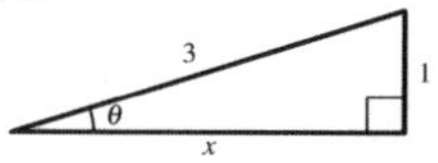

17. $4^2 + y^2 = 5^2 \Rightarrow y = 3$

$\cot\theta = \dfrac{4}{y} = \dfrac{4}{3}$

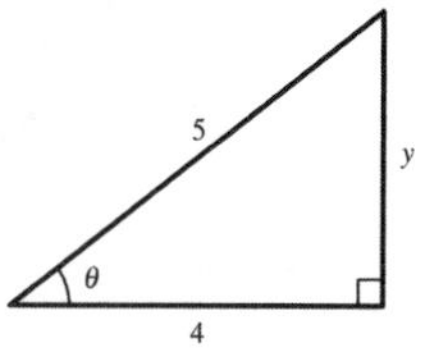

18. $5^2 + y^2 = 13^2 \Rightarrow y = 12$

$\csc\theta = \dfrac{13}{y} = \dfrac{13}{12}$

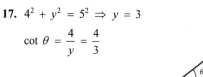

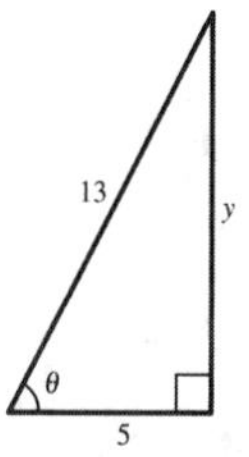

19. (a) $\sin 60° = \dfrac{\sqrt{3}}{2}$

$\cos 60° = \dfrac{1}{2}$

$\tan 60° = \sqrt{3}$

(b) $\sin 120° = \sin 60° = \dfrac{\sqrt{3}}{2}$

$\cos 120° = -\cos 60° = -\dfrac{1}{2}$

$\tan 120° = -\tan 60° = -\sqrt{3}$

(c) $\sin\dfrac{\pi}{4} = \dfrac{\sqrt{2}}{2}$

$\cos\dfrac{\pi}{4} = \dfrac{\sqrt{2}}{2}$

$\tan\dfrac{\pi}{4} = 1$

(d) $\sin\dfrac{5\pi}{4} - \sin\dfrac{\pi}{4} = -\dfrac{\sqrt{2}}{2}$

$\cos\dfrac{5\pi}{4} = \cos\dfrac{\pi}{4} = -\dfrac{\sqrt{2}}{2}$

$\tan\dfrac{5\pi}{4} = \tan\dfrac{\pi}{4} = 1$

20. (a) $\sin(-30°) = -\sin 30° = -\dfrac{1}{2}$

$\cos(-30°) = \cos 30° = \dfrac{\sqrt{3}}{2}$

$\tan(-30°) = -\tan 30° = -\dfrac{\sqrt{3}}{3}$

(b) $\sin 150° = \sin 30° = \dfrac{1}{2}$

$\cos 150° = -\cos 30° = -\dfrac{\sqrt{3}}{2}$

$\tan 150° = -\tan 30° = -\dfrac{\sqrt{3}}{3}$

(c) $\sin\left(-\dfrac{\pi}{6}\right) = -\sin\dfrac{\pi}{6} = -\dfrac{1}{2}$

$\cos\left(-\dfrac{\pi}{6}\right) = \cos\dfrac{\pi}{6} = \dfrac{\sqrt{3}}{2}$

$\tan\left(-\dfrac{\pi}{6}\right) = -\tan\dfrac{\pi}{6} = -\dfrac{\sqrt{3}}{3}$

(d) $\sin\dfrac{\pi}{2} = 1$

$\cos\dfrac{\pi}{2} = 0$

$\tan\dfrac{\pi}{2}$ is undefined.

21. (a) $\sin 225° = -\sin 45° = -\dfrac{\sqrt{2}}{2}$

$\cos 225° = -\cos 45° = -\dfrac{\sqrt{2}}{2}$

$\tan 225° = \tan 45° = 1$

(b) $\sin(-225°) = \sin 45° = \dfrac{\sqrt{2}}{2}$

$\cos(-225°) = -\cos 45° = -\dfrac{\sqrt{2}}{2}$

$\tan(-225°) = -\tan 45° = -1$

(c) $\sin\dfrac{5\pi}{3} = -\sin\dfrac{\pi}{3} = -\dfrac{\sqrt{3}}{2}$

$\cos\dfrac{5\pi}{3} = \cos\dfrac{\pi}{3} = \dfrac{1}{2}$

$\tan\dfrac{5\pi}{3} = -\tan\dfrac{\pi}{3} = -\sqrt{3}$

(d) $\sin\dfrac{11\pi}{6} = -\sin\dfrac{\pi}{6} = -\dfrac{1}{2}$

$\cos\dfrac{11\pi}{6} = \cos\dfrac{\pi}{6} = \dfrac{\sqrt{3}}{2}$

$\tan\dfrac{11\pi}{6} = -\tan\dfrac{\pi}{6} = -\dfrac{\sqrt{3}}{3}$

22. (a) $\sin 750° = \sin 30° = \dfrac{1}{2}$

$\cos 750° = \cos 30° = \dfrac{\sqrt{3}}{2}$

$\tan 750° = \tan 30° = \dfrac{\sqrt{3}}{3}$

(b) $\sin 510° = \sin 30° = \dfrac{1}{2}$

$\cos 510° = -\cos 30° = -\dfrac{\sqrt{3}}{2}$

$\tan 510° = -\tan 30° = -\dfrac{\sqrt{3}}{3}$

(c) $\sin\dfrac{10\pi}{3} = -\sin\dfrac{\pi}{3} = -\dfrac{\sqrt{3}}{2}$

$\cos\dfrac{10\pi}{3} = -\cos\dfrac{\pi}{3} = -\dfrac{1}{2}$

$\tan\dfrac{10\pi}{3} = \tan\dfrac{\pi}{3} = \sqrt{3}$

(d) $\sin\dfrac{17\pi}{3} = -\sin\dfrac{\pi}{3} = -\dfrac{\sqrt{3}}{2}$

$\cos\dfrac{17\pi}{3} = \cos\dfrac{\pi}{3} = \dfrac{1}{2}$

$\tan\dfrac{17\pi}{3} = -\tan\dfrac{\pi}{3} = \sqrt{3}$

23. (a) $\sin 10° \approx 0.1736$

(b) $\csc 10° \approx 5.759$

24. (a) $\sec 225° \approx -1.414$

(b) $\sec 135° \approx -1.414$

25. (a) $\tan\dfrac{\pi}{9} \approx 0.3640$

(b) $\tan\dfrac{10\pi}{9} \approx 0.3640$

26. (a) $\cot 1.35 \approx 0.2245$

(b) $\tan 1.35 \approx 4.455$

27. (a) $\cos\theta = \dfrac{\sqrt{2}}{2}$

$\theta = \dfrac{\pi}{4}, \dfrac{7\pi}{4}$

(b) $\cos\theta = -\dfrac{\sqrt{2}}{2}$

$\theta = \dfrac{3\pi}{4}, \dfrac{5\pi}{4}$

28. (a) $\sec\theta = 2$

$$\theta = \frac{\pi}{3}, \frac{5\pi}{3}$$

(b) $\sec\theta = -2$

$$\theta = \frac{2\pi}{3}, \frac{4\pi}{3}$$

29. (a) $\tan\theta = 1$

$$\theta = \frac{\pi}{4}, \frac{5\pi}{4}$$

(b) $\cot\theta = -\sqrt{3}$

$$\theta = \frac{5\pi}{6}, \frac{11\pi}{6}$$

30. (a) $\sin\theta = \frac{\sqrt{3}}{2}$

$$\theta = \frac{\pi}{3}, \frac{2\pi}{3}$$

(b) $\sin\theta = -\frac{\sqrt{3}}{2}$

$$\theta = \frac{4\pi}{3}, \frac{5\pi}{3}$$

31. $2\sin^2\theta = 1$

$$\sin\theta = \pm\frac{\sqrt{2}}{2}$$

$$\theta = \frac{\pi}{4}, \frac{3\pi}{4}, \frac{5\pi}{4}, \frac{7\pi}{4}$$

32. $\tan^2\theta = 3$

$$\tan\theta = \pm\sqrt{3}$$

$$\theta = \frac{\pi}{3}, \frac{2\pi}{3}, \frac{4\pi}{3}, \frac{5\pi}{3}$$

33. $\tan^2\theta = \tan\theta = 0$

$$\tan\theta(\tan\theta - 1) = 0$$

$\tan\theta = 0$ $\qquad$ $\tan\theta = 1$

$\theta = 0, \pi$ $\qquad$ $\theta = \frac{\pi}{4}, \frac{5\pi}{4}$

34. $2\cos^2\theta - \cos\theta - 1 = 0$

$$(2\cos\theta + 1)(\cos\theta - 1) = 0$$

$\cos\theta = -\frac{1}{2}$ $\qquad$ $\cos\theta = 1$

$\theta = \frac{2\pi}{3}, \frac{4\pi}{3}$ $\qquad$ $\theta = 0$

35. $\sec\theta\csc\theta - 2\csc\theta = 0$

$$\csc\theta(\sec\theta - 2) = 0$$

$(\csc\theta \neq 0$ for any value of $\theta)$

$$\sec\theta = 2$$

$$\theta = \frac{\pi}{3}, \frac{5\pi}{3}$$

36. $\sin\theta = \cos\theta$

$$\tan\theta = 1$$

$$\theta = \frac{\pi}{4}, \frac{5\pi}{4}$$

37. $\cos^2\theta + \sin\theta = 1$

$$1 - \sin^2\theta + \sin\theta = 1$$

$$\sin^2\theta - \sin\theta = 0$$

$$\sin\theta(\sin\theta - 1) = 0$$

$\sin\theta = 0$ $\qquad$ $\sin\theta = 1$

$\theta = 0, \pi$ $\qquad$ $\theta = \frac{\pi}{2}$

38. $\cos\left(\frac{\theta}{2}\right) - \cos\theta = 1$

$$\cos\left(\frac{\theta}{2}\right) = \cos\theta + 1$$

$$\sqrt{\left(\frac{1}{2}\right)(1 + \cos\theta)} = \cos\theta + 1$$

$$\left(\frac{1}{2}\right)(1 + \cos\theta) = \cos^2\theta + 2\cos\theta + 1$$

$$0 = \cos^2\theta + \left(\frac{3}{2}\right)\cos\theta + \left(\frac{1}{2}\right)$$

$$0 = \left(\frac{1}{2}\right)(2\cos^2\theta + 3\cos\theta + 1)$$

$$0 = \left(\frac{1}{2}\right)(2\cos\theta + 1)(\cos\theta + 1)$$

$\cos\theta = -\frac{1}{2}$ $\qquad$ $\cos\theta = -1$

$\theta = \frac{2\pi}{3}$ $\qquad$ $\theta = \pi$

$(0 = 4\pi/3$ is extraneous$)$

39. $(275 \text{ ft/sec})(60 \text{ sec}) = 16{,}500$ feet

$$\sin 18° = \frac{a}{16{,}500}$$

$$a = 16{,}500 \sin 18° \approx 5099 \text{ feet}$$

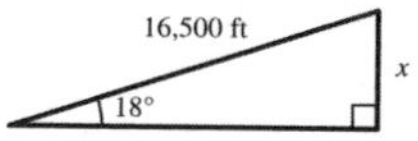

40. $\tan 3.5° = \dfrac{h}{13 + x}$ and $\tan 9° = \dfrac{h}{x}$

$(13 + x)\tan 3.5° = h \qquad x \tan 9° = h$

$13 \tan 3.5° + x \tan 3.5° = x \tan 9°$

$13 \tan 3.5° = x(\tan 9° - \tan 3.5°)$

$\dfrac{13 \tan 3.5°}{\tan 9° - \tan 3.5°} = x$

$h = x \tan 9° = \dfrac{13 \tan 3.5° \tan 9°}{\tan 9° - \tan 3.5°} \approx 1.295$ miles or 6839.307 feet

41. (a) Period: π

Amplitude: 2

(b) Period: 2

Amplitude: $\frac{1}{2}$

42. (a) Period: 4π

Amplitude: $\frac{3}{2}$

(b) Period: 6π

Amplitude: 2

43. Period: $\frac{1}{2}$

Amplitude: 3

44. Period: 20

Amplitude: $\frac{2}{3}$

45. Period: $\dfrac{\pi}{2}$

46. Period: $\frac{1}{2}$

47. Period: $\dfrac{2\pi}{5}$

48. Period: $\dfrac{\pi}{2}$

49. (a) $f(x) = c \sin x$; changing c changes the amplitude.

When $c = -2$: $f(x) = -2 \sin x$.

When $c = -1$: $f(x) = -\sin x$.

When $c = 1$: $f(x) = \sin x$.

When $c = 2$: $f(x) = 2 \sin x$.

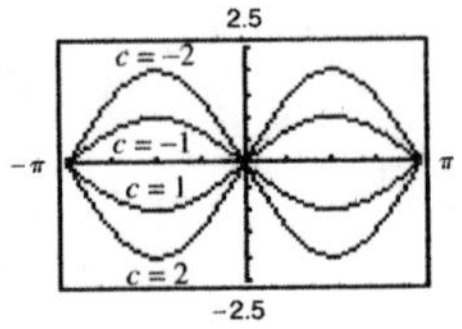

(b) $f(x) = \cos(cx)$; changing c changes the period.

When $c = -2$: $f(x) = \cos(-2x) = \cos 2x$.

When $c = -1$: $f(x) = \cos(-x) = \cos x$.

When $c = 1$: $f(x) = \cos x$.

When $c = 2$: $f(x) = \cos 2x$.

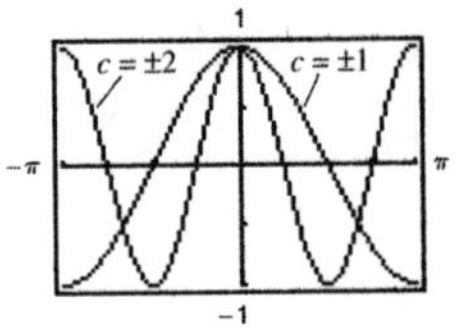

(c) $f(x) = \cos(\pi x - c)$; changing c causes a horizontal shift.

When $c = -2$: $f(x) = \cos(\pi x + 2)$.

When $c = -1$: $f(x) = \cos(\pi x + 1)$.

When $c = 1$: $f(x) = \cos(\pi x - 1)$.

When $c = 2$: $f(x) = \cos(\pi x - 2)$.

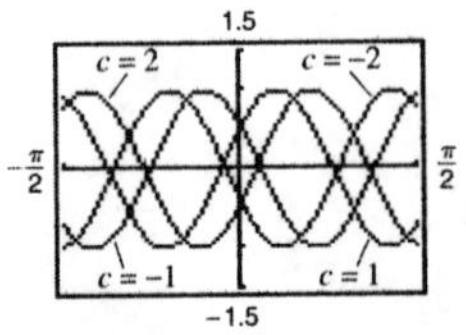

50. (a) $f(x) = \sin x + c$; changing c causes a vertical shift.

When $c = -2$: $f(x) = \sin x - 2$.

When $c = -1$: $f(x) = \sin x - 1$.

When $c = 1$: $f(x) = \sin x + 1$.

When $c = 2$: $f(x) = \sin x + 2$.

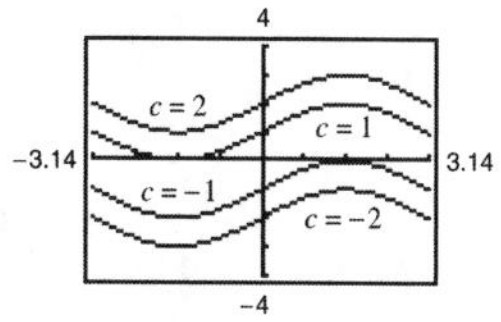

(b) $f(x) = -\sin(2\pi x - c)$; changing c causes a horizontal shift.

When $c = -2$: $f(x) = -\sin(2\pi x + 2)$.

When $c = -1$: $f(x) = \sin(2\pi x + 1)$.

When $c = 1$: $f(x) = \sin(2\pi x - 1)$.

When $c = 2$: $f(x) = -\sin(2\pi x - 2)$.

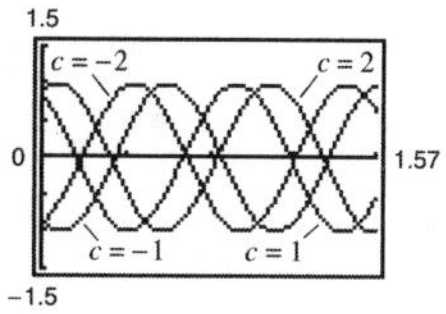

(c) $f(x) = c\cos x$; changing c changes the amplitude.

When $c = -2$: $f(x) = -2\cos x$.

When $c = -1$: $f(x) = -\cos x$.

When $c = 1$: $f(x) = \cos x$.

When $c = 2$: $f(x) = 2\cos x$.

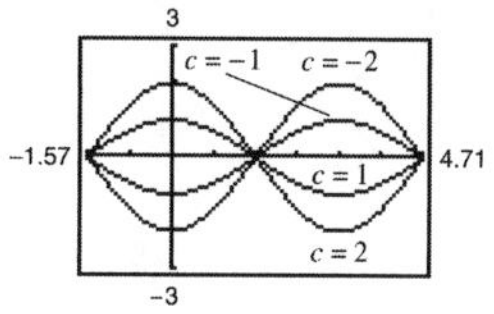

51. $y = \sin \dfrac{x}{2}$

Period: 4π

Amplitude: 1

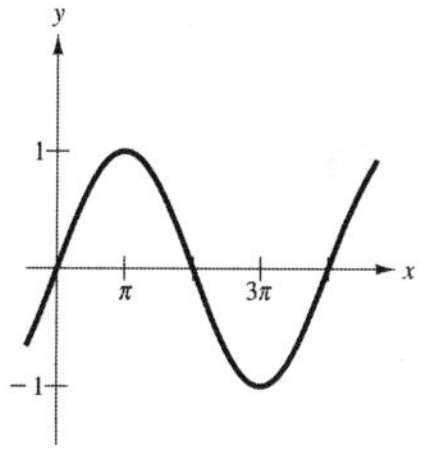

52. $y = 2\cos 2x$

Period: π

Amplitude: 2

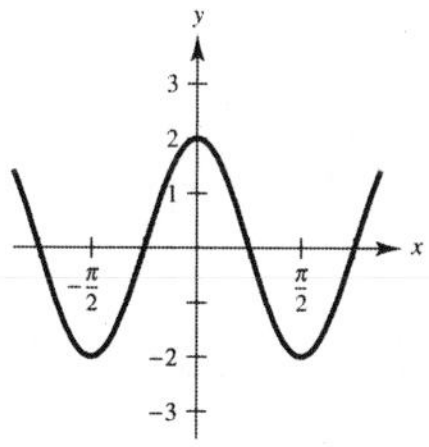

53. $y = -\sin \dfrac{2\pi x}{3}$

Period: 3

Amplitude: 1

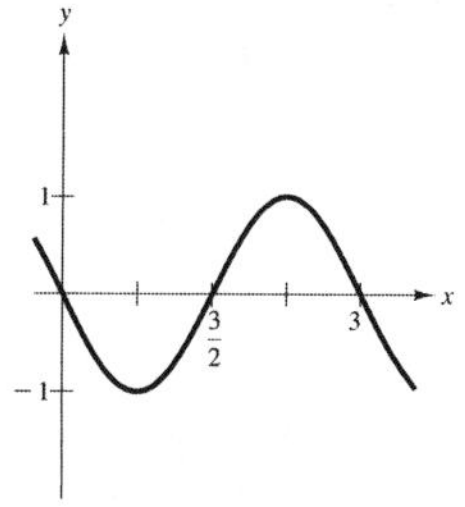

54. $y = 2\tan x$

Period: π

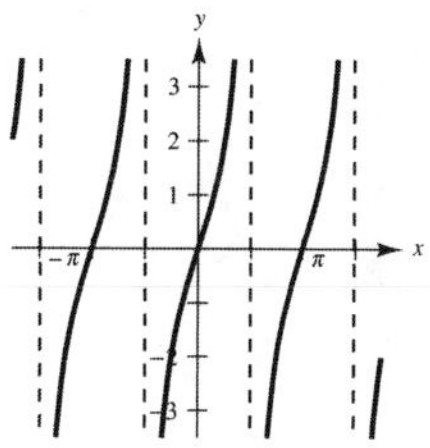

55. $y = \csc\dfrac{x}{2}$

Period: 4π

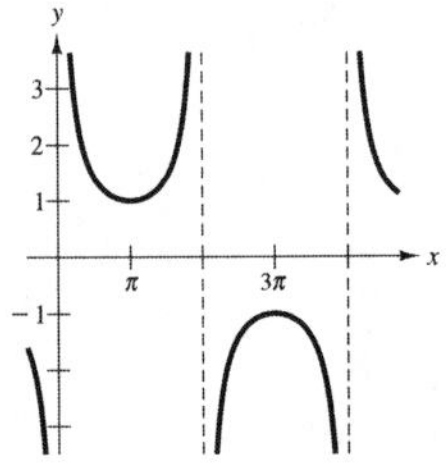

56. $y = \tan 2x$

Period: $\dfrac{\pi}{2}$

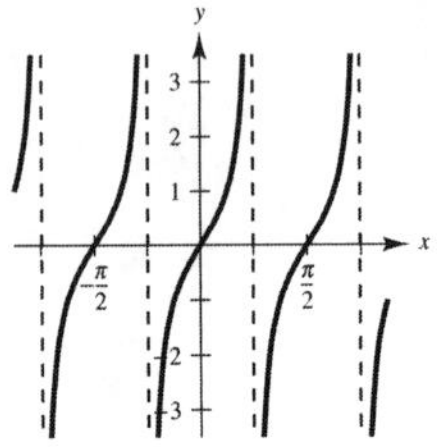

57. $y = 2\sec 2x$

Period: π

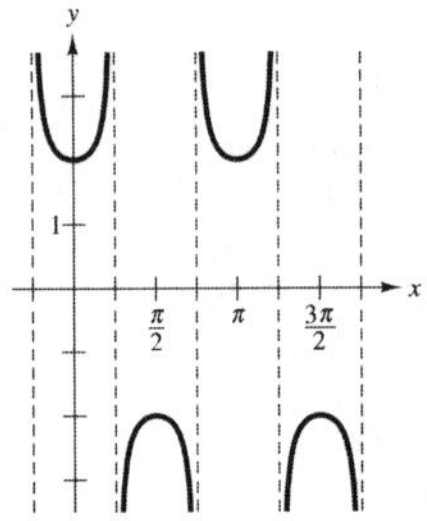

58. $y = \csc 2\pi x$

Period: 1

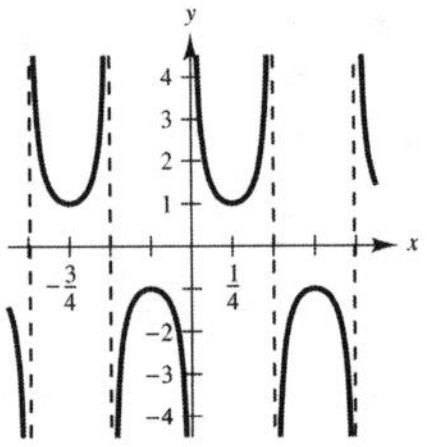

59. $y = \sin(x + \pi)$

Period: 2π

Amplitude: 1

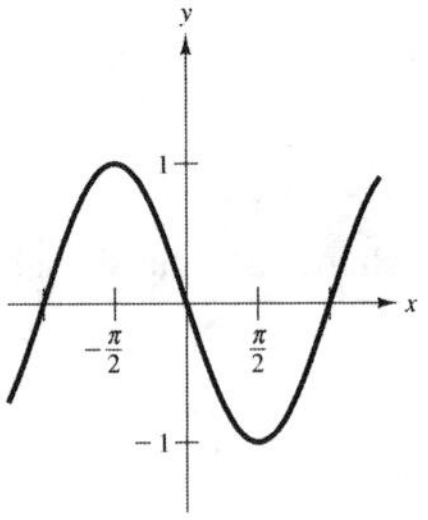

60. $y = \cos\left(x - \dfrac{\pi}{3}\right)$

Period: 2π

Amplitude: 1

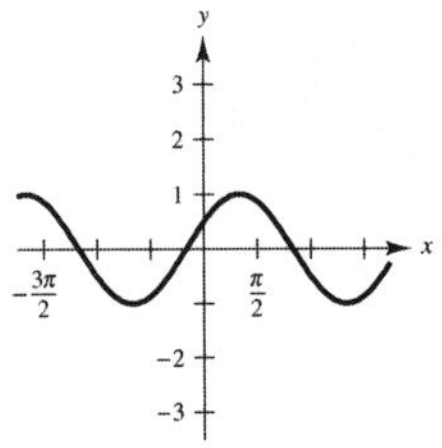

61. $y = 1 + \cos\left(x - \dfrac{\pi}{2}\right)$

Period: 2π

Amplitude: 1

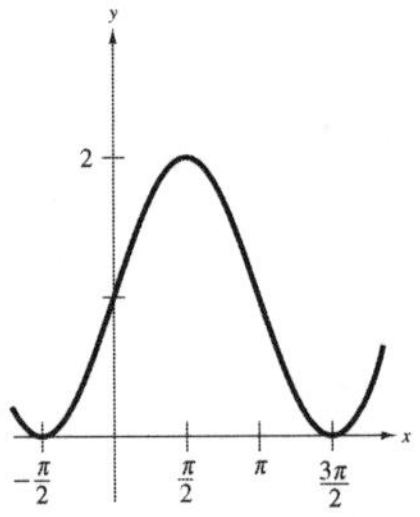

62. $y = 1 + \sin\left(x + \dfrac{\pi}{2}\right)$

Period: 2π

Amplitude: 1

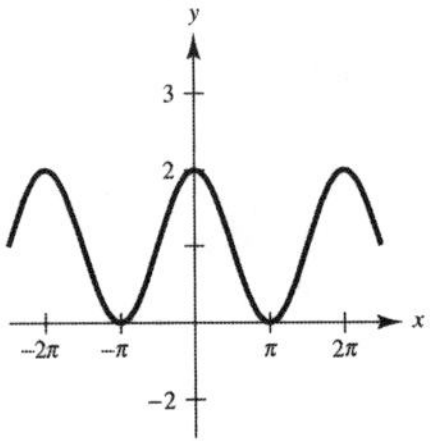

63. $y = a\cos(bx - c)$

From the graph, we see that the amplitude is 3, the period is 4π, and the horizontal shift is π. Thus,

$$a = 3$$

$$\frac{2\pi}{b} = 4\pi \Rightarrow b = \frac{1}{2}$$

$$\frac{c}{d} = \pi \Rightarrow c = \frac{\pi}{2}.$$

Therefore, $y = 3\cos\left[(1/2)x - (\pi/2)\right]$.

64. $y = a\sin(bx - c)$

From the graph, we see that the amplitude is $\frac{1}{2}$, the period is π, and the horizontal shift is 0. Also, the graph is reflected about the x-axis. Thus,

$$a = -\frac{1}{2}$$

$$\frac{2\pi}{b} = \pi \Rightarrow b = 2$$

$$\frac{c}{b} = 0 \Rightarrow c = 0.$$

Therefore, $y = -\frac{1}{2}\sin 2x$.

65. $f(x) = \sin x$

$g(x) = |\sin x|$

$h(x) = \sin|x|$

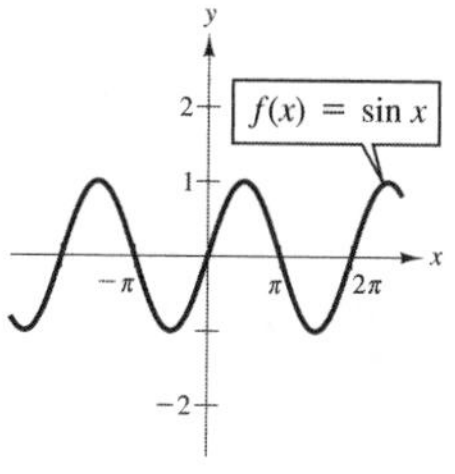

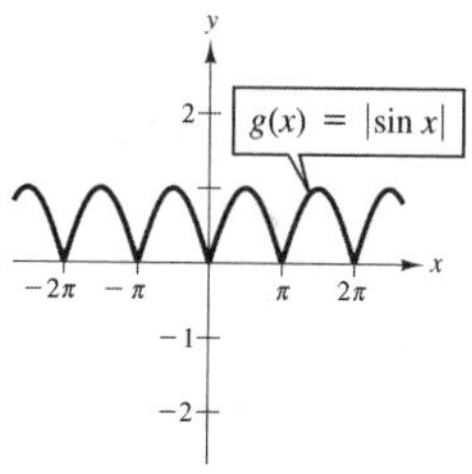

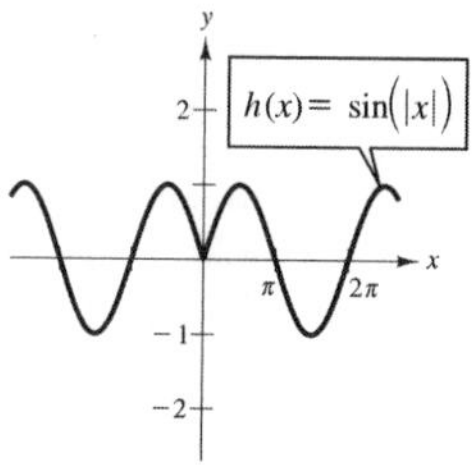

The graph of $|f(x)|$ will reflect any parts of the graph of $f(x)$ below the x-axis about the y-axis.

The graph of $f(|x|)$ will reflect the part of the graph of $f(x)$ to the right of the y-axis about the y-axis.

66. If $h = 51 + 50\sin\left(8\pi t - \frac{\pi}{2}\right)$, then $h = 1$ when $t = 0$.

67. $S = 58.3 + 32.5\cos\frac{\pi t}{6}$

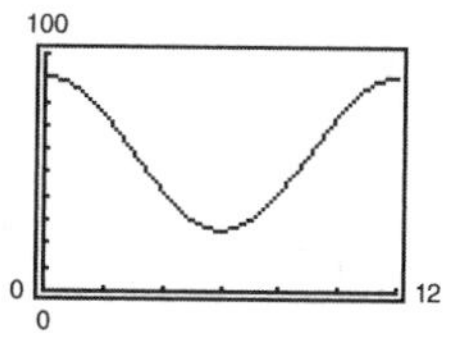

Sales exceed 75,000 during the months of January, November, and December.

68. (a) $5.35 - 2 = 3.35$

$5.35 + 2 = 7.35$

(b) $5.35 - 2(3) = -0.65$

(c) $13.35 = 5.35 + 2(4)$

$-4.65 = 5.35 - 2(5)$

True; because f and g have periods of 2 and intersect at $x = 5.35$, $f(13.35) = g(-4.65)$.

69. $f(x) = \frac{4}{\pi}\left(\sin \pi x + \frac{1}{3}\sin 3\pi x\right)$

$g(x) = \frac{4}{\pi}\left(\sin \pi x + \frac{1}{3}\sin 3\pi x + \frac{1}{5}\sin 5\pi x\right)$

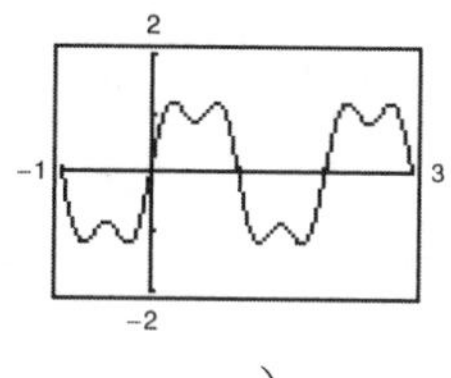

Pattern: $f(x) = \frac{4}{\pi}\left(\sin \pi x + \frac{1}{3}\sin 3\pi x + \frac{1}{5}\sin 5\pi x + \cdots + \frac{1}{2n-1}\sin(2n-1)\pi x\right)$, $n = 1, 2, 3\ldots$

70. $f(x) = \frac{1}{2} - \frac{4}{\pi^2}\left(\cos \pi x + \frac{1}{9}\cos 3\pi x\right)$

$g(x) = \frac{1}{2} - \frac{4}{\pi^2}\left(\cos \pi x + \frac{1}{9}\cos 3\pi x + \frac{1}{25}\cos 5\pi x\right)$

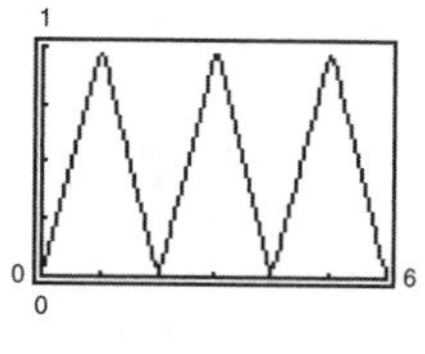

Pattern: $f(x) = \frac{1}{2} - \frac{4}{\pi^2}\left(\cos \pi x + \frac{1}{9}\cos 3\pi x + \frac{1}{25}\cos 5\pi x + \cdots + \frac{1}{(2n-1)^2}\cos(2n-1)\pi x\right)$, $n = 1, 2, 3\ldots$